THIOPHENE AND ITS DERIVATIVES

Part One

This is the Forty-Fourth Volume in the Series

THE CHEMISTRY OF HETEROCYCLIC COMPOUNDS

THE CHEMISTRY OF HETEROCYCLIC COMPOUNDS

A SERIES OF MONOGRAPHS

ARNOLD WEISSBERGER and EDWARD C. TAYLOR

Editors

THIOPHENE AND ITS DERIVATIVES

Part One

Edited by

Salo Gronowitz

University of Lund
Lund, Sweden

AN INTERSCIENCE® PUBLICATION

JOHN WILEY AND SONS

NEW YORK • CHICHESTER • BRISBANE • TORONTO • SINGAPORE

An Interscience® Publication
Copyright © 1985 by John Wiley & Sons, Inc.

Library of Congress Cataloging in Publication Data:

Main entry under title:

Thiophene and its derivatives.

(The Chemistry of heterocyclic compounds ; v. 44)
"An Interscience publication."
Includes index.
1. Thiophene. I. Gronowitz, Salo. II. Series.
QD403.T55 1985   547'.594   84-15356
ISBN 0-471-38120-9 (v. 1)

Printed in the United States of America

10 9 8 7 6 5 4 3 2 1

Contributors

L. I. BELEN'KII
N. D. Zelinsky Institute of Organic
  Chemistry
Academy of Sciences of the USSR
Moscow, USSR

P. H. BENDERS
Laboratory of Organic Chemistry
Twente University of Technology
Enschede, The Netherlands

F. BOHLMANN
Institut für Organische Chemie
Technische Universität
Berlin, Federal Republic of Germany

G. D. GALPERN
Institute of Petrochemical Synthesis
Academy of Sciences of the USSR
Moscow, USSR

YA. L. GOL'DFARB
N. D. Zelinsky Institute of Organic
  Chemistry
Academy of Sciences of the USSR
Moscow, USSR

SALO GRONOWITZ
Division of Organic Chemistry 1
Chemical Center
University of Lund
Lund, Sweden

ANITA HENRIKSSON-ENFLO
Institute of Theoretical Physics
University of Stockholm
Stockholm, Sweden

ALAIN LABLACHE-COMBIER
Laboratoire de Chimie Organique
  Physique

Université des Sciences et Techniques
  de Lille
Villeneuve d'Ascq, France

A. E. A. PORTER
Chemistry Department
University of Stirling
Stirling, Scotland

JEFFERY B. PRESS
Cardiovascular–CNS Research Section
American Cyanamid Company
Medical Research Division
Lederle Laboratories
Pearl River, New York.
Present affiliation:
Ortho Pharmaceutical Corporation
Raritan, New Jersey

MAYNARD S. RAASCH
Central Research and Development
  Department
Experimental Section
E. I. du Pont de Nemours and
  Company
Wilmington, Delaware

D. N. REINHOUDT
Laboratory of Organic Chemistry
Twente University of Technology
Enschede, The Netherlands

W. P. TROMPENAARS
Laboratory of Organic Chemistry
Twente University of Technology
Enschede, The Netherlands

C. ZDERO
Institut für Organische Chemie
Technische Universität
Berlin, Federal Republic of Germany

To the memory of

VICTOR MEYER

*great chemist and teacher
and founder of thiophene chemistry*

The Chemistry of Heterocylic Compounds

The chemistry of heterocyclic compounds is one of the most complex branches of organic chemistry. It is equally interesting for its theoretical implications, for the diversity of its synthetic procedures, and for the physiological and industrial significance of heterocyclic compounds.

A field of such importance and intrinsic difficulty should be made as readily accessible as possible, and the lack of a modern detailed and comprehensive presentation of heterocyclic chemistry is therefore keenly felt. It is the intention of the present series to fill this gap by expert presentations of the various branches of heterocyclic chemistry. The subdivisions have been designed to cover the field in its entirety by monographs which reflect the importance and the interrelations of the various compounds, and accommodate the specific interests of the authors.

In order to continue to make heterocyclic chemistry as readily accessible as possible new editions are planned for those areas where the respective volumes in the first edition have become obsolete by overwhelming progress. If, however, the changes are not too great so that the first editions can be brought up-to-date by supplementary volumes, supplements to the respective volumes will be published in the first edition.

ARNOLD WEISSBERGER

Research Laboratories
Eastman Kodak Company
Rochester, New York

EDWARD C. TAYLOR

Princeton University
Princeton, New Jersey

Preface

In 1952, in the first volume of *The Chemistry of Heterocyclic Compounds*, Howard D. Hartough described the state of research on the chemistry of thiophene and its derivatives up to 1950. Selenophene and tellurophene were also included in this monograph which, except for two chapters was written by Hartough alone. When this book was written, the explosive development triggered by the commercial process for thiophene from butane and sulfur, developed by Socony-Vacuum Oil Company in the 1940's, had just begun. The enormous amount of work carried out on this important aromatic five-membered heterocycle since 1950 makes it of course impossible for one person to cover all aspects, and an able group of specialists were assembled from all over the world to treat the entire field. This makes some minor overlaps between chapters unavoidable, but I think it is important to treat some topics from different angles of approach.

Because of the wealth of results and the rather large number of contributors, these volumes are not as strictly organized as some previous volumes in this series, but can be considered as a collection of topics on thiophene chemistry. Together, however, it is my hope that these chapters give as comprehensive a description as possible of the chemistry of thiophene and its monocyclic derivatives, based on the literature from 1950 to 1982. References to previous results, treated in Hartough's book, are also given when necessary.

The chapters fall in two categories: (1) those that treat syntheses, properties, and reactions of thiophenes, and (2) those that treat systematically functionalized simple thiophenes, such as alkylthiophenes, halothiophenes, aminothiophenes, thiophenecarboxylic acids, and so on. The latter chapters, as is customary in the Weissberger–Taylor series, contain tables of compounds with their physical properties, which should be very useful for all synthetic chemists. Part 1 of these volumes contains only chapters in category (1) and starts with a treatise on the preparation of thiophenes by ring-closure reactions and from other ring systems. It is followed by a chapter on theoretical calculations. Then, in two chapters, naturally occurring thiophenes in plants and in petroleum, shale oil, and coals are treated. The topic of the next chapter is the important field of pharmacologically active compounds. The synthetic use of thiophene derivatives for the synthesis of aliphatic compounds by desulfurization follows. Two chapters treat thiophenes modified at the sulfur, namely thiophene-1,1-dioxides and thiophene-1-oxides, and *S*-alkylation of thiophenes. In the last three chapters, the discussion on different reactivities of thiophenes starts with radical reactions of thiophenes, cycloaddition reactions, and photochemical reactions.

In the latter parts of this monograph, chapters on derivatives such as alkylthiophenes, halothiophenes, nitrothiophenes, aminothiophenes, hydroxy- and alkoxythiophenes, thiophenethiols, thiophenesulfonic acids, metal derivatives, formyl and acyl derivatives, thiophenecarboxylic acids, thienylethenes and

-acetylenes, arylthiophenes, and bithienyls are intercalated with chapters on physical and spectroscopical properties, electrophilic and nucleophilic substitution, and side-chain reactivity.

I wish to thank all the distinguished scientists who contributed chapters to these volumes for their splendid cooperation and my secretary Ann Nordlund for her invaluable help. I am also indebted to Dr. Robert E. Carter for correcting my chapter and those of some of the other authors whose native tongue is not English.

SALO GRONOWITZ

Lund, Sweden
August 1984

Contents

THIOPHENE AND ITS DERIVATIVES

Part One

This is the Forty-Fourth Volume in the Series

THE CHEMISTRY OF HETEROCYCLIC COMPOUNDS

CHAPTER I

Preparation of Thiophenes by Ring-Closure Reactions and from Other Ring Systems

SALO GRONOWITZ

Division of Organic Chemistry 1, Chemical Center, University of Lund, Lund, Sweden

1

I. INTRODUCTION

In this chapter, ring-closure reactions leading to various thiophene derivatives are discussed, as well as the use of di- and tetrahydrothiophenes and other heterocyclic compounds for the preparation of thiophenes. The aromatic nature of thiophene leads to the easy formation of the ring system under many different and sometimes unexpected conditions.

The various preparative methods can be systematized according to the number of compounds reacting in the ring-closure, as demonstrated by Meth-Cohn.[1] The different possibilities are indicated in Scheme 1; the transformations with solid lines are known.

The division is of course artificial, and it is a matter of taste whether certain related methods should be treated under $C_2 + C_2 + S$, $C_2S + C_2$, or $C_4 + S$. This is the case with the very versatile Gewald reaction, different modifications of which would fall in one of these classes. To provide a uniform treatment, they are all discussed under the $C_2S + C_2$ approach. Furthermore, the transformation of other rings into thiophenes is treated in a separate section and is not divided according to Scheme 1.

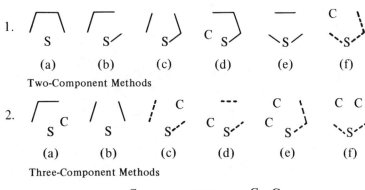

1.

(a) (b) (c) (d) (e) (f)

Two-Component Methods

2.

(a) (b) (c) (d) (e) (f)

Three-Component Methods

3.

(a) (b) (c)

Four-Component Methods

4.

Five-Component Methods

5.

(a) (b) (c)

One-Component Methods

Scheme 1

From the synthetic point of view, the two-component methods are by far the most important. The classical thiophene syntheses from disodium succinate and P_4S_{10} and the Paal–Knorr synthesis from 1,4-dicarbonyl compounds and P_4S_{10} fall under this heading. The modern commercial routes to thiophene from almost any four carbon units and a source of sulfur over a suitable catalyst at high temperature, and modern laboratory methods using diacetylenes and H_2S/HCl, are other examples of the $C_4 + S$ approach. The old Hinsberg reaction is a very useful example of the $C_2 + CSC$ technique. Modern methods like the Fiesselmann reaction, consisting of the condensation of thioglycolic esters with acetylenecarboxylic esters or other substrates with the same oxidation level, illustrates the $C_3 + CS$ procedure. The reaction of α-mercaptoketones with activated acetonitriles, constituting the original version of the Gewald reaction, illustrates the $C_2S + C_2$ approach. These two-component reactions are, therefore, treated first, followed by three- and one-component reactions.

It is an almost impossible task to include examples of all the reactions leading to thiophene derivatives. It is, however, my hope that the coverage of the *principal* routes to thiophenes is as complete as possible.

II. TWO-COMPONENT METHODS

1. $C_4 + S$ Methods

A.   *Thiophenes from Alkanes, Alkenes, and Alkadienes and Sulfur-Containing Reagents*

The discovery and development of the modern thiophene synthesis from butane, butadiene, or butenes with sulfur at Socony-Vacuum Oil Company in the early 1940's was described in detail in Hartough's monograph.[2] The reaction was carried out at 565°C and contact times between the hydrocarbon and sulfur vapor were about 2 sec. The most useful by-product of this reaction was 3-thiophenethiol[3] and a product believed to be 3,4-thiolanedithione but later shown to be 5-methyl-1,2-dithiol-3-thione.[4] Numerous patents describe the preparation of thiophene from butane, butenes, or butadiene and hydrogen sulfide using an Al_2O_3-Cr_2O_3 catalyst at about 600°C.[5-17] The reactions of C_6-hydrocarbons led to the formation of thiophene and 2- and 3-methylthiophene.[17a] Various catalyst compositions for this process have been described.[18-21] A study of the mechanism of the thiophene synthesis from *n*-butane and hydrogen sulfide over an alumina–lanthania–chromia-potassium oxide catalyst, using ^{14}C labeled intermediates, showed that the conversion of butane into butenes and butadiene took place during thiophene synthesis. In addition, this stepwise dehydrogenation is not the only possible route for thiophene formation, since both butane and butenes can react directly with hydrogen sulfide on the catalyst surface.[22] The use of piperylene[23-25] or pentadienes[26,27] in the reaction with hydrogen sulfide gave 2-methylthiophene, which was also obtained through the noncatalytic reaction of C_5-straight-chain hydrocarbons with sulfur.[28] From isoprene and hydrogen sulfide over a K_2O-promoted Al-Cr catalyst, 3-methylthiophene was prepared,[29] and from 2,3-dimethylbutadiene, 3,4-dimethylthiophene was obtained. 3-Ethylthiophene[30] has been prepared from the 2-ethylbutanol fraction from butanol manufacture by reaction with H_2S over the catalyst mentioned above.[31,32] Another alternative for the synthesis of thiophene is the reaction of C_4-hydrocarbons with sulfur dioxide over suitable catalysts.[33-42] 3-Vinylthiophene was obtained from the reaction of 3-methylpentane and sulfur dioxide.[43]

The influence of transition metal oxides on the catalytic activity of aluminum oxide in this reaction has been evaluated.[44] The equilibrium yields of thiophene in the reaction of C_4-hydrocarbons with sulfur dioxide or hydrogen sulfide were determined, and thermodynamic calculations on the synthesis of thiophene from butane, butenes, butadiene, and sulfur dioxide or hydrogen sulfide were carried out.[45] From 2,3-dimethyl-2-butene and 2,3-dimethyl-1,4-butadiene, 3,4-dimethylthiophene was obtained.[46] A review by Ryashentseva et al. on catalytic methods for obtaining thiophene and alkylthiophenes from C_2-C_6 hydrocarbons and from various organosulfur compounds[47,48] indicates that the reaction with hydrogen sulfide is the most promising one.[49] Sulfur atoms, S (^{1}D) atoms obtained from the gas-phase photodissociation of COS, reacted with 1,3-butadiene to give vinyl thiacyclopropane and thiophene.[50]

Thiophene has been prepared by passing C_4-molecules, such as butane, butene, n-butanol, or crotonaldehyde, together with carbon disulfide over a promoted chromium–aluminum oxide catalyst at about 500°C and with short contact times (5–8 sec).[51-53] With crotonaldehyde and butanol, yields of 78 and 42% of thiophene were obtained. From C_5-C_7 molecules, alkylthiophenes were obtained. Thus, 2-pentanol gave 70% of 2-methylthiophene, 1- and 3-hexanol yielded predominantly 2-ethylthiophene, 2-hexanol provided a 68% yield of 2,5-dimethylthiophene, and 2-heptanol gave an 84% yield of 2-ethyl-5-methylthiophene. 2-Pentanone was used for the preparation of 2-methylthiophene (80%) and 2-hexanone for the synthesis of 2,5-dimethylthiophene.[54] (2-Butyne-1,4-diol and butane-1,4-diol derived from it have also been used for the preparation of thiophene by reaction with H_2S[55,56] or sulfur[57] over Al_2O_3-Cr_2O_3 catalysts at high temperature.)

A method developed by Voronkov and coworkers especially for the synthesis of aryl-substituted thiophenes, consisted of heating alkenes and alkanes with at least four carbons in a straight chain with sulfur at 200–270°C[58] (Table 1). Usually, no solvent is used, although o-dichlorobenzene has been utilized in some cases.[59] Hydrogen sulfide is evolved, and the yields are often very low. When lower temperatures are used, the formation of mixtures of tetra- and dihydrothiophenes has been observed, as in the reaction of 2,6-dimethyl-octa-2,6-diene and sulfur at 140°C.[67] Much higher yields are obtained when 1,3-butadienes are used, and excellent yields (90–95%) have been reported for 2,5-diaryl- and 3,4-diarylthiophenes in the reaction of 1,4- and 2,3-diarylbutadienes with sulfur at about 200°C[68] (Scheme 2). 1,3-Dienes with various functional groups, such as carbethoxy, cyano, and carbon

TABLE 1.   REACTIONS OF ALKENES AND ALKANES WITH SULFUR TO GIVE

$$R_4 \underset{R_5}{\overset{R_3}{\left\langle\!\!\!\!\rule{0pt}{1em}\right.}}\underset{S}{\overset{R_2}{\rule{0pt}{1em}}}$$

Substrate	Temperature (°C)	Thiophene				Yield (%)	Reference
		R_2	R_3	R_4	R_5		
$C_6H_5CH=CHCH_2CH_3$	200–250	C_6H_5				5	58
$C_6H_5CH_2CH=CH_2$	220–230	C_6H_5				14.4	58
$C_6H_5CH_2-CH=CH-CH_3$	190–250	C_6H_5				15.3	58
$C_6H_5(CH_2)_3CH_3$	195–200	C_6H_5				5.1	58
$C_6H_5-CH-CH_2$ $\quad CH_3\ \ CH_3$	210–215		C_6H_5	C_6H_5		55	59
$C_6H_5CH-CH_2$ $H_5C_6-CH_2\ \ CH_3$ $\quad\ CH_3$	240–250	C_6H_5	C_6H_5			35	59
CH_2-CH_2 CH_2-CH_2	215–220	C_6H_5			C_6H_5	40	59
$C_6H_5CH_2,\ CH_2C_6H_5$ $C_6H_5-C=C((CH_3)_2)$ $\quad CH_3$	200–210		CH_3	C_6H_5		36	60
$C_6H_5-C=CH-C_2H_5$ $\quad\ CH_3$	210	CH_3		C_6H_5		26	61
$C_6H_5C=CH-CH_3$ $\quad C_2H_5$	210	CH_3	C_6H_5			15	61
$C_6H_5C=C-C_2H_5$ $\quad\ C_3H_7$	195–235	C_2H_5	C_6H_5		CH_3	7	61
$C_6H_5C=CHCH_2CH(CH_3)_2$ $\quad CH_3$	200–220	$(CH_3)_2CH$		C_6H_5		—	61
$C_6H_5-CH_2CH(CH_3)_2$	180–250		C_6H_5			44	62

![structure: p-CH₃OC₆H₄–CH(CH₃)–...–CH–CH₃] $p\text{-CH}_3\text{OC}_6\text{H}_4$	250	$p\text{-CH}_3\text{OC}_6\text{H}_4$		$p\text{-CH}_3\text{OC}_6\text{H}_4$	—	63
$C_6H_5CH_2CH=CHC_6H_4OCH_3\text{-}p$ (CH₃)	200	C_6H_5		$p\text{-CH}_3\text{OC}_6\text{H}_4$	21	64
$p\text{-CH}_3\text{OC}_6\text{H}_4CH_2CH=CH\text{-}C_6H_5$ (CH₃)	200	$p\text{-CH}_3\text{OC}_6\text{H}_4$		C_6H_5	41	64
$C_6H_5CH=CHCH_2CH_2CH_3$	210–235	CH_3		CH_3	12.5	65
$C_6H_5CH=CH\text{-}CHCH_3$ (CH₃)	210–220	C_6H_5		CH_3 (C_6H_5)	20.3	65
$C_6H_5\text{-}CH_2\text{-}CH_2\text{-}CH\text{-}CH_3$ (CH₃)	200–245	C_6H_5		CH_3	8.5	65
$CH_3C{=}CH_2CH_3 \;/\; C_6H_5CH$	210–220	C_6H_5	CH_3		11.6	65
[cyclopentanone structure]	200	CH_2CH_2CO		CH_3	10	66
[cyclopentanone structure]	200	CH_3–$C(CH_3){=}$ / CH_2–C–$C{=}O$		CH_3	4	66
[dithienyl-fused cyclopentanone structure]	200–210				1	66

7

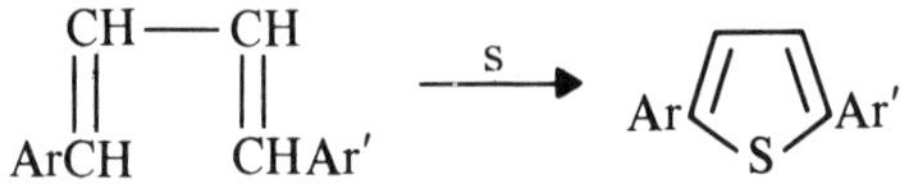

$$CH = CH$$

with structures as shown:

CH——CH
ArCH   CHAr' $\xrightarrow{S}$ Ar$\langle$S$\rangle$Ar'

Ar = Ar' = C_6H_5, p-$CH_3OC_6H_4$, 2-Th
Ar = C_6H_5;   Ar' = C_6H_4Cl (m, p), C_6H_4—C_6H_5

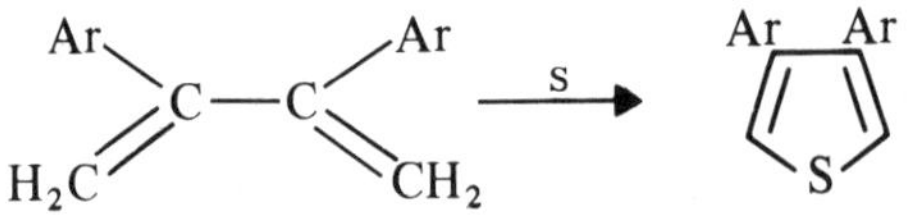

Ar$\diagdown$ $\diagup$Ar
        C—C $\xrightarrow{S}$ Ar   Ar
H$_2$C$\diagup$ $\diagdown$CH$_2$ $\langle$S$\rangle$

Ar = C_6H_5, $C_6H_4CH_3$-p, $C_6H_4OCH_3$-p

Scheme 2

groups, can also be used in this reaction. In the reaction of α,γ-diethylenic ketones, both thiophenes and 1,2-dithioles are formed[71, 72] (Scheme 3). With nitriles,[69, 70] Gewald-type reactions occur (see p. 42).

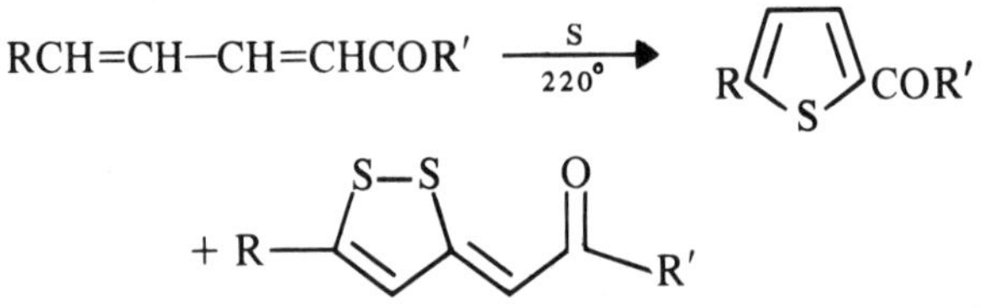

$$RCH=CH-CH=CHCOR' \xrightarrow[220°]{S} R\langle S\rangle COR'$$

$$+ R-\underset{S-S}{\overset{O}{\diagup\diagdown}}-R'$$

Scheme 3

Tetrachlorothiophene has been prepared by heating highly chlorinated C_4-type compounds with sulfur at temperatures between 150 and 300°C.[73, 74] The most convenient method for the synthesis of tetrachlorothiophene is the reaction of hexachlorobutadiene and sulfur at 205–240°C (Table 2).

The reaction of 2,3-diphenylbutadiene with bromine, followed by reaction of the mixture of dibromobutenes with sodium polysulfide in DMF, gave 3,4-diphenyl-thiophene as a dominant product, in addition to cyclic disulfides and tetrasulfides.[75] A convenient method for the synthesis of 3,4-di-t-butylthiophene consists of the reaction of 2,3-di-t-butylbutadiene, prepared by the copper-catalyzed reaction of t-butylmagnesium chloride with 1,4-dichloro-but-2-yne, with SCl_2 in CH_2Cl_2.[76]

A synthesis of the important penicillin side-chain, 3-thiophenemalonic acid, is shown in Scheme 4.[77–78a]

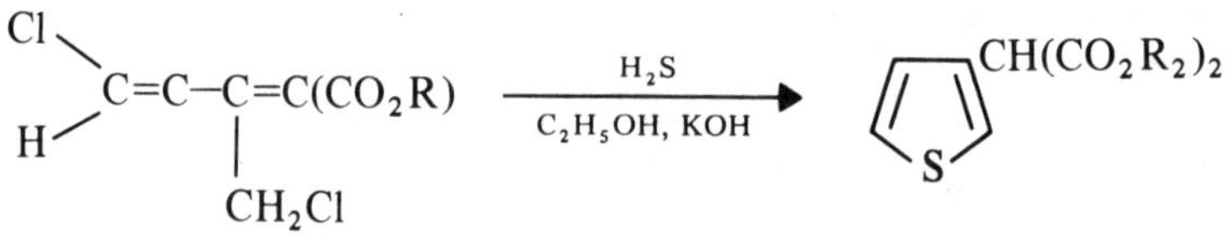

Cl$\diagdown$
        C=C-C=C(CO$_2$R) $\xrightarrow[C_2H_5OH, KOH]{H_2S}$ $\langle$S$\rangle$CH(CO$_2$R$_2$)$_2$
H$\diagup$ |
              CH$_2$Cl

Scheme 4

TABLE 2. REACTIONS OF 1,3-ALKADIENES WITH SULFUR TO GIVE

Substrate	Temperature (°C)	Thiophene				Yield (%)	Reference
		R_2	R_3	R_4	R_5		
H_3C (diene)	—	CH_3				36	160
H_3C —CH_3	—	CH_3			CH_3	28	160
$C_6H_5CH=CH—CH=CH_2$	212–217	C_6H_5				8.3	58
$CH_2=C—C=CH_2$ with CH_3, C_6H_5	210		Me	Ph		25	60
Cl-substituted diene	205–240	Cl	Cl	Cl	Cl	81	161
$C_6H_5CH=CH—CH=CHCOCH_3$	220	$COCH_3$			C_6H_5	10	72
$C_6H_5CH=CH—CH=CHCOC_6H_5$	220	COC_6H_5			C_6H_5	12[a]	72
$C_6H_5CH=CH—CH=CHCOC_6HOCH_3$-3p	220	$p\text{-}CH_3OC_6H_4CO$			C_6H_5	11[a]	72
$p\text{-}CH_3OC_6H_4CH=CH—CH=CHCOC_6H_5$	220	C_6H_5CO			$p\text{-}CH_3OC_6H_4$	17[a]	72
C_6H_5 ···COC_6H_5, $CO_2C_2H_5$	270	C_6H_5CO			C_6H_5	2	71
$p\text{-}CH_3OC_6H_4$ ···COC_6H_5, $CO_2C_2H_5$	270	C_6H_5CO			$p\text{-}CH_3OC_6H_4$	4.6	71
C_6H_5 ···$COCH_3{}^b$, $CO_2C_2H_5$	270	$COCH_3{}^b$			C_6H_5	1.3	71

9

TABLE 2. *Continued*

Substrate	Temperature (°C)	Thiophene R₂	R₃	R₄	R₅	Yield (%)	Reference
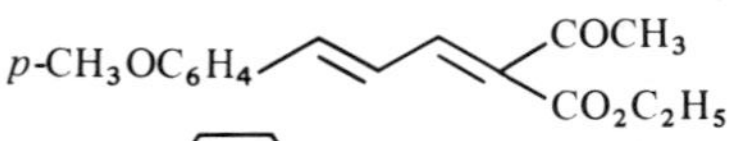 p-CH₃OC₆H₄ (—COCH₃, —CO₂C₂H₅)	270	COCH₃			p-CH₃OC₆H₄	2.7	71
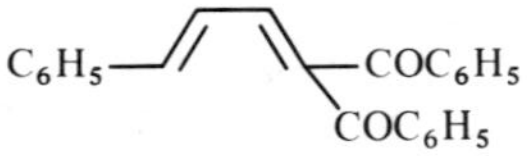 C₆H₅ (—COC₆H₅, COC₆H₅)	250	COC₆H₅			C₆H₅	2.5	71
 C₆H₅ (—COCH₃, COCH₃)	230	COCH₃			C₆H₅	2.1	71
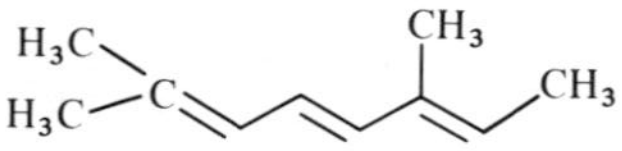 H₃C–C(CH₃)=... (CH₃, CH₃)	210	CH₃	CH₃		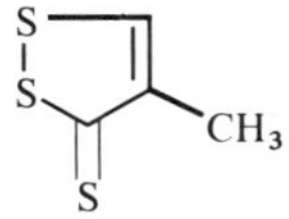 (dithiol structure) CH₃	10	162

^{a} α-(Dithiol-1,2-ylidene-3)ketones were also formed.
^{b} A total of 1.2% of ethyl 2-p-methoxyphenyl-5-thiophenecarboxylate was also formed.

10

B.   Thiophenes from the Reaction of Diacetylenes and Related Compounds with Hydrogen Sulfide

A convenient method for the synthesis of 2,5-disubstituted thiophenes is the reaction of conjugated diacetylenes with hydrogen sulfide in aqueous methanol, ethanol, or acetone at pH 9–10 and temperatures between 20 and 80°C.[79-81] It was shown recently that thiophene itself can be obtained in 94% yield, when diacetylene and hydrated sodium sulfide are reacted with potassium hydroxide in DMSO[81a] (Scheme 5). In some cases, glutathione or cystein was used as the source for hydro-

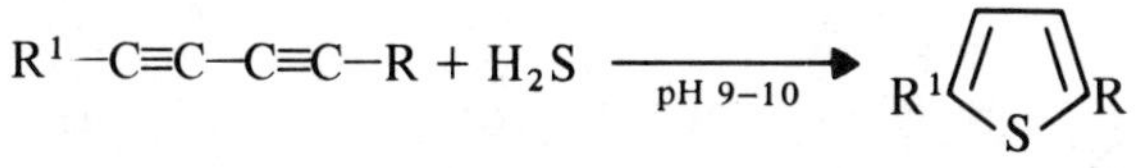

$$R^1-C\equiv C-C\equiv C-R + H_2S \xrightarrow[\text{pH 9-10}]{} R^1 \underset{S}{\diagdown} R$$

Scheme 5

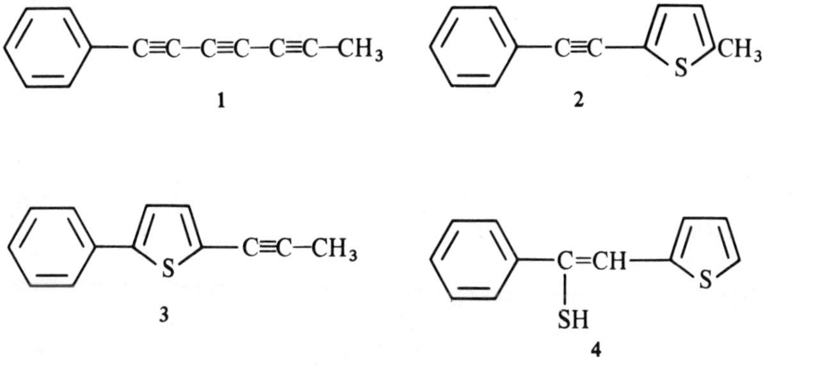

gen sulfide.[159] The diacetylene/H_2S reaction has been used for the synthesis of naturally occurring thiophenes.[82, 83] No systematic investigation of the influence of substituents on the rate of ring-closure has been carried out. It has been found, however, that di-(dialkylamino)diacetylenes ring-close very slowly, and a reaction time of 30–35 h at 80°C had to be used, whereas dialkoxymethyldiacetylenes ring-closed in 0.5 h.[84] In some cases, when the reaction was slow, refluxing butyl alcohol was used as solvent. Some interesting observations on the regiospecificity of the reaction have been made with unsymmetrical triacetylenes (Scheme 6).

Scheme 6

Thus, **1** gave only the phenyl thienylacetylene **2** and no phenylthiophene (**3**). Compound **4** was obtained as a by-product, indicating that the first intermediate is an enethiol upon addition of hydrosulfide to a triple bond.[85] Similarly, it is claimed that **5** gives only **6** and no **7** [84] (Scheme 7), (Table 3). A series of mono- and diglycosyl thiophenes has been synthesized recently by the reaction of mono- and diglycosylbutadiynes with sodium hydrosulfide.[84a]

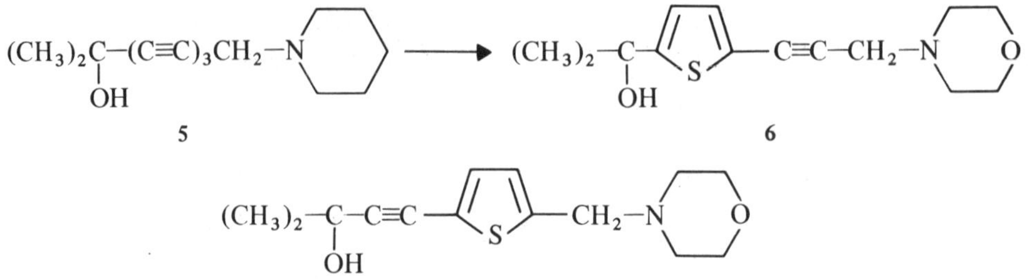

$$(CH_3)_2\underset{\underset{OH}{|}}{C}-(C\equiv C)_3CH_2-N\bigcirc \longrightarrow (CH_3)_2-\underset{\underset{OH}{|}}{C}-\underset{S}{\boxed{}}-C\equiv C-CH_2-N\bigcirc O$$

5   **6**

$$(CH_3)_2-\underset{\underset{OH}{|}}{C}-C\equiv C-\underset{S}{\boxed{}}-CH_2-N\bigcirc O$$

7

Scheme 7

TABLE 3.   THIOPHENES THROUGH THE REACTION BETWEEN DIACETYLENES AND
HYDROGEN SULFIDE

$$R_2C\equiv C-C\equiv C-R_2 + H_2S \longrightarrow \underset{R^1\ \ S\ \ R^2}{\boxed{}}$$

R_1	R_2	Yield (%)	Reference
H	H	20	80
H	CH_3	51	80
CH_3	CH_3	70	80
C_2H_5	C_2H_5	65	80
C_6H_5	C_6H_5	85	80, 79
COOH	C_6H_5	52	80
$C\equiv CC_6H_5$	C_6H_5	83	80, 79
$(C\equiv C)_2C_6H_5$	C_6H_5	75	80, 79
C_6H_5	H	56	82
2-Th[a]	C_6H_5	66	82
C_6H_5-2-Th	2-Th-C_6H_5	40	82
2-Fu[b]	2-Fu	68	82
3-Py[c]	C_3-Py	74	82
C_6H_5	$CH(OH)CH_3$	74	82
C_6H_5	$C(OH)(CH_3)_2$	33	82
$C_6H_5CH(OH)$	CH_3	64	82
$C_6H_5CH(OH)$	$CH(OH)C_6H_5$	14	82
$(CH_3)_2CH(OH)$	$C(OH)(CH_3)_2$	37	82
$C_2H_5(CH_3)C(OH)$	$C(OH)(CH_3)C_2H_5$	27	82
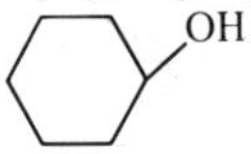 ⬡OH (cyclohexyl-OH)	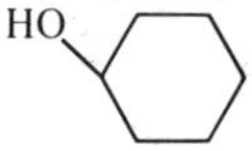 HO-⬡	55	82
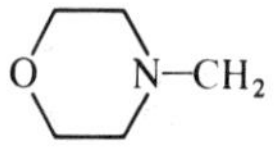 O⬡N-CH_2 (morpholino)	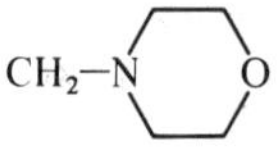 CH_2-N⬡O	32	82
C_6H_5-2-Th—	C_6H_5	24	82
C_6H_5	$CH=CH-CH_2OH$ t	60	82
C_6H_5CO	CH_3	54	82
$CH_3CH_2CH_2$	$CH\overset{t}{=}CHCH_2OH$	69	82
$CH_3CH_2CH_2$	$CH\overset{t}{=}CHCOOH$	36	82
$CH_3C\equiv C$	$HC\equiv CHCOOH$	42^d	82

TABLE 3. *Continued*

R₁	R₂	Yield (%)	Reference
2-Th—C≡C	$\overset{\text{OH}}{\underset{}{\text{CHCH}_3}}$	Good	83
o-O₂NC₆H₄OCH₂	CH₂OC₆H₄NO₂-o	85	163
m-O₂NC₆H₄OCH₂	CH₂OC₆H₄NO₂-m	89	163
p-O₂NC₆H₄OCH₂	CH₂OC₆H₄NO₂-p	91	163
H₅C₆OCH₂	CH₂OC₆H₅	95	164
o-ClC₆H₄OCH₂	CH₂OC₆H₄Cl-o	84	164
2,4-Cl₂C₆H₃OCH₂	CH₂OC₆H₃Cl₂-2,4	82	164
2,4,5-Cl₃C₆H₂OCH₂	CH₂OC₆H₂Cl₃-2,4,5	80	164
2,4,6-Cl₃C₆H₂OCH₂	CH₂OC₆H₂Cl₃-2,4,6	78	164
(H₃C)₂NCH₂	CH₂N(CH₃)₂	44	84
(H₅C₂)₂NCH₂	CH₂N(C₂H₅)₂	37	84
(piperidino)NCH₂	CH₂N(piperidino)	20	84
(CH₃)₂COH	CH₂N(piperidino)	60	84
(CH₃)₂COH	CH₂N(morpholino, with O)	33	84
H₃COCH₂	H₂COCH₃	55	84
H₅C₂OCH₂	CH₂OC₂H₅	53	84
(CH₃)₂CC≡C, OH	C(CH₃)₂, OH	18	84
(CH₃)₂C, OH	C≡C—CH₂N(morpholino, with O)	18	84
2-Th	2-Th	84	165
5-CH₃-2-Th	5-H₃C-2-Th	60	165
3-H₃C-2-Th	3-H₃C-2-Th	94.6	165
3-Th	3-Th	100	165
Fc[e]	CH₂OH	70	166
Fc	C₆H₅	51	166
Fc	Fc	80	166
H₃CFc	CH₂OH	55	167
H₃CFc	C₆H₅	53	167
H₃CFc	C₆H₄OCH₃-p	30	167
H₃CFc	C₆H₄Cl-p	80	167
Fc-C₆H₄OCH₂-p	p-CH₂OC₆H₄-Fc	58	168
Fc-C₆H₄O—CH₂-p	CH₂OH	52	168
Fc-C₆H₄OCH₂-p	C₆H₅	59	168

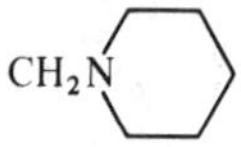

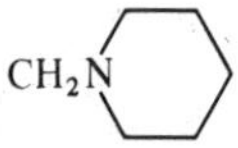

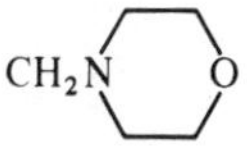

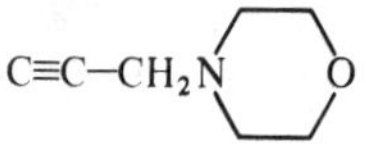

[a] Th means thienyl.
[b] Fu means furyl.
[c] Py means pyridyl.
[d] Glutathione was used instead of H₂S.
[e] Fc means ferrocenyl.

13

When **8** was heated with excess sodium hydroxide and hydrogen sulfide, the yield of **9** was increased to 50% and 7% of **10** was obtained, most probably by ring-closure of **11**[86] (Scheme 8). When tertiary alcohols such as **12** were reacted

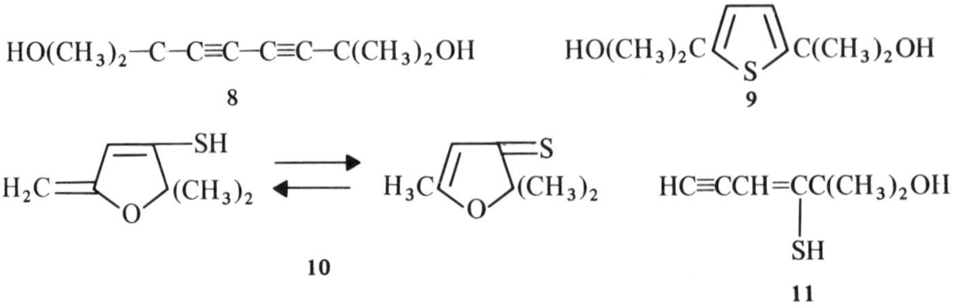
HO(CH3)2–C–C≡C–C≡C–C(CH3)2OH
8
HO(CH3)2C
S
C(CH3)2OH
9
SH
H2C
O
(CH3)2
H3C
O
S
(CH3)2
10
HC≡CCH=CC(CH3)2OH
SH
11

Scheme 8

with sodium methoxide in ethanol, dehydration occurred and the vinyl thiophenes (**13**) were obtained in 68% yield. Analogous results were obtained with the cyclohexyl derivative **14**, which gave **15** in 41% yield[87] (Scheme 9).

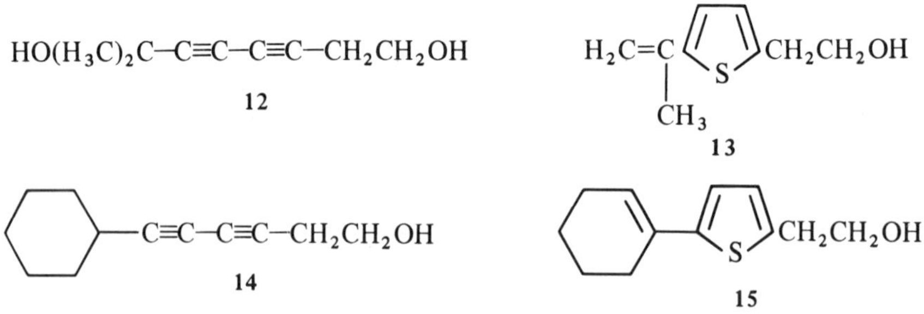
HO(H3C)2C–C≡C–C≡C–CH2CH2OH
12
H2C=C
CH3
S
CH2CH2OH
13
C≡C–C≡C–CH2CH2OH
14
S
CH2CH2OH
15

Scheme 9

Reaction of 5-chloro-3-penten-1-ynes with alkali sulfides or hydrosulfides at 30–100°C gave thiophene derivatives[88] (Scheme 10). The reaction of diphenyl-

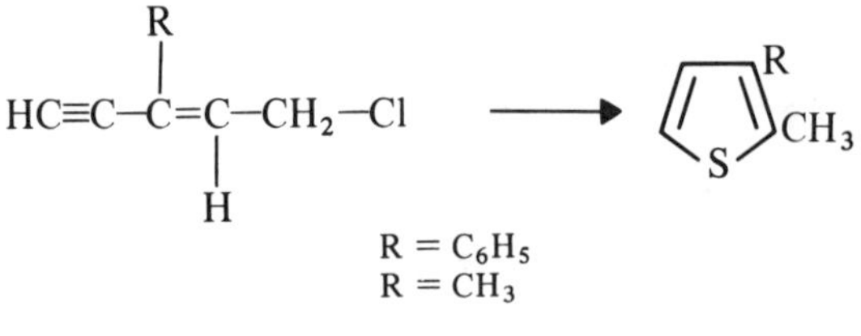
R
HC≡C–C=C–CH2–Cl
H
R
S
CH3
R = C6H5
R = CH3

Scheme 10

diacetylene with SCl_2 gave a 17% yield of 3,4-dichloro-2,5-diphenylacetylene.[88a] The reaction of vinyl acetylene with sodium hydrosulfide in aqueous DMSO gave 2,5-dihydrothiophene.[89] The reaction of vinyl acetylene with sulfur in an air–water–DMSO–KOH mixture gave thiophene as the main product.[89a] The reaction of 3-phenyl-butyn-1,3-ol with sulfur in ethyl benzoate at 210°C led to 3-phenylthiophene, and not to the expected dithiolethione.[90] The reaction of diarylbutadiynes with sulfur dichloride in benzene gave 3,4-dichlorothiophenes (Scheme 11).

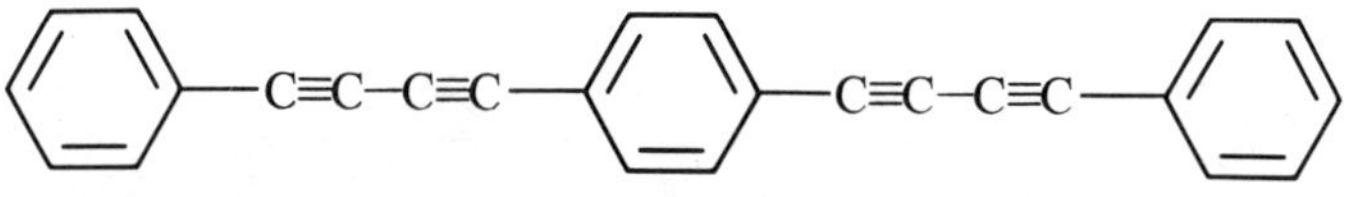

16

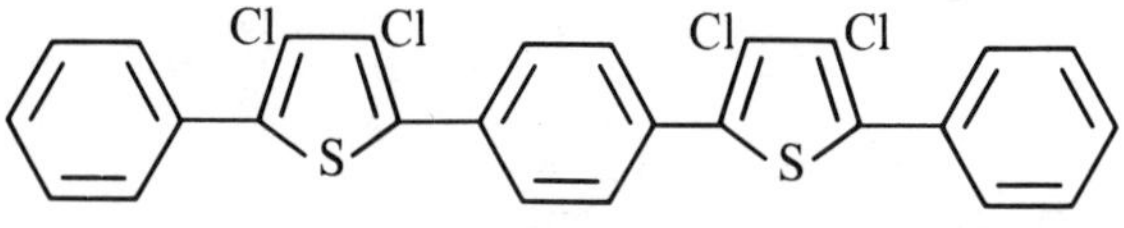

17

Scheme 11

From **16**, **17** was obtained in good yield.[91] Compounds prepared in this way are listed in Table 4.

TABLE 4.   THIOPHENES THROUGH THE REACTION BETWEEN DIACETYLENES AND SULFUR DICHLORIDE

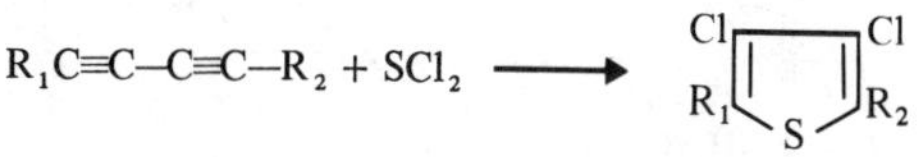

$$R_1C{\equiv}C{-}C{\equiv}C{-}R_2 + SCl_2 \longrightarrow$$

R_1	R_2	Yield (%)	Reference
C_6H_5	C_6H_5	80	91
$p\text{-}CH_3C_6H_4$	C_6H_5	43	91
$p\text{-}CH_3C_6H_4$	$p\text{-}CH_3C_6H_4$	55	91
$p\text{-}ClC_6H_4$	C_6H_5	16	91

The reaction of epoxides of acetylenes and vinylacetylenes with hydrogen sulfide in the presence of barium hydroxide at 50–80°C has been used as a general method for the preparation of alkyl, aryl, and vinyl thiophenes in yields between 30 and 75%[92-95] (see Table 5). It is believed that the reaction occurs via nucleophilic substitution at the epoxide ring, followed by dehydration and ring-closure[93] (Scheme 12). The reaction of adiponitrile with SCl_2 and NH_4Cl gave 3,4-dichloro-2,5-dicyanothiophene.[91a]

C.   Thiophenes from the Reactions of 1,4-Dicarbonyl- and Related Compounds with P_4S_{10}

The classical Paal-Knorr synthesis (Scheme 13) has been modified, by using tetralin as a solvent.[97] In some cases, the analogous furans were obtained as by-

TABLE 5. THIOPHENES THROUGH THE REACTION OF ACETYLENIC EPOXIDES WITH HYDROGEN SULFIDE IN THE PRESENCE OF BARIUM HYDROXIDE

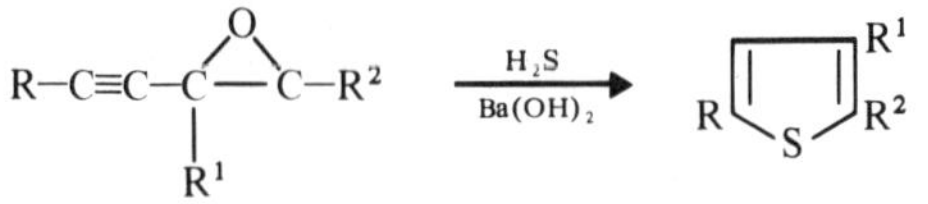

$$R-C\equiv C-\overset{\overset{\displaystyle O}{\diagdown}}{\underset{\underset{\displaystyle R^1}{|}}{C}}-C-R^2 \xrightarrow[\text{Ba(OH)}_2]{\text{H}_2\text{S}} R\underset{S}{\overset{R^1}{\boxed{}}}R^2$$

R	R¹	R²	Yield of Thiophene (%)	Reference
$\underset{}{\overset{\diagdown}{\diagup}}C=C\overset{\diagup H}{}$	CH₃	H	75	93, 92
$\underset{}{\overset{\diagdown}{\diagup}}C=C\overset{\diagup H}{}$	CH₃	CH₃		93, 92
CH₃	C₂H₅	H		93, 92
CH₃	CH₃	CH₃	73	93, 92
C₆H₅	CH₃	H		93
C₆H₅	C₂H₅	H		93
C₆H₅	CH₃	CH₃		93
(CH₃)₂CHOH	CH₃	H		93
cyclohexyl-OH	CH₃	H	91	95
cyclopentyl-OH	CH₃	H	51	95
H	C₂H₅CH[a]CH₃	H	50	96

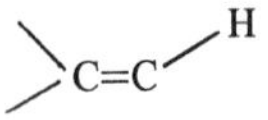

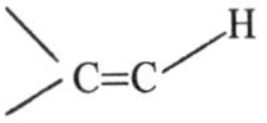

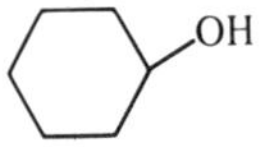

[a] Optical purity 51%.

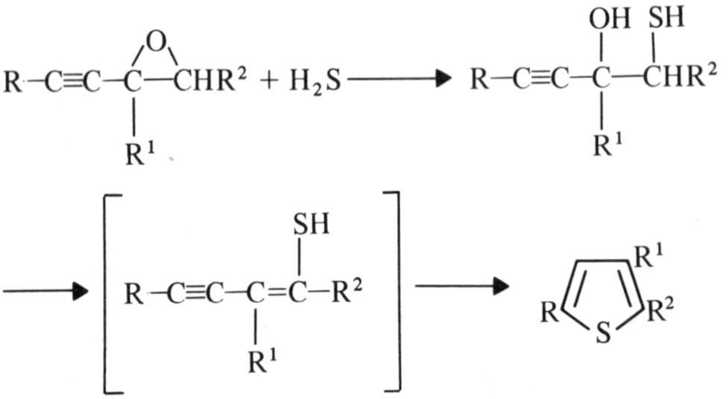

$$R-C\equiv C-\overset{\overset{\displaystyle O}{\diagdown}}{\underset{\underset{\displaystyle R^1}{|}}{C}}-CHR^2 + H_2S \longrightarrow R-C\equiv C-\overset{\overset{\displaystyle OH}{|}}{\underset{\underset{\displaystyle R^1}{|}}{C}}-\overset{\overset{\displaystyle SH}{|}}{C}HR^2$$

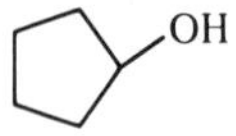

$$\longrightarrow \left[R-C\equiv C-\underset{\underset{\displaystyle R^1}{|}}{\overset{\overset{\displaystyle SH}{|}}{C}}=C-R^2 \right] \longrightarrow R\underset{S}{\overset{R^1}{\boxed{}}}R^2$$

Scheme 12

$$RCOCH_2CH_2COR^1 \xrightarrow{P_4S_{10}} R\underset{S}{\overset{}{\boxed{}}}R^1$$

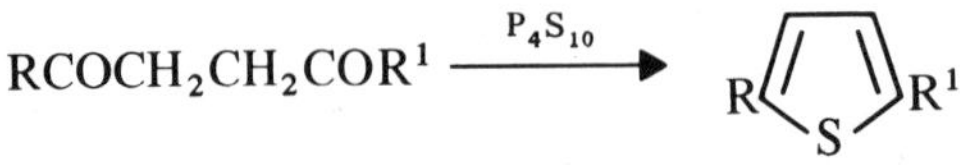

Scheme 13

products.[98,99] This reaction has also been applied to the synthesis of [8](2,5)-heterocyclophanes[100] (Scheme 14) and [n](2,4)-heterophanes.[101,102]

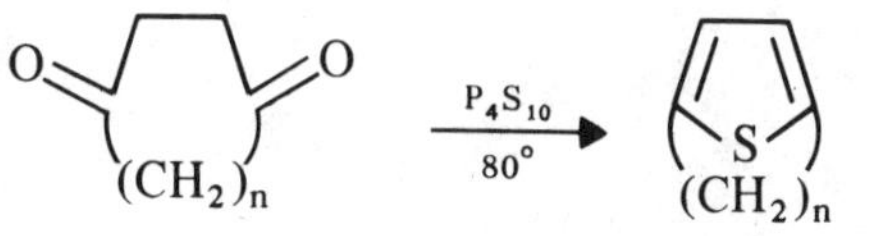

n	Yield (%)
8	51
11	37

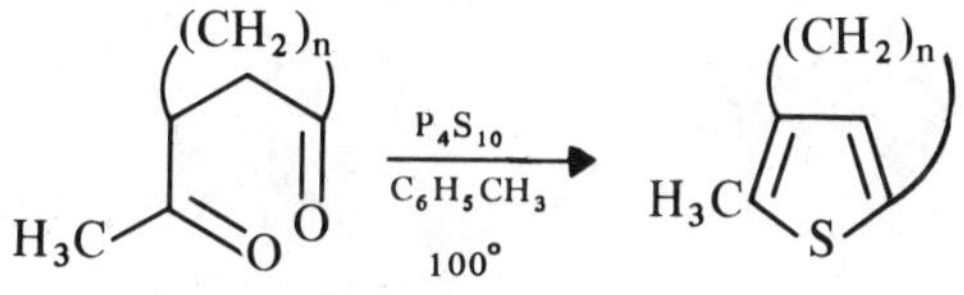

n	Yield (%)
6	61
7	51
9	70

Scheme 14

Strained systems[103] have also been prepared in this way, albeit in low yields (Scheme 15). The Paal–Knorr method has also been useful for the synthesis of nonclassical thienothiophenes[104,105] (Scheme 16). It is interesting to note that tetrabenzoylethane in xylene yields the *cis* and *trans* dihydro compounds, whereas

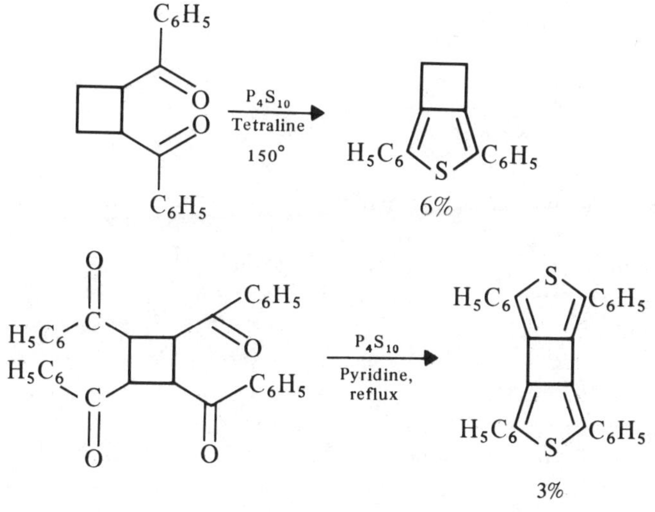

Scheme 15

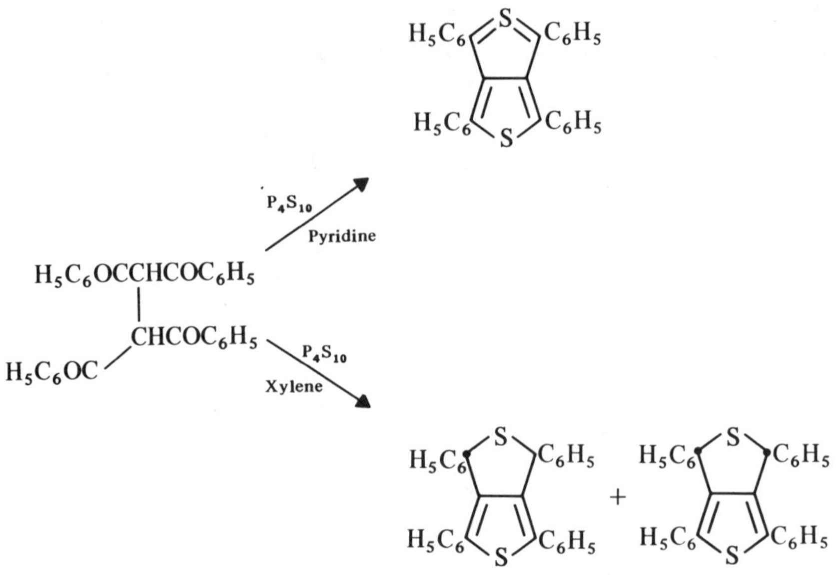

Scheme 16

in pyridine, the nonclassical compound was obtained.[106] Another example is the synthesis of the thienodiazole from 3,4-dibenzoyl 1,2,5-thiadiazole (53%)[107] (Scheme 17). More complex and functionalized 1,4-dicarbonyl compounds have

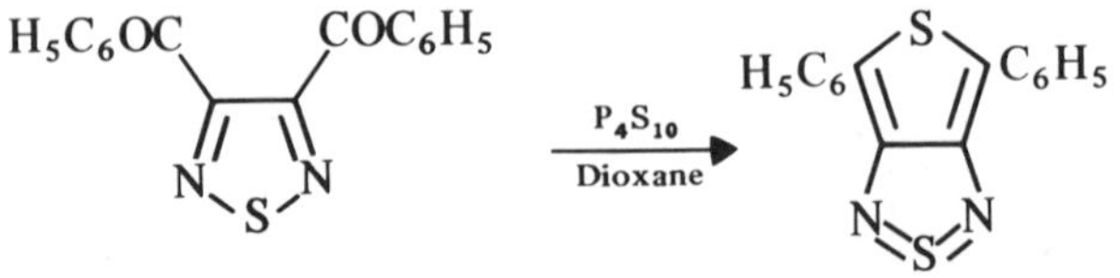

Scheme 17

also been used in the Paal–Knorr reactions, as for the synthesis of cyclopenta[b]-thiophenes,[108, 109] as shown in Scheme 18.

1,4-Dialdehydes or their acetals have successfully been used for the synthesis of 3,4-[110] and 2,4-thiophenedicarboxylic acid[111] (Scheme 19). However, when 2-ethoxalyl-4,4-diethoxybutyronitrile (**18**) was used in this reaction, only small amounts of 2,3-thiophenedicarboxylic acid were obtained and, instead, the isothiazole-fused compound (**19**) was the main product. However, the methyl homologue (**20**) gave 5-methyl-2,3-thiophenedicarboxylic acid in 24% yield.[112] In addition, 2,3,4- and 2,3,5-thiophenetricarboxylic acids have been obtained by this route[113] (Scheme 19). From **21**, ethyl 2-benzyl-4-thiophenecarboxylate has similarly been prepared[114] (Scheme 20). Minor amounts of carbothioate were also formed.

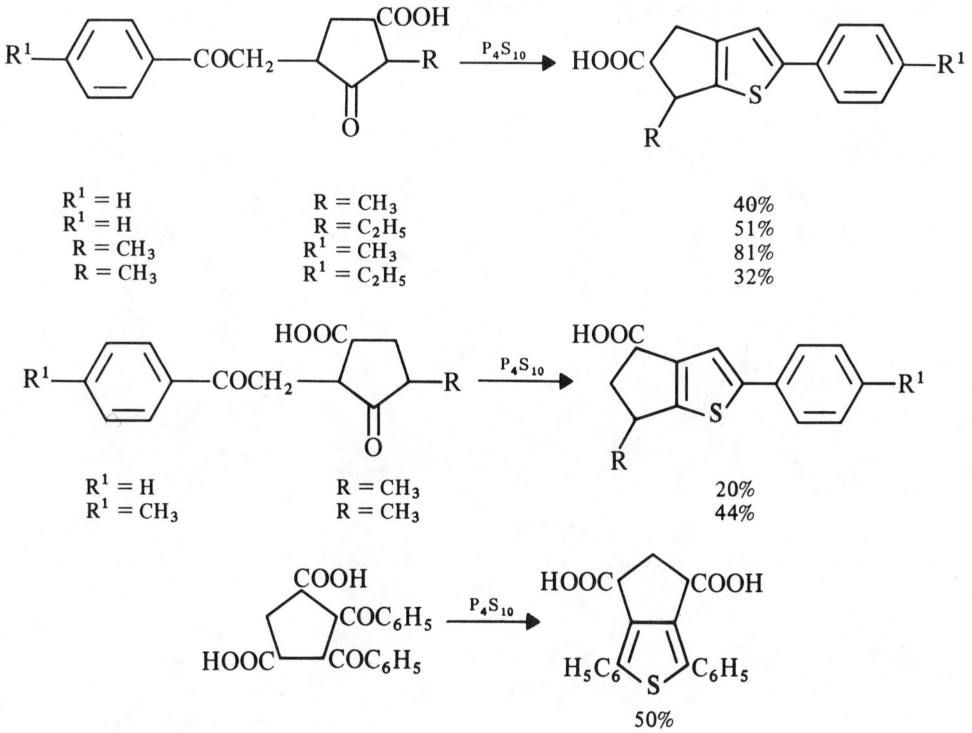

$R^1 = H$	$R = CH_3$	40%
$R^1 = H$	$R = C_2H_5$	51%
$R = CH_3$	$R^1 = CH_3$	81%
$R = CH_3$	$R^1 = C_2H_5$	32%

| $R^1 = H$ | $R = CH_3$ | 20% |
| $R^1 = CH_3$ | $R = CH_3$ | 44% |

50%

Scheme 18

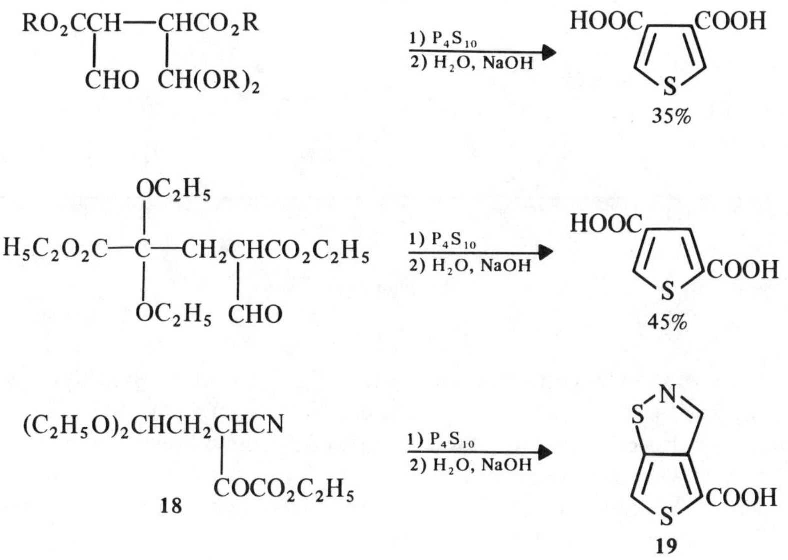

35%

45%

18

19

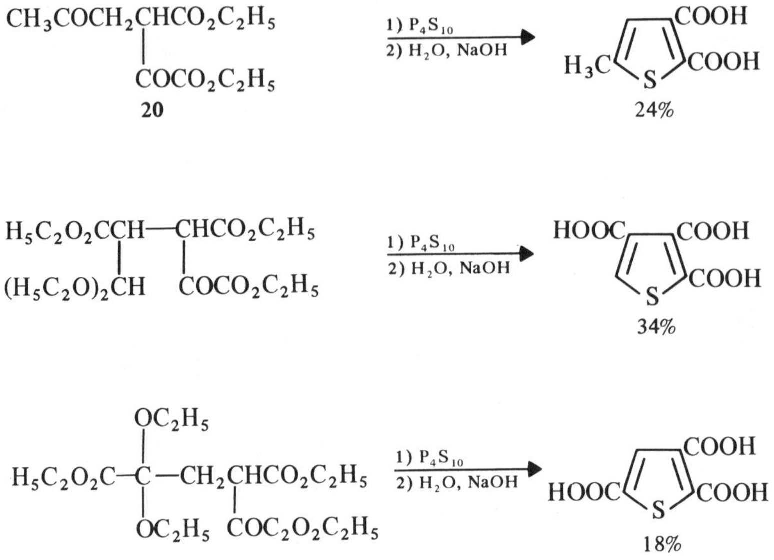

CH₃COCH₂CHCO₂C₂H₅ $\xrightarrow[\text{2) H}_2\text{O, NaOH}]{\text{1) P}_4\text{S}_{10}}$

$$\text{CH}_3\text{COCH}_2\overset{\displaystyle|}{\underset{\displaystyle COCO_2C_2H_5}{\text{CH}}}\text{CO}_2\text{C}_2\text{H}_5$$

20

24%

$$\text{H}_5\text{C}_2\text{O}_2\text{CCH}-\text{CHCO}_2\text{C}_2\text{H}_5$$
$$(\text{H}_5\text{C}_2\text{O})_2\text{CH} \quad \text{COCO}_2\text{C}_2\text{H}_5$$

$\xrightarrow[\text{2) H}_2\text{O, NaOH}]{\text{1) P}_4\text{S}_{10}}$

34%

$$\text{H}_5\text{C}_2\text{O}_2\text{C}-\overset{\displaystyle OC_2H_5}{\underset{\displaystyle OC_2H_5}{\text{C}}}-\text{CH}_2\overset{}{\text{CHCO}_2\text{C}_2\text{H}_5}$$
COC₂O₂C₂H₅

$\xrightarrow[\text{2) H}_2\text{O, NaOH}]{\text{1) P}_4\text{S}_{10}}$

18%

Scheme 19

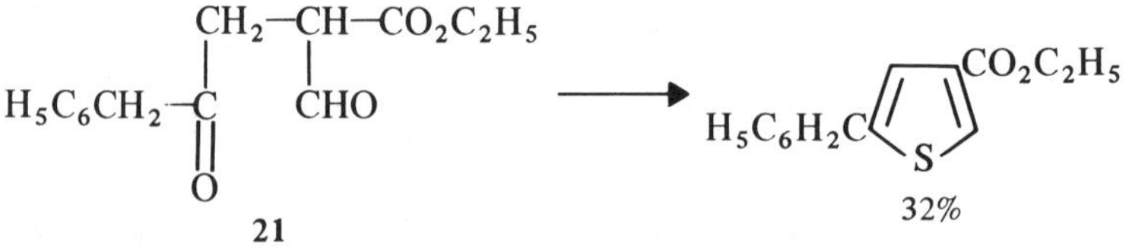

$$\text{H}_5\text{C}_6\text{CH}_2-\overset{}{\underset{\displaystyle O}{\text{C}}}\quad\overset{\displaystyle CH_2-CH-CO_2C_2H_5}{\underset{\displaystyle CHO}{}}$$

21

32%

Scheme 20

$$\text{R}_2\text{C}-\text{CH}_2'\quad\overset{\displaystyle R^1-C=O}{\underset{\displaystyle CH}{}}\text{CO}_2\text{C}_2\text{H}_5$$
O

$\xrightarrow[\text{Xylene}]{\text{P}_4\text{S}_{10}}$

22   **23**

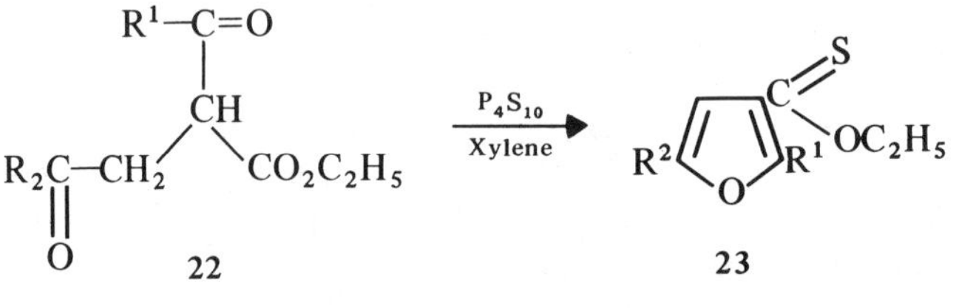

Scheme 21

The ester 1,4-diketone (**22**) reacted with P_4S_{10} in xylene to give 3-furancarbothioates (**23**)[115] in yields between 38 and 55% (Scheme 21). The Paal–Knorr reaction has also been used for the synthesis of 2,5-thiophenedi-β-propionic acid (**25**) from **24**, which was of interest for studying polymers[116] (Scheme 22). Recently, the yield of 2,5-disubstituted thiophenes from 1,4-diketones has been markedly increased by the use of Lawesson's reagents (**26**) (Table 6).

TABLE 6.   SOME SIMPLE ALKYL- AND ARYLTHIOPHENES THROUGH THE REACTION
OF 1,4-DICARBONYL COMPOUNDS WITH (a) P_4S_{10} AND (b) LAWESSON'S
REAGENT

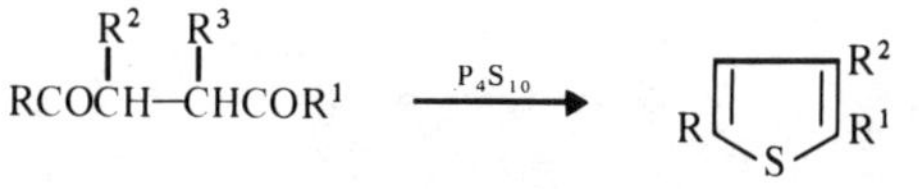

$$RCOCH\overset{R^2}{\underset{|}{|}}\text{—}CHCOR^1 \xrightarrow{P_4S_{10}} R\underset{S}{\overset{R^2}{\boxed{}}}R^1$$

Method	R	R^2	R^3	R^1	Yield (%)	Reference
a	H_3C	H	H	CH_3	58	125
					70	117
a	H_3C	H	H	$C(CH_3)_3$	53	169
a	$(H_3C)C_3$	H	H	$C(CH_3)_3$	84	170, 171
a	C_6H_5	H	H	C_6H_5	25	117
b	CH_3	H	H	CH_3	87	116a
b	$p\text{-}CH_3C_6H_4$	H	H	CH_3	86	116a
b	$p\text{-}CH_3OC_6H_4$	H	H	CH_3	90	116a
b	$p\text{-}BrC_6H_4$	H	H	CH_3	98	116a
b	C_6H_5	H	H	C_6H_5	80	116a
b	$p\text{-}CH_3C_6H_4$	H	H	$p\text{-}CH_3C_6H_4$	70	116a
b	C_6H_5	H	H	$p\text{-}CH_3OC_6H_4$	62	116a

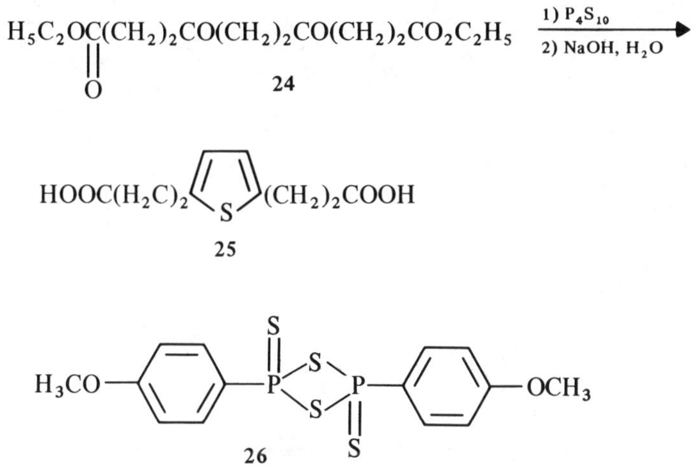

$$H_5C_2OC(CH_2)_2CO(CH_2)_2CO(CH_2)_2CO_2C_2H_5 \xrightarrow[\text{2) NaOH, } H_2O]{\text{1) } P_4S_{10}}$$

$$\underset{O}{\overset{||}{}}$$

24

25

26

Scheme 22

D.   Thiophenes from the Reactions of 1,4-Dicarbonyl- and Related Compounds with Hydrogen Sulfide and Hydrogen Chloride

A more convenient method than the Paal–Knorr reaction is, in many cases, the reaction of 1,4-dicarbonyl compounds with hydrogen sulfide and hydrogen chloride. Originally, in the synthesis of 2,5-diarylthiophenes, chloroform or benzene was used as solvent and a Lewis acid, such as anhydrous zinc chloride or stannic chloride, as catalyst.[117] However, it was found that the reaction is also fast in protic solvents, such as ethanol or methanol. From acetonylacetone, a 91% yield of 2,5-dimethylthiophene has been claimed.[118] A detailed investigation at temperatures about $-50°C$ showed that 2-mercapto-2,3-dihydrothiophenes and 2,5-dimercaptotetrahydrothiophenes were obtained as by-products, most probably formed via cyclizations of transiently formed *gem*-dithiols.[119, 120] Furans were also obtained, especially with phenyl-substituted 1,4-dicarbonyl derivatives, often as the main product. This was also true for the reaction of **27**, which gave **28** in 78–95% yield [121] (Scheme 23). The product distribution is shown in Table 7. The

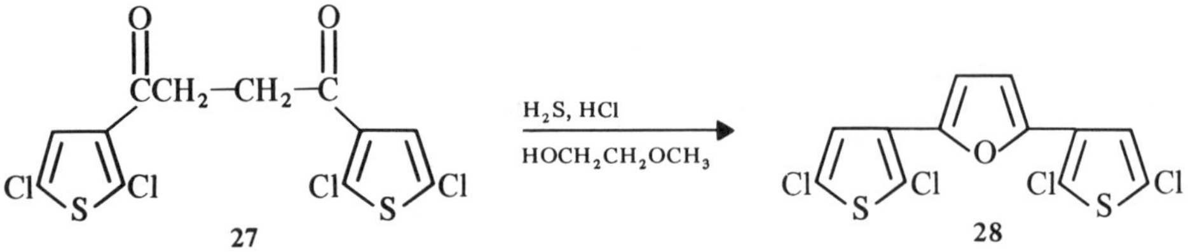

Scheme 23

TABLE 7.   PRODUCT COMPOSITION IN THE REACTION

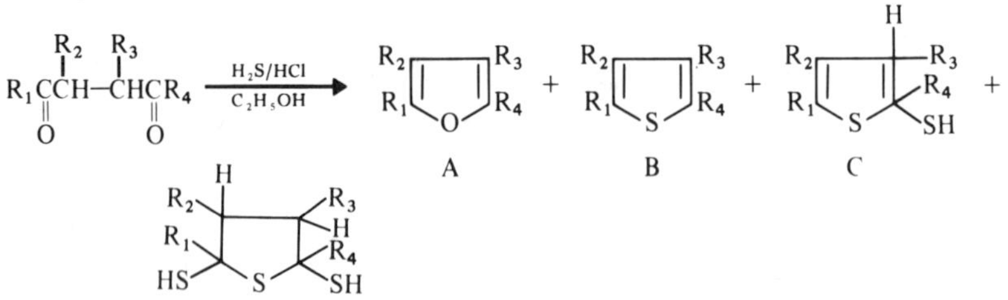

Substituents				Reaction Temperature ($°C$)	Yield of Products (%)			
R_1	R_2	R_3	R_4		A	B	C	D
CH_3	H	H	CH_3	-35	–	33	3	15
				< -50	–	49	0.9	13
$t\text{-}C_4H_9$	H	H	CH_3	-50	–	85	–	–
C_6H_5	H	H	CH_3	≈ -50	–	82	–	–
C_6H_5	H	H	C_6H_5	$-10 - +10^a$	53	29	–	–
				$-50 - 0^{a,b}$	33	53	–	–
CH_3	$CO_2C_2H_5$	H	CH_3	-40	–	53	20	–
CH_3	$CO_2C_2H_5$	CH_3	CH_3	-50	–	52	–	32
CH_3	$CO_2C_2H_5$	H	C_6H_5	-40	–	48	19	–
CH_3	$CO_2C_2H_5$	CH_3	C_6H_5	-40	–	73	–	–
C_6H_5	$C_2C_2H_5$	H	CH_3	-40	–	46	5	–
C_6H_5	$CO_2C_2H_5$	CH_3	CH_3	-40	–	48	12	–
C_6H_5	$CO_2C_2H_5$	H	C_6H_5	$-10 - +10^b$	80	–	–	–
C_6H_5	$CO_2C_2H_5$	CH_3	C_6H_5	$-10 - +10^b$	64	–	–	–

[a] The temperature rose to the maximum value when HCl was bubbled into the solution.
[b] The solvent was a mixture of C_2H_5OH and $CHCl_3$ (2:1).

TABLE 8.   2,5-DIARYLTHIOPHENES AND TERTHIENYLS THROUGH THE REACTION

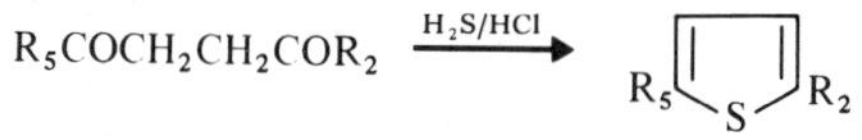

$$R_5COCH_2CH_2COR_2 \xrightarrow{H_2S/HCl} R_5 \!\! \diagdown\!\! \underset{S}{\boxed{}} \!\! \diagup\!\! R_2$$

Substituents				
R_5	R_2	Conditions	Yield (%)	Reference
C_6H_5	C_6H_5	$CHCl_3$, $ZnCl_2$	45	117
p-BrC_6H_4	p-BrC_6H_4	$CHCl_3$, $SnCl_4$	50, 70	117, 122
p-$CH_3OC_6H_4$	p-$CH_3OC_6H_4$	$CHCl_3$, $ZnCl_2$	60	117
2-Th	2-Th	CH_3OH, 50–55°	70	121
5-Cl-2-Th	5-Cl-2-Th	CH_3OH, 25°	90	121
2-Th	5-Cl-2-Th	CH_3OH, 25°	55	121
2-Cl-4-Th	2-Cl-4-Th	$C_2H_5OH/CHCl_3$	21	121

reaction is useful for the preparation of terthienyls[121] and diarylthiophenes,[117, 122] as listed in Table 8.

Some other examples of the reaction between 1,4-dicarbonyl compounds and H_2S/HCl are shown in Scheme 24. The hydroformylation of acetals of 2-substituted

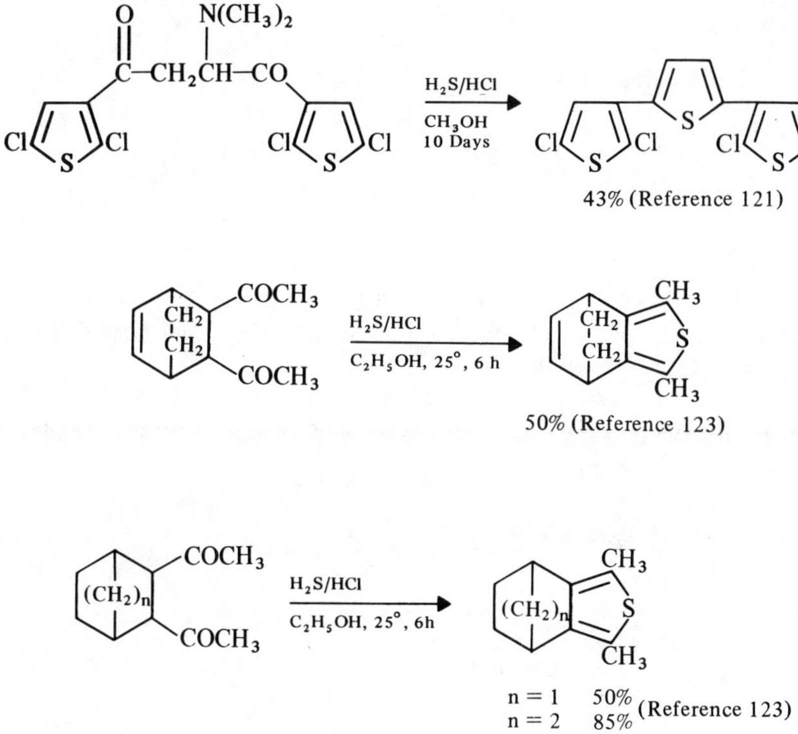

43% (Reference 121)

50% (Reference 123)

n = 1   50%
n = 2   85% (Reference 123)

Scheme 24

α,β-unsaturated aldehydes gives **29**, which are very useful starting materials for the synthesis of 3-alkylthiophenes (**30**)[124, 96] (Scheme 25). The yield of [11](2,5)-thiophenophane in the reaction of 1,4-cyclopentadecadione with H_2S/HCl was considerably lower than with P_2S_5.[126]

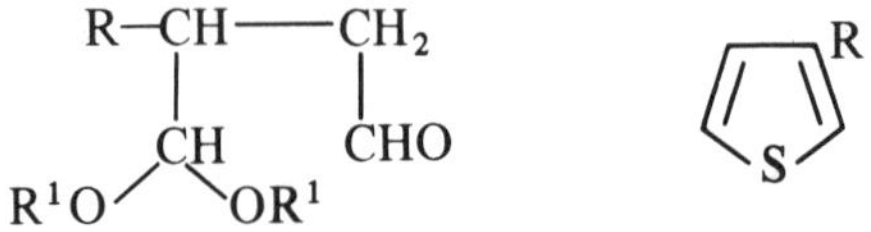

R = CHC$_2$H$_5$ 50–60%
              |
             CH$_3$
R = CH$_3$ 50–60%
R = C$_6$H$_5$ 50%

Scheme 25

The reaction of **31** with benzylmercaptan in dioxane/HCl gave 2,5-diphenyl-thiophene, probably via ring-closure of **32**[127] (Scheme 26).

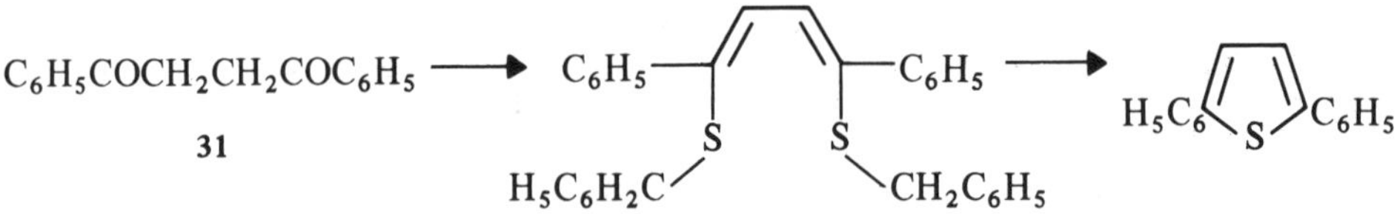

$C_6H_5COCH_2CH_2COC_6H_5$

31

Scheme 26

Tetraketones (**33**), readily available by Claisen condensation between diethyl oxalate and 2 moles of a methyl ketone, react with sulfur dichloride in $CHCl_3$ to give 2,5-diacyl-3,4-dihydroxythiophenes[128] (Scheme 27).

When acetonylacetone was reacted with H_2S over Al_2O_3 at 325°C, a 33% yield of 2,5-dimethylthiophene was obtained. The same reaction with γ-acetylpropyl alcohol gave 2-methyl-4,5-dihydrothiophene.[129]

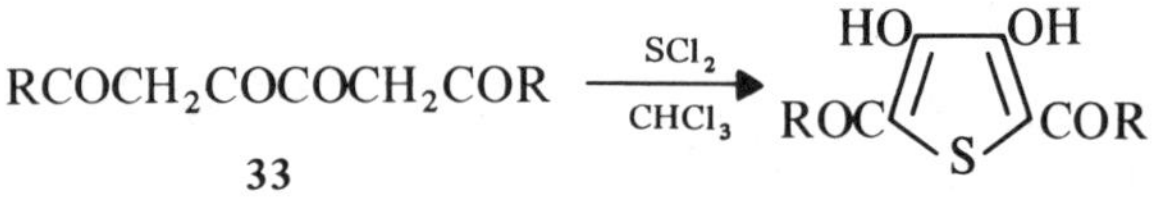

$$RCOCH_2COCOCH_2COR \xrightarrow[CHCl_3]{SCl_2}$$

33

R = C$_6$H$_5$ 33%
R = CH$_3$ 26%
R = CH$_2$CH$_2$CH(CH$_3$)$_2$ 20%
R = CH$_2$—C(CH$_3$)$_3$ 40%
R = C$_6$H$_5$; R = CH$_3$ 30%

Scheme 27

TABLE 9.   2-AMINOTHIOPHENES THROUGH THE REACTION OF β-KETONITRILES
WITH HCl/H$_2$S

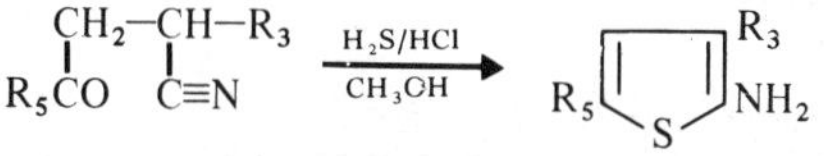

$$\underset{R_5CO \quad C\equiv N}{CH_2-CH-R_3} \xrightarrow[CH_3OH]{H_2S/HCl} R_5\underset{S}{\overset{R_3}{\bigcirc}}NH_2$$

R$_5$	R$_3$	Yield of Thiophene (%)	Melting Point (°C)	Reference
C$_6$H$_5$	H	95^a	Decomposing	129a
p-H$_3$CC$_6$H$_4$	H	89^a	Decomposing	129a
β-naphthyl	H	88^a	Decomposing	129a
CH$_3$	CO$_2$CH$_3$	75–82	116–117	129b
C$_6$H$_5$	CO$_2$CH$_3$	68–74	188–189	129b
C$_6$H$_5$	C$_6$H$_5$	77–85^a	192–195 dec.	129b
CH$_3$	C$_6$H$_5$	65–75^a	166–170	129b

^{a} Isolated as hydrochloride.

Some useful intermediates for the synthesis of 2-aminothiophenes are γ-ketonitriles, which are obtained by the reaction of Mannich bases from ketones with potassium cyanide, through the addition of HCN to α,β-unsaturated ketones, or through the reaction of α-haloketones with sodium ethyl cyanoacetate. If γ-ketonitriles are reacted at −10 to −5°C with anhydrous hydrogen sulfide and hydrogen chloride, 2-aminothiophenes are obtained in high yield.[129a,b] The compounds prepared in this way are given in Table 9.

E.   Thiophenes from γ-Keto Acids and Sulfurating Agents

The reaction of levulinic acid with P$_4$S$_{10}$, which yields the 5-methyl-2-hydroxy-thiophene system 34[130] has been reinvestigated. Depending upon the workup, the tautomeric form 35 or 36 can be obtained[131] (Scheme 28). The reaction gives

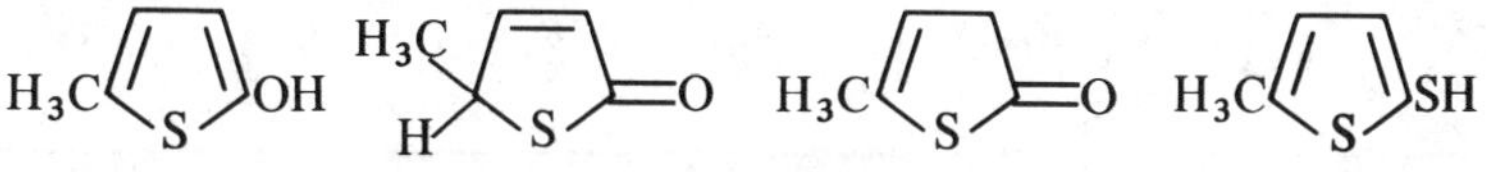

34   35   36   37

Scheme 28

irreproducible results and is very sensitive to the quality of P$_4$S$_{10}$ used. With some qualities, appreciable amounts of the thiol 37 were formed.[131] 2-Methylthiophene is also formed, which has been ascribed to the presence of P$_2$S$_3$ (P$_4$S$_7$) in the P$_4$S$_{10}$.[130] The reaction of P$_2$S$_3$ with levulinic acid, 2-methyllevulinic acid, and 3-methyllevulinic acid has been used for the synthesis of 2-methyl- (14%), 2,4-dimethyl- (34%) and 2,3-dimethylthiophene (3%).[132] The reaction of the sodium salt of optically active 38 with P$_4$S$_{10}$ gave 39 in 40% yield, together with 20% of an unidentified by-product,[133] and from the sodium salt of **40**, only 10% of **41** was obtained [134] (Scheme 29).

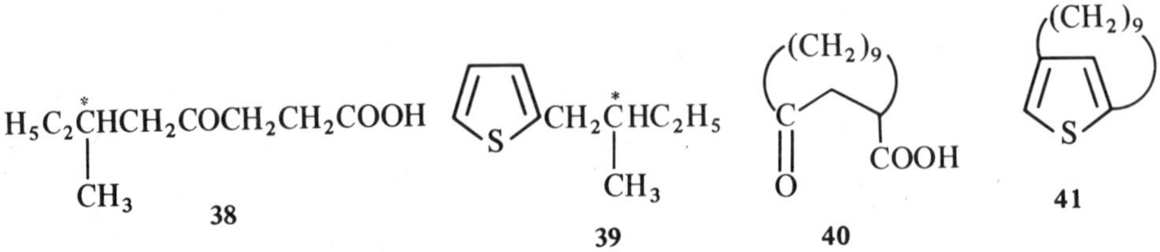

$$H_5C_2\overset{*}{C}HCH_2COCH_2CH_2COOH$$

CH$_3$ **38**

39

40

41

Ar—C(=O)—CH$_2$—H$_2$C—CO$_2$Et $\xrightarrow[\text{HCl}]{\text{H}_2\text{S}}$ Ar—[thiophene]—OC$_2$H$_5$

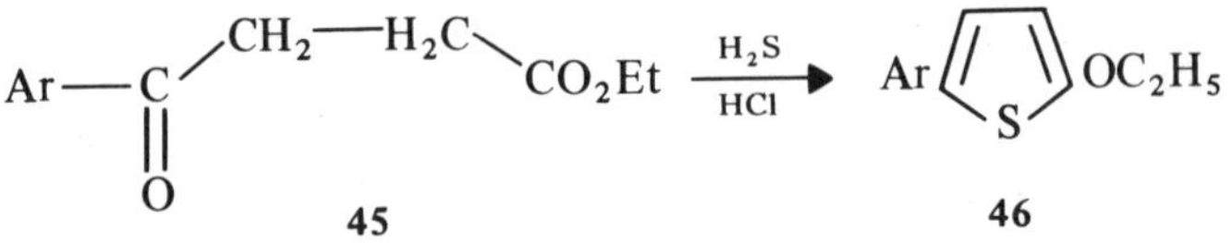

45 **46**

Ar = C$_6$H$_5$ 10%
Ar = p-ClC$_6$H$_4$ 20%
Ar = p-CH$_3$OC$_6$H$_4$ 15%
Ar = p-CH$_3$C$_6$H$_4$ 10%

Scheme 29

The reaction of aliphatic γ-oxo esters with H$_2$S/HCl gave a mixture of tautomers of the 5-alkyl-2-hydroxythiophene systems **42**, **43**, and 5-alkyl-5-mercaptothiolane-2-ones (**44**). The proportions were highly dependent on the bulkiness of the alkyl group [135] (see Table 10). Upon refluxing in pyridine, **44** could be transformed to a mixture of **42** and **43**, indicating that the γ-oxo ester is first converted to the *gem* dithiol or thioketone before ring-closure.[135] On the other hand, aromatic γ-oxo esters (**45**) gave 2-aryl-5-ethoxythiophenes (**46**) in the reaction with H$_2$S/HCl (Scheme 29).

TABLE 10.   PRODUCT DISTRIBUTION FROM THE REACTION OF γ-OXO ESTERS WITH H$_2$S/HCl

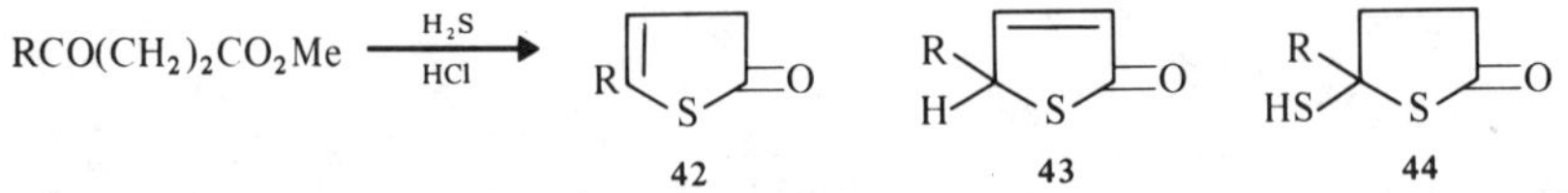

RCO(CH$_2$)$_2$CO$_2$Me $\xrightarrow[\text{HCl}]{\text{H}_2\text{S}}$

42   **43**   **44**

	Products (%)			
Esters	42	43	44	Yield (%)
R = CH$_3$	—	—	95	50
R = C$_2$H$_5$	9	18	73	45
R = C$_3$H$_7$	13	28	59	30
R = CH(CH$_3$)$_2$	25	75	0	30
R = C(CH$_3$)$_3$	23	70	0	25

The reaction of diethyl acetylsuccinate (47) with H_2S/HCl in ethanol led to a mixture of 48 ($\approx 60\%$), the thiophene derivatives (49) ($\approx 35\%$), and 50 ($\approx 5\%$), probably formed by cyclization of the nonobserved compounds 51. By modifying the experimental conditions, 49 could be isolated in 54% yield[136] (Scheme 30).

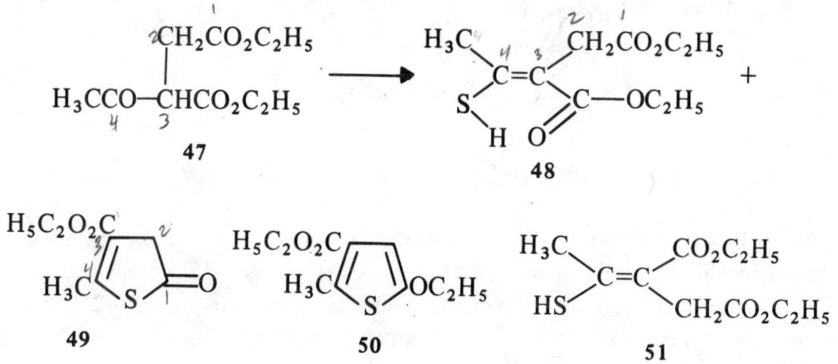

Scheme 30

F. Thiophenes from the Reaction of Succinic Acid Salts with P_4S_{10} and Other Sulfurating Agents

A modification of Victor Meyer's classical synthesis of thiophene from sodium succinate and P_4S_7 has been published[137] and patented.[170] Both $2\text{-}^{13}C\text{-}$ and $3\text{-}^{13}C$-enriched thiophenes have been prepared by this method from, appropriately, ^{13}C-enriched sodium succinate.[138] A detailed procedure for the synthesis of 3-methylthiophene in 52–60% yield from disodium methylsuccinate, using mineral oil as diluent, has also been published,[139] and 3-ethylthiophene was similarly obtained from disodium ethylsuccinate.[140] 3-Benzyl-3-α-methylbenzyl- and 3-α-dimethylbenzylthiophenes have been prepared from P_4S_7 and the sodium salts of the corresponding benzylsuccinic acids in 20–40% yield.[141] In connection with work on optically active (+)-S-3-sec.butylthiophene, the reaction of optically active sec.butylsuccinic acid with P_4S_{10} gave the desired compound in 40% yield and an optical purity of 43%.[96]

G. Thiophenes from Tetra- and Tricyanoethylenes and H_2S

The reaction of tetracyanoethylene (52) with hydrogen sulfide in the presence of pyridine yields 2,5-diamino-3,4-dicyanothiophene (53) in 92–95% yield.[172–175a] It has been shown that 53 is formed via tetracyanoethane (54) and can also be prepared by the reaction of 54 with sodium sulfide in water.[174] Substituted tetracyanoethanes can also be used and, from 55, a 64% yield of 56 was obtained.[176] Upon treatment with sodium hydroxide, 53 undergoes an interesting rearrangement to 2-amino-3,4-dicyano-5-mercaptopyrrole[174] (Scheme 31).

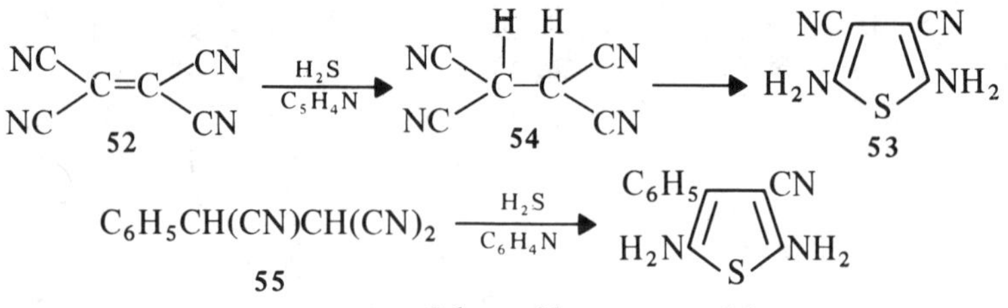

Scheme 31

H. Thiophenes Through the Willgerodt–Kindler Reaction

In 1965, four different research groups reported that aminothiophenes were formed in the reaction of 1-phenylbutanones,[177] 4-phenyl-3-butene-2-ones,[178,179] benzoyl acetone[180] and acetylphenylacetylene, and related compounds[181] with sulfur and morpholine at 100–135°C. Although 2-(4-morpholino)-5-phenylthiophene (**57**) was first obtained from the Willgerodt–Kindler reaction with benzalacetone in 1949, it was erroneously described as γ-phenylvinylthioacetomorpholide.[182] Similarly, 4-(4-morpholino)-2-phenylthiophene (**58**) was described as γ-phenylethynylthioacetomorpholide.[182] Depending upon the type of the four carbon units, 2- and/or 3-morpholine derivatives are obtained in various proportions (Table 11).

TABLE 11.   FORMATION OF 2-(4-MORPHOLINO-5-PHENYL)- AND 4-(MORPHOLINO-2-PHENYL)THIOPHENES IN THE WILLGERODT–KINDLER REACTION

Starting Material	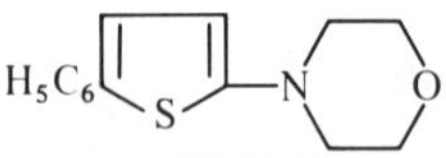 Yield (%)	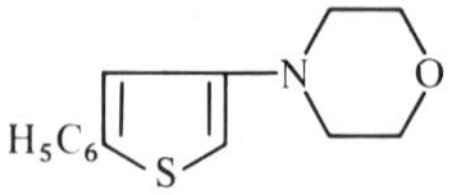 Yield (%)	Reference
$C_6H_5COCH_2CH_2CH_3$	a	–	177
$C_6H_5CH_2COCH_2CH_3$	a	–	177
$C_6H_5CH_2CH_2COCH_3$	a	–	177
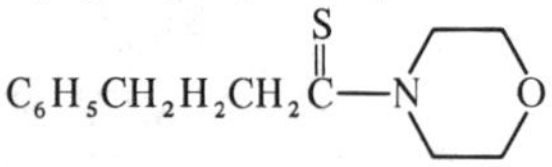 $C_6H_5CH_2H_2CH_2\overset{S}{\underset{}{C}}$—N O	a	–	177
$C_6H_5CH=CHCOCH_3$	~30	~5	181
$C_6H_5COCH_2COCH_3$	~30	~5	181
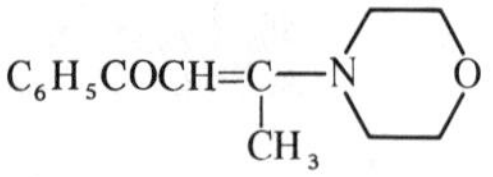 $C_6H_5COCH=C$—N O, CH_3	37	Traces	181, 183
$C_6H_5C≡C$—$COCH_3$	–	51	181
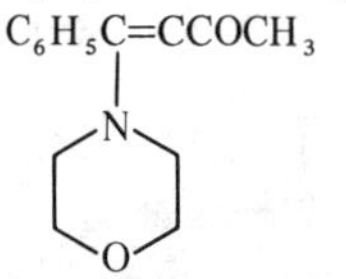 $C_6H_5C=CCOCH_3$	–	15	181[b], 183

[a] Only the 2-isomer, no yield given.
[b] Yield not given, 3-isomer main product containing 10% of 2-isomer.

The intermediate enamines also give aminothiophenes in 37% yield, and the reaction path in Scheme 32 has been suggested to explain the formation of the 2- and

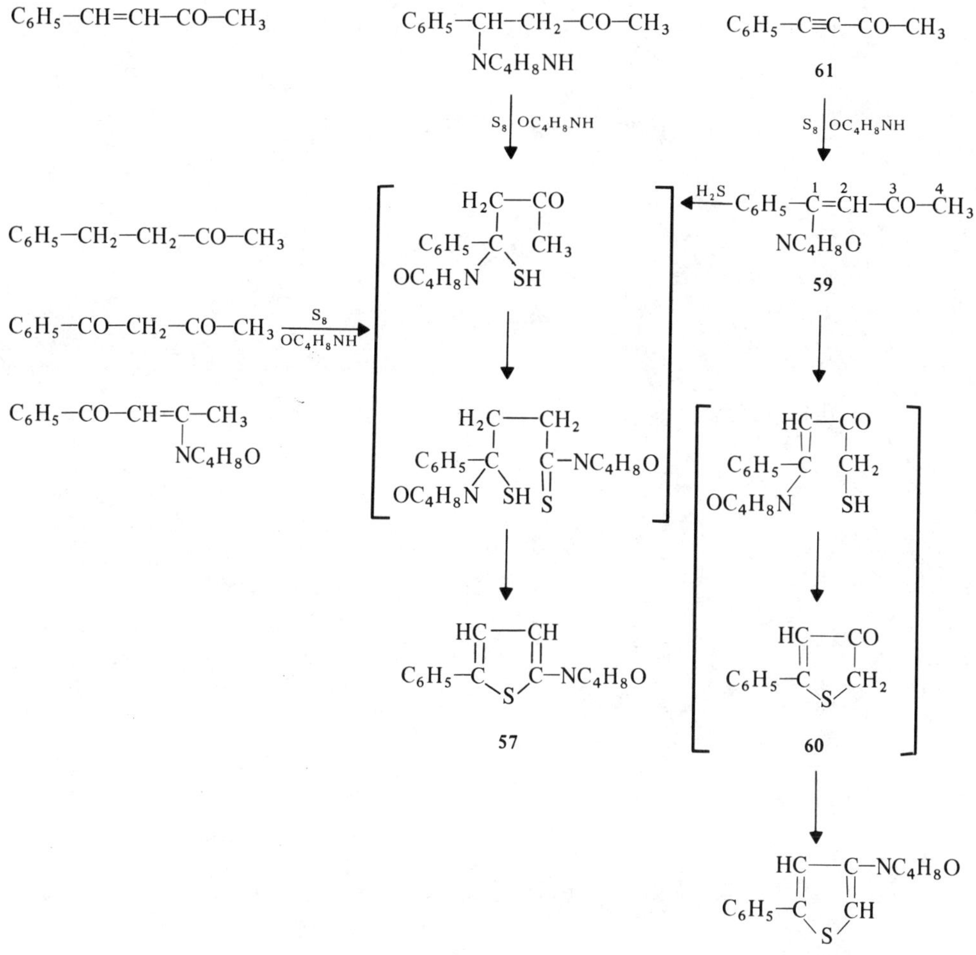

$C_6H_5-CH=CH-CO-CH_3$

$C_6H_5-CH-CH_2-CO-CH_3$
$\qquad\quad |$
$\qquad\quad NC_4H_8NH$

$C_6H_5-C\equiv C-CO-CH_3$
61

Scheme 32

3-morpholino intermediates.[183] The 3-derivative is assumed to be formed via **59** and the 5-phenyl-3-hydroxythiophene system (**60**). However, when the reaction of **59** or **61** is carried out in the presence of hydrogen sulfide to ensure a large excess of nucleophilic sulfur species from the beginning of the reaction, **57** is obtained as the main product, in addition to **58**.[183] From **62**, a 9% yield of **63** was obtained,[178] whereas **64** and **65** gave **66** in 5 and 24% yield, respectively[181] (Scheme 33). The reaction of **67** with sulfur and morpholine at 145°C gave **68** and **69** in 11 and 8% yield, respectively. The reaction of **70** with H_2S/HCl gave authentic **68** in 26% yield; its reaction with P_4S_{10} gave **69** in 31% yield[184] (Scheme 34).

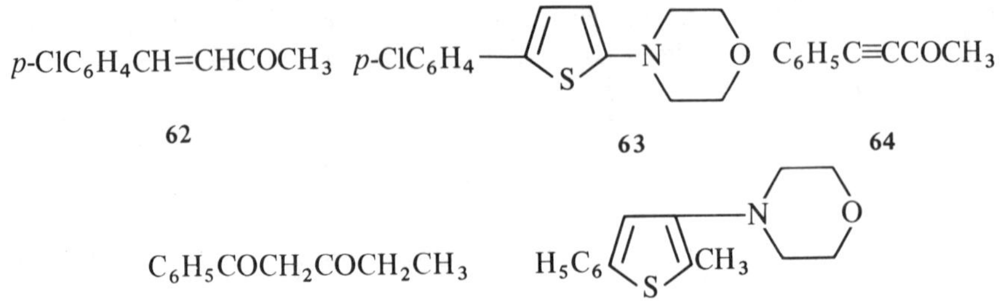

p-ClC$_6$H$_4$CH=CHCOCH$_3$ p-ClC$_6$H$_4$—[thiophene ring with S]—N[morpholine]O   C$_6$H$_5$C≡CCOCH$_3$

62   63   64

C$_6$H$_5$COCH$_2$COCH$_2$CH$_3$ H$_5$C$_6$[thiophene ring with S]CH$_3$ —N[morpholine]O

65   66

Scheme 33

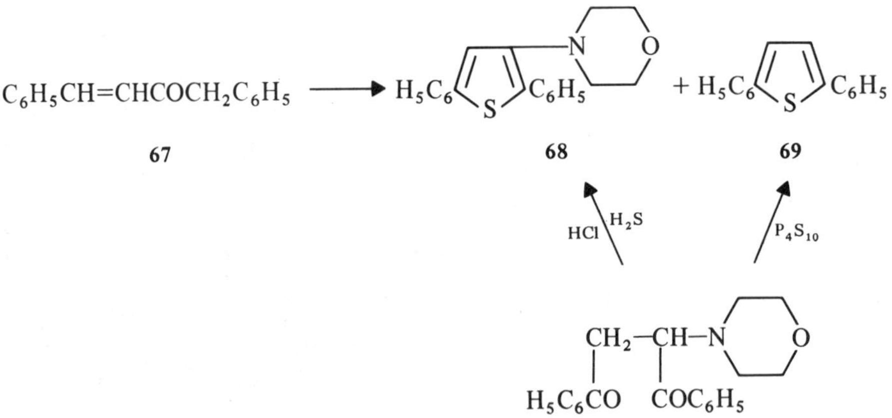

C$_6$H$_5$CH=CHCOCH$_2$C$_6$H$_5$ ⟶   H$_5$C$_6$[thiophene]C$_6$H$_5$ —N[morpholine]O   + H$_5$C$_6$[thiophene]C$_6$H$_5$

67   68   69

HCl ↖ $\diagdown$ H$_2$S $\diagup$ P$_4$S$_{10}$

CH$_2$—CH—N[morpholine]O

H$_5$C$_6$CO   COC$_6$H$_5$

70

Scheme 34

I.   Thiophenes by Various C$_4$ + S Methods

2,5-Thiophenedicarboxylic acid has been obtained through the reaction of the sodium salt of α,α'-dichloroadipic acid with Na$_2$S, followed by chlorination and dehydrochlorination.[142–145] Reaction of dimethyl α,α'-dichloroadipate with sulfur in benzotrichloride gave dimethyl 2,5-thiophenedicarboxylate.[146] Reaction of adipic acid (71) and related compounds 72 and 73 with thionyl chloride in pyridine gave the acid chlorides 74–76 in 63, 11, and 16% yield, respectively.[147] The follow-ing reaction sequence has been suggested for adipic acid[148] (Scheme 36).

Compounds of type 77, which are easily obtained by the alkylation of β-dicarbonyl compounds with propargyl bromides, react with H$_2$S/HCl in ethanol to give thiophenes (78)[149, 150] (Table 12). In a few cases, appreciable amounts of the corre-sponding furan were also formed. It is interesting to note that 79 reacts regio-specifically to give 80 and not 81 (Scheme 37). Compound 82 gives a 69% yield of

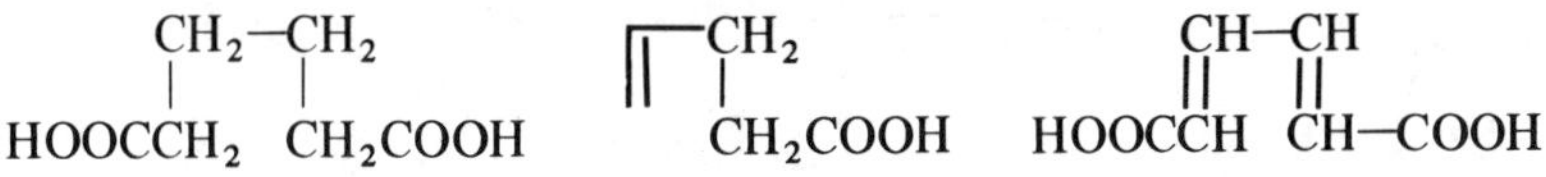

71 72 73

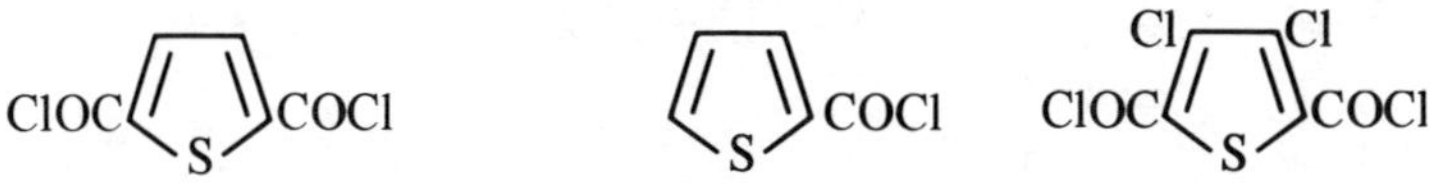

74 75 76

Scheme 35

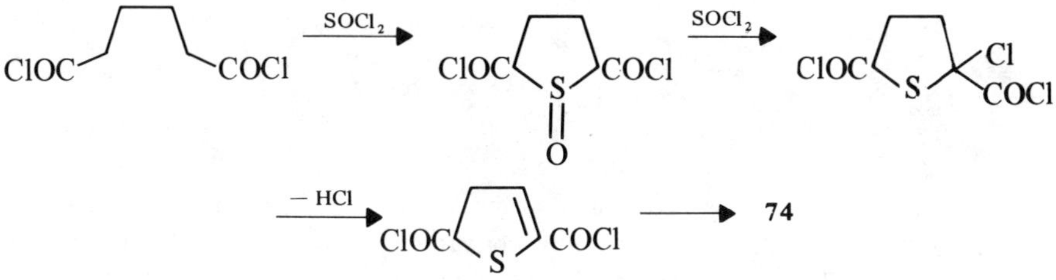

Scheme 36

TABLE 12. THE REACTION OF 1-ALKYN-5-ONES WITH H_2S/HCl[149,150]

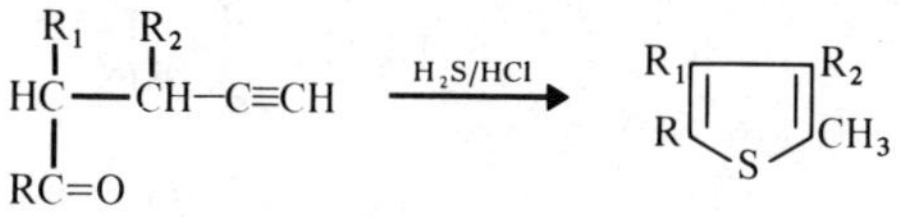

77 78

Substituents			Yield (%)
R	R_1	R_2	
CH_3	$CO_2C_2H_5$	H	69
n-C_3H_7	$CO_2C_2H_5$	H	80
C_6H_5	$CO_2C_2H_5$	H	65
2-Th	$CO_2C_2H_5$	H	2[a]
$(C_2H_5)_2CH$	$CO_2C_2H_5$	H	6[b]
CH_3	$CO_2C_2H_5$	C_6H_5	23[c]
CH_3	H	H	11
C_6H_5	H	H	71
2-Th	H	H	55
CH_3	C_6H_5	H	50[d]
CH_3	$COCH_3$	H	17
CH_3	COC_6H_5	H	6.5
C_6H_5	COC_6H_5	H	–[e]

[a] 70% of the corresponding furan was obtained.
[b] 92% of the corresponding furan was obtained.
[c] 7% of the corresponding furan was obtained.
[d] 5% of the corresponding furan was obtained.
[e] Polymerization.

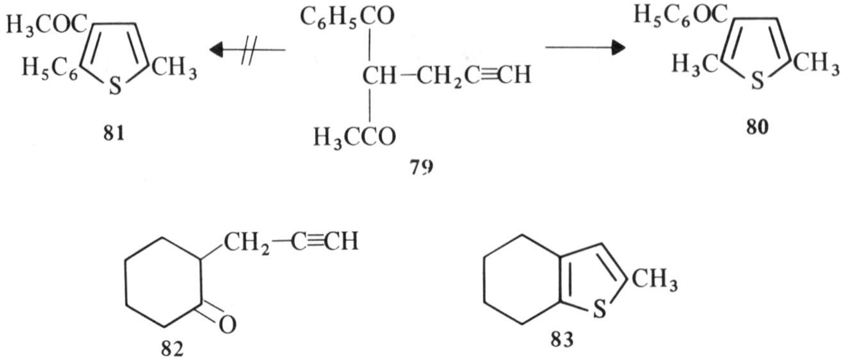

81 79 80 82 83

Scheme 37

83 with H_2S/HCl, whereas P_4S_{10} only yields 5%, together with 10% of the corresponding furan derivative.[150] Immonium derivatives such as **84–86** also react with H_2S to give **87**, **83**, and **88**, respectively (Scheme 38).

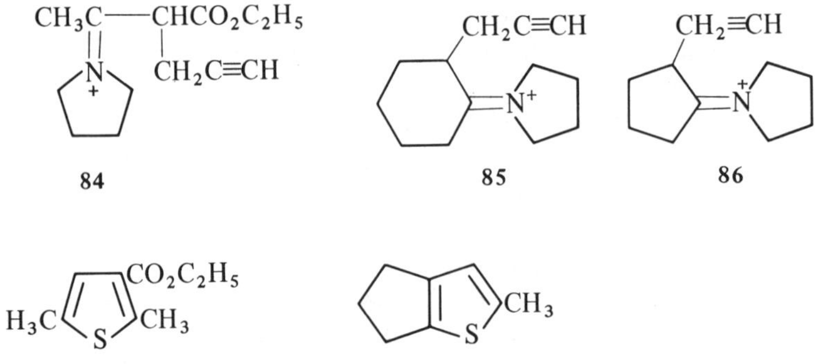

84 85 86 87 88

Scheme 38

The reaction of pentyn-4-carboxylic acid with P_4S_{10} yielded a mixture of 2-methylthiophene and **36**. The latter compound was also obtained through the reaction of the acid chloride of pentyn-4-carboxylic acid with KSH and H_2S in pyridine.[151]

β,γ-Dichloroketones, which can be prepared by the electrophilic addition of certain acid chlorides to allyl or methallyl chlorides, give 2- and 2,4-dialkylthiophenes upon reaction with P_4S_{10} in DMF, or dioxane,[152] or with K_2S/H_2S.[153] The key step after thionation is nucleophilic attack of the sulfur atom on the γ-carbon (Scheme 39), (Table 13).

The reaction of **89** with sodium sulfide gives **90**,[154] and **91** reacts with sodium hydrosulfide to give **92** (R = 1-methyl-2-benzimidazolyl) in 67% yield and

TABLE 13.   THE REACTION BETWEEN β,γ-DICHLOROKETONES AND SULFUR
           REAGENTS

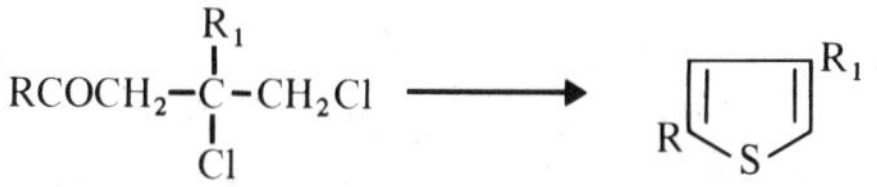

$$RCOCH_2-\overset{\overset{\displaystyle R_1}{|}}{\underset{\underset{\displaystyle Cl}{|}}{C}}-CH_2Cl \longrightarrow$$

Compounds		Yield (%)	Reference
R	R_1		
CH_3	H	53	153
C_2H_5	H	a	153
C_6H_5	H	a	153
Cyclohexyl	H	68	152
Cyclopentyl	H	70	152
1-Cl-Cyclohexyl	H	71	152
2-Cl-Cyclopentyl	H	64	152
4-Cl-Cyclohexyl	H	69	152
Cyclohexyl	CH_3	77	152
Cyclopentyl	CH_3	74	152
2-Cl-Cyclohexyl	CH_3	70	152
2-Cl-Cyclopentyl	CH_3	63	152
4-Cl-Cyclohexyl	CH_3	71	152

a Not stated in Chemical Abstract.

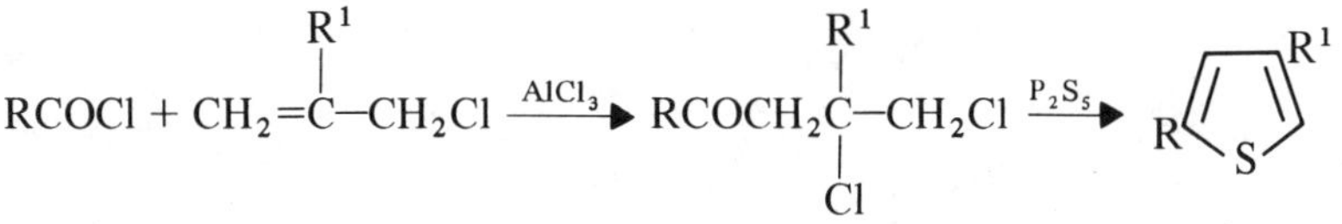

$$RCOCl + CH_2=\overset{\overset{\displaystyle R^1}{|}}{C}-CH_2Cl \xrightarrow{AlCl_3} RCOCH_2\overset{\overset{\displaystyle R^1}{|}}{\underset{\underset{\displaystyle Cl}{|}}{C}}-CH_2Cl \xrightarrow{P_2S_5}$$

Scheme 39

(R = 2-pyridyl) in 73% yield[155] (Scheme 40). The classical Benary reaction,[155a] consisting of the reaction of 2-chloroacetyl 3-aminocrotonate with potassium hydrosulfide, is useful for the synthesis of 3-hydroxy-4-carboethoxythiophenes[155b, 317]

Tetraphenylthiophene has been prepared by the reaction of dilithium tetraphenylbutadiene with SCl_2,[156] and in 90% yield by the reaction of 1,4-diiodotetraphenylbutadiene with lithium sulfide.[157] Dialkyl esters of **93** react with alkylsulfenyl

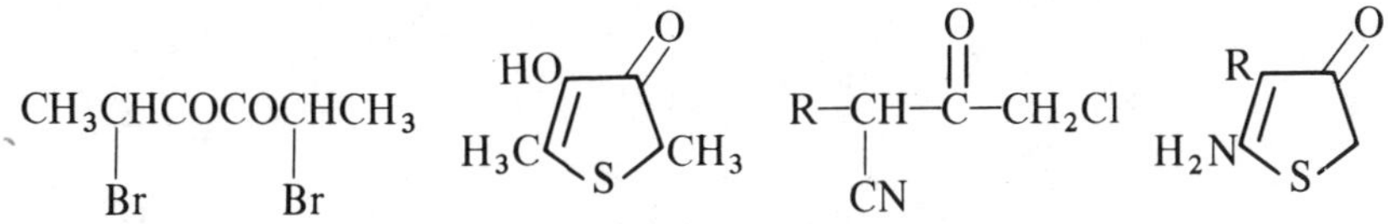

$$CH_3\underset{\underset{\displaystyle Br}{|}}{C}HCOCOCH\underset{\underset{\displaystyle Br}{|}}{C}H_3$$

89	**90**	**91**	**92**

$$R-\underset{\underset{\displaystyle CN}{|}}{C}H-\overset{\overset{\displaystyle O}{\|}}{C}-CH_2Cl$$

Scheme 40

chlorides at $-12-8°C$ in inert solvents to give dialkyl esters of thenylphosphonic acid **94**.[158] The reaction path is shown in Scheme 41. Reaction of the isomeric **95** with alkylsulfenyl chlorides leads to a mixture of **96** and **97**. Lower reaction temperatures favored the formation of **96**[158a] (Scheme 41). The reaction of 1,4-dicyanobutane with sulfur dichloride and ammonium chloride yielded 3,4-dichloro-2,5-dicyanothiophene.[158b]

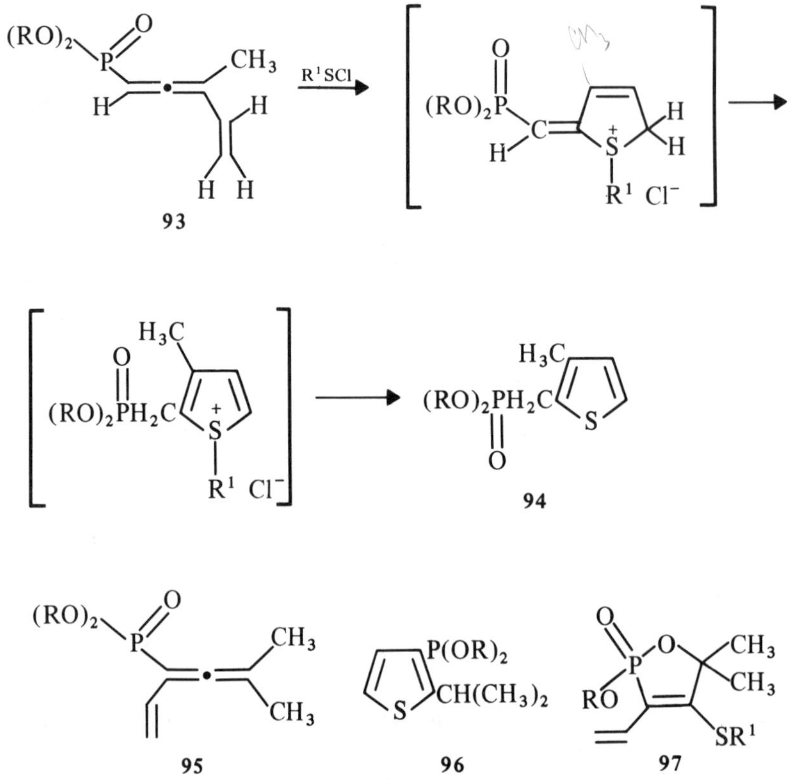

Scheme 41

2. C₂ + CSC Methods

A. The Hinsberg Reaction

The Hinsberg reaction[185] between diethyl thiodiacetate **98** and α-diketones (**99**) occurring under Claisen-type conditions, is very useful for the syntheses of 3,4-dialkyl- and 3,4-diarylthiophene-2,5-dicarboxylic acids, which give 3,4-dialkyl- and 3,4-diarylthiophenes upon decarboxylation (Table 14). By the use of conventional

TABLE 14. PREPARATION OF THIOPHENES BY THE HINSBERG REACTION

$$R_1COCOR_2 + R'O_2CCH_2SCH_2CO_2R' \longrightarrow$$

(thiophene product: $HOOC$–ring(S)–$COOH$ with R_1, R_2 substituents)

| Substituents | | | | Yield of Thiophenes | |
R₁	R₂	R'	Conditions	(%)	Reference
CH_3	C_2H_5	C_2H_5	$NaOC_2H_5/C_2H_5OH$ 0–5°	20	186
C_6H_5	H	C_2H_5	$NaOC_2H_5/C_2H_5OH$ 0–5°	57	187
C_6H_5	$p\text{-}H_{25}C_{12}C_6H_4$	C_2H_5	$NaOCH_3/CH_3OH$ C_6H_6, 25°	76	188
CH_3	CH_3	CH_3	$NaH/DMSO$ C_6H_6, 5°	43, 50[a]	189, 190
5-Cl-2Th	2,5-diCH₃=3Th	CH_3	CH_3CH_2ONa/C_2H_5OH 25°	77	121
$p\text{-}CH_3OC_6H_4$	$p\text{-}CH_3OC_6H_4$	CH_3	$t\text{-}BuOK/t\text{-}BuOH$ 60°	49[a]	191
$p\text{-}CH_3OC_6H_4$	$p\text{-}CH_3OC_6H_4$	C_2H_5	$t\text{-}BuOK/t\text{-}BuOH$ 60°	54[a]	191
C_6H_5	C_6H_5	$t\text{-}C_4H_9$	$t\text{-}BuOK/t\text{-}BuOH$ 60°	45[a]	191
$p\text{-}Cl\text{-}C_6H_4$	$p\text{-}Cl\text{-}C_6H_4$	C_2H_5	C_2H_5ONa/C_2H_5OH 20°	59	192
OC_4H_9	H	CH_3	$NaOCH_3/CH_3OH$	28[b]	193

[a] The monomethyl ester was obtained.
[b] The 3-hydroxythiophenes are obtained.

methods, as well as the ^{18}O-isotope technique, it has been shown that the Hinsberg reaction is a Stobbe-type condensation proceeding via a δ-lactone intermediate (100)[194] (Scheme 42). This offers the possibility of decarboxylating the initially

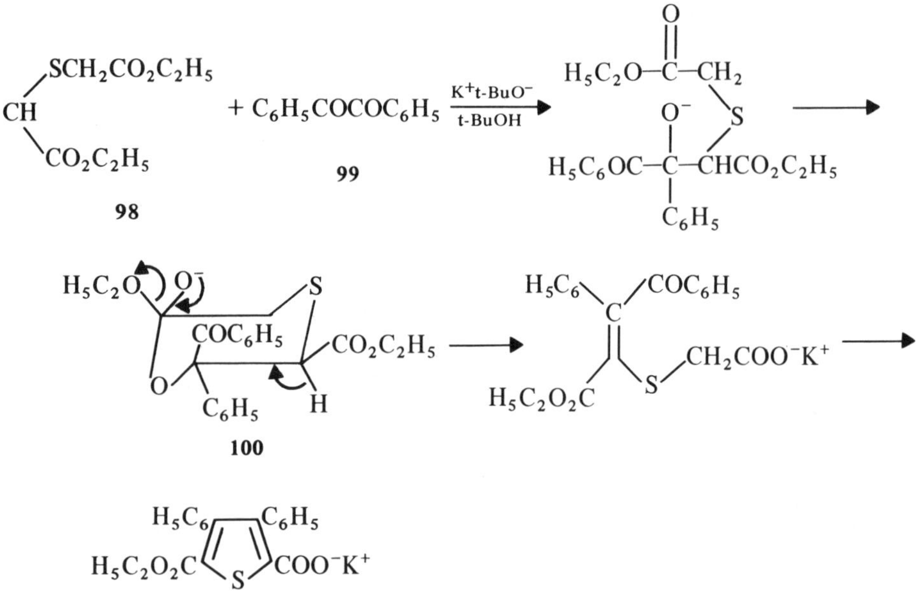

Scheme 42

obtained monoesters, leading to 3,4-substituted 2-thiophenecarboxylic acids. Phenyl glyoxal has also been used in the Hinsberg reaction for the synthesis of 3-phenyl-2,5-thiophenedicarboxylic acid.[187] 3,4-Bridged thiophene compounds (102) have been prepared from 101 in 36% yield[195] and from phenanthrene quinone; 103 was obtained in 33% yield[191] (Scheme 43).

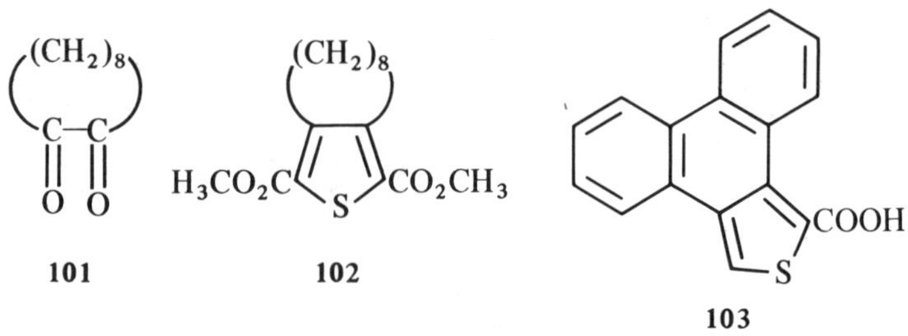

Scheme 43

The alkoxide-catalyzed condensation of ethyl phenylglyoxalate and methyl pyruvate (104) with thiodiacetate gives the 3-phenyl- and 3-methyl-4-hydroxy-thiophenedicarboxylates (105), indicating a normal Dieckmann-type mechanism and not a Stobbe-type condensation[196] (Scheme 44).

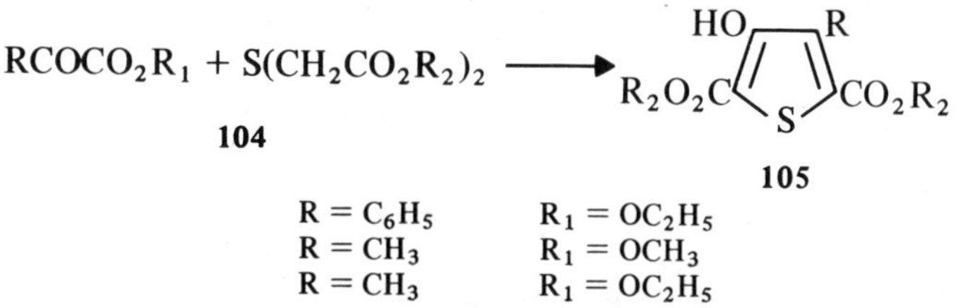

$$RCOCO_2R_1 + S(CH_2CO_2R_2)_2 \longrightarrow$$

104

[Structure **105**: thiophene ring with HO and R at top, R_2O_2C and CO_2R_2 on the ring, S at bottom]

105

R = C$_6$H$_5$	R$_1$ = OC$_2$H$_5$
R = CH$_3$	R$_1$ = OCH$_3$
R = CH$_3$	R$_1$ = OC$_2$H$_5$

Scheme 44

If ethyl oxalate is used as the α-dicarbonyl equivalent, diethyl 3,4-dihydroxy-2,5-thiophenedicarboxylate is obtained in 78% yield.[197, 198] Ethyl sulfinyl diacetate and ethyl sulfonyl diacetate have also been used in the Hinsberg-type condensation with ethyl oxalate to give the analogous thiophene-1-oxide and 1,1-dioxide, respectively.[198, 199]

The synthetic usefulness of the Hinsberg reaction has recently been increased by its application to diketosulfides, (RCOCH$_2$)$_2$S, leading to a facile synthesis of 2,5-diacylthiophenes.[200, 201] This method was first discovered in Fiesselmann's Laboratory in Erlangen, and was described in the Ph.D. theses of Erich Rose,[193] Siegfried Kroll,[193a] and Hugo Schädler.[193b]

The sodium methoxide-catalyzed condensation of glyoxal or phenylglyoxal[201] with various diketosulfides (**106**) gives 2,5-diacylthiophenes (**107**) directly. Biacetyl and benzil give 2,5-diacyltetrahydrothiophene-3,4-diols (**108**) in 80–90% yield under these conditions. The diols are cleanly converted in about 90% yield to the corresponding 2,5-diacylthiophenes upon treatment with thionyl chloride/pyridine[201] (Scheme 45). The diacylthiophenes prepared are given in Table 15. This

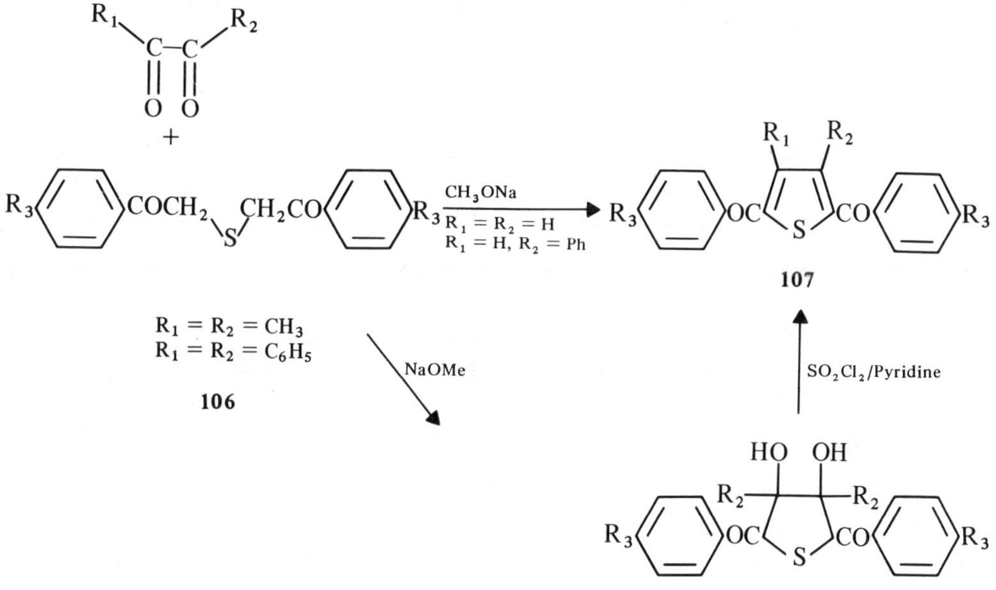

Scheme 45

TABLE 15. SYNTHESIS OF THIOPHENES BY MODIFIED HINSBERG REACTION

$$R_2COCH_2SCH_2COR_5 + RCOCOR_1 \longrightarrow$$

Thiophene product structure: R_4, R_3 on the ring positions; R_5OC and COR_2 attached, with S in the ring.

R_2	R_5	R	R_1	R_3	R_4	Yield (%)	mp (°C)	Reference
C_6H_5	OCH_3	H	H	H	H	29	70–71	193b
C_6H_5	OCH_3	H_3C	CH_3	CH_3	CH_3	50	175	193b
C_6H_5	OCH_3	OC_2H_5	OC_2H_5	OH	OH	53	140	193b
C_6H_5	OCH_3	C_6H_5	H	H	C_6H_5	59	112	193b
C_6H_5	OCH_3	CH_3	OC_2H_5	CH_3	OH	31	85	193
C_6H_5	OCH_3	$H_5C_2O_2C$	OC_2H_5	CO_2CH_3	OH	36, 58	94	193, 193b
C_6H_5	OC_2H_5	H	OC_4H_9	H	OH	52	71	193b
C_6H_5	OCH_3	H	OC_4H_9	H	OH	52	126–127	193
C_6H_5	OC_2H_5	C_6H_5	C_6H_5	C_6H_5	C_6H_5	82	186	193b
C_6H_5	OCH_3	C_6H_5	C_6H_5	C_6H_5	C_6H_5	32	184–185	193
CH_3	OCH_3	H	H	H	H	28	139	193a
CH_3	OCH_3	CH_3	CH_3	CH_3	CH_3	34	107	193a
CH_3	OCH_3	OC_2H_5	OC_2H_5	H	OH	41	167	193a
CH_3	OCH_3	OC_4H_9	OC_4H_9	OH	OH	30	136	193a
$C(CH_3)_3$	OCH_3	OC_2H_5	OC_2H_5	OH	OH			204
CH_3	CH_3	H	H	H	H	45, 50	172–173	193, 200
$CH_3(CH_2)_3$	$CH_3(CH_2)_3$	H	H	H	H	62	112–113	200
$C(CH_3)_3$	$C(CH_3)_3$	H	H	H	H	86	105.5–106	200
$CH_3(CH_2)_8$	$CH_3(CH_2)_8$	H	H	H	H	84	113–113.5	200
CH_3	CH_3	C_6H_5	H	C_6H_5	H	43	181–182	193
C_6H_5	C_6H_5	C_6H_5	H	C_6H_5	H	68, 73	106–108	193, 201
C_6H_5	C_6H_5	H	OC_4H_9	OH	H	50	136–137	193
C_6H_5	C_6H_5	$CO_2C_2H_5$	OC_2H_5	$CO_2C_2H_5$	H	36	123	193
C_6H_5	C_6H_5	OC_4H_9	OC_4H_9	OH	OH	25	184–185	193
C_6H_5	C_6H_5	H	H	H	OH	95	114–115	193, 200
$p\text{-}CH_3C_6H_4$	$p\text{-}CH_3C_6H_4$	H	H	H	H	96	143.5–144	200

p-ClC$_6$H$_4$	p-ClC$_6$H$_4$	H	H	H	91	244–245	200
p-BrC$_6$H$_4$	p-BrC$_6$H$_4$	H	H	H	95	254–255	200
p-O$_2$NC$_6$H$_4$	p-O$_2$NC$_6$H$_4$	H	H	H	74	192–194	200
p-CH$_3$OC$_6$H$_4$	p-CH$_3$OC$_6$H$_4$	H	H	H	92	192.5–193	200
3,5-(CH$_3$O)$_2$C$_6$H$_3$	3,5-(CH$_3$O)C$_6$H$_3$	H	H	H	95	96–96.5	200
2-Th	2Th	H	H	H	80	182.5–183	200
5-H$_5$C$_2$-2-Th	5-H$_5$C$_2$-2Th	H	H	H	86	114–114.3	200

approach has proven very useful for the synthesis of different thiophenophanes, such as **110** from **109**. Higher yields were obtained in many cases by using methanol instead of ethanol as solvent[202] (Scheme 46). Similarly, the furanothiophenophane **112** was prepared from **111**[203] (Scheme 47). Mixed sulfides, such as **113**, have also

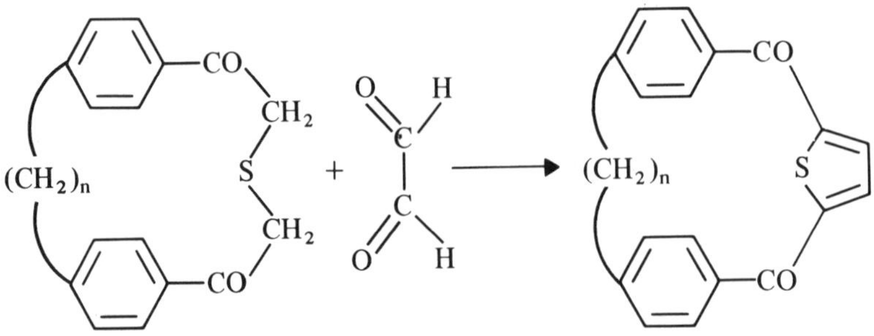

	109			**110**

n	Yield (%)
3	9
4	36
5 (CH$_3$OH)	34 (42)
6 (CH$_3$OH)	44 (61)
7	67
8 (CH$_3$OH)	58 (73)
10	91 (95)

Scheme 46

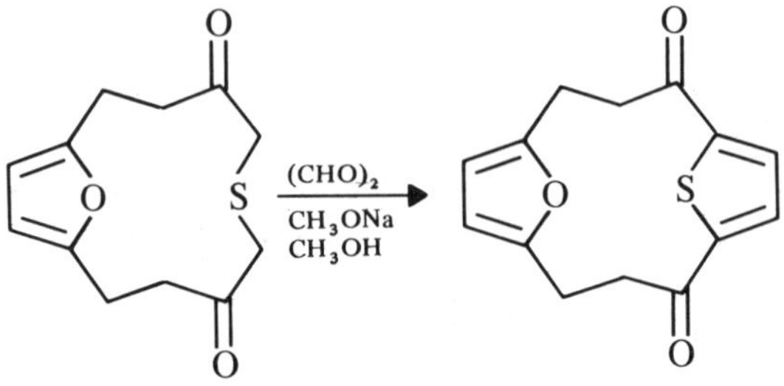

	111		**112**

Scheme 47

been used in the reaction with glyoxal, phenyl glyoxal, α-diketones, glyoxylates, mesaoxalates, and oxalates to give di- and trisubstituted thiophenes, 3-hydroxy-thiophenes, and 3,4-dihydroxythiophenes.[193–193b, 204] The reaction with unsymmetrical α-dicarbonyl compounds leads regiospecifically to only one isomer. Thus, the reaction of **113** (R = Ph) with phenylglyoxal, butylglyoxylate, or ethyl pyruvate gave only **114**, **115** (R = C$_6$H$_5$), and **116**, respectively. Similarly, **113** (R = CH$_3$) gave only **115** (R = CH$_3$) with butyl glyoxylate. In the case of **115** the structure was proven by authentic synthesis of the other possible isomer. In other

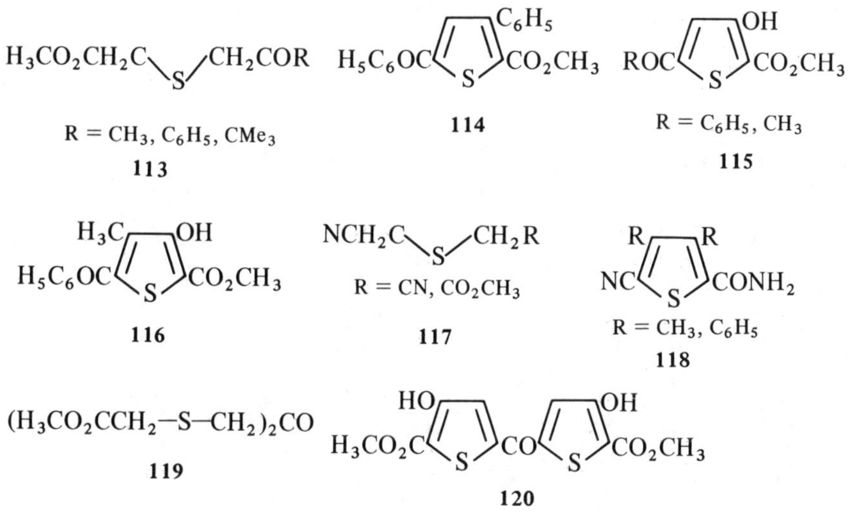

H_3CO_2CH_2C_S_/CH_2COR H_5C_6OC thiophene C_6H_5 / CO_2CH_3 ROC thiophene OH / CO_2CH_3

R = CH_3, C_6H_5, CMe_3   114   R = C_6H_5, CH_3
           113   115

H_5C_6OC thiophene H_3C / OH / CO_2CH_3 NCH_2C_S_/CH_2R NC thiophene R / R / $CONH_2$

           116   R = CN, CO_2CH_3   R = CH_3, C_6H_5
                                            117   118

(H_3CO_2CCH_2—S—CH_2)_2CO H_3CO_2C thiophene HO / CO thiophene OH / CO_2CH_3
           119
                                            120

Scheme 48

cases, the structure is assumed by analogy or by $FeCl_3$-color reactions. In the reaction of unsymmetrical sulfide (**113**, R = C_6H_5) with benzil, the reaction also stops at the tetrahydrothiophene-3,4-diol stage (see **108**), when carried out under the usual room temperature conditions. Heating to reflux leads to dehydration to give the thiophene, when the ethyl ester is used, instead of the methyl ester.[193b] Thiodiglycolic acid dinitrile (**117**, R = CN)[204a] and carbomethoxymethyl cyano-methyl sulfide (**117**, R = CO_2CH_3) can also be used in the Hinsberg reaction. Under the usual reaction conditions, however, one of the nitrile groups is sometimes converted to the amide. Thus, with diacetyl and benzil (**117**, R = CN), gave com-pounds **118**. The complex sulfide **119** reacts with butyl glyoxylate to give **120** (mp = 205–206°C) in 31% yield. The compounds prepared in Erlangen by modifi-cation of the Hinsberg reaction, together with their melting points, are given in Tables 15 and 16.

TABLE 16.   SYNTHESIS OF THIOPHENES BY MODIFIED HINSBERG REACTION

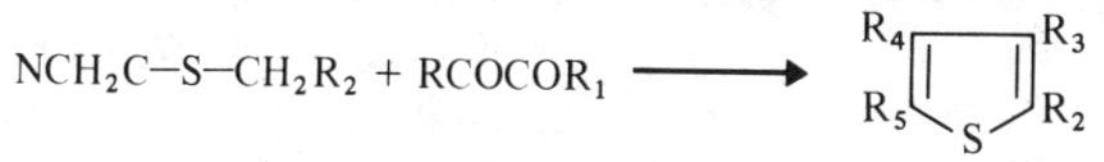

$NCH_2C—S—CH_2R_2 + RCOCOR_1 \longrightarrow$ R_4 thiophene R_3 / R_5 / R_2

R_2	R	R_1	R_3	R_4	R_5	Yield of thiophene (%)	mp (°C)	Reference
CN	CH_3	CH_3	H_3C	CH_3	$CONH_2$	66	192	193b
CN	C_6H_5	C_6H_5	C_6H_5	C_6H_5	$CONH_2$	80	261	193b
CO_2CH_3	OCH_3	OCH_3	OH	OH	CO_2CH_3	70	210	193b
CO_2CH_3	C_6H_5	C_6H_5	C_6H_5	C_6H_5	$CONH_2$	70	320	193b
CO_2CH_3	CH_3	OCH_3	OH	CH_3	CN	31	93	193b

B.   Thiophenes from α-Diketones and Wittig Reagents

The Wittig reagent (121) has been reacted with various α-diketones, such as 122, 123, and 124, in connection with the synthesis of very strained and unstable thiophene derivatives (125,[205] 126,[206, 207] and 127[208]), which were obtained in 14, 5, and 3.5% yield respectively (Scheme 49).

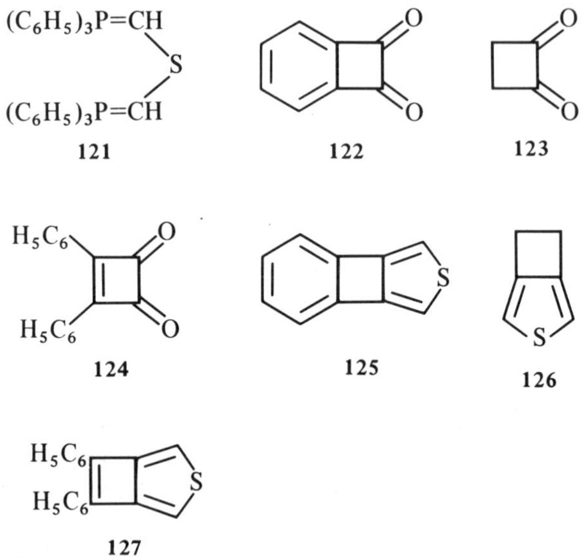

121   122   123

124   125   126

127

Scheme 49

3.   C₂S + C₂ Methods

A.   The Gewald Reaction

The Gewald reaction constitutes a very useful and versatile method for the synthesis of 2-aminothiophenes (130), with an electron-withdrawing substituent in the 3-position and alkyl and/or aryl groups in the 4- and 5-positions. In one version of the Gewald reaction, an α-mercaptoaldehyde or α-mercaptoketone (128) is reacted with an activated nitrile, such as malonitrile, methyl cyanoacetate, benzoylacetonitrile, or even p-nitrobenzylcyanide (129), in solvents such as ethanol or DMF, and with piperidine or triethylamine as catalyst, at about 50°C[209, 210] (Scheme 50). With malonitrile, water can be also used as a solvent. Recently, dioxane has been recommended as a most suitable solvent,[211, 212] since side reactions are suppressed. This methodology appears to be limited to aliphatic α-mercapto derivatives, which exist as dimers (1,4-dithianes) in the solid state, but react as monomers, since a dimer–monomer equilibrium is established in solution. Phenacyl

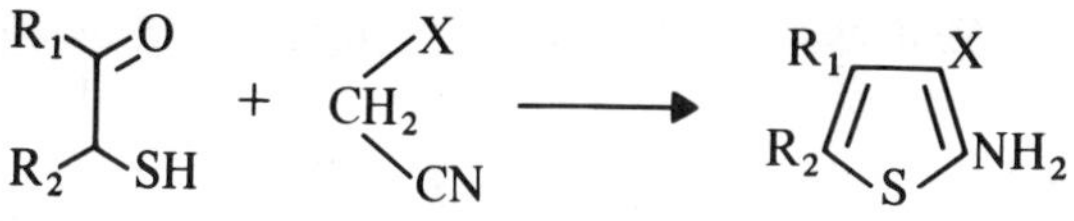

128 129 130

Scheme 50

mercaptane did not react under these mild conditions, nor do nonactivated nitriles such as benzyl cyanide or cyanoacetic acid. The detailed mechanism of the reaction has not been demonstrated, but it seems likely that the aldol-type condensation occurs first, followed by an attack of the thiolate on the cyano group. Thus, under certain conditions, using catalytic amounts of triethylamine in ethanol, α-mercapto-acetaldehyde (**131**) reacted with **132** to give **133**; larger amounts of triethylamine in DMF gave the amino ketone **134**[213] (Scheme 51). Similarly, the reaction of

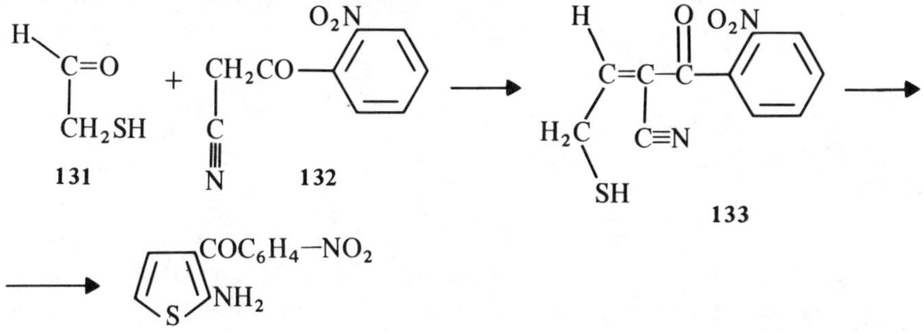

131 132 133

134

Scheme 51

2,4-dichlorophenacylcyanide with mercaptoacetaldehyde gave 2-amino-3-(2,6-dichlorobenzoyl)thiophene in 82% yield.[213a] Another way to obtain intermediates of type **133** is to condense α-chlorocyclohexanone (**135**) with an active nitrile (**136**) to **137** which, upon reaction with sodium hydrogen sulfide, yields **138**[210] (Scheme 52). Unfortunately, most other α-chloro ketones do not undergo the

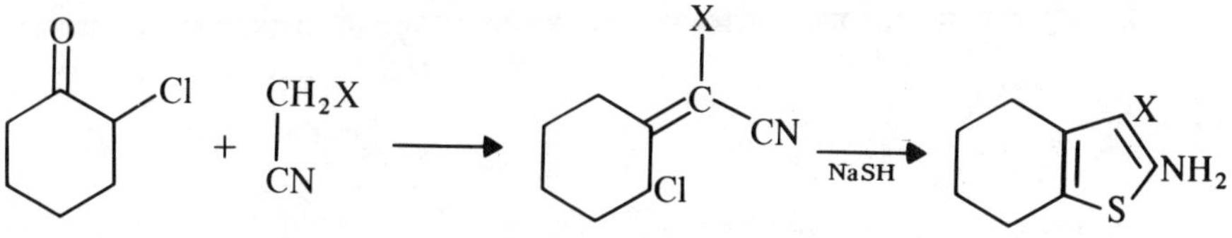

135 136 137 138

X = CN	95%
X = $CO_2C_2H_5$	80%
X = COC_6H_5	45%

Scheme 52

Knoevenagel–Cope condensation. Another way to arrive at compounds of type **137** is allylic bromination of **138a** to **139**, which, however, only gave useful yields of **140** with phenyl-substituted derivatives[210] (Scheme 53). A synthesis of 2-amino-

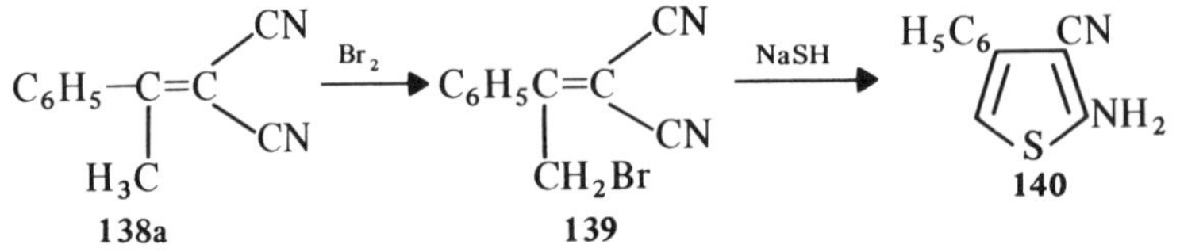

Scheme 53

4-carboethoxy-3-carbomethoxythiophene, through the reaction of ethyl γ-chloro-acetoacetate with sodium sulfide and methyl cyanoacetate in the presence of triethylamine in ethanol, has been described.[210a]

1,3-Dimercaptoacetone (**141**) reacts with malononitrile (**142**) in the presence of air to give **143** (Scheme 54). Compounds prepared according to this modification of the Gewald reaction are given in Table 17.

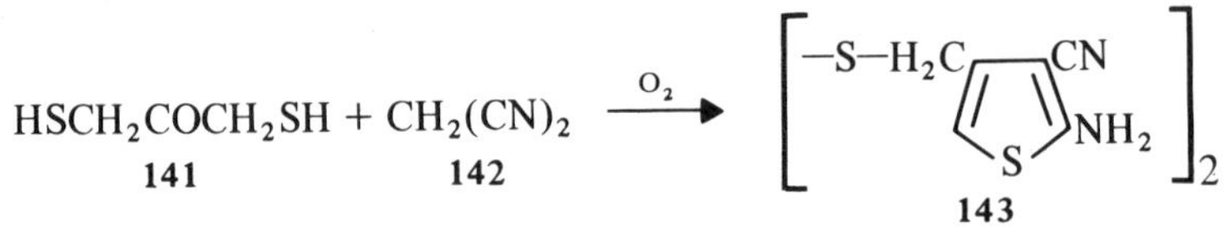

$$\text{HSCH}_2\text{COCH}_2\text{SH} + \text{CH}_2(\text{CN})_2 \xrightarrow{\text{O}_2}$$

141   142

Scheme 54

A more convenient modification of the Gewald reaction consists in the reaction at room temperature of aliphatic aldehydes, ketones, or β-dicarbonyl compounds (**144**) with active nitriles (**145**) and sulfur in the presence of amines.[221–222a] Excess ketone, alcohol, dioxane, or DMF can be used as solvents. Amines like diethyl- or triethylamine or morpholine have been employed and, in contrast to the modification in which catalytical amounts of amine were used, it is necessary to use 0.5–1.0 mole equivalents of amine, based on the amount of nitrile. The solvent can be of some importance to the outcome of the reaction. The reaction of methyl ethyl ketone and malononitrile to give 2-amino-3-cyano-4,5-dimethylthiophene was best catalyzed with morpholine, using excess ketone as solvent. With ethanol as the solvent, the product was only 2-butylidenemalononitrile[223] (Scheme 55). This

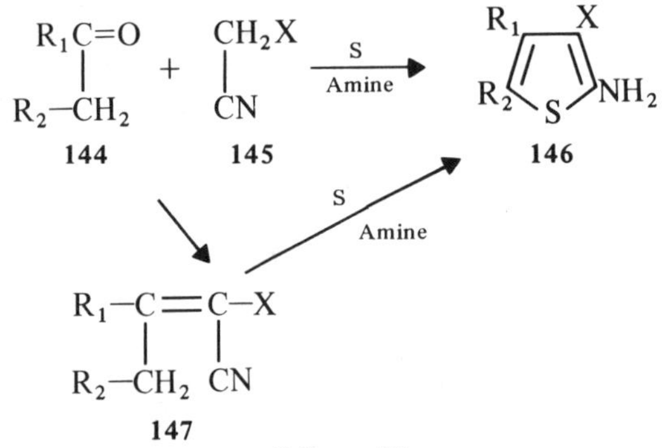

Scheme 55

TABLE 17.   GEWALD REACTIONS – MODIFICATION 1

$$R_1 \diagdown \diagup O \quad + \quad \begin{array}{c} CH_2-X \\ | \\ CN \end{array} \xrightarrow{Amine} \quad R_1 \diagup X$$

$$R_2 \diagup SH \qquad\qquad\qquad\qquad R_2 \diagdown_S \diagup NH_2$$

R_1	R_2	X	Yield of Thiophene (%)	Reference
H	H	CN	55	210
CH_3	H	CN	73	210
CH_3	CH_3	CN	70	210
C_2H_5	CH_3	CN	51	210
$-(CH_2)_4-$		CN	70	210
H	H	CO_2CH_3	46, 58	210, 214
CH_3	H	CO_2CH_3	75	210
CH_3	CH_3	CO_2CH_3	45	210
CH_3	H	$CO_2C_2H_5$	88	215
$-(CH_2)_4-$		$CO_2C_2H_5$	80	210
C_6H_5	H	$CO_2C_2H_5$	75	215
C_6H_5	CH_3	$CO_2C_2H_5$	97	215
H	H	$CONH_2$	60	216
CH_3	H	$CONH_2$	53	210
H	H	COC_6H_5	70	216
H	H	COC_6H_5	27	217
CH_3	H	COC_6H_5	40	210
H	H	$COC_6H_4CH_3\text{-}o$	60	216
H	H	$o\text{-}CH_3OC_6H_4CO\text{-}$	42	212
CH_3	H	$C_6H_4\text{--}NO_2\text{-}p$	34	210
H	H	$o\text{-}CH_3C_6H_4CO\text{-}$	73	212
H	H	$o\text{-}O_2N\text{--}C_6H_4CO\text{-}$	48	213
H	H	$m\text{-}O_2N\text{--}C_6H_4CO\text{-}$		211
H	H	$o\text{-}FC_6H_4CO\text{-}$	58	212
H	H	$o\text{-}F_3CC_6H_4CO\text{-}$	78	212
H	H	$3\text{-}O_2N\text{-}5\text{-}CH_3C_6H_3CO$	78	212
H	H	$3,5\text{-}Cl\text{--}C_6H_3CO$	74	212
H	H	$2,6\text{-}F\text{--}C_6H_3CO$	56	212
H	H	$o\text{-}H_3CSO_2C_6H_4CO$	79	212
H	H	$NCCH_2NHCO\text{-}$	44	218
H	H	$CSNH_2$	88	219
H	H	2-furyl-CO-	60	216
H	H	2-pyridyl-CO-	–	220
H	H	3-Me-2-pyridyl-CO·	–	220
H	H	2-ThCO-	70	216
CH_3	H	2-ThCO	48	217
$CH_2CO_2C_2H_5$	H	CO_2Me		210a
H	H	$2,6\text{-}di\text{-}Cl_2C_6H_4$	82	213a

one-pot procedure has been used very extensively for the preparation of numerous thiophenes (146), as is evident from Table 18.

Alternatively, a two-step procedure is preferable. The α,β-unsaturated nitrile 147 is first prepared by a Knoevenagel–Cope condensation and then reacted with sulfur and amine. Examples of the two-step procedure are given in Table 19.

TABLE 18. THE GEWALD REACTION – MODIFICATION 2

$$R_1\text{--CO} + \underset{\text{CN}}{\overset{\text{CH}_2X}{|}} + S \xrightarrow{\text{Amine}} \underset{R_2}{\overset{R_1}{\diagdown}}\!\!\boxed{}_{S}\!\!\underset{\text{NH}_2}{\overset{X}{\diagup}}$$
$$R_2\text{--CH}_2$$

R_1	R_2	X	Yield of Thiophene (%)	Reference
CH₃	CH₃	CN	42	222, 223
CH₃	CONHPh	CN	70	229[a]
	[(CH₂)₂SCH₂]	CN	61	237
	(CH₂)₄	CN	86	222, 223, 225, 237, 238
	—[(CH₂)₂N(CH₃)CH₂]—	CN	62	225
	CH—(CH₂)₃ CH₂N⟨morpholine⟩	CN	50	231
	—[CH(CH₃)(CH₂)₃]—	CN	45	231
	—(CH₂)₅—	CN	44	225, 238
	—[(CH₂)₂CH(CH₃)CH₂]—	CN	86[a]	225
	—[CH₂CH(CH₃)(CH₂)₂]—	CN	90[a]	225
	—[(CH₂)₂N(CH(CH₃)₂)CH₂]—	CN	74	239
	—[CH₂CH(CH₃)CH(CH₃)CH₂]—	CN	80[b]	225
	—(CH₂)₆—	CN	64	225
	—[(CH₂)₂N(C₄H₉)CH₂]—	CN	43	225
H₃C	C₆H₅CH₂	CN	17	224
H₃C	3,4-Cl₂C₆H₃CH₂	CN	18	224
CH₃	CONHC₆H₄Cl-p	CN	60	229[a]
CH₃	CONHC₆H₃Cl₂-2,5	CN	68	229[a]
	—[(CH₂)₂CH(C(CH₃)₃)CH₂]—	CN	79	225
H₃C	CONHC₆H₄—CH₃-p	CN	63	229[a]

[a] Another structure for these compounds is given in the publication.
[b] A cis/trans 5,7-Me₂ mixture.

R	R′	Z	Yield	Ref.
CH_3	$C_6H_5CH_2CH_2$	CN	29	224
CH_3	$3,4\text{-}Cl_2C_6H_3CH_2CH_2CH_2$	CN	24	224
$-(CH_2)_{10}-$		CN	42	225
$[(CH_2)_2CH(C_6H_5)CH_2]$		CN	63	237
$-[(CH_2)_2N(CH_2C_6H_5)CH_2]-$		CN	55	237
$-[(CH_2)_2N(CH_2C_6H_5)CH_2]-$		CN	71	225
$-[CH_2CH(C_6H_5)SCHC_6H_5]-$		CN	83	240
$-[CH_2CH(3,4\text{-}Cl_2C_6H_3)SCH(3,4\text{-}Cl_2C_6H_4)]-$		CN	85	240
$-[CH_2CH(p\text{-}F_3CC_6H_4)SCH(p\text{-}F_3CC_6H_4)]-$		CN	98	240
$-[CH_2CH(3,4\text{-}Cl_2C_6H_3)N(CH_3)CH_2(3,4\text{-}Cl_2C_6H_4)]-$		CN	58	240
$-[CH_2CH(C_6H_5)N(CH_3)CHC_6H_5]-$		CN	51	240
Cholestan-3-one		CN	30	238
2-Indanone		CN	41	237
2-Tetralone		CN	40	237
Tropinone		CN	50[c]	231
C_2H_5	CH_3	CO_2CH_3	40	222
CH_3	$COCH_3$	CO_2CH_3	31	222
H	CH_3	$CO_2C_2H_5$	42, 47	222, 241
H	C_2H_5	$CO_2C_2H_5$	75, 65	222, 241
CH_3	CH_3	$CO_2C_2H_5$	39	222
$-(CH_2)_3-$		$CO_2C_2H_5$	45	222
H	$CH(CH_3)_2$	$CO_2C_2H_5$	40	242
CH_3	CH_2COOH	$CO_2C_2H_5$	30	231
CH_3	$CO_2C_2H_5$	$CO_2C_2H_5$	32	222
$-(CH_2)_4-$		$CO_2C_2H_5$	82	222, 244
$-[CH_2N(CH(CH_3)_2)CH_2]-$		$CO_2C_2H_5$	61	223, 239
morpholine structure: $CH{-}(CH_2)_3$ with $N{-}CH_2{-}O$ ring		$CO_2C_2H_5$	50	231

[c] Probably contains some 2-amino-3-cyano-7-methyl-4,5,6,7-tetrahydrobenzo[b]thiophene.

TABLE 18. *Continued*

R₁	R₂	X	Yield of Thiophene (%)	Reference
	$-[CH_2CH(CH_3)(CH_2)_2]-$	$CO_2C_2H_5$	34	244
	$-[(CH_2)_2CH(CH_3)CH_2]-$	$CO_2C_2H_5$	70	244
	$-(CH_2)_5-$	$CO_2C_2H_5$	59	244
CH_3	$CONHC_6H_5$	$CO_2C_2H_5$	30	231
	$-[CH(C_6H_5)(CH_2)_3]-$	$CO_2C_2H_5$	50	231
	$-[CH_2CH(C_6H_5)NCH(C_6H_5)]-$ H	$CO_2C_2H_5$	35	231
	$-[CH_2CH(C_6H_5)NCH(C_6H_5)]-$ CH_3	$CO_2C_2H_5$	40	231
	$-[CH(CH_3)CH(C_6H_5)N-CHC_6H_5]-$ CH_3	$CO_2C_2H_5$	53	231
H	$-(CH_2)_4-$	$CONH_2$	61, 25	222, 245, 246
	C_6H_5	$CONH_2$	45	222
	$-[(CH_2)_2CH(OCOC_6H_5)CH_2]-$	$CONH_2$	42	246
	$-(CH_2)_4-$	$CONHCH_3$	29	246
	$-(CH_2)_4-$	$CONHC_2H_5$	35	231
	$-(CH_2)_4-$	$CONHC_6H_4CH_3$	62	247
	$-(CH_2)_4-$	$CONHC_6H_5$	41	231
H	CH_3	COC_6H_5		217, 248
CH_3	H	COC_6H_5		248
CH_3	CH_3	COC_6H_5		248
H	C_2H_5	COC_6H_5	70	249
CH_3	CH_3	COC_6H_5	72	249
i-C_3H_7	H	COC_6H_5	–	249
	$-(CH_2)_3-$	COC_6H_5	51	249
	$-(CH_2)_4-$	COC_6H_5	40	222
CH_3	n-C_3H_7	COC_6H_5	–	249
H	n-C_4H_9	COC_6H_5	–	249

H	—[CH₂C(CH₃)₂OCH₂]—	COC₆H₅	41	250
H	C₆H₅	COC₆H₅	86	217
H	CH₃	o-ClC₆H₄CO-	75	249
CH₃	C₂H₅	o-ClC₆H₄CO	61	249
H	CH₃	o-ClC₆H₄CO	59	249
CH₃	C₂H₅	o-ClC₆H₄CO	d	233
H	CH₃	p-ClC₆H₄CO	52	249
	i-C₃H₇	o-ClC₆H₄CO	–	249
	—(CH₂)₄—	p-ClC₆H₄CO	68	249
	—[CH₂C(CH₃)₂SCH₂]—	o-ClC₆H₄CO	46	250
	—[CH₂C(CH₃)₂OCH₂]—	o-ClC₆H₄CO	66	250
H	C₂H₅	o-BrC₆H₄CO	60	249
	—[CH₂C(CH₃)₂OCH₂]—	o-BrC₆H₄CO	50	250
	—[CH₂C(CH₃)₂OCH₂]—	p-BrC₆H₄CO	51	250
	C₂H₅	o-FC₆H₄CO	68	249
H	—(CH₂)₄—	m-F₃CC₆H₄CO	60	249
CH₃	CH₃	m-F₃CC₆HCO	–	249
	—[CH₂C(CH₃)₂OCH₂]—	m-O₂NO₂C₆H₄CO	46	250
H	C₂H₅	o-H₃COC₆H₄CO	73	249
	—(CH₂)₄—	o-CH₃OC₆H₄CO	71	249
H	C₂H₅	o-H₃CC₆H₄CO	58	249
	—[(CH₂)₂C*(CH₃)CH₂]—	C₆H₅CO	67	249
	—(CH₂)₅—	C₆H₅CO	58	249

d ^{35}S was used in the reaction.

TABLE 19. THE GEWALD REACTION – MODIFICATION 3

$$R_1\text{-}\overset{\displaystyle |}{C}=C\!\!\begin{array}{c}X\\ CN\end{array} \;\;\; R_2\text{-}CH_2 \;\; + \;\; S \;\;\xrightarrow{\text{Amine}}\;\; \begin{array}{c}R_1\\ R_2\end{array}\!\!\underset{S}{\boxed{}}\!\!\begin{array}{c}X\\ NH_2\end{array}$$

R₁	R₂	X	Yield of Thiophene (%)	References
CH_3	CH_3	CN	41	222
$PhCH_2$	CH_3	CN	34	224
$(CH_3)_2CH$	H	CN	–	226
	$-(CH_2SCH_2CH_2S)-$	CN	68	266
$(CH_3)_3C$	H	CN	48	226
	$-(CH_2)_4-$	CN	90	222
	$-[CH(CH_3)(CH_2)_3]-$	CN	45	231
	$-[CH_2C(CH_3)_2OCH_2]-$	CN	83	267
	$-[CH_2C(CH_3)_2SCH_2]-$	CN	92	267
	$-(CH_2)_5-$	CN	65	245
	$-(CH_2)_6-$	CN	65	245
$p\text{-}ClC_6H_4$	CH_3	CN	95	224
$p\text{-}ClC_6H_4$	C_2H_5	CN	86	224
	$-[CH(C_6H_5)(CH_2)_3]-$	CN	30	237
	$-(CH_2)_{10}-$	CN	60	245
	$-[CH(C_6H_5)(CH_2)_4]-$	CN	4	245
C_6H_5	C_6H_5	CN	95	224
	$-(CH_2)_{13}-$	CN	50	245
$C_6H_5CH_2$	C_6H_5	CN	98	224
1-Indanone		CN	25	237
1-Tetralone		CN	48	237
	$-[CH_2CH_2\text{-}S]-$	CO_2CH_3	41	227
C_2H_5	CH_3	CO_2CH_3	50	222
C_6H_{11}	H	CO_2CH_3	–	226
$2\text{-}Th$	H	$CO_2C_2H_5$	33	268
CH_3	H	$CO_2C_2H_5$	–	226

				Yield (%)	References
CH_3	$-(CH_2)_3-$	CH_3	$CO_2C_2H_5$	49	222, 268, 269
	$-(CH_2)_4-$		$CO_2C_2H_5$	52	222
	$-(CH_2)_5-$		$CO_2C_2H_5$	91	222
CH_3	$-(CH_2)_6-$	$CH_2CH_2OCOCH_3$	$CO_2C_2H_5$	85	245
C_6H_5	$-(CH_2)_6-$	H	$CO_2C_2H_5$	77	270
			$CO_2C_2H_5$	96	245
			$CO_2C_2H_5$	62	222, 268, 269
$p\text{-}O_2NC_6H_4$		H	$CO_2C_2H_5$	60	274
$p\text{-}FC_6H_4$		H	$CO_2C_2H_5$	68	269
C_6H_5		CH_3	$CO_2C_2H_5$	50	232
CH_3		C_6H_5	$CO_2C_2H_5$	38	222, 232, 269
$3,4,5\text{-}(CH_3O)C_6H_2$		H	$CO_2C_2H_5$	60	222
$p\text{-}CH_3C_6H_4$		H	$CO_2C_2H_5$	78	232, 269
$p\text{-}CH_3OC_6H_4$		H	$CO_2C_2H_5$	69	270
$2,4\text{-}(CH_3)_2C_6H_3$		H	$CO_2C_2H_5$	72	270
$2,5\text{-}(CH_3)_2C_6H_3$		H	$CO_2C_2H_5$	22	270
$2,4\text{-}(CH_3O)_2C_6H_3$		H	$CO_2C_2H_5$	60	270
$3,4\text{-}(CH_3O)_2C_6H_3$		H	$CO_2C_2H_5$	93	270
	$-[CH(C_6H_5)(CH_2)_3]-$	H	$CO_2C_2H_5$	50	231
	$-(CH_2SCH_2CH_2S)-$		$CO_2C_2H_5$	96	266
	$-(CH_2S(CH_2)_3S)-$		$CONH_2$		271
	$-(CH_2)_4-$		$CONH_2$	71, 75	222, 245
	$-(CH_2)_5-$		$CONH_2$	66	245
CH_3	$-(CH_2)_6-$	C_6H_5	$CONH_2$	58	222
	$-(CH_2)_{10}-$		$CONH_2$	80	245
	$-[CH(C_6H_5)(CH_2)_4]-$		$CONH_2$	48	245
			$CONH_2$	11	245
			C_6H_5CO	65	218
C_2H_5	$-[(CH_2)_2SCH_2]-$	CH_3	C_6H_5CO	56	217
	$-(CH_2)_4-$		C_6H_5CO	80	222
C_3H_7		C_2H_5	C_6H_5CO	79	217
C_6H_5	$-(CH_2)_6-$	H	C_6H_5CO	39	217, 226
			C_6H_5CO	—	217

TABLE 19. *Continued*

R₁	R₂	X	Yield of Thiophene (%)	Reference
	$-[(CH_2)_2N(CO_2C_2H_5)CH_2]-$	C_6H_5CO	72	217
	$-[(CH_2)_2N(COC_6H_5)CH_2]-$	C_6H_5CO	53	217
	$-(CH_2)_4-$	$C_6H_{11}CO$	95	217
	$-(CH_2)_4-$	$CONHC_6H_5$	41	231
	$-(CH_2)_4-$	$o-H_3CC_6H_4CO$	81	217
	$-(CH_2)_4-$	$o-ClC_6H_4CO$	37	217
	$-(CH_2)_4-$	$m-ClC_6H_4CO$	46	217
	$-(CH_2)_4-$	$p-CH_3OC_6H_4CO$	91	217
	$-(CH_2)_4-$	$m-CH_3OC_6H_4CO$	64	217
	$-(CH_2)_4-$	$o-FC_6H_4CO$	48	217
	$-(CH_2)_4-$	2-Furyl-CO-	77	217
	$-(CH_2)_4-$	2-Naphthyl-CO-	–	217
	$-(CH_2)_4-$	2-Thienyl-CO-	76	217, 272

In many cases, this third two-step reaction gives higher yields. Even more important is that certain ketones, such as alkylaryl ketones, do not give thiophenes in the one-pot modification,[220, 224] but give acceptable yields in the two-step technique. However, base-catalyzed polymerization of alkylidenemalonitriles is an annoying side-reaction; and with aralkylidenemalonitriles, aminothiophene formation was completely suppressed upon attempted reaction with amine and sulfur.[220]

Both the one- and two-step reactions have been carried out with numerous ketones and aldehyde or their Knoevenagel–Cope condensation products. In particular, cyclic and heterocyclic saturated ketones have been used extensively, as is evident from Tables 18 and 19. It was, however, observed that significantly lower yields (40–65%) were obtained with cyclic ketones from rings larger than six-membered. This trend is suggestive of increasing nonbounded repulsive interaction between methylene protons in middle- and large-sized rings fused to a planar five-membered ring.[225]

The mechanism of the two-pot reaction is not clear, but Gewald favours the route in which 147 is first formed and thiolated at the CH_2 group and then ring-closure occurs, as indicated in 148, and not the route via thiolation of the carbonyl derivative followed by condensation 149. The possibility that, when secondary amines such as morpholine or piperidine are used, enamines are formed from 147, appears unlikely, since triethylamine can also be used[140] (Scheme 56).

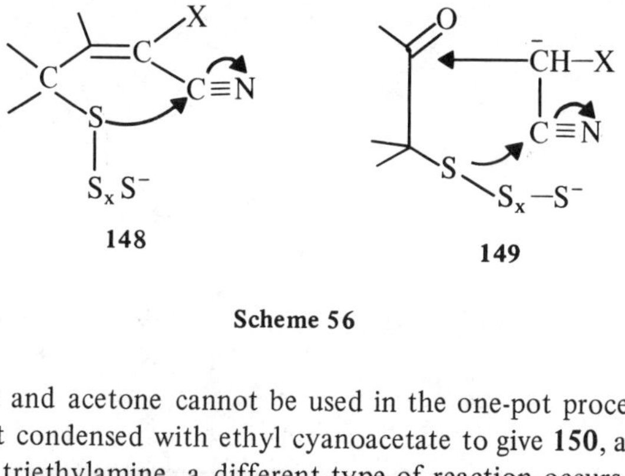

148   **149**

Scheme 56

Acetaldehyde and acetone cannot be used in the one-pot procedure. However, if acetone is first condensed with ethyl cyanoacetate to give 150, and then reacted with sulfur and triethylamine, a different type of reaction occurs and 2-amino-5-thiophenethiol (152) is formed, most probably via the *gem*-dithiol 151. The isolated product, in 71% yield, was the disulfide derived from 152[226] (Scheme 57).

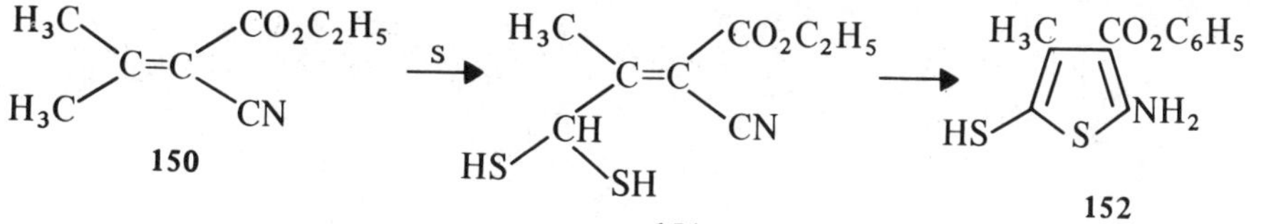

150   **151**   **152**

Scheme 57

Sterically crowded ketones with no methylene groups, such as methyl cyclohexyl, methyl isopropyl, or methyl *t*-butyl ketones, do not react with nitriles and sulfur. However, their condensation products with nitriles give the normal 2-amino-thiophenes with electron-withdrawing groups in the 3-position.[226] Methyl isopropyl and methyl *t*-butyl ketone give only condensation products with malonitrile. A ketone that reacts differently depending upon temperature is 3-thiacyclopentanone (153). In the reaction with methyl cyanoacetate and sulfur in the presence of diethylamine at 40°C in methanol, thiacyclopentanone gives the expected thiophene 154 in 30% yield.[227] However, when the reaction was carried out at room temperature or at 60°C, the sulfide 155 was obtained instead in 36% yield, and by treatment with hydrazine, it was reduced to 154. The same results, but with better yields, 60% of 155 and 47% of 154, were obtained if the condensation product 156 was used[227] (Scheme 58).

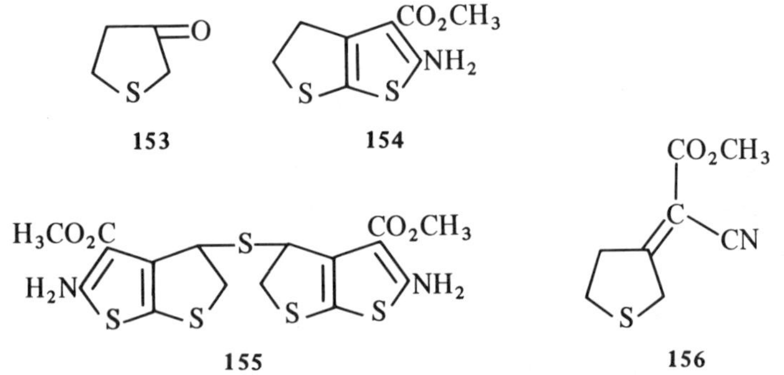

Scheme 58

With ketones that have only slightly different methylene groups. regiospecificity is seldom achieved. Thus, in the one-step procedure, 3-methylcyclohexanone (157) gave a mixture of 158 and 159 in which the 5-methyl isomer was assumed to be predominant, since it is sterically favored. The ratio between the two isomers was 9:1.[225] Surprisingly, the ylidene malonitrile from 1-phenylbutanone (160) gave the two aminonitriles, 161 and 162, in a 1:1 mixture.[224] It was expected that 161 would be the dominant isomer, since thiation of the benzylic position should have been favored. This is true for 2-tetralone (163), which only yields the angular isomer 164[224, 228] (Scheme 60).

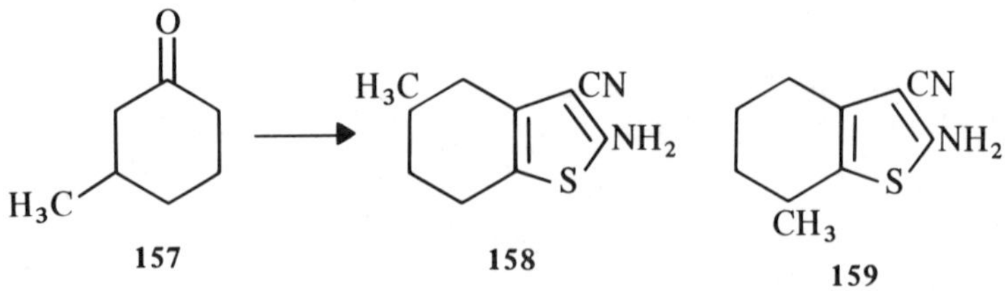

Scheme 59

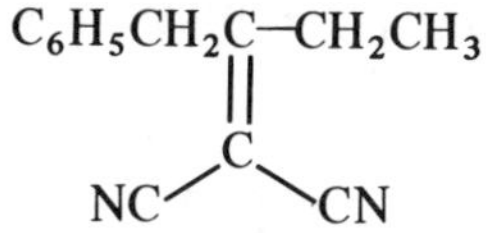

C6H5CH2C—CH2CH3   H5C2 CN   H5C6H2C CN

(structures 160, 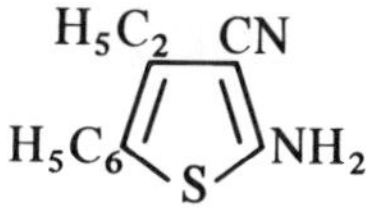 161, 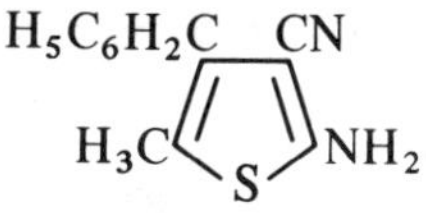 162)

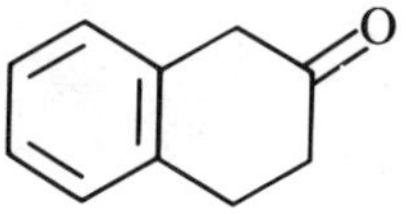

160   **161**   **162**

(structures 163, 164)

163

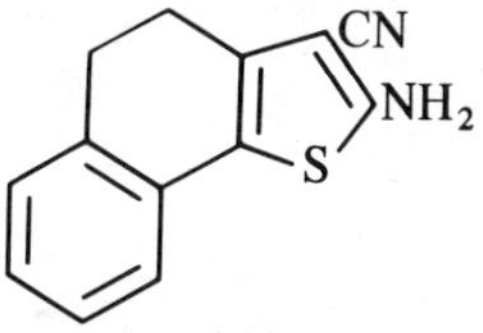

164

Scheme 60

As mentioned previously, β-dicarbonyl derivatives can also be used in the one-step procedure, although yields are only about 30%.[222] In the reaction of acetylacetone with methyl cyanoacetate and sulfur in the presence of diethylamine, both **165** and **166** are formed (Scheme 61). The latter is the sole product, if higher

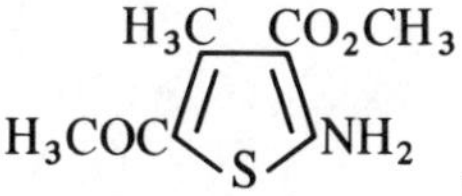

H3C CO2CH3   H3C CO2CH3

(structure 165)   NCHC=C (structure 166)
                              |
165   CH3

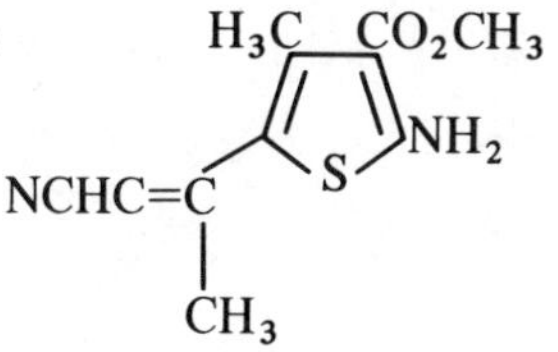

                         166

Scheme 61

temperature and excess methyl cyanoacetate is used. The formation of **166** depends upon an additional Knoevenagel condensation followed by ester cleavage, which must occur before the thiation, since **165** does not react under these conditions with methyl cyanoacetate to give **166**.[222] Also, anilides of acetoacetic acid (**167**) have been utilized in the reaction with malonitrile and sulfur, using ethanol as solvent and morpholine as the amine. It is stated in the publication that compounds (**168**) are formed, which must be a mistake, since the Gewald reaction can only lead to **169**[229] (Scheme 62). When cyclohexanone is reacted with cyanoacetohydrazide

ArNHCOCH2COCH3   ArNHOC CN

Ar = C6H5, 4-CH3C6H4—,   H3C (structure) NH2
     4-ClC6H4—,2,5-Cl2C6H3—

167   **168**

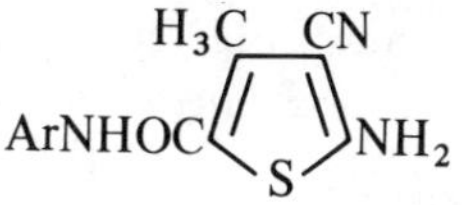

H3C CN

ArNHOC (structure) NH2

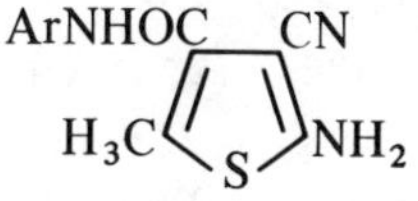

169

Scheme 62

(170) and sulfur in methanol, using morpholine as the amine, the thienopyrimidone derivative **171** is obtained in 40% yield (Scheme 63). Cycloheptanone reacts

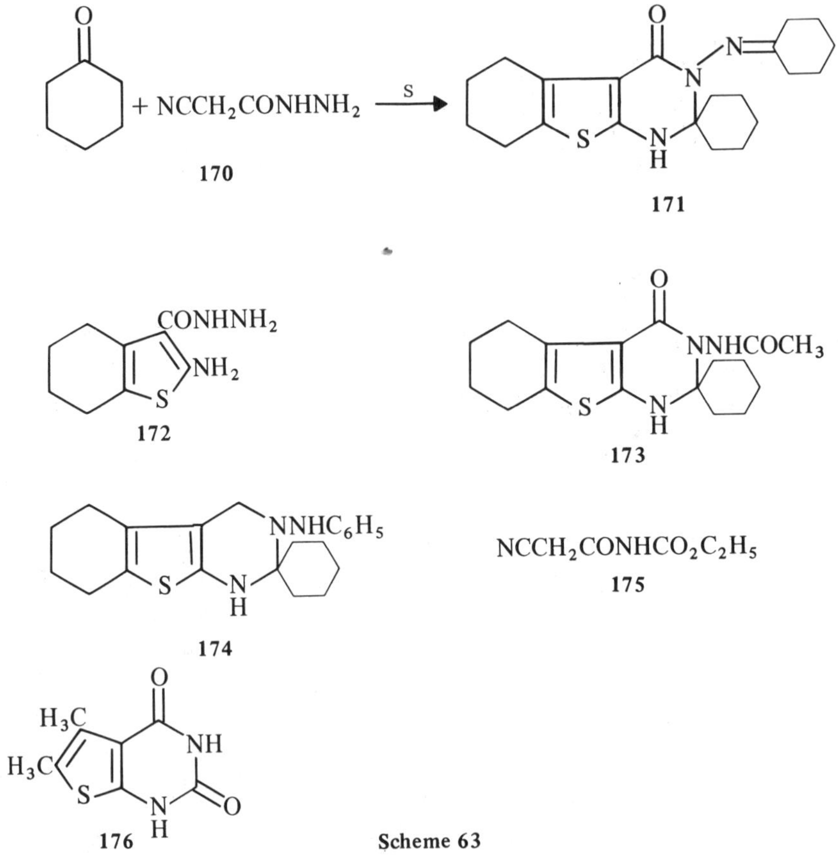

170

171

172

173

NCCH$_2$CONHCO$_2$C$_2$H$_5$

175

174

176

Scheme 63

analogously. Acid hydrolysis of **171** yields the hydrazide **172** in 70–75% yield.[230] The N-acetylcyanoacetohydrazide gives **173** in 57% yield. On the other hand, cyclopentanone yields no *spiro* compound, but gives the 2-amino-3-thiophene-carbonylhydrazide.[230] In a similar way, **174** was obtained from the reaction of N-cyanoacetylphenylhydrazine with cyclohexanone and sulfur.[231] The reaction of N-cyanoacetylurethane (**175**) with methyl ethyl ketone, sulfur, and triethylamine in boiling ethanol gave **176** in 80% yield.[232] A recent patent described the use of nitroacetonitrile in the Gewald reaction with α-mercaptoacetaldehyde or α-mercaptoketones, yielding 3-nitro-2-aminothiophene and 4-substituted 3-nitro-2-aminothiophenes.[232a] The reaction of aryl-substituted acetonitriles (Ar = C$_6$H$_5$, p-CH$_3$OC$_5$H$_4$, 2-pyridyl, 2-quinolyl) with alkylthioglycolates in refluxing pyridine in the presence of alkali metal oxides gave 3-aryl-2-amino-4(5H)-ketothiophenes.[232b] These reactions most probably proceed via Claisen condensation followed by ring-closure.

[35]S-labeled sulfur has been used in the Gewald reaction for the preparation of ^{35}S-labeled aminothiophenes.[233] Cyclohexanethione can be used instead of cyclohexanone in the reaction with malonitrile, sulfur, and triethylamine to give 2-amino-3-cyano-4,5-tetramethylenethiophene. Thiocamphor did not react.[273] A fourth minor modification in the synthesis of 2-amino-3-carbethoxythiophene (**178**) is to use enamines (**177**) instead of ketones in the one-step reaction,[232, 234, 227] but it offers no special advantage (Scheme 64). The imine **179** reacts with sulfur and methyl cyanoacetate to give **180** in 80% yield [235] (Scheme 65).

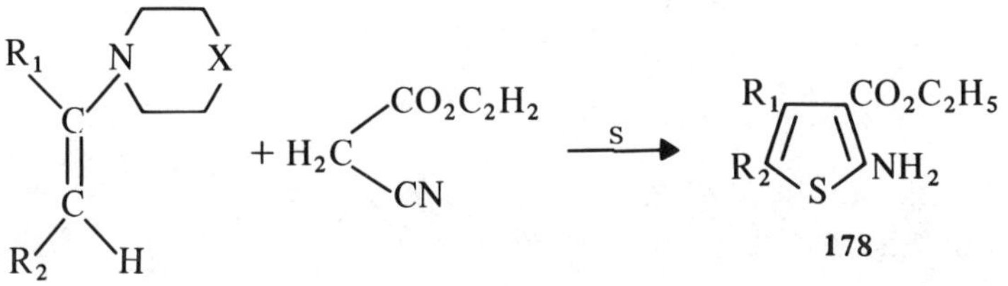

$$R_1 = R_2 = (CH_2)_4-, X = 0 \qquad 85\%$$
$$R_1 = H, R_2 = C_2H_5, X = CH_2 \qquad 46\%$$
$$R_1 = CH_3, R_2 = CO_2C_2H_5, X = 0 \qquad 33\%$$
$$R_1 = R_2 = C_6H_5, X = 0 \qquad 40\%$$

Scheme 64

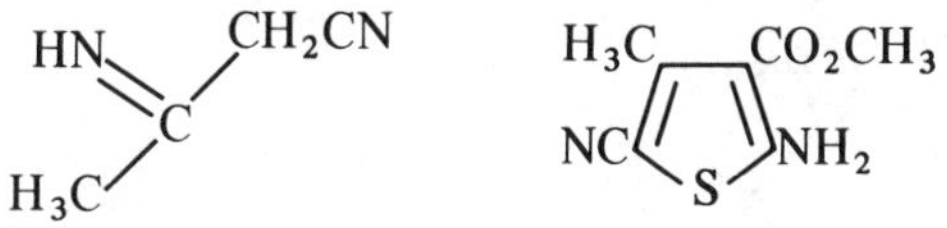

179   **180**

Scheme 65

It is also possible to prepare 2-aminothiophenes without electron-withdrawing groups in the 3-position, since an α,β-unsaturated nitrile such as **181** reacts with sulfur and diethylamine in ethanol to give **182** in 68% yield. The ethoxycarbonyl group is necessary to activate the CH_2 group for the thiation[236] (Scheme 66).

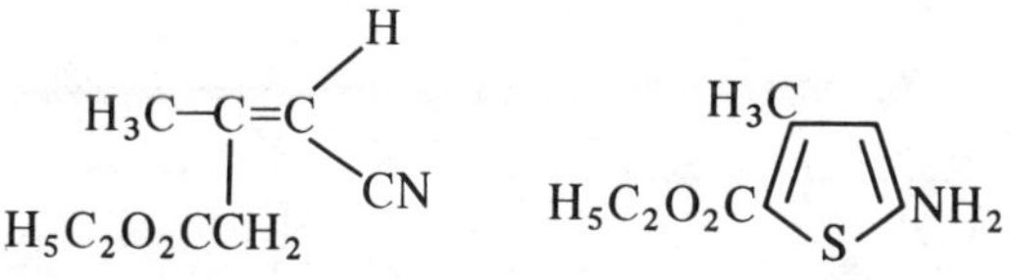

181   **182**

Scheme 66

Cinnamaldehyde (**183**) can also be used in a reaction similar to the Gewald reaction with ethyl cyanoacetate, sulfur, and triethylamine in DMF or, alternatively, the Knoevenagel condensation product (**184**) can be used according to the two-step route to the thioketones (**185**) (Scheme 67).

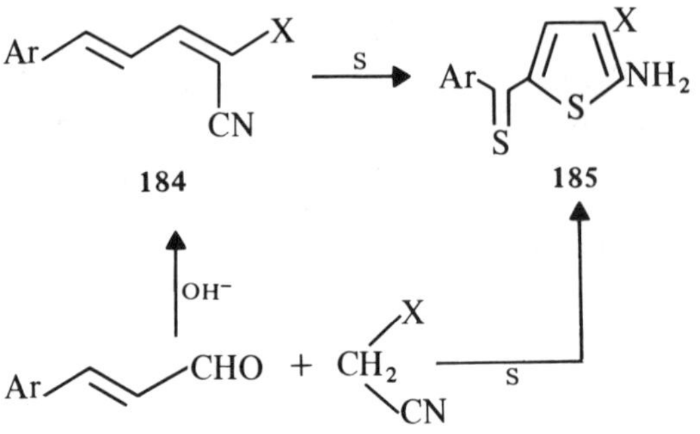

184   **185**

183

Scheme 67

A mechanism via **186** and **187** has been suggested (Scheme 68). The thiophenes prepared and the yields obtained are given in Table 20.

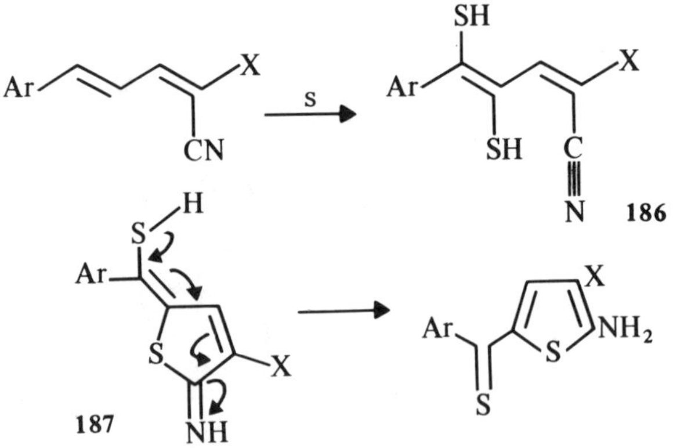

186

187

Scheme 68

TABLE 20.   GEWALD-TYPE REACTION OF PENTADIENENITRILES

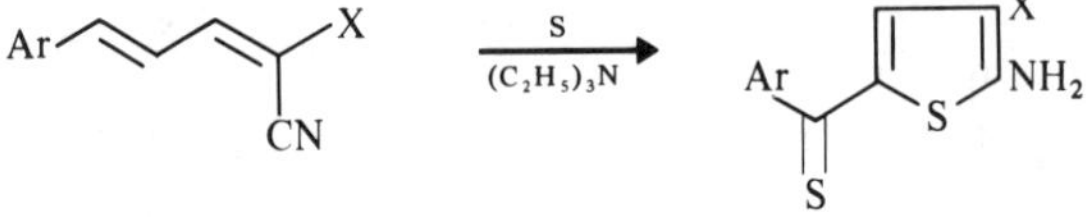

Ar	X	Yield of Thiophenes[a] (%)	Reference
C_6H_5	$CO_2C_2H_5$	40 (64)	69, 70
C_6H_5	COC_6H_5	21 (23)	69, 70
C_6H_5	$COC_6H_5CH_3$-p	26 (21)	69, 70
C_6H_5	$COC_6H_4OCH_3$-p	33 (24)	69, 70
p-$CH_3C_6H_4$	$CO_2C_2H_5$	16 (33)	69, 70
p-$CH_3C_6H_4$	COC_6H_5	18 (19)	69, 70
p-$CH_3OC_6H_4$	$CO_2C_2H_5$	30 (31)	69, 70
p-$CH_3OC_6H_4$	COC_6H_5	33 (18)	69, 70

[a] Yields in parenthesis are from the one-pot reaction of cinnamaldehydes with the active nitrile and sulfur. All yields were calculated on these aldehydes.

It may also be of interest to note what happens if active nitriles are reacted with sulfur and base in the absence of aldehyde or ketones that scavenge the reactive thiated intermediate, since these reactions are also of the $C_2 + C_2 + S$ type. If malonitrile is allowed to react with sulfur and triethylamine in DMF, the amino nitriles **188** and **189** are obtained in 30 and 20% yields, respectively.[251] Compound **189** is most probably formed via the aldol-type condensation product (**190**) followed by a normal Gewald reaction. Attempts to generalize this reaction by applying it to 2-amino-1-cyano-1,3-dicarboethoxypropane or 3-amino-2-cyano-3-benzylacrylate were unsuccessful. Compound **188** has also been obtained by the reaction of tetra-cyanoethylene or tetracyanoethane with hydrogen sulfide,[172-176] and in order to explain the formation of **188**, it is suggested that the thiation of malonitrile can lead to tetracyanoethane[251] (Scheme 69). It was recently shown[251a] that diethyl

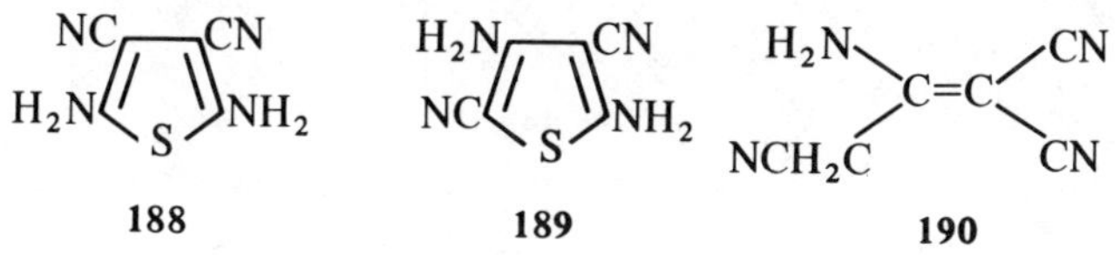

188   189   190

Scheme 69

2,5-diamino-3,4-thiophenedicarboxylate is formed in the reaction of ethyl cyano-acetate with sulfur in the presence of triethylamine and not diethyl 2,4-diamino-3,5-thiophenedicarboxylate, as previously claimed.[251] The first step in the reaction is most probably oxidative coupling to *sym.* diethyl dicyanosuccinate, which then reacts analogously to tetracyanoethane, The development of the Gewald reaction for the synthesis of 2-amino-3-cyano-, 2-amino-3-carbethoxy-, and 2-amino-3-carbonyl-substituted thiophenes has made possible the synthesis of an enormous number of thieno-fused heterocycles of great medicinal interest, such as thieno-pyrimidines and thienodiazepines.

B.   Various Reactions

A few examples of thiophene synthesis (**194**) in which the C_2S units are enethiolates (**191**) and the C_2 unit is an α-haloketone (**192**) or an acetylenic compound (**193**) have been described (Scheme 70). Enedithiolates, which have been used for this purpose, are the products of nitromethane and carbon disulfide (**195**). In aqueous solution at room temperature, they are reacted with α-chloro-acetaldehyde, α-chloroacetone, and phenacyl chloride, followed by methyl iodide and hydrochloric acid, to give the methylthio derivative **198** via **196** and **197**. With R = H, CH$_3$, and C$_6$H$_5$, 62, 33, and 27% of **198** was obtained, respectively[256] (Scheme 71). If methyl iodide is not added, the intermediate thiolate **197** can be oxidized to the disulfide **200**. It seems unlikely that the reaction proceeds via **199** to **198**.[256] The nitrodithiolate **195** has also been reacted with a number of

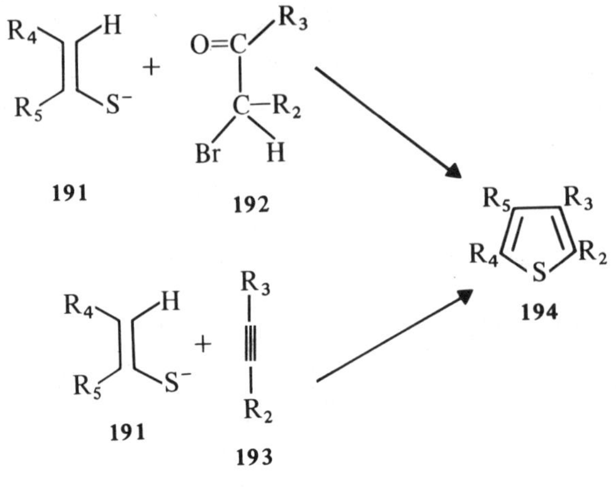

Scheme 70

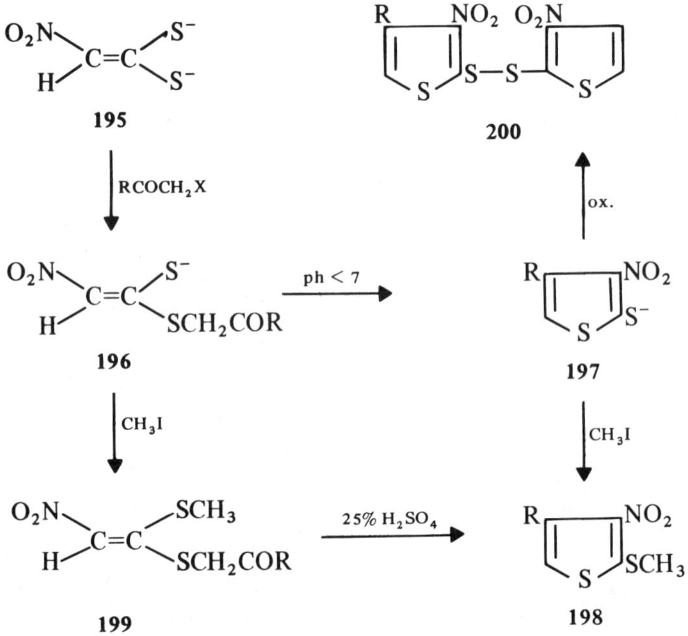

Scheme 71

chlorinated β-dicarbonyl derivatives to 3-nitrothienyl disulfides (**201**) in about 50% yield[252] (Scheme 72).

Another type of nitroenethiolate (**202**), prepared through the reaction of phenylisothiocyanate with nitromethane, gives 2-amino-3-nitrothiophenes (**203**) upon reaction with α-bromoketones[253] (Scheme 73).

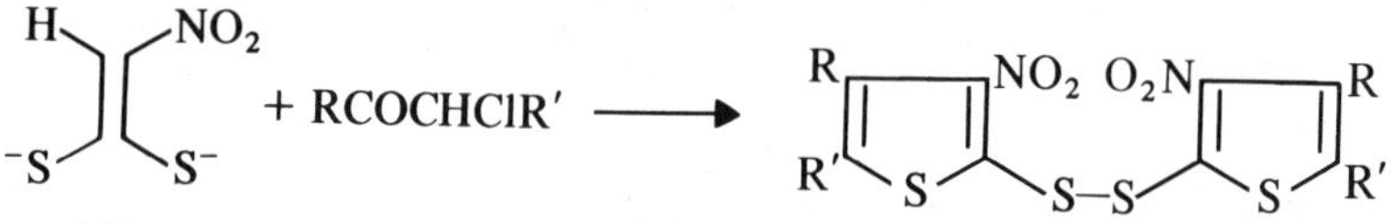

$$\text{195} + \text{RCOCHCIR}' \longrightarrow$$

R = H, R' = CO$_2$C$_2$H$_5$
R = CH$_3$, R' = CO$_2$C$_2$H$_5$
R = C$_6$H$_5$, R' = CO$_2$C$_2$H$_5$
R = CH$_3$, R' = COCH$_3$
R = C$_6$H$_5$, R' = COCH$_3$

Scheme 72

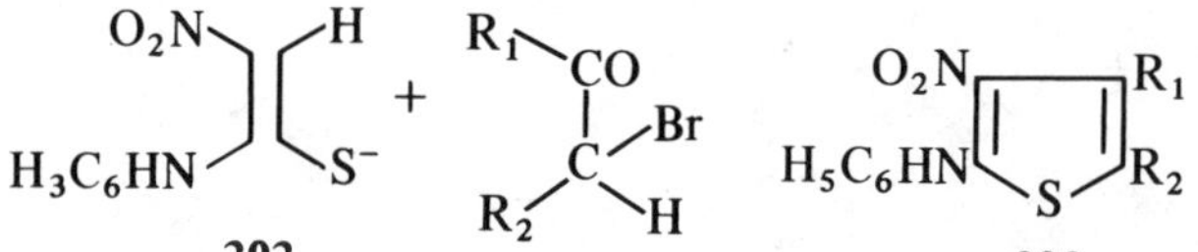

R$_1$ = C$_6$H$_5$, R$_2$ = H   63%
R$_1$ = CH$_3$, R$_2$ = H   58%
R$_1$ = R$_2$ = C$_6$H$_5$ 35%

Scheme 73

The possible ring-closure to a thiazole, according to Hantsch, does not occur.

Enedithiolates with CN, carbonyl. or ester functions, instead of the nitro group, are of even greater importance for the synthesis of thiophenes (compare the Gompper reaction), but react as a C$_3$S-building block with C-units and treated in the appropriate chapter.

The lithium salts of 2-arylethynethiolates react with alkyl phenylpropiolates to give alkyl 2,4-diaryl-3-thiophenecarboxylates in 65–82% yield; the corresponding potassium salts only give conjugate addition.[255a]

The enethiolate **204** reacts with dimethyl acetylenedicarboxylate in dimethoxyethane at −60°C in the presence of hydrogen chloride to give the dihydro derivative **205**[254] (Scheme 74). Refluxing α-chlorothioacetanilides (**206**) in methanol leads to

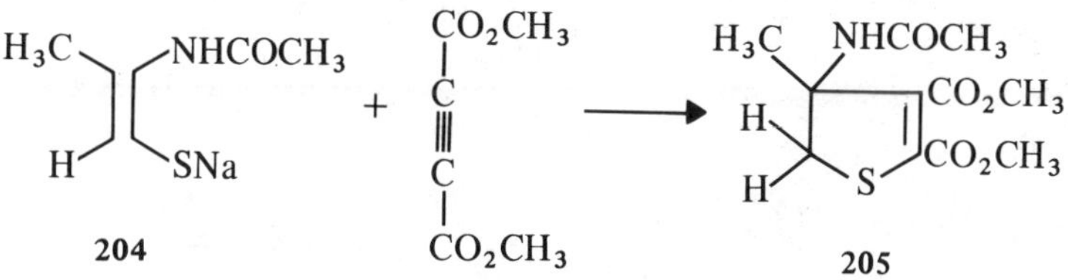

Scheme 74

the formation of 2,4-diaminothiophenes (**207**) in relatively low yields. The reaction is assumed to proceed via **208** and **209**, which undergoes C$_2$S + C$_2$ ring-closure to **209a**, which loses sulfur and chloride ion to give **207** after tautomerization[255] (Scheme 75).

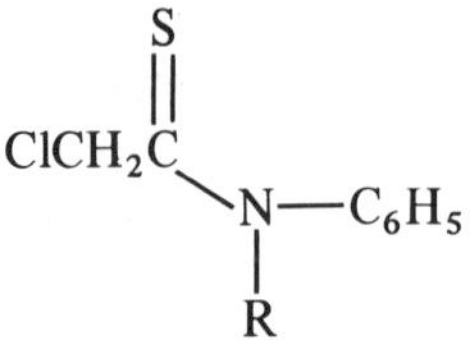

R = CH$_3$, C$_2$H$_5$, (CH$_3$)$_2$CH

206

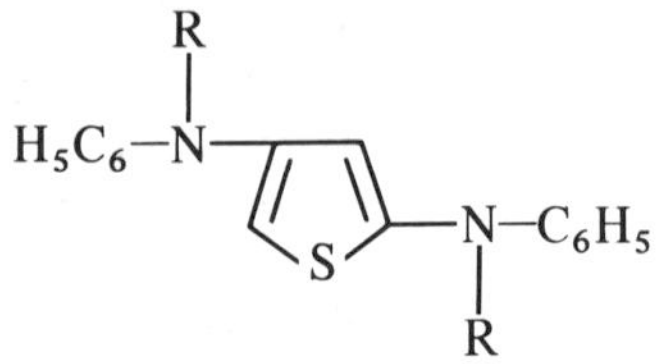

207

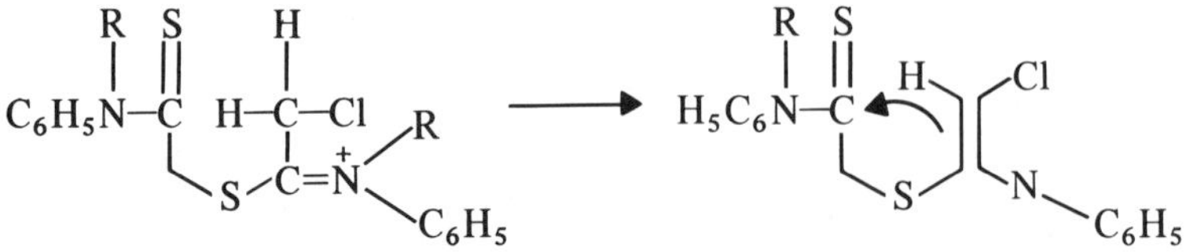

208   **209**

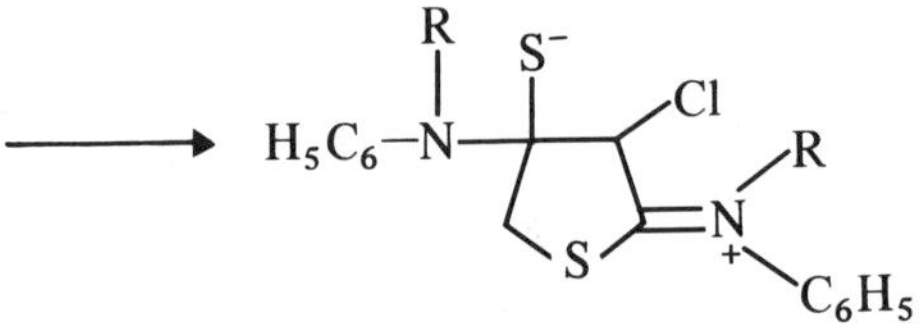

209a

Scheme 75

A very interesting route to 3-vinylthiophenes (**211**) utilizes α-mercapto aldehyde or ketones as the C$_2$S part and 1,3-butadienyltriphenylphosphonium salt (**210**) as the C$_2$-reagent. The reaction was best carried out in refluxing anhydrous picoline as solvent, although pyridine can be used, and with triethylamine as base[257] (Scheme 76). The mechanism of this reaction is not clear; usually, α-mercaptocarbonyl derivatives react with vinylphosphonium salts to give 2,5-dihydrothiophenes (see Section V).

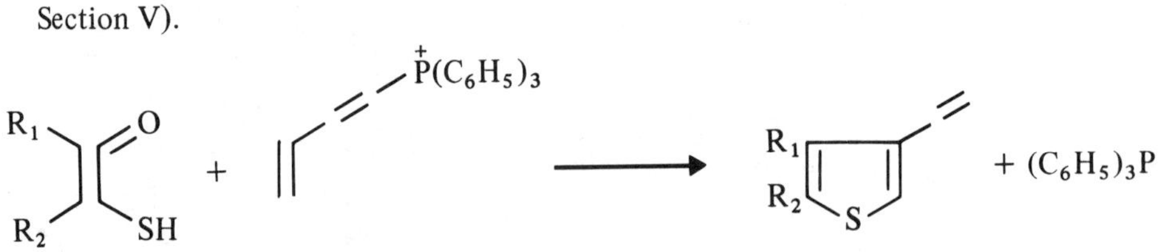

210   **211**

R$_1$ = R$_2$ = —(CH$_2$)$_4$—
R$_1$ = C$_2$H$_5$, R$_2$ = CH$_3$
R$_1$ = H, R$_2$ = C$_2$H$_5$

Scheme 76

In the reaction of 2-mercaptobutyraldehyde (**212**) with **210**, a second product was isolated and identified as 2,4-diethylthiophene (**215**). Apparently, the dimer **213** of **212** is dehydrated to **214**, followed by sulfur extrusion[257] (Scheme 77).

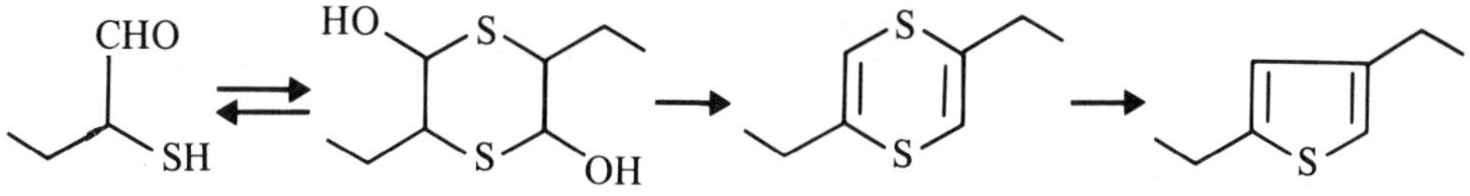

212   **213**   **214**   **215**

Scheme 77

Another C_2S unit that was used derives from β-mercaptoketones such as β-mercaptocyclohexanones (**216**), obtained by the addition of hydrogen sulfide to 2-cyclohexenones, and with α-dicarbonyl derivatives as C_2-reagents, leading to thiophenes of type **217**. As α-dicarbonyl reagents, 40% aqueous glyoxal, phenyl-glyoxal monohydrate, 40% aqueous pyruvic aldehyde, and 2,3-butandione have been used with benzene as cosolvent (Scheme 78). As catalyst, either the acidic

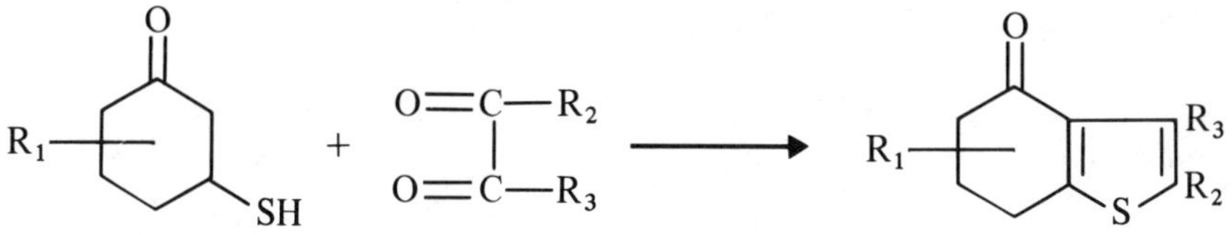

216   **217**

$R_1 = H, R_2 = R_3 = H$
$R_1 = 4\text{-}CH_3, R_2 = C_6H_5, R_3 = H$
$R_1 = 5\text{-}CH_3, R_2 = R_3 = CH_3$
$R_1 = H, R_2 = CH_3, R_3 = H$

Scheme 78

sulfonated ion-exchange resin Dowex 50W-XA (which made separation easy) or dodecylbenzenesulfonic acid (which gave a higher yield, but made isolation of pure product more difficult) was used.[258] Instead of the 3-mercaptocyclohexanones, their ethylene acetals, which are more stable, could also be used. The reaction of 3-mercaptocyclohexanone with pyruvic aldehyde was regiospecific. Only 3-methyl-6,7-dihydro-5H-benzo[b]thiophen-4-one was obtained, indicating that sulfur attacks at the aldehyde carbonyl. Yields are usually over 50%, except with butanedione, which even after a 48-h reaction only gave a 10% yield.[258]

Thiophenes are formed in the reaction of bis-aminodisulfides (**218**) with acetyl-enes at 140°C, and the following route for their formation has been suggested (Scheme 79). It is assumed that via a radicaloid pathway, the thiirene **219** is formed as an intermediate, which opens to the 1,3-dipoles, **220** and **221**, which react with the acetylene to give the thiophene **222** and **223**. The reaction shows high regio-specificity. Thus, with phenylacetylene ($R = C_6H_5$, $R' = H$), **222** ($R = C_6H_5$, $R' = H$) and **223** ($R = C_6H_5$, $R' = H$) were in a 74:26 ratio and a total yield of

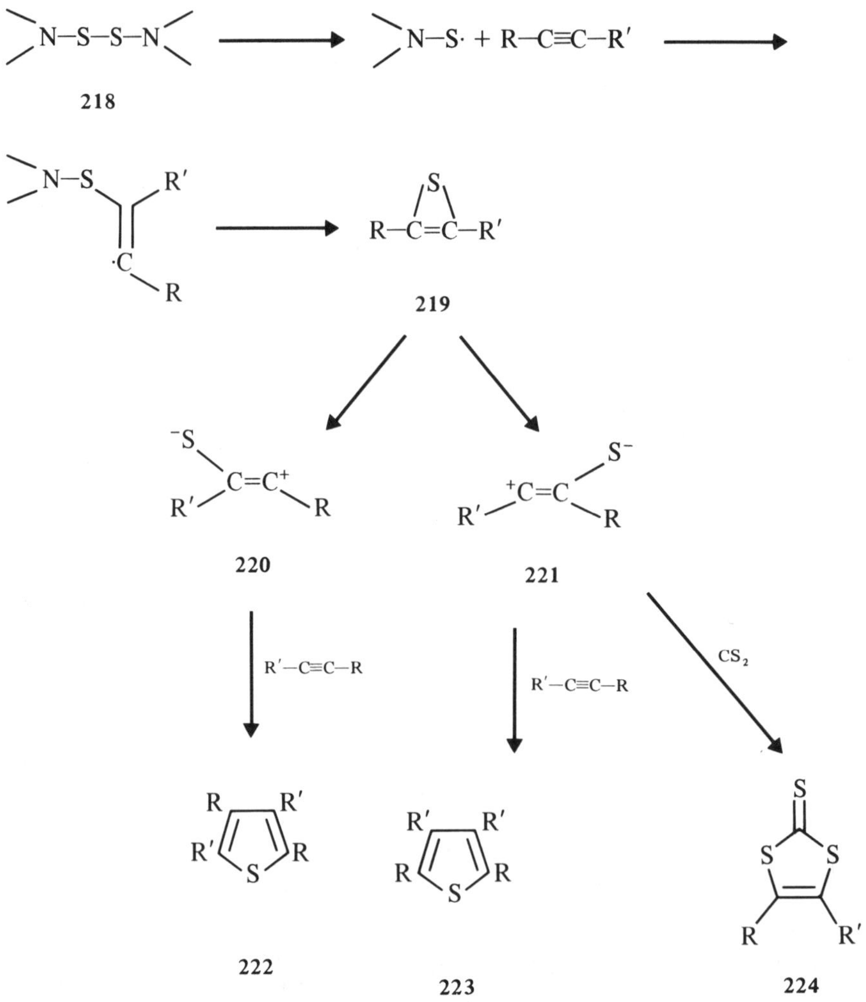

Scheme 79

52% was obtained. On the other hand, methyl propiolate ($R = CO_2CH_3$, $R' = H$) and ethyl phenyl propiolate ($R = CO_2C_2H_5$, $R' = C_6H_5$) gave exclusively 223 ($R = C_6H_5$, $R' = H$) and 223 ($R = CO_2C_2H_5$, $R' = C_6H_5$) in 48 and 67% yields, respectively. The regiospecificity was explained in terms of frontier molecular orbital theory, The intermediate dipoles could be trapped with carbon disulfide, resulting in the formation of trithiocarbonate (224).[259]

The reaction of nickel sulfide with acetylenes yields complexes of α,β-dithio-ketones such as 225, which react with diphenylacetylene at temperatures as low as 140°C to give tetraphenylthiophene in 78% yield, and with dimethyl acetylenedicarboxylate to give **226** in 84% yield.[260, 261] Another way of obtaining bis-dithio α-diketone complexes of nickel, platinum, and palladium is through the reaction of benzoins or acyloins with P_4S_{10}, followed by reaction with the metal salts.[262] The thiophenes are most probably formed via the 1,4-dithiines. The thioketo-carbenehexacarbonyl diiron complex (**227**, R = C_6H_5) also yields tetraphenyl-thiophene on heating. Upon reaction with diphenylacetylene, **227** (R = $CH_3OC_6H_4$) gives 2,3-bis-*p*-methoxyphenyl-4,5-diphenylthiophene. The simple complex (**227**, R = R' = H) gives thiophene upon heating and reaction with hexafluorobutyne-2 leads to the formation of 2,3-bis(trifluoromethyl)thiophene (Scheme 80). The

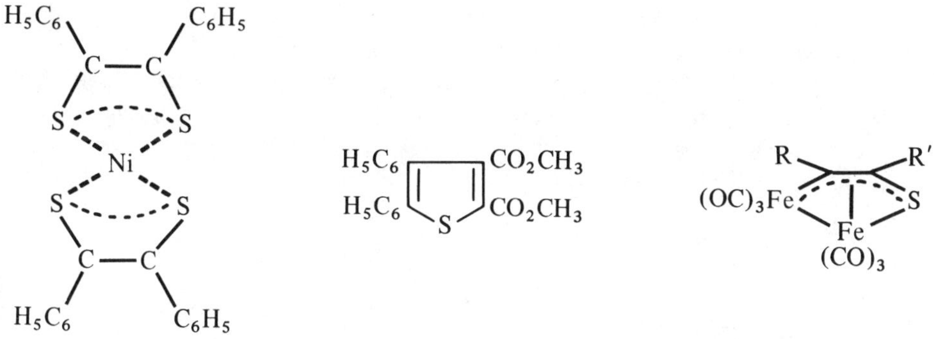

| 225 | 226 | 227 |

Scheme 80

thiophenes are formed via 1,3-dipolar addition of the thioketo carbene moiety to the acetylenes.[263] It is possible that transition metal complexes of the type discussed above are important in the classical synthesis of thiophene from acetylene and pyrite.[264] An example of a $C_2S + C_2$ reaction type at the thiophene-1,1-dioxide level is the reaction of thiirene 1,1-dioxides (**228**) with metallated nitriles that have no α-hydrogens. With the sodium salts of aryl-substituted nitriles, the 5-imino derivatives **229a** and **229b** were obtained in 85 and 42% yields, respectively, whereas the lithium salt of isobutyronitrile gave **230** in 30% yield (Scheme 81).

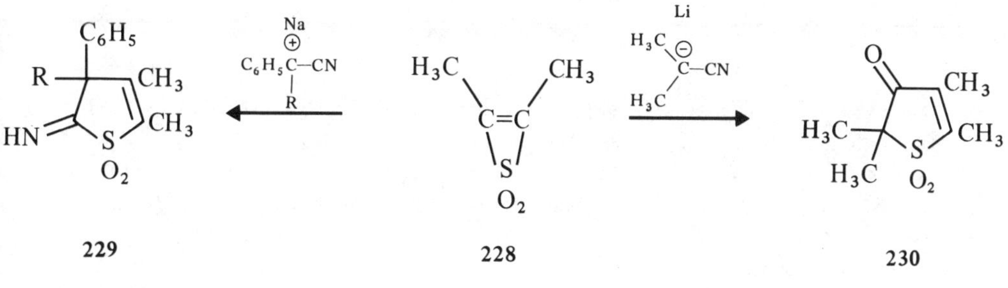

| 229 | 228 | 230 |

a: R = C_6H_5
b: R = CH_3

Scheme 81

Apparently, **229** is formed by attack of the aryl-substituted carbanion on the carbon atom of the thiirene, while the aliphatic carbanion attacked the sulfur atom to give **230**.[265]

4. $C_3S + C$ Methods

A.   The Gompper Reaction

Gompper et al. showed in 1962 that active methylene compounds can be condensed with carbon disulfide under basic conditions to give enedithiolates (**231**), which upon alkylation with α-halogenoacetic acid derivatives give ketene-mercaptals (**232**). These are easily ring-closed to 3-hydroxy- and 3-aminothiophene derivatives (**233**) upon treatment with a base (Scheme 82). Renewed ring-closure leads to thieno[2,3-b]thiophene derivatives (**234**). The reaction can also be carried out in a one-pot procedure, and stepwise alkylation of the enedithiolates can also be

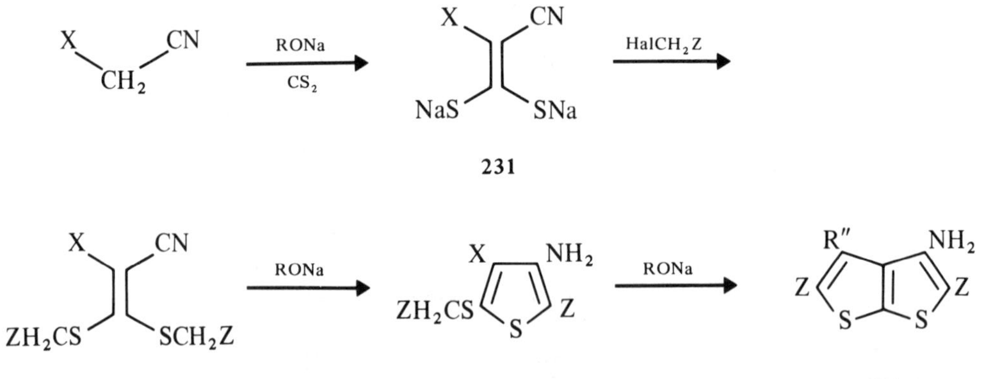

231

232   **233**   **234**

$$X = CN, CO_2R, CONH_2, C_6H_5$$
$$Z = CN, CO_2R, CONH_2$$
$$R'' = OH, NH_2$$

Scheme 82

achieved.[274, 275] This thiophene synthesis has been further developed. Other active methylene derivatives, such as aroyl acetonitriles[276, 277] and β-diketones[278, 279] have also been used, as well as 1-cyanomethylpyridinium chloride.[280] A recent example is the use of **234a**, prepared from benzylidene aminoacetonitrile which, upon reaction with methyl iodide and phenacyl bromide or bromo acetone, were ring-closed to **234b**[279a] (Scheme 83). As base in the condensation with carbon disulfide, sodium hydride or sodium hydroxide is usually used, but the ion-pair extraction technique using tetrabutylammonium salts has also been applied.[278] As alkylating agents, esters and amides of bromo- and chloroacetic acids, as well as α-chloro- or bromoketones,[278, 281] have been used. The alkylated products often ring-close spontaneously to the thiophenes, or catalytic amounts of sodium methoxide, ethoxide, or other bases are used to achieve the cyclization. Monoalkylation of 1,1-dithiolates with methyl chloroacetate or chloroacetamide led, in some cases, to the formation of 1,3-dithiolanone-4 (235).[275, 281] However, treatment of 235 with methyl iodide and sodium methoxide gave a mixture of 236 and 237, which was also obtained when treating the dithiolate with 1 mole each of methyl iodide, methyl chloroacetate, and sodium methoxide (Scheme 83). In the corresponding

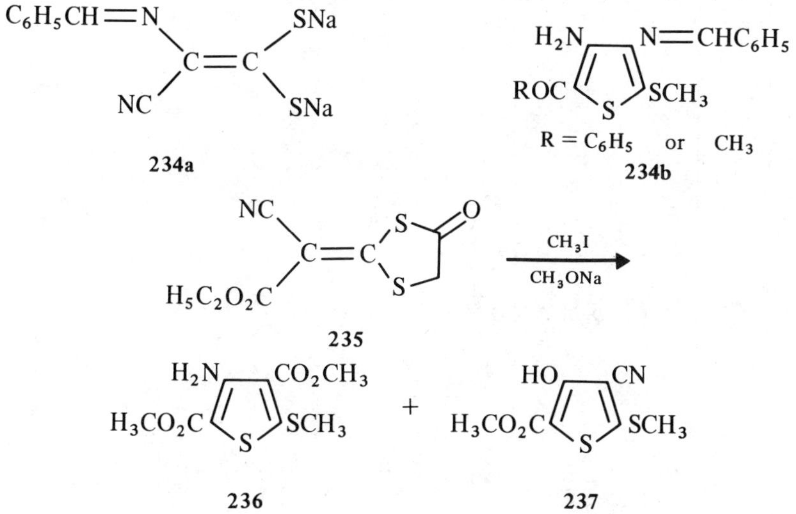

Scheme 83

reaction of the dithiolate (238) derived from aroylacetonitriles and CS_2, final ring-closure occurs selectively to the carbonyl group, giving 239[277] (Scheme 84).

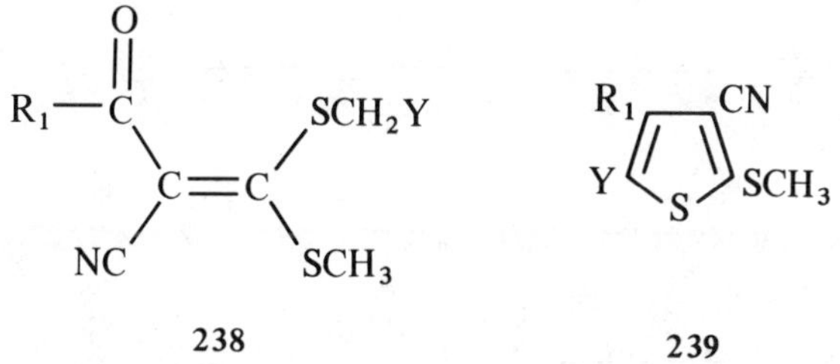

Scheme 84

The thiophene (240) formed from 1-cyanomethyl pyridinium chloride and CS_2, followed by alkylation, can be further transformed to 3,4-diaminothiophenes (241)[280, 280a] (Scheme 85). Compounds prepared by the Gompper reactions are given in Table 21. The 5-SCH_3 derivatives were obtained by stepwise alkylation with a halide having an active methylene group and then with methyl iodide.

TABLE 21. THE GOMPPER REACTION: $R_4CH_2Y + CS_2 + HalCH_2R_2 + (CH_3I) \longrightarrow$

Y	R_2	R_3	R_4	R_5	Yield of Thiophene (%)	Reference
CN	CO_2CH_3	NH_2	C_6H_5	$SCH_2CO_2CH_3$	94	275
CN	CN	NH_2	C_6H_5	SCH_2CN	81	275
CN	$CONH_2$	NH_2	C_6H_5	SCH_2CONH_2	82	275
CN	CO_2CH_3	NH_2	$CONH_2$	SCH_3	59	275
CN	CO_2CH_3	NH_2	CO_2CH_3	$SCH_2CO_2C_2H_5$	40	275
CN	CO_2CH_3	NH_2	C_6H_5	SCH_2CONH_2	65	275
CN	$CO_2C_2H_5$	NH_2	CN	$SCH_2CO_2C_2H_5$	87	275
CN	$CO_2C_2H_5$	NH_2	$CONH_2$	$SCH_2CO_2CH_3$	50	275
CN	$CO_2C_2H_5$	NH_2	$CONH_2$	$SCH_2CO_2C_2H_5$	—	275
CN	$CO_2C_2H_5$	NH_2	$CONH_2$	SCH_2COOH	33	275
CO_2CH_3	CO_2CH_3	OH	CO_2CH_3	$SCH_2CO_2CH_3$	—	275
CN	$CONH_2$	NH_2	CO_2CH_3	SH	56	284
C_6H_5CO	CO_2CH_3	C_6H_5	CN	$SCH_2CO_2CH_3$	14	276
CN	CO_2CH_3	NH_2	CO_2CH_3	SCH_3	—	281
$CO_2C_2H_5$	CO_2CH_3	OH	CN	SCH_3	—	281
CN	$p\text{-}C_6H_4CO$	NH_2	CN	$SCH_2COC_6H_4Br\text{-}p$	75	281
CH_3CO	$COCH_3$	CH_3	$COCH_3$	SH	45	278
CH_3CO	$COCH_3$	CH_3	$COCH_3$	SCH_3	52	277
CH_3CO	CO_2CH_3	CH_3	$COCH_3$	SH	47	277
CN	$CONH_2$	NH_2	$CO_2C_2H_5$	SCH_3	42	285
CN	$CONH_2$	NH_2	CN	SCH_3	78	285
CN	$-CONHCH_3$	NH_2	CN	SCH_3	71	285
CN	$CONH_2$	NH_2	$CONHCH_3$	SCH_3	65	285
CN	$CONHCH_3$	NH_2	$CONH_2$	SCH_3	—	285
CN	$CO_2C_2H_5$	NH_2	$CONH_2$	SCH_3	64	285
CN	$COCH_3$	NH_2	$CONH_2$	SCH_3	53	285
CN	COC_6H_5	NH_2	$CONH_2$	SCH_3	59	285
CN	$CONH_2$	NH_2	$CONHCH_3$	SCH_3	—	285

CN	CONHCH$_3$	NH$_2$	CONH$_2$	SCH$_3$	66	285
CN	CO$_2$CH$_3$	NH$_2$	CONH$_3$	SCH$_3$	58	285
C$_6$H$_5$CO	CN	C$_6$H$_5$	CN	SCH$_3$	45	277
p-BrC$_6$H$_4$CO	CN	p-BrC$_6$H$_4$	CN	SCH$_3$	41	277
p-ClC$_6$H$_4$CO	CN	p-ClC$_6$H$_4$	CN	SCH$_3$	37	277
3,4-Cl$_2$—C$_6$H$_3$	CN	3,4-Cl$_2$C$_6$H$_4$CO	CN	SCH$_3$	24	277
p-CH$_3$OC$_6$H$_4$	CN	p-CH$_3$OC$_6$H$_4$	CN	SCH$_3$	20	277
2-FurylCO	CN	2-Furyl	CN	SCH$_3$	23	277
2-ThCO	CN	2-Th	CN	SCH$_3$	35	277
C$_6$H$_5$CO	CO$_2$CH$_3$	C$_6$H$_5$	CN	SCH$_3$	45	277
p-BrC$_6$H$_4$CO	CO$_2$CH$_3$	p-BrC$_6$H$_4$	CN	SCH$_3$	60	277
p-ClC$_6$H$_4$CO	CO$_2$CH$_3$	p-ClC$_6$H$_4$	CN	SCH$_3$	52	277
3,4-Cl$_2$C$_6$H$_3$CO	CO$_2$CH$_3$	3,4-Cl$_2$C$_6$H$_3$	CN	SCH$_3$	44	277
p-CH$_3$OC$_6$H$_4$CO	CO$_2$CH$_3$	p-CH$_3$OC$_6$H$_4$	CN	SCH$_3$	36	277
2-ThCO	CO$_2$CH$_3$	2-Th	CN	SCH$_3$	21	277
C$_6$H$_5$CO	COCH$_3$	C$_6$H$_5$	CN	SCH$_3$	51	277
p-BrC$_6$H$_4$CO	COCH$_3$	p-BrC$_6$H$_4$	CN	SCH$_3$	33	277
p-ClC$_6$H$_4$CO	COCH$_3$	p-ClC$_6$H$_4$	CN	SCH$_3$	54	277
3,4-Cl$_2$C$_6$H$_4$CO	COCH$_3$	3,4-Cl$_2$—C$_6$H$_3$	CN	SCH$_3$	54	277
p-CH$_3$OC$_6$H$_4$	COCH$_3$	p-CH$_3$OC$_6$H$_4$	CN	SCH$_3$	30	277
2-FurylCO	COCH$_3$	2-Furyl	CN	SCH$_3$	63	277
2-ThCO	COCH$_3$	2-Th	CN	SCH$_3$	37	277
C$_6$H$_5$CO	CN	C$_6$H$_5$	CN	SCH$_2$CN	30	277
C$_6$H$_5$CO	CO$_2$CH$_3$	C$_6$H$_5$	CN	SCH$_2$CO$_2$CH$_3$	25	277
p-BrC$_6$H$_4$CO	CO$_2$CH$_3$	p-BrC$_6$H$_4$	CN	SCH$_2$CO$_2$CH$_3$	42	277
p-ClC$_6$H$_4$CO	CO$_2$CH$_3$	p-ClC$_6$H$_4$	CN	SCH$_2$CO$_2$CH$_3$	47	277
3,4-Cl$_2$—C$_6$H$_4$CO	CO$_2$CH$_3$	3,4-Cl$_2$C$_6$H$_3$	CN	SCH$_2$COCH$_3$	26	277
2-ThCO	COCH$_3$	2-Th	CN	SCH$_2$COCH$_3$	24	277
2-FurylCO	COCH$_3$	2-Furyl	CN	SCH$_2$COCH$_3$	24	277
C$_6$H$_5$CO	COC$_6$H$_5$	C$_6$H$_5$	CN	SCH$_2$COC$_6$H$_5$	20	277
C$_6$H$_5$CO	p-NO$_2$C$_6$H$_4$	C$_6$H$_5$	CN	SCH$_2$COC$_6$H$_4$NO$_2$-p	74	277
CN	CN	NH$_2$	CN (pyridinium N$^+$)	S$^-$	87	280

69

TABLE 21. *Continued*

Y	R_2	R_3	R_4	R_5	Yield of Thiophene (%)	Reference
CN	$CO_2C_2H_5$	NH_2	pyridinium	S^-	95	280
CN	COC_6H_5	NH_2	pyridinium	S^-	90	280
CN	$CONH_2$	NH_2	pyridinium	S^-	77	280
CN	COC_6H_5	NH_2	$-N=CHC_6H_5$	SCH_3	55	279a
CN	$COCH_3$	NH_2	$-N=CHC_6H_5$	SCH_3	37	279a

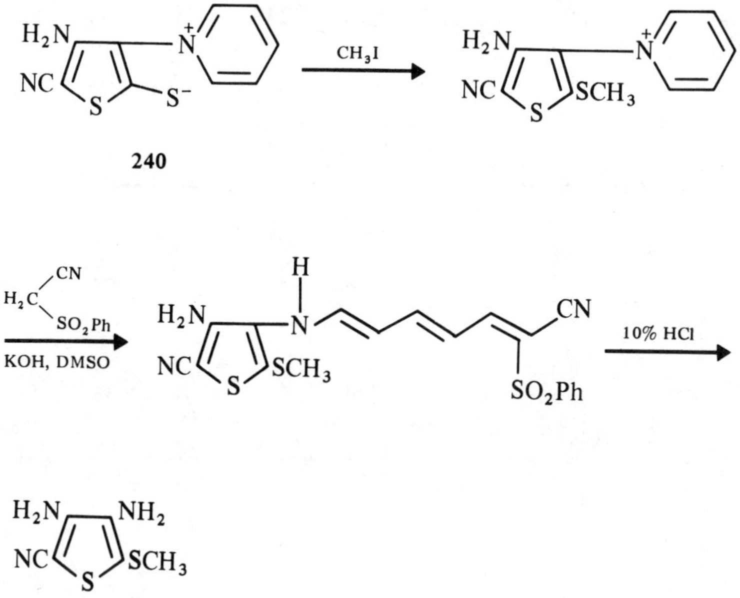

240

241

Scheme 85

Crude carbonyl sulfide can be used instead of carbon disulfide in condensation with malonitrile, and alkylation of the intermediate (**242**) with phenacyl bromide gives 3-amino 5-hydroxythiophene (**243**) in 64% yield[282] (Scheme 86). The gener-

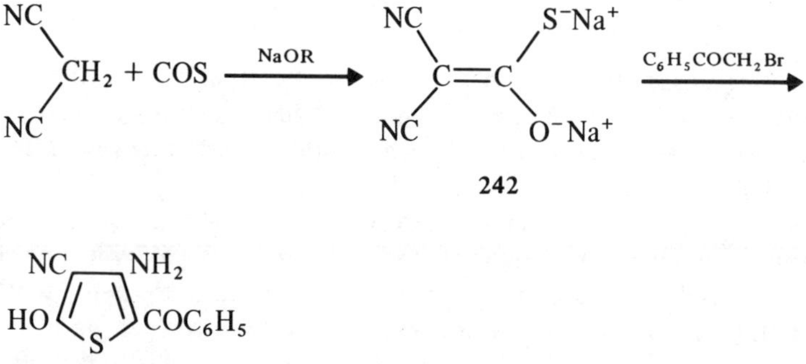

242

243

Scheme 86

ality of this reaction has not been demonstrated, and the use of ethyl cyanoacetate instead of malonitrile gave no thiophene.

A modification, also introduced by Gompper,[274] consists of the use of arylisothiocyanates instead of carbon disulfide in the condensation with active methylene derivatives, which leads to ketene S,N-acetals (245). By base treatment, these acetals can be ring-closed to thiophenes (246). From nitriles, 2,4-diamino derivatives are obtained (Scheme 87). However, depending upon reaction conditions,

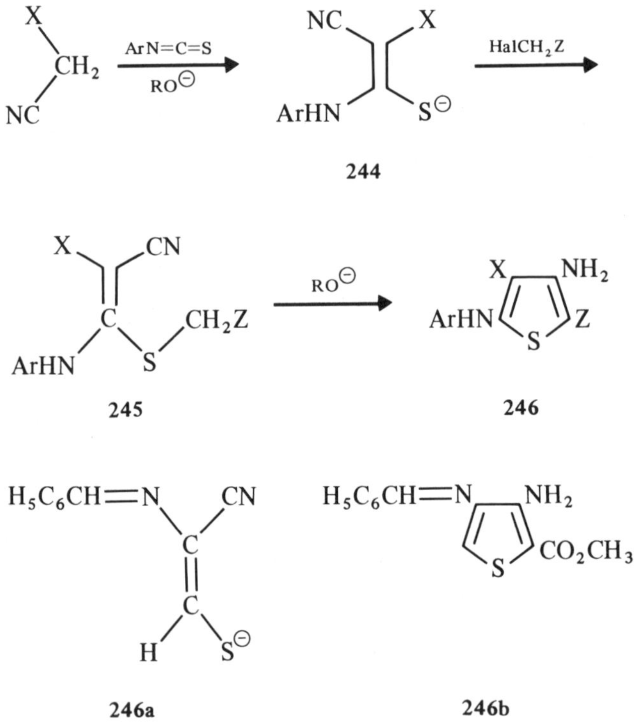

244

245   **246**

246a   **246b**

Scheme 87

alternative ring-closure to thiazolines may occur. Reaction of benzylidene aminoacetonitrile with ethyl thioformate gave the salt **246a**, which upon treatment with methyl chloroacetate followed by hydrogen chloride in wet ether gave **246b**, a key intermediate for biotin synthesis[274a] (Scheme 87).

The primary condensation products from cyanoacetates or cyanoacetamides give **247** upon alkylation with α-haloketones. In principle, **247** can ring-close to the thiophene **248** by attack of the carbanion derived from the α-position of the ketone on the cyano group, or by attack of the free electron-pair on nitrogen on the ketonic carbonyl to give the thiazoline **249** (Scheme 88). Reaction at low temperature in alcohol leads to **248**, whereas in refluxing ethanol, **249** is formed.[283] However, the structures of the components are also important. Thus, reaction of **250** with chloroacetone at low temperatures gave the expected thiophene (**251**), when R = C₆H₅, while the analogous alkyl derivative gave **252**.[283] Also, **253** gave the thiazolidone (**254**) at room temperature in ethanol (Scheme 89). In all cases when α-bromoketones were used instead of haloacetate as alkylating agents, the expected thiophenes were obtained.[277]

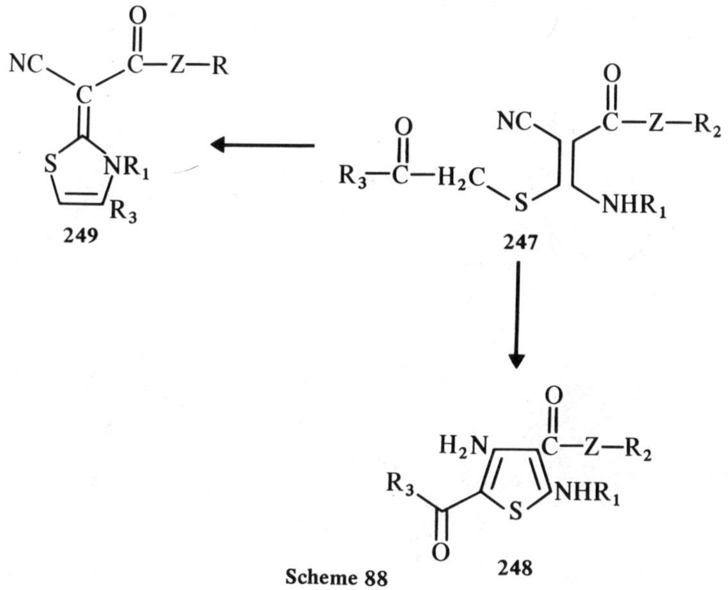

Scheme 88

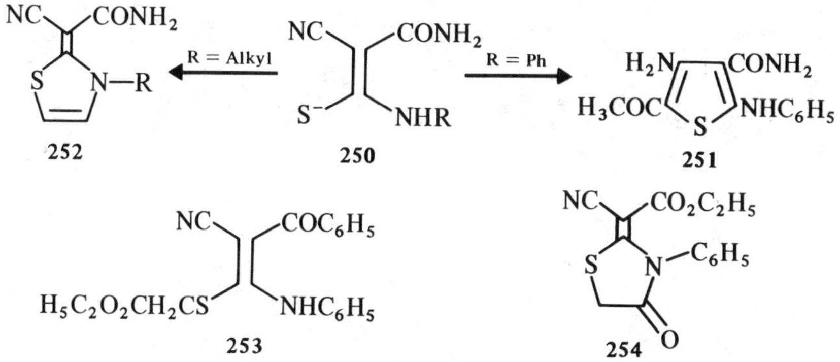

Scheme 89

The reaction of **255** with ethyl α-bromophenylacetate and 3-bromo-butan-2-one gives primarily **256** and **257**, which lost the carboethoxy group and acetyl group, respectively, to yield **258** and **259**. These products exist as stable imino forms[286] (Scheme 90) (Table 22).

Another way of using *S,N* or *S,O* keteneacetals for the synthesis of thiophenes starts from dialkyl monothione malonates (**260**), which are available from cyanoacetic esters via the imino ester hydrochlorides.[287] *S*-Alkylation with an α-haloester or α-haloketone gives the ketone acetal (**261**), which upon base treatment ringcloses to **262**. Compound **260** can also first be transformed to the amide **263**, which gives **264** upon ring-closure (Scheme 91).

TABLE 22. THE GOMPPER REACTION: $R_4CH_2Y + R\text{—}N{=}C{=}S + HalCH_2R_2 \longrightarrow$

$$\underset{R_5\;\;\;S\;\;\;R_2}{\overset{R_4\;\;\;\;R_3}{\text{(thiophene)}}}$$

Y	R	R_2	R_3	R_4	R_5	Yield of Thiophene (%)	Reference
CN	C_6H_5	C_6H_5CO	NH_2	$CO_2C_2H_5$	NHC_6H_5	60	283
CN	$C_6H_5CH_2$	CH_3CO	NH_2	$CO_2C_2H_5$	$NHCH_2C_6H_5$	20	283
CN	$C(CH_3)_3$	CH_3CO	NH_2	$CO_2C_2H_5$	$NHC(CH_3)_3$	40	283
CN	C_6H_{11}	CH_3CO	NH_2	$CONHCH_3$	NHC_6H_{11}	40	283
CN	CH_3	C_6H_5CO	NH_2	$CONHCH_3$	$NHCH_3$	40	283
CN	CH_3	CH_3CO	NH_2	$CONHCH_3$	$NHCH_3$	30	283
CN	C_3H_5	C_6H_5CO	NH_2	$CONHCH_3$	NHC_3H_5	30	283
CN	C_6H_5	C_6H_5CO	NH_2	$CONHCH_3$	NHC_6H_5	75	283
CN	C_6H_5	$p\text{-}ClC_6H_4CO$	NH_2	$CONCHCH_3$	NHC_6H_5	30	283
CN	C_6H_5	$C_6H_5\text{—}C_6H_4CO$	NH_2	$CONHCH_3$	NHC_6H_5	60	283
CN	$m\text{-}O_2NC_6H_4$	C_6H_5CO	NH_2	$CONHCH_3$	$NHC_6H_4NO_2\text{-}m$	45	283
CN	C_6H_5	C_6H_5CO	NH_2	$CONHC_6H_5$	NHC_6H_5	70	283
CN	C_6H_5	$C_6H_5\text{—}C_6H_5CO$	NH_2	$CONHC_6H_5$	NHC_6H_5	70	283
CN	$m\text{-}O_2NC_6H_4$	C_6H_5CO	NH_2	$CONHC_6H_5$	$NHC_6H_4NO_2\text{-}m$	60	283
CN	C_6H_5	CH_3CO	NH_2	$CONHC_6H_5$	NHC_6H_5	30	283
CN	C_6H_5	C_6H_5CO	NH_2	$CONHN{\langle}\text{piperidino}{\rangle}$	NHC_6H_5	80	283
CN	C_6H_5	C_6H_5CO	NH_2	$CONHN(CH_3)_2$	NHC_6H_5	60	283
C_6H_5CO	C_6H_5	CH_3CO	C_6H_5	CN	NHC_6H_5	75	277
C_6H_5CO	C_6H_5	C_6H_5CO	C_6H_5	CN	NHC_6H_5	79	277
C_6H_5CO	C_6H_5	$p\text{-}BrC_6H_4CO$	C_6H_5	CN	NHC_6H_5	95	277
C_6H_5CO	C_6H_5	$p\text{-}ClC_6H_4CO$	C_6H_5	CN	NHC_6H_5	95	277
C_6H_5CO	C_6H_5	$p\text{-}O_2NC_6H_4$	C_6H_5	CN	NHC_6H_5	21	277
CO	C_6H_5	$CO_2C_2H_5$	OH	$CONHC_6H_4OCH_3\text{-}p$	C_6H_5NH	62	289

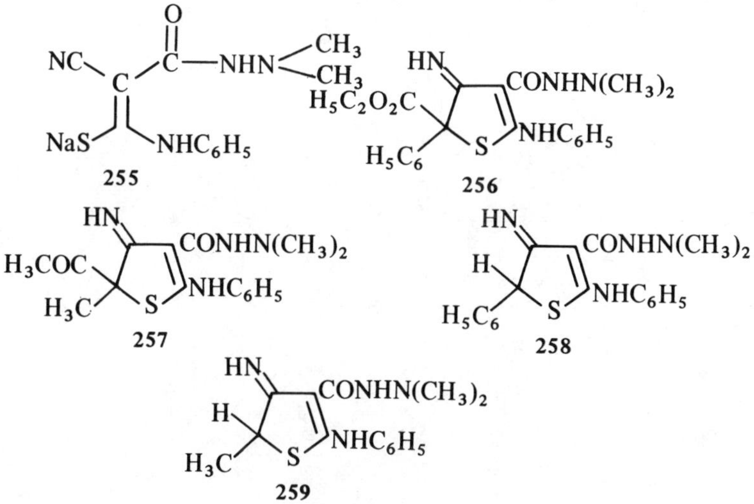

$$
\begin{array}{c}
\text{NC} \diagdown \underset{|}{\text{C}} - \overset{\overset{\text{O}}{\parallel}}{\text{C}} - \text{NHN} \diagup \overset{\text{CH}_3}{\underset{\text{CH}_3}{}} \\
\text{NaS} \diagup \overset{\|}{\text{C}} \diagdown \text{NHC}_6\text{H}_5
\end{array}
$$

255

$$
\begin{array}{c}
\text{HN} \\
\text{H}_5\text{C}_2\text{O}_2\text{C} \diagdown \quad \diagup \text{CONHN(CH}_3)_2 \\
\text{H}_5\text{C}_6 \diagdown_{\text{S}} \diagup \text{NHC}_6\text{H}_5
\end{array}
$$

256

$$
\begin{array}{c}
\text{HN} \\
\text{H}_3\text{COC} \diagdown \quad \diagup \text{CONHN(CH}_3)_2 \\
\text{H}_3\text{C} \diagdown_{\text{S}} \diagup \text{NHC}_6\text{H}_5
\end{array}
$$

257

$$
\begin{array}{c}
\text{HN} \\
\text{H} \diagdown \quad \diagup \text{CONHN(CH}_3)_2 \\
\text{H}_5\text{C}_6 \diagdown_{\text{S}} \diagup \text{NHC}_6\text{H}_5
\end{array}
$$

258

$$
\begin{array}{c}
\text{HN} \\
\text{H} \diagdown \quad \diagup \text{CONHN(CH}_3)_2 \\
\text{H}_3\text{C} \diagdown_{\text{S}} \diagup \text{NHC}_6\text{H}_5
\end{array}
$$

259

Scheme 90

$$
\underset{\textbf{263}}{\text{H}_5\text{C}_2\text{O}_2\text{CCH}_2\overset{\overset{\text{S}}{\parallel}}{\text{C}}\text{NR}_2} \quad \xrightarrow[\text{2) NaH}]{\text{1) HalCH}_2\text{CO}_2\text{C}_2\text{H}_5} \quad \underset{\textbf{264}}{\text{O}\diagdown\text{N}\diagdown\text{S}\diagup\overset{\text{OH}}{\diagdown}\text{CO}_2\text{C}_2\text{H}_5}
$$

$$
\uparrow \text{R}'_2\text{NH}
$$

$$
\underset{\textbf{260}}{\text{H}_3\text{CO}_2\text{CCH}_2\overset{\overset{\text{S}}{\parallel}}{\text{C}}\text{OCH}_3 + \text{HalCH}_2\text{COR}} \quad \xrightarrow{\text{NaH}} \quad \underset{\textbf{261}}{\text{H}_5\text{C}_2\text{O}_2\text{CCH}=\text{C}\diagup\overset{\text{SCH}_2\text{COR}}{\diagdown\text{OCH}_3}}
$$

$$
\xrightarrow{\text{NaH}} \quad \underset{\textbf{262}}{\text{H}_3\text{CO}\diagdown_{\text{S}}\diagup\overset{\text{OH}}{\diagdown}\overset{}{\underset{\overset{\|}{\text{O}}}{\text{C}}-\text{R}'}}
$$

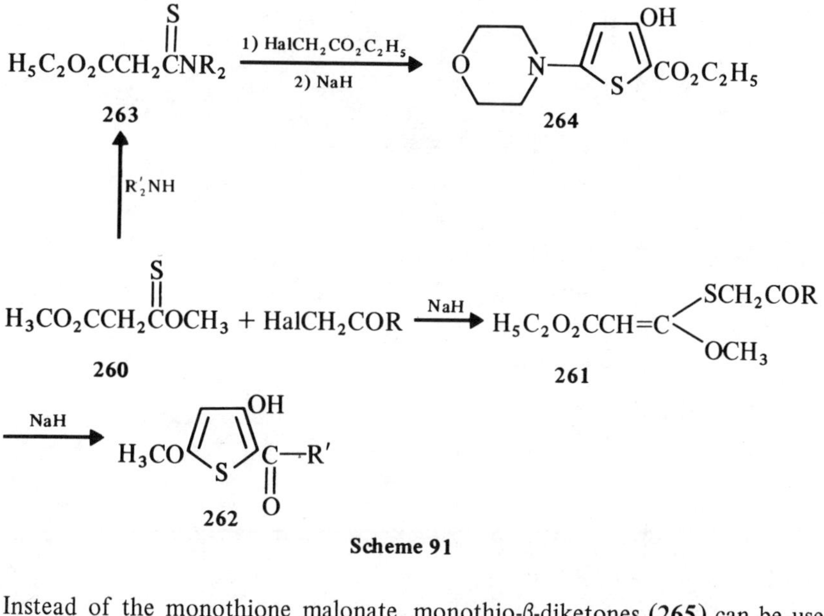

Scheme 91

Instead of the monothione malonate, monothio-β-diketones (**265**) can be used in the same way to give **266** [288] (Scheme 92). Compounds prepared in this manner are given in Table 23.

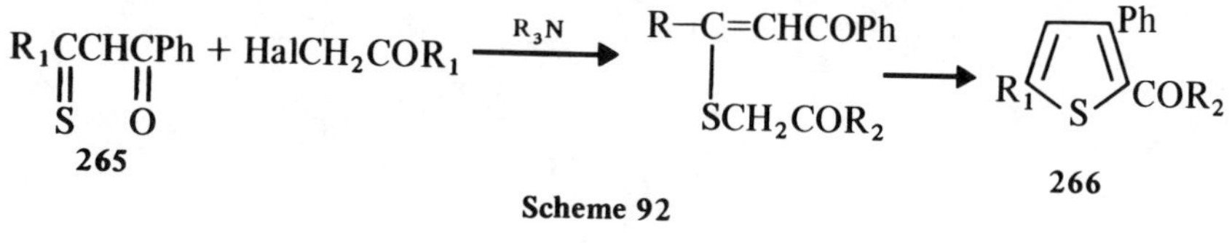

$$
\underset{\textbf{265}}{\text{R}_1\overset{\|}{\underset{\text{S}}{\text{C}}}\text{CHC}\overset{\|}{\underset{\text{O}}{}}\text{Ph} + \text{HalCH}_2\text{COR}_1} \quad \xrightarrow{\text{R}_3\text{N}} \quad \underset{\text{SCH}_2\text{COR}_2}{\text{R}-\text{C}=\text{CHCOPh}} \quad \longrightarrow \quad \underset{\textbf{266}}{\text{R}_1\diagdown_{\text{S}}\diagup\overset{\text{Ph}}{\diagdown}\text{COR}_2}
$$

Scheme 92

TABLE 23.   THIOPHENES FROM THIOESTERS AND RELATED COMPOUNDS AND α-HALOCARBONYL DERIVATIVES

$$R_1CH_2C(=S)R + HalCH_2R_2 \longrightarrow \text{(thiophene with } R_4, R_3, R_5, R_2, S)$$

R	R_1	R_2	R_3	R_4	R_5	Yield of Thiophene (%)	Reference
OCH_3	$CO_2C_2H_5$	CO_2CH_3	OH	H	OCH_3	90	287
OCH_3	$CO_2C_2H_5$	$CO_2C_2H_5$	OH	H	OCH_3	86	287
OCH_3	$CO_2C_2H_5$	COC_6H_5	OH	H	OCH_3	81	287
morpholino	$CO_2C_2H_5$	$CO_2C_2H_5$	OH	H	morpholino	61	287
C_6H_5	COC_6H_5	COC_6H_5	C_6H_5	H	C_6H_5	100	288
C_6H_5	COC_6H_5	$CO_2C_2H_5$	C_6H_5	H	C_6H_5	50	288
CH_3	COC_6H_5	COC_6H_5	C_6H_5	H	CH_3	60	288

Further development of this synthetic approach has been achieved by Hartke and Golz,[290-292] who condensed malonic acid derivatives or other active methylene derivatives with esters of thione or dithio acids in the presence of potassium alkoxides to give **267**, which was reacted with α-chlorocarbonyl derivatives to give thiophenes. From malonitrile, **268** ($R_1 = CH_3$, $R_2 = CN$) is obtained in 58% yield, and from diethyl malonate, 37% of **269** ($R_1 = CH_3$, $R_3 = CO_2C_2H_5$) was obtained. However, as observed previously with ketene S,S-acetals, if $R_2 = CO_2C_2H_5$ and $R_3 = CN$ in **267**, selectivity is not obtained and a mixture of the aminothiophene ·(**268**) and the hydroxythiophene (**269**) results[291] (Scheme 93). The thioester

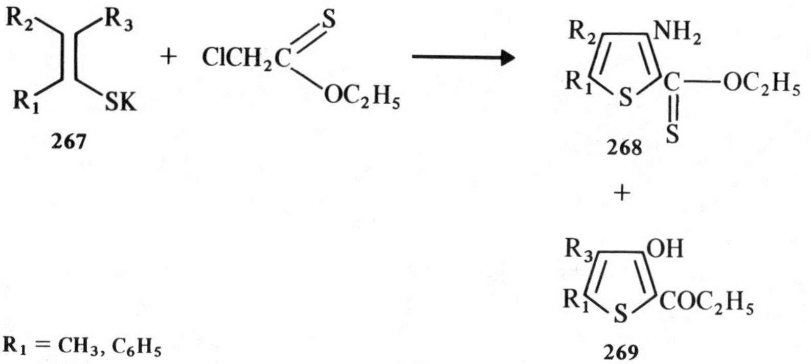

$R_1 = CH_3, C_6H_5$

Scheme 93

(**270**), prepared by alkylation of **269** ($R_1 = CH_3$, $R_3 = CO_2C_2H_5$), could again be condensed with malonitrile to **271** which, upon ring-closure with phenacyl bromide, gave the highly unsymmetrically substituted bithienyl (**272**)[292] (Scheme 94).

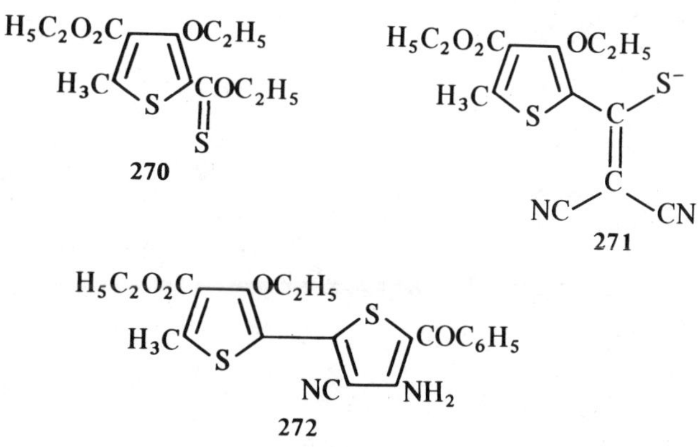

Scheme 94

Reaction of **267** ($R_1 = CH_3$) with chloromethyl methyl sulfide gives **273**, which is converted by reaction with trimethyloxonium fluoborate to the sulfonium salt (**274**). Upon cyclization by treatment with sodium cyanide, **274** gives **275**[292] (Scheme 95).

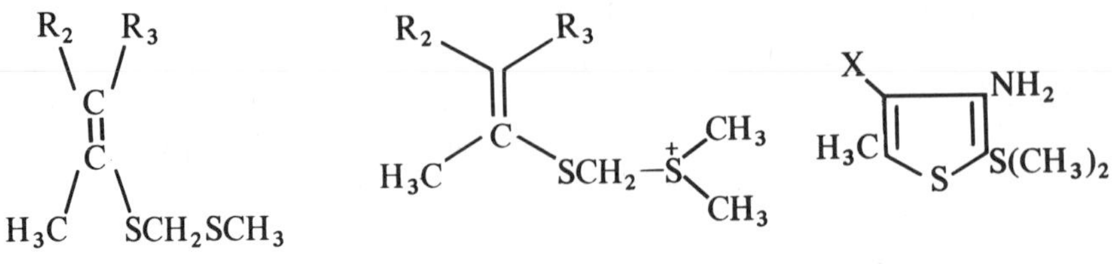

273

274

275
X = CN, CO$_2$C$_2$H$_5$

Scheme 95

B.   The Smutny, Rajappan, and Related Reactions

The discovery that 3-amino thioacrylamides and esters of type **276** are easily available from trithione and its alkylated salts, through reaction with primary or secondary amines, made a new synthesis of 5-alkylthio- or 5-amino-2-substituted thiophenes possible, through their reaction with α-halocarbonyl compounds in the presence of triethylamine. Thus, from methyl 3-morpholinodithioacrylate (**276**) and ethyl α-bromoacetate, 2-methylthio-5-carboethoxythiophene, **279** is obtained in almost quantitative yield. The reaction is assumed to proceed via S-alkylation to give **277** followed by sulfur ylide (**278**) formation and ring-closure[298] (Scheme 96). From 3-morpholino-thioacylmorpholide and α-bromo-p-nitro acetophenone,

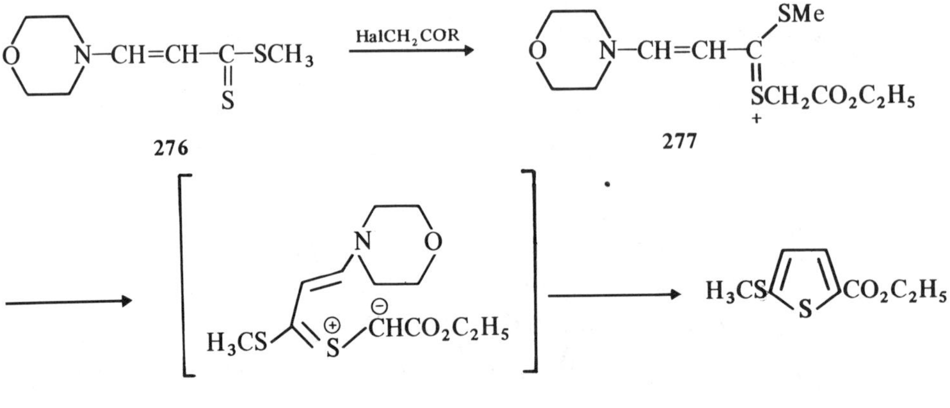

276

277

278

279

Scheme 96

2-morpholino-5-(p-nitrobenzoyl)thiophene was obtained in 70% yield. Furthermore, the electronegatively substituted dithioacrylate ester (**280**) gave the thiophene (**281**) with bromoacetone. However, **282** did not ring-close (Scheme 97).

Another route to 3-aminothioacrylamides consists of the reaction of enamines like **283** (easily obtained from ethyl acetoacetate) with isothiocyanate to **284**,

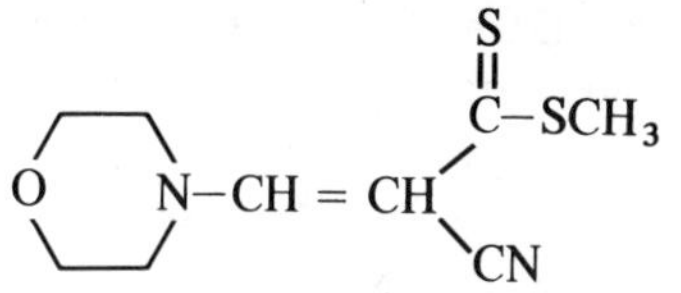

280

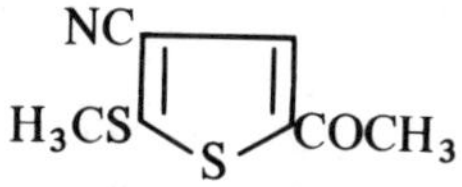

281

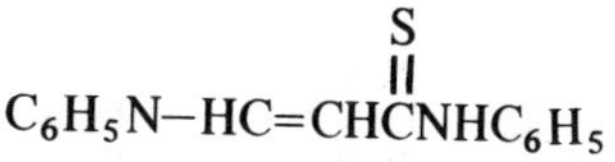

282

Scheme 97

which upon reaction with phenacyl bromide give **285** in over 50% yield. From **286**, **287** was obtained.[293-297] The presence of base is not necessary and ring-closure to thiophenes is achieved by refluxing in isopropanol. Ethyl α-chloroacetoacetate can

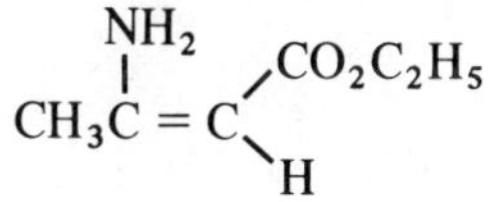

283

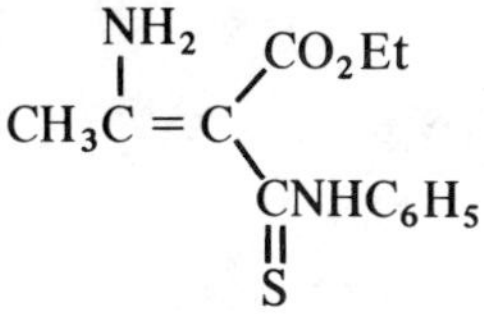

284

285

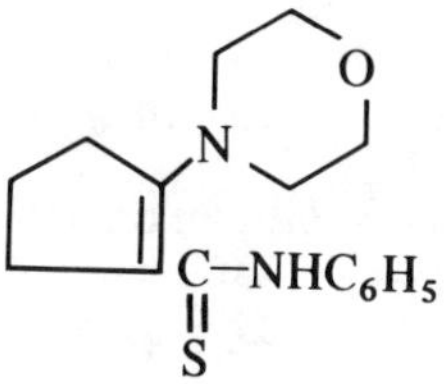

286

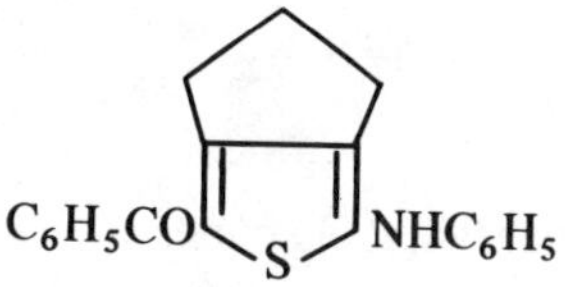

287

Scheme 98

be used instead of ethyl chloroacetate, as aromatization occurs after ring-closure with the loss of acetyl.[296] If bromonitromethane is used in the S-alkylation of **284**, instead of α-bromocarbonyl derivatives, 2-nitrothiophenes such as **290** are obtained via **288**. However, the yield is low because of competing isothiazole (**289**) formation, which, when the benzoyl isothiocyanate adduct is used, is the only product. Higher yields of thiophenes could be obtained by blocking the isothiazole route by dialkylation of the amino group.[299]

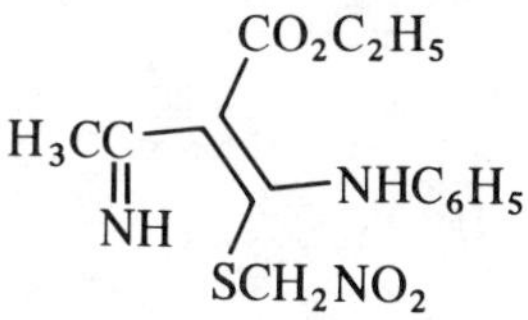

288

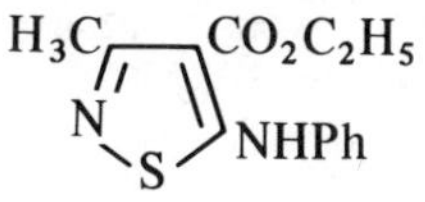

289

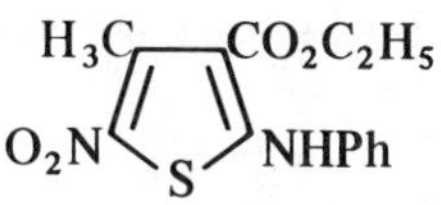

290

Scheme 99

The nitroketene aminals (291), first prepared by Gompper[284] through the condensation of nitromethane with carbon disulfide, alkylation, and reaction with secondary amines, have also been reacted with isothiocyanates to give 292, which, upon treatment with α-halocarbonyl compounds, gives 3-amino-4-nitrothiophenes (293)[300, 301] (Scheme 100). The reaction leads to thiophenes, except in the case of 294, which with phenacyl bromide gave a thiazole (295)[301] (Scheme 101).

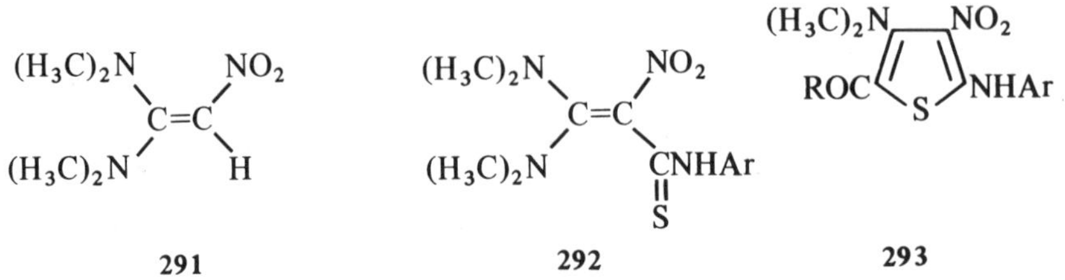

$(H_3C)_2N$⟍ NO_2
     $C=C$
$(H_3C)_2N$⟋ H

291

$(H_3C)_2N$⟍ NO_2
     $C=C$
$(H_3C)_2N$⟋ $\underset{\underset{S}{\parallel}}{C}NHAr$

292

$(H_3C)_2N$⟍⎯⎯⟍ NO_2
ROC⟍$_S$⟍$NHAr$

293

Scheme 100

$(H_3C)_2N$ NO_2
$(H_3C)_2N$ $\underset{\underset{S}{\parallel}}{C}NHCOC_6H_5$

294

$(H_3C)_2N$⟍   N⎯C_6H_5
$(H_3C)_2N$⟋ $_S$⟍COC_6H_5
     NO_2

295

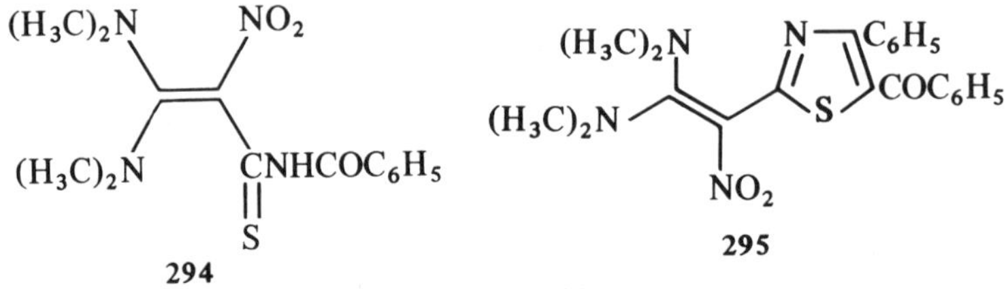

Scheme 101

Instead of 3-aminoacrylothioamides, propiolic thioamides, such as 296, can be used in the reaction with CH-acidic bromomethylene compounds, such as bromonitromethane, p-nitrobenzylbromide, or bromoacetonitrile, to give 297[302] (Scheme 102).

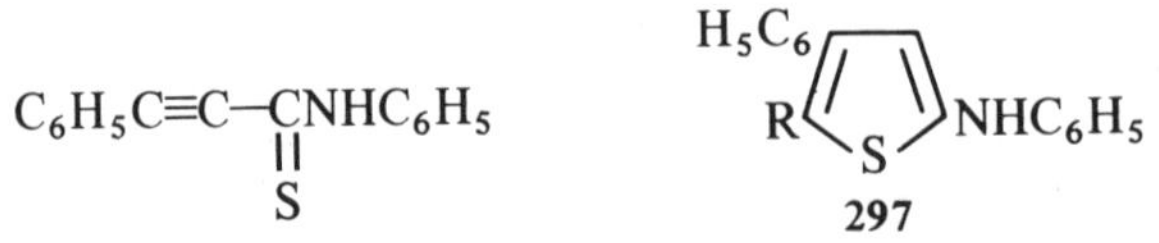

$C_6H_5C{\equiv}C-\underset{\underset{S}{\parallel}}{C}NHC_6H_5$

296

H_5C_6⎯⟍
R⟍$_S$⟍NHC_6H_5

297

R = NO_2 50%
R = $p\text{-}O_2NC_6H_4$ 45%
R = CN   83%

Scheme 102

Other compounds with the $>$N$-$C$=$C$-$C$=$S structural units are 2-aminovinyl thioketones (298), and their reaction with α-bromocarbonyl derivatives in the presence of triethylamine leads to 2-aryl-5-acyl thiophenes (299)[303-305] (Scheme 103). Compounds prepared by these reactions are listed in Table 24.

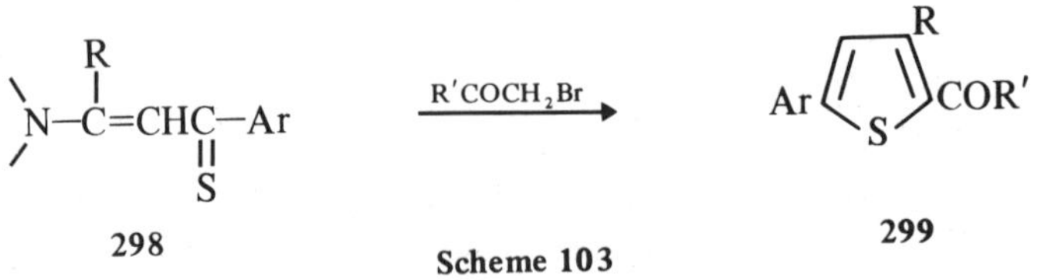

$\underset{/}{\overset{\backslash}{N}}-\underset{\underset{S}{\parallel}}{\overset{\overset{R}{|}}{C}}=CHC-Ar$

298

$\xrightarrow{R'COCH_2Br}$

Ar⟍⎯$\overset{R}{⟍}$
$_S$⟍COR'

299

Scheme 103

TABLE 24.   THIOPHENES FROM 3-AMINO THIOACRYLIC ACID DERIVATIVES AND α-HALO CARBONYL DERIVATIVES

$$\begin{array}{c} R \\ \diagdown \\ R_1 \end{array} N-CR_3=CR_4\underset{\underset{S}{\|}}{C}-R_5 + HalCH_2R_2 \longrightarrow \quad \underset{R_5}{\overset{R_4}{\diagup}}\!\!\!\Big\langle\!\!\underset{S}{\overset{R_3}{\diagdown}}\!\!R_2$$

R	R_1	R_2	R_3	R_4	R_5	Yield of Thiophene (%)	Reference
$(CH_2)_2O(CH_2)_2$		$CO_2C_2H_5$	H	H	SCH_3	Quantitative	298
$(CH_2)_2O(CH_2)_2$		$COCH_3$	H	H	SCH_3	70	298
$(CH_2)_2O(CH_2)_2$		COC_6H_5	H	H	SCH_3	70	298
$(CH_2)_2O(CH_2)_2$		$COC_6H_4NO_2\text{-}p$	H	H	SCH_3	83	298
$(CH_2)_2O(CH_2)_2$		$COC_6H_4NO_2\text{-}p$	H	H	morpholino	70	298
$(CH_2)_2O(CH_2)_2$		$COCH_3$	H	CN	SCH_3	—	298
H	H	COC_6H_5	CH_3	CO_2Et	$NHCOC_6H_5$	57	294
H	H	COC_6H_5	CH_3	CO_2Et	NHC_6H_5	88	294
H	H	COC_6H_5	CH_3	$COCH_3$	$NHCH_3$	30	294
H	H	$CO_2C_2H_5$	CH_3	$COCH_3$	$NHCH_3$	50	296
$(CH_2)_2O(CH_2)_2$		COC_6H_5	—(CH$_2$)$_3$—		NHC_6H_5	20	296
$(CH_2)_2O(CH_2)_2$		COC_6H_5	—(CH$_2$)$_4$—		NHC_6H_5	43	296
H	H	$CO_2C_2H_5$[a]	CH_3	$CO_2C_2H_5$	$NHCOC_6H_5$	3	296
H	H	$CO_2C_2H_5$[a]	CH_3	$CO_2C_2H_5$	NHC_6H_5	51	296
H	H	$CO_2C_2H_5$[a]	CH_3	$CO_2C_2H_5$	$NHC_6H_3Me_2\text{-}2,3$	23	296
H	H	$CO_2Bu\text{-}t$[a]	CH_3	$CO_2Bu\text{-}t$	$NHC_6H_3Me_2\text{-}2,3$	66	297
H	H	$CO_2Bu\text{-}t$[a]	CH_3	$CO_2Bu\text{-}t$	$NHC_6H_4CF_3\text{-}3$	25	297
H	H	NO_2	CH_3	$CO_2C_2H_5$	NHC_6H_5	15	299
H	H	NO_2	CH_3	$CO_2C_2H_5$	$NHC_6H_4OCH_3\text{-}p$	9	299
H	H	NO_2	CH_3	$CO_2C_2H_5$	NHC_6H_5	25	299
H	H	NO_2	CH_3	$COCH_3$	$NHCH_3$	12	299
H	H	NO_2	CH_3	$COCH_3$	$NHCH_3$	20	299
H	H	NO_2	CH_3	$CO_2C_2H_5$	NHC_6H_5	45	299
CH_3, —(CH$_2$)$_4$—, CH_3		NO_2	CH_3	$CO_2C_2H_5$	$NHCH_3$	42	299

TABLE 24. *Continued*

R	R_1	R_2	R_3	R_4	R_5	Yield of Thiophene (%)	Reference
CH_3	H	COC_6H_5	$NHCH_3$	NO_2	NHC_6H_5	47	300, 301
CH_3	CH_3	COC_6H_5	$(NCH_3)_2$	NO_2	$NHCH_3$	49	300, 301
CH_3	CH_3	COC_6H_5	$N(CH_3)_2$	NO_2	NHC_6H_5	48	300, 301
CH_3	CH_3	$COCH_3$	$N(CH_3)_2$	NO_2	$NHCH_2CH=CH_2$	20	300, 301
CH_3	H	$COCH_3$	$NHCH_3$	NO_2	NHC_6H_5	26	300, 301
CH_3	CH_3	COC_6H_5	$N(CH_3)_2$	NO_2	$NHC_6H_4-Cl\text{-}p$	57	301
CH_3	CH_3	COC_6H_5	$N(CH_3)_2$	NO_2	$NHC_6H_3(CH_3)_2\text{-}2,6$	68	301
H_3C	CH_3	COC_6H_5	$N(CH_3)_2$	NO_2	$NHC_6H_3(CF_3)_2\text{-}2,6$	17	301
H_3C	CH_3	COC_6H_5	$N(CH_3)_2$	NO_2	$NH-CH_2CH=CH_2$	47	301
$-(CH_2)_4-$		COC_6H_5	$N(CH_2)_4$	NO_2	NHC_6H_5	26	301
CH_3	CH_3	COC_6H_5	H	H	C_6H_5	56	303
CH_3	CH_3	$COC_6H_4Br\text{-}p$	H	H	C_6H_5	57	303
CH_3	CH_3	$COC_6H_4Br\text{-}p$	H	H	$C_6H_4CH_3\text{-}p$	45	303
CH_3	CH_3	COC_6H_5	H	H	$C_6H_4OCH_3\text{-}p$	55	303, 305
CH_3	CH_3	$COC_6H_4Br\text{-}p$	H	H	$C_6H_4OCH_3\text{-}p$	65	303
$-(CH_2)_5-$		$COCH_3$	H	H	C_6H_5	63	304, 305
$-(CH_2)_5-$		$COCH_3$	H	H	$C_6H_4Br\text{-}p$	71	304, 305
$-(CH_2)_5-$		$COCH_3$	H	H	$C_6H_4Cl\text{-}p$	55	304, 305
$-(CH_2)_5-$		$COCH_3$	H	H	$C_6H_4CH_3\text{-}p$	56	304, 305
$-(CH_2)_2O(CH_2)_2-$		$COCH_3$	H	H	$C_6H_4OCH_3\text{-}p$	40	304, 305
$-(CH_2)_5-$		COC_6H_5	H	H	$C_6H_4Cl\text{-}p$	63	305
$-(CH_2)_5-$		$COC_6H_4Br\text{-}p$	H	H	$C_6H_4Cl\text{-}p$	67	305
$-(CH_2)_5-$		COC_6H_5	H	H	2-Th	58	305
$-(CH_2)_5-$		$COC_6H_4Br\text{-}p$	H	H	2-Th	40	305
$-(CH_2)_5-$		COC_6H_5	C_6H_5	H	C_6H_5	66	305
$-(CH_2)_5-$		$COC_6H_4Br\text{-}p$	C_6H_5	H	C_6H_5	70	305

[a] Chloroacetoacetates were used as the reactive methylene compounds.

A special type of cyclic thioester, **300**, reacts with 2 eq. of α-bromocarbonyl derivatives in the presence of triethyl amine to give **301**, which upon treatment with potassium hydroxide gives **302**[306] (Scheme 104). A recent modification of

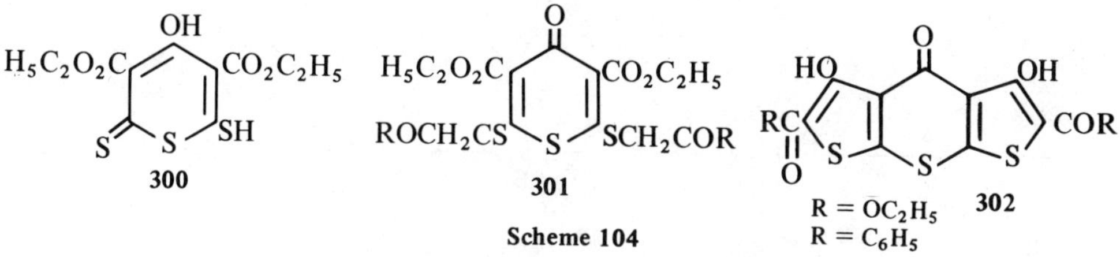

300 **301** **302**

R = OC$_2$H$_5$
R = C$_6$H$_5$

Scheme 104

the Rajappan method, which can be used for the synthesis of *N,N*-disubstituted 2-aminothiophenes consists in the reaction of thioamides with formamide chlorides to give **302a**, which upon alkylation is transformed to **302b**. When the alkylating agent contains an electron-withdrawing group, as in halomethylene ketones, *p*-nitrobenzylbromide or bromonitromethane ring-closure to **302c** occurs spontaneously or upon base-catalysis. Also, the free 3-amino- and 3-hydroxythioacrylamide prepared from **302a** could be used in this thiophene synthesis[306a] (see Table 25). Another recent synthesis of *N,N*-disubstituted 2-aminothiophenes consists in the *S*-alkylation of **302d** with propargyl bromide to **302e**, which upon heating in benzene, presumably via **302f** and **302g**, ring-closed to **302h**[306b] (Scheme 104a).

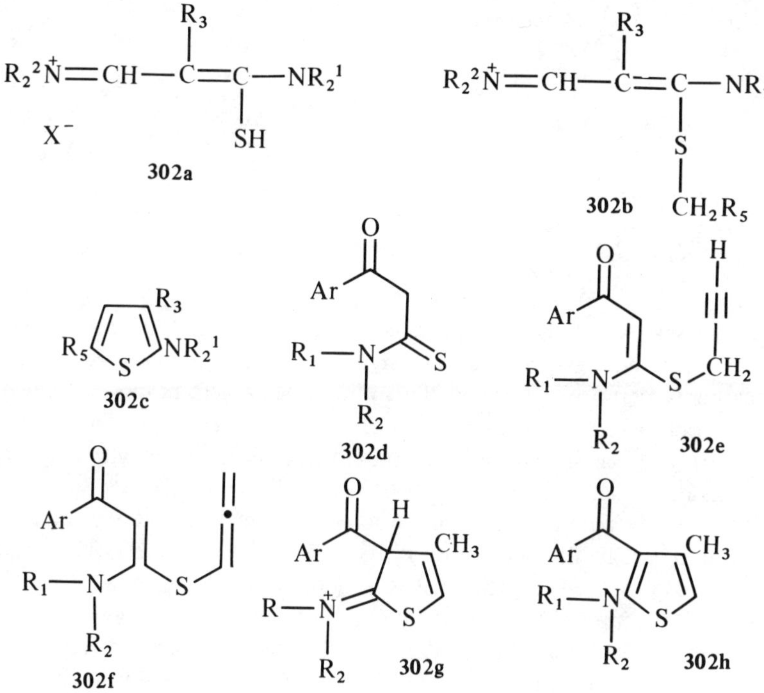

302a **302b**

302c **302d** **302e**

302f **302g** **302h**

Scheme 104a

TABLE 25.   *N,N*-DISUBSTITUTED 2-AMINOTHIOPHENES FROM 3-AMINOTHIOACRYL-
AMIDES AND ALKYLATING AGENTS[306a]

$$R_2^{2+}N=CH-\underset{\underset{SH}{|}}{\overset{\overset{R_3}{|}}{C}}=C-NR_2^1 + R_5CH_2X \longrightarrow \quad \underset{R_5}{\overset{}{\bigg[\bigg]}}\underset{S}{\overset{R_3}{\underset{}{}NR_2^1}}$$

R_3	NR_2^1	R_2^2	R_5	Yield of Thiophene (%)
C_6H_5	Morpholino	CH_3, CH_3	$p\text{-}NO_2C_6H_4$	81
C_6H_5	Morpholino	$-(CH_2)_4-$	CH_3CO	53
C_6H_5	Morpholino	CH_3, CH_3	C_6H_5CO	92
C_6H_5	Morpholino	$-(CH_2)_5-$	C_6H_5CO	78
C_6H_5	Morpholino	CH_3, C_6H_5	C_6H_5CO	26
C_6H_5	Piperidino	$-(CH_2)_5-$	C_6H_5CO	66
C_6H_5	Morpholino	CH_3, CH_3	$p\text{-}BrC_6H_4CO$	93
C_6H_5	Morpholino	CH_3, CH_3	$p\text{-}C_6H_5C_6H_4CO$	98
C_6H_5	Morpholino	CH_3, CH_3	NO_2	69
C_6H_5	Piperidino	$-(CH_2)_5-$	NO_2	11
$p\text{-}CH_3C_6H_4$	Morpholino	CH_3, CH_3	$p\text{-}BrC_6H_4CO$	84
$p\text{-}CH_3C_6H_4$	Morpholino	CH_3, CH_3	NO_2	72
$p\text{-}CH_3OC_6H_4$	Morpholino	CH_3, CH_3	C_6H_5CO	89
$p\text{-}CH_3OC_6H_4$	Morpholino	CH_3, CH_3	NO_2	90
$3,4\text{-}(CH_3O)_2C_6H_3$	Morpholino	CH_3, CH_3	$p\text{-}BrC_6H_4CO$	23
$p\text{-}ClC_6H_4$	Morpholino	CH_3, CH_3	C_6H_5CO	93
$p\text{-}ClC_6H_4$	Morpholino	CH_3, CH_3	NO_2	82
$p\text{-}C_6H_5C_6H_4$	Morpholino	$-(CH_2)_4-$	C_6H_5CO	94
α-Naphthyl	Morpholino	$-(CH_2)_4-$	$p\text{-}BrC_6H_4CO$	61
α-Naphthyl	Morpholino	$-(CH_2)_4-$	NO_2	85
CH_3	Morpholino	$-(CH_2)_4-$	$p\text{-}BrC_6H_4CO$	23
CH_3	Morpholino	CH_3, CH_3	NO_2	19

C.   The $C_3S + C$ Modification of the Fiesselmann Reaction

The sodium salt of thiocyclopentanonecarboxylic ester (**303**), prepared from the β-ketoester, hydrogen chloride, and hydrogen sulfide, was reacted with various α-halocarbonyl derivatives to yield the sulfide **304**, which, in most cases without isolation, ring-closed to thiophenes (**305**)[204] upon treatment with alkoxide (Scheme 105). It was shown that the product from the reaction with ethyl chloroacetate has the structure **305** and not **306**, as previously claimed.[307] Similarly, **307** and **308** were alkylated with α-halocarbonyl compounds, and the intermediate sulfide ring-closed to thiophene (**309**)[204] and **310**,[308] respectively. The analogous reaction of **311** gave **312**[309] (Scheme 106). More examples are given in Table 26, together with the mp's of the resulting thiophenes.

Thiolates, such as **303**, **307**, and **308**, were also reacted with α,α-dichloroacetone, and upon treatment with ethoxide gave **313** (mp 245°C),[204] **314** (mp 221–224°C), and **315** (mp 169°C)[308] in 27, 63, and 41% yield, respectively (Scheme 107).

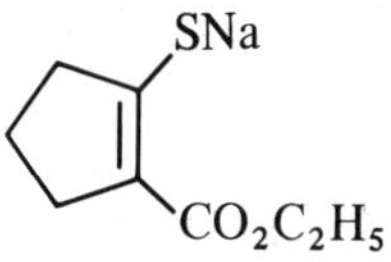

303

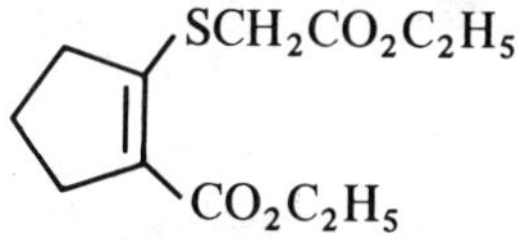

304

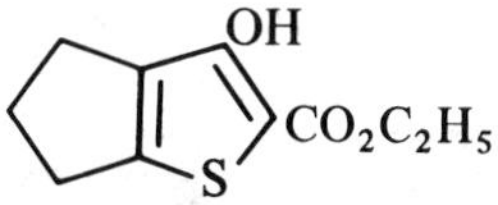

305

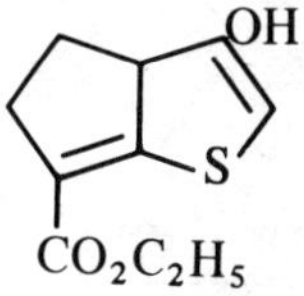

306

Scheme 105

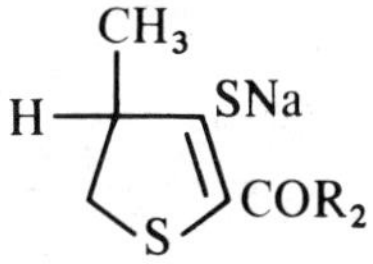

$R_2 = OC_2H_5, CH_3, C_6H_5$ $R = CH_3, Ph$ $R_2 = OC_2H_5, CH_3C_6H_5$

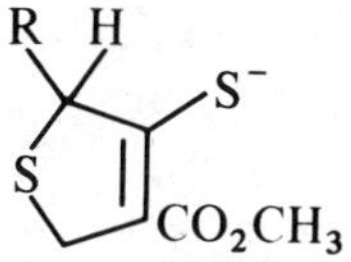

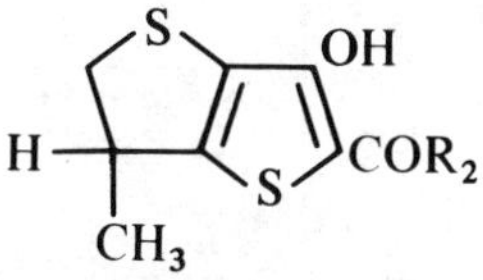

307   308   309

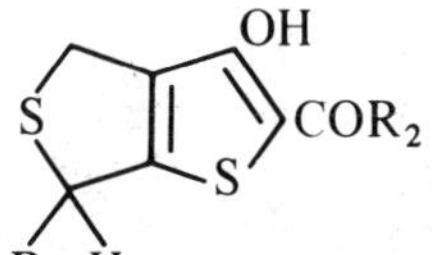

$R = CH_3, C_6H_5$
$R_2 = OCH_3, CH_3, C_6H_5$
310   311

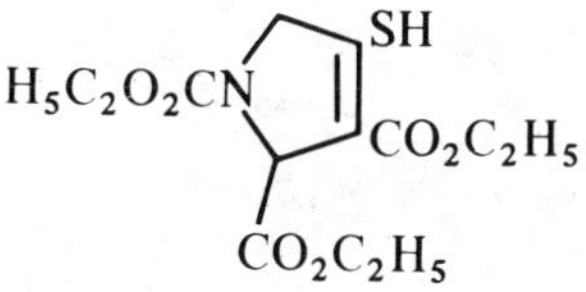

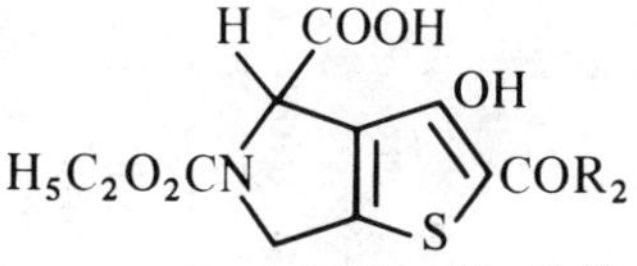

$R_2 = OCH_3, CH_3, C_6H_5$
312

Scheme 106

TABLE 26. THIOPHENES FROM β-THIOKETO ESTERS AND α-HALO CARBONYL DERIVATIVES

$$\underset{R_4}{\overset{R_5}{\bigvee}}\text{SNa} \quad + \text{HalCH}_2R_2 \longrightarrow \underset{R_5}{\overset{R_4}{\bigvee}}\text{OH} \atop R_2$$

R_4	R_5	R_2	Yield of Thiophene (%)	mp (°C)	Reference
(CH₂)₃		CO₂CH₃	28	53–53.5	204
(CH₂)₃		COC₆H₅	84	83	204
(CH₂)₃		COCH₃	58	51	204
(CH₂)₃		COC(CH₃)₃	87	90	204
(CH₂)₃		CONH₂	64	195–205	204
(CH₂)₃		CO₂C₂H₅	80	64.5	204
(CH₂)₄		COCH₃	23	47	204
(CH₂)₄		COC₆H₅	68	133–134	204
(CH₂)₄		CONH₂	85	186–188	204
(CH₂)₄		CO₂CH₃	87	54	204
[CH(CH₃)CH₂S]		COCH₃	86	72	204
[CH(CH₃)CH₂S]		COC₆H₅	92	69–70.5	204
[CH(CH₃)CH₂S]		COC(CH₃)₃	88	75	204
[CH(CH₃)CH₂S]		COCNH₂	85	190–198	204
[CH₂SCH(CO₂CH₃)]		CO₂CH₃	66	124	204
—[CH₂SCH(CO₂CH₃)]—		COCH₃	37	121	204
[CH(CH₃)SCH₂]		COC₆H₅	91	100	308
[CH(CH₃)SCH₂]		CN	81	170	308
[CH(CH₃)SCH₂]		COCH₃	63	67	308
[CH(CH₃)SCH₂]		CO₂CH₃	90	95	308
[CH(C₆H₅)SCH₂]		COC₆H₅	42	131	308
[CH(C₆H₅)SCH₂]		COCH₃	32	121	308
[CH(C₆H₅)SCH₂]		CO₂CH₃	47	150	308
[CH₂N(CO₂C₂H₅)CH₂]		CO₂C₂H₅	61	115	309
[CH₂N(CO₂C₂H₅)CH₂]		COCH₃	63	169	309
[CH₂N(CO₂C₂H₅)CH(CO₂C₂H₅)]		CO₂C₂H₅	51	95–96	309
[CH₂N(CO₂C₂H₅)CH(C₆H₅)]		CO₂CH₃	23	135–136	309

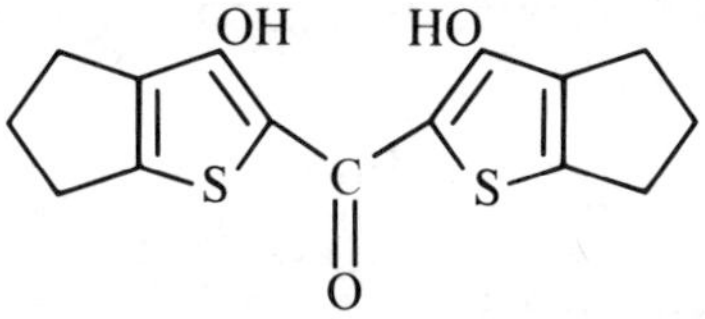

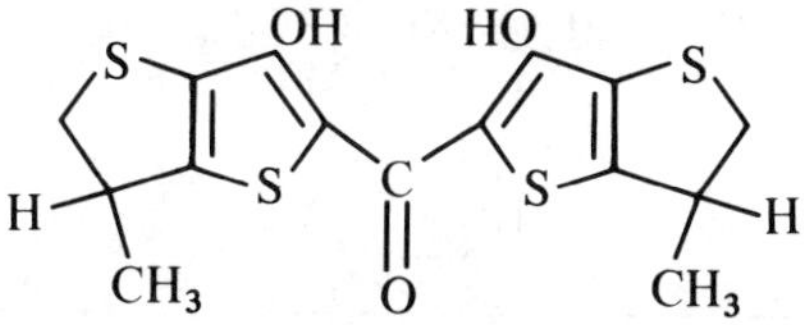

313

314

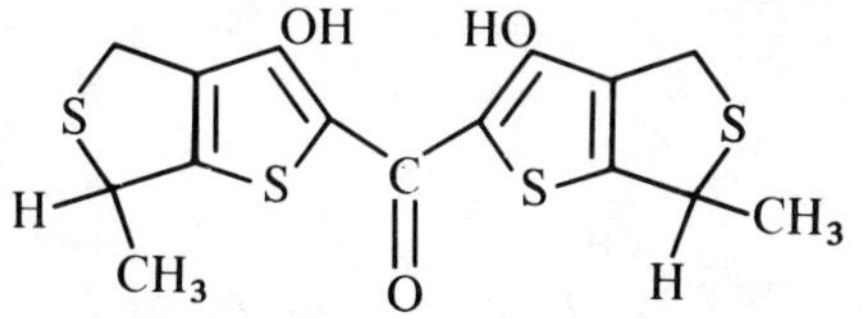

315

Scheme 107

After the development of the Vilsmeyer formylation of ketones to yield β-chloro-vinyl aldehydes (316), this thiophene synthesis became very important (Scheme 108) (Table 27). It consists of the reaction with sodium sulfide to give the salts of the corresponding enethiols (317), followed by reaction with α-bromocarbonyl compounds to give 318, which often spontaneously, or in some cases after treatment with alkoxides, ring-close to 319 in good yield.[310-312] It is also possible to use α-bromoesters of longer fatty acids such as α-bromopropionic acid or α-bromo-phenylacetic acid. Upon hydrolysis and distillation of the alkylated product (320), thiophenes (321) are obtained[312] (Scheme 109).

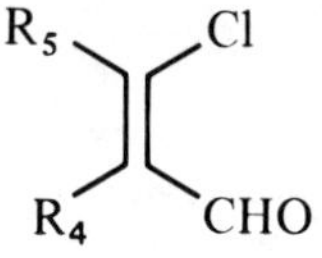

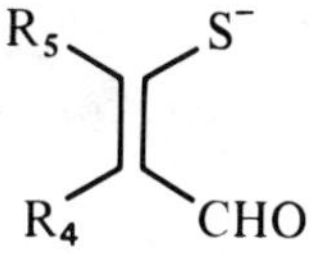

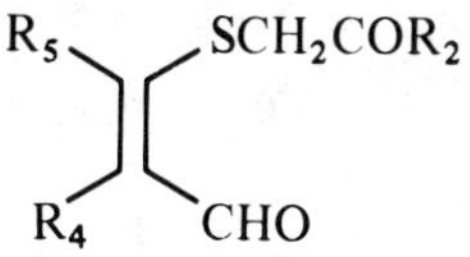

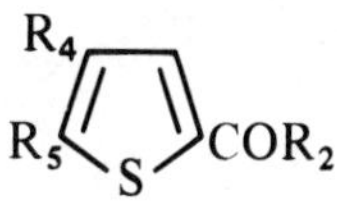

316

317

318

319

Scheme 108

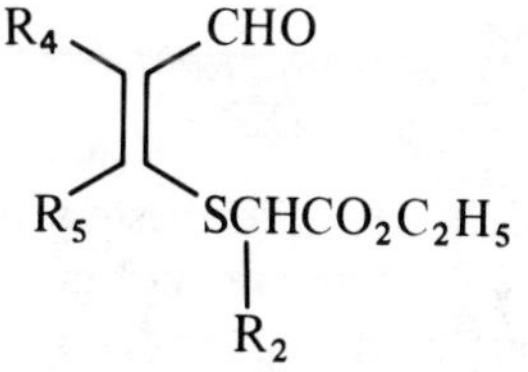

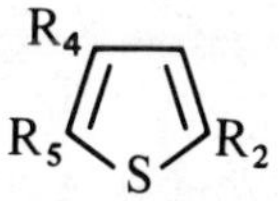

320

321

$R_5 = Ph, R_4 = CH_3, R_2 = CH_3$ 53%
$R_5 = R_4 = -(CH_2)_4^-, R_2 = CH_3$ 61%
$R_5 = R_4 = -(CH_2)_4^-, R_2 = Ph$ 50%

Scheme 109

TABLE 27.   THIOPHENES FROM β-CHLOROVINYL ALDEHYDES, SODIUM SULFIDE, AND α-HALOCARBONYL DERIVATIVES

$$\begin{array}{c} R_5 \diagdown \diagup Cl \\ \diagdown \diagdown \\ R_4 \diagup \diagdown CHO \end{array} + Na_2S + HalCH_2R_2 \longrightarrow \begin{array}{c} R_4 \\ R_5 \diagdown S \diagup R_2 \end{array}$$

R_4	R_5	R_2	Yield of Thiophene (%)	Reference
H	C_6H_5	COC_6H_5	–	310
C_6H_5	C_6H_5	COC_6H_5	86	311
C_6H_5	C_6H_5	$COCH_3$	60	311
C_6H_5	C_6H_5	CO_2CH_3	46	311
H	$t\text{-}C_4H_9$	$CO_2C_2H_5$	47	312
H	$t\text{-}C_4H_9$	NO_2	61	312
H	$t\text{-}C_4H_9$	COC_6H_5	65	312
CH_3	C_6H_5	$CO_2C_2H_5$	65	312
$-(CH_2)_4-$		$CO_2C_2H_5$	77	312
$-(CH_2)_4-$		COC_6H_5	91	312
$-(CH_2)_4-$		$COCH_3$	52	312
$-(CH_2)_4-$		CHO	45	312
$-(CH_2)_4-$		CN	30	312
$-(CH_2)_4-$		$C_6H_4NO_2\text{-}o$	76	312
$-(CH_2)_4-$		$C_6H_5NO_2\text{-}p$	75	312
$-(CH_2)_4-$		$C_6H_4CO_2C_2H_5\text{-}o$	27	312
$-(CH_2)_4-$		$CONH_2$	25	312

5.   C₃ + CS Methods

A.   The Fiesselmann Reaction

Hans Fiesselmann was born on October 23, 1909 in Erlangen and died on August 18, 1969. He studied in Erlangen and received his Ph.D. in 1936 on a thesis entitled "Indigo Farbstoffe der cis-Reihe." His habilitation in 1942 was "Diensynthesen mit Oxyprenen." His entire scientific career was at the University of Erlangen, where he became a professor.

a.   INTRODUCTION

Woodward and Eastman[313] showed in 1946 that esters of thioglycolic acid could be added to α,β-unsaturated esters in the presence of catalytic amounts of piperidine or sodium alcoholates. Upon Dieckmann condensation, these esters gave predominantly 4-carbomethoxy- or 2-carbomethoxy-3-ketotetrahydrothiophenes, depending upon reaction conditions. In 1954, Fiesselmann extended this reaction to α,β-acetylenic esters and found that this led to 3-hydroxy-2-thiophenecarboxylic acid derivatives.[314] In further work, he found that other compounds with the same oxidation level as acetylenes, such as β-keto esters, α,β-dihalo esters, and α- and β-halovinyl esters, could be used for the synthesis of 3-hydroxythiophenes. Using the corresponding nitriles instead of esters, numerous 3-aminothiophenecarboxylic acids have been obtained. Use of the corresponding aldehydes or ketones gave various substituted 2-thiophenecarboxylic acids. Instead of thioglycolates, it is possible to use α-mercaptoketones in the reaction with acetylenes, which yields 3-hydroxy-2-acylthiophenes.[193, 204] All of these are now called Fiesselmann reactions in the literature. They constitute one of the most useful and general thiophene syntheses. Unfortunately, Fiesselmann published only a few papers on this work and a number of patents. Most of the results are collected in Ph.D. theses from the University of Erlangen-Nürnberg* which, being not readily available, thus has led to the rediscovery of the Fiesselmann reaction during recent years.

b. THIOPHENES FROM ACETYLENES AND ALKYL THIOGLYCOLATES

The reaction of dimethyl acetylenedicarboxylate (322) with 2 eq. methyl thioglycolate in the presence of catalytic amounts of piperidine gives a thioacetal (323),[314, 315] which upon treatment with stronger base undergoes Dieckmann

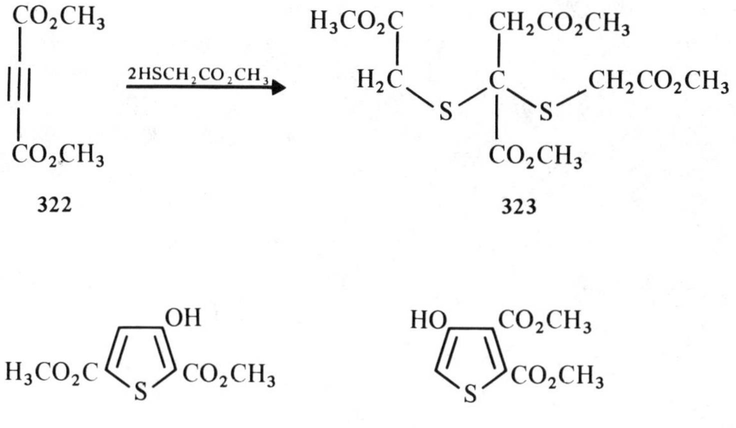

Scheme 110

cyclization and aromatization by the elimination of methyl thioglycolate to give 324 and can be carried out in a one-pot procedure by reacting the acetylene-dicarboxylate and alkyl thioglycolate with a suspension of sodium methoxide in benzene[314,317] or with alcoholic potassium hydroxide.[318] The reaction was later reinvestigated by Hendrickson et al. using acetylenedicarboxylate to prove that 324a was not formed.[319] Pyrroles and furans of this type are obtained in the reaction of dimethyl acetylenedicarboxylate with α-aminoketones and α-hydroxy-ketones, respectively.[319] It is claimed that phenylpropiolic acid and propiolic acid react similarly at the β-carbon to give 2-phenyl 3-hydroxy-2-thiophenecarboxylates

---

* I am much obliged to Professor H. J. Bestmann of this university for making these theses available to me. In this chapter, the results of these theses are therefore discussed in detail.

and 3-hydroxy-2-thiophenecarboxylates.[314,315,320] However, other authors later claimed that the reaction between esters of arylpropiolic acids, such as phenyl p-methoxyphenyl p-toloyl and 2-thienylpropiolic acids, and alkyl thioglycolates in the presence of a sodium methoxide suspension in benzene, leads to a mixture of alkyl 5-aryl-3-hydroxy-2-thiophenecarboxylate (325) and alkyl 5-aryl-3-hydroxy-4-thiophenecarboxylate (326), in which the latter compound often predominates.[115b,321-322a] This could indicate that the vinylic carbanion (327) formed in the Michael-type addition directly attacks the ester function in solvents like benzene to give 326.[323] It has been shown that in the presence of equivalent amounts of sodium methoxide, only 1 eq. alkyl thioglycolate can be added to dimethyl acetylenedicarboxylate.[314] Alternatively, as in the ethylene series,[313] the intermediate thioacetal could ring-close in either direction via 328a or 328b, depending upon the nature of the R-group and on the conditions used. It is claimed that under somewhat different conditions, using catalytic amounts of potassium t-butoxide in DMSO, only the unsaturated ether corresponding to 327 is formed from tetrolic ester and phenylpropiolate, in addition to 326.[323] When ethyl propiolate was reacted under the same conditions, a mixture of 50% of the cis-addition product, 33% of the trans-addition product, and 8% of 4-carbethoxy-3-hydroxythiophene was obtained. However, when an equivalent amount of t-butoxide in benzene was used, 2-carbomethoxy-3-hydroxythiophene was obtained in low yield[323] (Scheme 111). The 3-hydroxy-2-thiophenecarboxylates prepared in this way are given in Table 28.

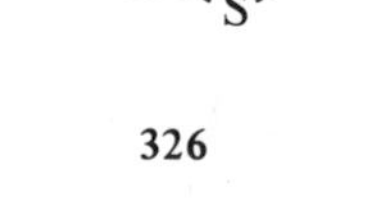

326 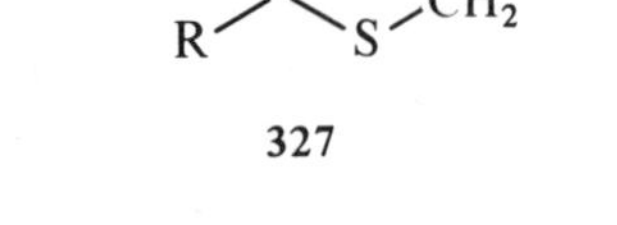   327

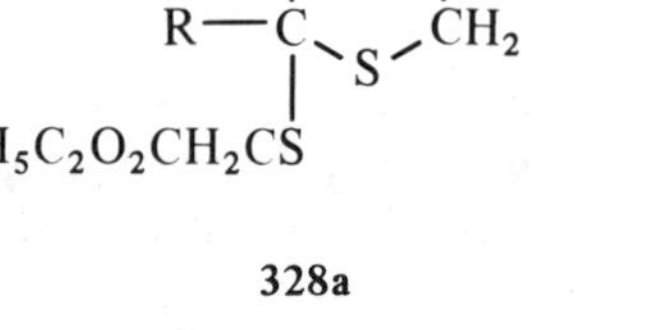

328a

328b   325

Scheme 111

TABLE 28.   FIESSELMANN SYNTHESIS FROM ACETYLENES AND ALKYL THIOGLY-
COLATES

$$R_1C{\equiv}C-R_2 + HSCH_2CO_2R_3 \longrightarrow$$

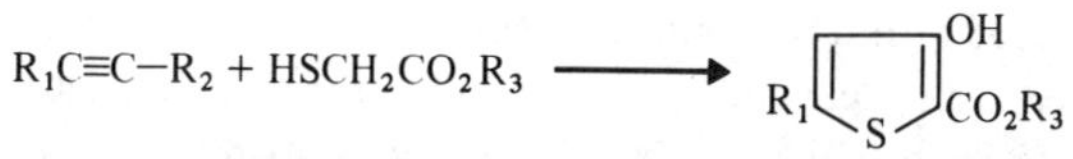

R_1	R_2	R_3	Yield of Thiophene (%)	Reference
CO_2CH_3	CO_2CH_3	CH_3	94	314
H	$CO_2C_2H_5$	C_2H_5	30	315
C_6H_5	$CO_2C_2H_5$	C_2H_5	36	315, 317
H	CO_2CH_3	CH_3	28	315, 323
C_6H_5	CO_2CH_3	CH_3	49	320
H	CO_2CH_3	CH_3	10	320
$(H_3C)_2CH$	$CO_2C_2H_5$	C_2H_5	46	317
$(CH_3)_3C$	$CO_2C_2H_5$	C_2H_5	77	317
α-Naphthyl	$CO_2C_2H_5$	C_2H_5	62	317
p-C_6H_5-C_6H_4	$CO_2C_2H_5$	C_2H_5	43	317

c.   THIOPHENES FROM β-KETO ESTERS AND ALKYL THIOGLYCOLATES

In 1954, Fiesselmann found that β-keto esters reacted in the cold with alkyl thioglycolates in the presence of catalytic amounts of acid to give a half mercaptol, which upon Dieckmann cyclization ring-closed to give alkyl 5-substituted 3-hydroxy-2-thiophenecarboxylates.[324] However, detailed investigations showed that the best method was to react the β-keto esters (329) with 2 eq. anhydrous or 80% thioglycolic acid in the presence of hydrogen chloride, which quantitatively gives the thioacetals (330), which are esterified in high yield to 331. Upon treatment of 331 with alcoholic potassium hydroxide or alcoholate, 5- as well as 4,5-substituted 3-hydroxy-2-thiophenecarboxylates (332) were obtained. The latter compounds are not available from acetylenes[325, 326] (Scheme 112). This

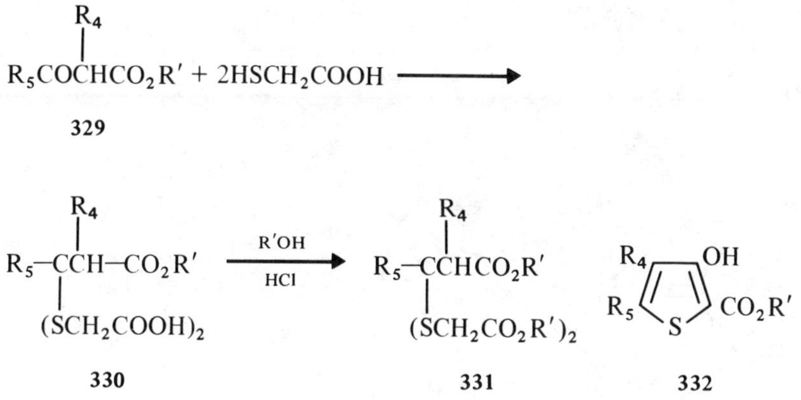

$$\underset{\textbf{329}}{R_5COCHCO_2R' + 2HSCH_2COOH \longrightarrow}$$

$$\underset{\textbf{330}}{R_5-CCH-CO_2R' \atop (SCH_2COOH)_2} \xrightarrow[HCl]{R'OH} \underset{\textbf{331}}{R_5-CCHCO_2R' \atop (SCH_2CO_2R')_2} \quad \underset{\textbf{332}}{}$$

Scheme 112

method of preparing the thioacetals (331) gave much better yields than the reaction of β-keto esters with alkyl thioglycolates, in the absence of solvents, which led to a mixture of 331 and unsaturated sulfide, if reaction times were not long enough. If the reaction is carried out in alcoholic solution, only unsaturated sulfides are obtained, and only those in which the ring-closing groups are *cis*-orientated can ring-close. This is of course the case when cyclic β-keto esters, such as cyclopentanone- or cyclohexanone-2-carboxylates, are used.[326] Later work has confirmed that only 332 and no 3-hydroxy-4-carboxy derivatives are formed from 331.[321, 322] It has been clearly proven that the ring-closure of 333 gives 334,[326] and not 335, as claimed by Chakrabarty and Mitra[327] (Scheme 113). The structure assignments given in Ref. 328 also appear to be erroneous.[326]

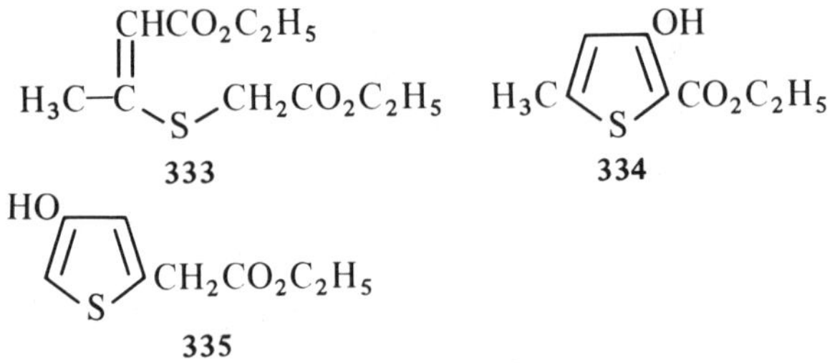

Scheme 113

The reaction of methyl α-formylphenylacetate with 2 eq. thioglycolic acid and esterification gave the thioacetal, which upon treatment with 2 N methanolic potassium hydroxide gave methyl 4-phenyl-3-hydroxythiophenecarboxylate (mp, 82°C) in 73% yield.[193a] The cyclic keto esters 336 and 337, obtained according to Ref. 313, upon reaction with alkyl thioglycolates or with thioglycolic acid followed by esterification, give the unsaturated sulfides 338 and 339, respectively, in which the ring-closing groups are *cis*-orientated. Treatment with alcoholate then gave the

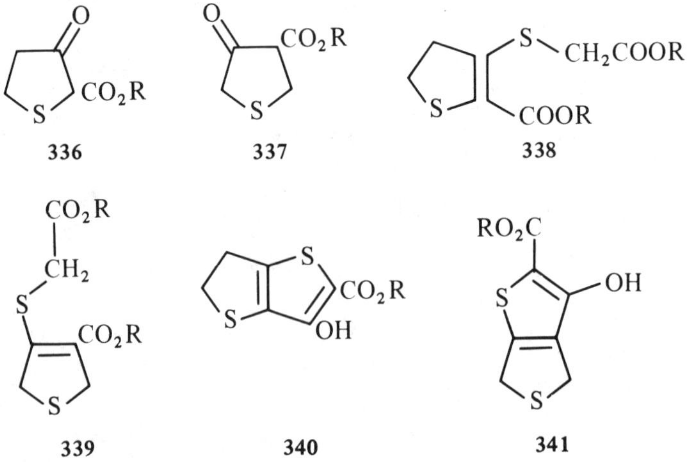

Scheme 114

thiophenes **340** and **341**, which could be aromatized to thienothiophenes.[329] In connection with work on thienopyrroles, the keto esters **342** and **343** were reacted

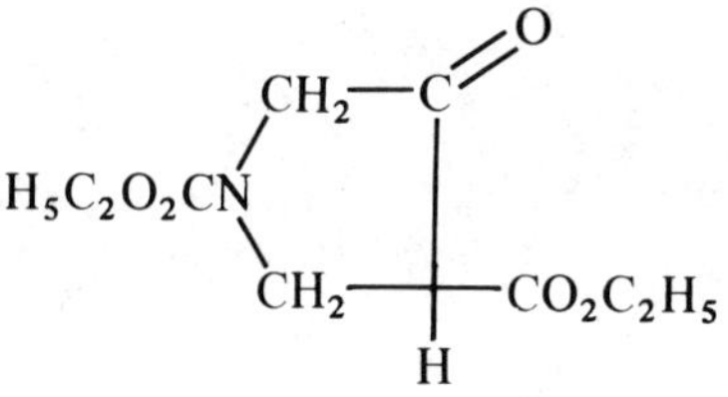

342 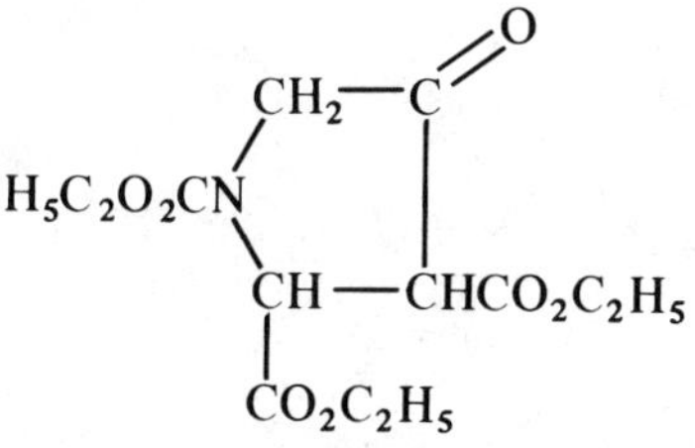 **343**

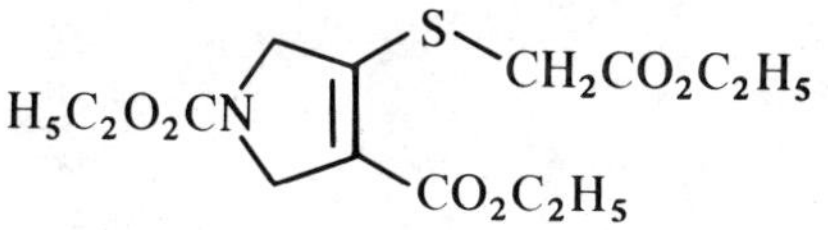

344

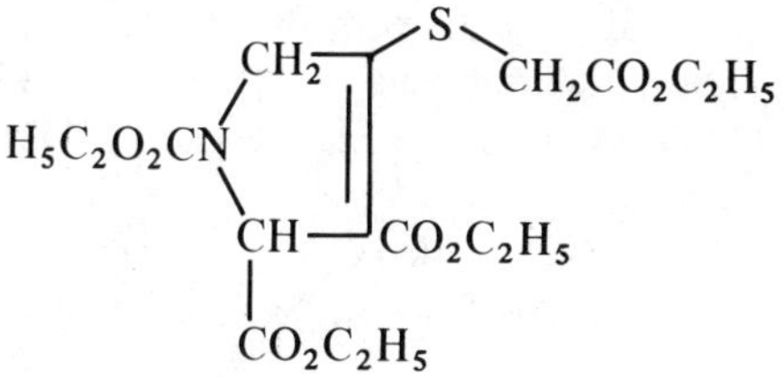

345

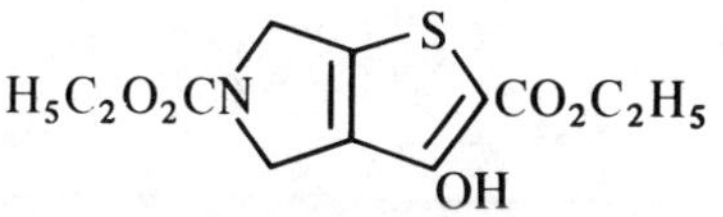

346 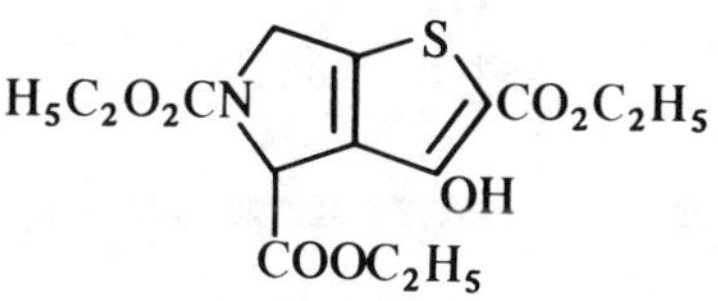 **347**

Scheme 115

with alkyl thioglycolate to give the unsaturated sulfides **344** and **345**, which upon ring-closure with sodium hydride in benzene gave higher yields of **346** and **347** than when ethoxide in ethanol was used[309] (Scheme 115).

Recently, a large number of piperidones **347a** has been transformed to **347b** by reaction with thioglycolic acid and esterification. In the ring-closure to **347c**, potassium carbonate in DMF was successfully used[332] (Scheme 115a).

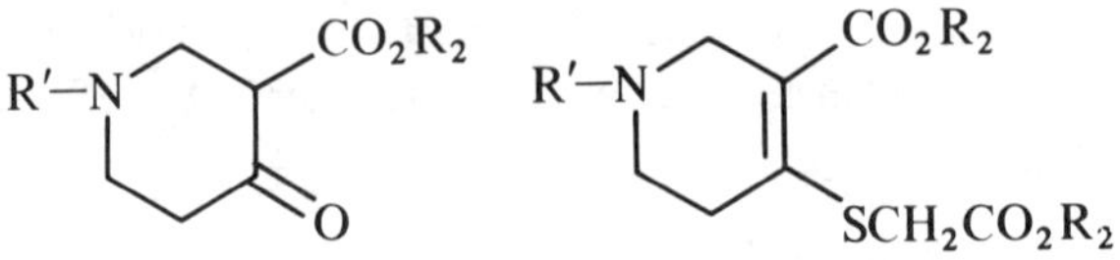

347a

347b

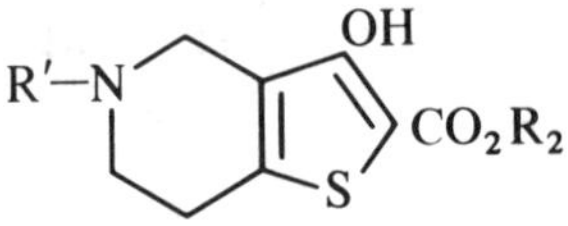

347c

Scheme 115a

The use of β-keto esters has been extended in various directions. Many β-keto esters functionalized in the 3-(α)-position with an acetoxy-,[330] alkylthio-, arylthio-, or phenacetyl amino group[331] can be transformed to thioacetals by the above methodology, and with the alkoxide ring-closed to 3,4-dihydroxythiophenes, 4-arylthio-3-hydroxythiophenes, and 4-amino-3-hydroxythiophenes.

Thiophene synthesis has also been attempted with 3-chloroacetylacetone, methyl α-chloroacetoacetate, and diethyl chlorooxaloacetate. From the latter compound (**347d**), diethyl 3,4-dihydroxythiophenedicarboxylate (**347f**) was obtained in 33% yield by reaction with ethyl thioglycolate in pyridine to **347e**, followed by ring-closure with sodium ethoxide in ethanol[193a] (Scheme 116). The compounds prepared are given in Table 29, together with their mp's, since many have not been described in the literature.

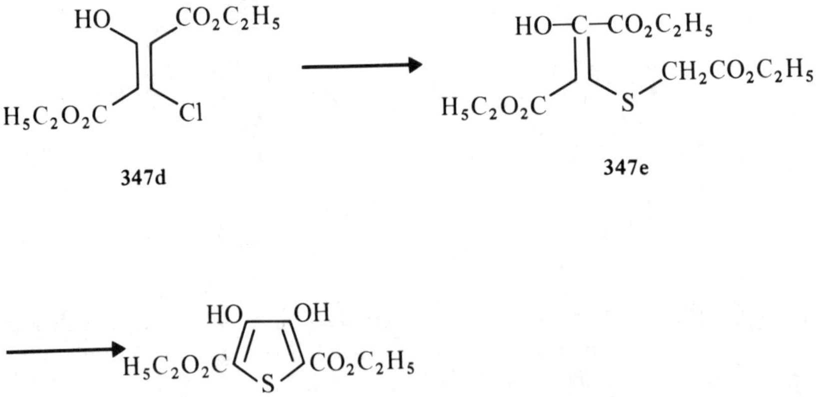

347d

347e

347f

Scheme 116

TABLE 29. FIESSELMANN SYNTHESES FROM β-KETOESTERS AND THIOGLYCOLIC ACID

$$R_5\overset{\overset{\textstyle O}{\|}}{C}\underset{\underset{\textstyle R_4}{|}}{C}H-CO_2R' + 2HSCH_2COOH \xrightarrow[R''OH]{H^+} \xrightarrow{NaOR'} \quad \overset{R_4}{\underset{R_5}{\Big|}}\!\!\!\overset{OH}{\underset{S}{\bigcirc}}\!\!CO_2R''$$

R_5	R_4	R'	R''	Yield of Thiophene (%)	mp (°C)	Reference
H_3C	H	CH_3	CH_3	79	53	325, 326
H_3C	H	C_2H_5	CH_3	81	53	325, 326
H_3C	H	C_2H_5	C_2H_5	75	23	325, 326
H_3C	C_2H_5	C_2H_5	CH_3	81	43–44	325, 326
H_3C	C_2H_5	C_2H_5	C_2H_5	73		325
H_5C_6	H	C_2H_5	CH_3	82	98	325, 326
H_5C_6	H	C_2H_5	C_2H_5	91	68	325, 326
$-(CH_2)_3-$		C_2H_5	C_2H_5	87	50–51	325, 326
H_3C	H_3C	C_2H_5	CH_3	79	52, 50	326, 333
H_3C	H_3C	C_2H_5	C_2H_5	76	50, 43	326, 333
$-(CH_2)_3-$		C_2H_5	CH_3	85	78, 68	326, 333
$-(CH_2)_4-$		C_2H_5	CH_3	88	62–63	326, 333
$-(CH_2)_4-$		C_2H_5	C_2H_5	76	34, 41	326, 333
$-(CH_2)_2S-$		C_2H_5	CH_3	76	170	329
$-CH(CH_3)CH_2S-$		C_2H_5	CH_3	74	53–54	329
$-CH_2CH(CH_3)S-$		C_2H_5	CH_3	71	75–76	329
$-CH_2SCH_2-$		C_2H_5	CH_3	100	139	329
$-CH(CH_3)SCH_2-$		C_2H_5	CH_3	63	95	329
$-CH(CH_3)SCH(CH_3)-$		C_2H_5	CH_3	41	114–115	329
$-CH_2SCH(CO_2CH_3)-$		C_2H_5	CH_3	48	124	329
H	CH_3	C_2H_5	CH_3			333
H	CH_3	C_2H_5	C_2H_5			333
C_5H_5	$OCOCH_3$	C_2H_5	CH_3	78[a]	146–147	330
C_6H_5	SC_2H_5	C_2H_5	CH_3	54	87	331
C_6H_5	SC_4H_9	C_2H_5	CH_3	47	94	331

TABLE 29. *Continued*

R_5	R_4	R'	R''	Yield of Thiophene (%)	mp (°C)	Reference
C_6H_5	SC_6H_5	C_2H_5	CH_3	46	101	331
CH_3	$OCOCH_3$	C_2H_5	CH_3	42	113	331
CH_3	$SCOCH_3$	C_2H_5	CH_3	50	47	331
CH_3	$NHCOCH_2C_6H_5$	C_2H_5	CH_3	40	202	331
CH_3	SC_2H_5	C_2H_5	CH_3	74		331
CH_3	SC_4H_9	C_2H_5	CH_3	81		331
CH_3	SC_6H_5	C_2H_5	CH_3	85	107	331
CH_3	C_6H_5	CH_3	CH_3	73	82	193a
H	$-CH_2N-CH_2-$ with $CO_2C_2H_5$ on N	C_2H_5	CH_3	61	117	309
	$-CH_2-N-CH(CO_2C_2H_5)-$ with $CO_2C_2H_5$ on N	C_2H_5	C_2H_5	51	99–99.5	309
	$-CH_2CH_2NHCH_2-$	C_2H_5	CH_3	60		332
	$-CH_2CH_2NHCH_2-$	C_2H_5	C_2H_5	96		332
	$-CH_2CH_2N(CH_3)CH_2-$	C_2H_5	C_2H_5	94		332
	$-CH_2CH_2N(CH_2C_6H_5)CH_2-$	CH_3	C_2H_5	78		332
	$-CH_2CH_2N(CH_2C_6H_4-Cl\text{-}o)CH_2-$	CH_3	CH_3	92		332
	$-CH_2CH_2N(CH_2C_6H_4Cl\text{-}o)CH_2-$	CH_3	C_2H_5	92		332
	$-CH_2CH_2N(CH_2C_6H_4Cl\text{-}o)CH_2-$	C_4H_9	C_4H_9	59		332
	$-CH_2CH_2N(CH_2C_6H_5)CH_2-$	C_2H_5	C_2H_5	91		332
	$-CH_2CH_2N(COC_6H_5)CH_2-$	CH_3	CH_3	48		332
	$-CH_2CH_2N(COC_6H_5)CH_2-$	C_2H_5	C_2H_5	98		332
	$-CH_2CH_2N(COC_6H_4CH_3\text{-}p)CH_2-$	CH_3	C_2H_5	67		332
	$-CH_2CH_2N(COC_6H_4H\text{-}o)CH_2-$	CH_3	C_2H_5	46		332
	$-CH_2CH_2N(COC_6H_4Cl\text{-}o)CH_2-$	CH_3	C_2H_5	30		332

a Isolated as dihydroxy derivative.

d. THIOPHENES FROM α,β-DIHALO ESTERS OR α-HALO α,β-UNSATURATED ESTERS AND ALKYL THIOGLYCOLATES

Esters derived from α,β-dihalocarboxylic acids (**347g**) also ring-close to give 3-hydroxy-2-thiophenecarboxylates (**348**) upon reaction with thioglycolates in the presence of $2N$ methanolic potassium hydroxides.[334-337] The reaction proceeds via elimination, to form α-haloacrylates (**347h**), and Michael addition of the thioglycolate to yield **347i**, followed by Dieckmann cyclization.[337] The reaction of α-haloacrylates was recently rediscovered.[338] Under different conditions, both

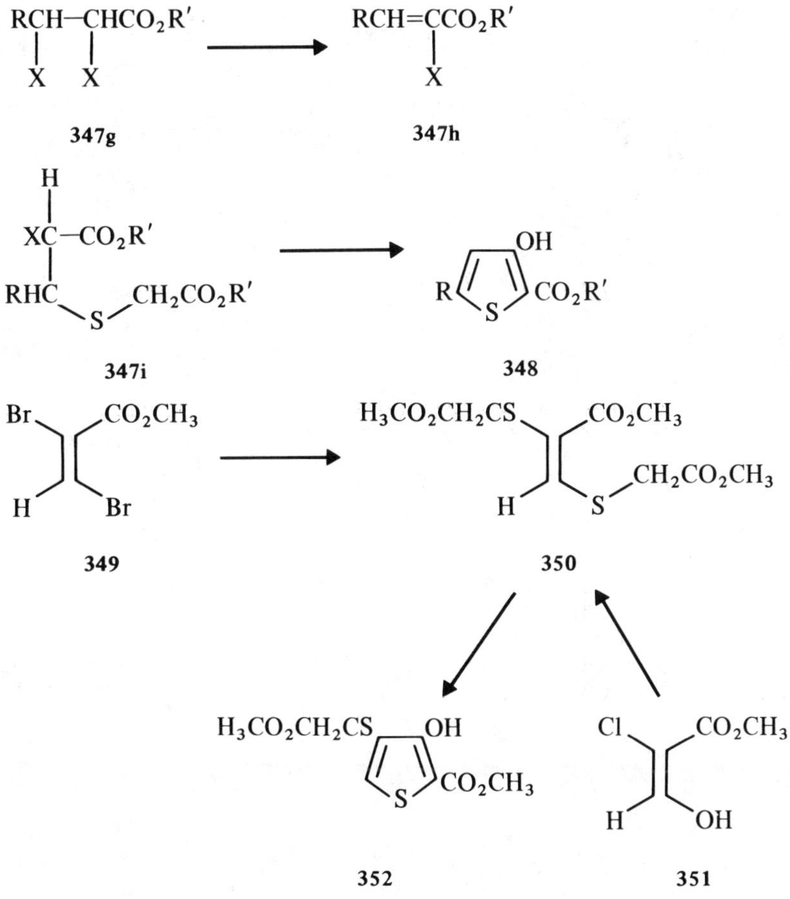

Scheme 117

347h and 347i could be isolated and separately ring-closed with the same high yield to 348. In pyridine with methyl thioglycolate, methyl-α,β-dibromoacrylate (349) gave the unsaturated dithio ether 350, which was also obtained from methyl formylchloroacetate (351). The reaction of 350 with methanolic sodium methylate gave **352** (mp 80°C) in 65% yield.[193a] Compound **350** has a great tendency to ring-close, which occurs even upon shaking with dilute sodium hydroxide solution, so the synthesis of **352** from **349** can be carried out in a one-pot procedure (Scheme 117). Dimethyl α,α'-dibromofumarate could also be reacted with methyl thioglycolate to give a mixture of **353** and **354**, which could be ring-closed to a mixture of **355** and **356**. Conditions for the preferential formation of **356** were described (Scheme 118). The compounds prepared are listed in Tables 30 and 31.

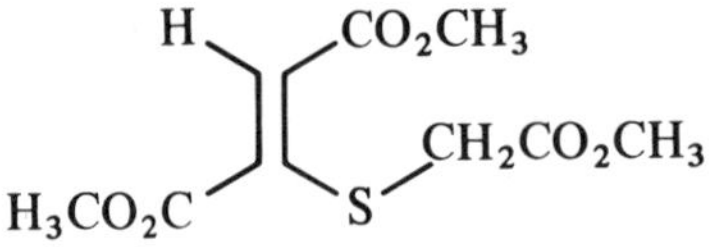

Scheme 118

TABLE 30.   FIESSELMANN SYNTHESES FROM α,β-DIHALOCARBOXYLATES WITH THIOGLYCOLATE

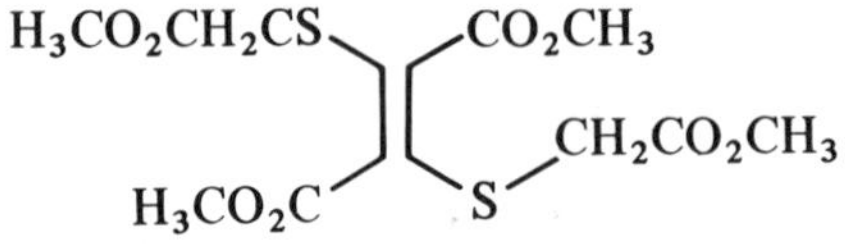

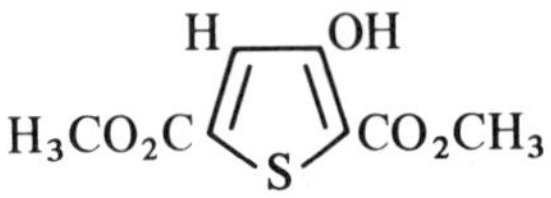

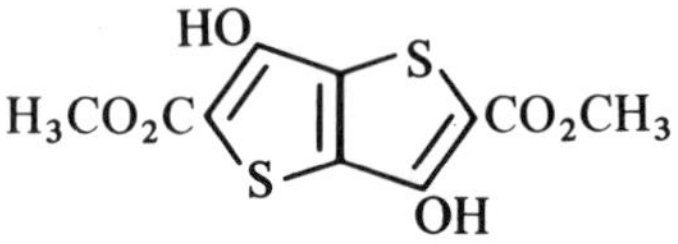

$$RCH-CHCO_2R' + HSCH_2CO_2R'' \xrightarrow{2\,N\,KOH} R\text{-thiophene-}^{OH}_{CO_2R''}$$

R	X	R'	R''	Yield of Thiophene (%)	mp (°C)	Reference
H	Cl	CH_3	CH_3	44, 71	43–44	335, 337
H	Cl	CH_3	C_2H_5	36		
CH_3	Cl	CH_3	CH_3	82, 85	53	335, 337
CH_3	Cl	CH_3	C_2H_5	30	–	335
CO_2CH_3	Br	CH_3	CH_3	60, 84	61	335, 337
C_6H_5	Br	CH_3	CH_3		98	334

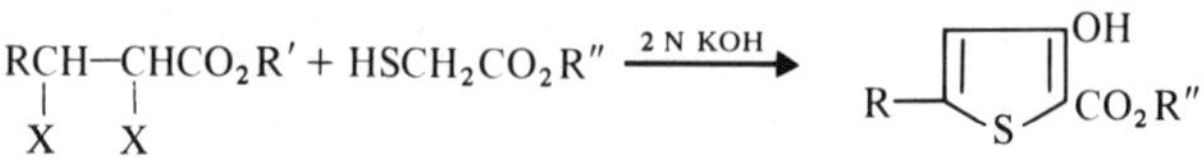

TABLE 31.   FIESSELMANN SYNTHESES FROM α-HALOACRYLATES WITH THIOGLYCOLATE

$$RCH = C - CO_2R' + HSCH_2CO_2R'' \longrightarrow R\text{-thiophene-}^{OH}_{CO_2R''}$$

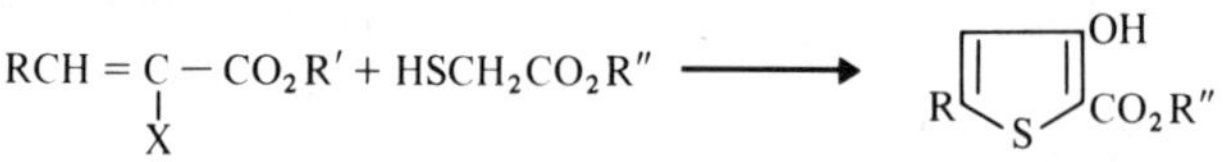

R	X	R'	R''	Yield of Thiophene (%)	Reference
H	Cl	CH_3	CH_3	75	337
CH_3	Cl	CH_3	CH_3	85	337
CO_2CH_3	Br	CH_3	CH_3	85	337

e. THIOPHENES FROM α,β-DIHALONITRILES AND β-CHLOROACRYLONITRILES AND ALKYL THIOGLYCOLATE

A synthetically useful form of the Fiesselmann reaction leading to thiophene analogues of anthranilic acid, namely 3-amino-2-thiophenecarboxylates, is the reaction of α,β-dihalonitriles and related compounds in the presence of base.[336,339-343] The best method is to use 2.5 eq. alcohol-free sodium alkoxide as base, with ether as solvent and with the addition of hydroquinone.[342] The α,β-dichloronitriles were obtained from α,β-unsaturated nitriles by chlorination in the presence of catalytic amounts of pyridine. However, α-phenyl-α,β-dichloro-dihydrocinnamic nitrile did not give thiophenes under these conditions.[343] Attempts to use α-formyl or β-keto-nitriles for the synthesis of 3-aminothiophene were not successful. Only 4-cyano-3-ketotetrahydrothiophene (357), prepared from acrylonitrile and methylthioglycolate, reacted with thioglycolic acid to give (358), which after esterification, even with catalytic amounts of sodium methoxide in methanol, spontaneously ring-closed to 359 (mp 157°C) in 92% yield owing to the advantageous *cis* stereochemistry[343] (Scheme 119). The compounds prepared are given in Table 32.

From β-chlorovinylaldehydes (360, X = Cl), the corresponding nitriles (361) are easily prepared via the oximes. In a one-pot procedure, they are ring-closed to 362 in very high yield by treatment with sodium methoxide in methanol[346] (Scheme 120). Alternatively, 360 (X = Cl) is transformed to 363 through reaction

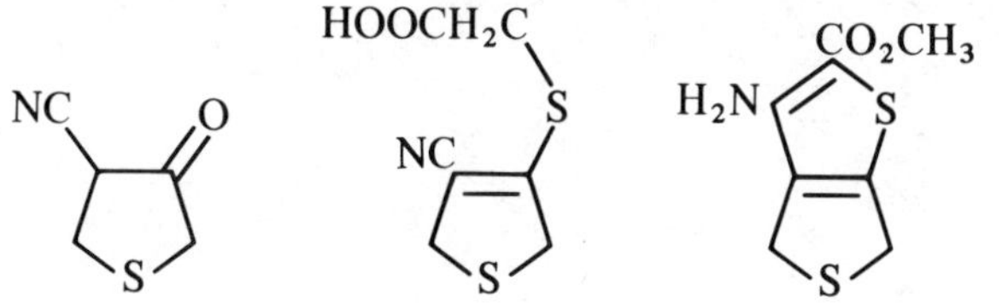

357   358   359

Scheme 119

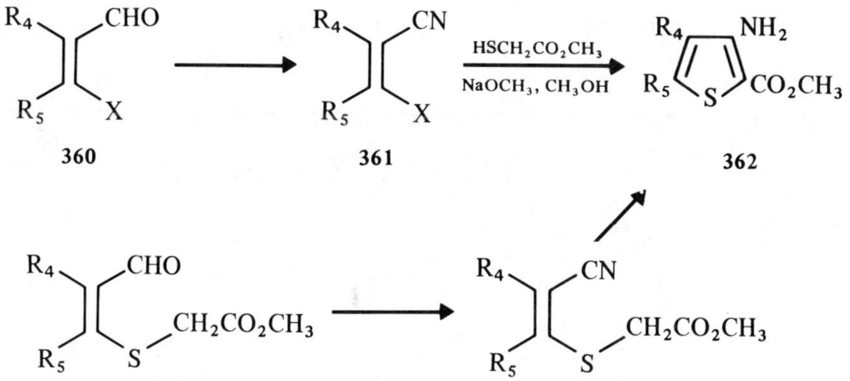

360   361   362

363   364

Scheme 120

TABLE 32.   FIESSELMANN SYNTHESES FROM α,β-DIHALONITRILES AND β-CHLORO ACRYLONITRILES AND THIOGLYCOLATE

$$R_5CH-CH-CN + HSCH_2CO_2R'' \longrightarrow$$
$$\overset{X\ \ X}{\underset{R_4}{|\ \ \ |}}$$

R_5	R_4	X	R''	Yield of Thiophene (%)	mp (°C)	Reference
H	H	Cl	CH_3	70, 43	65–66	342, 344
H	H	Cl	C_2H_5	44	42–43	342
CH_3	H	Cl	C_2H_5	37	59–61	342
C_6H_5	H	Cl	C_2H_5	25	104–105	342
H	CH_3	Cl	CH_3	54	85	343
H	C_6H_5	Cl	CH_3	2	186	343
H_3C	CH_3	Cl	CH_3	26	154	343
p-$H_3CC_6H_4$	H	Cl	CH_3	40–70	124	345
p-$H_3CC_6H_4$	H	Cl	C_2H_5	40–70	115	345
p-$CH_3OC_6H_4$	H	Cl	CH_3	40–70	185	345
p-$CH_3OC_6H_4$	H	Cl	C_2H_5	40–70	175	345

100

with the sodium salt of methyl thioglycolate in toluene, which is then transformed in the usual way to the nitrile (**364**) and ring-closed.[346] Recently, **364b** was prepared from **364a** and ethyl thioglycolate.[346a] Formylation of arylmethylcyanides gives **361** (X = OH), which upon reaction with methyl thioglycolate in the presence of concentrated hydrochloric acid, gave **362** (R_4 = Ar, R_5 = H) in 50–60% yield.[346b] Instead of the chloro derivative, alkoxy derivatives **364c** were recently used in the reaction with methyl or ethyl thioglycolate in alcoholic solvents, with potassium acetate as base, which gave 5-alkyl 3-amino-4-cyano-2-thiophenecarboxylate **364d** in 85–95% yield.[346c] The reaction of the analogous compounds **364e** with thioglycolates were not specific, and a mixture of the 3-hydroxy-4-cyano (**364f**) and 3-amino-4-carbethoxy (**364g**) derivatives was obtained in approximately equal amounts. In the case of **364e** [$R_5 = CH(CH_3)_2$],[346c] only the formation of **364f** [$R_5 = CH(CH_3)_2$] (18%) and thiazolidone (**364h**) was observed. The latter was the only product obtained from **364e** (R_5 = H). However, when $R_5 = C_6H_5$, **364f** was obtained in 77% yield[346c] (Scheme 120a).

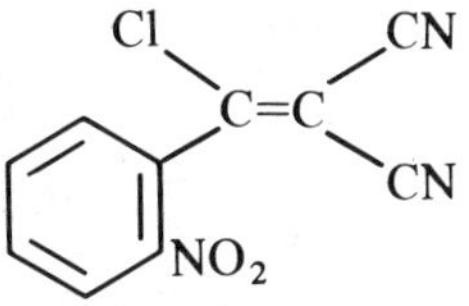

364a

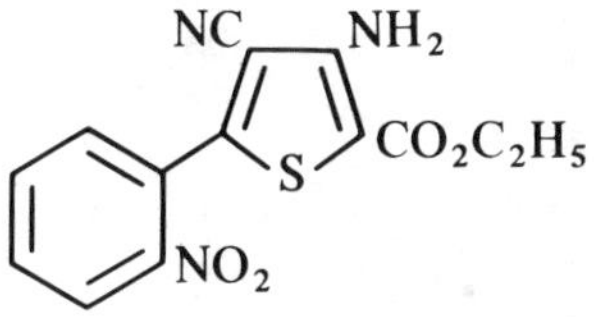

364b

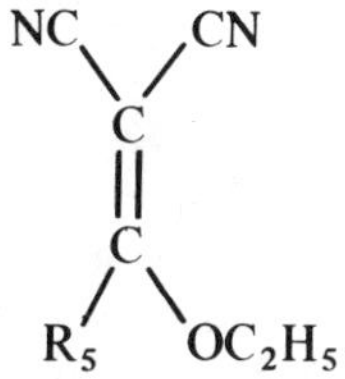

364c

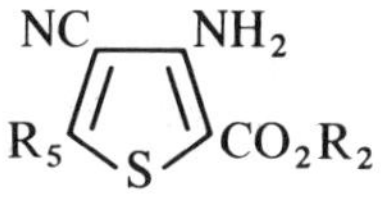

364d

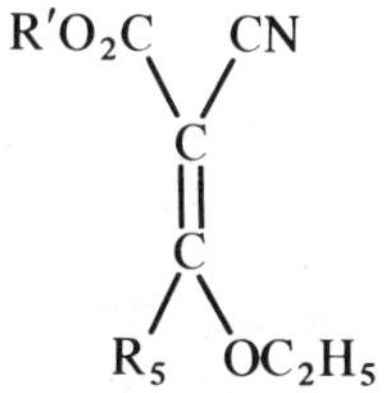

364e

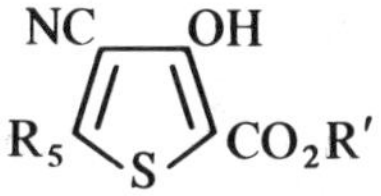

364f

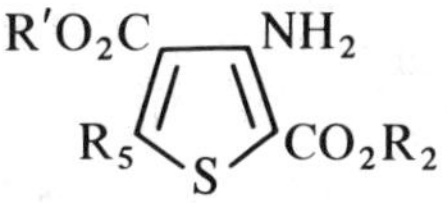

364g

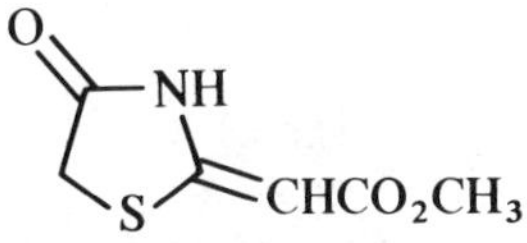

364h

Scheme 120a

Recently, methyl 3-amino-2-thiophenecarboxylate was prepared from α-chloroacrylonitrile.[338] The compounds prepared are listed in Table 33.

TABLE 33. FIESSELMANN SYNTHESES FROM β-CHLORO ACRYLONITRILES AND THIOGLYCOLATES

$$R_4\underset{R_5}{\overset{CN}{>}}X \; + \; HSCH_2CO_2R'' \longrightarrow R_4\underset{R_5}{\overset{NH_2}{\diamond}}{}_S\!CO_2R''$$

X	R_5	R_4	R''	Yield of Thiophene (%)	mp (°C)	Reference
Cl	H_3C	CH_3	CH_3	90	154	346
Cl	CH_3	C_2H_5	CH_3	93	76	346
Cl	CH_3	C_6H_5	CH_3	96		346
Cl	$(CH_2)_4$		CH_3	88	78	346
OH	H	C_6H_5	CH_3	50–60	69	346a
OH	H	$p\text{-}CH_3OC_6H_4$	CH_3	50–60–X	111	346a
OH	H	2-Benso[b]-furyl	CH_3	50–60	115	346a
OH	H	2-Th	CH_3	50–60	90	346a
OH	H	2,5-Di(CH_3)-3-Th	CH_3	50–60	136	346a
Cl	$2\text{-}O_2NC_6H_4$	CN	C_2H_5	70		346a
OC_2H_5	CH_3	CN	CH_3	97	202–204	346c
OC_2H_5	CH_3	CN	C_2H_5	88	140–141	346c
OC_2H_5	C_2H_5	CN	CH_3	83	152–154	346c
OC_2H_5	C_2H_5	CN	C_2H_5	87	111–113	346c
OC_2H_5	CH_3	CO_2CH_3	CH_3	37[a]	129–130	346c
OC_2H_5	CH_3	$CO_2C_2H_5$	C_2H_5	23[a]	75–76	346c
OC_2H_5	C_2H_5	CO_2CH_3	CH_3	25[a]	96–97	346c
OC_2H_5	C_2H_5	$CO_2C_2H_5$	C_2H_5	31[a]	72–73	346c

[a] Also, 3-hydroxy-4-cyanothiophenes were formed.

f. THIOPHENES FROM β-DIKETONES, KETOALDEHYDES AND RELATED COMPOUNDS, AND THIOGLYCOLATES

Habicht and Fiesselmann[347-349] showed that reactive and easily polymerizable β-ketoaldehydes (365), obtained by formylation of ketones and predominantly existing in their tautomeric hydroxymethylene form 366, reacted with thioglycolic acid to give thioacetals (367). A few drops of concentrated H_2SO_4 was used, instead of hydrochloric acid, in the thioacetalization and esterification to give 368. From alkyl thioglycolates and the ketoaldehydes, the unsaturated thio ethers (369) were formed (Scheme 121). Both thioacetals and thio ethers were obtained

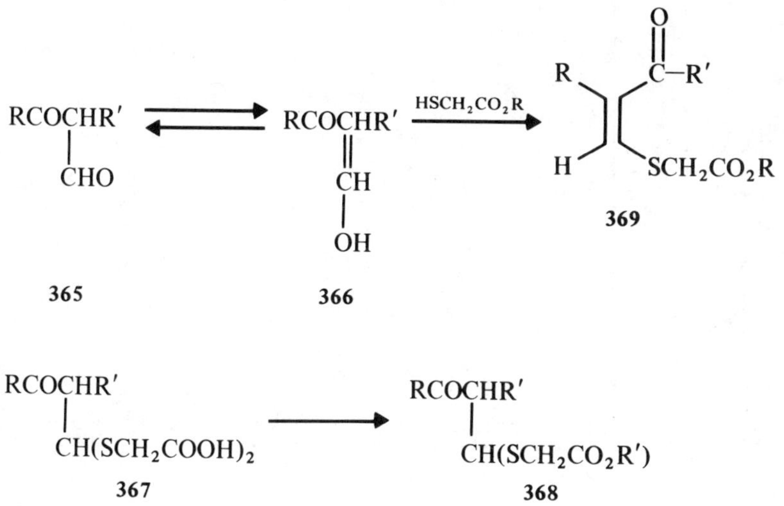

RCOCHR′   ⇌   RCOCHR′ $\xrightarrow{\text{HSCH}_2\text{CO}_2\text{R}}$
| ||
CHO   CH
|
OH

365   366   369

RCOCHR′   RCOCHR′
| |
CH(SCH₂COOH)₂   ⟶   CH(SCH₂CO₂R′)
367   368

Scheme 121

from acetals derived from the ketoaldehydes or from alkoxymethyleneketones via the reaction with thioglycolic acid or alkyl thioglycolate.

Both 368 and 369 could be ring-closed to substituted 2-thiophenecarboxylates with alcoholic potassium hydroxide or sodium alcoholate in ether, but preferentially by heating with sodium methoxide in methanol. The possibility of *cis–trans* isomerism in 369 does not affect the ring-closure, probably because of a low rotation barrier around the double bond. Compounds prepared by this reaction are given in Table 34. A simplified procedure consists of the direct reaction of the anhydrous sodium salt of hydroxymethylene acetophenone with methyl thioglycolate and hydrogen chloride at −5°C. After the removal of hydrogen chloride, ring-closure with sodium methylate gave a 29% yield of methyl 3-phenyl-2-thiophenecarboxylate.[350] In the reaction of formyl derivatives of β-diketones such as 370, the unsaturated sulfides (371) have equivalent carbonyl groups to which

TABLE 34.   FIESSELMANN SYNTHESES FROM 1,3-DICARBONYL COMPOUNDS AND DERIVATIVES WITH THIOGLYCOLATES

$$R_5COCH{-}COR_3 + HSCH_2COOH \xrightarrow[R''OH]{H^+} \overset{R_4}{\underset{R_5}{\bigsqcup}}\overset{R_3}{\underset{S}{}}CO_2R''$$

R_5	R_4	R_3	R''	Yield of Thiophene (%)	mp (°C)	Reference
H	CH_3	CH_3	CH_3	65	34	348
H	CH_3	C_2H_5	CH_3	68	25	348
H	$-(CH_2)_4-$		CH_3	64		348
H	$-(CH_2)_4-$		CH_3	58		348
H	CH_3	C_6H_5	CH_3	78		348
H	H	$p\text{-}CH_3C_6H_4$	CH_3	28		348
H	H	$p\text{-}CH_3OC_6H_4$	CH_3	25	102	348
H	H	C_6H_5	CH_3	18, 29	119	348, 350
H	H	CH_3	CH_3	49		348
H	H	CH_3	C_2H_5	45		348
H	H	$CH_2CH(CH_3)_2$	CH_3	61		348
H	$COCH_3$	CH_3	CH_3	63	69.5	348
H	$COCH_3$	CH_3	C_2H_5	55	89	348
H	CO_2CH_3	CH_3	CH_3	25	102	348
H	CO_2CH_3	C_6H_5	CH_3	62	99	348
H	$CO_2C_2H_5$	C_6H_5	C_2H_5	31	91	348
H	$COCH_3$	OH	CH_3	36	148	348
H	$COCH_3$	OH	C_2H_5	28	118	348
H	COC_6H_5	OH	C_2H_5	33	64	348
H	COC_6H_5	OH	CH_3	34	112	348
H_3C	$OCOCH_3$	CH_3	CH_3	33[a]	134	331
H_3C	SC_2H_5	CH_3	CH_3	65		331
H_3C	SC_4H_9	CH_3	CH_3	73		331
H_3C	C_6H_5	CH_3	CH_3	49		331
H	$OCOCH_3$	CH_3	CH_3	5[a]		331

C_6H_5	SC_2H_5	CH_3	CH_3	48		331
C_6H_5	SC_4H_9	CH_3	CH_3	41		331
C_6H_5	SC_6H_5	CH_3	CH_3	44	123	331
$-(CH_2)_4-$		CH_3	CH_3	31	43.5–45	351
$-(CH_2)_3-$		CH_3	CH_3	21	30	351

a Isolated as the 4-hydroxy derivative.

they can ring-close to give **372** (Scheme 122). However, the ethoxymethylene derivatives of methyl acetoacetate apparently gave two isomeric thioethers, **373** and **374**, which could be separated. Upon treatment of the mixture with base. ring-closure to 13% of **375** and 26% of **376** occurred (Scheme 123). The same yields

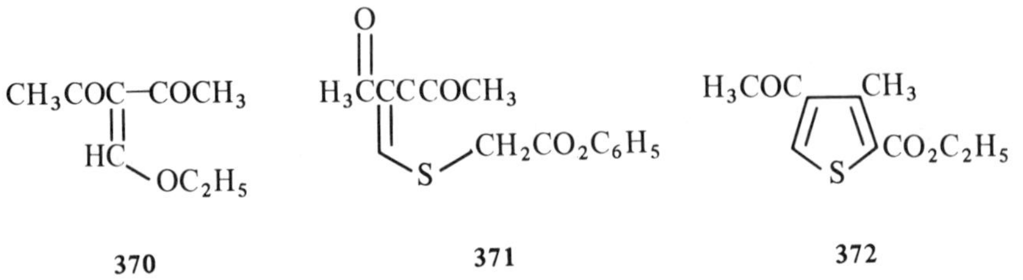

370   371   372

Scheme 122

373   374

375   376

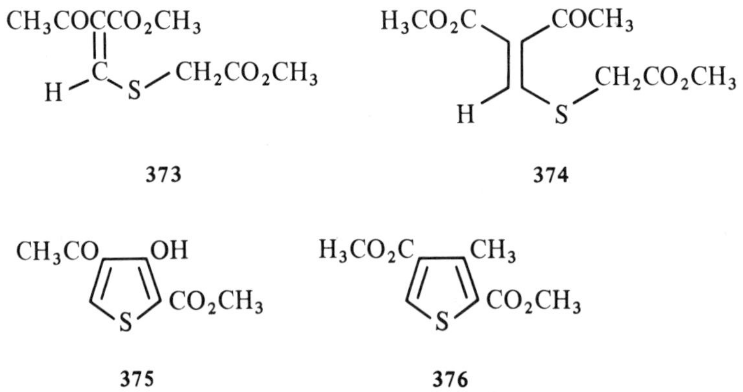

Scheme 123

were obtained if the ethoxymethylene derivative of methyl acetoacetate was reacted before ring-closure with 2 eq. methyl thioglycolate, or the mixture of **373** and **374** was reacted with a second equivalent of methyl thioglycolate under piperidine catalysis. In both cases, the thioacetal was formed as an intermediate.[348] If the crystalline form assumed to be **373** was reacted with 2 N methanolic potassium hydroxide, 33% of **375** and 18% of **376** were obtained, while with sodium methylate in ether, 36% of **375** and 25% of **376** were isolated. It is obvious that the stereochemistry at the double bond does not alone determine the outcome of ring-closure. Depending on either a low rotation barrier or addition of methoxide to the double bond before ring-closure, free rotation is achieved. The ethoxy-methylene derivatives of ethyl benzoylacetate (**377**) gave noncrystalline sulfides (**378**) with methyl and ethyl thioglycolate, and upon treatment with 2 N methanolic potassium hydroxide, only aldol condensation to give **379** in 62% yield was observed. On the other hand, reaction with sodium methylate in ether gave **379**

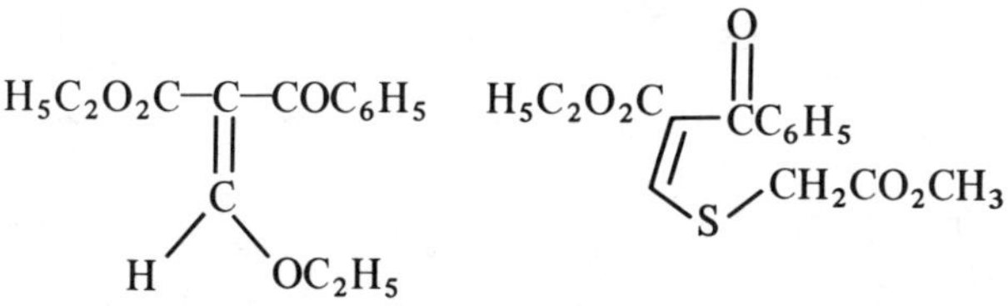

$H_5C_2O_2C-\overset{|}{\underset{||}{C}}-COC_6H_5$

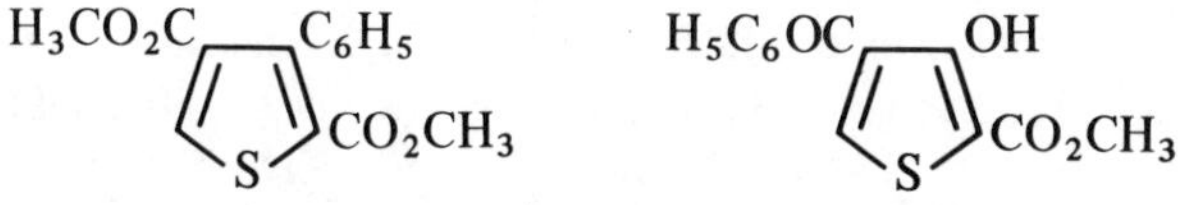

$\overset{H}{\diagup}\overset{}{\underset{OC_2H_5}{\diagdown}}C$

377

$H_5C_2O_2C\diagdown\overset{\overset{O}{||}}{\underset{\diagdown S\diagup}{C}}C_6H_5$ ... $CH_2CO_2CH_3$

378

$H_3CO_2C\diagdown\diagup C_6H_5$ / S / CO_2CH_3

379

$H_5C_6OC\diagdown\diagup OH$ / S / CO_2CH_3

380

Scheme 124

in 33% yield, together with 34% of **380**, formed via Dieckmann cyclization[348] (Scheme 124). With sodium ethoxide, the corresponding ethyl esters are prepared in the same ratio. 3-Heteroatom-substituted derivatives of acetylacetone (**381**), formed by hydrogen chloride catalyzed reaction with thioglycolic acid followed by esterification, can be transformed to **382**, which upon treatment with base leads

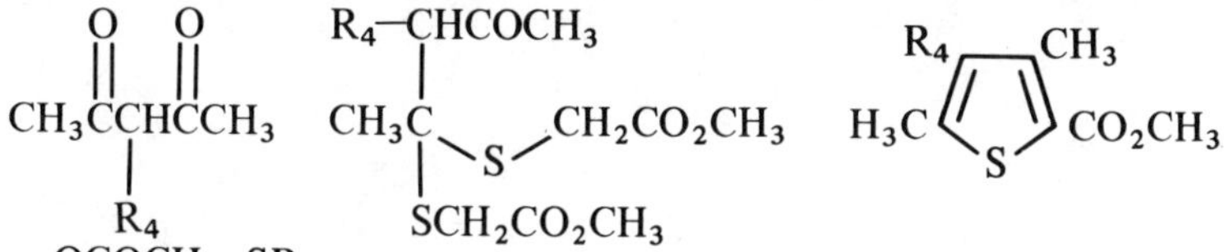

$\overset{O\quad O}{\underset{||\quad||}{CH_3CCHCCH_3}}$
$\underset{R_4}{|}$
$R_4 = OCOCH_3, SR$

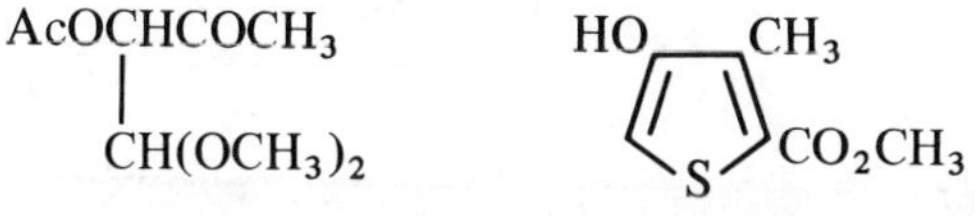

381

$\overset{R_4-CHCOCH_3}{\underset{|}{}}$
$CH_3C\diagdown_S\diagup CH_2CO_2CH_3$
$\underset{SCH_2CO_2CH_3}{|}$

382

$R_4\diagdown\diagup CH_3$ / H_3C S CO_2CH_3

383

$\overset{AcOCHCOCH_3}{\underset{|}{}}$
$CH(OCH_3)_2$

384

$HO\diagdown\diagup CH_3$ / S / CO_2CH_3

385

Scheme 125

to 4-hydroxy-, 4-alkylthio-, and 4-arylthio-thiophenes (**383**).[331] The reaction of **384** with methyl thioglycolate, followed by ring-closure, gave **385** in only 5% yield[331] (Scheme 125).

Unsymmetrical 1,3-diketones, like benzoyl acetones, give thioacetal formation at the carbonyl group bound to the aryl group upon treatment with thioglycolic acid and esterification (**386**).[331] Upon ring-closure, **387** is obtained (Scheme 126). Other unsymmetrical 1,3-diketones that have been investigated are acetylcyclo-

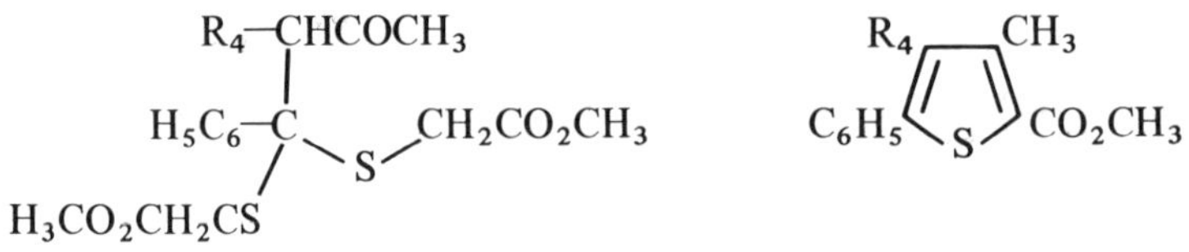

R$_4$—CHCOCH$_3$

H$_5$C$_6$—C—S—CH$_2$CO$_2$CH$_3$

H$_3$CO$_2$CH$_2$CS

R$_4$—CH$_3$

C$_6$H$_5$—S—CO$_2$CH$_3$

386   **387**

R$_4$ = R—S

Scheme 126

hexanone, acylcyclopentanone, and acyl-α-tetralone. Acetylcyclohexanone reacts at the cyclic keto group to give **388**, which upon treatment with sodium methoxide in methanol ring-closes to **389**. Acetylcyclopentanone gives the unsaturated sulfide (**390**), which is ring-closed to **391**[351] (Scheme 127). Acetyltetralone reacts very

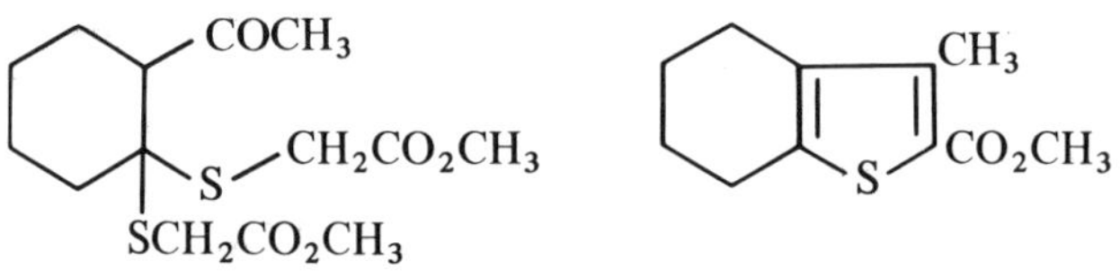

COCH$_3$

S—CH$_2$CO$_2$CH$_3$

SCH$_2$CO$_2$CH$_3$

CH$_3$

S—CO$_2$CH$_3$

388   **389**

COCH$_3$

S—CH$_2$CO$_2$CH$_3$

CH$_3$

S—CO$_2$CH$_3$

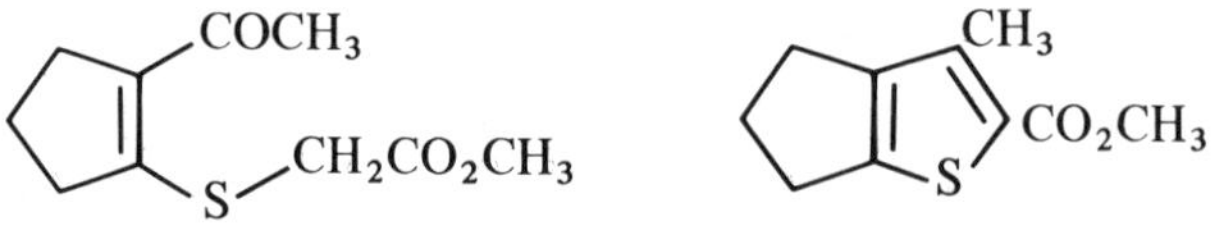

390   **391**

Scheme 127

slowly with thioglycolic acid at the acetyl group, giving a mixture of isomeric unsaturated thioethers, which could be separated and esterified to **392** and **393**. Upon treatment with sodium methoxide, both isomers give the thiophene **394** in about 85% yield, which confirms the facile isomerization during base-treatment (Scheme 128). An interesting deviation was observed when thiophene ring-closure was attempted with **395** or **396**. In both cases, the pyrone **397** was obtained[351] (Scheme 129).

α,γ-Diketocarboxylic acid derivatives have also been reacted with thioglycolic acid and alkyl thioglycolates, and the intermediate products ring-closed to thiophenes by base treatment. Thus, ethyl acetoneoxalate, upon reaction with thioglycolic acid probably gives **398**, which upon treatment with concentrated aqueous sodium hydroxide yields the dicarboxylic acid **399** in 38% yield. Ring-

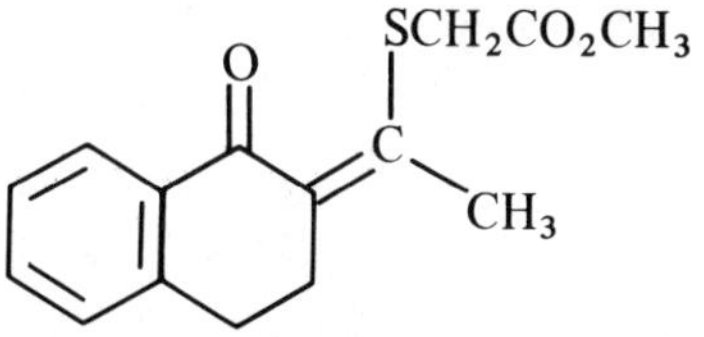

392

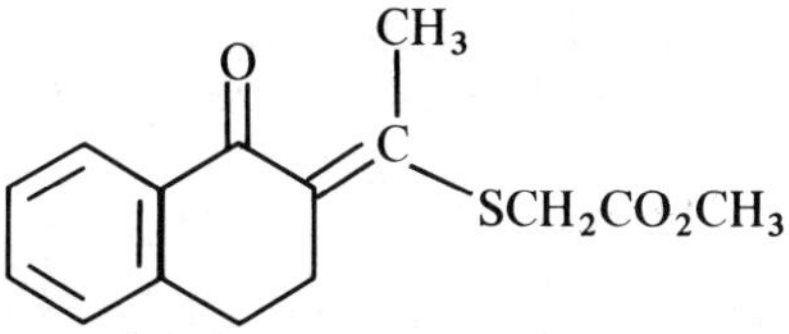

393

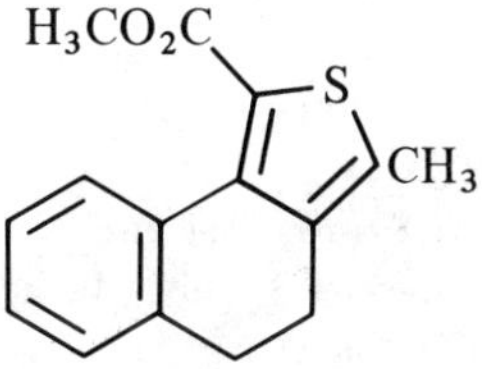

394

Scheme 128

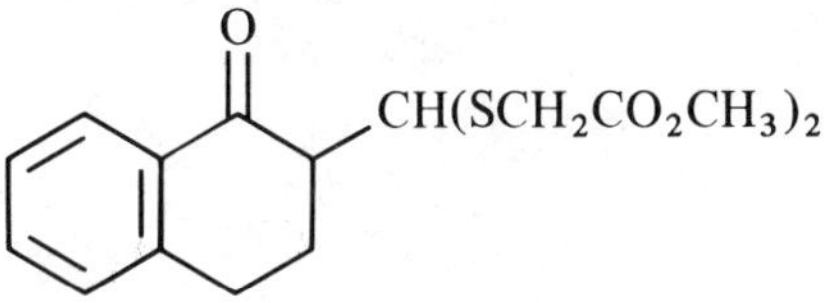

395

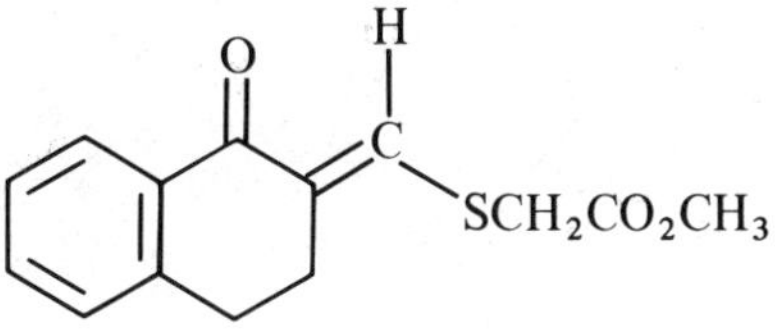

396

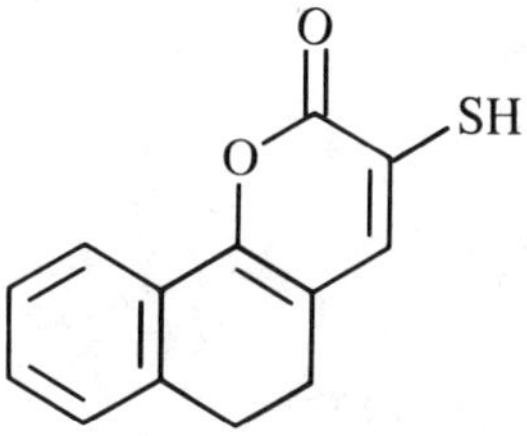

397

Scheme 129

closure with sodium ethoxide in ethanol was not successful (Scheme 130). However, the reaction of ethyl acetophenone oxalate with thioglycolic acid and

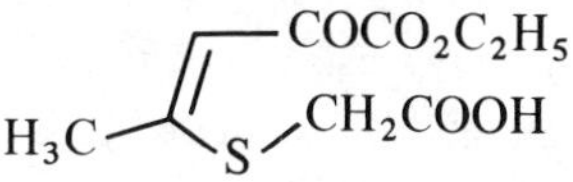

398

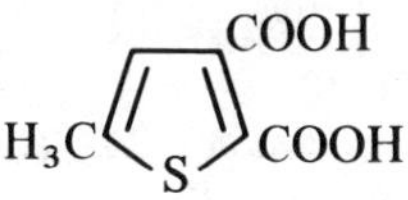

399

Scheme 130

hydrogen chloride in chloroform gave **400**, which upon treatment with concentrated sodium hydroxide gave the 2,5-thiophenedicarboxylic acid **401**[351] (Scheme 131). Ethyl cyclohexanoneoxalate reacts with thioglycolic acid at both carbonyl

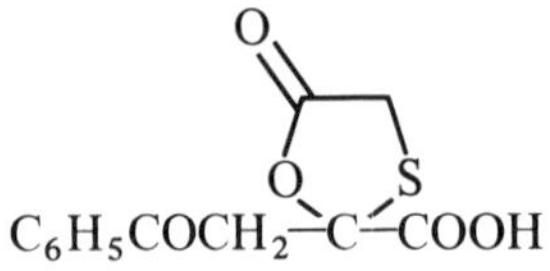

$C_6H_5COCH_2-C-COOH$

400

HOOC—[thiophene with C_6H_5]—COOH

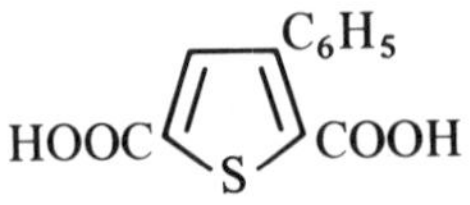

401

Scheme 131

groups, giving both **402** and **403**, since upon ring-closure both **404** and **405** were obtained, in 30 and 7% yield, respectively. The latter is derived by decarboxylation of the initially formed 2,3-thiophenedicarboxylic acid[351] (Scheme 132).

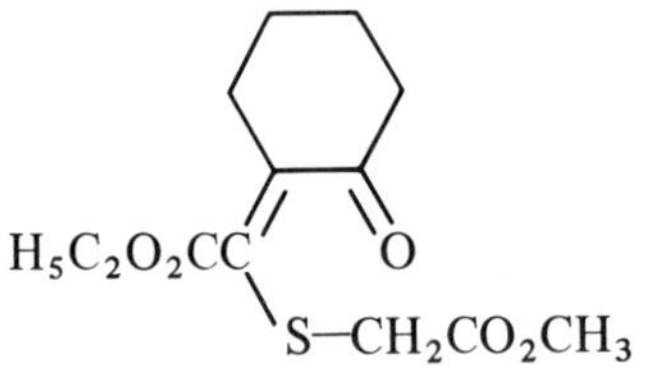

$H_5C_2O_2CC$ O
$S-CH_2CO_2CH_3$

402

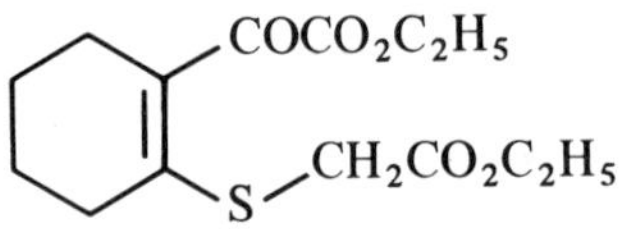

$COCO_2C_2H_5$
$S-CH_2CO_2C_2H_5$

403

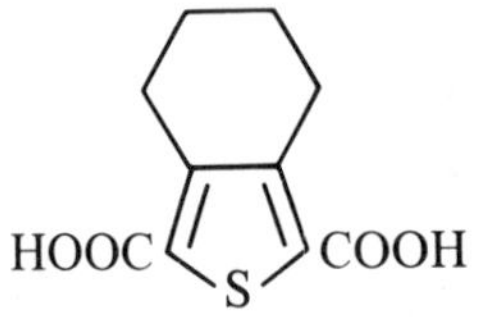

HOOC—[thiophene]—COOH

404

[thiophene]—COOH

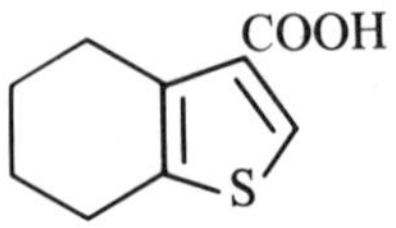

405

Scheme 132

g.   THIOPHENES FROM β-HALOVINYL ALDEHYDES AND KETONES WITH THIOGLYCOLATES

The discovery by Arnold and Zemlicka[352] that ketones with active methylene groups gave β-chlorovinyl aldehydes with Vilsmeier reagents led Fiesselmann to investigate the use of these compounds for the synthesis of thiophenes. The results are given in Schmidt's Ph.D. thesis.[346] He found a suitable one-pot procedure to prepare methyl 4,5-disubstituted 2-thiophenecarboxylate (**407**) in yields greater than 90%, by reacting the chlorovinylaldehyde **406** with 1.5 moles of methyl thioglycolate in methanol. An equivalent amount of sodium methoxide in methanol was added to achieve the condensation, and the temperature was kept at 10–15°C (Scheme 133). The only serious limitation is that the Vilsmeier formylation of unsym-

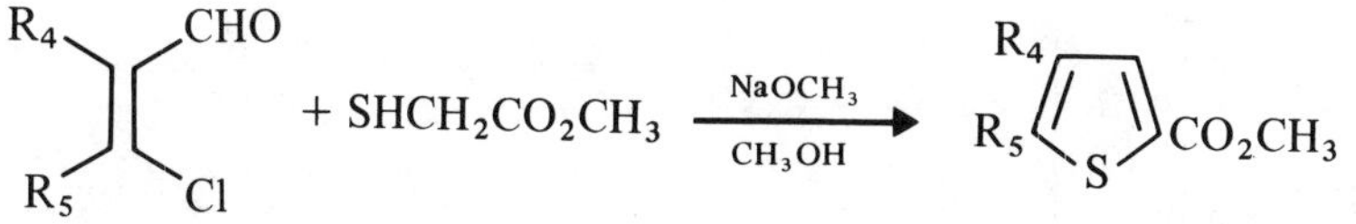

406   **407**

Scheme 133

metrically substitued ketones gives mixtures of isomeric chlorovinylaldehydes.[353] The reaction was extended to the unstable β-chlorovinylketones (408)[193a, 346] prepared through the aluminum chloride-promoted reaction of acid chlorides with acetylenes. Aliphatic compounds are especially unstable and polymerize easily. However, by careful reaction at 0°C with methyl thioglycolate in pyridine, the unsaturated thioethers (409) could be obtained. These were more stable than the starting material and were ring-closed by sodium methoxide treatment to 410 (Scheme 134). These convenient methods were later rediscovered by other

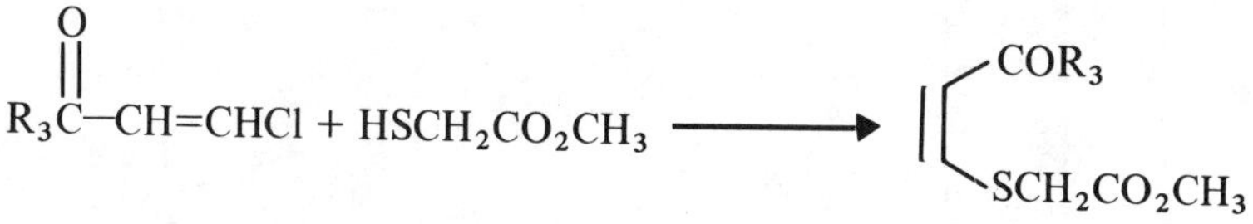

408   **409**

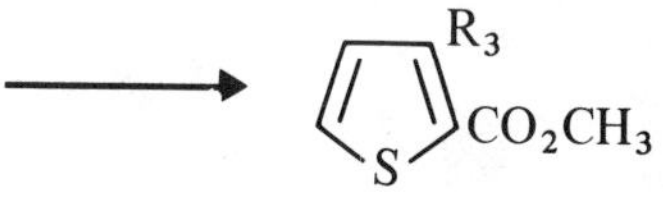

410

Scheme 134

workers.[345, 355–359] Compounds prepared according to this method are given in Table 35.

The reaction between β-halovinylcarbonyl compounds and thioglycolate is not stereospecific, and as was later found; it gives a mixture of unsaturated sulfides of type 411.[357] It was also determined that β-chlorovinylcarbonyl compounds react directly with 80% commercial thioglycolic acid upon treatment with pyridine-triethyl amine directly to 2-thiophenecarboxylic acids (413) and thiophenes (414).[356] Since 2-thiophenecarboxylic acids do not decarboxylate so easily, it appears reasonable that decarboxylation occurs at an intermediate stage (412),

TABLE 35. REACTION OF β-CHLOROVINYL ALDEHYDES AND KETONES WITH ALKYLTHIOGLYCOLATES

$$\begin{array}{c} R_3 \\ | \\ R_4 \diagdown C=O \\ \diagup \diagdown \\ R_5 \quad Cl \end{array} + HSCH_2CO_2R'' \longrightarrow \begin{array}{c} R_4 \diagup \diagdown \\ R_5 \diagdown S \diagdown CO_2R' \end{array}$$

R$_5$	R$_4$	R$_3$	R"	Yield of Thiophene (%)	mp (°C)	Reference
H$_3$C	H$_3$C	H	CH$_3$	88	32	346
H$_3$C	C$_2$H$_5$	H	CH$_3$	85		346
H$_3$C	C$_6$H$_5$	H	CH$_3$	95	64–65	346
C$_6$H$_5$	CH$_3$	H	CH$_3$	93		346
p-CH$_3$OC$_6$H$_4$	CH$_3$	H	CH$_3$	90		346
—(CH$_2$)$_3$—		H	CH$_3$	94	56–58	346
—(CH$_2$)$_4$—		H	CH$_3$	90		346
(naphthalene)		H	CH$_3$	90		346
C$_6$H$_5$	H	H	C$_2$H$_5$	40	29	354
p-ClC$_6$H$_4$	H	H	C$_2$H$_5$	56	83	354
p-BrC$_6$H$_4$	H	H	C$_2$H$_5$	74	79–80	354
C$_6$H$_5$	CH$_3$	H	C$_2$H$_5$	42	44	354
C$_6$H$_5$	C$_6$H$_5$	H	C$_2$H$_5$	38	77.5	354
—(CH$_2$)$_3$—		H	C$_2$H$_5$	50	37	354
—(CH$_2$)$_4$—		H	C$_2$H$_5$	64	30	354
—(CH$_2$)$_5$—		H	C$_2$H$_5$	70		354
—(CH$_2$)$_6$—		H	C$_2$H$_5$	70		357
(benzyl)		H	C$_2$H$_5$	65	83	354

112

Table (page rotated 90°). Structure at top left: a 1,8-disubstituted naphthalene bearing the bridge $-(CH_2)_3CO(CH_2)_2-$.

R¹	R²	R³	Yield (%)	m.p. (°C)	Ref.
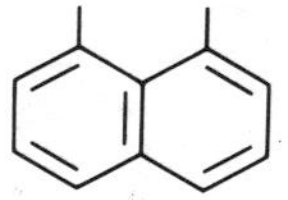 H	H	C_2H_5	87	98	354, 358
H	CH_3	CH_3	74	97	359
H	C_2H_5	H	63		193a
H	$n\text{-}C_3H_7$	H	71		193a
H	$n\text{-}C_5H_{11}$	H	63		193a
H	C_6H_5	CH_3	32, 50	118–119, 123	193a
H	$p\text{-}ClC_6H_4$	CH_3	40, 50	95–96, 98–99	357, 360
H	$p\text{-}CH_3OC_6H_4$	CH_3	48	101	357, 360
H	C_6H_5	C_2H_5	56	61.5	360
H	$p\text{-}BrC_6H_4$	C_2H_5	70	80–81	357
H	$p\text{-}BrC_6H_4$	CH_3	55	109–110	357
$-(CH_2)_5-$		CH_3	50	36	357
$-C-(CH_2)_2-C-$	$C(CH_3)_2$	CH_3	30	182–184	357

Structure (second section): bridge $-(CH_2)_4-$.

R¹	R²	R³	Yield (%)	m.p. (°C)	Ref.
H_5C_6	C_6H_5	CH_3	40	144–145	357
CH_3	H	H		139	356
C_6H_5	H	H		189	356
$p\text{-}CH_3C_6H_4$	H	H		216	356
$p\text{-}ClC_6H_4$	H	H		249	356
$p\text{-}BrC_6H_4$	H	H		251	356
CH_3	CH_3	H		186	356
C_2H_5	CH_3	H		113	356
C_6H_5	CH_3	H		147	356
C_6H_5	C_6H_5	H		261	356
H	H	C_6H_5		113	356
C_6H_5	C_6H_5	CH_3		189	356
H	H	H		196	356

TABLE 35. *Continued*

R₅	R₄	R₃	R″	Yield of Thiophene (%)	mp (°C)	Reference
	(o-CH₃-C₆H₄)S—CH₂—	H	H		237	356
	(o-CH₃-C₆H₄)O—CH₂CH₂—	H	H		240	356
	(o-CH₃-C₆H₄)S—CH₂CH₂—	H	H		234	356, 361
	(naphthyl, SCH₂—)	H	H		247	356

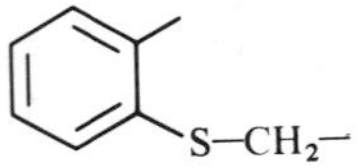

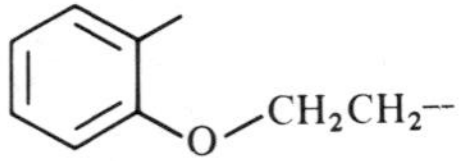

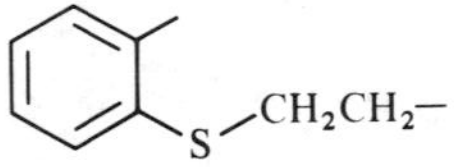

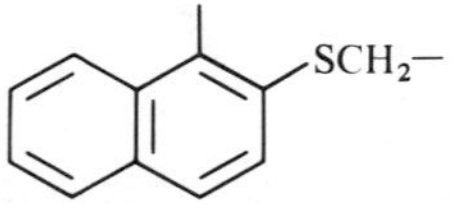

114

as indicated in Scheme 135.[356] The relative amounts of **413** and **414** depend upon the substituents, and the amount of **414** increases with temperature. The decarboxylative aromatization makes it possible to use other α-mercapto acids, which via **415** and **416** ring-close to 2-substituted thiophenes (**417**).[356] The

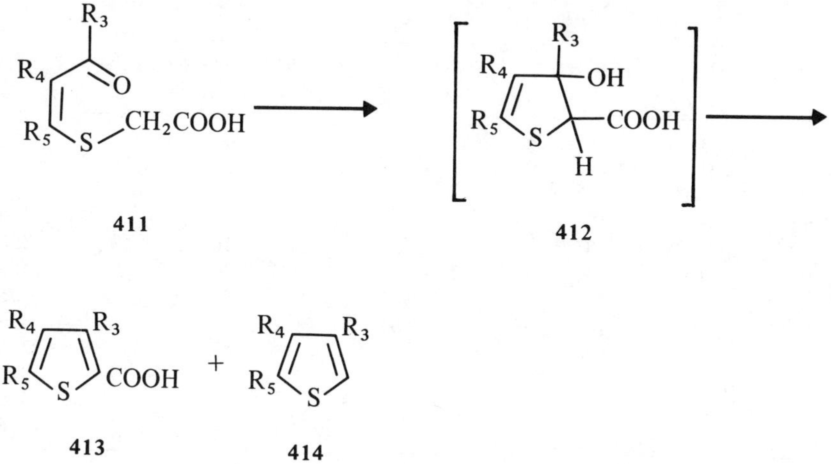

411   412

413   414

Scheme 135

thiophenecarboxylic acids (**417b**) have been prepared from **417a** by reaction with thioglycolic acid and 30% aqueous potassium hydroxide[359a] (Scheme 136). The compounds prepared are given in Table 36.

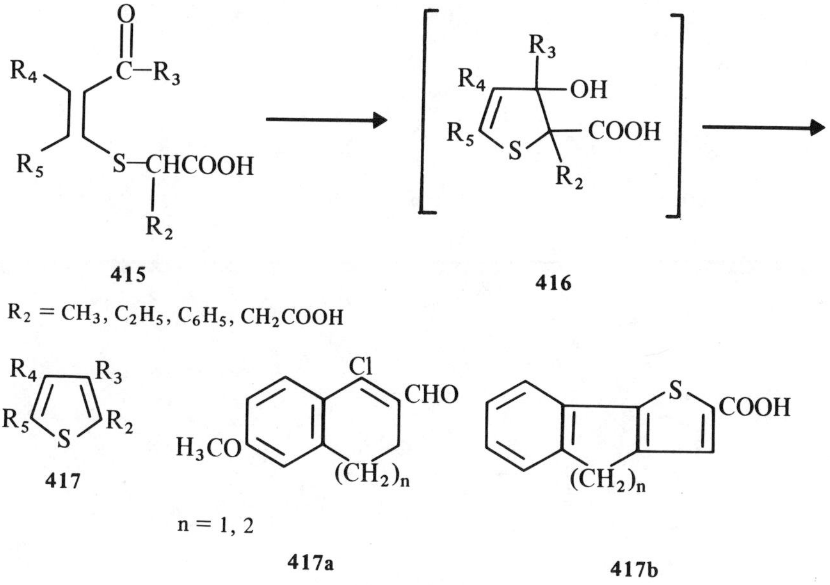

415   416

$R_2 = CH_3, C_2H_5, C_6H_5, CH_2COOH$

417

n = 1, 2

417a   417b

Scheme 136

TABLE 36.   FIESSELMANN SYNTHESES FROM β-CHLOROVINYLCARBONYL COMPOUNDS AND α-MERCAPTO ACIDS

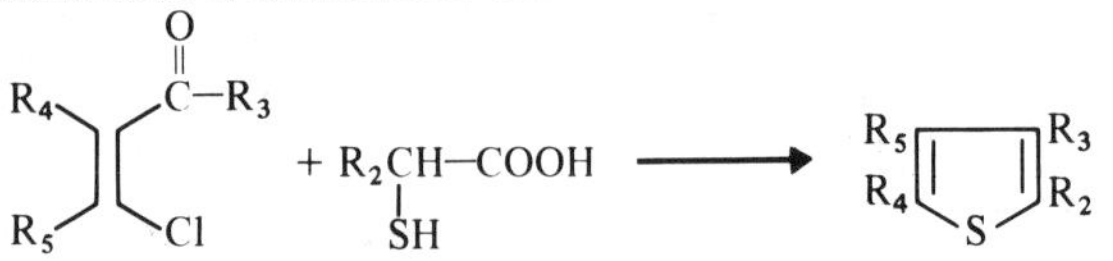

R_5	R_4	R_3	R_2	Yield of Thiophene (%)	Reference
C_6H_5	H	H	CH_3	63	356
C_6H_5	H	CH_3	CH_3	59	356
C_2H_5	H_3C	H	C_2H_5	32	356
C_6H_5	H	H	C_2H_5	45	356
C_6H_5	H	CH_3	C_2H_5	40	356
C_6H_5	H	CH_3	C_6H_5	34	356
C_2H_5	CH_3	H	CH_2COOH	80	356
C_6H_5	H	H	CH_2COOH	92	356
C_6H_5	H	CH_3	CH_2COOH	85	356
$-(CH_2)_4-$		H	CH_3		356
$-(CH_2)_4-$		H	C_2H_5		356
$-(CH_2)_4-$		H	C_6H_5		356

The reaction of β-chlorovinylaldehydes was also extended to the carbalkoxy derivative 418.[346] Upon treatment of 418 (R = CH₃) with sodium methoxide in methanol, only aldol condensation occurred, to give 419 in 70% yield; 418 (R = Ph) gave Dieckmann cyclization, leading to 420 in 71% yield[346] (Scheme 137).

From 421, 422 has been prepared[359] (Scheme 138).

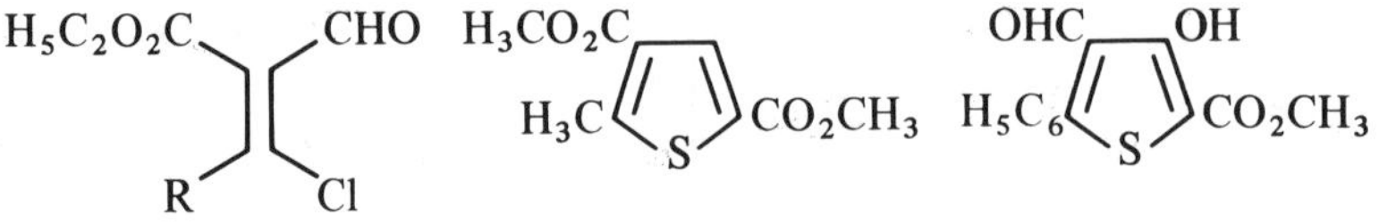

418   419   420

Scheme 137

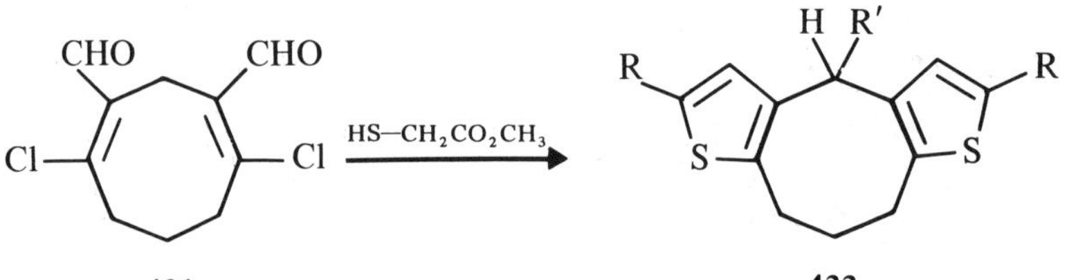

421   422

Scheme 138

h. THIOPHENES FROM α-BROMOVINYLKETONES AND THIOGLYCOLATES

The reaction of α-bromovinylketones **423** with methyl thioglycolate, followed by ring-closure from treatment with sodium methoxide in methanol, gave the corresponding 2-thiophenecarboxylates (**424**). Because of partial hydrolysis, it was in some cases more convenient to isolate the free acids. It was much more difficult to directly achieve thiophene ring-closure of the corresponding saturated dibromo derivatives, which are starting materials for the synthesis of **423**[336] (Scheme 139).

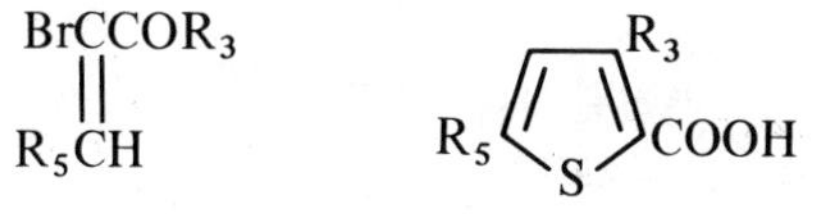

$$\begin{array}{cc} \underset{\mathrm{R_5CH}}{\overset{\mathrm{BrCCOR_3}}{\|}} & R_5\underset{S}{\diagup}\overset{R_3}{\diagdown}COOH \\ \mathbf{423} & \mathbf{424} \end{array}$$

Scheme 139

Instead of alkyl thioglycolates, different anilides of thioglycolic acid could be used. Low yields of thiophenes were obtained with α-halovinyl aldehydes such as α-bromoacrolein, α-chlorocrotonaldehyde, and α-bromocinnamic aldehyde.[348] The thiophene derivatives prepared in this way are listed in Table 37.

TABLE 37.   FIESSELMANN SYNTHESES FROM THE REACTION OF α-BROMOVINYL-KETONES AND THIOGLYCOLATES

$$\underset{\mathrm{R_5CH}}{\overset{\mathrm{BrCCOR_3}}{\|}} + \mathrm{HSCH_2\overset{O}{\overset{\|}{C}}-R'} \longrightarrow R_5\underset{S}{\diagup}\overset{R_3}{\diagdown}COR'$$

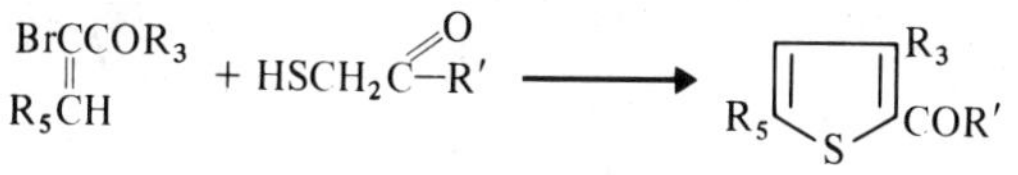

R_5	R_3	R'	Yield of Thiophene (%)	mp (°C)	Reference
C_6H_5	CH_3	H_3CO	78[a]	185	336
C_6H_5	CH_3	C_6H_5NH	62	174	336
C_6H_5	CH_3	p-$H_3CC_6H_4$	59	156	336
C_6H_5	C_6H_5	OCH_3	50[a]	225	336
H_3CO_2C	C_6H_5	OCH_3	75	110	336
H_3CO_2C	C_6H_5	C_6H_5NH	23	156	336
H_3CO_2C	$C_6H_4CH_3$-p	OCH_3	76	112–113	336
H_3CO_2C	α-Naphthyl	OCH_3	67	151	336
H_3CO_2C	CH_3	OCH_3	72	84	336
H	H	OCH_3	5[a]		348
CH_3	H	OCH_3	30		348
CH_3	H	OC_2H_5	26		348
C_6H_5	H	OCH_3	12	99	348
C_6H_5	H	OC_2H_5	28		348

[a] Isolated as the acid.

Other useful starting materials for thiophenes are 1,2-diaroyl-1,2-dihalogeno-ethylenes, which are prepared by the addition of bromine and chlorine to diaroyl-acetylenes. The bromo and chloro derivatives react differently with thioglycolate.[362] When sodium methoxide is used as condensing agent, the bromo derivatives (425) give 3-aryl-5-aroyl-2-thiophenecarboxylates (429) with no bromine in the 4-position. During the reaction, this must have been exchanged for hydrogen, with alkyl thioglycolate as the reducing agent. For example, the reaction of 1,2-diaroyl-1,2-dibromoethylene with methyl thioglycolate in pyridine gave 1,2-diaroyl-(carbomethoxymethylmercapto) ethylene as product. The main reaction path therefore probably proceeds via 426 to 428, as indicated in Scheme 140. In the ring-closure,

$$RCOCBr = CBrCOR + HSCH_2CO_2CH_3$$

425

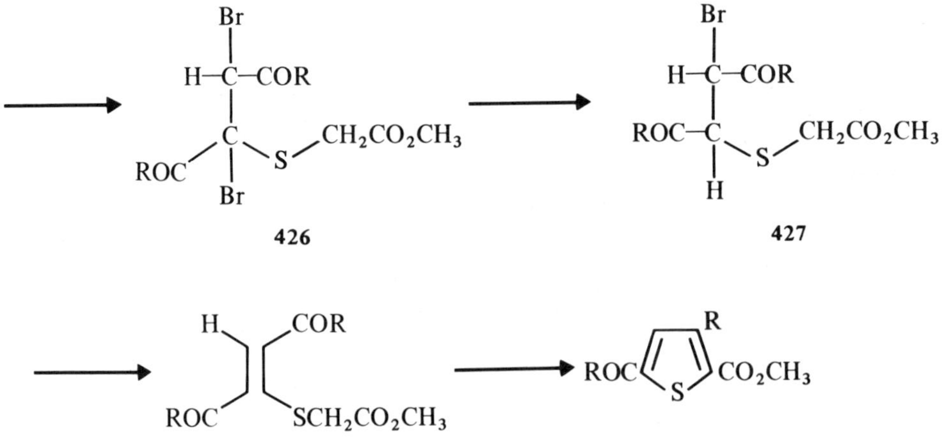

Scheme 140

small amounts of 3,6-diarylthieno[3,2-b]thiophene (433) are also formed as by-products. This is assumed to proceed through the intermediates (430–432) indicated in (Scheme 141).

When the dichloro compounds (434) were used, chlorine was not reduced by thioglycolate, and the intermediate (436) formed via 435 either ring-closed to the 4-chloro derivative (437) or reacted with methyl thioglycolate and ring-closed to the thieno[3,2-b]thiophene (439) via 438. In the phenyl case, 437 and 439 are formed in about equal amounts, and in the anisyl case, only 437 is formed; for the p-chloro compound, 437 and 439 are formed in 42 and 54% yield, respectively[362] (Scheme 142). The methyl 5-aroyl-3-arylthiophenes prepared in this way are given in Table 38.

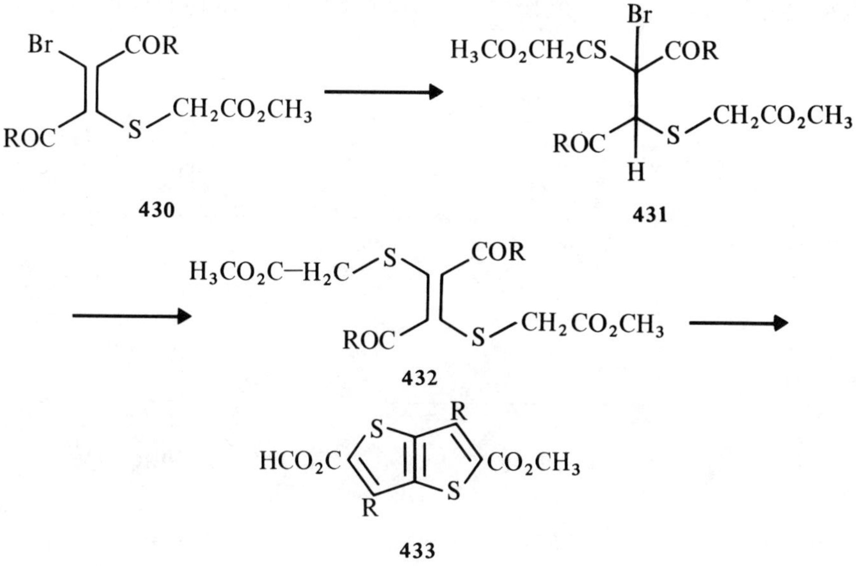
Br
COR
ROC
S
$CH_2CO_2CH_3$
430
$H_3CO_2CH_2CS$
Br
COR
ROC
H
S
$CH_2CO_2CH_3$
431
$H_3CO_2C{-}H_2C$
S
COR
ROC
S
$CH_2CO_2CH_3$
432
HCO_2C
S
R
S
CO_2CH_3
R
433

Scheme 141

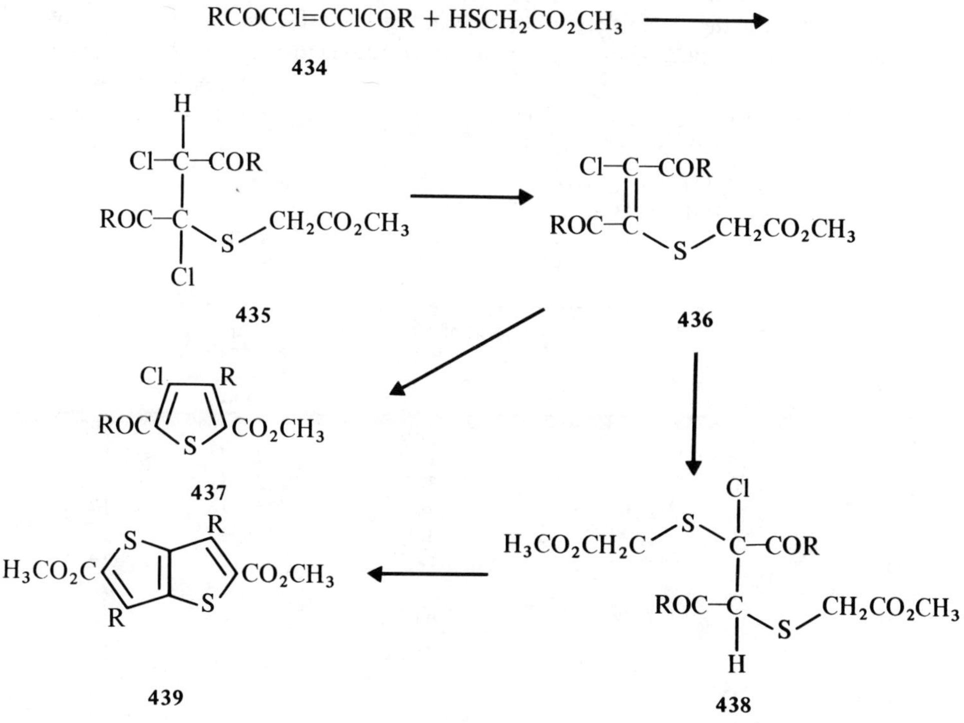
RCOCCl=CClCOR + $HSCH_2CO_2CH_3$
434
H
Cl—C—COR
ROC—C
Cl
S
$CH_2CO_2CH_3$
435
Cl—C—COR
ROC—C
S
$CH_2CO_2CH_3$
436
Cl
R
ROC
S
CO_2CH_3
437
$H_3CO_2CH_2C$
S
Cl
C—COR
ROC—C
H
S
$CH_2CO_2CH_3$
438
H_3CO_2C
S
R
S
CO_2CH_3
R
439

Scheme 142

TABLE 38.   FIESSELMANN SYNTHESES FROM 1,2-DIAROYL-1,2-DIHALOETHYLENES AND THIOGLYCOLATES

$$RCOC=C-COR + HS-CH_2CO_2CH_3 \longrightarrow$$

(with X X substituents on the ethylene carbons; product is a thiophene bearing R_4, R, ROC, CO_2CH_3 and S)

R	X	R_4	Yield of Thiophene (%)	mp (°C)	Reference
C_6H_5	Br	H	86	111	365
$p\text{-}CH_3OC_6H_4$	Br	H	90	136.5	362
$p\text{-}ClC_6H_4$	Br	H	58	149	362
C_6H_5	Cl	Cl	44	140	362
$p\text{-}CH_3OC_6H_4$	Cl	Cl	91	130	362
$p\text{-}ClC_6H_4$	Cl	Cl	42	158	362

i.   THIOPHENES FROM REACTIONS WITH OTHER α-MERCAPTO DERIVATIVES

Various changes in the SC-component are also tolerated in the Fiesselmann reaction. Thus, in addition to methyl and ethyl thioglycolate, amides and anilides of thioglycolic acid can also be used,[335–337] but the yields are often lower. Attempts to use the hydrazide of thioglycolic acid failed.[350]

The use of α-mercapto aldehydes and α-mercapto ketones, which exist in equilibrium with their stable 2,5-dihydroxy-1,4-dithiane forms, has proven to be of great synthetic value. Thus, α-mercapto acetone, α-mercapto acetaldehyde, and phenacyl mercaptane have been reacted with dimethyl acetylenedicarboxylate, methyl propiolate, and methyl phenyl propiolate[193] (Table 39) or α,β-dihalo esters,

TABLE 39.   FIESSELMANN SYNTHESES FROM ACETYLENIC DERIVATIVES AND α-MERCAPTOCARBONYL COMPOUNDS

$$R_5-C≡C-CO_2R' + HSCH_2R_2 \longrightarrow$$

(product is a thiophene bearing R_5, S, R_2, and OH)

R_5	R'	R_2	Yield of Thiophene (%)	mp (°C)	Reference
H_3CO_2C	CH_3	$COCH_3$	77	113	193
H_5C_6	CH_3	$COCH_3$	35	97–99	193
CH_3	CH_3	$COCH_3$	15	73–74	193
H	CH_3	$COCH_3$	30	50–51	193, 323
CH_3O_2C	CH_3	CHO	40	162–163	193
C_6H_5	CH_3	CHO	20^a	155–156	193
H_3CO_2C	CH_3	CHO	78	102	193
H	C_2H_5	COC_6H_5	46	64	204
C_6H_5	CH_3	COC_6H_5	41	123–124	204
H_3C	C_2H_5	COC_6H_5	41	58.5	204
$CH_3C≡C-$	CH_3	$COCH_3$		99.5–100	363
H_3CO_2C	CH_3	$CONH_2$	70	205	337

^{a} Methyl 2-phenyl-3-thiophenecarboxylate is also formed.

TABLE 40. FIESSELMANN SYNTHESES FROM α,β-DICHLORONITRILES AND α-MERCAPTOACETONE OR α-ACETYLMERCAPTOACETONITRILE

$$R_5CHClCClCN \xrightarrow{\text{HSCH}_2R_2} R_5 \underset{S}{\overset{}{\boxed{}}} \overset{NH_2}{\underset{R_2}{}}$$

R_5	R_4	R	R_2	Yield of Thiophene (%)	mp (°C)	Reference
H	H	H	$COCH_3$	69		343
H_3C	H	H	$COCH_3$	77	64	343
C_6H_5	H	H	$COCH_3$	78	137	343
H	CH_3	H	$COCH_3$	62	82	343
H_3C	CH_3	H	$COCH_3$	65	122	343
H	H	$COCH_3$	CN	25–30	44	337
H_3C	H	$COCH_3$	CN	62	103	337
C_6H_5	H	$COCH_3$	CN	75	128	337

α-halo-α,β-unsaturated esters,[337] and/or β-halo, α,β-unsaturated esters[204] to give 2-acyl 3-hydroxythiophenes. With α,β-dichloronitriles, 3-amino-2-acylthiophenes were obtained[343] (Table 40). It is also possible to use acetylated α-mercapto-carbonyl derivatives, which under the alkaline conditions of the Fiesselmann reaction give the thiolate, and from acetylmercaptoacetonitrile and α-halogen-α,β-unsaturated nitriles, 3-amino-2-cyanothiophenes were obtained[337] (Table 41). In general, the reactions with α-mercaptocarbonyl derivatives are much more sensitive to reaction conditions than those with the alkyl thioglycolates, and

TABLE 41. FIESSELMANN SYNTHESES FROM α,β-DIHALOCARBOXYLATES AND ACYLMERCAPTOCARBONYL AND ACYLMERCAPTOACETONITRILE DERIVATIVES

$$XCHCO_2CH_3 + ROCS-CH_2R_2 \longrightarrow R_5 \underset{S}{\overset{}{\boxed{}}} \overset{OH}{\underset{R_2}{}}$$
$$R_5CHX$$

R_5	X	R_2	Yield of Thiophene (%)	mp (°C)	Reference
H	Cl	CN	60	116	337
H_3C	Cl	CN	70	157	337
CO_2CH_3	Br	CN	70	183	337
C_6H_5	Br	CN	60	176	337
H	Cl	$COCH_3$	35	51	337
CH_3	Cl	$COCH_3$	45	74	337
CO_2CH_3	Br	$COCH_3$	70	113	337
H	Cl	CHO	10	86	337
CH_3	Cl	CHO	50	95	337
CO_2CH_3	Br	CHO	55	163	337
H	H	COC_6H_5	50	62	337
H	Cl	$CONH_2$	45	191	337
H_3C	Cl	$CONH_2$	60	220	337

temperature and base concentrations have to be strictly defined to avoid excessive polymerization and tar formation. The reaction of methyl phenylpropiolate (**440**) is of special interest, since the reaction with mercapto acetaldehyde and the following ring-closure proceeds in two directions, giving **441** and **442** in 11 and 20% yield, respectively[193] (Scheme 143). The intermediate unsaturated sulfide thus has

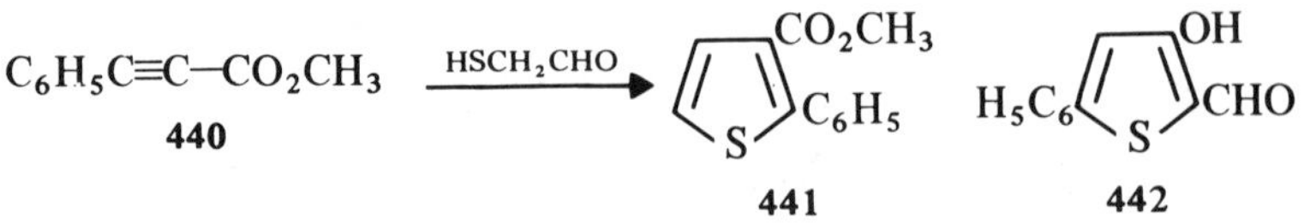
C6H5C≡C—CO2CH3
440
HSCH2CHO
CO2CH3
C6H5
S
441
OH
H5C6
CHO
S
442

Scheme 143

the possibility of undergoing aldol condensation to **441** or Claisen condensation to give **442**. As mentioned previously, alkyl thioglycolate shows a similar reaction pattern. It was found that the mode of ring-closure depends strongly on the reaction conditions. Thus, when catalytic amounts of potassium *t*-butoxide in DMSO are used, the reaction of propiolate with α-mercaptoacetone and α-mercaptocyclohexanone gives **443** and **444**, together with unsaturated thioethers.

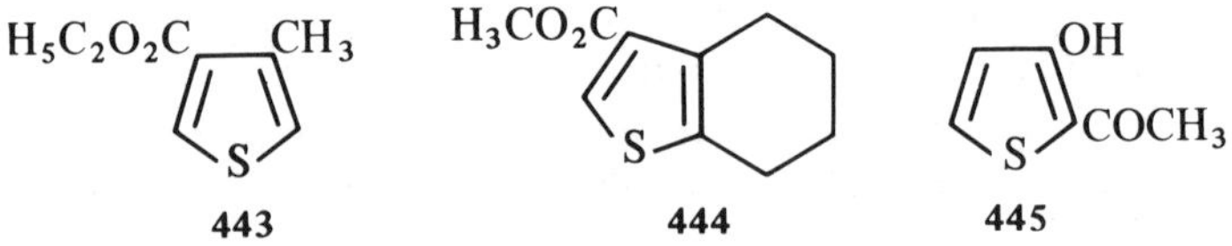
H5C2O2C
CH3
S
443
H3CO2C
S
444
OH
COCH3
S
445

Scheme 144

Similar results were obtained with ethyl tetrolate, ethyl phenylpropiolate, and ethyl *p*-nitrophenylpropiolate. Yet, when methyl propiolate was reacted with an equivalent amount of potassium *t*-butoxide in benzene, **445** was obtained.[323] The thiophenes obtained by aldol condensation are given in Table 42. Reaction of the dimer conrresponding to α,α-dimercaptoacetone (**446**) with acetylenic esters in a 1:2 ratio led upon treatment with base to double ring-closure, giving dithienyl ketones (**447**)[193] (Scheme 145). The reaction of acetylenic aldehydes (**448**) with

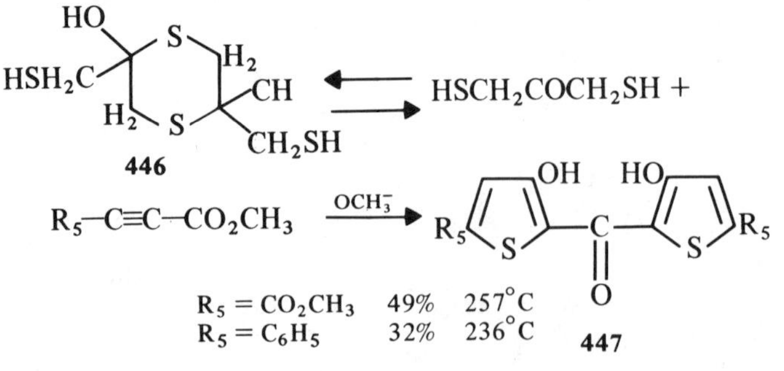
HO
S
H2
HSH2C
CH
H2
S
CH2SH
446
HSCH2COCH2SH +
R5—C≡C—CO2CH3
OCH3−
OH
HO
R5
S
C
O
S
R5
R5 = CO2CH3 49% 257°C
R5 = C6H5 32% 236°C
447

Scheme 145

TABLE 42. FIESSELMANN SYNTHESES FROM ACETYLENIC DERIVATIVES AND α-MERCAPTOCARBONYL COMPOUNDS

$$R_5C\equiv C-CO_2R + HS-CH{\Big<}{R' \atop R''} \longrightarrow RO_2C{\Big<}{} \overset{R'}{\underset{H}{C}}-R'' \longrightarrow$$

R_5	R	R'	R''	R_2	R_3	Yield of Thiophene (%)	Reference
H	CH_3	H	$COCH_3$	H	CH_3	43	323
H	CH_3	$-(CH_2)_4C{=}O$	$COCH_3$	$-(CH_2)_4-$	CH_3	16	323
H_3C	C_2H_5	H	$COCH_3$	H	CH_3	25	323
H_3C	C_2H_5	$-(CH_2)_4CO-$	$COCH_3$	$-(CH_2)_4-$	CH_3	58	323
C_6H_5	C_2H_5	H	$COCH_3$	H	CH_3	31	323
C_6H_5	C_2H_5	$-(CH_2)_4CO$	$COCH_3$	$-(CH_2)_4-$	CH_3	49	323
$p\text{-}NO_2C_6H_4$	C_2H_5	H	$COCH_3$	H	CH_3	22	323
C_6H_5	CH_3	H	CHO	H	H	11	323

mercaptoacetone or mercaptoacetaldehyde (449) has been used in the synthesis of
naturally occurring thiophenes such as junipal (450, $R_5 = H_3C-C\equiv C-$, $R' = H$)[364]
(Scheme 146).

$$R_5-C\equiv C-CHO + {}^-SCH_2COR' \longrightarrow R_5\underset{S}{\boxed{}}COR'$$

448   **449**   **450**

$R_5 = H$	$R' = CH_3$	48%
$R_5 = CH_3C\equiv C$	$R' = CH_3$	32%
$R_5 = CH_3C\equiv C$	$R' = H$	45%

Scheme 146

j. VARIOUS RING-CLOSURE REACTIONS ACCORDING TO FIESSELMANN

The reaction of methyl α,α,β-trichloropropionate with methyl thioglycolate
yields methyl 3-hydroxy-2-thiophenecarboxylate in 60% yield, and not the
expected 4-chloro-3-hydroxy derivative. The tribromopropionate reacts similarly.
Likewise, from dimethyl tribromosuccinate, a 35% yield of 3-hydroxy-2,5-dicarbo-
methoxythiophene was obtained.[330] A further development of the Fiesselmann
reaction is to use 1-chloropropenylidene-(3)-immonium salts (451), which are
obtained from formamide chlorides and substituted acetophenones. Compounds
(451) react with thioglycolates and substituted benzyl mercaptans in sodium
hydroxide or sodium alcoholate solution to give 5-substituted 2-thiophene-
carboxylates and 5-substituted 2-aryl thiophenes (452), respectively[365-367] (Scheme
147). Reaction of 451 with pyridine gave 453, which could also be ring-closed with
thioglycolate or benzyl mercaptans, using triethylamine as base. The compounds
prepared in this way are listed in Table 43.

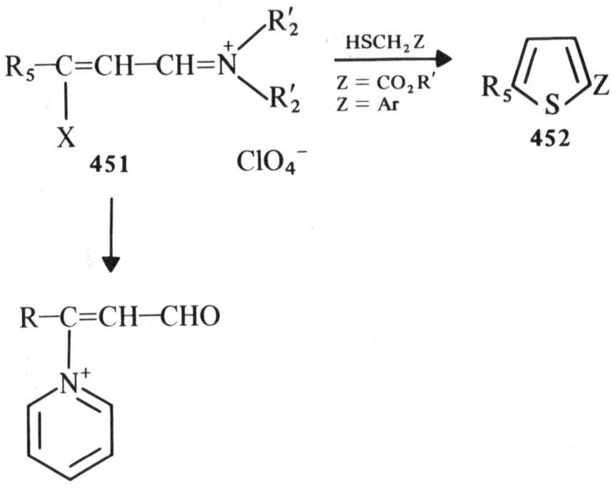

$$R_5-C=CH-CH=\overset{+}{N}\overset{R_2'}{\underset{R_2'}{\diagdown}} \quad \xrightarrow[\substack{Z = CO_2R' \\ Z = Ar}]{HSCH_2Z} \quad R_5\underset{S}{\boxed{}}Z$$

$$\underset{X}{|}$$

451 ClO_4^- **452**

$$R-C=CH-CHO$$

$$\underset{\overset{+}{N}}{|}$$

453

Scheme 147

TABLE 43.   PREPARATION OF 2-THIOPHENE CARBOXYLATES AND 2-ARYLTHIO-
PHENES FROM 1-CHLOROPROPENYLIDENE IMMONIUM SALTS AND
THIOGLYCOLATES AND BENZYL MERCAPTANS[367]

$$R_5\underset{\underset{Cl}{|}}{C}=CH-CH=\overset{+}{N}(CH_3)_2 + HSCH_2R_2 \longrightarrow R_5\overset{\fbox{}}{\underset{S}{}}R_2$$
$$ClO_4^-$$

R_5	R_2	Yield of Thiophene (%)
C_6H_5	CO_2CH_3	75
p-$H_3CC_6H_4$	CO_2CH_3	77
p-$H_3CC_6H_4$	$CO_2C_2H_5$	61
p-$O_2NC_6H_4$	CO_2CH_3	60
p-$H_5C_6C_6H_4$	CO_2CH_3	85
p-$H_3CO_2CC_6H_4$	CO_2CH_3	90
β-Naphthyl	CO_2CH_3	73
$C_6H_5CH=CH$	CO_2CH_3	43
p-$CH_3OC_6H_4CH=CH$	CO_2CH_3	55
o-$ClC_6H_4CH=CH$	CO_2CH_3	46
β-Naphthyl-$CH=CH$	CO_2CH_3	40
p-$HOC_6H_4CH=CH$	CO_2CH_3	58
p-$H_3CC_6H_4CH=CH$	CO_2CH_3	66
p-$ClC_6H_4CH=CH$	CO_2CH_3	55
$C_6H_5(CH=CH)_2$	CO_2CH_3	70
C_6H_5	p-$O_2NC_6H_4$	71
p-$H_3CC_6H_4$	p-$O_2NC_6H_4$	78
p-$H_3CC_6H_4$	p-NCC_6H_4	58
p-$H_3CC_6H_4$	p-$CH_3CONHC_6H_4$	60
p-$H_3CO_2CC_6H_4$	p-$O_2NC_6H_4$	65
β-Naphthyl	p-$O_2NC_6H_4$	84
$C_6H_5CH=CH$	p-$O_2NC_6H_4$	76
$C_6H_5(CH=CH)_2$	p-$O_2NC_6H_4$	80

The reaction of dimethyl acetylenedicarboxylate with tetramethyl thiourea in
dioxane gave about 20% yield of tetramethyl thiophenetetracarboxylate, together
with 28% of the tetracarbomethoxy derivative of 4-thiopyrone.[368]

B.   Thiophenes from Various C_3 + CS Ring-Closures

Phenyl isothiocyanate and 2-isothiocyanovinyl acetate react with ethyl γ-chloro
aceto acetate (454) in the presence of sodium hydride to give 2-amino-4-hydroxy-
thiophenes, which exist in the tautomeric keto forms (455)[369, 370] (Scheme 148).

$$H_5C_2O_2C-CH_2-CO \atop \underset{CH_2-Cl}{\diagdown} \longrightarrow H_5C_2O_2C\overset{\fbox{}}{\underset{RNH\ \ S}{}}O$$

454   455

$R = C_6H_5$
Scheme 148 $R = H_3CCO_2-CH=CH-$

It has been found that oxetanes (**456**) react with ethyl thioglycolate via **457** to give **458**, which upon heating to 90°C yields the 2,3-diaminothiophene **459**[371] (Scheme 149).

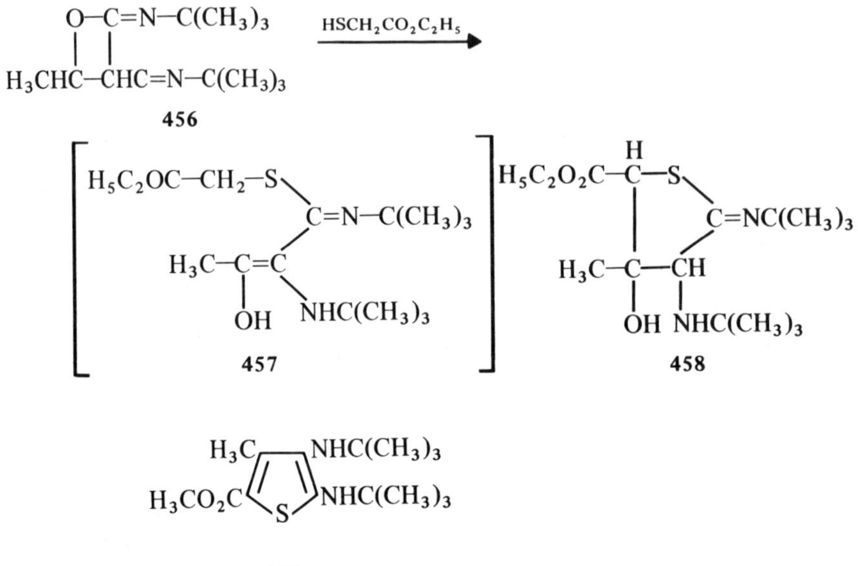

$$\text{O—C=N—C(CH}_3)_3$$
$$\text{H}_3\text{CHC—CHC=N—C(CH}_3)_3$$

$$\xrightarrow{\text{HSCH}_2\text{CO}_2\text{C}_2\text{H}_5}$$

456

$$\text{H}_5\text{C}_2\text{OC—CH}_2\text{—S}$$
$$\text{C=N—C(CH}_3)_3$$
$$\text{H}_3\text{C—C=C}$$
$$\text{OH} \quad \text{NHC(CH}_3)_3$$

457

$$\text{H}_5\text{C}_2\text{O}_2\text{C—C—S}$$
$$\text{H}$$
$$\text{C=NC(CH}_3)_3$$
$$\text{H}_3\text{C—C—CH}$$
$$\text{OH} \quad \text{NHC(CH}_3)_3$$

458

$$\text{H}_3\text{C} \quad \text{NHC(CH}_3)_3$$
$$\text{H}_3\text{CO}_2\text{C} \quad \text{NHC(CH}_3)_3$$

459

Scheme 149

Reaction of cyclopropenium salts with dimethylsulfonium methylide gives 2,3,4-substituted thiophenes in low yield, according to the reaction path indicated in Scheme 150.[372] Recently, it was found that treatment of 2-alkyne and allene derivatives with butyllithium and potassium *t*-butoxide in tetrahydrofuran, followed by carbon disulfide, *t*-butylalcohol, HMPTA, and methyl iodide, gave 3-substituted 2-methylthiothiophenes in reasonable yields. The only exception was *t*-butylallene, which, owing to steric reasons, gave 2-*t*-butyl-5-methylthio-thiophene.[372a]

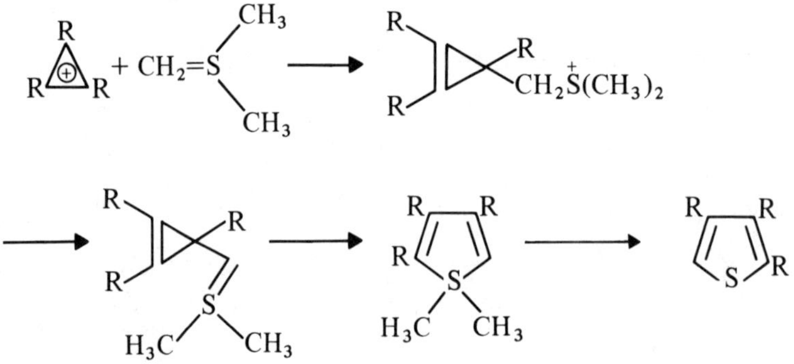

Scheme 150

III.   THREE- OR MORE-COMPONENT METHODS

1. $C_2 + C_2 + S$ Methods

A.   Thiophenes from 2 Moles of Acetylenes and Sulfur Reagent

The classical method of preparing thiophene from acetylene and hydrogen sulfide over Al_2O_3 at high temperature has been further studied.[373,374] Thiophene is also formed upon irradiation of hydrogen sulfide in the presence of acetylene.[375]

The reaction of diphenylacetylene with lithium, followed by sulfur dichloride, gives a 50% yield of tetraphenylthiophene. The reaction proceeds via 1,4-dilithio 1,2,3,4-tetraphenylbutadiene.[376,377] The reaction of diphenylacetylene with iron dodecacarbonyl leads to nonacarbonyl complexes, which upon photolysis in the presence of sulfur, give tetraphenylthiophene as the only product.[378] Decafluoro-tolane, upon heating with sulfur in benzene at 190°C, gives a 72% yield of tetra-kispentafluorophenylthiophene.[379] The reaction of potassium thiolacetate with diphenylacetylene at 140–160°C in dimethyl sulfoxide gave a 50–60% yield of tetraphenylthiophene. The reaction is assumed to proceed via addition of the thiolacetate to give **460**, which is hydrolyzed to **461**. This intermediate adds to another molecule of diphenylacetylene to give **462**, which under oxidative ring-closure gives **463** and dimethyl sulfide. The amount of dimethyl sulfide corre-sponded to the amount of **463** formed, and cis-cis bis(diphenylvinyl)sulfide did not give **463** under these conditions[380] (Scheme 151). The formation of 2,4-diphenyl-thiophene upon alkali treatment of β-phenylvinyl thiolacetate, prepared by the addition of thiolacetate to phenylacetylene, was observed previously.[381]

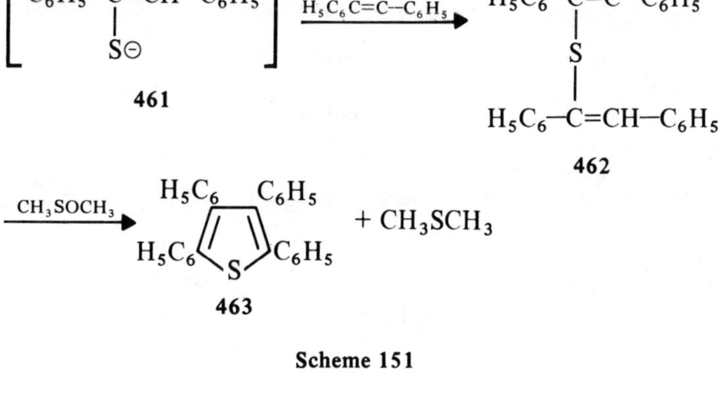

Scheme 151

Another interesting reaction of 2 mole of an acetylene with a sulfur reagent is the reaction of **464** with sodium sulfide in acetone to give **467**. The reaction is assumed to proceed via **465** and **466**[382] (Scheme 152). Through the reaction of **468**

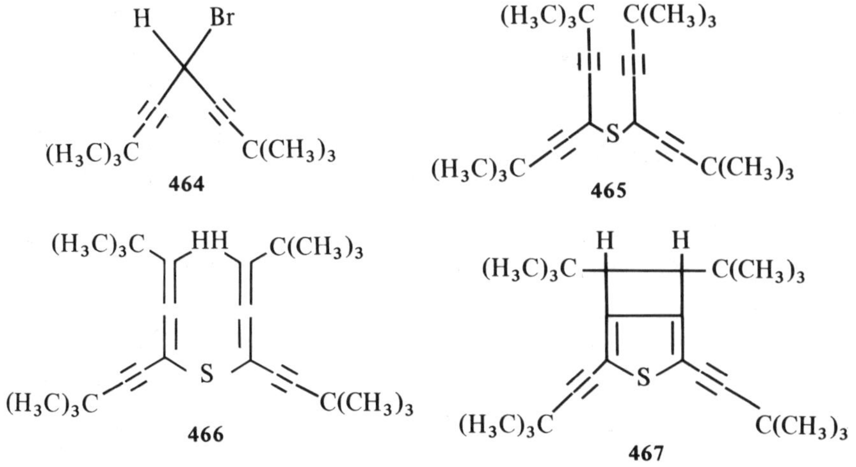

Scheme 152

(R = CH$_3$) with tris(triphenylphosphine)rhodium(I) chloride in boiling benzene, followed by reaction with sulfur, **469** (R = CH$_3$) was obtained in 50% yield. In the phenyl case, the nonolefinic ketone (**470**) could also be used, yielding 40% of **469** (R = C$_6$H$_5$)[383] (Scheme 153). Heating hexafluoro-2-butyne with sulfur and iodine at 180–200°C gave good yields of tetrakis(trifluoromethyl)thiophene.[384] Long heating of ethylthioacetylene at 90–100°C gave 2-ethylthiothiophene in 40% yield, together with much tar.[385]

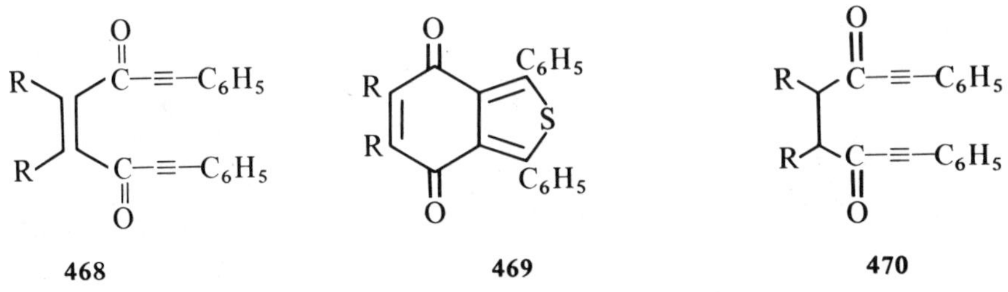

Scheme 153

B.   Thiophenes from 2 Moles of an Olefin and a Sulfur Reagent

A British patent describes the formation of thiophenes from ethylene and SO$_2$ in low yield.[386] The reaction of vinyl chloride with excess hydrogen sulfide at 530–550°C gives thiophene, in addition to vinyl hydrosulfide.[387] Tetrachlorothiophene is obtained by heating trichloroethylene or tetrachloroethylene with sulfur at 150–350°C.[388-390] The reaction of vinylidene chloride with hydrogen sulfide at 470–550°C gives 15% of 2-chloro- and 16% of 3-chlorothiophene as main products.[390a] Furthermore, the reaction of 1,2-dichloroethylene with hydrogen sulfide leads to 2- and 3-chlorothiophene, in addition to 2,4-dichlorothiophene.[391] The reaction of styrene with sulfur at higher temperatures yields 2,4-diphenylthiophene.[392,393] Catalytic amounts of 2-mercaptobenzothiazole increase the yield.[394] The mechanism of the reactions of hydrocarbons with sulfur to give thiophenes has been discussed.[395] 2,4-Diphenylthiophene has also been obtained from α- or β-bromostyrene and hydrogen sulfide,[396] as well as from 1- and 2-halo-1-phenylethanes.[397] α-Chlorostyrene with sulfur at 220°C gives 10–15% of 2,5-diphenylthiophene,[397] and from 1,2-diphenylethylene, 1,2-diphenylethane, and 1,2-diphenyl-1-chloroethane and 1,2-diphenyl-1,2-dichloroethane, tetraphenylthiophene has been obtained.[398] The reaction of acrylonitrile with sulfur at 160–200°C gives only 5% of 2,4-dicyanothiophene, but is still the most convenient method of obtaining this compound.[399,400] The reaction of methyl acrylate with sulfur gives a mixture of dimethyl 2,4-thiophenedicarboxylate.[401] Both dimethyl fumarate and dimethyl maleinate gave tetramethyl thiophenecarboxylate in about 20% yield.[401] The reaction of β-dimethylaminoacrylophenone with S_2Cl_2 gave 2,5-dibenzoyl-3-dimethylaminothiophene in 17% yield.[402]

The photolysis of enethiol esters leads to thiophenes.[403,404] From 471, 472, and 473, thiophenes 475, 476, and 477 were obtained in low yields, together with other products. The t-butyl derivative (474) gave no 3,4-di-t-butylthiophene[404] (Scheme 154). The following reaction path has been suggested: the enethiyl

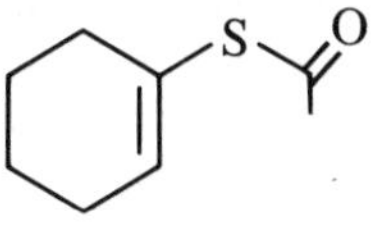

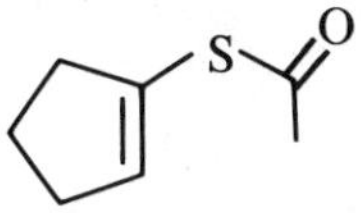

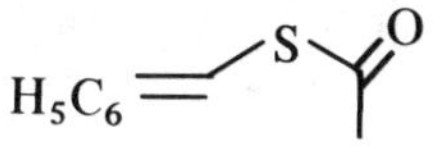

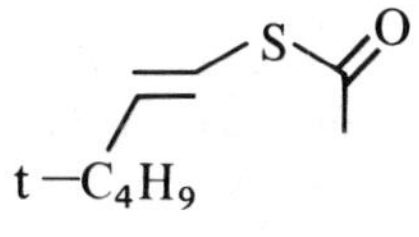

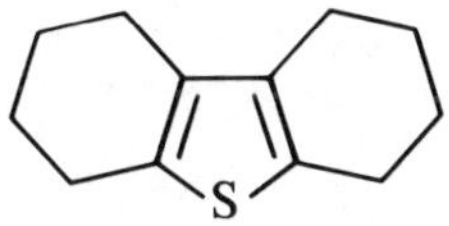

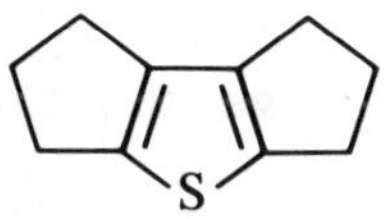

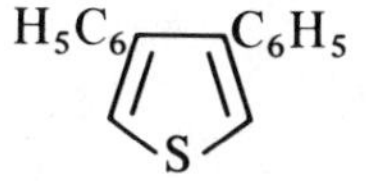

Scheme 154

radical (478) dimerizes by a head-to-head coupling to give 479 and the divinyl disulfide rearranges to 480 and ring-closes with the elimination of hydrogen sulfide to the 3,4-disubstituted thiophene (Scheme 155). However, from the photolysis of

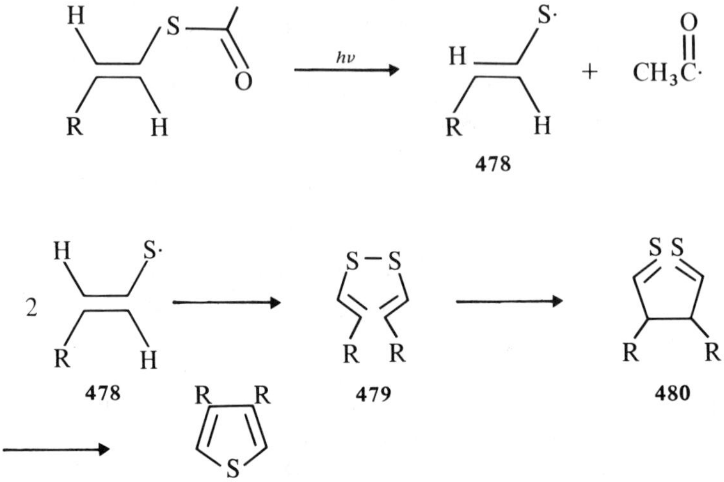

Scheme 155

S-(cis-1-propenyl)-L-cysteine, 2,4-dimethyl-, 3,4-dimethyl-, and 3-methylthiophene were obtained, and these authors suggest that the 2,4-isomer, which is the major product, is formed by head-to-tail coupling of 478 to 481, followed by ring-closure to 2,4-dimethylthiophene.[405,406] This reaction path was considered unlikely, and reaction via 482 and 483 was suggested instead[404] (Scheme 156).

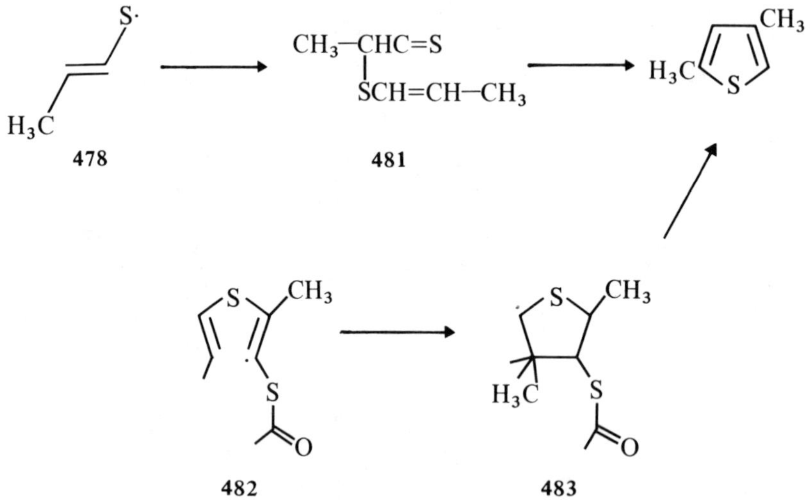

Scheme 156

C.   Thiophenes from Methyl Ketones and Sulfur Reagents

The reaction of acetophenone with sulfur and ammonia at room temperature leads to 2,4-diphenylthiophene in low yield.[407] It was found that the reaction of acetophenone anil with sulfur at 220–240°C, claimed to give pure 2,4-diphenyl-thiophene,[408] also results in the formation of some 2,5-diphenylthiophene.[409] 2,4-Di-(p-tolyl)thiophene and 2,4-di-(p-methoxyphenyl)thiophene have been prepared by this reaction.[409] The so-called anhydrotriacetophenedisulfides (formed upon the reaction of acetophenones with hydrogen chloride and hydrogen sulfide), upon heating with copper chromite in refluxing xylene, give 2,4-diarylthiophenes, such as 2,4-diphenyl-, 2,4-di-(p-fluoro)phenyl-2,4-di-(p-chloro)phenyl-, 2,4-di-(p-bromo)phenyl-, and 2,4-di-(p-iodo)phenylthiophene in about 80% yield.[410, 411] With p-(alkyl)phenyl methyl ketones, only trace amounts of the corresponding 2,4-diarylthiophenes were obtained.[411] The reaction of acetophenone with boron sulfide in refluxing benzene gave a 25% yield of 2,4-diphenylthiophene.[412] The reaction of propiophenone with thionyl chloride in pyridine gave 34% of 3,4-dibenzoylthiophene. With phenylpropyl ketone, only 7% of 2,5-dimethyl-3,4-dibenzoylthiophene was obtained. The mechanism of this reaction is not known, but has been suggested to proceed as indicted in Scheme 157.[413] The

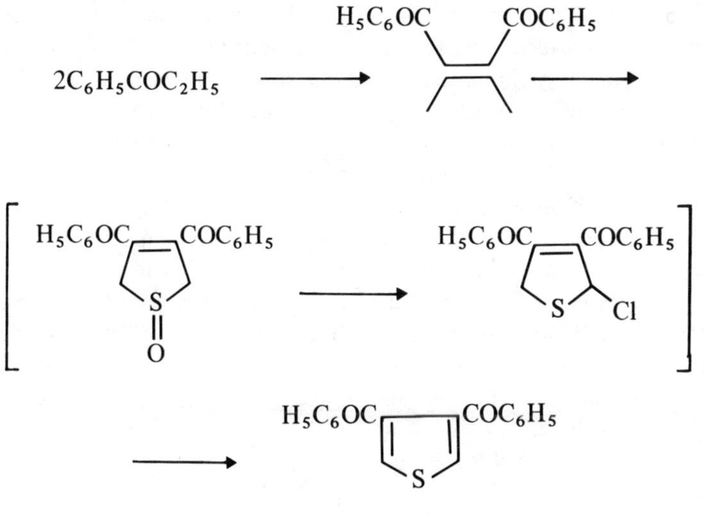

Scheme 157

reaction of acetophenones (**484**) with methoxycarbonylsulfenyl chloride (**485**) gives **486**, which upon treatment with 50% sulfuric acid gave 2,4-diarylthiophene (**487**)[414] (Scheme 158). The reaction of styrylmorpholine with sulfur in refluxing benzene gave 29% of 2,4-diphenyl- and 3% of 2,5-diphenylthiophene.[415]

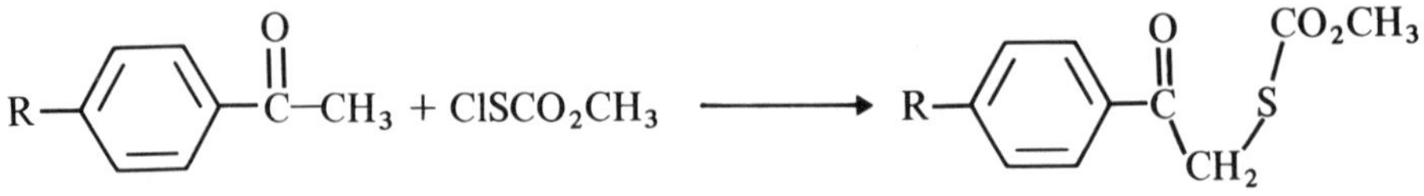

484 **485** **486**

R = H, CH_3, Br

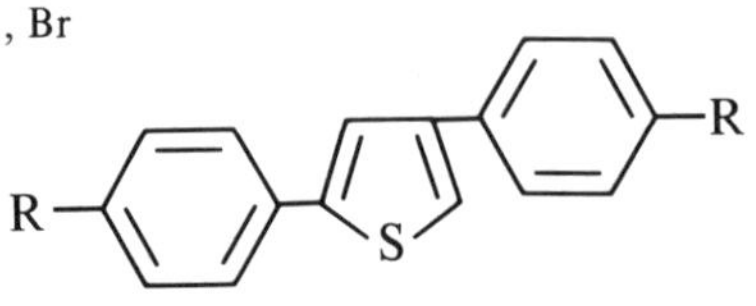

487

R = H	40%
R = CH_3	70%
R = Br	45%

Scheme 158

D. Various Reactions of $C_2 + C_2 + S$ Type and Reactions Involving Larger Numbers of Compounds

The reaction of ethylene oxide with hydrogen sulfide at 350–450°C in the presence of Al_2O_3 leads to the formation of thiophene, and from propylene oxide at 400°C, 2,4-dimethylthiophene was obtained.[415] Ethylene glycol, when reacted with hydrogen sulfide over Al_2O_3, also yields thiophene; the reaction most probably proceeds via ethylene oxide.[417] Thiophene is formed in the pyrolysis of ethyl mercaptane.[418] The reaction of benzyl chloride with sulfur at 200–240°C gives tetraphenylthiophene in 68% yield.[419–421] From *p*-chlorobenzyl bromide, tetrakis-(*p*-chlorophenyl)thiophene has similarly been obtained in 25–30% yield.[421] Tetra-(4-pyridyl)thiophene has been obtained in 90% yield from 4-picoline and sulfur.[422] The reaction of *p*-toluic acid and sulfur gives tetra-(*p*-carboxyphenyl)-thiophene, in addition to 4,4′-stilbendicarboxylic acid.[423,424] Thiophenes are formed upon the pyrolysis of substituted trithianes over Al_2O_3 or $SiO_2 \cdot Al_2O_3$ as catalyst;[425] they are also obtained on the pyrolysis of coal.[426] Heating α-D-glucose with hydrogen sulfide, ammonia, and water leads to a large number of volatile compounds, among them thiophenes.[427]

IV. ONE-COMPONENT METHODS

1. Ring-Closure of C_4-S Compounds to Thiophenes

Pyrolysis of dibutyldisulfide at 500°C in a nitrogen stream gives a 50% yield of thiophene.[427a] A series of 4-phenylbutylthio derivatives was cyclized to 2-phenyl-

hydrothiophenes by treatment with iodine or acid. Some hydrothiophenes dispro-
portionate to thiophenes and tetrahydrothiophenes during this ring-closure.[427b, 428a]

β-Aryl-α-mercaptoacrylic acids (488), prepared by alkaline hydrolysis of 5-aryl-
rhodanines, are useful starting materials for the preparation of 5-aryl-2-thiophene-
carboxylic acids through treatment with iodine or chlorine[428, 429] (Scheme 159).
It has been suggested that the reaction proceeds via a sulfenyl halide.[430]

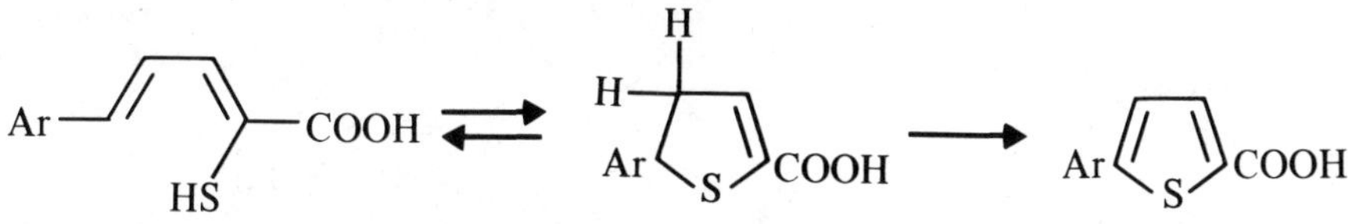

Scheme 159

488   489

The reaction has been extended to condensation products of 5-ethoxymethylene
rhodanine and β-dicarbonyl and β-cyanocarbonyl compounds (490), which upon
alkaline hydrolysis give thiophenes, probably via 491.[431] Thus, from 490 (R = R' =
COCH₃), owing to acid cleavage of the β-dicarbonyl derivative, 491 (R = COCH₃,
R' = H) is first formed, but could not be isolated, since it immediately ring-closed
to 5-methyl-2-thiophenecarboxylic acid. However, 52% of 493 was obtained from
492. Compounds 490 (R = CO₂C₂H₅, R' = CN) and 490 (R = COC₆H₅, R' = CN)
gave directly the aminothiophenes 494 and 495 in 90 and 53% yield, respectively.
The lower yield in the latter case was due to acid cleavage, as benzoic acid was
also isolated[431] (Scheme 160).

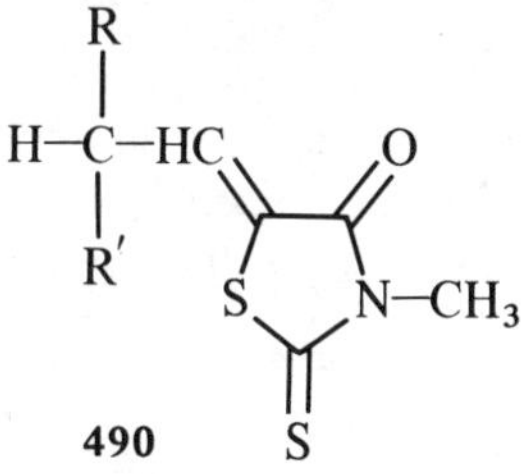

490 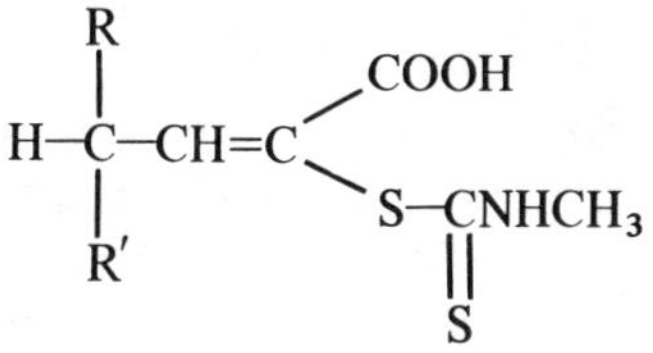   491

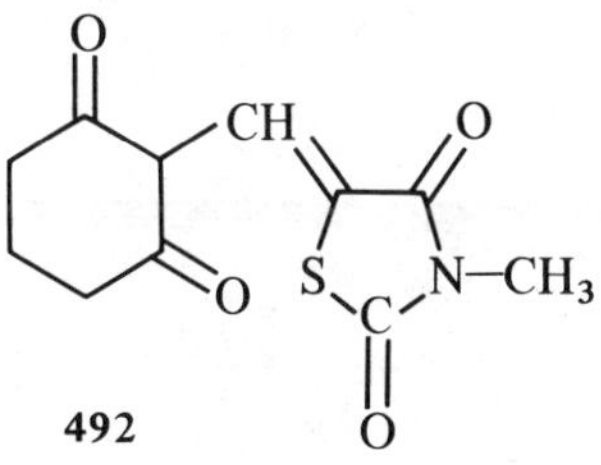

492 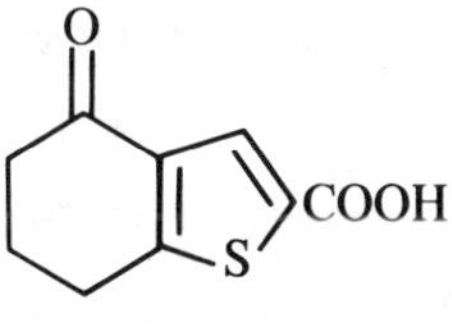   493

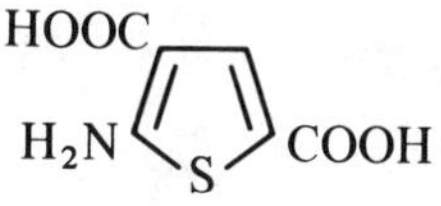

494 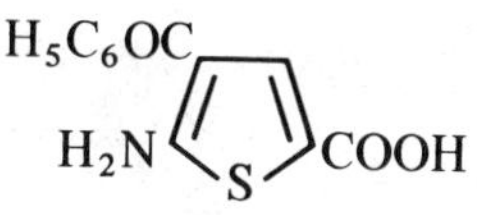   495

Scheme 160

Another example of ring-closure of γ-thiosubstituted nitriles is the preparation of 2-aminothiophene by treatment of γ-benzyl- or γ-benzhydrylthiocrotone nitriles with hydrogen chloride in ether.[432,433] The best yield (82%) was obtained from the benzhydryl *cis* isomer (496) and is supposed to proceed via 497[432,433] (Scheme 161). Treatment of 498 with 90% sulfuric acid for a few minutes gave the 2-amino-

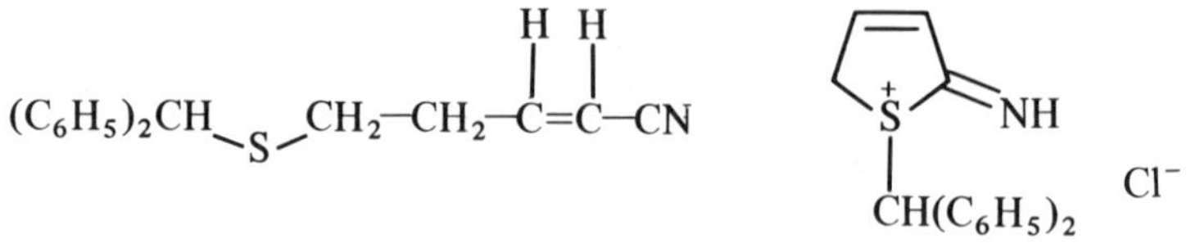

496

497

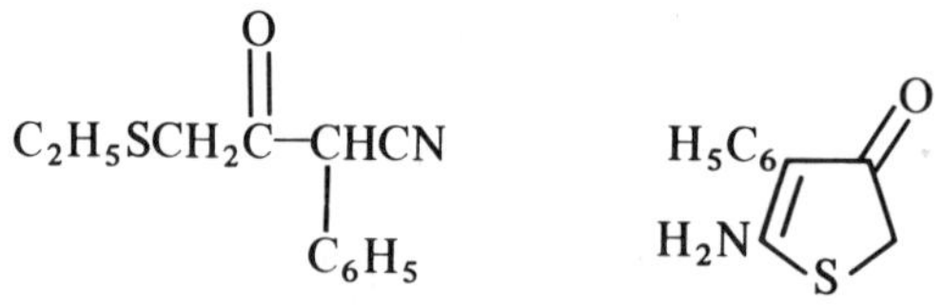

498

499

Scheme 161

4-hydroxythiophene (499). The *p*-chlorophenyl derivative was similarly prepared.[434] 1-(Alkyl- or phenylthio)-4-methylthiobutadienes (500) are metalated by butyllithium at $-40°C$ in THF in the presence of TMEDA to give the lithium derivative (501), which at $-40°C$ can be trapped as the methyl derivative upon reaction with methyl iodide. However, at 25–30°C, especially upon addition of HMPA, it rapidly ring-closes to 2-methylthiothiophene (502), which is further metalated by butyllithium in the 5-position[435] (Scheme 162).

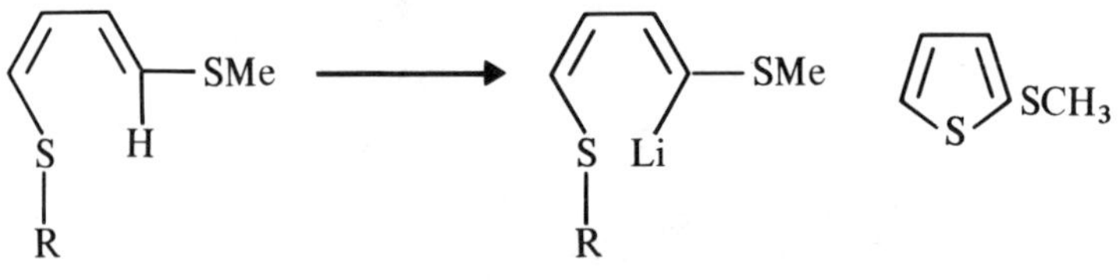

500

501

502

Scheme 162

The reaction of alkylthio substituted butatrienes, such as 503, with iodine in carbon tetrachloride gives a 63% yield of 504 (X = I) and isobutylene. The reaction with bromine leads to a mixture of 504 (X = Br) and 504 (X = H). The reaction of 505 gave one thiophene derivative, which is either 506 or 507[436] (Scheme 163). Treatment of the allenyldithio ester (508) with catalytic amounts of a strong base, such as sodium ethoxide in liquid ammonia,[438] or heating with traces of acid,[437]

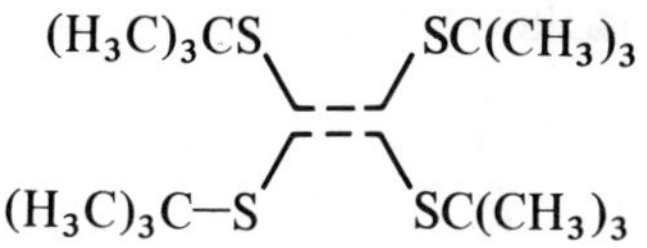

503

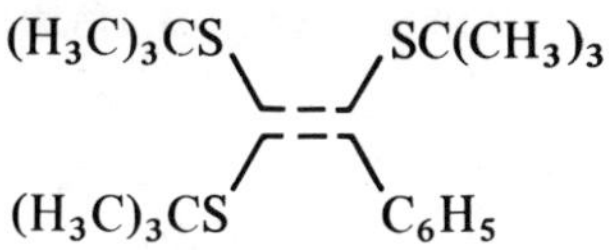

505

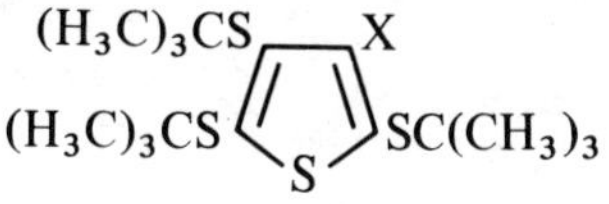

504

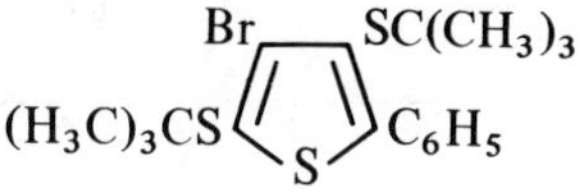

506

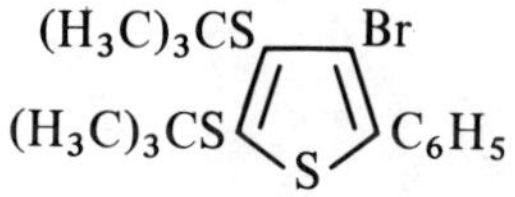

507

Scheme 163

leads to ring-closure to thiophenes (**510**) via **509**, as indicated in Scheme 164. If triethylamine is used as base, 2H-thiopyrans are formed instead. The allenic dithio esters (**508**) are formed by alkylation of the anion of dithio esters (**511**) with propargyl bromide to give **512**, followed by thermal rearrangement.[437] Thiophenes prepared in this way are listed in Table 44.

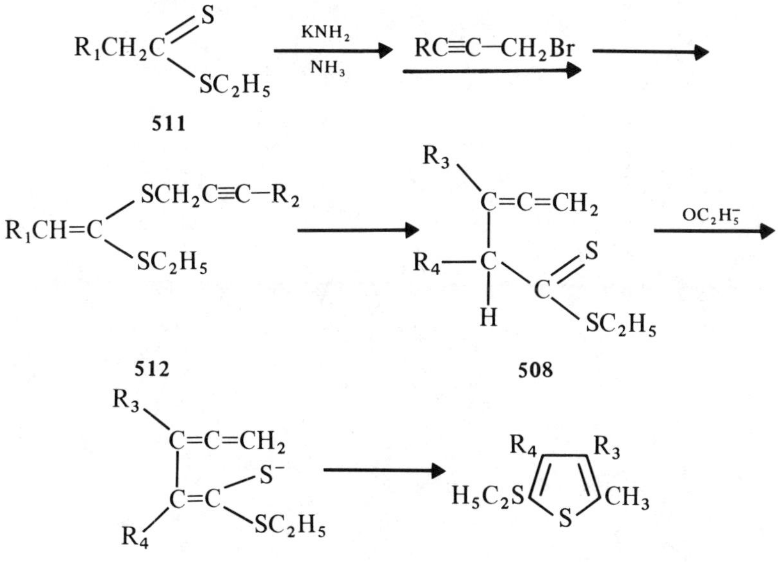

511

$R C \equiv C - CH_2 Br$

512

508

509

510

Scheme 164

TABLE 44.   THIOPHENES FROM ALLENIC DITHIO ESTERS

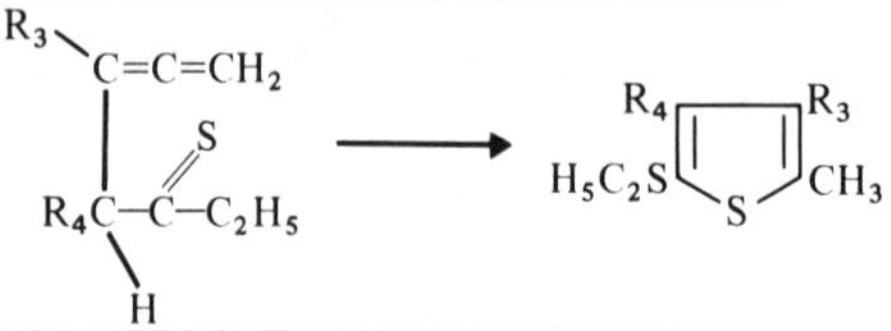

R_3	R_4	Yield of Thiophene[a] (%)	Yield of Thiophene[b] (%)
H	H		60
H	CH_3	54^c	65
H	C_2H_5		52
H	$CH(CH_3)_2$	42	37
H	$C(CH_3)_3$	40	
H	C_6H_5	38^c	
CH_3	H	82	53
CH_3	CH_3	75	54
CH_3	C_2H_5		62
CH_3	$CH(CH_3)_2$	73	35
CH_3	$C(CH_3)_3$	25	
CH_3	C_6H_5	50^c	

[a] According to Ref. 437, using an acidic catalyst.
[b] According to Ref. 438, using an ethoxide catalyst.
[c] Yield calculated on dithio ester[508].

Treatment of **513** with potassium *t*-butoxide in HMPTA gave the aminothiophenes (**516**) via **514–515**[439] (Scheme 165). The compounds prepared are collected

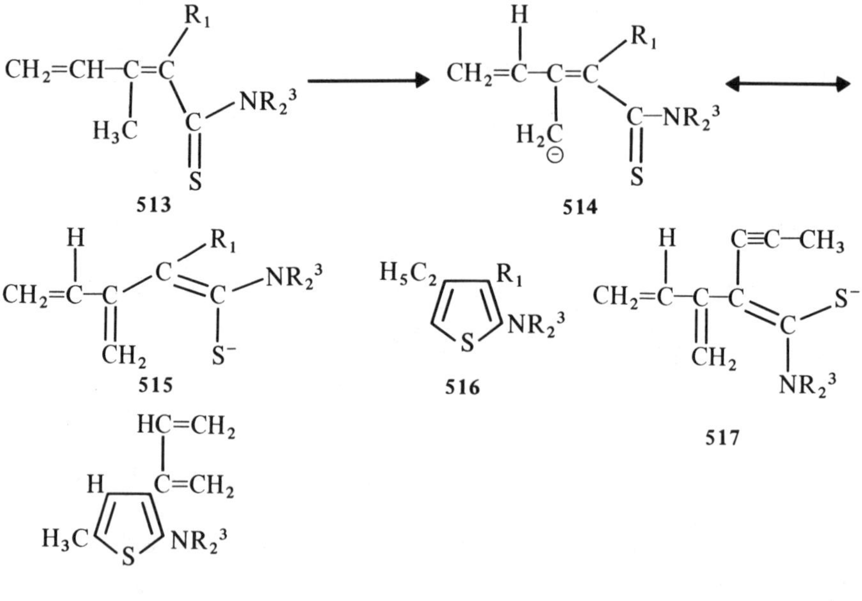

Scheme 165

TABLE 45. THIOPHENES FROM *N*,*N*-DIALKYL-2,4-PENTA-
DIENETHIOAMIDES[439]

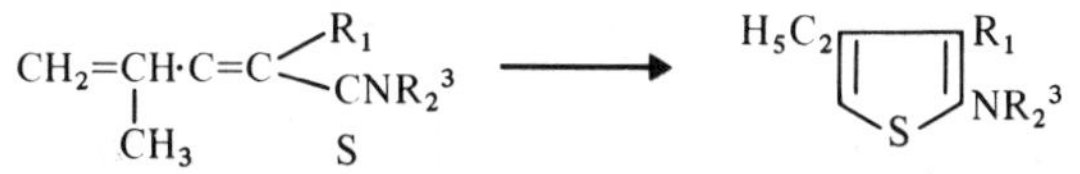

$$CH_2=CH\cdot C=C\underset{\underset{CH_3}{|}}{\overset{R_1}{\diagup}}\underset{S}{\overset{}{\diagdown}}CNR_2^3 \longrightarrow H_5C_2\underset{S}{\overset{}{\boxed{}}}\overset{R_1}{\underset{NR_2^3}{}}$$

R_1	R_2	Yield of Thiophene (%)
CH_3	C_2H_5	62
CH_3	C_3H_7	68
C_2H_5	C_2H_5	50
C_2H_5	C_3H_7	48
$CH_2=C(CH_3)$	C_2H_5	50
$CH_2=C(CH_3)$	C_3H_7	40

in Table 45. When **513** ($R_1 = C\equiv C-CH_3$) was reacted with a catalytic amount of potassium *t*-butoxide, the thiophene (**518**) was obtained in 40–45% yield. It is probably formed via **517** by the attack of sulfur on the acetylenic bond.[439] Acetylenic thioamides (**520**), prepared through [3,3]-sigmatropic rearrangement of 1-alkenyl allenyl sulfides (**519**) at room temperature to thioketenes and trapping with secondary amines, give thiophenes (**521**) upon treatment with *t*-butoxide in liquid ammonia or upon heating with HMPTA[400,441] (Scheme 166). Diacetylenes

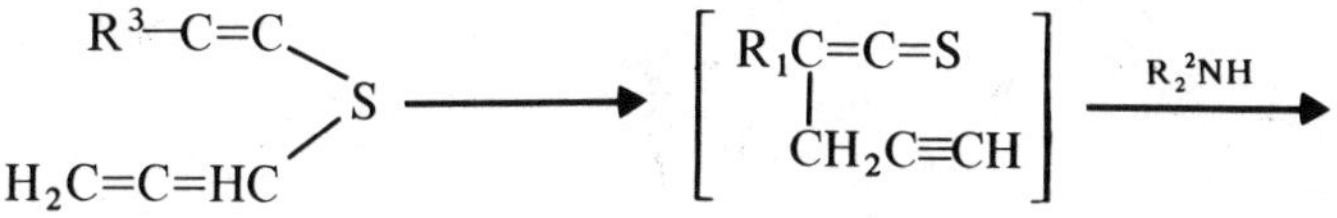

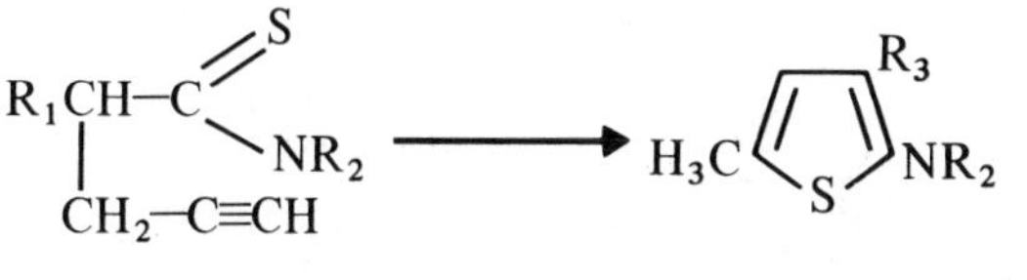

Scheme 166

(**522**) also undergo rearrangements to thioketenes (**523**), which can be trapped by secondary amines or by dialkylphosphines. Depending upon the nature of the substituents (R_1 and R_2) and upon the solvent, allenic thioamides (**523**), dienic thioamides (**524**), thiophenes (**525**), or mixtures of these types of compounds are formed (Scheme 167). If only a 20% excess of diethylamine is used, only a few percent of thiophenes are formed, and the main product is **524**. However, heating **522** with a 100% excess of amine in DMSO or HMPTA at 50°C, gave **525** as the only product. Similarly, with dialkylphosphines instead of diethylamine, **526** was obtained.[441] The 2-(dialkylamino)- and 2-(dialkylphosphino)thiophenes that were prepared are given in Table 46.

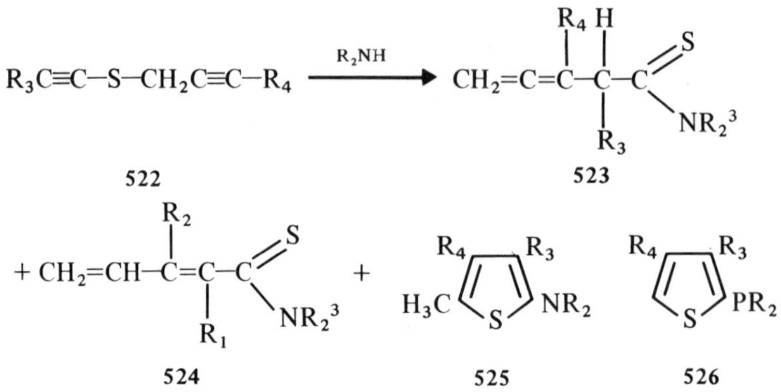

$$R_3C\equiv C-S-CH_2C\equiv C-R_4 \xrightarrow{R_2NH} CH_2=C=C-\underset{R_3}{\overset{R_4\;H}{\underset{|}{\overset{|}{C}}}}-C\overset{S}{\underset{NR_2{}^3}{\Bigg\langle}}$$

522   523

$$+\;CH_2=CH-\underset{R_1}{\overset{R_2}{\underset{|}{\overset{|}{C}}}}=C-C\overset{S}{\underset{NR_2{}^3}{\Bigg\langle}}\;+\; \underset{H_3C}{\overset{R_4}{\diagdown}}\!\!\overset{R_3}{\underset{S}{\diagup}}\!\!NR_2 \qquad \underset{}{\overset{R_4}{\diagdown}}\!\!\overset{R_3}{\underset{S}{\diagup}}\!\!PR_2$$

524   525   526

Scheme 167

TABLE 46.   THIOPHENES FROM 1-ALKYNYL-ALLENYLSULFIDES AND 1-ALKYNYL-2-ALKYNYLSULFIDES WITH SECONDARY AMINES AND PHOSPHINES

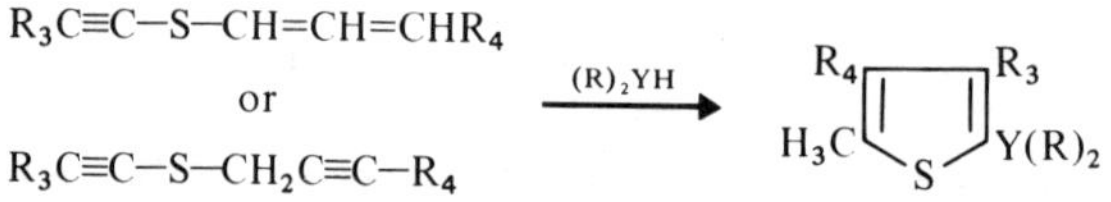

$$\begin{array}{c} R_3C\equiv C-S-CH=CH=CHR_4 \\ \text{or} \\ R_3C\equiv C-S-CH_2C\equiv C-R_4 \end{array} \xrightarrow{(R)_2YH} \underset{H_3C}{\overset{R_4}{\diagdown}}\!\!\overset{R_3}{\underset{S}{\diagup}}\!\!Y(R)_2$$

R_3	R_4	R	Y	Yield of Thiophene (%)	Reference
CH_3	H	C_2H_5	N	38, 59	440
CH_3	H	C_3H_7	N	41, 60	440
C_2H_5	H	C_2H_5	N	66, 58	440
C_2H_5	H	C_3H_7	N	72, 62	440
H_3C	CH_3	C_2H_5	N	50	441
H_3C	CH_3	C_3H_7	N	53	441
H_3C	C_2H_5	C_2H_5	N	54	441
H_3C	C_2H_5	C_3H_7	N	50	441
H_5C_2	CH_3	C_2H_5	N	50	441
H_5C_2	CH_3	C_3H_7	N	49	441
H_5C_2	C_2H_5	C_2H_5	N	51	441
H_5C_2	C_2H_5	C_3H_7	N	55	441
$CH_3C\equiv C$	CH_3	C_2H_5	N	46	441
$CH_3C\equiv C$	CH_3	C_3H_7	N	48	441
$CH_2=C(CH_3)$	CH_3	C_2H_5	N	a	441
$CH_2=C(CH_3)$	CH_3	C_3H_7	N	a	441
CH_3S	CH_3	C_2H_5	N	a	441
CH_3S	CH_3	C_3H_7	N	a	441
H_3C	CH_3	C_2H_5	P	41	441
CH_3	CH_3	C_3H_7	P	49	441
H_3C	C_2H_5	C_2H_5	P	39	441
H_3C	C_2H_5	C_3H_7	P	32	441
C_2H_5	CH_3	C_2H_5	P	35	441
C_2H_5	CH_3	C_3H_7	P	49	441
C_2H_5	C_2H_5	C_2H_5	P	36	441
C_2H_5	C_2H_5	C_3H_7	P	33	441

a Thiophenes were obtained together with **524**.

Mixtures of 2H-thiopyrans (**529**) and thiophenes (**528**) were obtained when propargyl (**527**) or allenic vinyl sulfides were heated in HMPTA or DMSO in the presence of amines[442,443] (Scheme 168).

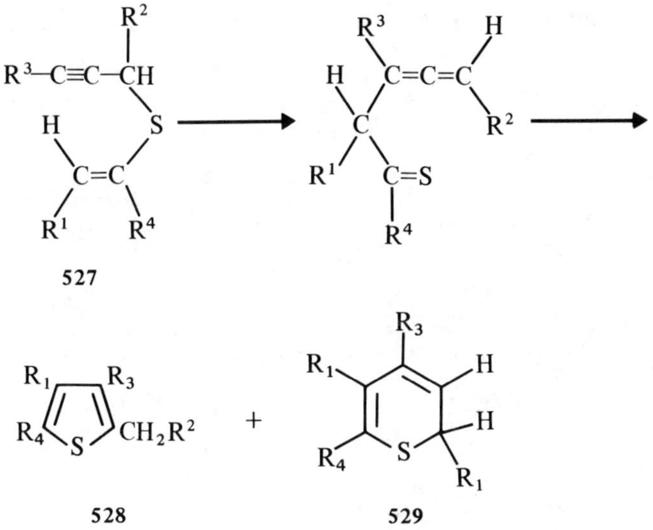

Scheme 168

In a few cases, as in the reaction with ethyl 3-propargylthiocinnamate in triethylamine, only the thiophene ethyl 2-phenyl-5-methyl-3-thiophenecarboxylate was obtained.[443] Propargyl vinyl sulfides have also been prepared by S-alkylation of thioketones with propargyl bromide. From methyl t-butyl thioketone and ethyl t-butylthioketone, 2-methyl-5-t-butylthiophene and 2,4-dimethyl-5-t-butylthiophene have been obtained.[444]

From cyanoalkynes and 2-propynethiol, the adduct (**530**) is obtained in high yield. Heating **530** leads to a (3,3)-sigmatropic rearrangement to **531**, which upon heating with diisopropylamine at 180°C in DMSO gives the thiophene **532**. If heated in silicone oil, 5-cyano-2H-thiopyrans are obtained[445] (Scheme 169).

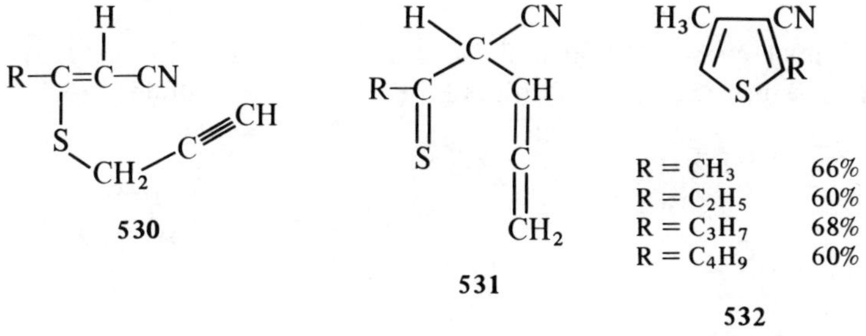

R = CH$_3$ 66%
R = C$_2$H$_5$ 60%
R = C$_3$H$_7$ 68%
R = C$_4$H$_9$ 60%

Scheme 169

Some more special cases of the C_4-S ring-closure have also been observed. Thus, the reaction of 2 moles of methyl propiolate with dimethyl sulfoxide at 125°C gives a 43% yield of **533**, which upon treatment with thionyl chloride gives dimethyl 2,4-thiophenedicarboxylate (**536**). The reaction is assumed to proceed via **534** and **535**[446] (Scheme 170). The reduction of **537** with LiAlH$_4$ gave **538**,

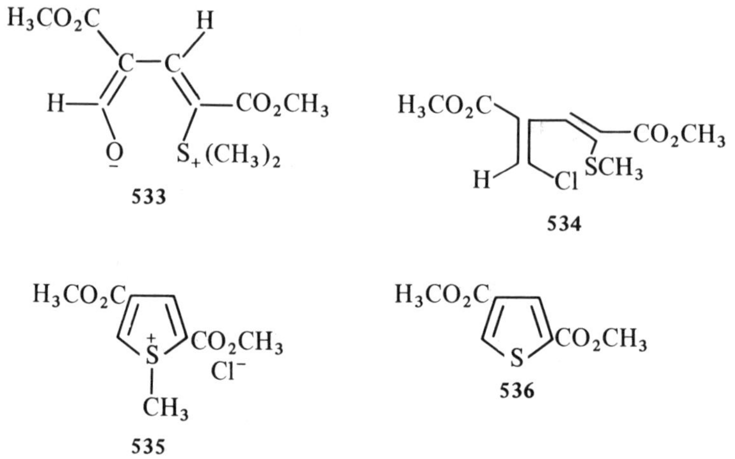

Scheme 170

which was dehydrated to **539** in the presence of P_2S_5 to avoid the elimination of hydrogen sulfide[447] (Scheme 171). The reaction of **540** with hydrogen sulfide first

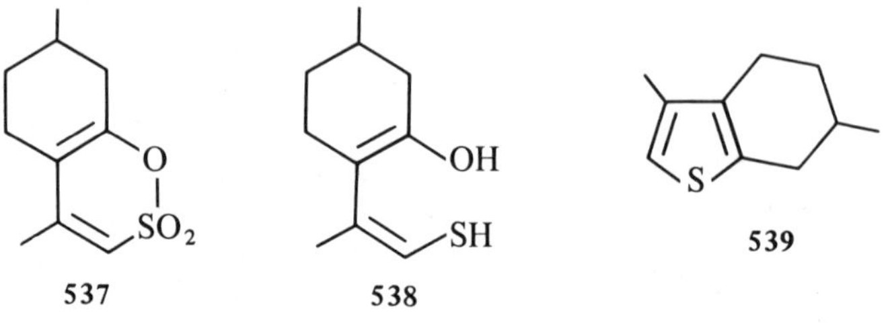

Scheme 171

described by Hantzsch,[448] and probably leading via **541** and **542** to 2-thiophene aldehyde, has been carried out with radioactive hydrogen sulfide[449] (Scheme 172).

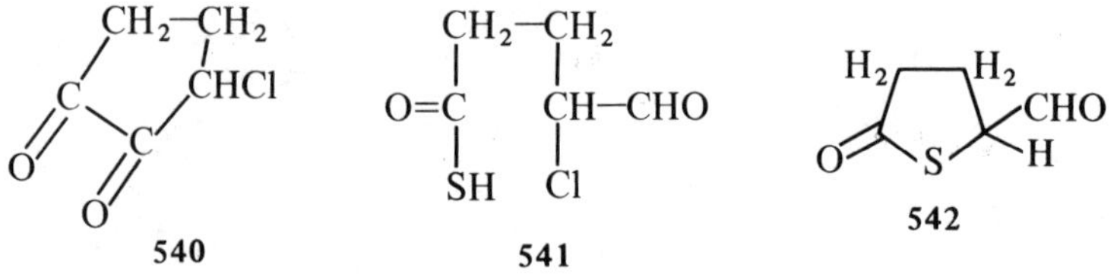

Scheme 172

Treatment of ethyl 2-amino-3-thiophenecarboxylates (543) with sodium ethoxide leads to ring-opening to 544 and ring-closure to 2-hydroxy-3-cyanothiophene (545)[450] (Scheme 173).

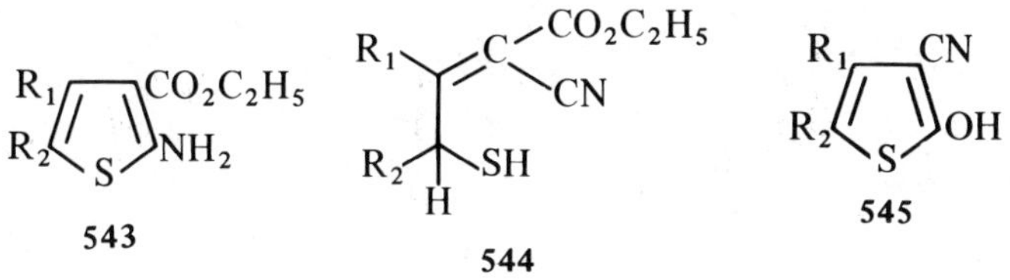

543

544

545

$R_1 = C_6H_5$, $R_2 = H$
$R_1 = CH_3$, $R_2 = H$
$R_1 = CH_3$, $R_2 = CO_2C_2H_5$
$R_1 = CH_3$, $R_2 = COCH_3$

Scheme 173

2.   Ring-Closure of C_3-S-C Compounds to Thiophenes

This type of ring-closure is in many cases related to the $C_3S + C$ reaction (see the Gompper reaction, Scheme 82)[275] and was in that section.

Enethiols (546) prepared from β-keto esters have been S-cyanomethylated with chloroacetonitrile to 547. Upon treatment of 547 with sodium ethoxide, ring-closure to 548 occurred[451] (Scheme 174). Alkylation of 546 with benzyl chloride

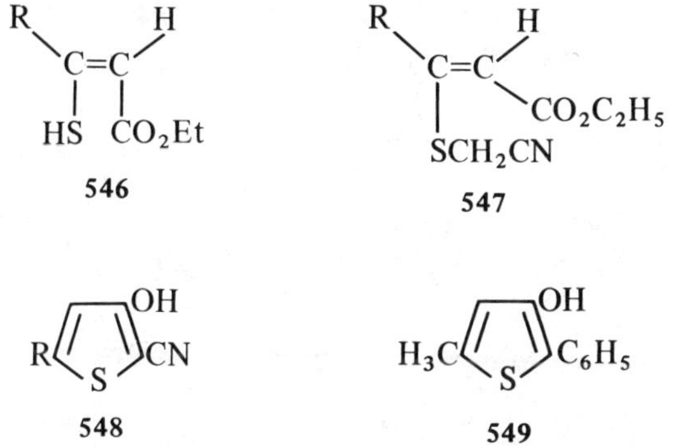

546

547

548

549

Scheme 174

or ethyl α-bromophenyl acetate followed by base-treatment gave (in the latter case, after elimination of the ester group) the hydroxythiophene (549).[451] S-Alkylation and ring-closure of monothio-β-ketones was discussed in Section II.4 and is shown in Scheme 92.[288] Competing Dieckmann and Thorpe cyclizations have been studied with 550 and 551, which led to a mixture of 3-hydroxy- (552) and 3-amino-thiophenes (553).[452] Cyclization without isomerization could not be achieved

(Scheme 175). The nature of the base and the conditions used had considerable influence on the proportions of **552** and **553** formed. Thus, sodium hydroxide in benzene gave a yield of 80% of **552** (R = $CO_2C_2H_5$) and 13% of **553** (R = $CO_2C_2H_5$)

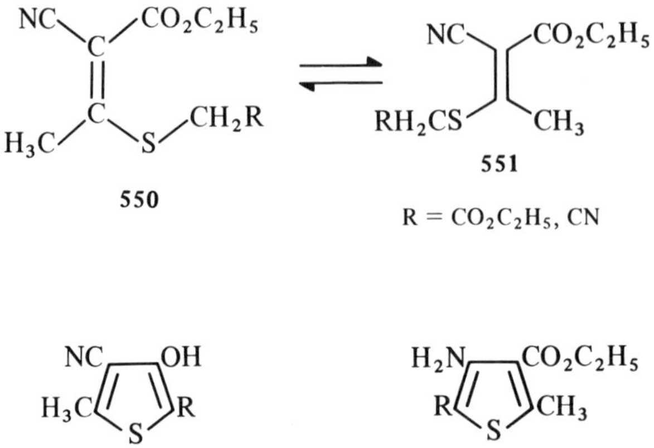

550 **551**

R = $CO_2C_2H_5$, CN

552 **553**

Scheme 175

after 5 h at room temperature; triethylamine in benzene gave a 32% yield of **552** (R = $CO_2C_2H_5$) and a 65% yield of **553** (R = $CO_2C_2H_5$) after 6 days at 75°C. With triethylamine in benzene at room temperature, as much as 95% of **553** (R = CN) was obtained, together with 3% of **552** (R = CN). The highest yield of **552** (R = CN) was obtained when Hünig-base was used, which gave 40% of **552** (R = CN) and 41% of **553** (R = CN) after 5 h in benzene at 20°C.[452] The thiophenes prepared are given in Table 47.

The regioselective deprotonation of *S,S*-dimethyl-α-oxoketene dithioacetals (**554**) with LDA at −78°C in HMPA proceeds via **555** to give 3,4-disubstituted

TABLE 47. 3-HYDROXYTHIOPHENES FROM *S*-ALKYLATED ENETHIOLATES

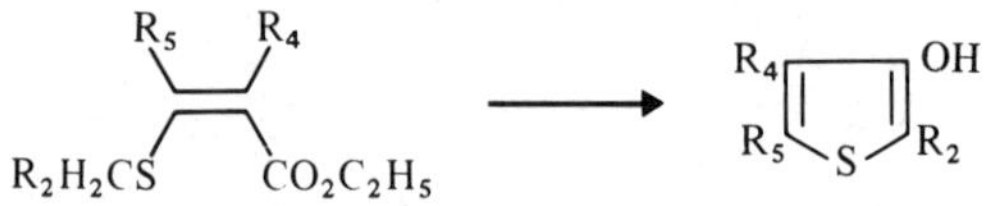

R	R_5	R_2	Yield of Thiophene (%)	Reference
H	CH_3	CN	78	451
H	$CH(CH_3)_2$	CN	57	451
H	$C(CH_3)_3$	CN	32	451
H	C_6H_5	CN	50	451
H	CH_3	C_6H_5	50	451
CN	CH_3	$CO_2C_2H_5$	80	452
CN	CH_3	CN	40	452

TABLE 48.   THIOPHENES FROM α-OXOKETENEDITHIOACETALS[454]

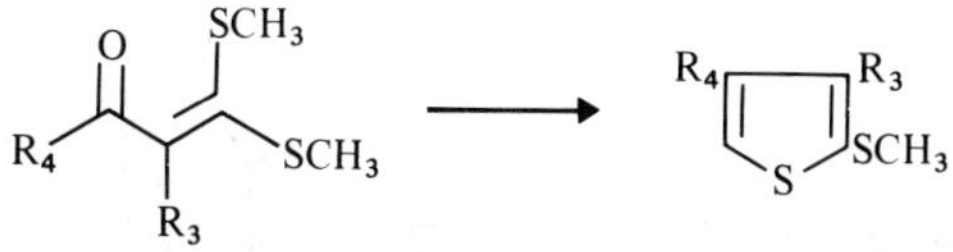

R$_3$	R$_4$	Yield of Thiophene (%)
C$_2$H$_5$	C$_6$H$_5$	55
H	C$_6$H$_5$	30
H	H$_3$COC$_6$H$_4$	30
OCH$_3$	C$_6$H$_5$	42
—(CH$_2$)$_3$C(CH$_3$)$_2$—		26
C$_6$H$_5$	H	22
C$_6$H$_5$	OC$_2$H$_5$	38

2-methylthiothiophenes (**556**).[453,454] The compounds prepared are listed in Table 48. Treatment of **556a** in an analogous way with LDA led to the 2,4-diamino-thiophene (**556b**), which exists as the imino tautomer.[454a,b] Compounds of type **556c**, prepared by the condensation of methyl aryl ketones with phenyl isothio-cyanates in the presence of sodium hydroxide followed by alkylation, gave the aminothiophenes (**556d**), if R$_4$ = COC$_6$H$_5$ or COCH$_3$ without additional base-catalysis. If R$_4$ = CO$_2$C$_2$H$_5$ or CO$_2$NH$_2$, **556e** was the product[454c] (Scheme 176).

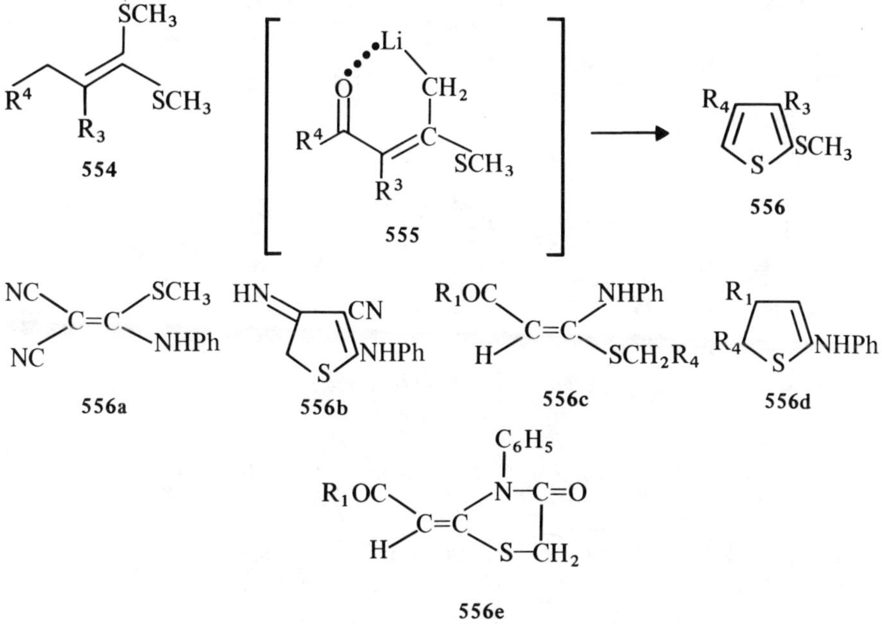

Scheme 176

The reaction of vinamidinium salts (557) with sodium amide in liquid ammonia gave the 2,4-diaminothiophene (558) in high yield, via the S-ylide. If R = C(CH$_3$)$_3$, a nonseparable mixture of allenes and thiophenes was obtained[455] (Scheme 177).

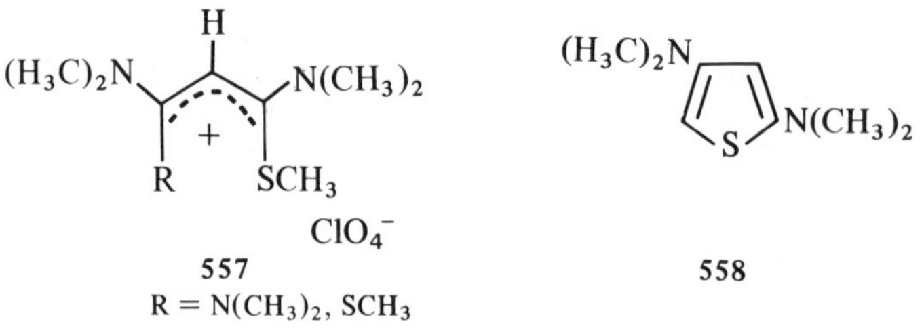

557
R = N(CH$_3$)$_2$, SCH$_3$

558

Scheme 177

The reaction of 559, prepared by the action of sulfur on ethyl cinnamate, with chloroacetic acid to give 2-phenyl-4-hydroxythiophene[456] has been modified. The intermediate (560) was isolated, and by heating with acetic anhydride and sodium acetate, it was cyclized, probably via 561, to give the acetate of 562[457] (Scheme 178).

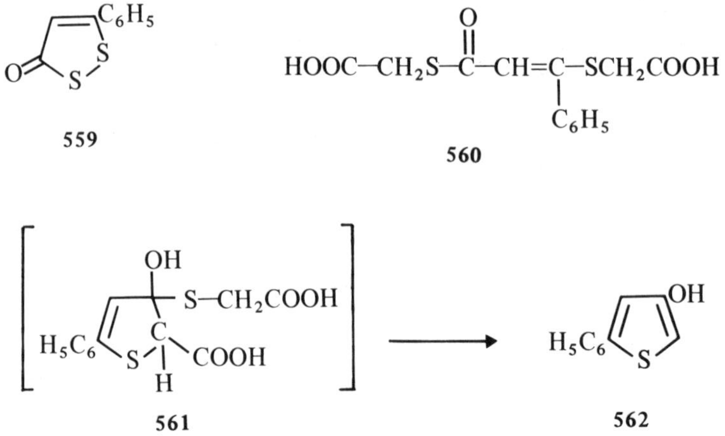

559

560

561

562

Scheme 178

3. Ring-Closure of C$_2$-S-C$_2$ Compounds to Thiophenes

Catalytic dehydrogenation of diethyl sulfide to thiophene in the presence of oxides and sulfides of various transition metals has been studied in detail.[458, 459] Thiophene has also been obtained by catalytic dehydrocyclization of diethyl disulfide, diethyl sulfoxide, and diethyl sulfone[460] and from ethyl vinyl and divinyl sulfides.[461, 462] From dehydrocyclization of di-n-propyl and diisopropyl sulfides on a Cr_2O_3–Al_2O_3–CuO–K_2O catalyst at 330–440°C, 2,4- and 2,5-dimethylthiophene were obtained as main products.[463] 3,4-Dialkylthiophenes have been obtained by heating divinyl sulfides at 150–200°C in the presence of potassium hydrogen sulfate.[464] The reaction is assumed to proceed via a [3,3]-sigmatropic rearrangement of **563** to **564**, which is cyclopolymerized to **565**, whereupon elimination catalyzed by potassium hydrosulfate yields 3,4-dimethylthiophene and 3,4-diethylthiophene in about 60% yield. Dialkyl disulfide, alkyl and 1-alkenyl monosulfide, and dialkyl trisulfide are formed as by-products. However, the possibility that 3,4-dialkylthiophenes are formed via enethiyl radicals (see Scheme 155) cannot be excluded (Scheme 179). From alkyl 1-propenyl sulfides,

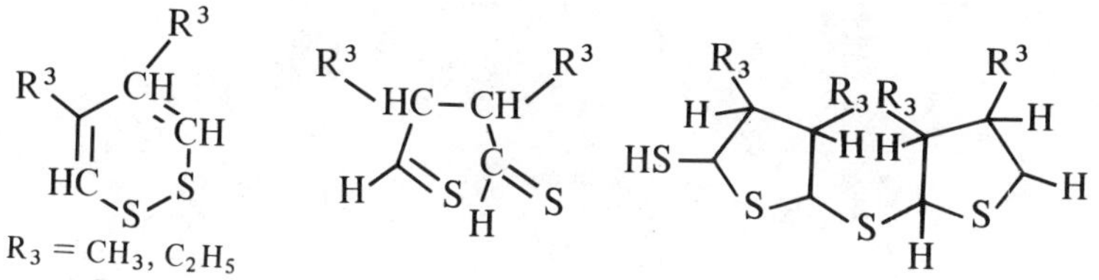

Scheme 179

3,4-dimethylthiophene was obtained upon heating at 150°C with aliphatic sulfur compounds.[464] Divinyl disulfides (**566**), obtained by oxidative coupling of anions derived from dithio esters with iodine, rearrange to bisdithio esters (**567**) upon heating in toluene at 100°C, followed by ring-closure via **568** to a mixture of ·**569** and **570** (Scheme 180). Heating **566** in toluene in the presence of 1 eq. of

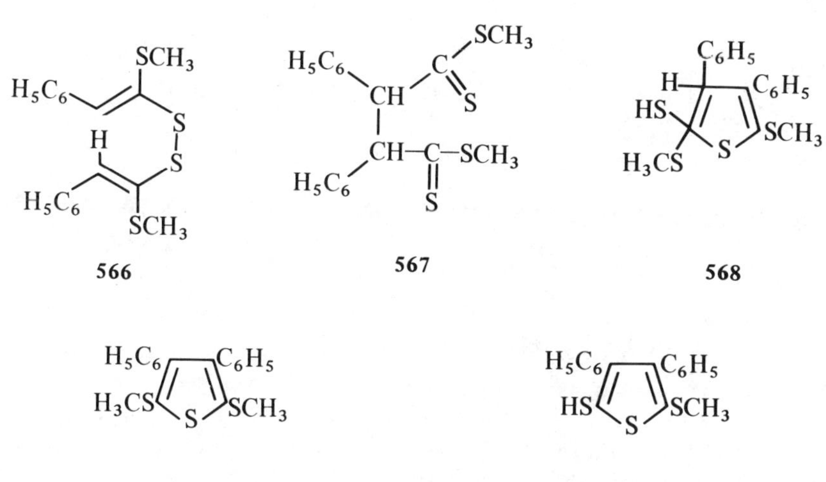

Scheme 180

potassium *t*-butoxide gave a quantitative yield of **570**.[465] The reaction of dimethyl acetylenedicarboxylate with disulfur dichloride using *N,N*-dimethylformamide as solvent and catalyst, gives **571** in 95% yield, which upon reaction with sodium thiophenolate yields **572**. Heating **572** for 5 min at 130–135°C gives **573**[466] (Scheme 181). Base-catalyzed rearrangement of bispropargyl sulfides leads to thio-

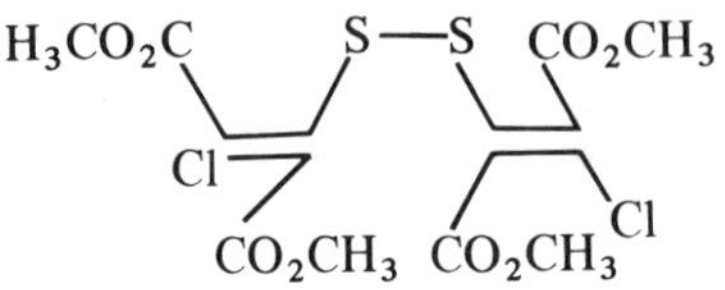

571 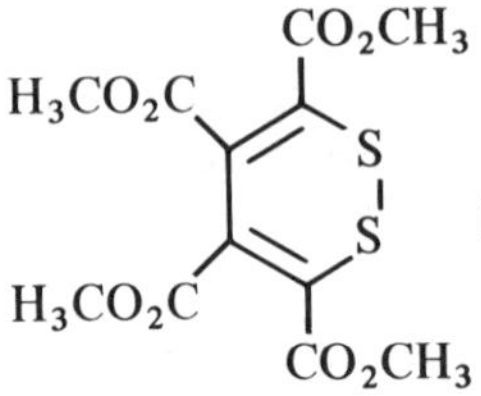   **572** 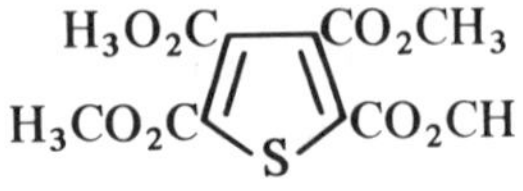   **573**

Scheme 181

phene derivatives. Thus, the reaction of **574** with potassium *t*-butoxide in THF at room temperature for 10 min gives a 73% yield of **578**.[467, 468] The reaction is assumed to proceed with the allene (**575**) and diradical (**576**) as intermediates (Scheme 182). The reaction of **579** (R = H) with potassium hydroxide in methanol

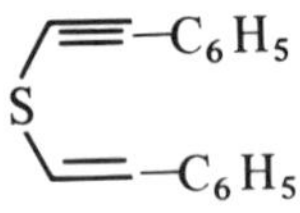

574 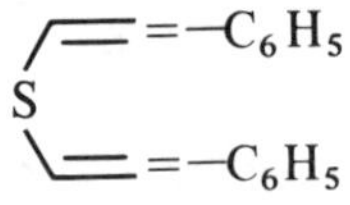   **575** 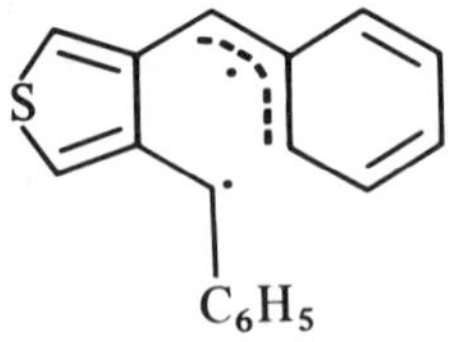   **576**

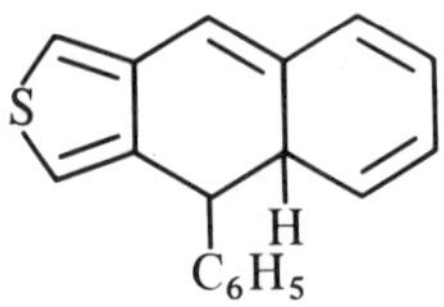

577 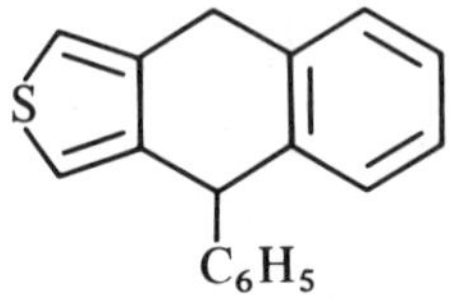   **578**

Scheme 182

gave the dimeric thiophene (**580**) in 12% yield, together with a trimeric derivative; **579** [R = C(CH$_3$)$_3$] gave a 51% yield of **581**.[467–470] The compounds are assumed to be formed via the dimethylene thiophene diradical (**582**), which could be trapped as the peroxide (**583**) with triplet oxygen.[469] Attempts to prepare diallene sulfide from γ-dimethylallenyllithium and sulfur dichloride led to the formation of **584** in low yield.[470] However, heating the easily available allenic sulfone (**585**) at 75°C gave a quantitative yield of **586**[471] (Scheme 183).

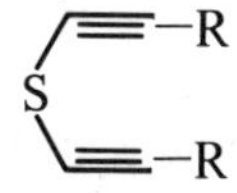
579

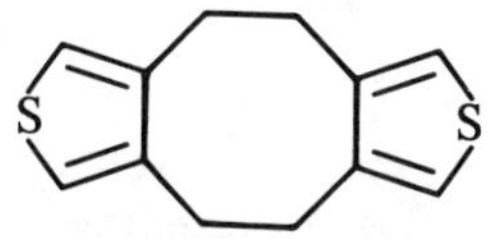
580

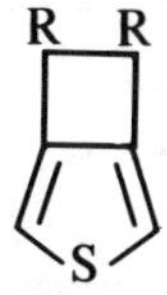
581

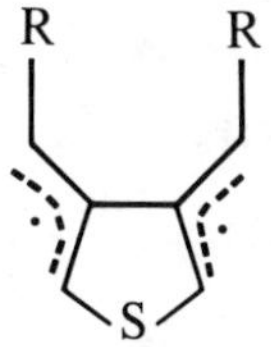
582

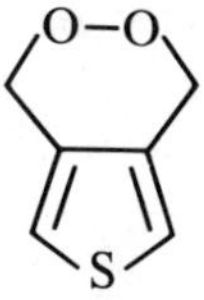
583

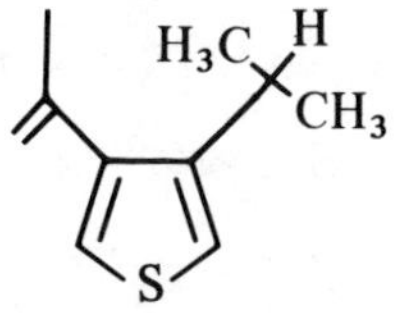
584

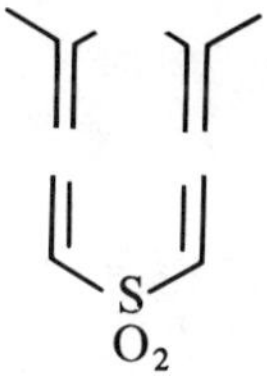
585

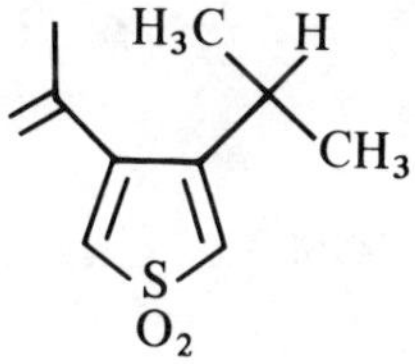
586

Scheme 183

The theoretically interesting thienocyclobutadiene (**590**) has been synthesized. It was formed in 5% yield in the flow pyrolysis of the *cis* isomer of **587** at 250°C, together with 30% of **588** and **589** in a 1:2 ratio. On standing at room temperature, it dimerized to **591**[472] (Scheme 184). *S*-α-Phenylphenacyl thiobenzoate gave, upon

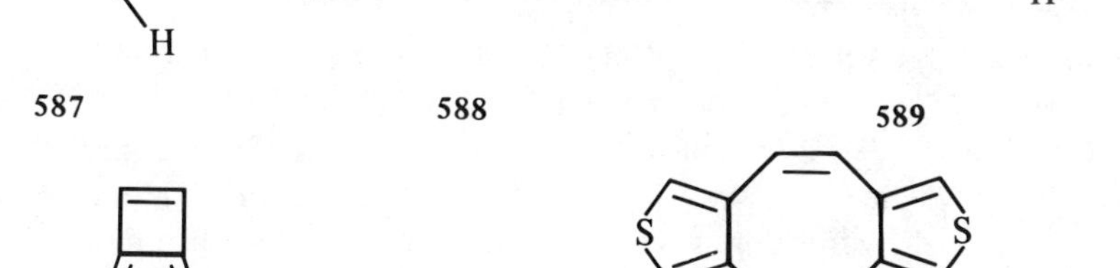
587 **588** **589**

590

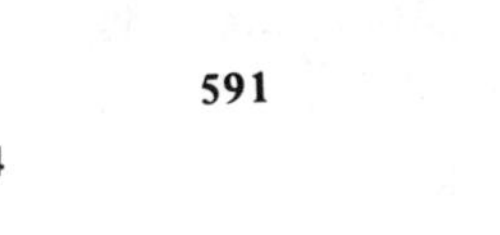
591

Scheme 184

attempted ring-closure to dithiolium salts, tetraphenylthiophene as the main product, even in the absence of hydrogen sulfide. It is assumed that the tetraphenyl-thiophene was formed via the α-mercaptodeoxybenzoin and the 1,4-dithiin.[473]

The reaction of phenacyl bromide with sodium dithioacetate in boiling ethanol led to ring-closure of the expected dithio ester (592) to the thiophenethiol (593)[473] (Scheme 185).

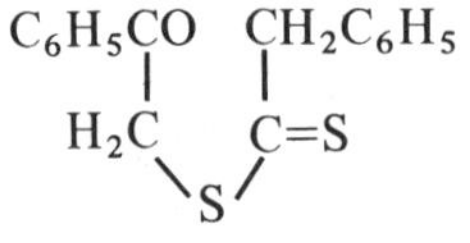

592 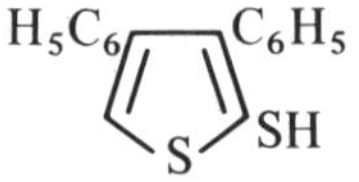   593

Scheme 185

V.   THIOPHENES FROM TETRA- AND DIHYDROTHIOPHENES

1.   Introduction

The reactions leading to di- and tetrahydrothiophenes could be systematized in the same way as those leading to thiophenes, according to the components undergoing the cyclization ($C_4 + S$, $C_2S + C_2$, $C_3S + C$, etc.). In addition, an aromatization step then leads to the thiophenes. However, this seems impractical, and instead these reactions are classified according to the type of reduced thio-phenes being aromatized. For a recent review on the syntheses of dihydrothio-phenes, see Ref. 473a.

2.   Dehydrogenation of Simple Tetrahydro- and Dihydrothiophenes

Dehydrogenation of tetrahydrothiophene and alkylated derivatives over various metal oxides gives thiophene and alkylthiophenes.[474-478a] Tetrahydrothiophene has also been aromatized by reaction with sulfur dioxide[479] and with sulfur under pressure.[480, 481] 3,3-Bithienyl and 3-thiophenethiol are probably by-products in the latter reaction.[480] Attempts to obtain 3,4-dihydroxythiophene by dehydrogenation of the tetrahydro derivative failed.[482] Chlorination of tetrahydrothiophene, with chlorine using iodine as catalyst, yields 2,3,4,5-tetrahydrotetrachlorothiophene, which could be dehydrochlorinated with sodium hydroxide in ethanol to a mixture of dichlorothiophenes.[481] In the chlorination of thiophene, chlorinated tetrahydro-thiophenes are also formed, which can be dehydrochlorinated to chlorothiophenes.[483] Aromatization, by treatment of highly chlorinated tetrahydrothiophenes with sulfur, gives disulfur dichloride as a by-product.[484] Reaction of 2,3-dichlorotetrahydro-thiophene (594), obtained by chlorination of tetrahydrothiophene, with Grignard reagents gives **595**. Upon treatment with potassium *t*-butoxide in DMSO, the saturated alkylthiophenes (**596**) are obtained in low yield.[485] 2-Alkenyl-1,2,2-trichlorovinylsulfides (**596a**, X = Cl) and 1,2-dichloro-1-propenylsulfides (**596a**, X = CH$_3$) give a mixture of **596b** (22–33%) and **596c** (0–22%) upon heating to 100–160°C. This involves rearrangement of the carbon skeleton and migration of chlorine. Treatment of **596b** (R$_2$ = H, X = Cl) with potassium *t*-butoxide led to the aromatization of **596d** via dehydrochlorination[485a] (Scheme 186).

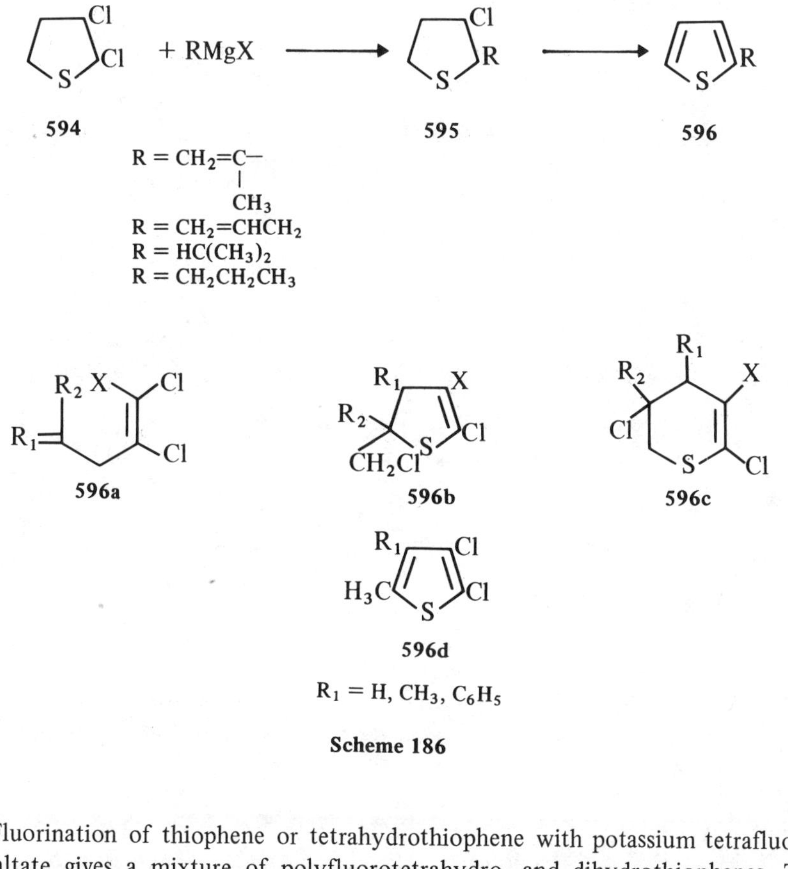

594 595 596

R = CH$_2$=C—
        |
        CH$_3$
R = CH$_2$=CHCH$_2$
R = HC(CH$_3$)$_2$
R = CH$_2$CH$_2$CH$_3$

596a **596b** **596c**

596d

R$_1$ = H, CH$_3$, C$_6$H$_5$

Scheme 186

Fluorination of thiophene or tetrahydrothiophene with potassium tetrafluorocobaltate gives a mixture of polyfluorotetrahydro- and dihydrothiophenes. The major product is **597**.[486] Treatment of **597** with molten potassium hydroxide at 250°C gave tetrafluorothiophene (**598**), whereas reaction with sodium methoxide in methanol gave **599**. From **600**, **601** and **602** have analogously been obtained[487] (Scheme 187). Pyrolysis of **600** also gave **601**.[488] Reaction of **602a** with S$_2$Cl$_2$ yields **603**, which upon treatment with zinc is claimed to give **598**.[489]

The first-order rate constant for the pyrolysis of 2,5-dihydrothiophene into thiophene and hydrogen has been determined.[490] Methyl or propyl tetrahydrothiophene-2,5-dicarboxylate was aromatized by refluxing with sulfur in trichlorobenzene.[491] Reaction of the dianion of ethyl acetoacetate with an episulfide (**604**)

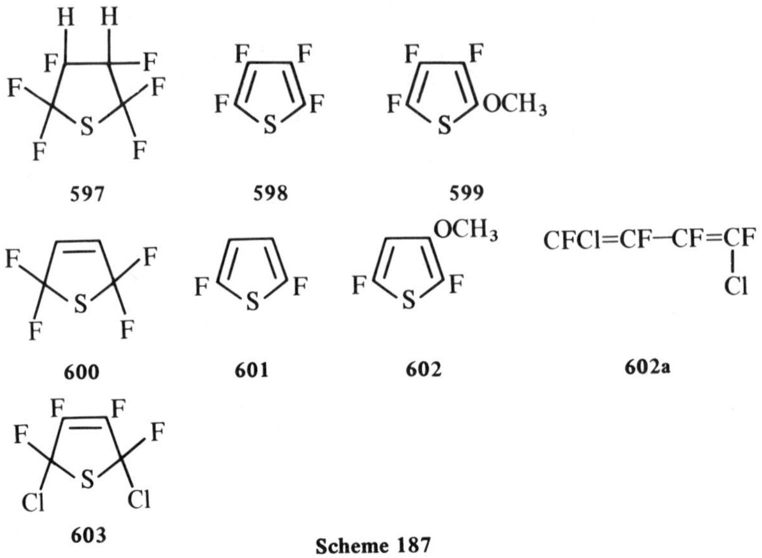

597   **598**   **599**

600   **601**   **602**   **602a**

603   **Scheme 187**

gave **605**,[492] which upon reaction with benzeneselenylbromide and oxidation gave **606** in 95% yield[493] (Scheme 188). The reaction of the 4,5-dihydrothiophenes

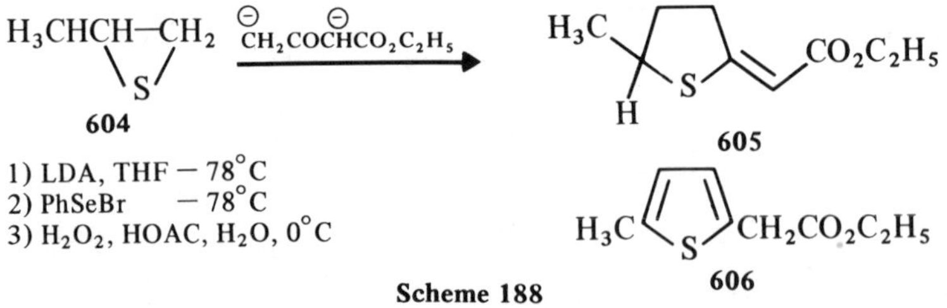

604

1) LDA, THF $-78°$C
2) PhSeBr $-78°$C
3) H_2O_2, HOAC, H_2O, $0°$C

605

606

Scheme 188

(**607**) with dimethyl diacetylenedicarboxylate proceeds via the cyclo adduct (**608**) to give dimethyl 2,3-thiophenedicarboxylates (**609**) and olefins[494] (Scheme 189).

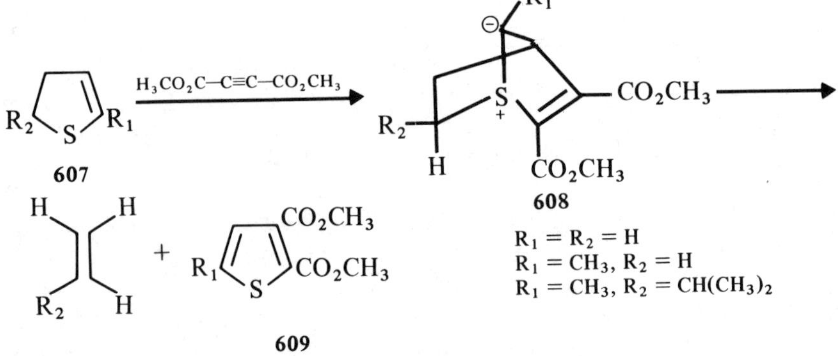

607

608

$R_1 = R_2 = H$
$R_1 = CH_3, R_2 = H$
$R_1 = CH_3, R_2 = CH(CH_3)_2$

609

Scheme 189

The reaction of succinic acid thioanhydride (610) with the phosphorane (611) gave a 49% yield of 612[495] (Scheme 190). Dichloromaleic acid thioanhydride (613),

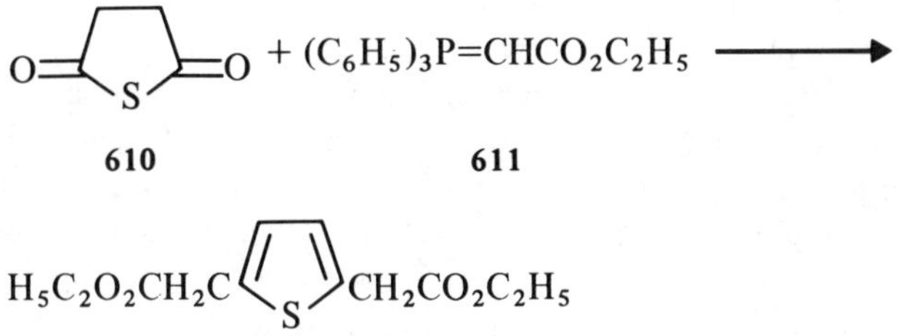

610   611

612

Scheme 190

obtained through reaction of tetrachlorothiophene with nitric acid, gives 614 upon reaction with zinc and acetic anhydride. Upon reaction with hydrogen sulfide, 615 is obtained via a dithiine as an intermediate[496] (Scheme 191).

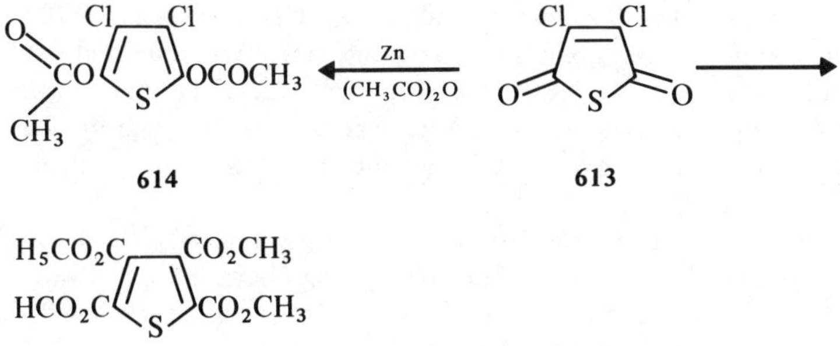

614   613

615

Scheme 191

Reaction of 616 with sodium sulfide in ethanol gave 617, which was aromatized to 618 by treatment with DDQ[497] (Scheme 192). The reaction of 1,2,3,5-tetra-O-acetyl-4-thio-D-ribofuranose with catalytic amounts of mercuric chloride and piperidine in DMF gave 46% of 4-acetoxy-2-(acetoxy-methyl)thiophene.[498]

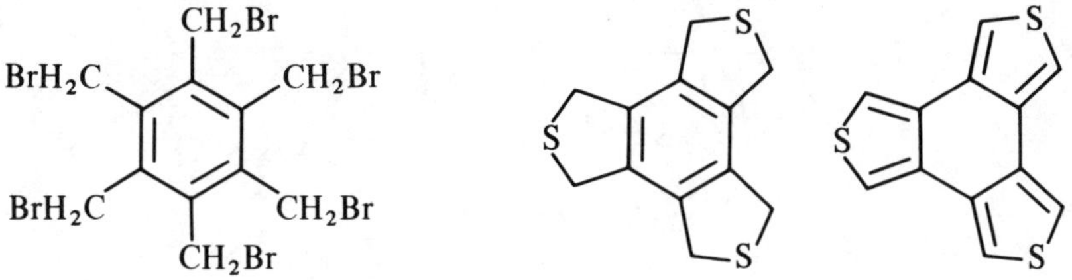

616   617   618

Scheme 192

3. Thiophenes from 3-Oxotetrahydrothiophenes

Some very useful starting materials for the synthesis of many thiophenes are 3-oxotetrahydrothiophenes, which were introduced by Woodward and Eastman.[313] They found that α,β-unsaturated esters added thioglycolic esters to give **619**, which, depending upon reaction conditions and substituents, could ring-close according to the C_3SC-principle to give **620**, or according to the C_2SC_2-principle to give **621**. Of course, if an α-substituted α,β-unsaturated acid (such as methacrylic acid) was used, only ring-closure, according to the C_3SC-principle, was obtained.[342] Because of the synthetic importance of **621**, modifications for its synthesis have been worked out.[329,342,429,500-502] The best way to obtain **621** appears to be the reaction with sodium methoxide in methanol.[500] Also **621** (R = H) is formed with about equal amounts of **620** (R = H) when cyclization is carried out at a lower temperature with sodium methoxide in ether. Recently, separation of **621** (R = H) from **620** (R = H) was achieved by flash column chromatography.[502] Complete regioselectivity in the Dieckmann cyclization could be achieved by using half-thiol diesters. Thus, treatment of **622** (prepared by the addition of methyl thioglycolate to ethyl thioacrylate) with sodium hydride in dry THF gave 74% of **620** (R = H); **623**, obtained from (2-ethoxycarbonylethylthio)acetyl chloride and lead ethylmercaptide, only gave **621** (R = H) in 70% yield under the same reaction conditions.[503] Compounds such as **619** can also be prepared through the reaction of α-halocarboxylates with β-mercaptocarboxylic acid derivatives and then ring-closed to **620** and **621**.[505,507,510,511]

If acrylonitrile is used instead of acrylate, ring-closure after Michael addition occurs according to the C_2SC_2-principle, giving 4-cyano-3-oxotetrahydrothiophene.[504] Instead of thioglycolate, α-mercaptoketones can also be used (Scheme 193). 3-Oxotetrahydrothiophenes that have been prepared are listed in Table 49.

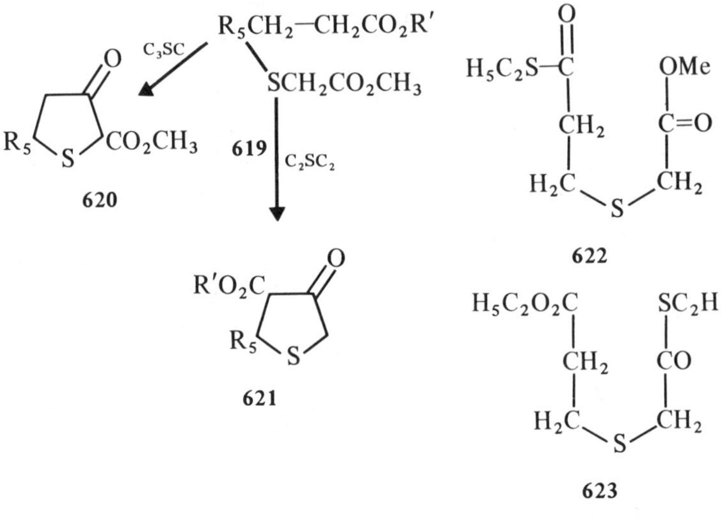

Scheme 193

TABLE 49. 3-OXOTETRAHYDROTHIOPHENES BY REACTION OF α,β-UNSATURATED ESTERS AND THIOGLYCOLIC ESTER

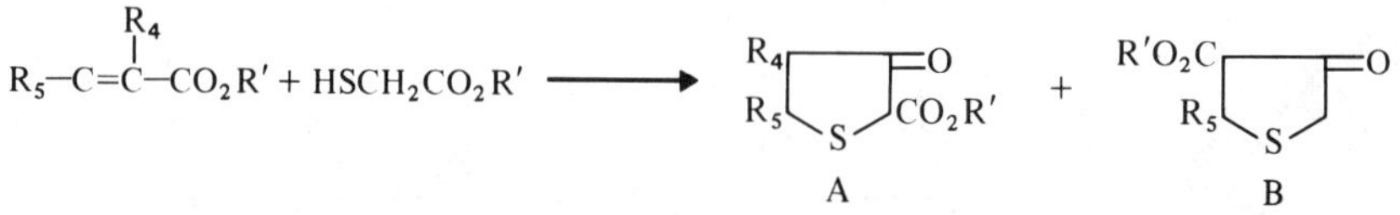

$$R_5-C(R_4)=C-CO_2R' + HSCH_2CO_2R' \longrightarrow A + B$$

A   B

R_5	R_4	R'	Yield of A (%)	Yield of B (%)	Reference
H	H	CH_3	45, 67	50[a]	333, 499, 500–502
H	CH_3	CH_3	60, 94		329, 335, 502
H	CH_3	C_2H_5	d	41[c]	335, 506, 507
CH_3	H	CH_3	41[a]	41[a]	502
CF_3	H	CH_3	b	b	508
CO_2CH_3	H	CH_3	–	76	309, 314, 509
C_6H_5	CO_2CH_3	CH_3	89	–	193b
C_6H_5	CH_3	CH_3	58	–	193b
CO_2CH_3	CH_3	CH_3	69	–	309
$(CH_2)_4CO_2CH_3$	H	CH_3	13	61	512, 513

[a] Separated by flash chromatography according to Ref. 502.
[b] Isomers not separated.
[c] According to Ref. 506.
[d] A mixture of A and B was obtained according to Ref. 507, when β-mercaptobutyric acid and chloroacetic acid were used.

Esters (620 and 621) can be hydrolyzed and decarboxylated to give 3-oxotetrahydrothiophenes.[313,505,507,514] Such compounds can also be obtained by the addition of thioglycolic acid to propenoic acids in dioxane in the presence of 1 eq. of triethylamine to give 624 in 70–90% yield. Ring-closure of the dicarboxylic acids to 625 was achieved by treatment with acetic anhydride, with lithium acetate as a catalyst, in 60–85% yield[515,516] (Scheme 194). The 3-tetrahydrothiophenes prepared in this way, which have been converted to various thiophenes, are given in Table 50.

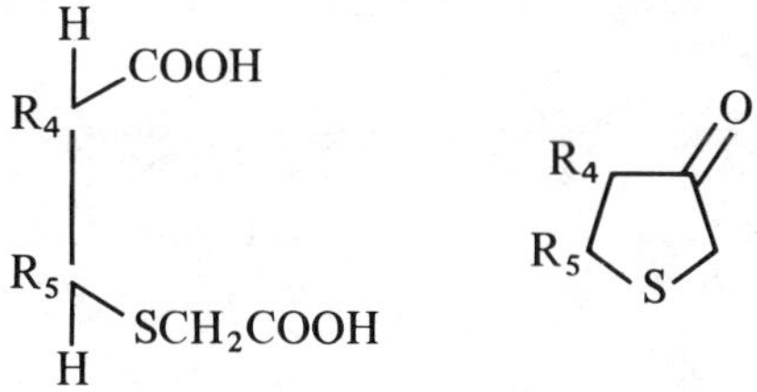

624   625

Scheme 194

3-Oxotetrahydrothiophenes have also been prepared from 626, which was obtained via reaction of diethyl maleate with hydrogen sulfide in the presence of triethylamine (626a) or when a mixture of diethyl maleate and ethyl acrylate

TABLE 50. 3-OXOTETRAHYDROTHIOPHENES FROM PROPENOIC ACIDS AND THIO-
GLYCOLIC ACID

$$R_5CH=CH-\underset{\underset{R_4}{|}}{C}H-COOH + HSCH_2COOH \longrightarrow$$

R_5	R_4	Yield of Product (%)	Reference
H	CH_3	91	515
CH_3	H	75	515
C_6H_5	H	80	515
C_6H_5	C_6H_5	70	515
$(H_3C)_2CH$	H	70	516
$(H_3C)_3C$	H	50	516
$p\text{-}H_3COC_6H_4$	H	70	516

was reacted in the same way (626b). This compound was obtained pure when
ethyl 3-mercaptopropionate was reacted with diethyl maleate in the presence of

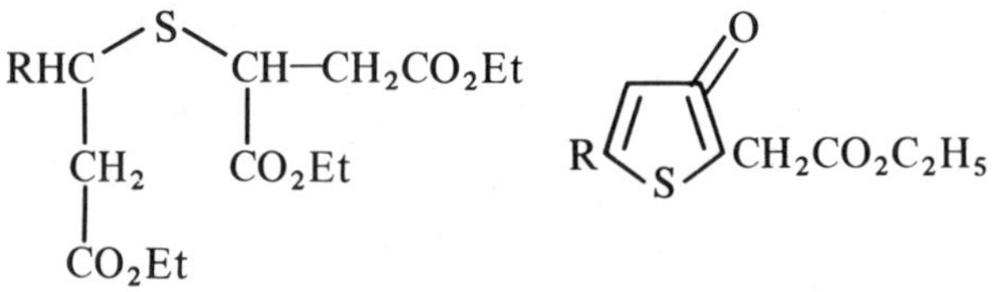

$$RHC \overset{S}{\diagdown} CH-CH_2CO_2Et$$

626

(a) R = CO_2Et
(b) R = H

627

(a) R = CO_2C_2H_5
(b) R = Me

Scheme 195

piperidine.[517] Ring-closure of 626 with magnesium ethoxide in xylene gave 627.
Pure 3-oxo-4-carbethoxy-tetrahydrothiophene was prepared through the reaction
of ethyl α-chloroacetoacetate with benzaldehyde and hydrogen chloride to give
628, which upon reaction with hydrogen sulfide in sodium ethylate solution gave
a 67% yield of 629[518] (Scheme 196). Instead of α,β-unsaturated carboxylic acids
and their esters, α,β-unsaturated ketones (630) of the corresponding Mannich
bases have also been used in the reaction with thioglycolate to give tetrahydro-
and dihydrothiophenes,[519,520] Primarily, 631 is formed upon ring-closure; this

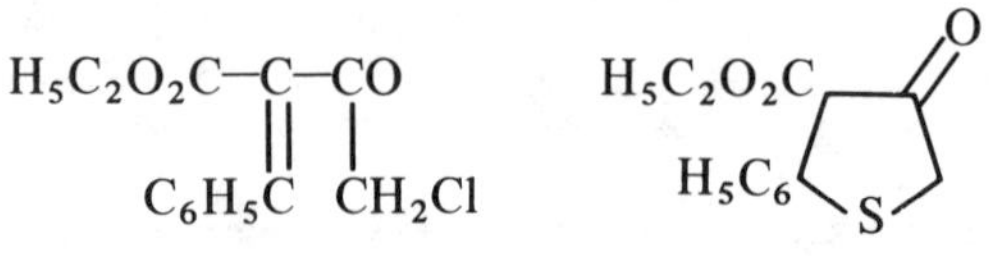

628

629

Scheme 196

compound, upon treatment with PPA, gives a mixture of **632**, **633**, and **634** owing to disproportionation during dehydration. The mixture would be aromatized to **634** by treatment with chloranil or diphenyl disulfide.[519, 520] Instead of thioglycolate, α-mercaptoaldehydes and α-mercaptoketones can be used.[308, 329] Thus, the reaction of 2-cyclohexen-1-one with mercaptoacetaldehyde diethylacetal gave **635**[521] which was transformed to **636** (Scheme 197).

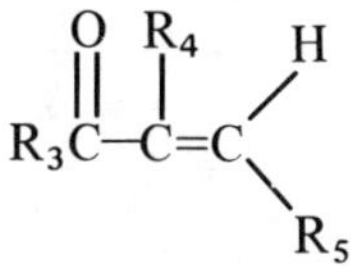

630

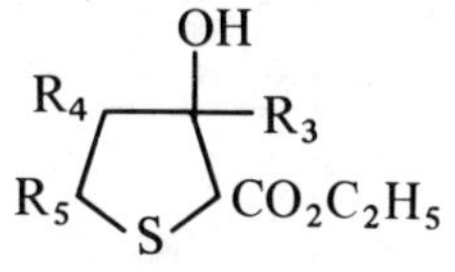

631

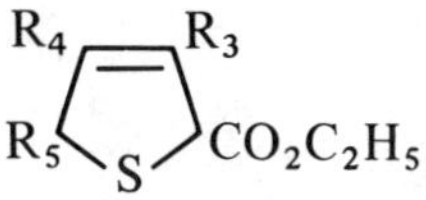

632

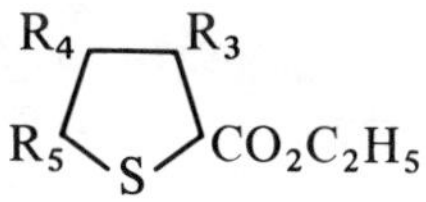

633

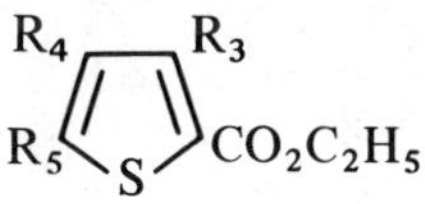

634

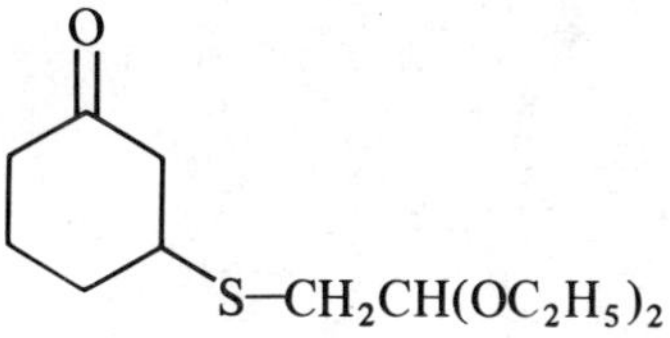

635

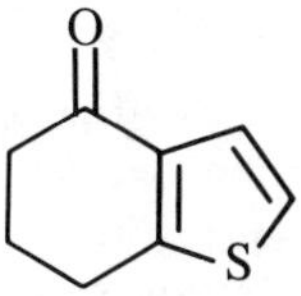

636

Scheme 197

The 3-oxotetrahydrothiophenes have been converted to thiophenes in many different ways. Direct aromatization of 3-oxotetrahydrothiophenes has been achieved by careful reaction with bromine[193b, 309, 335] or more conveniently with sulfuryl chloride.[501]

Sulfuryl chloride followed by pyridine was also used successfully for the transformation of **636a** to **636b**.[501a] Another new method of aromatization consists in the reaction with *N*-chlorosuccinimide followed by pyridine. This technique was used for the preparation of **636d** from **636c**[501b] (Scheme 197a) (Table 51).

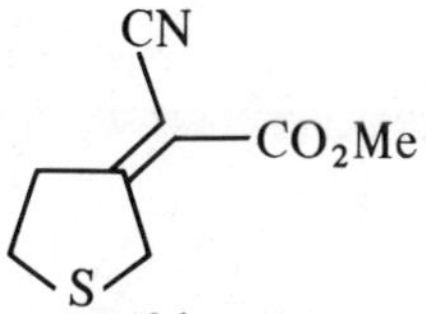

636a

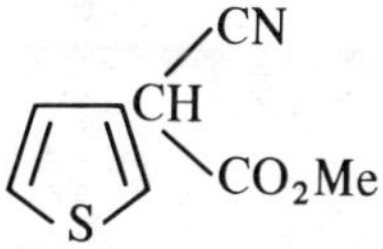

636b

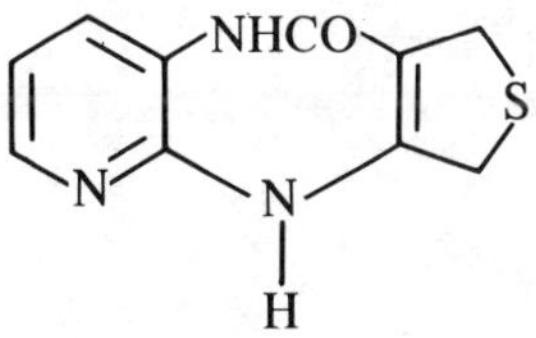

636c

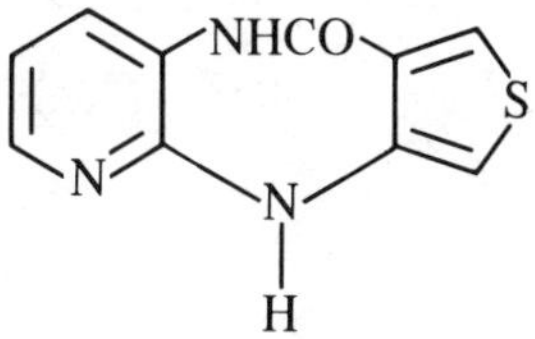

636d

Scheme 197a

TABLE 51.   AROMATIZATION OF 3-OXOTETRAHYDROTHIOPHENES TO 3-HYDROXYTHIO-
PHENES WITH (A) BROMINE AND (B) SO_2Cl_2

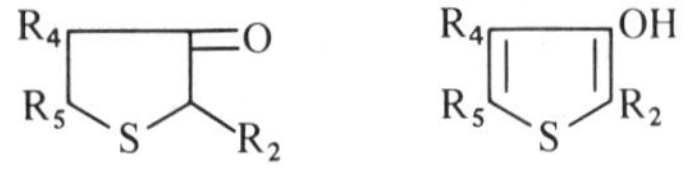

R_2	R_4	R_5	Yield of Thiophene (%)	Method	mp (°C)	Reference
CO_2CH_3	H	H	82	A, B		335, 501
CO_2CH_3	H	CH_3	75	A, B		335, 501
CO_2CH_3	CH_3	CO_2CH_3	15	A	107–108	309
CO_2CH_3	CO_2CH_3	C_6H_5	85	A	79	193b
H	CO_2CH_3	H	85	B		501
H	CO_2CH_3	CH_3	75	B		501
H	CN	H	77	B		501
H	CO_2CH_3	H	80	B		501
COC_6H_5	H	H	88	B	53–57	501
CN	H	H	76	B	88–92	501

In some cases, the 3-oxotetrahydrothiophene was first converted to the enol-acetate, enol methylether, or methylsulfonate before aromatization with sulfur chloride.[501,522] Treatment of 2- or 4-acyl-3-oxotetrahydrothiophenes with sulfur in DMF yields 2-acyl- or 4-acyl-3-hydroxythiophenes. The acyl-3-oxotetrahydro-thiophenes were prepared via the reaction of enamines derived from 3-oxotetra-hydrothiophenes with acid chlorides.[329] The 3-hydroxythiophenes prepared are listed in Table 52.

Another convenient method is aromatization with 40% H_2O_2[511] or 30% hydrogen peroxide in acetic acid.[506] Thus, 2-phenyl 4-carboethoxy-3-oxo-tetrahydrothiophene gave a 90% yield of ethyl 2-phenyl-3-hydroxy-4-thiophenecarboxylate.[511] Many different 3,4-disubstituted 2,5- and 4,5-dihydrothiophenes, derived from 3-oxo-tetrahydrothiophenes as indicated in Scheme 198 could be aromatized by 30%

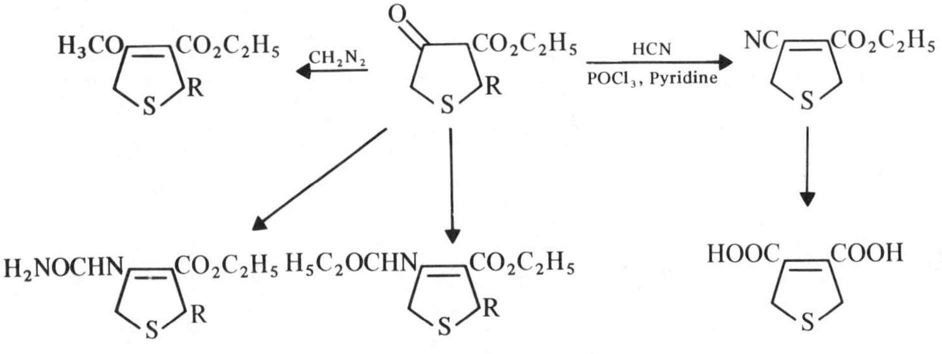

Scheme 198

TABLE 52. AROMATIZATION OF 2-ACYL-3-OXOTETRAHYDROTHIOPHENE
WITH SULFUR[329]

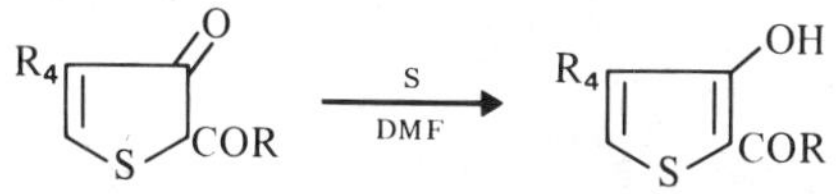

R	R_4	Yield of Thiophene (%)	mp (°C)
CH_3	H	47	50–51
CH_3	CH_3	25	36–37
C_2H_5	CH_3	50	40
C_6H_5	CH_3	50	77–78

hydrogen peroxide in acetic acid. If both β-substituents were electron-withdrawing, only aromatization was obtained, However, 3,4-dimethoxycarbonyl-2,5-dihydrothiophenes are exceptional and give a mixture of the thiophene and the 2,5-dihydrothiophene-1,1-dioxide. The 3,4-dicarboxylic acid also behaved curiously, probably giving the sulfoxide, which, upon heating in protic solvents, gave the thiophene. If one of the β-substituents is electron-donating and the other is electron-withdrawing, only sulfones are obtained. Oxidation with perbenzoic acid in chloroform gave the corresponding sulfone derivatives.[506] However, the use of iodosobenzene gave only thiophenes with all types of substituents.[523] A probable mechanism for the reaction with iodosobenzene is oxidation of the dihydrothiophene to the sulfoxide followed by dehydration. Thiophenes prepared from 2,5-dihydrothiophenes in these ways are collected in Table 53 and those from 4,5-dihydrothiophenes are given in Table 54.

Oxime formation followed by reaction with dry hydrogen chloride, most frequently in ether, gives 3-aminothiophenes in high yields,[508, 509, 513, 516, 524, 525] (Table 55).

Another route to 3-aminothiophenes consists of the reaction of 3-oxo-tetrahydrothiophenes (**637**) with ammonia to give the corresponding dihydrothiophenes (**638**), which can be aromatized to **639** by treatment with chloranil, bromine, or

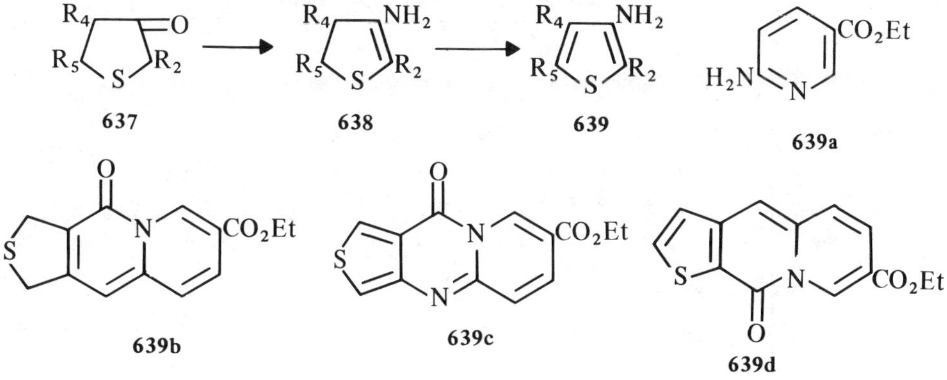

Scheme 199

TABLE 53. THE REACTION OF 2,5-DIHYDROTHIOPHENES WITH (A) 30% H_2O_2 IN ACETIC ACID, (B) PERBENZOIC ACID IN CHLOROFORM AND, (C) IODOSOBENZENE IN DIOXANE

$$R_4 \underset{S}{\overset{R_3,\,R_2}{\bigcirc}} \longrightarrow R_4 \underset{S}{\overset{R_3,\,R_2}{\bigcirc}} \;(\text{I}) \;+\; R_4 \underset{S\,O_2}{\overset{R_3,\,R_2}{\bigcirc}} \;(\text{II})$$

R_2	R_3	R_4	Method	Yield of I (%)	Yield of II (%)	Reference
H	COOH	COOH	A	68	0	506
H	COOH	COOH	B	–	84[a]	506
H	COOH	COOH	C	89	–	523
H_3C	COOH	COOH	A	90	–	506
H_3C	COOH	COOH	C	92	–	523
H	$CO_2C_2H_5$	CN	A	92	–	506
H	$CO_2C_2H_5$	CN	B	–	84	506
H	$CO_2C_2H_5$	CN	C	81	–	523
H_3C	$CO_2C_2H_5$	CN	A	92	Trace	506
H_3C	$CO_2C_2H_5$	CN	B	Trace	77	506
CH_3	$CO_2C_2H_5$	CN	C	81	–	523
H	CO_2CH_3	CO_2CH_3	A	23	53	506
H	CO_2CH_3	CO_2CH_3	B	–	95	506
H_3C	CO_2CH_3	CO_2CH_3	A	13	53	506
H_3C	CO_2CH_3	CO_2CH_3	B	–	75	506
H	$CO_2C_2H_5$	$NHCO_2C_2H_5$	A	–	69	506
H	$CO_2C_2H_5$	$NHCO_2C_2H_5$	B	–	86	506
H	$CO_2C_2H_5$	$NHCO_2C_2H_5$	C	72	–	523
CH_3	$CO_2C_2H_5$	$NHCO_2C_2H_5$	C	73	–	523
H	$CO_2C_2H_5$	$NHCONH_2$	A	–	67	506
H	$CO_2C_2H_5$	$NHCONH_2$	C	61	–	523
H_3C	$CO_2C_2H_5$	$NHCONH_2$	A	–	78	506
H_3C	$CO_2C_2H_5$	$NHCONH_2$	C	53	–	523
H	$CO_2C_2H_5$	OCH_3	A	–	52	506
H	$CO_2C_2H_5$	OCH_3	B	–	82	506
H	$CO_2C_2H_5$	OCH_3	C	54	–	523

[a] The sulfoxide was obtained in this case.

158

TABLE 54. THE REACTION OF 4,5-DIHYDROTHIOPHENES WITH (A) 30% H_2O_2 IN ACETIC ACID, (B) PERBENZOIC ACID IN CHLOROFORM, AND (C) IODOSOBENZENE IN DIOXANE

R_2	R_3	R_4	Method	Yield of I (%)	Yield of II (%)	Reference
H	COOH	COOH	A	97	–	506
H	COOH	COOH	C	94	–	523
HC	COOH	COOH	A	83	–	506
H_3C	COOH	COOH	C	76	–	523
H	CO_2CH_3	CO_2CH_3	A	10	60	506
H	CO_2CH_3	CO_2CH_3	B	–	97	
H	CO_2CH_3	CO_2CH_3	C	73^a	–	523

^{a} In addition, 11% of the monomethyl ester of 3,4-thiophenedicarboxylic acid was obtained.

sulfur in DMF.[329] The dihydro-derivatives (638) can also be obtained from ethylene sulfide and cyanoacetate[529] or by direct ring-closure of methacrylonitrile and methyl thioglycolate with sodium methoxide, which yields methyl 3-amino-4-methyl 4,5-dihydrothiophene-2-carboxylate in 65% yield. The aminothiophenes prepared in this way are given in Table 56.

TABLE 55. 3-AMINOTHIOPHENES FROM 3-OXOTETRAHYDROTHIOPHENES VIA OXIMES

R_2	R_4	R_5	Yield of Oxime (%)	Yield of Aminothiophene (%)	Reference
H	H	H		50	525
H_3C	H	H		49	525
H	CH_3	H		55	525
H	H	CH_3		55	525
H	CO_2CH_3	H	100	57	524
H	$CO_2C_2H_5$	CF_3	a		508
$CO_2C_2H_2$	H	CF_3	a		508
H	H	$(CH_3)_2CH$	95	65	516
H	H	$C(CH_3)_3$	85	80	516
H	CO_2CH_3	CO_2CH_3	83, 96	77, 85	509
$(CH_2)_4CO_2CH_3$	CO_2CH_3	H		96	513, 528
$(CH_2)_4CO_2C_2H_5$	$CO_2C_2H_5$	H			
$CH_3CH_2CH_2$	CO_2Et	H		46	526
H	CO_2CH_3	CH_3	91, 94	49, 83	193b, 527
CO_2CH_3	CH_3	H	95	85	193b
CO_2CH_3	CO_2CH_3	C_6H_5	80	75	193b
H	$CO_2C_2H_5$	C_6H_5	87	71	193b

^{a} See Ref. 528.

TABLE 56. 3-AMINOTHIOPHENES FROM 3-OXOTETRAHYDROTHIOPHENES AND AMINES

R_2	R_4	R_5	R'	R''	Yield of Enamine (%)	Yield of Thiophene (%)	Reference
H	H	H	$-(CH_2)_4-$		74	17	534
H	H	H	$-(CH_2)_5-$		74	20	534
H	H	H	$-(CH_2)_2O(CH_2)_2-$		78	20	534
H	CH_3	H	$-(CH_2)_4-$		80	30–50	515
H	H	CH_3	$-(CH_2)_4-$		77	30–50	515
H	H	C_6H_5	$-(CH_2)_4-$		78	30–50	515
H	C_6H_5	H	$-(CH_2)_4-$		80	30–50	515
CO_2CH_3	H	H	H	H	85	15	329
CO_2CH_3	CH_3	H	H	H	65	52	329
$CO_2C_2H_5$	H	CH_3	H	H	80		329
H	CO_2CH_3	CO_2CH_3	H	H	39		329
H	CH_3	$COCH_3$	H	H	39	80	329
H	CH_3	COC_6H_5	H	H	50	52	329

160

A route to secondary and tertiary aminothiophenes consists of the reaction of 3-oxotetrahydrothiophenes with amines, followed by aromatization of the intermediate enamine with chloranil,[530-533] diisopentyldisulfide of sulfur[309,515,516] (Table 56). Many derivatives have been obtained with substituted anilines in connection with work on pharmacologically active compounds. The compounds prepared, and their melting points, are given in Table 57.

Another example of this approach is the reaction of 637 ($R_2 = R_5 = H$, $R_4 = CO_2Me$) with 639a in the presence of a catalytic amount of p-toluene sulfonic acid, which gave 639b in 30% yield. Upon treatment with N-chlorosuccinimide in pyridine, 639b gave 639c in 55% yield. Applying the same reaction sequence to methyl tetrahydro-3-oxo-2-thiophene carboxylate gave 639d.[534a] The reaction of 3-oxo-4-carbomethoxytetrahydrothiophenes in ortho-phenylenediamine led to a mixture of 640, 641, and 642 in various amounts, which could be separated and aromatized with chloranil to the corresponding thiophenes.[535,536] Compounds (627) have been reacted with hydrazine to give 643, which could be converted to 644 in several steps.[517] Similarly, 4-carbomethoxy-3-oxotetrahydrothiophene, upon reaction with hydrazine, gave 645, which was aromatized to 646[537] (Scheme 200).

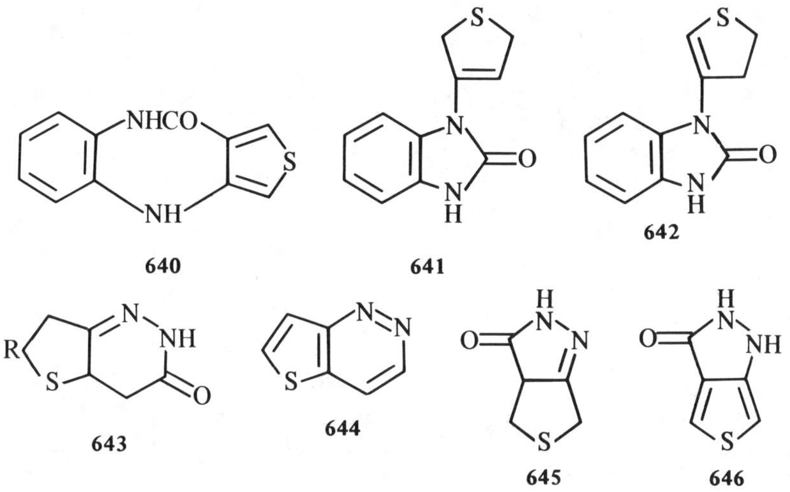

Scheme 200

Though reaction with H_2S-HCl, the 3-oxotetrahydrothiophenes were converted to the corresponding sulfur compounds (647, 648), which were then aromatized with sulfur,[204] or alkylated to 649 or acylated, and then aromatized by treatment with chloranil.[309,331] The 3-oxotetrahydrothiophenes could also be reacted with thioglycolic acid and hydrogen chloride to give 649 ($R = CH_2COOH$) directly, which then could be aromatized[329] or reacted with thiophenol or benzylmercaptan in the presence of hydrogen chloride to give 649 ($R = C_6H_5$ and 649 ($R = C_6H_5CH_2$), respectively.[533] These compounds could then be aromatized and further trans-

TABLE 57.   3-ARYLAMINOTHIOPHENES FROM 3-OXOTETRAHYDROTHIOPHENES AND SUBSTITUTED ANILINES

$$R_4\overset{O}{\underset{R_5}{\diagdown}}{}_S{\diagup}R_2 \;+\; ArNH_2 \;\longrightarrow\; R_4\overset{NHAr}{\underset{R_5}{\diagdown}}{}_S{\diagup}R_2 \;\longrightarrow\; R_4\overset{NHAr}{\underset{R_5}{\diagdown}}{}_S{\diagup}R_2$$

R₂	R₄	R₅	Ar	Yield of Dihydro (%)	Yield of Thiophene (%)	Thiophene mp (°C)	Reference
CO₂CH₃	CH₃	H	4-ClC₆H₄	78	58	106–108	533
CO₂CH₃	CH₃	H	C₆H₅	72	68	84	533
				(64)	(36)		309
CO₂CH₃	CH₃	H	3,4-di-Cl₂C₆H₃	82	51	102	533
CO₂CH₃	CH₃	H	2-H₃CC₆H₄	87	56	76–78	533
CO₂CH₃	CH₃	H	2,4-di-(H₃C)₂C₆H₃	75	54	46	533
H	CN	H	C₆H₅			88–89	531
H	CO₂CH₃	H	C₆H₅			77	531
H	CN	H	2,3-di-(H₃C)₂C₆H₃			107	530
H	CN	H	2,6-di-Cl₂C₆H₃			96	530
H	CN	H	2-Cl-6-CH₃C₆H₃			104	530
H	CN	H	2,6-di-(H₃C)₂C₆H₃			124	530
H	CN	H	4-(CH₃)₂CHC₆H₄			126	530
H	CN	H	4-C₂H₅OC₆H₄			114	530
H	CN	H	2,4,6-(CH₃)₃C₆H₂			78	530
H	CN	H	3,4-di-(CH₃)₂C₆H₃			69	530
H	CN	H	2-ClC₆H₄			135	530
H	CN	H	2,4,6-tri-Cl₃C₆H₂			162	530
H	CN	H	2,3-di-Cl₂C₆H₃			112	530
H	CN	H	2-Cl-5-CH₃C₆H₃			118	530
H	CN	H	2,4,6-Cl₃-5-CH₃C₆H			125	530
H	CN	H	2-Cl-3-CH₃C₆H₃			124	530
H	CN	H	2,4-Cl₂-3-CH₃C₆H₂			113	530
H	CN	H	2,6-Cl₂-4-C₂H₅OC₆H₂				530

H	CN	H	$2,6\text{-}Cl_2\text{-}4\text{-}C_2H_5O\text{-}5\text{-}CH_3C_6H$			136	530
H	CN	H	$2,4\text{-}Cl_2\text{-}5\text{-}CH_3C_6H$			157	530
H	CO_2CH_3	H	$2\text{-}O_2NC_6H_4$	86	96	120–122	532
H	CO_2CH_3	H	$4,5\text{-}Cl_2\text{-}O_2NC_6H_2$	75	67	162–163	532
H	CO_2CH_3	H	$4\text{-}CH_3O\text{-}2\text{-}O_2NC_6H_3$	84	90	170–171	532
H	CN	H	$4\text{-}Cl\text{-}2\text{-}O_2NC_6H_3$	80	99	213–214	532
H	CN	H	$4\text{-}CH_3S\text{-}2\text{-}O_2NC_6H_3$	57	95	167–169	532
H	CN	H	$4\text{-}CH_3O\text{-}2\text{-}O_2NC_6H_3$	72	88	149–150	532
H	CN	H	$4\text{-}F_3C\text{-}2\text{-}O_2NC_6H_3$	43	85	156–157	532
H	CN	H	$5\text{-}H_3C\text{-}2\text{-}O_2NC_6H_3$	34	62	127–128	532
H	CN	H	$2\text{-}Cl\text{-}3\text{-}CH_3C_6H_3$				504

TABLE 58. AROMATIZATION OF 4,5-DIHYDRO-3-ALKYLTHIOTHIOPHENES WITH (A) SULFUR, (B) BROMINE, (C) CHLORANIL, AND (D) SULFURYL CHLORIDE

R_2	R_4	R_5	R	Method	Yield of Thiophene (%)	mp (°C)	Reference
CO_2CH_3	CH_3	H	H	A	46		204
CO_2CH_3	H	H	$CH_2CO_2CH_3$	B	a^a	76	329
CO_2CH_3	CH_3	H	$CH_2CO_2CH_3$	B	b	58	329
H	CO_2CH_3	H	$CH_2CO_2CH_3$	B	c	54–55	329
CO_2CH_3	H	H	CH_3	C	75	73	331
CO_2CH_3	H	H	$COCH_3$	C	63	143	331
CO_2CH_3	CH_3	H	C_6H_5	C	61	102	533
CO_2CH_3	CH_3	H	$C_6H_5CH_2$	C	75	45	533
H	CN	H	C_6H_5	D	90	32–34	501

[a] The crude products gave methyl 3-hydroxythieno[3,2-b]-2-thiophenecarboxylates upon ring-closure in (a) 73%, (b) 78%, and (c) 87% yield.

formed to bicyclic and tricyclic heterocyclic systems. Most of this work is only available through the Ph.D. theses of Fiesselmann's students at Erlangen. The compounds prepared are given in Table 58.

2-Acetyl-3-oxotetrahydrothiophene reacted with methyl thioglycolate or thioglycolic acid in the presence of hydrogen chloride to give **653** (R = CH$_3$) and **653** (R = H) in 50 and 77% yield, respectively. Analogously, **654** was obtained from 2-benzoyl-3-oxotetrahydrothiophene in 38% yield. From 4-acetyl-3-oxotetrahydrothiophene, a 48% yield of **655** was obtained[329] (Scheme 201). The tosylate

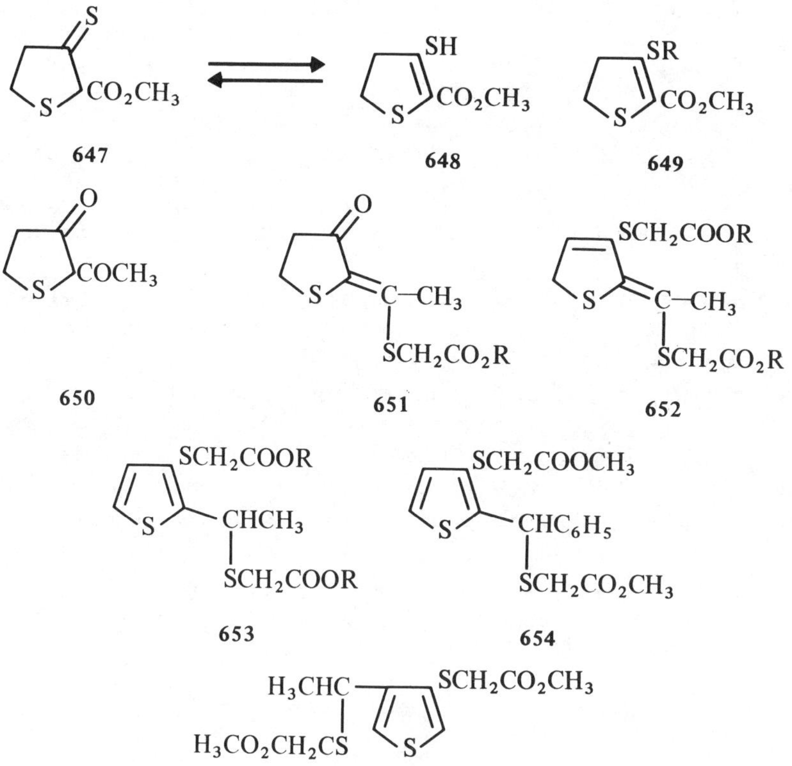

Scheme 201

of 4-carbomethoxy-3-oxotetrahydrothiophene (**656**) was reacted with sodium disulfide in acetone–methanol to give **657** in 85% yield, which, upon reaction with sulfuryl chloride, was aromatized to **658** in 90% yield.[501] Condensation of 4-carbomethoxy-3-oxotetrahydrothiophenes with aromatic aldehydes gave **659** (X = OCH$_3$), which, upon treatment with an arylamine in boiling xylene, gave the corresponding anilides (**659**, X = NHAr), usually with small amounts of the 3-hydroxythiophenes (**660**, X = NHAr), which were also readily obtained by acid- or base-catalyzed isomerization of **659** (X = NHAr). Treatment of **659** (X = OCH$_3$)

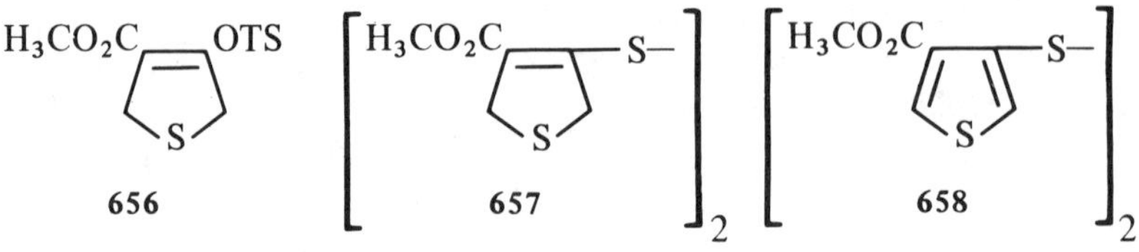

656

657₂

658₂

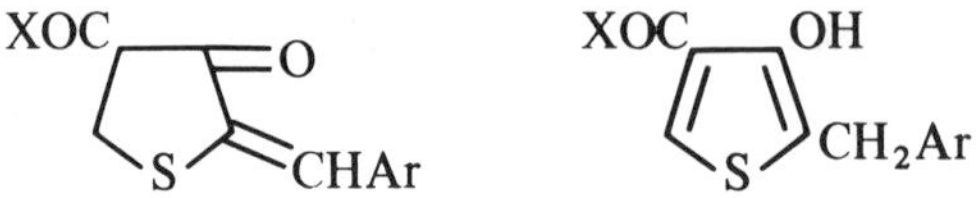

659

660

Scheme 202

with strong acid led to isomerization to the corresponding carbomethoxy-3-hydroxythiophenes.[538]

The reaction of **659** (X = OCH$_3$) with *p*-bromoaniline in picoline gave **660** (X = NHC$_6$H$_4$Br-*p*) in 51% yield. Some compounds prepared from **659** and from the *O*-methylation products of **659** are given in Table 59.

A very general method for the synthesis of 3-monosubstituted thiophenes (**663**) consists of the reaction of 3-oxotetrahydrothiophene **611** with Grignard reagents or organolithium compounds, followed by dehydratization and aromatization of the intermediate **662**.[539] Aromatization has been carried out with sulfur in DMF or with chloranil. 3-Substituted thiophenes prepared in this way are given in Table 60.

TABLE 59.   3-HYDROXYTHIOPHENES FROM 3-OXO-2-ARYLIDENE DERIVATIVES UPON TREATMENT WITH (A) HCl, (B) 10-CAMPHORSULFONIC ACID, AND (C) *p*-TOLUENESULFONIC ACID[538]

Ar	R	X	Method	Yield of Thiophene (%)
C$_6$H$_5$	H	*p*-BrC$_6$H$_4$NH	A	50
C$_6$H$_5$	H	*p*-BrC$_6$H$_4$NH	B	62
C$_6$H$_5$	H	*p*-CH$_3$C$_6$H$_4$NH	A	32
C$_6$H$_5$	H	*p*-CH$_3$C$_6$H$_4$NH	C	67
p-ClC$_6$H$_4$	H	*p*-ClC$_6$H$_4$NH	A	50
p-ClC$_6$H$_4$	H	*p*-ClC$_6$H$_4$NH	C	64
C$_6$H$_5$	H	OCH$_3$	B	78
p-ClC$_6$H$_4$	H	OCH$_3$	B	70
p-ClC$_6$H$_4$	CH$_3$	*p*-ClC$_6$H$_4$NH	B	62

TABLE 60. 3-SUBSTITUTED THIOPHENES FROM 3-OXOTETRAHYDROTHIOPHENE AND GRIGNARD REAGENTS FOLLOWED BY AROMATIZATION WITH (a) SULFUR AND (b) CHLORANIL

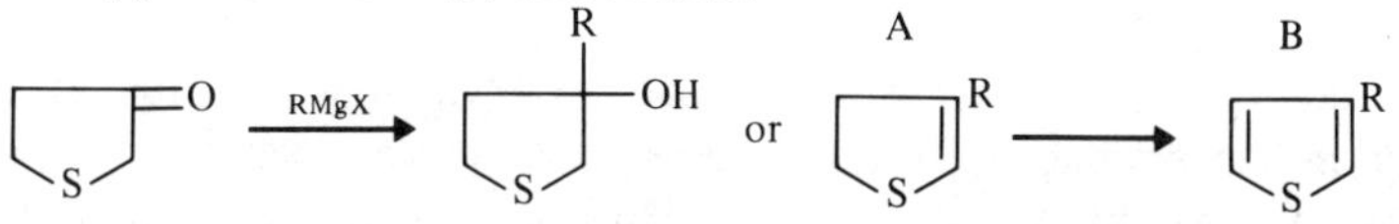

R	Yield of A (%)	Method	Yield of B (%)	Reference
2-Th	82	a	55	514
C_6H_5	86	a	88	514
C_6H_5		b	89	514
C_6H_5		a	57	514
5-C_2H_5-2-Th	72	b	70	514
2-Pyridyl	52–70	a	10[a]	539
3-Pyridyl	52–70	a	10[a]	539
4-Pyridyl	52–70	a	2[a]	539

[a] Overall yield.

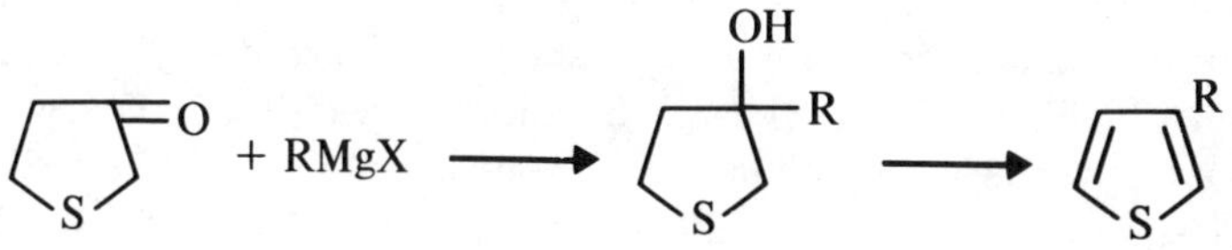

Scheme 203

However, difficulties were encountered in the dehydration of **662**, formed in the reaction of 3-ketotetrahydrothiophene with benzylic-type lithium derivatives, which were obtained by metalation of 2,4-lutidine and 2,3,4-trimethylpyridine[540] (Scheme 203).

Reduction of 4-carbomethoxy-3-ketotetrahydrothiophene to the corresponding 3-hydroxy derivative with aluminium amalgam, followed by treatment with sulfur, gave 3-thiophenecarboxylic acid after saponification.[499]

It has been found that **664**, upon heating with mineral acid, undergoes aromatization to **665**[541] (Scheme 204).

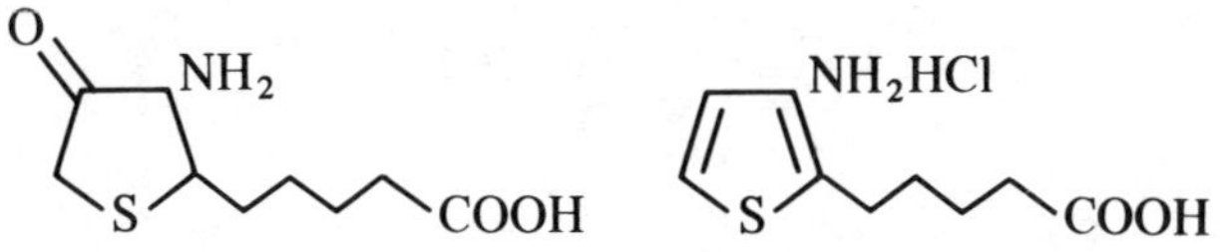

Scheme 204

4. Thiophenes from Dihydrothiophenes Obtained from Vinylphosphonium Salts

The reaction of α-mercaptoketones (666) with vinylphosphonium salts (667) in pyridine in the presence of triethylamine offers a convenient route to alkylated 2,5-dihydrothiophenes (668)[542] (Scheme 205). The use of α-mercaptoacetaldehyde

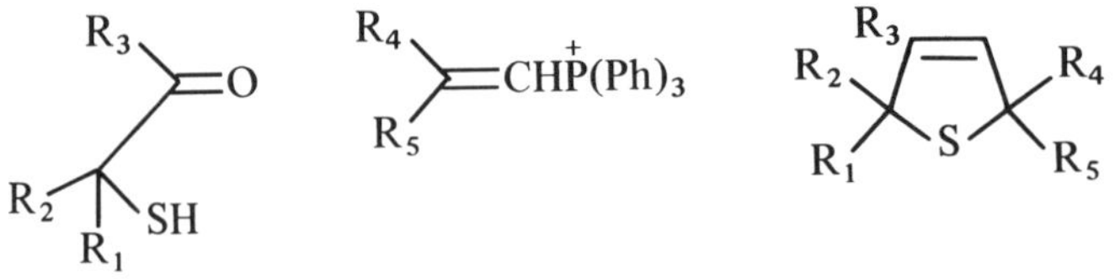

666   667   668

Scheme 205

and vinylphosphonium salts similarly gave 2,5-dihydrothiophenes not substituted in the 3,4-position.[543] It was also demonstrated that, preferentially, cis-2,5-dialkyl-2,5-dihydrothiophenes were obtained.[544]

Through the use of carbomethoxy vinylphosphonium salts (669) in the reaction with α-mercaptoketones and α-mercaptoaldehydes, 3-carbomethoxy-2,5-dihydrothiophenes (670) were obtained.[545, 546] If air was not rigorously excluded, thiophenes (671) were obtained as by-products, as in the preparation of 670 ($R_1 = R_3 = R_4 = H$, $R_2 = C_2H_5$, $R_5 = C_6H_5$) and 670 ($R_1 = R_4 = H$, $R_2 = R_3 = -(CH_2)_4-$, $R_5 = C_6H_5$)[545] (Scheme 206). The 2,5-dihydrothiophenes (668, $R_1 = R_4 = H$) could

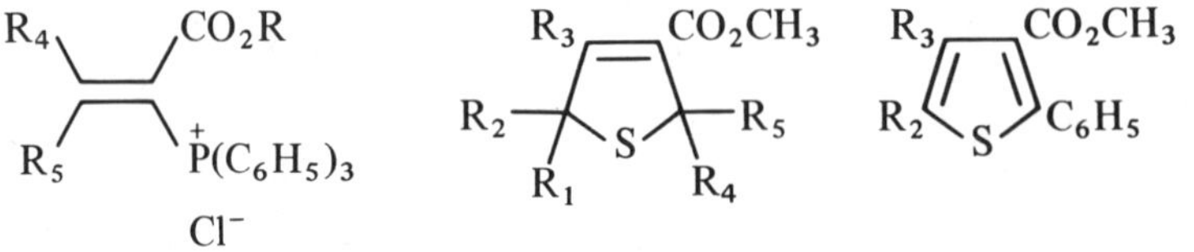

669   670   671

(a) $R_2 = C_2H_5$, $R_3 = H$
(b) $R_2 = R_3 = |CH_2|_4-$

Scheme 206

be dehydrogenated to thiophenes in high yield using chloranil in t-butanol or pyridine.[547] The results are given in Table 61.

5. Various Methods

Another route to 2,5-dihydrothiophenes consists of the reaction of Δ^3-1,3,4-thiadiazolines (672) with dimethyl acetylenedicarboxylate to give 673. Oxidation of 673 with m-chloroperbenzoic acid gave sulfoxides, which upon treatment with acetic anhydride gave 674[548, 549] (Scheme 207). The reaction of compounds of type

TABLE 61. THIOPHENES THROUGH THE REACTION OF α-MERCAPTOCARBONYL DERIVATIVES WITH VINYLPHOSPHONIUM SALTS FOLLOWED BY AROMATIZATION WITH CHLORANIL IN (A) t-BUTYLALCOHOL AND (B) PYRIDINE[547]

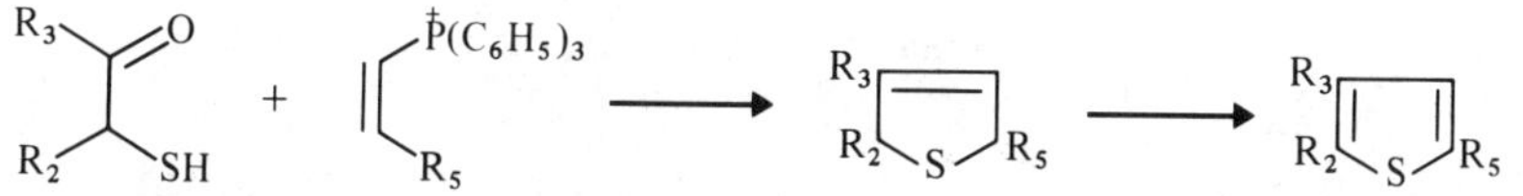

R_2	R_3	R_5	Yield from A (%)	Yield from B (%)	Yield from Vinylphosphonium Salt (%)
CH_3	CH_3	CH_3	87	83	72
C_2H_5	H	CH_3	82	80	57
CH_3	C_2H_5	CH_3	73	80	64
CH_3	C_2H_5	C_2H_5	89	82	57
$-(CH_2)_4-$		CH_3	90	86	82

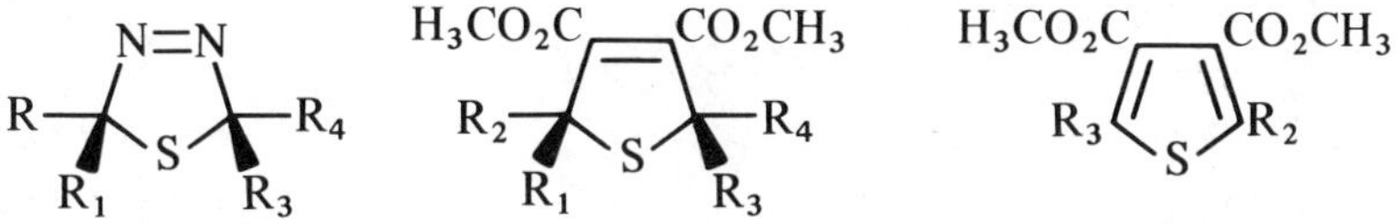

672   673   674

$R_1 = R_4 = H, R_2 = R_3 = + -C_4H_4$
$R_1 = R_4 = H, R_2 = R_3 = C_2H_5$

Scheme 207

675 with hydrogen chloride in alcohol gave 676, which could easily be aromatized[550,551] (Scheme 208). Thiirans (676a) react with malonitrile and sodium hydride to give 676b, which, after acylation were aromatized to 676c by NBS[551a] (Scheme 208).

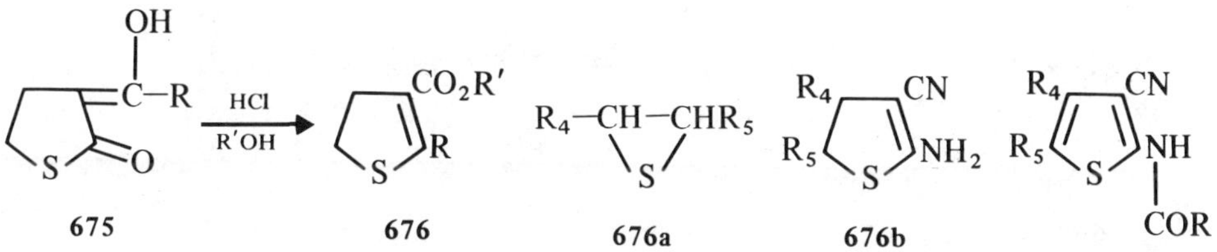

675   676   676a   676b   676c

	Yield (%)		Yield (%)		Yield (%)
$R_4 = R_5 = H$	57	$R = CH_3$	86	$R = C_6H_5$	99
$R_4 = H, \quad R_5 = CH_3$	62	$R = CH_3$	83	$R = C_6H_5$	92
$R_4 = C_6H_5, R_5 = H$	50	$R = CH_3$	86	$R = C_6H_5$	83

Scheme 208

VI. THIOPHENES FROM OTHER
HETEROCYCLIC COMPOUNDS

1. Thiophenes from Four-Membered Rings

Only a few cases of synthesis of thiophenes from four-membered rings are known. Thus, treatment of *trans*-2,4-diphenylthietane with potassium *t*-butoxide in DMF gave 2,3,5-triphenylthiophene together with several other products.[552] 3-Thietanone reacted rapidly with methanolic sodium hydrogen sulfide to give a mixture of 2,4-dimethyl-3-thiophenethiol, the corresponding disulfide, and some other compounds.[553] The reaction of **677** with sodium hydrogen sulfide in ethanol gave **678**, which upon heating at 100°C in vacuo, gave 4-methyl-2-hydroxythiophene (**679**) in 40% yield[554] (Scheme 209).

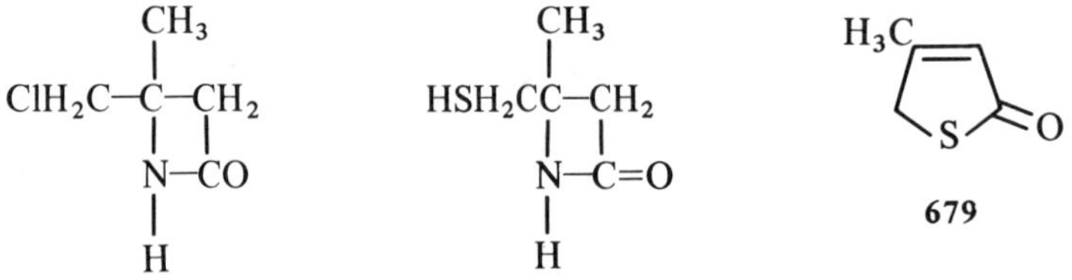

Scheme 209

2. Thiophenes from Five-Membered Rings

A. Thiophenes from Furans

Vapor-phase reactions of furans at 400–450°C with hydrogen sulfide over aluminium oxide have been utilized for the synthesis of thiophenes.[555–558] Tetrahydrofurans such as **680** have been converted to **681** by heating with P_2S_5 in toluene[559] (Scheme 210).

Most conveniently, thiophenes are obtained from furans by treatment with hydrogen sulfide in the presence of acids such as hydrogen chloride, perchloric acid,

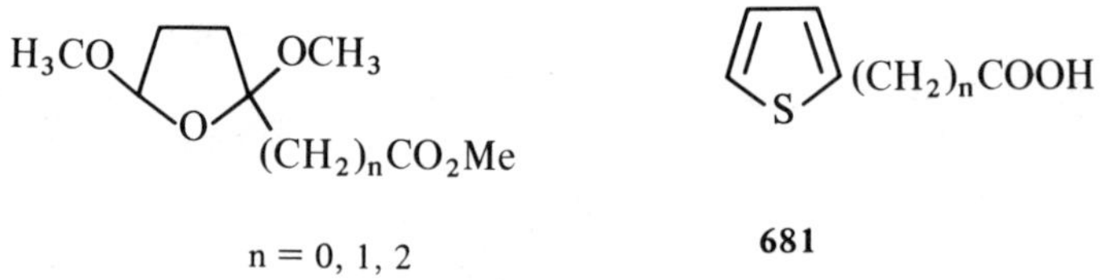

Scheme 210

TABLE 62.   THIOPHENES FROM FURANS[560, 561]

$$H_3C \overset{}{\underset{O}{\bigcap}} R \xrightarrow{H^+/H_2S} H_3C \overset{}{\underset{S}{\bigcap}} R$$

R	Yield of Thiophene
CH_3	80
C_2H_5	73
CH_2OH	50
$CH(OH)C_4H_9$	60
$(CH_2)_3OH$	63
$(CH_2)_2CH(OH)CH_3$	82
$C(CH_3)_2CH_2CH(OH)CH_3$	64
$(CH_2)_2COCH_3$	78
$C(CH_3)_2CH_2COCH_3$	74
$(CH_2)_2CO_2C_2H_5$	60

or trifluoroacetic acids, in solvents such as alcohols, ethers, or acetic acid at room temperature.[560, 560a, b] Instead of preparing 5-aryl-2-hydroxythiophenes (683b) by reacting 4-aryl-4-oxobutanoic acids with P_2S_5, it is considered better to first prepare the butenolide (683a) by treatment with acetic anhydride, followed by reaction with sodium hydrosulfide[560c] (Scheme 211). Some examples are given in Table 62. Thus, from 682, 683 was obtained.[203]

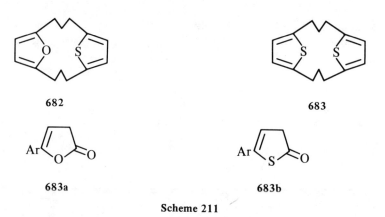

682

683

683a

683b

Scheme 211

B.   Thiophenes from 1,3-Oxathiolium Derivatives

2-Dialkylamino-1,3-oxathiolium salts (684, 685) have been prepared through the reaction of secondary amines with carbonyl sulfide, followed by alkylation with phenacyl bromide and ring-closure with concentrated sulfuric acid.[561, 562] 2-Aryl-1,3-oxathiolium salts can be prepared from N,N- disubstituted thioamides and phenacyl bromides,[563] or by the reaction of potassium thiolbenzoate and phenacyl bromide followed by cyclization with concentrated H_2SO_4[564] (Scheme 212). The reaction of **684** or the corresponding fluoroborates with the sodium salts of active

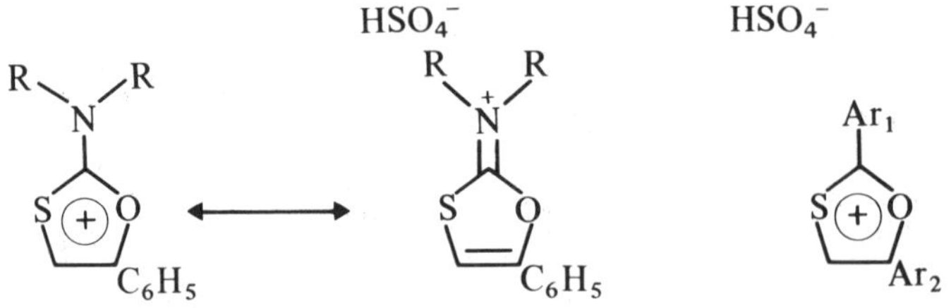

684   **685**   **686**

Scheme 212

methylene compounds (XCH_2Y) gives a mixture of products, such as **687–690**, under different conditions (Scheme 213). However, as is obvious from Tables 63 and

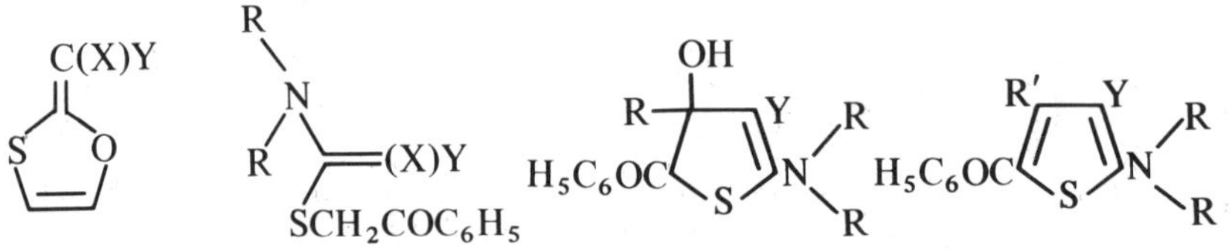

687   **688**   **689**   **690**

Scheme 213

64, conditions have been found that give the thiophenes (**690**) in high yields.[565-567] The mechanism indicated in Scheme 214 has been suggested: attack by the

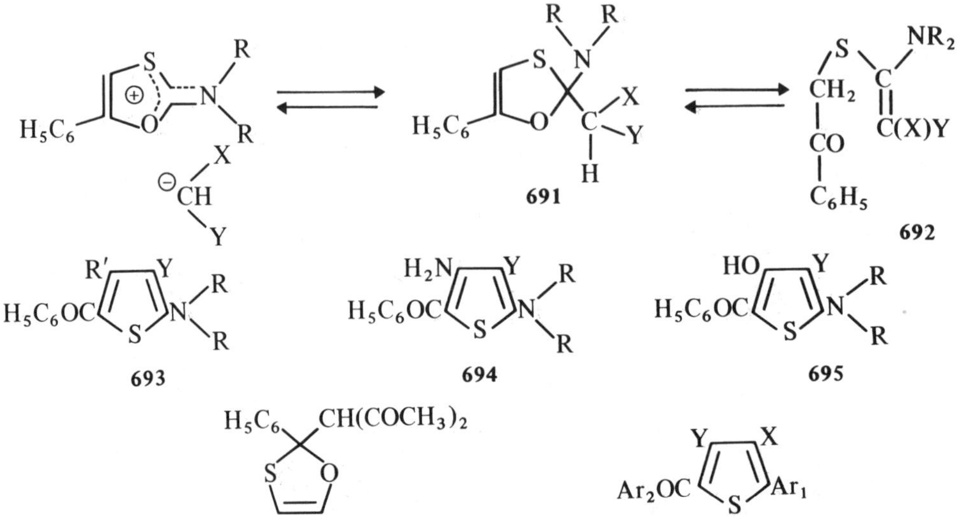

691   **692**

693   **694**   **695**

696   **697**

Scheme 214

TABLE 63. THIOPHENES FROM THE REACTION OF 5-PHENYL-N-(1,3-OXATHIOL-2-YLIDENE) TERNARY IMMONIUM SALTS WITH SALTS OF ACTIVE METHYLENE COMPOUNDS: (A) HYDROSULFATES IN METHYLENE CHLORIDE IN THE PRESENCE OF TRIETHYL-AMINE AND (B) FLUOROBORATES WITH THE SODIUM SALTS IN THF[565, 566]

Method	R_2N	X	Y	R_4	Yield of Thiophene (%)
A	$(CH_2)_5N$	CH_3CO	CH_3CO	CH_3	41
A	$(CH_2)_5N$	CH_3CO	C_6H_5CO	CH_3	33
A	$(CH_2)_5N$	CN	CN	NH_2	86
A	$(CH_3)_2N$	CN	CN	NH_2	94
A	$(CH_3)_2N$	CN	$CONH_2$	NH_2	17[a]
B	$(CH_2)_5N$	CH_3CO	CH_3CO	CH_3	73
B	$(CH_3)_2N$	CH_3CO	CH_3CO	CH_3	74
B	$(CH_2)_5N$	CH_3CO	C_6H_5CO	CH_3	82
B	$(CH_3)_2N$	CH_3CO	C_6H_5CO	CH_3	45
B	$(CH_2)_2O(CH_2)_2N$	CH_3CO	C_6H_5CO	CH_3	85
B	$(CH_2)_5N$	C_6H_5CO	C_6H_5	CH_3	41
B	$(CH_2)_2O(CH_2)_2N$	C_6H_5CO	C_6H_5	C_6H_5	67
B	$(CH_2)_5N$	CH_3CO	$CO_2C_2H_5$	CH_3	68
B	$(CH_3)_2N$	CH_3CO	$CO_2C_2H_5$	CH_3	74
B	$(CH_2)_5N$	C_6H_5CO	$CO_2C_2H_5$	C_6H_5	60
B	$(CH_2)_5N$	$CO_2C_2H_5$	CN	OH	50[a]
B	$(CH_3)_2N$	C_6H_5CO	$CO_2C_2H_5$	C_6H_5	65
B	$(CH_2)_5N$	C_6H_5CO	CN	C_6H_5	69

[a] Together with 25% of the corresponding 688.

TABLE 64.   THIOPHENES FROM 2-ARYL-SUBSTITUTED 1,3-OXATHIOLIUM SALTS AND SALTS OF ACTIVE METHYLENE DERIVATIVES

Ar	Ar'	X	Y	R_4	Yield of Thiophene (%)	Reference
C_6H_5	C_6H_5	CN	CN	NH_2	90	568
C_6H_5	p-BrC_6H_4	CN	CN	NH_2	100	568
p-$CH_3OC_6H_4$	C_6H_5	CN	CN	NH_2	75	568
p-$CH_3OC_6H_4$	p-ClC_6H_4	CN	CN	NH_2	100	568
p-$(CH_3)_2NC_6H_4$	C_6H_5	CN	CN	NH_2	90	568
C_6H_5	C_6H_5	CO_2CH_3	CN	OH	65	568
C_6H_5	p-BrC_6H_4	CO_2CH_3	CN	OH	72	568
p-$CH_3OC_6H_4$	C_6H_5	CO_2CH_3	CN	OH	45	568
p-$CH_3OC_6H_4$	p-ClC_6H_4	CO_2CH_3	CN	OH	90	568
p-$(CH_3)_2NC_6H_4$	C_6H_5	CO_2CH_3	CN	OH	95	568
C_6H_5	C_6H_5	CO_2CH_3	CO_2CH_3	OH	42	568
p-$CH_3OC_6H_4$	p-ClC_6H_4	CO_2CH_3	CO_2CH_3	OH	20	568
C_6H_5	C_6H_5	C_6H_5CO	C_6H_5	C_6H_5	40	564
C_6H_5	C_6H_5	$COCH_3$	$COCH_3$	CH_3	91	564
C_6H_5	C_6H_5	COC_6H_5	$COCH_3$	C_6H_5	20	564

[a] Together with 17% of 697.

carbanion on the 2-position gives **691**, which is reversibly ring-opened to **692**. Nucleophilic attack of the carbonyl oxygen of **692** on the ketene S,N-acetal carbon gives **687**. However, the main reaction path of **692** is ring-closure to thiophenes, according to the CSC$_3$-principle. When X = COR' and Y = COR', CO$_2$R', or CN, **693** is obtained. When X = Y = CN, the main product is **694**, and when X is CO$_2$R' and Y is CN or CONH$_2$, it is **695**. The corresponding reaction of **686** with carbanions ·from active methylene derivatives leads by a similar reaction path via **696** to compounds of type **697**.[564, 568] Reacting **698** with the carbanions from active methylene compounds gave compounds of type **699** in 50–80% yield[569] (Scheme 215).

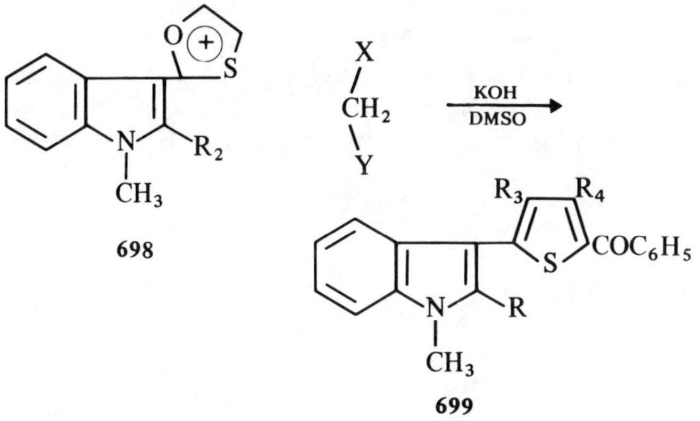

R_2 = Me	X = Y = CN	R_3 = CN,	R_4 = NH$_2$
R_2 = Ph	X = Y = CN	R_3 = CN,	R_4 = NH$_2$
R_2 = Ph	X = COCH$_3$, Y = CO$_2$C$_2$H$_5$	R_3 = CO$_2$C$_2$H$_5$,	R_4 = CH$_3$
R_2 = Ph	X = Y = COCH$_3$	R_3 = COCH$_3$,	R_4 = CH$_3$

Scheme 215

Reaction of thiobenzoic acid (**700**) with α-bromophenylacetyl chloride (**701**) in the presence of triethylamine gave the unstable **702**, which without isolation was reacted with dimethyl acetylenedicarboxylate to give **704** via **703** in low yield[571] (Scheme 216). This type of thiophene synthesis is of preparative interest, when

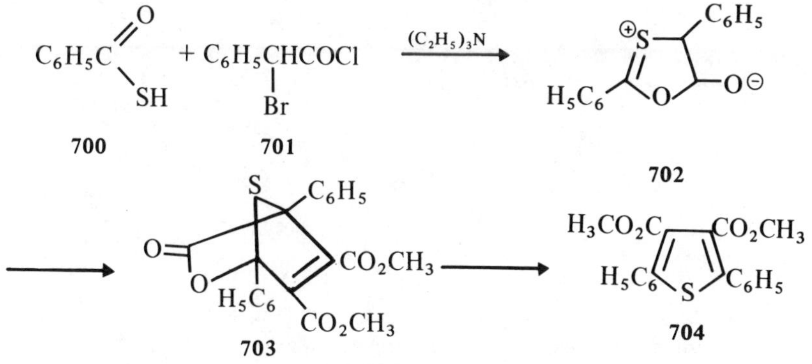

Scheme 216

other sydnones, like thiazolium and dithiolium compounds analogous to **702**, are used (see the following sections).

C.   Thiophenes from Thiazolium Derivatives

The anhydro 4-hydroxythiazolium hydroxide system (**708**) can be prepared by the *S*-alkylation of *N*-monosubstituted thioamides (**706**) with an α-halocarboxylic acid, followed by cyclodehydration with acetic anhydride–triethylamine[570] or, in some cases, when thioureas and carbamates are used, through the reaction of α-bromoacyl chlorides (**705**) with *N*-substituted thioamides (**706**) in the presence of triethylamine.[571] In this case, the reaction probably proceeds via **707**, although a ketene intermediate cannot be excluded. Reaction of **708** with acetylenic dipolarophiles, such as dimethyl acetylenedicarboxylate or other acetylenic

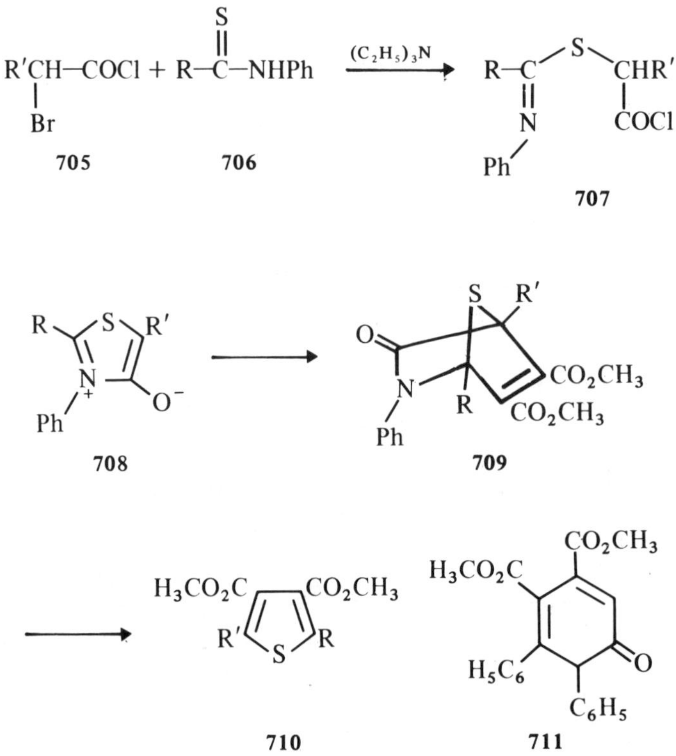

$$R'CH-COCl + R-\overset{\overset{\displaystyle S}{\|}}{C}-NHPh \xrightarrow{(C_2H_5)_3N} R-\overset{\overset{\displaystyle S}{C}}{\underset{\underset{\displaystyle Ph}{N}}{\|}}\!\!\diagdown\!\!\overset{\displaystyle S}{\underset{\displaystyle COCl}{CHR'}}$$

$$\underset{\displaystyle Br}{|}$$

          705   **706**   **707**

          708   **709**

          710   **711**

Scheme 217

derivatives, gives thiophenes (**710**) via the cycloaddition product (**709**). When R' = H, **709** yields **711** instead, through the elimination of sulfur[572] (Scheme 217). Some thiophenes prepared by this route are given in Table 65.

TABLE 65.   PREPARATION OF THIOPHENES THROUGH THE REACTION OF SYDNONE-LIKE COMPOUNDS WITH ACETYLENES

$$\underset{-O}{\overset{R_2}{\diagdown}}\!\!\underset{X}{\overset{S^+}{\diagup}}\!\!R_5 \;+\; R_3C\!\equiv\!C\!-\!R_4 \;\longrightarrow\; \underset{R_5}{\overset{R_4}{\diagdown}}\!\!\underset{S}{\diagup}\!\!\underset{R_2}{\overset{R_3}{\diagup}}$$

X	R_5	R_2	R_3	R_4	Yield of Thiophene (%)	Reference
N-C_6H_5	CH_3S	C_6H_5	CO_2CH_3	CO_2CH_3	67	571
N-C_6H_5	C_2H_5S	C_6H_5	CO_2CH_3	CO_2CH_3	63	571
N-C_6H_5	C_3H_7S	C_6H_5	CO_2CH_3	CO_2CH_3	46	571
N-C_6H_5	$(CH_3)_2N$	C_6H_5	CO_2CH_3	CO_2CH_3	69	571
N-C_6H_5	$(CH_3)_2N$	$CO_2C_2H_5$	CO_2CH_3	CO_2CH_3	16	571
S	p-$CH_3OC_6H_4$	C_6H_5	CO_2CH_3	CO_2CH_3	40	573, 574
S	p-$CH_3OC_6H_4$	H	CO_2CH_3	CO_2CH_3	47	573, 574
S	p-ClC_6H_4	H	CO_2CH_3	CO_2CH_3	44	574
S	C_6H_5	H	COC_6H_5	COC_6H_5	68	574
S	C_6H_5	C_6H_5	C_6H_5	H	91	575
S	C_6H_5	C_6H_5	C_6H_5CO	C_6H_5	78	575, 576
S	C_6H_5	C_6H_5	CO_2CH_3	C_6H_5	82	574, 575
S	C_6H_5	C_6H_5	CO_2CH_3	H	83	575, 576
S	p-$CH_3OC_6H_4$	C_6H_5	CO_2CH_3	CO_2CH_3	99	575, 576
S	p-$CH_3C_6H_4$	C_6H_5	CO_2CH_3	CO_2CH_3	90	576
S	C_6H_5	C_6H_5	CO_2CH_3	CO_2CH_3	65	575, 576
S	p-BrC_6H_4	CH_3	CO_2CH_3	CO_2CH_3	67	575, 576
S	p-$CH_3OC_6H_4$	C_6H_5	CO_2CH_3	CO_2CH_3	75	575
S	p-$(CH_3)_2NC_6H_4$	C_6H_5	CO_2CH_3	CO_2CH_3	90	575
S	C_6H_5	C_6H_5	CO_2CH_3	CO_2CH_3	88	575
S	p-$NO_2C_6H_4$	C_6H_5	CO_2CH_3	CO_2CH_3	70	575
S	C_6H_5	C_6H_5	$COCH_3$	C_6H_5	99	575

177

Compound **708** (R = R′ = Ph) has also been reacted with **712**. Both possible adducts, **713** and **714**, were formed in 48 and 33%, respectively. Upon pyrolysis of **713** and **714**, 2,5-diphenylthiophene and dimethyl-2,5-diphenyl-3,4-thiophene-dicarboxylate were formed in 73 and 84% yield, respectively, together with dimethyl-3,4-furandicarboxylate and furan[577] (Scheme 218). The reaction of the

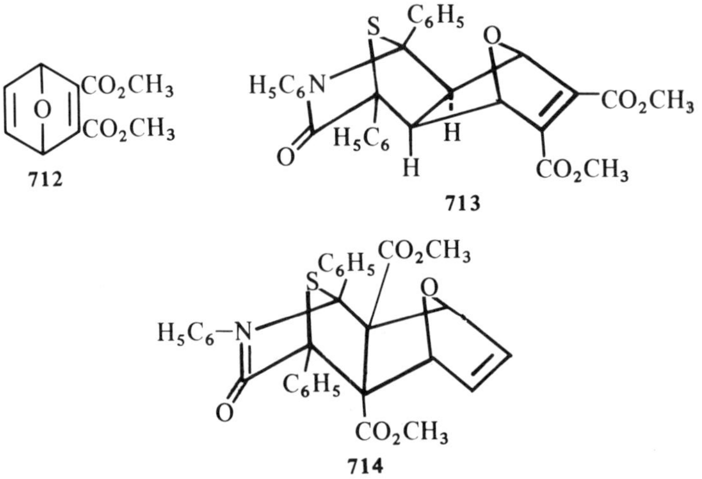

Scheme 218

N-arylsydnone imine (**715**) with dimethyl acetylenedicarboxylate gave **716**.[578] Thiazolium-5-thiolates (**717**) are easily available through the reaction of *N*-methyl-*N*-benzoyl phenylglycine with acetic anhydride and carbon disulfide. However, this compound shows no 1,3-dipolar activity, and the dimethylacetylene dicarboxylate, tetramethylthiophene tetracarboxylate is formed by an unknown route[579] (Scheme 219).

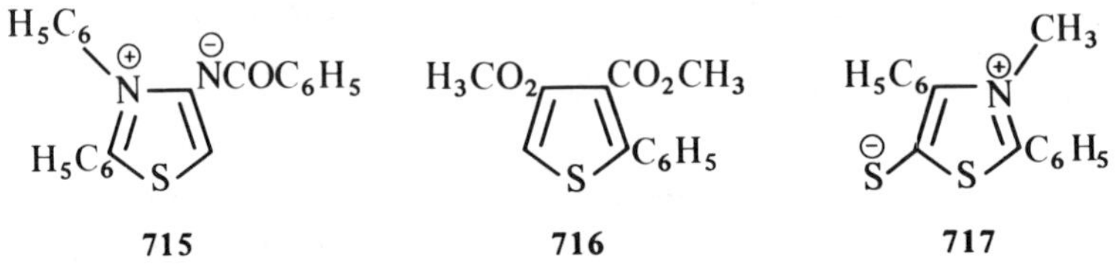

Scheme 219

α-Cyanothioamides (**718**) react with chloroacetonitrile to give 4-amino-2-cyano-methylene-4-thiazolines (**719**), which upon treatment with sodium ethylate yield 2,4-diaminothiophenes (**720**).[580] Treating **721** in the same way gave **722**,[580] and from **723**, **724** was obtained in 60–70% yield.[581] The reaction of the 2-thiazoline-5-thiones (**724a**) with tolane at 300°C gave tetraphenylthiophene as the main product; dimethyl acetylenedicarboxylate and methyl acetylenedicarboxylate gave 1,4-dithiafulvenes[581a] (Scheme 220).

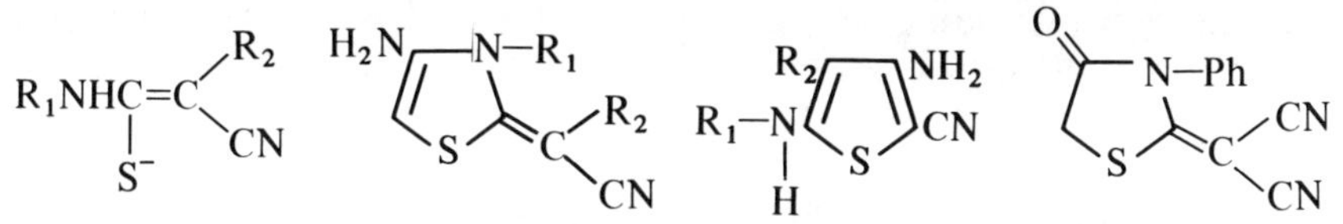

718   719   720   721

$R_1 = C_6H_5$ $R_2 = CO_2C_2H_5$
$R_1 = C_6H_5$ $R_2 = CN$
$R_1 = -CH_2-CH=CH$ $R_2 = CO_2C_2H_5$

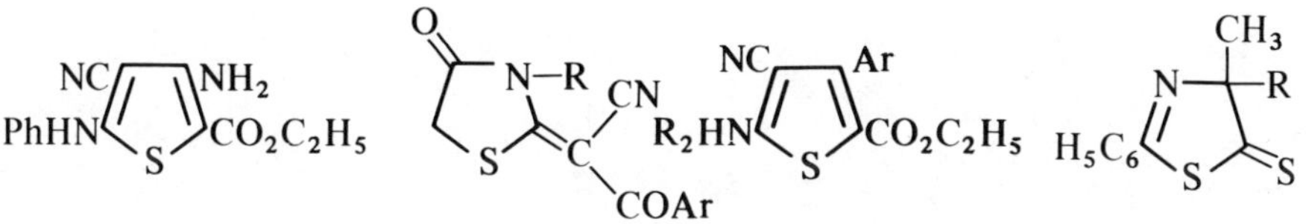

722   723   724   724a

$Ar = p\text{-}ClC_6H_4$ $R = C_6H_5$ $R = CH_3, C_6H_5$
$Ar = C_6H_5$ $R = p\text{-}ClC_6H_4$
$Ar = C_6H_5$ $R = p\text{-}CH_3C_6H_4$

Scheme 220

D. Thiophenes from 1,3-Dithiole Derivatives

The reaction of thiobenzoyl thioglycolic acids (725) with acetic anhydride leads to the anhydro-2-aryl-1,3-dithiolium 4-hydroxy system (726).[571,573-576,582] Sydnones of type 726 undergo 1,3-dipolar addition with acetylenes to give 727, which thermally eliminate carbonyl sulfide to give thiophenes (728) (Scheme 221). Tolane, however, did not react[575] (see Table 65).

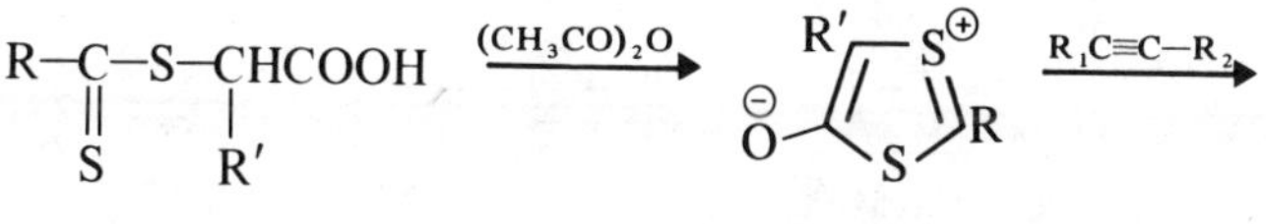

725   726

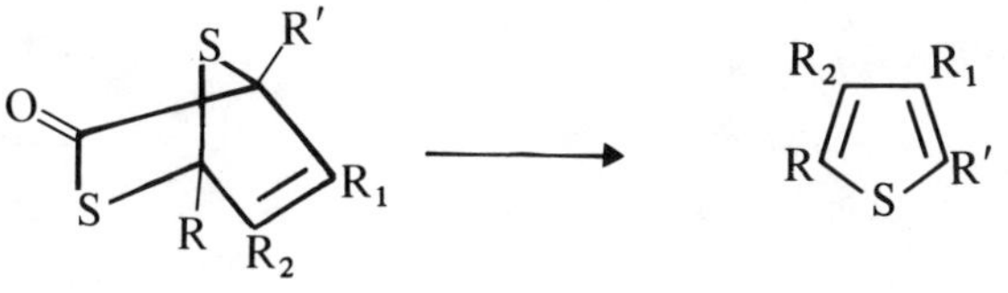

727   728

Scheme 221

Intramolecular cycloaddition has been carried out. Thus, heating **729** in xylene at 100°C for 15 min led directly to **731**, without isolation of the intermediate cycloadduct (**730**)[583] (Scheme 222).

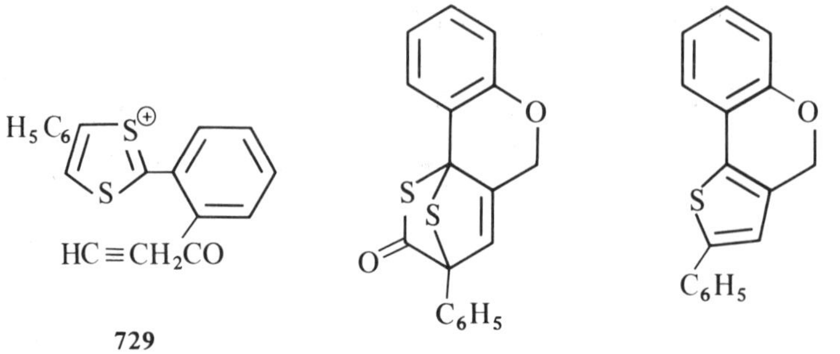

729 **730** **731**

Scheme 222

In an attempt to carry out cyclodehydration of **725** to **726** with dicyclohexyl carbodiimide or trifluoroacetic anhydride, dimers of **726** with the structure **732** were obtained. However, these dimers are in equilibrium with the monomers and give thiophenes upon reaction with dimethyl acetylenedicarboxylate.[592] The reaction of **712** with **726** (R = R′ = C_6H_5) occurred specifically across the less-substituted double bond to give **733** in 82% yield. Upon pyrolysis in refluxing xylene, 2,5-diphenylthiophene (92%) and dimethyl furan-3,4-dicarboxylate (91%) was obtained[577] (Scheme 223). Photolysis of **726** (R = R′ = Ph) leads via a dimer

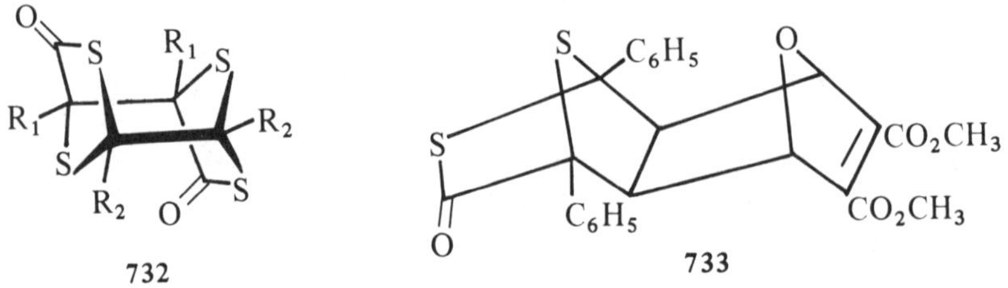

732 **733**

Scheme 223

to tetraphenyl-1,4-dithiin, tetraphenylthiophene, and some other products.[577a]

The reaction of **734** with **735** gave **736** in 20% yield (Scheme 224). 1-Pyrrolidine cyclopentene reacts differently, giving 1,3-dithionine-2-thiones.[584]

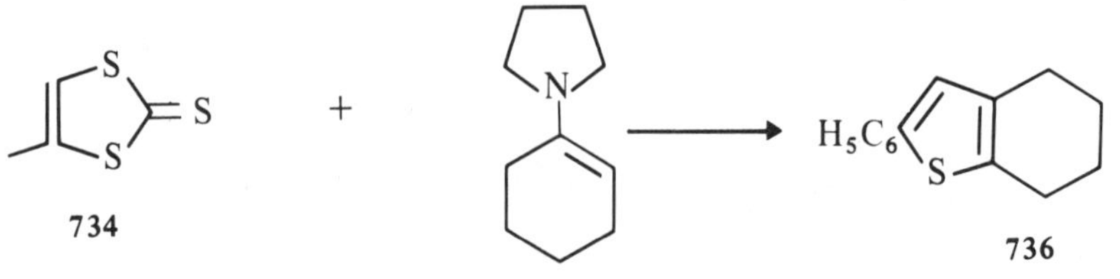

734 **735** **736**

Scheme 224

E. Thiophenes from Isothiazolium Salts

The reaction of isothiazolium salts (737), obtained by alkylation of isothiazoles, with the sodium salt of benzoylacetate gave 2-benzoylthiophenes (739). It is assumed that the reaction proceeds by nucleophilic attack of the phenacyl ion (actual or potential) on the ring-sulfur to give 738, followed by attack of the active methylene group and elimination of amine. The only exception is 737 ($R_5 = SCH_3$), in which case methanthiolate is preferentially lost and 740 is formed. The compounds that have been prepared are given in Table 66.

When 737 ($R_3 = R_5 = H$, $R = R_4 = C_6H_5$) and 737 ($R = R_4 = C_5H_5$, $R_5 = SMe$, $R_3 = H$) reacted with dimethylmethylene sulfurane, methylthiothiophenes 743 ($R_5 = H$) and 743 ($R_5 = SCH_3$) were obtained via 741 and 742 followed by oxidation. 2-Nitrobenzylidene sulfurane reacts somewhat differently with 737 ($R = R_4 = C_6H_5$, $R_3 = R_5 = H$), giving 2-p-nitrophenyl-4-phenylthiophene. In this case, the intermediate (744) loses dimethylsulfide[586] (Scheme 225).

The reaction of 745 with 746 gave 747[586] (Scheme 226).

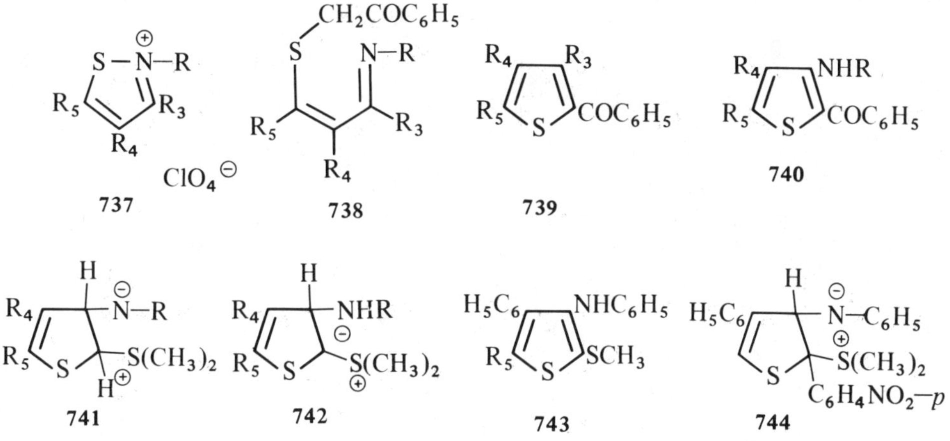

Scheme 225

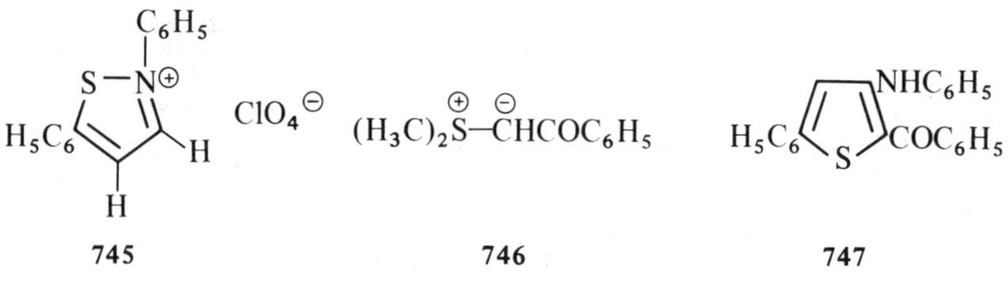

Scheme 226

TABLE 66.   THIOPHENES THROUGH THE REACTION OF ISOTHIAZOLIUM SALTS
WITH SODIUM BENZOYL ACETATE[585]

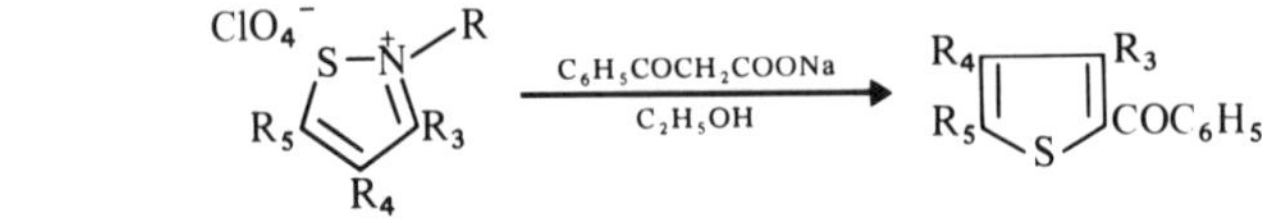

$$ \begin{array}{c} ClO_4^- \\ \underset{R_5}{S}-\overset{+}{\underset{R_3}{N}}-R \\ R_4 \end{array} \xrightarrow[C_2H_5OH]{C_6H_5COCH_2COONa} \underset{R_5}{\overset{R_4}{\underset{S}{|}}}\overset{R_3}{\underset{COC_6H_5}{}} $$

R	R_3	R_4	R_5	Yield of Thiophene (%)
CH_3	H	H	H	71
CH_3	H	H	C_6H_5	68
C_6H_5	H	H	C_6H_5	63
CH_3	H	C_6H_5	C_6H_5	68
C_6H_5	C_6H_5	H	C_6H_5	42
CH_3[a]	SCH_3	H	C_6H_5	38[b]
C_6H_5	H	C_6H_5	SCH_3	77[b]
CH_3	C_6H_5	H	SCH_3	74
C_6H_5	H	C_6H_5	H	40

[a] The iodide was used in this case.
[b] The product in this case was 5-phenyl-3-methylamino 2-benzoylthiophene.

F.   Thiophenes from 1,2-Dithiolium Derivatives

3-Amino-1,2-dithiolium salts (**748**), prepared from 3-chloro-1,2-dithiolium salts and secondary amines, react with nitromethane and sodium methoxide to give 2-nitro-3-aminothiophenes (**750**).[587] The reaction most probably proceeds by nucleophilic attack of the carbanion from nitromethane on the sulfur in position 1 to give **749**, followed by ring-closure on the thioamide group and elimination of H_2S (Scheme 227). The compounds prepared are listed in Table 67.

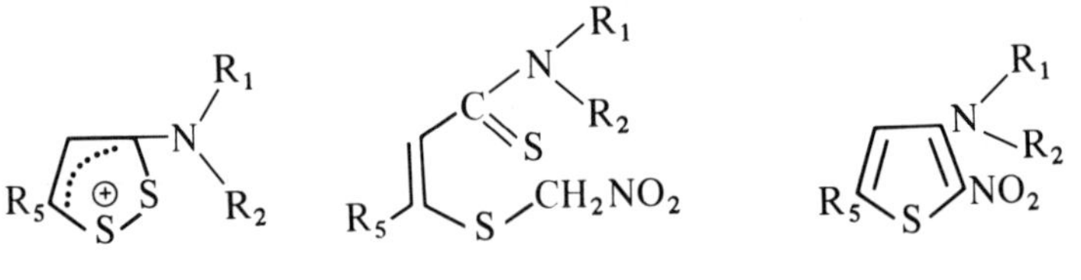

| 748 | 749 | 750 |

Scheme 227

The reaction of **751** with boiling potassium hydroxide in ethanol gives a dimorpholinodicyanothiophene in 71% yield, the structure of which was not determined.[588] The reaction of **752** (Ar = C_6H_5) with copper bronze gave a 50% yield of 2,5-diphenylthiophene together with 40% of a compound claimed to be

TABLE 67.   THIOPHENES FROM THE REACTION OF 3-AMINO-1,2-DITHIOLIUM SALTS WITH NITROMETHANE[587]

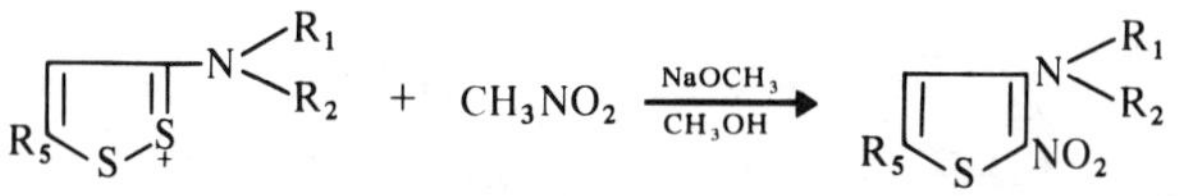

$$R_5 \overbrace{}_{S\,-\,\overset{+}{S}} N \overset{R_1}{\underset{R_2}{\big<}} + CH_3NO_2 \xrightarrow[CH_3OH]{NaOCH_3} R_5 \overbrace{}_{S} \overset{N \overset{R_1}{\underset{R_2}{\big<}}}{\underset{NO_2}{}}$$

R_1	R_2	R_5	Yield of Thiophene (%)
C_6H_5	CH_3	C_6H_5	52
C_6H_5	C_2H_5	C_6H_5	55
C_6H_5	$CH_2C_6H_5$	C_6H_5	57
C_6H_5	$CH_2CO_2C_2H_5$	C_6H_5	48
C_6H_5	C_6H_5	C_6H_5	43
$p\text{-}CH_3C_6H_4$	C_2H_5	C_6H_5	52
$p\text{-}CH_3OC_6H_4$	CH_3	C_6H_5	47
CH_3CH_2O	CH_2CH_3	C_6H_5	55

2,7-diphenylthiepin. From Ar $= p\text{-}CH_3OC_6H_4$, only 2,5-di-(p-anisyl)thiophene was obtained in 28% yield.[589] The reaction of 753 with alcoholic potassium hydroxide followed by acidification leads to the unstable (mercaptovinyl) thiophenethiol (754, R = H), which upon heating in pyridine ring-closes back to 753. These compounds can be methylated to give 754 (R = CH_3) with dimethyl sulfate and benzylated with benzyl chloride to give 754 (R = $C_6H_5CH_2$). The compound (754, R = CH_3) can also be prepared directly from 753 by reaction with sodium hydride and methyl iodide in DMSO[590, 591] (Scheme 227a). The compounds thus prepared are given in Table 68.

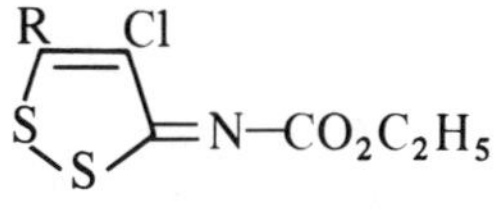

751

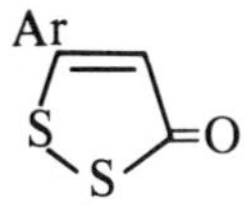

752

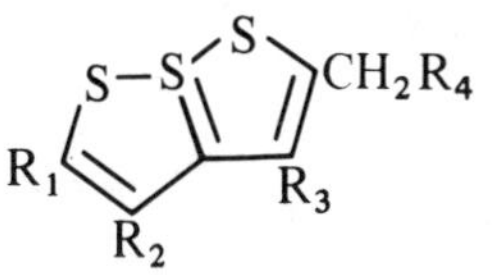

753

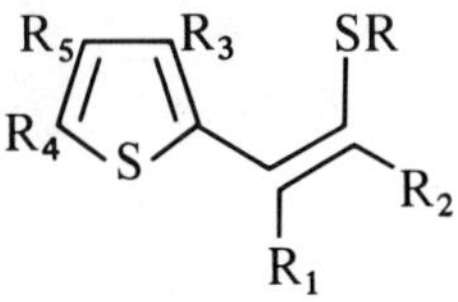

754

Scheme 227a

TABLE 68.   THIOPHENES BY BASE TREATMENT OF TRITHIA-1,6,6a,λ^4-PENTALENES[591]

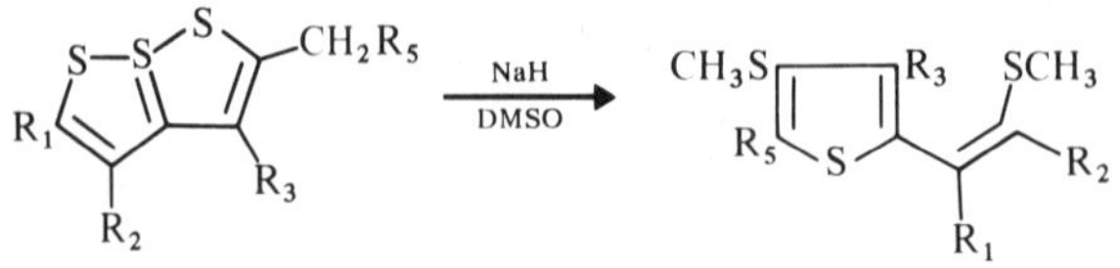

R_1	R_2	R_3	R_5	Yield of Thiophene (%)
H_3C	H	H	H	40
C_6H_5	H	H	H	65
p-$CH_3C_6H_4$	H	H	H	70
p-$CH_3OC_6H_4$	H	H	H	77
CH_3	—$CH_3CH_2CH_2$—		H	15
C_6H_5	H	H	CH_3	26
p-$CH_3OC_6H_4$	H	H	CH_3	10
C_2H_5	—CH_2—CH_2—CH_2—		CH_3	20
p-$CH_3OC_6H_4$	H	H	p-$CH_3OC_6H_4$	15

G.   Thiophenes from Thiadiazoles

Treatment of mesoionic 1,3.4-thiadiazoles (755) with dimethyl acetylene dicarboxylate gives 756 and S-cyanothioimidate derivatives (757). The reaction is assumed to proceed via 758 and 759[592] (Scheme 227b). It is possible that 756,

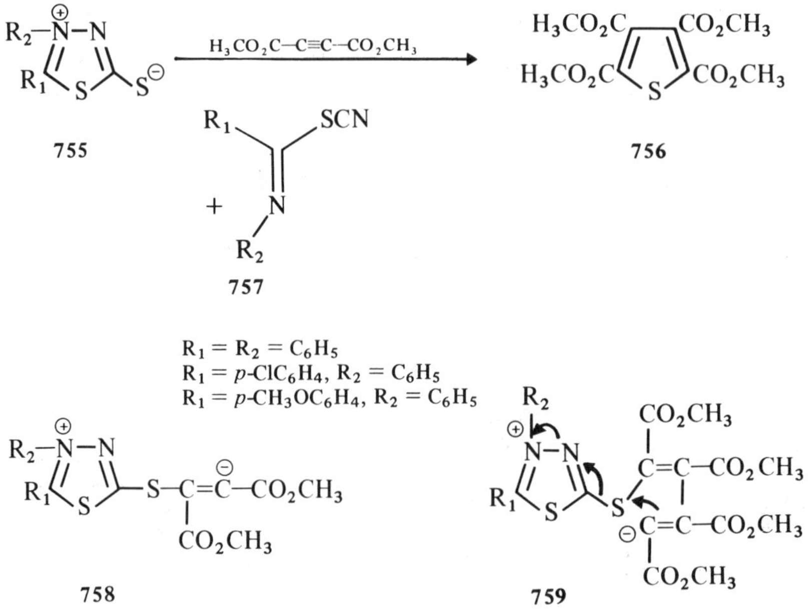

$R_1 = R_2 = C_6H_5$
$R_1 = p$-ClC_6H_4, $R_2 = C_6H_5$
$R_1 = p$-$CH_3OC_6H_4$, $R_2 = C_6H_5$

755

757

756

758

759

Scheme 227b

obtained from **717** in the reaction (which is discussed in regard to Scheme 219), is formed by a similar mechanism.

The photolysis of 1,2,3-thiadiazoles (**760**) has been investigated by several groups. Mixtures of 1,4-dithiines (**762**), *cis-trans* isomeric 1,3-dithiol derivatives (**763**), and symmetrical (**764**) and unsymmetrical thiophenes (**765**) are obtained, depending upon the nature of the substituents. The reaction is assumed to proceed via a 1,3-diradical (**761**)[593] (Scheme 227c). With aromatic substituted 1,2,3-thiadiazoles,

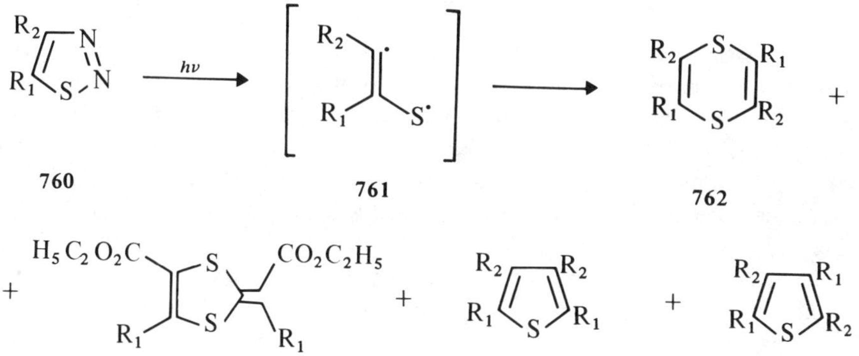

760   **761**   **762**

763   **764**   **765**

Scheme 227c

1,4-dithiafulvenes are also obtained,[594] whereas cycloalkeno-1,2.3-thiadiazoles also yield 1,2.4,5-tetrathianes, in addition to the previously mentioned classes of compounds.[595] Investigations of the gas-phase photolyses of 1,2,3-thiadiazole, and its 4- and 5-methyl derivatives, indicate that the reaction proceeds via thiirene intermediates, since trapping with hexafluoro-2-butyne gave 2,3-bis(trifluoromethyl)-thiophene with the parent compound and 5-methyl-2,3-bis(trifluoromethyl)-thiophene with *both* methyl 1,2,3-thiadiazoles.[596]

3.   Thiophenes from Six-Membered Heterocycles

A.   Thiophenes from Rings with One Heteroatom

The reaction of thiacyclohexane over acidic catalysts,[597] with triphenylcarbinol and trifluoroacetic acid[598] or nitromethane,[599] leads to 2-methylthiophene. Thiopyrylium iodide gives 2-thiophenecarboxaldehyde in 71% yield upon refluxing with MnO_2 in chloroform.[600]

Pyrolysis of 2H-thiopyrane derivatives (**765**) at 240–260°C leads to a mixture of thiophenes (**766** and **767**) together with other products. Compounds (**765**) are obtained by the reaction of aromatic thioketones and 1,3-butadiene. For the most part, thiofluorenone was used[601-603] (Scheme 228). The mechanism of this reaction is not completely understood.

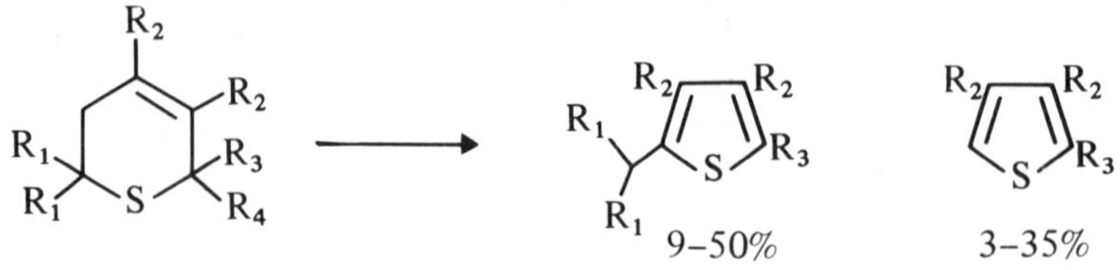

Scheme 228

The reaction of 2H-thiopyrane-2-ones (**768**, X = hydrogen or bromine) with piperidine or morpholine at about 70°C gives thiophene carboxamides (**770**) in 60–80% yield. The reaction probably occurs via the intermediate (**769**) and dehydrogenation (probably by oxygen) or dehydrobromination.[604] Similarly, the reaction of 3-bromo-4,5,6-triphenyl-2H-thiopyran-2-one with alcoholic potassium hydroxide gave 3,4,5-triphenyl-2-thiophenecarboxylic acid in 60% yield.[604] Triphenyl pyrylium fluoroborate (**771**) reacts with sodium disulfide to give an intermediate (**772**) similar to **769**, which upon oxidation with iodine or air gives **773** in 56 and 44% yield, respectively.[605] The reaction of cyclopenta[b]thiopyrans (**773a**)

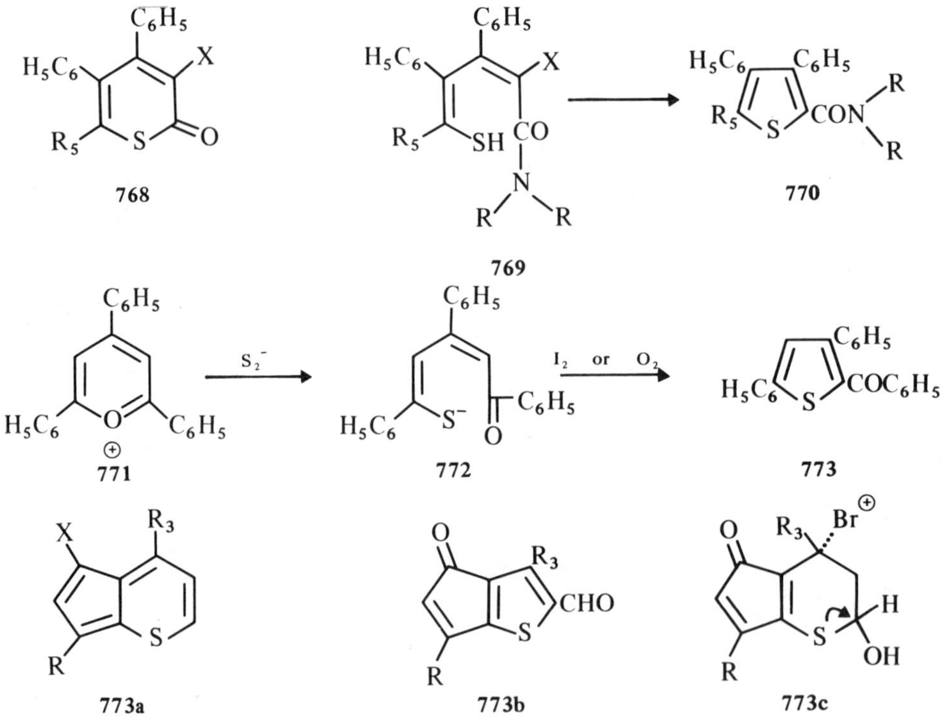

X = S-Bu-t, R = R$_3$ = t-Bu   58%
X = OCH$_3$, R = R$_3$ = t-Bu,   31%
X = S-Bu-t, R = t-Bu, R$_3$ = CH$_3$ 20%

Scheme 229

with 2 eq. of bromine led to 2-formylcyclopenta[b]thiophene-4-ones (773b), except with 773a (X = OCH$_3$, R = CH$_3$, R$_2$ = t-Bu), which gave the 5-bromo derivative of 773b (R = CH$_3$, R$_3$ = t-Bu). A ring-contraction mechanism via 773c as the key intermediate was suggested[605a] (Scheme 229).

The reaction of 774 with dimethyl oxosulfonium ylide gave a complex mixture of products, including 775 and 776[606] (Scheme 230).

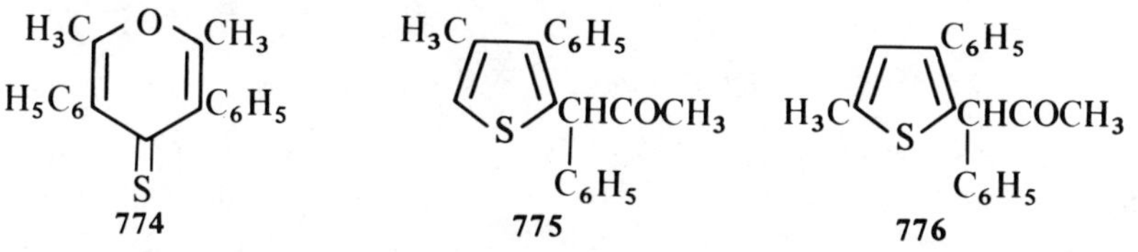

Scheme 230

Cycloaddition between phenyl azide and 777 to give 778 is followed by isomerization to 779. On heating of 779, a carbenoid rearrangement leads to 780, tautomeric with 3-hydroxy-4-anilino methylthiophene[607] (Scheme 231).

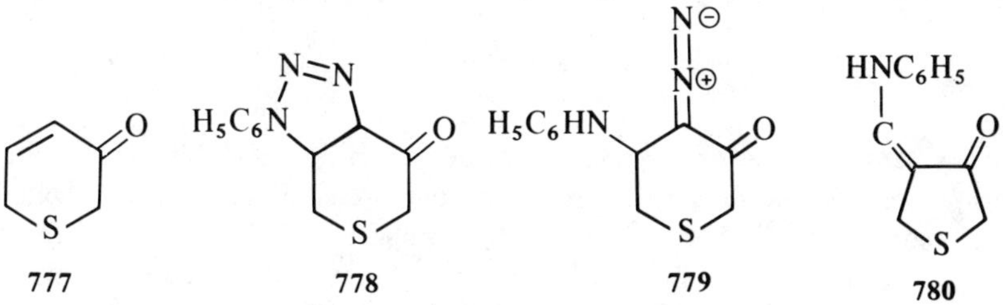

Scheme 231

The reaction of 780a with chloroacetone in the presence of sodium methoxide in methanol gave 780b in 44% yield[606a] (Scheme 232). The reaction of diphenyl-

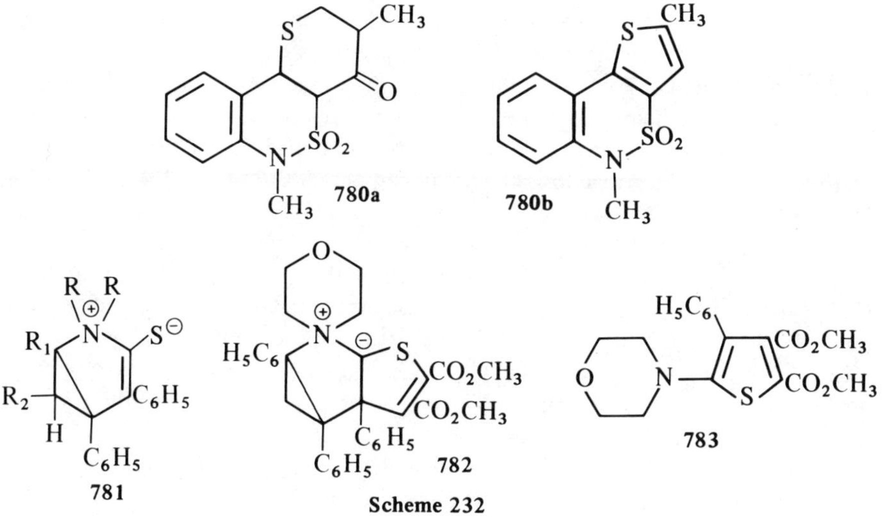

Scheme 232

cyclopropenethione with enamines leads to **781**. The reaction of **781** (R,R = morpholino, $R_1 = C_6H_5$, $R_2 = H$) with dimethyl acetylenedicarboxylate gave a 71% yield of **783**, probably via **782**, together with 66% of diphenylcyclopropene[608, 608a] (Scheme 232).

B.   Thiophenes from Rings with Two Heteroatoms

a.   THIOPHENES FROM 1,4-DITHIINS

Heating or oxidation of 1,4-dithiins (**784**) leads to the formation of thiophenes (**785**) with extrusion of sulfur[609-612] (Scheme 233). In particular, oxidation with

784   **785**

Scheme 233

hydrogen peroxide or peracid leads to rather unstable 1,4-dithiinoxides, which often are not isolated, and upon heating, decompose easily and in good yields to thiophenes.[609, 610, 613, 614] Hydrogen-poor Raney-nickel (W-7J) can also be used for the transformation of 1,4-dithiins to thiophenes. However, the reaction is accompanied by the formation of completely desulfurized products.[615] The ease of thiophene formation is, for instance, evident by the fact that 3,5-diphenyl 2-thiophene aldehyde was the product upon attempted Vilsmeyer formylation of 2,5-diphenyl-1,4-dithiin.[609] Upon nitration, the 2,5-diphenyl-3-nitro-1,4-dithiin is obtained, which even upon heating at 135°C gave 3,5-diphenyl-2-nitrothiophene (20%) and 3,5-diphenyl-4-nitrothiophene (11%).[610] From 2,5-diphenyl-3,6-dinitro-1,4-dithiin, a 25% yield of 2,4-diphenyl 3,5-dinitrothiophene was obtained upon oxidation with hydrogen peroxide.[612] From mono-bromo-2,5-diphenyl-1,4-dithiin, a low yield of 3-bromo-2,4-diphenylthiophene was obtained upon oxidation with hydrogen peroxide, whereas the dibromo derivative gave 2,4-diphenyl-3,5-dibromothiophene in 76% yield.[612] The reaction of 2,5-diphenyl-3-nitro-6-bromo-1,4-dithiin with hydrogen peroxide led regiospecifically to 2,4-diphenyl 3-nitro-5-bromothiophene in over 50% yield. The possible 3-bromo-5-nitro isomer was not observed.[612] Other unsymmetrical 1,4-dithiins have been prepared by the acidic hydrolysis of mixtures of Bunte salts derived from appropriate phenacyl halides. Oxidation of 2-p-methoxyphenyl-5-phenyl-1,r-dithiadiene with peracetic acid and decomposition gave a 39% yield of 2-phenyl-4-p-methoxyphenylthiophene and 49% of 2-p-methoxyphenyl-4-phenylthiophene.[613] Thermal decomposition of 2,5-dimethyl-1,4-dithiin gave a low yield of 2,4-dimethylthiophene.[616]

It has recently been found that, when 2,5-aryl 1,4-dithiins were heated in o-dichlorobenzene at 155°C, bis-(2,4-diaryl-3-thienyl)disulfide was formed as a by-product, together with the expected 2,4-diarylthiophene. On the basis of this result, and on kinetic data, the following reaction mechanism has been suggested[617] (Scheme 234).

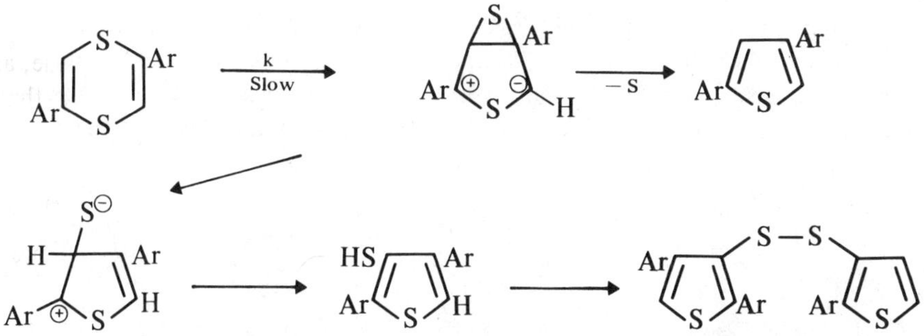

Scheme 234

The smooth decomposition of substituted 2,5-diphenyl-1,4-dithiinsulfoxide (786) in DMSO at 52°C has recently been used for the synthesis of a number of bromo- and nitro-2,4-diphenylthiophenes.[614] However, a detailed study of this reaction showed that its outcome is more complex in polar solvents like acetonitrile, in which (785) is formed together with (787). The amount of the latter compound increases with the presence of electron-donating substituents in the aryl group; in carbon tetrachloride, only 787 in formed, except when Ar = p-ClC$_6$H$_4$, when the proportion of 787 to 785 was 7:3.[618] Photolysis of 786 does not lead to thiophenes[618, 619] (Scheme 235).

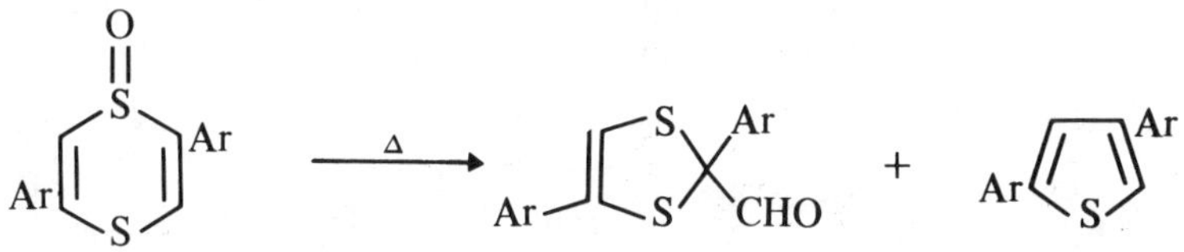

| 786 | 787a | 787b |

	CH$_3$CN	CCl$_4$	CH$_3$CN	CCl$_4$
	(%)	(%)	(%)	(%)
Ar = p-ClC$_6$H$_4$	3	68	97	32
Ar = C$_6$H$_5$	17	100	13	0
Ar = p-CH$_3$C$_6$H$_4$	51	100	49	0
Ar = p-CH$_3$OC$_6$H$_4$	61	100	39	0

Scheme 235

Tetracyano-1,4-dithiin (789) is easily available through the dimerization of sodium cyanodithioformate, prepared from sodium cyanide, carbon disulfide, and DMF, to 788, followed by reaction with thionyl chloride in 1,2-dimethoxyethane. Heating 789 over its melting point, or in 1,2-dimethoxyethane, gave tetracyano-thiophene in quantitative yield.[620-622] The extrusion of sulfur is catalyzed by cesium fluoride.[623] Oxidation of 791, prepared from chlorothiol maleic anhydride, with nitric acid gave 792.[624] From 793, prepared from 1,2-dithietenes and olefins, trifluoromethyl substituted thiophenes (794) have been obtained[625] (Scheme 236).

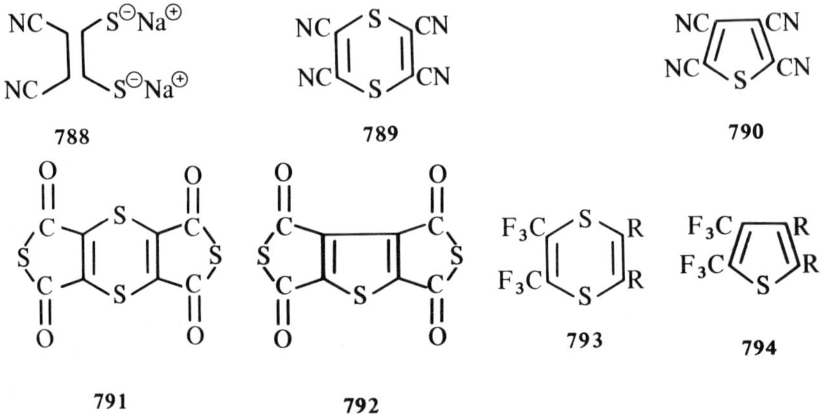

Scheme 236

Methylation of 795 with methyl iodide and silver fluoroborate gives predominantly 796, and in minor amounts the unstable 797, which, via the assumed intermediates 798 and 799, leads to 800. Under the reaction conditions, 800 is methylated to 801 which is obtained in 9% yield[626] (Scheme 237).

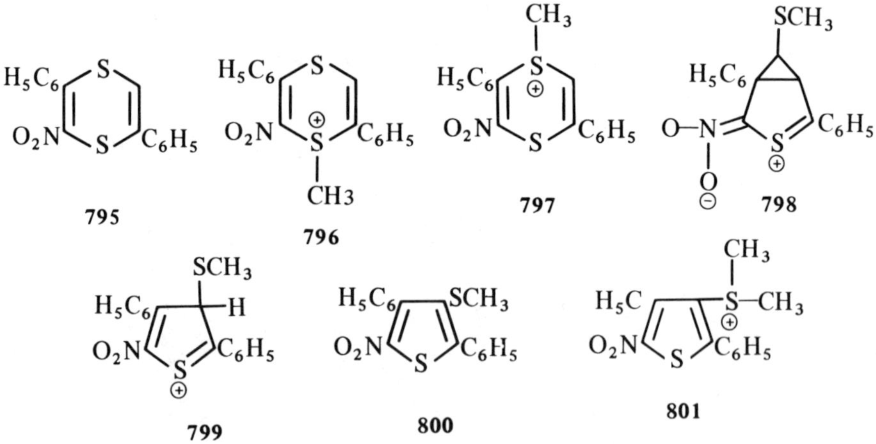

Scheme 237

The reaction of **802** with dimethyl acetylenedicarboxylate led via **803** to **804**, with elimination of phenylacetylene and sulfur dioxide[627] (Scheme 238).

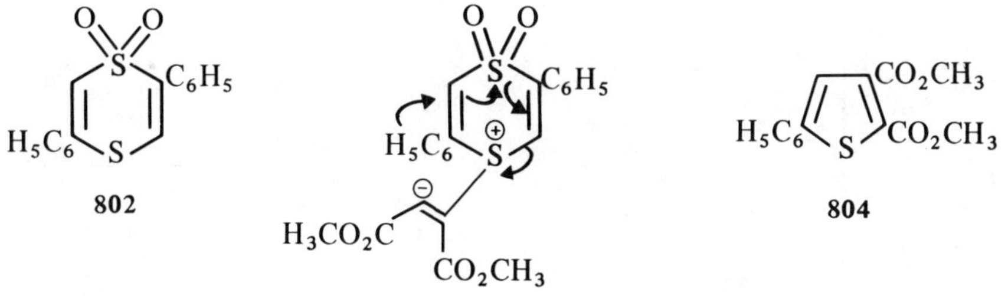

802

803

804

Scheme 238

Transition metal complexes such as **805** react with acetylenes to give 1,4-dithiins (**806**), which upon heating give thiophenes (**807**). These can also be obtained directly by carrying out the reaction between **805** and acetylenes at higher temperature[628] (Scheme 239).

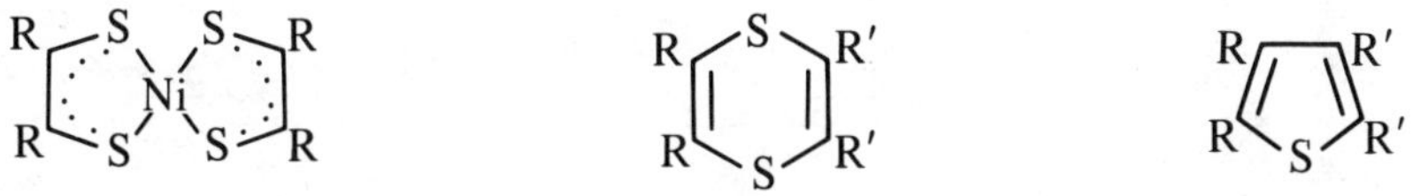

805

806

807

Scheme 239

Another route to 1,4-dithiins (**809**), and thus to thiophenes (**810**), consists of the reaction of 1,2-dithiacyclopentenone derivatives (**808**) with alcoholic alkoxide solution.[629-632] It has been suggested that the reaction occurs via **811** and **812** followed by dimerization to **809**. However, the possibility of dithiacyclobutene or thiacyclopropene intermediates cannot be excluded[632] (Scheme 240). Thiophenes prepared for 1,4-dithiins are given in Table 69.

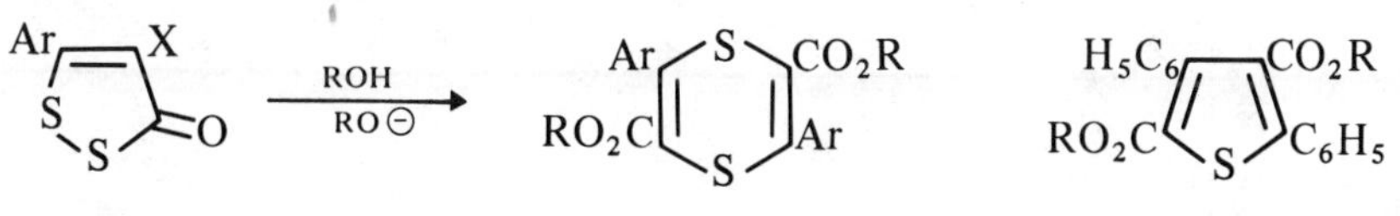

808

X = Cl or Br

809

810

811

812

Scheme 240

TABLE 69. THIOPHENES FROM 1,4-DITHIINES BY (a) HEATING, (b) OXIDATION WITH HYDROGEN PEROXIDE, OR (c) PERACETIC ACID

$$
\underset{\substack{R_4 \diagdown S \diagup R_3 \\ R_5 \diagup S \diagdown R_2}}{\text{(dithiine)}} \longrightarrow \underset{\substack{R_4 \diagdown R_3 \\ R_5 \diagup S \diagdown R_2 \\ \mathbf{A}}}{\text{(thiophene A)}} + \underset{\substack{R_5 \diagdown R_2 \\ R_4 \diagup S \diagdown R_3 \\ \mathbf{B}}}{\text{(thiophene B)}}
$$

Method	R_2	R_3	R_4	R_5	Yield of Thiophene A (%)	Yield of Thiophene B (%)	Reference
a	C_6H_5	H	C_6H_5	H	67		609
b	C_6H_5	H	C_6H_5	H	33		609
a	C_6H_5	H	C_6H_5	CHO	32		609
b	C_6H_5	H	C_6H_5	NO_2	31	46	610
b	C_6H_5	NO_2	C_6H_5	NO_2	25		612
b	C_6H_5	H	C_6H_5	Br	–	14	612
b	C_6H_5	NO_2	C_6H_5	Br	>50	–	612
a	C_6H_5	H	$p\text{-}CH_3OC_6H_4$	H	39	27	613
c	C_6H_5	H	$p\text{-}CH_3OC_6H_4$	H	39	49	613
a	$p\text{-}CH_3OC_6H_4$	H	$p\text{-}CH_3OC_6H_4$	H	76		613
a	CH_3	H	CH_3	H			616
a	CN	CN	CN	CN	100		622
a	$p\text{-}ClC_6H_4$	H	$p\text{-}ClC_6H_4$	CN			634
a	C_6H_5	CO_2CH_3	C_6H_5	CO_2CH_3	82		630
a	C_6H_5	$CO_2C_2H_5$	C_6H_5	$CO_2C_2H_5$	74		630
a	C_6H_5	$CO_2CH(CH_3)_2$	C_6H_5	$CO_2CH(CH_3)_2$	44		630
a	C_6H_5	C_6H_5	C_6H_5	C_6H_5	78		628
a	C_6H_5	C_6H_5	CO_2CH_3	CO_2CH_3	85		628
a	C_6H_5	C_6H_5	CH_3	CH_3	71		628
a	C_6H_5	C_6H_5	CF_3	CF_3	69		628
a	C_6H_5	C_6H_5	C_6H_5	H	42		628
a	α-Naphthyl	CO_2CH_3	α-Naphthyl	CO_2CH_3	82		632
a	β-Naphthyl	CO_2CH_3	β-Naphthyl	CO_2CH_3	80		632

b. THIOPHENES FROM OTHER RINGS

The 5-amino-1,2-dithiocyclopentenones (813) react differently, giving 1,2-dithiins (814), which upon treatment with copper powder or hydrogen-poor Raney-nickel give 815.[631, 633] The reaction of 814 (NR$_2$ = piperidino) with Grignard reagents gave 815 (NR$_2$ = piperidino), together with other products[631] (Scheme 241).

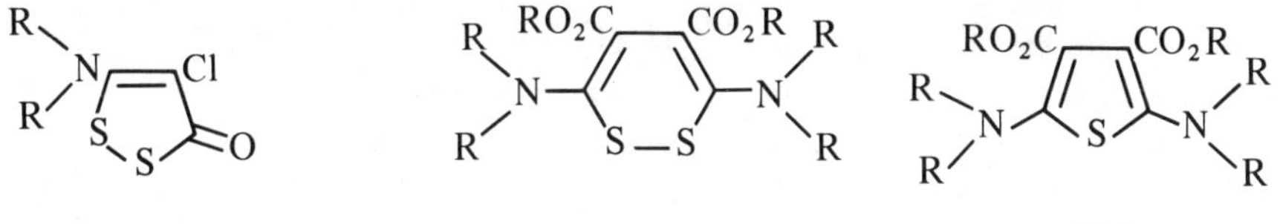

813

814

815

R$_2$N = Morpholino, piperidino

Scheme 241

Heating 816 in morpholine led to a mixture of 2,4- and 2,5-diphenylthiophene.[635] Heating 817 in refluxing toluene gave the thiophenethiol (818) in 40% yield. The diradicals (819 and 820) are suggested as intermediates.[636] The starting material (817) was obtained through the reaction of 2 moles of dimethyl ketenes with carbon disulfide, using triphenylphosphine as catalyst (Scheme 242).

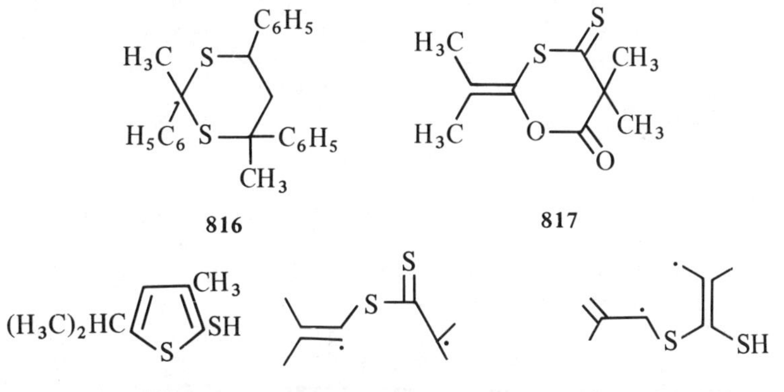

816

817

818

819

820

Scheme 242

REFERENCES

1. O. Meth-Cohn, in *Comprehensive Organic Chemistry*, Vol. 4, P. G. Sammes (Ed.), Pergamon Press, Oxford, 1979, p. 828.

2. H. D. Hartough, *Thiophene and Its Derivatives,* Interscience Publishers, New York, 1952.

3. P. D. Caesar and P. D. Branton, *Ind. Eng. Chem.*, **44**, 122 (1952).

4. S. Gronowitz and P. Moses, *Acta Chem. Scand.*, **16**, 105 (1962).

5. U.S. Patent No. 2,558,508; *Chem. Abstr.*, **46**, 1048 (1952).

6. U.S. Patent No. 2,558,507; *Chem. Abstr.*, **46**, 1047 (1952).

7. British Patent No. 632,306; *Chem. Abstr.*, **46**, 7592 (1952).

8. U.S. Patent No. 2,521,429; *Chem. Abstr.*, **44**, 11181 (1950).

9. British Patent No. 887,426; Chem. Abstr., **57**, 11169 (1962).

10. U.S. Patent No. 2,570,083; *Chem. Abstr.*, **46**, 5089 (1952).

11. U.S. Patent No. 2,694,074; *Chem. Abstr.*, **49**, 15974 (1955).

12. U.S.S.R. Patent No. 186,502: *Chem. Abstr.*, **66**, 115593 (1967).

13. U.S.S.R. Patent No. 229,543; *Chem. Abstr.*, **70**, 87556 (1969).

14. V. N. Kulakov, V. B. Abramovich, T. I. Selivanov, E. A. Bugai, and M. I. Akhmetshin, *Khim. Seraorg, Soedin., Soderzh. Neftyakh Nefteprod.*, **9**, 18 (1972); *Chem. Abstr.*, **79**, 126197 (1973).

15. U.S.S.R. Patent No. 380,658; *Chem. Abstr.*, **79**, 53173 (1973).

16. V. B. Abramovich, K. M. Vaisberg, M. F. Pankratova, B. I. Baglai, R. M. Masagutov, M. A. Ryashentzeva, and D. A. Khisaeva, *Neftekhimiya*, **14**, 289 (1974); *Chem. Abstr.*, **81**, 49495 (1974).

17. British Patent No. 603,103; *Chem. Abstr.*, **43**, 691 (1949).

17a. M. A. Ryasentsheva, E. P. Belanova, and Kh. M. Minachev, *Neftekhimija*, **22**, 231 (1982); *Chem. Abstr.*, **97**, 6089k (1982).

18. U.S. Patent No. 2,474,440; *Chem. Abstr.*, **43**, 6762 (1949).

19. U.S.S.R. Patent No. 186,393; *Chem. Abstr.*, **67**, 32583 (1967).

20. U.S.S.R. Patent No. 241,402; *Chem. Abstr.*, **71**, 64733 (1969).

21. U.S.S.R. Patent No. 199,103; *Chem. Abstr.*, **69**, 35928 (1968).

22. M. A. Ryashentseva, Kh. M. Minachev, A. A. Greish, G. V. Isagulyants, and U. A. Afanasieva, *Int. J. Sulfur Chem.*, **8**, 415 (1973).

23. V. N. Kulakov and V. B. Abramovich, *Khim. Seraorg. Soedin., Soderzh. Neftyakh Nefteprod.*, **8**, 28 (1968); *Chem. Abstr.*, **71**, 81063 (1969).

24. V. B. Abramovich, Ya. S. Amirov, I. Ya. Isyanov, M. F. Pankratova, V. P. Zemtov, and A. A. Shiryaeva, *Neftepererab. Neftekhim. (Moscow)*, 27 (1971); *Chem. Abstr.*, **76**, 14227 (1972).

25. V. B. Abramovich, V. N. Kulakov, and M. F. Pankratova, *Sb. Tr., Ufim. Neft. Inst.*, 154 (1971); *Chem. Abstr.*, **78**, 135977 (1973).

26. U.S.S.R. Patent No. 171,004, *Chem., Abstr.*, **63**, 13215 (1965).

27. V. B. Abramovich, V. N. Kulakov, M. F. Pankratova, and V. P. Zemtsov, *Kinet. Katal.*, **10**, 209 (1969); *Chem. Abstr.*, **70**, 118513 (1969).

28. U.S. Patent No. 3,350,408; *Chem. Abstr.*, **68**, 114429 (1968).

29. U.S.S.R. Patent No. 206,586; *Chem. Abstr.*, **69**, 43782 (1968).

30. V. B. Abramovich, V. N. Kulakov, and I. I. Tokarskaya, *Khim. Geterotsikl. Soedin., Sb.* **3**, 25 (1971); *Chem. Abstr.*, **78**, 71812 (1973).

31. U.S.S.R. Patent No. 178,809; *Chem. Abstr.*, **65** 2224 (1966).

32. V. B. Abramovich and V. N. Kulakov, *Zh. Vses. Khim. O-va im. D. I. Mendeleeva*, **11**, 480 (1966); *Chem. Abstr.*, **65**, 16926 (1966).

33. M. A. Ryashentseva, Y. A. Afanas'eva, and Kh. M. Minachev, *Geterogennyi Katal. Reakts. Poluch. Prevrashch. Geterosikl. Soedin.*, 221 (1971); *Chem. Abstr.*, **76**, 113002 (1972).

34. M. A. Ryashentseva, Yu. A. Afanas'eva, and Kh. M. Minachev, *Khim. Seraorg. Soedin.*,

Soderzh. Neftyakh Nefteprod., **8**, 22 (1968); *Chem. Abstr.*, **71**, 64678 (1969).

35. U.S.S.R. Patent No. 171,003; *Chem. Abstr.*, **63**, 9915 (1965).

36. U.S. Patent No. 2,558,716; *Chem. Abstr.*, **46**, 1048 (1952).

37. British Patent No. 627,247; *Chem. Abstr.*, **44**, 3030 (1950).

38. U.S. Patent No. 2,531,279; *Chem. Abstr.*, **45**, 2983 (1951).

39. U.S. Patent No. 2,557,678; *Chem. Abstr.*, **46**, 1047 (1952).

40. R. E. Conary, L. W. Devaney, L. E. Ruidisch, R. F. McCleary, and K. L. Kreuz. *Ind. Eng. Chem.*, **42**, 467 (1950).

41. R. C. Odioso, D. H. Parker, and R. C. Zabor, *Ind. Eng. Chem.*, **51**, 921 (1959).

42. C. R. Adams, *J. Catal.*, **11**, 96 (1968).

43. U.S. Patent No. 3,309,381; *Chem. Abstr.*, **67**, 21809 (1967).

44. M. A. Ryashentseva, Yu. A. Afanas'eva, and Kh. M. Minachev, *Izv. Akad. Nauk SSSR, Ser. Khim.*, 1067 (1970); *Bull. Acad. Sci. USSR, Chem. Sci.*, 1009 (1970).

45. Yu. A. Afanas'eva, M. A. Ryashentseva, Kh. M. Minachev, and I. I. Levitskii, *Izv. Akad. Nauk SSSR, Ser. Khim.*, 2012 (1970); *Bull. Acad. Sci. USSR, Chem. Sci.*, 1892 (1970).

46. Yu. K. Yur'ev and L. I. Khmel'nitskii, *Dokl. Akad. Nauk SSSR*, **92**, 101 (1953).

47. U.S.S.R. Patent No. 175,979; *Chem. Abstr.*, **64**, 7945 (1966).

48. T. S. Sukhareva, A. V. Mashkina, L. V. Shepel, and L. S. Zabrodova, *Katal. Sint. Org. Soedin. Sery,* 142 (1979); *Chem. Abstr.*, **93**, 220179 (1980).

49. M. A. Ryashentseva, Yu. A. Afanas'eva, and Kh. M. Minachev, *Khim. Geterotsikl. Soedin.*, 1299 (1971); *Chem. Heterocycl. Compd.*, 1215 (1971).

50. K. S. Sidhu, E. M. Lown, O. P. Strausz, and H. E. Gunning, *J. Am. Chem. Soc.*, **82**, 254 (1966).

51. M. A. Ryashentseva and Kh. M. Minachev, *Khim. Seraorg. Soedin., Soderzh. Neftyakh Nefteprod.*, **9**, 14 (1972); *Chem. Abstr.*, **79**, 126193 (1973).

52. German Offen. Patent No. 2,225,443; *Chem. Abstr.*, **78**, 111110 (1973).

53. F. Azizian and J. S. Pizey, *J. Chem. Tech. Biotechnol.*, **30**, 429 (1980).

54. F. Azizian and J. S. Pizey, *J. Chem. Tech. Biotechnol.*, **30**, 648 (1980).

55. Yu. K. Yur'ev, I. K. Korobitsyna, and E. K. Bridge, *Zh. Obshchei Khim.*, **20**, 744 (1950); *J. Gen. Chem. USSR*, **20**, 783 (1950).

56. I. Hirao and H. Hatta, *J. Pharm. Soc. Jpn.*, **74**, 446 (1954); *Chem. Abstr.*, **49**, 1696 (1955).

57. German Patent No. 1,224,749; *Chem. Abstr.*, **65**, 16942 (1966).

58. M. G. Voronkov, A. S. Broun, G. B. Karpenko, and B. L. Golshtein, *Zh. Obshchei Khim.*, **19**, 1357 (1949); *J. Gen. Chem. USSR*, **19**, 1357 (1949).

59. M. G. Voronkov, V. E. Udre, and A. O. Taube, *Khim. Geterotsikl. Soedin.*, 755 (1971); *Chem. Heterocycl. Compd.*, 703 (1971).

60. A. S. Broun, M. G. Voronkov, and F. I. Gol'dburt, *Nauchn. Byull. Leningrad Gos. Univ.*, No. 18, 14 (1947); *Chem. Abstr.*, **43**, 5392 (1949).

61. A. S. Broun and M. G. Voronkov, *Nauchn. Byull. Leningrad Gos. Univ.*, No. 20, 6 (1948); *Chem. Abstr.*, **43**, 5392 (1949).

62. A. S. Broun, M. G. Voronkov, and R. A. Shlyakhter, *Nauchn. Byull. Leningrad Gos. Univ.*, No. 18, 11 (1947); *Chem. Abstr.*, **43**, 5392 (1949).

62a. A. S. Broun and M. G. Voronkov, *J. Gen. Chem. USSR*, **17**, 1162 (1947); *Chem. Abstr.*, **42**, 1591 (1948).

63. J. Schmitt and A. Lespagnol, *Bull. Soc. Chim. Fr.*, 459 (1950).

64. W. E. Parham and E. T. Harper, *J. Am. Chem. Soc.*, **82**, 4936 (1960).

65. M. G. Voronkov and B. L. Golshtein, *J. Gen. Chem. USSR*, 1263 (1950).

66.   Y. Poirier, L. Legrand, and N. Lozac'h, *Bull. Soc. Chim. Fr.*, 1054 (1966).

67.   L. Bateman, R. W. Glazebrook, and C. G. Moore, *J. Chem. Soc.*, 2846 (1958).

68.   A. Tundo, *Boll. Sci. Fac. Chim. Ind. Bologna*, 18, 102 (1960).

69.   N. K. Son and Y. Mollier, *Compt. Rend.*, 273, 278 (1971).

70.   N. K. Son, R. Pinel, and Y. Mollier, *Bull. Soc. Chim. Fr.*, 471 (1974).

71.   N. K. Son, F. Clesse, H. Quiniou, and N. Lozac'h, *Bull. Soc. Chim. Fr.*, 3466 (1966).

72.   G. Pfister-Guillouzo and N. Lozac'h, *Bull. Soc. Chim. Fr.*, 153 (1963).

73.   U.S. Patent No. 3,278,553; *Chem. Abstr.*, 66, 2468 (1967).

74.   Yu. A. Ol'dekop and R. V. Kaberdin, *Vestsi Akad. Navuk Belarus. SSR, Ser. Khim. Navuk*, 131 (1972); *Chem. Abstr.*, 77, 164352 (1972).

75.   R. M. Dodson, V. Srinivasan, K. S. Sharma, and R. F. Sauers, *J. Org. Chem.*, 37, 2367 (1972).

76.   L. Brandsma, J. Meijer, H. D. Verkruijsse, G. Bokkers, A. J. M. Duisenberg, and J. Kroon, *J. C. S. Chem. Commun.*, 922 (1980).

77.   Japan Kokai Tokkyo Kohol Patent No. 79,24,867; *Chem. Abstr.*, 90, 186778 (1979).

78.   J. P. Clayton, A. W. Guest, A. W. Taylor, and R. Ramage, *J. C. S. Chem. Commun.*, 500 (1979).

78a. K. T. Veal and J. T. Gunter, European Patent 38,121; *Chem. Abstr.*, 96, 52175n (1982).

79.   K. E. Schulte, J. Reisch, and L. Hörner, *Angew. Chem.*, 72, 920 (1960).

80.   K. E. Schulte, J. Reisch, and L. Hörner, *Chem. Ber.*, 95, 1943 (1962).

81.   German Patent No. 1,202,796; *Chem. Abstr.*, 64, 6619 (1966).

81a. M. G. Voronkov, B. A. Trofimov, V. V. Kryuchkov, S. V. Amosova, Yu. M. Skvortsov, A. N. Volkov, A. G. Mal'kura, and R. Ya. Mushii, *Khim. Geterotsikl. Soedin.*, 1694 (1981).

82.   K. E. Schulte, J. Reisch, W. Herrmann, and G. Bohn, *Arch. Pharm.*, 296, 456 (1963).

83.   F. Bohlmann and P. Herbst, *Chem. Ber.*, 95, 2945 (1962).

84.   A. S. Zanina, G. N. Khabibulina, V. V. Legkoderya, and I. L. Kotlyarevskii, *Zh. Org. Khim.*, 8, 1527 (1972); *J. Org. Chem. USSR*, 8, 1556 (1972).

84a. J. M. J. Tronchet and A. P. Bonenfant, *Helv. Chim. Acta*, 64, 2322 (1981).

85.   K. E. Schulte and G. Bohn, *Arch. Pharm.*, 297, 179 (1964).

86.   A. N. Volkov, Yu. M. Skvortsov, Yu. V. Kind, and M. G. Voronkov, *Zh. Org. Khim.*, 10, 174 (1974); *J. Org. Chem. USSR*, 10, 176 (1974).

87.   A. S. Medvedeva, M. M. Demina, I. D. Kalikhman, and M. G. Voronkov, *Izv. Akad. Nauk SSSR, Ser. Khim.*, 1643 (1974); *Bull. Acad. Sci. USSR*, 1566 (1974).

88.   U.S. Patent No. 3,544,596; *Chem. Abstr.*, 74, 99857 (1971).

88a. T. J. Barton and R. G. Zika, *J. Org. Chem.*, 35, 1279 (1970).

89.   B. A. Trofimov, S. V. Amosova, G. K. Musorin, and M. G. Voronkov, *Zh. Org. Khim.*, 14, 667 (1978); *J. Org. Chem. USSR*, 14, 618 (1978).

89a. B. A. Trofimov, S. V. Amosova, G. K. Musorin, D. F. Kushnarev, and G. A. Kalabin, *Zh. Org. Khim.*, 15, 619 (1979).

90.   J. Teste and N. Lozac'h, *Bull. Soc. Chim. Fr.*, 442 (1955).

91.   K. E. Schulte, H. Walter, and L. Rolf, *Tetrahedron Lett.*, 4819 (1967).

91a. N. R. M. Smith and A. J. Banister, *Chem. Ind. London*, 907 (1982).

92.   F. Ya. Perveev and N. I. Kudryashova, *Zh. Obshchei Khim.*, 23, 976 (1953); *Chem. Abstr.*, 48, 8219 (1954).

93.   F. J. Perveejev and N. I. Kudrashova, *Dokl. Akad. Nauk SSSR*, 98, 975 (1954).

94. F. Ya. Perveev, *Vestn. Leningrad Univ.*, 10, No. 5, *Ser. Mat., Fiz. i Khim.*, No. 2, 145 (1955); *Chem. Abstr.*, 49, 8904 (1955).

95. F. Ya. Perveev and T. N. Kuren'gina, *Zh. Obshchei Khim.*, 25, 1619 (1955); *Chem Abstr.*, 50, 4900 (1956).

96. C. Botteghi, L. Lardicci, and R. Menicagli, *J. Org. Chem.*, 38, 2361 (1973).

97. E. Campaigne and W. O. Foye, *J. Org. Chem.*, 17, 1405 (1952).

98. H. Wynberg and U. E. Wiersum, *J. Org. Chem.*, 30, 1058 (1965).

99. W. H. Brown and G. F. Wright, *Can. J. Chem.*, 35, 236 (1957).

100. H. Nozaki, T. Koyama, and T. Mori, *Tetrahedron*, 25, 5357 (1969).

101. S. Fujita, T. Kawaguti, and H. Nozaki, *Tetrahedron Lett.*, 1119 (1971).

102. S. Hirano, T. Hiyama, S. Fujita, T. Kawaguti, Y. Hayashi, and N. Nozaki, *Tetrahedron*, 30, 2633 (1974).

103. P. J. Garratt and S. B. Neoh, *J. Org. Chem.*, 40, 970 (1975).

104. M. P. Cava, M. A. Sprecker, and W. R. Hall, *J. Am. Chem. Soc.*, 96, 1817 (1974).

105. M. P. Cava, M. Behforowz, G. E. M. Husbands, and M. Srinivasan, *J. Am. Chem. Soc.*, 95, 2561 (1973).

106. K. E. Potts and D. McKeough, *J. Am. Chem. Soc.*, 96, 4268 (1974).

107. J. D. Bower and R. H. Schlessinger, *J. Am. Chem. Soc.*, 91, 6891 (1969).

108. A. Ermili and L. Salamon, *Ann. Chim. (Italy)*, 59, 375 (1969).

109. S. Julia and J. M. Carulla, *An. Quim.*, 75, 904 (1979).

110. E. C. Kornfeld and R. G. Jones, *J. Org. Chem.*, 19, 1671 (1954).

111. R. G. Jones, *J. Am. Chem. Soc.*, 77, 4074 (1955).

112. R. G. Jones, *J. Am. Chem. Soc.*, 77, 4069 (1955).

113. R. G. Jones, *J. Am. Chem. Soc.*, 77, 4163 (1955).

114. M. Elliott, N. F. Janes, and B. C. Pearson, *J. Chem. Soc. (C)*, 2551 (1971).

115. C. Trebaul and J. Teste, *Bull. Soc. Chim. Fr.*, 2272 (1970).

116. N. Yoda, *Makromol. Chem.*, 55, 174 (1962).

116a. D. R. Shridhar, M. Jogibhukta, P. Shanthan Rao, and V. K. Handa, *Synthesis*, 1061 (1982).

117. E. Campaigne and W. O. Foye, *J. Org. Chem.*, 17, 1405 (1952).

118. U.S. Patent No. 3,014,923; *Chem. Abstr.*, 56, 8692 (1962).

119. F. Duus, *Acta Chem. Scand.*, 27, 466 (1973).

120. F. Duus, *Tetrahedron*, 32, 2817 (1976).

121. H. J. Kooreman and H. Wynberg, *Recueil*, 86, 37 (1967).

122. B. P. Das and D. W. Boykin, *J. Med. Chem.*, 20, 1219 (1977).

123. H. Wynberg and A. J. H. Klunder, *Recueil*, 88, 328 (1969).

124. C. Botteghi and L. Lardicci, *Chim. Ind.*, 52, 265 (1970).

125. M. W. Farrar and R. Levine, *J. Am. Chem. Soc.*, 72, 4433 (1950).

126. S. Gronowitz and T. Frejd, *Acta Chem. Scand.*, B30, 341 (1976).

127. W. Schroth, F. Billig, and G. Reinhold, *Z. Chem.*, 9, 229 (1969).

128. K. Balenović, A. Deljac, B. Gašpert, and Z. Štefanac, *Monatsh. Chemie*, 98, 1344 (1967).

129. Yu. K. Yur'ev and N. V. Makarov, *Zh. Obshchei Khim.*, 28, 885 (1958); *J. Gen. Chem. USSR*, 28, 857 (1958).

129a. W. Kees, "Die Bedeutung von Ketonmannichbasen zur Synthese von Thiophenabkömmlingen und Dithienothiazinen," dissertation, Friedrich Alexander Universität, Erlangen–Nürnberg, 1961.

129b. A. Klau, "Über die Darstellung und Umsetzungen von 3,5-disubstituierten 2-Aminothiophenen," dissertation, Friedrich Alexander Universität, Erlangen–Nürnberg, 1964.

130. W. Kues and C. Paal, *Chem. Ber.,* **19**, 555 (1886).

131. S. Gronowitz and R. A. Hoffman, *Ark. Kem.,* **15**, 499 (1960).

132. T. Hanzawa, H. Nishimura, and J. Mizutani, *Agric. Biol. Chem.,* **37**, 2393 (1973).

133. R. Menicagli, C. Botteghi, and M. Marchetti, *J. Heterocycl. Chem.,* **17**, 57 (1980).

134. S. Bradamante, R. Fusco, A. Marchesini, and G. Pagani, *Tetrahedron Lett.,* 11 (1970).

135. J. Brunet, D. Paquer, and P. Rioult, *Phosphorus and Sulfur,* **3**, 377 (1977).

136. F. Duus, *J. C. S. Perkin I,* 392 (1978).

137. I. Hirao, *J. Pharm. Soc. Jpn.,* **73**, 1023 (1953).

138. B. Bak, J. Christiansen, and J. T. Nielsen, *Acta Chem. Scand.,* **14**, 1865 (1960).

139. R. F. Feldkamp and B. F. Tullar, *Org. Synth.,* **34**, 73 (1954).

140. South African Patent No. 67,06,178; *Chem. Abstr.,* **70**, 96809 (1969).

141. E. C. Kooyman and J. B. H. Kroon, *Recueil,* **82**, 464 (1963).

142. Belgian Patent No. 611,897; *Chem. Abstr.,* **57**, 15075 (1962).

143. Swiss Patent No. 426,870; *Chem. Abstr.,* **68**, 87143 (1968).

144. British Patent No. 923,096; *Chem. Abstr.,* **61**, 643 (1964).

145. Polish Patent No. 61,515; *Chem. Abstr.,* **75**, 35704 (1971).

146. Czechoslovakian Patent No. 140,877; *Chem. Abstr.,* **77**, 61801 (1972).

147. S. Nakagawa, J. Okumura, F. Sakai, H. Hoshi, and T. Naito, *Tetrahedron Lett.,* 3719 (1970).

148. K. Oka, *Heterocycles,* **12**, 461 (1979).

149. K. E. Schulte, J. Reisch, and D. Bergenthal, *Angew. Chem.,* **77**, 1141 (1965).

150. K. E. Schulte, J. Reisch, and D. Bergenthal, *Chem. Ber.,* **101**, 1540 (1968).

151. K. E. Schulte and N. Jantos, *Arch Pharm.,* **292/64**, 16 (1959).

152. A. G. Ismailov, E. I. Mamedov, and V. G. Ibragimov, *Zh. Org. Khim.,* **13**, 2612 (1977); *J. Org. Chem. USSR,* **13**, 2424 (1977).

153. German Offen. Patent No. 2,006,277; *Chem. Abstr.,* **76**, 25099 (1972).

154. German Offen. Patent No. 2,214,540; *Chem. Abstr.,* **78**, 29615 (1973).

155. Yu. M. Volovenko and F. S. Babichev, *Khim. Geterotsikl. Soedin.,* 1425 (1977); *Chem. Heterocycl. Compd.,* 1146 (1977).

155a. E. Benary and A. Baravian, *Chem. Ber.,* **48**, 593 (1915).

155b. H. J. Jakobsen and S. O. Lawesson, *Tetrahedron,* **21**, 3331 (1966).

156. F. C. Leavitt, T. A. Manuel, and F. Johnson, *J. Am. Chem. Soc.,* **81**, 3163 (1959).

157. E. H. Braye, W. Hübel, and I. Caplier, *J. Am. Chem. Soc.,* **83**, 4406 (1961).

158. Ch. M. Angelov, M. Kirilov, K. V. Vachkov, and S. L. Spassov, *Tetrahedron Lett.,* **21**, 3507 (1980).

158a. C. M. Angelov and K. V. Vachkov, *Tetrahedron Lett.,* **22**, 2517 (1981).

158b. N. R. M. Smith and A. J. Banister, *Chem. Ind. London,* 907 (1982).

159. K. E. Schulte, J. Reisch, and W. Herrmann, *Naturwissenschaften,* 332 (1963).

160. B. A. Arbuzov and E. G. Katajev, *Dokl. Akad. Nauk SSSR,* **96**, 983 (1954).

161. E. J. Geering, *J. Org. Chem.,* **24**, 1128 (1959).

162. N. Lozac'h and Y. Mollier, *Bull. Soc. Chim. Fr.,* 1389 (1959).

163. A. G. Makhsumov, A. Safaev, and E. A. Mirzabaev, *Zh. Org. Khim.,* **5**, 1510 (1969); *J. Org. Chem. USSR,* **5**, 1472 (1969).

164. A. G. Makhsumov and E. A. Mirzabaev, *Khim. Geterotsikl. Soedin.,* **6**, 713 (1970); *Chem. Heterocycl. Compd.,* **6**, 661 (1970).

165. J.-P. Beny, S. N. Dhawan, J. Kagan, and S. Sundlass, *J. Org. Chem.*, **47**, 2201 (1982).

166. H. Egger and K. Schlögl, *Monatsh.*, **95**, 1750 (1964).

167. H. Falk, H. Lehner, and K. Schlögl, *Monatsh. Chemie.*, **101**, 967 (1970).

168. A. G. Makhsumov, T. Yu. Nasriddinov, and A. M. Sladkov, *Zh. Org. Khim.*, **7**, 1764 (1971); *J. Org. Chem. USSR*, **7**, 1833 (1971).

169. N. Messina and E. V. Brown, *J. Am. Chem. Soc.*, **74**, 920 (1952).

170. U.S. Patent No. 2,658,903; *Chem. Abstr.*, **48**, 13723 (1954).

171. R. Ramasseul and A. Rassat, *Bull. Soc. Chim. Fr.*, 3136 (1965).

172. T. L. Cairns, R. A. Carboni, D. D. Coffman, V. A. Engelhardt, R. E. Heckert, E. L. Little, E. G. McGeer, B. C. McKusick, and W. J. Middleton, *J. Am. Chem. Soc.*, **79**, 2340 (1957).

173. U.S. Patent No. 2,801,908; *Chem. Abstr.*, **52**, 1261 (1958).

174. W. J. Middleton, V. A. Engelhardt, and B. S. Fisher, *J. Am. Chem. Soc.*, **80**, 2822 (1958).

175. W. J. Middleton, *Org. Synth.*, **39**, 8 (1959).

175a. C. E. Nasakin, V. V. Alekseev, B. K. Promonekov, I. A. Abramov, and A. K. Bulai, *Zh. Org. Khim.*, **17**, 1958 (1981).

176. N. G. Sausen, V. A. Engelhardt, and W. J. Middleton, *J. Am. Chem. Soc.*, **80**, 2815 (1958).

177. F. Asinger and A. Mayer, *Angew, Chem.*, **77**, 812 (1965).

178. C. D. Slater and D. L. Heywood, *J. Heterocycl. Chem.*, **2**, 315 (1965).

179. G. Purrello, *Gazzetta*, **95**, 699 (1965).

180. F. Bottino and G. Purrello, *Gazzetta*, **95**, 1062 (1965).

181. T. Bacchetti, A. Alemagna, and B. Danieli, *Tetrahedron Lett.*, 2001 (1965).

182. D. Nightingale and R. A. Carpenter, *J. Am. Chem., Soc.*, **71**, 3560 (1949).

183. G. Purrello, *Gazzetta*, **97**, 549 (1967).

184. G. Purrello, *Gazzetta*, **97**, 557 (1967).

185. O. Hinsberg, *Berichte*, **43**, 901 (1910).

186. I. Ya. Postovskii, N. P. Bednyagina, and V. F. Kuznetsova, *Zh. Prikl. Khim.* **24**, 1071 (1951); *Chem. Abstr.*, **46**, 7563 (1952).

187. D. W. H. MacDowell and T. B. Patrick, *J. Org. Chem.*, **32**, 2441 (1967).

188. Netherland Patent No. 6,400,338: *Chem. Abstr.*, **62**, 4147 (1965).

189. D. J. Zwanenburg and H. Wynberg, *Recueil*, **88**, 321 (1969).

190. H. Wynberg and D. J. Zwanenburg, *J. Org. Chem.*, **29**, 1919 (1964).

191. D. J. Chadwick, J. Chambers. G. D. Meakins, and R. L. Snowden, *J. C. S. Perkin I*, 2079 (1972).

192. C. G. Overberger, H. J. Mallon, and R. Fine, *J. Am. Chem. Soc.*, **72**, 4958 (1950).

193. E. Rose, "Über Synthesen von Thiophenverbindungen," dissertation, Friedrich Alexander Universität, Erlangen–Nürnberg, 1961.

193a. S. Kroll, "Über neue Thiophenverbindungen." dissertation, Friedrich Alexander Universität, Erlangen–Nürnberg, 1961.

193b. H. Schädler, "Über 3-Aminothiophencarbonsäureester und über eine Variante der Hinsberg Synthese," dissertation, Friedrich Alexander Universität, Erlangen–Nürnberg, 1960.

194. H. Wynberg and H. J. Kooreman, *J. Am. Chem. Soc.*, **87**, 1739 (1965).

195. R. Helder and H. Wynberg, *Tetrahedron*, **31**, 2551 (1975).

196. P. P. Paranjpe and G. Bagavant, *Indian J. Chem.*, **11**, 313 (1973).

197. C. G. Overberger and J. Lal, *J. Am. Chem. Soc.*, **73**, 2956 (1951).

198. R. H. Eastman and R. M. Wagner, *J. Am. Chem. Soc.*, **71**, 4089 (1949).

199. C. G. Overberger, S. P. Lighthelm, and E. A. Swire, *J. Am. Chem. Soc.*, **72**, 2856 (1950).

200. Y. Miyahara, *J. Heterocycl. Chem.*, **16**, 1147 (1979).

201. Y. Miyahara, T. Inazu, and T. Yoshino, *Bull. Chem. Soc. Jpn.*, **53**, 1187 (1980).

202. Y. Miyahara, T. Inazu, and T. Yoshino, *Chem. Lett.*, 563 (1978).

203. Y. Miyahara, T. Inazu, and T. Yoshino, *Chem. Lett.*, 397 (1980).

204. H. Pfeiffer, "Über substituierte Hydroxy-furane-thiophene-thiophtene und Diamino-thiophtene," dissertation, Friedrich Alexander Universität, Erlangen–Nürnberg, 1961.

204a. D. A. Crombie, J. R. Kiely and C. J. Ryan, *J. Heterocycl Chem.*, **16**, 381 (1979).

205. P. J. Garratt and K. P. C. Vollhardt, *J. C. S. Chem. Commun.*, 109 (1970).

206. P. J. Garratt and D. N. Nicolaides, *J. C. S. Chem. Commun.*, 1014 (1972).

207. P. J. Garratt and D. N. Nicolaides, *J. Org. Chem.*, **39**, 2222 (1974).

208. P. J. Garratt and K. P. C. Vollhardt, *J. Am. Chem. Soc.*, **94**, 1022 (1972).

209. K. Gewald, *Angew, Chem.*, **73**, 114 (1961).

210. K. Gewald, *Chem. Ber.*, **98**, 3571 (1965).

210a. Nippon Kagaku Co. Ltd., Japanese Patent No. 81,143,245; *Chem. Abstr.*, **96**, 87016p (1981).

211. O. Hromatka, D. Binder, and P. Stanetty, *Monatsh. Chemie*, **104**, 104 (1973).

212. D. Binder, O. Hromatka, C. R. Noe, F. Hillebrand, and W. Veit, *Arch. Pharm.*, **313**, 587 (1980).

213. D. Binder, O. Hromatka, C. R. Noe, Y. A. Bara, M. Feifel, G. Habison, and F. Leierer, *Arch. Pharm.*, **313**, 636 (1980).

213a. D. Binder and P. Stanetty, *J. Chem. Res.*, 1073 (1981).

214. S. Gronowitz, J. Fortea-Laguna, S. Ross, B. Sjöberg, and N. E. Stjernström, *Acta Pharm. Suec.*, **5**, 563 (1968).

215. A. Cruceyra, V. Gomez-Parra, and R. Madroñero, *Anal. Quim.*, **73**, 265 (1977).

216. M. Robba, J. M. Lecomte, and M. Cugnon de Sevricourt, *Bull. Soc. Chim. Fr.*, 2864 (1974).

217. F. J. Tinney, J. P. Sanchez, and J. A. Nogas, *J. Med. Chem.*, **17**, 624 (1974).

218. K.-H. Weber and H. Daniel, *Lieb. Ann.*, 328 (1979).

219. German Patent No. 2,042,984; *Chem. Abstr.*, **76**, 140495 (1972).

220. O. Hromatka, D. Binder, P. Stanetty, and G. Marischler, *Monatsh. Chemie*, **107**, 233 (1976).

221. K. Gewald, *Z. Chem.*, **2**, 305 (1962).

222. K. Gewald, E. Schinke, and H. Böttcher, *Chem. Ber.*, **99**, 94 (1966).

222a. F. J. Tinney, W. A. Centenko, J. J. Kerbleski, D. T. Connor, R. J. Sorensen, and D. J. Herzig, *J. Med. Chem.*, **24**, 878 (1981).

223. W. O. Foye, J. Mickles, and G. M. Boyce, *J. Pharm. Sci.*, **59**, 1348 (1970).

224. A. Rosowsky, K. K. N. Chen, and M. Lin, *J. Med. Chem.*, **16**, 191 (1973).

225. A. Rosowsky, M. Chaykovsky, K. K. N. Chen, M. Lin, and E. J. Modest, *J. Med. Chem.*, **16**, 185 (1973).

226. K. Gewald and E. Schinke, *Chem. Ber.*, **99**, 2712 (1966).

227. K. Gewald and J. Schael, *J. Prakt, Chem.*, **315**, 39 (1973).

228. E. C. Taylor and J. G. Berger, *J. Org. Chem.*, **32**, 2376 (1967).

229. L. G. Sharanina and S. N. Baranov, *Khim. Geterotsikl. Soedin.*, 196 (1974).

230. K. Gewald and I. Hofmann, *J. Prakt, Chem.*, **311**, 402 (1969).

231. M. B. Devani, C. J. Shishoo, U. S. Pathak, S. H. Parikh, A. V. Radhakrishnan, and A. C. Padhya, *Indian J. Chem.*, **14B**, 357 (1976).

232. V. P. Arya and S. P. Ghate, *Indian J. Chem.*, **9**, 1209 (1971).

232a. Nippon, Kagaku Co. Ltd., Japanese Patent No. 81,100,780; *Chem. Abstr.*, **96**, 19954t (1981).

232b. Yu. M. Volovenko and F. S. Babichev, U.S.S.R. Patent No. 767,105; *Chem. Abstr.*, **95**, 24799e (1981).

233. M. Nakanishi, Y. Kato, T. Furuta, N. Arima, and H. Nishimine, *Yakugaku Zasshi*, **93**, 311 (1973).

234. V. I. Shvedov and A. N. Grinev, *Zh. Org. Khim.*, **1**, 2228 (1965).

235. German Patent No. 2,359,008; *Chem. Abstr.*, **81**, 137599 (1974).

236. K. Gewald, M. Hentschel, and R. Heikel, *J. Prakt. Chem.*, **315**, 539 (1973).

237. E. F. Elslager, P. Jacob, and L. M. Werbel, *J. Heterocycl. Chem.*, **9**, 775 (1972).

238. M. S. Manhas, V. V. Rao, P. A. Seetheraman, D. Succardi, and J. Pazdera, *J. Chem. Soc. C*, 1937 (1969).

239. A. S. Noravyan, A. P. Mkrtchyan, R. A. Akopyan, and S. A. Vartanyan, *Khim.-Farm. Zh.*, **14**, 37 (1980).

240. M. Chaykovsky, M. Lin, A. Rosowsky, and E. J. Modest, *J. Med. Chem.*, **16**, 188 (1973).

241. V. I. Shvedov, V. K. Ryzhkova, and A. N. Grinev, *Khim. Geterotsikl. Soedin.*, **3**, 1010 (1967).

242. E. B. Pedersen and D. Carlsen, *Tetrahedron*, **33**, 2089 (1977).

243. M. Perrissin, C. L. Duc, G. Narcisse, F. Bakri-Logeais, and F. Huguet, *Eur. J. Med. Chem. – Chim. Therap.*, **15**, 563 (1980).

244. M. Perrissin. C. L. Duc. G. Narcisse, F. Bakri-Logeais and F. Huguet, *Eur. J. Med. Chem. – Chim. Ther.*, **15**, 413 (1980).

245. V. P. Arya, *Indian J. Chem.*, **10**, 1141 (1972).

246. M. S. Manhas, M. Sugiura, and H. P. S. Chawla, *J. Heterocycl. Chem.*, **15**, 949 (1978).

247. F. Sauter, G. Reich, and P. Stanetty, *Arch. Pharm.*, **309**, 908 (1976).

248. O. Hromatka, D. Binder, C. R. Noe, P. Stanetty, and W. Veit, *Monatsh. Chemie*, **104**, 715 (1973).

249. M. Nakanishi, T. Tahara, K. Araki, M. Shiroki, T. Tsumagari, and Y. Takigawa, *J. Med. Chem.*, **16**, 214 (1973).

250. A. S. Noravyan, A. P. Mkrtchyan, I. A. Dzhagatspanyan, and S. A. Vartanyan, *Khim. -Farm. Zh.*, **11**, 62 (1977).

251. K. Gewald, M. Kleinert, B. Thiele, and M. Hentschel, *J. Prakt. Chem.*, **314**, 303 (1972).

251a. K. Gewald and A. Martin, *J. Prakt. Chem.*, **323**, 843 (1981).

252. G. Ronsisvalle, *Farm. Ed. Sci.*, **35**, 341 (1980).

253. H. Schäfer and K. Gewald, *Z. Chem.*, **15**, 100 (1975).

254. M. L. Petrov, N. A. Bunina, and A. A. Petrov, *Zh. Org. Khim.*, **14**, 2619 (1978).

255. J. P. Chupp, *J. Heterocycl. Chem.*, **7**, 285 (1970).

255a. L. S. Rodionova, M. L. Petrov, and A. A. Petrov, *Zh. Org. Khim.*, **17**, 2071 (1981).

256. L. Henriksen and H. Autrup, *Acta Chem. Scand.*, **24**, 2629 (1970).

257. J. M. McIntosh and F. P. Seguin, *Can. J. Chem.*, **53**, 3526 (1975).

258. R. P. Napier and C.-C. Chu, *Int. J. Sulfur Chem.*, *A*, **1**, 62 (1971).

259. F. M. Benitez and J. R. Grunwell, *Tetrahedron Lett.*, 3413 (1977).

260. G. N. Schrauzer and V. Mayweg, *J. Am. Chem. Soc.*, 84, 3221 (1962).

261. G. N. Schrauzer and V. Mayweg. *Z. Naturforsch.*, 19b, 192 (1964).

262. G. N. Schrauzer and V. P. Mayweg, *J. Am. Chem. Soc.*, 87, 1483 (1965).

263. G. N. Schrauzer and H. Kisch, *J. Am. Chem. Soc.*, 95, 2501 (1973).

264. W. Steinkopf and G. Kirchhoff, *Lieb. Ann.*, 403, 1 (1914).

265. Y. Yoshida, M. Komatzu, Y. Ohshiro, and T. Agawa, *J. Org. Chem.*, 44, 830 (1979).

266. V. P. Arya and S. P. Ghate, *Indian J. Chem.*, 9, 904 (1971).

267. A. S. Noravyan, A. P. Mkrtchyan, I. A. Dzhagatspanyan, R. A. Akopyan, N. E. Akopyan, and S. A. Vartanyan, *Khim.-Farm. Zh.*, 11, 38 (1977).

268. V. I. Shvedov and A. N. Grinev, *Khim. Geterotsikl. Soedin.*, 2, 515 (1966).

269. R. F. Koebel, L. L. Needham, and C. DeWitt Blanton, Jr., *J. Med. Chem.*, 18, 192 (1975).

270. V. I. Shvedov, V. K. Ryzhkova, and A. N. Grinev, *Khim. Geterotsikl. Soedin.*, 3, 239 (1967).

271. V. P. Arya, *Indian J. Chem.*, 10, 812 (1972).

272. U.S. Patent No. 3,558,606; *Chem. Abstr.*, 74, 141896 (1971).

273. Yu. A. Sharanin and L. G. Sharanina, *Khim. Geterotsikl. Soedin.*, 10, 1432 (1974).

274. R. Gompper and E. Kutter, *Angew. Chem.*, 74, 251 (1962).

274a. P. Rossy, F. G. M. Vogel, W. Hoffmann, J. Paust, and A. Nurrenbach, *Tetrahedron Lett.*, 22, 3493 (1981).

275. R. Gompper, E. Kutter, and W. Töpfl, *Lieb. Ann.*, 659, 90 (1962).

276. T. Liljefors and J. Sandström, *Acta Chem. Scand.*, 24, 3109 (1970).

277. M. Augustin, W.-D. Rudorf, and U. Schmidt, *Tetrahedron*, 32, 3055 (1976).

278. L. Daalgaard, L. Jensen, and S. O. Lawesson, *Tetrahedron*, 30, 93 (1974).

279. K. Clarke, W. R. Fox, and R. M. Scrowston, *J. C. S. Perkin 1*, 8944 (1980).

279a. W. D. Rudorf and M. Augustin, *Z. Chem.*, 22, 255 (1982).

280. Y. Tominaga, H. Fujito, Y. Matsuda, and G. Kobayashi, *Heterocycles*, 6, 1871 (1977).

280a. Y. Tominaga, H. Fujito, H. Norisue, A. Ushirogochi, Y. Matsuda, and G. Kobayashi, *Yakugaku Zasshi*, 99, 1081 (1979).

281. K. A. Jensen and L. Henriksen, *Acta Chem. Scand.*, 22, 1107 (1968).

282. H. Schäfer and K. Gewald, *J. Prakt. Chem.*, 317, 337 (1975).

283. R. Laliberté and G. Medawar, *Can. J. Chem.*, 48, 2709 (1970).

283a. R. Laliberté, U.S. Patent No. 3,506,669; *Chem. Abstr.*, 73, 3785c (1970).

284. R. Gompper and H. Schaefer, *Chem. Ber.*, 100, 591 (1967).

285. L. Henriksen and H. Autrup, *Acta Chem. Scand.*, 26, 3342 (1972).

286. R. Laliberté and G. Medawar, *Can. J. Chem.*, 49, 1372 (1971).

287. R. Raap, *Can. J. Chem.*, 46, 2255 (1968).

288. M. Takaku, Y. Hayasi, and H. Nozaki, *Bull, Chem. Soc. Jpn.* 43, 1917 (1970).

289. G. Barnikow, *J. Prakt. Chem.*, 34, 251 (1966).

290. K. Hartke and L. Peshkar, *Pharm. Zent.*, 107, 348 (1968); *Chem. Abstr.*, 70, 57761 (1969).

291. K. Hartke and G. Gölz, *Lieb. Ann.*, 1644 (1973).

292. G. Gölz and K. Hartke, *Arch. Pharm.*, 307, 663 (1974).

293. S. Rajappa and B. G. Advani, *Tetrahedron Lett.*, 5067 (1969).

294. S. Rajappa and B. G. Advani, *Indian J. Chem.*, 9, 759 (1971).

295. S. Rajappa and R. Sreenivasan, *Indian J. Chem.*, 9, 761 (1971).

296.   S. Rajappa and B. G. Advani, *Indian J. Chem.*, **12**, 1 (1974).

297.   S. Rajappa, B. G. Advani, and R. Sreenivasan, *Indian J. Chem.*, **12**, 4 (1974).

298.   E. J. Smutny, *J. Am. Chem. Soc.*, **91**, 208 (1969).

299.   S. Rajappa and R. Sreenivasan, *Indian J. Chem.*, **16B**, 752 (1978).

300.   S. Rajappa, B. G. Advani, and R. Sreenivasan, *Synthesis*, 656 (1974).

301.   S. Rajappa and R. Sreenivasan, *Indian J. Chem.*, **15B**, 301 (1977).

302.   W. Ried and L. Kaiser, *Synthesis*, 120 (1976).

303.   J. C. Meslin and H. Quiniou, *Compt. Rend. Ser. C.*, **273**, 148 (1971).

304.   J. C. Meslin, *Compt. Rend. Ser. C.*, **277**, 1391 (1973).

305.   J. C. Meslin, Y. T. N'Guessan, H. Quiniou, and F. Tonnard, *Tetrahedron*, **31**, 2679 (1975).

306.   W. Ried, B. I. Podkowik, and G. Oremek, *Lieb. Ann.*, 863 (1980).

306a.   J. Liebscher, B. Abegaz, and A. Areda, *J. Prakt. Chem.*, **325**, 168 (1983).

306b.   S. S. Bhattacharjee, H. Ila, and H. Junjappa, *Synthesis*, 410 (1983).

307.   C. K. Chandra, N. K. Chakrabarty, and S. K. Mitra, *J. Ind. Chem. Soc.*, **19**, 139 (1942).

308.   B. Hinkel, "Über die Synthese von Dihydro[3′,4′:3,2]thiopheno-thiophenen," dissertation, Friedrich Alexander Universität, Erlangen–Nürnberg. 1962.

309.   U. Tergau, "Über Thienopyrimidine und Thienopyrrole," dissertation, Friedrich Alexander Universität, Erlangen–Nürnberg, 1962.

310.   M. Pulst, M. Weissenfels, and L. Bayer, *Z. Chem.*, 287 (1973).

311.   M. Weissenfels and M. Pulst, *J. Prakt. Chem.*, **315**, 873 (1973).

312.   P. Cagniant and G. Kirsch, *Compt. Rend. Ser. C.*, **281**, 35 (1975).

313.   R. B. Woodward and R. H. Eastman, *J. Am. Chem. Soc.*, **68**, 2229 (1946).

314.   H. Fiesselmann and P. Schipprak, *Chem. Ber.*, **87**, 835 (1954).

315.   H. Fiesselmann, P. Schipprak, and L. Zeitler, *Chem. Ber.*, **87**, 841 (1954).

316.   H. Fiesselmann and W. Böhm, *Chem. Ber.*, **89**, 1902 (1956).

317.   A. Courtin, E. Class, and H. Erlenmeyer, *Helv. Chim. Acta*, **47**, 1748 (1964).

318.   H. Fiesselmann and P. Schipprak, *Chem. Ber.*, **89**, 1896 (1956).

319.   J. B. Hendrickson, R. Rees, and J. F. Templeton, *J. Am. Chem. Soc.*, **86**, 107 (1964).

320.   W. Böhm, "Über 3-Hydroxythiophen-carbonsäuren (2) 3-Hydroxy-thiophene und Bis-[thiophen-(2)]-indigos," dissertation, Friedrich Alexander Universität, Erlangen–Nürnberg, 1957.

321.   J. Brelivet, P. Appriou, and J. Teste, *Compt. Rend. Ser. C.*, **269**, 398 (1969).

322.   P. Appriou, J. Brelivet, and J. Teste, *Bull. Soc. Chim. Fr.*, 1497 (1970).

322a.   E. Larsson, *J. Prakt, Chem.*, **325**, 328 (1983).

323.   F. Bohlmann and E. Bresinsky, *Chem. Ber.*, **97**, 2109 (1964).

324.   H. Fiesselmann and G. Pfeiffer, *Chem. Ber.*, **87**, 848 (1954).

325.   H. Fiesselmann and F. Thoma, *Chem. Ber.*, **89**, 1907 (1956).

326.   F. Thoma, "Über die Einwirkung von Thioglycolsäure auf Ketone und β-Ketosäure-ester und über die Darstellung von Hydroxythiophen-carbonsäureestern," dissertation, Friedrich Alexander Universität, Erlangen–Nürnberg, 1957.

327.   M. K. Chakrabarty and S. K. Mitra, *J. Chem. Soc.*, 1385 (1940).

328.   K. Chandra, N. K. Chakrabarty, and S. K. Mitra, *J. Indian Chem. Soc.*, **19**, 139 (1942).

329.   R. Rippel, "Über Synthesen von Thienothiophenen und Thienodithiophenen," dissertation, Friedrich Alexander Universität, Erlangen–Nürnberg, 1965.

330.   H. Bourvé, "Über Hydroxythiophenverbindungen," dissertation, Friedrich Alexander Universität, Erlangen–Nürnberg, 1969.

331.   F. Schweigert, "Über die Darstellung substituierter Thiophencarbonsäuren-(2)," dissertation, Friedrich Alexander Universität, Erlangen–Nürnberg, 1964.

332.   J. P. Maffrand, D. Frehel, M. Miquel, and M. Roc, *Bull. Soc. Chim. Fr. II*, 48 (1978).

333.   J. Brelivet, P. Appriou, and J. Teste, *Bull. Soc. Chim. Fr.*, 1344 (1971).

334.   H. Fiesselmann, German Patent No. 1,020,641; *Chem. Abstr.*, **54**, 2357 (1960).

335.   F. Memmel, "Zur Darstellung von Oxythiophencarbonsäureestern und Oxythiophen-carbonsäureamiden bezw. -aniliden," dissertation, Friedrich Alexander Universität, Erlangen–Nürnberg, 1956.

336.   R. Knerr, "Synthese von Thiophencarbonsäuren," dissertation, Freidrich Alexander Universität, Erlangen–Nürnberg, 1958.

337.   H. Pirner, "Über Umsetzungen von α,β-Dihalogencarbonsäureestern mit α-Mercapto-carbonylverbindungen," dissertation, Friedrich Alexander Universität, Erlangen–Nürnberg, 1965.

338.   P. R. Huddleston and J. M. Barker, *Synth. Commun.*, 9, 731 (1979).

339.   Farbwerke Hoechst Akt.-Ges., British Patent No. 837,086; *Chem. Abstr.*, **54**, 24798 (1960).

340.   H. Fiesselmann, German Patent No. 1,055,077; *Chem. Abstr.*, **55**, 6497 (1961).

341.   H. Fiesselmann, German Patent No. 1,083,830; *Chem. Abstr.*, **55**, 17651 (1961).

342.   H. Meckl. "Über 3-Hydroxy-4-methyl-thiophencarbonsäure-(2)-methylester und über 3-Amino-thiophencarbonsäure-(2)-ester," dissertation, Friedrich Alexander Universität, Erlangen–Nürnberg. 1957.

343.   K.-H. Büttner, "Über 3-Aminothiophencarbonsäure-(2)-ester," dissertation Friedrich Alexander Universität, Erlangen–Nürnberg, 1961.

344.   S. Gronowitz, J. Fortea-Laguna, S. Ross, B. Sjöberg, and N. E. Stjernström, *Acta Pharm. Suec.*, **5**, 563 (1968).

345.   J. Brelivet and J. Teste, *Bull. Soc. Chim. Fr.*, 2289 (1972).

346.   K.-H. Schmidt, "Über die Umsetzung von β-Chlor-vinylaldehyden und β-Chlor-vinyl-nitrilen mit Thioglykolsäureester," dissertation, Friedrich Alexander Universität, Erlangen–Nürnberg, 1962.

346a. K. Gewald, U. Hain, and E. Schindler, German (East) Patent No. 146,952; *Chem. Abstr.*, **95**, 140416r (1981).

346b. G. Kirsch, D. Cagniant, and P. Cagniant, *J. Heterocycl. Chem.*, 19, 443 (1982).

346c. K. Sato, S. Kambe, A. Sakurai, and H. Midorikawa, *Synthesis,* 1056 (1982).

347.   H. Fiesselmann and H. Habicht, German Patent No. 1,092,929; *Chem. Abstr.*, **57**, 5894 (1962).

348.   H. Habicht, "Über substituierte Thiophencarbonsäure-2-ester," dissertation, Friedrich Alexander Universität, Erlangen–Nürnberg, 1958.

349.   H. Fiesselmann, German Patent No. 1,088,507; *Chem. Abstr.*, **56**, 456 (1962).

350.   M. Schneider, "Über das Thioglycolsäurehydrazid und Reaktionen des Hydrazins mit Thiophen- und 3-Hydroxythiophen-carbonsäure-2-ester," dissertation, Friedrich Alexander Universität, Erlangen–Nürnberg, 1960.

351.   U. Schüssler, "Über Umsetzungen von α,γ-Diketocarbonsäuren und Acylcyclanonen mit Thioglycolsäure und Thioglycolsäureestern," dissertation, Friedrich Alexander Universität, Erlangen–Nürnberg, 1967.

352.   Z. Arnold and J. Zemlicka, *Proc. Chem. Soc.*, 827 (1958).

353.   T. Frejd, J. O. Karlsson, and S. Gronowitz, *J. Org. Chem.*, **46**, 3132 (1981).

354.   S. Hauptmann, M. Weissenfels, M. Scholz, E.-M. Werner, H. J. Köhler, and J. Weisflog, *Tetrahedron Lett.*, 1317 (1968).

355.   S. Hauptmann, A. Hantschmann, and M. Scholz, *Z. Chem.*, 9, 22 (1969).

356. N. D. Trieu and S. Hauptmann, *Z. Chem.*, **13**, 57 (1973).

357. S. Hauptmann and E. M. Werner, *J. Prakt. Chem.*, **314**, 499 (1972).

358. S. Hauptmann, M. Scholz, H.-J. Köhler, and H.-J. Hofmann, *J. Prakt. Chem.*, **311**, 614 (1969).

359. J.-M. Magar, J.-F. Muller, and D. Cagniant, *Compt. Rend. Ser. C*, **286**, 241 (1978).

359a. M. Ghosh, R. Mukherjee, B. G. Chaterjee, and J. K. Ray, *Indian J. Chem. Sect. B.*, **20B**, 243 (1981).

360. H. Schaffer, "Über 3-Aminopyrrolcarbonsäure-(2)-ester und Pyrrolo(3, 2-d)pyrimidine," dissertation, Friedrich Alexander Universität, Erlangen–Nürnberg, 1964.

361. A. Ricci, D. Balucani, A. Fravolini, F. Schiaffella, and G. Grandolini, *Gazz. Chim. Ital.*, **107**, 19 (1977).

362. F. Horn, "Über Diaryl-thieno-(3, 2-b)-thiophene und Aroyl-aryl-thiophene," dissertation, Friedrich Alexander Universität, Erlangen–Nürnberg, 1965.

363. F. Bohlmann, H. Bornowski, and D. Kramer, *Chem. Ber.*, **96**, 584 (1963).

364. F. Bohlmann and E. Bresinsky, *Chem. Ber.*, **100**, 107 (1967).

365. J. Liebscher and H. Hartmann, German (East) Patent No. 97,205; *Chem. Abstr.*, **80**, 27091u (1964).

366. J. Liebscher and H. Hartmann, German (East) Patent No. 98,681; *Chem. Abstr.*, **80**, 70683a (1964).

367. J. Liebscher and H. Hartmann, *J. Prakt, Chem.*, **318**, 731 (1976).

368. E. Winterfeldt, *Chem. Ber.*, **100**, 3679 (1967).

369. A. W. Faull and R. Hull, *J. Chem. Res. (S)*, 240 (1979).

370. A. W. Faull and R. Hull, *J. C. S. Perkin 1*, 1078 (1981).

371. H.-J. Kabbe, *Chem. Ber.*, **104**, 2629 (1971).

372. B. M. Trost and R. Atkins, *Tetrahedron Lett.*, 1225 (1968).

372a. R. L. P. De Jong and L. Brandsma, *J. Organomet. Chem.*, **238**, C17 (1982).

373. M. A. Ryashentseva, Kh. M. Minachev, D. A. Sibarov, V. G. Barinov, and A. D. Kokurin, U.S.S.R. Patent No. 257,460; *Chem. Abstr.*, **72**, 132498q (1970).

374. D. A. Sibarov, M. A. Ryashentseva, V. G. Barinov, A. D. Kokurin, and K. M. Minachev, *Zh. Prikl. Khim. (Leningrad)*, **43**, 1767 (1970); *Chem. Abstr.*, **73**, 130826s (1970).

375. M. Tsukada, T. Oka, and S. Shida, *Chem. Lett.*, 437 (1972).

376. F. C. Leavitt, T. A. Manuel, F. Johnson, L. U. Matternas, and D. S. Lehman, *J. Am. Chem. Soc.*, **82**, 5099 (1960).

377. K. W. Hübel and E. H. Bray, U.S. Patent No. 3,149,101; *Chem. Abstr.*, **63**, 1819 (1965).

378. W. Hübel and E. H. Braye, *J. Inorg. Nucl. Chem.*, **10**, 250 (1959).

379. J. M. Birchall, F. L. Bowden, R. N. Haszeldine, and A. B. P. Lever, *J. Chem. Soc. (A)*, 747 (1967).

380. S. V. Amosova, B. A. Trofimov, N. N. Skatova, O. A. Tarasova, A. G. Trofimova, V. V. Takhistov, and M. G. Voronkov, *Dokl. Akad. Nauk SSSR*, **215**, 95 (1974).

381. H. Behringer, *Liebigs Annalen der Chemie*, **564** , 219 (1949).

382. H. Hauptmann, *Tetrahedron Lett.*, 3589 (1974).

383. J. Hambrecht, H. Straub, and E. Müller, *Tetrahedron Lett.*, 1789 (1976).

384. C. G. Krespan, U.S. Patent No. 3,052,691; *Chem. Abstr.*, **48**, 3449 (1963).

385. H. J. Boonstra and J. F. Arens, *Rec. Trav. Chim.*, **79**, 866 (1960).

386. Texaco Development Corp., British Patent No. 641,239; *Chem. Abstr.*, **45**, 7600 (1951).

387. M. G. Voronkov, E. N. Deryagina, H. A. Kuznetsova, and I. D. Kalikhman, *Zh. Org. Khim.,* **14**, 185 (1978).

388. R. N. Haszeldine, R. E. Banks, and J. M. Birchall, British Patent No. 1,069,943; *Chem. Abstr.,* **67**, 73518t (1967).

389. R. H. Goshorn and Th. E. Deger, U.S. Patent No. 3,350,410; *Chem. Abstr.,* **68**, 95668q (1968).

390. M. G. Voronkov, E. N. Deryagina, and V. I. Perevalova, *Khim. Geterotsikl. Soedin.,* 310 (1980).

390a. M. G. Boronkov, E. N. Deryagina, V. I. Perevalova, and O. B. Bannikova, *Zh. Org. Khim.,* **17**, 1103 (1981).

391. V. I. Perevalova, O. B. Bannikova, E. N. Deryagina, and M. G. Voronkov, *Zh. Org. Khim.,* **16**, 399 (1980).

392. L. R. Drake, U.S. Patent No. 2,538,722; *Chem. Abstr.,* **45**, 4269 (1951).

393. F. R. Mayo, *J. Am. Chem. Soc.,* **90**, 1289 (1968).

394. Th. K. Hanson and L. M. Kinnard, British Patent No. 696,439; *Chem. Abstr.,* **51**, 15587 (1957).

395. A. W. Morton, *J. Org. Chem.,* **14**, 761 (1949).

396. M. G. Voronkov, E. N. Deryagina, M. A. Kuznetsova, and V. I. Glukhikh, *Zh. Org. Khim.,* **16**, 2450 (1980).

397. M. G. Voronkov, V. E. Udre, and E. P. Popova, *Khim. Geterotsikl. Soedin.,* **3**, 1003 (1967).

398. M. G. Voronkov and V. E. Udre, *Khim. Geterotsikl. Soedin.,* **6**, 457 (1970).

399. Du Pont of Canada Ltd, Canadian Patent No. 672,716; *Chem. Abstr.,* **60**, 7999 (1964).

400. P. Pastour, P. Savalle, and P. Eymeri, *Compt. Rend.,* **260**, 6130 (1965).

401. H. Hopff and J. von der Crone, *Chimia,* **13**, 107 (1959).

402. R. Gompper, H. Euchner, and H. Kast, *Liebigs Annalen der Chemie,* **675**, 151 (1964).

403. J. R. Grunwell, *J. C. S. Chem. Commun.,* 1437 (1969).

404. J. R. Grunwell, D. L. Foerst, and M. J. Sanders, *J. Org. Chem.,* **42**, 1142 (1977).

405. H. Nishimura, T. Hanzawa, and J. Mizutani, *Tetrahedron Lett.,* 343 (1973).

406. H. Nishimura and J. Mizutani, *J. Org. Chem.,* **40**, 1567 (1975).

407. F. Asinger, M. Thiel, P. Püchel, F. Haaf, and W. Schäfer, *Liebigs Annalen der Chemie,* **660**, 85 (1962).

408. M. T. Bogert and P. P. Herrera, *J. Am. Chem. Soc.,* **45**, 238 (1923).

409. P. Demerseman, Ng. Ph. Buu-Hoi, R. Royer, and A. Cheutin, *J. Chem. Soc.,* 2688 (1954).

410. E. Campaigne, *J. Am. Chem. Soc.,* **66**, 684 (1944).

411. E. Campaigne, W. B. Reid, Jr., and J. D. Pera, *J. Org. Chem.,* **24**, 1229 (1959).

412. S. Jerumanis and J. M. Lalancette, *Can. J. Chem.,* **42**, 1928 (1964).

413. K. Oka, *Heterocycles,* **12**, 461 (1979).

414. W. Schroth, M. Hassfeld, W. Schiedewitz, and C. Pfotenhauer, *Z. Chem.,* **17**, 411 (1977).

415. G. Purrello and M. Piattelli, *Boll. Sedute Accad. Gioneia Sci. Nat. Catania,* 54 (1967); *Chem. Abstr.,* **70**, 3984e (1969).

416. Yu. K. Yuryev and K. Yu. Novitskii, *Zh. Obshch. Khim.,* **22**, 2187 (1952).

417. Yu. K. Yuryev, K. Yu. Novitskii, and E. V. Kukharskaya, *Dokl. Akad. Nauk SSSR,* **68**, 541 (1949).

418. J. L. Boivin and R. McDonald, *Can. J. Chem.,* **33**, 1281 (1955).

419. M. G. Voronkov and V. E. Uedre, U.S.S.R. Patent No. 165,470; *Chem. Abstr.*, **62**, 6461 (1965).

420. M. G. Voronkov and V. E. Udre, *Khim. Geterotsikl. Soedin.*, **1**, 683 (1965).

421. M. G. Voronkov and V. E. Udre, *Khim. Geterotsikl. Soedin.*, **1**, 148 (1965).

422. J. L. Keller, U.S. Patent No. 2,515,233. *Chem. Abstr.*, **44**, 8961 (1950).

423. W. G. Toland, Jr., J. B. Wilkes, and F. J. Brutschy, *J. Am. Chem. Soc.*, **75**, 2263 (1953).

424. W. G. Toland, Jr. and J. B. Wilkes, *J. Am. Chem. Soc.*, **76**, 307 (1954).

425. K. L. Kreuz, U.S. Patent No. 2,621,188; *Chem. Abstr.*, **47**, 1000q (1953).

426. L. I. Marnich, L. M. Ganzha, Zh. K. Lenkevich, and I. P. Shcherban, *Khim. Tverd. Topl.*, 127 (1970); *Chem. Abstr.*, **73**, 47315d (1970).

427. T. Shibamoto and G. F. Russell, *J. Agric. Food Chem.*, **25**, 109 (1977).

427a. M. G. Voronkov, E. N. Deryagina, and E. N. Sukhomazova, *Khim. Geterotsikl. Soedin.*, 565 (1981).

427b. E. Campaigne, R. L. White, and B. G. Heaton, *Int. J. Sulfur Chem. A*, **1**, 39 (1971).

428. E. Campaigne and R. E. Cline, *J. Org. Chem.*, **21**, 39 (1956).

429. P. M. Chakrabarty, N. B. Chapman, and K. Clarke, *Tetrahedron*, **25**, 2781 (1969).

430. P. M. Chakrabarty and N. B. Chapman, *J. Chem. Soc. (C)*, 914 (1970).

431. H. Behringer and K. Falkenberg, *Chem. Ber.*, **99**, 3309 (1966).

432. G. W. Stacy and D. L. Eck, *Tetrahedron Lett.*, 5201 (1967).

433. D. L. Eck and G. W. Stacy, *J. Heterocycl. Chem.*, **6**, 147 (1969).

434. S. Umio, K. Kariyone, and K. Tanaka, Japanese Patent No. 19,090('67), *Chem. Abstr.*, **69**, 10352e (1968).

435. R. H. Everhardus, H. G. Eeuwhorst, and L. Brandsma, *J. C. S. Chem. Commun.*, 801 (1977).

436. A. Roedig and G. Zaby, *Chem. Ber.*, **113**, 3342 (1980).

437. P. J. W. Schuijl. H. J. T. Bos, and L. Brandsma, *Rec. Trav. Chim.*, **88**, 597 (1969).

438. D. Schuijl-Laros, P. J. W. Schuijl, and L. Brandsma, *Rec. Trav. Chim.*, **88**, 1343 (1969).

439. J. Meijer and L. Brandsma, *Rec. Trav. Chim.*, **92**, 1331 (1973).

440. J. Meijer and L. Brandsma. *Rec. Trav. Chim.*, **91**, 578 (1972).

441. J. Meijer, P. Vermeer, H. J. T. Bos, and L. Brandsma. *Rec. Trav. Chim.*, **9**, 26 (1974).

442. D. Schuijl-Laros, P. J. W. Schuijl, and L. Brandsma, *Rec. Trav. Chim.*, **91**, 785 (1972).

443. L. Dalgaard and S.-O. Lawesson, *Tetrahedron*, **28**, 2051 (1972).

444. D. Barillier, L. Morin, D. Paquar, P. Rioult, M. Vazeux, and C. G. Andrieu, *Bull. Soc. Chim. Fr.*, 688 (1977).

445. R. A. van der Welle and L. Brandsma, *Rec. Trav. Chim.*, **92**, 667 (1973).

446. E. Winterfeldt and H.-J. Dillinger, *Chem. Ber.*, **99**, 1558 (1966).

447. W. Treibs, *Liebigs Annalen der Chemie*, **630**, 120 (1960).

448. A. Hantzsch, *Berichte*, **22**, 2838 (1889).

449. N. P. Buu-Hoi, *Bull. Soc. Chim. Fr.*, 1407 (1958).

450. K. Gewald, H. Jablokoff, and M. Hentschel, *J. Prakt. Chem.*, **317**, 861 (1975).

451. B. Hedegaard, J. Z. Mortensen, and S. O. Lawesson, *Tetrahedron*, **27**, 3853 (1971).

452. K. Hartke and F. Meissner, *Tetrahedron*, **28**, 875 (1972).

453. J. P. Marino and J. L. Kostusyk, *Tetrahedron Lett.*, 2489 (1979).

454   J. P. Marino and J. L. Kostusyk, *Tetrahedron Lett.*, 2493 (1979).

454a. M. Yokoyama, M. Kuranchi, and T. Inamoto, *Tetrahedron Lett.*, **22**, 2285 (1981).

454b. M. Yokoyama, M. Tohnishi, A. Kurihari, and T. Imamoto, *Chem. Lett.*, 1933 (1982).

454c. N. Ben Masour, W. D. Rudorf, and M. Augustin, *Z. Chem.*, 69 (1981).

455. R. Gompper and C. S. Scheider, *Synthesis*, 213 (1979).

456. P. Friedländer and S. Kielbasinski, *Berichte*, 45, 3889 (1912).

457. A. I. Kosak, R. J. E. Palchak, W. A. Steele, and C. M. Selwitz, *J. Am. Chem. Soc.*, 76, 4450 (1954).

458. A. V. Mashkina, T. S. Sukhareva, and V. I. Chernov, *Neftekhimiya*, 7, 110 (1967).

459. S. Trippler, A. F. Plate, T. A. Danilova, and M. A. Ryashentseva, *Neftekhimiya*, 8, 783 (1968).

460. A. V. Mashkina and T. S. Sukhareva, *Kinet. Katal.*, 5, 751 (1964); *Chem. Abstr.*, 61, 11883 (1964).

461. S. Trippler, T. Danilova, and A. F. Plate, *Neftekhimiya*, 10, 267 (1970).

462. M. G. Voronkov, E. N. Deryagina, S. V. Amosova, N. A. Kuznetsova, V. V. Kryuchkov, and B. A. Trofimov, *Khim. Geterotsikl. Soedin.*, 1579 (1975).

463. S. Trippler, T. A. Danilova, and A. F. Plate, *Vestn. Moskv. Izvest.*, 701 (1971).

464. H. Boelens and L. Brandsma, *Rec. Trav. Chim.*, 91, 141 (1972).

465. F. C. V. Larsson, L. Brandsma, and S. O. Lawesson, *Rec. Trav. Chim.*, 93, 256 (1974).

466. W. Ried and W. Ochs, *Chem. Ber.*, 105, 1093 (1972).

467. P. J. Garratt and S. Bin Neoh, *J. Am. Chem. Soc.*, 97, 3255 (1975).

468. P. J. Garratt and S. Bin Neoh, *J. Org. Chem.*, 44, 2667 (1979).

469. V. S. P. Cheng, E. Dominguez, P. J. Garratt, and S. Bin Neoh, *Tetrahedron Lett.*, 691 (1978).

470. S. Braverman, Y. Duar, and D. Segev, *Tetrahedron Lett.*, 3181 (1976).

471. S. Braverman and D. Segev, *J. Am. Chem. Soc.*, 96, 1245 (1974).

472. K. P. C. Vollhardt and R. G. Bergman, *J. Am. Chem. Soc.*, 95, 7538 (1973).

473. D. Leaver, W. A. H. Robertson, and D. M. McKinnon, *J. Chem. Soc.*, 5104 (1962).

473a. W. G. Blenderman and M. M. Joullie, *Heterocycles*, 19, 111 (1982).

474. R. D. Obolensev and A. V. Mashkina, *Khim. Seraorg. Soedin, Soderzh. Neft. Nefteprod. Akad. Nauk SSSR Bashkuzk, Fil.*, 4, 245 (1966); *Chem. Abstr.*, 57, 11139 (1962).

475. A. R. Kuzyev, *Katal. Sint. Org. Soedin. Sery*, 125 (1979); *Chem. Abstr.*, 94, 3890s (1981).

476. E. A. Viktorova, M. B. Vagabov, T. A. Danilova, and E. A. Karakhanov, *Katal. Sint. Org. Soedin, Sery.* 115 (1979); *Chem. Abstr.*, 94, 14834c (1981).

477. N. I. Shuikin and V. V. An, *Izv. Akad. Nauk SSSR, Otd. Khim. Nauk*, 1452 (1962).

478. G. Gardos, L. Hodossy, and T. K. Szabo, *Hung, J. Ind. Chem.*, 1, 115 (1973).

478a. M. V. Vagabov, *Vestn. Moskow Univ., Ser. 2. Khim.*, 23, 52 (1982); *Chem. Abstr.*, 96, 162465q (1982).

479. J. H. Blanc, J. Tellier, and C. Thibault, German Offen. Patent No. 2,062,587; *Chem. Abstr.*, 75, 63599w (1971).

480. W. Friedmann, *J. Inst. Pet.*, 37, 239 (1951).

481. F. Runge, E. Profft, and R. Drux, *J. Prakt. Chem.*, 279 (1955).

482. A. I. Kosak and R. L. Holbrook, *Ohio J. Sci.*, 53, 370 (1953).

483. H. L. Coonradt, H. D. Hartough, and G. S. Johnson, *J. Am. Chem. Soc.*, 70, 2564 (1948).

484. M. Hauptschein and V. Mark, U.S. Patent No. 3,364,233; *Chem. Abstr.*, 69, 10350c (1968).

485. J. P. Gouesnard and G. J. Martin, *Bull. Soc. Chim. Fr.*, 4452 (1969).

485a.   E. Nagashima, K. Suzuki, and M. Sekiya, *Chem. Pharm. Bull.*, **30**, 4384 (1982).
486.   J. Burdon, I. W. Parsons, and J. C. Tatlow, *J. Chem. Soc. (C)*, 346 (1971).
487.   J. Burdon, J. G. Campbell, I. W. Parsons, and J. C. Tatlow, *J. Chem. Soc. (C)*, 352 (1971).
488.   J. Burdon and I. W. Parsons, *J. Fluorine Chem.*, **13**, 159 (1979).
489.   E. M. Ilgenfritz and R. P. Ruh, U.S. Patent No.3,069,431; *Chem. Abstr.*, **58**, 10173h (1963).
490.   C. A. Wellington, T. L. James, and A. C. Thomas, *J. Chem. Soc. (A)*, 2897 (1969).
491.   J. Pirkl and C. Fisar, Czechoslovakian Patent No. 140,878. *Chem. Abstr.*, **77**, 75120n (1972).
492.   T. A. Bryson, *J. Org. Chem.*, **38**, 3428 (1973).
493.   C. A. Wilson, II and T. A. Bryson, *J. Org. Chem.*, **40**, 800 (1975).
494.   S. Fries and K. Gollnick, *Angew. Chem.*, **92**, 848 (1980).
495.   W. Flitsch and J. Schwietzer, *Lieb. Ann.*, 1967 (1975).
496.   O. Scherer and F. Kluge, *Chem. Ber.*, **99**, 1973 (1966).
497.   H. Hart and M. Sasaoka, *J. Am. Chem. Soc.*, **100**, 4326 (1978).
498.   R. L. Whistler and D. J. Hoffman, *Carbohydr. Res.*, **11**, 137 (1969).
499.   V. E. Kolchin and N. S. Vul'fson, *Zh. Obshch. Khim.*, **32**, 3731 (1962).
500.   O. Hromatka, D. Binder, and K. Eichinger, *Monatsh. Chemie*, **104**, 1520 (1973).
501.   Ph. A. Rossy, W. Hoffmann, and N. Müller, *J. Org. Chem.*, **45**, 617 (1980).
501a.   P. W. Raynolds, U.S. Patent No. 4,307,239; *Chem. Abstr.*, **96**, 142690f (1982).
501b.   J. B. Press, C. M. Hofmann, G. E. Wiegand, and S. R. Safir, *J. Heterocycl. Chem.*, **19**, 391 (1982).
502.   H. J. Liu and T. K. Ngooi, *Can. J. Chem.*, **60**, 437 (1982).
503.   Y. Yamada, T. Ishii, M. Kimura, and K. Hosaka, *Tetrahedron Lett.*, **22**, 1353 (1981).
504.   H. G. Alperman, H. Ruschig, and W. Meixner, *Arzneim. -Forsch.*, **22**, 2146 (1972).
505.   E. Larsson, *Sv. Kem. Tidskr.*, **57**, 24 (1945).
506.   T. Takaya, S. Kosaka, Y. Otsuji, and E. Imoto, *Bull. Chem., Soc. Jpn.*, **41**, 2086 (1968).
507.   E. Larsson and H. Dahlström, *Sv. Kem. Tidskr.*, **57**, 248 (1945).
508.   O. Hromatka, D. Binder, and K. Eichinger, *Monatsh. Chemie*, **105**, 127 (1974).
509.   D. Binder and P. Stanetty, *Synthesis*, 200 (1977).
510.   H. Schmid and E. Schnetzler, *Helv. Chim. Acta*, **34**, 896 (1951).
511.   A. P. Stoll and R. Suess, *Helv. Chim. Acta*, **57**, 2487 (1974).
512.   G. B. Brown, M. D. Armstrong, A. W. Moyer, W. P. Anslow Jr., B. R. Baker, M. Y. Querry, S. Bernstein, and S. R. Safir, *J. Org. Chem.*, **12**, 160 (1947).
513.   P. N. Confalone, G. Pizzolato, and M. R. Uskokovic, *Helv. Chim. Acta*, **59**, 1005 (1976).
514.   H. Wynberg, A. Logothetis, and D. VerPloeg, *J. Am. Chem. Soc.*, **79**, 1972 (1957).
515.   D. N. Reinhoudt, W. P. Trompenars, and J. Gewers, *Synthesis*, 368 (1978).
516.   D. N. Reinhoudt, J. Gewers, W. P. Trompenaars, S. Harkema, and G. J. van Hummel, *J. Org. Chem.*, **46**, 424 (1981).
517.   A. J. Poole and F. L. Rose, *J. Chem. Soc. (C)*, 1285 (1971).
518.   A. R. Surrey, H. F. Hammer, and C. M. Sutter, *J. Am. Chem. Soc.*, **66**, 1936 (1944).
519.   B. D. Tilak, H. S. Desai, and S. S. Gupte, *Tetrahedron Lett.*, 1609 (1964).
520.   B. D. Tilak and S. S. Gupte, *Indian J. Chem.*, **7**, 9 (1969).

521. R. P. Napier, H. A. Kaufman, P. R. Driscoll, L. A. Glick, Ch. Ch. Chu, and H. M. Foster, *J. Heterocycl. Chem.*, 7, 393 (1970).

522. J. B. Press, C. M. Hofmann, and S. R. Safir, *J. Org. Chem.*, 44, 3292 (1979).

523. T. Takaya, S. Hijikata, and E. Imoto, *Bull. Chem. Soc. Jpn.*, 41, 2532 (1968).

524. B. R. Baker, J. P. Joseph, R. E. Schaub, F. J. McEvoy, and J. H. Williams, *J. Org. Chem.*, 18, 138 (1953).

525. M. Murakami and M. Hikichi, Japanese Patent No. 69,12,895; *Chem. Abstr.*, 71, 101702b (1969).

526. P. N. Confalone, G. Pizzolato, M. R. Us'kokovic, and M. Rouge, U.S. Patent No. 4,317,915; *Chem. Abstr.*, 96, 199508g (1982).

527. D. Binder, C. R. Noe, and M. Zahora, *Arch. Pharm.*, 314, 557 (1981).

528. L. L. Chening and J. R. Piening, *J. Am. Chem. Soc.*, 67, 729 (1945).

529. C. H. R. Snyder and W. Alexander, *J. Am. Chem. Soc.*, 70, 217 (1948).

530. Netherlands Appl. Patent No. 6,604,742; *Chem. Abstr.*, 67, 21811p (1967).

531. O. Hromatka, D. Binder, and K. Eichinger, *Monatsh. Chemie*, 105, 1164 (1974).

532. J. K. Chakrabarti, J. Fairhurst, N. J. A. Gutteridge, L. Horsman, I. A. Ullar, C. W. Smith, D. J. Steggles, D. E. Tupper, and F. C. Wright, *J. Med. Chem.*, 23, 884 (1980).

533. K. Pöhlmann, "Über N-Substituierte 3-Aminothiophencarbonsäuren-(2)," dissertation, Friedrich Alexander Universität, Erlangen–Nürnberg, 1964.

534. F. A. Buiter, J. H. S. Weiland, and H. Wynberg, *Rec. Trav. Chim.*, 83, 1160 (1964).

534a. D. T. Connor, R. J. Sorenson, F. J. Tinney, W. A. Cetenko, and J. J. Kerbleski, *J. Heterocycl. Chem.*, 19, 1185 (1982).

535. O. Hromatka, D. Binder, and K. Eichinger, *Monatsh. Chemie*, 106, 375 (1975).

536. O. Hromatka, D. Binder, and K. Eichinger, *Monatsh. Chemie*, 106, 555 (1975).

537. M. Robba and N. Boutamine, *Bull. Soc. Chim. Fr.*, 1629 (1974).

538. R. Jaunin, *Helv. Chim. Acta*, 63, 1542 (1980).

539. H. Wynberg, T. J. Van. Bergen, and R. M. Kellogg, *J. Org. Chem.*, 34, 3175 (1969).

540. M. Alvarez, J. Bosch, R. Granados, and F. Lopez, *J. Heterocycl. Chem.*, 15, 193 (1978).

541. J. Vasilevskis, J. A. Gualtieri, S. D. Hutchings, R. C. West., J. W. Scott, D. R. Parrish, F. T. Bizzarro, and G. F. Field, *J. Am. Chem. Soc.*, 100, 7423 (1978).

542. J. M. McIntosh, H. B. Goodbrand, and G. M. Masse, *J. Org. Chem.*, 39, 202 (1974).

543. J. M. McIntosh and R. S. Steevenz, *Can. J. Chem.*, 52, 1934 (1974).

544. J. M. McIntosh and G. M. Masse, *J. Org. Chem.*, 40, 1294 (1975).

545. J. M. McIntosh and R. A. Sieler, *Can. J. Chem.*, 56, 226 (1978).

546. J. M. McIntosh and R. A. Sieler, *J. Org. Chem.*, 43, 4431 (1978).

547. J. M. McIntosh and H. Khalil, *Can. J. Chem.*, 53, 209 (1975).

548. J. Buter, S. Wassenaar, and R. M. Kellogg, *J. Org. Chem.*, 37, 4045 (1972).

549. R. M. Kellogg and W. L. Prins, *J. Org. Chem.*, 39, 2366 (1974).

550. F. Korte and K. H. Löhmer, *Chem. Ber.*, 90, 1290 (1957).

551. F. Korte and K. H. Büchel, *Chem. Ber.*, 93, 1021 (1960).

551a. K. Yamagata, Y. Tomioka, M. Yamazaki, T. Matsuda, and K. Noda, *Chem. Pharm. Bull. Tokyo*, 30, 4396 (1982).

552. R. M. Dodson and J. Yu Fan, *J. Org. Chem.*, 36, 2708 (1971).

553. B. Föhlisch and B. Czauderna, *Phosphorus & Sulfur*, 4, 167 (1978).

554. K. Glauss, *Tetrahedron Lett.*, 1271 (1974).

555. Pennsalt Chemicals Corp., Belgian Patent No. 623,801; *Chem. Abstr.*, 59, 8705g (1963).

556. Pennsalt Chemicals Corp., German Patent No. 1,228,273; *Chem. Abstr.*, **66**, 18665t (1967).

557. Yu. K. Yur'ev, *Vopr. Ispol'z. Pentozansoderzh. Syr'ya Tr. Vesoyuz. Soveshch., Riga*, 405 (1955); *Chem. Abstr.*, **53**, 14078 (1959).

558. Yu. K. Yur'ev, *Khim. i Prakt. Primen. Kremneorg. Soedin,. Tr. Konf. Leningrad*, 157 (1958); *Chem. Abstr.*, **53**, 17091 (1959).

559. K. Yu. Novitskii, N. K. Sadovaya, and A. N. Mentus, *Khim. Geterotsikl. Soedin.*, **3**, 48 (1971).

560. V. G. Kharchenko, I. A. Markushina, and T. I. Gubina, *Dokl. Akad. Nauk SSSR*, **255**, 1144 (1980).

560a. V. G. Kharchenko, I. A. Markushina, and T. L. Gubina, U.S.S.R. Patent No. 677,330; *Chem. Abstr.*, **95**, 24802a (1981).

560b. V. G. Kharchenko, T. I. Gubina, and I. A. Markushina, *Zh. Org. Khim.*, **18**, 394 (1982).

560c. G. A. Miller, N. D. Heindel, and J. A. Minatelli, *J. Heterocycl. Chem.*, **18**, 1253 (1981).

561. K. Hirai, *Tetrahedron Lett.*, 1137 (1971).

562. K. Hirai and T. Ishiba, *Chem. Pharm. Bull.*, **29**, 304 (1972).

563. H. Hartmann, *J. Prakt. Chem.*, **313**, 730 (1971).

564. K. Hirai and T. Ishiba, *Heterocycles*, **3**, 217 (1975).

565. K. Hirai and T. Ishiba, *Chem. Pharm. Bull.*, **19**, 2194 (1971).

566. K. Hirai and T. Ishiba, *Chem. Pharm. Bull.*, **20**, 2384 (1972).

567. K. Hirai, H. Sugimoto, and T. Ishiba, *J. Org. Chem.*, **45**, 253 (1980).

568. H. Hartmann, H. Schäfer, and K. Gewald, *J. Prakt. Chem.*, **315**, 497 (1973).

569. Y. Tominaga, Y. Matzuda, and G. Kobayashi, *Heterocycles*, **4**, 9 (1976).

570. K. T. Potts, E. Houghton, and U. P. Singh, *J. Org. Chem.*, **39**, 3627 (1974).

571. K. T. Potts, S. J. Chen, J. Kane, and J. L. Marshall, *J. Org. Chem.*, **42**, 1633 (1977).

572. K. T. Potts, E. Houghton, and U. P. Singh, *J. C. S. Chem. Commun.*, 1129 (1969).

573. K. T. Potts and U. P. Singh, *J. C. S. Chem. Commun.*, 569 (1969).

574. K. T. Potts, D. R. Choudbury, A. J. Elliott, and U. P. Singh, *J. Org. Chem.*, **41**, 1724 (1976).

575. H. Gotthardt, M. C. Weisshuhn, and B. Christl, *Chem. Ber.*, **109**, 753 (1976).

576. H. Gotthardt and B. Christl, *Tetrahedron Lett.*, 4747 (1968).

577. H. Matsukobo and H. Kato, *J. C. S. Perkin 1*, 2565 (1976).

577a. H. Kato, T. Shiba, N. Aoki, and H. Iijima, *J. C. S. Perkin Trans. 1*, 1885 (1982).

578. K. T. Potts and S. Husein, *J. C. S. Chem. Commun.*, 1360 (1970).

579. E. Funke, R. Huisgen, and F. C. Schaefer, *Chem. Ber.*, **104**, 1550 (1971).

580. K. Gewald and M. Hentschel, *J. Prakt. Chem.*, **318**, 343 (1976).

581. H. Dehne and P. Krey, *Pharmazie*, **33**, 687 (1978).

581a. Ch. Jenny, D. Obrecht, and H. Heimgartner, *Helv. Chim. Acta*, **65**, 2583 (1982).

582. H. Gotthard, C. M. Weisshuhn, O. M. Huss, and D. J. Brauer, *Tetrahedron Lett.*, 671 (1978).

583. H. Gotthardt and O. M. Huss, *Lieb. Ann.*, 347 (1981).

584. F. Ishii, R. Okazaki, and N. Inamoto, *Heterocycles*, **6**, 313 (1977).

585. D. M. McKinnon and M. E. Hassan, *Can. J. Chem.*, **51**, 3081 (1973).

586. D. M. McKinnon, M. E. R. Hassan, and M. Chauhan, *Can. J. Chem.*, **55**, 1123 (1977).

587. B. Bartho, J. Faust, R. Pohl, and R. Mayer, *J. Prakt. Chem.*, **318**, 221 (1976).

588. F. Boberg and R. Wiedermann, *Lieb. Ann.*, **734**, 164 (1970).

589.　M. A.-F. Elkaschef, F. M. E. Abdel-Megeid, and A. L. Elbarbary, *Tetrahedron,* **30,** 4113 (1974).

590.　F. Arndt and G. Traverso, *Chem. Ber.,* **89,** 124 (1956).

591.　A. Josse, M. Stavaux, and N. Lozac'h, *Bull. Soc. Chim. Fr.,* 1723 (1974).

592.　R. M. Moriarty and A. Chin, *J. C. S. Chem. Commun.,* 1300 (1972).

593.　K.-P. Zeller, H. Meier, and E. Müller, *Lieb. Ann.,* **766,** 32 (1972).

594.　W. Kirmse and L. Horner, *Lieb. Ann.,* **614,** 4 (1958).

595.　H. Bühl, U. Timm, and H. Meier, *Chem. Ber.,* **112,** 3728 (1979).

596.　J. Font, M. Torres, H. E. Gunning, and O. P. Strausz, *J. Org. Chem.,* **43,** 2487 (1978).

597.　A. K. Yus'kovich, T. A. Danilova, and L. M. Petrova, *Khim. Geterotsikl. Soedin.,* 713 (1973).

598.　L. M. Petrova, A. A. Freger, and E. A. Victorova, *Khim. Geterotsikl. Soedin.,* 1144 (1973).

599.　E. A. Victorova, A. A. Freger, A. V. Egorov, and L. M. Petrova, *Khim. Geterotsikl. Soedin.,* 141 (1973).

600.　I. Degani, R. Fonchi, and C. Vincenzi, *Gazz, Chim. Ital.,* **97,** 397 (1967).

601.　A. Schönberg and B. König, *Tetrahedron Lett.,* 3361 (1965).

602.　B. König, J. Martens, K. Praefcke, A. Schönberg, H. Schwarz, and R. Zeisberg, *Chem. Ber.,* **107,** 2931 (1974).

603.　K. Praefcke and C. Weichsel, *Lieb. Ann.,* 1604 (1980).

604.　I. El-Sayed El-Kholy, M. M. Mishrikey, and H. M. Fuid-Alla, *J. Heterocycl. Chem.,* **14,** 845 (1977).

605.　Ch. L. Pedersen, *Acta. Chem. Scand. B,* **29,** 791 (1975).

605a.　Y. Aso, M. Iyoda, and M. Nakagawa, *Tetrahedron Lett.,* **23,** 2473 (1982).

606.　H. Yamaoka, I. Mishima, and T. Hanafusa, *Bull. Chem. Soc. Jpn.,* **53,** 1763 (1980).

606a.　V. Cecchetti, A. Fravolini, and F. Schiaffella, *J. Heterocycl. Chem.,* **19,** 1045 (1982).

607.　K. Skinnemoen and K. Undheim, *Heterocycles,* **16,** 929 (1981).

608.　T. Eicher and S. Böhm, *Tetrahedron Lett.,* 3965 (1972).

608a.　Th. Eicher and S. Böhm, *Chem. Ber.,* **107,** 2238 (1974).

609.　W. E. Parham and V. J. Traynelis, *J. Am. Chem. Soc.,* **76,** 4960 (1954).

610.　W. E. Parham and V. J. Traynelis, *J. Am. Chem. Soc.,* **77,** 68 (1955).

611.　H. H. Szmant and L. M. Alfonzo, *J. Am. Chem. Soc.,* **79,** 205 (1957).

612.　W. E. Parham, I. Nicholson, and V. J. Traynelis, *J. Am. Chem. Soc.,* **78,** 850 (1956).

613.　W. E. Parham, E. T. Harper, and R. S. Berger, *J. Am. Chem. Soc.,* **82,** 4932 (1960).

614.　C. L. Gajurel and S. R. Vaidya, *Indian J. Chem.,* **19B,** 911 (1980).

615.　G. M. Badger, P. Cheuychit, and W. H. F. Sasse, *Aust. J. Chem.,* **17,** 353 (1964).

616.　W. E. Parham, G. L. O. Mayo, and B. Gadsby, *J. Am. Chem. Soc.,* **81,** 5993 (1959).

617.　K. Kobayashi, K. Mutai, and H. Kobayashi, *Tetrahedron Lett.,* 5003 (1979).

618.　K. Kobayashi, and K. Mutai, *Tetrahedron Lett.,* **22,** 5201 (1981).

619.　K. Kobayashi and T. Ohi, *Chem. Lett.,* 645 (1973).

620.　D. C. Blomstrom, U.S. Patent No. 3,207,728; *Chem. Abstr.,* **64,** 701 (1966).

621.　H. E. Simmons, U.S. Patent No. 3,400,134; *Chem. Abstr.,* **69,** 106715b (1968).

622.　H. E. Simmons, R. D. Vest, D. C. Blomstrom, J. R. Roland, and Th. L. Cairns, *J. Am. Chem. Soc.,* **84,** 4746 (1962).

623.　H. E. Simmons, R. D. Vest, S. A. Vladuchick, and O. W. Webster, *J. Org. Chem.,* **45,** 5113 (1980).

624. O. Scherer and F. Kluge, German Patent No. 1,276,048; *Chem. Abstr.,* **69**, 9021 (1968).

625. C. G. Krespan, U.S. Patent No. 3,073,844; *Chem. Abstr.,* **59**, 1650 (1963).

626. Th. E. Young and A. R. Oyler, *J. Org. Chem.,* **45**, 933 (1980).

627. K. Kobayashi and K. Mutai, *Tetrahedron Lett.,* 905 (1978).

628. G. N. Schrauzer and V. P. Mayweg, *J. Am. Chem. Soc.,* **87**, 1483 (1965).

629. F. Boberg, *Angew. Chem.,* **73**, 579 (1961).

630. F. Boberg, *Lieb. Ann.,* **679**, 118 (1964).

631. F. Boberg, H. Niemann, and J. Jovanović, *Lieb. Ann.,* **717**, 154 (1968).

632. A. Marei and M. M. A. El Sukkary, *U. A. J. Chem.,* **14**, 101 (1971).

633. F. Boberg. H. Niemann, and K. Kirchhoff, *Lieb. Ann.,* **728**, 32 (1969).

634. T. Bacchetti, A. Alemagna, and B. Danieli, *Tetrahedron Lett.,* 3569 (1964).

635. G. Purrello and A. Lo Vullo, *Boll. Sedute Accad. Gioenia Sci. Nat. Catania,* **9** [4], 46 (1967); *Chem. Abstr.,* **70**, 3690 (1969).

636. J. C. Martin, R. D. Burpitt, P. G. Gott, M. Harris, and R. H. Meen, *J. Org. Chem.,* **36**, 2205 (1971).

CHAPTER II

Theoretical Calculations on Thiophenes

ANITA HENRIKSSON-ENFLO

Institute of Theoretical Physics, University of Stockholm, Stockholm, Sweden

I. AN INTRODUCTION INTO QUANTUM CHEMICAL MOLECULAR ORBITAL METHODS

1. General

The quantum chemical methods[1] commonly in use today are all based on solving the nonrelativistic, time-independent Schrödinger equation (Eq. 1)

$$H\psi = E\psi \tag{1}$$

where H is the Hamilton operator of the system, consisting of the sum of a potential energy term, V, and a kinetic energy term, T;

$$H = T + V \tag{2}$$

and E is the eigenvalue of the state and ψ is the wave function that describes the spatial motion of all the particles of the system moving in the field of force specified by V.

Equation (1), although seemingly very simple, can, however, be solved exactly only for the one-electron case, that is, for H or H_2^+. For all other cases, approximate solutions are used.

Molecular systems have usually been treated in the Born–Oppenheimer approximation, which allows a separation of nuclear and electronic motions. The electronic system is then treated for a fixed geometry of the molecule.

In the electronic Hamiltonian, the kinetic energy operator has the form given in Eq. (3) and the potential energy operator has the form shown in Eq. (4).

$$T_{e1} = -\frac{h^2}{(8\pi^2 m)} \sum_i \nabla_i^2 \tag{3}$$

$$V_{e1} = -\sum_A \sum_i e^2 Z_A r_{Ai}^{-1} + \sum_{i<j} e^2 r_{ij}^{-1} \tag{4}$$

Thus, the Hamiltonian can be divided into a sum of one-electron parts, $H(i)$, and a sum of two-electron parts, $\sum e^2/r_{ij}$, describing the repulsion between the electrons (Eq. 5).

$$H = \sum_i H(i) + \sum_{i<j} e^2 r_{ij}^{-1} \tag{5}$$

For many-electron systems, the wave function ψ may be approximated by a product of one-electron functions, $\psi(i)$, each depending upon the coordinates of one electron only (Eq. 6).

$$\psi(1, 2, 3, \ldots, n) = \psi(1)\psi(2)\psi(3), \ldots, \psi(n) \tag{6}$$

It is convenient for molecular systems to build up the molecular wave function ψ as a linear combination of atomic orbitals, Φ, the LCAO–MO method (Linear Combination of Atomic Orbitals – molecular orbital)[2] (Eq. 7).

$$\psi = \sum C_i\Phi_i \tag{7}$$

The Schrödinger equation is then solved by the variational method, which yields a secular equation that must be solved. The calculated energy E is always higher than the true value (E_0); but for the case where the approximative wave function is equal to the true wave function, $E = E_0$.

In most molecules, the number of electrons are too large to be treated explicitly without computational difficulties. Since most chemical properties are determined by the outermost electrons, it can be convenient to treat only these explicitly, keeping the other electrons fixed. The advent of larger computers has allowed more electrons to be handled.

The calculations can be reduced by introducing certain parameters, often chosen to fit experimental data – the semiempirical methods. In the ab initio methods, no such parameters are introduced; all values, except the charges, are calculated.

2.   Semiempirical Methods

A.   The π-Electron Methods

a.   THE HÜCKEL METHOD

The quantum mechanical laws were applied early to chemical systems. The first calculations on organic molecules within the LCAO–MO formalism were made around 1930 by Hückel.[2] The Hückel method, HMO,[2,3] is a very crude one indeed, with many approximations. Thus, no interaction between the electrons is explicitly taken into account and the Hamiltonian, H, is reduced to a sum of one-electron contributions (Eq. 8)

$$H = \sum H(i) \tag{8}$$

Hückel also introduced the following approximations:

1. $H_{uu} = \int \phi_u H(i)\phi_u d\tau = \alpha$ (Coulomb integral)

2. $H_{uv} = \int \phi_u H(i)\phi_v d\tau = \beta$ (resonance integral)   for neighbors
$\qquad\qquad\qquad\qquad = 0$ for non-neighbors

3. $S_{uv} = \int \phi_u \phi_v d\tau = 1$ (overlap integral)   for $u = v$
$\qquad\qquad\qquad = 0$ for $u \neq v$

The Hückel method was useful for calculations on the π-electron part of planar conjugated organic systems, such as benzene. In this case, each atom contributes one orbital (a carbon $2p_z$) to the system. This leads to a secular determinant, Eq. (9), which can be handled easily.

$$|H_{uv} - \epsilon\delta_{uv}| = 0 \tag{9}$$

For π-electron systems containing heteroatoms, such as thiophene, the Hückel method can easily be modified by choosing variables α and β to fit the experimental data. The values of α and β were different for different types of experiments (e.g., UV spectra, ionization potentials, reduction potentials, oxidation potentials, etc.).

b.   THE PPP METHOD

The neglect of the electronic repulsion is a severe shortcoming of the Hückel method. The next step[4,5] in the development of more advanced quantum chemical methods was thus to include the electronic repulsion explicitly in the Hamiltonian (Eq. 10).

$$H = \sum_i H^c(i) + \sum_{i<j} e^2/r_{ij} \tag{10}$$

where r_{ij} is the distance between electrons i and j and H^c is the core Hamiltonian, consisting of the kinetic energy operator for an electron and the potential energy between an electron and all atomic cores of the molecule. In this case, the LCAO coefficients and energies are determined by the equations

$$\sum c_v(F_{uv} - \epsilon S_{uv}) = 0 \tag{11}$$

and

$$|F_{uv} - \epsilon S_{uv}| = 0 \tag{12}$$

where the elements in the Fock operator are given by the expressions in Eq. (13):

$$F_{uv} = H_{uv}^c + \sum\sum P_{\rho\sigma}(uv|\rho\sigma) - 0.5(u\rho|v\sigma) \tag{13}$$

in which

$$(uv|\rho\sigma) = \int \phi_u(1)\phi_v(1)\left(\frac{e^2}{r_{12}}\right)\phi_\rho(2)\phi_\sigma(2)d\tau \tag{14}$$

is called the Coulomb integral (not the same as for the Hückel method) and represents the repulsion between an electron distributed in space $\phi_u(1)\phi_v(1)$ and another in space $\phi_\rho(2)\phi_\sigma(2)$ and

$$(u\rho|v\sigma) = \int \phi_u(1)\phi_\rho(1)\left(\frac{e^2}{r_{12}}\right)\phi_v(2)\phi_\sigma(2)d\tau \tag{15}$$

is the exchange integral and represents the exchange between the electrons in the corresponding orbitals.

Finally, the bond order ($P_{\rho\sigma}$) is defined by

$$P_{\rho\sigma} = 2 \sum_k c_{k\rho} c_{k\sigma} \tag{16}$$

where the summation is extended over all occupied molecular orbitals.

Thus, to solve Eqs. (11) and (12), it is obvious that the bond order and the wave function must be known. This can be done by guessing an initial wave function, for example by using the results from a Hückel calculation as the initial guess. Equations (11) and (12) are then solved iteratively; that is, the calculations are repeated until the result differs from the input value by less than a certain limit. The orbitals thus obtained are called SCF orbitals (Self Consistent Field orbitals) and the technique used is the SCF method.

The equations to be solved are now much more complicated than those in the HMO method and require, in most cases, the help of electronic computers.

Another difficulty lies in the evaluation of the integrals, especially $(uv|\rho\sigma)$. To overcome these difficulties, some approximations can be made. In the most common method, the PPP method (Pariser–Parr–Pople method),[4,5] which also considers only π-electrons, the following approximations are made:

1. $(uv|\rho\sigma) = \delta_{uv}\delta_{\rho\sigma}\gamma_{u\rho}$

2. $S_{uv} = \delta_{uv}$

3. $H_{uv} = \beta_{uv}$ for neighbors
    $= 0$ for non-neighbors

The first two are called the ZDO approximation (Zero Differential Overlap). They reduce drastically the number of integrals, especially the two-electron integrals, keeping only one- and two-center integrals (γ_{uv}) different from zero:

$$(uu|vv) = \gamma_{uv} \text{ (Coulomb integral)}$$

Note that this is not the same as the Coulomb integral in the Hückel theory. However, approximations 2 and 3 together are the same as the HMO approximations.

The PPP method is developed in the π-electron theory, and the validity of the σ-π-separation has been studied thoroughly.[6] The evaluation of the integrals needed in the PPP method can be made in different ways, either theoretically by using Slater-type atomic orbitals or semiempirically by fitting the results of the calculations to experimental data. It was, however, found in the earliest calculations[4] that the use of nonempirical integrals gave poor results, and that the results were improved by the introduction of semiempirical parameters.

Different ways for obtaining these parameters have been proposed. In a critical evaluation of the approximations inherent in the PPP method, Fischer-Hjalmars[7] found that the ZDO approximation was equivalent to a formal transformation of the atomic orbital basis to an orthogonalized atomic orbital basis. A scheme for choosing the parameters, taking into account these considerations, has been proposed.[8]

B.   All-Valence Electron Methods

One serious shortcoming of the π-electron theories is just that they can handle only the π part of conjugated planar systems. For the treatment of saturated or nonplanar systems or the σ part in conjugated planar systems, other methods had to be developed. The next step was to include all valence shell electrons in the calculations. In the first half of the 1960's, two different methods were developed: the extended Hückel theory (EHT)[9] and the CNDO method (Complete Neglect of Differential Overlap).[10]

a.   THE EXTENDED HÜCKEL THEORY (EHT)

The extended Hückel theory is a direct extension of the Hückel theory. Thus, no electronic repulsion is taken into account explicitly. The Hamilton operator is a sum of one-electron Hamiltonians; hence, the procedure is not an SCF technique. The secular determinant is similar to the one for the Hückel method. Because the overlap between σ-orbitals is larger than between π-orbitals, all overlap integrals are calculated by the EHT method. Atomic orbitals were chosen as Slater-type orbitals. The diagonal elements (H_{ii}) were taken from experimentally measured atomic ionization potentials and the nondiagonal elements (H_{ik}) were calculated from Eqs. (17) or (18).

$$H_{jk} = 0.5K*S_{jk}*(H_{jj} + H_{kk}) \qquad (17)$$

or

$$H_{jk} = -K*S_{jk}*sqrt(H_{jj}H_{kk}) \qquad (18)$$

where the constant K was chosen to fit experimental data.

Extended Hückel calculations have been used for studies of conformation energies and orbital symmetries. Studies of orbital symmetries have been applied successfully to predictions of chemical reactivities.[11]

b.   THE CNDO METHOD

Another way to proceed is an extension of the PPP method, wherein the electronic repulsion is taken into account explicitly. The simplest version, the CNDO method, involves the ZDO approximation. Whereas the ZDO approximation can be justified for the PPP method,[7] this is not the case for the CNDO method,[12] because the overlap between σ-orbitals is much larger than between π-orbitals. To keep the results invariant under coordinate tranformation, the two-center Coulomb integrals (γ_{uv}) and resonance integrals (β_{uv}) were, in the original CNDO version, approximated to be integrals between s-type orbitals. The two-center Coulomb integrals can be calculated rather easily from Slater-type orbitals. The one-center integrals needed in the calculations were obtained from experimental data of atoms. Finally, the resonance integrals were chosen to fit results from minimal basis set ab initio calculations on diatomic molecules. Many other suggestions for parameterization have been made such as the method of Del Bene and Jaffe.[13] This technique, the CNDO/S-CI method, is parameterized to give good excitation energies.

c.   THE PNDO METHOD

To maintain invariant results under coordinate transformations, all two-center Coulomb integrals (γ_{uv}) were taken as integrals over s-type orbitals in the CNDO method. Another way to overcome this problem was suggested in the PNDO method (Partial Neglect of Differential Overlap), developed by Dewar and Klopman,[14] in which the integrals are evaluated after transformation to a set of symmetry axes between the two atoms in a bond. The one-center integrals are, however, still dependent on the coordinate axes. The parameters needed are chosen to give good agreement with experimental heats of formation, a quantity important for stabilities and reactivities of the molecules.

d.   THE INDO METHOD

As all one-center exchange integrals, owing to the ZDO approximation, are equal to zero, and all Coulomb integrals, owing to the invariance condition, are equal, the CNDO method gives just the average value of a configuration. Thus, for example, no difference between singlets and triplets can be obtained. In the INDO method (Intermediate Neglect of Differential Overlap),[10] the one-center exchange integrals were retained. There are several different modifications, parameterized to fit special types of experiments. The most common parameterization scheme, the original INDO, is an extension of the CNDO scheme. In the MINDO method (Modified Intermediate Neglect of Differential Overlap),[15] the parameters were chosen to give good heats of formation and have been successively refined: MINDO/1,[15] MINDO/2,[16] and MINDO/3.[17]

e.   THE PCILO METHOD

The PCILO method (Perturbative Configuration Interaction of Localized Orbitals)[18] chooses the most important configurations for interaction with the aid of localized orbitals. The parameters are based on the original CNDO parameters and the method has been extensively used for studies of conformational energies.

3.   Nonempirical Methods

A.   All-Electron Theories

For a complete description of a molecule, all electrons, valence, and inner shells are needed. The development of large computers has made it possible to treat many-electron systems with high accuracy. It should be pointed out, however, that some approximations still remain to be done. Thus, for example, electronic correlation cannot be accounted for by a single SCF calculation; but methods have been developed to treat this problem rather accurately by configuration interaction (CI).[19] Also, all calculations are done in the Born–Oppenheimer approximation, which states that the nuclei are in a fixed position.

In the ab initio methods, no empirical parameters, except the charge of the electron, are needed. The wave functions are expressed in terms of some basis functions, generally atomic orbitals, but other types of basis functions have also been used. The atomic orbitals can be expressed as Slater-type orbitals, which consist of the product of a radial and an angular part. The angular part consists of the spherical harmonics, whereas the radial part, $R(r)$, has the form of Eq. (19).

$$R(r) = N*P(r) \exp(-\alpha r) \qquad (19)$$

The integrals needed are all calculated. The number of integrals to be stored determines the limit of the computations and is defined by the size of the computer used.

In the practical calculations, it was easier to use Gaussian orbitals (Eq. 20) instead of Slater-type orbitals, although many Gaussian orbitals are needed to describe one Slater-type orbital.

$$R(r) = N*P(r) \exp(-\alpha r^2) \qquad (20)$$

When only one Gaussian orbital is utilized for the description of an orbital, the basis set used is said to be minimal. For two Gaussians per atomic orbital, the basis set is called double zeta, for three, triple zeta, and so on.

B.   Pseudopotential Methods

The inner-shell electrons generally do not take part in chemical reactions to any great extent. Thus, they can be treated as a fixed core. In the pseudopotential methods,[20] the inner-shell electrons are approximated by a potential. This method can be compared with the semiempirical all-valence electron methods, but no empirical parameters are needed.

4.   Applications of the Quantum Chemical Calculations

Through quantum chemical calculations, the wave function ψ for the electrons, the orbital energies, and the total energy of the system are obtained. These values give data of importance for determining the chemical and physical properties of a molecule. A great advantage of theoretical calculations is that calculations can be made on systems difficult or impossible to synthetisize or isolate. Thus, hypothetical and highly excited structures can also be studied.

A.   Total Energies

The total energy of a system gives a measure of the stability of the system: the lower the total energy, the more stable the system. From calculations of total energies for different geometries, reaction paths can be obtained. The reaction is said to proceed in the energy valley, that is, on a path where the energy is lowest.

Calculations on reaction paths have been a dream of many quantum chemists, but the calculations are in most cases very complicated. The number of the degrees of freedom are often very large; this leads to calculations on a huge amount of structures with different geometries. Other types of reactivity indices have thus been developed (see Section I.4.D).

Calculations of the total energy of a molecule with varying geometry give the most stable geometries and conformations. Moreover, vibrational frequencies can be obtained from these computations.

Calculations of the total energy of the equilibrium state give heats of formation and resonance energies. In many cases, the equilibrium geometry is known from experimental measurements or can be approximated from known bond lengths and bond angles.

Calculations of total energies of different electronic states give ionization potentials, electron affinities, and excitation energies. Such a procedure is often very tedious, and shorter ways to obtain these values have been developed (see Section I.4.B).

B.   Orbital Energies – Ionization Potentials and UV Spectra

According to Koopmans' theorem,[21] the orbital energies for closed-shell systems can be referred to the vertical ionization potential (Eq. 21).

$$\epsilon = - \text{IP} \qquad (21)$$

Thus, the orbital energies have been extensively used for predictions and assignments of ionization potentials, measured by photoelectron and X-ray electron spectroscopy. The energies of the unoccupied orbitals (virtual orbitals) give, in the same way, a measure of the electron affinities of a molecule.

Excitation energies can be obtained from orbital energies,[22] using the formulas in Eqs. (22) and (23),

$$E_{i-j} = \epsilon_j - \epsilon_i - J_{ij} + 2K_{ij} \quad \text{for singlets} \qquad (22)$$

$$E_{i-j} = \epsilon_j - \epsilon_i - J_{ij} \qquad \text{for triplets} \qquad (23)$$

where J_{ij} and K_{ij} are, respectively, the Coulomb and exchange integrals between orbitals i and j. In the Hückel theory, where no electron repulsion is taken into account and J and K thus are zero, the excitation energies are obtained from the difference between the orbital energies.

The virtual orbitals do not represent the excited state exactly, because the electronic repulsion is not correctly taken into account. The CI between the virtual orbitals gives a better description of the excited state.

C.   Charge Distributions

The wave function ψ represents the electronic distribution over the molecule.

The probability dP that an electron will be found in the space $d\tau$ is given by Eq. (24).

$$dP = \psi\psi*d\tau \tag{24}$$

In the LCAO–MO formalism, the electron density is given by Eq. (25).

$$\psi\psi*d\tau = \left(\sum c_v\Phi_v\right)^2 d\tau$$

$$= \sum c_v^2\phi_v^2 d\tau + \sum\sum c_u c_v \phi_u \phi_v d\tau \tag{25}$$

The net population on an atom r is defined by Eq. (26).

$$n(r) = \sum_j n_j c_{jr}^2 \tag{26}$$

where n_j is the occupation number of orbital j and the summation is taken over all occupied orbitals j. Similarly, the overlap population $n(r, v)$ is defined by Eq. (27)

$$n(r, v) = \sum c_{jr} c_{jv} S_{jrv} \tag{27}$$

In the ZDO approximation, where no overlap is involved, the population is thus given by Eq. (26) only. In other cases, different definitions for the populations can be made. The most common definition is from Mulliken.[23] Here, the net population is defined according to Eq. (26), and the gross population of an atom is defined as the net atomic population plus half the sum of all overlap populations related to that atom.

D. Reactivity Indices

For a complete description of reaction paths, accurate calculations of the total energy of the system at different geometries are needed. This is a very complicated procedure and is not yet possible to do, except for very small systems.

Simplified reactivity indices have been developed[3, 24] and most of these were applied to HMO calculations. They can also be used for other types of calculations, sometimes modified because of the approximations made.

The reactivity of a molecule can often be studied from purely electrostatic considerations. Thus, for example, if a certain region is highly positive or negative, an oppositely charged reactant is attracted to that position. Thus, the charges, defined by gross or net populations and the nuclear charges, are often used as reactivity indices.

Another index, which describes the electronic population in the bond region, is given by the bond order p_{rs}, defined by Eq. (27).

$$p_{rs} = \sum n_j c_{jr} c_{js} \tag{27}$$

The free valence index, F_r,[25] defined in Eq. (28)

$$F_r = N_{max} - N_r \tag{28}$$

in which N_r is the sum of all bond orders related to atom r and N_{max} is the maximum value of N_r, usually taken as $sqrt(3)$, gives an index of how much valence is left for a new bond to be built.

Because the highest occupied orbitals are supposed to be most affected by a reagent, it was suggested[26] that these orbitals should have more influence on reactivity than other lower lying orbitals. The energies and populations of the highest occupied orbitals (HOMO), can thus be used as reactivity indices. The lowest unoccupied orbitals (LUMO) are also of importance. In the concept of superdelocalizability defined by Eq. (29), the charges are weighted according to the energies (m_j) of the occupied orbitals.

$$S_r = 2\sum \frac{(c_{jr})^2}{m_j} \tag{29}$$

The reactivity indices mentioned above are all based on the assumption that the initial reactants only determine the reactivity. In chemical reactions, the reaction path is assumed to proceed generally through an activated complex to the final products. The energy barrier between the reactants and the activated complex may be the determining index for reactivity. For example, electrophilic substitution reactions are supposed to proceed by two mechanisms. Either the reagent binds weakly to the π-electronic system, forming a weak π-complex, or the conjugated system breaks at the reacting position and a rather strong σ-complex, a Wheland complex,[27] is formed as an intermediate. In the first case, reactivity indices based only on the static properties of the molecule, such as charges, bond orders, free valence, and superdelocalizabilities, are thought to be good. In the second case, the structure of the formed complex must also be considered.

The energy changes for reacting systems can be estimated rather easily in the HMO theory using Eq. (30), which gives the energy change ∂E, owing to changes in α and β.[28]

$$\delta E = q_r \delta \alpha_r + 2p_{rs} \delta \beta_{rs} + \text{higher-order terms} \tag{30}$$

Thus, for example, breaking a π-bond can be considered as a change in β_{rs} from a known value to zero. This gives the localization energy, a quantity used in studies of electrophilic aromatic substitution reactions based on the Wheland complex picture.

It can often be difficult to know which reactivity index should be used for a certain reaction. In cases where the activated complex is very similar to the reactants, the indices based only on the reactants (static indices) are applicable. In cases where the activated complex is very different from the reactants, both reactants and activated complex must be considered (e.g., in Eq. 30). Such an index is called a dynamic index.

If calculations of a certain index agree well with experimental data, this can give information about the reaction mechanism. For example, if the charge is a

good index, this suggests that the reaction is proceeding with only a small perturbation of the reactant. If the localization energy is a good index, the reaction
probably proceeds through a reaction intermediate.

Another important energy quantity is the resonance energy, obtained as the
difference between the π-electron energy of the system and the π-electron energy
of the corresponding localized double bonds.

E.   Properties

From quantum chemical data, many properties of a molecule can be calculated.
Examples of such properties are dipole moments, nuclear quadrupole moments,
and epr and nmr parameters.[1, 29] The equations involved are often very tedious and
the reader is referred to the special literature in this field.

II.   QUANTUM CHEMICAL CALCULATIONS ON THE THIOPHENE MOLECULE

1.   General

The thiophene molecule is a planar five-membered ring of C_{2v} symmetry, similar
to furan and pyrrole (see Fig. 1). The structure is well-known from microwave[30]
and electron diffraction studies.[31, 32]

There have been several quantum chemical calculations on thiophene and
thiophene-like compounds. All levels of approximation have been used, from
simple Hückel type,[33-54, 105-120] valence bond,[55] PPP,[54, 56-76, 121-131] EHT,[77, 78, 130]
CNDO/INDO,[79-89, 133-136] and MINDO[90-92, 137] to ab initio.[93-101, 138] A calculation[102] within the X_α-theory[103, 104] has also been performed.

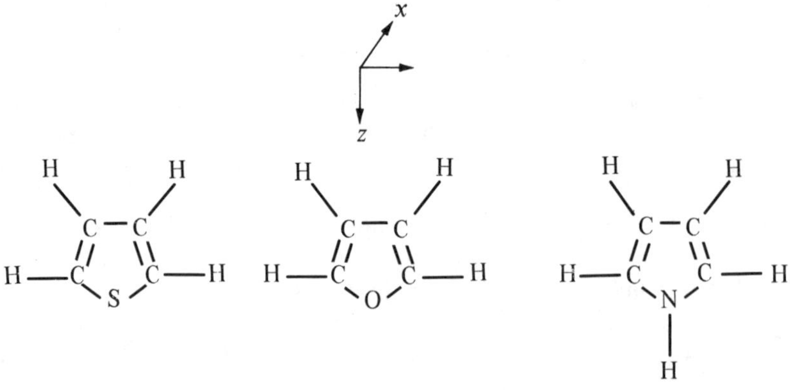

Fig. 1.   The structure of the iso-π-electronic molecules thiophenes, furan, and pyrrole, with
C_{2v} symmetry.

The calculations are listed in Table 1, with the name of the author(s), publication year, and application. Many different techniques are still in use today; it is not always necessary or meaningful to use the most sophisticated method. Depending on the problem to be studied, the method of calculation should be chosen.

2.   The Role of d-Orbitals on Sulfur

One of the earliest extensions of the HMO method, to include heteroatoms, was applied to the thiophene molecule.[33] In this work, Wheland and Pauling assumed that the sulfur atom contributes with a $3p_z$-orbital to the π-system. Thus, in the HMO approximation, the treatment of thiophene is similar to the treatment of pyrrole and furan, all molecules having six electrons in the π-system (as does benzene). The difference in the treatment of these three molecules is in the choice of α and β.

It is well-known that thiophene has more aromatic character than furan and pyrrole[139] and shows many similarities to benzene. Also, the electronegativity, according to Pauling,[140] is the same for sulfur and carbon, namely 2.5, whereas it is 3.0 for nitrogen and 3.5 for oxygen.

It was suggested that thiophene differs from furan by having low lying empty d-orbitals that could take part in the binding. This suggestion was first made by Schomaker and Pauling,[55] using the valence bond picture. From a consideration of bond lengths, resonance energies, and dipole moments, it was concluded that several types of structures were important. The relative weights of these structures were estimated to be 70% for structures without d-orbitals and 30% for structures including d-orbitals.

The effect of d-orbitals was studied in more detail with HMO calculations by Longuet-Higgins.[35] By employing $3p_z - 3d_{xz} - 3d_{yz}$-hybrid orbitals on the sulfur atom, the similarity with benzene was apparent. Other calculations[44] have shown, however, that the $3d$-orbitals are too high in energy to make any contribution.

There have been several calculations within the HMO theory. Some included d-orbitals,[35,37,38,40,41,48] others did not. Because of the semiempirical character, satisfactory agreement with experimental data can be obtained in both cases.

In the PPP-type calculations, inclusion of $3d$-orbitals leads to mathematical difficulties. Thus, in most of the calculations, the $3d$-orbitals are omitted. In the work of Bielefeld and Fitts,[62] a comparison between calculations with d-orbitals and without d-orbitals was made. It was found that the d-orbitals participate to only a small extent in the binding (0.05 electrons on $3d_{xz}$ on sulfur) (see Table 2), but that other properties, such as charge densities and the electronic spectrum, were markedly changed (Table 3). Similar calculations within the CNDO method also show the same trends.[79] The populations in the $3d_\sigma$ and $3d_\pi$ orbitals were 0.24 and 0.14 electrons, respectively. A further addition of $4s$ and $4p$ orbitals resulted in a population of only 0.04 electrons in these additional orbitals. On the other hand, the extra d-orbitals resulted in a change in the populations on all atoms (see Table 3). Thus, for example, the extra $3d$-orbital increased the sulfur population by 0.07 electrons, resulting in a change of charge from -0.03 to $+0.04$.

TABLE 1.

A. Survey over Theoretical Calculations on Thiophene

Method	Author	Reference	Year	Application
HMO	Wheland, Pauling	33	1935	Charges, reactivity
	Daudel et al.	34	1948	Reactivity
	Longuet-Higgins	35	1949	Effect of d-orbitals
	Nagakura, Hosoya	36	1952	Charges, bond orders, resonance energies
	Metzger, Ruffler	37	1954	Effect of d-orbitals
	de Heer	38	1954	Reactivity
	Melander	39	1955	Reactivity
	Kikuchi	40	1957	Charges
	Kreevoy	41	1958	Charges, free valences, reactivity
	Milazzo, DeAlti	42	1959	Electronic spectrum
	Maeda	43	1960	Effect of d-orbitals
	Mangini, Zauli	44	1960	Charges, UV spectra
	Pilar, Morris	45	1961	Reactivity, charges
	Eland	46	1969	Ionization potentials
	Decoret, Tinland	47	1971	Reactivity
	Hess, Schaad	48	1973	Resonance energies
	Rogers, Cammarata	49	1976	Diamagnetic susceptibility
	Aihara	50	1976	Resonance energies
	Osamura et al.	51	1976	C-13 Nmr chemical shifts
	Julg, Sabbah	52	1977	Comparison of thiophene and furan
	Duben	53	1978	Reaction mechanism of metal-catalyzed hydrodesulfurization
PPP	Parkanyi, Herndon	54	1978	Bond orders, bond lengths
	Parkanyi, Herndon	54	1978	Bond orders, bond lengths
	Santhamma	56	1956	UV Spectra, charges
	Sappenfeld, Kreevoy	57	1963	UV Spectra, charges, bond orders
	Pajol, Julg	58	1964	UV Spectra

228

	Author	No.	Year	Description
	Wachters, Davies	59	1964	UV Spectra
	Solony et al.	60	1965	UV Spectra, charges, bond orders, free valences
	Zweig, Hoffmann	61	1965	Orbitals, polarography, esr-parameters
	Bielefeld, Fitts	62	1966	Effect of d-orbitals
	Durr et al.	63	1966	Spectra
	Momicchioli et al.	64	1967	Spectra
	Billingsley, Bloor	65	1968	UV Spectra, ionization potentials
	Fabian et al.	66	1968	UV Spectra, ionization potentials, charges
	Chowdhury, Basu	67	1969	UV Spectra
	Phan-Tan-Luu et al.	68	1969	Charges, free valence
	Hammond	69	1970	UV Spectra
	Skancke, Skancke	70	1970	Parameterization, charges, UV spectra, geometries
ETH	Julg et al.	71	1970	Charges, bond orders, free valences
	Dewar, Trinajstic	72	1970	Heats of formation, geometries, charges
	Bodor et al.	73	1971	Geometry
	Wadt, Moomaw	74	1973	Effect of d-orbitals on spin-orbit coupling
	Håkansson et al.	75	1977	MCD Spectra
	Das Gupta, Birss	76	1980	Heat of atomization, resonance energy, charges
	Derrick et al.	77	1971	Ionization potentials
	Mårtensson, Chojnacki	78	1973	Electron density maps
CNDO/INDO	Clark	79	1967	Effect of d-orbitals on charges, nmr spectra
	Yonezawa et al.	80	1969	Spectra
	Tajiri et al.	81	1971	UV Spectra
	Galasso	82	1973	Effect of d-orbitals on nmr and esr parameters
		83	1975	Nmr Coupling constants
	Gold'farb et al.	84	1974	Reactivity
	Schulte, Schweig	85	1974	Effect of d-orbitals on spectra
	Epiotis et al.	86	1976	Furan, pyrrole, thiophene
	Andermann, Phillips	87	1976	Ionization potentials
	Chambers, Thomas	88	1977	Satellite peaks in PES
	Kluge, Scholz	89	1979	Ionization potentials
MINDO	Dewar	90	1975	Heat of formation, geometry, ionization potentials

TABLE 1. *(CONTINUED)*

Method	Author	Reference	Year	Application
	Dewar, Ford	91	1977	IR Spectra
	Defina, Andrews	92	1980	Geometry, ionization potentials
Ab initio	Clark, Armstrong	93	1970	Effect of *d*-orbitals
	Clark	94	1972	Effect of *d*-orbitals
	Gelius et al.	95	1972	Effect of *d*-orbitals, ionization potentials, charges, properties
	Palmer, Findlay	96	1972	Effect of *d*-orbitals
	Nitzsche et al.	97	1974	Populations
	von Niessen et al.	98	1976	Ionization potentials, charges, properties
	Bernardi et al.	99	1977	Geometry
	Kao, Radom	100	1979	Geometry
	Hilal	101	1980	Charges, reactivity
X_α	DeAlti, Decleva	102	1981	Ionization potentials

B. Calculations on Thiophene-like Molecules

Method	Author	Reference	Year	Application
HMO	deHeer	38	1954	Reactivity of thiophene and isothianaphtene
	Melander	39	1955	Reactivity of substituted thiophenes
	Kreevoy	41	1958	Charges, free valences, bond orders of thiophene and 1,4-dithiadiene
	Hess, Schaad	48	1973	Resonance energies of sulfur containing heterocycles
	Osamura et al.	51	1976	C-13 Nmr chemical shifts vs charges of substituted thiophenes
	Duben	53	1978	Reaction mechanism of metal-catalyzed hydrodesulfurization
	Parkanyi, Herndon	54	1978	Bond-order vs. bond lengths of sulfur-containing heterocycles
	Berthier, Pullman	105	1950	Charges and bond orders of sulfur containing heterocycles
	Melander	106	1955	Reactivity of nitrothiophenes

Author	Ref.	Year	Description
Koutecky	107	1959	Delocalization energies, charges, bond orders of thiopyrylium and related compounds
Zahradnik et al.	108	1959	Spectra and reactivity of sulfur containing heterocycles
	109	1963	Spectra of thiophenes
Fabian et al.	110	1965	Reactivity of substituted thiophenes
Östman	111	1965	
	112	1965	
	113	1968	
Gerdil, Lucken	114	1965	Esr and UV spectra and polarography of dibenzothiophene
Vincent et al.	115	1966	Reactivity of methyl-substituted thiophenes
Gusten et al.	116	1969	Reactivity of benzodithiophene
Klasinc, Humski	117	1970	Reactivity of substituted thiophene
Kamienski	118	1971	Nmr Chemical shifts vs. charges of substituted thiophenes
Scharf, Leismann	119	1973	UV Spectra of substituted maleic acid anhydrides
Janda et al.	120	1974	Reactivity and nmr spectra of methyl-substituted thiophenes
PPP — Parkanyi, Herndon	54	1978	Bond orders, bond lengths of sulfur containing heterocycles
Wachters, Davies	59	1964	Spectra of dithienyls
Momicchioli et al.	64	1967	Spectra of benzothiophene
Billingsley, Bloor	65	1968	Spectra of substituted thiophene
Fabian et al.	66	1968	Ionization potentials, UV spectra, charges, bond lengths of sulfur containing heterocycles
Chowdhury, Basu	67	1969	UV Spectra of thiazoles
Phan-Tan-Luu et al.	68	1969	Charges and free valences of substituted thiophenes
Skancke, Skancke	70	1970	Parameterization
Dewar et al.	72	1970	Heats of atomization, bond lengths, and ionization potentials of sulfur containing heterocycles
Bodor	73	1971	Rotation barriers of phenylthiazoles and protonated forms

TABLE 1. (CONTINUED)

Method	Author	Reference	Year	Application
	dasGupta, Birss	76	1980	Resonance energies of sulfur containing heterocycles
	Fabian et al.	121	1968	UV Spectra of iso-π-electronic heterocycles
	Fabian	122	1968	Spectra and charges of 2-dimethylamino-thiophene
	Clark	123	1968	Spectra of thienothiophenes
	van Reijendam	124	1970	Spectra of polyenyl substituted thiophenes
	Skancke	125	1970	Rotation barrier of bithienyl
	Corradi et al.	126	1973	Proton chemical shifts of sulfur containing heterocycles
	Hartmann et al.	127	1975	UV and 13-NMR data of amino- and carbonyl substituted thiophenes
	Mehlhorn et al.	128	1977	UV Spectra of sulfur containing heterocycles
		129	1978	Photoreactivity of sulfur containing heterocycles
	Fabian, Hartmann	130	1978	Spectra of sulfur containing heterocycles
	Fabian	131	1979	Spectra of sulfur containing heterocycles
EHT	Galasso, DeAlti	132	1971	Conformations of phenylthiophenes
CNDO/INDO	Tajiri et al.	81	1971	UV Spectra of thienothiophene
	Galasso	82	1973	Nmr and esr parameters of halothiophenes
		83	1975	
	Gold'farb et al.	84	1974	Reactivity of thienothiophene
	Rodmar	133	1971	F-Nmr Chemical shifts in fluorothiophenes
	Galasso, Trinajstic	134	1973	Hyperfine splitting constants in sulfur containing radicals
	Nagata et al.	135	1973	Conformations of carbonyl derivatives of thiophene
	de Jong et al.	136	1974	Charge and bond orders of thiophene dioxide
MINDO	Dewar	90	1975	Heats of formation, geometry, ionization potentials of sulfur containing heterocycles
	Galasso et al.	137	1981	Conformation of 3,3'-dithienylmethane

Ab initio	Clark	94	1972	Barrier of inversion of protonated thiophene and thiophene oxide
	Palmer, Findlay	96	1972	The effect of d-orbitals on sulfur containing heterocycles
	Bernardi et al.	99	1977	Geometry of thiophene and protonated thiophene
	Kao, Radom	100	1979	Conformations, stabilities, and charges of substituted thiophenes
	Galasso et al.	138	1981	Geometry of 2-(2-thienyl)-pyrrole

TABLE 2.   POPULATIONS IN THE $3d$-ORBITALS ON SULFUR IN THIOPHENE

Method	Basis on Sulfur	Population in $3d$, $4p$ on S			Reference
		$3d_\sigma$	$3d_\pi$	$4p$	
PPP	$3p_z$, $3d_{xz}$, $3d_{yz}$		0.050		62
	$3p_z$, $3d_{xz}$, $3d_{yz}$, $4p_z$		0.050	0.008	62
CNDO	$3s$, $3p$, $3d$	0.24	0.14		79
	$3s$, $3p$, $3d$, $4p$	0.24	0.14	0.04	79
Ab initio	Double zeta, $3d$	0.143	0.038		95
	Double zeta, $3d$, H_{2p}	0.143	0.037		95

Both the σ-system and the π-system were involved in this charge migration, and so approximately to the same amount. The good agreement with experimental dipole moments and coupling constants when the extra orbitals were neglected suggests, however, that the CNDO method overestimates these extra orbitals.

The effects of d-orbitals on the electronic spectrum are rather large.[62, 85] This case, where excited states are also included, is discussed in more detail in Section II.4.

In all semiempirical theories, the parameters are chosen to fit some experimental data. It is thus difficult to know how much a certain choice of parameters influences the results. In the ab initio calculations, no semiempirical parameters are needed. Thus, a final decision about the importance of the d-orbitals should be made by ab initio calculations. However, as long as the basis is not complete, variations in the choice of basis set can also influence the results.

In the careful ab initio calculations by Gelius, Roos, and Siegbahn,[95] the effects of basis set variations were studied. The Gaussian basis was essentially contracted to double zeta, and in some of the calculations additional basis functions, $2p$ on hydrogen and $3d$ on sulfur were added. Total energies for the different calculations are found in Table 4. The energy decrease caused by the addition of $2p$-functions on hydrogen was 0.023 (a.u.) and that caused by the addition of sulfur $3d$ was 0.030 (a.u.). In a similar study with a somewhat different basis set, Clark and Armstrong[93] found an energy lowering of 0.12 (a.u.) when d-functions on sulfur were added. The total energy was, however, lower in the calculations of Gelius et al. than in the calculations of Clark and Armstrong, showing that the former calculations started with a better basis set. The energy difference, 0.506 (a.u.), between these two calculations clearly shows that an improvement of the basis set is more important than the participation of d-orbitals. The role of d-orbitals in this case is just to improve the basis, as would any addition of other functions.

In Table 2, gross atomic populations from ab initio calculations are given.[95] The d-orbitals on sulfur have a small population, 0.14 electrons in d_σ and 0.04 in d_π, which are somewhat smaller than the values obtained from CNDO calculations[93] (Table 2).

The ab initio calculations also show rather large variations in the populations on the other atoms (Table 3). Thus, for example, the charge on sulfur varies from

TABLE 3. CHARGES ON THE DIFFERENT ATOMS IN THIOPHENE WITH d-ORBITALS AND WITHOUT d-ORBITALS ON SULFUR

Method	Basis on Sulfur	Gross Charges on the Atoms					Dipole Moment	Reference
		H_α	H_β	C_α	C_β	S		
PPP	$3p_z$			−0.06	0.04	0.21	0.18	62
	$3p_z, 3d_{xz}, 3d_{yz}$			−0.09	0.01	0.16		62
	$3p_z, 3d_{xz}, 3d_{yz}, 4p$			−0.08	0.01	0.15		62
CNDO	$3s, 3p$	0.03	0.03	−0.05	−0.07	0.13		79
	$3s, 3p, 3d$	0.05	0.03	−0.06	−0.02	−0.04		79
	$3s, 3p, 3d, 4s, 4p$	0.05	0.04	−0.05	−0.01	−0.05		79
Ab initio	Double zeta	0.27	0.24	−0.57	−0.18	0.48	0.96	95
	Double zeta, $3d$	0.25	0.23	−0.25	−0.24	0.02	0.61	95
	Double zeta, H_2p	0.15	0.11	−0.44	−0.06	0.49	0.97	95
	Double zeta, $3d$, H_2p	0.12	0.10	−0.11	−0.12	0.00	0.62	95
Ab initio	Minimal	0.17	0.26	−0.24	−0.17	0.14		96
	Minimal, $3d$	0.16	0.15	−0.17	−0.16	0.04		96
	Minimal, $3d, 3s$	0.16	0.15	−0.17	−0.17	0.03		96
Ab initio	Minimal	0.23	0.22	−0.40	−0.25			90
	Minimal, $3d$	0.22	0.21	−0.30	−0.24			90

TABLE 4.   TOTAL ENERGY OF THIOPHENE CALCULATED WITH AND WITHOUT
           d-ORBITALS ON SULFUR

Method	Basis on Sulfur	Energy (a.u.)	Reference
Ab initio	HF	− 551.077	98
Ab initio	Double zeta	− 550.923	95
	Double zeta, $3d$	− 550.976	95
	Double zeta, $3d$, H_{2p}	− 550.999	95
	Double zeta, H_{2p}	− 550.946	95
Ab initio	Minimal	− 550.075	96
	Minimal, $3d$	− 550.144	96
Ab initio	Minimal	− 550.417	93
	Minimal, $3d$	− 550.535	93

+ 0.482 with no polarization function to + 0.016 with d-orbitals on sulfur and to
+ 0.001 with both $2p$-orbitals on hydrogen and $3d$-orbitals on sulfur. These changes
in population, which are much larger than the d-orbital population on sulfur, are
found both in the σ- and the π-system. They clearly show that the extra d-orbital
on sulfur acts just as an extra polarization function, to improve an incomplete
basis set. The d-orbitals do not take part in the binding in thiophene, as is the case
where the sulfur is bonded to strongly electropositive atoms, as in SO_4^{2-}, SOF_2 and
SF_6.[141, 142] In the sulfate ion, for example, the $3d$-population on sulfur is as large
as 0.943.

The ab initio calculations performed have treated only the ground state. For
excited states, the d-orbitals are supposed to be more important. This case still
remains to be studied.

3.   Orbital Energies – Ionization Potentials

According to Koopmans' theorem, the orbital energy can be referred to as the
negative of the ionization potential. Thus, a comparison between the theoretically
calculated orbital energies and the experimentally measured ionization potentials
is a good check of the calculations, as well as a tool for interpretation of the
measurements.

Since the advent of the electron spectroscopic methods, photoelectron spectroscopy, and x-ray electron spectroscopy, an intensive study of ionization potentials
has started. Experimental studies on the thiophene molecule have been performed
in the photoelectron region by Eland[46] and Derrick et al.[77] and in the x-ray region
by Gelius et al.[95] Attempts to interpret the spectrum in terms of HMO calculations
gave rise to several questions.[46] A careful examination of vibrational structure,
comparison with spectra of similar molecules, especially furan, and theoretical
calculations (extended Hückel and ab initio) made it possible for Derrick et al.[77]
to assign all bands up to 25 eV (see Table 5). It was found that σ-orbitals lie between
the π-orbitals. Thus, an interpretation of the spectrum in terms of π-electron
theories is apparently unsatisfactory.

TABLE 5.   ORBITAL ENERGIES OBTAINED FROM AB INITIO AND ALL-VALENCE
ELECTRON   CALCULATIONS   COMPARED   WITH   EXPERIMENTAL
IONIZATION POTENTIALS (a.u.)^a

ETH (77)	CNDO (87)	Ab initio (95)	Ab initio (98)	Ab initio^b (98)	Assignment (98)	Experiment (77, 144)
−12.7	−13.2	−9.03	−9.23	−8.77	$1a_2(\pi)$	8.87
−12.2	−11.7	−9.3	−9.5	−9.0	$3b_1(\pi)$	9.52
−12.6	−13.1	−12.9	−13.1	−12.0	$11a_1$	12.1
−14.8	−22.0	−14.3	−14.3	−12.8	$2b_1(\pi)$	12.7
−13.7	−15.0	−14.2	−14.4	−13.3	$7b_2$	13.3
−13.7	−17.2	−14.7	−14.9	−13.4	$10a_1$	13.9
−14.0	−17.0	−15.9	−16.0	−14.5	$6b_2$	14.3
−16.1	−23.7	−19.1	−19.2	−17.5	$9a_1$	16.6
−17.5	−24.8	−20.4	−20.5	−18.7	$5b_2$	17.6, 17.8
−17.8	−27.2	−20.7	−20.8	−18.8	$8a_1$	18.3, 18.8
−23.2	−32.1	−26.8	−26.8		$7a_1$	22.1
−22.6	−34.8	−27.1	−27.0		$4b_2$	22.3
−27.9	−47.9	−32.2	−32.1		$6a_1$	26.1
		−181.4	−181.9		$1b_1$	
		−181.4	−181.9		$5a_1$	169.8
		−181.5	−182.0		$3b_2$	
		−244.2	−242.9		$4a_1$	
		−306.0	−306.4		$2b_2$	
		−306.1	−306.5		$3a_1$	290.3
		−306.6	−306.9		$1b_2$	
		−306.6	−306.5		$2a_1$	290.6
		−2502.6	−2503.6		$1a_1$	

^{a} The numbers given in parentheses are reference numbers.
^{b} Ab initio, many-body approach.

The assignments according to Derrick et al. have, however, been questioned in
some respects. On the basis of an intensity analysis of the ESCA spectrum and
ab initio calculations,[95,142] the ordering of the three outer orbitals were changed
(Table 5). The ordering of the orbitals from highest to lower energy was suggested
to be (in C_{2v}-symmetry): $1a_2(\pi)$, $3b_1(\pi)$, $11a_1$, $7b_2$, $2b_1(\pi)$, $10a_1$, $6b_2$, $9a_1$, $5b_2$,
$8a_1$, $7a_1$, $4b_2$, $6a_1$, $1b_1$, $5a_1$, $3b_2$, $4a_1$, $2b_2$, $3a_1$, $2a_1$, $1b_2$, $1a_1$. This ordering is also
supported by calculations with an ab initio many-body approach,[98] which includes
the effects of electron correlation and reorganization beyond the Hartree–Fock
approximation. Only in one respect is there a change in the ordering: the $7b_2$ and
$2b_1$ orbitals have changed place. They are, however, very close in energy in both
calculations.

The same ordering as obtained from the ab initio calculations of Gelius et al.
was also found by X_α calculations.[102] Experimental studies of the angular distribution in the photoelectron spectrum of thiophene[145] as well as in the Penning
electron spectrum[146] again agree with the ab initio many-body approach.[98]

Semiempirical calculations also give good numerical values for the ionization
potentials. In some cases, these results are due to the parameter choice, which is

fitted just to reproduce the experimental values. In other cases, for example, in the MINDO/3 calculations of Dewar,[90] where the parameters are chosen to give good heats of formation, an excellent agreement with experimental data was also obtained (calculated, 8.87 eV; observed, 8.9 eV).

4. Excitation Energies – UV Spectra

The experimental spectrum of thiophene has been measured in the vapor phase,[147,148] in solution,[149–153] (iso-octane, hexane, and ethanol), and in a stretched film.[75] It consists of three rather weak peaks at 5.16, 5.33, and 5.62 eV and a strong one at 6.6 eV. The existence of the 5.16 eV peak has, however, been questioned[153] and it is believed to be caused by impurities, probably benzene and toluene, which have about the same boiling points as thiophene. MCD Spectra of thiophene in stretched film[75] indicate at least two transitions of π-π^* type with different polarizations, one in plane at 5.26 eV and one out of plane at 5.64 eV. Any lower energy peak at about 5.16 eV was not found. Moreover, a triplet state at 3.9 eV has also been observed by direct excitation measurements.[154]

Quantum chemical calculations on the electronic spectrum can be made by separate calculations on the ground state and each excited state. Because they are rather tedious, most of the calculations on excited states are made with the aid of the virtual orbitals of the ground state, often by configuration interaction of the virtual orbitals.

For thiophene, most of the calculations of the excited states were done within the PPP approximation, but some CNDO- and INDO-type calculations were also done. The excited states have been treated by means of the virtual orbitals of the ground state in all cases. Until now, no ab initio calculation on the excited states of thiophene has been done.

In most of the calculations, the d-orbitals on sulfur were omitted. The results from such calculations are collected in Tables 6 and 7. The numerical values differ somewhat, depending on the different semiempirical parameters used, but both PPP and CNDO/INDO calculations seem to agree that the 5.33 eV peak can be assigned to an in-plane polarized π-π^* transition, whereas the 5.62 eV band can be assigned to an out-of-plane polarized π-π^* transition. The strong peak at 6.6 eV probably consists of two transitions, one in-plane polarized and one out-of-plane polarized.* Low-lying triplets of both A_1 and B_1 symmetries are also found in the calculations.

The effect of d-orbitals on the electronic spectrum has been studied in the PPP approximation by Bielefeld and Fitts[62] and in the CNDO approximation by Schulte and Schweig.[85] The results of these studies are given in Table 6.A. It can be found that the inclusion of the d-orbitals changes the ordering of the transitions and that the assignments may be changed. However, they do not agree any more with the experimentally found polarizations.[75]

---

* There is no quantum chemical evidence for the 5.16 eV transition, supposed to be due to impurities.

TABLE 6. PPP CALCULATED AND EXPERIMENTAL ELECTRONIC TRANSITIONS (eV)[a]

(57)	(59)	(60)	(62)	(65)	(67)	(69)	(71)	(75)	Assignment[b]	Experimental[c]
5.1	5.5	5.1	5.2	5.5	4.7	5.3	5.7	5.4	$1B_1$	(5.16)
5.7	6.9	5.4	5.7	5.7	5.4	5.8	6.1	5.5	$1A_1$	5.33
6.6	7.2	7.2	7.1		7.8		8.1	7.2	$1A_1$	5.62
	8.2	7.5	7.3		9.3		9.0	7.9	$1B_1$	6.6
		4.2			2.5				$3A_1$	3.9[d]
		3.4			3.8				$3B_1$	
		6.8			5.6				$3B_1$	
		5.5			5.9				$3A_1$	

[a] The numbers given in parentheses are reference numbers.
[b] Assignment according to Ref. 75.
[c] Refs. 147–153.
[d] Ref. 154.

TABLE 6A.　THE EFFECT OF d-ORBITALS ON THE ELECTRONIC SPECTRUM (ENERGY VALUES IN eV)

Method:	PPP (62)			CNDO (85)		Experimental[a]
Basis:	3p	3p, 3d	3p, 3d, 4p	3s, 3p	3s, 3p, 3d	
	5.2 $^1B_1$	4.9 $^1B_1$	5.3 $^1A_1$	5.4 $^1B_2$	5.0 $^1A_1$	5.33
	5.7 $^1A_1$	5.0 $^1A_1$	5.4 $^1B_1$	5.5 $^1A_1$	5.4 $^1B_2$	5.62
	7.1 $^1A_1$	6.6 $^1B_1$	6.7 $^1B_1$	6.9 $^1A_1$	6.2 $^1B_2$	6.6
	7.3 $^1B_1$	6.9 $^1A_1$	6.9 $^1A_1$	7.1 $^1B_2$	6.7 $^1A_1$	
	3.3 $^3B_1$	3.1 $^3A_1$	4.0 $^3A_1$			3.9[b]
	4.0 $^3A_1$	3.2 $^3B_1$	4.0 $^3B_1$			
	5.4 $^3A_1$	5.5 $^3A_1$	5.5 $^3A_1$			
	6.4 $^3B_1$	5.7 $^3B_1$	5.8 $^3B_1$			

[a] Ref. 147–153.
[b] Ref. 154.

The effect of d-orbitals upon the triplet spectra has been studied by Wadt and Moomaw[74] in the PPP approximation. They found that the contribution of individual $d\pi$ terms on spin-orbit coupling was of the order of 0.02–2.0 cm^{-1}, which is comparable in magnitude to other contributions.

As both PPP and CNDO calculations are parameterized to fit experimental measurements, it is difficult to judge how much the choice of parameters influences the results. A final assignment of the spectrum must thus await more sophisticated calculations on the nonempirical level. This is also the case for studies of the importance of d-orbitals on sulfur for the excited states.

5.　Charge Distributions

The electronic population on the different atoms can be deduced easily, in the ZDO approximations, from the coefficients of the atomic orbitals according to Eq. (26). In methods where overlap is included, the definition of charge is more complicated. The most common definition is from Mulliken,[23] according to whom the overlap populations are divided equally between the corresponding two atoms. It is, however, obvious that such a definition is not good for polar bonds.

The charges obtained from different types of calculations are listed in Table 8.

In the HMO calculations of Wheland and Pauling,[33] the effect of changes in the parameter α has been studied. It was found that in order to get more electrons on the α-carbon atom than on the β-carbon atom, the Coulomb integral on the α-carbon atom had to be changed with a value greater than 1/25 of the coulomb integral on sulfur.

As seen from Table 8, the charges on the atoms vary considerably from one calculation to another. Thus, for example, the charge on sulfur varies between 0.75 and 0.12 in the HMO calculations, between 0.36 and 0.09 in the PPP calculations, and between 0.48 and $-$ 0.001 in the ab initio calculations. In most of the

TABLE 7. CNDO/INDO CALCULATED AND EXPERIMENTAL ELECTRONIC TRANSITIONS (eV)[a]

INDO (80)	CNDO (81)	CNDO (85)				
		Parameter 1	Parameter 2	Parameter 3	Assignment[b]	Experimental[c]
5.5	5.4	5.4	5.1	5.2	$1B_2$	(5.16)
6.0	5.9	5.5	5.6	5.9	$1A_1$	5.3
7.7		6.9	7.1	7.2	$1A_1$	5.6
7.9		7.1	7.6	8.0	$1B_2$	6.6
			2.4		$3B_2$	3.9[d]
			3.4		$3A_1$	

[a] The number given in parentheses are reference numbers.
[b] Assignment according to Ref. 85. Note that the coordinate axes are different in the different calculations. Thus, the assignments have different symmetries.
[c] Refs. 147–153.
[d] Ref. 154.

TABLE 8. GROSS CHARGES AND BOND ORDERS IN THE THIOPHENE MOLECULE OBTAINED FROM DIFFERENT TYPES OF CALCULATIONS

Method	Reference	S	C_α	C_β	p(C–S)	p(C–C')	p(C–C)	Comments	
								α_S	α_C
HMO	33	0.62	−0.14	−0.08				0.5	0.0
		0.20	−0.08	−0.02				2.9	0.29
		0.12	0.02	−0.08				4.0	0.0
		0.13	−1.13	0.07				4.0	0.5
		0.31	−0.10	−0.06				2.0	0.25
		0.19	−0.13	0.04				3.0	0.375
	36	0.42	−0.10	−0.11	0.50	0.79	0.52		
	41	0.75	−0.16	−0.22	0.62	0.68	0.63		
	45	0.64	−0.17	−0.15	0.60	0.71	0.60		
PPP	57	0.19	−0.02	−0.07	0.34	0.88	0.39		
	59	0.14	−0.06	−0.01	0.31	0.86	0.47		
	60	0.09	−0.02	−0.03					
	62	0.21	−0.06	−0.04					
	70	0.17	−0.09	0.00					
	71	0.36	−0.13	−0.05					
	72	0.15	−0.05	−0.03					
CNDO	79	0.13	−0.05	−0.07					
MINDO	92	−0.13	0.09	−0.04					
Ab initio	95	0.48	−0.57	−0.18				Without d	
		0.02	−0.25	−0.24				With d	
	96	0.14	−0.24	−0.17				Without d	
		0.03	−0.17	−0.17				With d	
	98	0.00	−0.22	−0.17				Without d	
	99	0.26	−0.18	−0.08				Without d	
	100	0.23	−0.07	−0.05				Without d	
	101	0.26	−0.18	−0.08				Without d	

calculations the trend is, however, that the sulfur atom has a positive charge and the carbon atoms are negatively charged. In most cases, the α-carbon atom is slightly more negative than the β-carbon atom, but the opposite trend is also found.

The ab initio calculations show rather large variations in charge between the different calculations. This could be seen in Table 3, where it was found that a change in basis set caused rather large changes in the charges on the atoms. However, in the calculations with large basis sets with[95] or without[98] d-orbitals, the charges were about the same.

The charge of an atom in a molecule is difficult to measure experimentally, just as it is difficult to define theoretically. An estimate can be obtained from nmr chemical shifts.[158] Correlations between calculated charge densities and nmr chemical shifts[159] have been obtained by HMO[51, 118, 120] and PPP[126] calculations.

The charge distribution over the whole molecule can be measured more exactly through the dipole moment. In Table 9, calculated and experimental dipole moments are given. The dipole moments obtained in the most accurate calculations of von Niessen et al.,[98] 0.65 D, and of Gelius et al.,[95] 0.61 D, are in good agreement with the experimentally measured 0.55 D. Table 9 shows that the dipole moment is strongly dependent on the basis set, as long as the basis is not sufficient large. It should also be noted that the semiempirical calculations give dipole moments that are in good agreement with the experimental dipole moment in most cases, especially the PPP-type calculations.

An informative description of the charge distributions over a molecule can be obtained from electron density maps. Such maps have been made in the extended Hückel approximation by Mårtensson and Chojnacki[78] in different planes parallel and perpendicular to the molecular plane. Recently, such maps have also been obtained from ab initio calculations by Hilal.[101]

The thiophene molecule undergoes electrophilic substitution reactions preferentially at the α-position.[139] The first to discuss the reactivity of thiophene in

TABLE 9.   DIPOLE MOMENT OF THIOPHENE OBTAINED FROM DIFFERENT TYPES OF CALCULATIONS, TOGETHER WITH THE EXPERIMENTAL VALUE

Method	Reference	Dipole Moment	Comments
HMO	36	2.50	
PPP	57	0.50	
	59	-0.56	
	60	-0.62	
	62	0.18	
MINDO	92	2.22	
Ab initio	95	0.959	Double zeta basis
		0.606	Double zeta + S_{3d}
		0.619	Double zeta + S_{3d} + H_{2p}
		0.973	Double zeta + H_{2p}
	98	0.65	HF
	101	0.38	
Experimental	143	0.55	

quantum chemical terms were Wheland and Pauling,[33] who correlated the charges obtained from HMO calculations with the reactivity in a certain position.

In the ab initio calculations, the α-carbon atom is the most negatively charged; in the other types of calculation, either the α- or the β-position may be the most negatively charged. The differences in charge are rather small in all cases.

Useful information for studying the reactivity of a molecule, especially electrophilic substitution reactions, has been obtained from studies of molecular potentials.[157] The electrostatic interaction between the molecule and a positive test charge can be calculated and isopotential maps obtained. Such maps in planes parallel and perpendicular to the molecular plane of the thiophene molecule have been obtained in the ab initio calculations of Gelius et al.[95] A minimum was obtained at a distance of 1.74 Å from the molecular plane. In a plane parallel to the molecular plane, at the distance of 1.74 Å, three different minima were found, one at each double bond and one at the sulfur atom. From these calculations reaction mechanisms can be discussed. The reaction probably occurs through a weak π-complex; The attack happens at either the sulfur atom or the double bond. In the first case, the substitution takes place at the α-position, in the second, at the α- as well as the β-position. The charge in the frontier orbital, $1a_2$, favors the α-position.

6. Total Energies

The total energy of a molecule gives a measure of the stability of the molecule. From total energies, heats of formation, resonance energies, and localization energies can be deduced.

In the π-electron and the all-valence electron theories the total energy cannot be computed, since all electrons are not considered explicitly in the calculations. However, these methods can be parameterized to give good energy values, which is the case for the methods developed by Dewar.[14-17]

The total energy of thiophene calculated by ab initio methods is found in Table 10, as well as heats of formation and resonance energies obtained by semiempirical calculations and experimentally.

The geometry of a molecule can be obtained by minimizing the total energy as a function of the geometry. Such calculations have been done for thiophene on both the ab initio and the semiempirical level. The results are found in Table 11, where a good agreement between theory and experiment is illustrated.

In the PPP approximation, where the total energy cannot be obtained, the geometrical parameters are generally obtained from the bond orders, according to the formula in Eq. (31) derived by Coulson.[158]

$$r_{uv} = a - bp_{uv} \tag{31}$$

where the constants a and b are empirically found parameters. Bond lengths obtained in this way are also shown in Table 11. It can be noted that the results obtained by different techniques all yield rather good agreement with the experimental results, and that the results obtained form the semiempirical methods are especially good.

TABLE 10.   TOTAL ENERGIES, HEATS OF FORMATION, AND RESONANCE ENERGIES OBTAINED FROM DIFFERENT TYPES OF CALCULATIONS AND EXPERIMENTAL DATA

Method	Reference	Total Energy (Å)	Heat of Formation (kcal/mole)	Resonance Energy (kcal/mole)
HMO	36			26
VB	55			31
PPP	57			23
	72		100	
CNDO	79	− 507.148		
		− 516.149		
		− 517.388		
MINDO			32.5	
Ab initio	93	− 550.417		
		− 550.535		
	95	− 550.923		
		− 550.976		
	96	− 550.075		
		− 550.444		
	98	− 551.077		
	99	− 545.092		
	100	− 545.092		
	101	− 550.599		
Experimental	171		17.13	20

Calculations of the total energy as a function of the geometrical parameters result in a potential energy map from which vibrational frequencies can be deduced. This has been done within the MINDO approximation[91] using a method developed by McIver and Komornicki.[159] The results and the experimental values are collected in Table 12.

The most common way to calculate vibrational spectra is by harmonic force field calculations.[160] A complete vibrational assignment using this method was made by Cyvin et al.[161] and Scott.[162] The results of these are also found in Table 12. An

TABLE 11.   CALCULATED AND OBSERVED BOND LENGTHS (a.u.)

Method	Reference	$r(S-C)$	$r(C-C')$	$r(C-C)$
PPP	71		1.380	1.425
	72, 73	1.721	1.357	1.442
MINDO	90, 92	1.756	1.346	1.463
Ab initio	99	1.726	1.338	1.456
	100	1.732	1.335	1.454
Experimental				
Microwave	30	1.7140	1.3696	1.4232
Electrical diffraction	31	1.718	1.370	1.442
		1.716	1.366	1.442
	32	1.714	1.370	1.419

TABLE 12.   CALCULATED AND OBSERVED VIBRATIONAL FREQUENCIES (cm^{-1})

Assignment C_{2v}	Experimental (163)[a]	MINDO (91)	Force Field Calculations	
			(161)	(162)
a_1				
CH stretch	3126	3561	3127	3130
CH stretch	3098	3501	3104	3094
In-plane ring II	1409	1677	1427	1404
In-plane ring III	1360	1348	1364	1360
CH bend	1083	1058	1072	1090
CH bend	1036	988	1030	1034
In-plane ring IV	839	778	844	837
In-plane ring VII	608	460	642	603
a_2				
CH bend	903	802	893	901
CH bend	688	712	698	685
Out-of-plane ring I	567	435	565	565
b_1				
CH stretch	3125	3558	3129	3128
CH stretch	3086	3490	3089	3093
In-plane ring I	1504	1729	1532	1510
CH bend	1256	1100	1254	1263
CH bend	1085	970	1069	1081
In-plane ring V	872	774	857	889
In-plane ring VI	751	605	745	751
b_2				
CH bend	867	795	865	863
CH bend	712	699	712	709
Out-of-plane ring II	452	315	450	453

[a] Reference numbers given in parentheses.

excellent agreement with the experimental values[163] is obtained. Thus, the empirical methods give better results than the theoretical methods, which must be improved to yield good numerical data.

7. Properties

One-electron properties, such as dipole, second, quadrupole, third, and octupole moments, charge density at the nuclei, quadrupole coupling constants, and asymmetry parameters have been calculated at the ab initio level by Gelius et al.[95] and von Niessen et al.[98] In Table 13, those values for which experimental values were also available are collected. As seen from the table, very good agreement between theory and experiment is obtained.

The diamagnetic susceptibility has recently been correlated with the superdelocalizabilities calculated from Hückel theory[49] for a series of compounds. For thiophene, a correlation coefficient of 1.009 was obtained.

TABLE 13.   SOME MOLECULAR PROPERTIES OF THIOPHENE

Property	Calculated[a]	Calculated[b]	Experimental	Reference
		Dipole Moment (D)		
	0.61	0.65	0.55	143
		Quadrupole Moment (10^{-26} esu/cm^2)		
Q_{xx}	-7.85	-7.49	-8.3 ± 2.2	172
Q_{yy}	6.61	6.29	6.6 ± 1.5	
Q_{zz}	1.24	1.21	1.7 ± 1.6	
		Diamagnetic Susceptibility Tensor (10^{-6} erg/G^2 mole)		
$x_{xx}{}^4$	-439.70	-440.89	-438.1 ± 3.0	172
$x_{yy}{}^4$	-285.78	-286.57	-284.8 ± 3.0	
$x_{zz}{}^4$	-226.38	-227.34	-225.7 ± 3.0	
		Electric Field Gradient at Sulphur Nucleus (a.u.)		
$q_{xx}(S)$	1.51	1.56		
$q_{yy}(S)$	-1.92	2.02		
$q_{zz}(S)$	0.40	0.46		
		Potential at the Sulfur Nucleus (a.u.)		
$\Phi(S)$	-59.22	-59.20		

[a] According to Ref. 95.
[b] According to Ref. 98.

Nmr coupling constants in the CNDO formalism have been calculated by Clark,[79] using the Pople–Santry theory,[164] and in the CNDO/INDO formalism by Galasso,[82, 83] using the Blizzard–Santry theory,[165] which is based on perturbation calculations. In both cases, the effects of d-orbitals were studied. It was found that d-orbitals do not improve the results, which are listed in Table 14 together with experimental data.[166, 167]

III.   CALCULATIONS ON THIOPHENE-LIKE MOLECULES

1.   Trends in a Series of Similar Molecules

Because there are many approximations, especially in the semiempirical theories, no accurate results can be expected. In fact, in the simplest theory (the HMO theory), the energy values are expressed only as a factor of a certain parameter, β, whose value is not known. Yet such calculations are useful for many purposes. For example, certain trends in a series of similar molecules can be studied. By correlating experimental data with calculated values, predictions of substituent effects, for example, can be made. This type of correlation has also been used in studies of

TABLE 14.   CALCULATED   AND   EXPERIMENTAL   PROTON–PROTON   COUPLING
             CONSTANTS

Method	Basis on S	^3J $(\alpha\beta)$	^3J $(\beta\beta)$	^4J $(\alpha\beta')$	^4J $(\alpha\alpha)$	Reference
CNDO	$3s, 3p$	2.22	0.83	0.25	0.31	79
	$3s, 3p, 3d$	2.31	0.45	0.17	0.74	
	$3s, 3p, 3d, 4s$	2.41	0.47	0.15	0.82	
INDO-SS	$3s, 3p$	4.23	3.18	2.22	1.85	82
	$3s, 3p, 3d$	7.85	3.42	2.81	5.45	
INDO-S	$3s, 3p$	4.07	3.51	2.33	1.86	
	$3s, 3p, 3d$	4.41	3.51	2.38	2.25	
INDO-B	$3s, 3p$	2.75	1.94	− 0.40	1.53	
	$3s, 3p, 3d$	1.51	1.41	0.51	2.76	
CNDO	$3s, 3p$	2.22	0.83	0.25	0.31	83
	$3s, 3p, 3d$	2.31	0.45	0.17	0.74	
	$3s, 3p, 3d, 4s$	2.41	0.47	0.15	0.82	
Experimental		4.90	3.50	1.04	2.84	166
		5.15	3.45	1.04	2.75	167

substituent effects obtained by experimentally determined substituent constants, such as the Hammett constant. For thiophenes, this technique was used for correlating nmr chemical shifts of substituted thiophenes with experimentally determined substituent constants.[156]

Calculations of thiophene-like molecules are listed in Table 1. Most are made within the HMO and PPP theories, but some calculations of higher accuracy were performed. In the HMO calculations, ground-state properties, such as charges, bond orders, and reactivity indices were studied; whereas the PPP calculations very often consider excited states, that is, the electronic spectra. The all-valence electron methods are often used for studies of heats of formation, conformation energies, and esr and nmr data, as is also the case for the few ab initio calculations.

The electronic spectrum of a chemical compound has been easy to measure for a long time. Correlations with calculated values have been made in the HMO theory on, for example, sulfur containing heterocycles,[109–111] methyl substituted dibenzothiophenes,[114] and substituted maleic acid thioanhydrides[119] and in the PPP theory on sulfur containing heterocycles,[59, 66, 121, 131] dimethylamino thiophenes,[122] polyenyl-substituted thiophenes,[124, 130] amino- and carbonyl-substituted thiophenes,[127] methyl-, chlor-, and amino-substituted thiophenes,[68] and thienothiophenes,[123] which have also been studied in the CNDO approach.[81]

Polarographic reduction potentials have been correlated with HMO calculations on methyldibenzothiophenes,[114] ionization potentials with PPP-type calculations on methyl-, chloro-, and amino-substituted thiophenes,[68] and sulfur containing heterocycles.[66, 72, 76]

Nmr chemical shifts have been correlated with π-electron densities calculated within the HMO theory for substituted thiophenes[51, 118, 120] and within the PPP theory for benzothiophenes[126] and carbonyl-substituted thiophenes.[127] Also, the all-valence electron methods have been used for calculations on nmr data, for example, CNDO calculations on fluorothiophenes[133] and INDO calculations on halothiophenes.[82]

Esr-hyperfine coupling constants have been correlated with spin densities obtained from HMO calculations for dibenzothiophenes[114] and in the INDO approach for substituted thiopheneradicals.[134]

Generally, good correlation between theory and experiment has been found.

2. Reactivity of Thiophenes

The effects of substituents on the reactivity of thiophenes have mainly been studied with the aid of reactivity indices developed in the HMO theory. In the earliest studies, the reactivity of thiophene was discussed in terms of charges, free valences, and superdelocalizabilities.[33, 38] In a paper by Gronowitz and Rosenberg,[167] the effect of substituents on the reactivity of thiophene carboxylic acids and thiophenecarboxaldehydes was discussed using resonance structures described by Ingold.[168] The considerations were compared with IR–absorption measurements.

Static indices, such as charges, bond orders, and free valences, as well as dynamic indices, such as localization energies have been calculated by several authors using the HMO theory. Thus, Melander[39, 106] studied electrophilic, radical, and nucleophilic substitution reactions for thiophene and nitrothiophenes, using a model based on the d-orbital participation of Longuet-Higgins.[35] Renewed and refined calculations, with the substituents -CHO, -CN, and -NO$_2$ were performed by Östman,[112, 113] both with and without d-orbitals on sulfur.

Similar types of calculations have also been performed by Koutecky[107] on thiopyrylium and related compounds, by Janda[120] on methyl-substituted thiophenes, by Klasinc and Humski[117] on methyl-, methoxyl-, and thienyl-substituted thiophenes, and by Zahradnik et al.[109, 110] on a large number of thiophene-like compounds.

The results were compared with experimental data, such as alkylation, nitration, acylation, bromination, metalation, methylation, hydrogen–deuterium exchange, and hydrogen–tritium exchange. Generally, it was found that the dynamic index – the localization energy – was a good index. The static indices sometimes predicted the right order of reactivity, sometimes not, but the localization energy predicted the right reactivity in most cases. This suggests that the reaction mechanism probably proceeds through a Wheland-type transition state and not through a weak π-complex. This is also supported by the ab initio calculations of Gelius et al.[95]

Reactivity indices can of course also be studied by other methods than HMO. Thus, Gold'farb et al.[84] used the CNDO/2 method to calculate charge distributions and localization energies for a series of thienothiophenes that affect electrophilic substitutions. The localization energies fitted well to experimental data. What affects the charge distributions, good agreement with experiment and also with the calculated localization energies, was obtained for the reactivity of different positions in the same molecule, but not for the same position in different molecules.

Thus, even in this case, the localization energy seems to be an appropriate index.

Recently, an ab initio study was performed.[100] Charges and stabilization energies of a number of substituted thiophenes were discussed in terms of reactivity.

The reactivity indices mentioned above do not take into account the nature of the reagent. In the work of Decoret and Tinland,[47] the bromination of thiophene was studied using HMO theory by taking into account the reagent. The electrophilic reagent, Br^+, was assigned an orbital containing no electrons, whereas the thiophene molecule was treated as in ordinary HMO calculations. The site of bromination was correctly ascribed to the α-position.

The hydrodesulfurization of thiophenes is catalyzed by metal surfaces. A mechanism of this reaction has been proposed[170] that involves a metal bound to the sulfur atom. Duben[53] applied simple HMO theory to study this mechanism. The optimal situation requires that two metal d-orbitals interact with the thiophene π-orbitals. It was suggested that Mo^{3+} should be an effective catalyst. In the same study, the reactivities of methylsubstituted thiophenes were also considered. It was found that the theoretically calculated activation energies correlate well with those measured experimentally.

3. Geometries and Conformations

It is well-known experimentally that the thiophene molecule is planar. In protonated thiophene, the proton is supposed to bind to the sulfur atom through the π-system. Thus, the molecule is no longer planar. The structure and also the inversion barrier was computed with ab initio calculations and found to be 41 kcal/mole (39.8 kcal/mole with d-orbitals included) by Clark[94] and 33 kcal/mole by Bernardi et al.[99] The out-of-plane angle of hydrogen was 74°. Also, the sulfur atom was slightly out of the thiophene plane.

The geometry of thiophene-1-oxide is also supposed to be nonplanar; in this case, the oxygen atom is binding to the π-electron system. According to CNDO calculations,[94] the barrier of inversion was 24.5 kcal/mole.

There are several calculations on the conformation and rotational barriers of substituted thiophenes. For example, ab initio calculations of 2- and 3-substituted thiophenes, with the substituents -CHO, -OH, $-CHCH_2$, $-CH_3$, -CN, $-NO_2$, -F, and -Li have recently been performed[100] as well as calculations on 2-(2-thienyl)pyrrole.[138]

Quantum chemically, thiophene is rather big already. Thus, most of the calculations are semiempirical, mostly using the EHT, CNDO, and MINDO approximations, but PPP calculations have also been done.

In the PPP calculations, rotational barriers are calculated from

$$\beta = \beta_0 \cos \theta \qquad (32)$$

where the angle θ is the angle of rotation about a certain bond. Rotational barriers have been calculated in this way on 3,3'-bithienyl and 2,2'-bithienyl[125] and 2-phenylthiazole and its protonated form.[109]

Rotational barriers have been calculated in the EHT theory for methyl-substituted thiophenes,[132] in the CNDO theory on carbonyl derivatives of thiophene,[135] and in the MINDO approximation on 3,3'-dithienylmethane.[137]

Calculations of bond lengths in planar sulfur containing heterocycles in the MINDO approximation were made by Dewar.[90]

IV.   COMPARISON BETWEEN FURAN, PYRROLE, AND THIOPHENE

The molecules furan, pyrrole, and thiophene are all iso-π-electronic. Thus, in the π-electron theories, HMO and PPP, the three molecules should be treated the same way. The only difference lies in the choice of semiempirical parameters. A comparison between the properties of these three molecules thus seems appropriate. A complete review of all calculations is, however, out of the scope of this chapter.

According to Pauling,[140] the electronegativity of O, N, and S are 3.5, 3.0, and 2.5, respectively. This means that O binds electrons most strongly, followed by N, and then S. In fact, the electronegativity of sulphur is the same as that of carbon. Accordingly, in choosing the HMO parameters, the absolute values of α should be greatest for O and smallest for S. In the HMO calculations of Pilar and Morris,[45] the values of α were: $\alpha_O = \alpha_C + 0.5$, $\alpha_N = \alpha_C + 0.25$, and $\alpha_S = \alpha_C$. Also, the resonance integrals β were varied: $\beta_{CO} = 0.2\beta_{CC}$, $\beta_{CN} = 0.5\beta_{CC}$, and $\beta_{CS} = 0.6\beta_{CC}$.

Charges obtained from these calculations are listed in Table 15, together with corresponding values calculated by other HMO,[36] PPP,[60] MINDO,[92] and ab initio[98, 174, 101] methods. As shown in the table, the values obtained are widely spread, as was also found for calculations on charges. However, certain trends should be noted. In the π-electron theories, the oxygen is most negatively charged and sulfur is least, in accordance with considerations about electronegativity. In the all-valence electron and ab initio methods, this trend is no longer found. Now, the N atom is the most negatively charged. However, the N atom also binds a hydrogen in the σ-skeleton, and when this is taken into account, the trends are the same.[101]

Table 15 also includes dipole moments, both theoretical and experimental.[143, 173] The ab initio calculations are in good agreement with the experimental values, and the trends are correctly described.

Electron density maps have recently been produced with the aid of ab initio calculations.[101] An accumulation of negative charge was found around S and O, but not around N. The density in the heteroatom–carbon regions becomes more diffuse in the order: O–C < N–C < S–C.

The photoelectron spectra have been measured for all three molecules.[77, 175, 176] Calculated (MINDO,[92] ab initio[98]) and experimental ionization potentials for the six lowest ionizations are listed in Table 16. Good[92] and very good[98] agreement between experimental values and theoretically obtained values are found. In all cases, the first two ionizations occur from π-orbitals and the third from a σ-orbital. The fourth ionization is for thiophene from a π-orbital, whereas for pyrrole one σ-orbital, and for furan two σ-orbitals, are higher in energy than the corresponding π-orbital.

TABLE 15.   COMPARISON OF CHARGES FOR FURAN, PYRROLE, AND THIOPHENE

Molecule	HMO		PPP (60)	MINDO (92)	Ab initio		Experimental (143, 173)
	(45)[a]	(36)			(98, 174)	(101)	
Furan							
O	0.083	0.104	0.150	−0.311	−0.381	−0.20	
C_α	−0.018	−0.036	−0.040	0.229	0.016	0.05	
C_β	−0.023	−0.016	−0.035	−0.108	−0.230	−0.10	
H_α				0.008	0.188	0.08	
H_β				0.027	0.216	0.07	
$|\mu|$				0.20	1.29		0.72 D
Pyrrole							
N	0.438	0.255	0.225	0.081	−0.420	−0.28	
C_α	−0.107	−0.060	−0.061	−0.030	−0.082	0.02	
C_β	−0.112	−0.068	−0.052	−0.064	−0.215	−0.10	
H_α				0.021	0.177	0.06	
H_β				0.014	0.186	0.05	
H_N				0.035	0.288	0.21	
$|\mu|$				1.97	1.99		1.84 D
Thiophene							
S	0.641	0.422	0.094	−0.128	−0.001	0.26	
C_α	−0.168	−0.100	−0.020	0.088	−0.214	−0.26	
C_β	−0.152	−0.111	−0.027	−0.036	−0.174	−0.08	
H_α				0.004	0.199	0.07	
H_β				0.007	0.189	0.06	
$|\mu|$				2.22	0.65		0.55 D

[a] Numbers in parentheses are reference numbers.

TABLE 16.   COMPARISON OF IONIZATION POTENTIALS FOR FURAN, PYRROLE, AND THIOPHENE (eV)

Assignment	Furan			Pyrrole			Thiophene		
	Experimental (175)[a]	MINDO (92)	Ab Initio (174)	Experimental (176)	MINDO (92)	Ab Initio (98)	Experimental (77)	MINDO (92)	Ab Initio (98)
$a_2(\pi)$	8.88	8.38	8.87	8.21	8.20	8.17	8.87	8.92	8.77
$b_1(\pi)$	10.31	9.94	10.36	9.20	8.77	8.92	9.52	9.09	8.98
$a_1(\sigma)$	13.0	11.17	13.30	12.6	11.19	12.98	12.1	9.61	12.01
$b_1(\pi)$							12.7	13.11	12.88
$b_2(\sigma)$	14.4	11.80	14.69	13.0	11.44	13.39	13.3	10.43	13.29
$b_2(\sigma)$	15.1	12.92	15.17				14.3	12.32	14.52
$b_1(\pi)$	15.6	15.26	15.47	13.7	13.77	13.70			

[a] Number if parentheses are reference numbers.

TABLE 17.   COMPARISON OF UV SPECTRA FOR FURAN, PYRROLE, AND THIOPHENE
             (eV)

Assignment	Furan		Pyrrole		Thiophene	
	Experimental (177)[a]	PPP (60)	Experimental (148)	PPP (60)	Experimental (147–153)	PPP (60)
3-4	6.05	6.05	5.88	5.98	5.33	5.13
(2-4)–(3-5)	6.48	6.80				
2-4	7.40	7.40	6.77	6.56	5.62	5.42
3-5	7.80	7.96		7.64	6.6	7.18
2-5			7.21	7.19		7.47

[a] Numbers in parentheses are reference numbers.

The experimental UV spectra all consist of several peaks: for furan[77] at 6.1, 6.5, 7.4, and 7.8 eV, respectively; for pyrrole[148] at 5.9, 6.8, and 7.2 eV, and for thiophene[147–153] at 5.3, 5.6, and 6.6–7.7 eV (see Table 17). According to Solony et al.,[60] the longest wavelength transitions can all be assigned to a transition from the highest occupied orbital to the lowest unoccupied orbital. The "extra" low energy transition in furan at 6.48 eV may be due to a mixing of two states consisting of transitions from the second highest occupied orbital to the lowest unoccupied orbital and the highest occupied orbital to the second lowest unoccupied orbital. Hence, a transition follows for all the molecules from the second highest occupied molecular orbital to the lowest unoccupied, followed by transitions between highest occupied to second lowest unoccupied and second highest occupied to second lowest unoccupied orbitals.

V. SUMMARY

Results from theoretical calculations on thiophenes have been reviewed and discussed. Both semiempirical and nonempirical calculations have been performed. In both cases, good agreement with experiments has generally been obtained. All levels of approximations are still useful, depending on the problem to be solved.

REFERENCES

1.   See, for example, I. N. Levine, *Quantum Chemistry*, Allyn and Bacon, Boston, 1974; J. N. Murrell and A. Harget, *Semiempirical Self-Consistent Field Molecular Orbital Theory of Molecules*, Wiley-Interscience, New York, 1972; or any other textbook in quantum chemistry.
2.   E. Hückel, *Z. Phys.*, **70**, 204 (1931).
3.   A. Streitwieser, *Molecular Orbital Theory of Organic Chemists*, Wiley, New York, 1961.
4.   R. Pariser and R. G. Parr, *J. Chem. Phys.*, **21**, 466, 767 (1953).

5. J. A. Pople, *Trans. Faraday Soc.,* **49**, 1375 (1953).

6. P. G. Lykos and R. G. Parr, *J. Chem. Phys.,* **24**, 1166 (1956); **25**, 1301 (1956).

7. I. Fischer-Hjalmars, *J. Chem. Phys.,* **42**, 1962 (1965).

8. B. Roos and P. N. Skancke, *Acta Chem. Scand.,* **21**, 233 (1967).

9. (a) R. Hoffmann, *J. Chem. Phys.,* **39**, 1397 (1963); **40**, 2474 (1963). (b) R. Hoffmann and W. N. Lipscomb, *J. Chem. Phys.,* **36**, 2179 (1962); **37**, 177, 2872 (1962).

10. (a) J. A. Pople and G. A. Segal, *J. Chem. Phys.,* **44**, 3289 (1966); (b) J. A. Pople, D. L. Beveridge and P. A. Dobosh, *J. Chem. Phys.,* **47**, 2026 (1967).

11. R. B. Woodward and R. Hoffmann, *Die Erhaltung der Orbital symmetrie*, Verlag Chemie, 1970.

12. (a) H. A. B. Gray and A. J. Stone, *Theor. Chim. Acta,* **18**, 389 (1970); (b) K. R. Roby, *Chem. Letters,* **11**, 60 (1971); (c) D. B. Cook and R. McWeeny, *Chem. Phys. Lett.* **1**, 588 (1968).

13. J. Del Bene and H. H. Jaffe, *J. Chem. Phys.,* **48**, 1807, 4050 (1968); **49**, 1221 (1968).

14. M. J. S. Dewar and G. Klopman, *J. Am. Chem. Soc.,* **89**, 3089 (1967).

15. N. C. Baird and M. J. S. Dewar, *J. Chem. Phys.,* **50**, 1262 (1969).

16. M. J. S. Dewar and E. Haselbach, *J. Am. Chem. Soc.,* **92**, 590 (1970).

17. R. C. Bingham, M. J. S. Dewar, and D. H. Lo, *J. Am. Chem. Soc.,* **97**, 1285 (1975).

18. S. Diner, J. P. Malrieu, P. Claverie, and F. Jordan, *Chem. Phys. Lett.,* **2**, 319 (1968).

19. J. Paldus and J. Cizek, *Adv. Quant. Chem.,* **9**, 105 (1975).

20. J. C. Phillips and L. Kleinman, *Phys. Rev.,* **116**, 287 (1959).

21. T. Koopmans, *Physica,* **1**, 104 (1933).

22. C. C. J. Roothaan, *Rev. Mod. Phys.,* **23**, 69 (1951).

23. R. S. Mulliken, *J. Chem. Phys.,* **23**, 1833, 1841 (1955).

24. R. D. Brown, *Quart. Revs.,* **6**, 63 (1952).

25. C. A. Coulson, *Disc. Faraday Soc.,* **2**, 9 (1947); *J. Chim. Phys.,* **45**, 243 (1948).

26. K. Fukui, T. Yonezawa, and C. Nagata, *Bull. Chem. Soc. Jpn.,* **27**, 423 (1954); *J. Chem. Phys.,* **27**, 1247 (1957).

27. G. W. Wheland, *J. Am. Chem. Soc.,* **64**, 900 (1942).

28. C. A. Coulson and H. C. Longuet-Higgins, *Proc. Roy. Soc.,* **A191**, 39 (1947).

29. D. Neumann and J. W. Moskowitz, *J. Chem. Phys.,* **49**, 2056 (1968).

30. (a) B. Bak, D. Christensen, J. Rastrup-Andersen, and E. Tannenbaum, *J. Chem. Phys.,* **25**, 892 (1956); (b) B. Bak, D. Christensen, L. Hansen-Nygaard, and J. Rastrup-Andersen, *J. Mol. Spectrosc.,* **7**, 58 (1961).

31. W. R. Harshbarger and S. H. Bauer, *Acta Crystallogr. Sect.,* **B26**, 1010 (1970).

32. R. A. Bonham and F. A. Momany, *J. Phys. Chem.,* **67**, 2474 (1963).

33. G. W. Wheland and L. Pauling, *J. Am. Chem. Soc.,* **57**, 2086 (1935).

34. P. Daudel, R. Daudel, N. B. Buu-Hoi, and M. Martin, *Bull. Soc. Chim. Fr.,* 1202 (1948).

35. H. C. Longuet-Higgins, *Trans. Faraday Soc.,* **45**, 173 (1949).

36. S. Nagakura and T. Hosoya, *Bull. Chem. Soc. Jpn.,* **25**, 179 (1952).

37. J. Metzger and F. Ruffler, *J. Chim. Phys.,* **51**, 52 (1954).

38. J. deHeer, *J. Am. Chem. Soc.,* **76**, 4802 (1954).

39. L. Melander, *Ark. Kemi,* **8**, 361, 397 (1955).

40. K. Kikuchi, *Sci. Rep. Tohoku Univ.,* **40**, 133 (1957); **41**, 35 (1957).

41. M. M. Kreevoy, *J. Am. Chem. Soc.,* **80**, 5543 (1958).

42. G. Milazzo and G. DeAlti, *Gazz. Chim. Ital.,* **89**, 2479 (1959).

43. K. Maeda, *Bull. Chem. Soc. Jpn.,* **33**, 304 (1960).

44.   A. Mangini and C. Zauli, *J. Chem. Soc.,* 2210 (1960).

45.   F. L. Pilar and J. R. Morris II, *J. Chem. Phys.,* **34**, 389 (1961).

46.   J. H. D. Eland, *Int. J. Mass Spectrom. Ion Phys.,* **2**, 471 (1969).

47.   C. Decoret and B. Tinland, *Aust. J. Chem.,* **24**, 2679 (1971).

48.   B. A. Hess and L. J. Schaad, *J. Am. Chem. Soc.,* **95**, 3907 (1973).

49.   K. S. Rogers and A. Cammarata, *Experimentia,* **32**, 255 (1976).

50.   J. -i. Aihara, *J. Am. Chem. Soc.,* **98**, 2750 (1976).

51.   Y. Osamura, O. Sayanagi, and K. Nishimoto, *Bull. Chem. Soc. Jpn.,* **49**, 845 (1976).

52.   A. Julg and R. Sabbah, *C.R. Acad. Sci. Paris,* **285**, 421 (1977).

53.   A. J. Duben, *J. Phys. Chem.,* **82**, 348 (1978).

54.   C. Párkányi and W. C. Herndon, *Phosphorus and Sulfur,* **4**, 1 (1978).

55.   V. Schomaker and L. Pauling, *J. Am. Chem. Soc.,* **61**, 1769 (1939).

56.   V. Santhamma, *Proc. Nat. Inst. India,* **A22**, 204 (1956).

57.   D. S. Sappenfeld and M. Kreevoy, *Tetrahedron,* **19**, 157 (1963).

58.   L. Paujol and A. Julg, *Theor. Chim. Acta,* **2**, 125 (1964).

59.   D. J. H. Wachters and D. W. Davies, *Tetrahedron,* **20**, 2841 (1964).

60.   N. Solony, F. W. Birss, and J. B. Greenshields, *Can. J. Chem.,* **43**, 1569 (1965).

61.   A. K. Zweig and A. K. Hoffmann, *J. Org. Chem.,* **30**, 3997 (1965).

62.   M. J. Bielefeld and D. D. Fitts, *J. Am. Chem. Soc.,* **88**, 4804 (1966).

63.   F. Dürr, G. Hohlneicher, and S. Schneider, *Ber. Bunsenges. Phys. Chem.,* **70**, 803 (1966).

64.   F. Momicchioli and A. Rastelli, *J. Mol. Spectrosc.,* **22**, 310 (1967).

65.   F. P. Billingsley and J. E. Bloor, *Theor. Chim. Acta,* **11**, 325 (1968).

66.   J. Fabian, A. Mehlhorn, and R. Zahradnik, *J. Phys. Chem.,* **72**, 3975 (1968).

67.   C. B. Chowdhury and R. Basu, *J. Indian Chem. Soc.,* **46**, 779 (1969).

68.   R. Phan-Tan-Luu, L. Bouscasse, E. J. Vincent, and J. Metzger, *Bull. Soc. Chim. Fr.,* 1149 (1969).

69.   H. A. Hammond, *Theor. Chim. Acta,* **18**, 239 (1970).

70.   A. Skancke and P. N. Skancke, *Acta Chem. Scand.,* **24**, 23 (1970).

71.   A. Julg, M. Bonnet, and Y. Ozias, *Theor. Chim. Acta,* **17**, 49 (1970).

72.   M. J. S. Dewar and N. Trinajstić, *J. Am. Chem. Soc.,* **92**, 1453 (1970).

73.   N. Bodor, M. Farkas, and N. Trinajstić, *Croat. Chim. Acta,* **43**, 107 (1971).

74.   W. R. Wadt and W. R. Moomaw, *Mol. Phys.,* **25**, 1291 (1973).

75.   (a) R. Håkansson, B. Nordén, and E. W. Thulstrup, *Chem. Phys. Lett.,* **50**, 305 (1977); (b) B. Nordén, R. Håkansson, P. B. Pedersen, and E. W. Thulstrup, *Chem. Phys.,* **33**, 355 (1978).

76.   N. K. Das Gupta and F. W. Birss, *Tetrahedron,* **36**, 2711 (1980).

77.   P. J. Derrick, L. Åsbrink, O. Edqvist, B. -O. Jonsson, and E. Lindblom, *Int. J. Mass. Spectrom. Ion Phys.,* **6**, 177 (1971).

78.   O. Mårtensson and H. Chojnacki, *Acta Phys. Polonica,* **A44**, 259 (1973).

79.   (a) D. T. Clark, *Tetrahedron Lett.,* 5257 (1967); (b) D. T. Clark, *Tetrahedron,* **24**, 2663 (1968).

80.   T. Yonezawa, H. Konishi, and H. Kato, *Bull. Chem. Soc. Jpn.,* **42**, 1933 (1969).

81.   A. J. Tajiri, T. Asano, and T. Nakajima, *Tetrahedron Lett.,* **21**, 1785 (1971).

82.   V. Galasso, *Z. Naturforsch.,* **28a**, 1951 (1973).

83.   V. Galasso, *Chem. Phys. Lett.,* **32**, 108 (1975).

84. Ya. L. Gold'farb, V. P. Litvinov, G. M. Zhidomirov, I. A. Abronin, and R. Z. Zakharyan, *Chem. Scr.*, **5**, 49 (1974).

85. K. -W. Schulte and A. Schweig, *Theor. Chim. Acta*, **33**, 19 (1974).

86. (a) N. D. Epiotis, W. R. Cherry, F. Bernardi, and W. J. Hehre, *J. Am. Chem. Soc.*, **98**, 4361 (1976); (b) N. D. Epiotis and W. Cherry, *J. Am. Chem. Soc.*, **98**, 4365 (1976).

87. G. Andermann and D. R. Phillips, private communication cited in K. Taniguchi and B. L. Henke, *J. Chem. Phys.*, **64**, 3021 (1976).

88. S. A. Chambers and T. D. Thomas, *J. Chem. Phys.*, **67**, 2596 (1977).

89. G. Kluge and M. Scholz, *Z. Chem.*, **19**, 457 (1979).

90. M. J. S. Dewar, *Pure Appl. Chem.*, **44**, 767 (1975).

91. M. J. S. Dewar and G. P. Ford, *J. Am. Chem. Soc.*, **99**, 1685 (1977).

92. J. R. Defina and P. R. Andrews, *Int. J. Quant. Chem.*, **XVIII**, 797 (1980).

93. D. T. Clark and D. R. Armstrong, *Chem. Commun.*, 319 (1970).

94. D. T. Clark, *Int. J. Sulfur Chem. C*, **7**, 11 (1972).

95. U. Gelius, B. Roos, and P. Siegbahn, *Theor. Chim. Acta*, **27**, 171 (1972).

96. M. H. Palmer and R. H. Findlay, *Tetrahedron Lett.*, **41**, 4165 (1972).

97. L. E. Nitzsche and R. E. Christoffersen, *J. Am. Chem. Soc.*, **96**, 5989 (1974).

98. W. von Niessen, W. P. Kraemer, and L. S. Cederbaum, *J. Electron Spectrosc. Rel. Phenom.* **8**, 179 (1976).

99. F. Bernardi, A. Bottoni, and A. Mangini, *Gazz. Chim. Ital.*, **107**, 55 (1977).

100. J. Kao and L. Radom, *J. Am. Chem. Soc.*, **101**, 311 (1979).

101. R. Hilal, *J. Comput. Chem.*, **1**, 348, 358 (1980).

102. G. De Alti and P. Decleva, *Chem. Phys. Lett.*, **77**, 413 (1981).

103. J. C. Slater, *The Self-Consistent Field for Molecules and Solids*, Vol. 4., McGraw Hill, New York, 1974.

104. K. Johnson, *Adv. Quant. Chem.*, **7**, 143 (1973).

105. G. Berthier and B. Pullman, *Compt. Rend. Acad. Sci.*, **231**, 774 (1950).

106. L. Melander, *Acta Chem. Scand.*, **9**, 1400 (1955).

107. J. Koutecky, *Coll. Czech. Chem. Commun.*, **24**, 1608 (1959).

108. J. Koutecky, R. Zahradnik, and J. Paldus, *J. Chim. Phys.*, 455 (1959).

109. R. Zahradnik, C. Párkányi, V. Horák, and J. Koutecky, *Coll. Czech. Chem. Commun.*, **28**, 776 (1963).

110. R. Zahradnik and C. Párkányi, *Coll. Czech. Chem. Commun.*, **30**, 195 (1965).

111. J. Fabian, A. Mehlhorn, and R. Mayer, *Z. Chem.*, **5**, 22 (1965).

112. B. Östman, *J. Am. Chem. Soc.*, **87**, 3163 (1965).

113. B. Östman, *Acta Chem. Scand.*, **22**, 2765 (1968).

114. R. Gerdil and E. A. C. Lucken, *J. Am. Chem. Soc.*, **87**, 213 (1965); **88**, 733 (1966).

115. E. J. Vincent, R. Phan-Tan-Luu, and J. Metzger, *Bull. Soc. Chim. Fr.*, **11**, 3537 (1966).

116. H. Gusten, L. Klasinc, and O. Volkert, *Z. Naturforsch.*, **B24**, 12 (1969).

117. L. Klasinc and K. Humski, *Z. Naturforsch.*, **B25**, 324 (1970).

118. B. Kamienski and T. M. Krygoroski, *Tetrahedron Lett.*, **2**, 103 (1971).

119. H. -D. Scharf and H. Leismann, *Z. Naturforsch.*, **B28**, 662 (1973).

120. M. Janda, J. Šrogl, I. Stibor, P. Trška, and P. Vopatrná, *Coll. Czech. Chem. Commun.*, **39**, 3522 (1974).

121. J. Fabian, A. Mehlhorn, and R. Zahradnik, *Theor. Chim. Acta*, **12**, 247 (1968).

122. J. Fabian, *Z. Chem.*, **8**, 274 (1968).

123.   D. T. Clark, *J. Mol. Spectrosc.* **26**, 181 (1968).

124.   J. W. van Reijendam and M. J. Janssen, *Tetrahedron,* **26**, 1303 (1970).

125.   A. Skancke, *Acta Chem. Scand.,* **24**, 1389 (1970).

126.   E. Corradi, P. Lazzeretti, and F. Taddei, *Mol. Phys.,* **26**, 41 (1973).

127.   H. Hartmann and R. Radeglia, *J. Prakt. Chem.,* **317**, 657 (1975).

128.   A. Mehlhorn, B. Schwenzer, and K. Schwetlick, *Tetrahedron,* **33**, 1483 (1977).

129.   A. Mehlhorn, B. Schwenzer, H. -J. Bruckner, and K. Schwetlick, *Tetrahedron,* **34**, 481 (1978).

130.   J. Fabian and H. Hartmann, *Z. Chem.,* **18**, 145 (1978).

131.   J. Fabian, *Z. Phys. Chem.,* **260**, 81 (1979).

132.   V. Galasso and G. DeAlti, *Tetrahedron,* **27**, 4947 (1971).

133.   S. Rodmar, *Mol. Phys.,* **22**, 123 (1971).

134.   (a) V. Galasso and N. Trinajstic, *J. Chim. Phys. Physicochim. Biol.,* **70**, 1489 (1973); (b) V. Galasso, *Theor. Chim. Acta,* **34**, 137 (1974).

135.   S. Nagata, T. Yamabe, K. Yoshikawa, and H. Kato, *Tetrahedron,* **29**, 2545 (1973).

136.   F. deJong, A. J. Noorduin, T. Bouwman, and M. J. Janssen, *Tetrahedron Lett.,* **13**, 1209 (1974).

137.   V. Galasso, E. Montoneri, and G. C. Pappalardo, *J. Mol. Struct.,* **76**, 43, 48 (1981).

138.   V. Galasso, L. Klasinc, A. Sabljic, N. Trinajstic, G. C. Pappalardo, and W. Steglich, *J. Chem. Soc. Perkin Trans.,* **2**, 127 (1981).

139.   H. D. Hartough, *Thiophene and Its Derivatives.* Interscience, New York, 1952.

140.   L. Pauling, *The Nature of the Chemical Bond*, Cornell University Press, Ithaca, NY, 1948.

141.   B. Roos and P. Siegbahn, *Theor. Chim. Acta,* **21**, 368 (1971).

142.   U. Gelius, B. Roos, and P. Siegbahn, *Theor. Chim. Acta,* **23**, 590 (1971).

143.   B. Harris, R. J. W. LeFe've, and E. P. A. Sullivan, *J. Chem. Soc.,* 1622 (1953).

144.   U. Gelius, C. J. Allan, G. Johansson, H. Siegbahn, D. A. Allison, and K. Siegbahn, *Phys. Scr.,* **3**, 237 (1971).

145.   J. A. Sell and A. Kuppermann, *Chem. Phys. Lett.,* **61**, 355 (1979).

146.   T. Munakata, K. Kuchitsu, and Y. Harada, *J. Electron Spectrosc. Rel. Phenom.,* **20**, 235 (1980).

147.   W. C. Price and A. P. Walsh, *Proc. Roy. Soc. (London),* **A179**, 201 (1941).

148.   (a) G. Milazzo, *Spectrochim. Acta,* **2**, 245 (1944); (b) G. Milazzo, *Gazz. Chim. Ital.,* **78**, 835 (1948); (c) G. Milazzo, *Gazz. Chim. Ital.,* **83**, 392 (1953).

149.   F. S. Boig, G. W. Costa, and I. Osvar, *J. Org. Chem.,* **18**, 775 (1953).

150.   G. Leandrix, A. Mangini, F. Montanari, and R. Passerini, *Gazz. Chim. Ital.,* **85**, 769 (1955).

151.   S. Gronowitz, *Ark. Kemi,* **13**, 239 (1958).

152.   J. Sice', *J. Phys. Chem.,* **64**, 1572 (1960).

153.   G. Horva'th and A. I. Kiss, *Spectrochim. Acta,* **A23**, 921 (1967).

154.   M. R. Padhye and S. R. Desai, *Proc. Phys. Soc. (London),* **A65**, 298 (1952).

155.   (a) M. Karplus and J. A. Pople, *J. Chem. Phys.,* **38**, 2803 (1963); (b) J. B. Stothers, *Carbon-13-NMR-Spectroscopy*, Academic Press, New York, 1972.

156.   S. Gronowitz and R. A Hoffmann, *Ark. Kemi,* **13**, 279 (1958).

157.   R. Bonaccorsi, C. Petrolongo, E. Scrocco, and J. Tomasi, *Quantum Aspects of Heterocyclic Compounds in Biochemistry,* Vol. 2, E. Bergmann and B. Pullman (Eds.), Academic Press, New York, 1970, p. 181.

158. C. A. Coulson, *Proc. Roy. Soc.,* **A169**, 413 (1939).

159. (a) J. W. McIver and A. Komornicki, *Chem. Phys. Lett.,* **10**, 303 (1971); (b) *J. Am. Chem. Soc.,* **94**, 2625 (1972).

160. E. B. Wilson, Jr., J. C. Decius, and P. C. Cross, *Molecular Vibrations,* McGraw Hill, New York, 1955.

161. (a) B. N. Cyvin and S. J. Cyvin, *Acta Chem. Scand.,* **23**, 3139 (1969); (b) S. J. Cyvin, B. N. Cyvin, and G. Hagen, *Acta Chem. Scand.,* **23**, 3407 (1969).

162. D. W. Scott, *J. Mol. Spectr.,* **31**, 451 (1969).

163. T. Shimanouchi, Tables of Molecular Vibrational Frequencies, National Standard Reference Data Series, Washington, DC, 1972.

164. J. A. Pople and D. P. Santry, *Mol. Phys.,* **8**, 1 (1964).

165. A. C. Blizzard and D. P. Santry, *J. Chem. Phys.,* **55**, 950 (1971).

166. J. M. Read, Jr., C. T. Mathis, and J. H. Goldstein, *Spectrochim. Acta,* **21**, 85 (1965).

167. S. Gronowitz and A. Rosenberg, *Ark. Kemi,* **8**, 23 (1956).

168. C. K. Ingold, *Structure and Mechanism in Organic Chemistry,* Cornell University Press, Ithaca, NY, 1953.

169. B. Östman, *Acta Chem. Scand.,* **22**, 2754 (1968).

170. J. M. J. G. Lipsch and G. C. A. Schuit, *J. Catal.,* **15**, 179 (1969).

171. S. Sunner, *Acta Chem. Scand.,* **9**, 847 (1955).

172. D. H. Sutter and W. H. Flygare, *J. Am. Chem. Soc.,* **91**, 4063 (1969).

173. A. D. Buckingham, B. Harris, and R. J. W. Le Fevre, *J. Chem. Soc.,* 1626 (1953).

174. W. von Niessen, L. S. Cederbaum, and G. H. F. Diercksen, *J. Am. Chem. Soc.,* **98**, 2066 (1976).

175. P. J. Derrick, L. Åsbrink, O. Edqvist, B. -O. Jonsson, and E. Lindholm, *Int. J. Mass. Spectrom. Ion Phys.,* **6**, 161 (1971).

176. P. J. Derrick, L. Åsbrink, O. Edqvist, B. -O. Jonsson, and E. Lindholm, *Int. J. Mass. Spectrom. Ion Phys.,* **6**, 191 (1971).

177. K. Watanabe and T. Nakayama, *J. Chem. Phys.,* **29**, 48 (1958).

CHAPTER III

Naturally Occuring Thiophenes

F. BOHLMANN and C. ZDERO

Institut für Organische Chemie, Technische Universität, Berlin, Federal Republic of Germany

I. INTRODUCTION

Previous reviews of thiophenes and their derivatives mentioned only simple, naturally occurring thiophenes from coal tar and shale oil and only one compound from Compositae, terthienyl (isolated from *Tagetes erecta L*),[1] was described. Since then the number of thiophene derivatives from this family has risen to more than 150, and a few have been isolated from fungi. Most probably, all of these compounds are derived from acetylenic precursors; consequently, most of them still have acetylenic bonds, but a few are shown that do not contain these bonds. However, in many cases, feeding experiments with labeled precursors have shown that they also are formed in the plants from acetylenes. The key step is the addition of H_2S or its biochemical equivalent to conjugated triple bonds. Though the details of this reaction are still not known, it is most likely a two-step reaction, because the thiophenes often cooccur with thioenolethers, which are also formed in a two-step reaction, as shown by feeding experiments.[2] The following scheme is proposed:

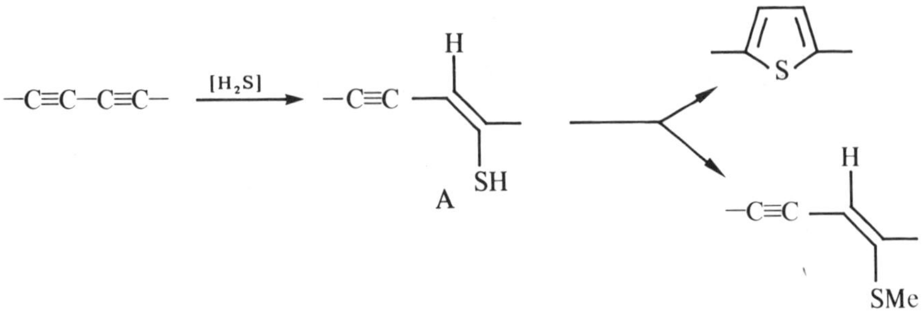

The thiophenes isolated so far from Compositae occur in the roots as well as in the aerial parts. Normally, the concentration is low (10^{-2}–$10^{-4}\%$), but may rise to about 1% in special cases. These compounds normally are extracted with ether–petrol mixtures from the roots or the aerial parts. Separation by column and thin layer chromatography then affords the single thiophene derivatives, which can be detected by their UV maxima. These are sometimes characteristic and allow the elucidation of the chromophoric system. Most important, however, are the ^1H nmr spectra, which often lead directly to the structures. In a few cases, chemical transformations and degradations and further ^1H nmr studies are necessary to decide between two possible structures. ^{13}C nmr data, though characteristic, have not been used for structure elucidation. Of course, high resolution mass spectroscopy, especially to establish the molecular formula, is important. The structures of several naturally occurring thiophenes have been established further by synthesis. Since probably all compounds are derived from acetylenic precursors, the known thiophenes will be discussed in terms of biogenetic considerations.

II. NATURALLY OCCURRING THIOPHENES

1. Thiophenes Derived from C_{10}-Acetylenes

Most of these thiophenes (Scheme 1) are formed from the isomeric dehydromatricariaesters 1 and 2 by the addition of H_2S or its biochemical equivalent. Accordingly, these thiophenes were isolated mainly from members of the tribe Anthemideae (family Compositae), where their precursors, together with corresponding methylthioenolether, are widespread.[4] Starting with cis- and trans-dehydromatricariaester (1 and 2), four thiophenes, 3,[4] 4,[5] 5,[6] and 6,[6] are possible, since all were isolated several times.[4] Two more thiophenes (7[7] and 8[7]) are formed from 5 and 6 by oxidative elimination of C-10, a reaction that is very common in the field of naturally occurring acetylenes.[4] Also, the lactone 9[8] is derived from 6

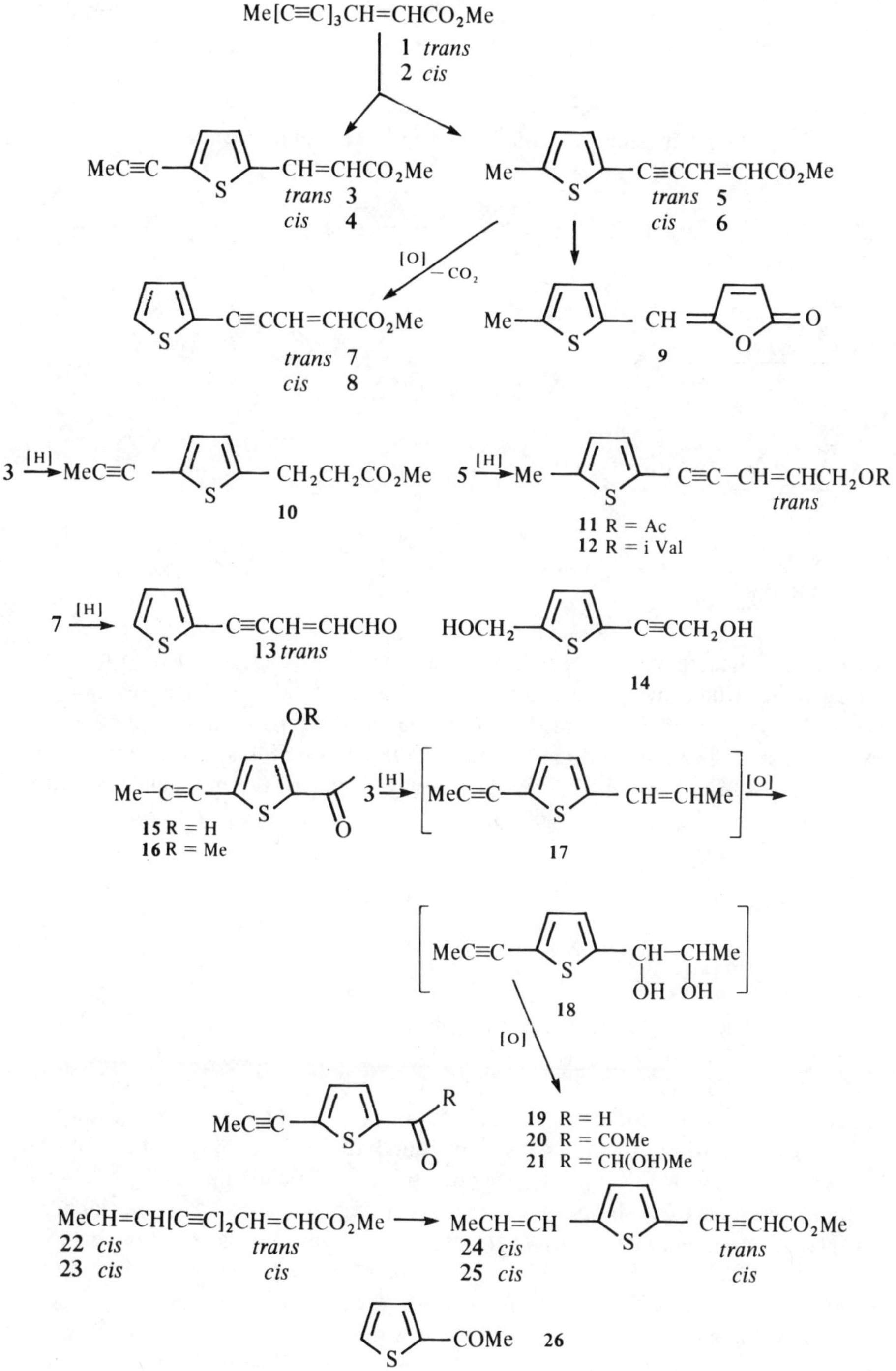

Me[C≡C]₃CH=CHCO₂Me

1 *trans*
2 *cis*

MeC≡C—⟨S⟩—CH=CHCO₂Me
trans **3**
cis **4**

Me—⟨S⟩—C≡CCH=CHCO₂Me
trans **5**
cis **6**

[O] — CO₂

⟨S⟩—C≡CCH=CHCO₂Me
trans **7**
cis **8**

Me—⟨S⟩—CH=⟨O⟩=O
9

3 →[H] MeC≡C—⟨S⟩—CH₂CH₂CO₂Me
10

5 →[H] Me—⟨S⟩—C≡C—CH=CHCH₂OR
trans
11 R = Ac
12 R = i Val

7 →[H] ⟨S⟩—C≡CCH=CHCHO
13 *trans*

HOCH₂—⟨S⟩—C≡CCH₂OH
14

OR
Me—C≡C—⟨S⟩—C(=O)
15 R = H
16 R = Me

3 →[H] [MeC≡C—⟨S⟩—CH=CHMe] →[O]
17

[MeC≡C—⟨S⟩—CH—CHMe]
                      | |
                     OH   OH
18

[O]

MeC≡C—⟨S⟩—C(=O)R
19 R = H
20 R = COMe
21 R = CH(OH)Me

MeCH=CH[C≡C]₂CH=CHCO₂Me → MeCH=CH—⟨S⟩—CH=CHCO₂Me
22 *cis*   *trans*   **24** *cis*   *trans*
23 *cis*   *cis*   **25** *cis*   *cis*

⟨S⟩—COMe **26**

Scheme 1

by addition of the free carboxyl group to the 4.5-triple bond. The structures **3–9** were elucidated by spectroscopic methods and were established by synthesis. The lithiated thiophene **27**, in reaction with dimethyl formamide, afforded the aldehyde **28**, which by Wittig reaction was transformed to **3**.[9] Similarly, **5** and **6** were obtained starting with methylethinylthiophene (**31**), prepared via **30** from the

$$MeC{\equiv}C\!-\!\!\underset{27}{\overset{S}{\text{[thiophene]}}}\!\!-Li + HCONMe_2 \longrightarrow MeC{\equiv}C\!-\!\!\underset{28}{\overset{S}{\text{[thiophene]}}}\!\!-CHO$$

$$\xrightarrow{Ph_3P\,=\,CHCO_2Me} 3 \quad Me\!-\!\!\underset{29}{\overset{S}{\text{[thiophene]}}}\!\!-COMe \xrightarrow{PCl_5} Me\!-\!\!\underset{30}{\overset{S}{\text{[thiophene]}}}\!\!-C(Cl_2)Me$$

$$\xrightarrow{NaNH_2} Me\!-\!\!\underset{31}{\overset{S}{\text{[thiophene]}}}\!\!-C{\equiv}CH \xrightarrow[HCONMe_2]{EtMgBr} Me\!-\!\!\underset{32}{\overset{S}{\text{[thiophene]}}}\!\!-C{\equiv}CCHO$$

$$\xrightarrow{Ph_3P\,=\,CHCO_2Me} \quad 5 + 6$$

ketone **29**. Reaction with ethylmagnesium bromide and dimethylformamide afforded **32**, which by Wittig reaction gave **5** and **6**.[6] Similar Wittig reaction of thienylpropargylaldehyde afforded **7**.[6] Compounds **3** and **4** also have been obtained by addition of sodium sulfide to *cis*- and *trans*-dehydromatricariaester.[10] The lactone **9** could be synthesized by aldol condensation of the butenolide **34** with **33**.[8] Compound **10**, a 2,3-dihydro derivative of **3**, has been isolated from the

$$Me\!-\!\!\underset{33}{\overset{S}{\text{[thiophene]}}}\!\!-CHO + \underset{34}{\overset{O}{\text{[butenolide]}}}{=}O \longrightarrow 9$$

Artemisia species;[11] its structure clearly followed from the ^{1}H nmr data, as shown by the clear difference in the chemical shifts of the thiophene methyl and the acetylene methyl signals. Compounds **11** and **12**, present in *Osmitopsis* species,[12, 13] are obviously also products of the biochemical reduction of **5**, followed by esterification; **12** has also been synthesized. Compound **13**[14] is most likely formed in this way from **7**, though it could as well arise from compounds like **267** or **294** by degradation, since these thiophenes cooccur in *Anthemis* species.[4] The C$_8$-thiophene **14** present in *Centaurea ruthenica* is only formed by the degradation of C$_{13}$-thiophene. The structures of **15** and **16** are somewhat unusual. They have been isolated from *Artemisia*[15] as well as from *Liatris*[16] and *Ageratina*[17] species. Nothing is known about their biogenesis, but surely acetylenic precursors must be assumed. The structures followed clearly from their ^{1}H nmr spectra and were further established by synthesis.[18] Addition of the anion of mercapto acetone to **35** afforded, in the presence of alkali through the intermediate **36**, the hydroxy ketone **15**, which on methylation gave **16**. So far, it is not clear, if the thiophenes **19–21**

$$Me[C{\equiv}C]_2CO_2Me \xrightarrow{^{\ominus}SCH_2COMe} \left[MeC{\equiv}C{-}\overbrace{\underset{S-CH_2COMe}{}}^{CO_2Me} \right] \longrightarrow 15 \xrightarrow{MeI} 16$$

35
36

isolated from the Basidiomycete *Daedelea juniperea*[19, 20] are formed from **3**. If the latter would be reduced to **17**, hydroxylation could lead to **18**, which by further oxydation could be transformed to the natural compounds. Compound **19** also has been synthesized,[21] using a biomimetic synthesis by addition of H_2S to a triyne as starting material; MnO_2-oxidation afforded **19**. The thiophenes **24** and **25**

$$Me[C{\equiv}C]_3CH_2OH \xrightarrow{H_2S,\ OH^{\ominus}} MeC{\equiv}C{-}\overbrace{\underset{S}{}}{-}CH_2OH \xrightarrow{MnO_2} 19$$

37

obviously are formed by addition of H_2S to the isomeric matricariaesters **22** and **23** respectively. Thus far, these thiophenes have been isolated only from *Anthemis austriaca*.[22, 23] The structure of **25** clearly followed from the ^1H nmr data. The couplings $J_{2,3}$ and $J_{8,9}$ showed that both double bonds are *cis*-configurated, whereas the ^1H nmr data of **24** indicate a 2,3-*trans*-double bond. The structures were further established by synthesis.[24] Starting with 2,5-diiodothiophene, reaction with the cuprous salt of propyne, followed by reaction with the salt of methyl propiolate (**39**), afforded the diyne **40**, which on partial hydrogenation gave **25**. 2-Acetyl-thiophene (**26**) has been isolated from a *Bidens* species;[25] it is not known if it is also formed by degradation.

$$MeC{\equiv}C{-}\overbrace{\underset{S}{}}{-}I + CuC{\equiv}CCO_2Me \longrightarrow$$

38   39

$$MeC{\equiv}C{-}\overbrace{\underset{S}{}}{-}C{\equiv}CCO_2Me \xrightarrow{H_2} 25$$

40

2. Thiophenes Derived from C_{13}-Acetylenes

A. Monothiophenes Derived from Tridecapentaynene

Most of the known, naturally occurring thiophenes are derived from the wide-spread tridecapentaynene **41**. Feeding experiments have established this assumption for several thiophenes.[26] Addition of 1 eq. H_2S would lead to four different monothiophenes. So far, however, only two, **42**[27] and **43**,[28] have been isolated from many species of the Compositae.[4] Both compounds are further transformed to a large number of derivatives (see Scheme 2). The UV spectra of **42** and **43** are very

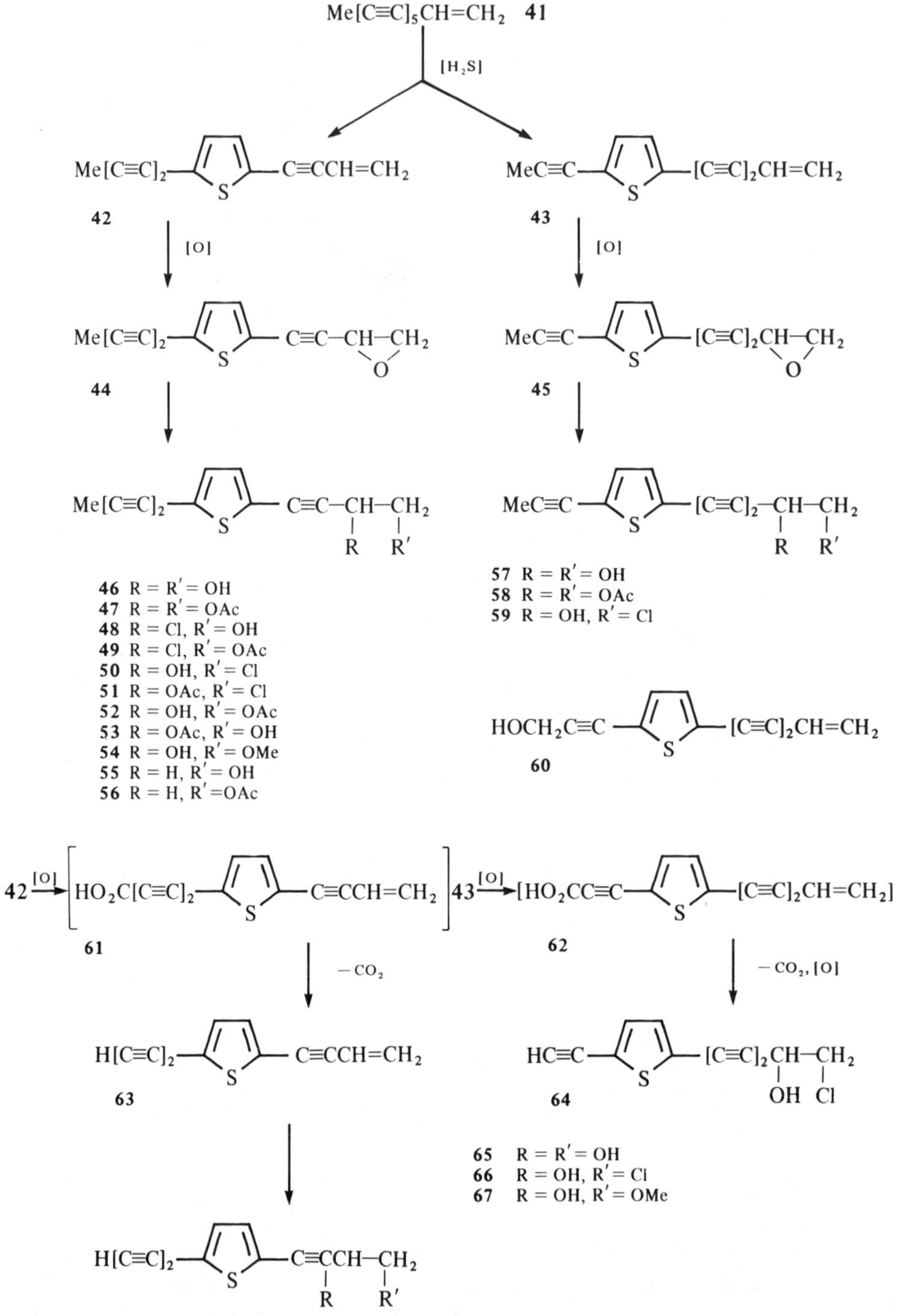

$Me[C{\equiv}C]_5CH{=}CH_2$ **41**

[H₂S]

42 $Me[C{\equiv}C]_2$—[thiophene]—$C{\equiv}CCH{=}CH_2$

43 $MeC{\equiv}C$—[thiophene]—$[C{\equiv}C]_2CH{=}CH_2$

[O]

44 $Me[C{\equiv}C]_2$—[thiophene]—$C{\equiv}C{-}CH{-}CH_2$ (epoxide O)

45 $MeC{\equiv}C$—[thiophene]—$[C{\equiv}C]_2CH{-}CH_2$ (epoxide O)

$Me[C{\equiv}C]_2$—[thiophene]—$C{\equiv}C{-}CH{-}CH_2$ with R, R'

46 R = R' = OH
47 R = R' = OAc
48 R = Cl, R' = OH
49 R = Cl, R' = OAc
50 R = OH, R' = Cl
51 R = OAc, R' = Cl
52 R = OH, R' = OAc
53 R = OAc, R' = OH
54 R = OH, R' = OMe
55 R = H, R' = OH
56 R = H, R' = OAc

$MeC{\equiv}C$—[thiophene]—$[C{\equiv}C]_2{-}CH{-}CH_2$ with R, R'

57 R = R' = OH
58 R = R' = OAc
59 R = OH, R' = Cl

$HOCH_2C{\equiv}C$—[thiophene]—$[C{\equiv}C]_2CH{=}CH_2$
60

$42 \xrightarrow{[O]}$ [$HO_2C[C{\equiv}C]_2$—[thiophene]—$C{\equiv}CCH{=}CH_2$]   **61**

$43 \xrightarrow{[O]}$ [$HO_2CC{\equiv}C$—[thiophene]—$[C{\equiv}C]_2CH{=}CH_2$]   **62**

$-CO_2$

$-CO_2, [O]$

$H[C{\equiv}C]_2$—[thiophene]—$C{\equiv}CCH{=}CH_2$
63

$HC{\equiv}C$—[thiophene]—$[C{\equiv}C]_2CH{-}CH_2$ with OH Cl
64

65 R = R' = OH
66 R = OH, R' = Cl
67 R = OH, R' = OMe

$H[C{\equiv}C]_2$—[thiophene]—$C{\equiv}CCH{-}CH_2$ with R, R'

Scheme 2

similar, and the ^{1}H nmr data are only slightly different. For structure elucidation, therefore, the hydrogenated thiophenes are prepared and their structures established by comparing them with the synthetic dialkyl thiophenes. Furthermore, both structures are established by synthesis.[29] A convenient route is the method of Stevens and Castro,[30] elaborated further by Atkinson and Curtis.[16] 2,5-Diiodothiophene (68) afforded 70, which on reaction with the cuprous salt of butenyne yielded 42. Similarly, reaction of 68 with the cuprous salt of propyne, followed by reaction with the salt of hexadiynene (73), gave 43.[29]

$$I-\overset{\displaystyle\langle\!\langle\;\;\rangle\!\rangle}{S}-I + Cu[C{\equiv}C]_2Me \longrightarrow$$

$$\textbf{68} \qquad\qquad \textbf{69}$$

$$Me[C{\equiv}C]_2-\overset{\displaystyle\langle\!\langle\;\;\rangle\!\rangle}{S}-I + CuC{\equiv}CCH{=}CH_2 \longrightarrow \textbf{42}$$

$$\textbf{70} \qquad\qquad \textbf{71}$$

$$MeC{\equiv}C-\overset{\displaystyle\langle\!\langle\;\;\rangle\!\rangle}{S}-I + Cu[C{\equiv}C]_2CH{=}CH_2 \longrightarrow \textbf{43}$$

$$\textbf{72} \qquad\qquad \textbf{73}$$

Only the separation of unreacted iodo-compounds from the substituted thiophenes can sometimes be a problem.

The biogenetic pathway from 42 and 43, respectively, to 44–59 is obvious. First, epoxidation occurs and then the epoxides 44 and 45 are opened, leading to the natural products; only the reductive cleavage of 44 leading to 55 and 56 is an exception. Though their biosynthesis is not established, this pathway could be demonstrated in a very similar case (97). The epoxide 44 was first isolated from *Echinops* species,[27] its structure could be elucidated clearly from the UV maximum and the ^{1}H nmr data. Though the UV spectra of type 44 compounds are slightly different only from those of type 45, small differences are characteristic and allow the assignment of the chromophoric system. The nature of the oxygen function of 44 can be deduced easily from the typical signals of the epoxide protons (3.50 t and 2.84 d/J = 3.2 Hz). Also, 46–49 are present in several *Echinops* species;[4,27] the structures followed from the UV and ^{1}H nmr spectra. As 48 on acetylation afforded 49, the relative position of chlorine and hydroxyl could be established easily. Furthermore, treatment of 48 or 49 with diluted alkali afforded 44, which could be transformed with diluted acid to 46. Periodate oxidation of 46 gave the aldehyde 74; its ^{1}H nmr data further supported the proposed distribution of the acetylenic

$$\textbf{46} \xrightarrow{\;HIO_4\;} Me[C{\equiv}C]_2-\overset{\displaystyle\langle\!\langle\;\;\rangle\!\rangle}{S}-C{\equiv}CCHO \qquad \textbf{74}$$

bonds. From the roots of *Eclipta erecta*, 50–52 were isolated;[31] 50 and 51 were isomers of 48 and 49, respectively. Consequently, dilute alkali transformed both into 44. As expected, the chemical shifts of H-1 and H-2 in 48 and 50 are very

$$52 \xrightarrow{\text{MnO}_2} \text{Me[C} \equiv \text{C]}_2 - \underset{S}{\boxed{\text{thiophene}}} - \text{C} \equiv \text{CCOCH}_2\text{OAc} \quad 75$$

similar and, therefore, are not conclusive for structural assignments. Compound **52** afforded the ketoacetate **75** on oxidation with manganese dioxide, thus establishing the structure of the hydroxy acetate **52**. The isomer **53** is present in the roots of *Pluchea suaveolens*;[32] the methoxy derivative **54** has been isolated from *Cullumia setosa*.[33] The 2-desoxy compounds **55** and **56** also are present in *Echinops* species.[27] Their structures followed directly from the spectral data. Most of the thiophenes (**44, 46–56**) have been synthesized too. The diiodide **68** was used again as starting material. Reaction with the cuprous salt **76** afforded **77**, which gave **55** with the cuprous salt of pentadiyne.[34] Compounds **44, 46–49,** and **52** were obtained simi-

$$68 + \text{CuC} \equiv \text{CCH}_2\text{CH}_2\text{OH} \longrightarrow \text{I} - \underset{S}{\boxed{\text{thiophene}}} - \text{C} \equiv \text{CCH}_2\text{CH}_2\text{OH} +$$

$$\quad\quad\quad\quad\quad 76 \quad\quad\quad\quad\quad\quad\quad\quad\quad\quad\quad\quad 77$$

$$\text{Cu[C} \equiv \text{C]}_2\text{Me} \longrightarrow 55$$

$$69$$

larly.[34] Reaction of **70** with the cuprous salt **78** afforded **79**, which could be transformed to the natural compounds by the usual methods.

$$70 + \text{CuC} \equiv \text{CCHCH}_2\text{OAc} \longrightarrow \text{Me[C} \equiv \text{C]}_2 - \underset{S}{\boxed{\text{thiophene}}} - \text{C} \equiv \text{CCHCH}_2\text{OAc}$$

78 ... 79

$$\longrightarrow 44, 46–49, 52$$

The isomeric epoxide **45** and the diacetate **58** have so far been isolated only from a *Cullumia* species,[35] while the diol **57** is present in *Platycarpha glomerata*.[37] The chlorohydrin **59** was isolated from the roots of *Pterocaulon virgatum*[38] together with the corresponding desmethyl compound **64**. The relative position of chlorine can be deduced from the mass spectra, as the 1-chloro compounds always show a strong fragment formed by the elimination of CH₂Cl. The propargylic oxidation product of **43** is the alcohol **60**, which is present in a *Rudbeckia* species.[39] The structure clearly followed from the UV and ¹H nmr spectra and has been established by synthesis.[34] Using the method of Atkinson and Curtis,[16] reaction of 2,5-diiodo thiophene with the cuprous salt **80** afforded the acetal **81**, which on reaction with the cuprous salt of hexadiynene gave **82**. Hydrolysis afforded **60**, which is identical with the natural compound:

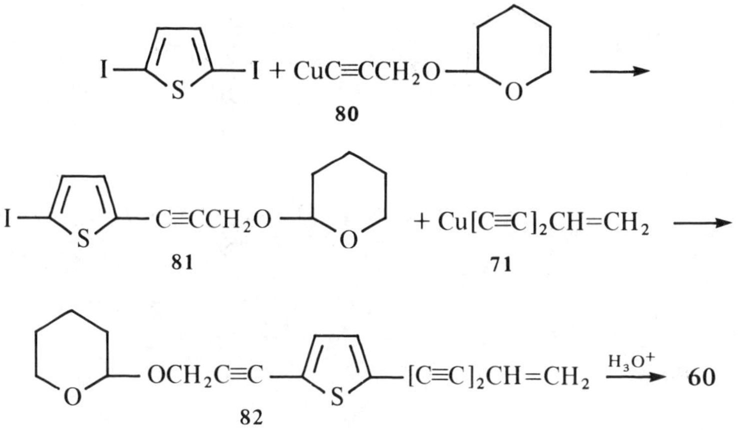

From *Eclipta erecta* **63** and **66** were isolated;[31] *Berkheya maritima* afforded **65**[36] and *Cullumia setosa* gave **54**[31] and **67**.[33] The presence of monosubstituted acetylenes clearly followed from the characteristic IR band and from the ^{1}H nmr spectra. If we look at the possible biogenetic pathways for these compounds, the cooccurrence of **60** and **64** may support the possibility that, in these cases, the terminal methyl group of **42** and **43** is first oxygenated to **61** and **62**, respectively, and then, by the loss of CO_2, transformed to **63** and **64** respectively. Of course, these steps could also occur subsequent to the modification of the vinyl group. Similar pathways have been established with several dithiophenes.[4]

B.   Bithienyls Derived from Tridecapentaynene

As outlined in Section II.A, tridecapentaynene can be transformed to **42** and **43** by the addition of H_2S or its biochemical equivalent. Further addition leads to **83** and **92** respectively (see Scheme 3). Compounds **83** and **92** have been isolated from several species, most of them belonging to the tribe Heliantheae.[4] The structures followed from the ^{1}H nmr data, which clearly indicated the presence of methyl-substituted thiophenes [2.42 d (J = 0.8) and 6.32 dq (J = 3.5, 0.8)]. The typical signals of a vinylacetylene were almost identical with those of the bithienyl **91** [6.02 dd (J = 17.5, 11), 5.71 dd (J = 17.5, 2), 5.54 dd (J = 11, 2)]. The UV maximum of **83** is shifted by 6 nm to a longer wavelength, if compared to that of **91**. Furthermore, the structure of **83** has been established by synthesis.[40] Starting with 2-methyldithienyl, introduction of iodine in the 2'-position, and then reaction with the cuprous salt of vinyl acetylene, afforded **83**, which was identical with the natural compounds.

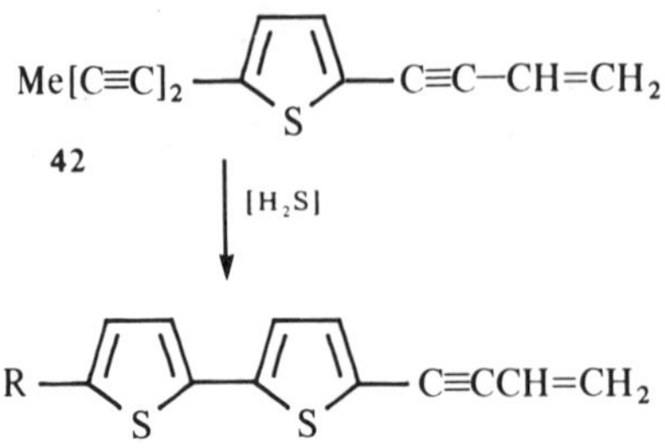

Me[C≡C]₂——⟨S⟩——C≡C—CH=CH₂   MeC≡C——⟨S⟩——[C≡C]₂CH=CH₂

42   **43**

[H₂S] ↓   [H₂S] ↓

R——⟨S⟩——⟨S⟩——C≡CCH=CH₂   MeC≡C——⟨S⟩——⟨S⟩——CH=CH₂

83	R = Me	**88**	R = CH₂OTigl
84	R = CH₂OH	**89**	R = CH₂OSen
85	R = CH₂OAc	**90**	R = CHO
86	R = CH₂OiBu	**91**	R = H
87	R = CH₂OAng		

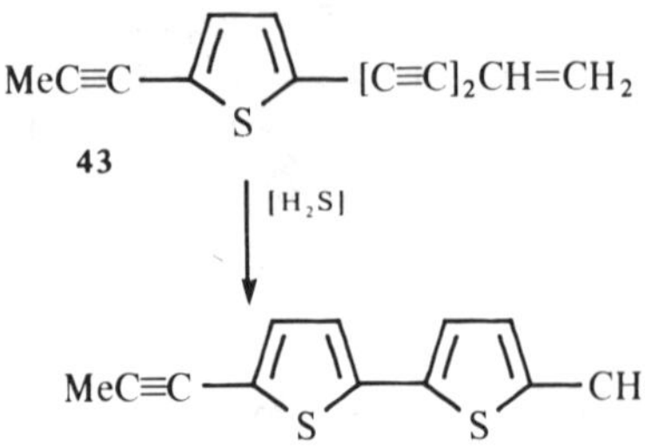

92

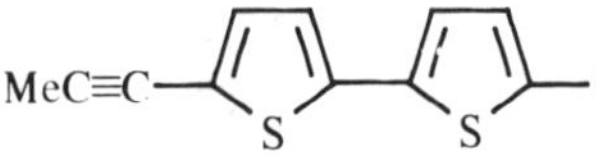

MeC≡C——⟨S⟩——⟨S⟩——CH₂CH₂OAc   **93**

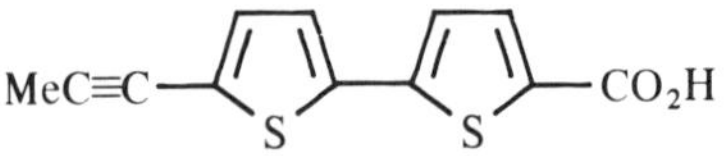

MeC≡C——⟨S⟩——⟨S⟩——CO₂H   **94**

Me——⟨S⟩——⟨S⟩——C≡CCH—CH₂
                               R   R'

⟨S⟩——⟨S⟩——C≡CCH—CH₂
                          R   R'

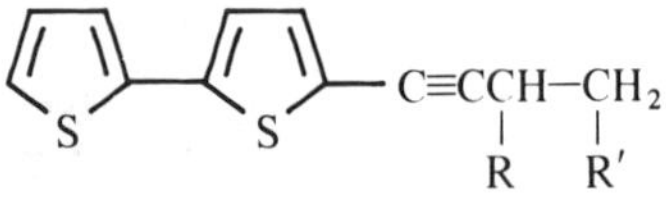

99	R = R' = OH
100	R = R' = OAc
101	R = OH, R' = OAc
102	R = OAc, R' = OH
103	R = OH, R' = OiVal
104	R = OH, R' = Cl
105	R = H, R' = OH
106	R = H, R' = OAc
107	R = H, R' = OiVal

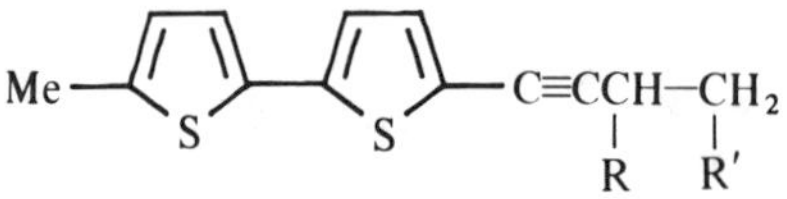

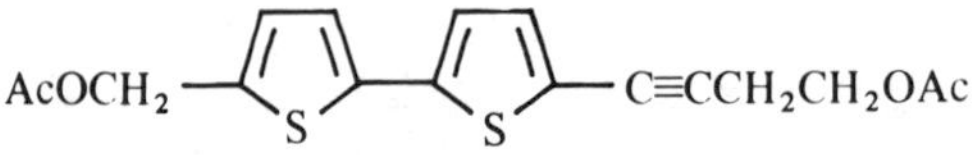

95	R = R' = OAc
96	R = OH, R' = Cl
97	R = H, R' = OH
98	R = H, R' = OAc

AcOCH₂——⟨S⟩——⟨S⟩——C≡CCH₂CH₂OAc

108

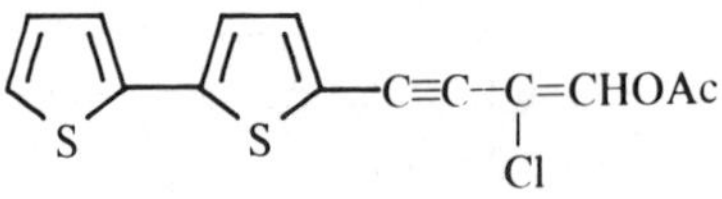

⟨S⟩——⟨S⟩——C≡C—C=CHOAc
                             Cl

109 *trans*
110 *cis*

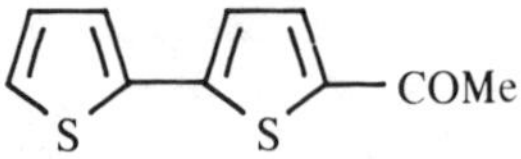

⟨S⟩——⟨S⟩——COMe

111

⟨S⟩——⟨S⟩——C≡C——☐——C≡C——⟨S⟩——⟨S⟩

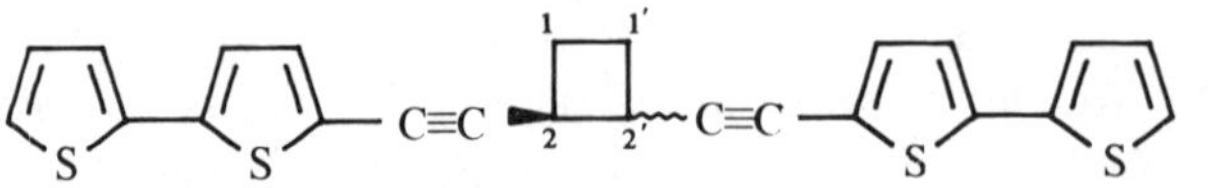

112 (2,2'-*cis*)
113 (2,2'-*trans*)

Scheme 3

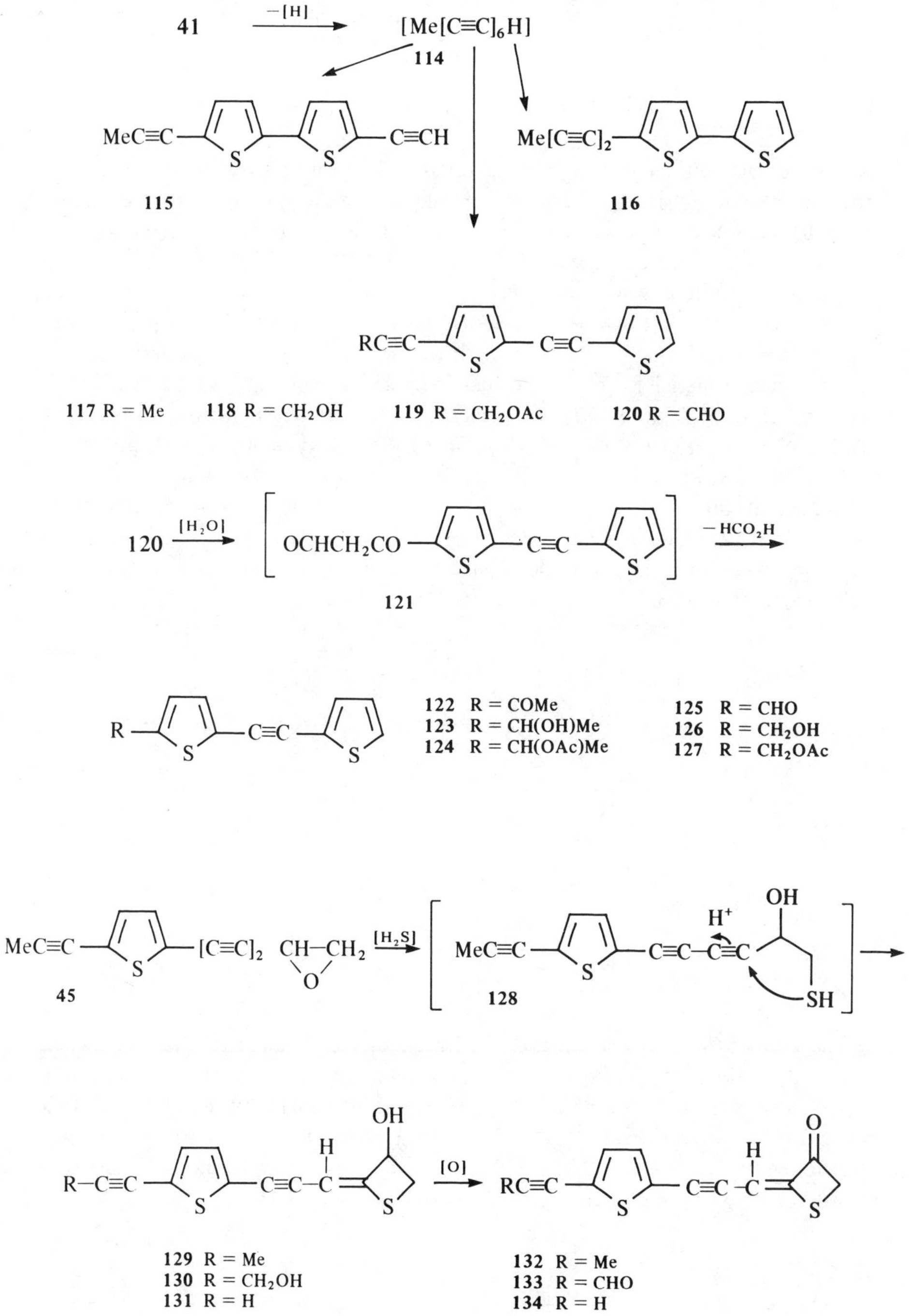

$$41 \xrightarrow{-[H]} [Me[C{\equiv}C]_6H]$$

114

MeC≡C—⟨S⟩—⟨S⟩—C≡CH   Me[C≡C]₂—⟨S⟩—⟨S⟩

115   **116**

RC≡C—⟨S⟩—C≡C—⟨S⟩

117 R = Me   **118** R = CH₂OH   **119** R = CH₂OAc   **120** R = CHO

$$120 \xrightarrow{[H_2O]} \left[OCHCH_2CO-\langle S \rangle-C{\equiv}C-\langle S \rangle \right] \xrightarrow{-HCO_2H}$$

121

R—⟨S⟩—C≡C—⟨S⟩

122 R = COMe	**125** R = CHO
123 R = CH(OH)Me	**126** R = CH₂OH
124 R = CH(OAc)Me	**127** R = CH₂OAc

MeC≡C—⟨S⟩—[C≡C]₂ CH—CH₂ $\xrightarrow{[H_2S]}$ [MeC≡C—⟨S⟩—C≡C—C≡C—CH(OH)—CH₂—SH]
                               \O/

45   **128**

R—C≡C—⟨S⟩—C≡C—C=⟨ OH / S⟩ $\xrightarrow{[O]}$ RC≡C—⟨S⟩—C≡C—C=⟨ O / S⟩

129 R = Me	**132** R = Me
130 R = CH₂OH	**133** R = CHO
131 R = H	**134** R = H

Scheme 3 (Continued)

$$Me\text{---}\underset{S}{\boxed{}}\text{---}\underset{S}{\boxed{}}\text{---}R + CuC\equiv CCH=CH_2 \longrightarrow 83$$

135 R = H
136 R = I

71

The isomeric bithienyl **92** has been isolated first from a *Guizotia* species;[39] this structure also followed from the ^{1}H nmr data. The chemical shift of the methyl group (2.06 s) is typically different from that of **83**. Of course, the signals of the vinyl protons are shifted downfield [6.67 dd (J = 17.11), 5.06 d (J = 17), 5.08 d (J = 11)] in comparison with those of **83**. Also the chemical shifts of the β-thiophene protons are slightly different. The UV maxima of **92** has a pronounced shift to longer wavelength compared with that of **83**.[39] A number of derivatives of **83** have been isolated from Compositae species. The corresponding alcohol (**84**) was first isolated from *Echinops* species[27] and its structure was established by synthesis.[41] Vilsmeier reaction of **91** afforded the aldehyde **90**, which is present in a *Dyssodia* species;[42] reduction with sodium boranate gave **84**, which is identical with the natural compound. Acetylation gave **85**, which was isolated first from *Flaveria repanda*;[41] **84** has also been prepared starting with **137**.[40] Reaction with the cuprous salt of vinyl acetylene gave the ester **138**, which on reaction with lithium alanate yielded **84**.

$$MeO_2C\text{---}\underset{S}{\boxed{}}\text{---}\underset{S}{\boxed{}}\text{---}I + CuC\equiv CCH=CH_2 \longrightarrow$$

137

$$MeO_2C\text{---}\underset{S}{\boxed{}}\text{---}\underset{S}{\boxed{}}\text{---}C\equiv CCH=CH_2 \longrightarrow 84$$

138

The esters **86–89** cooccur in the aerial parts of *Eclipta erecta*,[31] their separation being extremely difficult. From the ^{1}H nmr data, the nature of the ester residues, could, however, be easily deduced. The most widespread bithienyl is **91**, first isolated from *Tagetes erecta*.[43] The structure was established by synthesis, starting wih dithienyl acetylene (**139**).[44] Grignard reaction with dimethylformamide afforded **140**, which by Wittig reaction was transformed to **91**:

$$\underset{S}{\boxed{}}\text{---}\underset{S}{\boxed{}}\text{---}C\equiv CH \xrightarrow[DMF]{EtMgBr} \underset{S}{\boxed{}}\text{---}\underset{S}{\boxed{}}\text{---}C\equiv CCHO \xrightarrow{Ph_3P=CH_2} 91$$

139   **140**

A second synthesis starts with 2-iodobithienyl.[40] Reaction with the cuprous salt of vinylacetylen then afforded **91**, obviously formed by oxidative decarboxylation of **83**. This has been established indirectly by feeding labeled **41**.[26] However, **41** could be transformed to **91** through **42** and **83**. Further transformations of **83** and **91** also are very common among Compositaes. Probably all compounds isolated so far are formed through the epoxides of **83** and **91**. The alcohol **97** and the corresponding acetate **98** have been isolated from *Dyssodia setifolia*.[45] The structures clearly followed from the ^{1}H nmr data. The chlorohydrin **96** is present in *Pterocaulon virgatum*.[38] The position of the hydroxyl clearly followed from the mass spectrum, where a strong fragment $M\text{-}^{\cdot}CH_2Cl$ is visible. The diacetate **95** was present again in the roots of *Dyssodia setifolia*.[45] An oxidation product of **98**, the diacetate **108**, has been isolated from *Porophyllum ruderale*;[46] its structure followed from the ^{1}H nmr data, which are very similar to those of **98**, with only the methyl signal being replaced by the signals of the CH_2OAc-group. The dithiophenes **99–107** are derived from the epoxide of **91**. Most of these have been isolated first from *Echinops* species,[27] but **105** and **106** are also typical for *Tagetes* and have been isolated from many other genera also. The structures could usually be deduced directly from the ^{1}H nmr data and only a few chemical transformations were necessary to establish the positions of functional groups. Compound **107**, the isovalerate of **105**, has been isolated from *Haploesthes greggii var. texana*.[42] The diol **99**, the diacetate **100**, and the isomeric hydroxyacetates **101** and **102** are present in *Echinops* species.[27] Although the structures of **99** and **100** followed directly from the ^{1}H nmr data, the position of the hydroxy group in **101** was established by manganese dioxide oxidation to the corresponding ketone. The hydroxy isovalerate **103** is present in the aerial parts of *Dyssodia setifolia*.[45] The ^{1}H nmr data of the corresponding acetate could be assigned easily, while those of **103** showed, in part, overlapping multiplets. The chlorohydrin **104** has been isolated thus far only from *Pterocaulon virgatum*.[38] The fragmentation in the mass spectrum clearly showed again that the chlorine was at C-1. Most probably, **109** is also derived from **91**. From *Berkheya adlamii*, only one of the possible isomers was isolated.[47] A synthesis was therefore necessary to establish the configuration of the double bond.[47] Reaction of 2-iodo bithienyl (**141**) with the cuprous salt **142** gave the acetate **143**, which, after hydrolysis and manganese dioxide oxidation, afforded **144**. Reaction with **145** in the presence of diisopropyl ethylamine gave the isomers **109** and **110**:

The chemical shifts of H-1 in **109** and **110** allowed the assignment of the stereochemistry. In the *cis*-isomers **110**, this proton is deshielded by the acetylenic bond. This isomer was later isolated also.[36] Biogenetically, these compounds may be formed via the isomer of **104**, which, after oxidation to the corresponding aldehyde **146**, may be transformed to enol-acetates:

$$\left[\text{146} \quad \text{—C≡CCHCHO} \atop \text{Cl} \right] \longrightarrow 109 + 110$$

2-Acetyl dithienyl (**111**) only has been isolated from *Haploesthes greggii var. texana*.[42] Its structure easily followed from the ^{1}H nmr data. Furthermore, it is identical with synthetic material.[48] Surely, **111** is a degradation product. However, nothing is known about the biogenetic pathway. Also, **94** is undoubtedly formed by biochemical oxidation of **92**; it has been isolated so far only from *Arctium lappa*.[49] Two isomeric dimers of **91** have been isolated together with **91** from *Cardopatium corymbosum*.[50] From the ^{1}H nmr data, the presence of *cis, trans*-isomeric cyclobutane derivatives could be deduced. Several thiophenes most probably are formed via tridecahexayne (**114**), which may be formed by dehydrogenation of **41**. Addition of 2 eq. H_2S leads to **115**, **116**, and **117**. A further possibility, 1,4- and 9,12-addition leading to **179**, is realized in the adducts **190** and **191** (see Scheme 4) and is supported by the isolation of **115** from *Tagetes erecta*.[39] The structure followed from the ^{1}H nmr data and the IR-band of the monosubstituted acetylene; furthermore, the structure was established by synthesis.[51] The reaction of diiodo bithienyl (**147**) with the cuprous salt of propargylaldehyde acetal (**148**) afforded **149**, which on partial hydrolysis gave **150**. Boranate reduction and hydrolysis yielded **151**, which can be degraded with alkali to **152**. Transformation to **115** was achieved via the bromide by alanate reduction. This somewhat unusual route was taken because the separation of the reaction products without oxygen functions is extremely difficult.

The isomer **116** has been isolated from *Dyssodia papposa*.[42] It shows a broad UV maximum at 348 nm, whereas **115** displays two maxima (353 and 347 nm). The ^{1}H nmr data are of course very different. From the spectrum of **116**, the presence of a monosubstituted thiophene is obvious [7.24 dd (J = 5 and 1), 7.01 dd (J = 5 and 3.8), 7.18 dd (J = 3.8 and 1)]. Compound **117** is only present in a *Berkheya* species.[52] Its UV spectrum is different from that of **115** and **116** (λ_{max} = 359, 353, and 341.5 nm). Since the ^{1}H nmr data did not directly establish the structure, the triple bonds were hydrogenated, leading to **154**; its ^{1}H nmr and mass spectra clearly led to the proposed structure:

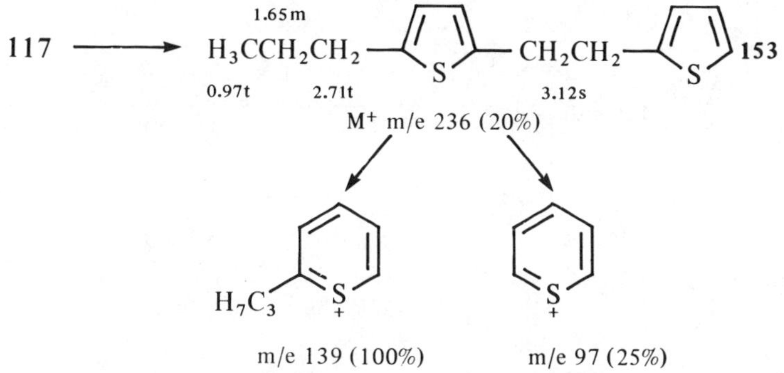

Also, in *Berkheya* species, the derivatives **118–120** and the degraded compounds **122–127** are present.[52] The formation of these compounds from **120** would require hydration of the triple bond, leading to **121**, which by acid splitting gives **122**. Further transformation could lead to **123–127**. The structures **117–127** have been established by synthesis.[53, 54] Reaction of the iodocompound **81**[34] with the cuprous salt **154** afforded **155**, which after hydrolysis gave **118**. Manganese dioxide oxidation afforded **120**, and acetylation yielded the acetate **119**. The tosylate **156** could be transformed to **117**. All compounds synthesized were identical with the natural compounds:[53]

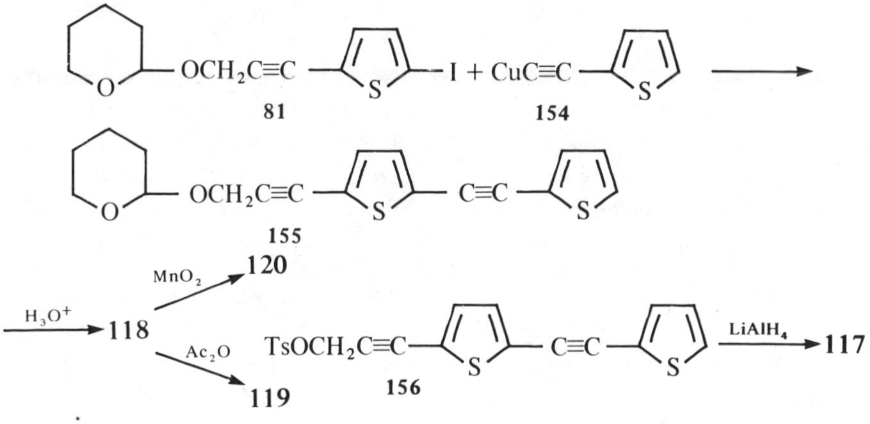

For the synthesis of **125–127**, 2,5-diiodothiophene was transformed to the mono-Grignard compound **157**, which on reaction with orthoformiate gave **158**. Hydrolysis yielded 5-iodothiophenealdehyde (**159**), which on reaction with the cuprous salt **154** gave the aldehyde **125**; this could easily be transformed to **126** and **127**.[54]

Compounds **122–124** were prepared similarly, starting with 5-iodo-2-acetyl-thiophene (**160**), which on reaction with **154** gave **122**. This compound was transformed readily to the carbinol **123** and the corresponding acetate **124**:[54]

The unusual thiophenes **129–134** are only present in *Berkheya* and *Cullumia* species.[54–56] Most probably, **129** is biogenetically formed via the epoxide **45**, which by reaction with H_2S may give **128**. Nucleophilic addition of the thiol to the terminal triple bond would then give **129**. This pathway is supported by feeding with labeled **43**, which is obviously the precursor of **45**.[57] The structure elucidation of these compounds caused some difficulties. The ^1H nmr data of **132** were in agreement with **172**, a structure whose existence was very likely, based on bio-genetic considerations. However, a synthesis of **172** ruled out this possibility. Reaction of acetylene dicarboxylate with mercaptoacetic ester afforded **161**, which was transformed by saponification and decarboxylation to **163**. Acetylation and reaction with thionyl chloride gave the acetate **164**, which after addition of the ylene **165** gave the betaine **166**. Thermal elimination of triphenylphospine oxide yielded **167**, following the method of Märkl.[58] Saponification afforded the acid **168**, which after reacetylation could be decarboxylated to **169**. On reaction with **72**, its cuprous salt gave the acetate **170**. Saponification yielded **171**, which was in equlibrium with **172**. Indeed, the ^1H nmr data were close to those of the natural compound, but slightly different. Furthermore, the natural compound showed no equilibrium with an enol form. The only other possible structure, therefore, was

132, though the presence of a thiethanone was not very likely. Finally, this structure was established by synthesis.[55] Reaction of 72 with the cuprous salt of propargyl-aldehyde acetal gave 173, which after hydrolysis to 174 was condensed with 3-thiethanone, yielding the isomeric ketones 132 and 175:

$$\text{MeC}\equiv\text{C}-\underset{S}{\overset{}{\boxed{}}}-\text{I} + \text{CuC}\equiv\text{CCH(OR)}_2 \longrightarrow$$

72

$$\text{MeC}\equiv\text{C}-\underset{S}{\overset{}{\boxed{}}}-\text{C}\equiv\text{CCH(OR)}_2 \overset{\text{H}_3\text{O}^+}{\longrightarrow}$$

173

$$\text{MeC}\equiv\text{C}-\underset{S}{\overset{}{\boxed{}}}-\text{C}\equiv\text{CCHO} + \underset{S}{\overset{O}{\boxed{}}} \longrightarrow$$

174

$$\textbf{132} + \text{MeC}\equiv\text{C}-\underset{S}{\overset{}{\boxed{}}}-\text{C}\equiv\text{C}-\text{C}\overset{\text{C}-\text{S}}{\underset{\text{H}}{\diagdown}}$$

175

Boranate reduction of **132** afforded **129**. Comparing the ^{1}H nmr data of **129** with those of the reduction product of **175** also allowed assignment of the stereochemistry of the double bond. The corresponding desmethyl compounds **131** and **134** are synthesized similarly. The aldehyde **150**[51] was decarbonylated by treatment with alkali, followed by hydrolysis to **176**; its condensation with 3-thiethanone gave **134** and boranate reduction yielded **131**:

$$\text{OHCC}\equiv\text{C}-\underset{S}{\overset{}{\boxed{}}}-\text{C}\equiv\text{CCH(OR)}_2 \overset{\substack{1)\ \text{OH}^\ominus \\ 2)\ \text{H}_3\text{O}^+}}{\longrightarrow}$$

150

$$\text{HC}\equiv\text{C}-\underset{S}{\overset{}{\boxed{}}}-\text{C}\equiv\text{CCHO} + \underset{S}{\overset{O}{\boxed{}}} \longrightarrow \textbf{134}$$

176

C.   Terthienyls Derived from Tridecapentaynene

As mentioned, terthienyl (**189**) was the first naturally occurring thiophene derivative and was isolated from the Compositae member *Tagetes erecta.*[1] Several derivatives of **189** have been isolated also (**181–188** and **195**). The biogenetic origin of **189** from tridecapentaynene (**41**) has been established by feeding with labeled **41**.[26] However, some questions concerning the sequence of the single steps had to be answered. Obviously, a dehydrogenation step was necessary for the introduction

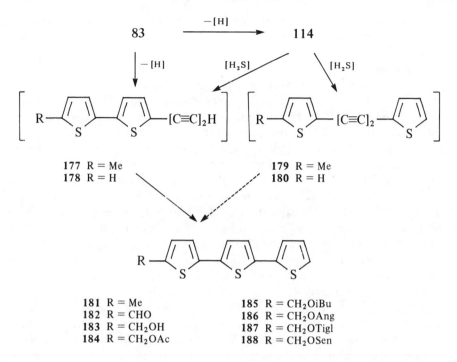

181 R = Me
182 R = CHO
183 R = CH₂OH
184 R = CH₂OAc

185 R = CH₂OiBu
186 R = CH₂OAng
187 R = CH₂OTigl
188 R = CH₂OSen

189 R = H

190 *trans* SMe
191 *cis*

192

193

194 R = H
195 R = Me

Scheme 4

of the required 1.2-acetylene bond, which could be achieved from **41** to **114**, from **83** to **178**, from **91** or the monothiophene **42** to the corresponding dehydrocompounds. Theoretically, even more possibilities can be proposed, and the question, at what stage is the terminal methyl eliminated by oxidative decarboxylation, also needs to be considered. Further feeding experiments with labeled **177–180, 83,** and **91** have shown[59] that, at least in *Tagetes erecta*, the most likely precursors for **181** and **189** are **177** and **178**, respectively. However, **179** and **180** were also transformed to **181** and **189**, but the incorporation was lower, Even less incorporation was observed by feeding with labeled **83** and **91**, which, however, were transformed to **99** and **101** with high incorporation. Most probably, the required enzymes for the addition of H_2S or its biochemical equivalent, as well as those for the oxidative decarboxylation, are not very specific, since **179** and **180** are not present in *Tagetes.* Although **189** is relatively widespread in Compositae, **181** has been isolated so far only from *Eclipta, Dyssodia,* and *Tagetes* species. Its structure followed from the molecular formula and the ^1H nmr data, which show the typical upfield shift of the β-thiophene proton at C-11. Compounds **182–188** have been isolated from *Eclipta erecta,*[31] and **182** was also present in a few other genera. The structure of **182** clearly followed from the spectroscopic data. The UV maximum is shifted to 389 nm by the additional carboxyl group, which caused a strong downfield shift of H-11 in the ^1H nmr spectrum; the chemical shifts of the thiophene protons were influenced only slightly. The esters **184–188** caused some difficulties during their separation, but the nature of the ester residues can be deduced easily from the typical ^1H nmr signals. Thus far, **195** is the only β-substituted terthienyl derivative isolated. Its structure followed from the spectroscopic data and has been isolated from *Dyssodia anthemidifolia*[45] only. The UV maximum is 370 nm, clearly indicating a bathochromic shift when compared with the maximum of terthienyl. The ^1H nmr spectrum in C_6D_6 at 270 MHz could be fully interpreted and the position of the methoxy group could be clearly assigned. Because of the influence of the methoxy group, the chemical shifts of the protons H-1 through H-3 and H-10 through H-13 are slightly different. Compound **195** may be formed as shown in Scheme 4, with **180** as the precursor via **192** and **193**, which could be cyclized to **194**. The thioenol ethers **190** and **191**, which have been isolated from a *Berkheya* species together with terthienyl (**189**),[56] are obviously derived from **180**.

The ^1H nmr data would agree also with an isomeric 7-mercaptomethyl derivative. Therefore, these thiophenes were synthesized, starting with **180**, which was obtained by Glaser-coupling of the thienyl acetylene (**196**).[56] Addition of methyl mercaptide afforded the *cis, trans*-isomeric thienol ethers **190** and **191**, which were identical with the natural compounds.

Peracid oxidation leads to the isomeric sulfones **197** and **198**, and their ^1H nmr data allow the assignment of the configurations, as the *cis*-orientated olefinic proton is always more deshielded than the *trans*-orientated. Furthermore, the position of the methyl sulfone group followed from the ^1H nmr data. The cooccurrence of **189** and **190** indicates that, at least in this species, terthienyl may be formed via **180**, which supports the formation of terthienyl by different routes.

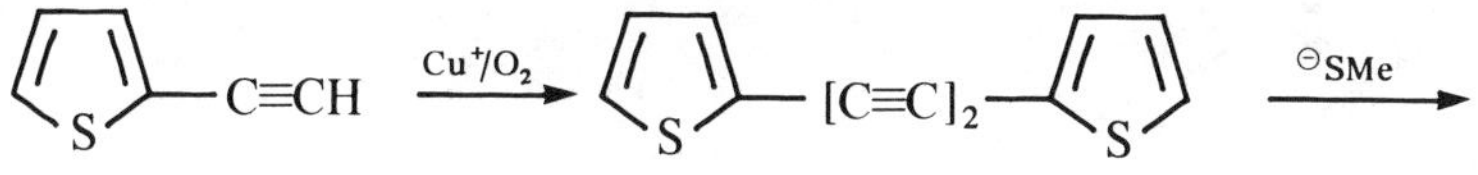

196   180

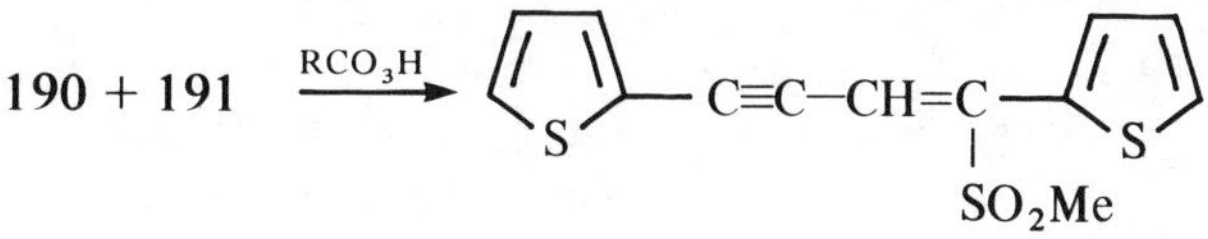

190 + 191 $\xrightarrow{RCO_3H}$

197   trans
198   cis

D. Thiophenes Derived from Trideca-1.11-Dien-3.5.7.9-Tetrayne

Several thiophenes are derived from the tetrayne **199**, although the number is much smaller than those derived from **41**, which, however, is much more widespread (Scheme 5). Only two of the three possible additions of H_2S or its biochemical equivalent are used in Compositae, 5,8-Addition leads to **200**, first isolated from *Lasthenia aristata*, together with the derivatives **201–203**.[28] These structures could also be deduced from the spectroscopic data. The UV maxima of **200–202** are typically different from those of type **42** ($\lambda_{max} = 336$ nm); the ^{1}H nmr data clearly indicate the presence of a disubstituted *trans*-double bond and vinyl group. The UV maxima of the aldehyde **203** is shifted to 360.5 nm. The structures **201–203** are further established by synthesis. 2,5-Diiodothiophene was coupled with the cuprous salt of vinylacetylene and then with **213**, yielding the tetrahydropyranyl-ether **214**. Hydrolysis afforded **201**, acetylation gave **202**, and manganese dioxide oxidation afforded **203**.[24] The spectral data of these compounds were identical with

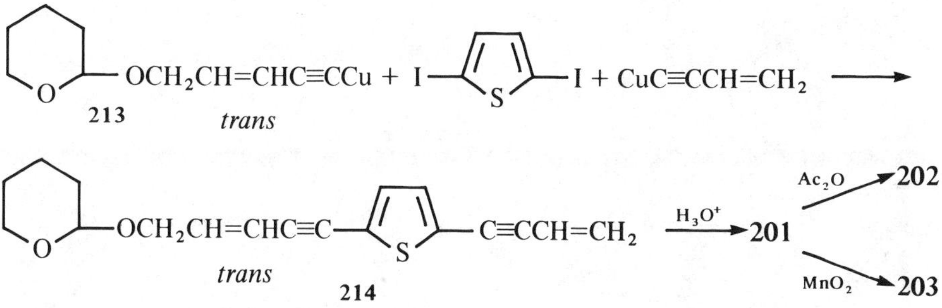

213   trans

trans
214

those of the natural products; **208** and **209** are obviously formed via the corresponding epoxides of **200**. Both diols have been isolated from the roots of *Serratula pectinata*,[60] but could not be separated. The structures followed from the ^{1}H nmr data and from the results of periodate splitting, which gave the aldehydes **211** and **215** respectively. The ^{1}H nmr data of **211** and **215** established

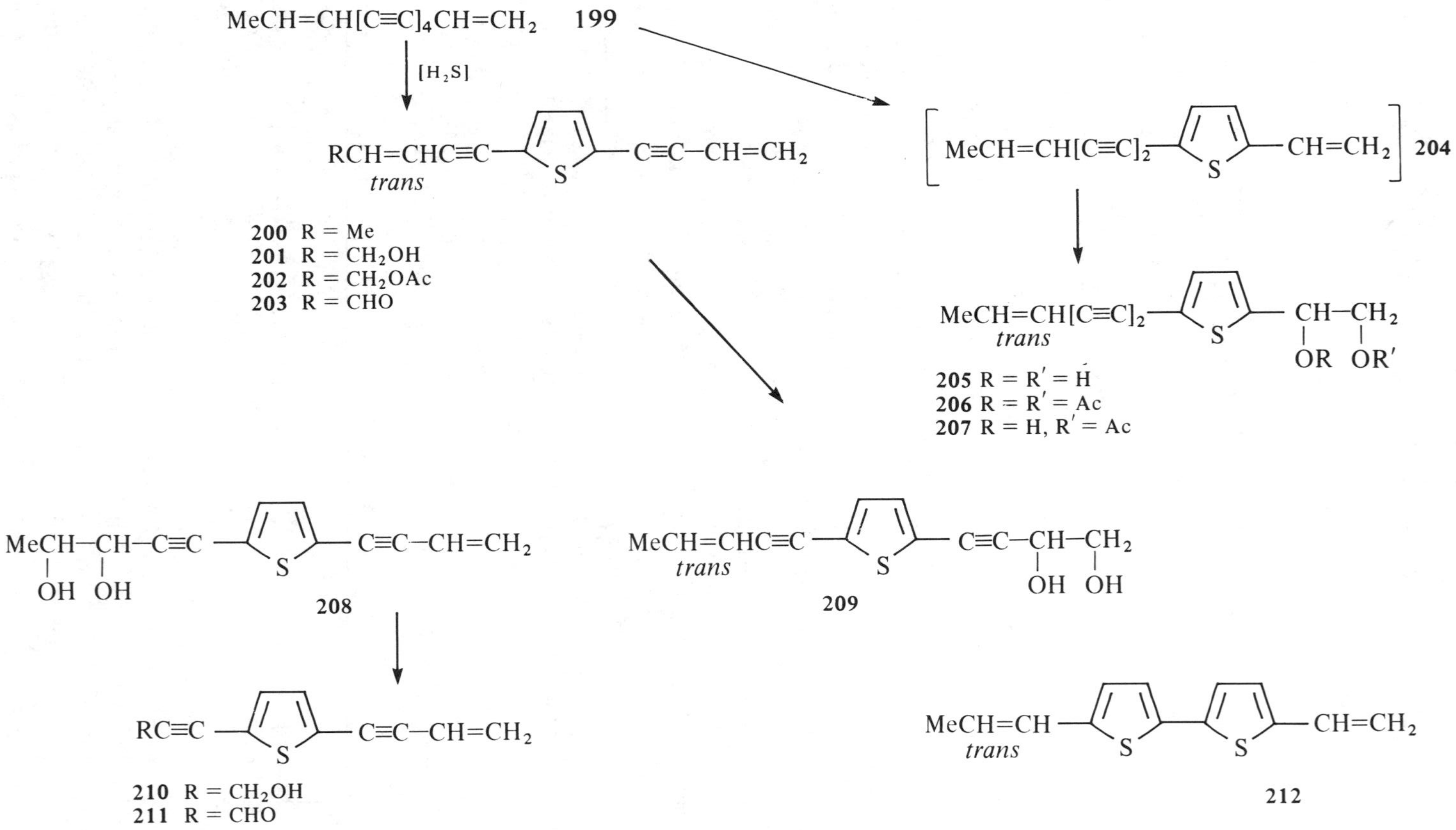

MeCH=CH[C≡C]₄CH=CH₂ 199

[H₂S]

RCH=CHC≡C—⟨S⟩—C≡C—CH=CH₂
trans

200 R = Me
201 R = CH₂OH
202 R = CH₂OAc
203 R = CHO

$[$ MeCH=CH[C≡C]₂—⟨S⟩—CH=CH₂ $]$ 204

MeCH=CH[C≡C]₂—⟨S⟩—CH—CH₂
                        | |
                        OR OR'
trans

205 R = R' = H
206 R = R' = Ac
207 R = H, R' = Ac

MeCH=CHC≡C—⟨S⟩—C≡C—CH—CH₂
                          | |
                          OH OH
trans 209

MeCH—CH—C≡C—⟨S⟩—C≡C—CH=CH₂
   | |
   OH OH
208

RC≡C—⟨S⟩—C≡C—CH=CH₂

210 R = CH₂OH
211 R = CHO

MeCH=CH—⟨S⟩—⟨S⟩—CH=CH₂
trans
212

Scheme 5

282

OHCC≡C—[thiophene]—C≡CCH=CH₂   MeCH=CHC≡C—[thiophene]—C≡CCHO
                                                    trans

211   **215**

the structures of the diols; **211** was present in the same plant[60] as was the alcohol **210**, which on manganese dioxide oxidation gave **211**. The structures of **210** and **211** were established by synthesis.[34] Reaction of 2,5-diiodothiophene first with the cuprous salt of propargylaldehyde acetal and then with that of vinylacetylene gave **217**, which on hydrolysis yielded **211**. Boranate reduction gave **210**. The aldehyde **215** was obtained similarly, by reaction of **216** with the cuprous salt of *trans*-pentenyne, which gave **218**. However, a *cis*, *trans*-mixture was formed. Hydrolyses, therefore, also gave a mixture, from which the *trans*-isomer **215** could be separated by crystallization. The drastic conditions of this reaction obviously lead to this kind of *cis*, *trans*-isomerization, which is a disadvantage of this method.

$(RO)_2CHC≡CCu + I$—[thiophene]—I ⟶

$(RO)_2CHC≡C$—[thiophene]—I **216**

CuC≡CCH=CH₂   CuC≡CCH=CHMe
                                    trans

$(RO)_2CHC≡C$—[thiophene]—C≡CCH=CH₂   MeCH=CHC≡C—[thiophene]—$C≡CCH(OR)_2$
                                                      trans
217   *cis*   **218**

H₃O⁺   H₃O⁺

211 ⟶ **210**   **215**

A 3,6-addition of H_2S to **199** would lead to **204**, which is so far not isolated from natural sources. However, **205–207** are obviously derived from this thiophene. All three compounds were isolated from the roots of *Serratula radiata*.[61] The structures followed from the ¹H nmr data and a few chemical transformations. Whereas periodate splitting of **205** yielded the aldehyde **219**, manganese dioxide oxidation of **207** gave the ketoacetate **220**. Saponification of **206** yielded **205**, and boranate reduction and saponification of **220** also gave **205**. These results clearly showed that all compounds were closely related, differing only in the nature of the oxygen functions. As in other cases, the intensity of the acetylenic band is the IR spectrum of **219** is a good indication of whether the triple bond is conjugated to the carbonyl group or not. In the second case, a strong band near $2200\,cm^{-1}$ can be observed.

$$205 \xrightarrow{\text{HIO}_4} \text{MeCH=CH—C≡C—}\underset{S}{\text{[thiophene]}}\text{—CHO} \quad 219$$

trans

$$207 \longrightarrow \text{MeCH=CHC≡C—}\underset{S}{\text{[thiophene]}}\text{—COCH}_2\text{OAc} \quad 220$$

trans

The bithienyl derivative **212** is obviously formed by further addition of H_2S to **204**. It has been isolated from the aerial parts of *Bidens connatus*.[62] Its UV spectrum is typically different from that of other bithienyl derivatives ($\lambda_{max} = 366$, 262, and 254 nm).

E.   Thiophenes Derived from Trideca-1.3-Diene-5.7.9.11-Tetrayne

Some naturally occurring thiophenes are derived from trideca-1.3-diene-5.7.9.11-tetrayne (Scheme 6) (**221**). Again, only two of the three possible modes of H_2S-addition are realized in the compounds isolated so far. The only simple mono-thiophene of this type, **222**, has been isolated from the roots of *Rudbeckia amplexicaulis*.[28] Its structure was deduced for the ^{1}H nmr data and from the structure of the hydrogenation product, which was identical with a synthetic sample. Therefore, an isomeric structure with a pentadiyne side chain (**231**) could be excluded. The latter is obviously the precursor of **223**, a bithienyl derivate first isolated from a *Bidens* species.[63] The data, however, were not fully in agreement with those of a synthetic sample.[64] Later, **223** was isolated from *Rudbeckia triloba*.[39] The spectral data were identical with those of the synthetic product. The synthesis of **223** starts with 2-methyl bithienyl, which was transformed by a Vilsmeier reaction to **238**. Wittig reaction with **239** gave **223**, which was characterized as its maleic anhydride adduct.

$$\text{Me—}\underset{S}{\text{[thiophene]}}\text{—}\underset{S}{\text{[thiophene]}} \longrightarrow \text{Me—}\underset{S}{\text{[thiophene]}}\text{—}\underset{S}{\text{[thiophene]}}\text{—CHO +}$$

237   238

$$\text{Ph}_3\text{P=CHCH=CH}_2 \longrightarrow 223$$

239

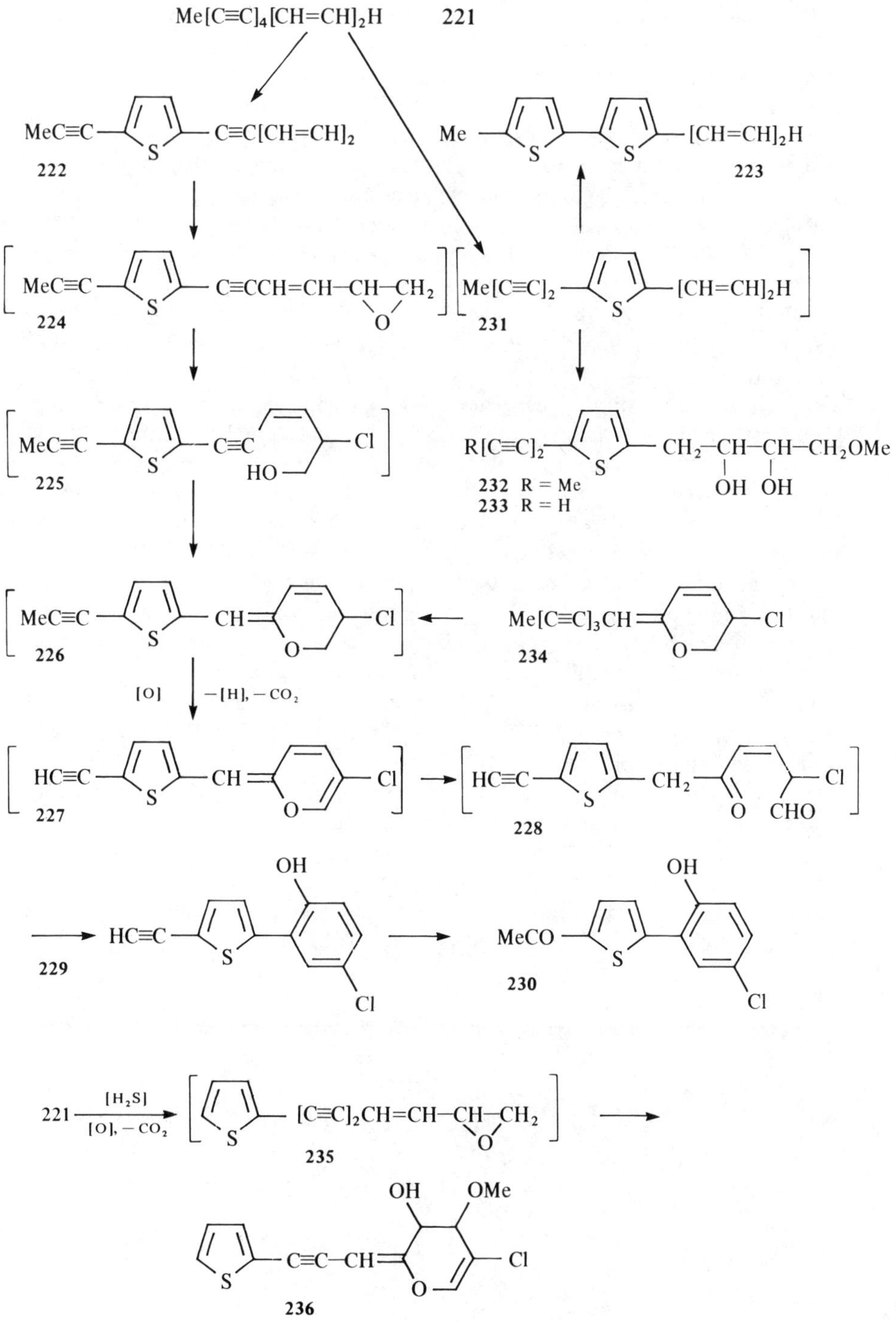

Scheme 6

Compound **231** is most probably the precursor also of **232** and **233**, both isolated from *Cullumia setosa*.[45] The position of the methoxy group was deduced from the mass spectra, which in both cases showed elimination of ·CH_2OMe. The arrangement of the hydroxy groups followed from the corresponding ^1H nmr data. The assignment was possible by spin decoupling. The side chain of **232** and **233** is an unusual one, but it is formed most probably via oxidation of the diene side chain of **231**, while the methyl group is eliminated as usual by oxidative decarboxylation. Further unusual thiophenes have been isolated from *Helichrysum* species, the chlorophenols **229** and **230** and the chloroenolether **236**.[65] The molecular formula, estimated by high resolution mass spectroscopy, clearly indicated the presence of chlorine compounds; the IR spectra showed that **229** was a monosubstituted acetylene (3310, 2110 cm^{-1}), **230** was a conjugated ketone, and **236** was an enol ether with a hydroxy group (1650 and 3610 cm^{-1}). The ^1H nmr data of **229** and **230** indicated trisubstituted benzene derivatives with an α,α'-disubstituted thiophene part, and the terminal groups obviously were C≡CH (3.42 s) and MeCO (2.54 s), respectively. The substitution pattern of the aromatic ring, however, could be deduced by comparing the chemical shifts with those of the corresponding acetates. Even then, the assignment was not completely sure and, therefore, was established by synthesis of **229** and **230**.[66] The necessary phenolic acetylene (**243**) was obtained via the aldehyde **240**, which on reaction with carbontetrabromide and triphenylphosphine gave **241**. After protection of the phenolic hydroxyl, elimination of HBr afforded **243**. Coupling with but-1-yn-3-ol gave **244**, which on reaction with sodium sulfide gave the thiophene **245**. Manganese dioxide oxidation followed by hydrolysis yielded **230**; its spectra data was fully identical with the natural compound. Similarly, **229** was obtained by starting with **243**, which was coupled with propinol. Reaction with sodium sulfide gave the thiophene **247**, which was transformed via the aldehyde to the acetylenic derivate **229**; its spectral data was also identical with those of the natural thiophene.

The ^1H nmr data of **236**, and those of the corresponding acetate, showed the presence of a monosubstituted thiophene, whereas the other down field signals could be assigned by spin decoupling. As the double doublet of the H-9 signal was shifted downfield in the spectrum of the acetate, the relative position of the hydroxy and methoxy group was established. Though the biogenesis of these chloro compounds has not been established, the formation via the epoxides **224** and **235**, respectively, is very likely. Upon ring opening with chloride ion (a typical reaction in the field of acetylenic compounds), **224** would give **225**, which could be cyclized to **226**. A similar compound, **234**, is present in *Anaphalis* species[67] and, therefore, also could be transformed to **226** by the formal addition of H_2S. Dehydrogenation and oxidative decarboxylation would then give **227**, which on hydrolysis could lead to **228**. Aldol condensation transforms **228** to **229**, which on hydration would lead to **230**. Similarly, the epoxide **235** could be transformed to **236**. The cooccurrence of **229** and **236** supports this assumption.

F.   Thiophenes Derived from C_{13}-Triynes

Only a few thiophenes are derived from C_{13}-triynes like **250, 251**, and **266**, which are closely related to each other (**249** is the precursor)[4] (Scheme 7). Obviously, **254** and the corresponding acetate **255** are formed via **251** by the formal addition of H_2S, followed by oxidation and decarboxylation of the thus far unknown thiophene **253**. Both **254** and **255** are present in the aerial parts of *Glossopappus macrotis*.[68] The structures clearly followed from the UV spectra ($\lambda_{max} = 337$ and 316 nm), the molecular formula, and the ^1H nmr data, from which the nature of the side chain can be easily deduced [**255**: 4.09 t ($J = 6.5$), 2.48 dt ($J = 6.5, 6.5$), 5.74 dt ($J = 15, 6.5$), 6.19 dd ($J = 15, 10$), 6.60 dd ($J = 15.5, 10$), and 6.59 d ($J = 15.5$)]. In the spectrum of **254**, the H-1 triplet is shifted upfield to 3.61 ppm. Oxidation of **250** leads to **252**, which by elimination of water can be transformed to a furan, a compound isolated from different genera of the tribe Anthemideae. The formal addition of H_2S then leads to **256**, a thiophene, which has not been isolated. However, the derivatives **257–262** are known and all occur in representatives of the tribe Anthemideae. The compounds **257–259** and the *cis, trans*-isomeric aldehydes **260** and **261** have been isolated from *Santolina chamaecyparissus*,[69, 70] which also contains the *cis, trans*-isomeric furans **263** and **264**, which are obviously formed by oxidative decarboxylation of **256**; **263**, however, is relatively widespread in the tribe Anthemideae.[4] The structure of **259** followed from the ^1H nmr data, which clearly showed that a disubstituted thiophene was present (7.00 d and 6.93 d); further downfield, signals were typical for an α-substituted furan [7.30 dd ($J = 1.8, 0.5$, and 0.5), 6.37 ddd ($J = 4, 1.8$, and 0.5), and 6.87 d ($J = 4$)]. Furthermore, from the coupling of the olefinic protons, the presence of the *cis*-double bond could be deduced, and the nature of the terminal group followed from the typical signals. Because the chemical shifts of the β-proton thiophene acetylenes and thiophene olefines are different, the position of the triple bond also can be deduced from the ^1H nmr data. In the spectra of **261** and **262**, a considerable downfield shift of the signal of the thiophene proton clearly indicates the position of the aldehyde group (7.58 and 7.54 d, respectively), while the stereochemistry of the double bond followed from the couplings observed. The UV maxima of **261** (370, 302, and 280 nm) are of course at longer wavelengths than those of the corresponding desformyl compound **263** (357 nm). Similarly, the structures of **263** and **264** also followed from the ^1H nmr data. They are further confirmed by synthesis.[6] Wittig reaction of thienylpropargylaldehyde (**271**) with the phosphoran **272** afforded the thiophene **263**, which is identical with the natural product.

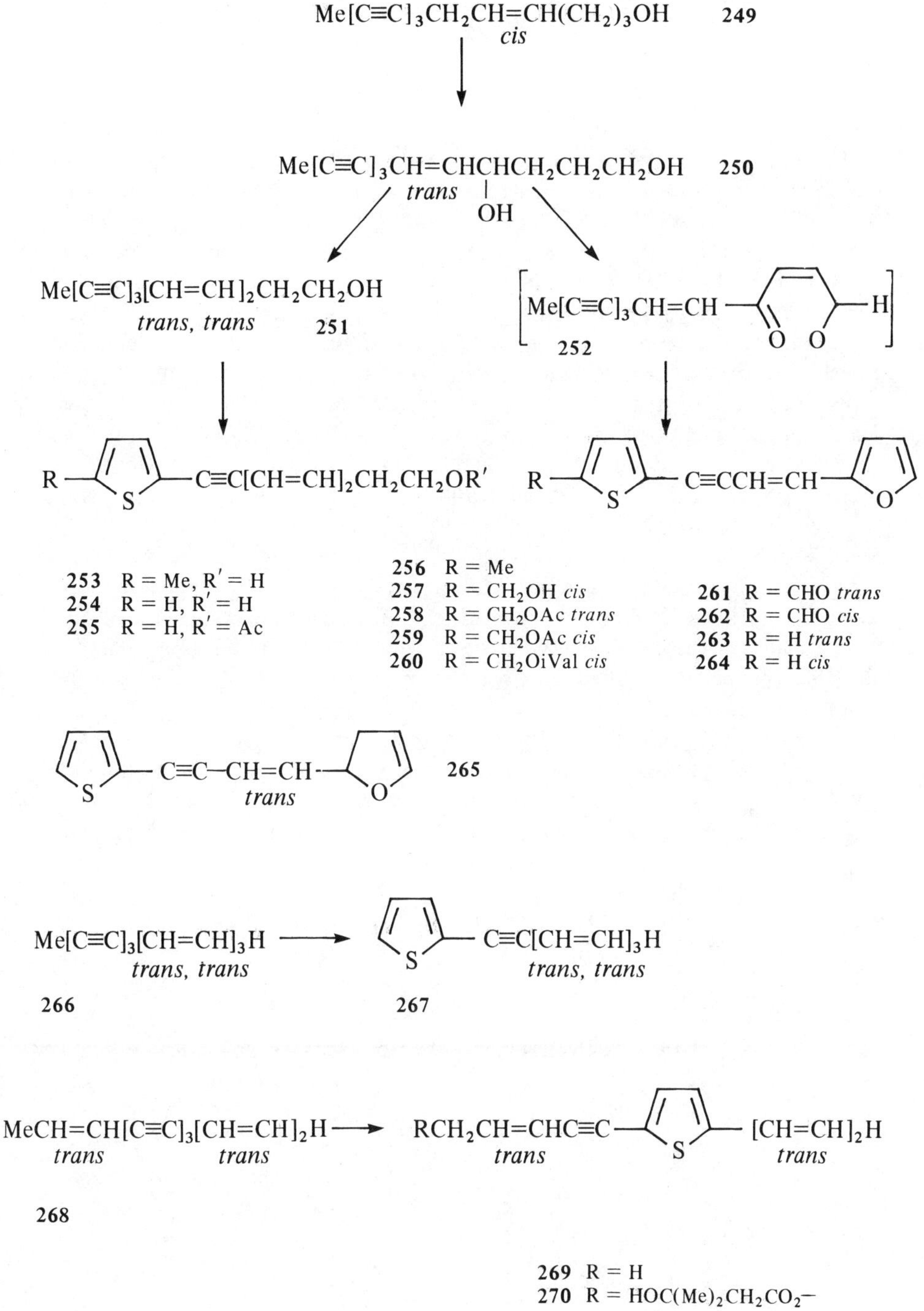

Me[C≡C]₃CH₂CH=CH(CH₂)₃OH   **249**
cis

Me[C≡C]₃CH=CHCHCH₂CH₂CH₂OH   **250**
trans |
OH

Me[C≡C]₃[CH=CH]₂CH₂CH₂OH
trans, trans   **251**

[Me[C≡C]₃CH=CH—⟨⟩—H]
        O O
   252

R—⟨S⟩—C≡C[CH=CH]₂CH₂CH₂OR′

R—⟨S⟩—C≡CCH=CH—⟨O⟩

253 R = Me, R′ = H
254 R = H, R′ = H
255 R = H, R′ = Ac

256 R = Me
257 R = CH₂OH *cis*
258 R = CH₂OAc *trans*
259 R = CH₂OAc *cis*
260 R = CH₂OiVal *cis*

261 R = CHO *trans*
262 R = CHO *cis*
263 R = H *trans*
264 R = H *cis*

⟨S⟩—C≡C–CH=CH–⟨O⟩   **265**
      trans

Me[C≡C]₃[CH=CH]₃H ⟶ ⟨S⟩—C≡C[CH=CH]₃H
trans, trans   *trans, trans*

266   **267**

MeCH=CH[C≡C]₃[CH=CH]₂H ⟶ RCH₂CH=CHC≡C—⟨S⟩—[CH=CH]₂H
trans   *trans*   *trans*   *trans*

268

269 R = H
270 R = HOC(Me)₂CH₂CO₂−

Scheme 7

$$\text{(271)} \quad \text{—C≡CCHO} + \text{Ph}_3\text{P=CH—(272)} \longrightarrow \text{263}$$

271   272

The isovalerate **260** has been isolated with **259** and **262** from the roots of *Santolina rosmarinifolia.*[71] The structure clearly followed from the ¹H nmr data, which were similar to those of **259**. Also, the nature of the ester group can be easily deduced from the typical signals. The dihydrofuran derivative **265** has been isolated from *Glossopappus macrotis*, which also contains **254** and **255**. Compound **265** is closely related to the corresponding triyne isolated from similar species.[4] The structure followed from the molecular formula, and UV spectrum, and especially from the ¹H nmr data, which showed the presence of a dihydrofuran derivative, while the signals of the thiophene protons were similar to those of **263**. The triene **267** surely is derived from **266** by the formal addition of H_2S, followed by oxidative decarboxylation. This thiophene was isolated first from *Matricaria* species,[28,72] but has been isolated from several other members of the tribe Anthemidae and from *Xeranthemum* species.[4] The structure was deduced from the spectroscopic data and from some chemical transformations and was established by synthesis.[28] Starting with the Grignard compound of 2-ethinyl thiophene, reaction with hexa-2,4-dien-1-al afforded **273**. Elimination of water yielded **267**. In a second synthesis,[72] Zincke-aldehyde (**274**) was used as starting material. The aldehyde obtained (**275**) on reaction with methylmagnesium iodide and elimination of water also gave **267**.[72]

Compound **269** is the formal H_2S-addition product of the isomeric triyne **268**. It has been isolated from a *Centaurea* species[4] and the structure followed from the UV and ¹H nmr spectra. The only derivative of **269** is the unusual ester **270**, which was isolated from a *Coreopsis* species.[71] The mass spectrum indicated the presence

$$\text{—C≡CMgBr} + \text{OHC[CH=CH]}_2\text{Me} \longrightarrow$$

$$\text{—C≡CCH(OH)[CH=CH]}_2\text{Me} \xrightarrow{\text{H}^+} \text{267}$$

273

$$\text{—C≡CMgBr} + \text{PhN(Me)[CH=CH]}_2\text{CHO} \longrightarrow$$

274

$$\text{—C≡C[CH=CH]}_2\text{CHO} \xrightarrow[-\text{H}_2\text{O}]{\text{MeMgI}} \text{267}$$

275

of a derivative of **269** by the fragment m/e 199 ($C_{13}H_{11}S$), while the molecular ion corresponds with the molecular formula $C_{18}H_{20}O_3S$. Saponification afforded the corresponding alcohol **281**; acetylation gave an acetate. Manganese dioxide oxidation of the alcohol yielded an aldehyde. The ^1H nmr spectra led to the structure **270**, which was established by synthesis.[73] The iodo compound **276**, on reaction with the cuprous salt **213**, afforded the acetate **277**. Saponification and manganese dioxide oxidation gave the aldehyde **278** and Wittig reaction with **279** led to **280**. Hydrolysis afforded **281**, which was transformed to the bromide **282**. This bromide reacted with the silver salt **283** to yield **270**, which is identical with natural compound.

$$\text{—OCH}_2\text{CH=CHC≡CCu} \quad + \quad I\text{—}\underset{S}{\underset{\|}{\diagdown}}\text{—CH}_2\text{OAc} \longrightarrow$$

213   *trans*   276

$$\text{— OCH}_2\text{CH=CHC≡C —}\underset{S}{\underset{\|}{\diagdown}}\text{—R} \xrightarrow[\text{Ph}_3\text{P=CHCH=CH}_2]{\text{MnO}_2}$$

277   R = CH_2OAc
278   R = CHO   279

$$\text{RCH}_2\text{CH=CHC≡C —}\underset{S}{\underset{\|}{\diagdown}}\text{—[CH=CH]}_2\text{H} + \text{AgO}_2\text{CCH}_2\text{C(Me)}_2\text{OH} \longrightarrow 270$$

trans   283

280   R =
281   R = OH
282   R = Br

3.   Thiophenes Derived from C_{14}-Acetylenes

A few thiophenes are derived from acetylenic compounds with a C_{14} chain. The precursor of **287** and **288** is obviously the triyne **284** (Scheme 8). Formal addition of H_2S would lead to **286**, which by oxidative decarboxylation can be transformed to **287** and **288**. Both compounds were isolated from *Xeranthemum cylindraceum*.[28] Saponification of **288** afforded the alcohol **287**, which was further characterized as its azo benzoate. The ^1H nmr spectrum of the latter established the structure and stereochemistry of the homo derivatives of **254** and **255**. These

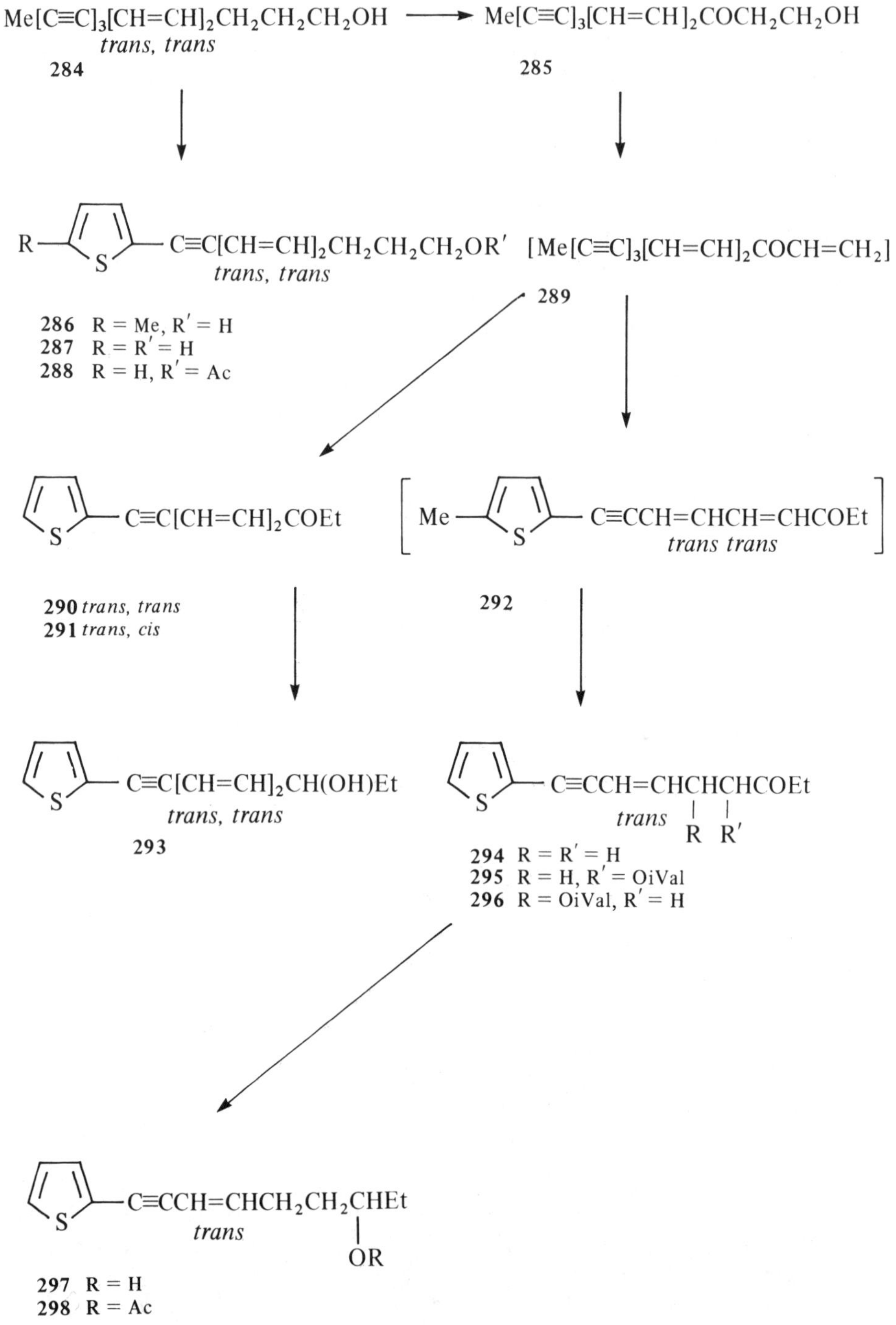

$Me[C{\equiv}C]_3[CH{=}CH]_2CH_2CH_2CH_2OH \longrightarrow Me[C{\equiv}C]_3[CH{=}CH]_2COCH_2CH_2OH$

trans, trans

284   **285**

$R-\underset{S}{\text{⟨thiophene⟩}}-C{\equiv}C[CH{=}CH]_2CH_2CH_2CH_2OR'$ $[Me[C{\equiv}C]_3[CH{=}CH]_2COCH{=}CH_2]$

trans, trans

289

286 R = Me, R′ = H
287 R = R′ = H
288 R = H, R′ = Ac

$\underset{S}{\text{⟨thiophene⟩}}-C{\equiv}C[CH{=}CH]_2COEt$ $\left[Me-\underset{S}{\text{⟨thiophene⟩}}-C{\equiv}CCH{=}CHCH{=}CHCOEt \right]$

                                                         trans trans

290 *trans, trans*   **292**
291 *trans, cis*

$\underset{S}{\text{⟨thiophene⟩}}-C{\equiv}C[CH{=}CH]_2CH(OH)Et$ $\underset{S}{\text{⟨thiophene⟩}}-C{\equiv}CCH{=}CHCHCHCOEt$

                     trans, trans   *trans*   R R′

293   **294** R = R′ = H
                                                  295 R = H, R′ = OiVal
                                                    296 R = OiVal, R′ = H

$\underset{S}{\text{⟨thiophene⟩}}-C{\equiv}CCH{=}CHCH_2CH_2CHEt$

                     trans |
                                          OR

297 R = H
298 R = Ac

Scheme 8

Me[C≡C]₂CH=<chemical structure>

299

→ <chemical structure>—CH=<chemical structure>

300

Me[C≡C]₃—<chemical structure>

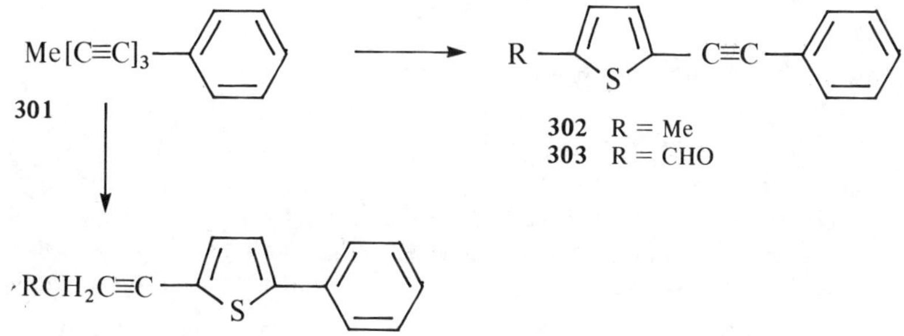

301

→ R—<chemical structure>—C≡C—<chemical structure>

302 R = Me
303 R = CHO

·RCH₂C≡C—<chemical structure>

304 R = H
305 R = OAc

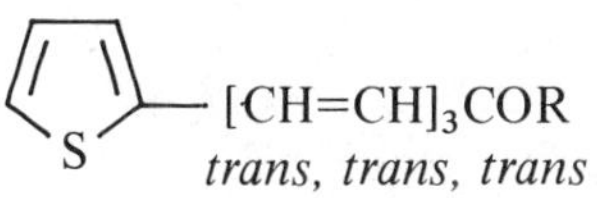

<chemical structure>—[CH=CH]₃COR
trans, trans, trans

306 R = NHCH₂CHMe₂

307 R =  N<chemical structure>

308 R = 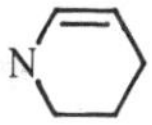 N<chemical structure>

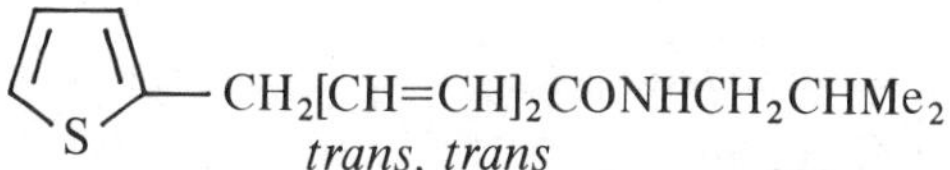

<chemical structure>—CH₂[CH=CH]₂CONHCH₂CHMe₂
trans, trans
                                    309

Scheme 8 (Continued)

thiophenes, however, are present only in members of the tribe Anthemideae, while *Xeranthemum* belongs to the Cynareae. Allylic oxidation of **284** followed by elimination of water would lead to **289**, the precursor of a widespread ketone[4] formed by hydrogenation of **289**. Two compounds, **290** and **294**, are derived from these precursors by formal addition of H₂S and oxidative decarboxylation. The first, **290**, has been isolated from a *Matricaria* species;[72] the structure follows from the spectroscopic data and has been established by synthesis.[72] The thiophene aldehyde **275**, on reaction with ethylmagnesium bromide, gave **293**, which afforded **290** (identical with the natural compound) on oxidation. Originally, **293** was not

$$\text{[thiophene]} - C \equiv C[CH=CH]_2CHO + EtMgBr \longrightarrow$$

275

$$\text{[thiophene]} - C \equiv C[CH=CH]_2CHEt \xrightarrow{MnO_2} 290$$

293 $\overset{|}{OH}$

known as a natural compound. It was isolated from an *Anthemis* species[14] with the *trans, cis*-isomers of **290**. The position of the *cis* double bond followed from the chemical shift of H-6, which is strongly deshielded by the ketogroup. UV isomerization transformed **291** to **290**, which further established the structure. The dihydro derivative of **290**, the ketone **294**, has also been isolated for the first time from a *Matricaria* species.[72] Its structure was established by synthesis.[74] Cadiot-coupling of ethinyl thiophene with the bromo acetylene **310** gave the carbinol **311**, which on reduction with lithium alanate afforded **297**. Jones oxidation gave **294**, which is identical with the natural compound. At the time, **297** was not known as a natural

$$\text{[thiophene]} - C \equiv CH + BrC \equiv CCH_2CH_2CH(OH)Et$$

310

$$\longrightarrow \text{[thiophene]} - [C \equiv C]_2CH_2CH_2CH(OH)Et$$

311

$$\xrightarrow{LiAlH_4} \text{[thiophene]} - C \equiv CCH=CHCH_2CH_2CH(OH)Et \xrightarrow{CrO_3} 294$$

trans

297

compound, but was isolated later from *Anthemis saguramica*.[14] The corresponding acetate is present in *Matricaria caucasica*.[75] The isomeric isovalerates **295** and **296** have also been isolated from *Anthemis saguramica*.[14] The structures followed from the spectral data. The UV spectra indicated the nature of the chromophoric system, and the IR spectra showed that keto esters were present. The molecular formulae were deduced from the mass spectra, and the ^{1}H nmr data allowed the assignment the positions of the functional groups. In the spectrum of **296** the downfield signal of the proton under the ester group is a threefold doublet at 5.64 ppm; the corresponding signal in the spectrum of **295** is a double doublet at 4.98 ppm, clearly indicating that the ester group is positioned α- to the keto group in the latter and α- to the double bond in the former. All other signals were fully in agreement with the proposed structures. Finally, the structure of **296** was established by synthesis. 2-Iodothiophene afforded **312** on reaction with the cuprous salt **213**; its hydrolysis gave **313**, which on oxidation yielded **13**. Reaction with the lithio compound **314** gave the amide **315**, which, after protection of the hydroxyl group, could be transformed to **317**. Hydrolysis and esterification finally led to **296**.

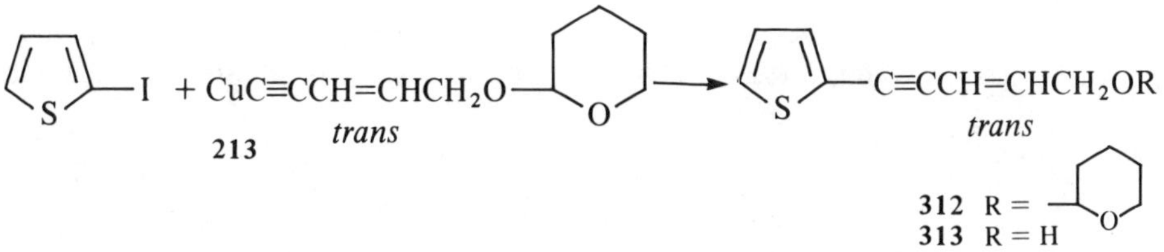

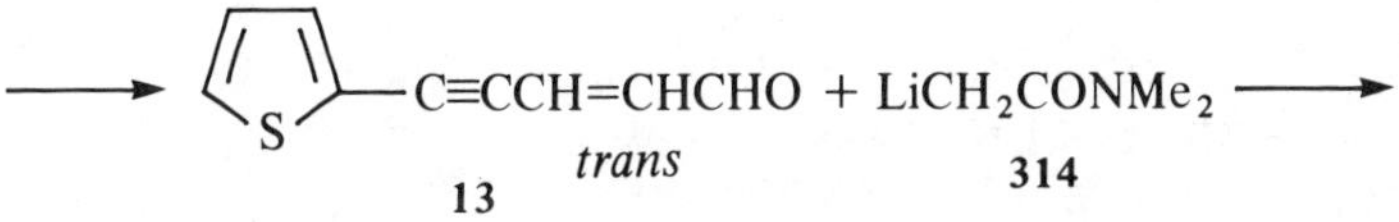

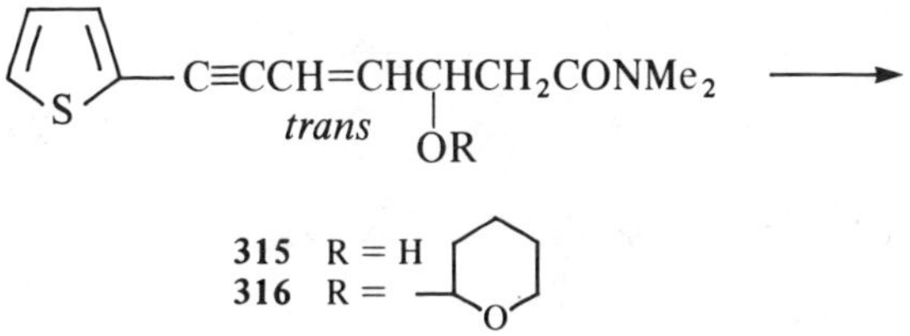

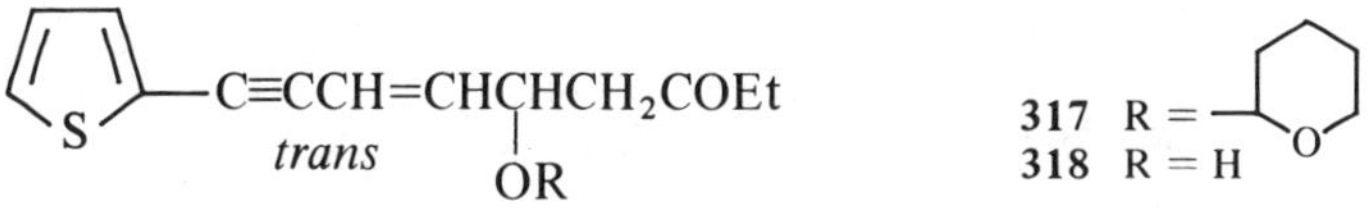

$$\text{(thiophene)}-\text{I} + \text{CuC}\equiv\text{CCH}=\text{CHCH}_2\text{O}-\text{(oxane)} \longrightarrow \text{(thiophene)}-\text{C}\equiv\text{CCH}=\text{CHCH}_2\text{OR}$$

213   *trans*

trans

312 R = (oxane)
313 R = H

$$\longrightarrow \text{(thiophene)}-\text{C}\equiv\text{CCH}=\text{CHCHO} + \text{LiCH}_2\text{CONMe}_2 \longrightarrow$$

13   *trans*   **314**

(thiophene)—$\text{C}\equiv\text{CCH}=\text{CHCHCH}_2\text{CONMe}_2 \longrightarrow$

trans OR

315 R = H
316 R = (oxane)

(thiophene)—$\text{C}\equiv\text{CCH}=\text{CHCHCH}_2\text{COEt}$

trans   OR

317 R = (oxane)
318 R = H

Though **300** has no triple bond, it is obviously derived from the widespread spiroenolether **299**. It was isolated first from *Artemisia ludoviciana*[6] and then from several other species, all belonging to the Anthemideae.[4] The structure was deduced from the spectral data and some chemical degradations. Permanganate oxidation afforded thiophene carboxylic acid, while partial hydrogenation and ozonization gave the lactone **320**, whose structure followed from the molecular formula and the IR band, which indicated the presence of a butyrolactone. In combination with the ^{1}H nmr data, therefore, only the structure **300** was possible, which could be established by synthesis.[74]

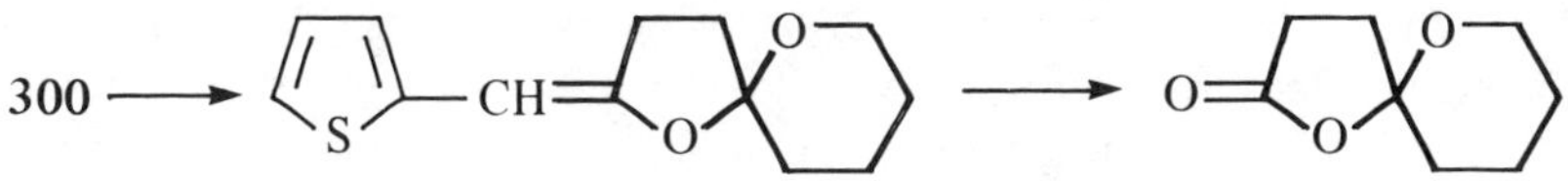

300 $\longrightarrow$ (thiophene)—CH=(spiro lactone) $\longrightarrow$ O=(spiro lactone)

319   **320**

Reaction of furfuryl chloride with the Grignard compound **321** afforded **322**, which after hydrogenation yielded **323**. Lithiation and reaction with chloromethyl thiophene gave **324**, which after hydrolysis and addition of bromine in methanol in the presence of potassium acetate gave **326**. Thermal elimination of methanol finally afforded **300**, whose spectral data were identical with those of the natural compound.

Some compounds are derived from the phenyl triyne **301**, which itself is formed from a C_{14}-triyne.[4] Although **304** and **305** are formed by formal 4,7-addition of H_2S, a 2,5-addition leads to **302** and oxidation would give **303**, which has been isolated from a *Coreopsis* species.[76] In the same genus, the isomeric acetate **305** is present.[77] The isomer **327** was obtained by boranate reduction of the aldehyde **303**, also isolated from a *Coreopsis* species,[39] followed by acetylation. The ^{1}H nmr data of **302** clearly led to the structure, but those of **305** were not conclusive. However, a clear decision is possible using the mass spectra (e.g., in the spectrum of **327**, M-⁻OAc leads to the base peak **328**), but the corresponding fragment of **304** is - less intensive. The structure of **305** was also established by synthesis.[77] Reaction of the cuprous salt **80** with the iodo compound **329** gave **330**, which after hydrolysis and acetylation gave **305** (identical with the natural compound).

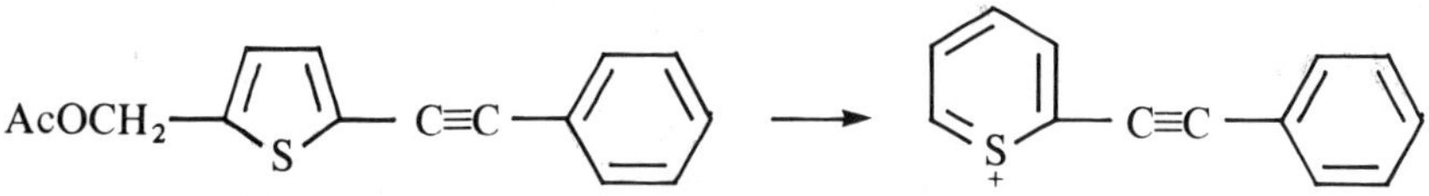

327 → 328

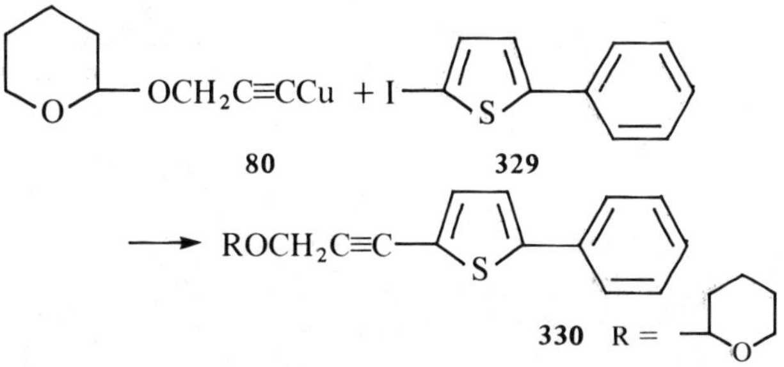

80 + 329

→ ROCH$_2$C≡C— 330 R =

A few thiophenes isolated from Compositae are amides. Three such compounds, **306–308**, have been isolated from *Otanthus maritimus*,[78] which differs only in the nature of the amide group. The unknown chromophoric system was that of a thienyl triene amide, with UV maxima around 360 and 347 nm. The configuration of the double bonds clearly followed from the ^{1}H nmr data after addition of Eu(fod)$_3$ as a shift reagent. Also, the presence of monosubstituted thiophenes and the nature of the amide groups could be deduced directly from the ^{1}H nmr spectra. However, in the case of **308**, the spectra were complicated owing to the existence of two conformers. Thus **306–308** most probably are derived from the unknown diyne **332**, which itself is derived from **331**.[4] β-Oxidation and hydrolysis could lead to **334**, which is known as an amide.[4] Transformation to **335** followed by dehydrogenation would then give **306–308**. The pathway through **332** has been established for the corresponding amides of type **334**.[79]

$$MeCH_2CH_2[C≡C]_2CH_2CH=CH(CH_2)_3CO_2R \qquad \textbf{331}$$
$$cis$$

$$[MeCH_2CH_2[C≡C]_2CH_2CH_2CH_2[CH=CH]_2CO_2R] \qquad \textbf{332}$$

[O]

$$[MeCOCH_2[C≡C]_2CH_2CH_2[CH=CH]_2COR] \qquad \textbf{333}$$

$$Me[C≡C]_2CH_2CH_2[CH=CH]_2COR \qquad \textbf{334}$$

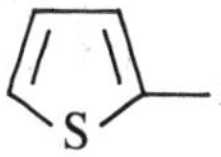

—CH$_2$CH$_2$[CH=CH]$_2$COR $\xrightarrow{-[H]}$ 306–308

335

The structure of **306** has been established by synthesis.[80] Reaction of 2-thiophene aldehyde with the Grignard compound **336** afforded the carbinol **337**, which after partial hydrogenation and acid treatment yielded the dienal **339**. Wittig reaction with the phosphorane **340** gave **306**, which was identical with the natural amide. Similar reaction of **339** with the phosphorane **341** afforded the ester **342**, which could be transformed to **307** and **308** through the acid chloride **344** by reaction with piperidine and Δ^2-piperidein, respectively. The compounds obtained were

$$\text{S} \text{—CHO} + \text{BrMgC} \equiv \text{CCH} = \text{CHOMe} \longrightarrow \text{S} \text{—CH(OH)C} \equiv \text{CCH} = \text{CHOMe} \xrightarrow{H_2}$$

336   337

$$\text{S} \text{—CH(OH)[CH=CH]}_2\text{OMe} \xrightarrow{H_3O^+} \text{S} \text{—[CH=CH]}_2\text{CHO} \quad 339$$

338   *trans, trans*

339 + Ph$_3$P=CHCONHCH$_2$CHMe$_2 \longrightarrow$ 306

340

339 + Ph$_3$P=CHCO$_2$Me $\longrightarrow$ S—[CH=CH]$_3$COR $\longrightarrow$ 307, 308

341   *trans, trans, trans*

                              342 R = OMe
                              343 R = OH
                              344 R = Cl

again identical with the natural amides. A further amide, the dienamide **309**, has been isolated for the first time from *Argyranthemum frutescens*.[81] Permanganate oxidation gave thiophene carboxylic acid and the isobutylamide of oxalic acid, while diluted alkali transformed **309** to the thienyldiene **345** – its Diels–Alder adduct only has the UV spectrum of a monosubstituted thiophene. Together with the ^{1}H nmr data, therefore, the structure of this amide could be assigned.

309 $\xrightarrow{OH^\ominus}$ S—[CH=CH]$_2$CH$_2$CONHCH$_2$CHMe$_2$ 345

Compound **309** is derived most probably from **346**, in a manner similar to the pathway described for **306** through **335**, though again the latter itself has so far not been isolated.

MeCH$_2$CH$_2$[C≡C]$_2$CH$_2$CH=CH(CH$_2$)$_2$CO$_2$R $\longrightarrow$

346   *cis*

Me[C≡C]$_2$CH$_2$[CH=CH]$_2$COR $\longrightarrow$ 309

347   *trans, trans*

4.   Dithiocompounds Derived from C_{13}-Acetylenes

A few red-colored compounds, **348–351**, have been isolated from several Compositae, together with the corresponding thiophenes[4] (Scheme 9). These

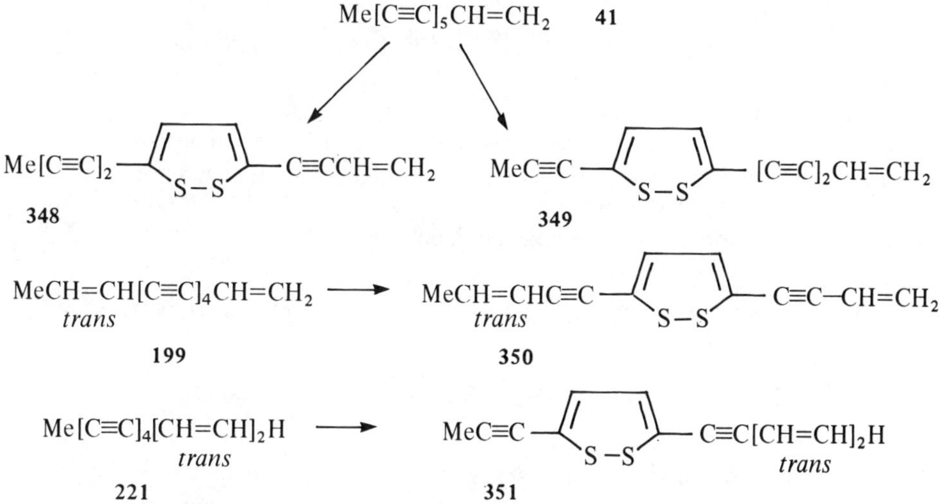

$$Me[C{\equiv}C]_5CH{=}CH_2 \quad \textbf{41}$$

Me[C≡C]₂— (thiophene ring, S–S) —C≡CCH=CH₂   MeC≡C— (thiophene ring, S–S) —[C≡C]₂CH=CH₂

348   **349**

MeCH=CH[C≡C]₄CH=CH₂ ⟶ MeCH=CHC≡C— (thiophene ring, S–S) —C≡C–CH=CH₂
trans   *trans*

199   **350**

Me[C≡C]₄[CH=CH]₂H ⟶ MeC≡C— (thiophene ring, S–S) —C≡C[CH=CH]₂H
trans   *trans*

221   **351**

Scheme 9

compounds are transformed easily by the loss of sulphur to thiophenes by TLC or short heating. Therefore, thiophenes could be formed through these compounds by the addition of disulfide to corresponding acetylenes, followed by loss of the sulphur:

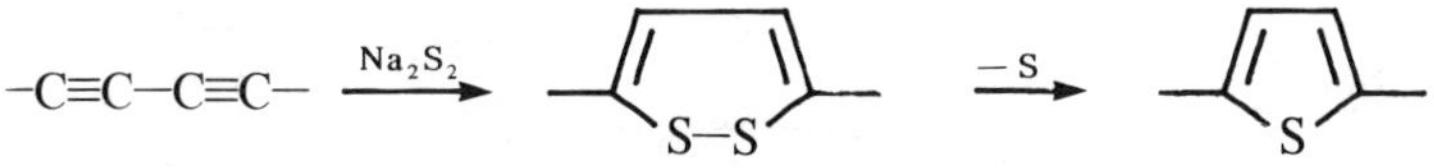

$$-C{\equiv}C-C{\equiv}C- \xrightarrow{Na_2S_2} \text{(disulfide ring, S–S)} \xrightarrow{-S} \text{(thiophene ring, S)}$$

However, under laboratory conditions, this addition leads to five-membered disulfides of type **352**[82] and, therefore, this assumption is very unlikely:

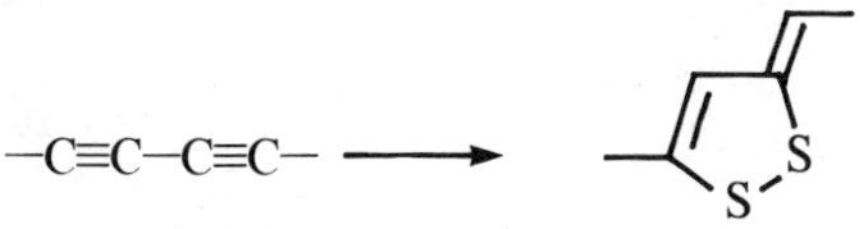

$$-C{\equiv}C-C{\equiv}C- \longrightarrow \text{(ring structure, S–S)} \quad \textbf{352}$$

The structure elucidation of **348–351** caused some difficulties. The occurrence of **348** from *Eriophyllum caespitosum* was reported first by Mortensen and Sörensen,[83] who also discussed an isomeric structure. The easy transformation of **348** to the corresponding thiophene **42** and the ¹H nmr data excluded such a structure;[84] however, the red color of these compounds favors the existence of an equilibrium:

353

The n→π*-transitions around 490 nm are similar to those of thioketones and, therefore, would support such an assumption. Thus far, however, no synthesis of these
compounds has been achieved. The only known compound of this type is **353**,[86]
which, however, is yellow. Boranate reduction of **349** afforded an unstable dihydro
compound, which also supports the presence of an equilibrium, since disulfides
normally are not reduced by boranate. The structures **349–359**[84, 85] followed from
the transformation to the corresponding thiophenes.

III. SPECTROSCOPIC DATA OF ACETYLENIC THIOPHENES

1. UV Spectra

Although most of the polyacetylenes show very characteristic UV spectra,[4] the
acetylenic thiophenes normally display broad UV maxima, which of course are
less typical. However, in many cases, a choice between different chromophoric
systems is possible. In Table 1, the maxima of the main types are listed. Though

TABLE 1.   UV MAXIMA OF DIFFERENT TYPES OF ACETYLENIC THIOPHENES IN
ETHER[a]

Type	Compound	Maxima (nm)
I	$-C\equiv C-$	274
II	$-[C\equiv C]_2$	312, 300
III	$-C\equiv C-CH=CH-$	309, 294
IV	$-C\equiv C[CH=CH]_2-$	338, 316
V	$-C\equiv C[CH=CH]_3H$	355, 332
VI	$-[C\equiv C]_2CH=CH-$	(346.5), 323, (316), 268 258, 253.5

TABLE 1. *Continued*

Type	Compound	Maxima (nm)
VII	$-C\equiv C-$ [thiophene] $-[C\equiv C]_2-$	340.5, 320, 245, 233
VIII	$MeC\equiv C-$ [thiophene] $-[C\equiv C]_2CH=CH_2$	357.5, 338, (324), 274, 264, 258
IX	$Me[C\equiv C]_2-$ [thiophene] $-C\equiv C-CH=CH_2$	357, 339, 332, 253
X	$-CH=CH-C\equiv C-$ [thiophene] $-C\equiv C-$	(333), 316
XI	$MeCH=CH-C\equiv C-$ [thiophene] $-C\equiv C-CH=CH_2$	(355), 333.5
XII	$-CH=CH-C\equiv C-$ [thiophene] $-[CH=CH]_2H$	351
XIII	$MeC\equiv C-$ [thiophene] $-C\equiv C-[CH=CH]_2H$	336, 315
XIV	[bithiophene] $-C\equiv C-$	332, 325, 239
XV	[bithiophene] $-[C\equiv C]_2-$	348
XVI	$Me-$ [bithiophene] $-C\equiv CCH=CH_2$	350, 251
XVII	$MeCH=CH-$ [bithiophene] $-CH=CH_2$	366, 262, 254
XVIII	$MeC\equiv C-$ [thiophene] $-C\equiv C-$ [thiophene]	359, 353, 341.5, 334.5, 329, 320.5, 255
XIX	$HOCH_2-$ [terthiophene]	354 (terthienyl 350)
XX	$Me-$ [thiophene] $-C\equiv C-CH=CH-CO_2R$	338 (*cis*), 332 (*trans*)
XXI	$MeC\equiv C-$ [thiophene] $-CH=CH-CO_2R$	338
XXII	[thiophene] $-C\equiv C-CH=$ [furanone] $=O$	371, 281, 237

TABLE 1.   *Continued*

Type	Compound	Maxima (nm)
XXIII		418, 393, 306.5, 294
XXIV		357, 332 (*trans*), 339 (*cis*)
XXV		363, 348, (332), 255
XXVI		(342), 327 (315)
XXVII		312
XXVIII		(320), 305, 241, 228

a Numbers in parentheses indicate shoulders in the spectra.

the maxima are shifted to longer wavelengths by increasing degrees of conjugation, a clear assignment of the chromophoric system sometimes is doubtful. All compounds with maxima around 350 nm often cause difficulties, though small differences in the spectra can be conclusive. As an example, the thiophenes of type VIII and IX may be mentioned. While the long wavelength maxima of IX show more fine structure, VIII has a characteristic maximum at 274 nm, which is not present in IX. Therefore, the whole spectrum must always be compared to identify a compound and, in most cases, ^{1}H nmr and mass spectra are also necessary for clear identification. Of course, the UV spectra are very useful for detecting acetylenic thiophenes in the crude fraction, as these spectra are the most sensitive ones.

2. ^{1}H Nmr Spectra

In many cases, the most important method for structure elucidation of acetylenic thiophenes is ^{1}H nmr spectroscopy. Because of the characteristic chemical shifts and couplings of thiophene protons, their detection in a ^{1}H nmr spectrum is normally easy, though in some cases complications may arise from the overlapping of signals, especially in bithienyl derivatives. Highfield ^{1}H nmr spectroscopy in such cases is very useful. The degree of substitution usually follows from the spectra, of course, allowing differentiation between mono- and disubstituted thiophenes. Small differences in the coupling $J_{2,3}$ and $J_{3,4}$ allow the assignment of the signals. The interpretation of shielding and deshielding effects is more difficult. This can be visualized from the examples listed in Table 2. Though not rigorously

TABLE 2.   ¹H NMR DATA OF SOME ACETYLENIC THIOPHENES

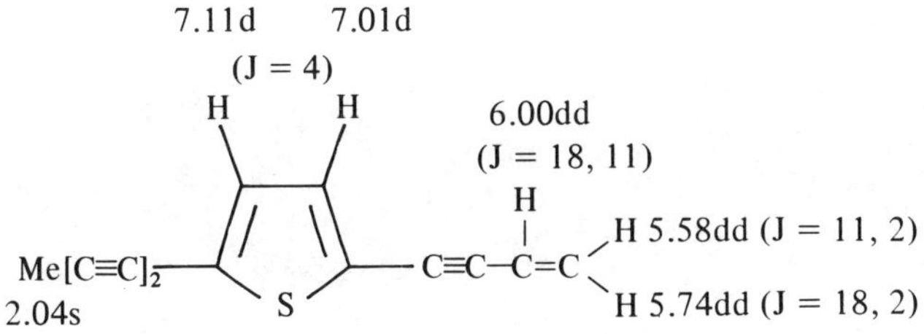

7.11d   7.01d
(J = 4)

H   H   6.00dd
(J = 18, 11)

H
| H 5.58dd (J = 11, 2)
42   Me[C≡C]₂⁻   C≡C−C=C
2.04s   S   H 5.74dd (J = 18, 2)

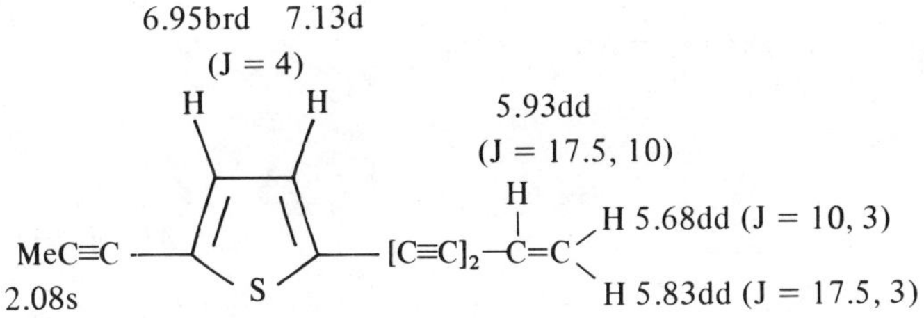

6.95brd   7.13d
(J = 4)

H   H   5.93dd
(J = 17.5, 10)

H
| H 5.68dd (J = 10, 3)
43   MeC≡C ⋅⋅⋅   [C≡C]₂−C=C
2.08s   S   H 5.83dd (J = 17.5, 3)

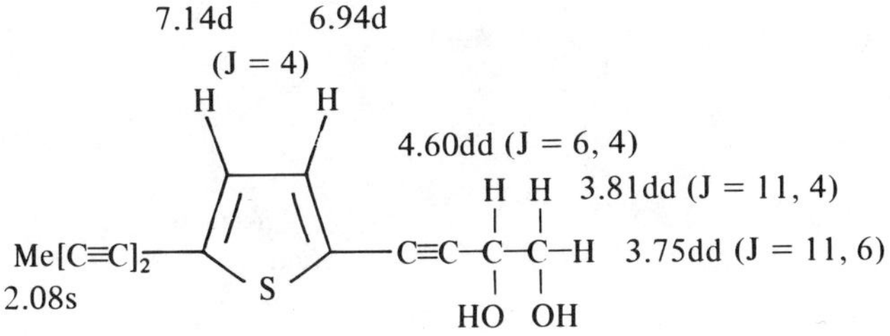

7.14d   6.94d
(J = 4)

H   H   4.60dd (J = 6, 4)
H H 3.81dd (J = 11, 4)
| |
46   Me[C≡C]₂⁻   C≡C−C−C−H   3.75dd (J = 11, 6)
2.08s   S | |
HO OH

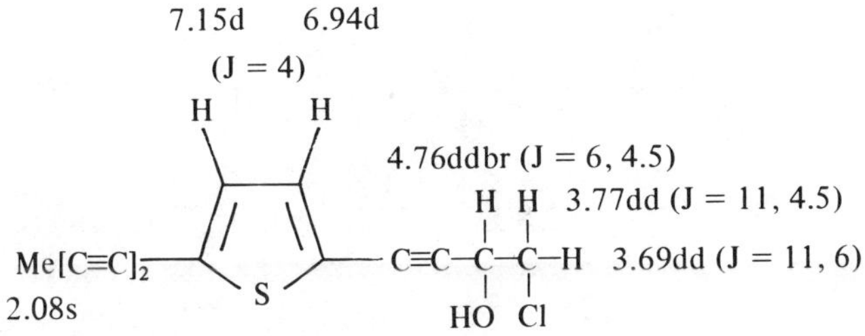

7.15d   6.94d
(J = 4)

H   H   4.76ddbr (J = 6, 4.5)
H H 3.77dd (J = 11, 4.5)
| |
50   Me[C≡C]₂⁻   C≡C−C−C−H   3.69dd (J = 11, 6)
2.08s   S | |
HO Cl

7.17d   7.11d
(J = 4)

H   H   4.60dd (J = 6, 4)
H H 3.83dd (J = 10, 4)
| |
65   H[C≡C]₂⁻   C≡C−C−C−H
3.39s   S | |
HO OH   3.76dd (J = 10, 6)

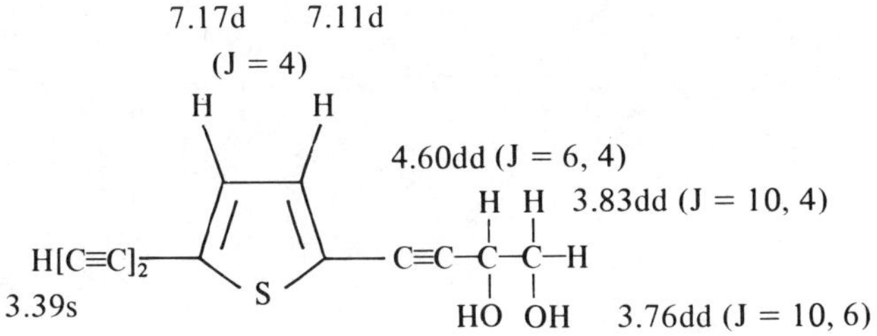

TABLE 2. *Continued*

67 H[C≡C]₂—
3.38s

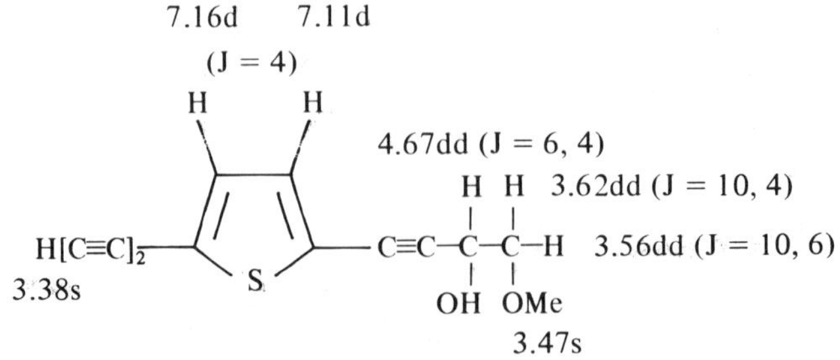

7.16d 7.11d
(J = 4)
H H

—C≡C—C—C—H
        | |
       OH OMe
           3.47s

4.67dd (J = 6, 4)
H H 3.62dd (J = 10, 4)
3.56dd (J = 10, 6)

91 7.23dd H—
    (J = 5, 1)

7.01dd 7.18dd 7.03d 7.10d
(J = 5, 4) (J = 4, 1) (J = 3.8)
H H H H

—C≡C—C=C
         H
H 5.54dd (J = 11, 2)
H 5.71dd (J = 17.5, 2)

6.02dd
(J = 17.5, 11)
H

97 7.22dd H—

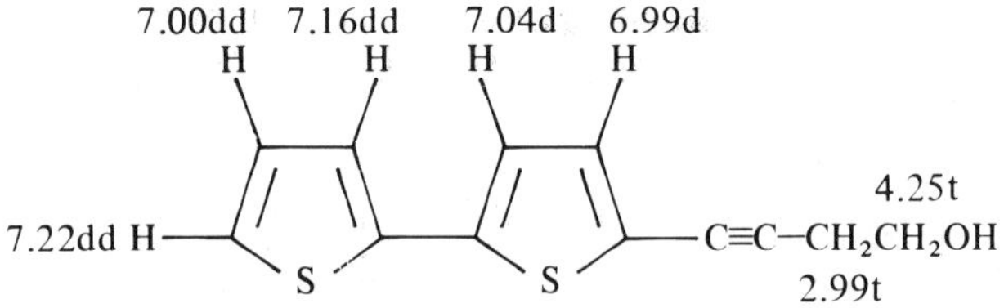

7.00dd 7.16dd 7.04d 6.99d
H H H H

—C≡C—CH₂CH₂OH
4.25t
2.99t

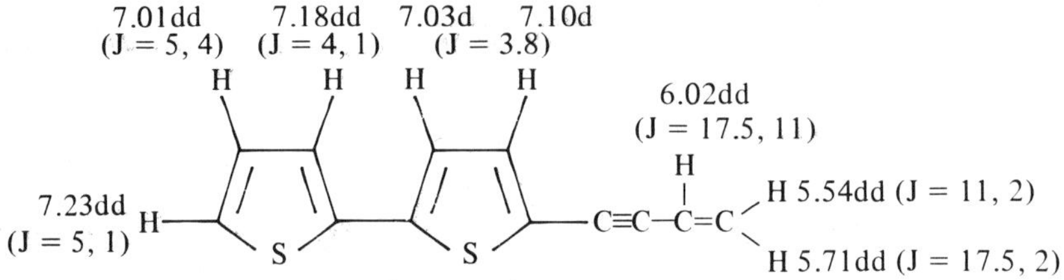

109 7.28dd H—

7.05dd 7.23dd 7.09d 7.23d
H H H H

—C≡C—
Cl
—H s7.77
OAc s2.26

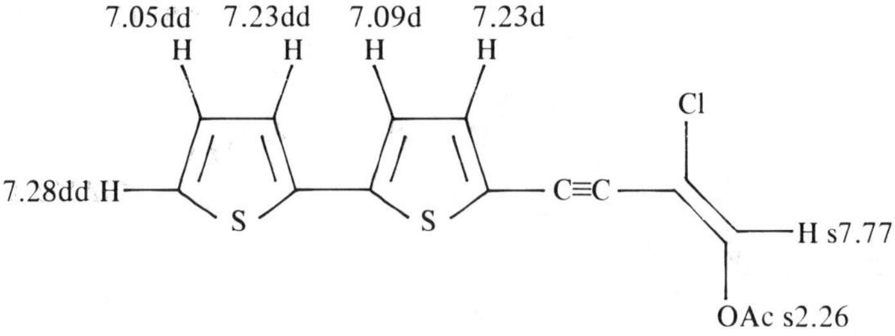

110

Cl
—OAc s2.29
H 7.85s

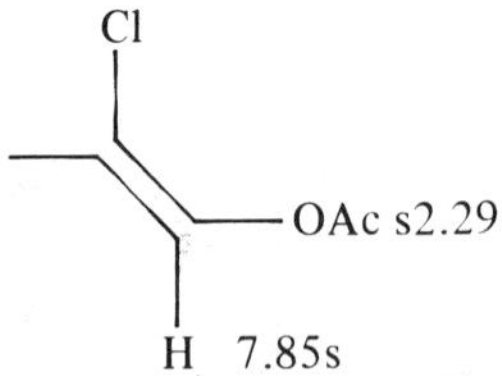

TABLE 2.   *Continued*

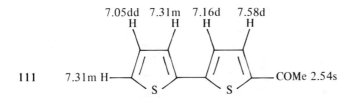

111   7.31m H

7.05dd 7.31m 7.16d 7.58d
H   H   H   H

COMe 2.54s

116   Me[C≡C]₂
2.05s

7.17d   7.00d   7.18dd
(J = 3.7)   (J = 4, 1)
H   H   H   H 7.01dd (J = 5, 4)

H 7.24dd (J = 5, 1)

181   Me
2.50d

6.59dq

189   7.22dd H
(J = 5, 1)

7.02dd   7.17dd 7.07s
(J = 5, 3.7) H   H   H

195   6.69dd H

6.76dd   6.69s   3.32s 7.26dd
6.99dd H   H   H   OMe H   H 6.80dd

H 6.90dd

$J_{1,2} = J_{11,12} = 5$ $J_{1,3} = J_{10,12} = 1.4$ $J_{2,3} = J_{10,11} = 3.4 \, (C_6D_6)$

TABLE 2.   *Continued*

established, the given assignments show that a diyne group has a stronger deshielding effect than a monoacetylene. Acetylenic methyl groups normally display a signal around 2.05 ppm; the thiophene methyl signal is always at lower fields (ca. 2.5 ppm). The differences between chemical shifts of the methyl group in mono- and diyne derivatives are very small and, therefore, are not conclusive. The signals of acetylenic protons are shifted downfield in thiophene acetylenes, owing to the low electronic density on the terminal carbon. The long-range deshielding effect of the thiophene ring can be visualized too, if the spectra of **42** and **43** are compared. The shielding effect of an α-methyl group is also obvious (see chemical shifts in **181** and **189**). As usual, the deshielding effect of α-carbonyl is always present. However, it can be transferred over two triple bonds, as can be seen from the data of **354**, which is obtained by oxidation of **57**.

3. ^{13}C Nmr Spectra

As mentioned previously, ^{13}C nmr data have not been used very much in structure elucidation of acetylenic thiophenes. However, some data are available for such compounds[3] (see Table 3). The C,H-couplings were also estimated, and these are listed in Table 3 for one compound. The overall picture of these data shows that a second thienyl residue has a small deshielding effect on the carbons of the first one only. The largest effect is that on the α-carbon (ca. 10 ppm). The couplings are influenced only a little by substituents. As in the benzene derivatives the coupling C—H—β is always larger than C—H—α. The chemical shift of the acetylene carbons clearly shows a deshielding effect of the thiophene ring. The data presented show that some aspects may be useful in structure elucidation. However, ^{13}C nmr data alone are not sufficient.

TABLE 3. ¹³C NMR DATA OF SOME ACETYLENIC THIOPHENES

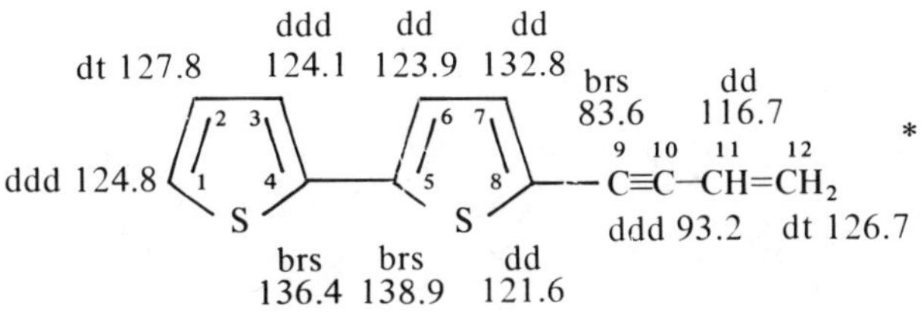

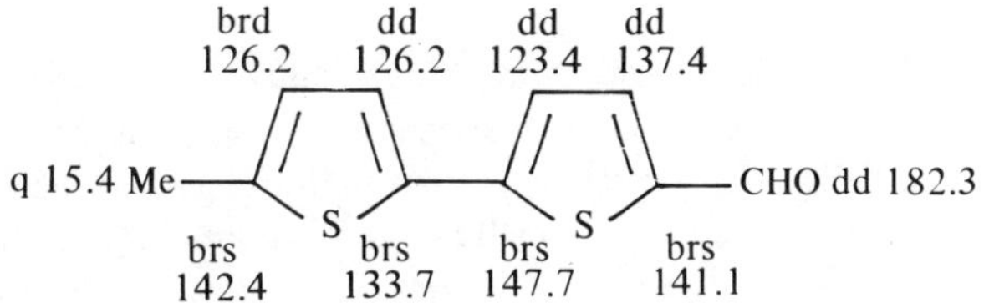

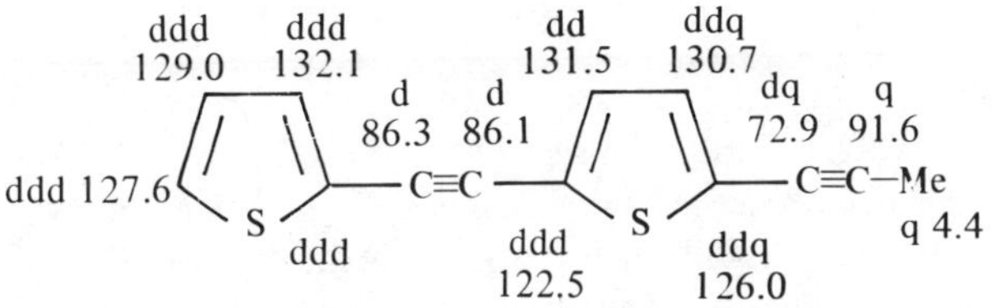

ddd   dd   dd
dt 127.8   124.1 123.9 132.8   brs   dd
                                   83.6   116.7   *
                                   9   10   11   12
ddd 124.8 |2 3\ /6 7\   —C≡C—CH=CH₂
           |1 4/ \5 8/
           \S/   \S/   ddd 93.2   dt 126.7
         brs   brs   dd
        136.4 138.9 121.6

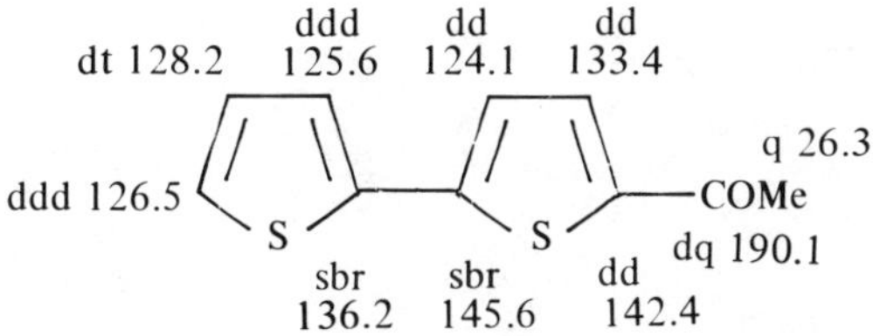

ddd   dd   dd
dt 128.2   125.6 124.1 133.4

ddd 126.5 / \ / \   q 26.3
           \S/ \S/ —COMe
         sbr   sbr   dd   dq 190.1
        136.2 145.6 142.4

brd   dd   dd   dd
126.2   126.2   123.4 137.4

q 15.4 Me— / \ / \ —CHO dd 182.3
          \S/ \S/
        brs   brs   brs   brs
       142.4 133.7 147.7 141.1

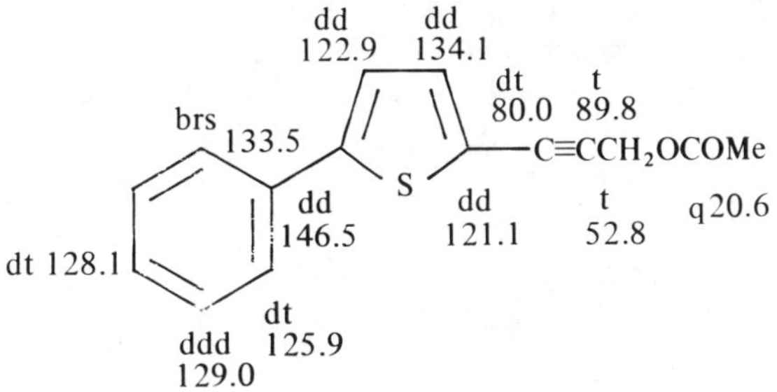

dd   dd
122.9   134.1
                  dt   t
                 80.0   89.8
brs 133.5 / \ —C≡CCH₂OCOMe
           \S/
         dd   dd   t   q 20.6
        146.5   121.1   52.8
dt 128.1

    dt
ddd 125.9
129.0

ddd   ddd   dd   ddq
129.0   132.1   131.5 130.7
          d   d   dq   q
         86.3 86.1   72.9   91.6
ddd 127.6 / \ —C≡C— / \ —C≡C—Me
           \S/   \S/   q 4.4
          ddd   ddd   ddq
         122.5   126.0

* J(Hz): C-1, H-1 = 188; C-1, H-2 = 6; C-1, H-3 = 10; C-2, H-1 = C-2, H-3 = 5; C-2, H-2 = 168;
C-3, H-1 = 9; C-3, H-2 = 6; C-3, H-3 = C-6, H-6 = 168; C-6, H-7 = C-7, H-6 = 5; C-7, H-7 =
170; C-8, H-6 = 11; C-8, H-7 = 6; C-10, H-11 = 2; C-10, H-12 = 9; C-11, H-11 = 168; C-11,
H-12 (cis) = 6; C-11, H-12 (trans) < 1; C-12, H-11 = 6; C-12; H-12 = 163.

dt   ddd   dd
127.8   123.7   124.3

ddd 124.6

S   S   S
brs   brs
137.6   136.3

dt   ddd   d   S   OMe 958.9
128.0   123.5 113.5 153.1   ddd   dt
                              122.8   127.0

ddd 124.4   ddd 123.9

S   brs   brs   brs
brs   134.8 124.8 132.8
137.5

4. Mass Spectra and Other Spectroscopic Data

Mass spectra are mainly important for establishing the molecular formula. However, in some cases, the fragmentation patterns can also be useful in structure elucidation. In several cases, the nature of end groups can often be established by mass spectroscopy, especially in the case of chlorohydrins, where a clear differentiation of the two possible structures is easy. Furthermore, the strong tendency of forming the thiapyrryliumion can be used (see **327**). Most fragmentations do not show any special features. As usual, the picture is determined by functional groups and the fragmentation of the thiophene ring itself is not pronounced.

The IR data of course are necessary to establish the nature of functional groups, but do not deserve a detailed discussion. Because of the weak polarization of a triple bond by the thiophene ring, the intensity of the acetylenic vibration normally is low. Strong acetylenic bands are visible only in cases with a conjugated carbonyl group. Thus far, nothing is known about the absolute configuration of the chiral acetylenic thiophenes. Also, no CD or ORD measurements are available.

IV. DISTRIBUTION OF ACETYLENIC THIOPHENES

Thiophenes have been isolated, except from a single microorganism, only from Compositae. In Table 4, the occurrence of these compounds is listed, following the tribal arrangement of the family. Only a few species belonging to the tribes Vernonieae and Eupatorieae contain thiophenes. In most cases, only the compounds derived from tridecapentaynene (**41**) were isolated. Two exceptions are the thiopheneketones **15** and **16**. No thiophenes have been isolated from members of the tribe Astereae. In the tribe Inuleae, several genera contain different types of thiophenes, most of which are derived from **41**. By far, most genera that contain thiophenes belong to the tribe Heliantheae. Therefore, in the table, their

TABLE 4.   DISTRIBUTION OF ACETYLENIC THIOPHENES

Species	Compound	Reference
	Vernonieae	
Ethulia conyzoides	42	4
Pseudostifftia kingii	200	100
Vernonia anisochaetoides	43	87
V. grandiflora	43	35
V. saltensis	42	88
	Eupatorieae	
Ageratina glabrata	16	17
Liatris pycnostachya	15	16
L. scariosa	15	16
L. spicata	15	16
Mikania scandens	42, 348	4
	Inuleae	
Athrixia arachnoidea	189	89
Bellida graminea	91	35
Blumea lacera	43	90
B. viscosa	43	35
Buphthalmum grandiflorum	83–85	91
B. salicifolium	83, 85	91
Calocephalus citreus	42	92
Helichrysum acutatum	236	93
H. panduratum	236	65
H. polycladum	189	94
H. populifolium	43	94
H. splendidum	189	95
H. tenuifolium	229, 230	65
H. trilineatum	100, 229	94
Lasiopogon muscoides	43	35
Leontonyx squarrossus	46	96
Macowania cf. hamata	42	97
Pluchea camphorata	43	35
Pluchea dioscorides	42, 46, 48, 49	4
P. foetida	42, 43	98
P. indica	42, 46, 48, 49	4
P. odorata	42, 43, 44, 60	99
P. suaveolens	42, 43, 53	32
P. tomentosa	42	35
Pterocaulon virgatum	50, 59, 64, 96, 104	38
Schoenia cassiniana	43, 85, 91	35
Sphaeranthus indicus	43	4
Stoebe vulgaris	42	96
Tessaria absinthioides	42	101
T. integrifolia	91, 189	101
Brachylaena discolor	43	35
Tarchonanthus camphoratus	43	28
T. trilobus	43	102

TABLE 4. *Continued*

Species	Compound	Reference
	Heliantheae	
Ambrosiinae		
Ambrosia artemisiifolia	43, 349	84
A. eliator	43, 349	84
A. chamissonis	349	103
A. cumanensis	43, 349	103
A. trifida	42, 43, 349, 351	35, 84
A. trifoliata	42, 43, 349	84
Iva xanthifolia	43, 349	84
Melampodiinae		
Melampodium divaricatum	222, 351	104
M. longifolium	222, 351	84
M. paludosum	222	4
M. rhomboideum	222	4
Milleriinae		
Guizotia abyssinica	200	4
G. oleifera	200	39
Milleria quinquefolia	348	4
Rudbeckiinae		
Rudbeckia amplexicaulis	83–85, 222, 223	28, 39
R. bicolor	222, 351	84
R. fulgida	42, 43, 48	4
R. hirta	222, 351	84
R. laciniata	222	105
R. newmannii	222, 351	4
R. nitida	222	105
R. speciosa	222, 351	84
R. sullivantii	222, 351	4
R. triloba	43, 60	39
Zaluzaniinae		
Zaluzania discoidea	44	35
Ecliptinae		
Aspilia eggersii	200	35
A. montevidensis	43	4
A. parvifolia	43	35
Eclipta alba	42, 91, 183	31, 35, 106
E. erecta	42, 50–52, 63, 66, 83–89, 104, 181–189	31
Engelmannia pinnatifida	46	35
Flourensia cernua	42	107
F. resinova	42	35

TABLE 4. *Continued*

Species	Compound	Reference
Oyedaea boliviana	42, 348	108
Podachaenium eminens	43, 57	109
Steiractinia sodiroi	348	35
Verbesina alata	42, 348	4
V. alternifolia	42	111
V. boliviana	42	110
V. cinerea	42, 348	110
V. latisquamata	43, 349	111
V. occidentalis	43, 349	112
Wedelia forsteriana	42, 83–85	113
W. grandiflora	43	114
W. paludosa	43	4
W. triloba	43	114
Zexmenia hispida	42, 348	115
Helianthinae		
Viguiera stenoloba var. chihuahense	116	116
Neurolaeninae		
Calea pilosa	83, 85, 93	117
Coreopsidinae		
Bidens connata	200, 212	62
B. ferulaefolia	223	118
B. frondosa	212	4
B. maximowicziana	200	119
B. pilosa	26	25
B. radiata	212, 223	118
Coreopsis bigelowii	203	4
C. grandiflora	203, 303, 304	39, 76
C. nuecensis	304, 305	77
C. parvifolia	42, 43	120
C. saxicola	203	35
C. verticillata	270	121
Thelesperma simplicifolium	200	35
Coulterellinae		
Coulterella capitata	43	35
Pectidinae		
Dyssodia acerosa	90, 105, 106, 116, 181, 182, 189	42
D. anthemidifolia	84, 85, 91, 105, 106, 181, 195	45
D. decipiens	83, 91, 106, 181, 189, 195	122
D. papposa	106, 189, 195	42
D. setifolia	85, 91, 95, 97–100, 103, 105, 106	45
D. setifolia var. setifolia	91, 105, 106, 189	35
Hymenantherum tenuifolium	91, 105, 106	122
Porophyllum lanceolatum	91, 105, 106, 189	122

TABLE 4.   *Continued*

Species	Compound	Reference
P. ruderale	91, 98, 100, 108, 189	46, 122
Tagetes coronopifolia	91, 189	35
T. elliptica	91, 189	35
T. erecta	91, 115, 189	39, 44
T. filifolia	91, 189	35
T. glandulifera	91, 104–106	4
T. gracilis	91, 105, 106, 181, 189	122
T. indica	91, 106, 189	4
T. lemmoni	91, 189	35
T. lucida	91, 106, 189	4
T. microglossa	105, 189	123
T. minuta	91, 189	35
T. patula	91, 92, 106, 189	122
T. pauciloba	43	4
T. signata	91, 106	4
T. tenuifolia	91, 189	35
T. terniflora	91, 105, 106, 189	122
T. zypaquirensis	91	122
Thymophylla tenuiloba	91, 105, 106	122
Flaveriinae		
Flaveria australasica	84, 85, 91	4
F. bidentis	91, 99, 189	35
F. campestris	85, 91, 100, 189	35
F. chloraefolia	85, 91, 99, 100, 189	124
F. pringlei	85, 91, 100, 106, 189	35
F. repanda	85, 189	41
F. trinervata	85, 189	4
Madiinae		
Achyrachaena mollis	42	28
Hemizonia corymbosa	43	28
Layia elegans	43	28
L. platyglossa	43	28
Baeriinae		
Eriophyllum caespitosum	42, 348	84
E. lanatum	42, 348	125
E. staechadifolium	43, 349	125
Lasthenia aristata	200–203	28
L. chrysostoma	200–203, 350	85
L. coronaria	200–203, 350	85
L. glaberrima	200	28
L. maritima	200–203	4
Chaenactidinae		
Arnica sachalinensis	42, 91	4
Chaenactis glabriuscula	42, 348	4
Palafoxia hookeriana	43, 349	4

TABLE 4. *Continued*

Species	Compound	Reference
P. texana	**349**	4
Picradeniopsis woodhousei	**200, 350**	42
Schkuhria abrotanoides	**43, 349**	4
S. advena	**43, 349**	84
S. multiflora	**43, 349**	126
S. pinnata	**43, 349**	127
S. advena	**43, 349**	127

Gaillardiinae

Gaillardia pulchella	**189**	4
Haploesthes greggii	**91, 105–107, 111, 189**	42
Helenium tenuifolium	**43**	128
Hymenoxys robusta	**43**	35

Anthemideae

Achillea sudetica	**263**	129
Anacylus depressus	**267**	4
A. radiatus	**6**	130
A. tomentosus	**306**	78
Anthemis arvensis	**267**	131
A. austriaca	**24, 25**	22
A. brachycentros	**4**	131
A. cupaniana	**4**	132
A. ersula	**25**	35
A. melanolepis	**5**	35
A. monantha	**3, 5, 6**	35
A. montana	**4**	131
A. punctata	**4**	131
A. punctata var. sicula	**4**	4
A. saguramica	**13, 290, 291, 293–297**	14
A. scariosa	**6**	131
Argyranthemum foeniculaceum	**309**	4
A. frutescens	**309**	81
A. gracile	**309**	4
Artemisia absinthium	**10**	11
A. arborescens	**10, 15, 16**	11, 15
A. austriaca	**6**	4
A. canariensis	**10**	133
A. koidzumii	**300**	134
A. ludoviciana	**300**	135
A. orientalis	**300**	4
A. princeps	**297**	4
A. purshiana	**300**	4
Artemisia reptans	**300**	35
A. sieversiana	**10**	133
A. stelleriana	**300**	4
A. stolonifera	**300**	35
A. superba	**264**	4
A. verlotorum	**5, 6, 9**	4, 35

TABLE 4.   *Continued*

Species	Compound	Reference
A. vulgaris	5, 6	4. 35
Balsamita major	300	136
Chamaemelum fuscatum	7, 8, 25, 294	4, 7, 35, 137
C. nobile	5, 6	8
C. nobile var. discoides	6, 9	4
C. oreades	267, 294	75
C. santolinoides	5, 6	4
Chamomilla recutita	267	4
Chrysanthemum coronarium	309	139
Cladanthus arabicus	6	78
Coleostephus myconis	263	4
Cotula reptans	3	138
Diotis martima (*Otanthus*)	306–308	78
Glossopappus macrotis	254, 255, 265	68
Heteranthemis viscade hirta	300	136
Lepidophorum repandum	263	35
Matricaria caucasica	267, 290, 294, 297, 298	75
M. maritima	5, 267, 290	35
M. maritima ssp. subpolare	294	35
M. tetragonosperma	267, 290, 291, 294	35
M. trichophylla	7, 267, 290, 294, 297	75
Osmitopsis asteriscoides	11, 12	13
Pentzia suffructicosa	4	35
Plagius flosculosus	264	136
Santolina africana	264	4
S. chamaecyparissus	257–259, 261, 262, 264	69
S. pectinata	259, 264	35
S. pinnata	263, 264	4
S. rosmarinifolia	259, 260, 262–264	71
S. scariosa	264	4
Tanacetum boreale	4	136
T. odessanum	300	140
T. vulgare	4	4
Tripleurospermum decipiens	267, 290, 294	35
T. grandiflorum	267, 294	75
T. hookeri	267, 294	35
T. inodorum	267, 290, 294	75
T. melanolepis	294, 297	35
T. oreades	267, 294	75
T. oreades var. tschihatchewii	267, 294, 297	35
T. sevanense	5, 294	35
T. subpolare	267, 294	35
T. tenuifolium	5, 267, 290, 294, 297	75
T. transcaucasicum	5, 294	35

Libeae

Erato polymnoides	189, 200	141
Philoglossa mimuloides	200	142

TABLE 4.   *Continued*

Species	Compound	Reference
	Senecioneae	
Senecio deppeanus	**43**	143
	Arctotheae	
Berkheya adlamii	91, 109, 189	47
B. angustifolia	91, 131, 134, 189	56
B. armata	117–119, 122–126	52
B. barbata	100, 129, 131, 132, 134, 189	52
B. bergeriana	91, 100, 109, 110, 189	36
B. bipinnatifida	43, 91, 105, 117, 118, 189	36
B. carduodes	189–191	56
B. cirsifolia	91, 189	36
B. coriacea	91, 100, 105, 106	56
B. debilis	91, 100, 105	36
B. decurrens	117, 119, 120	52
B. echinata	91, 100, 109, 110, 117, 189	36
B. erysithales	91, 100, 106, 117, 118	36
B. fruticosa	100, 102, 106, 189	52
B. herbacea	117–120, 125–127	52
B. heterophylla var. radiata	91, 100, 105, 106, 117, 118	52
B. insignus	91, 100, 105, 106	36
B. macrocephala	91, 100, 189	52
B. maritima	46, 65, 100, 109, 110, 189	36
B. multijuga	91, 100, 105, 106	36
B. onopordifolia	91, 99, 100, 105, 106, 109, 189	52
B. pannosa	105, 106, 189	36
B. purpurea	117, 119, 125	52
B. radula	91, 100, 109, 189	52
B. rhapontica ssp. aristosa	99, 100, 105, 106, 109, 189	36
B. rhapontica ssp. platyptera	100, 106, 189	36
B. rhapontica ssp. rhapontica	91, 100, 105, 106, 189	36
B. rigida	117, 118, 122, 125 –	35
B. robusta	91, 110, 189	36
B. setifera	117–119, 189	36
B. speciosa	91, 100, 109, 189	36
B. umbellata	46, 117, 118, 189	36
B. ssp. novum aff. bipinnatifida	91, 99, 106, 117–119	36
Cullumia bisulca	42, 91, 117	52
C. decurrens	43, 45, 58, 91, 100, 106	35
C. setosa	44, 46, 54, 67, 232, 233	33, 52
C. squarrosa	43, 91, 100, 105, 117, 129–134, 189	56
C. sulcata	91, 100, 105, 117, 189	56
Cuspidia cernua	91, 100, 105, 106, 129, 131, 189	52
Didelta carnosa	91, 100, 105, 106, 117, 189	52
D. spinosa	100, 106, 189	52
Platycarpha glomerata	42, 43, 57, 117–119	37

TABLE 4. *Continued*

Species	Compound	Reference
	Cynareae	
Arctium lappa	94	49
Cardopatium corymbosum	91, 100, 105, 106, 112, 113, 189	35, 50
Carthamus tinctorius	267	35
Centaurea aggregata	200	35
C. angustifolia	200	144
C. bella	200	4
C. carduiformis	91	35
C. cataonica	200	4
C. cineraria	200	105
C. cristata	49	144
C. dealbata	200	4
C. depressa	91, 189	35
C. indurata	200	4
C. jacea	200	105
C. jacea ssp. angustifolia	200	35
C. jacea ssp. jacea	200	4
C. jacea ssp. jacea var. pectinata	200	4
C. jacea ssp. macroptilon	200	4
C. kotschyi	189	35
C. leucophylla	200	4
C. macroptilon	200	4
C. nicaensis	200	4
C. nigra	200	105
C. phrygia ssp. pseudophrygia	200	4
C. pulcherrima	200	105
C. ruthenica	14	145
C. sadleriana	200	35
C. sevanensis	200	35
C. somchetica	200	35
C. transalpina	200	4
C. uliginosa	200, 269	4
Chartolepis glastifolia	200	27
Echinops bannaticus	84, 85, 91, 99, 100, 105, 106, 189	27
E. chamtavicus	42, 44, 48, 49, 55, 56, 84, 85, 91, 99–102, 105, 106, 189	4
E. commutatus	42, 44, 48, 49, 55, 56, 84, 85, 91, 99–102, 105, 106, 189	27
E. cornigerus	84, 85, 91, 99, 100, 105, 106, 189	27
E. dahuricus	42, 44, 48, 49, 55, 56, 84, 85, 91, 99–102, 105, 106, 189	27
E. exaltus	42, 44, 48, 49, 55, 56, 84, 85, 91, 99–102, 105, 106, 189	4
E. horridus	42, 44, 48, 49, 55, 56, 84, 85, 91, 99–102, 105, 106, 189	27
E. humilis	42, 44, 48, 49, 55, 56, 84, 85, 91, 99–102, 105, 106, 189	4
E. niveus	84, 85, 99, 100, 105, 106, 189	27
E. persicus	42, 44, 48, 49, 55, 56, 84, 85, 91, 99–102, 105, 106, 189	27

TABLE 4.   *Continued*

Species	Compound	Reference
E. ritro	42, 44, 48, 49, 55, 56, 84, 85, 91, 99–102, 105, 106, 189	27
E. sphaerocephalus	42, 44, 48, 49, 55, 56, 84, 85, 91, 99–102, 105, 106, 189	27
E. spinosissimus	91, 189	35
E. strigosus	42, 44, 48, 49, 55, 56, 84, 85, 91, 99–102, 105, 106, 189	27
E. viscosus	42, 44, 48, 49, 55, 56, 84, 85, 91, 99–102, 105, 106, 189 ——	4
Rhaponticum carthamoides	205–207	35
Saussurea pectinata	200–203, 205, 206, 208–211	60
Serratula radiata	200–202, 205–207	61
S. xeranthemoides	200–202	35
Xeranthemum annuum	254, 255, 267	28
X. cylindraceum	254, 255, 267	28
X. foetidum	254, 255, 267	4
X. inapertum	267, 287, 288	4
Tricholepis radicans	43	35
	Mutisieae	
Mutisia coccinea	91, 100	146
M. homoeantha	91, 106, 189	147

placements in the subtribes is also listed. For the subtribe Ambrosiinae, the cooccurrence of 43 and the corresponding dithio compound 349 seems to be characteristic; in the Melampodiinae, 222 is widespread. In the Milleriinae, only *Guizotia* contains 200, whereas *Milleria* afforded the dithio compound 348. For the Rudbeckiinae, the occurrence of 222 seems to be typical, though compounds derived from tridecapentaynene are also present. The subtribe Ecliptinae can be characterized by the thiophenes 42 and 43 and its derivatives, often accompanied by the corresponding dithio compounds 348 or 349. From the Helianthinae, only the rare bithienyl 116 has been isolated, yet the Coreopsidinae are rich in thiophenes. Most are derived from trideca-1,11-diene-3,5,7,9-tetrayne (199), but the phenylthiophenes 303–305 are also present in *Coreopsis* species. The small subtribe Coulterellinae contains 43. The Pectidinae, previously placed in the tribe *Helenineae*, are rich in the thiophenes. Here, the occurrence of bi- and terthienyls seems to be characteristic. They have been isolated from the genera *Dyssodia, Hymenantherum, Porophyllum,* and *Tagetes.* The same is true for the subtribe Flaveriinae; its placement was also in question. The chemistry clearly supports, however, incorporation into the Heliantheae near Pectidinae. The Madiinae again contain thiophenes derived from 41, while the Baeriinae mainly contain those derived from 199 with a few from 41, which are present in *Eriophyllum.* The Chaenactidinae, previously placed in the tribe *Helenieae*, can also be characterized by the occurrence of thiophenes derived from 41; dithio compounds are also frequent. The placement of this group in the Heliantheae, therefore, is strongly supported by the chemistry. The same is true for the Gaillardiinae, where again thiophenes derived from **41** are present. This group previously was placed in the Helenieae. Thus far, thiophenes have not been isolated only from some subtribes belonging to the Heliantheae. Most of these are smaller groups and many of the genera have not been investigated. However, the chemotaxonomic relevance of the thiophenes for this tribe is obvious. This is also true for the tribe Anthemideae. Here, thiophenes derived from **41** or **199** have never been isolated. They are replaced by those derived from C_{10} and C_{13} acetylenes. The C_{10} compounds especially are typical for this tribe, since they have never been isolated from other tribes. As shown in the table, some thiophenes may also be useful markers for certain genera. Together with other acetylenes, therefore, the occurrence of these compounds is chemotaxonomically important. A typical example is the displacement of the genus *Osmitopsis* from the Astereae to the Anthemideae. The few species investigated from the Liabeae indicate that thiophenes derived from **199** are present. Thiophenes in the tribe Arctotheae are widespread. Most are derived from **41** and **114**, respectively. In many genera, every species investigated contains these compounds. The placement of *Platycarpha* in this tribe is strongly supported by the chemistry, since compounds like **117** have never been isolated from members of other tribes. Therefore, the proposed placement in the tribe Cynareae can also be excluded. No thiophenes have been isolated from the tribe Calenduleae and only one species from the Senecioneae has so far afforded a thiophene. In the tribe Cynareae, mainly thiophenes derived from **199** were isolated. However, from the subtribe Echinopeae, which sometimes has been separated from the Cynareae as a new tribe, thiophenes derived from **41** have been isolated. Bi- and terthienyl derivatives are also widespread. In the subtribe Carlineae, thiophenes derived from C_{13}-triynes are widespread. *Cardopatium*, however, contains bi- and terthienyls and, therefore, may be better placed in the Echinopeae.

In the tribe Mutisieae, thiophenes derived from **41** or **114** have been isolated from two species. Both belong to the genus *Mutisia*. No thiophenes seems to be present in the Lactuceae, which, however, is often separated as a new family. The overall picture of the distribution of thiophenes in the Compositae shows that the ability to yield thiophene ring formations seems to be fundamental for this family. Most probably, loss of the required enzyme during evolution is the reason that these compounds are missing in some tribes.

V. CONCLUSIONS

If we look at the structures of all the thiophenes isolated from nature, it is very likely that they are all derived from compounds that have at least a conjugated diyne grouping. The diyne, however, must be activated by conjugation with a carbonyl group; those compounds with three and more conjugated triple bonds seem to be sufficiently reactive for the addition of an obviously nucleophilic sulphur compound, though its nature still is not clear. Most probably, an enzyme is present that liberates $^{\ominus}SH$. The cooccurrence of thiophenes and methylthioenol ethers, both formed from the same precursor, is a strong indication that compounds like A (p. 262) are intermediates in the formation of thiophenes.

The structural requirements of the corresponding acetylenic compound is obviously the reason that the thiophenes have been isolated only from the Compositae, where such acetylenes are widespread. In the Umbelliferae, acetylenes are also widespread, but so far no triynes or tetraynes have been isolated. The diynes, however, normally are not activated by conjugation, which may explain why thiophenes are missing in the family.

Several thiophenes isolated from the Compositae show biological activity, especially antibiotic, antinematodic activity against special mites. This means that at least some of the thiophenes are useful as protectors for the plants. However, only a few compounds have been tested on a broad scale, and the reason why the Compositae produce these thiophenes is difficult to ascertain. In principle, this question cannot be answered since during evolution the metabolic ability of the species changed, mainly because of mutations, and now they are forced to produce all these compounds. Perhaps these compounds have contributed to the survival of certain species, whereas other species, which were forced by mutation to produce useless or even injurious compounds, have died out. We can only hope to find out what may be the advantage of having the ability to produce these compounds. Surely several compounds are useless, but very probably many are useful in some way. At present, we know that the acetylenic thiophenes are useful as chemotaxonomical markers. Further investigation should provide a clearer picture of their total purpose.

REFERENCES

1. L. Zechmeister and J. Sease, *J. Am. Chem. Soc.*, **69**, 270 (1947).

2. F. Bohlmann, U. Hinz, A. Seyberlich, and J. Repplinger, *Chem. Ber.*, **97**, 809 (1964).

3. R. Zeisberg and F. Bohlmann, *Chem. Ber.*, **108**, 1040 (1975).

4. F. Bohlmann, T. Burkhardt, and C. Zdero, *Naturally Occurring Acetylenes*, Academic Press, London and New York, 1973.

5. E. Guddal and N. A. Sörensen, *Acta Chem. Scand.*, **13**, 1185 (1959).

6. F. Bohlmann, H. Bornowski, and H. Schönowsky, *Chem. Ber.*, **95**, 1733 (1962).

7. F. Bohlmann, W. von Kap-herr, L. Fanghänel, and C. Arndt, *Chem. Ber.*, **98**, 1411 (1965).

8. F. Bohlmann and C. Zdero, *Chem. Ber.*, **99**, 1226 (1966).

9. L. Skatteböl, *Acta Chem. Scand.*, **13**, 1460 (1959).

10. K. E. Schulte, private communication.

11. H. Greger, *Phytochemistry*, **17**, 806 (1978).

12. F. Bohlmann and C. Zdero, *Chem. Ber.*, **105**, 1919 (1972).

13. F. Bohlmann and C. Zdero, *Chem. Ber.*, **107**, 1409 (1974).

14. F. Bohlmann, K. M. Kleine, and C. Arndt, *Chem. Ber.*, **99**, 1642 (1966).

15. F. Bohlmann, K. M. Kleine, and H. Bornowski, *Chem. Ber.*, **95**, 2934 (1962).

16. R. Atkinson and R. Curtis, *Phytochemistry*, **10**, 454 (1971).

17. F. Bohlmann, J. Jakupovic, and M. Lonitz, *Chem. Ber.*, **110**, 301 (1977).

18. F. Bohlmann, H. Bornowski, and D. Kramer, *Chem. Ber.*, **96**, 1229 (1963).

19. J. Birkenshaw and P. Chaplen, *Biochem. J.*, **60**, 255 (1955).

20. R. T. Curtis and J. A. Taylor, *J. Chem. Soc. [C]*, 1813 (1969).

21. K. E. Schulte and N. Jantos, *Arch. Pharm.*, **292**, 536 (1959).

22. F. Bohlmann, K. M. Kleine, and C. Arndt, *Liebigs Ann. Chem.*, **694**, 149 (1966).

23. F. Bohlmann, C. Zdero, and H. Schwarz. *Chem. Ber.*, **107**, 1074 (1974).

24. F. Bohlmann, P. H. Bonnet, and H. Hofmeister, *Chem. Ber.*, **100**, 1200 (1967).

25. F. Bohlmann, H. Bornowski, and K. M. Kleine, *Chem. Ber.*, **97**, 2135 (1964).

26. F. Bohlmann and U. Hinz, *Chem. Ber.*, **98**, 876 (1965).

27. F. Bohlmann, C. Arndt, K. M. Kleine, and H. Bornowski, *Chem. Ber.*, **98**, 155 (1965).

28. F. Bohlmann, K. M. Kleine, and C. Arndt, *Chem. Ber.*, **97**, 2125 (1964).

29. F. Bohlmann and E. Bresinsky, *Chem. Ber.*, **100**, 1209 (1967).

30. R. D. Stevens and C. E. Castro, *J. Org. Chem.*, **28**, 3313 (1963).

31. F. Bohlmann and C. Zdero, *Chem. Ber.*, **103**, 834 (1970).

32. F. Bohlmann, J. Ziesche, R. M. King, and H. Robinson, *Phytochemistry*, **19**, 969 (1980).

33. F. Bohlmann and K. H. Knoll, *Phytochemistry*, **18**, 1060 (1979).

34. F. Bohlmann, P. Blaskiewicz, and E. Bresinsky, *Chem. Ber.*, **101**, 4163 (1968).

35. F. Bohlmann and coworker, unpublished.

36. F. Bohlmann, N. Le Van, T. Van Cuong Pham, A. Schuster, V. Zabel, and W. H Watson, *Phytochemistry*, **18**, 1831 (1979).

37. F. Bohlmann and C. Zdero, *Phytochemistry*, **16**, 1832 (1977).

38. F. Bohlmann, W. R. Abraham, R. M. King, and H. Robinson, *Phytochemistry*, **20**, 825 (1981).

39. F. Bohlmann, M. Grenz, M. Wotschokowsky, and E. Berger, *Chem. Ber.*, **100**, 2518 (1967).

40. E. Atkinson, R. F. Curtis, and G. T. Phillips, *J. Chem. Soc. (C)*, 2011 (1967).

41. F. Bohlmann and K. M. Kleine, *Chem. Ber.*, **96**, 1229 (1963).

42. F. Bohlmann, C. Zdero, and M. Grenz, *Phytochemistry*, **15**, 1309 (1976).

43. J. H. Uhlenbrock and J. D. Bijloo, *Rec. Trav. Chim. Pays-Bas*, **78**, 382 (1959).

44. F. Bohlmann and P. Herbst, *Chem. Ber.*, **95**, 2945 (1962).

45. F. Bohlmann and C. Zdero, *Chem. Ber.*, **109**, 901 (1976).

46. F. Bohlmann, J. Jakupovic, H. Robinson, and R. M. King, *Phytochemistry*, **19**, 2760 (1980).

47. F. Bohlmann, C. Zdero, and W. Gordon, *Chem. Ber.*, **100**, 1193 (1967).

48. S. Gronowitz and R. Ekman, *Ark. Kemi*, **17**, 90 (1961).

49. S. Obata, M. Yoshikura, and T. Washino, *Nippon Nogei Kagaku*, **44**, 437 (1970); **74**, 108136.

50. A. Selva, A. Arnone, R. Mondelli, V. Prio. L. Ceranlo, S. Petruso, S. Plescia, and L. Lamartina, *Phytochemistry*, **17**, 2097 (1978).

51. F. Bohlmann and J. Kocur, *Chem. Ber.*, **108**, 2149 (1975).

52. F. Bohlmann and C. Zdero, *Chem. Ber.*, **105**, 1245 (1972).

53. F. Bohlmann, C. Zdero, and H. Kapteyn, *Chem. Ber.*, **106**, 2755 (1973).

54. F. Bohlmann and J. Kocur, *Chem. Ber.*, **107**, 2115 (1974).

55. F. Bohlmann and W. Skuballa, *Chem. Ber.*, **107**, 497 (1973).

56. F. Bohlmann and A. Suwita, *Chem. Ber.*, **108**, 515 (1975).

57. R. Brunke, Dissertation Technische Universität, Berlin, 1977.

58. G. Märkl, *Chem. Ber.*, **94**, 3005 (1961).

59. R. Jente and C. Olatunji, unpublished.

60. F. Bohlmann and C. Zdero, *Chem. Ber.*, **100**, 1910 (1967).

61. F. Bohlmann and E. Waldau, *Chem. Ber.*, **100**, 1206 (1967).

62. F. Bohlmann, C. Arndt, K. M. Kleine, and M. Wotschokowsky, *Chem. Ber.*, **98**, 1228 (1965).

63. S. Jensen and N. A. Sörensen, *Acta Chem. Scand.*, **15**, 1885 (1961).

64. L. Skatteböl, *Acta Chem. Scand.*, **15**, 2047 (1961).

65. F. Bohlmann and W. R. Abraham, *Phytochemistry*, **18**, 839 (1979).

66. F. Bohlmann, W. Knauf, and L. Misra, Tetrahedron (in press).

67. F. Bohlmann, C. Arndt, and C. Zdero, *Chem. Ber.*, **99**, 1648 (1966).

68. F. Bohlmann and C. Zdero, *Chem. Ber.*, **108**, 739 (1975).

69. F. Bohlmann and C. Zdero, *Chem. Ber.*, **101**, 2062 (1968).

70. F. Bohlmann and C. Arndt, *Chem. Ber.*, **99**, 135 (1966).

71. F. Bohlmann and C. Zdero, *Chem. Ber.*, **106**, 845 (1973).

72. N. A. Sörensen, *Pure Appl. Chem.*, **2**, 569 (1961).

73. F. Bohlmann and P. D. Hopf, *Chem. Ber.*, **106**, 3621 (1973).

74. F. Bohlmann, H. Jastrow, G. Ertinghausen, and D. Kramer, *Chem. Ber.*, **97**, 801 (1964).

75. F. Bohlmann, H. Mönch, and P. Blaskiewicz, *Chem. Ber.*, **100**, 611 (1967).

76. J. S. Sörensen and N. A. Sörensen, *Acta Chem. Scand.*, **12**, 771 (1958).

77. F. Bohlmann and C. Zdero, *Chem. Ber.*, **101**, 3243 (1968).

78. F. Bohlmann, C. Zdero, and A. Suwita, *Chem. Ber.*, **107**, 1038 (1974).

79. F. Bohlmann and E. Dallwitz, *Chem. Ber.*, **107**, 2120 (1974).

80. F. Bohlmann and C. Hühn, *Chem. Ber.*, **110**, 1183 (1977).

81. E. Winterfeldt, *Chem. Ber.*, **96**, 3349 (1963).

82. F. Bohlmann and E. Bresinsky, *Chem. Ber.*, **100**, 107 (1967).

83. J. T. Mortensen and N. A. Sörensen, *Acta Chem. Scand,*, **18**, 2392 (1964).

84. F. Bohlmann and K. M. Kleine, *Chem. Ber.*, **98**, 3081 (1965).

85. F. Bohlmann and C. Zdero, *Phytochemistry*, **17**, 2032 (1978).

86. H. J. Barber and S. Smiles, *J. Chem. Soc.*, 1141 (1928).

87. F. Bohlmann, G. Brindöpke, and R. C. Rastogi, *Phytochemistry*, **17**, 475 (1978).

88. F. Bohlmann, P. K. Mahanta, and L. Dutta, *Phytochemistry*, **18**, 289 (1979).

89. F. Bohlmann and C. Zdero, *Phytochemistry*, **16**, 1773 (1977).

90. F. Bohlmann and C. Zdero, *Tetrahedron Lett.*, **2**, 69 (1969).

91. F. Bohlmann and E. Berger, *Chem. Ber.*, **98**, 883 (1965).

92. J. S. Sörensen, J. T. Mortensen, and N. A. Sörensen, *Acta Chem. Scand.*, **18**, 2182 (1964).

93. F. Bohlmann and W. R. Abraham, *Phytochemistry*, **18**, 1754 (1979).

94. F. Bohlmann, C. Zdero, W. R. Abraham, A. Suwita, and M. Grenz, *Phytochemistry*, **19**, 873 (1980).

95. F. Bohlmann and A. Suwita, *Phytochemistry*, **18**, 885 (1979).

96. F. Bohlmann and A. Suwita, *Phytochemistry*, **17**, 1929 (1978).

97. F. Bohlmann and C. Zdero, *Phytochemistry*, **16**, 1583 (1977).

98. F. Bohlmann and P. K. Mahanta, *Phytochemistry*, **17**, 1189 (1978).

322   F. Bohlmann and C. Zdero

99.   F. Bohlmann and C. Zdero, *Chem. Ber.*, **109**, 2653 (1976).

100.   F. Bohlmann, C. Zdero, R. M. King, and H. Robinson, *Phytochemistry*, **19**, 2669 (1980).

101.   F. Bohlmann, C. Zdero, and M. Silva, *Phytochemistry*, **16**, 1302 (1977).

102.   F. Bohlmann and A. Suwita, *Phytochemistry*, **18**, 677 (1979).

103.   F. Bohlmann, C. Zdero, and M. Lonitz, *Phytochemistry*, **16**, 575 (1977).

104.   F. Bohlmann and N. Le Van, *Phytochemistry*, **16**, 1765 (1977).

105.   R. E. Atkinson and R. F. Curtis, *Tetrahedron Lett.*, **5**, 297 (1965).

106.   N. R. Krishaswamy, T. R. Seshadri, and B. R. Sharma, *Tetrahedron Lett.*, **6**, 4227 (1966).

107.   F. Bohlmann and M. Grenz, *Chem. Ber.*, **110**, 295 (1977).

108.   F. Bohlmann and C. Zdero, *Phytochemistry*, **18**, 492 (1979).

109.   F. Bohlmann and N. Le Van, *Phytochemistry*, **16**, 1304 (1977).

110.   F. Bohlmann, M. Grenz, R. K. Gupta, A. K. Dhar, M. Ahmed, R. M. King, and H. Robinson, *Phytochemistry*, **19**, 2391 (1980).

111.   F. Bohlmann and M. Lonitz, *Chem. Ber.*, **111**, 254 (1978).

112.   F. Bohlmann and M. Lonitz, *Phytochemistry*, **17**, 453 (1978).

113.   F. Bohlmann and C. Zdero, *Chem. Ber.*, **104**, 958 (1971).

114.   F. Bohlmann and N. Le Van, *Phytochemistry*, **16**, 579 (1977).

115.   F. Bohlmann and M. Lonitz, *Chem. Ber.*, **111**, 843 (1978).

116.   F. Bohlmann, C. Zdero, and P. K. Mahanta, *Phytochemistry*, **16**, 1073 (1977).

117.   F. Bohlmann, U. Fritz, R. M. King, and H. Robinson, *Phytochemistry*, **20**, 743 (1981).

118.   S. L. Jensen and N. A. Sörensen, *Acta Chem. Scand.*, **15**, 1885 (1961).

119.   F. Bohlmann and C. Zdero, *Chem. Ber.*, **108**, 440 (1975).

120.   F. Bohlmann and C. Zdero, *Chem. Ber.*, **110**, 468 (1977).

121.   F. Bohlmann and C. Zdero, *Chem. Ber.*, **103**, 2095 (1970).

122.   F. Bohlmann and C. Zdero, *Phytochemistry*, **18**, 341 (1979).

123.   A. Castro and C. Castro, *Rev. Latinoam. Quim.* **9**, 204 (1978).

124.   F. Bohlmann, M. Lonitz, and K. H. Knoll, *Phytochemistry*, **17**, 330 (1978).

125.   F. Bohlmann, C. Zdero, J. Jakupovic, H. Robinson, and R. M. King, *Phytochemistry*, **20**, 2239 (1981).

126.   F. Bohlmann, J. Jakupovic, H. Robinson, and R. M. King, *Phytochemistry*, **19**, 881 (1980).

127.   F. Bohlmann and C. Zdero, *Phytochemistry*, **16**, 780 (1977).

128.   F. Bohlmann, K. M. Rode, and C. Zdero, *Chem. Ber.*, **100**, 537 (1967).

129.   F. Bohlmann and C. Zdero, *Chem. Ber.*, **106**, 1328 (1973).

130.   F. Bohlmann and K. M. Kleine, *Chem. Ber.*, **96**, 588 (1963).

131.   F. Bohlmann, K. M. Kleine, C. Arndt, and S. Köhn, *Chem. Ber.*, **96**, 1616 (1965).

132.   F. Bohlmann, C. Arndt. H. Bornowski, and K. M. Kleine, *Chem. Ber.*, **96**, 1485 (1963).

133.   H. Greger, *Planta Med.*, **35**, 84 (1979).

134.   F. Bohlmann and C. Zdero, *Phytochemistry*, **19**, 149 (1980).

135.   F. Bohlmann, H. Bornowski, and H. Schönowski, *Chem. Ber.*, **95**, 1733 (1962).

136.   F. Bohlmann, C. Arndt, H. Bornowski, K. M. Kleine, and P. Herbst, *Chem. Ber.*, **97**, 1179 (1964).

137.   F. Bohlmann and C. Zdero, *Chem. Ber.*, **103**, 2856 (1970).

138.   J. S. Sörensen, B. Ve, T. Anthonsen, and N. A. Sörensen, *Austr. J. Chem.*, **21**, 2037 (1968).

139.   A. Romo de Vivar, F. Montiel, and W. Diaz, *Rev. Latinoam. Quim.,* **5**, 32 (1974); **81**, 60839.

140.   F. Bohlmann and K. H. Knoll, *Phytochemistry,* **17**, 319 (1978).

141.   F. Bohlmann and M. Grenz, *Phytochemistry,* **18**, 334 (1979).

142.   F. Bohlmann, M. Grenz, and C. Zdero, *Phytochemistry,* **16**, 285 (1977).

143.   F. Bohlmann, K. H. Knoll, C. Zdero, P. K. Mahanta, M. Grenz, A. Suwita, D. Ehlers, N. Le Van, W.-R. Abraham, and A. A. Natu, *Phytochemistry,* **16**, 965 (1977).

144.   F. Bohlmann, K. M. Rode, and C. Zdero, *Chem. Ber.,* **99**, 3544 (1966).

145.   R. Jente, F. Bohlmann, and S. Schöneweiß, *Phytochemistry,* **18**, 829 (1979).

146.   F. Bohlmann and C. Zdero, *Phytochemistry,* **16**, 239 (1977).

147.   F. Bohlmann, C. Zdero, and N. Le Van, *Phytochemistry,* **18**, 99 (1979).

CHAPTER IV

Thiophenes Occurring in Petroleum, Shale Oil, and Coals

G. D. GALPERN

Institute of Petrochemical Synthesis, Academy of Sciences of the USSR, Moscow, USSR

I. INTRODUCTION

Thiophene and its derivatives are among the typical heteroatomic components of the majority of fossil fuels and numerous derivatives thereof.

Thiophene, the parent compound of the heterocyclic series of interest is obtained primarily from light oil in the production of coke-oven benzene. The bulk of information regarding thiophenes was provided, however, by studies of the composition of crude oils and petroleum products, in which thiophenes are widely represented, particularly in higher fractions and residues.

In contrast to coals and the organic matter of shales, crude oils displaying a high degree of reduction are characterized by the presence not only of thiophenes (frequently in high concentrations), but also of a plurality of diverse perhydrothiophenes (thiophanes or thiolanes) and thiacyclohexane derivatives (thianes), which are all closely related structurally.

These two series of saturated heterocyclic compounds have much in common with cycloalkanes (naphthenes), as evidenced by the predominance of alkyl derivatives with one or several methyl or ethyl groups and one, or in rare instances two, long alkyl chain(s). In the majority of cases, molecular mass growth results in increasing the number of rings in molecules, and higher fractions are noted for the presence of polycyclic compounds in which catacondensed compounds are predominant.

It is essential that the functional derivatives of hydrocarbons present as admixtures in all crudes bear, for the most part, one functional group; two such groups are rare. In an analogous manner, heteroatomic compounds contain one (or rarely two)

325

heteroatoms per molecule. This is true of the majority of heterocyclic compounds and, first and foremost, of thiophenes and thiacycloalkanes of interest.

The number of isomeric forms increases sharply on transition from lower to higher homologues, thereby complicating investigation of higher fractions. In the case of heterocyclic compounds, these difficulties are particularly pronounced, because the aforementioned peculiarity of the composition of heterocyclic compounds causes, on transition to higher fractions, a marked drop in the share of hydrophilic functional groups and heteroatoms in relation to the hydrophobic hydrocarbon skeleton. It is evident that the sensitivity of functional analysis decreases drastically on transition from light distillates to residuum and that the resolving power of the methods for separating complicated mixtures, in terms of the functional attributes of mixture components, diminishes.

A distinctive aspect of the molecular mass distribution of petroleum components is that the share of the majority of heterocyclic components grows as the temperature of cuts increases. The predominance of saturated hydrocarbons in light and middle fractions is observed in typical crude oils, whereas the higher fractions, and particularly residues, contain the concentrates of aromatic hydrocarbons and heteroatomic components. In the first approximation, the hydrocarbon moiety of petroleum distillates may be regarded as a system of compounds practically devoid of interatomic interactions. On the contrary, the heteroatomic components are characterized by diverse intermolecular interactions. Of prime importance are the various types of hydrogen bonds, since the interactions in question result in the formation of higher molecular-weight associates from low molecular compounds and hence upset the linearity of molecular mass distribution. Many of the low-molecular functional derivatives are absent in light fractions, but feature prominently in higher fractions and residues. This situation occurs because the latter fractions, unlike the light fractions, are non-Newtonian liquids. The interaction with aromatic hydrocarbons, especially with the bi- and polycyclic aromatics concentrating in higher petroleum fractions, plays a prominent role for heterocyclic compounds.

The multitude of fossil fuel components and the functional proximity of their selected groups necessitated the development of specific, structural-group methods of separating and characterizing polycomponent systems. The progress in this field relied for a long time on a laborious trial-and-error approach, but the mid-20th century witnessed a vigorous advance in instrumentation techniques. These developments made it possible to characterize the composition of sophisticated systems and the structure of their components, using relatively small amounts of test substances that required a short period of time for analysis and provided for a higher precision of separation and analysis. In the area under consideration, the methods used are chromatography, thermodiffusion, mass spectrometry, IR and near-UV spectroscopy, proton and ^{13}C magnetic resonance employed in combination with modern automated and recording computer equipment. Complexing and some extractive techniques also play an essential role.

II. THIOPHENES OCCURRING IN PETROLEUM

Prior to 1948, only thiophene, thiophane, 2- and 3-methylthiophanes, and *cis*- and *trans*-dimethylthiophanes were reliably identified in crude oils. In 1948, the American Petroleum Institute (API) in cooperation with the U.S. Bureau of Mines initiated API Research Project 48, which was devoted to an investigation of the organic sulfur compounds in crude oils, and the report[1] published in 1972 summarizes this systematic 20-year study. It presents in an exhaustive manner the results of studies performed in compliance with Research Project 48 and describes in detail numerous techniques used for the isolation, separation, analysis, and characterization of organic sulfur compounds in crude oils from Wasson, Texas and, to a lesser extent, those from Wilmington, Deep River, Agha Jari. The methods for synthesizing sulfur compounds were developed both before and during these investigations.

Previous knowledge and identification reliability of thiophenes, thiophanes, and thiacyclohexanes at the commencement of Research Project 48 can be seen from Table 1, which was compiled on the basis of Smith's data.[2]

The data obtained in the course of studying the composition of sulfur compounds in crude oils are most complete for Wilmington and Wasson crudes.[1] Table 2 lists such data for thiophenes and thiacycloalkanes.

The Wilmington crude oil distillate boiling up to 250°C is characterized by the presence of thiophene and alkylthiophenes, whereas the corresponding Wasson crude oil distillate is free of thiophene, but contains benzo[b]thiophene and alkylbenzothiophenes. Benzothiophenes are amply represented in the higher fractions of crude oils containing organosulfur compounds with a high degree of cyclization.

The characterization of sulfur-containing components of crude oils and the establishment of their individual composition and structure are feasible only for the

TABLE 1. THIOPHANES AND THIACYCLOALKANES IDENTIFIED BEFORE 1948[2]

	Number of Carbon Atoms					
	4	5	6	7	8	Total
Thiophenes						
Theoretically possible	1	2	7	13	30	53
Identified	1	2	6	2	5	16
Thiophanes						
Theoretically possible[a]	1	2	12	20	30	65
Identified	1	2	8	7	7	25
Thiacyclohexanes						
Theoretically possible[a]	0	1	3	15	15	34
Identified	0	1	3	0	0	4

[a] The number of optical isomers was not estimated.

TABLE 2.   THIOPHENES AND THIACYCLOALKANES IDENTIFIED IN WILMINGTON AND WASSON CRUDE OILS[1]

Wilmington Crude Oil (1.38% S)		Wasson Crude Oil (1.85% S)	
Alkylthiophenes	*Thiophanes*	*Thiophanes*	*Benzothiophenes*
2-Methyl-	2-Methyl-	Thiophane	Benzo/b/-
3-Methyl-	3-Methyl-	2-Methyl-	2-Methyl-
2-Ethyl-	2,2-Dimethyl-	3-Methyl	3-Methyl-
3-Ethyl-	*trans*-2,5-Dimethyl-	2,2-Dimethyl-	4-Methyl-
2,5-Dimethyl-	*cis*-2,5-Dimethyl-	*trans*-2,5-Dimethyl-	7-Methyl-
2,4-Dimethyl-	*trans*-2,4-Dimethyl-	*cis*-2,5-Dimethyl-	2-Ethyl-
2,3-Dimethyl-	*cis*-2,4-Dimethyl	*trans*-2,4-Dimethyl-	3-Ethyl
3,4-Dimethyl-	*trans*-2,3-Dimethyl	*cis*-2,4-Dimethyl-	2,3-Dimethyl-
2-*n*-Propyl-	*cis*-2,3-Dimethyl-	3,3-Dimethyl-	2,4-Dimethyl-
2-Isopropyl	*trans*-3,4-Dimethyl-	*cis-trans*-3,4-Dimethyl-	
	cis-3,4-Dimethyl-	2,2,5-Trimethyl-	*Thiaindans*
Thiaindans	2,2,5-Trimethyl-	2-Ethyl-	Thiaindan
Thiaindan	2-Ethyl-	2.2.5,5-Tetramethyl	2-Methyl-
2-Methyl-	2,2,5,5-Tetramethyl-	*trans*-2,5-Dimethyl-	3-Methyl-
3-Methyl-	2,4,4-Trimethyl-	*cis*-2,5-Dimethyl	5- and/or 7-Methyl-
3,3-Dimethyl-			2,2-Dimethyl-
2,3-Dimethyl-	*Thiacyclohexanes*	*Thiacyclohexanes*	3,3-Dimethyl-
2,2-Dimethyl-	Thiacyclohexane	Thiacyclohexane	2,5/2,7/5,7-Dimethyl-
3,5- or 3,7-	2-Methyl-	2-Methyl-	2,3-Dimethyl-
Dimethyl	3-Methyl-	3-Methyl-	2,4-Dimethyl-
3,6-Dimethyl-	4-Methyl-	4-Methyl-	2,6-Dimethyl-
2,5- or 2,7-			3,5/3,7-Dimethyl-
Dimethyl-			3,6-Dimethyl-
2-Ethyl- and			2-Ethyl-
2,6-dimethyl-			2-Methyl-2-ethyl-
2-Methyl-3-ethyl-			2-Methyl-3-ethyl-
2,4-Dimethyl-			2,2,4-Trimethyl-
2,2,5- or 2,2,7-			2,2,5-Trimethyl-
Trimethyl-			
2,3,4-Trimethyl or			
3-ethyl-4-methyl-			
2,2,4-Trimethyl-			
2-Methyl-2-ethyl-			

compounds that boil under 250°C and contain up to 9 or 11 carbon atoms per molecule. Here, use is made of multistep separation and group isolation flow sheets, involving diverse chemical and physical methods of analysis at each step. The outline of Wasson crude oil processing to separate organosulfur compounds from the 111–150°C (Fig. 1) and 200–250°C (Fig. 2) fractions[1] are presented below.

These fractions are practically devoid of nitrogen compounds; oxygen compounds are presented therein as traces of phenols and volatile acids. The ultimate identification of compounds was performed using gas–liquid chromatography, gel chromatography, microdesulfurization JR spectroscopy, and low-resolution mass spectrometry.

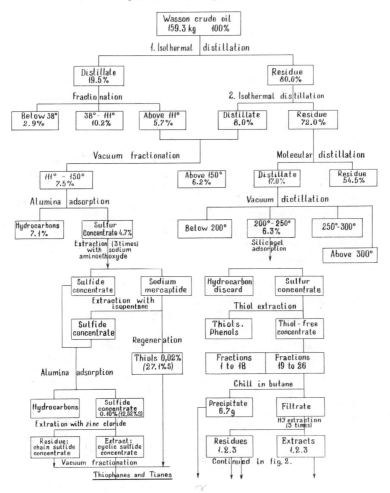

Fig. 1. Processing of Wasson crude oil to produce sulfur compounds concentrates (Thompson and cow. (1)].

Identification procedures are described in great relevant detail in Ref. 1, which contains an extensive review of relevant literature and a list of 58 publications devoted to the problems of separation and identification that were investigated during API Research Project 48. In 1980, Thompson presented a brief summary of this review.[3]

The realization of this research program necessitated prolonged work by a large team of experts in diverse fields, namely, organic, physical, and analytical chemistry. The results obtained made it possible to develop less sophisticated techniques of analyzing the composition of crude oils and bitumoids. Studies carried out in compliance with API Research Project 48 unequivocally showed that in some instances elemental sulfur and disulfides are inherent in native crudes and do not at all times appear as a result of secondary oxidation of H_2S and thiols.

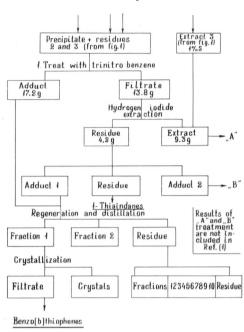

Fig. 2. Processing of Wasson 200° to 250° fractions to produce benzo[b]thiophene and 1-thiaindane [Thompson and cow. (1)].

The findings of earlier investigations involving the products of sulfuric acid treatment of petroleum distillates were corroborated in the majority of cases; this was particularly true in the case of data reported by Birch and associated in the 1950's.[4] Unfortunately, the data in question provide no basis for the quantitative estimate of the content of individual thiophenes and thiacycloalkanes. As can be seen from a number of analyses in Ref. 1, the concentrations of individual alkylthiophenes on a crude oil basis are equal to 1.10^{-5}–1.10^{-6} wt.%, and the concentrations of alkylcycloalkanes are greater by 2 orders of magnitude, so that in the study of organosulfur compounds in crude oil the primary task comprises obtaining sulfur concentrates, as shown in Figs. 1 and 2. Birch and McAllan[5] examined in detail the applicability of an aqueous solution of mercuric acetate for the separation of sulfides from light fractions and also of thiophenes unsubstituted at the alpha-position (practically no reaction occurs between 2,5-dimethylthiophene and mercuric acetate). The method was used in the investigation of kerosines from Agha Jari, Iran,[4] and Midwest crude oils.[6] To separate alkyl sulfides, Wilson[7] resorted to ligand-exchange chromatography, in which the stationary phase comprised deactivated silica gel impregnated with a solution of mercuric acetate in acetic acid, and 50% aqueous acetic acid was the mobile phase. Kaimai and Matsunaga[8] employed ligand-exchange thin-layer chromatography with mercuric acetate on acidic alumina, the latter adsorbent being more active than silica gel. Two-dimensional chromatography in combination with photoelectric scanning and mass-spectrometric analysis enabled these workers to establish, with an accuracy of ± 2–4% in the 280–440°C boiling range fraction of Kuweit crude oil, the presence of alkylated thiophenes (C_9H_{19}–C_{17}–H_{35} alkyls) and thiamonocycloalkanes (C_5H_{11}–$C_{13}H_{27}$ alkyls), using n-hexane as the developer. Sulfides and alkylthiophenes were separated quantitatively from dibenzothiophene. In the crude oil, the content of sulfides and thiophenes equals from 7 to 20% of the total amount of organosulfur compounds, while the respective content of dibenzothiophenes varies from 93 to 80%. Escalier et al.[9] analyzed Basra heavy crude oil by complex formation with mercuric acetate and elution chromatography on silica gel, followed by gas–liquid chromatography using a flame-ionization and a flame-photometric detector. They identified in the crude oil 3-methylthiophene, dimethylthiophene, 2-propylthiophene, 2,3,5-trimethylthiophene, 2,3,4-trimethylthiophene, methyl-n-propylthiophene, diethylthiophene, two dimethyl-ethylthiophenes, and tetramethylthiophene. The overall concentration of thiophenes was approximately 40–47 ppthsd, the concentration of each component being 1–8 ppthsd, except for 2,3,5-trimethylthiophene, which was present in a concentration of up to 23–25 ppthsd. The acidimetric method of nonthiophene sulfur determination suggested by Lamathe[10,11] for heavy petroleum products up to asphaltenes is based on the acid-back titration of sulfide and mercaptan complexes with mercuric acetate. This technique was further refined by Bardina et al.[12,13] who succeeded in determining the sulfide, sulfoxide, and sulfone functions simultaneously present in a test sample.

Vogh and Dooley[14] developed an unorthodox modification of ligand-exchange chromatography for sulfide separation from aromatic concentrates, which is adapted for the treatment of petroleum heavy ends. This technique is based on high-pressure liquid chromatography using BioRex ion-exchange resin (200–400 mesh), the Cu form of the ion-exchanger being prepared by substituting Na ions for Cu ions (from copper sulfate). Aromatic hydrocarbons and thiophenes are sharply separated from saturated and aromatic sulfides. To separate aliphatic sulfides from cyclic sulfides (Fig. 1), extraction with zinc chloride was utilized.

Chertkov and associates[15,16] in their study of a 150–250°C fraction of Arlan crude oil suggested a two-stage extraction of organic sulfur compounds by 86 and 91% sulfuric acid (at each extraction stage, the recommended acid-to-fraction ratio is 1:5, at an extraction ratio equal to 2). This method found extensive application in investigations carried out by chemists in the USSR. Lyapina and associates[17] performed a comparative investigation of two separative procedures — an acid and a liquid chromatographic pathway (Figs. 3 and 4). The quantitative ratios of the groups of compounds obtained by both pathways are close to each other, but the second procedure provides for somewhat higher yields, owing to decreased resinification and associated losses. The particulars of the separative procedure presented in Fig. 3 are given in Ref. 18. The isolation of sulfides by silver nitrate and sulfide separation from thiophenes and hydrocarbons described in Ref. 19 was based on the findings reported by Pailer et al.,[20] and an impetus to the employment of a dual adsorbent (silica gel and alumina) was given by Hirsch et al.[21] Bondarenko and associates[22] recommended a two-stage extraction with phenol and furfurol as an alternative to sulphuric acid extraction of middle crude oil distillates for the

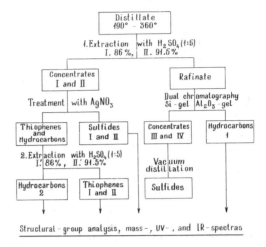

Fig. 3.   Processing of 190° to 360° petroleum distillates to produce thiophenes and sulfides by the H_2SO_4 pathway [Ljapina, Gal'pern and cow. 18, 52].

process separation of sulphides and thiophenes. This method makes it possible to obtain the concentrates of refined sulfides and also the concentrates of poly-aromatics and thiophenes, as well as of their benzologues and naphthenologues. The feasibility of a sulfide extraction process was evaluated.[23] Extracting sulfides and thiophenes as an admixture from diesel oil distillates was proposed on a pilot scale using spent sulfuric acid from alkylation.[24]

With high-boiling petroleum fractions, the number of compounds and isomers thereof increases to such an extent that, in most cases, identification is confined to various modifications of group and structural-group analyses. For practical purposes, it is essential to effect separation as regards functional attributes, that is,

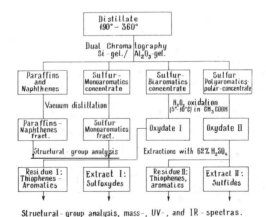

Fig. 4.   Processing of 190°–360° petroleum distillate to produce thiophenes and sulfides by chromatographic pathway [Ljapina, Gal'pern and cow. 18, 52].

into polar and nonpolar compounds. Acids, bases, sulfur, nitrogen, and oxygen-containing neutral compounds fall into the polar group, whereas the group of nonpolar compounds includes hydrocarbons of various homologous series (cyclic and acyclic with diverse degrees of cyclization and hydrogen deficiency), in particular, aromatic hydrocarbons. These attributes are also employed for characterizing further the groups of polar compounds. As noted, the share of functional groups relative to the hydrocarbon moiety of heteroatomic compounds in crude oil (relative to the number of C atoms) drops with the increasing molecular mass of compounds. It is evident that here the structure of the hydrocarbon moiety acquires a decisive role and affects the acid–base and redox properties of the functional groups. At this level, the significance attached to the analysis of polycomponent systems of organic compounds is governed by the fact that monofunctional subsystems, provided certain structural limitations are observed, may frequently find application along with, and preferably over, individual compounds. This situation is illustrated later in conjunction with the discussion of thiophene and thiophane derivatives.

It by no means follows from the foregoing discussion that studies of individual compounds of multicomponent systems present no interest. Two fields of research necessitate progress in the analysis of specific compounds of fossil fuels at this level too. Mention should be made in this connection primarily of genetically typical compounds that are essential for geochemistry, since their study clarifies the origin of natural compounds of interest. Unfortunately, nearly all organosulfur compounds have so far escaped investigation in this direction. The second field is concerned with the biological activity of numerous fossil fuels and petroleum and bitumoid fractions that have already found application in medical practice. Although the search for the active principles of such natural formations has long since been initiated, the results are far from being exhaustive. The role of this research trend is demonstrated herein by investigations of the so-called fish shales. It is also appropriate to draw attention to the part played by polar components in the formation of aqueous emulsions of crude oils and petroleum products. The phenomena of the formation and destruction of such emulsions are known to be promoted to a greater extent by the systems of compounds than by individual components of the systems in question. That the search among the group of petroleum acid components yielded interesting results is demonstrated by Seifert's investigations.[25]

The systematic study of heavy ends of crude oils and the characterization of their sulfur-containing components were initiated by the U.S. Bureau of Mines and the American Institute of Petroleum (Research Project 60). Five crude oils of different types, from the following oil fields, were studied: Prudhoe Bay (Alaska), Gach Saran (Iran), Swan Hills (Alberta, Canada), Wilmington (California, USA), and Recluse (Wyoming, USA). The first results reported in Ref. 26 were followed by summaries of the investigation.[27–30] An outline of more detailed but less available reports[31–34] is presented in Refs. 35 and 36.

In the study under consideration, fractions boiling in the 370–535°C range were processed according to a general flow sheet; the steps included taking off the distillate and subjecting it to the extraction of acid and base components by liquid chromatography on an anionite and a cationite, respectively; the neutral nitrogen compounds were removed by ferrous chloride treatment. The refined distillates were obtained in a yield of about 95%, except for Recluse crude oil, which because of its specific composition required an additional dewaxing step that was responsible for decreasing the refined fraction yield to 77%. The paraffin was shown by mass spectrometry to contain 65% of n-alkanes, 19% of alkylmonocycloalkanes, and about 10% of di-, tri-, and polycycloalkanes, with an admixture of monocycloaromatics (about 1%). The refined feed stock was separated using liquid chromatography with a column packed with a dual-adsorbent silica and alumina gels, the fractions of saturated hydrocarbons, mono-, and diaromatics, and polyaromatics (polar) being collected. Direct mass-spectrometric analysis showed that in the fraction of saturated compounds, the number of rings increased continuously from 0 to 6 and aromatic hydrocarbons were absent. The other three fractions were subjected to further gel-permeation chromatography, followed by the serial mass-spectrometric analysis of subfractions. The results cited in Ref. 35 for mono- to tetracyclic derivatives of thiophene and thiophane (to be more exact, thiamonocycloalkanes) are listed in Table 3.

As Table 3 illustrates, in the crude oil fractions investigated, the content of thiacycloalkanes is insignificant and varies from 0 to 1 mass %.

The presence of alkylthiophenes and their naphthenologues was established in Wilmington crude oil only; benzothiophene and its derivatives are typical for all crude oils studied. The bulk of thiophenes, however, is represented by condensed polycyclic systems, preferably with one thiophene ring; the dominant compounds are benzo- and dibenzothiophenes. Characteristic of all crude oils is the presence of homologous series forming a nearly regular sequence from C_nH_{2n-18} to C_nH_{2n-30}.

Increased concentrations of alkylbenzo- and alkyldibenzothiophenes appear to be typical of the majority of higher petroleum fractions. This feature was clearly demonstrated earlier by Drushel and Sommers,[37] who studied the sulfur compounds in vacuum gas oil, the Safania (Saudi Arabia) crude oil fraction boiling from 425 to 455°C. The multistage separative procedure employed included chromatography using a silica gel-packed column, oxidation with hydrogen peroxide in acetic acid, and chromatographic resolution of the oxidate, also on silica gel. The benzene eluate contained predominantly aromatic hydrocarbons, whereas in the 1,4-dioxane eluate sulfones were present (the principal product to be investigated), which were reduced with zinc in hydrochloric acid to obtain initial sulfur compounds. The concentration maxima were found to correspond to the homologous series $C_nH_{2n-10}S$ (benzothiophenes) and $C_nH_{2n-16}S$ (dibenzothiophenes). In the initial fraction, the content of thiaindanes, $C_nH_{2n-10}S$, is appreciable, whereas the concentration of alkylthiophenes is insignificant, since no alkylthiophenes are present in higher fractions.

In the Zapadnosurgut crude oil fraction boiling in the 410–450°C range, Lyapina et al. showed thiamonocycloalkanes to be present and found that the alkylated thiophenes:benzothiophene:dibenzothiophene ratio equals 1:33:10.5. The observed relatively high concentration of alkylthiaindane is nearly five times as

TABLE 3. THIACYCLOALKANES AND THIOPHENES IN 370–535°C FRACTIONS OF CRUDE OILS[a]

Homologous Series	Oil Fields				
	Prudhoe Bay	Gach Saran	Swan Hills	Wilmington	Recluse
Thiamonocycloalkanes					
$C_nH_{2n}S$	Traces	0.61	–	0.10	0.88
Thiabicycloalkanes					
$C_nH_{2n-2}S$	Traces	0.14	–	0.04	0.91
Thiatricycloalkanes					
$C_nH_{2n-4}S$	–	–	–	0.06	0.99
Thiatetracycloalkanes					
$C_nH_{2n-6}S$	–	–	–	0.04	0.76
Alkylthiophenes					
$C_nH_{2n-4}S$	–	–	–	0.44	–
Cycloalkylthiophenes					
$C_nH_{2n-6}S$	–	–	–	0.25	–
Dicycloalkylthiophenes					
$C_nH_{2n-8}S$	–	–	–	0.41	–
Alkylbenzothiophenes					
$C_nH_{2n-8}S$	1.91	3.39	1.17	2.15	0.64
Cycloalkylbenzothiophenes					
$C_nH_{2n-10}S$	0.23	2.50	0.88	1.52	0.9
Dicycloalkylbenzothiophenes					
$C_nH_{2n-12}S$	0.09	1.06	–	1.22	–
Overall content of thiophene compounds in 370–535°C fraction	14.42	21.07	14.80	16.73	7.90

[a] According to data in Ref. 35; mass % based on crude oil.

great as that of alkylthiophenes. The results obtained for the 360–400°C fractions from various crude oils[38-40] are in agreement with the data listed above, as can be readily seen from Table 4.

The specificity of composition of Usa and Yareg crude oil fractions manifests itself in the total absence of saturated thiacycloalkanes, in contrast to the presence of thiaindans (benzothiophanes). The content of alkylthiophenes is under 10%, which was also typical of Orenburg crude oil. Gas condensate from the Urtabulak field has a high content of alkylthiophenes and is unique among other crude oils under investigation in that its content of alkylthiophenes is four times as great as that of dibenzothiophenes. Zapadnosurgut, Siberia crude oil is noted for its content of benzothiophenes, exceeding the content of alkylthiophenes by a factor of nearly seven. Urtabulak crude oil is further characterized by a content of thiamonocyclo-alkanes equal to nearly 90% of the sum total of thiacycloalkanes. It should be emphasized that Vyakhirev and associates established[41] as early as 1965 the presence of low concentrations of thiophene, 2-methyl, 3-methyl, and 2,5-dimethyl thiophenes in gasoline from the Ramashkino field.

TABLE 4. THIACYCLOALKANES AND THIOPHENES IN 200–360°C FRACTIONS IN SELECTED SOVIET CRUDE OILS[a]

Homologous Series	Oil Field						
	Arlan	Zapadnosurgut	Samatlor	Urtabulak	Orenburg	Yarega	Usa
Thiamonocycloalkanes $C_nH_{2n}S$	48.0 (3.59)	59.7 (4.07)	56.5 (1.28)	89.7 (0.50)	46.1 (0.39)	— (—)	— (—)
Thiabicycloalkanes $C_nH_{2n-2}S$	17.4 (1.30)	16.4 (1.11)	20.6 (0.46)	21.0 (0.10)	22.2 (0.18)	— (—)	— (—)
Thiatricycloalkanes $C_nH_{2n-4}S$	8.4 (0.62)	5.3 (0.36)	8.4 (0.19)	—	7.9 (0.07)	— (—)	— (—)
Thiaindans $C_nH_{2n-8}S$	8.2 (0.61)	5.5 (0.38)	4.6 (0.10)	12.7 (0.05)	22.5 (0.18)	5.9 (0.32)	17.0 (1.04)
Alkylthiophenes $C_nH_{2n-4}S$	25.5 (1.54)	11.3 (0.64)	21.2 (0.47)	62.6 (0.30)	3.7 (0.12)	6.4 (0.35)	9.8 (1.76)
Alkylbenzothiophenes $C_nH_{2n-8}S$	46.0 (2.78)	73.5 (4.14)	42.3 (0.93)	14.2 (0.04)	30.5 (1.03)	24.4 (1.30)	28.8
Cycloalkylbenzothiophenes $C_nH_{2n-10}S$	21.2 (1.28)	1.4 (0.08)	16.3 (0.36)	3.5 (0.01)	9.3 (0.33)	16.2 (0.87)	13.5
Dicycloalkylbenzothiophenes $C_nH_{2n-12}S$	3.8 (0.23)	7.4 (0.42)	8.2 (0.18)	2.8 (0.02)	20.0 (0.69)	10.2 (0.55)	12.8
Dibenzothiophenes $C_nH_{2n-16}S$	—	0.7 (0.04)	4.6 (0.10)	5.2 (0.02)	36.5 (1.25)	15.4 (0.83)	14.4

[a] According to data in Refs. 38–40; mass % based on the sum of cycloalkanes or thiophenes (in brackets – based on the fraction mass).

The data considered above invariably show the concentrations of individual thiophene and alkylthiophenes in crude oils to be generally not high, although rare exceptions to this rule do occur. In contrast, the concentrations and amounts of "thiophenogens," primarily alkylthiocycloalkanes, are substantially greater. Thiophane and homologues thereof undergo relatively facile catalytic dehydrogenation and yield thiophenes; thianes (thiacyclohexanes) likewise lend themselves readily to dehydrogenation and form alkylthiophanes and then thiophenes. This trend of research is exemplified by studies[42-45] concerned with thiamono- and dicycloalkanes and benzothiacycloalkanes. A further group of "thiophenogens" includes exhaustively hydrogenated condensed polycyclic systems incorporating one thiophene ring. Subjecting such systems to pyrolysis results in thiophene cycle "dealkylation" and the formation of thiophene and its homologues. That this is indeed the case can be seen from analytical studies relying on the use of pyrolytic chromatography for the characterization of the high-molecular moiety of crude oils and kerogen.[46,47] Analytical pyrolysis is discussed in detail in Ref. 48. The inverse problem, namely, the conversion of a part of thiophenes present in petroleum distillates, is dealt with in Ref. 49, and is concerned with the ionic hydrogenation of thiophene concentrates from petroleum middle distillates with a view toward converting these concentrates into sulfide concentrates. The reaction was carried out using trifluoroacetic acid and triethylsilane in the presence of diethoxyboron trifluoride as catalyst (5–10°C, 2 h); the reactions products were small amounts of thiacycloalkanes, thiabicycloalkanes, and a little benzothiophane. Benzothiophenes were found to undergo hydrogenation much slower than do thiophenes, presumably owing to the presence of benzothiophenes substituted in position 2. Studies carried out in the Soviet Union with regard to the conversion of middle distillate sulfide consisting primarily of thiocycloalkanes into sulfoxides and sulfones prompted investigations in the field of selective hydrogenation, as reported in Ref. 49.

Research in the oxidation field of organic S compound of petroleum fractions was initiated and continued for many years in our laboratory and in the laboratories of the Bashkir branch of The USSR Academy of Sciences, and at the Institute of Organic Physical Chemistry at the Kazan branch.

We showed the oxidation potentials of organic sulfides to be markedly distinct from the oxidation potentials of hydrocarbons and other groups of sulfur compounds.[50] This work ran parallel with the finding that, in petroleum fractions, the oxidation of sulfides with hydrogen peroxide to sulfoxides is likely to provide a facile route for sulfide separation from other petroleum fraction components.[51] The employment of the technique of petroleum distillate oxidation in the procedure for separating and characterizing sulfides in the middle fraction of the Yuzhnouzbekistan crude oil was summarized in Refs. 52–54. The sulfides were found to comprise 61.1% alkylthiamonocycloalkanes and 12.3% alkylthiabicycloalkanes, thiaalkyltri-, and polycycloalkanes (hydrogen deficiency, Z, from 4 to 8) with an admixture of thiaindanes and its derivatives. The subfractions of alkylthiamonocycloalkanes contained thiophane and thiane derivatives; 2-n-alkyl-4,5-dimethylthiophanes apparently constituted the predominant group. A more detailed investigation of the composition of this crude oil was carried out under the guidance of I. U. Numanov.[55] To control the content of sulfide sulfur in the starting distillates, an oxidative method of iodatometric analysis was elaborated in detail.[56] Laboratory investigations led to the development of techniques for obtaining sulfoxides from the distillates of sulfur-containing and high-sulfur crude oils and also from the concentrates of sulfur compounds in these crudes.[57] Several variants of this reaction were described: (1) oxidation with hydrogen peroxide in the presence of perchloric acid, followed by precipitating the perchlorates of sulfoxides and subjecting them to hydrolysis with aqueous soda;[58] (2) oxidation with hydrogen peroxide in the presence of sulfuric acid, followed by hydrolysis of the aqueous layer;[59] (3) oxidation under the conditions of foam and emulsion formation in the presence of acetic acid, followed by extracting the target product also with acetic acid and recovering the acid;[60] (4) oxidation with organic peroxides.[61] Variant (3) of the process for the preparation of sulfoxides was elaborated to the stage of technical tests on a pilot unit and was patented.[62] The oxidation of sulfide-containing petroleum fractions in the crude oil was conducted in Ufa, on a pilot unit in a reactor using the foam-emulsion reaction system.[63] Introducing some modifications in the process conditions (adding a mixture of sulfuric and acetic acids, doubling the amount of hydrogen peroxide used, and elevating slightly the reaction temperature) makes it possible to obtain sulfones from the same feed stock.[64] Parallel studies in this field were made in Ufa.[65] The latter investigations resulted in the elaboration of a unique process for the oxidation of sulfide concentrates with hydrogen peroxide, which dispenses with the use of acids and polar solvents, but calls for distilling of a part of the water.[66]

The technique of selective oxidation of sulfides to sulfoxides in an acetic anhydride was suggested in Ref. 67. An earlier comparison of sulfide oxidation in acetic acid and acetic anhydride[68] showed that the oxidation of sulfides to sulfoxides by hydrogen peroxide experienced essential acceleration because of the combined effect of acetic anhydride and perchloric acid, both under homogeneous and heterogeneous (emulsion system) conditions; the depth of aromatic sulfide oxidation increases nearly threefold, as compared to oxidation in acetic acid.

The method of sulfide oxidation to sulfoxides by hydrogen peroxide proved successful for the purification of petroleum nitrogen concentrates from the admixtures of organic sulfur compounds.[69] The oxidates thus obtained are then dissolved in the 10- to 15-fold amount of acetic anhydride and the solution is treated with a large-pore kationite (KU-1) to completely remove the sulfur compounds from the nitrogen compounds.

The Arlan crude oil 190–360°C fraction and the Romashkino crude oil 200–315°C fraction subjected to stepwise oxidation by hydrogen peroxide (at temperatures of 25, 60, and 100°C) and chromatography each yielded four groups of sulfones,[70] as demonstrated by the polarographic, EPR,[71] and mass spectrometric[72] techniques. The results obtained made it possible to assess the content of 1,1-dioxobenzothiophenes in petroleum sulfones as 60–95% (S percentage based on the sum of S in the starting fractions). This value is substantially higher than the findings reported earlier and is ascribed to the possible presence of essential amounts of tetrasubstituted thiophenes (and 2,3-disubstituted benzothiophenes). Relevant tests showed that the "petroleum sulfones" displayed high activity in the treatment of cattle trichophytosis, microsporia in cats and dogs, and other skin diseases in animals. "Petroleum sulfones" are noted for their low toxicity for warm blooded animals; LD_{50} for white mice is 3.84 mg/kg.

It is noteworthy that thiophene falls under the category of petroleum components that inhibit completely *Pseudomonosis* and *Mycobacterium* cultures adapted to withstand the effects of petroleum products.[73]

Petroleum sulfoxides prepared from the middle fractions of sulfur-containing crude oils consist mostly of thiamono and bicycloalkanes, among which the derivatives of tetrahydrothiophene (thiophane) are predominant. The sulfoxides obtained from the Arlan crude oil diesel fuel were tested as a possible top dressing for cereal crops.[74] The yields of rye and spring wheat were found to increase by 12–20% as a result of petroleum sulfoxide application at a rate of 1 litre per hectare; grain quality characteristics are simultaneously improved. Sulfoxides from Romashkino, Surgut, and Arlan crude oil distillates displayed a repellent action comparable with that of hexamide toward fleas and house flies.[75] Petroleum sulfoxides were found to possess a broad spectrum of pest control activity.[76] They were tested[77] as acaricides on cotton crops and found to be significantly superior to nitraphen; no mite resistance to sulfoxides was observed.

The extractive power of cyclic sulfides and sulfoxides of petroleum origin toward noble metals was investigated.[78] The employment of these agents for the extraction of niobium and tantalum was described,[79] and petroleum sulfoxides (PSO) as extractants were reported to outperform the currently employed agents (e.g., tributyl phosphate, methyl isopropyl ketone, and cyclohexanone). The application of sulfoxides as extractants of inorganic compounds, primarily rare-earth elements, appears to hold much promise.[80] Sulfoxides have been investigated as extractants for gallium from hydrochloric solutions[81] and also for organic acids.[82]

Saturated sulfoxides derived from petroleum were studied under the conditions of the Pummerer reaction,[83] and preliminary investigations involved di-*n*-hexyl and diisoamyl sulfoxides,[84] thiophane sulfoxide, and 6-methyl-1-thiachroman sulfoxide.[85] Thiophane sulfoxide yields acetoxythiophane and the dimer of dihydrothiophene, whereas 6-methyl-1-oxotetralin reacts with acetic anhydride to form 6-methyl-2-acetoxy-1-thiatetralin (thiachromane) as the principal product, 6-methyl-1-thiaoctalin (4H) being a minor impurity. 2-Methyl-1-oxothiadecalin was converted into 2-methyl-2,3-octalin and the yield was 57%. The conversion of $(C_5-C_6)_2$ sulfoxides yields predominantly a mixture of *cis*- and *trans*-isomers of 1,2-unsaturated sulfides, but the reaction is complicated by the sulfone–sulfide redox rearrangement. Sulfoxides produced from the distillates of Romashkino (average composition, $C_{14.1}H_{28.6}SO_{1.1}$ and $C_{12.3}H_{24.0}SO_{1.2}$) and Yuzhnouzbekistan (average composition, $C_{13.5}H_{24.5}SO_{1.2}$) crude oils reacted up to 90% with acetic anhydride at 100°C. As a result of the reaction, the C_{12} and C_{13} sulfoxides lost all oxygen; in the C_{14} compounds, the content of oxygen varied from 0.3 to 1.0 atom per mole, the sulfoxide group being totally absent. The C_{14} and C_{12}

compounds yielded the products that contain from 2 to 4 atoms per molecule less than do the starting sulfoxides, presumably because of the dehydrogenation of thiabicycloalkane sulfoxides. In the case of C_{14} compounds, partial preservation of the acetoxy derivate appears to be feasible, and the process is accompanied by the partial oligomerization of petroleum sulfoxide and resinification.

Of greater promise is the production of arylsulfonium salts derived from sulfoxides[86] and sulfides,[87] which are expected from sulfur-bearing crude oil distillates. The following aspects of this problem have been investigated:[88] (1) PSO condensation with phenol and anisole in the presence of perchloric acid and phosphorus oxychloride to obtain the perchlorates of p-hydroxy and p-methoxyphenylthionylcycloalkanes; (2) conversion of perchlorates of p-hydroxyphenylthionylcycloalkanes into sulfonium bases (hydrated zwitterions) under the effect of anionites; (3) dehydration and oligomerization of these zwitterions; (4) the conversion of sulfonium bases into chlorides by hydrochloric acid. As shown earlier,[89] 1-oxothiophanes and to a lesser degree 1-oxothianes with a free hydrogen atom at position 2 react with phenol. From mass-spectroscopic data, it is reasonable to infer that the petroleum sulfoxides predominantly participating in the reaction are the perchlorates of dimethylethyl, methylpropyl, and dimethylpropyl-1-oxothiophanes. The sulfonium bases prepared using an anionite undergo rapid dimerization and, for the most part, yield dimers containing a cleaved heterocycle. Thiophane derivatives only participated in this reaction.

It was established that individual hydroxy- and methoxyphenylthiacyclanyl sulfonium salts[90] in the presence of amines undergo C—S bond cleavage and, accordingly, heterocycle opening to form 1-hydroxy(1-methoxy)-aryl-4 (5)-aminoalkyl sulfides. The chloro(bromo) sulfonium salts of 2,5-dihydroxyarylthianes, those of petroleum origin inclusive, were recently shown[87] to form 2,5-acetoxyarylhaloalkyl sulfides, via heterocycle opening, when boiled with acetic anhydride. The reactions of p-hydroxyphenyl derivatives proceed in an analogous manner.

The reactions of sulfonium salt decyclization discussed above can be represented by the following scheme:

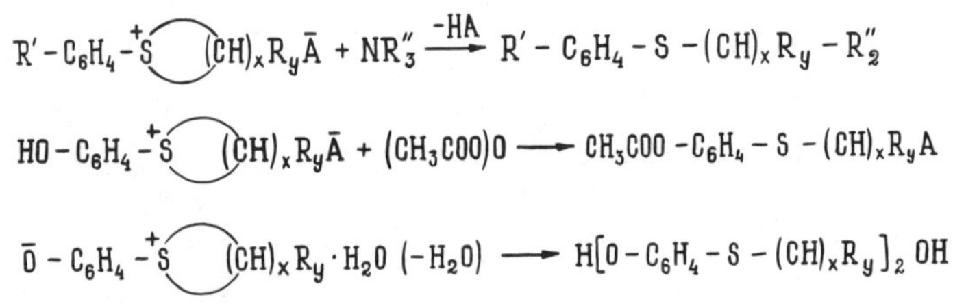

$$R'-C_6H_4-\overset{+}{S}\underset{}{\bigcirc}(CH)_xR_y\bar{A} + NR''_3 \xrightarrow{-HA} R'-C_6H_4-S-(CH)_xR_y-R''_2$$

$$HO-C_6H_4-\overset{+}{S}\underset{}{\bigcirc}(CH)_xR_y\bar{A} + (CH_3COO)O \longrightarrow CH_3COO-C_6H_4-S-(CH)_xR_yA$$

$$\bar{O}-C_6H_4-\overset{+}{S}\underset{}{\bigcirc}(CH)_xR_y\cdot H_2O\ (-H_2O) \longrightarrow H[O-C_6H_4-S-(CH)_xR_y]_2OH$$

R = H or Alk ;   R' = HO-, H₃CO-, CH₃COO- ;   R'' = NH₂ or Nalk or N-cycloalk or ArN-heterocycles ;   HA = H₂SO₄ or HClO₄ or HCl ;   x = 4 or 5 ;   y = 5 or 4.

A number of these compounds may be assumed to possess biological activity and to provide a convenient source of intermediates for fine organic synthesis.

III.   THIOPHENES OCCURRING IN OIL SHALES AND COALS

The organic matter of oil shales and coals has much in common as regards composition complexity and the principal classes of components; the main differences are confined to the ratio of higher and lower components, the latter being typical of crude oils. Petroleum heavy ends and residues resemble most closely the organic compounds from shales and coals. In these fossil fuels, an essential role is played by aromatic compounds, including those containing thiophene rings condensed to a greater or smaller extent with benzene rings. The enhanced content of polar compounds is likewise typical.

Because of the distinctive features of the organic moiety of solid fossil fuels, general schemes developed for the analysis of petroleum heavy ends and residuum have of late been applied in the analysis of the organic moiety in question, as exemplified by Ref. 92, which contains a chapter devoted to the methods of analysis of coal and coal products.[93] Unfortunately, the available data on the structure of native compounds of solid fossil fuels are scarce and until recently have been confined to the determination of mineral (pyrite and sulfate) and organic sulfur (overall content). In solid fossil fuel processing, the primary recourse was to high-temperature techniques, so that the volatile components formed were typical pyrogenic products containing unsaturated and aromatic hydrocarbons and admixtures of heterocyclic compounds of sulfur and nitrogen. Among sulfur compounds, the products identified were methyl- and polymethylthiophenes, some lower homologues thereof, and, in much higher concentrations, benzothiophenes with short side chains. From a practical point of view, commercial separation has been limited only to unsubstituted precursors of these compound series, namely, thiophene proper and benzothiophene.

Pilot plant production of synthetic liquid fuel from coal and shales generally relies on high-temperature processes involving, for the most part, the addition of free or combined hydrogen. The resultant product contains a minimum amount of organic sulfur compounds, predominantly polybenzologues of thiophenes noted for their enhanced thermal stability. Saturated heterocyclic compounds are represented in trace amounts; organic sulfur compounds in somewhat greater concentrations occur in the products of thermal or thermocatalytic processing of sour crude oil residues. Thus, Stekhun[94] investigated liquid products from coking the residues of Zapadnosibir crude oils and found, by the functional analysis method, that the sulfur compounds in the distillates collected from the boiling point to 190°C, from 190 to 360°C, and above 360°C had the overall sulfur content of 0.66, 1.71, and 2.10% by mass, respectively. Accordingly, they are composed to the extent of 78.8, 82.4, and 78.6% of compounds characterized by the presence of "residual sulfur," that is, predominantly of the derivative of thiophene and benzelogues thereof.

Galegos[95] resorted to gas chromatography in combination with mass spectrometry and found that the products of fluidized bed, catalytic cracking of crude oil contained thiophene and its alkyl derivatives; all of the 8 theoretically possible methyl-, dimethyl-, and ethylthiophenes; 9 of the 12 C_3-substituted thiophenes;

and 7 of the 31 C_4-substituted thiophenes. The latter were characterized by mass numbers without establishing their structure, since no pertinent characteristics of individual compounds were available to the author.[95]

As early as 1964, Wingerter and Prinzler[96] investigated sulfur compounds in light oil produced from Böhlen brown coal (GDR). Acid and base components were preliminarily removed from the light oil; the neutral oil thus obtained containing 3.33 mass % sulfur. As a result of gas–liquid chromatography on silica gel, the light oil was separated into six fractions of the following composition: fractions 1 and 2 — naphthenes and paraffins; fraction 3 — olefins and aromatic hydrocarbons; fraction 4 — aromatic hydrocarbons and thiophenes; fraction 5 — thiophenes and sulfides; fraction 6 — sulfides and nitrogen and oxygen compounds. Fractions 3–6 were subjected to low-temperature extraction with a solution of sulfur dioxide in butane and the resultant extracts were chromatographed on silica gel. After vacuum distillation and chromatographic purification, thiophenes were precipitated by mercuric chloride from fractions 4 and 5, and the complexes formed were treated with $6N$ hydrochloric acid and distilled with steam. The sum of thiophenes underwent fine rectification and then was separated by gas–liquid chromatography. As a result, the following compounds were identified: 2- and 3-methylthiophenes, 2- and 3-ethylthiophenes, 2,3-, 2,4-, 2,5-, and 3,4-dimethylthiophenes, thiophene, and 2-methylthiophane. The overall content of thiophenes was 4.2%, based on the neutral oil. 2-Methylthiophene is the major component (1.56%), and the sum of thiophenes with alkyl substituents at position 2 equals 3.2%. The minor components are 3-ethylthiophene (0.01%), thiophene (0.03%), and 3-methylthiophene (0.05%).

The information regarding sulfur in the products of coal desulfurization, the methods of coal analysis, and the origin of sulfur in coals were summarized in Ref. 92, in which Chakrabarti notes that in coals formed under fresh water conditions, the content of sulfur is low compared with that in coals formed in saline water or in estuaries. The content of sulfur in coals varies over a wide range, from under 1% to over 10%. For example, Rasa coal from Istria (Yugoslavia) contains about 11% sulfur. It is further noted that in Assam coals (India), organic sulfur compounds are represented, according to Ref. 97, by thiols, alkyl sulfides, and disulfides, and only in part by cyclic compounds. The content of cyclic sulfur compounds appears to be underrated, owing to the absence of modern methods of investigation.

Dooley et al.[93] cite a detailed scheme of investigating the composition of liquid products produced from coals, which comprises a modification of the scheme developed earlier for the study of heavy ends and residues of crude oils.[29] The results obtained are indicative of the presence of benzo(b)thiophene and higher benzologues of thiophene. However, no data about alkyl- and cycloalkylthiophenes are presented. Recent publications[98,99] devoted to the analytical characterization of liquid products of coal processing, synthetic fuels, and the products of shale processing emphasize the absence of alkyl- and cycloalkylthiophenes in the specimens tested, in which benzo(b)thiophene and its benzologues are the predominant sulfur compounds. That the absence of thiophenes with long alkyl chains results in many instances from their thermal beta-degradation, which involves the formation of alkenes and methylthiophenes passing into low-yield light fractions, cannot be ruled out. The concentration of thiophenes in the Colorado shale oil fraction boiling below 200°C was studied by Dinneen and presented in a report[100] summarizing research in this field carried out at the U.S. Bureau of Mines in the 1952–1961 period. This fraction contained thiophene (0.001%), methyl-, dimethyl-, and trimethylthiophenes, 2-ethylthiophene, 2-isopropylthiophene, and a number of di- and trisubstituted thiophenes; the total number of alkylthiophenes identified was 17. Among alkylthiophenes, 13 compounds have alkyls in the alpha-position to the sulfur atom, including a number of 2,5-disubstituted compounds. 2-Methyl-5-ethylthiophene and 2-methyl-5-isopropylthiophene are present in the highest concentrations, 0.024 and 0.021%, respectively. The concentration of 3-methylthiophene is the lowest (0.0008%) and benzothiophene accounts for 0.005%. Thiophenes, benzothiophenes, and higher benzologues and naphthenologues of thiophene are contained in gas oil with a mean molecular mass of 335.

Eisen and Rang investigated organic sulfur compounds in Esthonian shale gasoline[101] produced in tunnel ovens from shales of the Kiviyli deposit. The 60–150°C fraction was found to contain 1.19% of thiophenes, among them 0.06% of thiophene itself, 0.28% of 2,5-dimethylthiophene, 0.22% of 2-ethylthiophene, 0.13% of 3-methylthiophene, and 0.04% of 1,4-dimethylthiophene. Qualitative analysis showed that the 150°C fraction contained 2-n-propyl-, 3-isopropyl-, and 2-methyl-5-ethylthiophenes.

Klein reports[102] that the Colorado shale oil contains all possible monoalkyl derivatives of thiophene with C_1–C_4 alkyls and points out that the structure of organic sulfur compounds in the shale oil has little relevance to the parent kerogen, insofar as these compounds appear because pyrolysis of the compounds yields various degrees of thermal stability. Some of these compounds evolve hydrogen sulfide at the early stages of pyrolysis; other compounds undergo degradation at higher conversion degrees and yield thermostable heterocyclic compounds.

Clugston et al.[103] studied sulfur compounds in the gas oil fraction of the Atabaska and Cold Lake oil-bearing sands and the heavy oil from the Loadminster and Medicine River deep-seated limestone deposits. The dominant sulfur compounds in aromatic hydrocarbons with which they are associated were separated by distillation-simulating gas chromatography and gas chromatography with subsequent mass-spectroscopic characterization. The predominant form of organic sulfur compounds comprised alkyl derivatives of benzo- and dibenzothiophene with short side chains, among which some isomers are most prominent. No alkyl- and cyclo-alkylthiophenes were discovered.

Studies carried out by Pailer and associates[104-107] on the products of low-temperature carbonization of oil shale from Seefeld (Tirol) are of unique and unmatched interest. Oil shale raffinate (designated as Friedrich 111) was submitted by the Austrian Ichtyol Society. The raffinate, freed of acids and bases, was distilled, and the residue boiling above 140°C/11 torr and comprising 26.2% of the raffinate charged was fractionated.

The selected fractions were separated into components by the following six-stage procedure: (1) column chromatography on alumina, (2) column chromatography on silica gel, (3) column chromatography on silica gel impregnated with 10% AgNO$_3$, (4) column chromatography on silica gel impregnated with 10% picric acid, (5) thin-layer chromatography on silica gel impregnated with picric acid, and (6) preparative gas chromatography.

The identification of the separated components was performed using nmr, IR spectroscopy, and near-UV spectroscopy.

In fraction 2 (sulfur content, 19.25%), benzothiophene, 2-methyl-5-phenylthiophene, and 5-methyl-2,2'-thienyl were identified. Fraction 4 (53 g), boiling in the 97–100°C/1 torr range, was separated into nine subfractions, among which subfraction 4 contained thiophenes with long alkyl chains not discussed below. Subfraction 5 contains diethylbenzo(b)thiophene, benzothiophenes with C$_9$–C$_{10}$ alkyl chains, diethylphenylthiophene, and benzothiophenes with C$_4$–C$_5$ alkyl chains. The separation and identification procedures are described in detail in Ref. 104.

Fraction 7 was separated on alumina into 1000 eluates combined to form 25 subfractions; the eluents were cyclohexane with progressively increasing additions of benzene. The eluates were combined using gas chromatography data, and chromatographic characteristics served as a basis for selecting 12 subfractions to be subjected to detailed analysis. The sum of subfractions 5 and 6 contains 20 isomeric homologues of thiophene with C$_{10}$–C$_{12}$ alkyl side chains, with some of the isomers having a long n-alkyl (or slightly branched) chain. The resolution of isomers presented difficulties because of their large number. Subfraction 10 contains phenylthiophene, presumably of the beta series, 2-(1'-methyl)indanyl-5-methylthiophene and n-butyl, isoamyl, n-heptyl, and heptyl benzo(b)thiophenes. Of particular interest is the presence of 2,2'-dithienylmethane derivatives I–III. Subfractions 12–14, apart from benzo- and thienothiophenes, contain 2-phyenyl-5-isoamylthiophenes and a series of 2,2'-dithienyl derivatives (IV–IX). In subfraction 18, compounds X–XV occur.

XIV.

XIII.

XV.

XVI.

XVII.

XVIII.

XIX.

XX.

XXI.

XXII.

XXIII.

XXIV.

XXV.

Subfractions 11 and 12 contain exclusively condensed dibenzo- and naphtho-thiophenes, compounds XVI and XVII and also oxygen compounds (dibenzofuran and 2,6-dimethyldibenzodioxane 1,4) (see Ref. 105).

Subfractions 1–4 are almost completely devoid of sulfur. Subfractions 5 and 6 presumably contain thiophenes with long alkyl chains that undergo separation before polyalkylated thiophenes and/or benzothiophenes containing six or seven C atoms in side chains. Compound XVIII and diethyl- and methylethyldithienyls with the unidentified position of alkyl groups are also present. The presence of 2,3-dimethyl-5-(1-methylindanyl)-thiophene (XXV) and of dibenzothiophenes was established in subfraction 10; subfraction 11 was found to contain 2-phenyl-benzothiophene, dithienyls with C_4 and C_5 side chains, phenylbenzothiophene, and methylphenylbenzothiophene. Compound III and a mixture of dithienylethanes and propanes, which were not investigated further, and also 1,2-napthothiophene and its 5-methyl derivative and compound XIX were also found in subfraction 11. In subfraction 14, a series of dithienylmethane and 1,2-dithienylethane derivatives (XX–XXV) are observed (see Ref. 106).

The fraction of neutral shale oil boiling in the 80–190°C/3.5 torr range was further separated into 14 fractions, from which subfraction 12 was selected for detailed investigation.

Subfraction 12 (168–175°C/3.5 torr) was found to contain mono-, di-, and trialkylthiophenes. The monosubstituted compounds are characterized by $C_{15}H_{31}$–$C_{18}H_{37}$ alkyls, whereas in the trisubstituted thiophenes, the sum of three alkyl groups varies in the same range. Also identified were the series of mono- and trisubstituted 2-phenyl-thiophenes, disubstituted 2,2-dithienyls, monosubstituted 2,2-di-thienylmethanes (C_3–C_8 alkyls), thienylbenzothiophenes, and benzothiophenyl- and thiophenylmethanes, as well as a number of tricyclic condensed compounds having one or two thiophene rings. Details are given in Ref. 107.

Pailer's studies clearly demonstrated the presence in oil shale of complicated sulfur compounds that might be regarded as precursors of lower thiophenes in the products of rigid thermolysis of shales and shale oils.

It is reasonable to assume that a careful examination of the organic matter of other shales and coals having different metamorphism grades — provided recourse is made to mild methods of extraction and extract investigation — would bring us closer to elucidation of the nature of organic compounds in fossil fuels, their metamorphism under native conditions, and transformation under the conditions of thermal and thermocatalytic processing. Such studies would provide a better insight into the biological activity of a plurality of organic sulfur compounds obtained for the first time from natural sources. It should be emphasized that these investigations involve the synthesis and resynthesis of numerous, heretofore unknown, individual compounds, the study of which is fundamental to the progress of organic chemistry.*

REFERENCES

1. H. T. Rall, C. J. Thompson, H. J. Coleman, and R. L. Hopkins, *Sulfur Compounds in Crude Oil*, U.S. Bureau of Mines Bulletin 659, Washington, DC, 1972, p. 187.
2. H. M. Smith, *Crude Oil: Qualitative and Quantitative Aspects. The Petroleum World*, U.S. Bureau of Mines Information Circular 8286, Washington, DC, 1966, p. 41.
3. C. J. Thompson, 9th International Symposium on Organic Sulfur Chemistry, Abstracts, Riga, 1980, p. 32.
4. S. F. Birch, *J. Inst. Petrol*, **39**, 185 (1953).
5. S. F. Birch and D. T. McAllan, *J. Inst. Petrol.*, **37**, 443 (1951).
6. S. F. Birch, T. V. Cullum, R. A. Dean, and R. L. Denuer, *Ind. Eng. Chem.*, **47**, 240 (1955).
7. L. Wilson Orr, *Anal. Chem.*, **38**, 1559 (1966).
8. T. Kaimai and A. Matsunaga, *Anal. Chem.*, **50**, 268 (1978).
9. J. Escalier, J. P. Massone, and M. Mariche, *Analysis*, **7**, 58 (1979).
10. M. J. Lamathe, *Compt. Rend. Acad. Sci.* (*Paris*), **263**, 872 (1966).
11. M. J. Lamathe, *Chim. Anal.*, **49**, 119 (1967).
12. T. A. Bardina, E. N. Karaulova, N. N. Bezinger, and G. D. Galpern, *Zh. Anal. Khim.* (*Moscow*), **35**, 2045 (1980).

* In this connection, it appears pertinent to recommend the monograph *Chromatography in Petroleum Analysis*,[108] which might be useful to workers engaged in the study of thiophenes in fossil fuels and their products.

13. T. A. Bardina, E. N. Karaulova, and G. D. Galpern, 9th International Symposium on Organic Sulfur Chemistry, Abstracts, Riga, 1980, p. 213.

14. J. W. Vogh and J. E. Dooley, *Anal. Chem.*, **47**, 816 (1975).

15. Ya. B. Chertkov, V. G. Spirkin, and V. N. Demishev, *Neftekhimiya*, **5**, 741 (1965).

16. V. G. Spirkin and Ya. B. Chertkov, *Neftekhimiya*, **8**, 453 (1968).

17. N. K. Lyapina, M. A. Parfenova, T. S. Nikitina, A. A. Valtsova, and V. S. Nikitina, *Neftekhimiya*, **19**, 921 (1979).

18. V. S. Nikitina, N. K. Lyapina, F. G. Sattarova, N. S. Lyubopytova, and M. A. Parfenova, *Neftekhimiya*, **11**, 264 (1971).

19. V. S. Nikitina, N. K. Lyapina, and A. D. Ulendeeva, *Neftekhimiya*, **10**, 594 (1970).

20. M. Pailer, W. Oesterreicher, and E. Simonitsch, *Mh. Chem.*, **96**, 48 (1965).

21. D. E. Hirsch, A. L. Hopkins, H. J. Coleman, F. A. Cotton, and C. J. Thompson, *Anal. Chem.*, **44**, 915 (1972).

22. M. F. Bondarenko, M. A. Pais, and Z. I. Abramovich, *Neftekhimiya*, **17**, 904 (1977).

23. M. A. Pais, V. S. Bogdanov, M. F. Bondarenko, and G. V. Portnova, *Neftekhimiya*, **20**, 607 (1980).

24. D. I. Kondakov, N. K. Lyapina, V. S. Nikitina, A. A. Smarkalov, Yu. E. Nikitin, and M. A. Parfenov, *Khimiya i Fizika Nefti i Neftekhimicheskii Sintez* (Chemistry and Physics of Petroleum and Petrochemical Synthesis), Inst. Khim. Bashkir. Fil. Akad. Nauk SSSR, Ufa, 1976, p. 224.

25. W. K. Seifert, in *Progress in the Chemistry of Organic Natural Products*, Serial No. 32, Springer Verlag, New York, 1975, p. 49.

26. H. J. Coleman, J. E. Dooley, D. E. Hirsch, and C. J. Thompson, *Anal. Chem.*, **45**, 1724 (1973).

27. C. J. Thompson, J. E. Dooley, D. E. Hirsch, and C. C. Ward, *Hydrocarbon Process.*, **52**(9), 123 (1973).

28. J. E. Dooley, C. J. Thompson, D. E. Hirsch, and C. C. Ward, *Hydrocarbon Process.*, **53**(4), 93 (1974).

29. J. E. Dooley, E. E. Hirsch, and C. J. Thompson, *Hydrocarbon Process.*, **53**(18), 141 (1974).

30. C. J. Thompson, J. E. Dooley, J. W. Vogh, and D. E. Hirsch, *Hydrocarbon Process.*, **53**(8), 93–98 (1974).

31. J. E. Dooley, R. L. Hopkins, D. E. Hirsch, H. J. Coleman, and C. J. Thompson, U.S. Bureau of Mines Report 7770, Washington, DC, 1973, p. 25.

32. J. E. Dooley, D. E. Hirsch, H. J. Coleman, and C. J. Thompson, U.S. Bureau of Mines Report 7821, Washington, DC, 1973, p. 30.

33. D. E. Hirsch, J. E. Dooley, H. J. Coleman, and C. J. Thompson, U.S. Bureau of Mines Report 7893, Washington, DC, 1974, p. 28.

34. C. J. Thompson, J. E. Dooley, J. W. Vogh, and D. E. Hirsch, U.S. Bureau of Mines Report 7945, Washington, DC, 1974, p. 25.

35. J. E. Dooley, D. E. Hirsch, C. J. Thompson, and C. C. Ward, *Hydrocarbon Process.*, **53**(11), 187 (1974).

36. C. J. Thompson, C. C. Ward, and J. C. Ball, Bartlesville Energy Research Center Report 76/8, 1976, p. 28.

37. H. V. Drushel and A. L. Sommers, *Anal. Chem.*, **39**, 1819 (1967).

38. M. A. Parfenova, V. S. Nikitina, K. K. Lyapina, T. S. Nikitina, A. A. Smarkalov, and R. A. Shaigardanova, *Khimiya i Fizika Nefti i Neftekhimicheskii Sintez* (Chemistry and Physics of Petroleum and Petrochemical Synthesis), Inst. Khim. Bashkir. Fil. Akad. Nauk SSSR, Ufa, 1976, pp. 19–27.

39.   L. A. Melnikova, N. K. Lyapina, and L. R. Karmanova, *Neftekhimiya,* **20,** 612 (1980).

40.   N. V. Agadzhanova, N. K. Lyapina, R. B. Alieva, V. S. Shmakov, and M. A. Parfenova, *Neftekhimiya,* **23,** 424 (1983).

41.   D. A. Vyakhirev, L. E. Reshetnikova, G. Ya. Malkova, and N. I. Malyugina, in *Khimiya Seroorganicheskikh Soedinenii Soderzhashchikhsya v Nefti i Nefteproduktakh* (Chemistry of Organosulfur Compounds Contained in Petroleum and Petroleum Products), Vol. 9, Vysshaya Shkola Publishers, Moscow, 1972, p. 373.

42.   A. K. Yushkovich, T. A. Danilova, and E. A. Victorova, *Neftekhimiya,* **22,** 689 (1982).

43.   T. Yu. Filipova, Kh. M. Minachev, Ya. I. Isakov, M. G. Vagabov, and E. A. Karakhanov, *Vestn. Mosk. Gos. Univers., Ser. Khim.,* **22,** 511 (1982).

44.   M. V. Vagabov, S. K. Dzamalov, E. A. Karakhanov, and E. A. Victorova, *Neftekhimiya,* **21,** 64 (1981).

45.   S. K. Dzhamalov, M. V. Vagabov, E. A. Victorova, and E. A. Karakhanov, *Vestn. Mosk. Gos. Univers., Ser. Khim.,* **19,** 225 (1978).

46.   A. Giraud and M. A. Bestongeff, *J. Gas Chromatogr.,* 464 (1967).

47.   M. A. Bestongeff and D. Joly, *Proc. 7th World Petrol. Congr. Mexico City, 1967,* **9,** 1929 (1968).

48.   P. A. Quinn, J. Swanson, H. U. C. Mekzelaar, and P. G. Kistenmaker, in *Analytical Pyrolysis,* (C. E. R. Jones and C. A. Cramers, Eds.), Elsevier, Amsterdam, 1977, p. 408.

49.   N. K. Lyapina, Z. N. Parnes, G. A. Tolstikov, M. A. Parfenova, and V. S. Smarkalov, *Neftekhimiya,* **22,** 693 (1982).

50.   V. G. Lukuaniza and G. D. Galpern, *Izvest. Acad. Nauk SSSR, Otdel Khim. Nauk,* 130, (1956).

51.   E. K. Karaulova and G. D. Galpern, *Khim. Topl.,* (9) 39, (1956).

52.   G. D. Galpern, T. A. Bardina, L. A. Barykina, T. S. Bobruyskaya, E. S. Brodskii, and E. N. Karaulova, *Seroorganicheskie Soedineniya* (Organosulfur Compounds), Vol. 1, (Report at the 12th Scientific Session on the Organosulfur Compounds in Petroleum, Riga, 1971), Zinatne, Riga, 1976, p. 42.

53.   G. D. Galpern, *Int. J. Sulfur Chem. B,* 6(2), 115 (1971).

54.   G. D. Galpern, T. S. Bobruyskaya, E. S. Brodskii, T. A. Bardina, E. N. Karaulova, R. A. Khmeltskii, and A. A. Polyakova, *Neftekhimiya,* **10,** 743 (1970).

55.   I. U. Numanov and G. P. Nasyrov, *Geteroatomnye Componenti Neftei iz Tadzhikskoi Vpadini* (Heteroatomic Components in Crude Oils from Tadzhik Depression), Donish, P. H., Dushanbe, 1973, p. 259.

56.   G. D. Galpern, G. P. Girina, and V. G. Lukyaniza, in *Metody Analiza Organicheskikh Soedinenii Nefti, ikh Smesei i Proizvodnikh* (Methods of Analysis of Petroleum Organic Compounds, Their Mixtures and Derivatives), Acad. Nauk SSSR P.H. Moscow, 1960, p. 58.

57.   G. D. Galpern, E. N. Karaulova, T. A. Bardina, and T. S. Bobruyskaya, 3rd Organic Sulfur Symposium, Abstracts, University of Caen, France, 1967, p. 66.

58.   E. N. Karaulova, G. D. Galpern, and T. A. Bardina, USSR Inventor's Certificate 186,454 (1966); priority, November 27, 1965.

59.   E. N. Karaulova, G. D. Galpern, T. A. Bardina, and A. S. Kharitonov, USSR Inventor's Certificate 206,579 (1967); priority, January 9, 1967.

60.   T. P. Burmistrova, T. A. Bardina, G. D. Galpern, E. N. Karaulova, N. A. Luchai, N. N. Terpilovskii, and A. A. Khitrik, USSR Inventor's Certificate 322,996 (1971); priority, September 13, 1971.

61.   T. P. Burmistrova, A. A. Khitrik, N. N. Terpilovskii, G. D. Galpern, E. N. Karaulova, and T. A. Bardina, USSR Inventor's Certificate 392,687 (1973); priority, May 7, 1973.

62.   T. P. Burmistrova, T. A. Bardina, G. D. Galpern, E. N. Karaulova, N. A. Luchai, N. N.

Terpilovskii, and A. A. Khitrik, US Patent No. 3,792,095 (February 12, 1974); British Patent No. 1,339,318 (August 9, 1971); French Patent No. 2,131,209 (September 10, 1972); BRD Patent No. 2,140,293 (December 5, 1974); DDR Patent No. 91,025 (June 5, 1972); priority, September 11, 1971.

63. L. M. Sagryatskaya, R. M. Masagutov, A. Ch. Shapiro, M. F. Bondarenko, T. P. Burmistrova, R. Sh. Latypov, A. A. Khitrik, and Z. A. Kireeva, *Neftekhimiya*, **14**, 765 (1974).

64. T. P. Burmistrova, A. A. Khitrik, G. D. Galpern, E. N. Karaulova, P. Sh. Latypov, N. N. Terpilovskii, N. Z. Gilmanshina, and A. Kh. Shapiro, USSR Inventor's Certificate 469,326 (January 17, 1975); priority, September 14, 1972.

65. Yu. E. Nikitina, A. P. Kapina, Yu. I. Murinov, V. G. Ben'kovskii, and N. K. Lyapina, *Khimiya i Fizika Nefti i Neftekhimicheskii Sintez* (Chemistry and Physics of Petroleum and Petrochemical Synthesis), Inst. Khim. Bashkir. Fil. Akad. Nauk, SSSR, Ufa, 1976, p. 28.

66. Yu. E. Nikitin, N. K. Lyapina, V. G. Ben'kovski, N. I. Antipov, K. I. Boldov, L. S. Vlasov, A. K. Ivanov, V. V. Bulantsev, and V. A. Mikhailov, USSR Inventor's Certificate 397,514 (1974).

67. A. B. Sviridova, V. I. Laba, and E. N. Prilezhayeva, *Zh. Org. Khim.* (*USSR*), **7**, 2480 (1971).

68. E. N. Karaulova, T. A. Bardina, G. D. Galpern, and T. S. Bobruyskaya, *Neftekhimiya*, **6**, 480 (1966).

69. N. N. Besinger, M. A. Abdurakhmanov, and G. D. Galpern, *Neftekhimya*, **1**, 149 (1961).

70. F. N. Mazitova, N. A. Iglamova, and R. P. Kondrat'eva, *Neftekhimiya*, **16**, 631 (1976).

71. N. A. Iglamova, F. N. Mazitova, A. A. Vafina, and A. V. Iliasova, *Neftekhimiya*, **19**, 264 (1979).

72. N. A. Iglamova, F. N. Mazitova, and E. S. Brodskii, *Neftekhimiya*, **22**, 407 (1982).

73. S. N. Litvinenko, G. P. Grigor'eva, N. G. Sanina, A. M. Malkov, and I. F. Tikhonov, *Khim. Tekhnol. Topl. Masel*, **15**(7), 18 (1970).

74. G. E. Radzeva, N. N. Ryakhovskaya, A. Kh. Shapiro, and L. M. Zagryadskaya, in *Tezisi Dokladov na 14 Nauchnoi Sessii po Khimii i Technologii Organicheskikh Soedinenii Sery i Sernistykh Neftei* (Abstracts of Papers Presented at the 14th Scientific Session on Chemistry and Technology of Organosulfur Compounds and Sulfur-Bearing Crude Oils), Zinatne, Riga, 1976, p. 103.

75. G. A. Kashafutdinov, F. N. Mazitova, and N. A. Iglamova, in *Tezisi Dokladov na 14 Nauchnoi Sessii po Khimii i Technologii Organicheskikh Soedinenii Sery i Sernistykh Neftei* (Abstracts of Papers Presented at the 14th Scientific Session on Chemistry and Technology of Organosulfur Compounds and Sulfur-Bearing Crude Oils), Zinatne, Riga, 1976, p. 70.

76. T. V. Garipov and D. K. Chervyakov, in *Tezisi Dokladov na 14 Nauchnoi Sessii po Khimii i Technologii Organicheskikh Soedinenii Sery i Sernistykh Neftei* (Abstracts of Papers Presented at the 14th Scientific Session on Chemistry and Technology of Organosulfur Compounds and Sulfur-Bearing Crude Oils), Zinatne, Riga, 1976, p. 68.

77. I. U. Numanov, V. P. Chayko, B. A. Borovkov, V. G. Kovalenkov, P. M. Stepanov, and N. T. Radzhabov, in *Tezisi Dokladov na 13 Nauchnoi Sessii po Khimii i Tekhnologii Organicheskikh Soedinenii Sery i Sernistykh Neftei* (Abstracts of Papers Presented at the 13th Scientific Session on Chemistry of Organosulfur Compounds and Sulfur-Bearing Crude Oils), Zinatne, Riga, 1974, p. 67.

78. V. A. Pronin, M. V. Usolzeva, S. M. Shostakovskii, N. S. Nikolskii, and S. I. Syunyaev, in *Tezisi Dokladov na 13 Nauchnoi Sessii po Khimii i Tekhnologii Organicheskikh Soedinenii Sery i Sernistykh Neftei* (Abstracts of Papers Presented at the 13th

Scientific Session on Chemistry of Organosulfur Compounds and Sulfur-Bearing Crude Oils), Zinatne, Riga, 1974, p. 267.

79.   A. I. Nikolaev and A. G. Babkin, in *Tezisi Dokladov na 13 Nauchnoi Sessii po Khimii i Tekhnologii Organicheskikh Soedinenii Sery i Sernistykh Neftei* (Abstracts of Papers Presented at the 13th Scientific Session on Chemistry of Organosulfur Compounds and Sulfur-Bearing Crude Oils), Zinatne, Riga, 1974, p. 268.

80.   A. G. Babkin and A. I. Nikolaev, in *Tezisi Dokladov na 13 Nauchnoi Sessii po Khimii i Tekhnologii Organicheskikh Soedinenii Sery i Sernistykh Neftei* (Abstracts of Papers Presented at the 13th Scientific Session on Chemistry of Organosulfur Compounds and Sulfur-Bearing Crude Oils), Zinatne, Riga, 1974, p. 270.

81.   Yu. V. Itkin, A. M. Reznik, M. Ya. Shpirt, and L. D. Iwchenko, *Zh. Prikl. Khim.* (*Leningrad*), **48**, 1510 (1975).

82.   Yu. E. Nikitin, N. L. Egutkin, and Yu. I. Murinov, *Zh. Prikl. Khim.* (*Leningrad*), **48**, 51 (1975).

83.   R. Pummerer, *Bericht,* **43**, 411 (1910).

84.   E. N. Karaulova, G. D. Galpern, V. D. Nikitina, L. R. Barykina, I. V. Cherepanova, D. K. Zhestkov, F. V. Kozlova, and G. Yu. Pek, *Neftekhimiya,* **10**(4), 599 (1970).

85.   E. N. Karaulova, G. D. Galpern, C. D. Nikitina, I. V. Cherepanova, and L. R. Barykina, *Neftekhimiya,* **12**(1), 104 (1972).

86.   E. N. Karaulova, USSR Inventor's Certificate 335,941 (published 1973); priority, December 1970.

87.   T. S. Bardina, E. N. Karaulova, and G. D. Galpern, USSR Inventor's Certificate 681,054 (1979); priority, May 31, 1976.

88.   E. N. Karaulova, T. N. Bobruyskaya, G. D. Galpern, V. D. Nikitina, and L. R. Barykina, *Neftekhimiya,* **23**(2), 259 (1983).

89.   E. N. Karaulova, G. D. Galpern, V. D. Nikitina, T. A. Bardina, and L. M. Petrova, *Khim. Geterozikl. Soed.* (*Riga*), 1479 (1973).

90.   E. N. Karaulova, G. D. Galpern, T. S. Bobruyskaya, and V. D. Nikitina, *Dokl. Akad. Nauk SSSR,* **216**, 91 (1974).

91.   T. A. Bardina, L. R. Barykina, I. N. Dehtyareva, and G. D. Galpern, USSR Inventor's Certificate 836,008 (1981); priority, July 2, 1979.

92.   Cl. Karr, Jr. (Ed.), *Analytical Methods for Coal and Coal Products,* Vol. 1, Academic Press, New York, 1978, p. 580.

93.   J. E. Dooley, C. I. Thompson, and S. E. Sheppele, in *Analytical Methods for Coal and Coal Products* (Cl. Karr, Jr., Ed.), Academic Press, pp. 467–498.

94.   A. I. Stekhun, *Khim. Tekhnol. Topl. Masel,* **20**, 25 (1975).

95.   E. J. Gallegos, *Anal. Chem.,* **47**, 1150 (1975).

96.   K. H. Wingerter and H. P. Prinzler, *Chem. Technol.,* **16**, 473 (1964).

97.   J. K. Chowdhlury, P. B. Datta, and S. R. Ghosh, *J. Sci. Indian Res.,* **11**B, 150 (1952).

98.   Ch. Willey, M. Iwas, R. N. Castle, and M. L. Lee, *Anal. Chem.,* **53**, 400 (1981).

99.   D. W. Later, M. L. Lee, K. D. Bartle, R. C. Kong, and D. L. Wassilaros, *Anal. Chem.,* **53**, 1612 (1981).

100. G. U. Dinneen, *Proc. Am. Petrol. Inst.,* **42**(VIII), 41 (1962).

101. O. G. Eisen and S. A. Rang, in *Khimiya Seroorganicheskikh Soedinenii Soderzhashchikhsya v Nefti i Nefteproducktakh* (Chemistry of Organosulfur Compounds Contained in Petroleum and Petroleum Products) Vol. 6, Khimiya Publications, USSR, 1964, p. 121.

102. R. F. Klein, in *Developments in Petroleum Science,* Vol. 5, Teh Tie Yen and G. V. Chilingarian, Eds.), Elsevier, Amsterdam, 1976.

103. D. M. Clugston, A. E. George, D. S. Montgomery, G. T. Smiley, and H. Sawatsky in *Advances in Chemistry,* No. 151, American Chemical Society, Washington, DC, 1976, pp. 11–27.

104. M. Pailer and H. Bergretter, *Mh. Chem.,* **104,** 297 (1973).

105. M. Pailer and H. Grünhaus, *Mh. Chem.,* **104,** 312 (1973).

106. M. Pailer and L. Berner-Fenz, *Mh. Chem.,* **104,** 339 (1973).

107. M. Pailer and V. Hozek, *Mh. Chem.,* **106,** 1259 (1975).

108. K. H. Altgelt and T. H. Gown (Eds.), *Chromatography in Petroleum Analysis,* Vol. 15 in Chromatographic Sciences Series, Marcel Dekker Inc., New York, 1979.

CHAPTER   V

Pharmacologically Active Compounds and other Thiophene Derivatives

JEFFERY B. PRESS

Cardiovascular–CNS Research Section, American Cyanamid Company,
Medical Research Division, Lederle Laboratories, Pearl River, New York

Present affiliation of Dr. Press: Ortho Pharmaceutical Corporation, Raritan, New Jersey.

I. INTRODUCTION

The role of thiophene derivatives in the design and synthesis of pharmacologically important molecules has grown enormously since 1950. The advent of economical commercial sources of thiophene at that time led to increased study of its chemical properties. The development of new synthetic methods beginning in the 1960's led to new procedures for the synthesis of novel thiophene derivatives. Concurrent with this increase in the knowledge of thiophene chemistry, medicinal chemistry began to mature with rational drug design. As a consequence, many new thiophene derivatives were synthesized, and numerous insights into the modes of action of biological agents have been gained since the first major review of the pharmacology of thiophene and its derivatives.[1]

During the last three decades, numerous reviews of thiophene, including some coverage of biological activity, have appeared. Martin-Smith and Reid[2] wrote a detailed account of progress in the 1950's, which is relied upon heavily in this review. Additionally, surveys by Nord,[3] Nobles,[4,5] and Böhm and Zieger[6] provide useful updates of progress in this area. Gronowitz has also published several reviews.[7-11]

There are a variety of reasons why thiophene derivatives are interesting to the pharmaceutical chemist. One of the oldest is based upon the concept of bioisoterism, as developed by Erlenmeyer.[12,13] Thiophene, with its six-π electron aromaticity, is electronically and sterically similar to benzene (as well as furan and pyrrole). As a result, thiophene analogues of biologically active benzene derivatives may well exhibit similar activities. At the same time, the presence of a heteroatom, or the lower resonance energy in thiophene, may alter its metabolic fate; thus, the thiophene derivative may have less toxic effects and/or a better therapeutic profile.

A second reason that thiophene derivatives are of interest in medicinal chemistry lies in the development of structure–activity relationships. Many examples exist in the literature where various electron donating or withdrawing substituents on a benzene ring system are required to maximize activity. Since thiophene behaves as a electron-rich aromatic system, it may be superior to substituted benzenes in certain cases. In addition, the thiophene derivatives often provide chemical novelty and, thus, patentability to systems known to have pharmacological activity. This patent novelty is sometimes sufficient for the pharmaceutical industry to develop these agents.

Thiophene is used in two principal ways by the medicinal chemical investigator. The most interesting chemically is the use of thiophene either as the central ring or as part of a central fused ring system. These compounds clearly provide the greatest challenge to the synthetic organic chemist and the utmost insight into the effects thiophene causes upon biological activity. Many biological agents of this type have been prepared, based upon active benzene derivatives. In recent years, molecules have been prepared to investigate inherent properties and have been based less upon benzene isosteres. The second use of thiophene to the medicinal chemist is the replacement of pendent aromatic rings on biologically important molecules with thiophene. Clearly, this is a less challenging problem chemically, but is important to the development of structure–activity relationships in aryl-substituted systems.

This chapter attempts to cover reports of biologically active thiophene derivatives since the summary by Blicke[1] in the early 1950's through 1981. The magnitude of such a project places some limitations upon its scope. In general, unless a thiophene derivative has obvious biological importance that is not reported in the routine chemical literature, patent coverage is omitted. This is a reasonable omission, because frequently chemical and, more importantly, biological information is incompletely revealed in the patent literature, and thus little useful information for the purposes of this chapter is provided. It is the intent here to report all thiophene systems with confirmed biological activity or thiophene isosteres of known active benzene systems. Clearly, such a project is difficult, and this author apologizes in advance for omissions of pertinent information.

The major topic areas in this chapter follow the format used by the American Chemical Society in the highly instructive *Annual Reports in Medicinal Chemistry* series. Some areas of biological investigation overlap and placement of these subjects is at the whim of the author. Furthermore, some agents have more than one type of activity but, in general, will be mentioned only in one place as a result of space considerations.

Several observations and conclusions have been made previously and should be mentioned in advance. Thiophene derivatives in general have been more toxic than their benzene counterparts and have had lower activity.[2,3,5] This observation seemed to slow reports of thiophene medicinal chemical research in the 1960's and early 1970's. As thiophene derivatives have become less tied to benzene counterparts, this "increased toxicity" seems to be of less concern. Second, the 3-thiophene derivatives are more active and less toxic than the 2-isomers.[2,4,14] These comparisons were made in the late 1950's when a sufficient number of 2- and 3- isomers of active compounds had been prepared. The difficulty of producing 3-substituted thiophenes has frequently prevented the formation of the various isomers for direct comparison, but clearly all thiophene isomers should be prepared in order to arrive at valid conclusions about the effects of thiophene on biological activity and toxicity.

II.   CENTRAL NERVOUS SYSTEM THERAPY

Effective psychotherapeutic drug development paralleled the growth in importance of thiophene in pharmaceutical chemistry. The discovery of chlorpromazine, imipramine, and chlorodiazepoxide as useful therapeutic agents for the treatment of schizophrenia, depression, and anxiety, respectively, led to the era of psychopharmacology in the 1950's. Several excellent reviews have appeared on these subjects.[15,16] Not long after these discoveries, thiophene derivatives were already being investigated for their role in the treatment of psychic disorders.

1. Antipsychotic Agents

The development of chlorpromazine in 1952 as an effective treatment for schizophrenia demonstrates the vagaries and frustrations of drug research. Seemingly minor modifications of the antihistaminic drug phenergan led to chlorpromazine, which caused unexpected pharmacological effects in laboratory animals. Chlorpromazine was placed into clinical trials as an antiemetic agent, because of its anticholinergic properties. It was only in the clinic that the drug was found to alleviate psychoses without inducing sedation; this antischizophrenic action has also been termed "neuroleptic" or "major tranquilizing" activity. As is the case for many chemotherapeutic agents, chlorpromazine causes drug-induced side effects, such as Parkinsonism, which are sometimes more disturbing than the disease state being treated. Drug research in the area of antipsychotic agents is currently directed at minimization of these untoward side effects.

Thiophene analogues of promazine derivatives were prepared and the importance of examining all of the thiophene positional isomers[17] was demonstrated. Of the three promazine isosteres, the 2,3- and 3,4- annelated derivatives 1 and 2 showed

1, $R = CH_2CH_2CH_2N(CH_3)_2$

4, $R = CH_2CH_2CH_2N \underset{\_\_}{\frown} NCH_2CH_2OH$

2 $R = CH_2CH_2CH_2N(CH_3)_2$

5 $R = CH_2CH_2CH_2N \underset{\_\_}{\frown} NCH_2CH_2OH$

3 $R = CH_2CH_2CH_2N(CH_3)_2$

6 $R = CH_2CH_2CH_2N \underset{\_\_}{\frown} NCH_2CH_2OH$

7 $R^1 = H$
8 $R^1 = Alkyl$

9 $R^1 = H$
10 $R^1 = CH_3$

activity in the rat, whereas the 3,2- annelated derivative 3 was inactive. These compounds were tested for catalepsy, ptosis, and sedation and for amphetamine antagonism as a secondary means of evaluation. The order of activity of the 2-substituents (X) in 1 and 2 was the same as for other neuroleptic agents, that is,

2-CF$_3$ > 2-Cl > H. The bis-thienothiazine analogues could not be prepared, presumably because of their instability.[18]

In later work, the same research group prepared the more potent hydroxyethylpiperazinylpropyl derivatives 4, 5, and 6. In these cases, 4 and 5 were much more potent than 6, with substituent effects similar to those already noted. These studies measure the influence of neuroleptic drugs on dopamine metabolism by examining increases in homovanillic acid concentrations in rat brain.[19] These workers concluded that "thienobenzothiazines differ quantitatively, but not qualitatively, from the phenothiazines in their effect of dopamine metabolism."

During the course of developing phenothiazine drugs as neuroleptics, other related ring systems such as dibenzothiepine, dibenzoxazepine, and dibenzodiazepine also showed neuroleptic activity. One of the most interesting discoveries was that clozapine showed good antipsychotic effects without causing the worrisome extrapyramidal side effects observed for previously used drugs.[20] This discovery led to the preparation of some thiophene isosteres. Three methods to synthesize 10-alkylamino-4H-thieno[3,4-b][1,5]benzodiazepines (7) were reported and the compounds showed expected biological properties.[21] The most active compounds had an N-alkylpiperazine or N-hydroxyethylpiperazine substituent at C-10 (R^2) and various substituents on the benzene ring (X). Little structure–activity correlation was found by varying X.

Unfortunately, the promising neuroleptic activity found in these compounds, as evidenced by motor activity effects or amphetamine antagonism, was accompanied by catalepsy in rats. Thus, derivatives of 7 appeared to behave as classical neuroleptics and not as clozapine-like agents. Interestingly, reductive alkylation at N-4 gave 8, which was active not only as a neuroleptic but also as an antidepressant in laboratory animals. Thienobenzodiazepine (8) represents a unique type of mixed-action CNS agent.

The alternative 3,4- annelated clozapine isostere 9 has also been prepared.[22] Derivatives of 9 gave 10, which had mixed antidepressant/neuroleptic activity, as predicted from results for 8, but which also had reduced potency.

The related 2,3- annelated 11 and 12 were synthesized using some standard

11   12

procedures.[23] As noted earlier, piperazinyl derivatives were the most active. Halogen (Cl or F) substitution on C-7 of the phenyl ring enhanced activity, but positional isomers had diminished activity. Short-chain alkyl substitutents (R^2 = CH$_3$–,

CH$_3$CH$_2$–, or (CH$_3$)$_2$CH–) at the 2-position of thiophene also seemed to increase activity. As in the 3,4-series, derivatives of **11** were far more active than isomer **12**, which was inactive in these tests. Many examples of **11** showed a profile of activity similar to clozapine in that they did not cause catalepsy in rats at doses that blocked conditioned avoidance.

In another paper, the same group prepared the 3,2- annelated isomers **13** as well

13

as several 3,4- derivatives of **7**, using procedures different from those previously reported.[24] In contrast to the results for the three promazine isosteres, **1**, **2**, and **3**, derivatives of **13** were as active as their **11** counterparts. In contrast to **11**, alkyl substitution on the thiophene eliminates activity. These workers also found that **7** caused less cataleptic liability in their animal models than had been previously observed.[21]

A further example of clozapine isosteres has recently been reported.[25] In this case, both benzene rings were replaced by heterocycles, namely, by 3,4- annelated thiophene and pyridine (**14**, R^1 = H, CH$_3$–, R^2 = H, CH$_3$–, and HOCH$_2$CH$_2$–). In this case, no interesting pharmacological activity was observed.

Although clozapine provides the best clinical agent to model because of its low side-effect liabilities, thiophene analogues of other neuroleptic agents have also been prepared. The 3,4- annelated thiophene isosteres of loxapine (**15**) and clothiepine (**16**) were synthesized and tested for their effects on motor activity in rats or their antiamphetamine effects in mice.[22] Both the oxygen- and sulfur-containing central ring compounds were potent neuroleptic agents with clothiepine isostere **16** an order of magnitude more potent than the oxygen analogue **15**. When **16** was substituted on the benzene ring (Y = 7-Cl or 7,8-dimethyl), neuroleptic activity disappeared. When **16** was halogenated on the thiophene ring (Z = Cl), antidepressant activity was observed. In comparison, **15** showed more typical neuroleptic substitution effects with Y = 7-substitutents preserving, and other positional substituents eliminating, activity. In the case of halogenation of the thiophene (**15**, Z = Cl), neuroleptic activity was maintained and potential antidepressent activity developed.

The thiophene isostere of perathiepin (**17**, X = Y = H) was prepared using a nine-step synthetic procedure and was shown to have a high degree of central depressant activity with relatively low toxicity. Some mild cataleptic activity of **17** ("peradithiepin") was noted. This activity was enhanced by chlorination at the 2-

14   **15**   X = O   **17**
             16   X = S

position (**17**, X = H, Y = Cl)[27] or by the introduction of unsaturation across the 4,5- bridge.[28] The effect of fluorine substitution (**17**, H = F, Y = H) was also investigated.[29] This compound had similar acute toxicity, higher central depressant activity, and very high catalepsy liability. Interestingly, this derivative also was inactive toward apomorphine-induced stereotypes and hence was concluded to be not a "true neuroleptic." As a consequence, the 2-chloro-8-fluoro derivative **17** (X = F, Y = Cl) was prepared.[30] This compound was found to be the most interesting of the series, because it demonstrated the character of a tranquilizer of long duration without cataleptic side effects.

Continuing in the search for atypical neuroleptic agents, thiophene annelated [3,2-c] and [2,3-c][1]benzazepines **18** and **19** were prepared in an extensive investigation.[31] Classical neuroleptic pharmacological effects in both systems were found when there was no substitution ($R^1 = R^2 = H$). When thiophene was substituted ($R^2 = H$, $R^1 = 2$-Cl or 2-CH$_3$), these effects were enhanced, as measured by cataleptogenic effects, antagonism of apomorphine-induced gnawing, and *in vitro* receptor-bonding studies. This enhancement was more marked for **18** as compared to **19**. When **18** and **19** were disubstituted ($R^1 \neq H$, $R^2 \neq H$), distinctly weaker neuroleptic effects but strong bonding affinities were observed. The substitution at the 7- position (R^2) produced "atypical" neuroleptics resembling clozapine. In general, **18** exhibited stronger neuroleptic effects than **19**.

18   **19**

Apparently, the most important thiophene derivative to be prepared as a "major tranquilizer" was developed from this study.[31] Clinical trials were carried out on NT 104-252 (**18**, $R^1 = H$, $R^2 = 7$-Cl, $R^3 = CH_3-$) as a consequence of its interesting pharmacological profile. The drug proved to be efficacious in the treatment of schizophrenia, with rare observations of extrapyramidal side effects. Unfortunately, NT 104-252 caused proconvulsive effects that led to seizures and clinical trials were abandoned.

2.   Anxiolytics/Anticonvulsants

The discovery of chlordiazepoxide as the first psychotherapeutic agent of the 1,4-benzodiazepine type changed the concepts of treating clinical forms of anxiety. Until its introduction, anxiety was treated by sedatives such as meprobamate. Pharmacological studies of chlordiazepoxide and later of diazepam showed that benzodiazepines were anticonvulsants, muscle relaxants, hypnotics, and taming agents. Numerous improved benzodiazepines have been prepared since the 1950's, with concomitant publications often appearing. Sternbach has reviewed and updated this area of research frequently.[32,33] For a thorough review of the current state of anxiolytic research, see Fielding and Lal.[34]

Among the modifications of the diazepam molecule designed to improve efficacy and lower side-effect liability, many thiophene isosteres have been prepared. The 2,3- annelated thieno[1,4]diazepinones **20** and **21** have been investigated

$$20 \qquad R^1 = CH_3$$
$$21 \qquad R^1 = H$$

in great detail.[35,36] Structure–activity relationships reveal that alkyl substitution at the 2- or 3- position (R^2 or R^3) of thiophene enhances activity, with the ethyl derivative ($R^2 = CH_3CH_2-$) having the best effect. Substituent effects on the pendent phenyl ring showed that best effects were achieved when X was ortho-chlorine or fluorine. The effects of N-substitution were not predictable but **20** (N-methyl) was superior to the unsubstituted **21**. One compound selected for further evaluation on the basis of its pharmacological profile, which included potency two to three times that of diazepam, was clotiazepam (Y-6047, **20**, $R^2 = CH_3CH_2-$, $R^3 = H$, X = o-Cl).[36,37]

The Binder group has also been pursuing thienodiazepines. In a number of short papers, they have reported the preparation of all three thieno isosteres of 1,4-benzodiazepinones. The 2,3- annelated diazepines, **20** and **21**, were developed to investigate various substitution effects. In particular, thiophene monosubstitution ($R^2 = NO_2$, Cl, or CO_2R),[38,40] thiophene disubstitution ($R^2 = $ Cl and $R^3 = $ Cl or

NO_2),[39] and the unsubstituted parent system[41] were prepared. More recent reports of substitution on the pendent phenyl group of **20** and **21** include biological testing results.[42,43] In these cases, o-substitution was found to give superior activity (analogous to reports by other workers) and a series of ortho-nitro derivatives ($X = o$-NO_2) was developed in detail. These compounds demonstrated expected activity as anticonvulsants, as measured by strychnine-induced seizures, electroshock, and pentylenetetrazol-induced convulsions. The pyridyl substituted system **20** has also been prepared.[44]

The 3,2- annelated systems were also prepared by this group. Thus, the parent unsubstituted compound **22** ($R^2 = R^3 = H$)[45] and several substituted thiophene systems **22** ($R^2 = H$, $R^3 = CF_3$),[46] **22** ($R^2 = NO_2$, $R^3 = H$),[47] and **22** ($R^3 = Cl$, $R^2 = H$)[48] were prepared. No biological effects were reported.

The other isosteric system (**23**) was prepared using methodology other than

22

23

that used to prepare **21** and **22**. The effects of substitution on the thiophene of these 3,4- annelated compounds was investigated by the preparation of trifluoromethyl (**23**, $R^3 = CF_3$, $R^2 = H$),[49] chlorine, or nitro derivatives (**23**, $R^2 = Cl$ or NO_2, $R^3 = H$).[50] Other compounds were also prepared.[50–52] Here, again, the results of these molecular alterations on the biological effects were not reported.

Another variation on this diazepam-type structure was repeated at about the same time[53] by a Parke Davis research group. In this case, the 2,3- annelated thiophene system was substituted in the 2- and 3- position with a saturated four-carbon methylene bridge. Detailed structure–activity analysis showed this methylene bridge to produce the greatest activity. The most active compounds were those where the pendent ring was unsubstituted or was ortho-fluorinated (**24**, $X = H$ or

24

o-F). The aryl group could also be replaced by 2-thienyl. Methyl substitution at nitrogen (**24**, R = CH$_3$—) enhanced pharmacological activity. From these results, bentazepam (CI-718, **24**, R = H, X = H) was chosen as a clinical lead and its effects in man were studied.

In related work, heterocyclic analogues of **24** were prepared, wherein an oxygen (**25**) or sulfur (**26**) was introduced into the saturated six ring.[54] The most active compound of the series was the unsubstituted oxygen derivative **25** (Y = H). Phenyl-substituted **25** or the sulfur system **26** had little pharmacological activity as anticonvulsant agents.

In the more recent years of research on [1,4]benzodiazepines, triazolo-fused derivatives have revealed interesting anxiolytic properties. As a consequence, thiophene isosteres of this active series have also been prepared. A series of papers appeared describing the development of the clinical candidate Y-7131 (**27**, R =

25   X = O
26   X = S

27

CH$_3$CH$_2$—).[55-59] Pharmacological activity was assessed by anti-pentylenetetrazole effects and inhibition of fighting episodes in mice. The *o*-chlorophenyl derivative was the most active of the series, in contrast to earlier observations wherein *o*-fluoro substitution imparted the most activity to benzodiazepines. Alkyl substituents on thiophene enhanced activity as noted earlier. Finally, methyl or ethyl introduced on the triazole ring also gave superior results. Y-7131 was the most potent compound of these studies.

Similar compounds have been reported by another group.[60] In this research, **27** (R = Br) was the most potent possible anxiolytic agent. Chlorine and ethyl substitution were investigated, but were found to be less effective. Alkyl substitution on the triazole ring was further examined and cyclohexyl proved to be an effective replacement for methyl. These workers also reported that the *o*-chlorophenyl moiety gave the best effect. Based on these studies, brotizolam (WE-941 **27**, R = Br) and WE-973 (**27**, methyl replaced by cyclohexyl) were placed in clinical trial.[61]

Imidazole-fused thieno[1,4]diazepines **28** and **29** were synthesized at approximately the same time as the derivatives mentioned above. Using procedures that have become standard for the preparation of this type of thiophene compound, **27** and **28** (R = H, CH$_3$) were reported, but no biological activity was mentioned.[62]

Prior to the discovery of benzodiazepines that had remarkable anxiolytic effects with very few side effects, sedatives and hypnotics were prescribed to relieve the

28   29

symptoms of clinical anxiety. These compounds, such as meprobamate, were frequently amides and numerous thiophene derivatives were prepared. These compounds generally were less interesting chemically, but were active either as sedatives
or muscle relaxants.

French workers recently reported a series of compounds (30) bearing a 3-thienyl
substituent.[63] Disubstituted amides (30, $R^1 = R^2 = CH_3CH_2CH_2-$ or R^1 and R^2
form a 5- or 6-methylene ring) were more active than α-monosubstituted amides.
These compounds were anticonvulsants comparable to meprobamate in efficacy,
as measured in laboratory animals. Neurotropic anticonvulsant activity was found
for a number of acyl derivatives of 2-aminobenzenesulfonamide. In particular,
thiophene 31 was among the most active compounds prepared.[64] Methaqualone
analogues 32[65] and 33[6] have been prepared and found to have hypnotic activity.
Neither of these compound types showed any superiority compared to the parent
compound.

30

31   32   33

Hydantoins and barbituric acid derivatives have also received a great deal of
modification in attempts to improve their anticonvulsant properties and reduce side
effects. Research in these areas has fallen off, diazepine compounds have proven

their superiority in the treatment of anxiety. A large series of 5,5-diphenylhydantoin isosteres was prepared. The spectrum of activity of the 2-thienyl analogue **34** (Ar^1 = 2-thienyl, Ar^2 = phenyl) was similar to the parent compound, but its therapeutic profile was better.[66] Interestingly, the 3-thienyl compound **34** (Ar^1 =

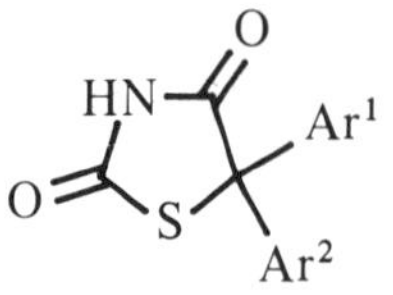

34

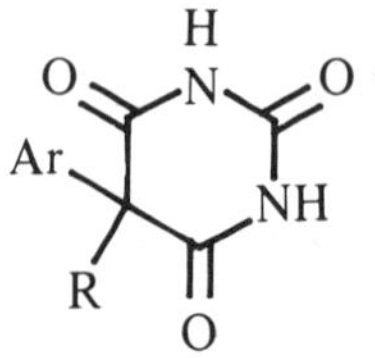

35

3-thienyl, Ar^2 = phenyl) had higher anticonvulsant activity and lower toxicity.[67] Anticonvulsant activity was also reported for the bis-thiophene analogue **34** (Ar^1 = Ar^2 = 2-thienyl).[2] Monothiophene-substituted hydantoins (**34**, Ar^1 = 2-thienyl, Ar^2 = H) also had anticonvulsant activity, but less than that of other interesting series.[68,69]

Several thienyl barbituric acid derivatives also have been reported. The 2-thienyl isostere of phenobarbitol (**35**, Ar = 2-thienyl, R = CH_2CH_3—) was found to be as active as the known drug.[70] The 3-thienyl compound **35** (Ar = 3-thienyl, R = CH_2CH_3—) produced different and unexpected results in biological tests. This compound produced sedation without hypnotic side effects.[71]

Several 2- and 3-substituted thiophene analogues of propoxyphene have been prepared.[72,73] Compound **36** showed an interesting combination of sedative and stimulant properties. A series of thiophene isosteres of protoberberine alkaloids, which were CNS depressants, was recently reported.[74] D-ring (**37**), A-ring (**38**), and A/D-ring (**39**) analogues were prepared, but no biological data were reported.

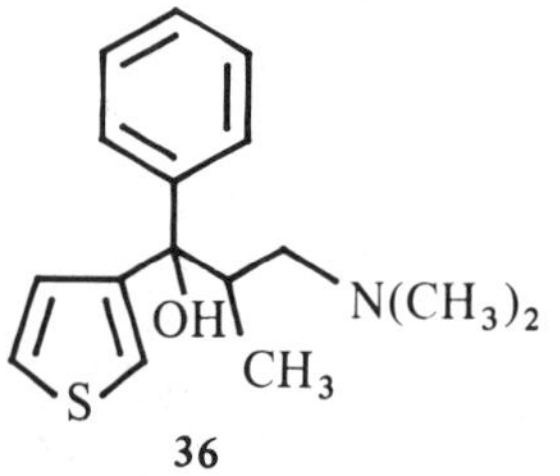

36

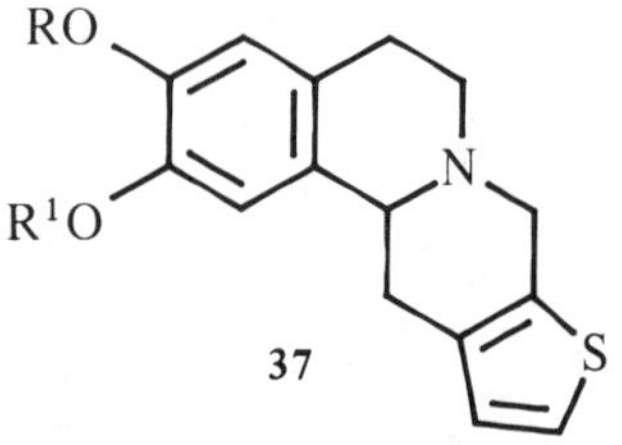

37

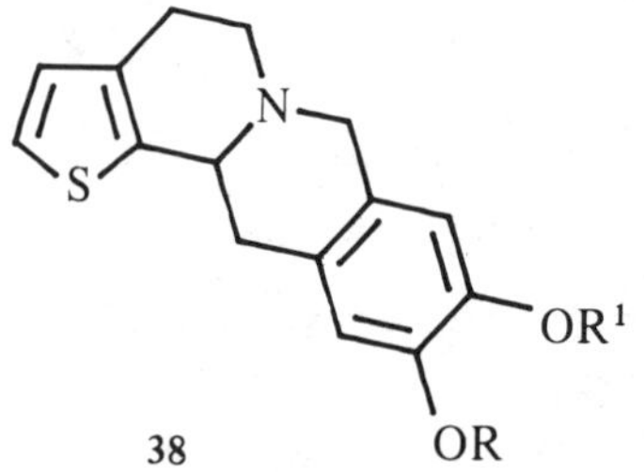

38

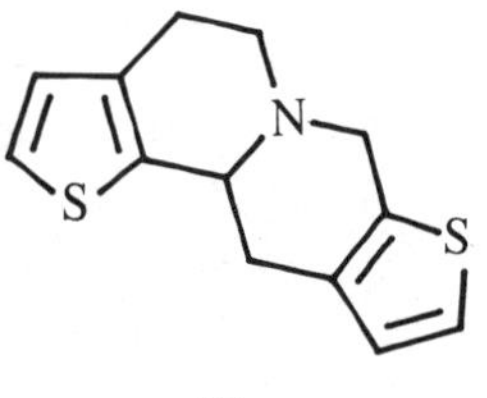

39

3.   Antidepressants

As was the case for the serendipitous discoveries of the previous classes of psychotherapeutic agents, the antidepressant imipramine was initially prepared to improve upon the profile of the antipsychotic chlorpromazine. Once the properties of imipramine were discovered, new animal models were established to measure antidepressant effects and to differentiate them from the actions of neuroleptic agents. As this development implies, many antidepressant drugs are tricyclic compounds related to major tranquilizers and are discussed in detail in a recent review.[75] With intense interest in this research area beginning in the mid-1950's, thiophene isosteres of known drugs soon appeared.

Pizotylin (BC-105, **40**) went into clinical trials in the late 1960's.[76-78] This compound was found to be an antagonist of biogenic amines such as 5-hydroxytryptamine, histamine, and acetylcholine. Besides this "classic antidepressant" activity, the drug caused the typical anticholinergic side effects of dry mouth, increased appetite, and nausea. Clinical trials showed **40** to be effective in the treatment of depression and the control of migraine episodes. It was comparable to melleril and imipramine in its mode of action.

The very similar drug **41**, with a dimethylaminopropylidene side chain attached

to the tricycle instead of the methylpiperidinylidene group, was placed in the clinic at approximately the same time.[79-81] In spite of the similarities of **40** and **41** (code IB-503), their profile of biological activity differed. Studies in laboratory animals showed **41** to have sedative and neuroleptic properties greater than amitriptyline but less than chlorpromazine. In the clinic, **41** was found effective and preferred for the treatment of mania and manic-melancholic psychoses. Unfortunately, the structure–activity development of this family of compounds has not appeared and other thiophene annelations have not been reported.

More recently, thiophene analogues of amitriptyline, nortriptyline, and chloro-amitriptyline were prepared.[82] In this case, bis-thiophene isosteres were considered with varying modes of annelation and antidepressant effects were ascertained by measurement of noradrenalin and 5-hydroxytryptamine re-uptake *in vitro*. Compound **42** (R = R^1 = H) was the most active of these derivatives and was as active as nortriptyline. Using *in vivo* tests, **42** (R = H, R^1 = CH$_3$–) and **43** (R^1 = CH$_3$–) were almost as potent as **42** (R = R^1 = H). The chloro-substituted derivatives (**44**)

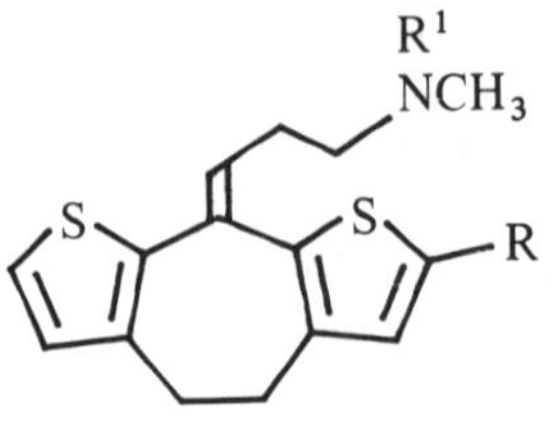

42

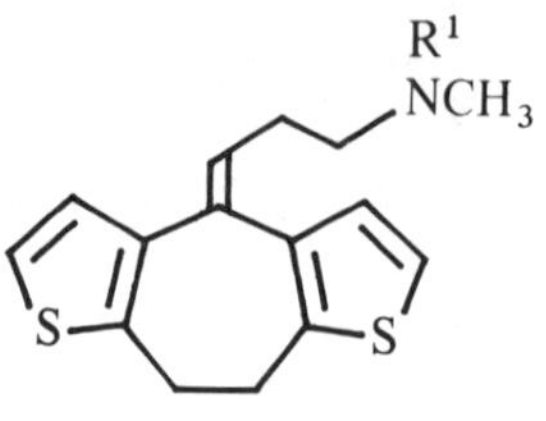

43

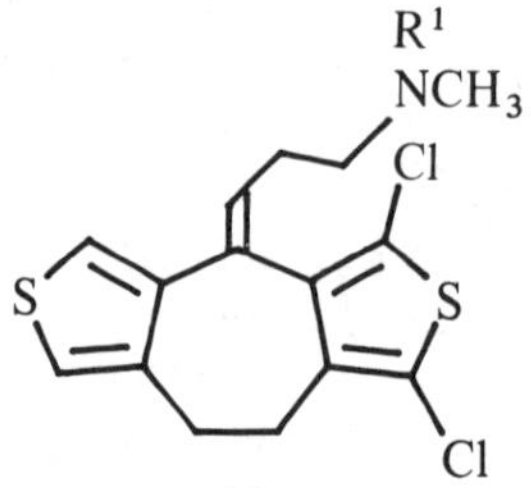

44

were barely active in both test systems. When these compounds were tested for antireserpine effects, they were less active than amitriptyline and nortriptyline. The change of annelation pattern once again was found to have significant effects on activity. Isomer **42** (R = R^1 = H) was 10 times more active than **43** (R^1 = H) as an inhibitor of noradrenaline uptake, but was one-tenth as active for the inhibition of 5-hydroxytryptamine uptake.

Thiophene isosteres of imipramine and clomipramine were prepared in an attempt to find antidepressants that had fewer side effects.[83] The 3,2- annelated (**45**) and 2,3- annelated (**46**) thiophene derivatives were synthesized, as well as their

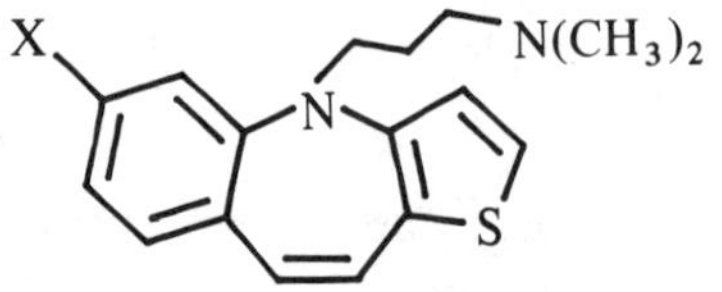

45

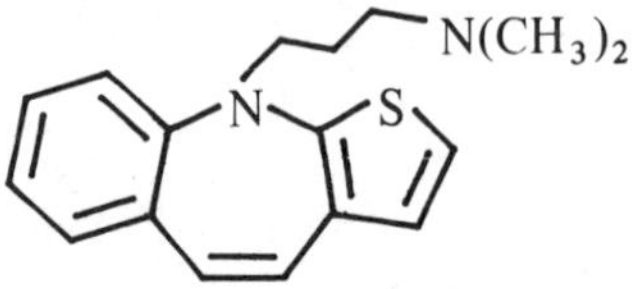

46

dihydro counterparts. Although no biological data was presented, unsaturated derivatives **45** (X = H) was chosen for further evaluation as an antidepressant. This compound, RU-15687, was the most interesting in animal experiments, because its pharmacological profile was similar to imipramine, but with a lesser degree of cardiac toxicity.

4.   Analgetics and Related Agents

In comparison to the first three areas of central nervous system (CNS) disease therapy, attempts to control pain with various natural and synthetic materials dates to significantly earlier times than the mid-1950's. Historically, these agents may be separated into the categories of strong (powerful, centrally-acting agents) and weak (locally active, with or without some activity in the CNS). The former category, which includes the narcotic compounds (i.e., morphine and codeine), has strong sedation and addiction liabilities. The obvious drawback to the weaker agents is that they are ineffective in treating severe pain. Research continues to investigate newer strong analgesics that do not have addiction potential and non-narcotic agents with fewer side effects.

An important thiophene-containing analgesic (47) was reported as a result of

47

an extensive structure–activity investigation.[84] This compound (Tinordin, Y-3642) and its hydrochloride salt were compared advantageously to known analgesic and anti-inflammatory agents such as acetylsalicyclic acid, phenylbutazone, and indo-methacin. It was superior to aminopyrine in alleviating pain (benzoquinone writh-ing test in mice), similar to acetylsalicyclic acid in preventing edema (induced by various adjuvants), and similar to aminopyrine as an antipyretic.[84,85] This com-pound was used in clinical trials because of its interesting biological activity.

Amide derivatives of the tetrahydrobenzo[b]thiophene 48 that are related to the tetrahydropyridine 47, were recently prepared.[86] A variety of substituents (48:

48

Z = H, 5-CH_3, 6-CH_3; R = alkyl, aryl, piperazinoalkyl, and morpholinoalkyl) were investigated in basic pharmacological screens. Most of the derivatives were minor depressants; six (48, n = 1, R = aryl) were found to be more potent than acetyl-salicylic acid or oxyphenbutazone as analgesic agents. Several related 2-amino-thiophenes have also shown minor analgesic activity[87] and others will be mentioned as anti-flammatory agents.

An area of analgesic research that has developed over the last three decades is the study of benzomorphans. These compounds have strong analgesic activity, but have lessened addiction liability. Several hetero-morphans have been prepared including the three thieno- annelated systems 49, 50, and 51. The first such system reported

was the 2,3- annelated derivative **50**, because of the relative ease of synthesis using
well-established routes.[88] In particular, **50** (R^1 = H; R^2 = H, CH_3; R^3 = H; R^4 =
H, Br, $COCH_3$) was prepared and tested for analgetic activity. As a class of compounds, **50** had a high degree of toxicity and only a few derivatives exhibited
analgetic effects; this mode of thiophene annelation was concluded to produce
uninteresting results.

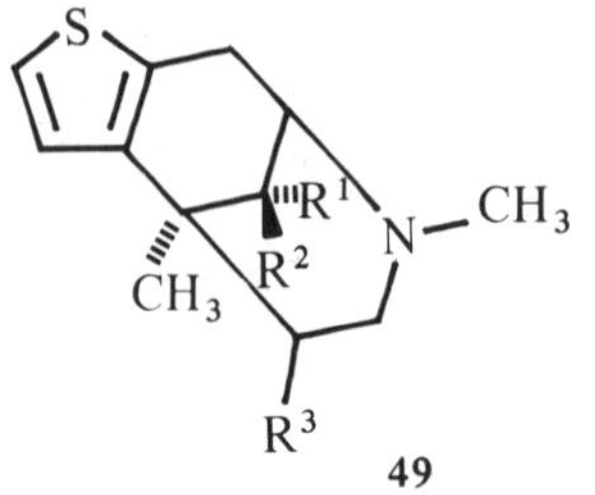

49

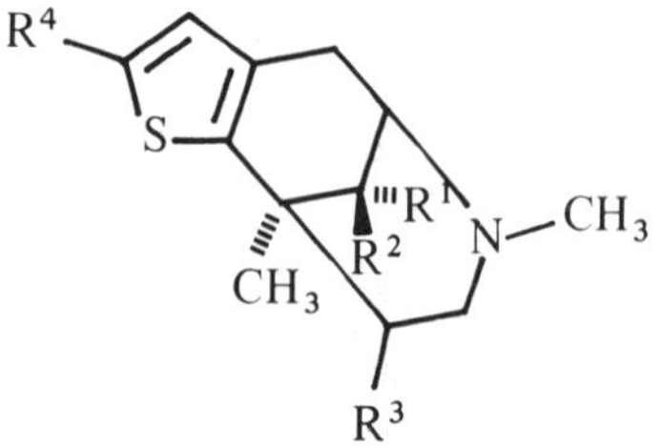

50

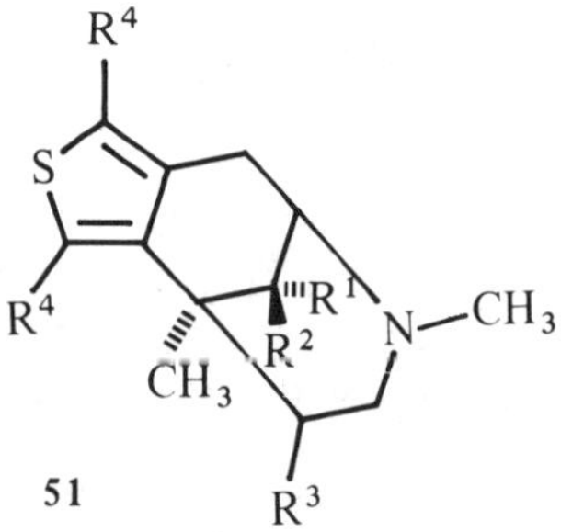

51

The 3,2- annelated isosteres (**49**) of benzomorphan were subsequently synthesized.[89,90] Using alternative synthetic procedures, **49** (R^1 = R^2 = R^3 = H;
R^1 = CH_3, R^2 = R^3 = H; R^1 = R^2 = H, R^3 = CH_3) was prepared, but unfortunately had no biological evaluation. Finally, the 3,4- annelated systems (**51**) were
prepared,[91,92] as well as new examples of **49** and **50**. Analgetic evaluation of these
compounds showed some interesting results, but no data was reported.

A new class of narcotic antagonists, (2-*exo*-3-*endo*)-2-aryltropane-3-carboxylic
esters,[93] and hypoglycemic agents with analgesic activity, (*exo,exo*)-2-aryltropane-
3-carboxylic esters,[94] recently appeared. Tropane esters (**52**) revealed an interesting

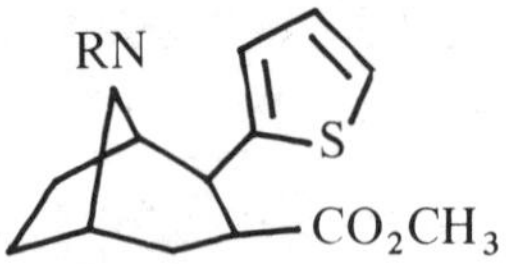

52

diversity of biological activities. The thiophene analogues were found to be substantially more active than the phenyl derivatives as hypoglycemic agents with
moderate analgesic properties; the two types of activity could not be separated,

although extensive substituent effects were studied. The most interesting compound (52, R = CH$_3$) showed apparent analgesic effects with little addiction liability and some mild respiratory depression. Further work was abandoned on this system. Related thiophene-substituted 3,8-diazabicyclo[3,2,1]octanes were endowed with high analgesic activity.[95]

Fentanyl, a well-known analgesic with high potency, rapid onset, and short duration of action, was the subject of structural modifications. As part of this study, 2-thienylethyl- or 2-phenylethyl- substitution was found to afford extremely potent compounds. One of the most interesting compounds, 53, was selected for

53

further pharmacological investigation.[96] This compound (R30,730) was 4521 times more potent than morphine and had a relatively short duration of action and an unusually high safety margin. Because of the powerful morphine-like activity and low toxicity of 53 (363 times safer than morphine), it was chosen for additional study.[97] The related thenoyl esters of 1-methyl-4-piperidinol were devoid of analgetic activity.[98]

A powerful class of analgesic agents was discovered while modifying diphenylallylamines that were known to have local anesthetic and antihistaminic properties. Thiambutenes (54) possess strong analgesic activity, but have typical addiction liability.[99,100] The hydrogenated derivatives of 54 have less analgesic activity.[101] One derivative has found use as a veterinary analgesic and is discussed in another section. Other derivatives have been used clinically and were found to induce drowsiness and nausea similar to morphine.[102] The similarities between 54 and methadone cannot be overlooked, both in terms of structure and activity. Monothiophene isosteres 55 (Ar = phenyl)[103] and dithiophene isosteres 55 (Ar = 2-thienyl)[104,105] of methadone have been prepared. Other similar compounds are also reviewed.[106,107]

Other modifications of this series of compounds change their primary CNS activity to that of antitussive agents. Compound 56 was reported to have a more

54
56 R = CH$_3$, NR^{1}R^2 = N⟨ ⟩

55

potent antitussive action than morphine, with the dextro form twice as active as the racemic mixture.[108] Compounds **57**, modeled on **56** and ephedrine, also had antitussive activity.[109] A great deal of analogue work has been reported.[110,111] A clinical agent, bitioden (**58**), was developed as a cough suppressant as a result of this research.[112,113] Quinuclidine analogue **59** was recently reported among other compounds and was found to have cough-suppressing activity inferior to bitiodin, with similar toxic effects.[114]

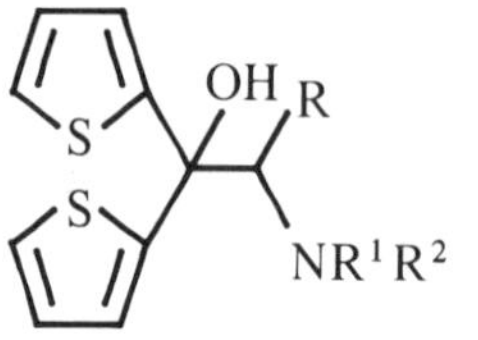

57

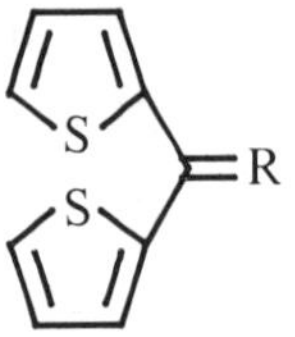

58 R =

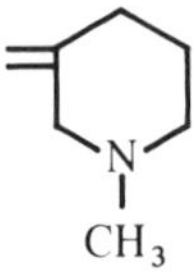

59 R =

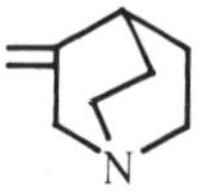

Returning to compounds primarily with analgetic activity, numerous agents with aspirin-like effects have been reported. A 2-thiophene carboxylic acid derivative (**60**) was found to have interesting analgesic and antipyretic properties.[115] A 2-thienylchalcone derivative was also found to have intermediate analgesic activity.[116] Dihydrothienocoumarins (**61**) have received some attention recently. Compound **61**

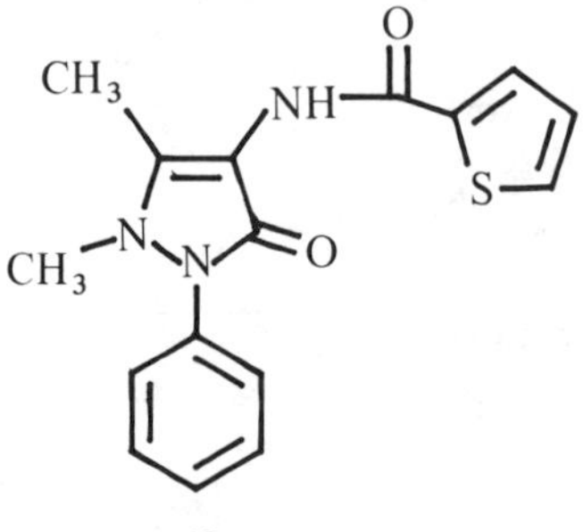

60

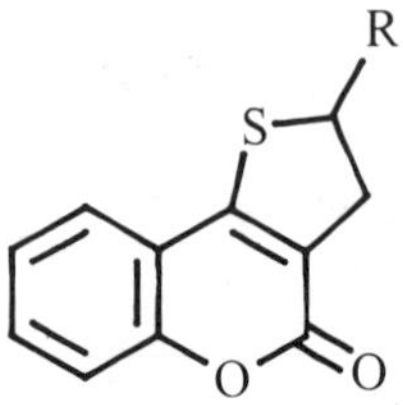

61

(R = 2-thienyl, 2-pyridyl, substituted phenyl) showed aspirin-like activity in the acetic acid-induced writhing test in mice.[117] Some related thienobenzopyranones modeled on known analgetic agents were prepared; no activity was given.[118] A series of pyridoxine derivatives with central analgetic activity was synthesized.

Of numerous analogues prepared, only thiophene **62** showed potency slightly greater than morphine.[119]

Reports of thiophene-containing local anesthetics have also appeared. A compound similar to lidocaine in its effects (**63**) has been used clinically, especially in the field of dentistry.[120] A series of 2-thienyl-β-piperidinylethyl ketones containing alkyl-substituted thiophene was prepared and found active as local anesthetics.[121] Additionally, some ω-(*N,N*-dialkylamino)alkyl thiophenes showed this type of activity.[122]

         62   63

5. Stimulants and Mood-Altering Drugs

Certainly no discussion of CNS-active agents would be complete without a discussion of drugs that are used illicitly. Amphetamines, tetrahydrocannabinol, and phencyclidine all may be included in this category and several reports of thiophene isosteres of these agents have appeared. Analogues of chloramphetamine (**64**) were prepared and found to decrease 5-hydroxytryptamine concentration in rat brain.[123] Chlorine substitution at C_4 or C_5 (**64**, X or Y = Cl) caused a potent

                              64

and long-lasting depletion of 5-HT; dichloro-derivative **64** (X = Z = Cl) was a 5-HT uptake blocker, rather than a depleter. Dichloro **64** (X = Y = Cl) was most similar to chloramphetamine in effect and duration, as measured in behavioral studies in mice.

Additional heterocyclic analogues of amphetamine and various acyl derivatives were prepared.[124] Of the many derivatives in the study, **64** (X = Y = H, Z = CH₃) caused a significant increase in papillary muscle contractile force, and the desmethyl derivative **64** (X = Y = Z = H) showed behavioral effects similar to amphetamine. Compound **65** was prepared as a potential nonstimulant anorexic agent.[125] About half of this series of compounds produced significant anorexia, with only one of

the active compounds causing increased motor activity. The best compounds were
65 (X = 4-Cl, R = R^1 = H; X = 4-Cl, R = isopropyl, R^1 = H; X = 4-F, R = isopropyl, R^1 = H).

An extensive study of cannabinoids and heterocyclic analogues of tetrahydrocannabinol led to several thiophene derivatives. Compounds **66** (R = 1,2-dimethyl-
heptyl, R^1 = H) and **67** (R = 1,2-dimethylheptyl) were found equipotent in the
Dopa potentiation test, but **66** showed analgesic properties as well.[126] Compound
66 (R = 1,2-dimethylheptyl, R^1 = CH$_3$) was less active than its methyl isomer in
sedative–hypnotic potency.[127] In spite of the extensive research in this area, the
goal of separating the various components of cannabinoid biological activity has
continued to be elusive.

Of the many analogues of phencyclidine prepared, thiophene isostere **68** has
been the subject of extensive study. Conformational studies and effects on activity
have been reported.[128,129] Molecules with an axial aryl group and equatorial piperidine moiety are more active as measured by the roto-rod test. Compound **68** was
identified as an illicit drug in 1972 and was entered into Schedule I of the Controlled Substances Act in 1975.[130] It is reportedly more potent than its phenyl
isostere.

65

66

67

68

III. PHARMACODYNAMIC AGENTS

Cardiovascular drugs represent the second largest class of therapeutic agents
marketed in the world. Antihypertensives and diuretics are the most frequently
prescribed drugs to treat cardiovascular disease, but progress in the understanding
of the sympathetic and parasympathetic nervous systems as well as the role of
biogenic amines has led to many new drug therapies. Several biogenic amines affect
more than vascular smooth muscles; cholinergic mechanisms, for example, are

involved in the cardiovascular, gastrointestinal, and central nervous systems. Drug effects in this area of research are often difficult to define to one specific target. The complexity of the autonomic nervous system and greater awareness of side-effect liabilities for long-term drug therapies have created the need for more specific and safer drugs. Additionally, the introduction of the spontaneously hypertensive rat as a model for hypertension has played a major role in the direction of cardiovascular research for the past several decades.

1.   Anticholinergic Agents

The development of thiophene analogues of antispasmodic agents has been an active research area since Blicke's review.[1] The systematic modification of atropine and local anesthetics and the development of new agonists and antagonists of biogenic amines that are active in the autonomic nervous system have been fruitful areas of discovery. Thiophene derivatives have been reported as adrenergic, para-sympathetic (selective as to musculotropic or neurotropic activity), and ganglionic blocking agents.

Historically, basic esters of α-substituted thienyl glycolic, thienyl acetic, or thienyl propionic acids were found to be antispasmodics and are represented by general structure 69. One early clinical agent was penthienate methobromide (70), which was developed as an isostere of active phenyl derivatives.[131] This compound was more active than atropine in preventing acetylcholine-stimulated intenstine contractions in rabbits and anesthetized dogs.[132,133] It greatly reduced gastric secretions and intenstinal motility.[134] The cyclohexyl analogue of 70 also has been extensively studied and showed mydriatic activity.[135,136] The effects of thiophene substitutional position were studied in 71 and 72. The 2- isomer 71 was more active than 72 and almost as active as atropine.[137] When the diethylamine portion of 72 was replaced by piperidine, the highly active spasmolytic agent P.P.T. was produced.[137,138] Several closely related derivatives have also been prepared.[131,139]

71	2-Thienyl
72	3-Thienyl

Compounds related to the thiambutene analgesics have been reported as anticholinergic agents. Among these are the tertiary amino butanol derivative **73** and its related propanol analogue.[140,141] Resolution of **74** showed that the L form was a more active anticholinergic than its optical antipode.[142] Simple thienyl alkyl amines related to the biogenic amines showed musculotropic activity, as evidenced by inhibition of barium chloride-induced spasms in intenstinal strips.[143] Bisdialkylaminopropane derivative **75** was 10-fold more active than papaverine as an antispasmodic agent.[144]

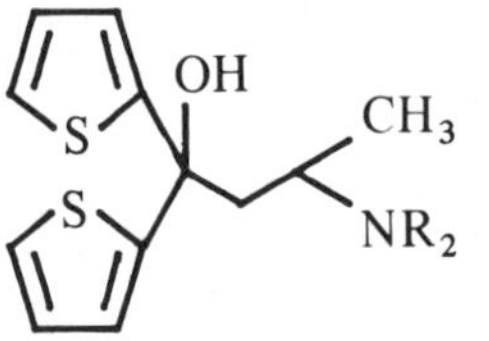

73 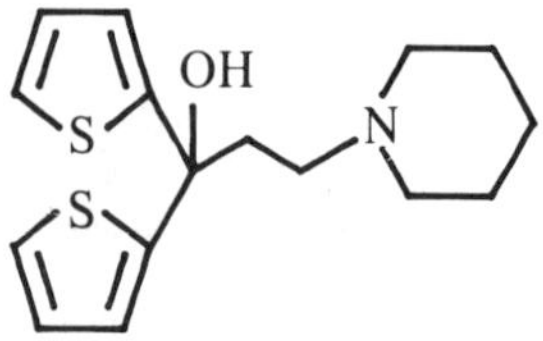   **74**

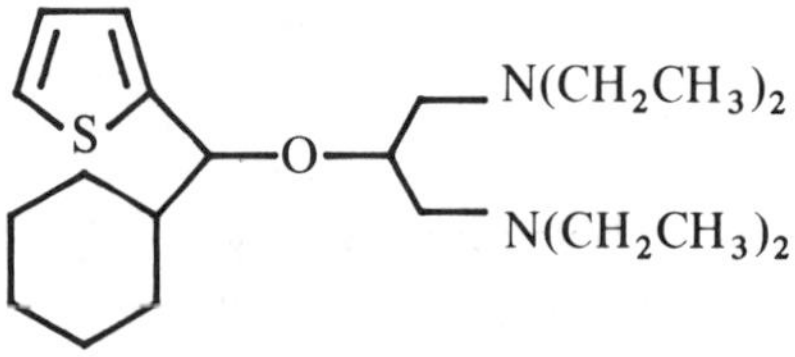

75

Glycolic acid derivatives received extensive investigation in the 1960's and a renaissance of interest occurred in the mid-1970's. Anticholinergic activity was studied for a variety of compounds, including an acetate derivative of a thienyl glycolate.[145] A series of 3-piperidinyl glycolates, including the 2-thienyl derivative **76**, was prepared. The anticholinergic activity of **76** was superior to that of its

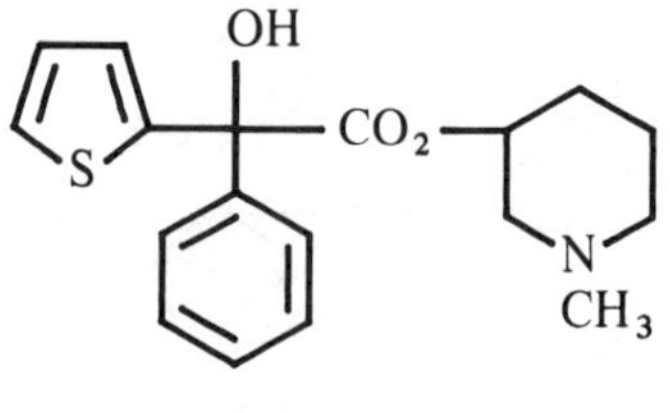

76

phenyl isostere and was the most active of the study.[146] Examination of the CNS effects of these derivatives also illustrated that **76** was the most potent psychotomimetic with the shortest duration of action.[147] Additional studies measuring the effects of these anticholinergic agents on motor activity and body temperature have appeared.[148] Related quinuclidinyl analogues of **76** were also potent anticholinergic agents with pronounced behavioral effects.[149] Other thienyl derivatives of **76** have also been prepared.[150,151] Psychotomimetic effects of compounds including **70** have been reported.[152]

In related studies, the development of a clinical vasodilator with action similar to that of papaverine was reported. Cetiedil (**77**) was prepared as a result of extensive structure–activity investigations on a series of spasmolytic agents first synthesized in 1965.[153] The compounds had been prepared initially based upon earlier observations.[154,155] Modifications of structure **78** showed the following: (1) cyclo-

77

78

pentyl or cyclohexyl substituents (R) were the most interesting, with cyclohexyl compounds showing more neurotropic activity and musculotropic activity;[156] (2) the highest ratio of desired musculotropic/neurotropic activity was seen for cyclic amine derivatives, with R_2^2 as part of a seven-ring having the best therapeutic ratio;[153] (3) branching ($R^1 = CH_3$) reduced the desired activity.[157] Pharmacological[158] and clinical investigation of cetiedil, comparing it to papaverine and aminophylline,[159] revealed **77** to be a strong papaverine-like and weak atropine-like anticholinergic agent. Among its interesting properties, **77** produces peripheral vasodilation without causing changes in other cardiovascular parameters. It enhances the activity of β-adrenergic stimulants and also affects the role of calcium in membranes.

Amide derivatives of phenylthienylhydroxyacetic acids (**79**) were prepared to investigate their spasmolytic properties. The effects of N-alkylation (R) or variation of the basic amine (NR^1R^2) were not predictable;[160] the most interesting compound was **79** (R = CH_3, NR^1R^2 = pyrrolidinyl). Disubstituted 2-thiopheneacetonitriles **80** were also prepared as anticholinergic/antispasmodic agents. In the

79

80

first report, thieno aminobutyronitriles (**80**, n = 2) showed *in vitro* musculotropic activity and the best compound (**80**, n = 2, NR^1R^2 = 4-methylpiperidyl, R^3 = isopropyl) had sixfold improvement in potency compared to papaverine.[161] Subsequently, additional butyronitriles (**80**, n = 2) and valeronitriles (**80**, n = 3) were investigated.[162] Allyl- and benzyl-substituted butyronitriles (**80**, n = 2, $R^1 = CH_3$,

R^2 = benzyl) had peripheral vasodilator activity 5 times greater than papaverine, whereas valeronitrile derivatives (**80**, n = 3, R = nC_4H_9— or allyl, NR^1R^2 = pyrrolidine) had spasmolytic properties 8–16 times greater than papaverine. This differentiation of spasmolytic and vasodilatory action, noted previously during the development of cetiedil, demonstrates the importance of detailed structure–activity relationships and of new pharmacological testing procedures.

Interest in thiambutene-like compounds has reappeared. Tiemonium methylsulfate (**81**), which is related to **73** and **74**, has an interesting pharmacological

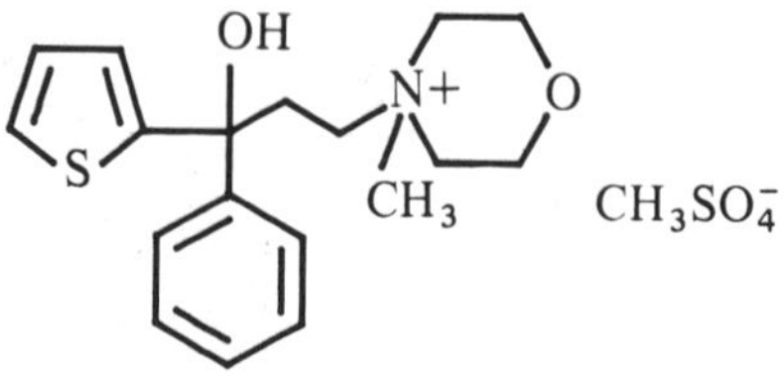

81

profile.[163] It antagonizes acetylcholine stimulation (atropine-like) as well as musculotropic stimulation (papaverine-like), but binds with histamine H_1 receptors and does not effect α-adrenergic receptors unlike papaverine. The "polyvalent antispasmodic" properties of **81** were concluded to be unlike atropine, since much of its action occurs at the muscular membrane. Even more related to the thiambutene series of analgesics, **82** and the corresponding alcohols were reported to have

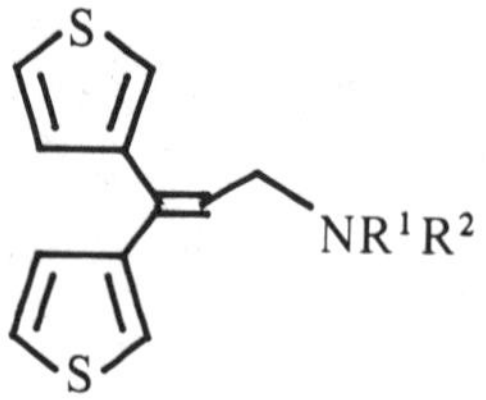

82

musculotropic-type spasmolytic activity.[164] Some derivatives in this study also showed peripheral vasodilatory effects as well as some expected analgetic properties.

In very similar work, a different research group combined the concepts derived from the study of oxyfedrin and the work described above to prepare the cerebrally active clinical agent **83** (tinofedrin, D-8950). As part of the development of **83**,

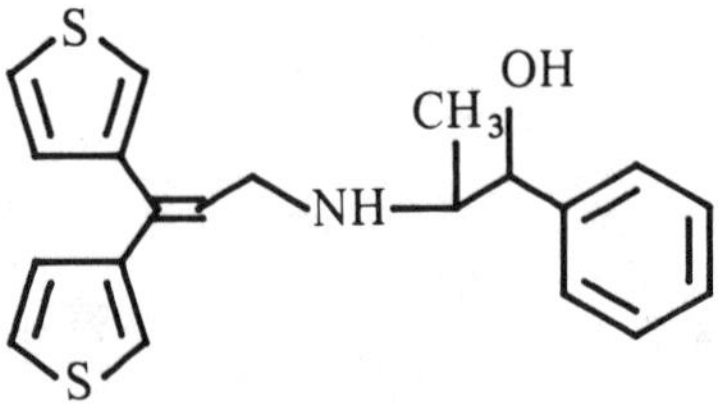

83

thienyl derivatives were found to be superior to their phenyl isosteres in activity and were less toxic.[165] Isomeric 2-thienyl derivatives were less interesting than **83**. Hydrated or hydrogenated compounds related to **83** showed significant activity as well.[165] New syntheses and optical resolution of **83** have recently appeared.[166] Pharmacological studies in dogs showed that tinofedrin was remarkably superior to standard drugs in improving cerebral and peripheral blood flow.[167] The oximino analogues of these compounds were also prepared and were active cardiotropic agents. The diphenyl isostere was the most active at preventing spasmogen-induced intestinal contractions, yet thiophene containing **84** also had significant activity.[168]

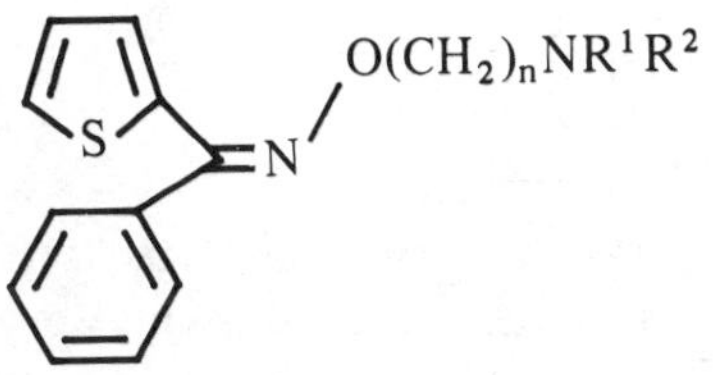

84

The ∈-isomer was determined to be the more potent form.[168, 169]

Other anticholinergic or antispasmodic agents with significantly variant structures have also appeared. The ganglion-blocking antihypertensive trimethaphan camsylate (**85**) has been an important clinical agent for some time.[170] Recent

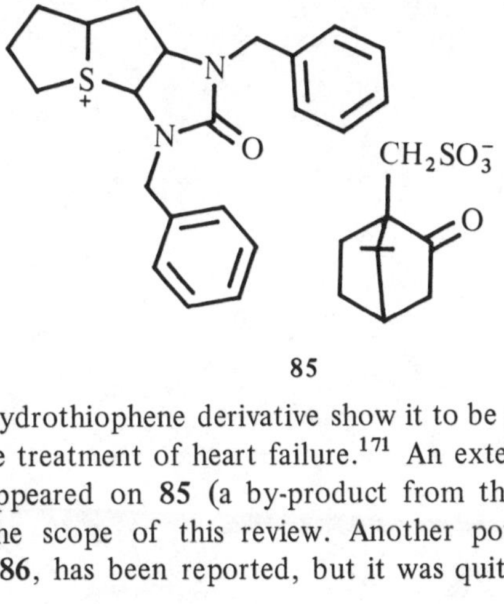

85

reports on this tetrahydrothiophene derivative show it to be a vasodilator that may be used for the acute treatment of heart failure.[171] An extensive listing of all the literature that has appeared on **85** (a by-product from the synthesis of biotin) would be beyond the scope of this review. Another potential antispasmodic, sulfonium derivative **86**, has been reported, but it was quite toxic.[172] A series of

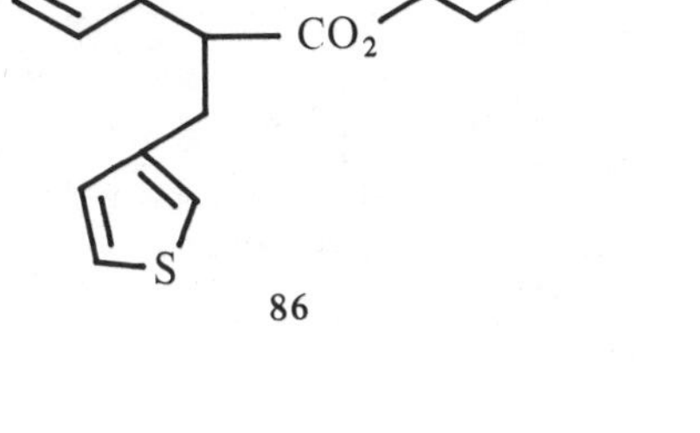

86

3-tropanyl-2,3-diarylacrylates was prepared which had atropine-like activity, but increased toxicity. Among the derivatives reported, **87** was slightly less active and more toxic than its phenyl isostere.[173] Pyranothiophene compounds were prepared to investigate antispasmodic activities and **88** was found to be superior to others synthesized.[174]

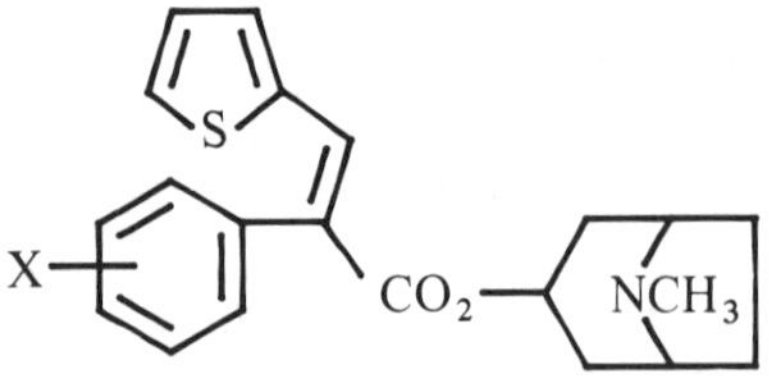

87

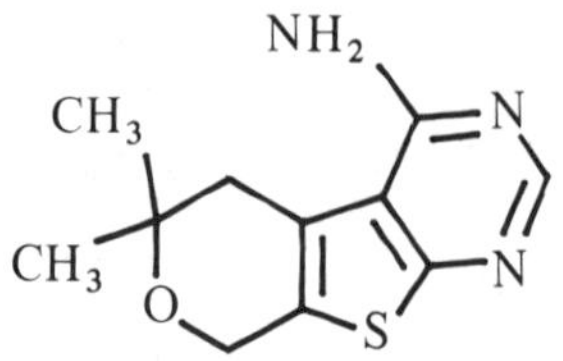

88

Timepidium bromide (**89**) is a recently marketed antispasmodic agent closely related to diphemanil sulfate. This derivative is orally active, in part because it is easily absorbed through the gastrointestinal tract. It has the activity of atropine with fewer side effects.[175,176] A series of rigid bicyclic analogues of these compounds has also received recent attention. Diarylmethylene quinolizidine quaternary ammonium salts **90**, **91**, and **92** (1-, 2-, and 3- isomers, respectively) showed significant inhibition of constriction caused by acetylcholine in isolated ileum and appeared to be more active than **89**.[175] Thienyl isosteres (Ar = 2-thienyl) were more potent than the phenyl derivatives with activities 2- > 3- > 1 substitution effects. New methods of synthesizing these[177] and related compounds with anticholinergic activity have appeared.[178] Related series of 2- and 1-diarylmethylene indolizidinium halides **93** and **94** have also been synthesized. As with quinolizidine

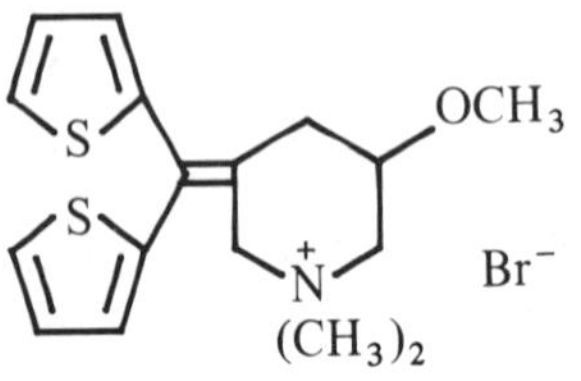

89

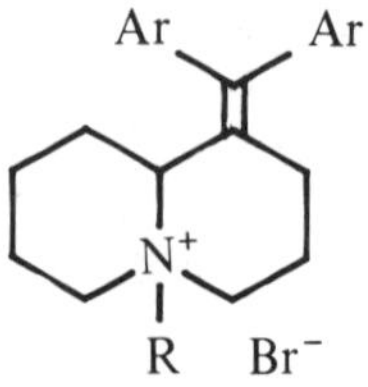

90

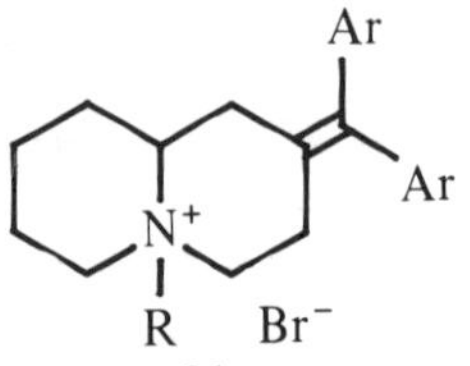

91

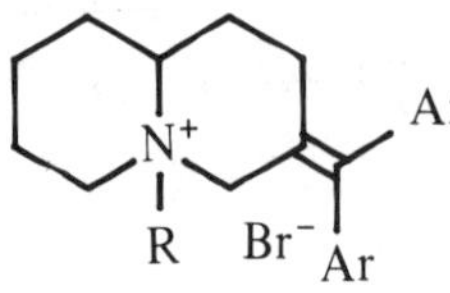

92

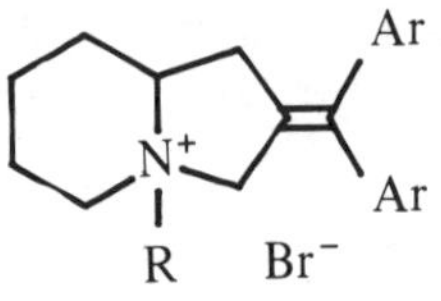

93

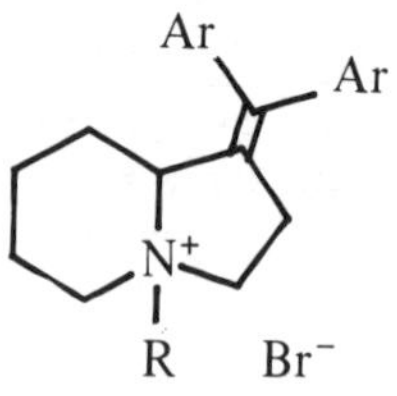

94

derivatives, the 2- substituted **93** were superior to the 1- substituted **94**, but in contrast to the earlier work, no differences between the thiophene and benzene isosteres (Ar) were observed. These compounds were more active anticholinergic agents than their monocyclic analogues.[179]

2.   Adrenergic Receptor Blockers

Agents that affect the sympathetic nervous system have gained great importance in the treatment of cardiovascular disease. These drugs have antihypertensive properties as potent as the ganglionic blocking agents discussed in the previous section, without serious parasympathetic side effects. Drugs such as clonidine, which acts centrally on the α-adrenergic system, have been used extensively as hypotensive agents in the clinic. In addition, β-adrenergic receptor blockers, such as propranolol, are effective in treating angina pectoris, arrythmias, and hypertension. Recently, these drugs have been effective in preventing second heart attacks. Some of these therapeutic areas have been reviewed;[180] they differ from the topic considered in the previous section in that these therapies have only become areas of intense research interest in the last two decades. Thiophene analogues of the adrenergic amines epinephrine[181] and ephedrine[182] have been prepared.

Tiamenidine (Hoe-440, **95**) was synthesized as part of an extensive structure–activity investigation based upon clonidine. This derivative lowered blood pressure in renal hypertensive and spontaneously hypertensive rats at threefold higher doses than those required for clonidine.[183] There were fewer sympathomimetic side effects for **95**, as compared to the model. At elevated doses, **95** had vasoconstrictor activity.[183] Tiamenidine was far less active than clonidine either as a sedative agent or in other CNS side effects, such as locomotor activity, conditioned avoidance, or barbital potentiation.[184] A pharmacokinetic study of this antihypertensive agent, which activates the α-adrenoreceptors in the CNS, has appeared.[185]

95

Other thiophene derivatives have also been reported to affect the sympathetic nervous system. Thiophene isosteres of the sympathicolytic alkaloid sendaverine (**96**, R = H) were synthesized.[186] The methyl ethers **96** (R = CH$_3$) also were pre-

96

pared, but no biological activity was studied. The synthesis and pharmacological properties of thiophene analogues of pilocarpine have been reported. These derivatives had no parasympathetic activity as measured in guinea pig ileum.[187]

More research activity has been focused on agents that affect the β-adrenoreceptor. A series of thienylethanolamines (**97**) was prepared to investigate their potential as antihypertensive agents.[188,189] The isostere of dichloroisoproterenol (**97**, $R^1 = R^2 = Cl$, $R^3 = H$, $R^4 = $ isopropyl) had acute blood pressure-lowering

97

properties, whereas the phenyl derivative was inactive. Substitution at R^1 and R^2 was essential for activity, as was the thiophene 2- substitution of the alkylamine side chain.[189] Bromo- substitution of thiophene also produced effective hypotensive agents. Many of the compounds showed biological effects resulting from β-blockade as well as inhibition at α-receptors. Compound **98** was prepared in the same study as a possible vasodilator.[189]

98

Another research group also synthesized **97** as a potential β-adrenoreceptor blocking agent.[190] All of the dichlorinated thiophene derivatives, and especially **97** ($R^1 = R^2 = Cl$, $R^3 = H$, $R^4 = $ isopropyl), exhibited nonspecific spasmolytic activity. All compounds except **97** ($R^1 = R^2 = R^3 = H$, $R^4 = $ isopropyl) decreased the resting heart rate in test animals, but to a smaller degree than propranolol. Chlorine- substitution at positions 4 and 5 (R^2, R^1) favored blockade predominately at the β-adrenergic receptors. Derivatives with isopropyl- or *tert*-butyl-nitrogen substitution (R^4) were chosen for further study.

Additional dichloro derivatives of **97** were subsequently prepared;[191] QM-5119 (**97**, $R^1 = R^2 = Cl$, $R^3 = H$, $R^4 = $ *tert*-butyl) was found to be a more effective antagonist on myocardial β-receptors than on those in bronchial smooth muscle and was the most interesting compound in the study.[191] Further evaluation of this compound in anesthesized cats showed that it was effective against induced arrythmias and that it had local anesthetic activity. It compared favorably with propranolol and alprenolol.[192] A more detailed study of structure–activity relationships, including replacement of chlorine by bromine, followed.[193] Both types of halogenation on **97** produced similar pharmacological effects. Monohalogenation at the 4- or 5- positions (R^2 or R^1) produced agents with moderate blocking activity while mono- substitution at the 3- position (R^3) eliminated activity. The highest activity was observed for derivatives dihalogenated at the 4- and 5- positions, with β-adrenergic effects comparable to propranolol.[193]

During a search for drugs affecting the peripheral autonomic nervous system, **99** was prepared.[194] This novel aryl thioether, in contrast to more typical aryloxy derivatives, antagonized the cardiovascular effects of isoproternol. It was five times more potent than propranolol, significantly longer in duration of effect, and virtually devoid of other pharmacological actions.

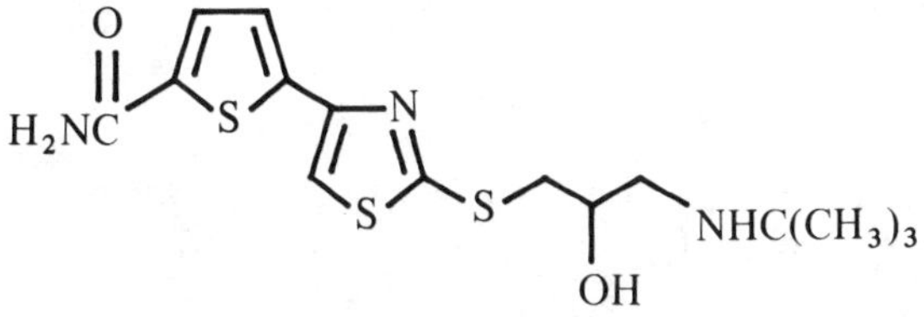

99

3.   Other Cardiovascular Agents

Several thiophene derivatives have been reported to be cardioactive agents with less clearly defined mechanisms of action than those previously discussed. An anti-arrythmic agent designated CH-200 (**100**) was recently reported.[195] The compound

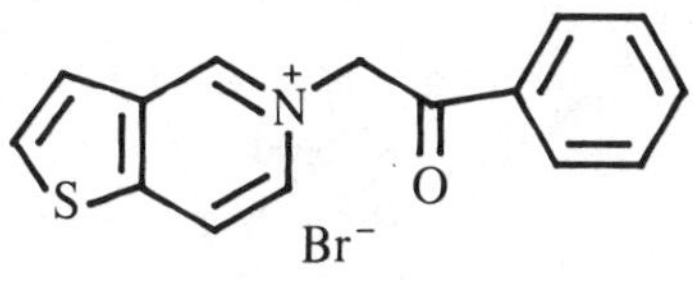

100

was found to produce antiarrythmic effects more effectively than procainamide or lidocaine in beagle dogs.[196,197] Its slow onset and long duration of action were qualitatively similar to procainamide, but **100** did not cause severe acute hypotension. Studies indicated that **100** may be useful in treating arrythmia accompanied by hypotension.

A thiophene analogue of dipyrimadole has been reported as a hypotensive agent.[198] Compound **101** (VK-774) inhibits noradrenaline-induced platelet aggregation in rabbits. Of the 16 derivatives prepared, **101** was the most potent compound,

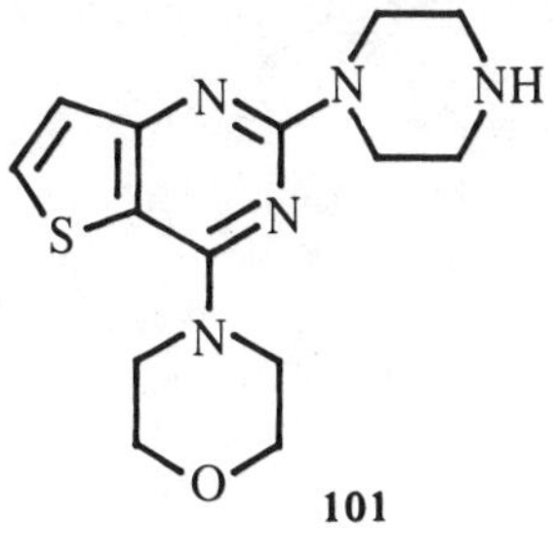

101

although no structure–activity relationship could be defined. The compound was an effective antithrombitic agent, with no fibrinolytic activity.[198]

A series of compounds was prepared to investigate hypotensive activity in normotensive rats.[199] The phenyl isosteres of **102** were the most active, and most

102

thiophene derivatives were devoid of activity; compound **102** was the only thienyl derivative with pharmacological activity. Compounds that inhibit the enzyme carbonic anhydrase were investigated as possible cerebrovasodilators. Based upon similarities to acetazolamide, which is a carbonic anhydrase inhibitor diuretic, and some 1,4-benzenesulfonamides, which are potent anticonvulsants, **103** was prepared

103

as a novel sulfonamide with potential anticonvulsant cerebral vasodilation activity. Sulfone derivatives were superior to sulfoxides and sulfides (**103**, $n = 2 > 1 > 0$) and 3- or 4- halogen substitution on the aryl groups produced the best activity. The most interesting compound, **103** (Ar = 4-fluorophenyl, $n = 2$, R = H), had anti-convulsant properties and caused increased cerebral blood flow without diuresis.[200]

4.   Diuretics

Diuretics are the cornerstone of antihypertensive therapy. Two independent mechanisms operate when diuretics are used to control blood pressure: (1) reduction in extracellular fluid by excretion and (2) alteration of vascular responsiveness to catecholamines. Levels of sodium and potassium excreted are controlled to some extent by the area of the kidney affected by the drug. Loop-type diuretics such as furosemide and ethacrynic acid, aldosterone antagonists such as spironolactone, and agents affecting cortical dilution such as the thiazides are all used clinically in the treatment of edema and essential hypertension and yet operate in different ways.

With interest in diuretics dating back 70 or more years to mercurial agents, it is not surprising that thiophene derivatives have received attention during the history of diuretic development. Based on the dramatic activity of chlorothiazide and hydrochlorothiazide, a number of thiophene disulfonamides were prepared.[201] The most active member of the series (**104**, R = CH_3CH_2-, R^1 = H) had one-fifth the activity of hydrochlorothiazide and caused excessive sodium excretion. The chloro-derivative had moderate activity, and the bromo- compound was inactive (**104,**

$$SO_2NHR^1$$

R^{1}NHO$_2$S   S   R

104

R = Cl or Br). The effects of positional substitution on thiophene were examined in **104** (R = CH$_3$, R^1 = H), **105,** and **106.** Methyl derivative **105** was moderately active, similar to the 5-methyl **104,** but sulfonamide isomer **106** was only weakly active. As was the case for the phenyl isosteres, nitrogen substitution (**104,** R^1) destroyed diuretic activity.[201]

CH$_3$ H$_2$NO$_2$S   CH$_3$

H$_2$NO$_2$S   S   SO$_2$NH$_2$ S   SO$_2$NH$_2$

105   **106**

Based on the thiadiazole diuretic diamox and some reports that the thiophene sulfonamides were inhibitors of carbonic anhydrase,[202] 12 sulfonamides were examined for their diuretic activity.[203] Some of these compounds had been pre-pared previously.[201,204] The most interesting compounds in the study were **104** (R = R^1 = H) and **104** (R = CH$_3$, R^1 = H). Bisthiophene compounds and mono-sulfonamide derivatives showed only marginal activity. Unfortunately, these com-pounds caused elevated levels of potassium excretion and, consequently, little future work was planned. A 2-thienyl-substituted thiazolo[2,3-b]-1,3,4-thiadiazole exhibited significant diuretic activity.[205] A series of β-substituted aryl ethylamines, including a thiophene derivative, caused diuresis in rats.[206]

Investigation of the diuretic "T.S.S." (**107**) was reported within the last decade.

S   S   SO$_3^-$Na$^+$

S

S

107

The drug caused saluresis accompanied by increased urinary volume; potassium excretion and the Na/K ratio also increased.[206] The effectiveness of the compound was judged to be intermediate between chlorothiazide and furosemide. Its effects were caused by the inhibition of sodium reabsorption in the ascending limb of the loop of Henle.[207] Further studies on **107** established that the site of action of the drug was the ascending limb and that T.S.S. had a mechanism of action similar to furosemide and ethacrynic acid.[208]

The important clinical diuretic, tienilic acid (**108**), is a thiophene derivative.[209]

108

The compound was prepared as a member of a series of heteroaroyl-substituted phenoxyacetic acids and was the most interesting derivative of the series. The 3-thienyl isomer of this compound was more toxic and caused more potassium excretion. Substitution on the thiophene of **108** or replacement by furan also reduced biological activity. Although these compounds were extensions of previous work done on ethacrynic acid, the diuretic profile of **108** showed a strong hypo-uricemic activity unlike other diuretics. Extensive secondary evaluation of **108** was also reported.[209] Unfortunately, the drug produced adverse reactions, with about 40 cases of liver damage reported in late 1979. This toxicity caused the FDA to have tienilic acid removed from the market. Although **108** is still sold as a diuretic in Europe (the benefits outweigh the risk of hepatic damage), close medical supervision is now required during clinical use.

A series of vinylogous analogues of tienilic acid were reported recently.[210] As was the case for tienilic acid, **109** was the best compound of the 20 derivatives prepared, and the 3-thiophene isomer was less active. The diuretic activity of **109** was half that of ethacrynic acid; **109** was also half as toxic.

109

Finally, annelated analogues of tienilic acid have also been reported.[211] A variety of aroyl substituents were investigated and the best compound was thenoyl derivative **110**. When the carbonyl was reduced to a carbinol or a methylene group,

110 CO_2H

natriuretic activity was greatly reduced. When the carboxylic acid moiety of **110** was replaced by an ester, aldehyde, hydrazide, or acyl guanidine, considerable activity was retained. The (+) enantiomer of **110** was twice as natriuretic as the racemate and 80 times more potent that tienilic acid. Interestingly, whereas the diuretic and saluretic activity was affected by the (+) enantiomer, the (−) enantiomer of **110** caused uricosuric activity.[211,212]

IV. METABOLIC DISEASE THERAPY

Control of metabolic diseases using synthetic agents has been a subject of intense investigation since 1899, when acetylsalicyclic acid was first used clinically. In addition to salicylates, steroids and arylacetic acids have been used as anti-flammatory, antipyretic, and antiarthritic drugs. These agents also have some efficacy in the treatment of allergies, as do antihistamines. During the long course of investigation of these drugs, a large number of thiophene derivatives were prepared that found use in the clinic. Other areas of metabolic disease research include investigations of diabetes mellitus, hyperlipidemia, atherosclerosis, and several prostaglandin mediated processes. Many of the latter areas of investigation developed recently since the discoveries in microbiology and biochemistry began to elucidate their metabolic pathways. As a result, a great deal of biological research evolved contemporaneously with increased utilization of thiophene derivatives in medicinal chemistry.

1. Anti-Inflammatory Agents

Antiphlogistic or anti-inflammatory drugs frequently have analgesic properties. Several thiophene derivatives already mentioned in the CNS section of this chapter have useful antirheumatoid effects. Both steroidal and nonsteroidal agents have been clinically efficacious in treating inflammatory diseases, but steroids have severe side effects that reduce their utility. Nonsteroidal anti-inflammatory drugs, such as aspirin, and the newer arylacetic acid derivatives, such as indomethacin, have received intensive investigation in attempts to improve potency and minimize their side effects (especially in regards to gastrointenstinal disturbances).

Suprofen (R-25, 061, **111**) was first reported as a potent inhibitor of prostaglandin biosynthesis.[213] This analogue of ibuprofen was the most interesting compound of a series of heteroaroyl benzeneacetic acids and showed a marked anti-writhing activity. The compound was subsequently found to be a peripheral analgesic and to be active against adjuvant-induced inflammation in rats. Additional studies demonstrated that **111** had nonnarcotic analgesic properties and was equipotent to new peripheral analgesic drugs such as zomepirac and diflusinal.[214] Suprofen appeared to be a tissue-selective inhibitor of prostaglandin synthesis and continues to be an interesting pharmacological agent.[214] The isomeric tiaprofenic

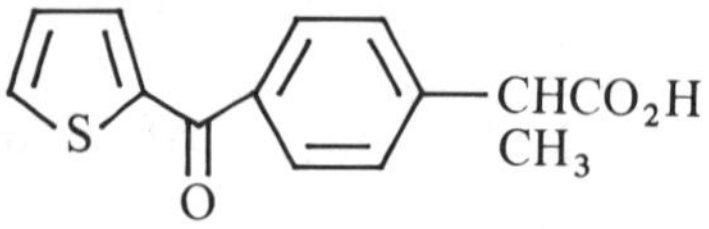

111

acid (**112**, RU-15060) was prepared by a different research group. This compound was equally as active as indomethacin for treating carrageenin-induced inflammation.

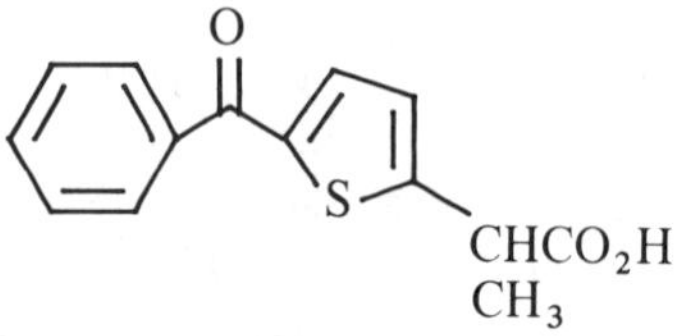

112

The activity of **112** was less than that of indomethacin on sustained models of inflammation, while **112** had a tendency to induce gastric ulcers.[215] The pharmacokinetics of **112** in humans and dogs has been reported.[216] Other 2-benzoyl derivatives of **112** were reported, as well as some additional isomeric thiophene compounds. In general, these compounds were equally efficacious both as analgesic and as anti-inflammatory agents.[217] Structurally rigid analogues of suprofen were prepared to investigate conformational requirements of the drug receptor site. Of the three thenoyl derivatives prepared, **113** was the best anti-inflammatory agent, using carrageenin-induced foot edema in rats as the measure of activity.[218]

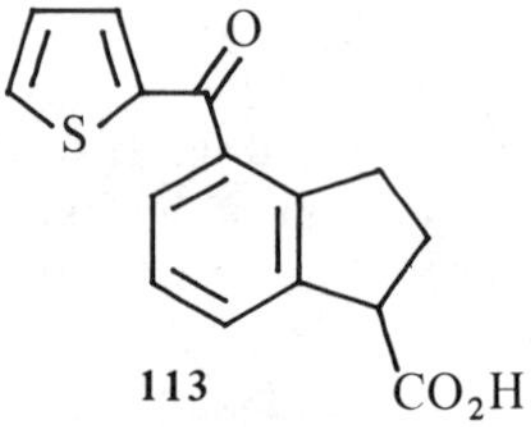

113   CO$_2$H

The thiophene isosteres of indomethacin were recently prepared. The series of compounds **114** (R = H, CH$_3$; R^1 = H, CH$_3$, CH$_3$CH$_2$–) was synthesized via Fischer-indole cyclization, but was found to have significantly weaker antiphlogistic

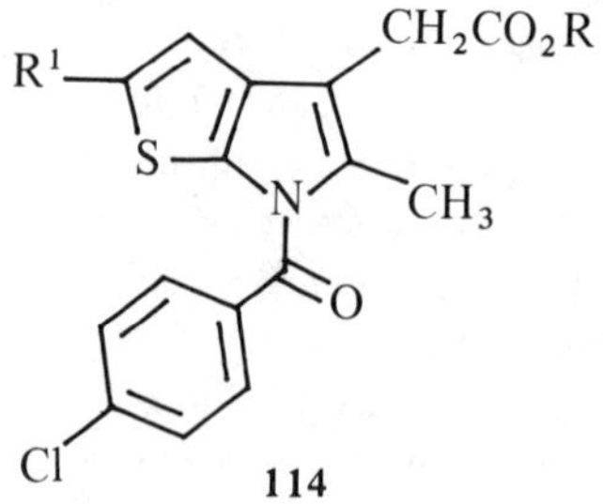

114

and analgesic effects than the phenyl analogue.[219] A brief report on the synthesis of **115** (X = O, R = H) appeared, wherein **115** was compared to its active benzene isostere as well as to its furan counterpart. Compound **115** (X = O, R =

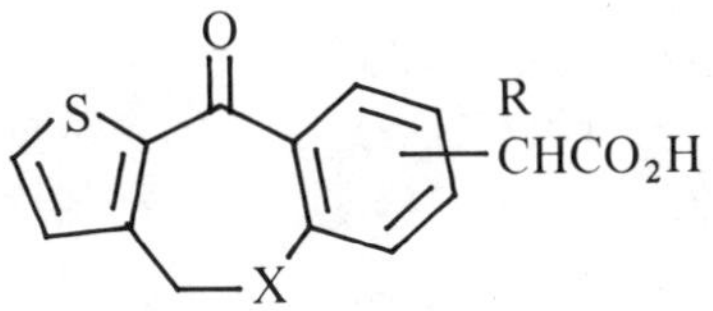

115

H) was twice as potent as its benzene isostere and 10-fold more active than the furan compound. The therapeutic ratio (anti-inflammatory activity/gastric irritation liability) of this thiophene derivative was 25 times that of indomethacin.[220]

In a related but more detailed study, a different research group prepared some derivatives and isosteres of dibenz[b,c]oxepin-3-acetic acids. Two annelated thiophene derivatives, **115** and **116**, were synthesized, as well as numerous dibenz-

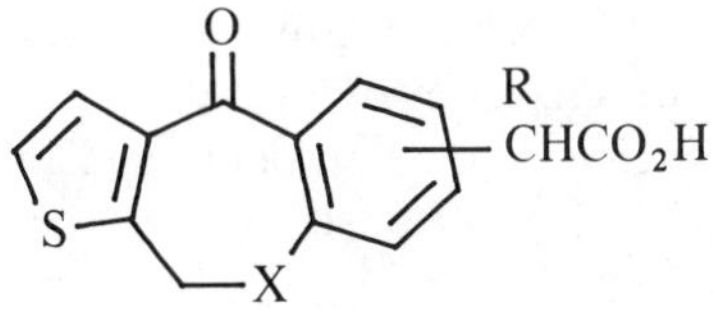

116

oxepins. Different side chains (R = H, CH$_3$) and varying side-chain positions and heteroatoms (X = O, S) were investigated. None of the thiophene derivatives were as active as the dibenzoxepins; **116** (X = S, R = H) had the best activity (0.59× indomethacin) of these isosteres.[221] Additional synthetic work directed at **115** (X = O, S; R = H, CH$_3$) and **116** (X = S, R = H) was reported, but no additional pharmacological activity was revealed.[222] Other arylthiophene acetic acids had highly potent anti-inflammatory properties.[223]

Fenamic acid analogues and other thiophene amino acid derivatives have also been prepared as antiphlogistic agents. 4-Anilino-3-thiophene carboxylic acid derivatives (**117**) were reported to be equipotent with indomethacin and flufenamic acid.[224] Additional development led to adantate (HOE-473, **118**) which was a

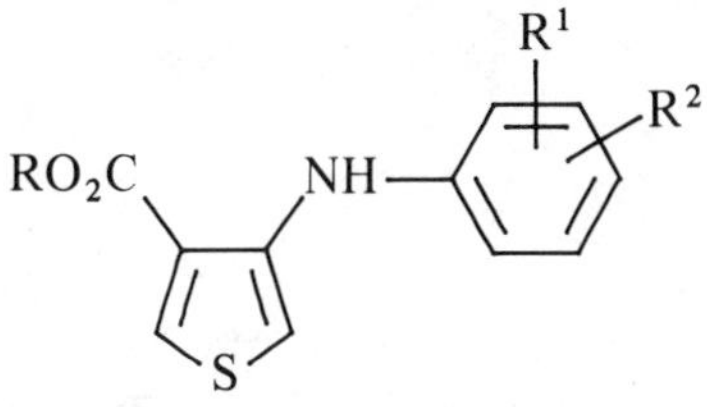

117	R = H
118	R = CH$_2$O$_2$CCH$_3$, R^1 = 2-Cl, R^2 = 3-CH$_3$

highly active antiphlogistic agent. The compound was compared favorably to phenylbutazone and indomethacin and did not adversely affect the gastric mucosa. This compound was the subject of further investigation.[225] Other thiophene acid derivatives have been found to be anti-inflammatory agents. Tinordin (47), discussed earlier as an analgesic agent, also has good anti-inflammatory activity.[84] A closely related thienopyrimidine (119) was prepared and showed good activity

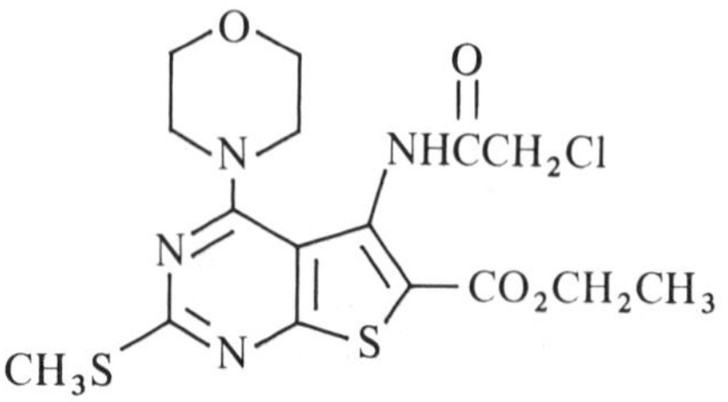

119

against induced rat paw edema.[226] Other related thiophene-containing substances that have reported anti-inflammatory properties are 120 and 121. Compounds of the former type that had appreciable inhibition of carrageenin-induced edema in mice were aryl derivatives of 120 (R = 4-CH$_3$C$_6$H$_4$— or 4-FC$_6$H$_4$—).[227] The second series of compounds had analgesic, anticonvulsant, and antimicrobial activity, with 121 having the best anti-inflammatory activity.[228]

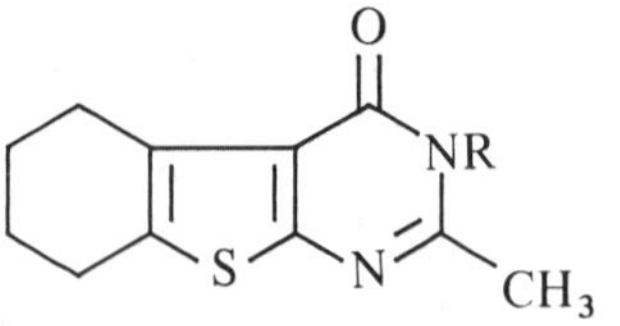

120

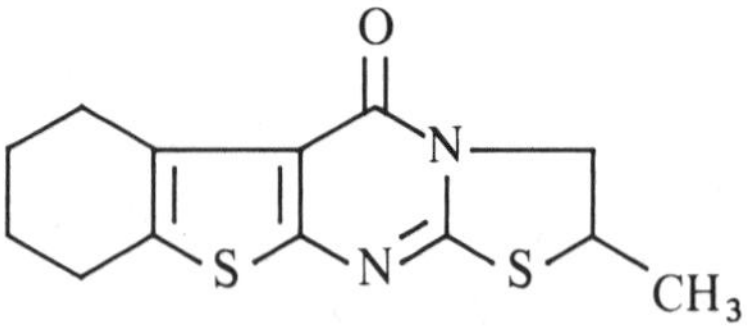

121

A variety of structures other than those reported above have shown antiphlogistic properties. Aroyl acetonitriles and related enamine derivatives were reported to be potential antiarthritic agents. Among the most interesting compounds were the 2-thiophene compound 122 and its 3-substituted isomer. The 2-substituted derivative

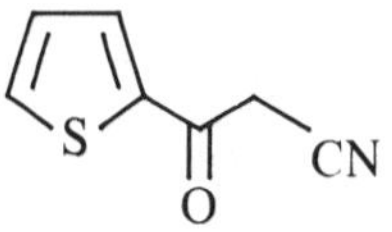

122

was more active in the rat adjuvant arthritis model and none of the active compounds in this study acted via prostaglandin synthetase inhibition.[229] A series of 2-hydroxyethyl and carboxyalkyl ethers of aromatic oximes had pronounced activity against carrageenin-induced edema in rats.[230] Some 4H-1,4-benzothiazines were synthesized and investigated for anti-inflammatory activity. The phenyl isosteres of **123** were effective in preventing carrageenin-induced edema in rats, but thienyl derivative **123** was completely ineffective.[231] Several 6-thenoylbenz-

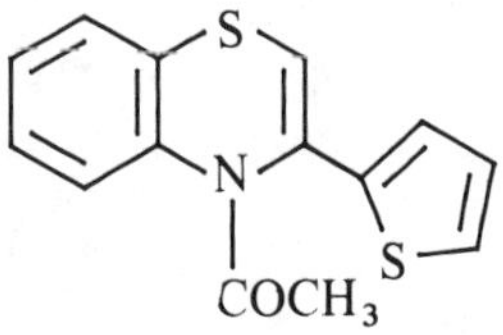

123

oxazolines had anti-inflammatory and analgesic properties.[232,233] A series of 3-substituted pyrroles were anti-inflammatory agents; the 2-thienyl derivative was 25% as active as the best phenyl compounds in the study.[234] Several derivatives of sulfolane were developed as a new class of anti-inflammatory agents. Many of the derivatives of **124** compared favorably to phenylbutazone, hydrocortisone, and brufen.[235] Other aryl sulfoxides have also been found to have anti-inflammatory activity.[236]

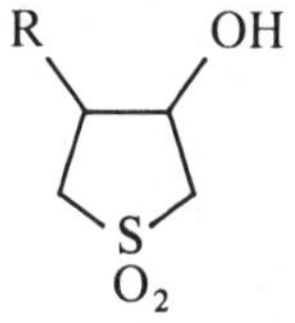

124

2. Prostaglandin Analogues

Prostaglandins occur in nearly all organs of the body and cause diverse physiological effects. Prostaglandins, as well as other products from the arachadonic acid cascade such as thromboxanes and prostacyclins, have been implicated in the processes of vasoconstriction and vasodilation, platelet aggregation, fertility control, and inflammatory processes. Many anti-inflammatory agents mentioned in the previous section are believed to interfere with prostaglandin synthetase as their mode of action. It was inevitable that agents with so many powerful biological effects would be the target of organic chemical syntheses. Thiophene analogues of various prostaglandins have appeared during the last decade.

The preparation of thiaprostaglandin analogues has been one area of interest. Several reports have appeared on the synthesis of **125**, but no biological activity

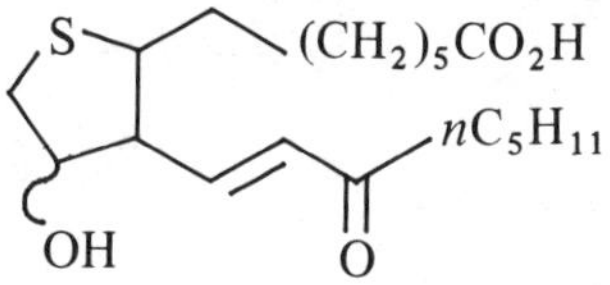

125

was mentioned.[237,238] The report of the structure and action of prostacyclin in 1976 led almost immediately to the synthesis of sulfur analogue **126** as a stable physiological mimic of the naturally occurring material.[239,240] The free acid **126** (R = H) was comparable to natural prostacyclin in platelet aggregation inhibition but, in contrast, **126** caused vasoconstriction. The compound is stable for several hours in neutral solution.[239] The S-oxide and S,S-dioxide of related compounds were also prepared.[241]

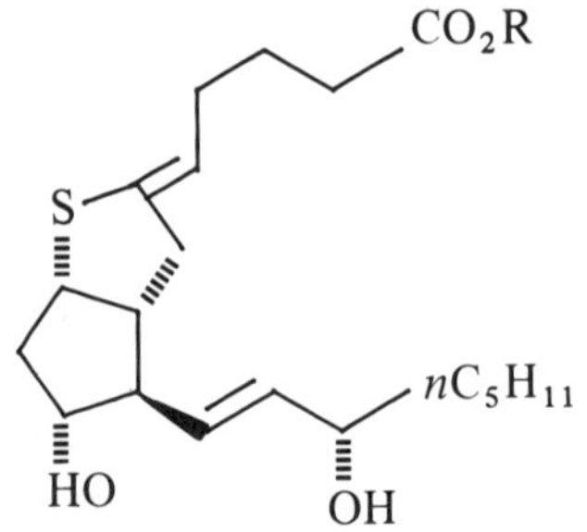

126

Thiophene-substituted prostaglandins have also been prepared in order to maximize the desirable effects of these agents. Several derivatives of 11-desoxy-prostaglandin $F_{2\alpha}$ were reported to have superior luteolytic activity, compared to $PGF_{2\alpha}$. The all-hydrogen (**127**, X = H) and the 5-chloro derivative (**127**, X = Cl)

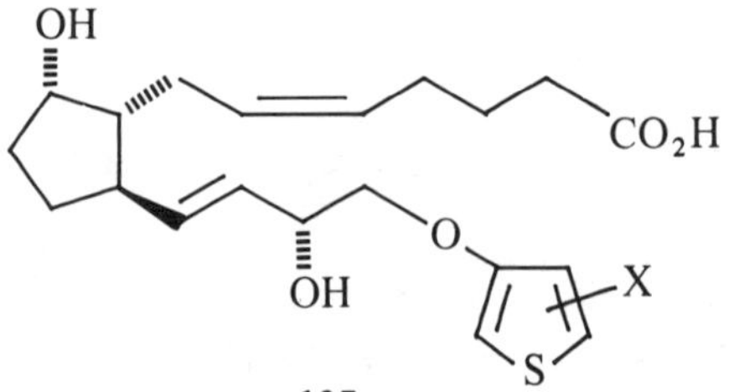

127

were 25 times more potent than the natural product.[242] Using the general synthetic procedures of Corey, prostaglandin analogues of $PGF_{2\alpha}$ with a 16-thienyloxy side chain (**128**) were prepared. A variety of substituted thiophenes (**128**, R^1, R^2 = CH_3, CF_3, Cl, and Br), as well as positional isomers of thiophene, were investigated. The most interesting compound (designated HR-847) was a 3-thienyl isomer that

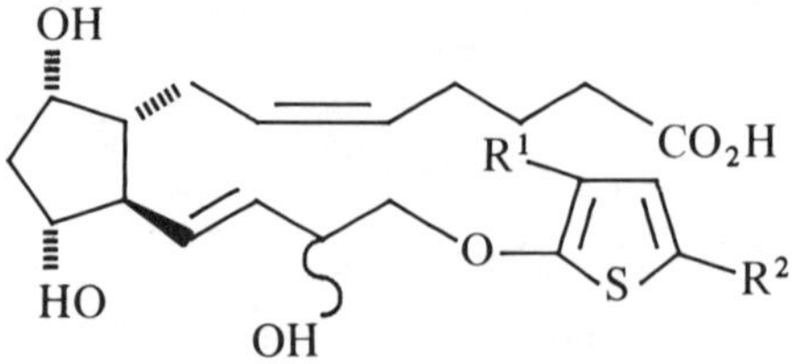

128

was 100 times more active than natural $PGF_{2\alpha}$ in terminating pregnancies in hamsters.[243,244] The compounds were also more potent than the desoxy analogues 127.[243,245] Acetylenic thiophene derivatives of prostaglandins, wherein a structurally rigid moiety replaced the naturally occurring *n*-pentyl group, also had luteolytic activity.[246]

3.   Steroid Analogues

Steroids, like prostaglandins, have been implicated in a variety of physiological processes. Besides sexual differentiation, these compounds affect inflammatory reactions, reproduction, and tumor development. Unlike prostaglandins, which have only been subjects of intense investigation for the past 15 years, steroid research has been an active area for decades. Numerous thiophene analogues have appeared in the literature.

One of the earliest reports of a thiophene-containing steroid is the synthesis of 129, an analogue of 3-desoxyisoequilenin; no estrogenic activity was reported.[247]

A recent synthesis of thiophene analogues of equilenin has been reported.[248] Steroidal [2,3-c]thiophenes 130 and 131 were prepared by several multistep syntheses. Substituents (X) on thiophene were varied during the course of synthesis.

Biological activity of these androstane analogues was not specified, but other papers were cited.[249] Another thiasteroid (132) was prepared in order to study optical rotatory dispersion phenomena.[250]

An extensive study of androgenic thiasteroids has been the subject of numerous papers. In the first of a series of reports, **133** ($R^1 = CH_3CO_2-$, $R^2 = R^3 = H$, n = 0) was the most active compound, with potency equal to that of testosterone.[251] Later, additional derivatives of **133** (n = 0, 1, 2) were prepared. The sulfoxide and

133

sulfones were inactive androgenic agents. This series of compounds measured the effects of 7α-methyl, 19-nor-, and 17α-alkyl modifications upon anabolic–androgenic activity and the results indicated that steric effects are the important factor in these drug–receptor interactions.[252] Additional related androstane derivatives have also been prepared.[253] Recently, some 7β-methyl analogues of **133** were reported and tested for antitumor activity. The compounds did not reduce breast tumor weight significantly, and the 7β-methyl substituent decreased myotrophic–androgenic activity.[254]

Several papers have appeared reporting the synthesis of A-nor-3-thiasteroidal systems represented by **134**. No biological activity is revealed;[255,256] a review of

134

heteroatom-containing steroids included these compounds.[257] A total synthesis of A-nor-3-thiaestra-1,5(10)-dien-17β-ol was recently published, but no biological activity was reported for this derivative of **134**.[258,259] The perhydrogenated analogue of **134** has also been prepared.[260] Other partially hydrogenated systems related to **134** have been reported, but pharmacological activity was not studied.[261] A chiral synthesis of **135** using a biomimetic acid-catalyzed ring closure has been accomplished.[262] Racemic **135** had previously been prepared.[263,264] The thiophene isostere of 16-thia-D-homoestrogen was prepared in the hope of improving the antifertility and serum lipid-lowering properties of the steroid. Compound **136** was synthesized using a multistep sequence, but no pharmacological data were reported.[265]

135

136

Estrogenic activity and antihormonal action of compounds related to diethyl-stilbestrol were the subjects of several reports in the 1950's and early 1960's. Many thiophene isosteres of stilbenes have been prepared, but they have no activity.[266-268] Dithiophene derivatives of **137** also had no estrogenic activity.[269] The thiophene isostere of diethylstilbestrol (**138**) was prepared; interestingly, **138** (X = H) was

137

138

active as an estrogen, whereas its phenyl counterpart was not.[270] When the thio-phene of **138** was oxygen-substituted (X = OH, OCH$_3$), the compounds were marginally estrogenic.[271]

4. Antihistamines

Histamine is a naturally occurring amine synthesized by biorganisms via decar-boxylation of histidine. It is a potent vasodilator, but causes strong stimulation of bronchial muscle and gastric acid secretion. Early trials attempting to desensitize individuals to the effects of histamine led to the development of the ethylene-diamine and aminoalkylether series of antihistamines in the mid-1940's. Subsequent research revealed these early antihistamines to be H$_1$ antagonists that were effective in treating allergic and bronchial reactions. These compounds strongly interfere with histamine, but exert local anesthetic, sedative, and spasmolytic actions in clinical usage. A second type of antihistamine, designated an H$_2$ antagonist, was defined in the early 1970's and has been found effective in the treatment of peptic ulcer. Cimetidine is the most notable example of this type of antihistamine to date.

Among the most clinically important thiophene derivatives are those that were prepared in the 1940's and 1950's as antihistaminic agents. The large number of published reports on compounds such as thenalidine, thenyl diamine, metha-phenaline, and chlorothen precludes a detailed review in this chapter, but the articles by Blicke[1] and Martin-Smith and Reid[2] give a good account of most of this work. Methapyrilene (**139**) has had extensive clinical use as both a prescription and an over-the-counter medication. It was recently removed from the market because of potential chronic toxicity. Structure–activity studies on **139** demonstrated the effects of 2- vs. 3- substitution on thiophene as compared to the phenyl

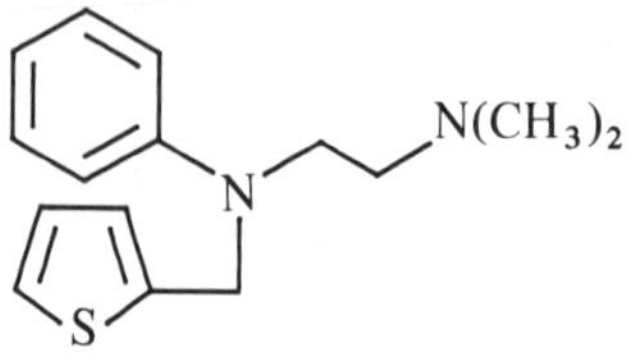

139

isostere. The 3-substituted thiophene isomer is three times more active than **139**,[272] and **139** is two times more active than the phenyl derivative.[273,274] A molecular structure determination of **139** supported the premise that antihistaminic activity was the result of competitive binding at the receptor site.[275]

Several other simple thiophene derivatives have been reported as antihistamines. Oximoesters of 2-benzoylthiophene (**140**) were prepared and had good activity. The best compound of the study was **140** (R = CH_3CH_2-), which had antihistam-

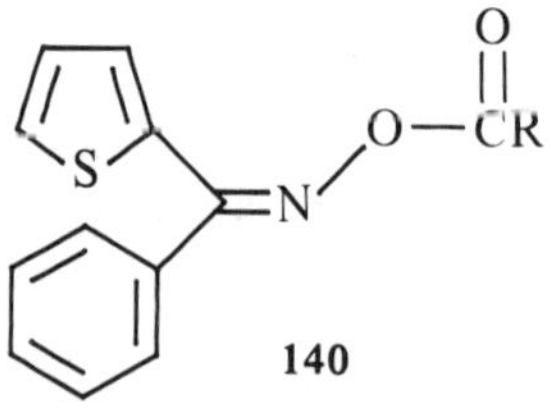

140

inic but no anticholinergic activity.[276] Amides and esters of substituted thienyl-acrylic acids were prepared. Compounds **141** (X = H, CH_3, CH_3CH_2-, Cl) showed good antihistaminic activity in isolated guinea pig ileum.[277] Several thenoyl 2-chromones were recently found to have slight antiallergy activity.[278] A group of sulfolanes were also reported to have antiallergic/decongestant properties.[279]

As mentioned, tricyclic neuroleptic drugs were discovered as a result of research on tricyclic antihistamines. Pizotyline (BC-105, **40**) is a thiophene-fused tricyclic derivative having antihistaminic properties as well as antidepressant effects. Ketotifen (HC-20-511, **142**) was recently reported as a new, orally active antianaphylactic

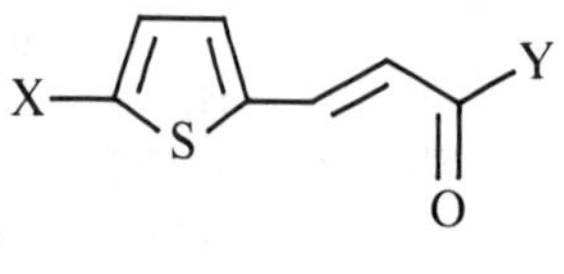

141

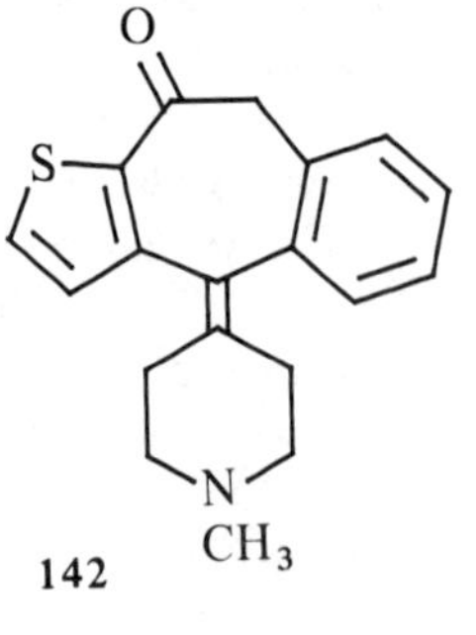

142

agent.[280] Compound **142** inhibited histamine release from mast cells, inhibited cAMP phosphodiesterase in various organs, and proved to be an effective antihistamine. Its effects were equal to clemastine and were of long duration with minimal side effects.[280] Several clinical studies showed **142** to be superior to known drugs.[281–283] Various aryl-substituted derivatives of **142**, as well as the isomeric ketone and its derivatives, were reported prior to the detailed pharmacological profile mentioned above;[284] no biological effects were discussed in this paper.

Dithiadene (**143**) was developed as an antihistaminic and antiallergic agent, based upon the activity of the analogous dibenzo[b,e]thiepin system.[285] This

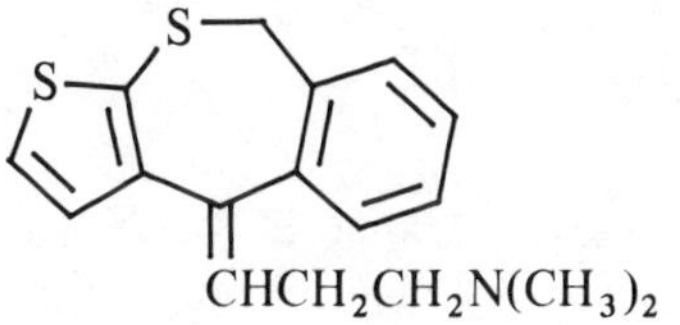

143

compound also had antidepressant properties. A subsequent report demonstrated that the two side-chain geometrical isomers of **143** do not differ in degree of antihistaminic activity. Other compounds related to **143** were also prepared, but all had significantly reduced antihistaminic activity.[286] The piperidyl analogue of **143**, as well as the sulfone and sulfoxide derivatives, were synthesized, but these compounds had only weak antihistaminic activity with increased toxicity.[287] The thieno[3,2-c]-isomer of **143** has been prepared more recently. Compound **144**

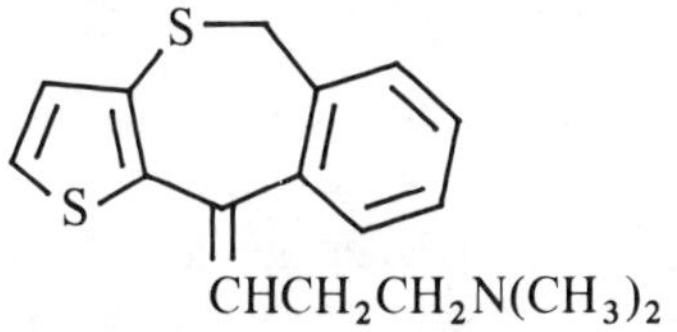

144

displayed intense antihistamine effects in guinea pigs, had 50% of the activity of dithiadene in histamine detoxification tests, and did not show significant antidepressant effects. The derivative was concluded to be a potent antihistamine.[288] The related compounds **145** have been prepared, but had little activity of interest;[287] other related compounds also were inactive.[289]

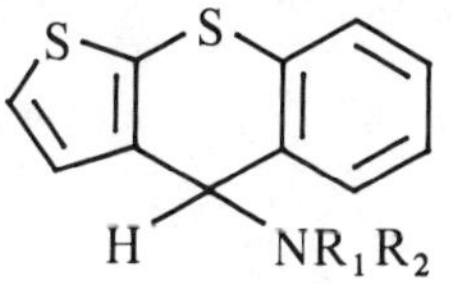

145

A small series of dialkyl-substituted thieno[2,3-d]pyrimidines was prepared to investigate their pharmacological profile. Of these, the most interesting was QM-1143 (146, R = CH₃), which had antihistaminic potency similar to that of dephen-

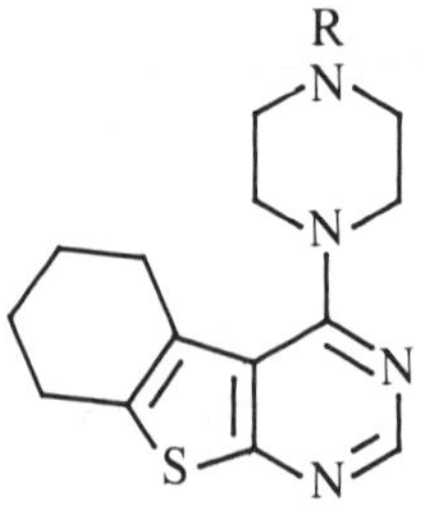

146

hydramine. The compound also had some CNS effects typical of other antihistaminic agents.[290] A detailed report on 4-oxothieno[2,3-d]pyrimidines (147) appeared recently. These compounds showed weak oral activity as antiallergic agents, as

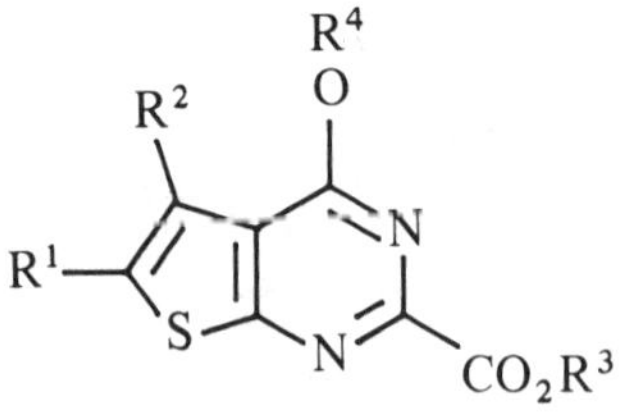

147

measured in the rat passive cutaneous anaphylaxis test. The best compounds had H or CH_3 at the 5-position (R^2) and lower alkyl at the 6-position ($R^1 = R^3 = CH_3CH_2—$; $R^2 = R^4 = H$).[291] The related pyrido[1,2-a]-fused systems (148) also had antiallergy activity. Substitution of lower alkyl at R^1 and an acidic functional

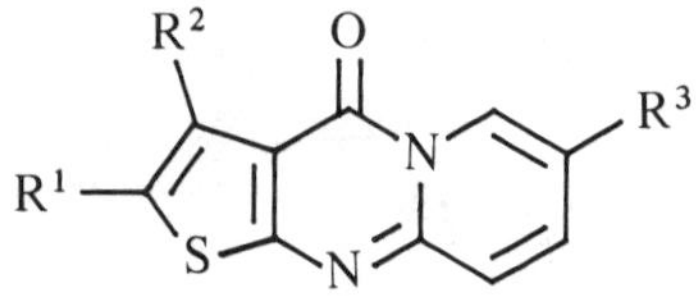

148

group at R^3 gave compounds with potent activity. The best compounds (147, $R^1 = CH_3CH_2—$; $R^2 = H$; $R^3 = CO_2H$ or tetrazol) were orally active and were 10 times more active than doxantrazole.[292] Thienyl-substituted imidazo[2,1-b]-quinoxalin-5(10H)-ones were discovered to be a new class of bronchodilating agent.[293]

5. Hypolipidemic Agents

Attempts to control the levels of circulating cholesterol and plasma triglycerides recently gained clinical importance, since epidemiological studies correlated elevated serum cholesterol levels with an increased propensity toward heart attack. Currently, studies indicate that the ratio of high density lipoprotein (HDL) to low density lipoprotein (LDL) may even be more important in this regard. Clofibrate and procetofene are two clinical agents used to control cholesterol and triglyceride levels.

The thiophene analogues of clofibrate and procetofene were recently synthesized. The former, **149** (X = O, R = Cl), had only weak hypolipidemic activity as

149

compared to clofibrate. When oxygen was replaced by sulfur (X = S), the resultant compound was far more potent than clofibrate as hypocholesteremic and hypotriglyceridemic agent. Even the deschloro analogue **149** (X = S, R = H) had potent activity. The thiophene isostere of procetofene (**149**, X = S, R = —COC$_6$H$_4p$Cl) was a hypolipidemic agent, but had less potent cholesterol-lowering activity.[294] A related derivative (**150**) was prepared as part of a series of alkyloxyarylcarboxylic

150

acids. In this study, branched p-alkoxy side chains had inferior biological properties and both thiophene and furan analogues were superior to their phenyl isosteres. These compounds lowered blood lipids and inhibited fatty acid synthesis.[295] Related thienyl-substituted phenyl derivatives also had hypocholesterolemic activity.[296]

Some benzo[b]thiophene-2-carboxylic acids, as well as their thienyl isosteres, were prepared and their hypolipidemic activity was measured. The mode of annelation of thiophene had great importance, since the thieno[3,2-b] system (**152**) showed activities exceeding those of clofibrate, but the thieno[2,3-b] isomers (**151**)

151

152

were less active. Chlorine substitution enhanced, while fluorine or methyl substitution reduced, activity. The best compound in the series was 152 (R = Cl, R^1 = H), which was far more potent than clofibrate in reducing triglyceride and cholesterol serum levels.[297]

Based on reports of the hypolipidemic properties of a series of 5-sulfamoylbenzoic acids, thiophene isosteres (153) were prepared. Although cholesterol levels were not reduced in test animals, 153 (R = CH$_3$) reduced triglycerides in a manner similar to the phenyl isostere and somewhat superior to clofibrate. Compound 153 (R = H) showed properties similar to those of clofibrate.[298]

HO$_2$C

Cl   S   SO$_2$N

R

R

153

6.   Hypoglycemic Agents

Diabetes mellitus is a metabolic disease wherein the body poorly utilizes blood glucose. One major clinical treatment of the disease involves the use of prescription insulin to enhance natural insulin secretion; this is the only successful method to treat juvenile diabetes, in which insulin synthesis and secretion is absent. In adult-onset diabetes, many mild cases may be controlled by dietary restriction of sugar intake and by agents that reduce levels of blood sugar.

Thiophene isosteres of the sulfonylurea group of hypoglycemic agents (154) have been the subject of an ongoing research program in Egypt. A series of thio-

X

Y
||
S   SO$_2$NHCNHR

154

phene sulfonylureas,[299,300] thioureas,[301] semicarbazides and semicarbazones,[302] thiosemicarbazides,[303] and 2,4-thiophenesulfonylureas[304] have been prepared and their effects on circulating blood sugar levels were measured. All of these derivatives generally were inferior to monosubstituted phenyl derivatives, and N-phenyl compounds showed the greatest hypoglycemic activity, as evidenced by prolongation of glucose-induced hyperglycemia in test animals.

A series of hypoglycemic aralkyl lactamimides was prepared to further develop the structure–activity relationship of derivatives previously found active. The compounds were tested in glucose-primed rats, and compound 155 as well as several furan and phenyl isosteres had activities as potent as tolbutamide. The size of

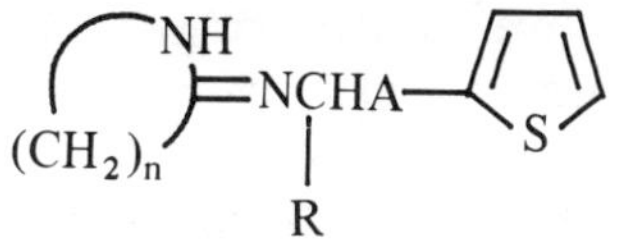

155

the lactam ring did not significantly affect activity, but alkyl substitution (R) enhanced activity. The best compounds of the study were **155** (n = 7, A = CH$_2$, R = CH$_3$CH$_2$—; n = 5, A = CH$_2$, R = allyl; and n = 5, R = cyclopropyl).[305]

Several other thiophene derivatives have potential for use in controlling blood sugar. The three possible thiophene isosteres of the artificial sweetner saccharin (**156–158**) have been prepared and the effects of thiophene annelation were demon-

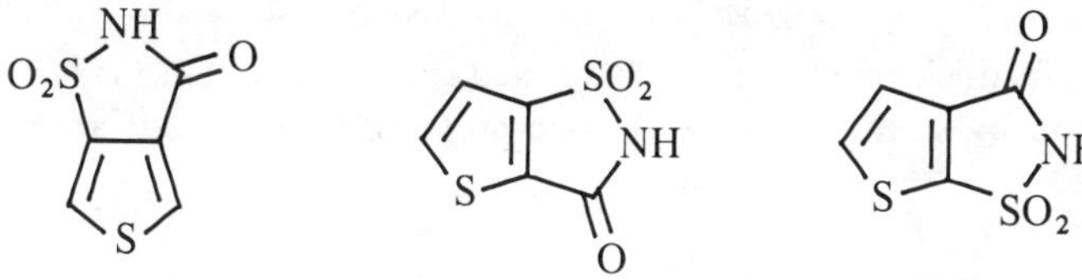

156   **157**   **158**

strated to be important. The [3,4-d] isomer **156** was 1000 times sweeter than sucrose and did not have the bitter metallic aftertaste associated with saccharin.[306] Considerable hypoglycemic activity for **159** (R = o-hydroxyphenyl) was reported as part of a study of 1,5-benzodiazepines.[307]

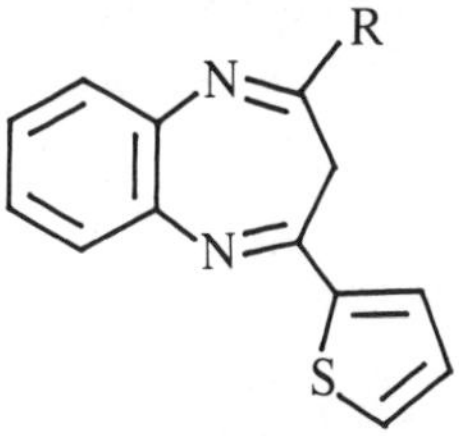

159

7. Platelet Aggregation Inhibitors/Antithrombitics

Agents that enhance or attenuate coagulation of the blood have important clinical use. Prevention of thrombitus and, thus, the prevention of strokes and heart attacks have been the target of much research; heparin and coumarin derivatives are used for this treatment. Coagulants, on the other hand, are used to treat several less-known metabolic diseases, such as hypoprothrombinemia, as well as overdoses of the anticoagulants mentioned above. These drugs probably affect prostaglandin synthetic pathways to produce their effects. Certainly, it is clear that thromboxane A$_2$, part of the arachidonic metabolic cascade, causes platelet aggregation.

Drugs that interfere with prostaglandin synthesis such as aspirin were reported to have anticoagulant properties. Hence, many of the compounds mentioned earlier as antiarthritic or analgetic agents affect platelet aggregation to some degree. The carboxylic acid derived from adantate (118), as well as other 4-amino-3-thiophene carboxylic acid derivatives, have both fibrinolytic and aggregation-inhibiting effects.[308] Another report of the clinical application of substituted N-phenyl-4-aminothiophene carboxylic acids noted that o-chloro substitution of phenyl produced the best antiaggregation effects. This study also found that N-(2-chloro-3-methylphenyl)-4-aminothiophene-3-carboxylic acid was the most active. These compounds were the first heterocyclic compounds observed to exhibit fibrinolytic activity.[309, 310]

Tinordin (Y-3642, 47), previously discussed as an analgetic agent, was reported also as a platelet aggregation inhibitor.[311] Alkanoic acid derivatives that are analogues of nonsteroidal anti-inflammatory agents also have platelet aggregation inhibitory effects. A number of 5-substituted 2-thienylacetic acids (161), as well as isomeric phenyl acetic acids (160), were prepared to investigate antiaggregatory

effects. The most potent compound of the study, 160 (R = CH_3), was more effective than aspirin at preventing collagen-induced platelet aggregation. These compounds behaved like nonsteroidal anti-inflammatory agents such as indomethacin in other tests.[312]

Thiophene derivatives of warfarin, 162, and their alcohol counterparts have

been prepared. The alcohols have enhanced anticlotting effects as compared to the ketonic derivatives.[313] A series of imidazo[1,2-a]thienopyrimidin-2-ones were found to have blood platelet aggregation inhibitory activity. Thieno[2,3-d]-isomers (163), as well as the [3,2-d]- and [3,4-d]- annelated derivatives, were prepared.[314]

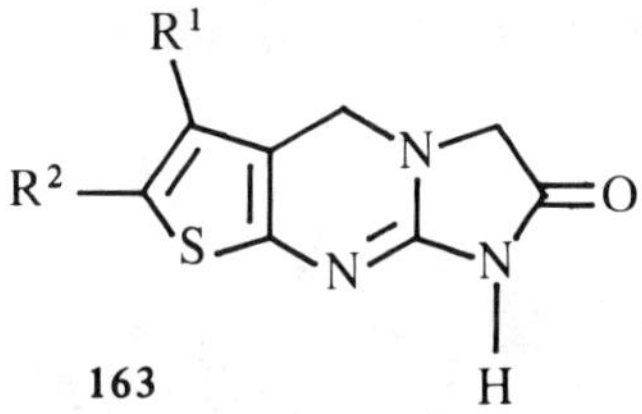

R¹ structure 163

163 H

Substitutional changes at R^1 and R^2 showed that increased lipophilic character around the thiophene nucleus enhanced inhibitory action and that the [2,3-d]-isomer **163** ($R^1, R^2 = -(CH_2)_4-$) was the most interesting compound. Compound **163** showed highly potent inhibitory activity against various aggregating agents.[315,316]

Ticlopidine (**164**) was found to be a potent blood platelet aggregation inhibitor

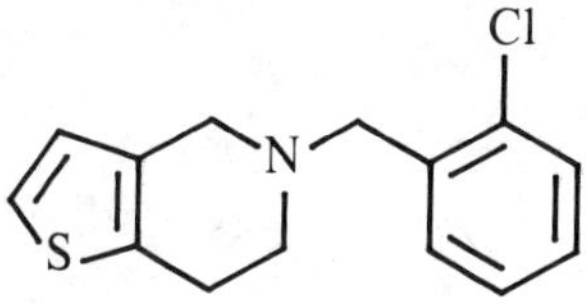

164

and antithrombitic agent.[317] A series of furan and thiophene compounds was prepared, but no structure–activity correlation could be established. Thiophenes were superior to their furan isosteres.[318] There have been alternative syntheses of **164** reported recently.[319-321] (5-Methoxy-2-thienyl)thioacetic acid (**165**) had

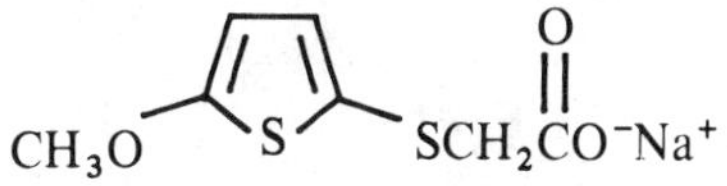

$$CH_3O \diagup S \diagdown SCH_2\overset{O}{\overset{\|}{C}}O^-Na^+$$

165

corticosteroid-like effects in test animals, including decreases in blood lymphocytes.[322] Finally, another thiophene isostere of dihydropyridamole was reported to have pharmacological activity. In this case, compound VK-744 decreased aggregation of platelets and was a fibrinolytic agent.[323,324] Several 3-thienylthiazolo[3,2-b][2,4]-benzodiazepines inhibited ADP-induced platelet aggregation in plasma.[325]

8. Other Agents

Several thiophene derivatives having antifertility properties in addition to analogues of steroids, were prepared. A series of thieno[2,3-d]pyrimidines including **166** were prepared and caused mild inhibition of pregnancy. Several also had some anti-inflammatory activity.[326] Compounds related to antihistaminic agent

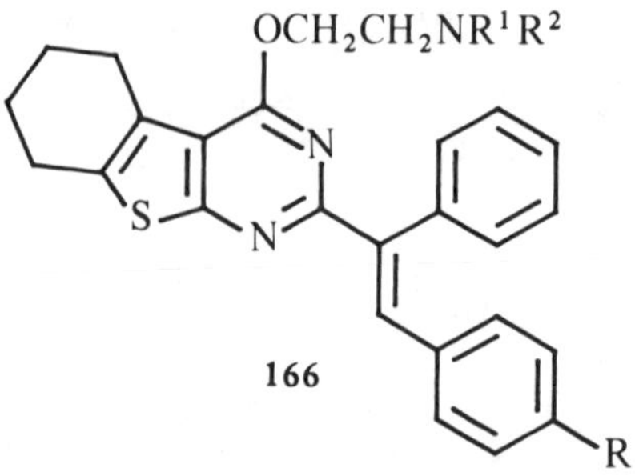

$$OCH_2CH_2NR^1R^2$$

166

R

142 were found to have ovulation-inhibiting properties; a series of compounds including **167** was prepared. Methyl substitution on nitrogen and on the seven-

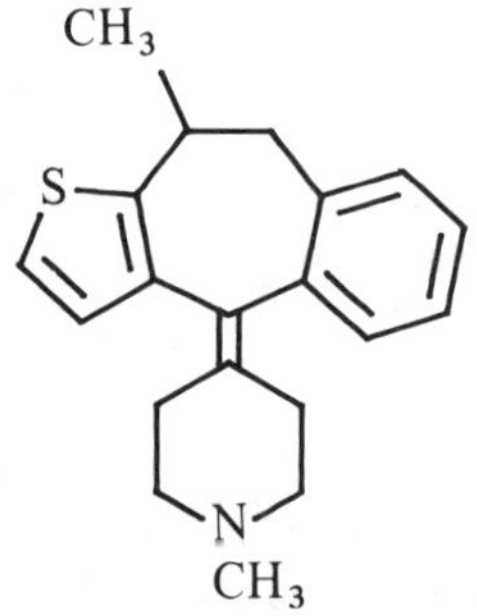

CH₃

167

ring enhanced activity and the exocyclic double bond was required. Derivative **167**, designated research number 26-921, was the best compound at inhibiting ovulation and secretion of the luteinizing hormone.[327-329]

A number of reports have appeared on the clinical utility of 2-thiophene-carboxylic acid and various substituted derivatives. The compounds have been found to lower serum calcium and inhibit other sclerotic processes,[330,331] inhibit bone resorption,[332,333] reduce blood sugar[334,335] and act as hypolipidemic agents.[336] Other studies on the biological effects of these compounds are too numerous to mention here.

V.   INFECTIOUS DISEASE THERAPY

The concept of chemotherapy (treating infections with drugs) was proposed by Paul Erlich nearly 80 years ago. Only a few agents, such as mercury, ipecac, and quinine, were utilized for these purposes prior to Erlich's time. Synthetic drugs were discovered that treated infections, albeit only those caused by protozoa. It was not until the 1930's that sulfonamide drugs were developed that effectively treated bacterial infection. Drug-induced toxic effects were also discovered in these agents, and the challenge to develop better anti-infectives with considerably lower toxicity was taken by the medicinal synthesis chemists. Fleming discovered penicillin and its antibacterial properties at about the same time. Later, semisynthetic penicillins and cephalosporins were developed, which advanced the art of treating infections still further. A number of other agents, such as macrolide antibiotics, streptomycin-like compounds, and tetracyclines, that showed greater specificity for certain types of infection were developed.

The concept of treating cancer with chemotherapeutic agents developed from ideas concerning anti-infectives. Cancer treatment attempts to interfere with RNA or DNA synthesis in the aberrant cancer cells with therapeutic agents which thus destroy the diseased cells.

1.   β-Lactam Antibiotics

The development of effective anti-infective agents is probably the single greatest contributor to the increased life expectancy of 20th-century man. Minor infections that could kill 100 years ago are now treated easily with little or no inconvenience to patients. The continuing challenge presented to medicinal chemists consists of preparing new antibiotics that have fewer side effects, better stability, and better activity against several infective strains that remain immune to chemotherapy. Resistant strains also pose an ongoing problem, since bacteria may develop resistance to some chemotherapeutic agents as time progresses.

Semisynthetic penicillins with thiophene incorporated into the molecules have been reported. Ticarcillin (BRL-2288, **168**) is the thiophene isostere of the clinically

168

important carbenicillin and has also been used in the clinic. A series of reports on this α-carboxy-3-thienylmethyl penicillin have shown it to be an efficacious antibiotic. In particular, **168** was the most active against *H. influenzae* and meningococci, as compared to benzylpenicillin, ampicillin, rifampin, and cephalothin.[337] It was two- to fourfold more active than carbenicillin against a majority of *Pseudomonas* strains, but was inactive toward *Klebsiella pneumoniae*.[338] Compound **168** was active against a variety of gram-positive and gram-negative bacteria and was bactericidal.[339] Human trials showed **168** to be effective against *Pseudomonas*, with no increase in resistance developed over 5 years of clinical use.[340] Other reports on ticarcillin and ticarcillin cresyl sodium have also appeared.[341-343]

Thiophene derivatives related to cloxacillin and oxacillin have also been prepared. The purpose of these syntheses was to determine if the thiophene analogues
of **169** would retain the penicillinase-stable features of the known drugs. Compounds **169** (R^1 = H, CH_3; R^2 = CH_3, C_6H_5—; R^3 = CH_3) showed good activity

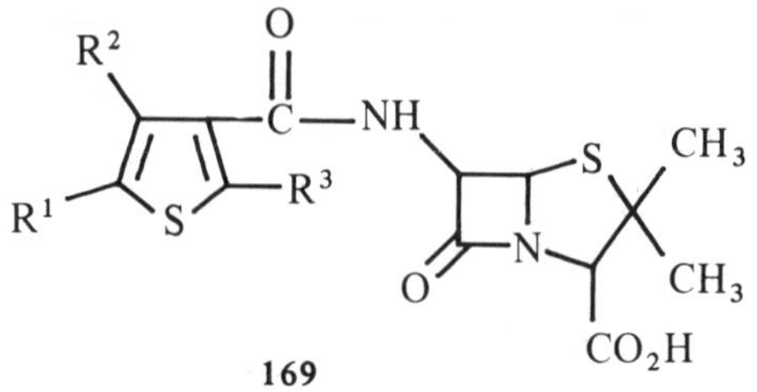

169

against penicillinase-producing staphylococci and were as active as oxacillin. When
the substituent R^1 was a t-butyl group, activity decreased. The derivatives that had
the highest activity were less stable to acid than oxacillin and, therefore, would
probably not be well-absorbed orally.[334] Ampicillin analogues containing an α-
thienylglycol side chain have also been prepared,[345] and ketenimino penicillins
have been reported.[346]

Bisnorisopenicillin (**170**) was prepared to investigate whether this structural
type might exhibit the increased antibacterial spectrum characteristic of cephalosporins. Compound **170**, as well as its monocyclic precursor, exhibited activity

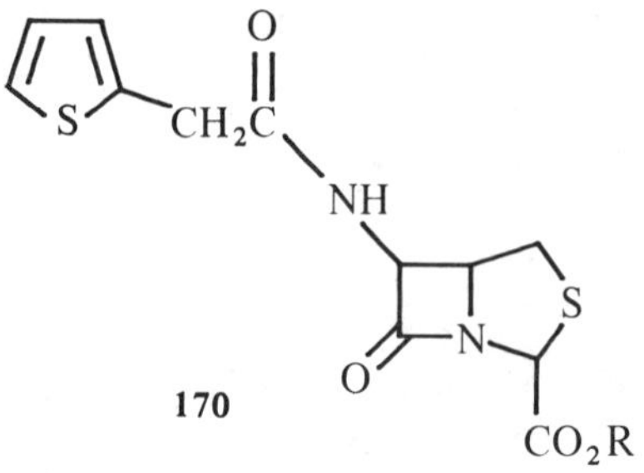

170

against gram-negative bacteria comparable to thienylpenicillin (**171**). The compound was substantially less active than **171** against *Staphylococcus aureus* or
Streptococcus faecalis.[347] Thienylpenicillin (**171**) had been prepared earlier as part

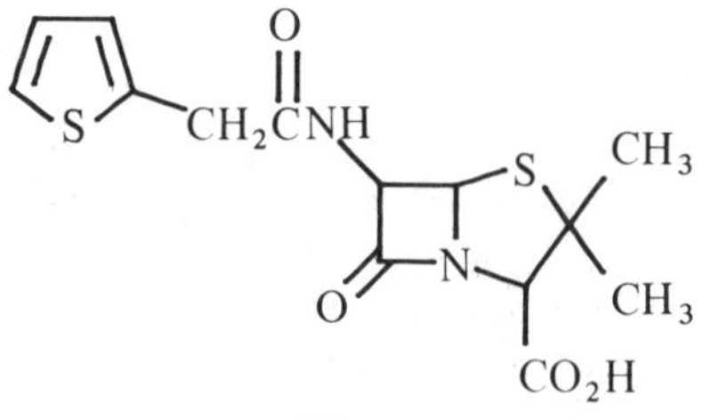

171

$$\begin{array}{ll} 172 & R^1 = R^2 = H, \ R = O_2CCH_3 \\ 173 & R^1 = R^2 = H \\ 174 & R^1 = H, \ R^2 = OCH_3, \ R = O_2CCH_3 \\ 175 & R, R^1 = O, \ R^2 = H \\ 176 & R, R^1 = CH=CHCH_2X \\ 178 & R^1 = R^2 = H, \ R = \\ \\ 179 & R^1 = R^2 = H, \ R = \end{array}$$

of an extensive synthetic program that led to the preparation of the clinically important cephalosporin antibiotic cephalothin (172).[348]

Compound 172 was the first marketed cephalosporin and was the most interesting agent of the new class of semisynthetic cephalosporins. These antibiotics as a class were superior to previous drugs in that they were very nontoxic, were acid- and penicillinase-stable, and had a broad spectrum antibiotic activity against both gram-positive and gram-negative microorganisms. The 3-thienyl isomer of cephalothin, as well as other heterocyclic analogues, had significantly less activity.[348] Detailed reports of the biological activity of 172 showed it to be similar to the actions of ampicillin combined with a penicillinase-resistant penicillin. The low toxicity of the drug allowed the use of large doses in the clinic, which facilitated the treatment of infection.[349] Other derivatives closely related to cephalothin have also been prepared. Compounds 173 [R = $(CH_3)_3CCO_2-$, $CH_3OCH_2CO_2-$, $(CH_3)_3CO_2C-NHCH_2CO_2-$, and $NH_2CH_2CO_2-$] were all prepared using an improved procedure for acylation of desacetylcephalothin. None of these derivatives were superior to the drug.[350] The oxime of cephalothin, designated number 10485, was prepared and evaluated in humans, but gave unexpectedly low serum levels and urinary recoveries.[351]

7-Methoxycephalosporins demonstrated increased gram-negative activity. The 7-methoxy derivative of cephalothin (174) was prepared and the 7-methoxy substituent was found to have no pronounced effect on the reactivity of the β-lactam. This was in contrast to an expected three- to fivefold decrease in reactivity, as was found in methoxypenicillins.[352] The aldehyde cephalosporin 175 has also been prepared.[353,354] Aldehyde 175 had antibiotic properties and was used to prepare a series of 3-(3-substituted prop-1-enyl)cephalosporins (176). The vinylogue of cephalothin (176, X = CH_3CO_2-) had biological effects very similar to 172.[354]

Cefoxitin (177) is a semisynthetic cephamycin whose synthesis was reported in 1972.[355] Cefoxitin was similar in activity to cephalothin against sensitive gram-

177

negative strains and was more effective against cephalothin-resistant indole-producing *Proteus* strains. Compound **177** was remarkably stable to β-lactamase.[356] Studies comparing **177** to cephalothin[357–359] and to cephaloridine[360, 361] showed the compound was active *in vivo* against a wide variety of bacteria, including penicillin-resistant staphylococci. The compound was superior to eight other cephalosporins against cephalothin-resistant strains of *Escherichia coli*, *Klebsiella*, and *Proteus mirablis*.[360] The effects of 7-α-methoxy substitution on cephalosporins including **177** have been reported.[362] The standard cephamycin drug used today is cefoxitin.

Another thiophene-containing cephalosporin that has been used clinically is cephaloridine (**178**). This compound was not studied as extensively as the previously mentioned drugs. Cephaloridine has much greater potential nephrotoxicity and is less resistant to staphylococcal penicillinase than cephalothin.[363–365] Another cephalosporin used in the clinic, **179** (FR-10024), was prepared by several groups.[354, 366] This compound was a broad spectrum antibiotic that compared favorably to cephaloridine and cephalothin. FR-10024 had a unique biliary excretion pattern, as compared to the known cephalosporins.[366] The discovery that thienylureido cephalosporin **180** (R = H), with its C-7 side chain with an

180

L-configuration, was more active against gram-negative organisms that its side-chain isomer[367] led to the preparation of the 7α-methoxy derivatives **180** (R = CH$_3$O—). It was anticipated that the 7α-methoxylation would increase the stability of the drug to β-lactamases. The D-form of **180** (R = CH$_3$O—) was found to have a broad spectrum of antibacterial activity, particularly against β-lactamase producing organisms, and was designated SQ-14,349.[368] Further research to compare the methoxy derivative to the unsubstituted **180** (R = H) has been reported.[369]

Several reports of the preparation and activity of 3-trifluoromethyl cephalosporin derivatives have appeared. These compounds were prepared because electron-withdrawing substituents at the 3-position of cephalosporins facilitates nucleophilic cleavage of the β-lactam, which results in enhanced antimicrobial action. Compound 181 was prepared by total synthesis, but its biological effects were not reported.[370] Subsequently, 181 was prepared enzymatically and was found to be twice as active

181   X = CF$_3$

182   X =

183   X = —CH=CH—

as the synthesized racemic material. The compound was active against *S. aureus* and *E. coli*.[371] A large series of 3-trifluoromethylcephalosporins, including the thienyl derivatives of 181, were synthesized and some structure–activity studies were done. In this series, introduction of a methoxy group at the 7-position decreased activity.[372] In somewhat related work, derivatives of 182 containing substituted thiadiazoles at the 3-position have also been prepared. The activity of 182 (R = Na) against gram-positive organisms was similar to that of cephalothin, and it was superior against gram-negative organisms.[373] The same workers prepared 3-vinyl cephalosporin derivatives of 183; the thiophene-substituted 183 was one-fourth as active as the best compound of the study.[374] Several thienyl bis-cephalosporin derivatives have been prepared.[375,376]

Other studies involving modifications on cephalosporin antibiotics include the preparation of 4-aldehyde derivatives,[377] 3-nitrile substituted cephems,[378] C-4 carboxyl modified compounds,[379] and 3-azidomethyl derivatives.[380] Alkoxy-methyl[381] and hindered acyloxymethyl substituents[382] and halogens have also been placed at the 3-position of thienyl cephalosporins.[383] Derivatives with 2-methyl, 2-methylene,[384] and 2-thiomethyl substituents have been reported.[385] 3-Thenoylthio substitution[386] and 3-cyanomethyl substitution[387] have also been investigated. Sulfocephalosporins have shown strong activity against *Pseudomonas* and other gram-negative bacteria.[388]

Some nuclear analogues of cephalosporins have recently been prepared. The 3-acetoxymethyl-Δ^3-*O*-2-isocephem system (184) was synthesized and resolved

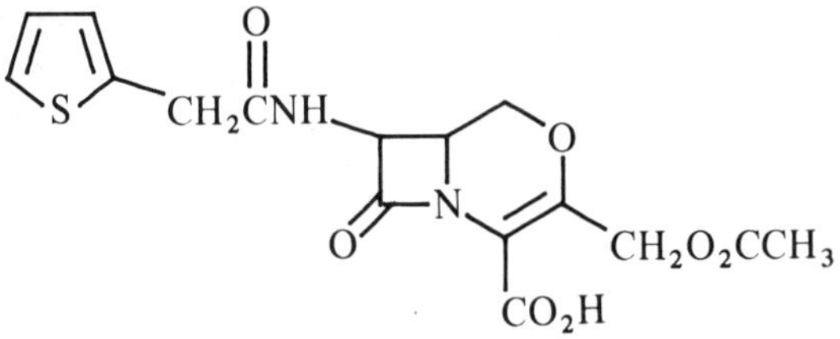

184

into its optical antipodes. The activity of this cephalothin analogue was found to reside primarily in the dextrorotatory isomer, (+)-**184**. The activity of this isomer compared favorably to **172**.[389] The 2-oxodesthiocephalosporin **185** was prepared, as well as some nonthiophene-containing systems. No activity was reported.[390] This system was also prepared as part of the syntheses of carbon analogues of cefoxitin and 7α-methoxydeacetylcephalothin (**186**). The bioactivity spectra of

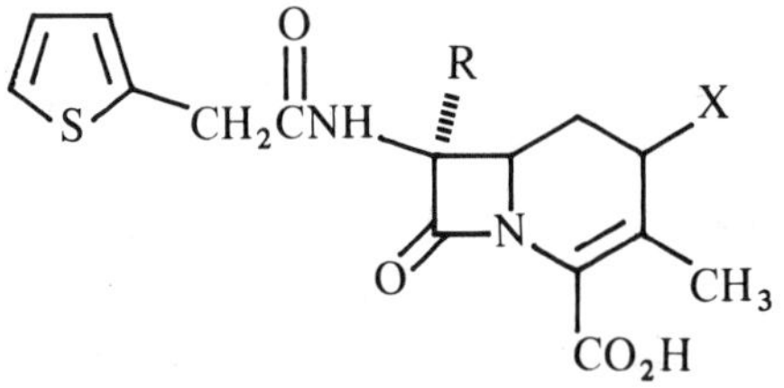

185   X = O_2CCH_3, R = H
186   X = H, R = OCH_3

these compounds against numerous microorganisms were similar to those of the natural sulfur-containing systems.[391] Finally, a series of novel thiophene-substituted β-lactams were prepared as potential precursors to chemotherapeutic agents. Compound **187** had no reported biological activity.[392]

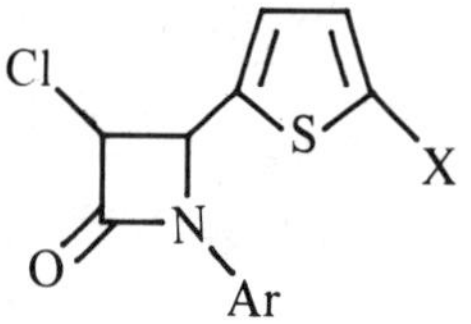

187

2.   Other Anti-Infectives

Other antimicrobial agents have been used as anti-infective compounds in addition to the β-lactams. Besides tetracyclines and aminoglycosides for which there are few reported thiophene analogues, sulfonamides, chloramphenicol, nitrofurans, and nalidixic acid are useful clinical agents that have numerous related thiophene derivatives. A number of such thiophene compounds reported to have germicidal activity have already been mentioned by Blicke.[1]

A great deal of research investigating thiophene derivatives related to chloramphenicol was reported in the 1950's. This work culminated in the synthesis of the DL-*threo* isostere 188[393,394] as well as the *erythro* isostere 188.[395] As in the

188

case of chloramphenicol, the compound with the *threo* configuration had no antibiotic activity, while the *erythro* isomer was active; it had only 20% of the activity of the known drug.[395] Other related compounds have also been prepared, but have minimal antibiotic activity.[396-399] The superior antibacterial effectiveness of the thiophene isostere of sulfonilamide (189) and of the related nitro analogue (190) have been reported.[400]

A large number of 5-nitrothiophene derivatives have been prepared, based upon the antibacterial effects of isosteric furan compounds. Simple nitrothiophene derivatives such as acetals (191),[401] oximes (192),[402] semicarbazones (193),[402] thiosemicarbazones,[403] and glyoxal derivatives (194)[404] have all been prepared and found to have fungicidal and bacteriostatic activity. The 4-nitrothiophene analogues of 194 were reported at the same time and were more active and less toxic than 194.[404] A series of 5-nitrothiophene vinyl derivatives has also been prepared. Thus,

189		
190	X = $-SO_2NH_2$	
191	X = $-CH(OR)_2$	
192	X = $-CH=NOH$	
193	X = $-CH=NNHCONH_2$	
194	X = $-COCH=NR$	

reports of vinyl amides (195), vinyl esters (196),[405] hydrazones (197),[406] and nitrones[407] have appeared. In the case of the former agents, poor solubility made antibacterial evaluation difficult, while derivatives of 197 were less active than the desvinyl compounds 194. The position of the nitro group was critical for analogues of 194; 5-nitro compounds were active antibacterials, but 4-nitro derivatives were not.[406] Several vinylsulfonamides (198) have also been prepared and found to have antibacterial effects comparable to nitrofurantoin against a variety of microbes.[408] Other thenoyl derivatives have been investigated for antimicrobial effects.[409-411]

$$O_2N \overset{\overset{\displaystyle \text{[thiophene ring]}}{}}{\underset{S}{}} \diagup CH=CH \diagup X$$

195	X = —CONHAr
196	X = —CO$_2$R
197	X = —C(CH$_3$)=NR
198	X = —SO$_2$NR^{1}R^2

Several di(nitrothienyl)sulfides were prepared in the hope of obtaining compounds with reduced toxicity while maintaining broad spectrum antibacterial activity. Compound **199** was the best of the series of isomers; all of the derivatives

$$O_2N \overset{\text{[thiophene]}}{\underset{S}{}} S \overset{\text{[thiophene]}}{\underset{S}{}} NO_2$$

199

prepared in the study were more active than their phenyl isosteres.[412,413] It is interesting that the related acetamido analogues of **199** had antitumor activity.[414] Thienylthiazoles (**200**)[415,416] as well as thienylquinazolines (**201**)[415] were syn-

200   R = [thiazole ring] —NHR^1

201   R = [quinazoline ring with NR^{2}R^3]

thesized, but had only marginal biological effects. Guanyl derivatives **200** [R^1 = —C(NH)NH$_2$] had a broad spectrum antimicrobial activity as well as effects against *Mycobacterium tuberculosis*.[417] A series of thioamides (**202** and **203**) were pre-

$$[\text{thiophene}] \diagup (CH=CH)_n \overset{S}{\underset{\|}{C}}NH_2$$

202	n = 1
203	n = 0

pared wherein the vinylogues (**202**) were significantly more active against tubercule bacillus than **203**. The thiophene derivatives prepared in this series were the most active of the heterocycles studied.[418] In a search for compounds active against pox viruses, a series of thiosemicarbazones was prepared. Compound **204** (Hoe-105) was active against vaccinia viruses and was an effective tuberculostat.[419–421]

$$NC\!-\!\overset{\displaystyle S}{\underset{\displaystyle S}{\bigcirc}}\!-\!CH\!=\!NNH\overset{\displaystyle S}{\underset{\displaystyle \parallel}{C}}NH_2$$

204

A more detailed study of **204** showed it to be as efficacious as thiaacetazone.[421] Several 5-arylthiophene-2-carboxylic acid derivatives showed little interesting antitubercular activity.[422] Some thenoyl amides and hydrazides were prepared, but had limited activity.[423] Other simple 5-methylsulfinyl thiophene derivatives (**205**) were recently prepared as potential antitubercular agents, but were no more active than their phenyl isosteres.[424]

$$CH_3\overset{\displaystyle O}{\underset{\displaystyle \parallel}{S}}\!-\!\overset{\displaystyle}{\underset{\displaystyle S}{\bigcirc}}\!-\!CH\!=\!NNHR$$

205

Thiophene as well as other aromatic hydrazides of disubstituted glycolic acids demonstrated high antitubercular activity.[425,426] A large number of imine, semicarbazone, and thiosemicarbazone derivatives of 2-thiophenecarboxaldehyde were reported, with **206** having the best antimicrobial action of the series against both sensitive and resistant strains.[427] The bithiophene analogue of atophan (**207**) as

206   **207**

well as a number of carboxylic acid derivatives were prepared, but few had significant activity.[428] Another study of bithiophenes found **208** to be the most active of the azomethine derivatives prepared.[429] A similar set of nonnitrated analogues of **208** also had weak antimicrobial activity.[430] Several thienylacetylenenic derivatives were prepared as part of a large study of the microbiological effects of acetylene compounds.[431] 2-(1-Naphthyl)thiophene-5-carboxaldehyde and the corres-

208

ponding carboxylic acid had moderate activity against gram-positive bacteria and high activity in relation to dermatophyte fungi.[432] Thienyl-substituted Mannich bases showed good broad-spectrum antibacterial activity.[433]

Nalidixic and oxolinic acids have proven themselves as clinically useful chemotherapeutic agents active against gram-negative bacteria. Thiophene isosteres of these compounds have been synthesized. Initially, **209** ($R^1 = CH_3$, $R^2 = H$, $R^3 = CH_3CH_2-$) was reported as the best compound of several derivatives prepared.[434–436] Compound **209** was active against *Proteus sp.* and *E. coli*, but inactive

209

against *Pseudomonas aeruginosa* and *Enterococci* strains. Resistance to **209** could be induced quite easily.[436] Other derivatives of **209** where substituent R^3 was larger than ethyl had significantly less activity.[434] These compounds were prepared independently by another research group.[437] Subsequently, *N*-alkoxy derivatives of **209** ($R^3 = -OCH_3$, $-OCH_2CH_3$, etc.) were prepared and **209** ($R^1 = CH_3$, $R^2 = H$, $R^3 = -OCH_3$) was found to have antimicrobial activity similar to nalidixic acid.[436] The thiophene analogues of 2-(5-nitro-2-furyl)cinchonic acid derivatives were devoid of significant antibacterial activity.[438]

Benzilidene derivatives **210** showed good antimicrobial activity against gram-

210

positive organisms as well as antifungal properties. Compounds wherein the 4-benzilidene moiety was absent had reduced activity.[439] In somewhat related work, the 2-thienyl derivatives of magnesidin and many other similar compounds were synthesized; the thiophene derivatives were as active as any in this report.[440] Several thieno[3,4-d]imidazoles (**211**) were prepared, but none had better antimicrobial

211

action than the standard [2-(α-hydroxybenzyl)-benzimidazole].[441] Several other fused thiophene systems, including **212**, were reported to be potential antibacterial agents. Compound **212** (R = PhCH$_2$−, R^1 = CH$_3$) is isosteric with the known

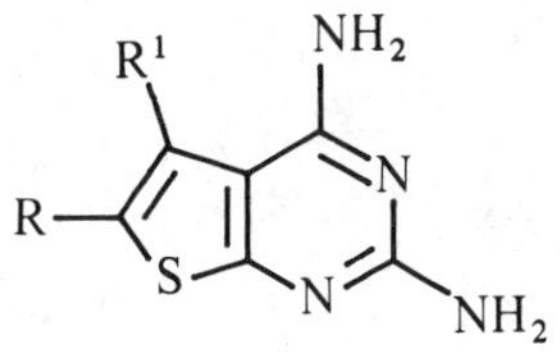

212

potent antibacterial pyrido[2,3-d]pyrimidine system and **212** was found to be 10–50 times more potent than this system at low pH.[442] Several thienoazepinones were prepared, but all were less active than the isosteric furan derivatives.[443] Thienodiazepines (**213**) have appeared in several reports, but no activity was mentioned.[444,445]

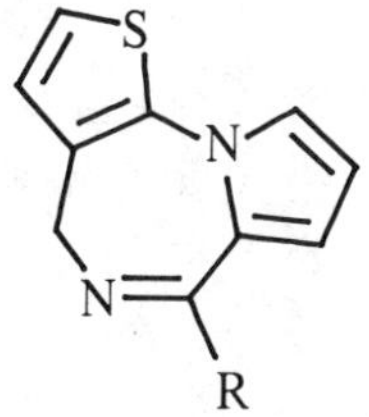

213

Thiophene derivatives containing boron have been subjects of interest in recent years. 4,5-Borazarothieno[2,3-c]pyridines (**214**) and thiophene isomer **215** were prepared and found to have biological activity.[446−448] In particular, **215** (X =

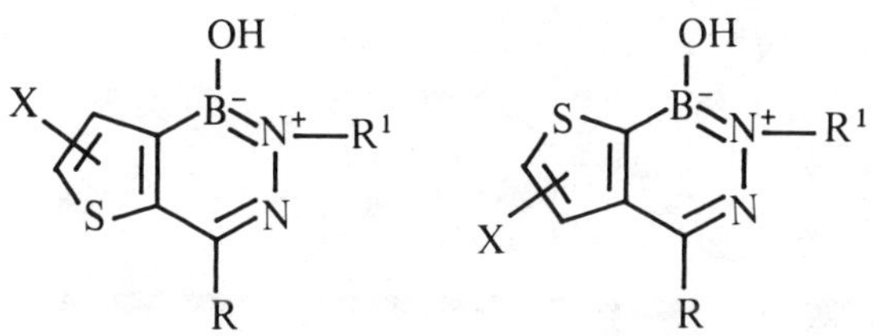

214   **215**

R = H, R^1 = SO$_2$Ar) showed high antibacterial activity against gram-negative organisms.[446,447] The cyclic nature of these derivatives was important for activity. In a more detailed structure–activity study, 2-alkyl substitution on thiophene (X = 2-CH$_3$−, 2-CH$_3$CH$_2$−) gave an increased antibacterial effect.[446,447] A study on the mode of action of these agents, as well as ICI-65,468 (**216**) and ICI-74,704 (**217**, X = H), showed them to be bacteriostats. The most active compound was ICI-75,188 (**217**, X = Cl).[449] A recent report on **214** (X = 2-CH$_3$, R^1 = −CO$_2$CH$_2$CH$_2$CH$_3$) as an antibacterial has appeared.[450]

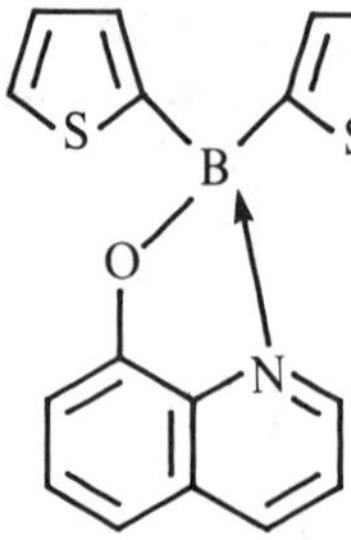

216 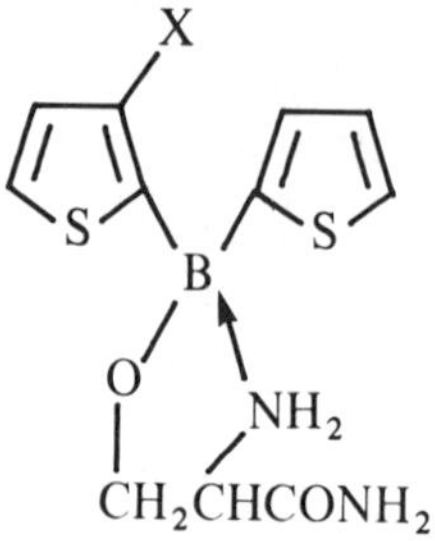   217

Specific antitubercular compounds were actively investigated in the 1950's. These compounds frequently were thiosemicarbazones,[451] acetamides,[452] ketones,[453] and aldehydes[454] of relatively simple structures. Thiosemicarbazone **218** was reported

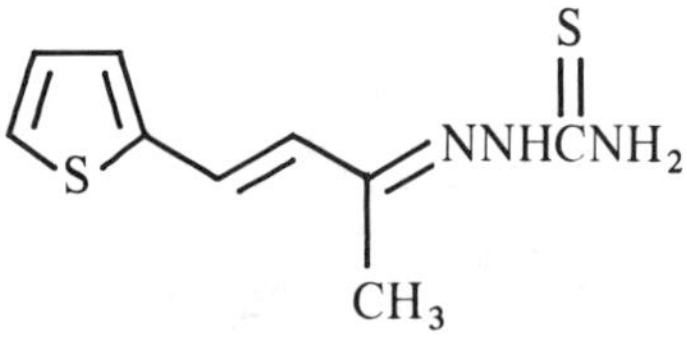

218

to completely inhibit growth of *M. tuberculosis* with the 3-thiophene isomer having greater activity.[455] Several carboxylic acid hydrazides were found to have moderate antitubercular activity.[456-459] The isonicotinyl hydrazone **219** and related compounds were highly active against several varieties of tuberculosis bacteria.[460,461]

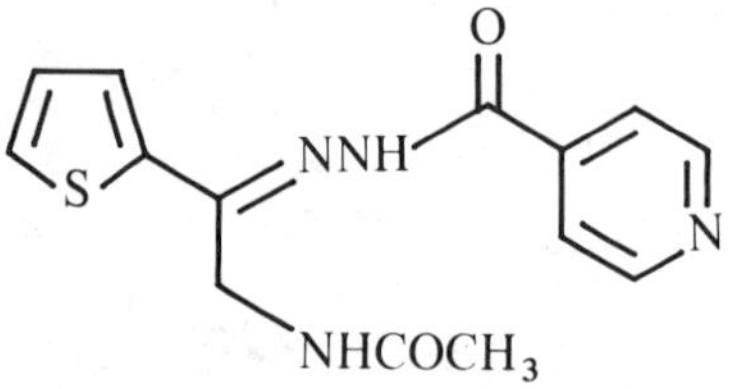

219

A variety of compounds of simple structures, similar to those mentioned above, were reported to have specific antifungal activity. Of these, several additional nitrothiophene derivatives,[462] as well as thienylmercury compounds[463,464] were highly active. Several 3-thienylrhodanine derivatives also have been prepared and have demonstrated fungistatic activity of moderate potency.[465]

3.   Antiparasitics/Antiprotozoals

Parasitic worm infections are a major cause of disease to humans and livestock in most areas of the world. Schistosomiasis is a major disease in underdeveloped countries. Protozoan infections such as malaria are still the most common cause of morbidity and mortality in the world. Some thiophene derivatives have been prepared as active agents for the treatment of these diseases. Several of the best anthelminths reported are thiophene derivatives and are discussed in Section VI.

A variety of thiophene derivatives have been reported as amebicides and antiprotozoal agents. Thenoyl derivative **220** has been reported as an agent active

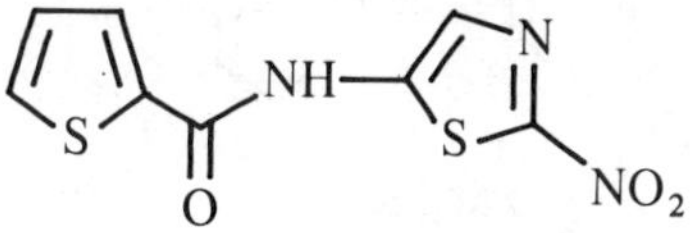

220

against Trichomonas. This compound, known as atrican, has had brief mention in the literature, although it has been used clinically.[466] Thienyl-substituted benzoheterocycles (**221**) were found to be potent inhibitors of Trichomonas *in vitro*.

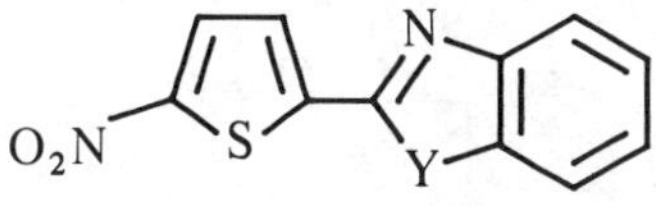

221

Other derivatives with furan and pyrrole substitution were also prepared, but the only derivative that was active *in vivo* was **221** (Y = NH). Several of these compounds were also effective against pinworm infections.[467] A series of 5-nitrothienylidene derivatives of indanones, benzofuranones, and chromanones was found to have amebicidal as well as antibacterial activity. Compound **222** had remarkable activity against *Trichomonas vaginalis*.[468,469] Related heteroarylidine carbostyril derivatives had some antitrichomonal activity.[470,471]

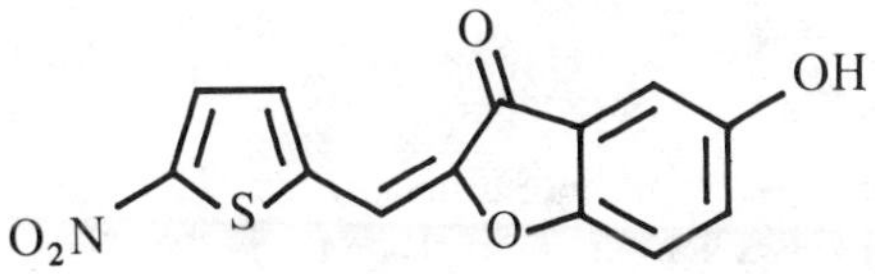

222

Considerable antiprotozoal activity was discovered for 5-nitrothienylidenes (**223**). In this system, thiophene derivatives were superior to their furan isosteres.[472]

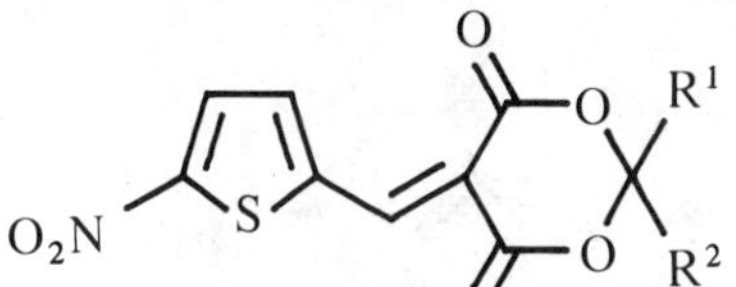

223

Thenoyl ester **224** as well as its furan and phenyl isosteres were potent antiamebic derivatives; the furan derivative was superior in efficacy, had lower toxicity, and was selected for field trials.[473] A series of thiophene derivatives related to **192, 196,**

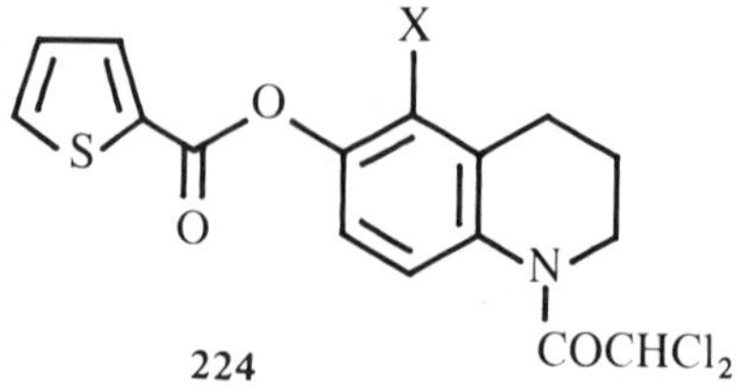

224

and **220** was found to have activity against Trichomonas, amebae, and helminths. The furan isosteres were once again superior.[474] Several 2,5-bis(4-guanylphenyl)thiophenes as well as their pyrrole isosteres were prepared. Although none of the derivatives showed antimalarial activity, **225** was the best compound against trypanosomes and produced cures in mice at dose levels of ca. 1 mg/kg.[475]

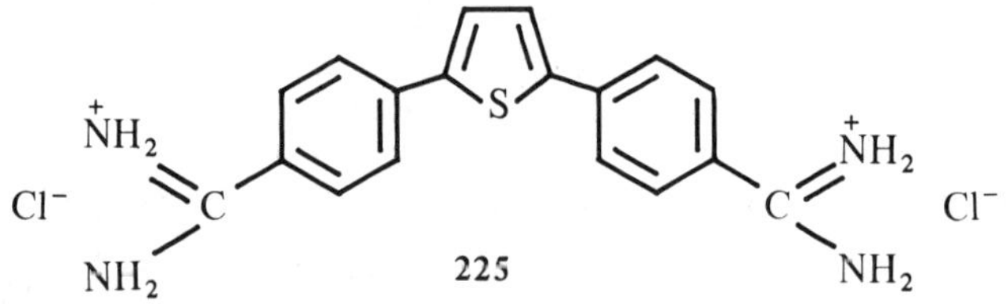

225

5-Nitrothiophenecarboxaldehyde hydrazone derivatives related to antibacterial agents reported earlier have been synthesized and found to be more active than the corresponding furan derivatives against schistosomiasis. Compound **226** reduced

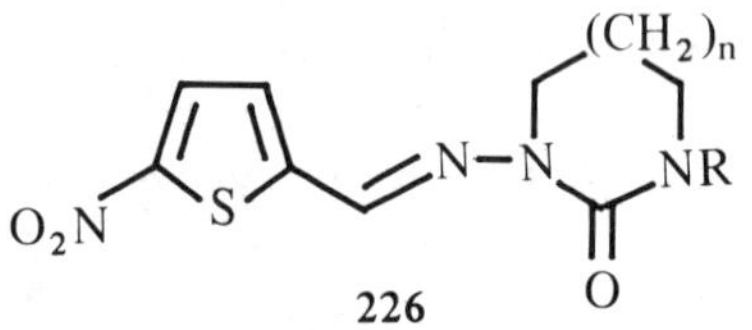

226

production of viable eggs and killed adult worms. The most interesting compound (**226**, n = 1, R = H) was selected for further study.[476] Several vinyl nitrothiophene derivatives related to **226** were previously reported to have weak schistosomicidal activity.[477] Vinyl sulfonamides (**198**) previously mentioned as antibacterial agents were also found to be anthelminths.[408] Several thiosemicarbazones had good activity against *Trypanosoma cruzi*.[478] Trichlorothiophene carboxamides (**227**)

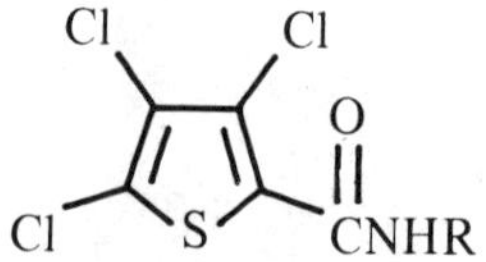

227

were prepared as antiparasitic agents for agricultural use, but were found to have anthelminthic properties. None of these compounds was active enough for human clinical trials.[479] Thienyl acrylamides and pyrazolines have been prepared as potential schistosomicidal agents.[480]

A series of papers reporting the synthesis and biological properties of substituted 2,4-diaminothieno[2,3-d]pyrimidines related to antibacterial agent 212 have appeared. These compounds were prepared as analogues of pyrimethamine and found to be antifolates and specifically antimalarials.[481] Compound 228 (Y =

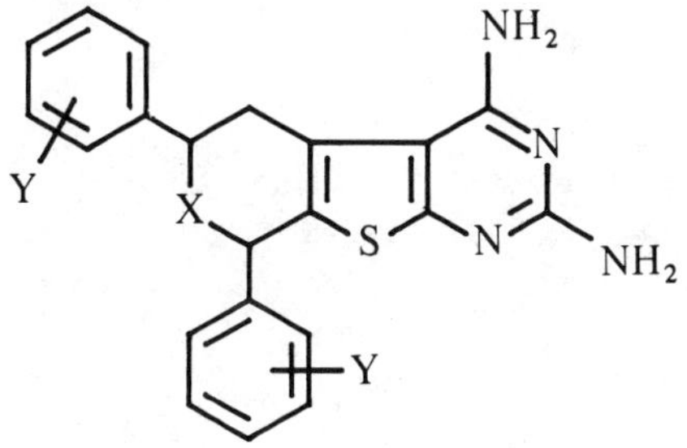

228

p-CF$_3$, X = S) was active against *Plasmodium berghei* at a dose of 640 mg/kg.[482] Tetracyclic analogues of 228 were also prepared, and several of these compounds showed activity similar to that mentioned above. Little further work was planned.[483] Thienyl-substituted quinoline derivatives[484] and oxazoline derivatives[485] were effective antimalarial agents.

4. Cancer Chemotherapy/Immunosuppressives

Increased awareness of the possibility to "cure" cancer using an aggressive regimen of surgery, radiation, and chemotherapy has given a strong impetus to the search for new, efficacious and less toxic chemical agents. These compounds may operate in several ways to combat cancer; they might interfere with RNA replication, activate or inactivate the immune response system of the host, or interfere with metabolic pathways. Not surprisingly, thiophene derivatives have been synthesized as active compounds in this area of drug research.

Thieno[2,3-b]azepine-4-ones (229) were prepared as potential antineoplastic

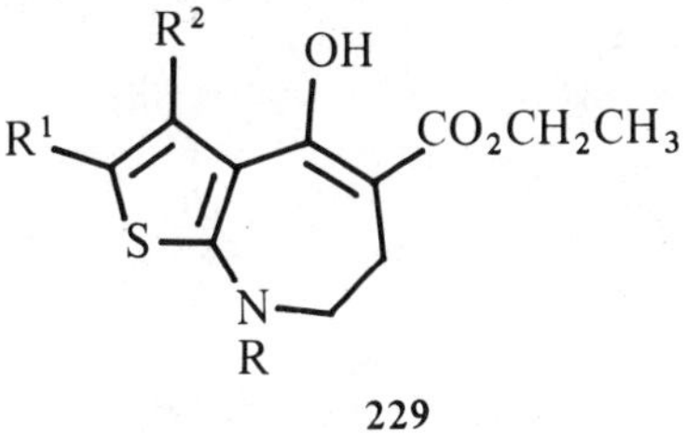

229

agents. Although the phenyl isostere of **229** was found to have good activity against Crocker's sarcoma, the thiophene derivatives had little interesting activity.[486] Several reports on the synthesis of monothiouronium (**230**) and dithiouronium (**231**) salts of 2,5-disubstituted thiophenes have appeared. Compound **230** was

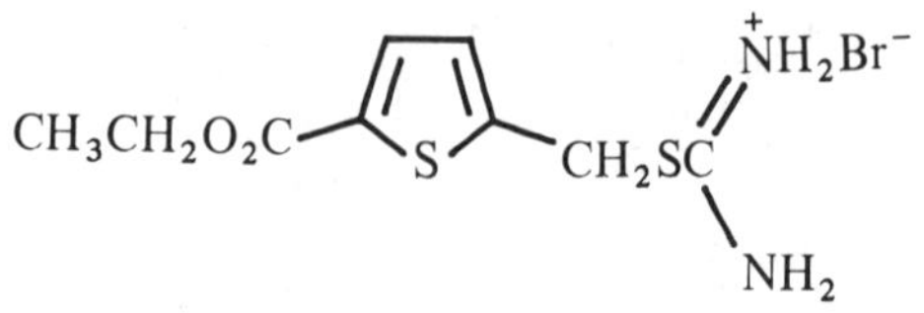

230

reported to be highly active against Yoshida sarcoma in rats, but was inactive against fibrosarcoma in mice.[487] Bis-salt **231** was active but toxic.[488] In both cases,

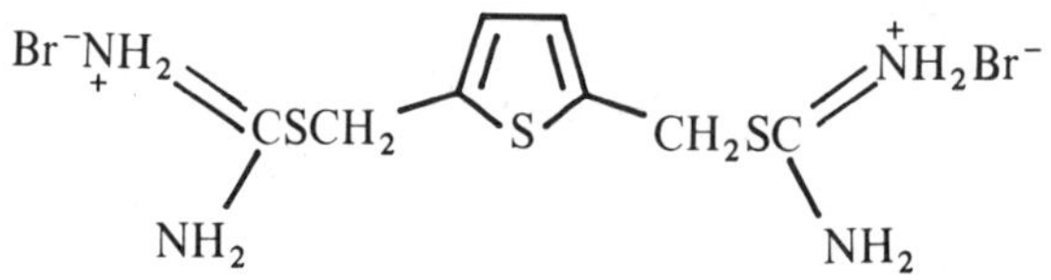

231

no further work was planned. Several thiophenecarboxylic acids and esters also were investigated for their anticancer effects.[488-490] Thiophene acetylene **232** was prepared as part of a large study of acetylenic carbamates that were oncolytic agents. Compound **232** had only modest activity.[491-493] A series of 5-nitrothio-

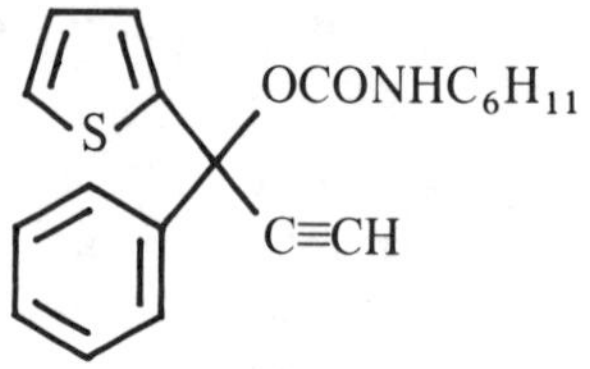

232

phene-2-vinyl substituted heterocycles as well as their furan counterparts had activity against Erlich ascites carcinoma in mice. Antitumor activity seemed related to chemical structure in a manner similar to that noted earlier for antibiotic activity.[494]

Citioline (**233**) has been used to protect against radiation damage caused by

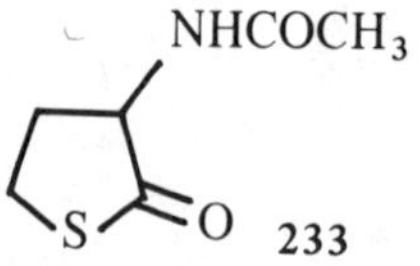

233

cancer therapy.[495] Antiviral agents have been used with limited success for anti-tumor effects. Some simple derivatives of thiophenecarboxaldehydes and ketones were found to have antiviral effects.[496-499]

Immunosuppressive compounds have also played a role in cancer chemotherapy, as well as in diseases of the autoimmune system. A clinical cytotoxic agent, ICI-47.776 (234), was developed from a series of 5-arylidene-2-oxodihydrothio-

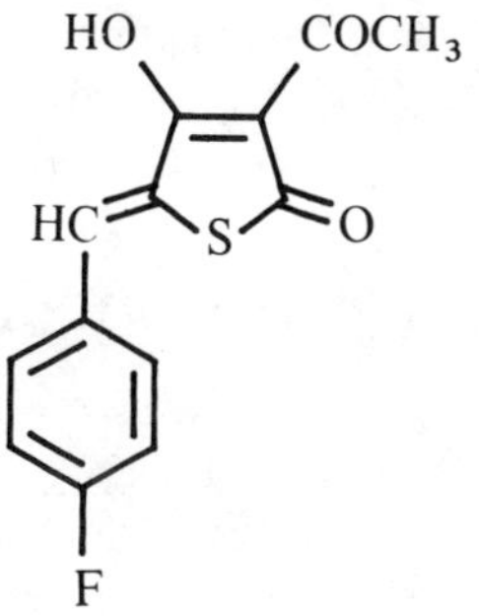

234

phenes.[500] Compound 234 disrupts the supply of adenosine triphosphate in the cell by inhibiting oxidative phosphorylation in the mitochondria.[501,502] ICI-47,776 had effects similar to other cytotoxic drugs, such as cyclophosphamide, metho-trexate, and 6-mercaptopurine. The compound was a potent inhibitor of antibody synthesis.[503] The thiophene analogue of cinanserin was a potent inhibitor of primary immune reponse as well as of serotonin.[504]

Several semisynthetic anticancer agents have been reported. 4'-Demethyl-epipodophyllotoxin-β-D-thenylidene glucoside (VM-26, 235) has been used in

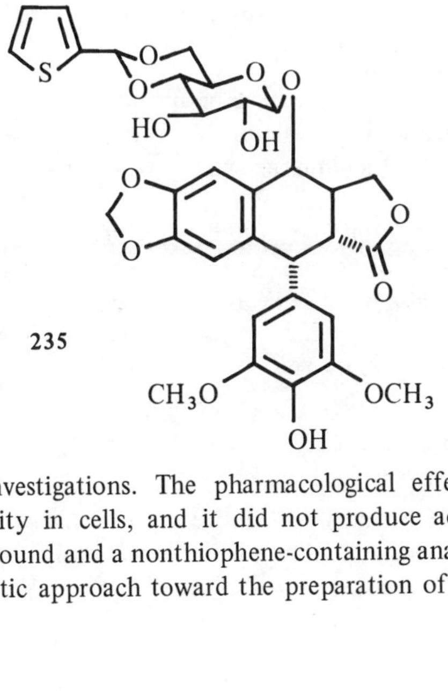

235

numerous clinical investigations. The pharmacological effects of 235 included high cytostatic activity in cells, and it did not produce accumulation of meta-phases.[505] This compound and a nonthiophene-containing analogue were developed as part of a systematic approach toward the preparation of more potent and less toxic podophyllotoxin derivatives.[506] VM-26 has been reported to treat malignant intracranial neoplasms[507] and lymphocytic leukemia,[508] among other malignant diseases. A thenoyl derivative of Damavaricin C inhibited the growth of virus-transformed cells and human-derived cancer cells. The compound also was active against ascites and solid Sarcoma 180 *in vivo*.[509]

N-Benzoyl-β-2-thienyl-DL-alanine and its 3-thienyl isomer were inhibitors in a microbial antitumor screen. The 2-thienyl isomer was 50% more active than the 3-isomer.[510] Several synthetic dehydrodipeptides, including several thienyl derivatives, rendered macrophages cytolytic for several tumor cells.[511] Antimalarial compounds **238** were also antifolates and had some antitumor effects.[481–485]

VI. VETERINARY AND AGRICULTURAL AGENTS

The control of disease and growth in food grains and domestic animals is a principal concern for veterinary and agricultural scientists. As the world population continues to increase dramatically, great pressures are placed on farm productivity. Attempts to promote crop growth and to protect crops from the ravages of insect infestation have been reasonably successful. As more is learned about insecticide toxicity in the ecosystem, more selective and less toxic pesticides are required. Increased insect resistance to known agents also creates the need for new pesticidal compounds. Domestic animals suffer from the parasitic infections mentioned previously for man. Besides the treatments discussed earlier, several anthelminths have been developed and used successfully for treating pets and livestock. Although many agents containing thiophene are active in these areas of investigation, much relevant research lies obscured in the patent literature.

1. Anthelminths

Nematode infections are a significant problem in pets and livestock. Several of the most potent anthelminths in clinical use today are the thiophene derivatives pyrantel, morantel, and thenium closylate. Other active thiophene-containing anthelminths are discussed in Section V.

A series of 2-arylethyl- or 2-arylvinylimidazolines and tetrahydropyrimidines was discovered to have broad-spectrum anthelmintic activity. Based on the nematodic activity of **236** discovered previously,[512,513] this series was developed to

236

produce pyrantel (**237**), which was active against nematodes that infest dogs, horses, livestock, and man.[514] Structure–activity studies revealed that various aryl systems decrease in potency in the order: 2-thienyl > 3-thienyl > phenyl >

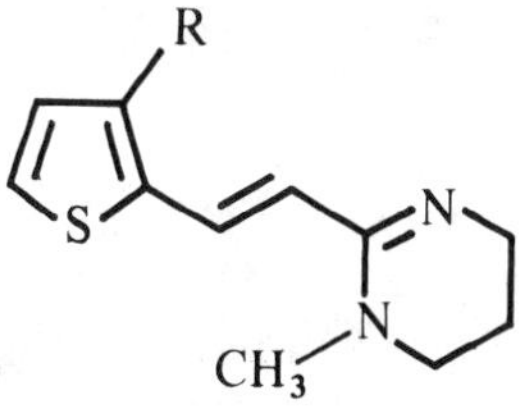

237 R = H
241 R = CH$_3$

2-furyl. The arylethyl analogues of **237** were usually less potent. Although *N*-alkyl substitution enhanced activity, substitution at the other positions reduced activity.[514] Test results for the activity of pyrantel in various hosts have been recorded.[515]

In related work, 1-(2-arylvinyl)pyridinium salts also were found to have anthelmintic properties. Compound **238** was among the best compounds of this report, with a decrease in potency among various aryl substituents as noted for pyrantel.[516]

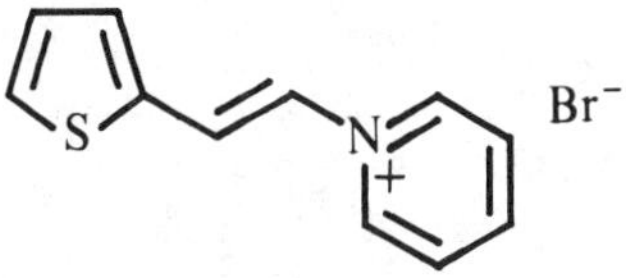

238

Noncyclic amidines (**239**) related to pyrantel were prepared and found to have interesting activity. Disubstitution on N and no substitution on N' was required for activity; 2-thienyl derivatives were once again more potent than the 3-thienyl

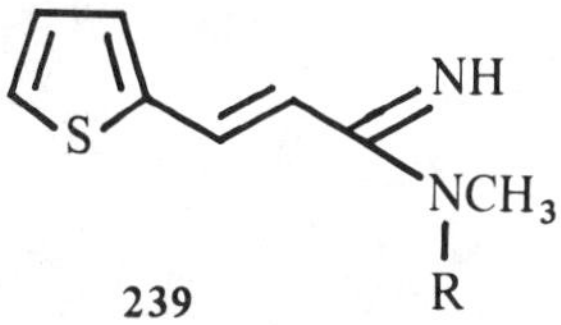

239

isomers.[517] The 2-thiazoline (**240**, n = 2) and 2-dihydrothiazine (**240**, n = 3) analogues of pyrantel were synthesized and several were highly active against roundworm infections. Only a few thiazoline derivatives were active, but many dihydrothiazine derivatives had good potency. Pyrantel had a better spectrum of activities than any of the compounds **240** that were prepared.[518]

Finally, the last of the series of papers concerning this area of research reported

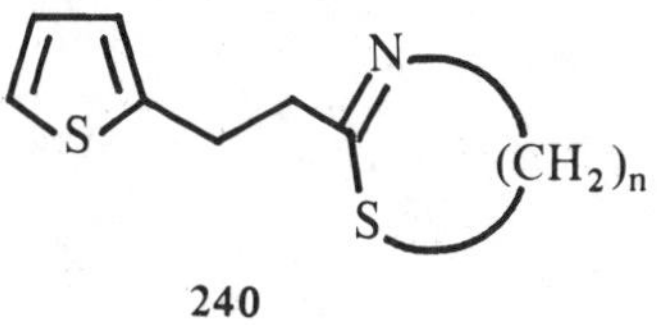

240

the activity of morantel (**241**), which is another clinically important anthelminth. In addition to reviewing structure–activity relationships already noted for compounds **237–240**, *ortho-* substitution on the aryl systems was found to promote activity. Morantel and pyrantel were the most active against major nematode infections of sheep.[519] Test results in morantel in other species were noted.[520]

Another clinically important thiophene-containing anthelminth was developed earlier. Thenium closylate (**242**) was developed from concepts derived from

242

the active bephenium series of compounds. Compound **242** was more active against hookworms and roundworms in dogs and cats and was less emetic than bephenium.[521,522] The compound was ineffective against tapeworms.[522] Recently, a combination of thenium closylate and piperazine phosphate was found to be effective against *Toxocera canis* in young dogs.[523]

Several reports of compounds that have limited antiparasitic activity have appeared. Nitrothiophenes (**226**) mentioned earlier were found to treat schistosomiasis in infected mice.[524] Another nitrothiophene, RO-11-0761, was an effective anthelminth.[525] The acetamide derivative of 5-nitro-2-thienylamine, Sch-1773, was reported to have antihistomonas activity in turkey poults, but was not as active as ipronidazole.[526] Azindole **243** had anthelmintic activity, but its 6-azindole

243

isomer did not. Neither **243** nor other derivatives in this report had interesting broad-spectrum activity.[527] Hetero-triarylmethane derivatives also have interesting activity in this regard.[528] Several 5-nitro-2-thienyl-substituted quinazoline derivatives had interesting anthelmintic and antimicrobial properties.[529,530]

2.   Anesthetics

Veterinary medicine deals with animals that are far larger in size than man. As a consequence, larger doses of more powerful drugs are required to immobilize these animals for treatment. Thiambutene (**54**), which was discussed as an analgesic in Section II, has been used in vetinary practice as an anesthetic and an immobilization drug.[531] Numerous reports of the drug's use in horses, sheep, and large, wild animals appeared in the late 1960's.[532-537]

Tiletamine (CI-534, **244**) was developed more recently as an agent of potency

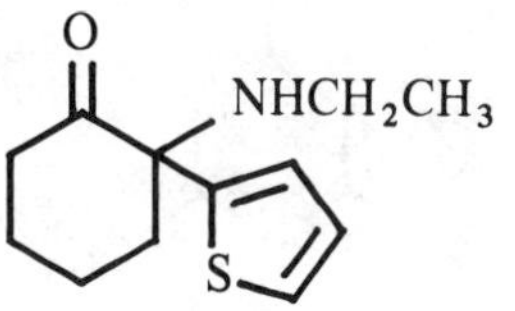

244

intermediate between phencyclidene and ketamine. The compound acts principally on the CNS and was found to produce catalepsy in all animal species tested.[538] The drug has been used as a surgical anesthetic in rabbits[539] and cats[540] and as a taming agent in cats.[541] Compound **244** caused some convulsions in animals and, as a consequence, this compound was combined with the tranquilizer muscle-relaxant zolazepam. This combination, designated CI-744 (telazol), has been used in monkeys,[542] sheep,[543] other nonhuman primates,[544,545] and wild carnivores such as lions and leopards.[546,547] Telazol produced rapid onset of anesthesia for surgical procedures with a wide safety margin.

3. Pesticides

Pesticides are used in agriculture to control insects and undesirable weeds. Blicke reviewed numerous thiophene analogues of DDT.[1] In general, these compounds were less active than DDT, but had the same mode of action.[548] A more recent report has included several of these derivatives.[549] The thiophene analogues of 2,2,2-trichloro-1-(p-chlorophenyl)ethanol (**245**) were prepared and found to be inactive against house flies.[550]

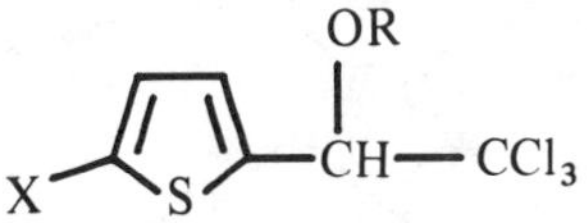

245

Several derivatives of thieno[2,3-d]pyrimidines have been prepared as potential pesticides. Compound **246** (X = Cl) was active against pigweed and wild mustard. Several piperazine derivatives prepared from **246** (X = Cl) had fungicidal activity.[551]

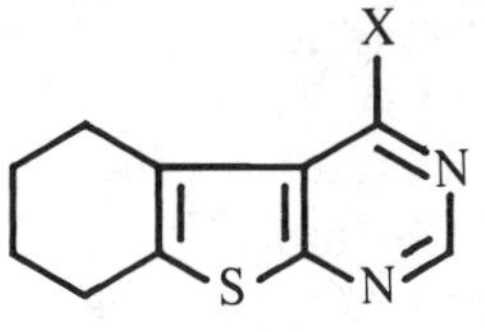

246

In a later report, **246** (X = —NHN=CHR) showed some herbicidal activity against pigweed, velvet leaf, and red millet.[552] *N*-Arylthieno[2,3-d]pyrimidin-4-amines (**247**) had some pesticidal activity.[553] Several derivatives of 2-amino-3-carboethoxy-

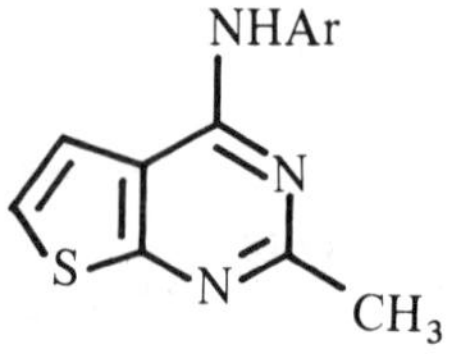

247

thiophene (**248**) were prepared and had some activity against *Botrytis cinera* as well as several microorganisms.[554]

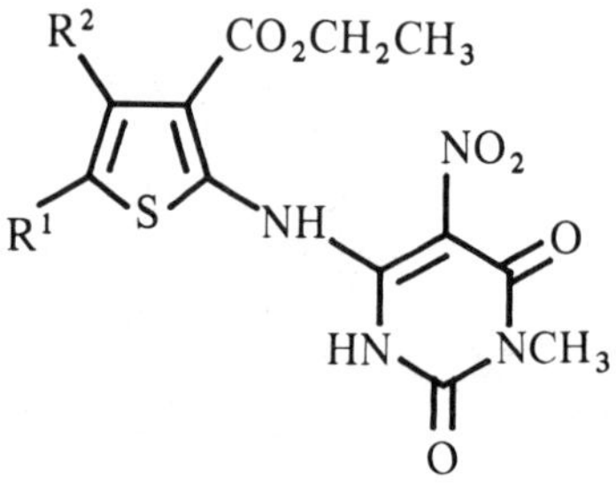

248

Carbamate esters of 3-cyano-4-hydroxythiophenes (**249**) and their pyrrole isosteres have been synthesized as potential insectides. Methyl- (**249**, R = H) and dimethyl- (**249**, R = CH$_3$) carbamates were evaluated for inhibition of house fly

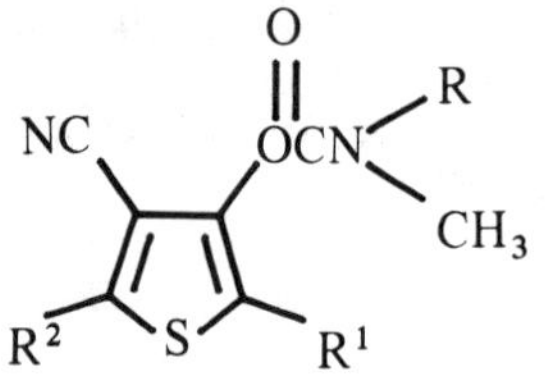

249

acetylcholinesterase and were active. Several compounds also had aphicidal activity.[555] Thienyl aromatic sulfones (**250**, n = 0, X = 2) were prepared and found to have little ovicidal activity. The pyridyl isosteres of this series were the most interesting compounds reported.[556] A subsequent paper described the vinylogues (**250**, n = 1). In this report, the sulfides (**250**, n = 1, X = 0) were the most active, with ovicidal activity against the eggs of red spider mites.[557] Related thiomethyl thiocyanates were fungicidal and were active against powdery mildew.[558]

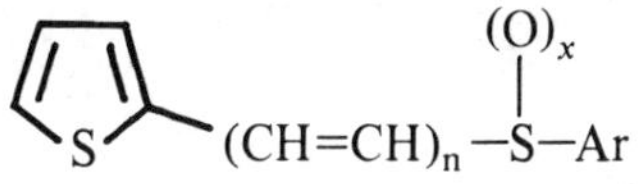

$$\text{(thiophene)} - (CH=CH)_n - \overset{(O)_x}{\underset{|}{S}} - Ar$$

250

4. Growth Regulators

Only a few thiophene-containing growth regulators have appeared in the literature. Sulbenox (**251**) was reported to be a novel growth stimulant in sheep. The
compound was not estrogenic, androgenic, or goitrogenic in rat tests and did not
function as an antibacterial agent. The evaluation of **251** as a growth promoter for

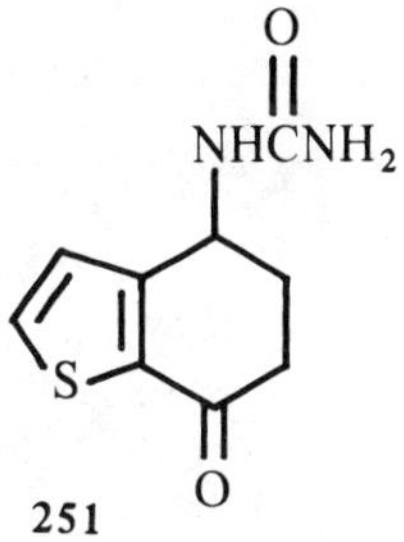

$$\overset{O}{\underset{\|}{NHCNH_2}}$$

251

feed cattle was under investigation.[559] A compound that affects hormone-steroid
metabolism, 5-(4-chloro-5-sulfamoyl-2-thienylaminophenyl)-tetrazole, has been
designated BM-02001 and evaluated for growth regulation in animals.[560]

Chemangro 8728 (**252**) was studied as a growth stimulant for snap beans. The
compound increased yields of beans by increasing pod production. Compound **252**
did not affect seed or fiber development, color, or trace metal content of the
harvested beans.[561]

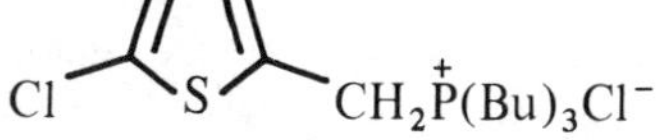

$$Cl - \text{(thiophene)} - CH_2\overset{+}{P}(Bu)_3Cl^-$$

252

VII. MISCELLANEOUS PROPERTIES OF THIOPHENE DERIVATIVES

Thiophene-containing compounds affect biological systems in ways other than
those described. Derivatives have been prepared that are "antimetabolites" and/or
that interfere with metabolic pathways in plants and animals. Thiophene analogues
of vitamins have also been prepared. The carcinogenicity and toxicity of thiophene
derivatives have been reported.

The 2-thienyl (**253**) and 3-thienyl isosteres of phenylalanine have been prepared

NH$_2$

253

and investigated for their antiviral and anticancer properties.[14,462,563] The 3-isostere was a more active antagonist of β-phenylalanine[3,14] and activity was found only in the L form of the compound.[564] Early investigations of DL-253 showed it to have antiviral activity[565-569] as well as some anticancer activity.[570-572] No extremely interesting effects were observed.[573] Derivative 253 inhibited protein synthesis in buds[574] and formation of virus RNA.[575,576] The compound inhibited the progression of HeLa cells into mitosis when phenylalanine was absent in the cell culture.[577] It also competed for uptake and blocked utilization of D-tryptophan in *E. coli*.[578]

The 3-thienyl analogue of phenylalanine has been reported to inhibit tumor growth and antibody production.[579-583] The compound inhibited protein and *n*-RNA synthesis.[584,585] The immunosuppressive properties of this isomer have been noted for rats,[586] mice,[587,588] and *in vitro*. The derivative also prevented ethionine pancreatitis in rats and rabbits.[589] *N*-Benzoyl derivatives of β-3-thienyl-DL-alanine and DL-253 were active as inhibitors in a microbial antitumor screen.[510] Nitrofuran derivatives containing thienyl- substituted amino acid residues have had antimicrobial activity.[590]

Synthetic peptides derived from thiophene-containing amino acids have been prepared as antitumor agents and antimetabolites. A number of peptides containing thienyl-DL-alanine units have been reported.[591-595] Early reports found these peptides to have growth stimulation and inhibition effects on *E. coli* that were no greater than simple amino acids.[591-593,596,597] Later studies showed that some of these peptides had greater effects than the corresponding free amino acids.[511,595] Analogues of thyroliberin(TRF)[598] and oxytocin containing[599] thiophene units have also been prepared. Both of these analogues maintained some of the activity of the naturally occurring peptides. Some *N*-acetyldehydro-3-(2-thienyl)alanyl amino acid derivatives caused necrogenic activity.[600]

Thiophene analogues of other naturally occurring biologically active molecules have been synthesized. Thiomuscarine iodide (254) and its isomers have been

OH

CH$_3$ CH$_2$Ṅ(CH$_3$)$_3$ I$^-$
254

reported[601,602] and have been investigated for pharmacological effects. Among retinoic acid analogues investigated for anticancer effects, 255 was active.[603] Thiophene-containing cyclic analogues of juvenile hormone have been prepared to investigate the structure–activity relationships of the naturally occurring materials.[604] Several nucleosides containing thiosugars such as 4'-thiocordycepin (256)[605] and analogues of 1-β-D-ribofuranosyl-1,2,4-triazole-3-carboxamide[606] have

been reported. No attempt is made here to categorize all reports of thioribofurano-sides. Studies investigating the effects of 5-(2-thienyl)valeric acid on biotin syn-thesis have been reported; the compound inhibited an intermediate enzyme process in biotin biosynthesis.[607] Several thienyl-substituted 4,8-dihydrotoxoflavin deriva-tives have been synthesized.[608]

Other thiophene derivatives interfere with enzymatic and metabolic pathways. The thiophene analogue of oxidenone (257) was prepared as a possible tyrosine hydroxylase inhibitor.[609] Compound 258 represents a recently discovered class of

potent cyclic AMP phosphodiesterase inhibitors.[610] N-Aralkyl dithiocarbamates inhibited dopamine-β-hydroxylase.[611] Thenoylcyclopentadienyl manganese tri-carbonyl (259) was reported to inhibit mitochondrial respiration more effectively than the methyl-substituted derivative.[612] Compound **234**, previously discussed as an immunosuppressive agent, has been found to inhibit electron transport and uncouple photophosphorylation in chloroplasts.[613] Anilinothiophenes (**260**)

260

deactivated reactions in the watersplitting enzyme system by photosynthesis.[614] Later studies on these compounds demonstrated their effects on photooxidation of cytochrome B-559 in chloroplast fragments.[615] Nitrothiophene derivatives, as well as other nitroheterocyclic sensitizers, reacted with cytochrome-C and thus interfered with mitochondrial energy metabolism.[616] Thiophene dicarboxylates inhibited tumor growth by interfering with the hexose monophosphate pathway[617,618] Trimethaphan camsylate (**85**) previously discussed as a ganglionic blocker was found to inhibit plasmid cholinesterase.[619]

A large number of studies of the chelating agent 2-thenoyltrifluoroacetone (**261**) have been published. Compound **261**, initially believed to specifically inhibit

261

mitochondrial succinate dehydrogenase[620,621] was found to inhibit malate dehydrogenase.[622] This compound inhibited oxidations of NADH, succinate, and malate in mitochondria.[623] A specific site of interaction of **261** in the mitochondrial respiratory chain has been proposed.[624]

Thiophene derivatives have also been used to investigate electronic and structural requirements of enzyme reactions. Sulfonamide (**262**) was used as part of a QSAR study of sulfonamide binding to carbonic anhydrase.[625] Biotin derivatives **263** were

262

263

prepared as potential affinity labels to investigate the carbon dioxide transfer reaction of biotin.[625] Compounds **263** were used to probe for the active site in the biotin-dependent enzyme, acetyl-CoA carboxylase from *E. coli* B.

The toxic effects of thiophene and its derivatives have been the subjects of numerous investigations. Methapyrilene (**139**), an antihistamine removed from the market because of hepato-toxicity, was investigated and found to nitrosate in the presence of nitrite. The formation of these nitrosamines was presumed to cause the reported incidence of liver tumors in rats.[626] Other metabolic studies on the drug were reported.[627] Thiophene analogues of the known carcinogen, 4-aminobiphenyl, were found to have the expected mutagenic effects *in vitro*. Some doubt as to their capability of eliciting tumors *in vivo* was expressed.[628] Compound **264**, designated

264

VR-6 and prepared as a urinary tract antiseptic, was found to have mutagenic and carcinogenic properties.[629] 2-Nitrothiophene was more mutagenic than the 3-isomer.[630] Several 5-nitrothiophene derivatives that had anthelmintic properties caused cancer in rats.[631-634] Similar results were obtained with other nitrothiophenes.[635] Thienylanthracenes were synthesized to study their potential carcinogenicity.[636]

The general biological effects of thiophene were thoroughly summarized by Blicke.[1] Additional studies of the toxic effects of thiophene, thiophane, and certain 2-substituted derivatives were reported. Both unsubstituted compounds were quite toxic and caused behavioral and liver disorders.[637] Chronic inhalation effects of thiophene on animals were also reported.[638] Inhalation toxicity of sulfolane caused convulsions in rats and chronic convulsions and death in monkeys.[639] The toxic effects of thiophene on Purkinje cells have been studied.[640-641] The metabolism of thiophene has been investigated in rabbits.[642] The metabolism of 2-thiophenecarboxylic acid has been studied.[643]

VIII.   CONCLUSIONS

This chapter demonstrates the explosion of research on biologically active thiophene derivatives that has occurred since the review prepared by Blicke. A thiophene compound has been prepared for therapy for almost all disease states. In addition, thiophene agents have been used effectively in veterinary medicine and in agriculture. Studies of other thiophene derivatives have led to a greater understanding of metabolic and enzymatic processes.

As basic biological research continues to unravel the molecular causes for disease and toxic effects, continued investigation of thiophene and its derivatives will almost certainly continue to be a source of new and useful agents of therapy. Based upon the enormous progress demonstrated over the last several decades, the conclusions of Martin-Smith and Reid that "work [now] could profitably be concentrated on . . . other ring systems" were obviously premature. We can only expect further important advances for thiophene research in the future.

IX.   ADDENDUM

Since the completion of the first draft of this chapter, several reports of biologically active thiophene derivatives appeared in 1982 and early 1983 that are worthy of note. In the field of CNS research, additional investigations into the structure–activity relationships of neuroleptic agent 11 revealed that compound 265 had the best therapeutic index of the series.[644] Like clozapine, 265 had sig-

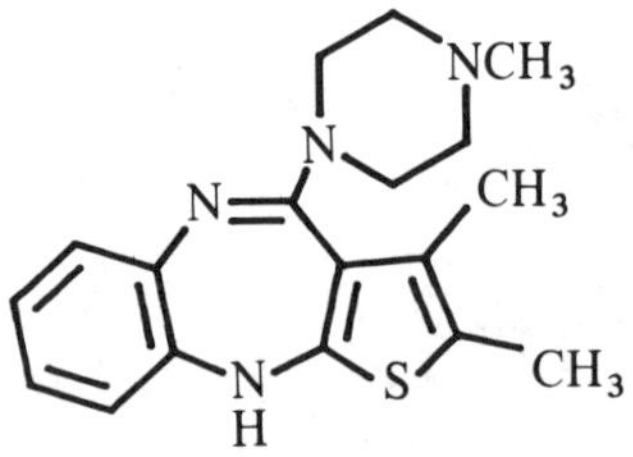

265

nificant anticholinergic effects that probably are a result of the 3-methyl substituent's interference of the rotation of the piperazine moiety. As noted earlier,[23] a 2-methyl substituent was essential for antidopaminergic activity.

Additional work on diazepine research has been reported. The mechanism of action of brotizolam (27, R = Br) has been studied.[645] In a more novel result, structural modifications have completely altered the biological activity of benzodiazepines. Tifluadom (KC5103, 266) did not have minor tranquilizer effects,

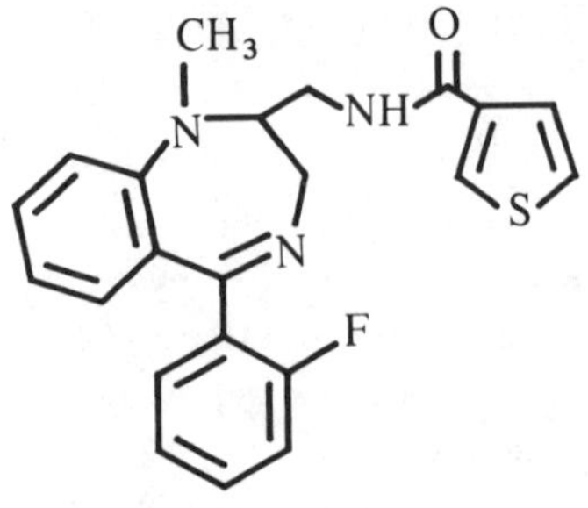

266

but rather behaved as an analgesic agent. Its effects were reversed by naloxone but not by typical benzodiazepine antagonists; these results indicated that **266** was an opiate analgesic operating on K-receptors.[646]

Several additional pharmacodynamic agents have also appeared in the recent literature. A complete review of the synthesis and pharmacological properties of the clinical antispasmodic agent tiquizium bromide (**92**, Ar = 2-thienyl) has been reported.[647] The activity of the cerebral vasodilator UK-17,022 (**103**, Ar = 4-fluorophenyl, n = 2, R = H) has also been summarized.[648] Oxazolidine **267**, a

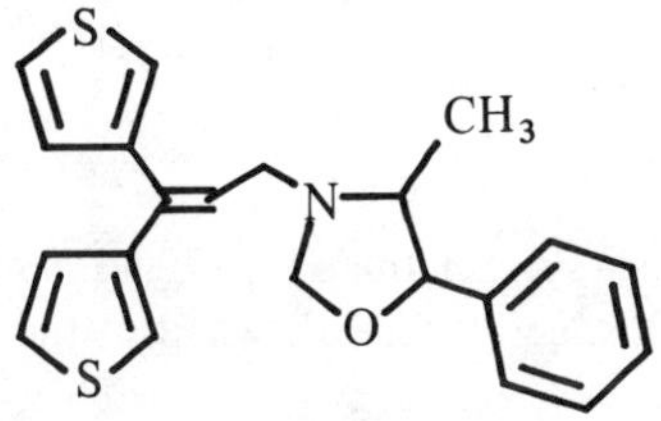

267

derivative of tinofedrin (**83**), was synthesized as another potential cerebral vaso-dilator.[649]

Several novel analogues of prazosin have been reported as antihypertensive agents. Both the 2-substituted and 3-substituted thiophene derivatives of **268** showed similar potent activity in the spontaneously hypertensive rat, but the

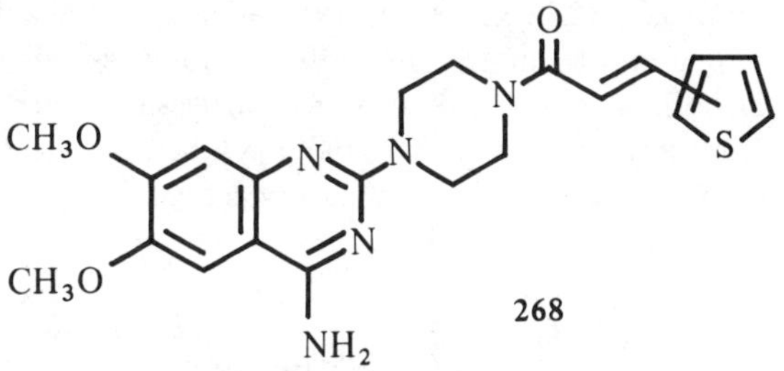

268

phenyl and furan analogues had better activity in renal hypertensive rats and were the subjects of more detailed investigation.[650] These compounds showed α-adreno-receptor blocking effects. Thiophene analogues of the β-adrenoceptor blocking agent toliprolol have been prepared.[651] Derivative **269** exhibited inhibitory activity of the same order as that of propanolol. Only the 5-methyl analogues had proper-

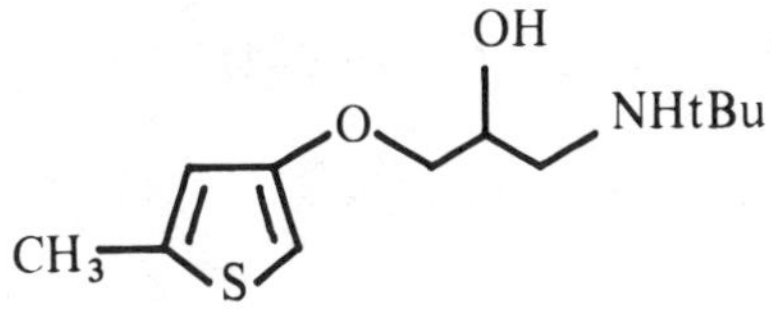

269

ties worthwhile for further study. A new clinical diuretic agent has been studied.[652] Azosemide (270) was found to be an effective diuretic whose mode of action may

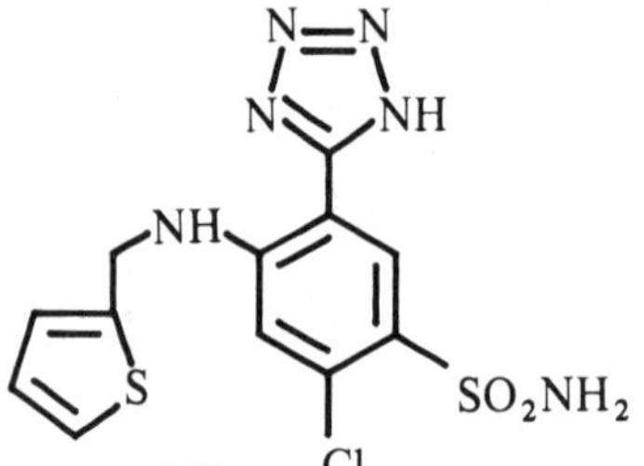

270

involve the renin–angiotensin system. In an investigation of thieno[2,3-d]pyrimidine derivatives, 271 was the most active diuretic agent of the series with activity similar to that of chlorthiazide.[653]

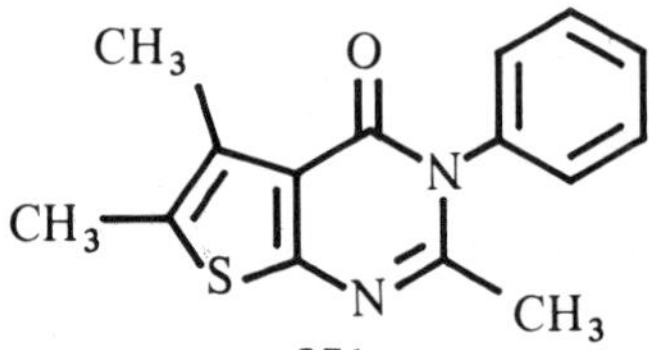

271

New thiophene derivatives have also been recently reported as metabolic disease therapy agents. Tenoxicam (272) has been reviewed as a clinical anti-inflammatory agent.[654] The compound was more potent that diclofenac and indomethacin in animal models of inflammation and also had analgesic properties. Tenoxicam seemed to have little influence on prostaglandin synthesis. Tetrazole analogues of 271 were reported to have some anti-inflammatory activity.[655]

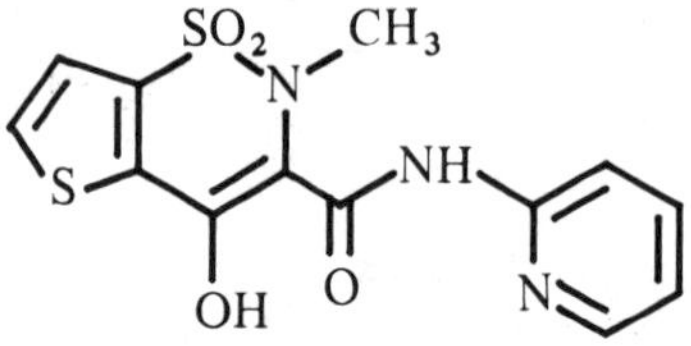

272

Additional references to the synthesis of thiaprostaglandin (126) have appeared.[656,657] The isomer of 125, namely 273, and the ketone analogues have also been prepared.[658] Additionally, new thiasteroids have been prepared and

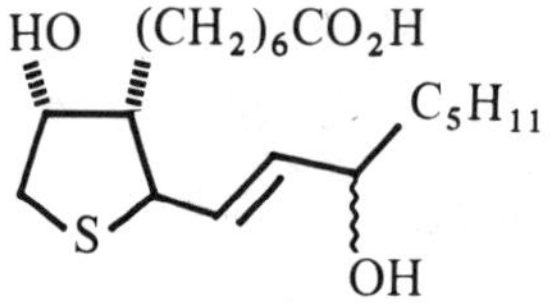

273

reviewed.[659] Compound **274**, the 1-thiasteroid analogue of **134**, has been synthesized.[660] Several novel polyhetero steroid analogues, including **275**, were the subjects of synthetic investigations.[661]

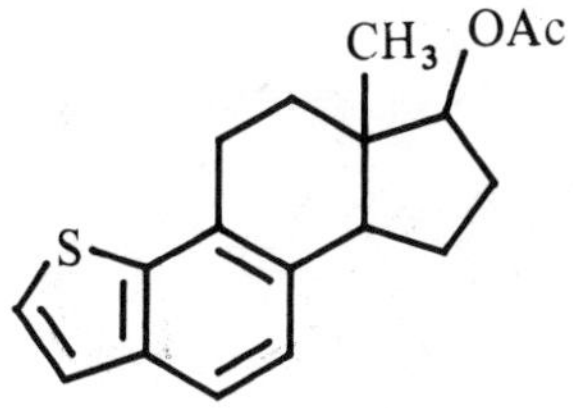

274

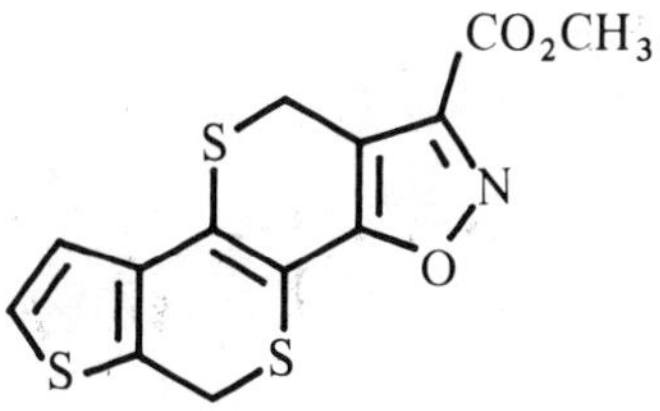

275

Research on antihistaminic agents has also continued. The therapeutic index of ketotifen (**142**), as well as several other H_1-antagonists, has been determined.[662] A new antiaminic compound related to **142** and dithiadene (**143**) has recently been reported.[663] Pipethiadene (**276**), the piperidine analogue of dithiadene, had outstanding antihistamine, antiserotonin, antireserpine, and anticataleptic activity. Pipethiadene was investigated clinically as an agent to treat migraines.

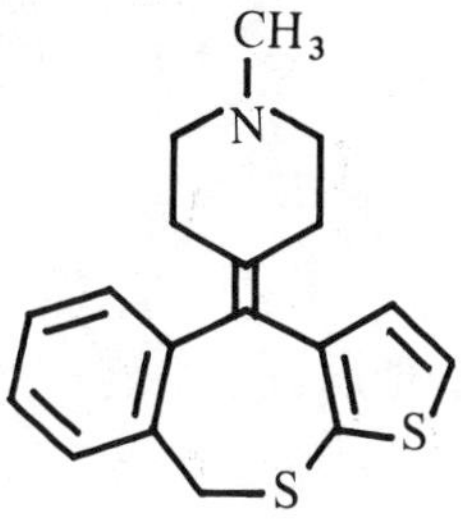

276

Antibiotic development has led to several important thiophene derivatives. The penicillin temocillin sodium (**277**, BRL-17,421) had activity against a wide range of bacteria, including a large number of gram-negative organisms.[664] Compound **277**, the methoxy analogue of ticarcillin (**168**), had little activity against

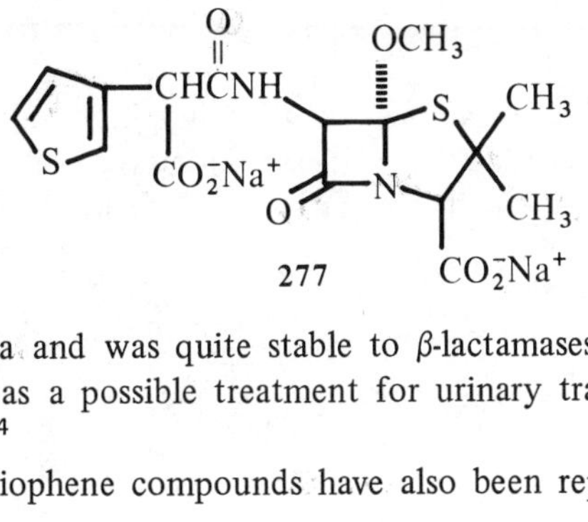

277

gram-positive bacteria and was quite stable to β-lactamases. The compound was tested in the clinic as a possible treatment for urinary tract, lower respiratory, and skin infections.[664]

Other β-lactam thiophene compounds have also been reported. Cephalosporin

278, the isostere of cephaloridine (**178**), has been prepared from penicillin precursors.[665] Several N^7-hydroxy and N^7-alkoxy cephalosporins have recently been

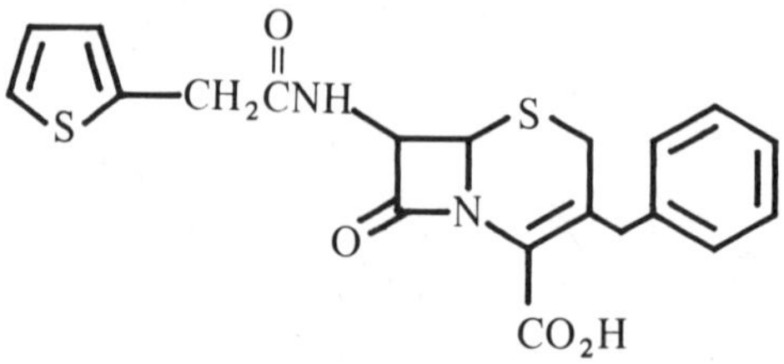

278

prepared.[666] These compounds, based upon thienamycin-type structures, show significant antibacterial activity; **279** (R = CH$_3$) showed superior activity as compared with **279** (R = H). The amido hydrogen of β-lactam antibiotics was con-

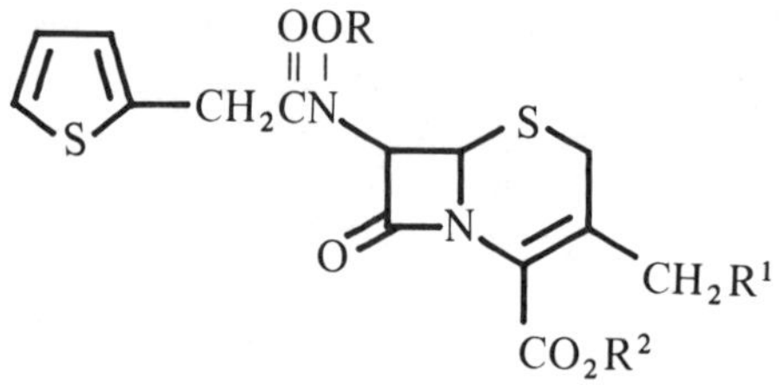

279

cluded to be unnecessary for antibacterial activity.[666] Finally, in a study of monosulfactams, **280** was prepared and shown to have antibacterial activity, although this compound was not the most potent in the series.[667]

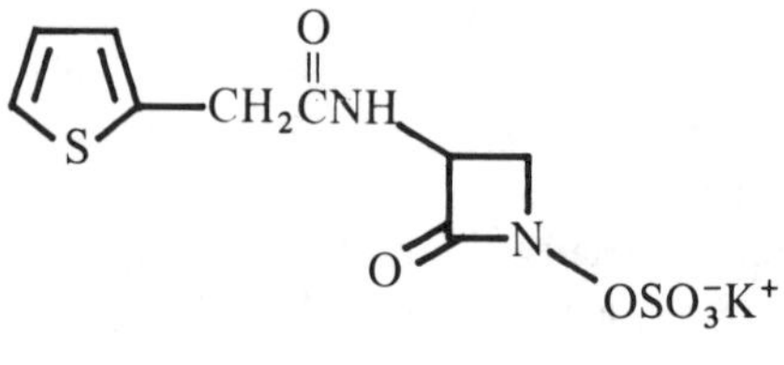

280

Non-β-lactam antibiotics containing the thiophene nucleus have also been reported. Thiolactomycin (**281**, R = CH$_3$) and thiotetromycin (**281**, R = CH$_3$CH$_2$—) have been isolated and have exhibited a wide spectrum of antibacterial activity.[668, 669]

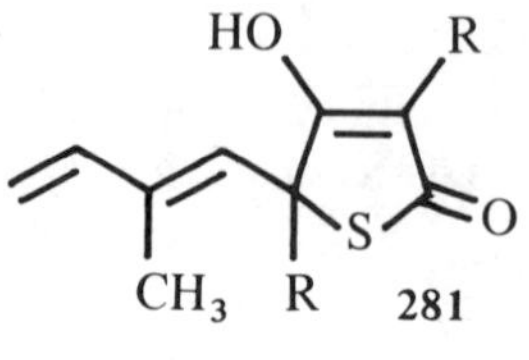

281

Both compounds had low acute toxicity. Reports of 2-thiophenesulfonamide derivatives[670] and 2-thiophenecarboxamide derivatives[671] as weak antibacterial agents have appeared.

Thiophene-containing compounds have been prepared for other approaches to infectious disease therapy. A series of substituted pyrimidine derivatives including **282** have been reported to amplify the antibacterial effects of phleomycin. Thio-

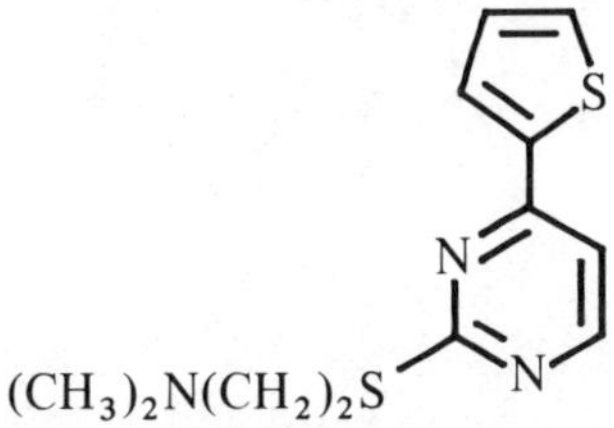

$(CH_3)_2N(CH_2)_2S$

282

phene **282** was more active than its thiazole analogues.[672] Thiacage compounds **283** (n = 1, 2) possessed antiviral activity similar to their furanyl counterparts, but suffered from undesired CNS side effects.[673] A new clinical antifungal agent with

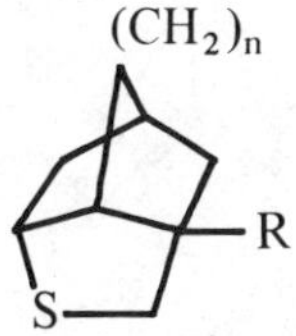

$(CH_2)_n$

283   R = $-NH_2$, $-CH_2NH_2$

activity comparable to miconazole has been developed.[674] Compound **284** was tested *in vitro* and *in vivo* and was shown to be an active fungicide rather than a fungostat. It exhibited good oral activity, unlike miconazole.[674]

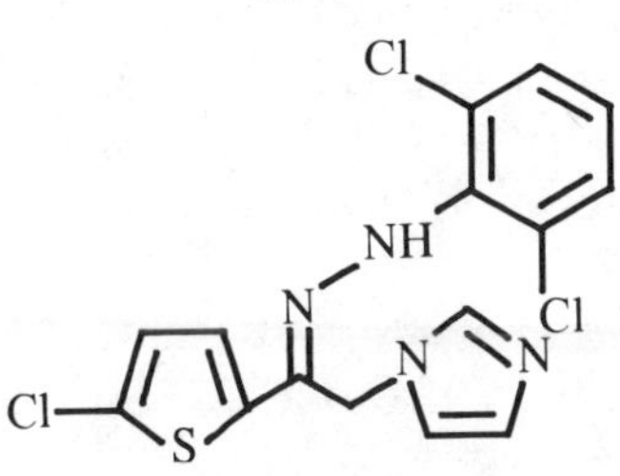

284

Other thiophene derivatives with potential utility include the thio-sugar compounds **285**. These suitably functionalized sugars were prepared for the synthesis

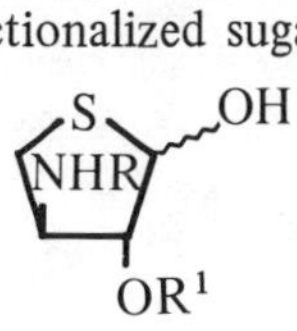

285

of adriamycin analogues.[675] Antianoxic activity was reported for a series of vinyl-sulfoxides and vinylsulfones.[676] Among the numerous thienyl derivatives prepared, **286** was the most potent derivative and it provided the greatest survival time of test

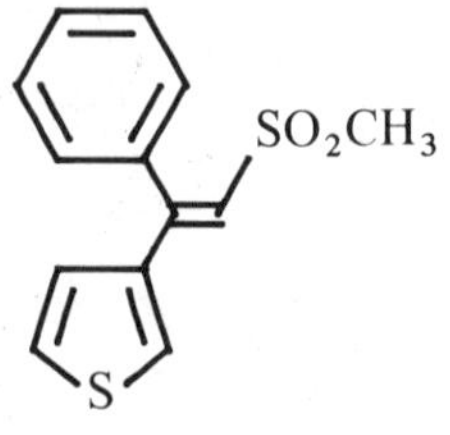

286

animals as compared to controls. The activity of **286** was conjectured to arise from inhibition of enzymes such as cytochromoxidase and succinode-hydrogenase in the respiratory chain.[676] Thieno[3,2-d]pyrimidine nucleosides such as **287**, the isostere of inosine, have been prepared as possible cellular function activators.[677]

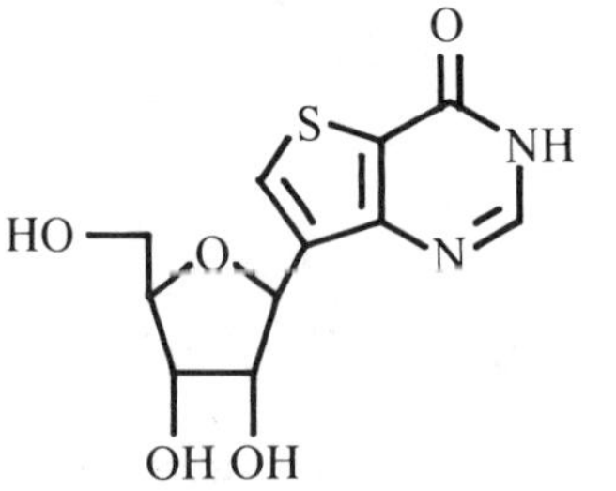

287

As noted in the conclusion, combined advances in synthetic organic chemical methodology and in the understanding of the underlying causes of various disease states have led to an enormous increase in the number of thiophene derivatives prepared for medical investigations. With such rapidly expanding areas of know-ledge, new, useful derivatives of thiophene will most certainly continue to be synthesized and developed.

REFERENCES

1. F. F. Blicke, in *Thiophene and Its Derivatives*, (H. D. Hartough, Ed.), Interscience, New York, 1952, pp. 29–45.

2. M. Martin-Smith and S. T. Reid, *J. Med. Pharm. Chem.*, **1**, 507 (1959).

3. F. F. Nord, A. Vaitiekunas, and L. J. Owen, *Fortschr. Chem. Forsch.*, **3**, 309 (1955).

4. W. L. Nobles, in *Pharmaceutical Sciences: Fourth Annual Visiting Lecture Series*, College in Pharmacy, University of Texas, Austin, 1961, pp. 149–185; *Chem. Abstr.*, **58**, 409h (1963).

5. W. L. Nobles and C. D. Blanton, Jr., *J. Pharm. Soc.*, **53**, 115 (1964).

6. R. Böhm and G. Zieger, *Pharmazie*, **35**, 1 (1980).

7. S. Gronowitz, in *Advances in Heterocyclic Chemistry*, Vol. 1., (A. R. Katritsky, Ed.), Academic Press, New York, 1963, pp. 1–124.

8. S. Gronowitz, in *Organic Compounds of Sulfur, Selenium and Tellurium*, Vol. 2, (D. H. Reid, Ed.), The Chemical Society, London, 1973, pp. 352–496.

9. S. Gronowitz, in *Organic Compounds of Sulfur, Selenium and Tellurium*, Vol. 3, (D. H. Reid, Ed.), The Chemical Society, London, 1975, pp. 400–493.

10. S. Gronowitz, in *Organic Compounds of Sulfur, Selenium and Tellurium*, Vol. 4, (D. R. Hogg, Ed.), The Chemical Society, London, 1977, pp. 244–299.

11. S. Gronowitz, in *Organic Compounds of Sulfur, Selenium and Tellurium*, Vol. 5, (D. R. Hogg, Ed.), The Chemical Society, London, 1979, pp. 247–305.

12. H. Erlenmeyer, E. Berger, and M. Leo, *Helv. Chim. Acta*, **16**, 733 (1933).

13. H. Erlenmeyer, *Bull. Soc. Chim. Biol. (Paris)*, **30**, 792 (1948).

14. E. Campaigne, *J. Am. Pharm. Assoc.*, **46**, 129 (1957).

15. M. Gordon (Ed.), *Psychopharmacological Agents*, Vol. III, Academic Press, New York, 1974.

16. E. Usdin and I. S. Forrest (Eds.), *Psychotherapeutic Agents. Part II: Applications*, Marcel Dekker, Inc., New York, 1977.

17. C. J. Grol and H. Rollema, *J. Med. Chem.*, **18**, 857 (1975).

18. C. J. Grol and J. S. Faber, *Rec. Trav. Chim. Pays-Bas*, **89**, 68 (1970).

19. C. J. Grol, H. Rollema, D. Dijkstra, and B. H. C. Westerink, *J. Med. Chem.*, **23**, 322 (1980).

20. H. Gross and E. Langner, *Wien. Med. Wochenschr.*, **116**, 814 (1966).

21. J. B. Press, C. M. Hofmann, N. H. Eudy, W. J. Fanshawe, I. P. Day, E. N. Greenblatt, and S. R. Safir, *J. Med. Chem.*, **22**, 725 (1979).

22. J. B. Press, C. M. Hofmann, N. H. Eudy, I. P. Day, E. N. Greenblatt, and S. R. Safir, *J. Med. Chem.*, **24**, 154 (1981).

23. J. K. Chakrabarti, L. Horsman, T. M. Hotten, I. A. Pullar, D. E. Tupper, and F. C. Wright, *J. Med. Chem.*, **23**, 878 (1980).

24. J. K. Chakrabarti, J. Fairhurst, N. J. A. Gutteridge, L. Horsman, I. A. Pullar, C. W. Smith, D. J. Steggles, D. E. Tupper, and F. C. Wright, *J. Med. Chem.*, **23**, 884 (1980).

25. J. B. Press, C. M. Hofmann, G. E. Wiegand, and S. R. Safir, *J. Heterocycl. Chem.*, **19**, 391 (1982).

26. M. Rajšner, J. Metyšová, and M. Protiva, *Farmaco Ed. Sci. (Pavia)*, **23**, 140 (1968).

27. M. Rajšner, J. Metyšová, and M. Protiva, *Coll. Czech. Chem. Comm.*, **35**, 378 (1970).

28. K. Šindelář, J. Metyšová, and M. Protiva, *Coll. Czech. Chem. Comm.*, **36**, 3404 (1971).

29. M. Rajšner, F. Mikšik, J. Metyšová, and M. Protiva, *Coll. Czech. Chem. Comm.*, **44**, 2997 (1979).

30. Z. Polívka, J. Holubek, E. Svátek, J. Metyšová, and M. Protiva, *Coll. Czech. Chem. Comm.*, **46**, 2222 (1981).

31. F. Hunziker, R. Fischer, P. Kipfer, J. Schmutz, H. R. Bürki, E. Eichenberger, and T. G. White, *Eur. J. Med. Chem. Chim. Ther.*, **16**, 391 (1981).

32. L. H. Sternbach, L. O. Randall, and S. R. Gustafson, in *Psychopharmacological Agents*, Vol. I, (M. Gordon, Ed.), Academic Press, New York, 1964, pp. 137–224.

33. L. O. Randall, W. Schallek, L. H. Sternbach, and R. Y. Ning, in *Psychopharmacological Agents*, Vol. III, (M. Gordon, Ed.), Academic Press, New York, 1974, pp. 175–282.

34.   S. Fielding and H. Lal (Eds.), *Anxiolytics,* Futura Publishing Co., Mount Kisco, NY, 1979.

35.   M. Nakanishi, T. Tahara, K. Araki, M. Shiroki, T. Tsumagari, and Y. Takigawa, *J. Med. Chem.,* **16**, 214 (1973).

36.   M. Nakanishi, T. Tsumagari, Y. Takigawa, S. Shuto, T. Kenjo, and T. Fukuda, *Arzneim.-Forsch.,* **22**, 1905 (1972).

37.   M. Nakanishi and M. Setoguchi, *Arzneim.-Forsch.,* **22**, 1914 (1972).

38.   O. Hromatka and D. Binder, *Monatsh. Chem.,* **104**, 704 (1973).

39.   O. Hromatka, D. Binder, and P. Stanetty, *Monatsh. Chem.,* **104**, 709, 920 (1973).

40.   O. Hromatka, D. Binder, and W. Veit, *Monatsh. Chem.,* **104**, 973 (1973).

41.   O. Hromatka, D. Binder, C. R. Noe, P. Stanetty, and W. Veit, *Monatsh. Chem.,* **104**, 715 (1973).

42.   D. Binder, O. Hromatka, C. R. Noe, F. Hillebrand, W. Veit, and J. E. Blum, *Arch. Pharm.,* **313**, 587 (1980).

43.   D. Binder, O. Hromatka, C. R. Noe, Y. A. Bara, M. Feifel, G. Habison, F. Leierer, and J. E. Blum, *Arch. Pharm.,* **313**, 636 (1980).

44.   O. Hromatka, D. Binder, P. Stanetty, and G. Marischler, *Monatsh. Chem.,* **107**, 233 (1976).

45.   O. Hromatka, D. Binder, and G. Pixner, *Monatsh. Chem.,* **104**, 1348 (1973).

46.   O. Hromatka, D. Binder, and K. Eichinger, *Monatsh. Chem.,* **105**, 138 (1974).

47.   O. Hromatka and D. Binder, *Monatsh. Chem.,* **104**, 1105 (1973).

48.   O. Hromatka, D. Binder, and G. Pixner, *Monatsh. Chem.,* **106**, 1103 (1975).

49.   O. Hromatka, D. Binder, and K. Eichinger, *Monatsh. Chem.,* **105**, 135 (1974).

50.   O. Hromatka, D. Binder, and K. Eichinger, *Monatsh. Chem.,* **104**, 1513, 1599 (1973).

51.   O. Hromatka, D. Binder, and K. Eichinger, *Monatsh. Chem.,* **105**, 123, 135 (1974).

52.   K. Grohe and H. Heitzer, *Liebigs Ann. Chem.,* 1947 (1977).

53.   F. J. Tinney, J. P. Sanchez, and J. A. Nogas, *J. Med. Chem.,* **17**, 624 (1974).

54.   A. S. Noravyan, A. P. Mkrtchyan, I. A. Dzhagatspanyan, and S. A. Vartanyan, *Khim.-Farm. Zh.,* **11**(10), 62 (1977).

55.   T. Tahara, K. Araki, M. Shiroki, H. Matsuo, and T. Munakata, *Arzneim.-Forsch.,* **28**, 1153 (1978).

56.   T. Tsumagari, A. Nakajima, J. Fukuda, S. Shuto, T. Kenjo, Y. Morimoto, and Y. Takigawa, *Arzneim.-Forsch.,* **28**, 1158 (1978).

57.   M. Setoguchi, S. Takehara, A. Nakajima, T. Tsumagari, and Y. Takigawa, *Arzneim.-Forsch.,* **28**, 1165 (1978).

58.   M. Nakanishi, T. Tsumagari, S. Shuto, T. Kenjo, T. Fukuda, and M. Setoguchi, *Jpn. J. Pharmacol.,* **24** 113 (1974).

59.   Y. Kato and H. Nishimine, *Arzneim.-Forsch.,* **28**, 1170 (1978).

60.   K. H. Weber, A. Bauer, A. Langbern, and H. Daniel, *Liebigs Ann. Chem.,* 1257 (1978).

61.   A. N. Nicholson and C. M. Wright, *Neuropharmacology,* **19**, 491 (1980).

62.   R. I. Fryer, J. V. Early, and A. Walser, *J. Heterocycl. Chem.,* **15**, 619 (1978).

63.   P. de Cointet, P.-J. Grossi, C. Pigerol, M. Broll, and P. Eymard, *Eur. J. Med. Chem. Chim. Ther.,* **15**, 223 (1980).

64.   L. Raffa, M. DiBella, L. Dibella, and G. Conti, *Farmaco Ed. Sci. (Pavia),* **19**, 425 (1964).

65.   T. Hisano, M. Ichikawa, A. Nakagawa, and M. Tsuji, *Chem. Pharm. Bull.,* **23**, 1910 (1975).

66.   E. A. Swinyard, W. C. Brown, and L. S. Goodman, *J. Pharmacol.,* **106**, 319 (1952) and references therein.

67. E. E. Campaigne and H. L. Thomas, *J. Am. Chem. Soc.*, 77, 5365 (1955).

68. J. J. Spurlock, *J. Am. Chem. Soc.*, 75, 1115 (1953).

69. L. M. Long, C. A. Miller, and G. J. Chen, *J. Am. Chem. Soc.*, 71, 669 (1949).

70. F. F. Blicke and M. F. Zienty, *J. Am. Chem. Soc.*, 63, 2945 (1941).

71. E. E. Campaigne and R. L. Patrick, *J. Am. Chem. Soc.*, 77, 5425 (1955).

72. H. A. Luts and W. L. Nobles, *J. Pham. Sci.*, 51, 1173 (1962).

73. F. C. Rogers and W. L. Nobles, *J. Pharm. Sci.*, 51, 273 (1962).

74. S. Jeganathan and M. Srinivasan, *Phosphorus and Sulfur*, 11, 125 (1981).

75. S. Fielding and H. Lal (Eds.), *Antidepressants*, Futura Publishing, Mount Kisco, NY, 1975.

76. F. Sicuteri, B. Anselmi, and P. L. Del Bianco, *Drug. Dig.*, 3, 299 (1967/1968).

77. A. Gehring, P. Blaser, R. Spiegel, and W. Pöldinger, *Arzneim.-Forsch.*, 21, 15 (1971).

78. W. V. Krumholz, J. A. Yaryura-Tobras, and L. White, *Curr. Ther. Res.*, 10, 342 (1968).

79. H. Gross and E. Kaltenbäck, *Drug. Dig.*, 3, 61 (1967/1968).

80. E. Vencovsky, Vl. Šedivec, E. Peterová, and P. Baudis, *Arzneim.-Forsch.*, 19, 491 (1969).

81. E. Messmer, *Arzneim.-Forsch.*, 19, 735 (1969).

82. B. Yom-Tov, S. Gronowitz, S. B. Ross, and N. E. Stjernström, *Acta Pharm. Suec.*, 11, 149 (1974).

83. J. Guillaume, L. Nédélec, M. Cariou, and A. Allais, *Heterocycles*, 15, 1227 (1981).

84. M. Nakanishi, H. Imamura, and Y. Maruyama, *Arzneim.-Forsch.*, 20, 998 (1970) and references therein.

85. M. Nakanishi, H. Imamura, K. Ikegami, and K. Goto, *Arzneim.-Forsch.*, 20, 1004 (1970).

86. M. Perrissin, C. L. Duc, G. Narcisse, F. Bakri-Logeais, and F. Huguet, *Eur. J. Med. Chem. Chim. Ther.*, 15, 413 (1980).

87. M. B. DeVani, C. J. Shishoo, U. S. Pathak, S. H. Parikh, A. V. Radhakrishnan, and A. C. Padhya, *Indian J. Chem.*, 14B, 357 (1976).

88. T. A. Montzka and J. D. Matiskella, *J. Heterocycl. Chem.*, 11, 853 (1974).

89. J. Bosch, R. Granados, and F. Lopez, *J. Heterocycl. Chem.*, 12, 651 (1975).

90. J. Bosch, M. Alvarez, and R. Granados, *An. Quim. C-Org. Bioquim.*, 77, 346 (1981).

91. M. Alvarez, J. Bosch, R. Granados, and F. Lopez, *J. Heterocycl. Chem.*, 15, 193 (1978).

92. M. Ban. Y. Baba, K. Muira, Y. Kondo, K. Suzuki, and M. Hori, *Chem. Pharm. Bull.*, 24, 1679 (1976).

93. R. L. Clarke, A. J. Gambino, A. K. Pierson, and S. J. Daum, *J. Med. Chem.*, 21, 1235 (1978).

94. R. L. Clarke, M. L. Heckeler, A. J. Gambino, S. J. Daum, H. R. Harding, A. K. Pierson, D. G. Teiger, J. Pearl, L. D. Shargel, and T. J. Goehl, *J. Med. Chem.*, 21, 1243 (1978).

95. L. Fontanella, E. Occelli, and E. Testa, *Farmaco Ed. Sci. (Pavia)*, 30, 742 (1975).

96. P. G. H. Van Daele, M. F. L. DeBruyn, J. M. Boey, S. Sanczuk, J. T. M. Agten, and P. A. J. Janssen, *Arzneim.-Forsch.*, 26, 1521 (1976).

97. W. F. M. Van Bever, C. J. E. Niemegeers, K. H. L. Schellekens, and P. A. J. Janssen, *Arzneim.-Forsch.*, 26, 1548 (1976).

98. J. A. Waters, *J. Med. Chem.*, 20, 1496 (1977).

99. D. W. Adamson and A. F. Green, *Nature*, 165, 122 (1950).

100. H. Isbell and H. G. Fraser, *J. Pharmacol.*, 109, 417 (1953).

101. D. W. Adamson, W. M. Muffin, and A. F. Green, *Nature*, 167, 153 (1951).

102.   H. F. Fraser, T. L. Nash, G. D. Vanhorn, and H. Isbell, *Arch. Int. Pharmacodyn.*, **98**, 443 (1954).

103.   D. J. Brown, A. H. Cook, and I. Heilbron, *J. Chem. Soc.*, 113 (1949).

104.   L. O. Randall and G. Lehmann, *J. Pharmacol.*, **93**, 314 (1948).

105.   E. A. Schildknecht and E. V. Brown, *J. Am. Chem. Soc.*, **77**, 954 (1955).

106.   M. Sander, *Arzneim.-Forsch.*, **4**, 375 (1954).

107.   T. Yabuuchi, *Chem. Pharm. Bull.*, **8**, 169 (1960).

108.   R. Kimura, M. Ogawa, and T. Yabuuchi, *Chem. Pharm. Bull.*, **7**, 171 (1959).

109.   R. Kimura and T. Yabuuchi, *Chem. Pharm. Bull.*, **6**, 159 (1958).

110.   R. Kimura, T. Yabuuchi, and Y. Tamura, *Chem. Pharm. Bull.*, **8**, 103 (1960).

111.   Y. Kasé, T. Yuizono, and M. Muto, *J. Med. Chem.*, **6**, 118 (1963).

112.   Y. Kasé, T. Yuizono, T. Yamasaki, T. Yamada, S. Tamuja, and I. Condo, *Chem. Pharm. Bull.*, **7**, 372 (1959).

113.   Y. Sasaki, J. Sugihara, A. Watanabe, M. Sakuma, M. Otsuka, and Y. Sato, *J. Pharm. Soc. Jpn.*, **89**, 345 (1969).

114.   E. E. Mikhlina, V. Y. Vorob'eva, N. A. Komarova, J. M. Sharanov, A. I. Polezhaeva, M. D. Mashkovskii, and L. N. Yakhontov, *Khim. Farm. Zh.*, **10**(11), 56 (1976).

115.   G. Linari and R. Spanò, *Chim. Ther.*, **5**, 138 (1970).

116.   J. D. Couquelet, J. M. Couquelet, M. Payard, F. Fauran, and A. Thihault, *Ann. Pharm. Fr.*, **36**, 151 (1978).

117.   J. Xicluna, J. E. Ombetta, J. Novarro, J. F. Robert, and J. J. Panouse, *Eur. J. Med. Chem. Chim. Ther.*, **14**, 523 (1979).

118.   N. D. Heindel and J. A. Minatelli, *J. Heterocycl. Chem.*, **13**, 669 (1976).

119.   H. G. Kraft, L. Fiebig, and R. Hotovy, *Arzneim.-Forsch.*, **11**, 922 (1961).

120.   J. E. Winther and B. Nathalang, *Scand. J. Dent. Res.*, **80**, 272 (1972).

121.   E. Profft, *Chemikerzeitung*, **82**, 295 (1958).

122.   M. H. Kim and R. D. Schuetz, *J. Am. Chem. Soc.*, **74**, 5102 (1952).

123.   S. Conde, R. Madronero, M. P. Fernandez-Tome, and J. del Rio, *J. Med. Chem.*, **21**, 978 (1978).

124.   W. O. Foye and S. Tovivich, *J. Pham. Sci.*, **68**, 591 (1979).

125.   R. I. Mrongovius, P. Ghosh, A. G. Bolt, and B. Ternai, *Arzneim.-Forsch.*, **31**, 1718 (1981).

126.   R. K. Razdan, B. Z. Lerris, G. R. Handrick, H. C. Dalzell, H. G. Pars, J. F. Howes, N. Plotnikoff, P. Dodge, A. Dren, J. Kyncl. L. Shoer, and W. R. Thompson, *J. Med. Chem.*, **19**, 549 (1976).

127.   R. K. Razdan, G. R. Handrick, H. C. Dalzell, J. F. Howes, M. Winn, N. P. Plotnikoff, P. W. Dodge, and A. T. Dren, *J. Med. Chem.*, **19**, 552 (1976).

128.   J. P. Vincent, B. Kartalowski, P. Geneste, J. M. Kamenka, and M. Lazdunski, *Proc. Natl. Acad. Sci.*, **76**, 4678 (1979).

129.   P. Geneste, J. M. Kamenka, S. N. Ung, P. Herrman, R. Goudal, and G. Trouiller, *Eur. J. Med. Chem. Chim. Ther.*, **14**, 301 (1979).

130.   A. J. Shulgin and D. E. MacLean, *Clin. Toxicol.*, **9**, 553 (1976).

131.   C. H. Tilford, L. A. Doerle, M. G. Van Campen, and R. S. Shelton, *J. Am. Chem. Soc.*, **71**, 1705 (1949).

132.   F. P. Ludueña and A. M. Lands, *J. Pharmacol.*, **110**, 282 (1954).

133.   R. M. Clark and B. R. Clark, *Arch. Int. Pharmacodyn.*, **112**, 458 (1957).

134.   F. Stiegman and R. A. Dolehide, *Am. J. Dig. Dis.*, **22**, 37 (1955).

135.   A. M. Lands, K. Z. Hooper, H. M. McCarthy, and R. F. Feldkamp, *Proc. Soc. Exp. Biol. NY*, **66**, 452 (1947).

136. A. Flickenstein, R. Mushaweck, and F. Bohlinger, *Arch. Exp. Path. Pharmakol.,* **211,** 132 (1950).

137. E. E. Campaigne and R. C. Burgeois, *J. Am. Chem. Soc.,* **75,** 2702 (1953).

138. F. H. Meyers and B. E. Abreu, *J. Pharmacol.,* **104,** 387 (1952).

139. J. H. Biel, E. P. Sprengler, H. A. Leiser, J. Hormer, A. Drukker, and H. L. Friedman, *J. Am. Chem. Soc.,* **77,** 2250 (1955).

140. D. W. Adamson and A. F. Green, *Nature,* **165,** 122 (1950).

141. J. Cymerman-Craig and R. J. Harrisson, *Aust. J. Chem.,* **8,** 378 (1955).

142. J. P. Long, F. P. Luduēna, B. F. Fullar, and A. M. Lands, *J. Pharmacol.,* **117,** 29 (1956).

143. F. F. Blicke and F. Leonard, *J. Am. Chem. Soc.,* **74,** 5105 (1952).

144. H. G. Morren, R. Denager, S. Trolin, H. Strubbe, R. Linz, G. Dony, and R. Collard, *Ind. Chim. Belge,* **20,** 733 (1955).

145. P. Duchêne-Marullaz and J. Vocher, *Compt. Rend. Soc. Biol.,* **156,** 1634 (1962).

146. L. G. Abood, A. Ostfield, and J. H. Biel, *Arch. Int. Pharmacodyn.,* **120,** 186 (1959).

147. L. G. Abood, *J. Med. Pharm. Chem.,* **4,** 469 (1961).

148. V. C. Lipman, P. S. Shurrager, and L. G. Abood, *Arch. Int. Pharmacodyn.,* **146,** 174 (1963).

149. L. Albanus, *Acta. Pharm. Tox.,* **28,** 305 (1970).

150. A. Meyerhöffer and O. Wahlberg, *Acta Chem. Scand.,* **27,** 868 (1973).

151. G. P. Nilles and R. D. Schuetz, *J. Med. Chem.,* **13,** 1249 (1970).

152. M. Cohen, *Arch. Int. Pharmacodyn.,* **169,** 412 (1967).

153. M. Robba and R. C. Moreau, *Ann. Pharm. Fr.,* **23,** 103 (1965).

154. F. Leonard, *J. Am. Chem. Soc.,* **74,** 2915 (1952).

155. F. Leonard and L. Simet, *J. Am. Chem. Soc.,* **77,** 2855 (1955).

156. M. Robba and Y. LeGuen, *Chim. Ther.,* **1,** 238 (1966).

157. M. Robba and Y. LeGuen, *Chim. Ther.,* **2,** 120 (1967).

158. J. R. Bossier, M. Aurousseau, J. F. Giridicelli, and D. Duval, *Arzneim.-Forsch.,* **28**(II), 2222 (1978).

159. J. A. Simaan and D. M. Aviado, *J. Pharm. Exp. Ther.,* **198,** 176 (1976).

160. F. Clemence, O. LeMartret, F. Fournex, G. Plassard, and M. Dagnaux, *Chim. Ther.,* **7,** 14 (1972).

161. H. Tron-Loisel, P. Brossier, O. Campagnon, B. Grozean, P. L. Compagnon, and D. Branceni, *Eur. J. Med. Chem. Chim. Ther.,* **12,** 379 (1977).

162. H. Tron-Loisel, P. Brossier, P. L. Compagnon, and D. Branceni, *Eur. J. Med. Chem. Chim. Ther.,* **13,** 351 (1978).

163. C. Labrid, G. Dureng, H. Bert, and P. Duchene-Marulloz, *Arch. Int. Pharmacodyn.,* **223,** 231 (1976).

164. M. Robba and D. Duval, *Chim. Ther.,* **8,** 22 (1973).

165. K. Thiele, K. Posselt, H. Offermans, and K. Thiemer, *Arzneim.-Forsch.,* **30**(I), 747 (1980).

166. A. Kleeman, J. Heese, and J. Engel, *Arzneim.-Forsch.,* **31**(II), 1178 (1981).

167. K. Thiemer, F. Stroman, I. Szelenyi, and A. V. Schlichtegroll, *Arzneim.-Forsch.,* **28**(II), 1343 (1978).

168. W. G. Haney, R. G. Brown, E. I. Isaacson, and J. N. Delgado, *J. Pharm. Sci.,* **66,** 1602 (1977).

169. J. Laforest and G. Thullier, *J. Heterocycl. Chem.,* **14,** 793 (1977).

170. L. O. Randall, W. G. Peterson, and G. Lehman, *J. Pharm. Exp. Ther.,* **97,** 48 (1949).

171. E. Braunwald, *New Engl. J. Med.,* **297,** 331 (1977).

172. H. A. Luts, W. A. Zuccarello, J. F. Grattan, and W. R. Nobles, *J. Pharm. Sci.,* **53,** 840 (1964).

173. H. C. Caldwell, J. A. Finkelstein, P. P. Goldman, A. J. Swak, J. Schlosser, C. Pelikan, and W. G. Groves, *J. Med. Chem.,* **13,** 1076 (1970).

174. A. S. Noravyan, A. P. Mkrtchyan, I. A. Dzhagatspanyan, R. A. Akonyan, N. E. Akonyan, and S. A. Vartanyan, *Khim. Farm. Zh.,* **38** (1977).

175. N. Kawazu, T. Kanno, S. Saito, and H. Tamaki, *J. Med. Chem.,* **15,** 914 (1972).

176. T. Meshi, S. Nakamura, and T. Kanno, *Chem. Pharm. Bull.,* **21,** 1709 (1973).

177. E. Koshinaka, N. Ogawa, S. Kurata, K. Yamagishi, S. Kubo, I. Matsubara, and H. Kato, *Chem. Pharm. Bull.,* **27,** 1454 (1979).

178. E. Koshinaka, N. Ogawa, K. Yamagishi, H. Kato, and M. Hanaoka, *Yakugaku Zasshi,* **100,** 88, 100 (1980).

179. H. Kato, E. Koshinaka, N. Ogawa, K. Yamagishi, K. Mitani, S. Kubo, and M. Hanaoka, *Chem. Pharm. Bull.,* **28,** 2194 (1980).

180. J. H. Laragh (Ed.), *Topics in Hypertension,* Yorke Medical Books, Dun-Donnelly Publishers Inc., New York, 1980.

181. E. D. Bergmann and Z. Goldschmidt, *J. Med. Chem.,* **11,** 1121 (1968).

182. J. M. Barker, D. J. Byron, and P. R. Huddleston, *J. Chem. Soc. (C),* 2183 (1969).

183. E. Lindner and J. Kaiser, *Arch. Int. Pharmacodyn.,* **211,** 305 (1974).

184. P. Simon, R. Chermat, and J. R. Boissier, *Therapie,* **30,** 855 (1975).

185. H. G. Eckert, S. Baudner, K. E. Weimer, and H. Wissman, *Arzneim.-Forsch.,* **31(I),** 419 (1981).

186. J. M. Barker and P. R. Huddleston, *Org. Prep. Proc. Int.,* **13,** 429 (1981).

187. H. Y. Aboul Enein, A. A. Al Badr, S. E. Ibrahim, and M. Ismail, *Pharm. Acta Helv.,* **55,** 228 (1980).

188. J. F. Bagli and E. Ferdinandi, *Can. J. Chem.,* **53,** 2598 (1975).

189. J. F. Bagli, W. D. Mackay, E. Ferdinandi, M. N. Cayen, I. Vavra, T. Pugsley, and W. Lippmann, *J. Med. Chem.,* **19,** 876 (1976).

190. C. Corral, V. Darias, M. P. Fernández-Tomé, R. Madronéro, and J. del Río, *J. Med. Chem.,* **16,** 882 (1973).

191. V. Darias, R. Madroñero, and J. del Río, *Arzneim.-Forsch.,* **24,** 1751 (1974).

192. V. Darias and J. del Río, *Arzneim.-Forsch.,* **24,** 1756 (1974).

193. S. Conde, C. Corral, R. Madroñero, A. Sanchez Alvarez-Insúa, M. P. Fernández-Tomé, J. del Río, and M. Santos, *J. Med. Chem.,* **20,** 970 (1977).

194. Y. Hara, E. Sato, A. Miyagishi, A. Aisaka, and T. Hibino, *J. Pharm. Sci.,* **67,** 1334 (1978).

195. D. Aubert, G. Barthelemy, and A. Bernat, *J. Pharmacol.,* **6,** 364 (1975).

196. K. Hashimoto, T. Tsukada, H. Matsuda, and S. Imai, *Eur. J. Pharmacol.,* **45,** 185 (1977).

197. J. Castañer, *Drugs of the Future,* **3,** 271 (1978).

198. P. J. Roberts, *Drugs of the Future,* **3,** 477 (1978).

199. W. V. Curran and A. Ross, *J. Med. Chem.,* **17,** 273 (1974).

200. I. T. Barnish, P. E. Cross, R. P. Dickinson, M. J. Perry, and M. J. Randall, *J. Med. Chem.,* **24,** 959 (1981).

201. G. deStevens, A. Halamandaris, S. Ricca, Jr., and L. H. Werner, *J. Med. Pharm. Chem.,* **1,** 565 (1959).

202. G. J. Martin, C. P. Balant, S. Avakian, and J. M. Beiler, *Arch. Int. Pharmacodyn.,* **98,** 286 (1954).

203. A. Buzas, J. Frossard and J. Leste, *Ann. Pharm. Fr.,* **19**, 31 (1961).

204. A. Buzas and J. Teste, *Bull. Soc. Chim. Fr.,* 793 (1960).

205. M. M. Kochhar, M. Salahi-Asbahi, and B. B. Williams, *J. Pharm. Sci.,* **62**, 336 (1973).

206. C. Fauran, J. Eberle, D. Berthon, G. Huguet, M. Servant, B. Pourrias, G. Raynaud, M. Garrivet, and V. Hecaen, *Chim. Ther.,* **6**, 453 (1971).

207. B. Dartigues, J. Roquebert, J. Canellas, and C. Peyraud, *Therapie,* **27**, 491 (1972).

208. B. Dartigues, J. Roquebert, J. Canellas, and C. Peyraud, *Therapie,* **27**, 501 (1972).

209. G. Thuillier, J. LaForest, B. Cariou, P. Bessin, J. Bonnet, and J. Thuillier, *Eur. J. Med. Chem. Chim. Ther.,* **9**, 633 (1974).

210. A. Nuhrich, C. LaBlanche, G. Devaux, A. Carpy, P. Dufour, C. Nguyenba, and J. Roquebert, *Eur. J. Med. Chem. Chim. Ther.,* **16**, 551 (1981).

211. W. F. Hoffman, O. W. Woltersdorf, Jr., F. C. Novello, E. J. Cragoe, Jr., J. P. Springer, L. S. Watson, and G. M. Fanelli, Jr., *J. Med. Chem.,* **24**, 865 (1981).

212. G. M. Fanelli, Jr. and D. L. Bohn, *Pharmacologist,* **21**, 275 (1979).

213. P. G. H. Van Daele, J. M. Boey, V. K. Sipido, M. F. L. DeBruyn, and P. A. J. Janssen, *Arzneim.-Forsch.,* **25**, 1495 (1975).

214. R. J. Capetola, D. A. Shriver, and M. E. Rosenthale, *J. Pharm. Exp. Ther.,* **214**, 16 (1980).

215. H. Fujimura, K. Tsurumi, Y. Hiramatsu, Y. Tamura, S. Kokuba, and M. Yanagihara, *Oyo Yakuri,* **9**, 715 (1975); through *Chem. Abstr.,* **83**, 188321 (1975).

216. J. Pottier, D. Berlin, and J. P. Raynaud, *J. Pharm. Sci.,* **66**, 1030 (1977).

217. F. Clémence, O. LeMartret, R. Fournex, G. Plassard, and M. Dagnaux, *Eur. J. Med. Chem. Chim. Ther.,* **9**, 390 (1974).

218. T. Aono, M. Imanishi, Y. Kawano, S. Kishimoto, and S. Noguchi, *Chem. Pharm. Bull.,* **26**, 2475 (1978).

219. D. Binder, C. R. Noe, G. Habison, and J. Chocholous, *Arch. Pharm.,* **312**, 169 (1979).

220. D. E. Aultz, A. R. McFadden, and H. B. Lassman, *J. Med. Chem.,* **20**, 456 (1977).

221. T. Yoshioka, M. Kitagawa, M. Oki, S. Kubo, H. Tagawa, K. Ueno, W. Tsukada, M. Tsubokawa, and A. Kasahara, *J. Med. Chem.,* **21**, 633 (1978).

222. H. Tagawa and K. Ueno, *Chem. Pharm. Bull.,* **26**, 1384 (1978).

223. J. S. Kaltenbronn and T. O. Rhee, *J. Med. Chem.,* **17**, 654 (1974).

224. H. G. Alperman, *Arzneim.-Forsch.,* **20**, 293 (1970).

225. H. G. Alperman, H. Ruschig, and W. Meixner, *Arzneim.-Forsch.,* **22**, 2146 (1972).

226. A. Santulli, D. H. Kim, and S. V. Wanser, *J. Heterocycl. Chem.,* **8**, 445 (1971).

227. M. S. Manhas, S. D. Sharma, and S. G. Amin, *J. Med. Chem.,* **15**, 106 (1972).

228. M. B. Devani, C. J. Shishoo, U. S. Pathak, S. H. Parikh, G. F. Shah, and A. C. Padhya, *J. Pharm. Sci.,* **65**, 660 (1976).

229. D. N. Ridge, J. W. Hanifin, L. A. Harten, B. D. Johnson, J. Menschik, G. Nicolou, A. E. Sloboda, and D. E. Watts, *J. Med. Chem.,* **22**, 1385 (1979).

230. J. van Dijk and J. M. A. Zwagemakers, *J. Med. Chem.,* **20**, 1199 (1977).

231. F. DeSimone, A. Dini, R. A. Nicolaus, E. Ramundo, M. DiRosa, and P. Persico, *Farmaco Ed. Sci. (Pavia),* **35**, 333 (1980).

232. J.-P. Bonte, D. Lesieur, C. Lespagnol, J.-C. Cazin, and M. Cazin, *Eur. J. Med. Chem. Chim. Ther.,* **9**, 497 (1974).

233. J.-P. Bonte, D. Lesieur, C. Lespagnol, M. Plat, J.-C. Cazin, and M. Cazin, *Eur. J. Med. Chem. Chim. Ther.,* **9**, 491 (1974).

234. K. Sakai, M. Suzuki, K. Nunami, N. Yoneda, Y. Onoda, and Y. Iwasawa, *Chem. Pharm. Bull.,* **28**, 2384 (1980).

444   J. B. Press

235.   G. A. Tolstikov, N. N. Novitskaya, B. V. Flekhter, D. N. Lazareva, V. A. Davydova, and E. G. Kamalova, *Khim.-Farm. Zh.*, **12**(12), 33 (1978).

236.   M. Miocque, H. Moskowitz, J. Blanc-Guenee, A. M. Saint-Marc, G. Raynaud, J. Thomas, C. Gouret, and B. Rourrias, *Chim. Ther.*, **7**, 283 (1972).

237.   I. Vlattas, L. DellaVecchia, and A. O. Lee, *J. Am. Chem. Soc.*, **98**, 2008 (1976).

238.   I. Vlattas and L. DellaVecchia, *Tetrahedron Lett.*, 4267, 4459 (1974).

239.   K. C. Nicolaou, W. E. Barnette, G. P. Gasic, and R. L. Magolda, *J. Am. Chem. Soc.*, **99**, 7736 (1977).

240.   M. Shibasaki and S. Ikegami, *Tetrahedron Lett.*, 559 (1978).

241.   K. C. Nicolaou, R. L. Magolda, and W. E. Barnette, *J. Chem. Soc. Chem. Comm.*, 375 (1978).

242.   J. Buendia and J. Schalbar, *Tetrahedron Lett.*, 4499 (1977).

243.   W. Bartmann, G. Beck, U. Lerch, H. Teufel, and B. Schölkens, *Prostaglandins*, **17**, 301 (1979).

244.   J. Sandow, W. V. Rechenberg, B. Schölkens, and U. Weithmann, *Acta Endocrin.*, **87**, Supp. 215, 45 (1978).

245.   K. U. Weithmann, W. Bartmann, G. Beck, U. Lerch, E. Konz, and B. A. Schölkens, *Thromb. Haemostasis*, **42**, 119 (1979).

246.   D. G. Fletcher, K. H. Gibson, H. R. Moss, D. R. Sheldon, and E. R. H. Walker, *Prostaglandins*, **12**, 493 (1976).

247.   R. J. Collins and E. V. Brown, *J. Am. Chem. Soc.*, **79**, 1103 (1957).

248.   S. R. Ramadas and P. S. Srinivasan, *Steroids*, **30**, 213 (1977).

249.   H. Kaneko, Y. Yamato, and M. Kurokawa, *Chem. Pharm. Bull.*, **16**, 1200 (1968).

250.   P. Laur, H. Häuser, J. E. Gurst, and K. Mislow, *J. Org. Chem.*, **32**, 498 (1967).

251.   M. E. Wolff and G. Zanati, *J. Med. Chem.*, **12**, 629 (1969).

252.   M. E. Wolff, G. Zanati, G. Shanmugasundarum, S. Gupte, and G. Aadahl, *J. Med. Chem.*, **13**, 531 (1970).

253.   G. Zanati, G. Gaare, and M. E. Wolff, *J. Med. Chem.*, **17**, 561 (1974).

254.   W. H. Chiu and M. E. Wolff, *J. Med. Chem.*, **22**, 1257 (1979).

255.   I. R. Trehan, D. K. Sharma, and D. V. Rewal, *Indian J. Chem.*, **11**, 827 (1973).

256.   I. R. Trehan, R. Inder, and D. V. Rewal, *Indian J. Chem.*, **14B**, 210 (1976).

257.   V. F. Shner, V. A. Rulin, and N. N. Suvorov, *Khim.-Farm. Zh.*, **12**(4), 22 (1978).

258.   P. S. Jogdeo and G. V. Bhide, *Steroids*, **35**, 133 (1980).

259.   P. S. Jogdeo and G. V. Bhide, *Steroids*, **33**, 601 (1979) and references therein.

260.   C. M. Cimarusti, F. F. Giarrusso, P. Grabowich, and S. D. Levine, *Steroids*, **26**, 359 (1975).

261.   M. Kishi and T. Komeno, *Tetrahedron*, **27**, 1527 (1971).

262.   A. A. Macco and H. M. Buck, *J. Org. Chem.*, **46**, 2655 (1981).

263.   A. A. Macco, R. J. deBrouwer, and H. M. Buck, *J. Org. Chem.*, **42**, 3196 (1977).

264.   A. A. Macco, R. J. deBrouwer, M. M. P. Nossin, E. F. Godefroi, and H. M. Buck, *J. Org. Chem.*, **43**, 1591 (1978).

265.   T. Terasawa and T. Okada, *Steroids*, **37**, 445 (1981) and references therein.

266.   Ng. Ph. Buu-Hoï and Ng. Hoan, *J. Org. Chem.*, **17**, 350 (1952).

267.   R. J. Collins and E. V. Brown, *J. Am. Chem. Soc.*, **79**, 1103 (1957).

268.   Ng. H. Nam, Ng. Ph. Buu-Hoï, and Ng. D. Xuong, *J. Chem. Soc.*, 1690 (1954).

269.   J. Sicé and M. Mednick, *J. Am. Chem. Soc.*, **75**, 1628 (1953).

270.   W. R. Biggerstaff and O. L. Stafford, *J. Am. Chem. Soc.*, **74**, 419 (1952).

271. W. R. Biggerstaff, H. Arzoumanian, and K. L. Stevens, *J. Med. Chem.*, **7**, 110 (1964).

272. E. Campaigne and W. M. LeSuer, *J. Am. Chem. Soc.*, **71**, 333 (1949).

273. H. M. Lee, W. G. Dinwiddie, and K. K. Chen, *J. Pharmacol.*, **90**, 83 (1947).

274. A. M. Lands, J. O. Hoppe, O. H. Siegmund, and F. P. Luduena, *J. Pharmacol.*, **95**, 45 (1949).

275. G. R. Clark and G. J. Palenik, *J. Am. Chem. Soc.*, **94**, 4005 (1972).

276. S. L. Lee, B. B. Williams, and M. M. Kochlar, *J. Pharm. Sci.*, **56**, 1354 (1967).

277. B. Mazière, M. Mazière, J. C. Bovay, and N. Dat-Xuong, *Chim. Ther.*, **4**, 265 (1969).

278. M. Payard, P. Tronche, J. Bastide, P. Bastide, and G. Chavernac, *Eur. J. Med. Chem. Chim. Ther.*, **16**, 453 (1981).

279. L. C. Weaver, W. M. Alexander, and B. E. Abbieu, *Arch. Int. Pharmacodyn.*, **156**, 414 (1965).

280. U. Martin and D. Römer, *Arzneim.-Forsch.*, **28**(I), 770 (1978).

281. K. Kuokkanen, *Acta Allergol.*, **30**, 73 (1975).

282. J. P. Girard and M. Cuevas, *Acta Allergol.*, **32**, 27 (1977).

283. K. Kuokkanen, *Acta Allergol.*, **32**, 316 (1977).

284. E. Waldvogel, G. Schwarb, J.-M. Bastian, and J.-P. Bourquin, *Helv. Chim. Acta* **59**, 866 (1976).

285. M. Protiva, M. Rajšner, E. Adlerová, V. Seidlová, and Z. J. Vejdělex, *Coll. Czech. Chem. Comm.*, **29**, 2161 (1964).

286. M. Rajšner, E. Svátek, J. Metyš, and M. Protiva, *Coll. Czech. Chem. Comm.*, **39**, 1366 (1974).

287. M. Rajšner, J. Metyš, and M. Protiva, *Coll. Czech. Chem. Comm.*, **32**, 2854 (1967).

288. M. Rajšner, J. Metyš, B. Kakaï, and M. Protiva, *Coll. Czech. Chem. Comm.*, **40**, 2905 (1975).

289. M. Rajšner and M. Protiva, *Coll. Czech. Chem. Comm.*, **33**, 1846 (1968).

290. V. Darias, M. P. Fernandez-Tome, R. Madroñero, J. Del Rió, and A. Vila-Coro, *Chim. Ther.*, **7**, 224 (1972).

291. D. L. Temple, J. P. Yevich, R. R. Covington, C. A. Hanning, R. J. Seidehamel, H. K. Mackey, and M. J. Baitek, *J. Med. Chem.*, **22**, 505 (1979).

292. F. J. Tinney, W. A. Cetenko, J. J. Kerbleski, D. J. Connor, R. J. Sorenson, and D. J. Herzig, *J. Med. Chem.*, **24**, 878 (1981).

293. G. E. Hardtmann, G. Koletar, O. R. Pfister, J. H. Gogerty, and L. C. Iorio, *J. Med. Chem.*, **18**, 447 (1975).

294. S. Gronowitz, R. Svenson, G. Bondesson, O. Magnusson, and N. E. Stjernström, *Acta Pharm. Suec.*, **15**, 361 (1978).

295. R. A. Parker, T. Kariya, J. M. Grisar, and V. Petrow, *J. Med. Chem.*, **20**, 781 (1977).

296. S. Yurugi, A. Miyake, M. Tomimoto, H. Matsumura, and Y. Imai, *Chem. Pharm. Bull.*, **21**, 1885 (1973).

297. S. Gronowitz, M. Herslöf, R. Svenson, G. Bondesson, O. Magnusson, and N. E. Stjernström, *Acta Pharm. Suec.*, **15**, 368 (1979).

298. B. Dafgård, S. Gronowitz, G. Bondesson, O. Magnusson, and N. E. Stjernström, *Acta Pharm. Suec.*, **11**, 309 (1974).

299. A. A. Abou Ouf, M. M. El-Kerdawy, W. A. Abdulla, and H. A. Selim, *J. Drug Res.*, **2**, 71 (1969).

300. A. A. Abou Ouf, M. M. El-Kerdawy, and H. A. Selim, *J. Drug. Res.*, **6**, 123 (1974).

301. M. M. El-Kerdawy and H. A. Selim, *J. Drug. Res.*, **5**, 135 (1973).

302. A. A. Abou Ouf, M. M. El-Kerdawy, and H. A. Selim, *J. Drug. Res.*, **5**, 127 (1973).

303.   M. M. El-Kerdawy and H. A. Selim, *Bull. Fac. Pharm. Cairo Univ.*, **12**, 235 (1973).

304.   F. A. El-Telbany, B. Abdel-Fattah, and M. Khalifa, *Egypt J. Pharm. Sci.*, **16**, 397, 403 (1975).

305.   J. M. Grisar, G. P. Claxton, and N. L. Wiech, *J. Med. Chem.*, **19**, 365 (1976).

306.   P. A. Rossy, W. Hoffmann, and N. Müller, *J. Org. Chem.*, **45**, 617 (1980).

307.   S. U. Kulkarni and K. A. Thakar, *J. Ind. Chem. Soc.*, **53**, 279 (1976).

308.   K. N. von Kaulla and D. Thilo, *Klin. Wschr.*, **48**, 668 (1970).

309.   K. N. von Kaulla and D. Thilo, *Thromb. Diath. Haemorrh. Supp.* **42**, 345 (1970).

310.   D. Thilo and K. N. von Kaulla, *J. Med. Chem.*, **13**, 503 (1970).

311.   M. Nakanishi, M. Imamura, and K. Goto, *Biochem. Pharmacol.*, **20**, 2116 (1971).

312.   S. Yoshimura, S. Takahashi, A. Kawamata, K. Kikugawa, H. Suehiro, and A. Aoki, *Chem. Pharm. Bull.*, **26**, 685 (1978).

313.   E. Boschetti, D. Molho, J. Chabert, M. Grand, and L. Fontaine, *Chim. Ther.*, **7**, 20 (1972).

314.   H. Yamaguchi and F. Ishikawa, *J. Heterocycl. Chem.*, **18**, 67 (1981).

315.   H. Yamaguchi and F. Ishikawa, *Chem. Pharm. Bull.*, **28**, 3172 (1980).

316.   F. Ishikawa, A. Kosasayama, H. Yamaguchi, Y. Watanabe, J. Saegusa, S. Shibamura, K. Sakuma, S. Ashida, and Y. Abiko, *J. Med. Chem.*, **24**, 376 (1981).

317.   J. J. Thebault, C. E. Blatrix, J. F. Blanchard, and E. A. Panak, *Clin. Pharmacol. Ther.*, **18**, 485 (1975).

318.   M. Podesta, D. Aubert, and J. C. Ferrand, *Eur. J. Med. Chem.*, **9**, 487 (1974).

319.   J.-P. Maffrand and R. Borgegrain, *Heterocycles*, **12**, 1479 (1979).

320.   J.-P. Maffrand and D. Frehel, *Bull. Soc. Chim. Fr. 2*, 48 (1978).

321.   K. Satake, T. Imai, M. Kimura, and S. Morosawa, *Heterocycles*, **16**, 1271 (1981).

322.   P. Staben, A. S. Bhargava, C. Schobel, F. Siegmund, and P. Gunzel, *Arzneim.-Forsch.*, **31**(II), 1735 (1981).

323.   U. Horch, R. Kadatz, Z. Kopitar, J. Ritschard, and H. Weisenberger, *Thromb. Diath. Haemorr. Supp.* **42**, 253 (1970).

324.   H. Gastpar, *Thromb. Diath. Haemorrh. Supp.*, **42**, 291 (1970).

325.   E. F. Elslager, J. R. McLean, S. C. Perricone, D. Potoczak, H. Veloso, D. F. Worth, and R. H. Wheelock, *J. Med. Chem.*, **14**, 397 (1971).

326.   M. S. Manhas, S. G. Amin, S. D. Sharma, B. Dayal, and A. K. Bose, *J. Heterocycl. Chem.*, **16**, 371 (1979).

327.   J. M. Bastian and M. Marko, *Experimentia*, **32**, 413 (1976).

328.   E. Waldvogel, G. Schwarb, J.-M. Bastian, and J. P. Bourquin, *Helv. Chim. Acta*, **59**, 866 (1976).

329.   M. Marko and E. Flückiger, *Experimentia*, **32**, 491 (1976).

330.   C. T. Chan, H. Wells, and D. M. Kramsch, *Circ. Res.*, **43**, 115 (1978).

331.   C. T. Chan, D. M. Kramsch, and H. Wells, *Fed. Proc.*, **35**, 599 (1976).

332.   V. S. Fang, C. Minkin, and P. Goldhaber, *Science*, **172**, 163 (1971).

333.   W. Lloyd, C. Minkin, M. Bresnahan, P. Baer, and H. Wells, *J. Dent. Res.*, **54**, Special Issue B, B87 (1975).

334.   V. S. Fang, *Arch. Int. Pharmacodyn.*, **176**, 193 (1968).

335.   V. S. Fang, *Arch. Int. Pharmacodyn.*, **178**, 315 (1969).

336.   P. Lechert, M. Freyss-Beguin, E. Van Brussel, and N. Mathieu-Levy, *Therapie*, **26**, 831 (1971).

337.   L. D. Sabath, L. L. Stumpf, S. J. Wallace, and M. Finland, *Antimicrob. Agent Chemother.*, **1**, 53 (1970).

338. H. C. Neu and E. B. Winshell, *Antimicrob. Agent Chemother.*, 1, 385 (1970).

339. R. Sutherland, J. Burnett, and G. N. Rolinson, *Antimicrob. Agent Chemother.*, 1, 390 (1970).

340. H. C. Neu and G. J. Garvey, *Antimicrob. Agent Chemother.*, 7, 457 (1975).

341. T. Mita, T. Obe, N. Ito, M. Sugimoto, O. Matsumoto, and J. Ishigami, *Chemotherapy*, 25, 2833 (1977).

342. R. D. Libke, J. T. Clarke, E. D. Ralph, R. P. Luddy, and W. M. M. Kirby, *Clin. Pharm. Ther.*, 17, 441 (1975).

343. V. Rodriguez, J. Inaki, and G. P. Bodney, *Antimicrob. Agent Chemother.*, 4, 31 (1973).

344. S. Gronowitz, J. Rehnö, K. Titlestad, M. Vadzis, B. Sjöberg, P. Bamberg, B. Ekström, and U. Forsgren, *Acta Pharm. Suec.*, 9, 381 (1972).

345. M. Hatanaka and T. Ishimaru, *J. Med. Chem.*, 16, 978 (1973) and references therein.

346. A. W. Taylor and G. Burton, *Tetrahedron Lett.*, 3831 (1977).

347. W. F. Huffman, R. F. Hall, J. A. Grant, and K. G. Holden, *J. Med. Chem.*, 21, 413 (1978).

348. R. R. Chauvette, E. H. Flynn, B. G. Jackson, E. R. Lavagnino, R. B. Moun, R. A. Mueller, R. P. Pioch, R. W. Roeske, C. W. Ryan, J. L. Spencer, and E. Van Heyningen, *J. Am. Chem. Soc.*, 84, 3401 (1962).

349. P. Nauman, *Arzneim.-Forsch.*, 16, 1099 (1966).

350. D. A. Berges, *J. Med. Chem.*, 18, 1264 (1975).

351. C. H. O'Callaghan, *Antimicrob. Agent Chemother.*, 13, 628 (1978).

352. P. P. K. Ho, R. D. Towner, J. M. Indelicato, W. J. Wilham, W. A. Spitzer, and G. A. Koppel, *J. Antibiot.*, 26, 313 (1973).

353. J. W. Chamberlin and J. B. Campbell, *J. Med. Chem.*, 10, 966 (1967).

354. P. J. Beeby and J. A. Edwards, *J. Med. Chem.*, 20, 1665 (1977).

355. S. Karady, S. H. Pines, L. M. Weinstock, F. E. Roberts, G. S. Brenner, A. M. Hoinowski, T. Y. Cheng, and M. Sletzinger, *J. Am. Chem. Soc.*, 94, 1410 (1972).

356. J. M. T. Hamilton-Miller, D. W. Kerry, and W. Brumfitt, *J. Antibiot.*, 27, 42 (1974).

357. H. Wallick and D. Hendlin, *Antimicrob. Agent Chemother.*, 5, 25 (1974).

358. H. Russell, D. R. Daoust, S. B. Zimmerman, D. Hendlin, and E. O. Stapley, *Antimicrob. Agent Chemother.*, 5, 38 (1974).

359. W. Brumfit, J. Kosmidis, J. M. T. Hamilton-Miller, and J. N. G. Gilchrist, *Antimicrob. Agent Chemother.*, 6, 290 (1974).

360. L. Verbist, *Antimicrob. Agent Chemother.*, 10, 657 (1976).

361. A. K. Miller, E. Celozzi, Y. Kong, B. A. Pelak, D. Hendlin, and E. O. Stapley, *Antimicrob. Agent Chemother.*, 5, 33 (1974).

362. N. A. C. Curtis, G. W. Ross, and M. G. Boulton, *J. Antimicrob. Chemother.*, 5, 391 (1979).

363. *Drugs of Today*, 1, 92 (1965).

364. R. M. Atkinson, J. P. Curie, B. Davis, D. A. H. Pratt, H. M. Sharpe, and E. G. Tomick, *Toxicol. Appl. Pharmacol.*, 8, 398 (1966).

365. R. M. Atkinson, J. D. Carsey, J. P. Curie, J. R. Middleton, D. A. H. Pratt, H. M. Sharpe, and E. G. Tomick, *Toxicol. Appl. Pharmacol.*, 8, 407 (1966).

366. M. Nishida, T. Murakawa, N. Okada, S. Fukada, H. Sakamoto, S. Nakamoto, Y. Yokota, and Y. Kono, *Antimicrob. Agent Chemother.*, 11, 51 (1977).

367. H. Breuer, W. D. Treuner, H. J. Schneider, M. G. Young, and H. I. Basch, *J. Antibiot.*, 31, 546 (1978).

368. H. E. Applegate, C. M. Cimarusti, J. E. Dolfini, W. H. Koster, M. A. Ondetti, W. A.

Slusarchyk, M. G. Young, H. Breuer, and U. D. Treuner, *J. Antibiot.*, **31**, 561 (1978).

369.   H. H. Gadebusch, H. I. Basch, P. Lukaszow, B. Remsburg, and R. Schwind,*J. Antibiot.*, **31**, 570 (1978).

370.   T. Watanabe, Y. Kawano, T. Tanaka, T. Hashimoto, M. Nagano, and T. Miyadera, *Tetrahedron Lett.*, 3053 (1977).

371.   N. Serizawa, K. Nakagawa, S. Kamimura, T. Miyadera, and M. Arai, *J. Antibiot.*, **32**, 1016 (1979).

372.   Y. Kawano, T. Watanabe, J. Sakai, H. Watanabe, M. Nagano, T. Nishimura and T. Miyadera, *Chem. Pharm. Bull.*, **28**, 70 (1980).

373.   T. Sugawara, H. Masuya, T. Matsuo and T. Miki, *Chem. Pharm. Bull.*, **28**, 2116 (1980).

374.   T. Hashimoto, Y. Kawano, S. Natsumē, T. Tanaka, T. Watanabe, M. Nagano, S. Sugawara, and T. Miyadera, *Chem. Pharm. Bull.*, **26**, 1803 (1978).

375.   P. Mazzeo and F. Segnalini, *Farmaco Ed. Sci. (Pavia)*, **36**, 916 (1981).

376.   R. Reiner, U. Weiss, and P. Angehrn, *Eur. J. Med. Chem. Chim. Ther.*, **10**, 10 (1975).

377.   P. J. Beeby, *J. Med. Chem.*, **20**, 173 (1977).

378.   J. L. Fahey, R. A. Firestone, and B. G. Christensen,*J. Med. Chem.*, **19**, 562 (1976).

379.   T. Jen, B. Dienel, J. Frazee, and J. Weisbach,*J. Med. Chem.*, **15**, 1172 (1972).

380.   D. Willner, A. M. Jelenevsky, and L. C. Cheney,*J. Med. Chem.*, **15**, 948 (1972).

381.   J. A. Webber, G. W. Huffman, R. E. Koehler, C. F. Murphy, C. W. Ryan, E. M. Van Heyningen, and R. T. Vasileff,*J. Med. Chem.*, **14**, 113 (1971).

382.   S. Kukolja,*J. Med. Chem.*, **13**, 1114 (1970).

383.   R. R. Chauvette and P. A. Pennington,*J. Med. Chem.*, **18**, 403 (1975).

384.   I. G. Wright, C. W. Ashbrook, T. Goodson, G. V. Kaiser, and E. M. Van Heyningen, *J. Med. Chem.*, **14**, 420 (1971).

385.   G. V. Kaiser, C. W. Ashbrook, T. Goodson, I. G. Wright, and E. M. Van Heyningen, *J. Med. Chem.*, **14**, 426 (1971).

386.   J. M. Essery, U. Corbin, V. Sprancmanis, L. B. Crast, Jr., R. G. Graham, P. F. Misco, Jr., D. Willner, D. N. McGregor, and L. C. Chaney,*J. Antiobiot.*, **27**, 573 (1974).

387.   J. A. Webber and R. T. Vasileff, *J. Med. Chem.*, **14**, 1136 (1971).

388.   H. Nomura, I. Minami, T. Hitaka, and T. Fugono,*J. Antiobiot.*, **29**, 928 (1976).

389.   T. T. Conway, G. Lim, J. L. Douglas, M. Menard, T. W. Doyle, P. Rivest, D. Horning, L. R. Morris, and D. Cimon, *Can. J. Chem.*, **56**, 1335 (1978).

390.   A. Martell, T. W. Doyle, and B. Y. Luh, *Can. J. Chem.*, **57**, 614 (1979).

391.   R. A. Firestone, J. L. Fahey, N. S. Maciejewicz, G. S. Patel, and B. G. Christensen, *J. Med. Chem.*, **20**, 551 (1977).

392.   M. S. K. Youssef and K. M. Hassan, *Rev. Roum. Chim.*, **26**, 81 (1981).

393.   G. Carrara and G. Weitnauer, *Gazz. Chim. Ital.*, **81**, 142 (1951).

394.   E. C. Hermann and A. Kreuchunas, *J. Am. Chem. Soc.*, **74**, 5168 (1952).

395.   C. F. Huebner, P. A. Diassi, and C. R. Scholz,*J. Org. Chem.*, **18**, 21 (1953).

396.   H. Keskin, C. D. Mason, and F. F. Nord, *J. Org. Chem.*, **16**, 1333 (1951).

397.   O. Dann and B. Gotz, *Z. Naturforsch. B.*, **12**, 191 (1957).

398.   M. C. Rebstock, C. D. Stratton, and L. L. Bambas,*J. Am. Chem. Soc.*, **77**, 24 (1955).

399.   M. C. Rebstock and C. D. Stratton, *J. Am. Chem. Soc.*, **77**, 3082 (1955).

400.   J. Bulkacz, M. A. Apple, J. C. Craig, and A. R. Naig,*J. Pharm. Sci.*, **57**, 1017 (1968).

401.   P. M. Theus and W. Weuffen, *Arch. Pharm.*, **300**, 629 (1967).

402.   P. M. Theus and H. Tiedt, *Arch. Pharm.*, **301**, 424 (1968).

403.   D. M. Wiles and T. Suprunchuk, *J. Med. Chem.*, **14**, 252 (1971).

404. F. M. Riccieri, F. Gualtieri, and B. Babuchieri, *Farmaco Ed. Sci. (Pavia)*, **20**, 707 (1965).

405. R. Kimura, T. Yabuuchi, and M. Hisaki, *Chem. Pharm. Bull.*, **10**, 1232 (1962).

406. D. A. Kulikova, Y. D. Churkin, and L. V. Panfilova, *Khim.-Farm. Zh.*, **14**(4), 36 (1980).

407. H. K. Kim and R. E. Bambury, *J. Med. Chem.*, **14**, 366 (1971).

408. D. R. Shridhar, C. V. Riddy Sastry, K. B. Lal, A. K. Marwah, G. S. Reddi, K. K. Bhopale, H. N. Tripathi, R. S. Khokhar, K. Tripathi, and G. S. T. Sai, *Indian J. Chem.*, **20B**, 234 (1981).

409. Ng. D. Xuong and Ng. Ph. Buu Hoi, *Compt. Rend.*, **253**, 3115 (1961).

410. H. Raffa, M. DiBella, L. DiBella, and G. M. Lolli, *Farmaco Ed. Sci. (Pavia)*, **20**, 786 (1965).

411. M. Likar, P. Schauer, M. Japelj, M. Globokar, M. Oklobdzija, A. Povše, and V. Šunjic, *J. Med. Chem.*, **13**, 159 (1970).

412. G. Ronsisvalle and G. Blandino, *Farmaco Ed. Sci. (Pavia)*, **36**, 785 (1981).

413. E. Winkelmann, W. Raether, and W.-H. Wagner, *Arzneim.-Forsch.*, **26**, 1543 (1976).

414. G. Ronsisvalle and G. Pappalardo, *Farmaco Ed. Sci. (Pavia)*, **32**, 678 (1977).

415. C. Y. Wang, C. W. Chiu, K. Muraoka, P. D. Michie, and G. T. Bryan, *Antimicrob. Agent Chemother.*, **8**, 216 (1975).

416. V. A. Smirnov, A. E. Lipkin, and T. B. Ryskina, *Khim.-Farm. Zh.*, **6**(6), 24 (1972).

417. V. P. Arya, F. Fernandez, and V. Sudarsanam, *Indian J. Chem.*, **10**, 598 (1972).

418. G. Pappalardo, B. Tornetta, P. Condorelli, and A. Bernardini, *Farmaco Ed. Sci. (Pavia)*, **22**, 808 (1967).

419. E. Winkelmann and H. Rolly, *Arzneim.-Forsch.*, **22**, 1704 (1972).

420. V. Hochstein-Mintzel, H. Stickl, and H. Rolly, *Arzneim.-Forsch.*, **22**, 1717 (1972).

421. W. H. Wagner and E. Winkelmann, *Arzneim.-Forsch.*, **22**, 1713 (1972).

422. V. I. Shvedov and O. A. Safonova, *Khim.-Farm. Zh.*, **12**(11), 53 (1978).

423. M. Likar, P. Schauer, M. Japelj, M. Globokar, M. Oklobdžija, A. Povše, and V. Šunjic, *J. Med. Chem.*, **13**, 159 (1970).

424. E. Kesler and S. Gronowitz, *Monatsh. Chem.*, **111**, 119 (1980).

425. J. S. Berdinskii, T. A. Sakulina, L. D. Orlova, G. N. Pershin, and T. N. Zykova, *Khim.-Farm. Zh.*, **11**(1), 83 (1977).

426. J. S. Berdinskii, O. O. Makeeva, and G. N. Pershin, *Khim.-Farm. Zh.*, **2**(7), 33 (1968).

427. F. Fujikawa, K. Hirai, O. Sawada, H. Toyoshima, S. Tamura, M. Naito, and S. Tsukuma, *Yakugaku Zasshi*, **82**, 1681 (1962).

428. M. N. Zemtsova, P. L. Trakhtenberg, A. E. Lipkin, and T. B. Ryskina, *Khim.-Farm. Zh.*, **7**(8), 13 (1973).

429. K. I. Vakhreeva, A. E. Lipkin, T. B. Ryskina, and N. I. Skachkova, *Khim.-Farm. Zh.*, **7**(3), 24 (1973).

430. K. I. Vakhreeva, M. G. Viderker, P. I. Buchin, A. E. Lipkin, and T. B. Ryskina, *Khim.-Farm. Zh.*, **6**(1), 24 (1972).

431. J. Reisch, W. Spitzner, and K. E. Schulte, *Arzneim.-Forsch.*, **17**, 816 (1967).

432. N. V. Stulin, A. E. Lipkin, D. A. Kulikova, and E. A. Rudzim, *Khim.-Farm. Zh.*, **9**(11), 20 (1975).

433. D. Ducher, J. Couquelet, R. Cluzel, and J. Couquelet, *Chim. Ther.*, **8**, 552 (1973).

434. P. M. Gilis, A. Haemers, and W. Bollaert, *Eur. J. Med. Chem. Chim. Ther.*, **13**, 265 (1978).

435. P. M. Gilis, A. Haemers, and W. Bollaert, *Eur. J. Med. Chem. Chim. Ther.*, **15**, 185 (1980).

436. P. M. Gilis, A. Haemers, and S. R. Pattyn, *Antimicrob. Agent Chemother.*, **13**, 533 (1978).

437. M. A. Khan and A. E. Guarçoni, *J Heterocycl. Chem.*, **14**, 807 (1977).

438. I. Lalezari, F. Ghabgharan, and R. Maghsoudi, *J. Med. Chem.*, **14**, 465 (1971).

439. G. D. Rees, J. K. Sugden, and N. J. Van Abbe, *Pharm. Acta Helv.*, **50**, 451 (1975).

440. S. V. Bhat, H. Kohl, B. N. Ganguli, and N. J. de Souza, *Eur. J. Med. Chem. Chim. Ther.*, **12**, 53 (1977).

441. H. Berner and H. Reinshagen, *Monatsh. Chem.*, **107**, 299 (1976).

442. B. Roth, *J. Med. Chem.*, **12**, 227 (1969).

443. R. Royer, G. LaMotte, J.-P. Bochelet, P. Demerseman, R. Cavier, and J. Lemoine, *Eur. J. Med. Chem. Chim. Ther.*, **11**, 221 (1976).

444. S. Rault, M. C. de Sevricourt, and M. Robba, *C.R. Acad. Sci. Ser. C,* **285**, 381 (1977).

445. S. Rault, M. C. de Sevricourt, H. El Khashef, and M. Robba, *C.R. Acad. Sci. Ser. C,* **290**, 169 (1980).

446. S. Gronowitz, T. Dahlgren, J. Namtvedt, C. Roos, B. Sjöberg, and U. Forsgren, *Acta Pharm. Suec.*, **8**, 377 (1971).

447. S. Gronowitz, T. Dahlgren, J. Namtvedt, C. Roos, G. Rosēn, B. Sjöberg, and U. Forsgren, *Acta Pharm. Suec.*, **8**, 623 (1971).

448. D. Florentin, B.-P. Rogues, J. M. Metzger, and J.-P. Collin, *Bull. Soc. Chim. Fr.*, 2620 (1974).

449. P. J. Bailey, G. Cousins, G. A. Snow, and A. J. White, *Antimicrob. Agent Chemother.*, **17**, 549 (1980).

450. G. Högenauer and M. Woisetschläger, *Nature*, **293**, 662 (1981).

451. Ng. Ph. Buu-Hoi, Ng. Hoan, and D. Lavit, *J. Chem. Soc.*, 4590 (1952).

452. J. Craig-Cymerman and D. Willis, *J. Chem. Soc.*, 1071 (1955).

453. D. Liberman, M. Moyeux, A. Rouaix, J. Maillard, L. Hengel, and J. Himbert, *Bull. Soc. Chim. Fr.*, 957 (1953).

454. Ng. Ph. Buu-Hoi, D. Lavit, and Ng. D. Xuong, *J. Chem. Soc.*, 1581 (1955).

455. W. L. Nobles, *J. Am. Chem. Soc.*, **77**, 6675 (1955).

456. D. Liberman, N. Rist, F. Grumbach, M. Moyeux, B. Gauthier, A. Rouaix, J. Maillard, J. Himbert, and S. Cals, *Bull. Soc. Chim. Fr.*, 1430 (1954).

457. Ng. Ph. Buu-Hoi, Ng. D. Xuong, R. Royer, and D. Lavit, *J. Chem. Soc.*, 547 (1953).

458. R. I. Meltzer, A. D. Lewis, F. H. McMillan, J. D. Genzer, F. Leonard, and J. A. King, *J. Am. Pharm. Assoc.*, **42**, 594 (1953).

459. Ng. Ph. Buu-Hoi, Ng. D. Xuong, F. Binon, and Ng. H. Nam, *C.R. Acad. Sci.*, **235**, 329 (1952).

460. H. C. Byerman, J. S. Bontekoe, W. J. Vander Burg, and W. L. C. Veer, *Rev. Trav. Chim. Pays-Bas,* **73**, 109 (1954).

461. G. Carrara, V. D'Amato, G. Rolland, and E. Fusapoli, *Gazz. Chim. Ital.*, **83**, 459 (1953).

462. A. Vecchi and G. Melone, *J. Org. Chem.*, **22**, 1636 (1957).

463. Y. Inoue and C. Tomizawa, *Bot. Kab.*, **18**, 33 (1953).

464. G. N. Mahapatra and M. K. Rout, *J. Indian Chem. Soc.*, **34**, 653 (1957).

465. F. C. Brown, C. K. Bradsher, E. C. Morgan, M. Tetenbaum, and P. Wilder, *J. Am. Chem. Soc.*, **78**, 384 (1956).

466. H. Tuchmann-Duplessis and L. Mercier-Parot, *C.R. Acad. Sci.*, **258**, 5103 (1964).

467. G. L. Dunn, P. Actor, and V. J. DiPasto, *J. Med. Chem.*, **9**, 751 (1966).

468. R. Albrecht, H. J. Kessler, and E. Schröder, *Chim. Ther.*, **6**, 352 (1971).

469. R. Albrecht, H. J. Kessler, and E. Schröder, *Arzneim.-Forsch.*, **21**, 127 (1971).

470. D. R. Shridhar, C. V. Reddy Sastry, A. K. Mehrotra, R. Nagarajan, B. Lal, and K. K. Bhopale, *Indian. J. Chem.*, **19B**, 59 (1980).

471. D. R. Shridhar, C. V. Reddy Sastry, N. K. Vaidya, S. R. Moorty, G. S. Reddi, G. S. Thapar, and S. K. Gupta, *Indian J. Chem.*, **16B**, 704 (1978).

472. U. Herzog and H. Reinshagen, *Eur. J. Med. Chem. Chim. Ther.*, **11**, 415 (1976).

473. D. M. Bailey, E. M. Mount, J. Siggins, J. A. Carlson, A. Yarinsky, and R. G. Slighter, *J. Med. Chem.*, **22**, 599 (1979).

474. R. Cavier, J. Cenac, R. Royer, and L. Rene, *Chim. Ther.*, **5**, 270 (1970).

475. B. P. Das and D. W. Boykin, *J. Med. Chem.*, **20**, 1219 (1977).

476. R. M. Lee, M. W. Mills, and G. S. Sach, *Experimentia*, **33**, 198 (1977).

477. D. W. Henry, V. H. Brown, M. Cory, J. G. Johansson, and E. Bueding, *J. Med. Chem.*, **16**, 1287 (1973).

478. H. R. Wilson, G. R. Revankar, and R. L. Tolman, *J. Med. Chem.*, **17**, 760 (1974).

479. D. Pillon, S. Trinh, and R. Cavier, *Chim. Ther.*, **5**, 32 (1970).

480. M. M. El-Kerdawy, A. A. Samour, and A. A. El-Agamey, *Pharmazie*, **30**, 76 (1975).

481. A. Rosowsky, M. Chaykovsky, K. K. N. Chen, M. Lin, and E. J. Modest, *J. Med. Chem.*, **16**, 185 (1973).

482. M. Chakovsky, M. Lin, A. Rosowsky, and E. J. Modest, *J. Med. Chem.*, **16**, 188 (1973).

483. A. Rosowsky, K. K. N. Chen, and M. Lin, *J. Med. Chem.*, **16**, 191 (1973).

484. J. P. Schaefer, K. S. Kulkarni, R. Costin, J. Higgins, and L. M. Honig, *J. Heterocycl. Chem.*, **7**, 607 (1970).

485. T. R. Herrin, J. M. Pauvlik, E. V. Schuber, and A. O. Geisler, *J. Med. Chem.*, **18**, 1216 (1975).

486. R. F. Koebel, L. L. Needham, and C. DeWitt Blanton, Jr., *J. Med. Chem.*, **18**, 192 (1975).

487. V. N. Gogte, B. D. Tilak, K. N. Gadekar, and M. B. Sahasrabudhe, *Tetrahedron*, **23**, 2443 (1967).

488. V. N. Gogte, L. G. Shah, B. D. Tilak, K. N. Gadekar, and M. B. Sahasrabudhe, *Tetrahedron*, **23**, 2437 (1967).

489. M. B. Sahasrabudhe, M. K. Nerurkar, M. V. Nerurkar, B. D. Tilak, and M. D. Bhavasar, *Br. J. Cancer*, **14**, 547 (1960).

490. M. B. Sahasrabudhe, M. V. Nerurkar, L. B. Kotnis, B. D. Tilak, and M. D. Bhavasar, *Nature*, **184**, 202 (1959).

491. R. D. Dillard, G. A. Poore, N. R. Easton, M. J. Sweeney, and W. R. Gibson, *J. Med. Chem.*, **11**, 1155 (1968).

492. R. D. Dillard, G. A. Poore, D. R. Cassady, and N. R. Easton, *J. Med. Chem.*, **10**, 40 (1967).

493. W. P. Purcell and J. M. Clayton, *J. Med. Chem.*, **11**, 199 (1968).

494. H. Katae, H. Iwana, Y. Takase, and M. Shimizu, *Arzneim.-Forsch.*, **17**, 1030 (1967).

495. I. L. Lodē, *Therapiewoche*, **29**, 4343 (1979).

496. E. E. Campaigne, P. A. Monroe, B. Arnwine, and W. L. Archer, *J. Am. Chem. Soc.*, **75**, 988 (1953).

497. R. L. Thompson, S. A. Minton, J. E. Officer, and G. H. Hitchings, *J. Immunol.*, **70**, 229 (1953).

498. S. A. Minton, J. E. Officer, and R. L. Thompson, *J. Immunol.*, **70**, 222 (1953).

499. M. Weintraub and W. G. Kemp, *Can. J. Microbiol.*, **1**, 549 (1955).

500. D. M. O'Mant, *J. Chem. Soc. C*, 1501 (1968).

452   J. B. Press

501.   T. J. Franklin, B. B. Newbould, D. M. O'Mant, A. I. Scott, G. J. Stacy, and G. E. Davies, *Nature,* **210,** 638 (1966).

502.   T. J. Franklin and B. Higginson, *Biochem. J.,* **102,** 705 (1967).

503.   G. E. Davies, *Immunology,* **14,** 393 (1968).

504.   J. Krapcho, R. C. Millonig, C. F. Turk, and B. J. Amrein, *J. Med. Chem.,* **12,** 164 (1969).

505.   H. Stähelin, *Eur. J. Cancer,* **6,** 303 (1970).

506.   C. Keller-Juslen, M. Kuhn, A. von Wartburg, and H. Stähelin, *J. Med. Chem.,* **14,** 936 (1971).

507.   B. D. Sklansky, R. S. Mann-Kaplan, A. F. Reynolds, Jr., M. L. Rosenblum, and M. D. Walker, *Cancer,* **33,** 460 (1974).

508.   G. Rivera, R. J. Aur, G. V. Dahl, C. B. Pratt, A. Wood, and T. L. Avery, *Cancer,* **45,** 1286 (1980) and references therein.

509.   K. Onodera, Y. Aoi, and K. Sasaki, *Agr. Biol. Chem.,* **40,** 2209 (1976).

510.   T. T. Otani and M. R. Briley, *J. Pharm. Sci.,* **68,** 260 (1979).

511.   H.-U. Schorlemmer, W. Opitz, E. Etschenberg, D. Bitter-Suermann, and U. Hadding, *Cancer Res.,* **39,** 1847 (1979).

512.   J. E. Lynch and B. Nelson, *J. Parasitol.,* **45,** 659 (1959).

513.   W. J. Farrington, *Aust. J. Chem.,* **17,** 230 (1964).

514.   J. W. McFarland, L. H. Conover, H. L. Howes, Jr., J. E. Lynch, D. R. Chisholm, W. C. Austin, R. L. Cornwell, J. C. Daniliwicz, W. Courtney, and D. H. Morgan, *J. Med. Chem.,* **12,** 1066 (1969).

515.   See Refs. 21 and 22 in Ref. 514.

516.   J. W. McFarland and H. L. Howes, Jr., *J. Med. Chem.,* **12,** 1079 (1969).

517.   J. W. McFarland and H. L. Howes, Jr., *J. Med. Chem.,* **13,** 109 (1970).

518.   J. W. McFarland, H. L. Howes, Jr., L. H. Conover, J. E. Lynch, W. C. Austin, and D. H. Morgan, *J. Med. Chem.,* **13,** 113 (1970).

519.   W. C. Austin, R. L. Cornwell, R. M. Jones, and M. Robinson, *J. Med. Chem.,* **15,** 281 (1972).

520.   See Refs. 6–20 in Ref. 519.

521.   R. B. Burrows, P. Clapham, D. A. Rawes, F. C. Copp, and O. D. Standen, *Nature,* **188,** 945 (1960).

522.   R. B. Burrows and W. G. Lillis, *Am. J. Vet. Res.,* **23,** 77 (1962).

523.   R. M. Corwin and T. A. Miller, *Am. J. Vet. Res.,* **39,** 263 (1978).

524.   R. M. Lee, *Br. Soc. Parasitol. Proc.,* **75,** 18 (1977).

525.   H. R. Stöhler, *Tropenmed. Parasitol.,* **28,** 276 (1977).

526.   R. E. Bradley and L. Panitz, *J. Parasitol.,* **62,** 643 (1976).

527.   M. H. Fisher, G. Schwartzkopf, Jr., and D. R. Hoff, *J. Med. Chem.,* **15,** 1168 (1972).

528.   F. Sauter, P. Stanetty, W. Macka, and A. Mesbah, *Monatsh. Chem.,* **107,** 495 (1976).

529.   R. J. Alaimo and C. J. Hatton, *J. Med. Chem.,* **15,** 118 (1972).

530.   R. J. Alaimo and H. E. Russell, *J. Med. Chem.,* **15,** 335 (1972).

531.   E. Walton, *Rep. Prog. Appl. Chem.,* **40,** 386 (1955).

532.   C. J. Cansfield, *Vet. Rec.,* **83,** 475 (1968).

533.   M. J. Hayes, *Vet. Rec.,* **83,** 528 (1968).

534.   J. E. Keen, *Vet. Rec.,* **83,** 502 (1968).

535.   A. M. Harthoon, *Vet. Rec.,* **84,** 151 (1969).

536.   G. M. Massey, *Aust. Vet. J.,* **49,** 207 (1973).

537. W. D. Harbison, R. F. Slocombe, S. J. Watts, and G. A. Stewart, *Aust. Vet. J.*, **50**, 543 (1974).

538. G. Chen, C. R. Ensor, and B. Bohner, *J. Pharmacol. Exp. Ther.*, **168**, 171 (1969).

539. G. Chen, *Am. J. Vet. Res.*, **29**, 869 (1968).

540. R. R. Bennett, *Am. J. Vet. Res.*, **30**, 1469 (1969).

541. R. Chen, *Am. J. Vet. Res.*, **29**, 863 (1968).

542. J. A. McNamara, D. L. Sly, and B. J. Cohen, *Am. J. Vet. Res.*, **35**, 1089 (1974).

543. G. H. Conner, R. W. Coppock, and C. C. Beck, *Vet. Med.*, **69**, 479 (1974).

544. F. E. Eads, *Vet. Med.*, **71**, 648 (1976).

545. M. Bush, R. Custer, J. Smeller, and L. M. Bush, *J. Am. Vet. Med. Assoc.*, **171**, 866 (1977).

546. W. J. Boever, J. Holden, and K. K. Kane, *Vet. Med.*, **72**, 1722 (1977).

547. J. M. King, B. C. R. Bertram, and P. H. Hamilton, *J. Am. Vet. Med. Assoc.*, **171**, 894 (1977).

548. R. B. March, R. L. Metcalf, and L. L. Lewaller, *J. Econ. Ent.*, **45**, 851 (1952).

549. R. L. Metcalf and T. R. Fukuto, *Bull. World Health Org.*, **38**, 633 (1968); *Chem. Abstr.*, **69**, 51154 (1968).

550. R. C. Blinn, F. A. Gunther, and R. L. Metcalf, *J. Am. Chem. Soc.*, **76**, 37 (1954).

551. V. J. Ram, *Arch. Pharm.*, **312**, 19 (1979).

552. V. J. Ram, H. K. Pandey, and A. J. Vlietinck, *J. Heterocycl. Chem.*, **18**, 1277 (1981).

553. K. E. Nielsen and E. B. Pedersen, *Chem. Scripta*, **18**, 245 (1981).

554. V. J. Ram, *Arch. Pharm.*, **312**, 726 (1979).

555. I. J. Kay and N. Punja, *J. Chem. Soc. C*, 2409 (1970).

556. P. A. Van Zwieten, J. Meltzer, and H. O. Huisman, *Rec. Trav. Chim.*, **81**, 616 (1962).

557. W. P. Trompen and H. O. Huisman, *Rec. Trav. Chim.*, **85**, 175 (1966).

558. A. G. M. Willems, A. Tempel, D. Hamminga, and B. Stork, *Rec. Trav. Chim.*, **90**, 97 (1971).

559. G. Asato and R. D. Wilbur, *Experimentia*, **35**, 1458 (1979).

560. H. C. Erbler, *Z. Versuchstierkd.*, **21**, 112 (1979).

561. D. R. Tompkins, W. A. Sistrunk, and J. W. Fleming, *Hortic. Sci.*, **6**, 393 (1971).

562. K. Dittmer, *J. Am. Chem. Soc.*, **71**, 1205 (1949).

563. R. Garst, E. Campaigne, and H. G. Day, *J. Biol. Chem.*, **180**, 1013 (1949).

564. M. F. Ferger and V. du Vigneaud, *J. Biol. Chem.*, **174**, 241 (1948).

565. K. Dittmer, *Ann. NY Acad. Sci.*, **52**, 1274 (1950).

566. M. E. Rafelson, H. E. Pearson, and R. J. Winzler, *Arch. Biochem. Biophys.*, **29**, 69 (1950).

567. G. C. Brown, *J. Immunol.*, **69**, 441 (1952).

568. R. T. Cushing and H. R. Morgan, *Proc. Soc. Exp. Biol. NY*, **79**, 497 (1952).

569. U. Rothfels, *J. Exp. Zool.*, **125**, 17 (1954).

570. J. A. Jacquez, R. K. Barclay, and C. C. Stock, *J. Exp. Med.*, **96**, 499 (1952).

571. J. A. Jacquez, C. C. Stock, and R. K. Barclay, *Cancer, NY*, **6**, 828 (1953).

572. H. R. Morgan, *J. Exp. Med.*, **99**, 451 (1954).

573. *Cancer Res. Suppl.*, **3**, (1955).

574. A. Szweykowska and J. Schneider, *Acta Soc. Bot. Pol.*, **36**, 735 (1967); *Chem. Abstr.*, **68**, 86303 (1968).

575. N. V. Kaverm, G. A. Galegov, and I. V. Tsvetkova, *Vop. Virusol.*, **13**, 168 (1968); *Chem. Abstr.*, **69**, 17107 (1968).

576. A. Veckenstedt and I. Eisenhuth, *Acta Virol.*, **15**, 192 (1971); *Chem. Abstr.*, **75**, 60305 (1971).

577. D. N. Wheatley and M. S. Inglis, *Exp. Cell Res.*, **107**, 191 (1977).

578. J. Kuhn, *Antimicrob. Agent Chemother.*, **12**, 322 (1977).

579. R. W. Wissler, E. Frazier, K. Soules, P. Barker, and E. C. Bristow, III, *Am. Med. Assoc. Arch. Pathol.*, **62**, 62 (1956).

580. Z. Hruban and R. W. Wissler, *Cancer Res.*, **20**, 1530 (1960).

581. E. C. Bristow, III and R. W. Wissler, *Lab. Invest.*, **10**, 31 (1961).

582. Z. Hruban, R. W. Wissler, and A. Slelers, *Lab. Invest.*, **11**, 382 (1962).

583. A. Misefari and M. F. LaVia, *Infect. Immunol.*, **3**, 810 (1971).

584. B. Hotham-Iglewski and M. F. LaVia, *Proc. Soc. Exp. Biol. Med.*, **131**, 895 (1969).

585. M. F. LaVia, T. Leavitt, and S. K. Page, *Proc. Soc. Exp. Biol. Med.*, **139**, 399 (1972).

586. M. F. LaVia, S. A. Uriu, N. D. Barber, and A. E. Warren, *Proc. Soc. Exp. Biol. Med.*, **104**, 562 (1960).

587. A. Misefari and M. F. LaVia, *Infect. Immunol.*, **4**, 240 (1971).

588. A. Misefari, G. Costa, and A. L. Costa, *Pharmacol. Res. Comm.*, **9**, 785 (1977).

589. G. L. Monto and R. A. Guillan, *Clin. Res.*, **25**, 111A (1977).

590. E. Szarvasi, L. Fontaine, and A. Betheder-Matibet, *J. Med. Chem.*, **16**, 281 (1973).

591. F. W. Dunn and K. Dittmer, *J. Biol. Chem.*, **188**, 263 (1951).

592. F. W. Dunn, *J. Biol. Chem.*, **227**, 575 (1957).

593. F. W. Dunn, J. M. Ravel, and W. Shive, *J. Biol. Chem.*, **219**, 809 (1956).

594. F. W. Dunn, J. Humphreys, and W. Shive, *Arch. Biochem. Biophys.*, **71**, 475 (1957).

595. J. T. Hill and F. W. Dunn, *J. Med. Chem.*, **12**, 737 (1969).

596. F. W. Dunn, *Science*, **120**, 146 (1954).

597. F. W. Dunn, *J. Org. Chem.*, **21**, 1525 (1956).

598. A.-M. Bellocq, S. Castensson, and H. Sievertsson, *Biochem. Biophys. Res. Comm.*, **74**, 577 (1977).

599. C. W. Smith, G. Skala, and R. Walter, *J. Med. Chem.*, **21**, 115 (1978).

600. W. Opitz, M. Schwiertz, S. Raddatz, and P. R. Imberge, *Arzneim.-Forsch.*, **31**(I), 402 (1981).

601. C. H. Eugster and K. Allner, *Helv. Chim. Acta*, **45**, 1750 (1962).

602. M. Giannella, M. Pigini, P. Rüedi, and C. H. Eugster, *Helv. Chim. Acta*, **62**, 2329 (1979).

603. A. M. Jetten and M. E. R. Jetten, *Nature*, **278**, 180 (1979).

604. K. Hejno and F. Šorm, *Coll. Czech. Chem. Comm.*, **41**, 479 (1976).

605. R. G. S. Ritchie, D. M. Vyas, and W. A. Szarek, *Can. J. Chem.*, **56**, 794 (1978).

606. M. V. Pickering, J. T. Witkowski, and R. K. Robins, *J. Med. Chem.*, **19**, 841 (1976).

607. Y. Izumi, H. Fukuda, Y. Tani, and K. Ogata, *Agric. Biol. Chem.*, **42**, 579 (1978).

608. F. Yoneda and T. Nogamatsu, *Chem. Pharm. Bull.*, **23**, 2001 (1975).

609. T. Tsujikawa and M. Hayashi, *Chem. Pharm. Bull.*, **25**, 3147 (1977).

610. E. A. Harrison, Jr., K. C. Rice, and M. E. Rogers, *J. Heterocycl. Chem.*, **14**, 909 (1977).

611. W. O. Foye and J. P. Speranza, *Eur. J. Med. Chem. Chim. Ther.*, **9**, 177 (1974).

612. N. Autissier, B. Gautheron, P. Dumas, J. Brosseau, and A. Loireau, *Toxicology*, **8**, 125 (1977).

613. R. P. F. Gregory, *Biochem. Biophys. Acta*, **368**, 228 (1974).

614. G. Rengor, *Biochim. Biophys. Acta*, **256**, 428 (1972).

615. J. Maroc and J. Garnier, *Biochim. Biophys. Acta*, **548**, 374 (1979).

616. C. L. Greenstock, J. E. Biaglow, and R. E. Durand, *Br. J. Cancer Suppl. III*, **37**, 11 (1978).

617. Ref. 489 and references contained therein.

618. M. B. Sahasrabudhe, M. V. Nerurkar, and L. B. Kotnis, *Acta Union Int. Cent. Le Cancer*, **20**, 221 (1964).

619. A. H. Anton, S. Czinn, J. Jazwa, L. Tam, and L. Amaranath, *Res. Comm. Chem. Path. Pharmacol.*, **22**, 375 (1978) and references therein.

620. T. E. King, *Adv. Enzymol.*, **28**, 115 (1966).

621. J. B. Warshaw, K. W. Lam, and D. R. Sanadi, *Arch. Biochem. Biophys.*, **115**, 312 (1966) and references therein.

622. M. Gutman and E. Hartstein, *Biochim. Biophys. Acta*, **481**, 33 (1977).

623. P. R. Rich, A. L. Moore, and W. D. Bonner, Jr., *Biochem. J.*, **162**, 205 (1977) and references therein.

624. W. J. Ingledew and T. Ohnishi, *Biochem. J.*, **164**, 617 (1977).

625. H. Flaster and H. Kohn, *J. Heterocycl. Chem.*, **18**, 1425 (1981).

626. W. J. Mergens, F. M. Vane, S. R. Tannenbaum, L. Green, and P. L. Skipper, *J. Pharm. Sci.*, **68**, 827 (1979).

627. R. Ziegler, B. Ho, and N. Castagnoli, *J. Med. Chem.*, **24**, 1133 (1981).

628. J. Ashby, J. A. Styles, D. Anderson, and D. Paton, *Br. J. Cancer*, **38**, 521 (1978).

629. M. Baumeister and L. Lützen, *Arch. Toxicol.*, **42**, 259 (1979).

630. C. E. Voogd, *Mut. Res.*, **38**, 117 (1976).

631. S. M. Cohen and G. T. Bryan, *Fed. Proc.*, **32**, 825 (1973).

632. C. Y. Wang, C. W. Chiu, and G. T. Bryan, *Fed. Proc.*, **34**, 828 (1975).

633. S. M. Cohen, E. Erturk, and G. T. Bryan, *J. Natl. Cancer Inst.*, **57**, 277 (1976).

634. C. Y. Wang, C. W. Chiu, and G. T. Bryan, *Biochem. Pharmacol.*, **24**, 1563 (1975).

635. C. Y. Wang, K. Muraoka, and G. T. Bryan, *Cancer Res.*, **35**, 3611 (1975).

636. F. A. Vingiello, S.-G. Quo, P. Polss, and P. Henson, *J. Med. Chem.*, **7**, 832 (1964).

637. G. A. Mikhailets, *Toksiol. Seraorgan. Soedin. Ufa, Sb.*, 4 (1964); through *Chem. Abstr.*, **63**, 7550 (1965).

638. G. A. Mikhailets, I. B. Mikhailets, and D. G. Pel'ts, *Toksiol. Seraorgan. Soedin. Ufa, Sb.*, 21 (1964); through *Chem. Abstr.*, **63**, 7558 (1965).

639. M. E. Anderson, R. A. Jones, R. G. Mehl, T. A. Hill, L. Kurlansiks, and L. J. Jenkins, Jr., *Toxicol. Appl. Pharmacol.*, **40**, 463 (1977).

640. R. Albrechtsen, N. H. Diemer, and M. H. Nielson, *Acta Path. Microbiol. Scand.*, **82A**, 791 (1974).

641. P. Bradley and M. Berry, *Neuropath. App. Neurobiol.*, **5**, 9 (1979).

642. H. G. Bray and F. M. B. Carpanini, *Biochem. J.*, **109**, 11P (1968).

643. R. E. Cripps, *Biochem. J.*, **134**, 353 (1973).

644. J. K. Chakrabarti, T. M. Hotten, S. E. Morgan, I. A. Pullar, D. M. Rackham, F. C. Risius, S. Wedley, M. O. Chaney, and N. D. Jones, *J. Med. Chem.*, **25**, 1133 (1982).

645. J. Ishiko, C. Inagaki, and S. Takaori, *Neuropharmacology*, **22**, 221 (1983).

646. D. Römer, H. H. Büscher, R. C. Hill, R. Maurer, T. J. Petcher, H. Zeugner, W. Benson, E. Finner, W. Milkowski, and P. W. Thies, *Nature*, **298**, 759 (1982).

647. *Drugs of the Future*, **7**, 655 (1982).

648. *Drugs of the Future*, **7**, 279 (1982).

649. J. Engel, A. V. Schlichtegroll, and W. S. Sheldrick, *Arzneim.-Forsch.*, **32**(I), 475 (1982).

650.   T. Sekiya, H. Hiranuma, S. Hata, S. Mizogami, M. Hanazuka, and S. Yamade, *J. Med. Chem.*, **26**, 411 (1983).

651.   S. Conde, C. Corral, J. Lissavetsky, V. Darias, and D. Martin, *Eur. J. Med. Chem.-Chim. Ther.*, **18**, 151 (1983).

652.   Y. Suzuki, M. Ito, and T. Komura, *Nippon Yakurigaku Zasshi*, **80**, 395 (1982); *Chem. Abstr.*, **98**, 27561 (1983).

653.   M. J. Kulshreshtha, S. Bhatt, M. Pardasani, and N. M. Khanna, *J. Ind. Chem. Soc.*, **58**, 982 (1981).

654.   *Drugs of the Future*, **7**, 493 (1982).

655.   C. J. Shishoo, M. B. Devani, M. D. Karvekar, G. V. Ullas, S. Ananthan, V. S. Bhadti, R. B. Patel, and T. P. Gandhi, *Indian J. Chem.*, **21B**, 666 (1982).

656.   K. C. Nicolaou, W. E. Barnetti, and R. L. Magolda, *J. Am. Chem. Soc.*, **103**, 3472 (1981).

657.   H. Yokomori, Y. Torisawa, M. Shibasaki, and S. Ikegami, *Heterocycl. Spec. Issue*, **18**, 251 (1982).

658.   I. T. Harrison, R. J. K. Taylor, and J. H. Fried, *Tetrahedron Lett.*, 1165 (1975).

659.   S. R. Ramadas, P. C. Chenchaiak, N. S. Chandrakumar, M. V. Krishna, P. S. Srinivasan, V. V. S. K. Sastry, and J. A. Rao, *Heterocycles*, **19**, 861 (1982).

660.   T. Komeno, H. Iwakura, and K. Takeda, *Heterocycles*, **10**, 207 (1978).

661.   S. R. Ramadas and N. S. Chandrakumar, *Phosphorus and Sulfur*, **13**, 79 (1982).

662.   R. L. Rojas Martinez, O. Yodu Ferral, N. Yodu Ferral, and J. Aquirre Fernandez, *Rev. Cubana Med.*, **21**, 251 (1982); *Chem. Abstr.*, **97**, 174797 (1982).

663.   Z. Polivka, M. Rajšner, J. Metyš, J. Holubek, E. Svátek, M. Ryska, and M. Protiva, *Coll. Czech. Chem. Comm.*, **48**, 623 (1983).

664.   *Drugs of the Future*, **7**, 273 (1982).

665.   B. R. Cowley, D. C. Humber, B. Laundon, A. G. Long, and A. L. Lynd, *Tetrahedron*, **39**, 461 (1983).

666.   D. Hagiwara, K. Sawada, T. Ohnami, and M. Hashimoto, *Chem. Pharm. Bull.*, **30**, 3061 (1982).

667.   E. M. Gordon, M. A. Ondetti, J. Pluscec, C. M. Cimarusti, D. P. Bonner, and R. B. Sykes, *J. Am. Chem. Soc.*, **104**, 6053 (1982).

668.   H. Sasaki, H. Oishi, T. Hayashi, I. Matsuura, K. Ando, and M. Sawada, *J. Antibiot.*, **35**, 396 (1982).

669.   S. Omura, Y. Iwai, A. Nakagawa, R. Iwata, Y. Takahashi, H. Shimizu, and H. Tanaka, *J. Antibiot.*, **36**, 109 (1983).

670.   A. M. El-Naggar, F. S. M. Ahmed, A. M. Abd El-Salam, and T. M. Ibrahim, *Egypt. J. Chem.*, **23**, 273 (1980).

671.   A. M. El-Naggar, M. N. Aboul-Enein, and A. A. Makhlouf, *J. Ind. Chem. Soc.*, **59**, 783 (1982).

672.   D. J. Brown, W. B. Cowden, and L. Strekowski, *Aust. J. Chem.*, **35**, 1209 (1982).

673.   S. Inokuma, A. Sugie, K. Moriguchi, H. Shimomura, and J. Katsube, *Heterocycles*, **19**, 1909 (1982).

674.   R. L. Dyer, G. J. Ellames, B. J. Hamill, P. W. Manley, and A. M. S. Pope, *J. Med. Chem.*, **26**, 442 (1983).

675.   J. O. Jones and R. S. McElhinney, *J. Chem. Res.*, 1368 (1982).

676.   M. Madesclaire, D. Roche, A. Carpy, and A. Boucherle, *Arch. Pharm.*, **315**, 741 (1982).

677.   W.-Y. Ren, M.-I. Lim, B. A. Otter, and R. S. Klein, *J. Org. Chem.*, **47**, 4633 (1982).

CHAPTER VI

Reduction and Desulfurization of Thiophene Compounds

L. I. BELEN'KII and Ya. L. GOL'DFARB

N. D. Zelinsky Institute of Organic Chemistry, Academy of Sciences of the USSR, Moscow, USSR

I. INTRODUCTION

The reductive processes are essentially a special part of thiophene chemistry. This is due to the poisoning effect of the bivalent sulfur compounds on most heterogeneous catalysts. Also, reductive cleavage of thiophene rings with retention or removal of sulfur atoms gives various compounds of the aliphatic, cycloaliphatic, and heterocyclic series. The whole reaction sequence, including the synthesis of thiophene or its homologues by reactions of hydrocarbons with sulfur, SO_2, or H_2S, the introduction of various substituents in the thiophene nucleus, and, finally, the ring cleavage may be regarded as a route of functionalization of simple C_4-C_6 petroleum hydrocarbons.

Some aspects of the reduction of thiophene derivatives and the synthesis of various compounds by reductive cleavage of the thiophene ring are the subject of many reviews and monographs. The main data concerning the problems under discussion are summarized in Ref. 1.

Numerous synthetically potent reductive processes are described for thiophene compounds. In principle, thiophenes are capable of undergoing the same reductive transformations as other aromatic compounds; furthermore, they may be subjected to specific reactions of hydrogenolysis with retention or removal of sulfur atoms. Many reductive processes in the thiophene series proceed more easily and are more selective than similar transformations of benzene analogues. This is due to the activating effect of the heteroatom on the α-positions of the thiophene ring and to the ability of the sulfur atom to take part in delocalization of not only the positive but also the negative charge and the odd electron. But there are serious limitations to these processes in the presence of heterogeneous catalysts, which are often poisoned by thiophene compounds.

II. REDUCTION OF SUBSTITUTENTS WITHOUT AFFECTING THE THIOPHENE RING

Many reducing reagents used in the aromatic series allow the reduction of substituents, without affecting the thiophene ring. In particular, zinc and tin in acid media, lithium aluminum hydride, sodium borohydride, hydrazine, and others are used for this purpose. In this section, several examples are given to illustrate the application of these reagents for the reduction of thiophenes. In some cases, reagents used for the reduction of substituents can reduce or break up the thiophene ring itself. For instance, sodium in alcohol or in liquid ammonia, sodium amalgam, and heterogeneous catalysts, including Raney nickel, have such properties. Data related to the use of such reagents and also to electrochemical reduction are considered below in more detail.

1. Reduction of Oxygen-Containing Substituents

The reduction of aldehydes and ketones plays an important role in various transformations of thiophenes and is a subject that has been thoroughly investigated. For example, the reduction of the CO group in acylthiophenes into CH_2 is the principal route to many alkylthiophenes. The Clemmensen reduction,[2-4] the Wolf–Kishner reduction, and the modern modification of the latter by Huang–Minlon[5-9] are the most important reactions used for preparative purposes. Water, heated almost to the boiling point may also be used in some cases as solvent[10,11] (Scheme 1).

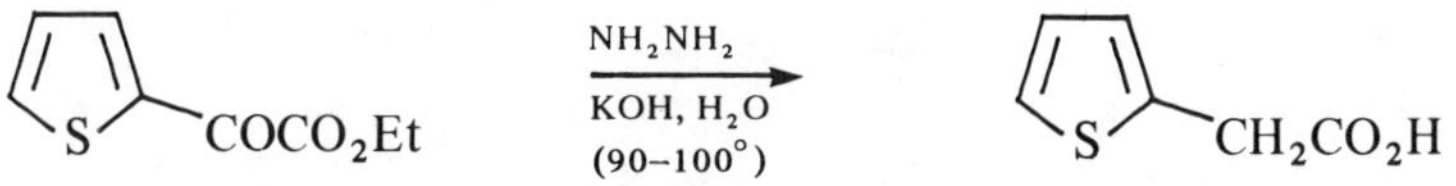

$$\text{(thiophene)–}COCO_2Et \xrightarrow[\substack{KOH, H_2O \\ (90-100°)}]{NH_2NH_2} \text{(thiophene)–}CH_2CO_2H$$

Scheme 1

Catalytic reduction of the CO group of acylthiophenes into CH_2 was achieved in the presence of palladium sulfide[12] or rhenium heptasulfide.[12,13] The transformation of 2-thiophenecarboxaldehyde and 2-acetylthiophene into 2-methylthiophene and 2-ethylthiophene, respectively, by means of synthesis-gas in the presence of a cobalt catalyst is also described.[14,15]

The reduction of aldehydes and ketones is widely used for the preparation of alcohols and pinacols. Various general methods may be used for these purposes. With the exception of catalytic hydrogenation, the reduction conditions for aldehydes and ketones of the thiophene series are generally not different from those used for the benzene analogues. In particular, the Meerwein–Ponndorf reaction gives good results in the case of acylthiophenes.[16] There are also no difficulties when lithium aluminum hydride is utilized for the reduction of ester and carbonyl groups. The use of lithium aluminum deuteride complex with (−) quinine for the transformation of 2-thiophenecarboxaldehyde into optically active carbinol (1) is an interesting example of such a reduction[17] (Scheme 2).

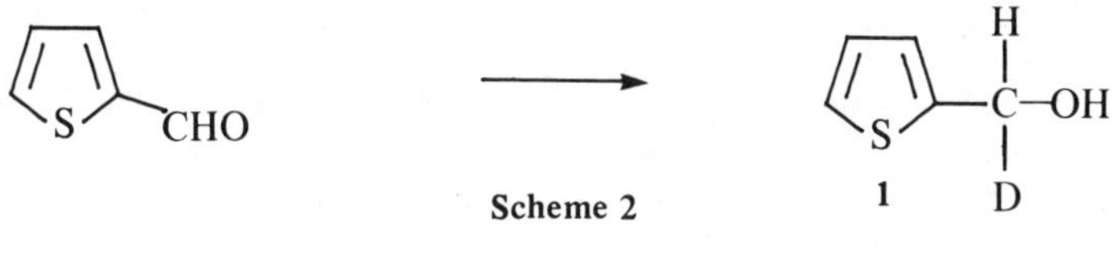

$$\text{(thiophene)–}CHO \longrightarrow \text{(thiophene)–}\underset{\underset{D}{|}}{\overset{\overset{H}{|}}{C}}-OH$$

Scheme 2   1

The traditional reducing reagents have also been used successfully. The reduction of 2-thienylglyoxylic acid into 2-thienylglycolic acid with sodium amalgam was described first in 1886.[18] For the reduction of aldehydes, the use of metals is especially effective. Thus, 2-thiophenecarboxaldehyde gives 2-thenyl alcohol in high yield by the action of zinc powder in alkaline alcohol solution[19] or in aqueous acetic acid.[20] The reduction of 2-thiophenecarboxaldehyde into 2-thenyl alcohol in conditions of modified Cannizzaro reaction should also be noted.[21] According to Caullet et al.,[22, 23] the ketones of the thiophene series are successfully reduced into alcohols using the electrochemical method. for instance, phenyl-2-thienylcarbinol was obtained from 2-benzoylthiophene in more than 80% yield.

Pinacols are available from thiophene ketones using the action of zinc in acetic acid.[24] The first attempts to prepare pinacol from 2-acetylthiophene by electrochemical reduction failed.[22, 25] But more careful choice of preparative electrolysis conditions gave 2,3-di-(2-thienyl)-2,3-butanediol from 2-acetylthiophene in up to 70% yield[22, 26] and 1,2-di-(2-thienyl)-1,2-ethanediol from 2-thiophenecarboxaldehyde in up to 90% yield.[27, 28] The study of the influence of supporting electrolyte composition on the stereochemistry of products made it possible to work out the conditions of stereoselective condensation; the ratio of *d,l*- (2a) and *meso*-forms (2b) in the case of 2-thiophenecarboxaldehyde reduction was 8:1.[28]

2a   2b

In the course of studying the preparative reduction of 2-thiophenecarboxaldehyde, it was found that this process yields not only pinacols (2a, b) but also aldehydes of the 2,2'-dithienylmethane series (3 and 4) from products of the

3   4

"head-to-tail" coupling of radical anions (5a, b)[29, 30] (Scheme 3). The formation of these products is suppressed by adding a small amount of lithium perchlorate to the supporting electrolyte – tetrabutylammonium perchlorate.

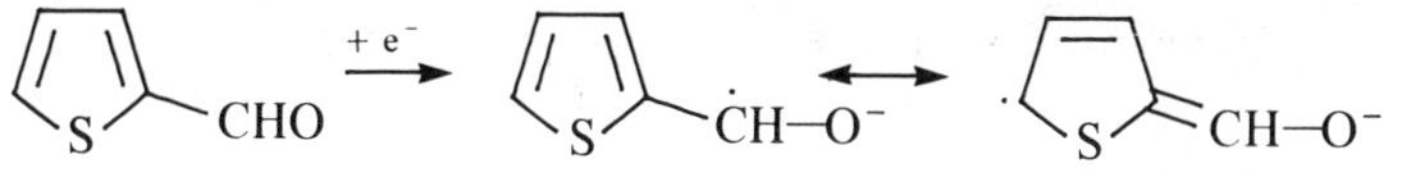

Scheme 3

The reduction of the CHOH group of alkylthienylcarbinol into CH_2 is usually carried out by the action of lithium aluminum hydride in the presence of aluminum chloride.[31] Lithium aluminum hydride is also often used to obtain thienylalkanols from thiophenecarboxylic acids and their derivatives.[32, 33] The possibilities of alumohydride reduction and the ways of CO group protection may be illustrated by the preparation of unambigeous samples of 4- (**6**) and 5-hydroxymethyl-2-acetyl-thiophenes (**7**)[33] (Schemes 4 and 5).

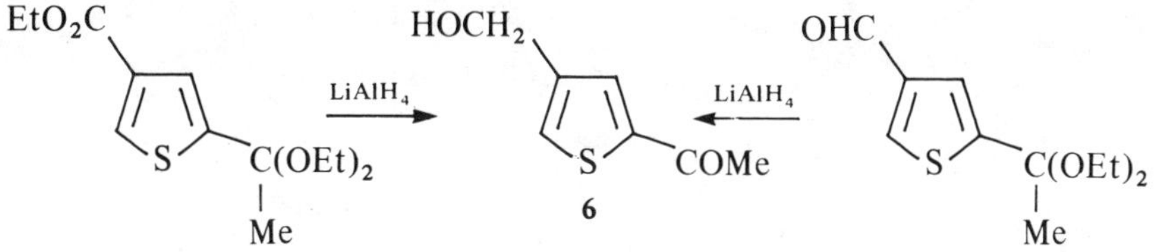

Scheme 4

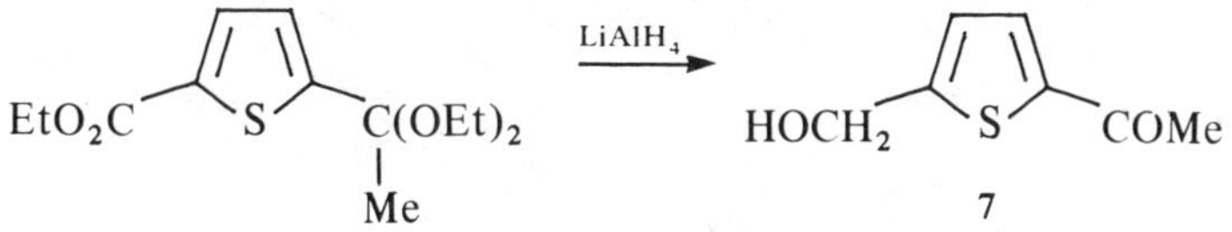

Scheme 5

For the transformation of 2- and 3-thiophenecarboxylic acids into the corresponding methylthiophenes, a very unusual method was used: the action of trichlorosilane and tertiary amine on the acid, followed by the cleavage of formed thenyltrichlorosilane in alkaline media.[34] Unfortunately, the yields of the products were only 20 and 11%, respectively.

2.   Reductive Dehalogenation of Thiophene Compounds

Reduction of the halogen is of great importance in the preparative chemistry of thiophene. This process is often used for the synthesis of β-substituted derivatives which are not accessible. The halogen atom becomes a protecting group for the α-position. The reduction of 2,3,5-tribromothiophene, which leads to the formation of 3-bromothiophene, should be mentioned first, because it is an important intermediate in syntheses of various β-substituted thiophenes. This process can be carried out by a number of procedures, of which the action of zinc in acetic acid according to Gronowitz[35] is apparently the most convenient procedure.

Recently, a simplified procedure was proposed that allows the "one-pot" transformation of corresponding unhalogenated heterocycles into 3-bromothiophene, its methyl homologues, and also 3-bromoselenophene.[36] This synthesis comprises bromination in the presence of sodium acetate followed by the addition of the metallic zinc. The hydrogen bromide formed during debromination liberates acetic acid from sodium acetate and thus the reaction conditions are close to those described in Ref. 35, but yields in the modified procedure are lower (~ 50%).

Some compounds having acyl or hydroxyalkyl groups in the β-position undergo rearrangement during debromination by zinc in acetic acid[37] (Schemes 6 and 7).

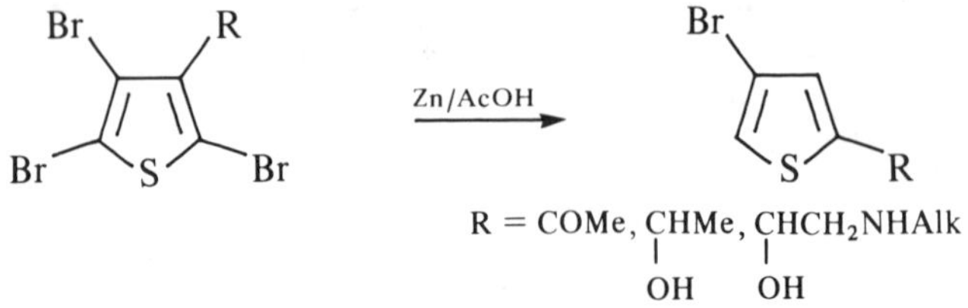
Br
R
Br
S
Br
Zn/AcOH
Br
S
R
R = COMe, CHMe, CHCH2NHAlk
OH OH

Scheme 6

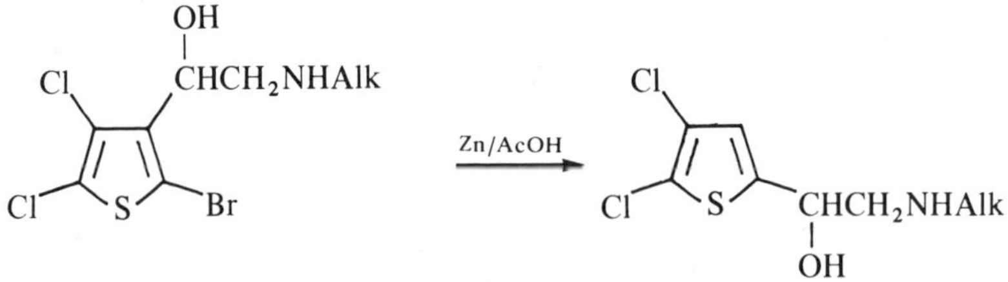
OH
Cl
CHCH2NHAlk
Cl
S
Br
Zn/AcOH
Cl
Cl
S
CHCH2NHAlk
OH

Scheme 7

Reductive dehalogenation of 2- and 3-bromothiophenes and isomeric chlorothiophenes with calcium in methanol is also known,[38] but this process requires a great quantity of calcium metal (its molar quantity should be 15 times greater than that of dehalogenated compound). Moreover, it does not always proceed to the end, especially in the case of chloride reduction, though it lasts 6–14 h. Zinc in aqueous dioxane was used successfully for the transformation of a chloromethyl group into a methyl.[39]

There are also cases of reductive debromination of thiophenes with Raney nickel,[40] but careful dosage of the catalyst is neccessary, because of the possibility of reductive desulfurization. From other heterogeneous catalysts, palladium–charcoal is used for the substitution of chlorine by hydrogen in thiophenes.[41,42] Recently an immobilized palladium complex was proposed as a catalyst. It allows the smooth replacement of chlorine or bromine atoms by hydrogen without affecting the thiophene ring or carbonyl substituents including the formyl group.[43,44] The respective α- and β-, mono- and dideuterated thiophenes are easily obtained using gaseous deuterium instead of hydrogen.[44]

In some cases, when it is desired to remove selectively one of the two α-bromine atoms, one can use the action of 1 eq. of butyllithium, followed by decomposition of the lithium–halogen exchange product formed with water. The transformation of 2,3,5-tribromothiophene into 2,4-dibromide is carried out in this way.[45, 46] The action of metallic copper in organic acids (usually in propionic acid) is a specific method of α-halogen reduction in the presence of −I−M− substituents in α'- or β-position.[47−49]

The replacement of α-halogen by hydrogen was observed during the action of sodium alkoxide in dimethylsulfoxide.[50] Gronowitz and his coworkers, using isomeric bromoiodothiophenes as models, demonstrated that sodium methylate causes the halogen disproportionation (the so-called "halogen dance") leading in all cases to the same mixture of 3-bromo-, bromoiodo-, diiodobromo-, and triiodobromothiophenes.[51] This transformation takes place in various solvents — methanol, pyridine, and hexamethylphosphortriamide.

3.   Reduction of Nitrogen-Containing Substituents

In the thiophene series, the reduction of nitro compounds is comparatively seldom used. In particular, the classical Zinin reaction with benzene derivatives is restricted in thiophene series because of the instability of the corresponding amines, which can be isolated only as salts.[52, 53] The reductive acylation of nitro-substituted thiophenes provides interesting possibilities. This reaction is carried out by the action of iron in acetic acid in the presence of acetic anhydride.[54−57]

2-Acetylaminothiophene can be obtained by hydrogenation of 2-nitrothiophene under pressure with rhenium heptasulfide as catalyst in the presence of acetic anhydride.[58] Also, the hydrogenation of NO_2 groups of thiophene nitro compounds in the presence of palladium catalyst should be noted.[41] A rather unusual method of reductive acylation consists of the interaction of Raney nickel with nitro-substituted thiophenes in acetic anhydride.[59, 60]

Reduction of oximes of the thiophene series with sodium amalgam usually gives low yields,[61, 62] which are probably due to the formation of thiophene ring reduction products.[63] The use of zinc in hydrochloric acid is preferable; the yields of the corresponding aminoalkylthiophenes are 30–50%.[63, 64] The use of stannous chloride in hydrochloric acid allows selective reduction of the oximino group in the presence of the keto group in ω-isonitroso-2-acetylthiophene (8). The yield of 2-aminoacetylthiophene hydrochloride (9) is 48%[65] (Scheme 8).

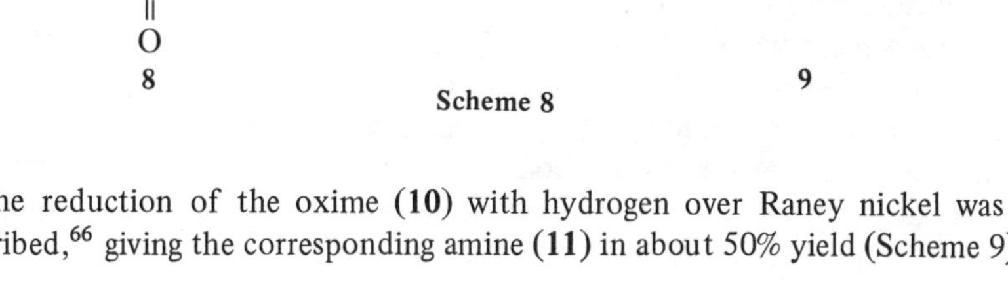

Scheme 8

The reduction of the oxime (10) with hydrogen over Raney nickel was also described,[66] giving the corresponding amine (11) in about 50% yield (Scheme 9).

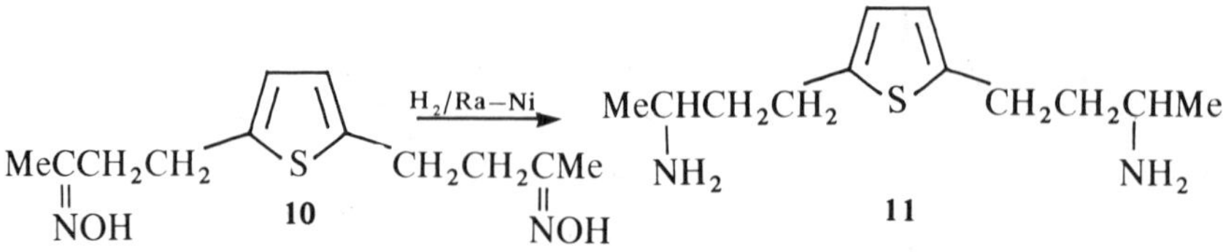

Scheme 9

Numerous amino acids of the thiophene series have been prepared by the reduction of ketoacid oximes. This transformation is carried out by the action of amalgamated aluminum[67, 68] or tin in hydrochloric acid,[69] or better with zinc in aqueous ammonia.[70]

The reduction of Schiff bases into the corresponding aminomethylthiophenes is achieved by the action of sodium borohydride[71, 72] or by hydrogen in the presence of rhenium heptasulfide as a catalyst.[73]

If there are electron-withdrawing substituents in the thiophene ring, the obtained amines are sufficiently stable and there is no need for acylation of the amino group. For instance, by the reduction of ethyl 5-nitro-2-thiophenecarboxylate with amalgamated aluminum, ethyl 5-amino-2-thiophenecarboxylate is obtained in 78% yield.[74] Substituted 4-nitro- (12) and 5-nitro-2-thiophenecarboxamides (13) are smoothly reduced with freshly precipitated iron (II) hydroxide into the corresponding aminothiophenes (14 and 15)[75] (Schemes 10 and 11).

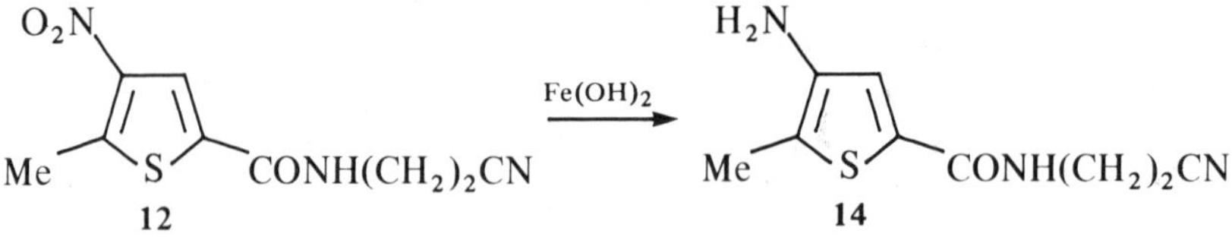

Scheme 10

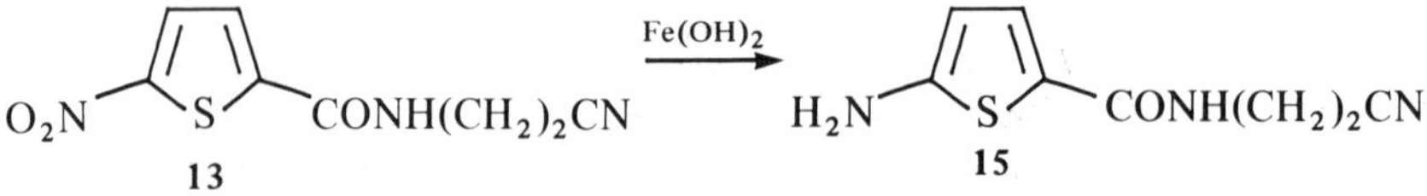

Scheme 11

Catalytic reduction of 4- and 5-nitro-2-thiophenesulfonamides in the presence of Raney nickel leads to 4- and 5-amino-substituted 2-thiophenesulfonamides.[76] It is possible that the presence of the sulfonamide group stabilizes the molecule and prevents poisoning of the catalyst.

Condensed derivatives with an amino group in the benzene ring are also quite stable. They can be obtained from the corresponding nitro-substituted compounds by various methods, including the reduction with hydrazine hydrate[77] and hydrogenation[78] in the presence of Raney nickel. Japanese authors described the preparative electrochemical reduction of the cyano group in 3-benzo[b]thiophen-acetonitrile over copper cathode covered with nickel black.[79]

4.   Reduction of Multiple Bonds in Side Chain

α,β-Unsaturated bonds activated by electron-withdrawing substituents can be reduced using various reagents while the thiophene nucleus remains unchanged. For example, β-(2-thienyl)propionic is smoothly obtained from β-(2-thienyl)acrylic acid by reduction with sodium amalgam.[65] Lithium aluminum hydride also reduces the carboxylic group, transforming the same acid into 3-(2-thienyl)propanol.[80] The latter is available more easily from the corresponding ester.[81] Similarly, the alumo-hydride reduction of thienyl nitroolefines (16) is a convenient method for pre-paring thienylalkylamines (17)[82-84] (Scheme 12).

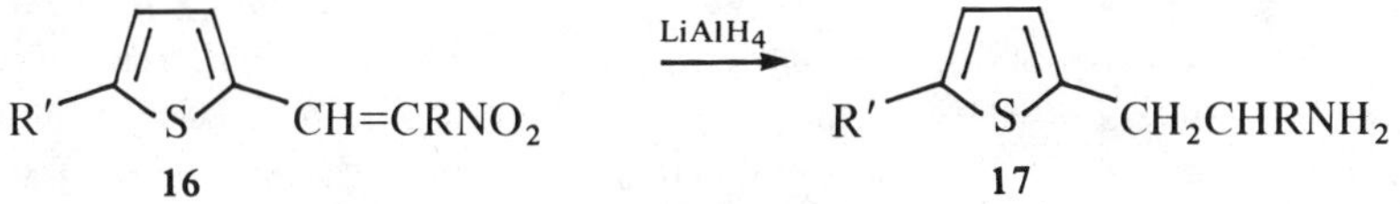

Scheme 12

For the catalytic hydrogenation of side-chain double bonds, various catalysts can be used. Strange as it is, this process can be carried out even in the presence of Raney nickel (see, for example, Ref. 85). It appears that the Wilkinson catalyst is the most suitable one for these purposes.[86] A modified chiral rhodium catalyst allows homogeneous hydrogenation to be carried out with high optical purity[87] (Scheme 13).

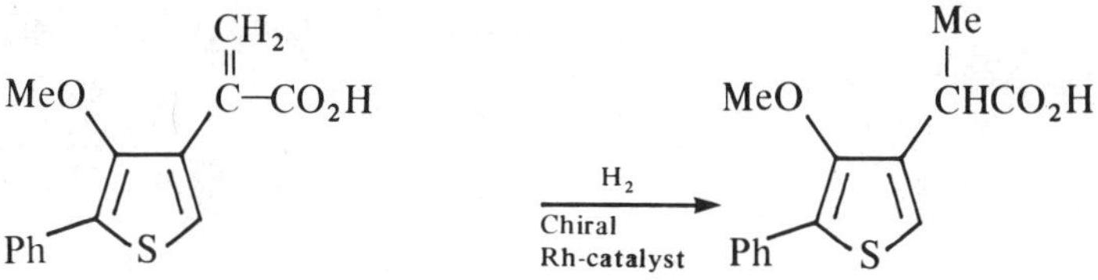

(+)-Isomer: Yield 97%,
optical purity 88%

Scheme 13

5.   Reduction of Sulfur-Containing Substituents

In some cases, standard procedures can be used for the reduction of sulfur-containing substituents in thiophenes. In particular, the preparation of 2-thio-phenethiol from 2-thiophenesulfonic acid under the action of zinc has been described.[88] The reduction of thiophene sulfones can be demonstrated by the transformation of thionessal sulfone into thionessal with zinc in acetic acid in the presence of hydrochloric acid, a process described by Ginsberg.[89] Under the same conditions, benzo[b]thiophenesulfone is reduced into benzo[b]thiophene,[90] but with lithium aluminum hydride, one obtains 2,3-dihydrobenzo[b]thiophene.[90]

There are data concerning the reduction of thiocyanothiophenes into corresponding thiophenethiols or disulfides by the action of lithium aluminum hydride and sodium borohydride,[91] or by cyclooctatetraene dipotassium salt.[92] The latter reagent,[92] as well as lithium aluminum hydride,[93] may be used for the transformation of thiophene disulfides into mercaptans. Diselenides of the benzo[b]thiophene series are reduced into corresponding selenides by the action of sodium borohydride.[94]

Specific difficulties in the reduction of sulfur-containing substituents are often connected with hydrogenolysis of C—S bonds. Therefore, it is neccessary in such cases to select experimental conditions very carefully, so as not to effect thiophene ring. In this connection, the reductive cleavage of alkylthienylsulfides leading to the corresponding mercaptans, a process studied by Gol'dfarb and Kalik, demonstrates the situation described above very convincingly. It is carried out in conditions very close to Birch reduction — by the action of sodium in liquid ammonia, but in the absence of alcohols that are often used as proton donors in the reduction of the thiophene ring (see Section IV.1). The yields of thiophenethiols (18) formed by the reduction of alkyl-2-thienylsulfides (19) with sodium in liquid ammonia are up to 80%[95] (Scheme 14).

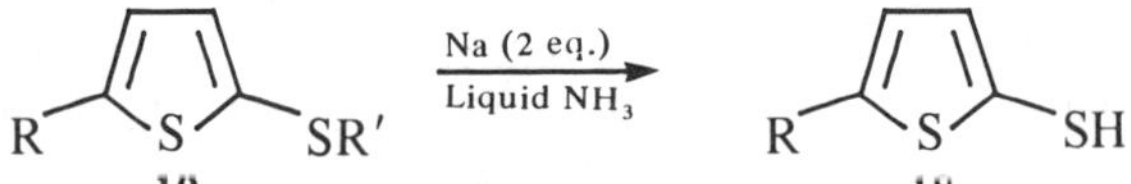

Scheme 14

Dealkylation of bis-sulfides (20) with 4 eq. of sodium leads to bis-mercaptothiophenes (21),[96] whereas with 2 eq. of sodium, mercaptosulfides (22) are obtained.[97] In all cases, the yields are 60–80% (Scheme 15).

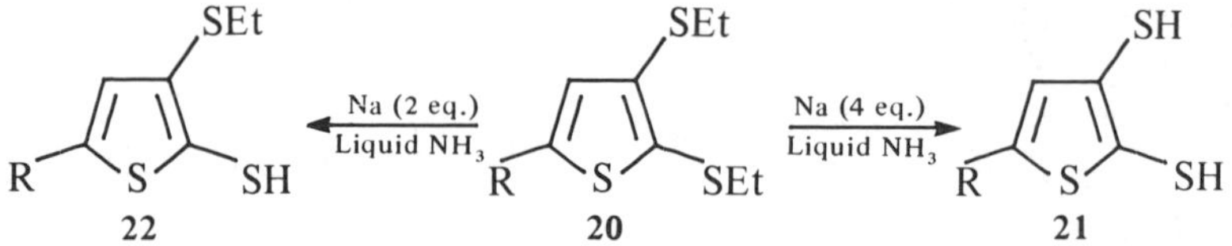

R = H, Et
Scheme 15

By the reductive cleavage of alkyl thienylsulfides, the corresponding thiophenethiolate anions are formed directly. They are stable toward alkali metal in liquid ammonia, but when starting sulfide has an electron-withdrawing substituent, sodium is not used only for the cleavage of the side chain C—S bond. For instance, 2-ethylthio-5-ethyl-3-thiophenecarboxylic acid is readily transformed under the action of 2 eq. sodium into the corresponding mercaptoacid. However, in the case of an excess of sodium (up to 10 eq.), cleavage of the thiophene ring with sulfur elimination occurs.[98]

The reductive cleavage of alkylthio thiophenecarboxaldehydes and their acetals under the action of sodium in liquid ammonia was widely used for the synthesis of so-called mercaptoaldimines (24) (see Refs. 99 and 100a), which are able to form chelates. These aldimines exist mainly in another tautomeric form, namely 5-alkyl-3-(aminomethylene)-4-thiolene-2-thiones (24a)[101] (Scheme 16).

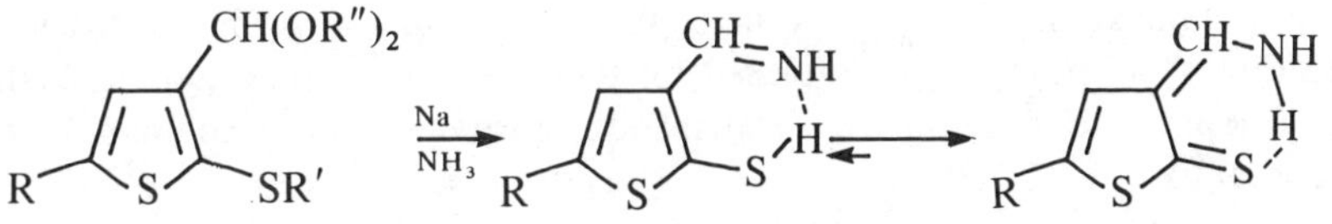

Scheme 16

Along with mercaptoaldimines of the described type above (24), [R = H or alkyl as well as R = $(CH_2)_n NAlk_2$, $(CH_2)_2OH$, $(CH_2)_2OEt$[102-104]], there were also obtained analogous derivatives having four coordinating centers in one thiophene ring (25)[105] or in two cycles divided by a methylene bridge (26).[106]

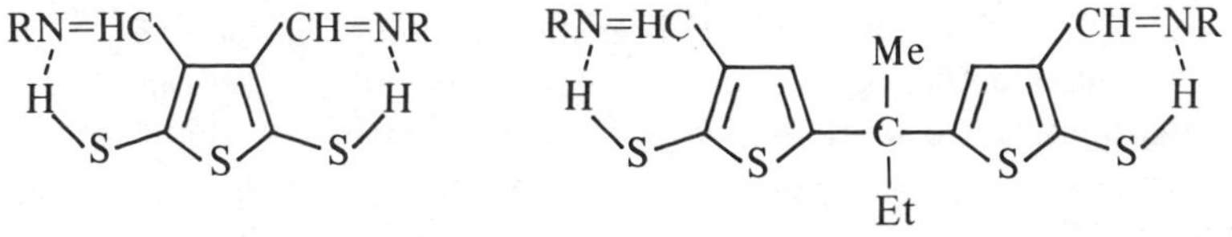

Mercaptoaldimines can also be prepared in the benzo[b]thiophene[107] and thieno-[3,2-b]thiophene series[108] by the reductive cleavage of acetals obtained from the corresponding aldehydo sulfides. Mercaptoaldimines (aminomethylenethiolene-thiones) are not formed by any other mutual arrangement of functions in the starting alkylmercapto acetal. For instance, in the case of 5-ethylthio-2-thiophene-carboxaldehyde acetal, it is neccessary to use 6 eq. sodium to finish reduction. The sodium is used not only for C–S bond cleavage, but also for the transformation of the acetal group into methyl.[109] The same is true for acetal of 3-ethylthio-2-thiophenecarboxaldehyde (27),[110] but if the alkali solution is acidified after the additon of 2 eq. of sodium, it is possible to isolate the product, which may be considered as a tautomeric mixture of 28a and 28b[111] (Scheme 17).

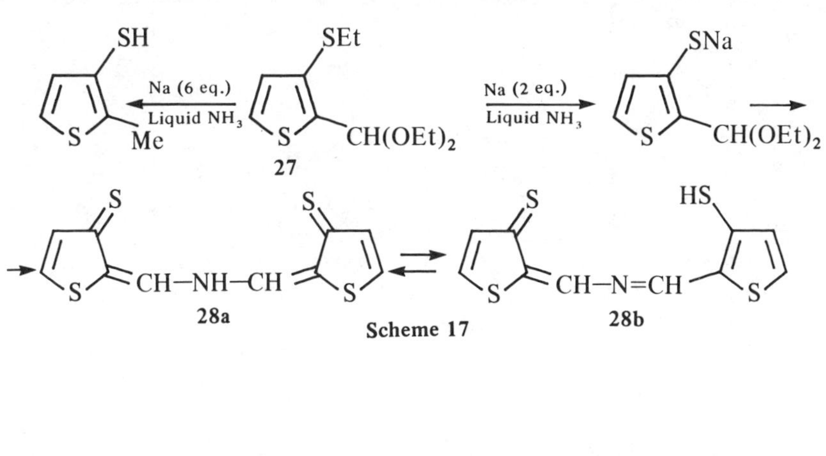

Scheme 17

In the case of benzyl thienylsulfides, cleavage of C—S bond adjacent to thiophene ring may be observed. Thus, by the action of sodium in liquid ammonia on the diethylacetal of 2-benzylthio-5-ethyl-3-thiophenecarboxaldehyde, mercaptoaldimine and 5-ethyl-3-thiophenecarboxaldehyde were obtained.[102] It is apparent that the elimination of the exocyclic sulfur atom results in a comparatively low yield (56%) of the mercaptan formed by the reduction of 5-benzylthio-2-ethylthiophene in liquid ammonia, which demands a greater amount of sodium (4.6 eq.) than usual.[95]

III. HYDROGENATION OF THIOPHENES INTO DIHYDRO- AND TETRAHYDROTHIOPHENE DERIVATIVES

1.   Catalytic Hydrogenation

The possibilities of catalytic hydrogenation of thiophenes are very restricted by the poisoning action of thiophenes on catalysts and also by ring destruction under hydrogenation conditions (see Ref. 112). In particular, thiophene itself is often used as a model sulfur-containing substance when studying the poisoning of catalysts (see Section IV.2) and such widely used catalysts as Raney nickel, which causes the reductive desulfurization of thiophenes (see Section IV.3).

Only few catalysts are suitable for hydrogenation of the thiophene ring, but their use is often limited by the high cost, drastic conditions, and low selectivity of the process. To some extent, palladium catalyst is an exception, because it acts at moderate temperature and pressure.[41] But the poisoning action of thiophene compounds on this catalyst makes it necessary to use large amounts, almost equal to the quantity of the substance under hydrogenation. Therefore, the process is warranted only in special cases, for example, in stereospecific synthesis of biotin (29) via the corresponding thiophenic precursor (30)[113, 114] (Scheme 18).

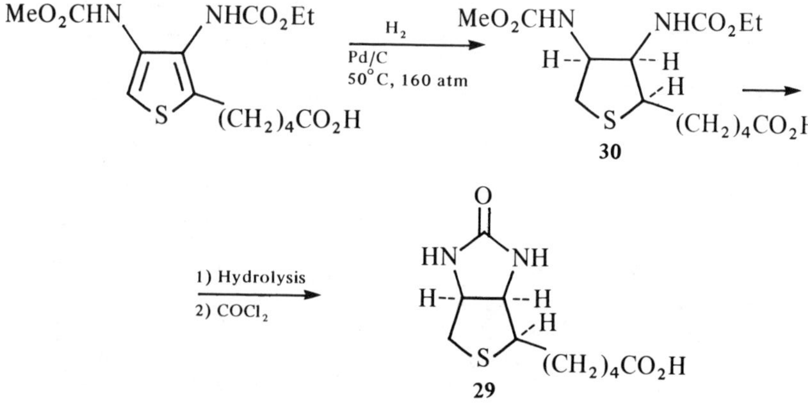

Scheme 18

Claeson and Jonsson[115] have carried out the hydrogenation of 5-(2-thienyl)-valeric acid over palladium–charcoal and resolved the obtained racemic acid in enantiomers through cinchonidine salts. Determination of the absolute configuration of these isomers made possible the stereochemical correlation of biotin, since the stereochemistry of the so-called (−)-thiophane-valeric alcohol − the product of (+)-biotin degradation − is the same as that of (−)-tetrahydro-5-(2-thienyl)valeric acid.

Other metals of the platinum group are not used for hydrogenation of thiophenes. For example, the rate of thiophene hydrogenation over rhodium catalyst falls sharply to zero upon 5–6% of conversion.[116] However, data exist that homogeneous rhodium complex catalysts are able to reduce thiophene and benzo[b]-thiophene.[117] tetrahydrothiophenes can be obtained in high yields using rhenium heptasulfide as catalyst,[58, 118, 119] but drastic conditions (250–500°C, 100–300 atm) and the high price of the catalyst limit its use. Apparently, rhodium disulfide does not have any advantage over rhodium heptasulfide.[120] The activity of rhodium chlorides is lower than that of sulfides in thiophene hydrogenation.[121]

Hydrogenation of the thiophene ring also takes place with such catalysts as sulfides of molybdenum, tungsten, cobalt, and nickel,[122, 123] but reactions over these catalysts proceed at very drastic conditions, leading to poor yields of tetrahydrothiophenes and also to the formation of destruction products. For instance, when cobalt polysulfide is used for hydrogenation of acylthiophenes,[122] conversion does not exceed 30% and, as a rule, mixtures of alkylthiophenes and the corresponding tetrahydro derivatives, with the former predominating, are obtained. By hydrogenation of thiophene and alkylthiophenes over molybdenum sulfide on alumina, the yields of tetrahydrothiophenes are about 50%. The moderate yields probably are caused by hydrogenolysis. This is the main process in hydrorefining of oil products carried out over similar catalysts (see Section IV.2).

Of some interest is the hydrogenation of thiophenes over cobalt catalysts in the presence of carbon monoxide (synthesis-gas is used in these reactions).[124,125] For the preparation of the catalyst, cobalt carbonate with addition of dicobalt octacarbonyl or cobalt hydrocarbonyl was used. It is supposed that these additives act as homogeneous hydrogenation catalysts.[124] But this method is limited by the difficulty of retaining oxygen-containing functions. For instance, hydrogenation of 2-thenyl alcohol leads to 2-methyltetrahydrothiophene and hydrogenation of 2-acetylthiophene leads to 2-ethyltetrahydrothiophene.[124] An exception to this is the carboxy group, which is retained in the product.[125]

2. Ionic Hydrogenation

There are now quite convenient methods for the reduction of the thiophene nucleus that do not use heterogeneous catalysts; electrophilic ionic hydrogenation is one. Parnes, Bolestova, Belen'kii, and Kursanov were the first to use ionic hydrogenation in the thiophene series.[126, 127] This reaction (see Refs. 128 and 129) consists of consecutive reversible protonation and irreversible addition of hydride–

ion and is carried out by means of trialkylsilane in proton acid medium (usually in CF$_3$COOH). When applied to thiophenes, the mechanism may be represented by Scheme 19.[130]

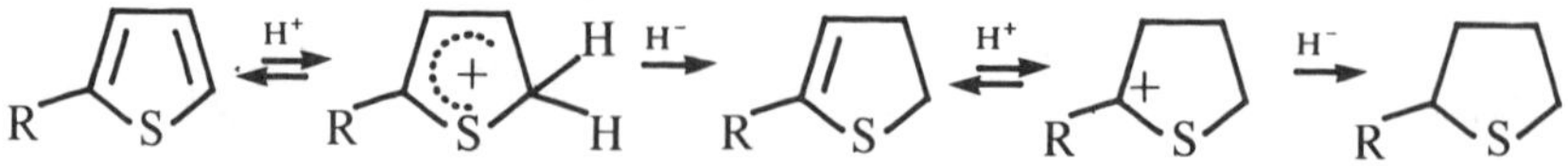

Scheme 19

Various thiophenes bearing electron-releasing substituents enter into this reaction. Acylthiophenes (31) under the conditions of ionic hydrogenation give alkylthiophenes (33), which are then reduced into tetrahydrothiophenes (34)[130] (Scheme 20). 2-Thenyl alcohol (32), R = H, obtained from 2-thiophenecarboxaldehyde, is easily dehydrated and octahydrodithenyl ether (35) is the final product of the reduction.[130]

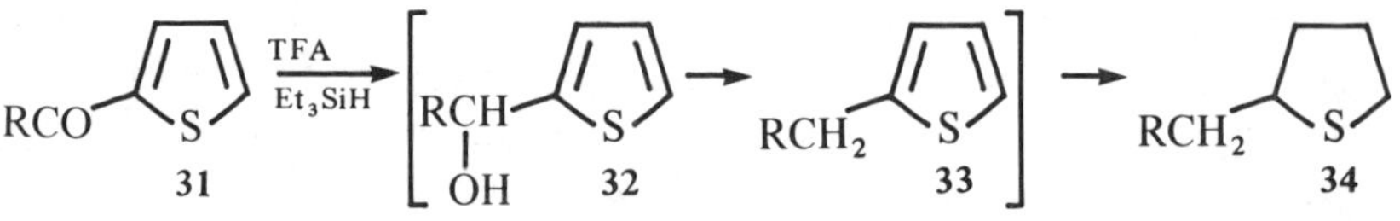

Scheme 20

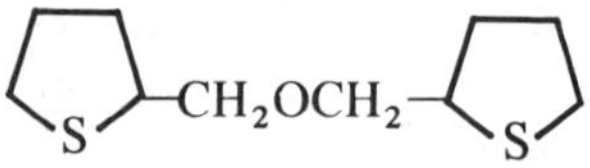

35

In the case of 2-halogen- and 2-alkylthio-substituted thiophenes, hydrogenolysis of the bond between the substituent and the ring takes place as well as hydrogenation of the ring (Schemes 21 and 22). Hydrogenolysis of the C—S bond in the side chain is observed also during the reduction of 2-thenylsulfides (Scheme 23).[130]

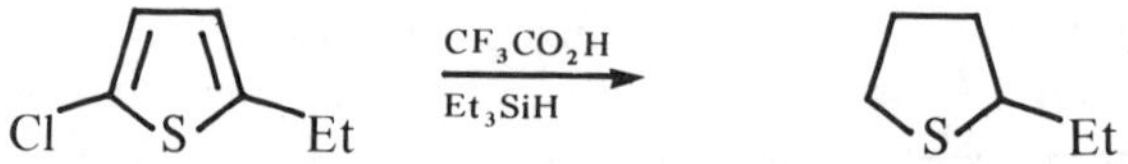

Scheme 21

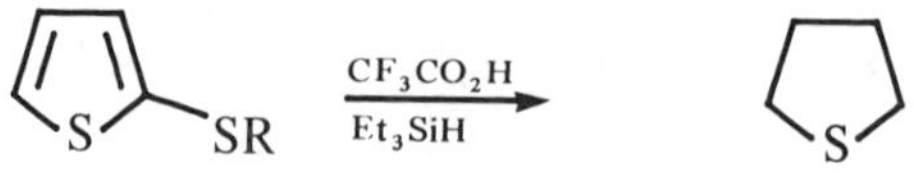

Scheme 22

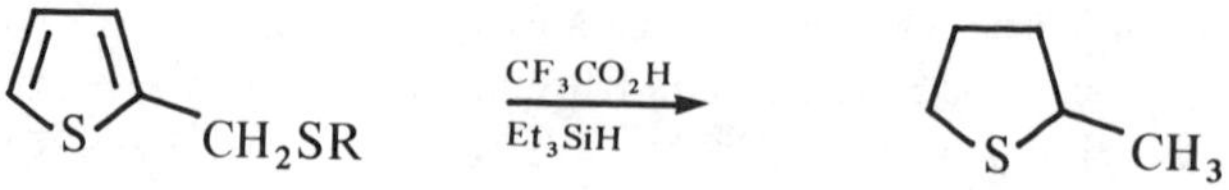

Scheme 23

Through ionic hydrogenation, alkyltetrahydrothiophenes with various terminal functional groups are obtained, for example, with carboxy[130] and dialkylamino groups.[131] Mono- and diphenylthiophenes,[132] 2,2'-bithiophen,[133] and benzo[b]-thiophene and its homologues[134] also undergo ionic hydrogenation.

In some cases, ionic hydrogenation meets with difficulties caused, in the opinion of the authors,[132] by steric hindrances. Thus, in contrast to other isomers, it proved impossible to hydrogenate 2,5-diphenylthiophene at ordinary conditions of the reaction. 2,5-Di(2-thenyl)thiophene (36) is transformed by ionic hydrogenation into the partial reduced product (37); the middle nucleus is not affected[130] (Scheme 24).

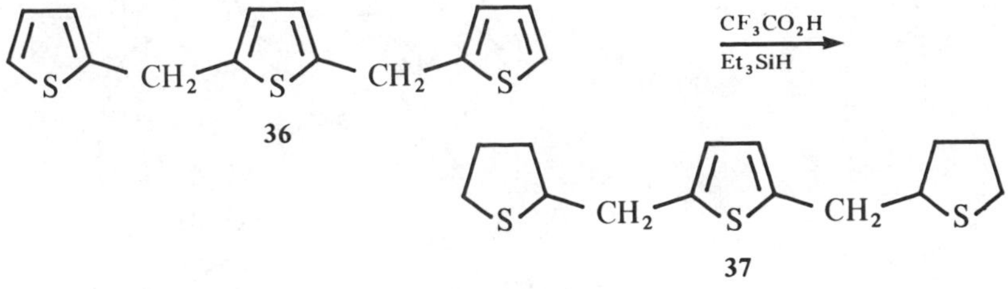

Scheme 24

Tetradehydrobiotin is hydrogenated with difficulty and the yield of cis-biotin is only 10%.[135] As shown by Zav'yalov and Dorofeeva, the process is strongly facilitated in the case of some methyl-substituted tetradehydrobiotins (38) (Scheme 25). The authors explain this fact by the activating influence of the methyl groups.[136]

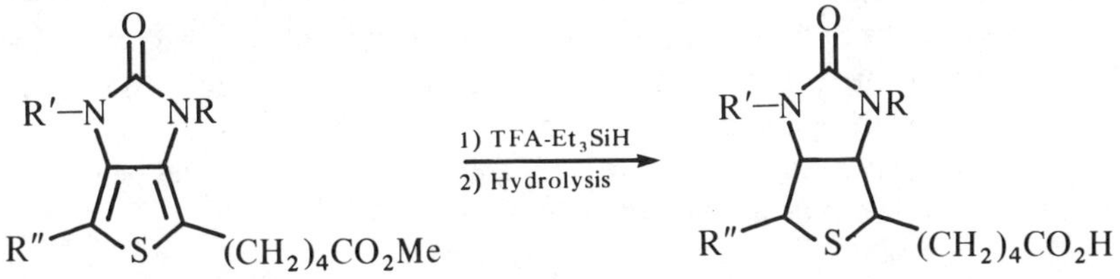

	Yield
(a) R = R'' = H, R' = Me	82%
(b) R = R' = H, R'' = Me	72%
(c) R = R' = Me, R'' = H	Low
(d) R = R' = R'' = H	Traces

Scheme 25

Under ordinary conditions (triethylsilane in CF_3COOH, 50°C), ionic hydrogenation of thiophenes proceeds very slowly (as a rule, it requires 20–60 h), and this long time is the principal inconvenience of the reaction, when applied to thiophenes. Parnes, Kursanov, and coworkers have shown that the rate of the reaction is increased considerably in the presence of boron trifluoride etherate,[137, 138] p-toluenesulfonic acid, lithium tosylate, or lithium perchlorate.[139] With such additives (1–3% by weight), the hydrogenation of the thiophene ring proceeds at room temperature and is usually complete in 0.5–2 h. The influence of additives on ionic hydrogenation is consistent with the mechanism of the reaction described above. This mechanism was confirmed[130] by the reduction of 2,5-dimethylthiophene with the aid of deuteriosilane, which resulted in tetrahydro-2,5-dimethylthiophene having deuterium in one α- and one β-position of the ring (39) (Scheme 26).

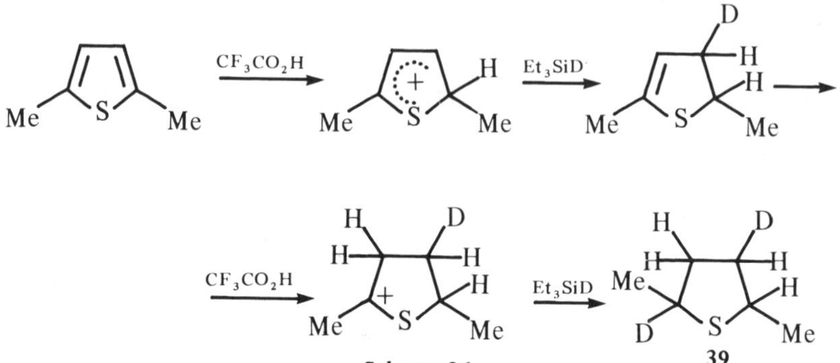

Scheme 26

The first step of the reaction is α-C-protonation of thiophene, its homologues, and some of its derivatives. It proceeds quantitatively under the action of aluminum chloride and HCl in dichloromethane or 1,2-dichloroethane on thiophenes, and σ-complexes formed are quite stable in solution and can be retained for a long time at room temperature.[140] A study of the hydrogenation of such stable complexes (40), formed by the protonation of 2,5-dimethylthiophene and 2,3,5-trimethylthiophene, led to the elaboration of the novel variant of thiophene ionic hydrogenation with the system triethylsilane-HCl-AlCl$_3$ (Scheme 27); it should be noted that aluminum chloride can be used in catalytic amounts.[141,142]

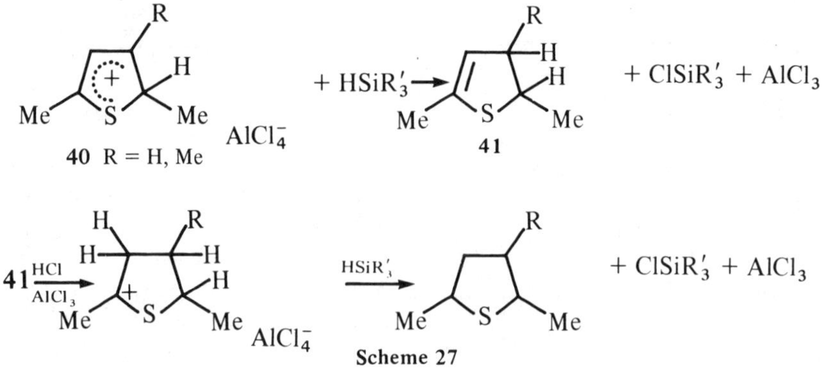

Scheme 27

The use of the triethylsilane-HCl-AlCl$_3$ system not only accelerates the ionic hydrogenation as it takes place when the additives mentioned above are used, but also allows the hydrogenation of some compounds that are stable under the usual conditions of the reaction (triethylsilane in trifluoroacetic acid), for example, 2,5-diphenylthiophene.[142] The yields of tetrahydrothiophenes are up to 80%. At the same time, none of the variants of ionic hydrogenation considered allows tetrahydrothiophene from unsubstituted thiophene to be obtained; in all cases, a mixture of di- and tetrahydrothiophenes is formed, where the former predominates. This situation may be explained by the low concentration of secondary carbenium ions (42), which should be formed during the second stage of hydrogenation.

42

It is interesting to note that with the use of superacid HF–TaF$_5$, the process of ionic hydrogenation leads to tetrahydrothiophene in 80% yield.[143] Under these conditions, isopentane serves as a hydride ion donor.

Recently, a new reduction reaction of thiophenes was proposed, leading mainly to 2,5-dihydrothiophenes with an admixture of tetrahydrothiophenes.[144] The authors suppose that it consists of consecutive transfers of an electron and a proton to thiophenium ions. Zinc metal is used as the source of electrons and trifluroacetic acid is the proton source.

3. Electrochemical Reduction

Preparative electroreduction of thiophenes has now been developed considerably. It should be emphasized that direct reduction is practically unapplicable to thiophene and its homologues because of their high reduction potentials. However, the use of indirect reduction with the aid of electrochemically generated radical anions of other organic substances able to transfer an electron to a molecule, the reduction of which is very difficult, creates favorable possibilities, as shown by Mairanovskii et al.[145] In particular, electrolysis of dimethylformamide solution of thiophene in the presence of diphenyl as a "carrier" and 10% water as a proton donor leads to the formation of dihydro- (53.8%) and tetrahydrothiophene (42.8%), the conversion being 55%.[146] The yield of reduction products is affected not only by the water content, but also by the presence of Zn^{2+} ions, which according to Ref. 147, accelerates the protonation of intermediate thiophene radical anions by water.

Direct electrochemical reduction has been carried out successfully with thiophenecarboxylic acids that have an electron-withdrawing substituent stable under the conditions of the process and this substituent lowers considerably the reduction potential of the thiophene ring. It allows the easy transformation of 2-thiophenecarboxylic acid and its derivatives (43 and 44) into the corresponding 2,5-dihydro derivatives (45 and 46), the yields being more than 90%[148, 149] (Schemes 28 and 29).

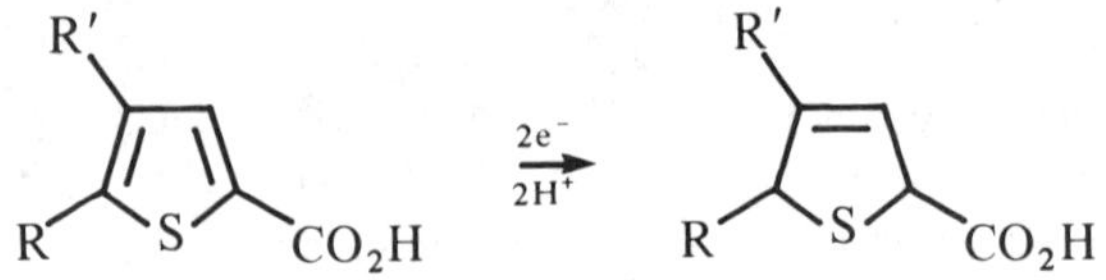

Scheme 28

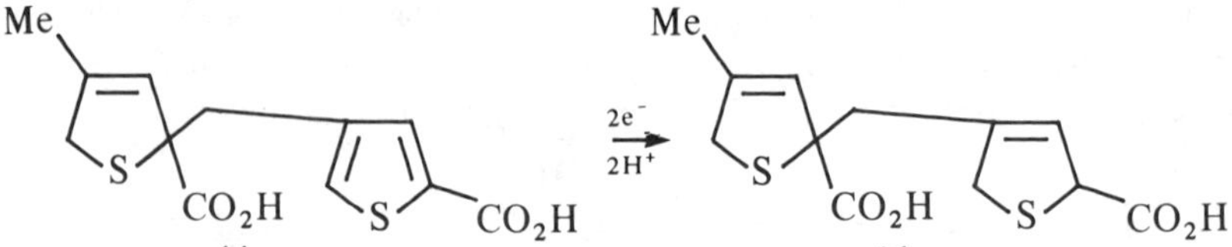

Scheme 29

Since dihydrothiophenes are readily desulfurized as such (for instance, in photolysis conditions[150] or by thermolysis of corresponding sulfones[151, 152]), the transformations considered create a new, interesting route from thiophenes to isoprenoids.

In contrast to 2-thiophenecarboxylic acid[148] and 2-furancarboxylic acid,[153] which are readily transformed into 2,5-dihydro derivatives, 2-selenophenecarboxylic acid gives three dimeric acids by electroreduction. Of these three, only one (47) was sufficiently stable to be isolated,[154]

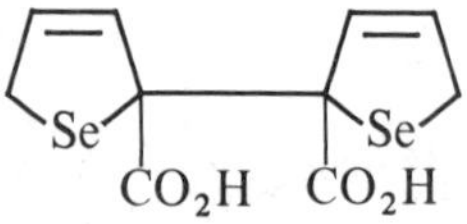

Recently, an interesting new reaction was found by Gul'tyai et al. — the reductive electrochemical carboxylation of methyl 2-thiophenecarboxylate (48), which gives a 2,3-dihydro-2,5-thiophenedicarboxylic acid half ester (49) with an admixture of products arising from the so-called "secondary catalysis," which results in the introduction of two CO_2 molecules.[155, 156] If the reaction is carried out at a low concentration of starting ester (0.1 M), That is, under conditions of suppression of "secondary catalysis," the yield of the dihydro acid increases up to 80%. The mechanism of the main process may be represented by Scheme 30.[156]

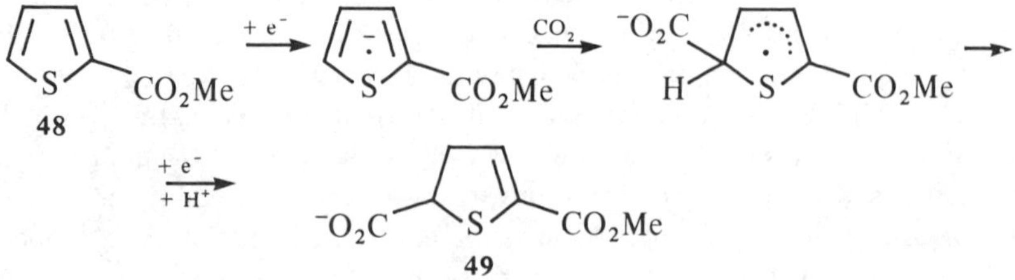

Scheme 30

4.   Other Methods

In some cases, reduction of the thiophene ring may be carried out with sodium amalgam. Thus, 2-thiophenecarboxylic acid is transformed by the action of the excess of 2.5% sodium amalgam in aqueous potassium hydroxide into tetrahydro-2-thiophenecarboxylic acid (the yield is 60%) and 2-thiophenecarboxaldehyde oxime becomes 2-aminomethyltetrahydrothiophene.[63]

Some dihydrothiophenes can be obtained by the reduction of the corresponding thiophenes with alkali metals in liquid ammonia; but as a rule, these reactions are complicated by subsequent transformations leading to acyclic compounds. For this reason, they are considered in the following section, which is devoted to reductive cleavage of C—S bonds.

IV.   REDUCTIVE CLEAVAGE OF THE THIOPHENE RING WITH THE RUPTURE OF ONE OR TWO C—S BONDS — SOME TRANSFORMATIONS OF RING-OPENING PRODUCTS

In this section the reactions of thiophenes leading to compounds of other classes are considered. Depending on the reagents used and reaction conditions, the sulfur atom of the thiophene ring may be removed, or retained in the products as a mercapto group, which can be transformed into other functional groups. In the course of these transformations, partial or complete hydrogenation of unsaturated bonds is observed as well as modification of substituents. Generally, the methods considered below are of great synthetic potential. The most important are the systematically studied reductive desulfurization with Raney nickel and the reductive cleavage by the action of alkali metals in liquid ammonia, which are investigated extensively. Catalytic hydrodesulfurization is of great importance for the technology of oil refining, but has very limited preparative use. Therefore, it is not considered in detail.

1.   Reduction with Alkali Metals in Liquid Ammonia and Related Processes

Reduction of thiophenes with alkali metals in liquid ammonia may lead to various products: dihydrothiophenes, unsaturated mercaptans, and sulfur-free compounds. The reaction conditions — the nature of the metal and the presence of a proton donor and its efficiency — substantially influence the results of the process.

Birch and McAllan, who first applied this reaction to thiophene, showed that by the action of sodium in liquid ammonia, in the presence of methanol as a proton donor, thiophene is converted into 2,3- and 2,5-dihydrothiophenes along with the products of subsequent reduction — butenthiols, butenes, and hydrogen sulfide.[157,158] They succeeded in isolating 2,3-dihydrothiophene (12.3%) and 2,5-dihydrothiophene (26.2%) from the mixture formed.[158] The reduction of

3-methylthiophene proceeds similarly; the yield of 2,5- and 4,5-dihydro-3-methyl-thiophene mixtures is ~ 45%.[159] However, in the case of 2-methylthiophene, the yield of dihydro-2-methylthiophenes decreased to 7.5% and the unsaturated thiol became the main product.[159] During the reduction of 2,5-dimethylthiophene, it was found impossible to isolate dihydro-2,5-dimethylthiophene. At the same time, small amounts of 2-pentanone and 2-hexanone were isolated in the above cases mentioned after work-up of the reaction mixtures.[159]

During the reduction of thiophene with sodium in liquid ammonia, in the presence of ammonium bromide as a proton donor, Krug and Tocker[160] obtained the same results, that Birch and McAllan had observed in the presence of methanol. However, Hückel and Nabih[161] did not notice even traces of dihydrothiophenes when they applied the modified procedure of thiophene hydrogenation; in contrast to the method described in Ref. 160 the proton donor was added after decoloration of the sodium solution in liquid ammonia. Butylmercaptan was obtained in low yield and the great bulk of thiophene, as the authors assume, was converted into butenes or butane since most of the sulfur was found in the form of sodium sulfide.

It was noted that unsaturated thiols are formed during the action of sodium in liquid ammonia on thiophene and its homologues. The formation of mercaptan from benzo[b]thiophene proceeds quite easily. As Hückel and Nabih demonstrated,[161] the reaction of benzo[b]thiophene with sodium in liquid ammonia leads to o-ethylthiophenol in 78% yield. According to the same authors, dibenzothiophene is not changed under the conditions of Birch reduction (in the presence of methanol) or Krug and Tocker reduction (in the presence of ammonia salt as proton donor). At the same time, the use of the Hückel procedure (see above) leads to 1,4-dihydrodibenzothiophene,[161] but not to the cleavage product. However, Birch and Nasipuri[162] later described the transformation of dibenzothiophene into 2-cyclohexenylthiophenol with undetermined position of the olefinic double bond. Hückel and coworkers[163] showed that the direction of reaction is determined by the sequence in which the reagents are mixed and that 1,4-dihydrodibenzothiophene is formed when a proton donor is not used. This result is consistent with the data of Gilman and Jacobi[164] and is explained in Ref. 163 by an intermediate formation of a bis-metallorganic compound.

To understand the reasons for the different results obtained in the action of alkali metals in liquid ammonia on thiophenes, one should consider briefly the reaction mechanism discussed in many publications, in particular, in the paper of Birch and Nasipuri[162] (see also Ref. 165). The reduction mechanism consists in consecutive additions of solvated electrons and protons to the aromatic or hetero-aromatic substance (Scheme 31).

$$ArH + e^- \rightleftharpoons ArH^{\overline{\cdot}} \xrightarrow{\text{ROH}} ArH_2^{\cdot} \xrightarrow{e} ArH_2^- \xrightarrow{\text{ROH}} ArH_3$$

Scheme 31

If the proton donor is absent, the course of the process may become complicated. In particular, the possibility of the formation of a dianion of the starting aromatic compound (which is in fact the same as the bis-organometallic compound and can be transformed into a dihydro derivative during treatment of the reaction mixture with the proton donor) is discussed in Ref. 162 (Scheme 32).

$$\text{ArH}^{\bar{}} + e^- \rightleftharpoons \text{ArH}^{2-} \xrightarrow{\text{2 ROH}} \text{ArH}_3$$

Scheme 32

At the same time, by the use of ammonium salts as proton donors, solvated electrons can be spent for the reduction of the ammonium ion according to Scheme 33.[160]

$$2\text{NH}_4^+ + 2e^-(\text{NH}_3)_x \longrightarrow \text{H}_2 + (2x + 2)\text{NH}_3$$

Scheme 33

The higher the concentration of NH_4^+ ions, the more probable is this undesirable reaction.

Lithium metal is a more active reducing reagent than sodium,[160] and this is consistent with values of their oxidizing potentials in liquid ammonia.[166,167]

According to the reaction mechanism, the presence of electron-withdrawing substituents should facilitate reduction. In Ref. 165, unpublished data of Slobbe is cited, which describes the formation of β,γ-unsaturated δ-mercaptoacids in nearly quantitative yield as a result of the action of 5 eq. lithium on 2-thiophenecarboxylic acid and 5-methyl-2-thiophenecarboxylic acid in liquid ammonia in the presence of methanol. The stereochemistry of these δ-mercapto acids was not established. Gol'dfarb and coworkers[168] demonstrated that by the reduction of 2-thiophene-carboxylic acid with 3 eq. lithium in liquid ammonia in the presence of methanol, 2,5-dihydro-2-thiophenecarboxylic acid is formed, but the main product is 5-mercapto-cis-3-pentenoic acid. When 5 eq. lithium are used, this mercapto acid is the only product. Blenderman et al.[169] carried out the reduction of 2-thiophene-carboxylic acid without an alcohol as proton donor and obtained a mixture of products in which the dihydro acid mentioned above was already the main component. But its isolation offered difficulties. In the case of 3- and 5-methyl-2-thiophenecarboxylic acids (used in this reaction as lithium salts), the action of 2 eq. lithium in liquid ammonia (without an alcohol) leads to the corresponding 2,5-dihydro-2-thiophenecarboxylic acids in 75 and 50% yields[170] (Scheme 34).

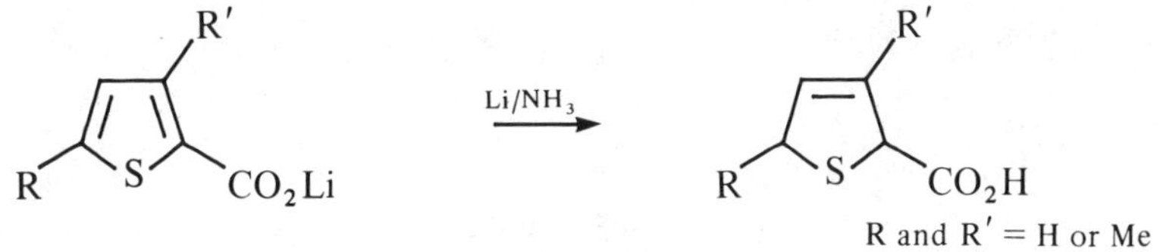

R and R$'$ = H or Me

Scheme 34

When studying the reactions of 2-thiophenecarboxylic acid and its 3-, 4-, and 5-methyl-substituted derivatives with lithium in liquid ammonia, Gol'dfarb et al.[168,171] demonstrated regio- and stereospecific reductive cleavage that resulted in Z-isomers of the corresponding β,γ-unsaturated δ-mercapto acids (50) (Scheme 35). This result is in good agreement with the reaction mechanism (Scheme 36) proposed in Ref. 171.

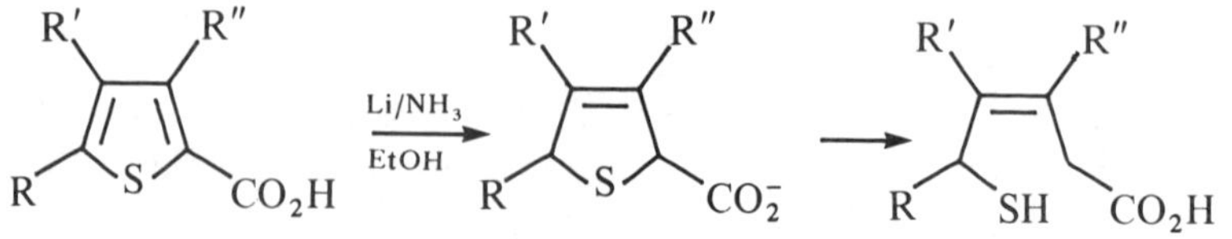

(a) R = R' = R'' = H　　(c) R = R'' = H, R' = Me
(b) R = Me, R' = R'' = H　(d) R = R' = H, R'' = Me

Scheme 35

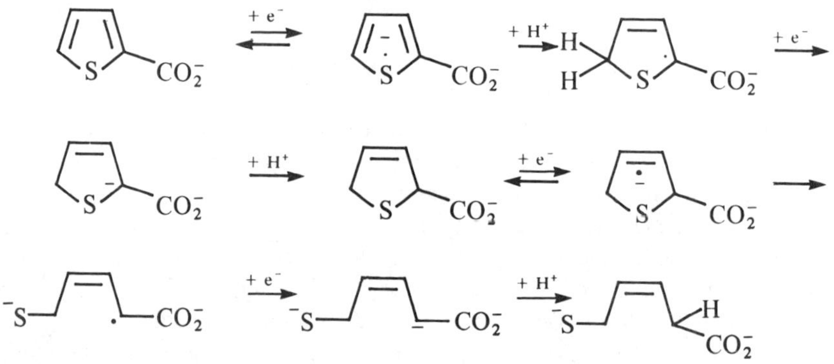

Scheme 36

In light of the mechanism described above, it is quite easy to understand the recently published results[172] on the reductive alkylation of 2-acylthiophenes (51). In this case, the action of sodium in liquid ammonia on 2-acylthiophenes as well as on 2-thiophenecarboxylic acid was carried out in the presence of a proton donor (ethanol) and was followed by treatment of the mixture with ammonium chloride and then with 2–4.5 eq. of an alkylating reagent (Scheme 37). According to the authors,[172] the first stage of the reaction is the formation of a cyclic dianion.

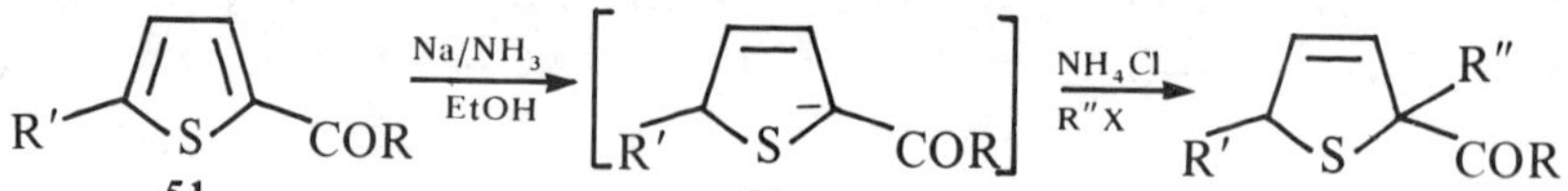

R = OH, alkyl, cycloalkyl; R' = H, alkyl; R'' = Me, Bu, CH_2Ph, $CH_2CH=CH_3$

Scheme 37

Protonation of the latter with ammonium chloride leads to the cyclic allyl anion (52). Owing to the effect of the ring sulfur and acyl substituent, the negative charge in this anion is localized in position 2, at which the attack of the alkylating reagent (alkyl halide) occurs. Oxidation of the resulting 2,5-dihydrothiophenes (53) into sulfones and pyrolysis of the latter made it possible to prepare a number of alkylalkadienylketones.[172]

Substituted allylmercaptans formed during the cleavage of the thiophene ring, which were considered above, can be reduced under the action of solvated electrons with the formation of hydrogen sulfide and unsaturated substances that do not contain sulfur.[160] Furthermore, allylmercaptans, or more precisely, the corresponding thiolate anions (54) formed during the interaction of thiophenes and alkali metals in liquid ammonia are rather labile and are capable of nonreductive transformations with retention or loss of the sulfur atom. These transformations, which take place during the treatment of reaction mixtures, are discussed below.

Reduction of thiophenes with alkali metals in liquid ammonia in the presence of an alcohol studied in detail by Gol'dfarb and Zakharov. They demonstrated in particular that 2-alkylthiophenes give α,β-unsaturated sulfides (55 and 56) by the action of lithium in liquid ammonia, followed by treatment with alkyl halide or ethylene oxide[173-175] (Scheme 38).

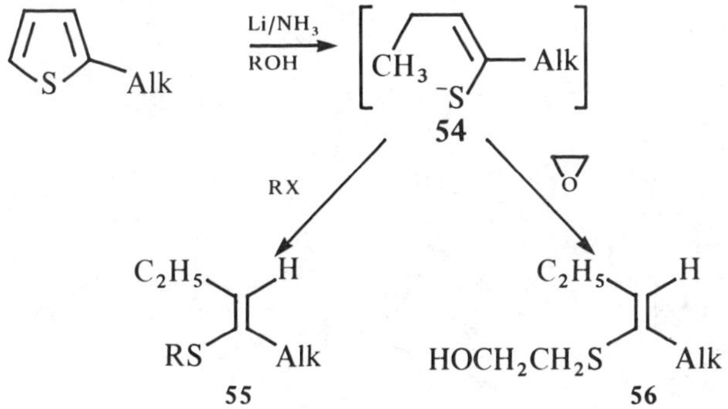

Scheme 38

When the mixture formed by the action of lithium in liquid ammonia on alkyl-thiophenes (57) underwent hydrolysis, dialkylketones (58) were obtained in 45–75% yields.[176] In the case of thiophene itself, a small amount of butyraldehyde, identified as its semicarbazone, was formed.[176] Under the same conditions, ω-(2-thienyl)alkanoic acids are converted into aliphatic ketoacids (59)[176,177] and ω-(2-thienyl)alkanols[178] and 2-(ω-dialkylaminoalkyl)thiophenes[179] are converted into ketoalcohols (60) and aminoketones (61) in 70–80% yields (Scheme 39).

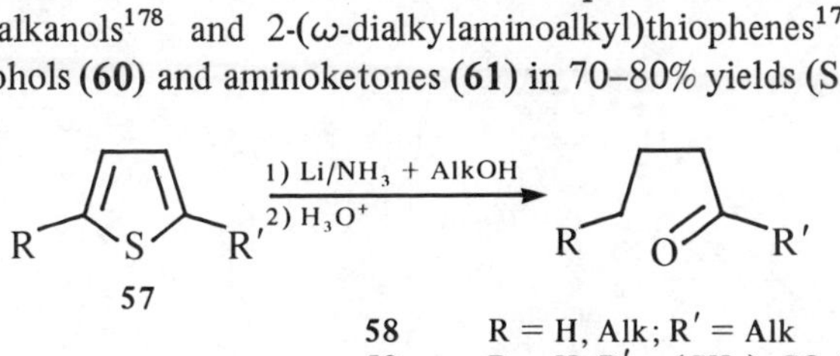

58	R = H, Alk; R' = Alk
59	R = H, R' = $(CH_2)_n CO_2H$, n = 3–5, 9
60	R = H, R' = $(CH_2)_n OH$, n = 2–6
61	R = H, R' = $(CH_2)_n NEt_2$, n = 2–6;
	R = H, R' = $CH_2 NMe_2$

Scheme 39

In some cases, under conditions of reductive cleavage, substituted thiophene-carboxylic acids and thienylcarbinols undergo hydrogenolysis of the C–S bond and products that do not contain sulfur are formed. Such transformations are described by Semenovskii and Emel'yanov[180,181] for the tri- and tetracyclic acids 62 and 63 when lithium was used in liquid ammonia at 25°C under pressure (Schemes 40 and 41).

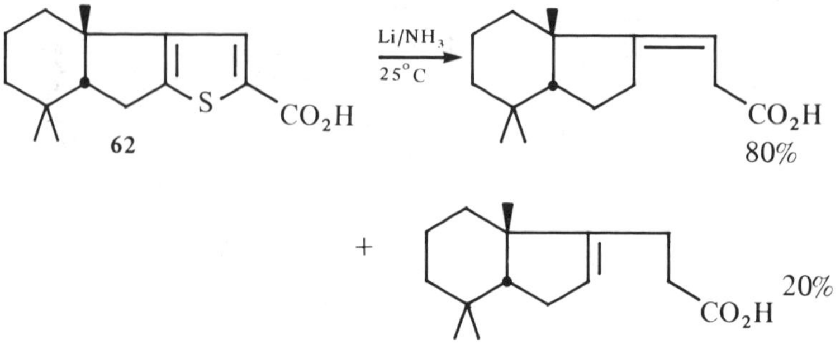

Scheme 40

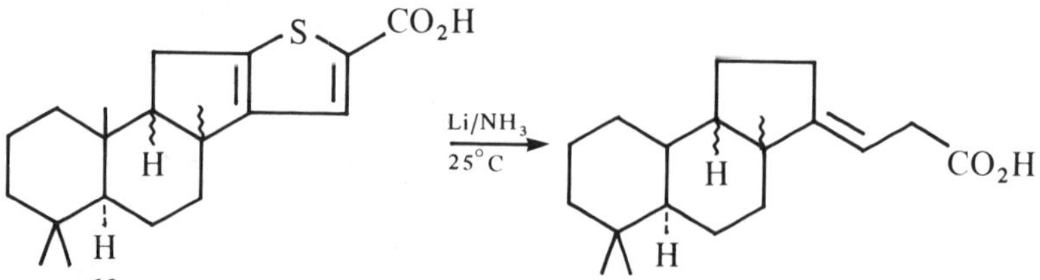

Scheme 41

In the case of reductive cleavage of 5-dialkylaminoalkyl-2-thiophenecarboxylic acids, Zakharov, Gol'dfarb, and Stoyanovich[182] observed regio- and stereoselective formation of E,E-alkadienecarboxylic acids (64) (Scheme 42).

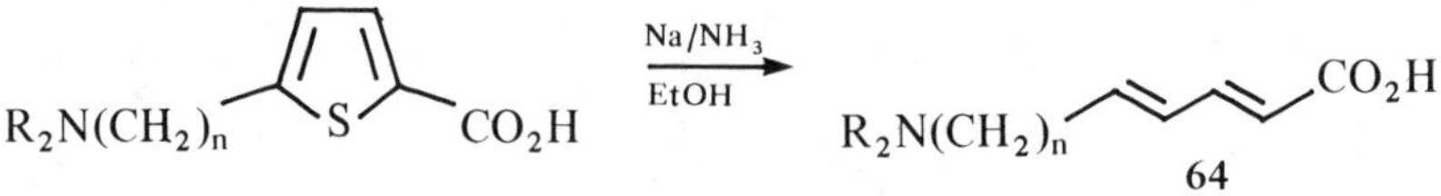

R = Me, n = 1
R = Et, n = 2

Scheme 42

For the stereoselective synthesis of E-homoallyl carbinols (65) from thienyl-(66) and dihydrothienylmethanols (67), Lozanova, Moiseenkov, and Semenov-skii[183,184] used successfully the action of lithium in ethylamine (Scheme 43).

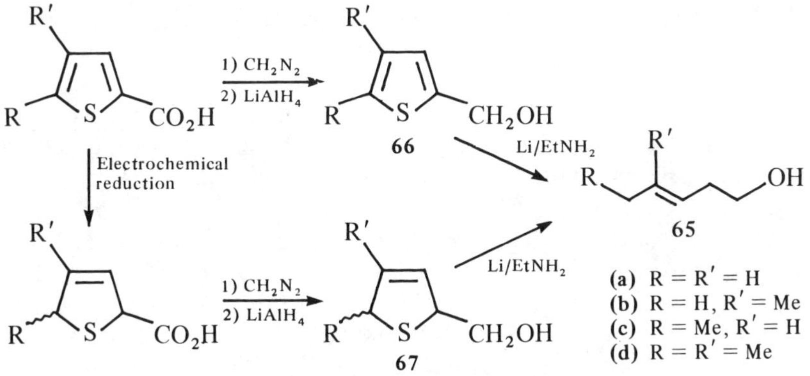

Scheme 43

The use of the same reagent allowed E,E-homofarnesol (**72**) to be prepared via the thiophene derivatives **68–71** (Scheme 44)[185] and homogeraniol and some of its homologues (**74**) to be obtained from the dithienylmethane derivative (**73**) (Scheme 45).[186]

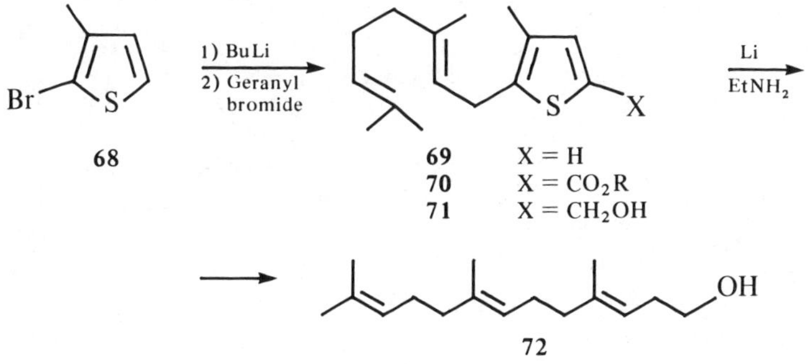

Scheme 44

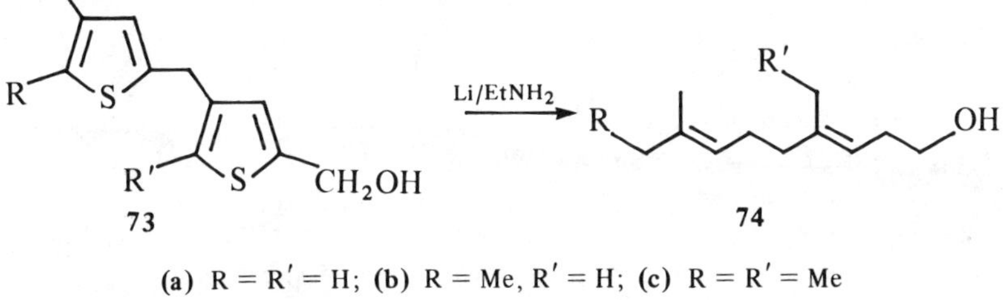

(a) R = R′ = H; (b) R = Me, R′ = H; (c) R = R′ = Me

Scheme 45

As shown by the data discussed above, the reductive cleavage of thiophenes with alkali metals in liquid ammonia leads to various products, including polyfunctional derivatives. Since this method has not been studied for very long, we can expect new and interesting results in this field in the near future.

To complete this section, some data should be mentioned concerning reductive cleavage of thiophenes in conditions related to the action of alkali metals in liquid ammonia. The reductive cleavage of some sulfur-containing compounds, including condensed thiophenes under the action of $Ca(NH_3)_6$, is described in Ref. 187. Under these conditions, benzo[b]thiophene is converted almost quantitatively (yield 96%) into o-ethylthiophenol and dibenzothiophene gives o-cyclohexenyl-thiophenol (yield 76.7%) with the admixture of phenylcyclohexane (19.8%). At the same time, 2,3-dimethylbenzo[b]thiophene and 1,3-dihydrobenzo[c]thiophene are transformed in high yields into products of desulfurization.[187]

It is well known that reductive cleavage of benzo[b]thiophene with sodium in alcohol leads to a mixture of o-ethylthiophenol (yield 34%) and 2,3-dihydrobenzo-[b]thiophene.[188] The reaction probably proceeds via the latter substance.[189,190] Under the same conditions, thiophene is converted into butylmercaptan (yield 20–25%)[161] and thieno[3,2-b]thiophene into 3-mercapto-2-ethylthiophene.[191] The action of sodium in dioxane on dibenzothiophene causes the cleavage of one or two C—S bonds, depending on the quantity of the metal, with the formation of 2-mercaptobiphenyl or biphenyl, respectively.[192]

Recently, some methods of desulfurization using metallic sodium were described;[193,194] however, they are of no preparative importance.

2.   Catalytic Hydrodesulfurization

Transformation of thiophene into aliphatic or arylaliphatic compounds takes place in the process of catalytic hydrodesulfurization, which plays a very important role in oil refining. Thiophene, and its homologues and annelated thiophenes are often used as model substances for the study of conditions and catalysts of hydro-desulfurization (see, for instance, Refs. 195–201). However, the preparative use of such transformations is apparently of no importance for a number of reasons. The high temperatures of hydrodesulfurization (from 150 to 600°C for different catalysts, mostly 300–400°C[112]) make it difficult to retain functional groups. Furthermore, under the conditions of this process, even thiophene and its homo-logues are converted into mixtures of hydrocarbons. This is caused by the complex mechanism of desulfurization and the possibility of simultaneous cracking.[112]

Hydrodesulfurization of thiophene and of 10 mono- and dialkylthiophenes over an Al—Co—Mo sulfide catalyst at 300–310°C was studied in detail.[202] Unsaturated hydrocarbons with different positions of double bonds, as well as substances formed as the result of dealkylation, were detected in the products of hydro-desulfurization, along with paraffins. In particular, C_{10}-hydrocarbons obtained by hydrogenolysis of 2-ethyl-5-butylthiophene at 410°C contain only about 12% n-decane and the main part of them was almost an equilibrium mixture of 1-, 2-, and 3-decenes. Besides, the formation of 2-butyl- (3.5%), 2-ethyl- (5.9%), and 2,5-diethylthiophene (1.1%) was noted under the conditions mentioned above.

Hydrodesulfurization of benzannelated thiophenes apparently proceeds more uniformly, leading to the corresponding alkylbenzenes or biphenyls (see, for instance, data concerning conversions of benzo[b]thiophene and dibenzothiophene over the same catalyst.[203,204]) At the same time, the desulfurization conditions for condensed and polyarylsubstituted thiophenes are more drastic than for thiophene and alkylthiophenes.[198,199] For example, hydrodesulfurization of dibenzothiophene and tetraphenylthiophene does not proceed over the Ni–W sulfide catalyst at 220°C and 200 atm H_2.[197]

The data concerning the conditions of the process and the nature of the catalysts used in hydrodesulfurization are summarized in a number of special publications (see Refs. 112, 205, and 206–208). It should be noted that the preparative possibilities of desulfurization under conditions of heterogeneous catalysis are not sufficiently known at present. For instance, this can be seen from the results in Ref. 209, in which reductive desulfurization over 0.5% palladium on alumina was followed by determination of products, with the aid of g.l.c., and was used for the analysis of sulfur-containing organic substances. It was demonstrated that at 150–250°C the process proceeded without by-products and that the formation of cracking and isomerization products only occurred at about 400°C. When desulfurization is carried out under optimal conditions, it gives saturated hydrocarbons that retain the skeleton of the starting compound. In the case of benzo[b]thiophene and its homologues, the benzene ring undergoes hydrogenation and the corresponding alkylcyclohexanes are obtained.[209]

It was demonstrated[210] that in the presence of Ni–W sulfide catalyst under hydrogen pressure of 1500 psi at 300°C, ketones (75) undergo reductive desulfurization with simultaneous reduction of the keto group to CH_2, leading to the corresponding hydrocarbons (76), which are often obtained in high yields (Scheme 46).

$$RCO\text{—}\underset{\textbf{75}}{\underset{S}{\diagdown}}\longrightarrow\underset{\textbf{76}}{R(CH_2)_4CH_3}$$

R = Me, Et, Pr, $n\text{-}C_{17}H_{35}$, Ph

Scheme 46

Thus, with the proper choice of conditions, the catalytic desulfurization may be used for the transformation of thiophenes into hydrocarbons. However, it is not suitable for preserving functional compounds. The latter is achieved by the use of reductive desulfurization with Raney nickel, which is considered in the following section.

3.   Reductive Desulfurization of Thiophenes with Raney Nickel and Related Reagents

Reductive desulfurization under the action of Raney nickel plays an exceptionally important role as a synthetic route, making it possible to obtain compounds of various classes from thiophene derivatives, and also as a method of structure elucidation of the latter. Bougault, Cattelain, and Chabrier were the first to observe desulfurization of various organic compounds, including thiophenes under the action of Raney nickel. They reported their results on May 28, 1938 in the French Chemical Society. The summary of this report was published in 1939[211] and a detailed paper appeared the next year.[212] At the same time, they proposed the action of Raney nickel as a method of refining benzene and toluene to free them from thiophene and methylthiophene.[213] Soon after, the use of this reaction by duVigneaud and coworkers[214] for biotin structure elucidation attracted general attention to desulfurization with Raney nickel.

Beginning with the publication of Blicke's and Sheets' results[215] in 1948, desulfurization of thiophene derivatives became widely used for the synthesis of various compounds of other classes. The synthetic importance of the reaction under discussion is determined by some distinctive features. Owing to the aromatic nature of the thiophene, various substituents can be introduced into its ring and these substituents are retained or changed in the process of reductive desulfurization. Sulfur elimination is followed as a rule by complete saturation of the four-carbon fragment of the thiophene ring. In the process of reductive desulfurization, elimination of sulfur from sulfur-containing substituents also takes place and as a result complete loss of substituents such as alkylthio-groups may occur. From substituents that do not contain sulfur and are transformed under the conditions of reductive desulfurization, one should note the carbonyl group that is reduced to the hydroxy group. Transformations of other substituents are observed more rarely, and they are mentioned in the discussion of appropriate examples. Conversion of thiophenes (77) in the process of reductive desulfurization, which may be demonstrated by Scheme 47, permits various substances with straight or branched chains

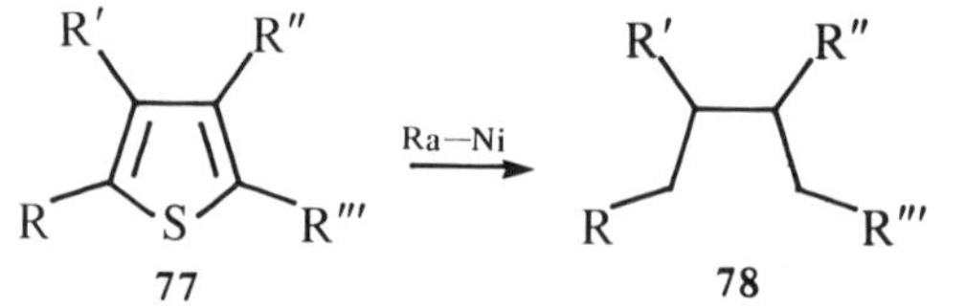

Scheme 47

to be obtained (78), in particular carbo- and heterocyclic compounds (if two atoms of the thiophene ring are linked by a chain) and the lengthening of aliphatic chains by four methylene units.

Some aspects of the mechanism of reductive desulfurization were discussed in Refs. 100b and 216–220. Within the limits of this section, it is essential to note that the mechanism of this process probably includes the cleavage of C–S bonds with free radical formation (sulfur is bound in the form of nickel sulfide) and subsequent saturation of free valences and double bonds with hydrogen adsorbed by Raney nickel. When the reaction is carried out in the presence of specially degassed or "aged" Raney nickel, with an insufficient amount of this catalyst, or at temperatures promoting the removal of adsorbed hydrogen, the products of radical recombination are detected. However, under ordinary conditions (at 20–80°C in ethanol or methanol with 5–10 times more Raney nickel in weight than thiophene compound without any additional supply of hydrogen), desulfurization proceeds as a rule without difficulties and provides retention of the carbon skeleton. Complications may be associated with incomplete hydrogenation of double bonds and sometimes with nonselective transformations of substituents. Modifications of Raney nickel W-2,[221] W-6, and W-7[222] are used more often, other modifications are described in Refs. 222 and 223.

There are some cases when desulfurization does not take place, and this is usually explained by steric shielding of the sulfur atom (see, for instance, Refs. 224 and 225). These data should, however, be regarded with certain scepticism. The steric hindrances undoubtedly retard the reaction; however, the failure of the attempts to desulfurize some thiophene compounds is caused either by insufficient activity of Raney nickel or by nonoptimal reaction conditions. Examples illustrating this situation are given below. On the other hand, there are real limitations to Raney nickel desulfurization. For example, the method[226] proposed for quantitative determination of sulfur using reductive desulfurization of various compounds with Raney nickel seems not to be a universal one, since during its elaboration, only thiophene and benzo[b]thiophene were studied.

At present, a number of reviews and monographs consider the Raney nickel desulfurization of organic sulfur compounds including thiophenes.[100b-d, 216-220, 227] One of the most well-known reviews of the subject[219] notes that by 1960 desulfurization of about 190 thiophene compounds was described. Since then, the number of such examples has at least doubled. The available material is considered below, grouped according to related compounds formed as a result of reductive desulfurization. To avoid repetition, the substances of different classes can be given as one group, if for the synthesis of starting compounds, similar methods are used and the reaction proceeds on the same lines. Along with substituted thiophenes, the corresponding condensed systems are also considered.

A.   Synthesis of Aliphatic and Aromatic Hydrocarbons

It is advisable to consider the preparation of hydrocarbons and also of other substances of the aliphatic and aromatic series from thiophene derivatives together, since in these cases, the methods of synthesis of the starting thiophenes do not differ very much and benzene rings as a rule are not hydrogenated in reductive desulfurization conditions.

Reductive desulfurization of thiophene and 2-methylthiophene was described in the papers of Bougault et al.[212, 213] Conversion of these compounds into butane and pentane has been suggested as a method of purifying benzene and toluene. Blicke and Sheets were the first to use this reaction for the synthesis of hydrocarbons. They obtained ethylbenzene from benzo[b]thiophene or 3-hydroxybenzo[b]thiophene in high yields (75–86%) and also biphenyl from dibenzothiophene.[228] Later, other authors studied reductive desulfurization of various condensed systems possessing thiophene rings. Transformation of dibenzothiophene into biphenyl is described particularly in Refs. 229 and 230. Elimination of selenium atom from dibenzoselenophene, dibenzoselenophene-5-oxide and -5,5-dichloride proceeds in the same way.[229] Biphenyl is also formed by reductive desulfurization of 3-bromo-dibenzothiophene.[192] The action of Raney nickel does not cause the hydrogenolysis of the C—F bond; reductive desulfurization of octafluorodibenzothiophene leads to octafluorobiphenyl in high yield (74%).[231]

Carruthers[225] reported an unsuccessful attempt to desulfurize 4,6-dimethyl-dibenzothiophene, but later he and Douglas[232] described desulfurization of the not less hindered 2,4,6,8-tetramethyldibenzothiophene and also of 2,3,6,7- as well as 2,3,6,8-tetramethyldibenzothiophene and 3-ethyl-6,8-diethylnaphtho[1,2-b]thiophene, which occurred in distillates of Kuwait oil. Kruber and Raeithel, while studying coal-tar, carried out desulfurization of naphtho[1,2-b]thiophene that they had isolated from anthracene oil.[233] Carruthers prepared this substance independently and also proved its structure by hydrogenolysis into 2-ethylnaphthalene.[234]

Banfield et al.[235] failed to carry out desulfurization of naphtho[2,1-b]thiophene with Raney nickel in boiling ethanol, but were successful when they carried out this reaction in boiling ethyleneglycol. Desulfurization of some synthesized substituted benzo[b]thiophenes, benzo[c]thiophenes, naphtho[1,2-b]thiophenes, naphtho-[2,1-b]thiophenes,[236-239] and also of 4,5,6,7-tetrahydro-1,4,4-trimethylbenzo[b]-thiophene isolated from kerosene fractions of Middle East oil[240] was undertaken to prove the structures of these compounds. Kao, Tilak, and Venkataraman[241] obtained 1- and 2-phenylnaphthalenes as the result of hydrogenolysis of isomeric benzothiafluorenes.

Wynberg et al.[242] described reductive desulfurization of 4,5-dihydro-3-phenyl-thiophene — the intermediate in the synthesis of 3-phenylthiophene. To prove the structures of dimers formed from benzo[b]thiophene-1,1-oxide, their reductive desulfurization was carried out.[243-246] In this connection, it should be noted that the photodimer undergoes dissociation under the conditions of the reaction (boiling with Raney nickel in the mixture of ethanol with acetone), resulting in the formation of ethylbenzene, which is caused by desulfurization of benzo[b]thiophene-1,1-dioxide.[244]

The synthesis of hydrocarbons that are hard to prepare was not the object of the experiments mentioned above. In most cases, the authors did not even mention the yields of the products. The paper of Gol'dfarb and Kondakova published in 1952,[247] where the reductive desulfurization of 2,5-dibenzhydrylthiophene (79) leading to 1,1,6,6-tetraphenylhexane (80, R = Ph, R' = H) was described, may be regarded as the first publication devoted to the synthesis of hydrocarbons from synthetically prepared substituted thiophenes. Sometime later, Gol'dfarb and Korsakova[248] prepared 2,7-dimethyl-2,7-diphenyloctane (80, R = R' = Me) and 2,2,7,7-tetraphenyloctane (80, R = Ph, R' = Me) (Scheme 48) in the same way;

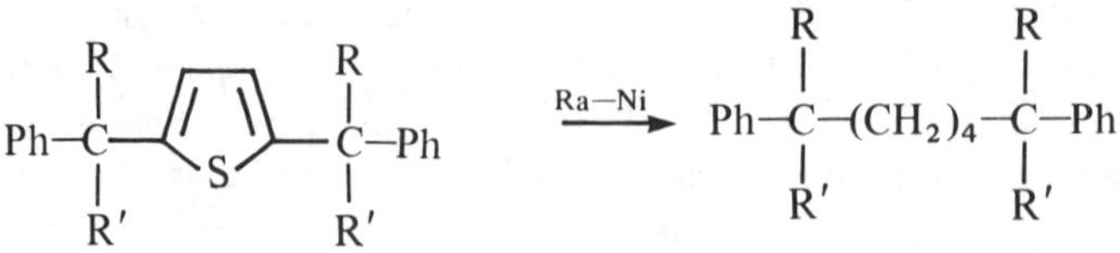

Scheme 48

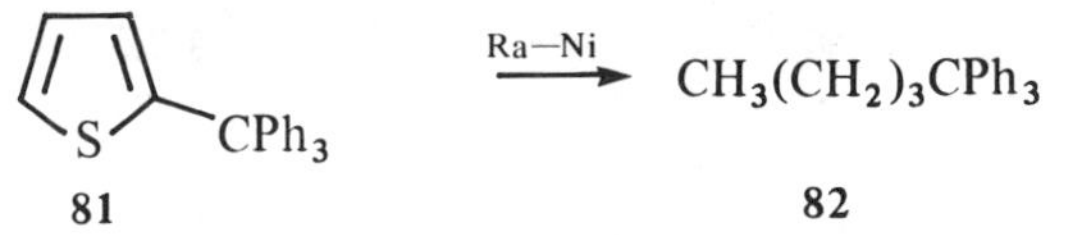

Scheme 49

2-tritylthiophene (81) was transformed into 1,1,1-triphenylpentane (82) (Scheme 49). The yields of hydrocarbons are 70–80% and the starting compounds are easily obtained by alkylation of thiophene with the corresponding carbinols.

A curious transformation useful for the synthesis of substituted cyclobutanes was described by Baker and coworkers.[249] They reported that reductive desulfurization of dibenzo[b,f]thieno[3,2-b]thiophene (83) with Raney nickel in boiling methanol gave dibenzyl, but in boiling ethanol, 1,2,3,4-tetraphenylcyclobutane was formed (84). However, Badger et al.[250] failed to get the same results and obtained dibenzyl as the major or single product in all cases. When less active Raney nickel was used (particularly not fresh, but kept for 3–18 months), dibenzyl was accompanied by stilbene and by the product of partial desulfurization — 2-phenyl-benzo[b]thiophene (85) (Scheme 50).

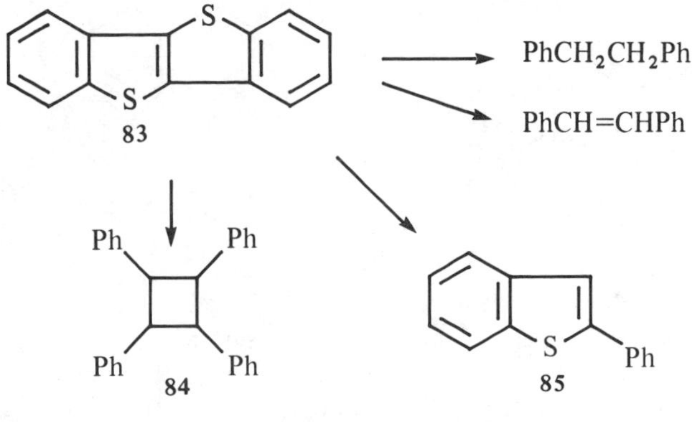

Scheme 50

By the action of Raney nickel on 2-benzylthiophene, n-amylbenzene is formed.[210] From 3-*tert*-butylbenzo[b]thiophene, 2,2-dimethyl-3-phenylbutane is obtained,[251] which contains the product of incomplete hydrogenation of the side chain as an admixture (5–10%). To obtain the individual compound, an additional hydrogenation of the desulfurization product was carried out. Dann and Hauck[252] prepared 2,3-diphenylbutane from 3,4-diphenylthiophene (Scheme 51). This reaction gives predominantly the meso-form (yield 55%), which is also more easily isolated.

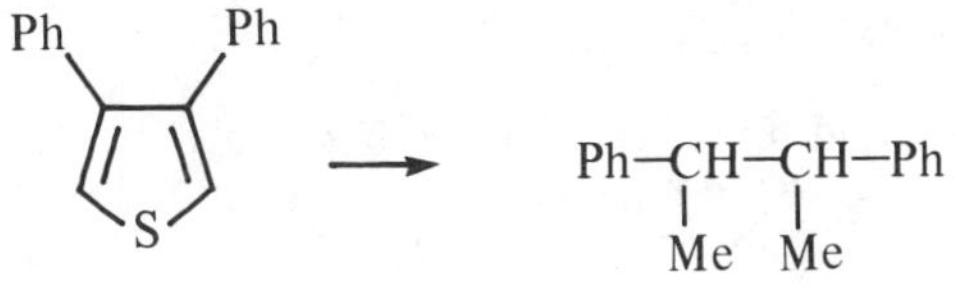

Scheme 51

Badger et al.[253] transformed tetraphenylthiophene into 1,2,3,4-tetraphenylbutane (Scheme 52) and flavophene (**86**) into dibenzoperylene (**87**) (Scheme 53). In the two latter cases, the reductive desulfurization was carried out at 100°C in xylene or mesitylene with yields of about 30%.

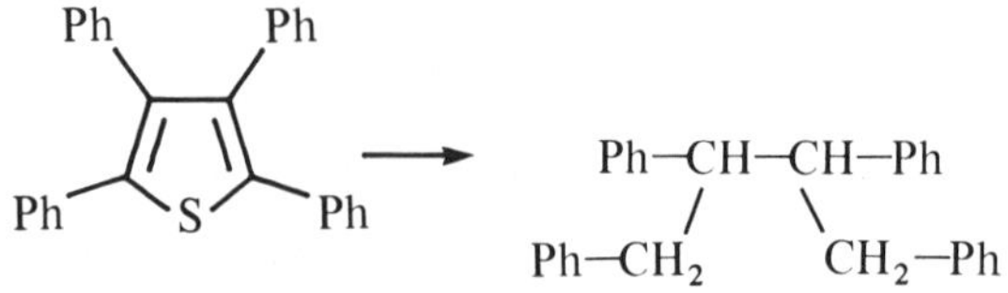

Scheme 52

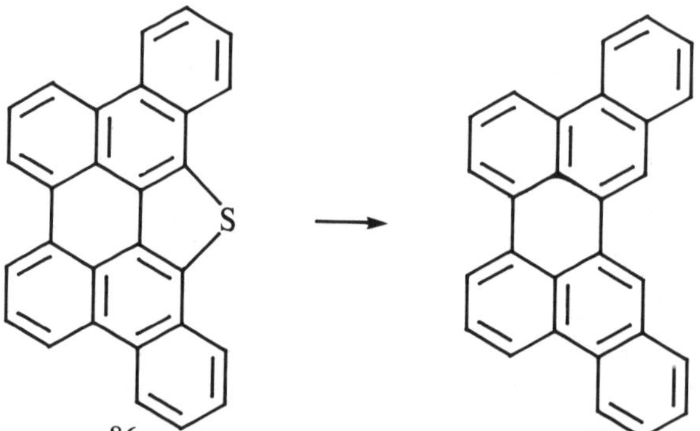

Scheme 53

As mentioned above, benzene rings are not affected by the reductive desulfurization. However, in the case of some polycyclic systems, a partial hydrogenation of benzene rings occurs. Davies and Porter[254] showed that under the action of Raney nickel on naphtho[2,1-a]dibenzothiophene (**88**), 1-phenyl-1,2,3,4,5,6,7,8-octahydrophenanthrene (**89**) was formed (Scheme 54).

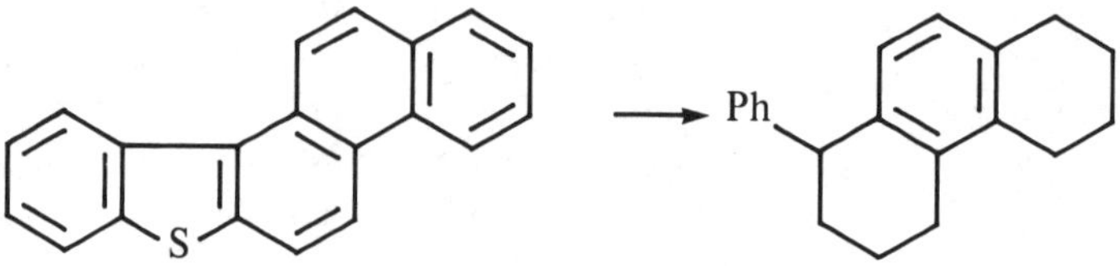

Scheme 54

By the reductive desulfurization of the dibenzofulvene derivative (**90**, X = S), obtained from fluorene and thiophene-2,5-dicarboxaldehyde, Pastour and Barrat also observed hydrogenation of benzene rings; the reaction of the selenophene analogue (**90**, X = Se) proceeds in the same way (Scheme 55).[255]

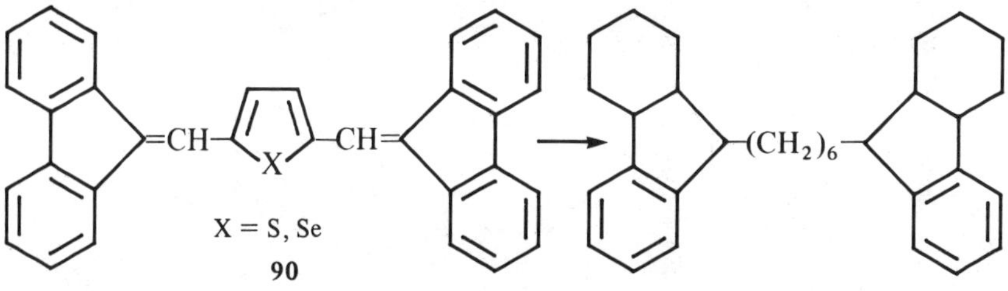

Scheme 55

The specific route to the alkylaromatic hydrocarbons consists in reductive desulfurization of arylthienylcarbinols of type **91**. According to Badger et al.,[256] this transformation includes hydrogenolysis of the C—O bond in the starting carbinol with the formation of arylthienylalkane (**92**), which is then converted into the hydrocarbon **93**. In most cases, the formation of alkylarylcarbinol and its dehydration product is not observed (Scheme 56). 9-(2-Thienyl)fluorenol-9 (**94**) behaves similarly (Scheme 57). However, in the case of diphenyl(2-thienyl)carbinol (**95**) and phenyldi(2-thienyl)carbinol (**96**), the corresponding alkylarylcarbinol (**97** or **98**) and arylalkene (**99** or **100**) were obtained (Schemes 58 and 59).

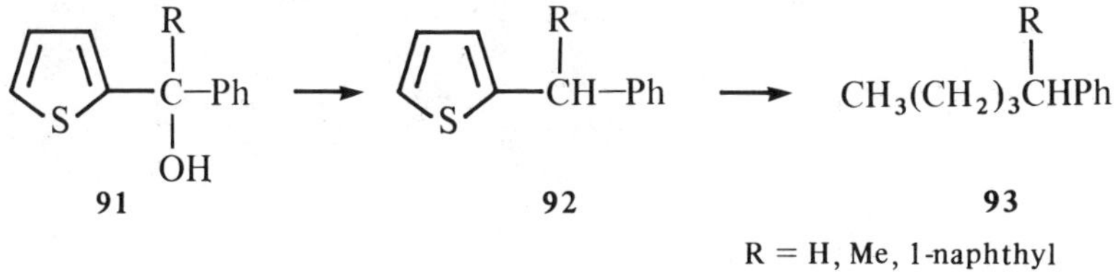

R = H, Me, 1-naphthyl

Scheme 56

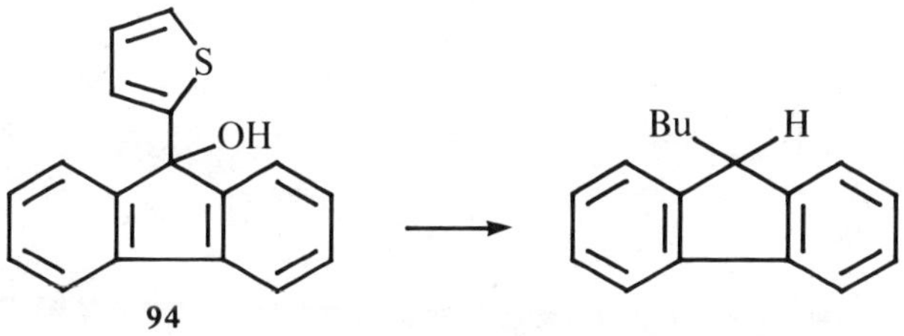

Scheme 57

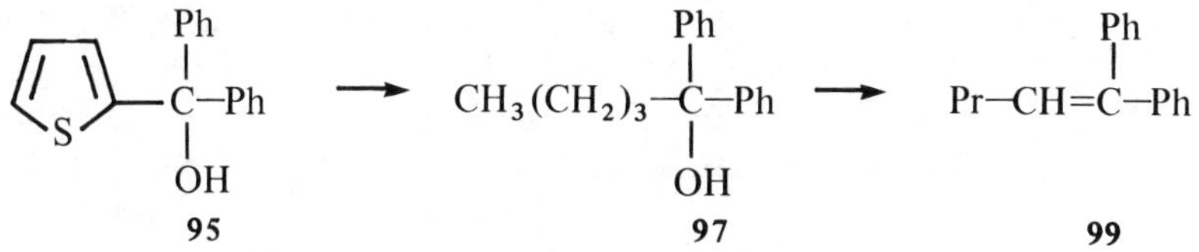

Scheme 58

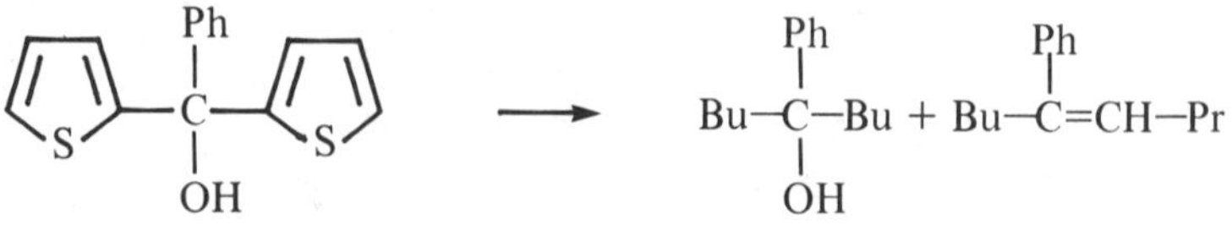

Scheme 59

This behavior of the carbinols is in accordance with the data on the transformation of acetophenone and methylphenylcarbinol into ethylbenzene under the action of Raney nickel[228] and also of methylpentylphenylcarbinol and dibutylphenylcarbinol into 2-phenylheptane and 5-phenylnonane, respectively.[256]

The possibility of synthesizing higher aliphatic hydrocarbons from thiophene derivatives was proved in 1956 by Gol'dfarb and Danyushevskii,[7] who described the transformation of di-(2-thienyl)thiophene (101) into n-tetradecane (Scheme 60), and also by Wynberg and Logothetis,[257] who obtained n-eicosane from 2-cetylthiophene (102) (Scheme 61) and 4-methyloctane by dehydration and reductive desulfurization of methylpropyl(2-thienyl)carbinol (103) (Scheme 62).

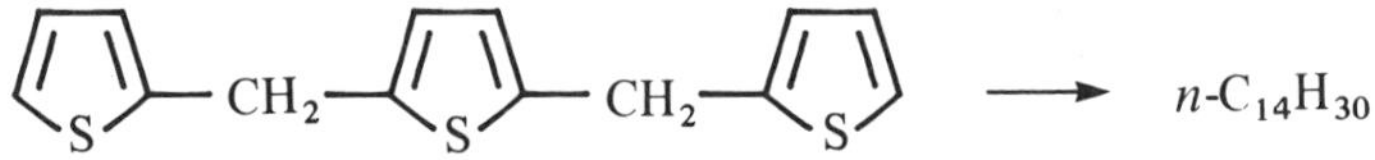

Scheme 60

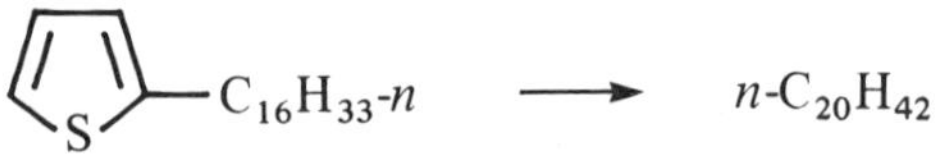

Scheme 61

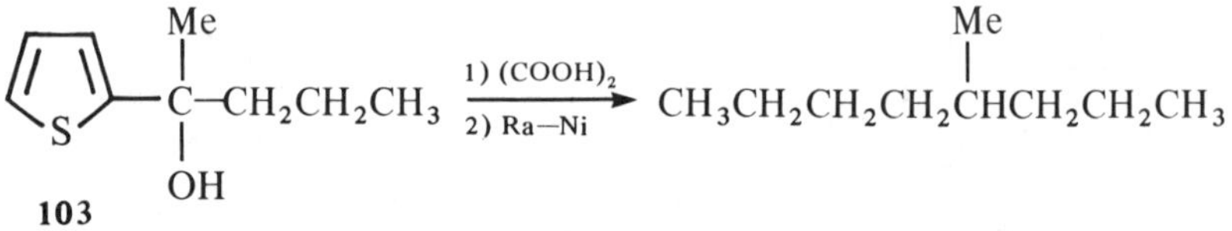

Scheme 62

Reductive desulfurization helped to solve an important theoretical problem about the optical rotation value of compounds having four different alkyls at the asymmetric carbon atom. Wynberg et al.[258] transformed optically active 3-(5-ethyl-2-thienyl-3-(5-methyl-2-thienyl)hexane (104, R = R''' = Et, R' = Pr, R'' = Me), into ethylpropylpentylhexylmethane (105) which proved to be optically inactive within the limits of the precision of the method. Methylpropylbutylhexylmethane obtained in the same way by Hulshof and Wynberg[259] had very low optical rotation ($[M]_{578} + 0.12°$). Sy et al.[260, 261] described the synthesis of several tetraalkylmethanes of type 105 by reductive desulfurization of di(2-thienyl)alkanes

(104) obtained by the condensation of ketones with thiophene. It should be noted that simultaneous hydrogenolysis of the C—Cl bond may occur: 1-chloro-2,2-di(2-thienyl)propane (104, R = Me, R' = CH$_2$Cl, R" = R"' = H) is converted into 5,5-dimethylnonane (Scheme 63).

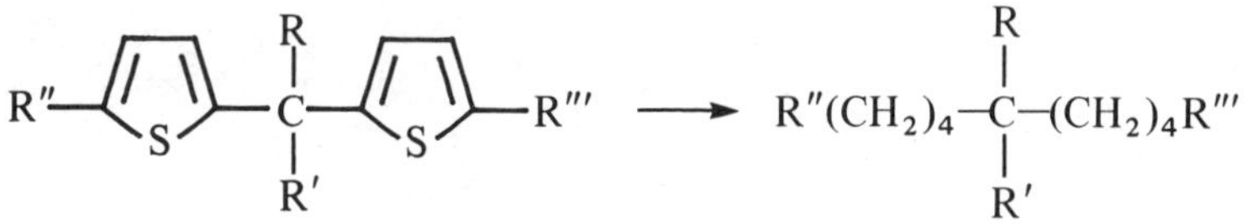

Scheme 63

B. Synthesis of Carboxylic Acids

Reductive desulfurization of acids of the thiophene series is a most widely developed method for the synthesis of aliphatic, arylaliphatic, and aromatic compounds. This is due to the importance of acids and undoubtedly to their easy identification, as well as to the stability of the carboxy group under the reaction conditions and the great possibility of varying these conditions. This process can be carried out not only in organic solvents, but also in water, reductive desulfurization being combined in one stage with the action of alkali on the Ni—Al alloy.

The first synthesis of carboxylic acids using reductive desulfurization were described by Blicke and Sheets.[215,228] These authors studied the action of Raney nickel on some acids of the benzo[b]thiophene and thiophene series and obtained the corresponding arylaliphatic and aliphatic carboxylic acids in yields varying from 75 to 98%.

Results obtained by Papa, Schwenk, and Ginsberg[262] were very important for the development of the synthesis of carboxylic acids from thiophenes. They demonstrated that desulfurization of acids may be carried out not only with Raney nickel, but also with an Ni—Al alloy in aqueous alkali. Later, this very convenient modification of the reaction was often used by other investigators and was termed the Papa–Schwenk method. Among the acids whose desulfurization was studied in Ref. 262 are 2-thiophenecarboxylic acid, its 4- and 5-methyl-substituted homologues, and 4-(2-thienyl)butyric and 3-(2-thienyl)acrylic acids (Schemes 64–66).

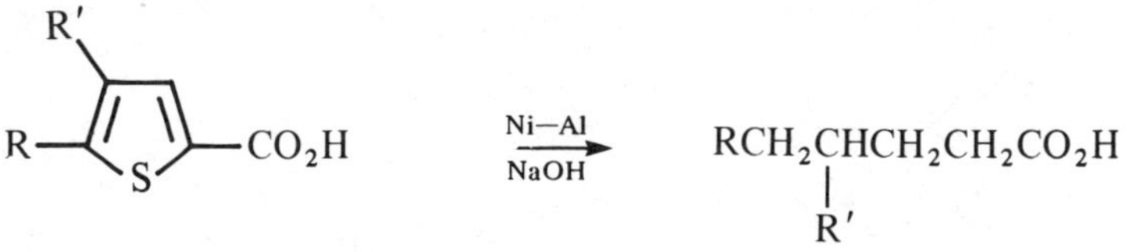

(a) R = R' = H; (b) R = Me, R' = H; (c) R = H, R' = Me

Scheme 64

$$\text{(thiophene)}-(CH_2)_3CO_2H \longrightarrow CH_3(CH_2)_6CO_2H$$

Scheme 65

$$\text{(thiophene)}-CH=CHCO_2H \longrightarrow CH_3(CH_2)_5CO_2H$$

Scheme 66

Using reductive desulfurization, Modest and Szmuszkovicz[263] obtained alkyl-substituted derivatives of some aromatic anhydrides from the corresponding condensed systems possessing thiophene rings. The synthesis of the starting compounds mentioned above was developed by the same authors.[264]

The synthetic potentialities of reductive desulfurization were very quickly appreciated by many authors. In 1954, a paper by Hansen[265] appeared which, using acetic acid (106, R = Me) as an example, demonstrated that it was possible to obtain, by the use of thiophene, aliphatic acids (107) with five carbon atoms more than the starting acid (Scheme 67).

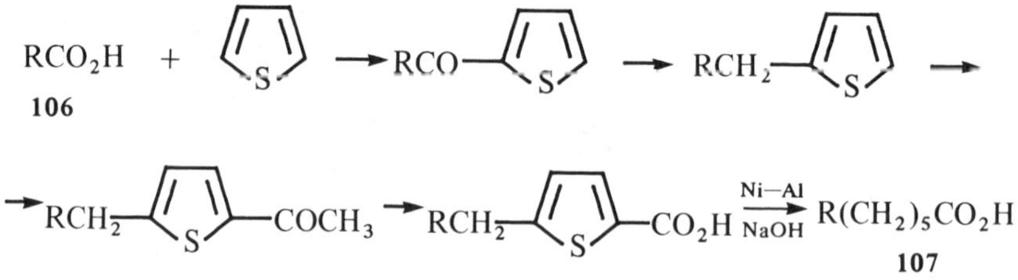

$$RCO_2H + \text{(thiophene)} \longrightarrow RCO-\text{(thiophene)} \longrightarrow RCH_2-\text{(thiophene)} \longrightarrow$$
106

$$\longrightarrow RCH_2-\text{(thiophene)}-COCH_3 \longrightarrow RCH_2-\text{(thiophene)}-CO_2H \xrightarrow[\text{NaOH}]{\text{Ni-Al}} R(CH_2)_5CO_2H$$
107

Scheme 67

Practically at the same time, many papers by Badger, Buu-Hoi, Gol'dfarb, Wynberg, and their groups appeared in which the possibilities mentioned were widely used and applied to new classes of compounds. Badger, Rodda, and Sasse[266] proposed a scheme based on the use of 2,5-dimethylthiophene and dicarboxylic acids (Scheme 68). These authors also demonstrated that long-chain carboxylic acids

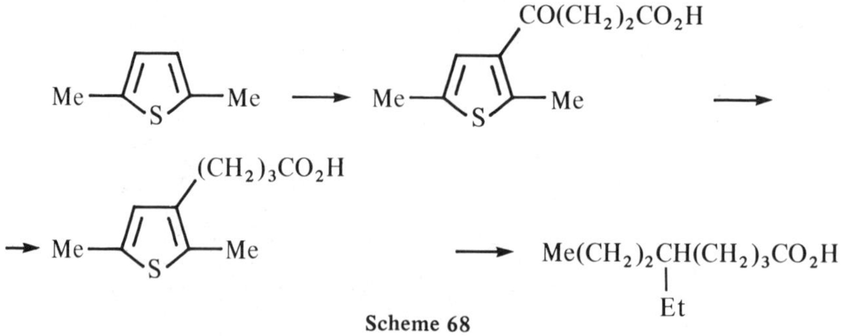

$$Me-\text{(thiophene)}-Me \longrightarrow Me-\text{(thiophene with }CO(CH_2)_2CO_2H)-Me \longrightarrow$$

$$\longrightarrow Me-\text{(thiophene with }(CH_2)_3CO_2H)-Me \longrightarrow Me(CH_2)_2\underset{\underset{Et}{|}}{CH}(CH_2)_3CO_2H$$

Scheme 68

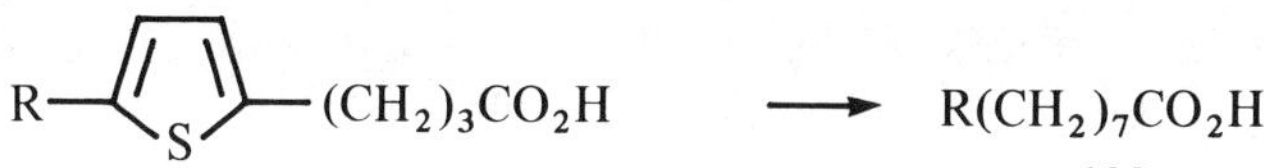

$$R-\underset{S}{\boxed{}}-(CH_2)_3CO_2H \longrightarrow R(CH_2)_7CO_2H$$

108

R = H, Et, n-heptyl, 5-ethyloctyl

Scheme 69

(**108**) can be obtained from thiophene and its homologues (Scheme 69) and dicarboxylic acids (**109**), can be obtained from 2,2-di(2-thienyl)propane (Scheme 70).[267] The yields at the stage of desulfurization vary from 51 to 93%.

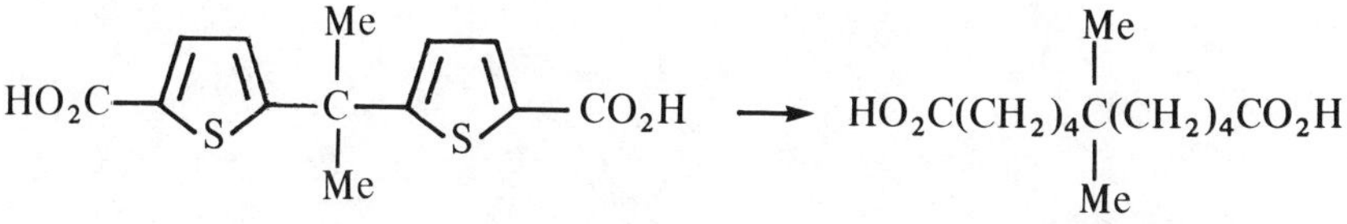

$$HO_2C-\underset{S}{\boxed{}}-\underset{\overset{Me}{|}}{\underset{\underset{Me}{|}}{C}}-\underset{S}{\boxed{}}-CO_2H \longrightarrow HO_2C(CH_2)_4\underset{\overset{Me}{|}}{\underset{\underset{Me}{|}}{C}}(CH_2)_4CO_2H$$

Scheme 70

109

Sy, Buu-Hoi, and Xuong[268] described the total synthesis of d,l-tuberculostearic (**110**), d,l-11-methyllauric (**111**), and arachidonic (**112**) acids, which is illustrated in Scheme 71 for the first acid. The acids **111** and **112** are obtained in the same way, and for them only the stages of reductive desulfurization are shown (Schemes 72 and 73).

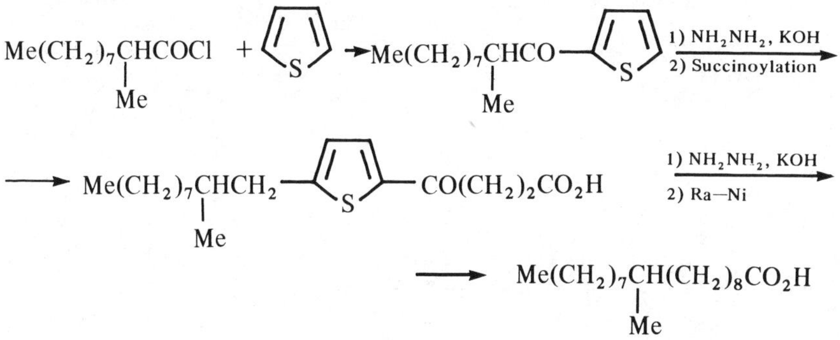

$$Me(CH_2)_7\underset{\underset{Me}{|}}{CHCOCl} + \underset{S}{\boxed{}} \rightarrow Me(CH_2)_7\underset{\underset{Me}{|}}{CHCO}-\underset{S}{\boxed{}} \xrightarrow[\text{2) Succinoylation}]{\text{1) NH}_2\text{NH}_2\text{, KOH}}$$

$$\longrightarrow Me(CH_2)_7\underset{\underset{Me}{|}}{CHCH_2}-\underset{S}{\boxed{}}-CO(CH_2)_2CO_2H \xrightarrow[\text{2) Ra—Ni}]{\text{1) NH}_2\text{NH}_2\text{, KOH}}$$

$$\longrightarrow Me(CH_2)_7\underset{\underset{Me}{|}}{CH}(CH_2)_8CO_2H$$

110

Scheme 71

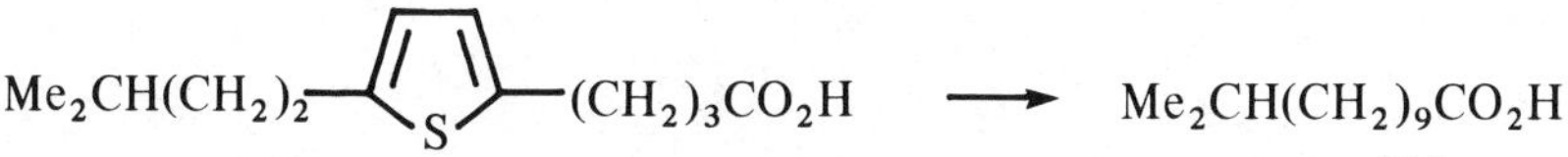

$$Me_2CH(CH_2)_2-\underset{S}{\boxed{}}-(CH_2)_3CO_2H \longrightarrow Me_2CH(CH_2)_9CO_2H$$

111

Scheme 72

$$Me(CH_2)_{11}-\underset{S}{\boxed{}}-(CH_2)_3CO_2H \longrightarrow Me(CH_2)_{18}CO_2H$$

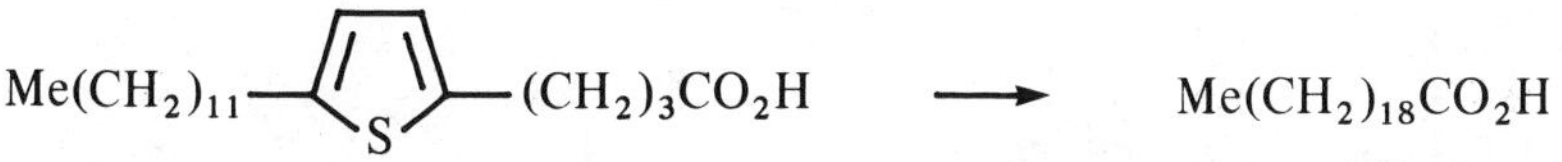

112

Scheme 73

The same authors also carried out the synthesis of some branched monocarboxylic acids (113 and 114)[6,269] (Schemes 74 and 75). Spaeth and Germain[270] also obtained some branched acids when proving the structure of acetylation products of 3-isopropylthiophene.

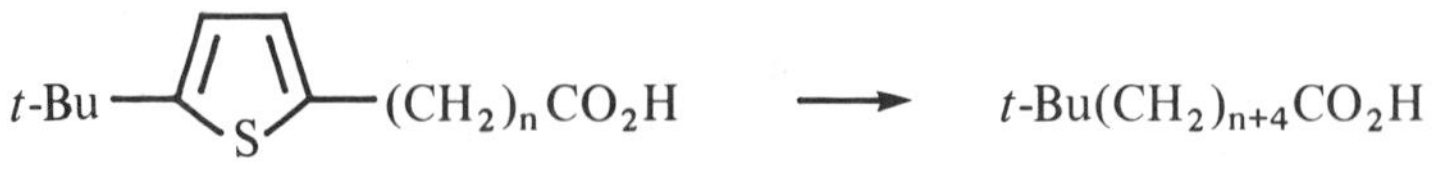

$$t\text{-Bu}\underset{S}{\diagdown}\!\!\!-\!(CH_2)_n CO_2H \quad\longrightarrow\quad t\text{-Bu}(CH_2)_{n+4}CO_2H$$

$$\mathbf{113} \quad n = 0, 3$$

Scheme 74

$$R\underset{S}{\diagdown}\!\!\!-\!R \overset{CO_2H}{} \quad\longrightarrow\quad R(CH_2)_2\underset{CH_2R}{\overset{|}{C}}HCO_2H$$

$$\mathbf{114} \quad R = Me, Pr$$

Scheme 75

When studying the synthesis of various long-chain compounds, Wynberg and Logothetis[257] obtained n-tricosanoic and n-pentadecanoic acids by the reductive desulfurization of corresponding 5-alkyl-2-thiophenecarboxylic acids (115) (Scheme 76). Starting from the ketocarboxylic acids (116) of the thiophene series, McGhie et al.[271] have synthesized several branched-chain acids and one straight-chain acid.

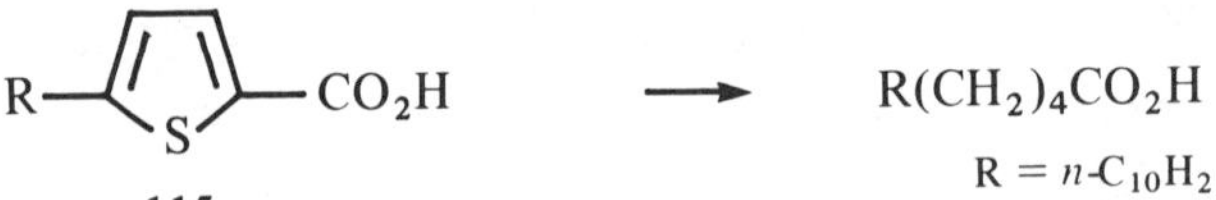

$$R\underset{S}{\diagdown}\!\!\!-\!CO_2H \quad\longrightarrow\quad R(CH_2)_4CO_2H$$

$$\mathbf{115} \qquad R = n\text{-}C_{10}H_{21}, n\text{-}C_{18}H_{37}$$

Scheme 76

But the unusual sequence of the reactions used by the authors (Scheme 77) is probably inconvenient for carboxylic acid preparation, since it includes two additional steps: reduction of the reaction mixture after hydrogenolysis and oxidation of an aliphatic hydroxy acid before the Kishner reduction.

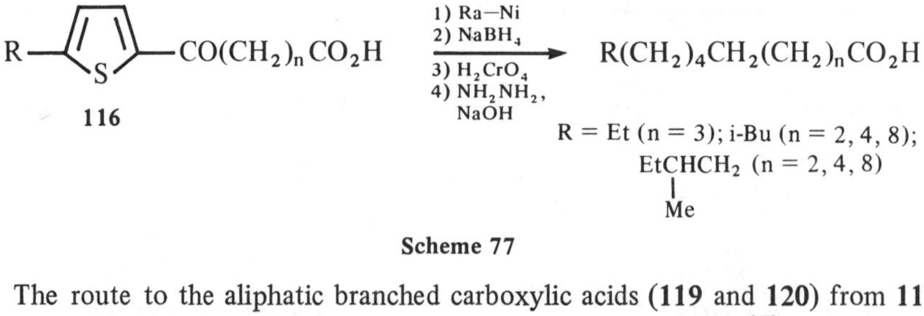

$$R\underset{S}{\diagdown}\!\!\!-\!CO(CH_2)_n CO_2H \xrightarrow[\substack{3)\ H_2CrO_4 \\ 4)\ NH_2NH_2, \\ NaOH}]{\substack{1)\ Ra\text{—}Ni \\ 2)\ NaBH_4}} R(CH_2)_4CH_2(CH_2)_n CO_2H$$

$$\mathbf{116}$$

$$R = Et\ (n = 3);\ i\text{-Bu}\ (n = 2, 4, 8);$$
$$EtCHCH_2\ (n = 2, 4, 8)$$
$$\underset{Me}{\overset{|}{}}$$

Scheme 77

The route to the aliphatic branched carboxylic acids (119 and 120) from 117, 118 and 3,3-dimethylglutaric anhydride, used by Buu-Hoi et al.,[272] is doubtlessly more rational (Schemes 78 and 79).

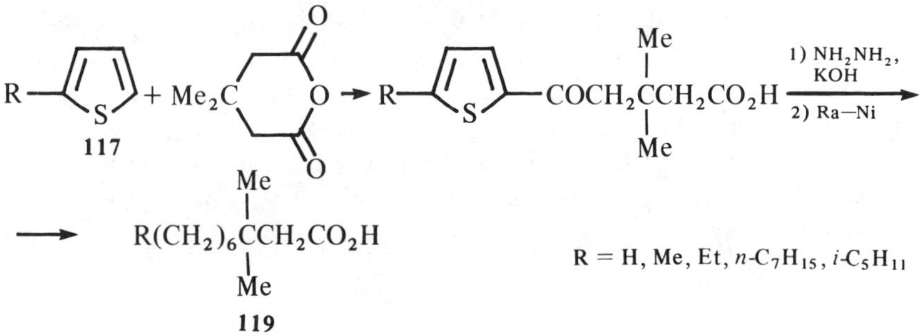

Scheme 78

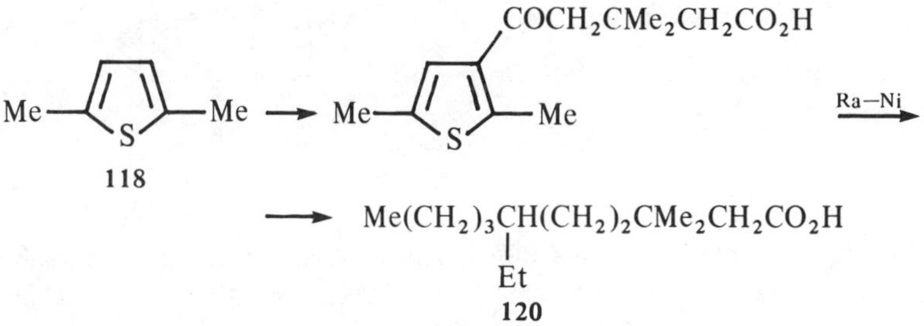

Scheme 79

Buu-Hoi et al.[273,274] demonstrated the possibility of synthesizing the higher aliphatic dicarboxylic acids **121** and **122** from thiophene (Schemes 80 and 81).

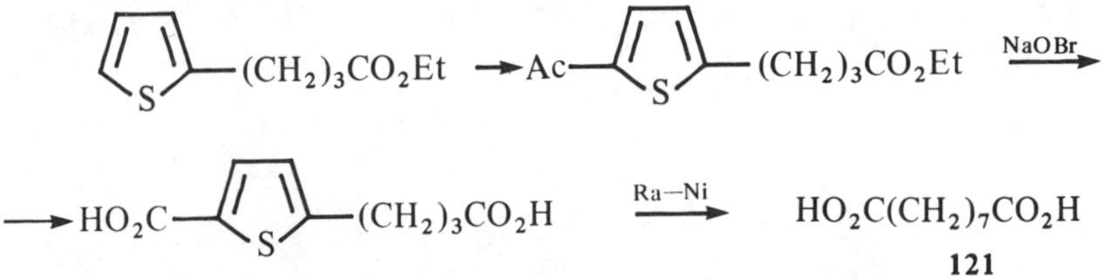

Scheme 80

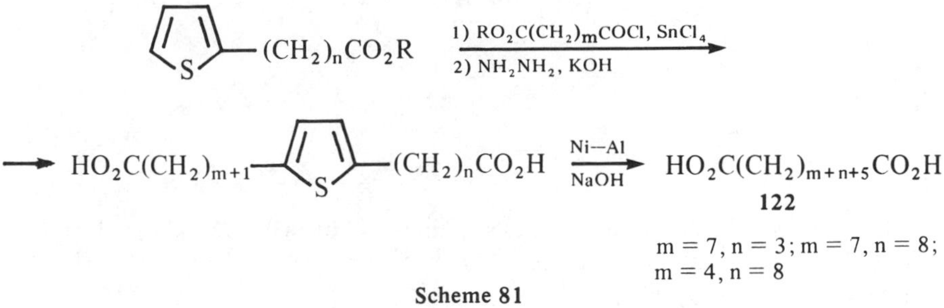

$$m = 7, n = 3; m = 7, n = 8;$$
$$m = 4, n = 8$$

Scheme 81

The attempt of Dann and Hauck[252] to carry out the reductive desulfurization of 3,4-diphenyl-2,5-thiophenedicarboxylic acid (123, R = H) was unsuccessful. However, starting from the diester of the same acid (123, R = Me), these authors succeeded in obtaining dimethyl meso-3,4-diphenyladipate (124). Buu-Hoi, Sy, and Xuong[275] previously described hydrogenolysis of the diethyl ester (123, R = Et) (Scheme 82) and similar transformation of thiophenetetracarboxylic acid into 1,2,3,4-tetracarboxybutane.

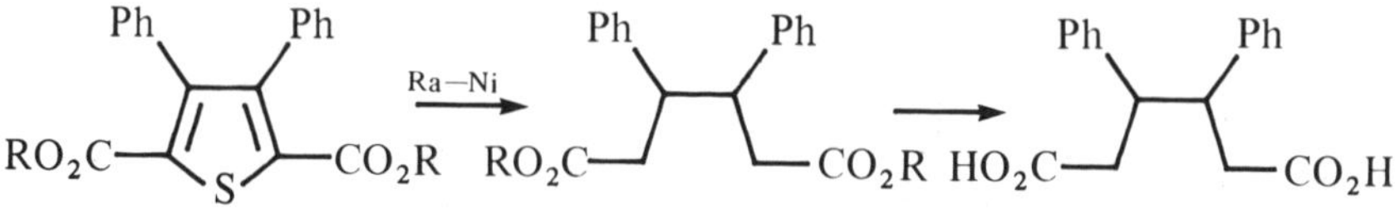

123
R = Me, Et

124

Scheme 82

Gol'dfarb et al. obtained some mono- and dicarboxylic acids of the aliphatic series to prove the structure of the products of several reactions.[276-278] Scheme 83 shows the synthesis of one of these acids (125).

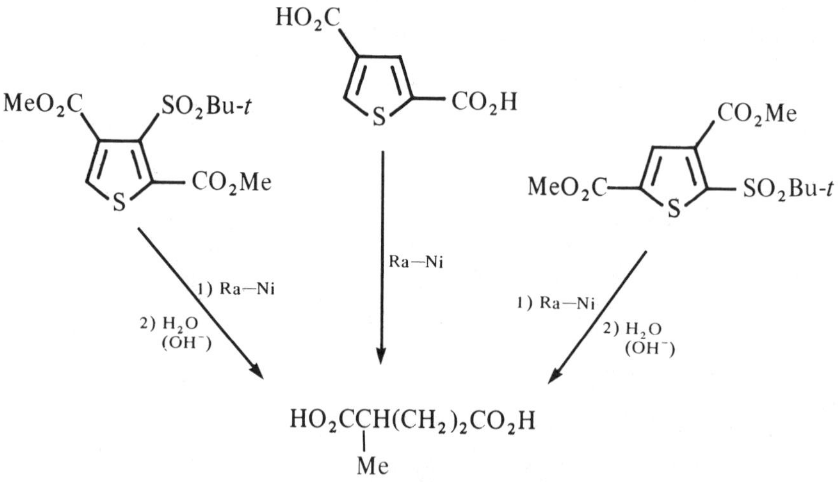

$HO_2CCH(CH_2)_2CO_2H$
|
Me
125

Scheme 83

Buu-Hoi et al. described the synthesis of mono- (126 and 127) and dicarboxylic acids (128) from phenylthienylalkanes (129)[263,273,279] (Scheme 84) and 2-aryl-3-(2-thienyl)acrylonitriles (130)[280,281] (Scheme 85).

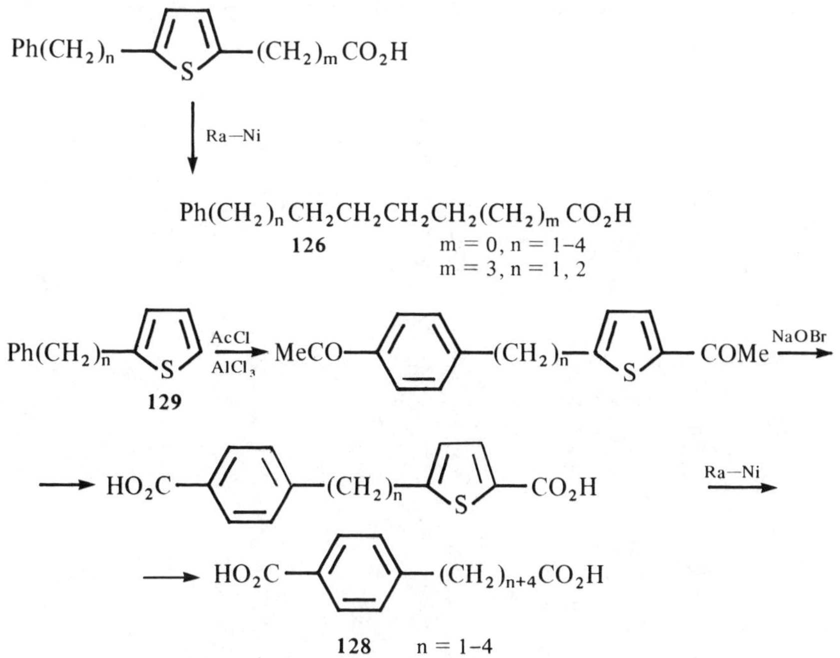

Ph(CH$_2$)$_n$ —[thiophene]— (CH$_2$)$_m$CO$_2$H

$\downarrow$ Ra—Ni

Ph(CH$_2$)$_n$CH$_2$CH$_2$CH$_2$CH$_2$(CH$_2$)$_m$CO$_2$H

126   m = 0, n = 1–4
             m = 3, n = 1, 2

Ph(CH$_2$)$_n$ —[thiophene]— $\xrightarrow[\text{AlCl}_3]{\text{AcCl}}$ MeCO—[benzene]—(CH$_2$)$_n$—[thiophene]—COMe $\xrightarrow{\text{NaOBr}}$

129

$\longrightarrow$ HO$_2$C—[benzene]—(CH$_2$)$_n$—[thiophene]—CO$_2$H $\xrightarrow{\text{Ra—Ni}}$

$\longrightarrow$ HO$_2$C—[benzene]—(CH$_2$)$_{n+4}$CO$_2$H

128   n = 1–4

Scheme 84

R—[thiophene]—CHO + PhCH$_2$CN $\xrightarrow{\text{(OH}^-\text{)}}$ R—[thiophene]—CH=CCN $\xrightarrow{\text{Hydrolysis}}$
                                                                |
                                                               Ph

130

$\longrightarrow$ R—[thiophene]—CH=CCO$_2$H $\xrightarrow[\text{NaOH}]{\text{Ni—Al}}$ R(CH$_2$)$_5$CCO$_2$H
                                    | |
                                   Ph   Ph

127

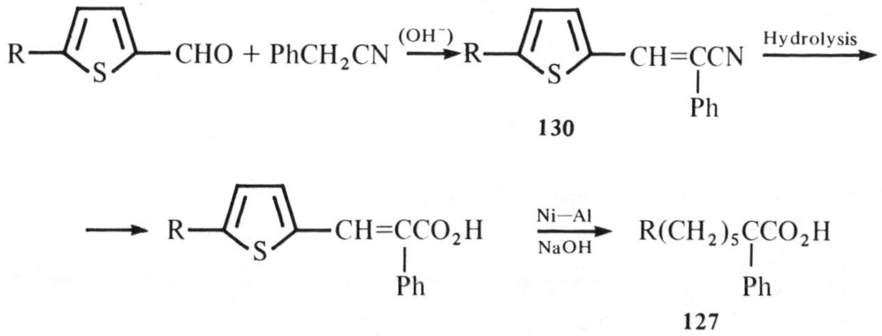

R = H, Et, Ph(CH$_2$)$_k$ (k = 1–3)

Scheme 85

The use of di-(2-thienyl)methane (**131**) is especially effective for the synthesis of long-chain mono- and dicarboxylic acids. It was studied in great detail by Buu-Hoi, Sy, and Xuong[273,274] as well as by Gol'dfarb and Kirmalova.[282–285] The first group of authors showed that starting from di-(2-thienyl)methane, one can obtain 1,9-nonanedicarboxylic acid[273] and 1,14-tetradecanedicarboxylic acid[274] (Scheme 86).

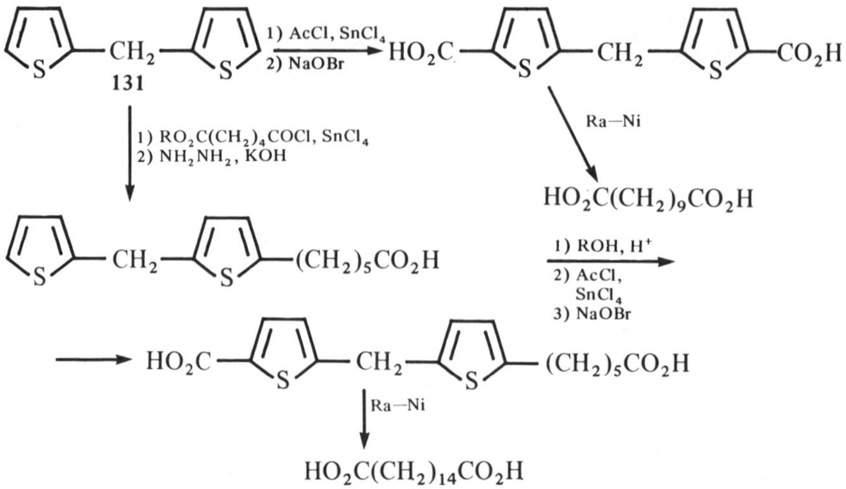

Scheme 86

The second group of authors showed that acylation of the two free α-positions of di-(2-thienyl)methane with chlorides of dicarboxylic acid monoesters opened a shorter route to higher dicarboxylic acids with an odd number of carbon atoms, such as 1,15-pentadecanedicarboxylic and 1,17-heptadecanedicarboxylic acids[282,284] (Scheme 87). The long-chain monocarboxylic acids (**132** and **133**) can easily be

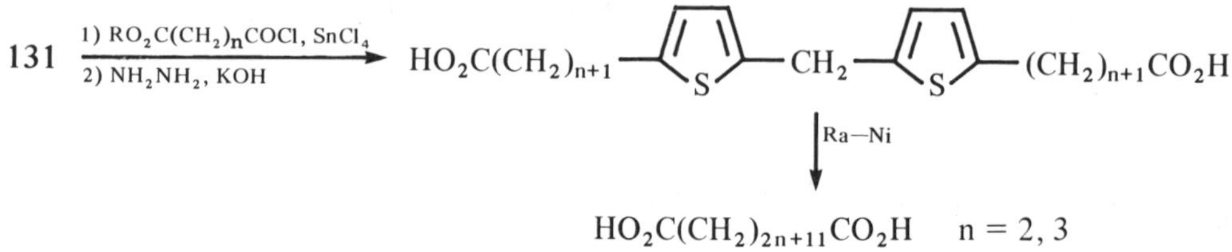

Scheme 87

obtained using monometalation and carboxylation or monoacylation of di-(2-thienyl)methane and its monoalkyl-substituted derivatives, with subsequent transformations similar to those described above[283–285] (Scheme 88).

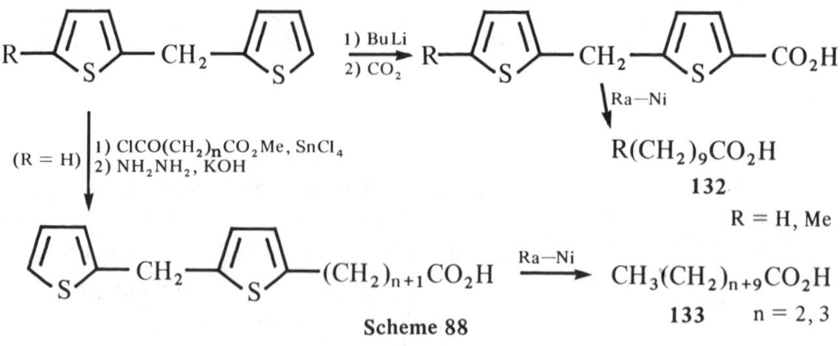

Scheme 88

Schemes 89 and 90 show interesting examples of the synthesis of branched aliphatic carboxylic acids, starting from 2,2-di-(2-thienyl)propionic acid (134)[262] and 2,5-di-(2-thienyl-2-propyl)thiophene (135).[275]

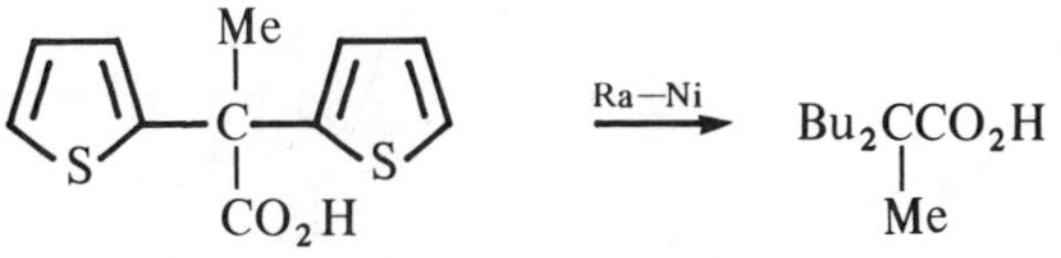

Scheme 89

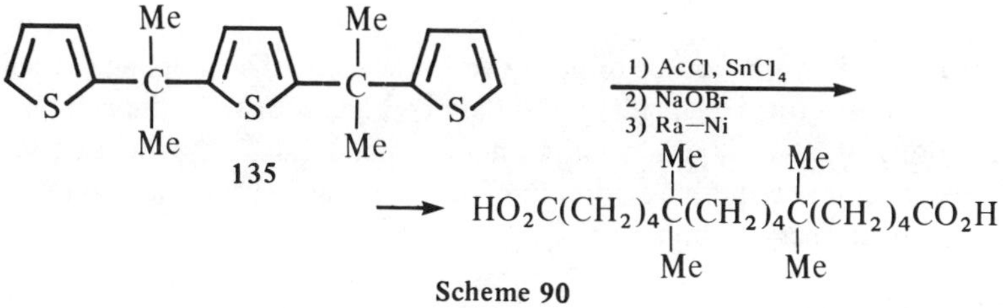

Scheme 90

There are data about reductive desulfurization of acids derived from 2,2'- and 2,3'-bithiophenes (136–141).[286,287] These transformations, which usually proceeded in high yields, as well as the hydrogenolysis of derivatives of thieno[2,3-b]thiophene (142), thieno[3,2-b]thiophene (143),[288] thieno[3,4-b]thiophene (144–146),[289] and dithieno[3,2-b;2,3-d]thiophene (147),[290] were performed to prove the structures of the corresponding bi- and tricyclic compounds.

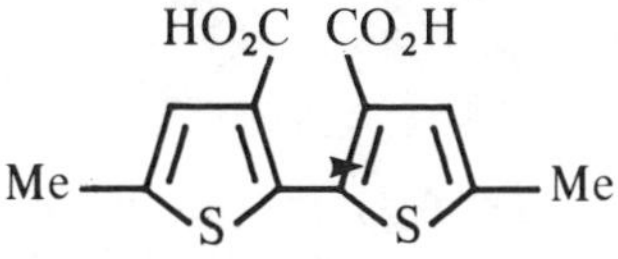

136

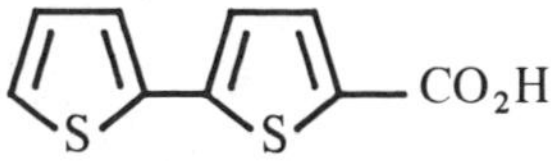

137

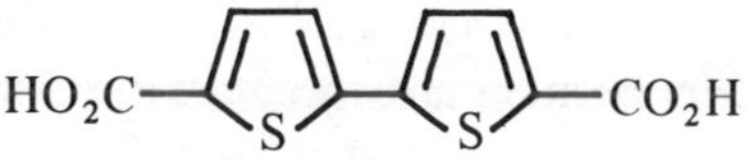

138

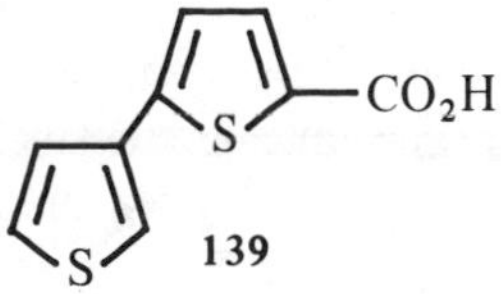

139

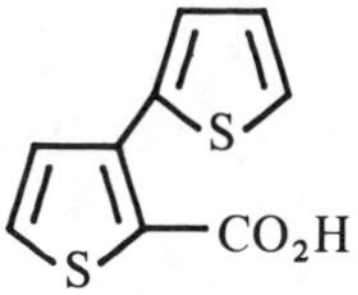

140

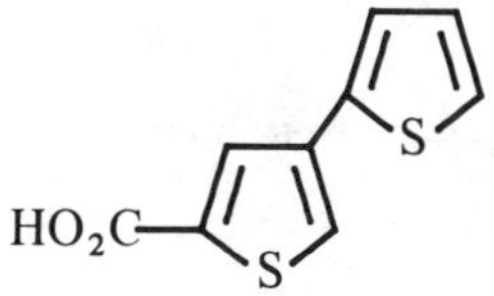

141

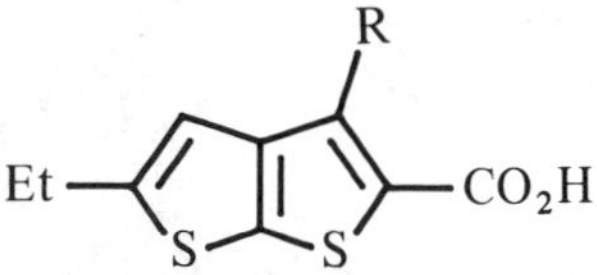

142

R = H, Me

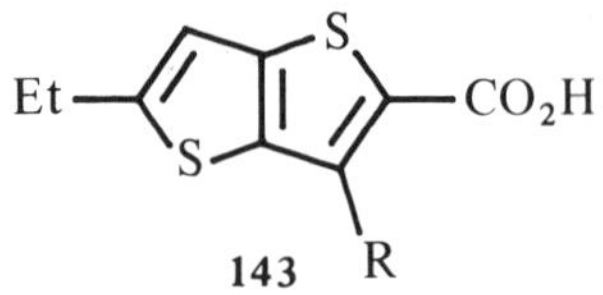

143　R
R = H, Me

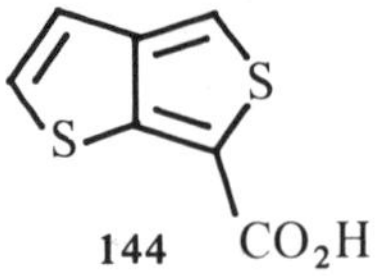

144　CO_2H

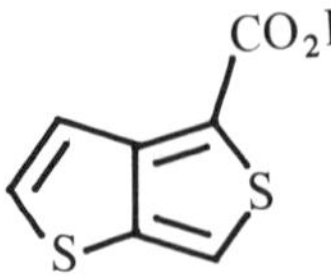

145

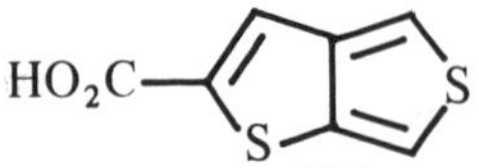

146

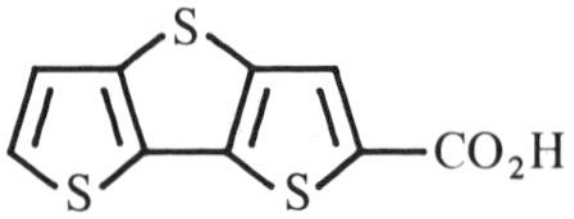

147

Reductive desulfurization of optically active acids of the thiophene series (**148** and **149**) was used for the elucidation of their configurations and the correlation of configurations of some aliphatic and aromatic compounds[291,292] (Scheme 91) and also in connection with the synthesis of optically active hydrocarbons with an asymmetric quaternary carbon atom[259] (Scheme 92).

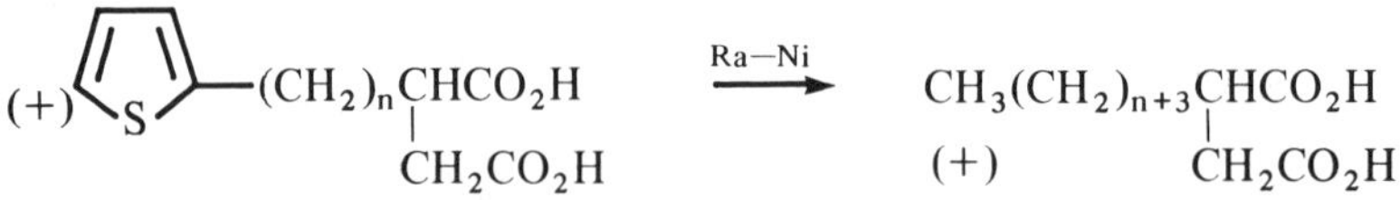

148　n = 0, 1

Scheme 91

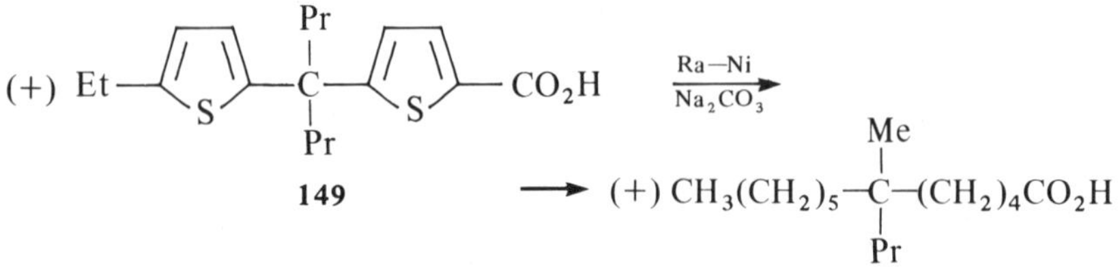

149

Scheme 92

The use of the synthesis of aliphatic substances via thiophene derivatives made it easy to introduce a ^{14}C label in definite chain position and to obtain 9-^{14}C-stearic (**150**, R = n-C_9H_{19}), -palmitic (**150**, R = n-C_7H_{15}), and -capric (**150**, R = Me) acids[293] (Scheme 93).

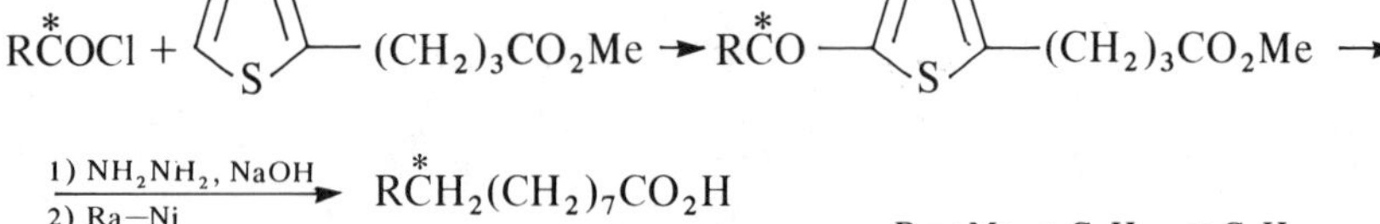

150

R = Me, n-C_7H_{15}, n-C_9H_{19}

Scheme 93

Buu-Hoi et al. proposed an exceptionally simple method for obtaining aliphatic carboxylic acids labeled with deuterium or tritium,[294-297] which consists of treating the carboxylic acids of the thiophene series with Ni–Al alloy in heavy or super-heavy water in the presence of NaOD or NaOT (Schemes 94–97). Probably the only drawback of this method is that the hydrogen isotope is simultaneously introduced into several positions of the molecule, though these positions are easily determined by PMR spectra.[298]

$$R \text{—}\underset{S}{\overset{}{\diagdown}}\text{—}(CH_2)_nCO_2H \xrightarrow[\text{NaOT, T}_2\text{O}]{\text{Ni–Al}} RCT_2(CHT)_2\text{—}CT_2(CH_2)_nCO_2H$$

(a) n = 0, R = H; (b) n = 1, R = H; (c) n = 7, R = $n\text{-}C_6H_{13}$;

(d) n = 3, R = Me(CH$_2$)$_7$CHCH$_2$
$\qquad\qquad\qquad\qquad$ |
$\qquad\qquad\qquad\qquad$ Me

Scheme 94

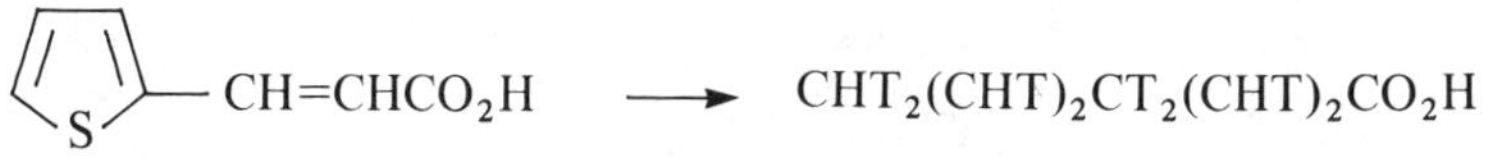

$$\underset{S}{\overset{}{\diagdown}}\text{— CH=CHCO}_2H \longrightarrow CHT_2(CHT)_2CT_2(CHT)_2CO_2H$$

Scheme 95

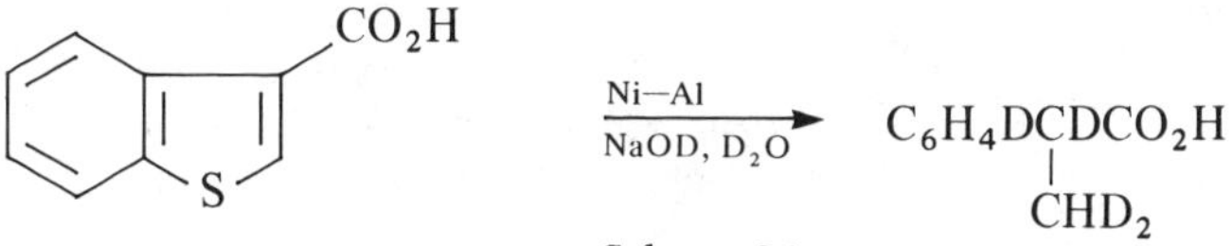

$$\xrightarrow[\text{NaOD, D}_2\text{O}]{\text{Ni–Al}} C_6H_4DCDCO_2H$$
$$\qquad\qquad\qquad\qquad |$$
$$\qquad\qquad\qquad\qquad CHD_2$$

Scheme 96

$$\longrightarrow C_6H_4DCDCH_2CHCO_2H$$
$$\qquad\qquad\quad |\qquad\quad |$$
$$\qquad\qquad\quad CHD_2\quad R$$

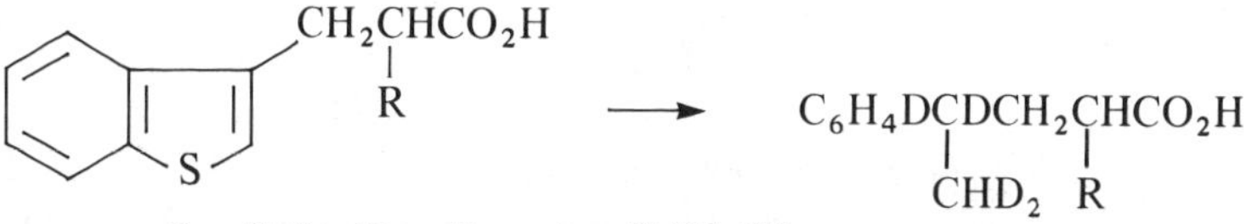

R = H, Et, i-Pr, s-Bu, n-Am, PhCH$_2$CH$_2$

Scheme 97

In the cases described above, acids containing one or two atoms of the hydrogen isotope at the same carbon atom were obtained. An interesting method is used for preparing acids having trideuteriomethyl or tritritiomethyl groups.[297] Acids **151** and **152** are used as starting compounds in this process, in which the free α-positions of the thiophene ring are substituted by chlorine or bromine atoms and the latter are exchanged for deuterium or tritium during reductive desulfurization (Schemes 98 and 99). The yields of the labeled acids are 70–90%.

$$X\text{—}\underset{S}{\overset{}{\diagdown}}\text{—CO}_2H \longrightarrow CD_3CHDCHDCD_2CO_2H$$

151

X = Cl, Br $\qquad\qquad$ **Scheme 98**

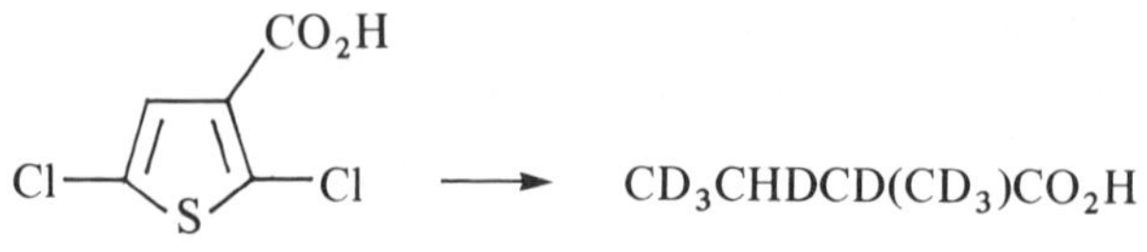

$$Cl{-}\underset{S}{\overset{CO_2H}{\|}}{-}Cl \longrightarrow CD_3CHDCD(CD_3)CO_2H$$

152

Scheme 99

C. Synthesis of Ketones, Alcohols, Phenols, Ethers, and Acetals

In one of their papers,[228] Blicke and Sheets described the transformation of 2-benzoylthiophene under the action of Raney nickel, leading to valerophenone in 75% yield. The same reaction was reproduced in Ref. 299. Later, reductive desulfurization of ketones of the thiophene series was studied by many authors and it was found that the process was not so simple as in the case of carboxylic acids. Studying the reductive desulfurization of 2-acetylthiophene in xylene at 100°C, Hurd and Rudner[300] observed that along with 2-hexanone (76.5% yield), the formation of ethanol (7.6%) and acetaldehyde (10%) also took place. Badger, Rodda, and Sasse[267] found that the result of reductive desulfurization of 2-acetyl-thiophene was determined by the Raney nickel activity. Thus, by the use of nickel W-6, they obtained not only 2-hexanone and acetaldehyde (the latter arose in the authors opinion[267,300] as a result of thiophene ring cleavage), but also the product of radical recombination, 2,11-dodecanedione. However, using nickel W-7, 2-hexanone was not isolated at all, and only 2,11-dodecanedione and acetaldehyde were obtained.[267]

Wynberg and Logothetis,[257] as well as Gol'dfarb and Konstantinov,[301,302] studied the reductive desulfurization of some thiophene ketones and showed that in this process the transformation of the keto group into a hydroxy group took place to a certain extent. Thus, 5-n-octadecyl-2-acetylthiophene (**153**, R = Me, R$'$ = n-C$_{18}$H$_{37}$) is transformed into the corresponding secondary alcohol (**154**) in 83% yield and a similar conversion of n-pentadecyl 2-thienyl ketone (**153**, R = n-C$_{15}$H$_{31}$, R$'$ = H) occurred in 90% yields[257] (Scheme 100). Reductive desulfurization of 5-methyl-2-thienyl 2-thienyl ketone (**155**) leads to butyl pentyl carbinol[301] (Scheme 101).

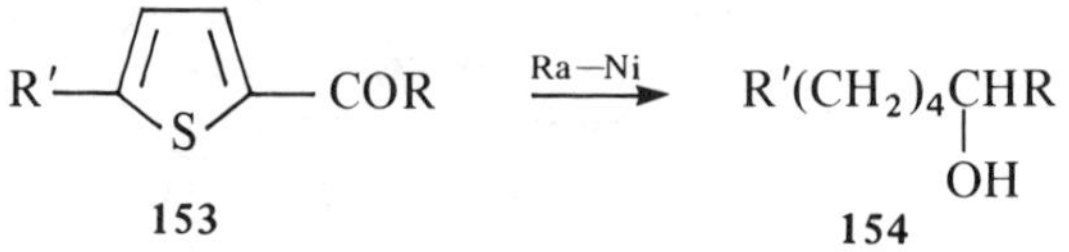

$$R'{-}\underset{S}{\|}{-}COR \xrightarrow{\text{Ra--Ni}} R'(CH_2)_4\underset{\underset{OH}{|}}{C}HR$$

153   **154**

(a) R = Me, R$'$ = n-C$_{18}$H$_{37}$; (b) R = n-C$_{15}$H$_{31}$, R$'$ = H

Scheme 100

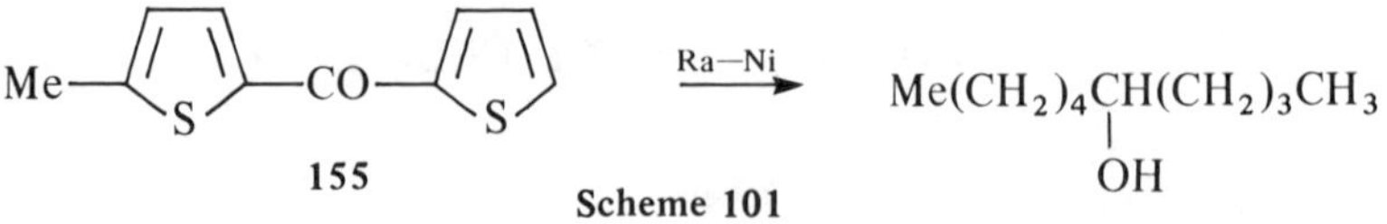

$$Me{-}\underset{S}{\|}{-}CO{-}\underset{S}{\|} \xrightarrow{\text{Ra--Ni}} Me(CH_2)_4\underset{\underset{OH}{|}}{C}H(CH_2)_3CH_3$$

155

Scheme 101

To obtain aliphatic ketones, it was proposed in Ref. 257 to oxidize the mixture formed by desulfurization of ketones with chromium trioxide. In this way, 2,11-dodecanedione was obtained (91%) from 5,5'-diacetyl-2,2'-bithiophene (**156**, R = Me, R' = MeCO) and *n*-octyl *n*-heptadecyl ketone from 5-stearoyl-2,2'-bithiophene (**156**, R = *n*-C$_{17}$H$_{35}$, R' = H) (Scheme 102).

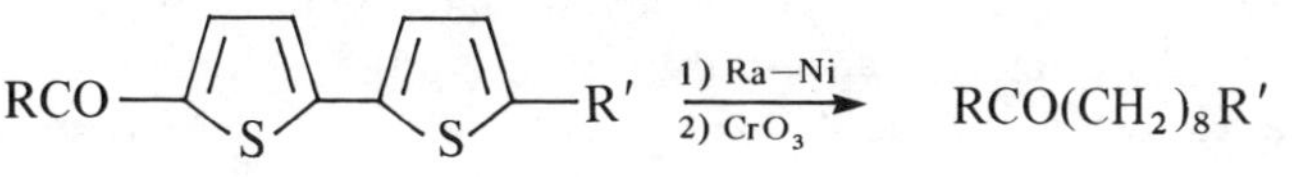

156

(**a**) R = Me, R' = Ac; (**b**) R = *n*-C$_{17}$H$_{35}$, R' = H

Scheme 102

Kao, Tilak, and Venkataraman[241,303] studied the reductive desulfurization of thioindigo and some related substances. One could expect that the products of this reaction should be diphenacyl and its derivatives. In fact, however, subsequent reduction takes place, leading to the formation of 2,4-diphenyl-7-butanol and 1,3-diphenylbutane from 2,3'-bis-thionaphthene indigo (**157**) and of 1,4-diphenylbutane from thioindigo (**158**)[241] (Schemes 103 and 104). At the same time, acetophenone was obtained from thioindoxyl (**159**)[241] (Scheme 105).

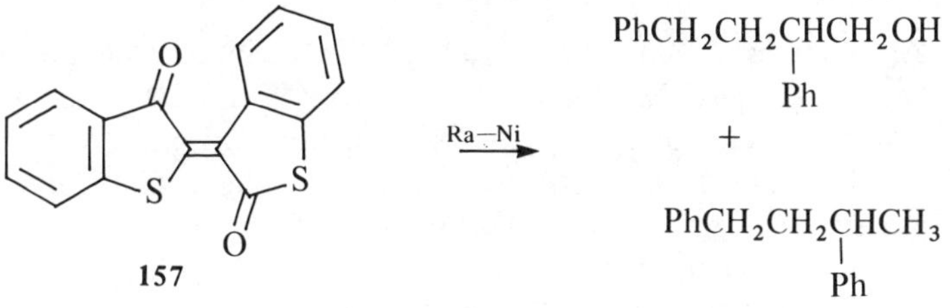

157

PhCH$_2$CH$_2$CHCH$_2$OH
|
Ph
+
PhCH$_2$CH$_2$CHCH$_3$
|
Ph

Scheme 103

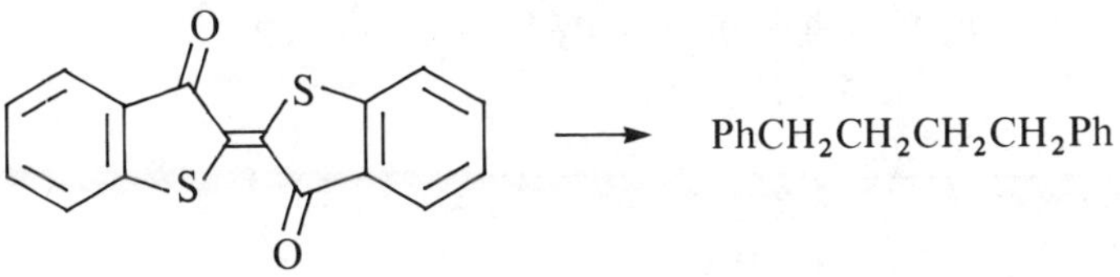

158

PhCH$_2$CH$_2$CH$_2$CH$_2$Ph

Scheme 104

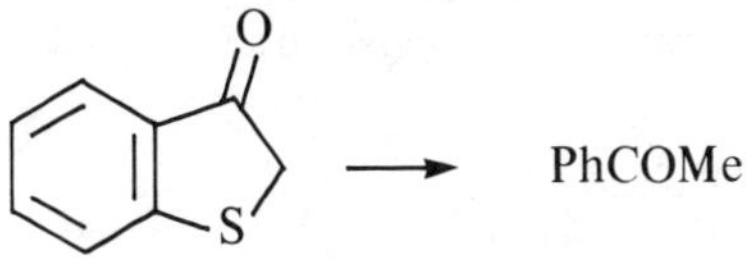

159

PhCOMe

Scheme 105

Taking on insufficient amount of Raney nickel, one can avoid reduction of the keto group, but in this case, the reaction does not proceed to the end and considerable amounts of diketones are formed (products of recombination of the intermediate radicals). Thus, Badger and Sasse,[230] using freshly prepared Raney nickel (from 125 g of alloy, i.e., about 60 g of catalyst) on 30 g of 2-benzoyl thiophene in methanol, isolated 10 g of starting ketone and obtained 10 g of valerophenone and 3.5 g of 1,8-dibenzoyloctane. In the case of 2-propionylthiophene, 20 g of starting ketone was recovered and 4.4 g of 3-heptanone and 0.066 g of dimeric product were obtained. The yield of 2-acetylthiophene desulfurization products was very low (probably because of the cleavage mentioned above), but as in previous cases, only one-third of the starting ketone was recovered. About the same amounts of 5-ethyl-2-acetylthiophene and 3-acetylbenzo[b]thiophene were recovered in the desulfurization of these ketones under similar conditions, without any identification of diketones.[230]

Gol'dfarb and Konstantinov[302] showed that in some cases steric hindrances promote the preservation of keto groups. Thus, they succeeded in obtaining butyl *tert*-butyl ketone without any admixture of corresponding alcohol from 2-pivaloyl-thiophene (Scheme 106), but 2-butyrylthiophene and di-(2,5-dimethyl-3-thienyl)

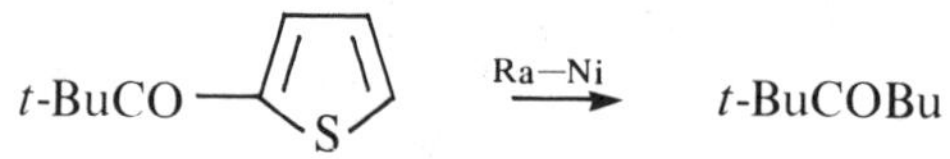

$$ t\text{-BuCO} \underset{S}{\langle\!\!\!\square\!\!\!\rangle} \xrightarrow{\text{Ra-Ni}} t\text{-BuCOBu} $$

Scheme 106

ketone gave mixtures of ketones and alcohols. The attempts to desulfurize highly hindred ketones (3-acetyl-2,5-di-*tert*-butylthiophene[301] and di-(2,5-di-*tert*-butyl-3-thienyl) ketone[302] were completely uncussessful. Desulfurization of 5,5'-diacetyl-di-(2-ethyl-3-thienyl)methane (160) was described by Gol'dfarb and Vol'ken-shtein[304] (Scheme 107) and the same reaction of 5,5'-diacetyl-di-(2-thienyl)methane (161) was studied by a group of Japanese authors[305] (Scheme 108). The yields of the corresponding diketones were 70–90%. It is interesting to note that the authors of Ref. 305 did not succeed in the desulfurization of either di-(2-thienyl)methane itself or its 5,5'-dinitro- and dichloro derivatives.

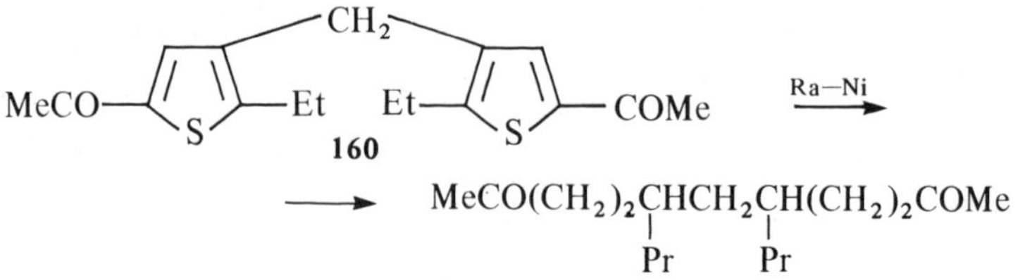

$$ \text{MeCO} \underset{S}{\langle\!\!\!\square\!\!\!\rangle} \text{Et} \quad \text{Et} \underset{S}{\langle\!\!\!\square\!\!\!\rangle} \text{COMe} \xrightarrow{\text{Ra-Ni}} $$

160

$$ \longrightarrow \quad \text{MeCO(CH}_2)_2\text{CHCH}_2\text{CH(CH}_2)_2\text{COMe} $$
$$ \qquad\qquad\qquad\qquad\quad \overset{|}{\text{Pr}} \qquad \overset{|}{\text{Pr}} $$

Scheme 107

$$ \text{MeCO} \underset{S}{\langle\!\!\!\square\!\!\!\rangle} \text{CH}_2 \underset{S}{\langle\!\!\!\square\!\!\!\rangle} \text{COMe} \xrightarrow{\text{Ra-Ni}} \text{MeCO(CH}_2)_9\text{COMe} $$

161

Scheme 108

5-Methyl-5-acetylnonane was obtained by desulfurization of di-(2-thienyl)-2-butanone (162)[261] (Scheme 109). Reductive desulfurization of 2-(perfluorobutyryl)-thiophene (163) leads to aliphatic ketone 164 with the admixture of carbinol (165)[306] (Scheme 110). It is important to emphasize that fluorine atoms are retained during this reaction.

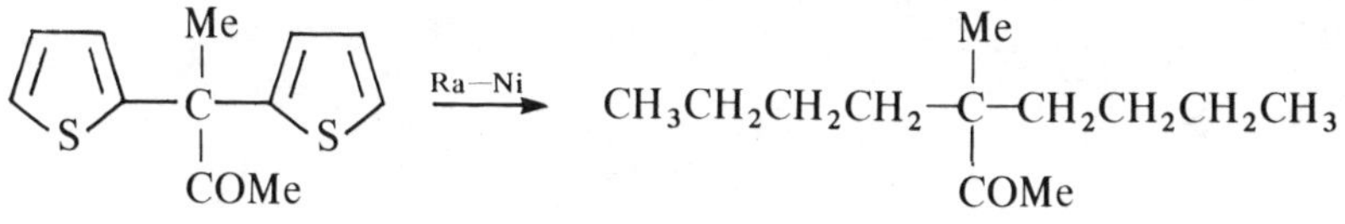

$$\text{Me}$$

$$162 \xrightarrow{\text{Ra—Ni}} CH_3CH_2CH_2CH_2-\underset{\underset{\text{COMe}}{|}}{\overset{\overset{\text{Me}}{|}}{C}}-CH_2CH_2CH_2CH_3$$

162

Scheme 109

$$C_3F_7CO-\text{[thiophene]} \xrightarrow{\text{Ra—Ni}} C_3F_7COC_4H_9 + C_3F_7\underset{\underset{\text{OH}}{|}}{CH}C_4H_9$$

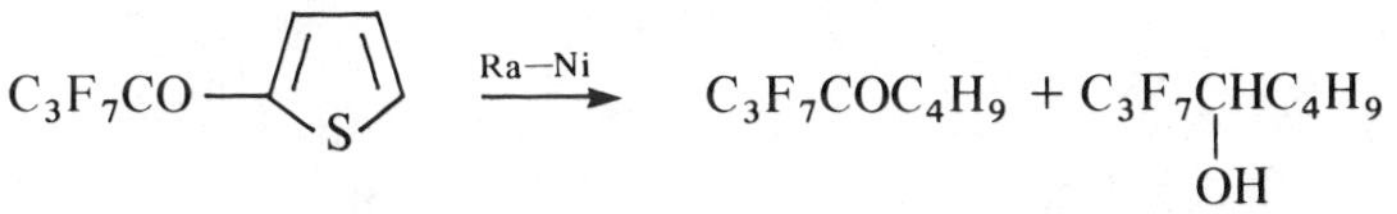

163   164   165

Scheme 110

In some cases, the keto group can be retained if the reductive desulfurization is carried out in the presence of lower aliphatic ketones (mostly acetone). This method was successfully used in the synthesis of macrocyclic ketones (see Section IV.3.I). It should be noted here that Stetter and Rajh,[307] when carrying out desulfurization in the presence of methyl ethyl ketone, obtained several aliphatic and arylaliphatic diketones (166 and 167) in 50–80% yields (Schemes 111 and 112).

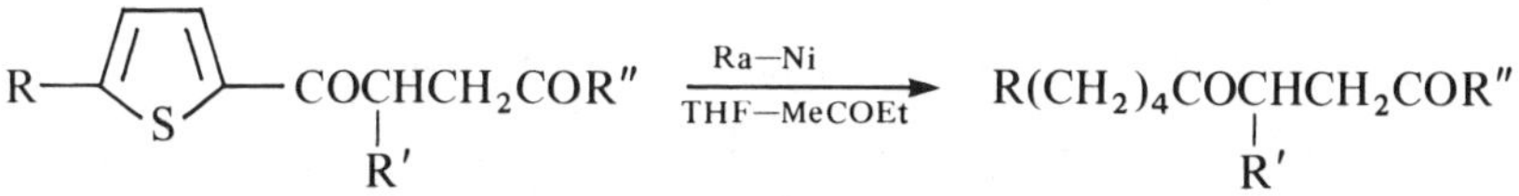

$$R-\text{[thiophene]}-\underset{\underset{R'}{|}}{CO}CHCH_2COR'' \xrightarrow[\text{THF—MeCOEt}]{\text{Ra—Ni}} R(CH_2)_4\underset{\underset{R'}{|}}{CO}CHCH_2COR''$$

166

(a) R = R′ = H, R″ = Me; (b) R = H, R′ = Ph, R″ = Me;
(c) R = H, R′ = R″ = Ph

Scheme 111

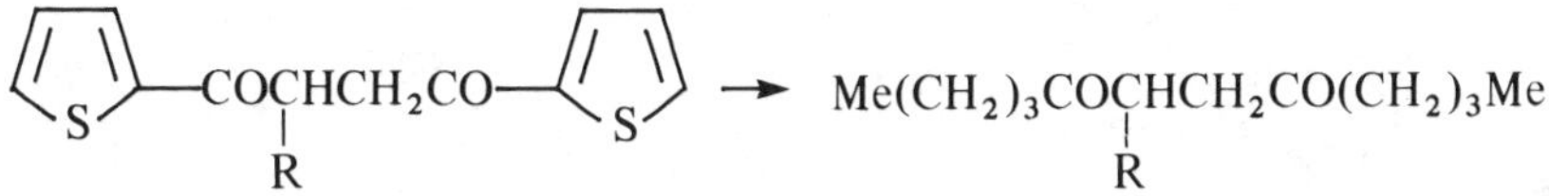

$$\text{[thiophene]}-\underset{\underset{R}{|}}{CO}CHCH_2CO-\text{[thiophene]} \longrightarrow Me(CH_2)_3\underset{\underset{R}{|}}{CO}CHCH_2CO(CH_2)_3Me$$

167   R = H or Ph

Scheme 112

Reductive desulfurization was used also to prove the structures of some ketones of the thiophene series. In particular, Gol'dfarb, Kalik, and Kirmalova[276] identified 2-hexanone and 2,11-dodecanedione in the desulfurization products of 2-acetyl-5-

methylthiothiophene (Scheme 113). Wynberg and Bantjes[287] obtained 2,15-hexadecanedione from diacetylterthiophene (168). Challenger et al. studied reductive desulfurization of acetylation products of thieno[3,2-b]thiophene (169) and thieno[2,3-b]thiophene (170).[308, 309]

$$MeCO-\underset{S}{\text{⟨thiophene⟩}}-SMe \xrightarrow[\text{2) } CrO_3]{\text{1) Ra—Ni}} MeCO(CH_2)_3CH_3 + MeCO(CH_2)_8COMe$$

Scheme 113

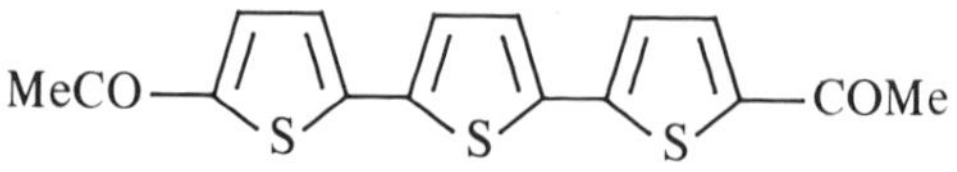

$$MeCO-\underset{S}{\text{⟨⟩}}\underset{S}{\text{⟨⟩}}\underset{S}{\text{⟨⟩}}-COMe$$

168 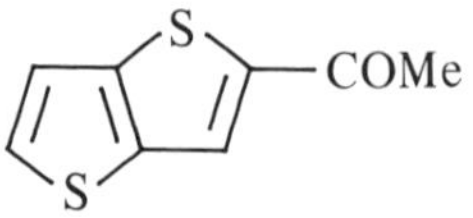   169

Reductive desulfurization of aldehydes 171–173 of the thiophene[301] and the dithienylmethane series[7] is a very convenient method for the synthesis of primary alcohols (Schemes 114–116). The yields of aliphatic alcohols are up to 65%.

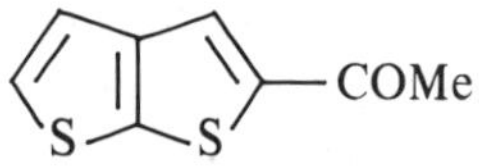

$$\underset{S}{\text{⟨⟩}}\underset{S}{\text{⟨⟩}}-COMe \qquad R-\underset{S}{\text{⟨⟩}}-CHO \xrightarrow{\text{Ra—Ni}} R(CH_2)_4CH_2OH$$

170 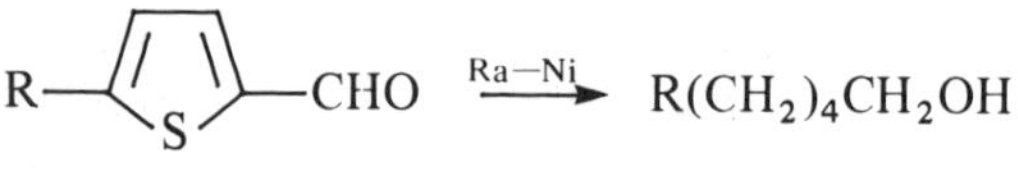   171   R = Bu, t-Bu

Scheme 114

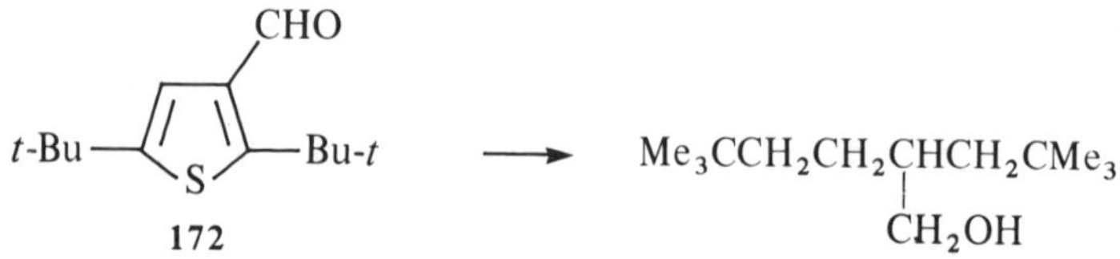

$$t\text{-Bu}-\underset{S}{\text{⟨⟩}}-\text{Bu-}t \longrightarrow Me_3CCH_2CH_2\underset{|}{C}HCH_2CMe_3$$
$$\qquad\qquad\qquad\qquad CH_2OH$$

172

Scheme 115

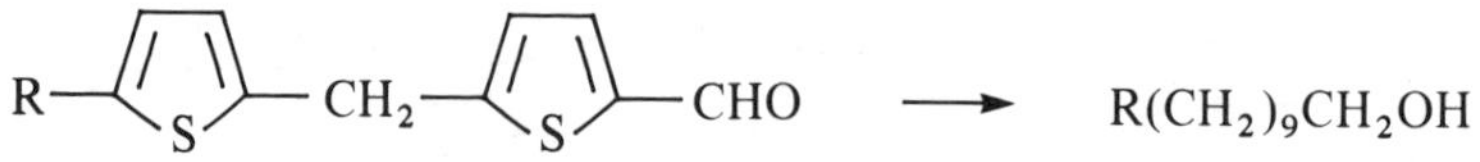

$$R-\underset{S}{\text{⟨⟩}}-CH_2-\underset{S}{\text{⟨⟩}}-CHO \longrightarrow R(CH_2)_9CH_2OH$$

173   R = H, Me

Scheme 116

In light of the data mentioned above, it is no wonder that desulfurization of primary alcohols of the thiophene series leads without difficulty to aliphatic alcohols. Synthesis of higher alcohols from β-hydroxyethyl-substituted derivatives of dithienylmethane (174), described by Gol'dfarb and Kirmalova[283, 285] (Scheme 117), is of interest. The starting compounds (174) are obtained smoothly by the

174 R = H, Me, CH$_2$CH$_2$OH

Scheme 117

action of ethylene oxide on the corresponding lithium derivatives. The yields of alcohols and glycols reach 80%. Desulfurization of some hydroxymethyl derivatives (**175**) also proceeds very easily[252] (Scheme 118). The hydroxy group may be retained even in desulfurization of tertiary alcohols (Scheme 119). Unfortunately, in the paper by Wynberg and Logothetis,[257] the yield of the formed aliphatic carbinol (**176**) is not given.

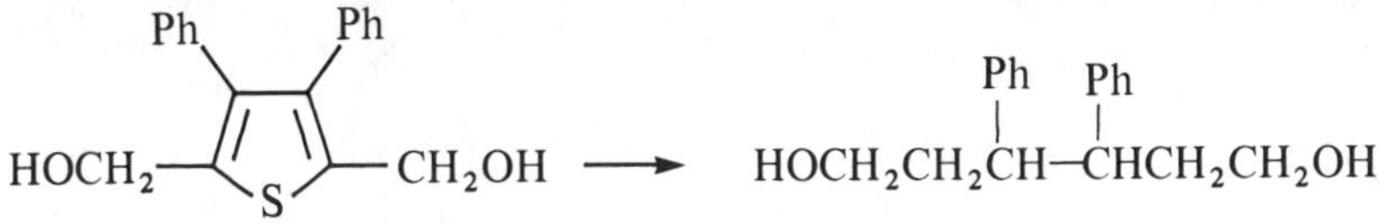

175

Scheme 118

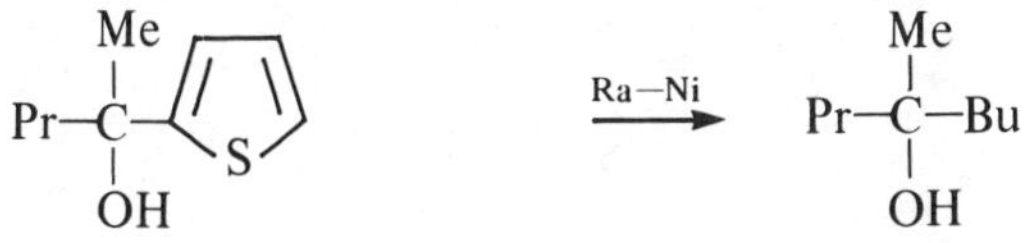

176

Scheme 119

Reductive desulfurization of some hydroxy derivatives of benzothiophene gives alkylphenols[262] (Scheme 120). Hydrogenolysis of tetrahydrothiophene derivative **177** was used by Gassman and Amick[310] in the interesting synthesis of *ortho*-substituted phenols (Scheme 121).

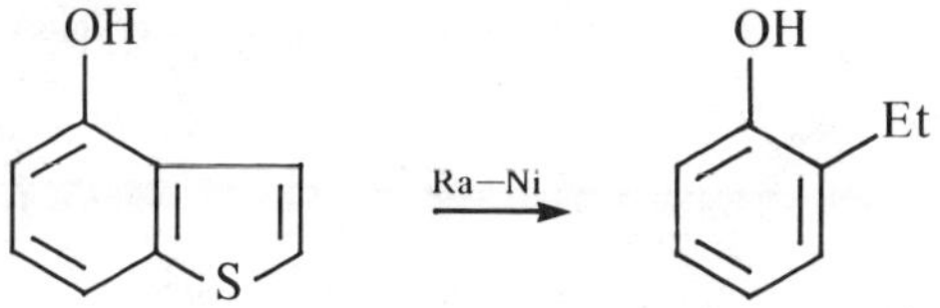

Scheme 120

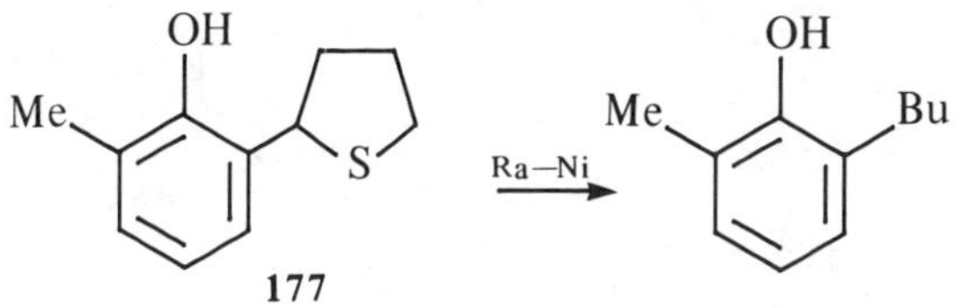

177

Scheme 121

The ether linkage does not change under the action of Raney nickel and ethers of the thiophene series are transformed under the conditions of reductive desulfurization into the corresponding aliphatic or aromatic ethers. Thus, Gol'dfarb and Konstantinov[311] obtained some aliphatic ethers (**178–180**) in yields of 50–60% (Schemes 122–124). An attempt to desulfurize the sterically hindered 2,5-di-*tert*-butyl-3,4-dimethoxythiophene (**181**, R = *tert*-Bu) was unsuccessful.

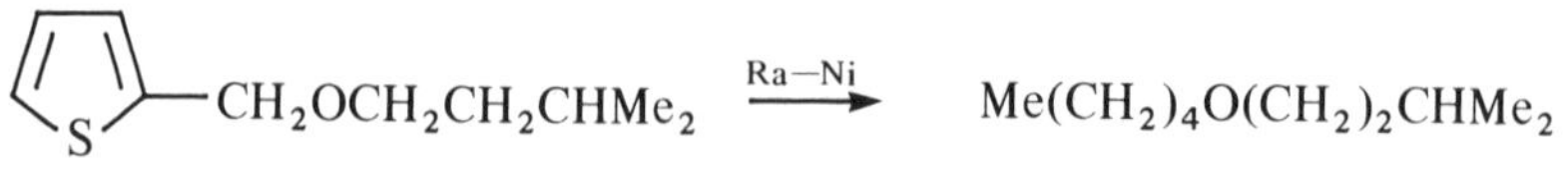

178

Scheme 122

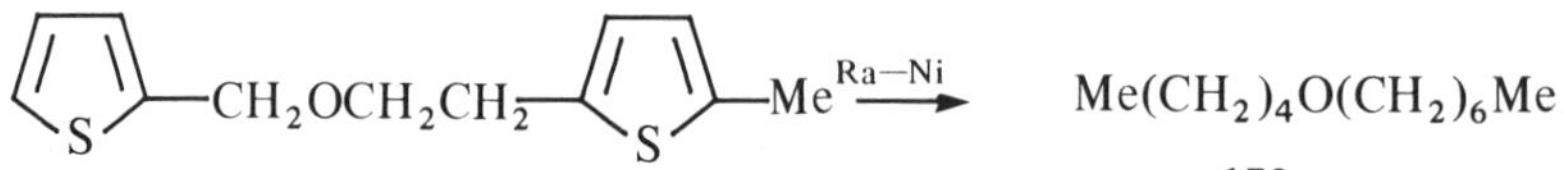

179

Scheme 123

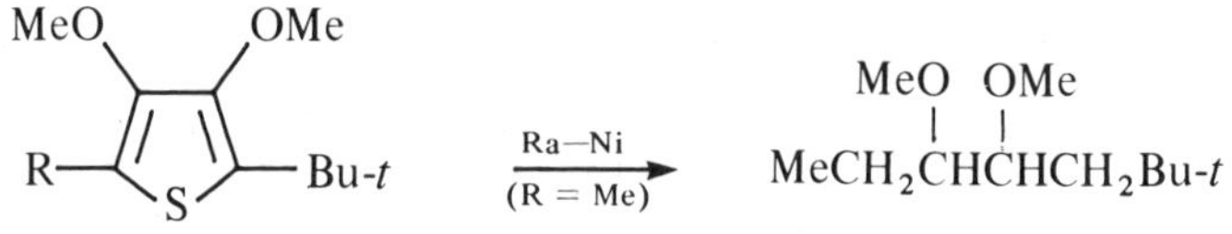

181 **180**

Scheme 124

Several arylaliphatic ethers were synthesized from the ethers of thiophene[252] (Scheme 125) and benzo[b]thiophene series.[235,238,241,312–314] The synthesis of the ether (**182**)[238] is shown in Scheme 126.

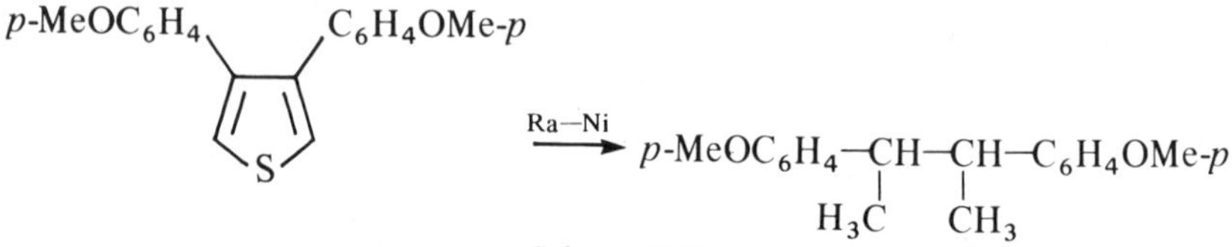

Scheme 125

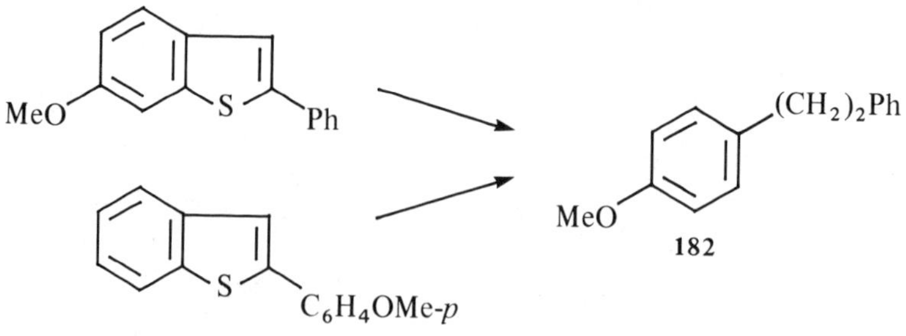

182

Scheme 126

Under the action of Raney nickel on the diethylacetal of 5-butyl-2-thiophene-carboxaldehyde (183), the aliphatic acetal (184) is formed together with ethyl nonyl ether, the product of hydrogenolysis of one of the C—O bonds (Scheme 127). In the case of the diethylacetal of 2-thiophenecarboxaldehyde, only the

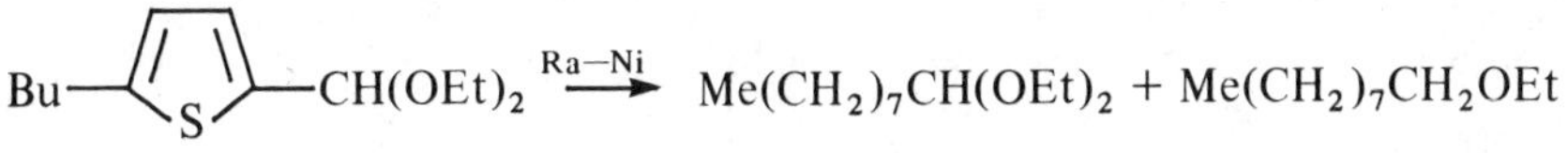

Bu—[thiophene]—CH(OEt)$_2$ $\xrightarrow{\text{Ra—Ni}}$ Me(CH$_2$)$_7$CH(OEt)$_2$ + Me(CH$_2$)$_7$CH$_2$OEt

183   184

Scheme 127

ethylacetal of valeraldehyde was isolated[311] (Scheme 128). Reductive desulfurization of the ethylene acetals 185 and 186 leads to the corresponding 2-alkyl-1,3-dioxolanes in yields of 40–50%[302] (Schemes 129 and 130); that is, transformation of an aldehyde into the cyclic acetal makes it possible to retain the aldehyde group under conditions of reductive desulfurization.

[thiophene]—CH(OEt)$_2$ $\longrightarrow$ Me(CH$_2$)$_3$CH(OEt)$_2$

Scheme 128

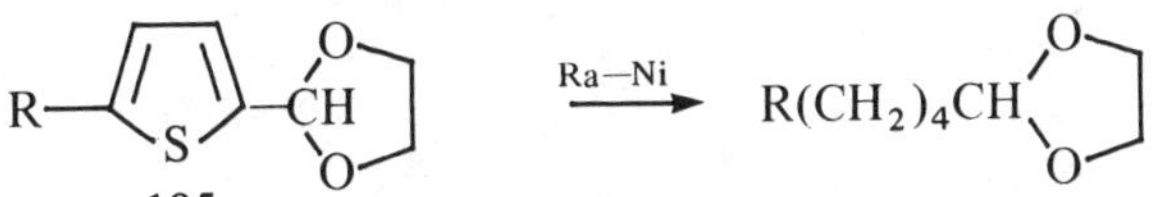

R—[thiophene]—CH(O–O) $\xrightarrow{\text{Ra—Ni}}$ R(CH$_2$)$_4$CH(O–O)

185

R = H, Bu

Scheme 129

[thiophene]—C(R')(R)—[thiophene]—CH(O–O) $\longrightarrow$ Me(CH$_2$)$_3$C(R')(R)(CH$_2$)$_4$CH(O–O)

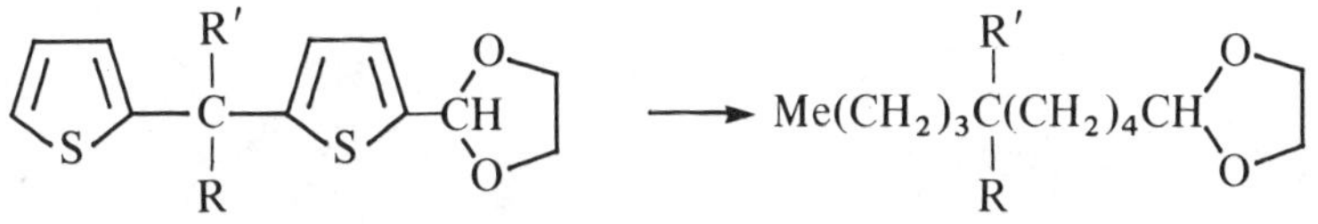

186

(a) R = R' = H; (b) R = Me, R' = Et

Scheme 130

D. Synthesis of Hydroxy-, Alkoxy-, and Ketocarboxylic Acids

From the data considered in the previous two sections, it is evident that, in principle, there is a possibility of obtaining aliphatic carboxylic acids with additional oxygen-containing functions from thiophene derivatives. Practically, the success of such synthesis is determined by the availability of starting compounds and the choice of desulfurization conditions.

The reductive desulfurization of thiophenic precursors containing an hydroxy or alkoxy group proceeds quite smoothly. The synthesis and desulfurization of one such precursor was described by Gol'dfarb and Kirmalova.[282] These authors obtained hydroxy acid (188) as the result of the action of 2 eq. butyllithium on 5-(β-hydroxyethyl)-di-(2-thienyl)methane (187), followed by carboxylation. Desulfurization of this acid led to 12-hydroxydodecanoic acid in 94% yield (Scheme 131).

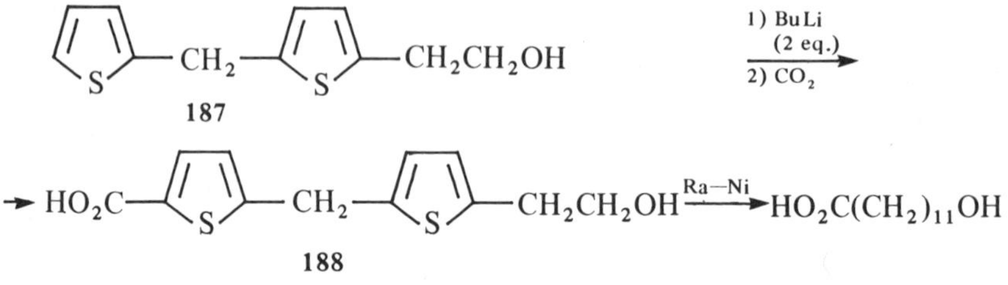

Scheme 131

Miller, Haymaker, and Gilman[315] suggested another scheme for the synthesis of long-chain hydroxycarboxylic acids (189, R = n-alkyl C_2–C_{17}, n = 2–8), which includes the preparation of 2-alkylthiophene (190), its transformation into an ester of the ketoacid (191), reduction of the latter into hydroxy acid (192), and desulfurization (Scheme 132). The yields at all stages are sufficiently high, including the hydrogenolysis, which gives 65–80%.

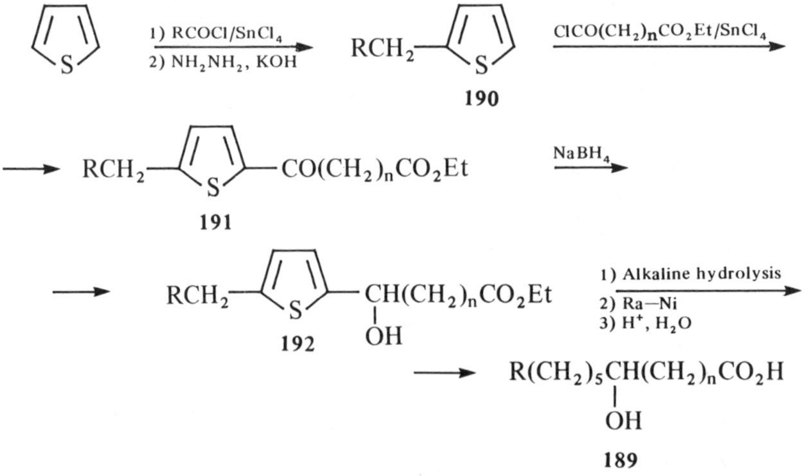

Scheme 132

Sicé[316] and Gronowitz[317] described reductive desulfurization of isomeric 5- and 3-methoxythiophenecarboxylic acids, leading, respectively, to 5-methoxyvaleric acid (82% yield) and 3-methoxyvaleric acid (63% yield). In the first case, desulfurization was carried out under the action of Raney nickel in the aqueous solution of sodium bicarbonate, in the second, according to the Papa–Schwenk method. The synthesis of carboxylic acid derivatives bearing the methoxy group in the benzene ring is described by Dann and Hauck[252] and also by Modest and Szmuszkovicz.[263]

Hydrogenolysis of optically active 2-thienylglycolic acid was used by Grono-witz[318] for the correlation of the configuration of mandelic acid with that of hydroxy acids of the aliphatic series (Scheme 133).

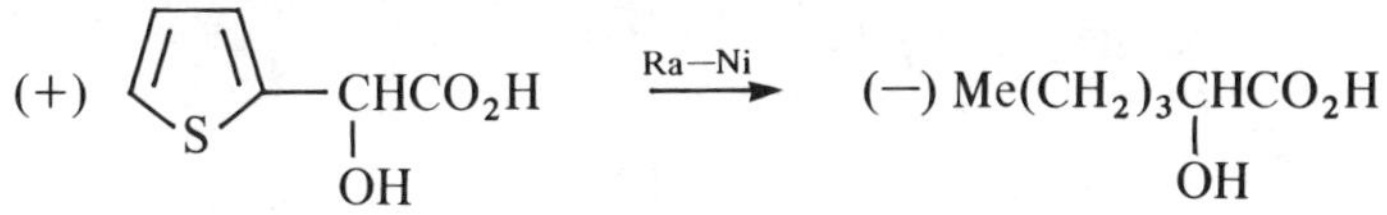

$$(+) \quad \underset{S}{\boxed{}}-\underset{\underset{OH}{|}}{CHCO_2H} \quad \xrightarrow{Ra-Ni} \quad (-) \, Me(CH_2)_3\underset{\underset{OH}{|}}{CHCO_2H}$$

Scheme 133

Reductive desulfurization of keto acids can lead to aliphatic keto acids, hydroxy acids, or their mixtures. Keto groups in the benzene ring can be reduced into CH_2 during hydrogenolysis.[319] Papa, Schwenk, and Ginsberg[262] obtained 10-hydroxy-myristic acid (Scheme 134) by the action of Ni–Al alloy on 9-(2-thenoyl)-pelargonic acid (**193**). They also obtained a mixture of 4-ketocaprylic acid (**195**) and γ-caprylolactone (**196**) from β-(2-thenoyl)propionic acid (**194**) (Scheme 135). Increasing the amount of alloy, they obtained only the lactone (**196**).

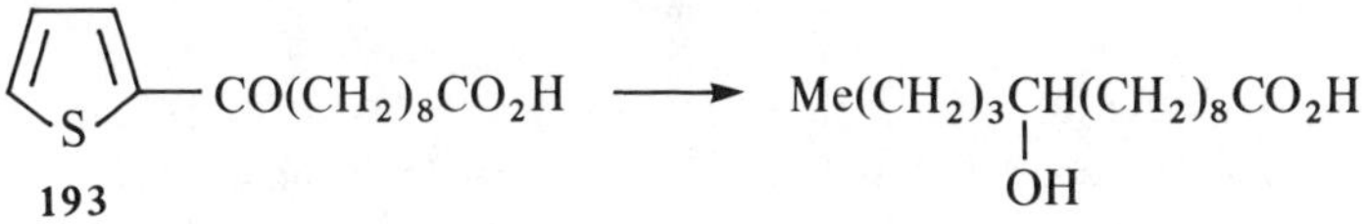

$$\underset{\underset{\textbf{193}}{}}{\underset{S}{\boxed{}}}-CO(CH_2)_8CO_2H \quad \longrightarrow \quad Me(CH_2)_3\underset{\underset{OH}{|}}{CH}(CH_2)_8CO_2H$$

Scheme 134

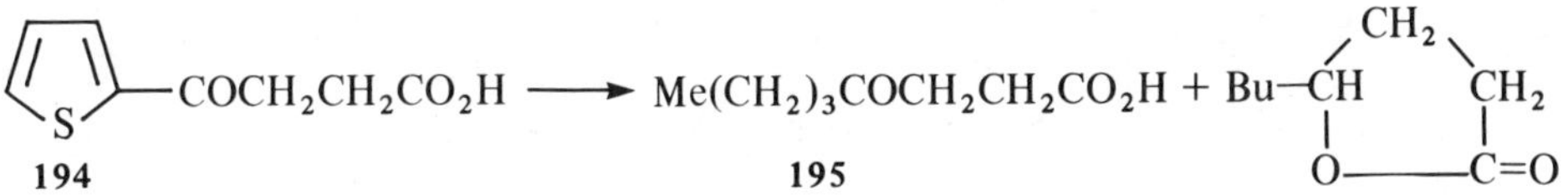

$$\underset{\underset{\textbf{194}}{}}{\underset{S}{\boxed{}}}-COCH_2CH_2CO_2H \quad \longrightarrow \quad \underset{\textbf{195}}{Me(CH_2)_3COCH_2CH_2CO_2H} + \underset{\textbf{196}}{Bu-\overset{\overset{\displaystyle CH_2}{}}{\underset{\underset{O\underline{}C=O}{|}}{CH}}\overset{}{\underset{}{CH_2}}}$$

Scheme 135

Badger, Rodda, and Sasse investigated the influence of the method of Raney nickel preparation and of the quantity of this catalyst on the course of desulfur-ization.[267] They obtained 4-ketocaprylic acid (**195**) in 57% yield from acid (**194**) and from β-(5-bromo-2-thenoyl)propionic acid (**197**) the same acid (**195**) was obtained in the lower yield (25%) due to the formation of the dimerization product (**198**) (Scheme 136).

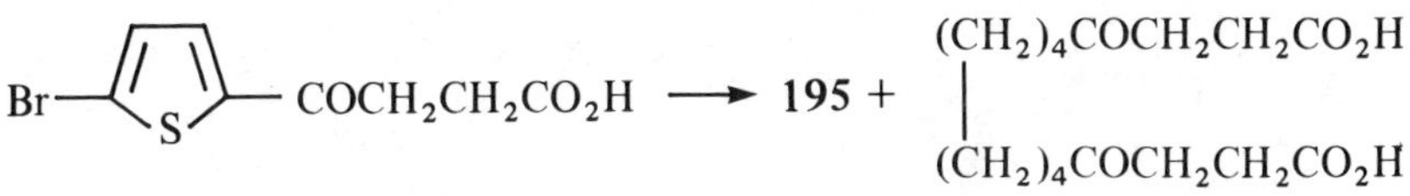

$$Br-\underset{\underset{\textbf{197}}{}}{\underset{S}{\boxed{}}}-COCH_2CH_2CO_2H \quad \longrightarrow \quad \textbf{195} + \underset{\textbf{198}}{\underset{(CH_2)_4COCH_2CH_2CO_2H}{\overset{(CH_2)_4COCH_2CH_2CO_2H}{|}}}$$

Scheme 136

Badger and Sasse[230] transformed β-(2-thenoyl)propionic acid (**194**) into γ-caprylolactone (**196**) and together with this produt a small amount of dihydroxydicarboxylic acid (**199**) was isolated. The branched ketoacid (**201**) was obtained by desulfurization of 3-(β-carboxypropionyl)-2,5-dimethyl-4-ethylthiophene (**200**)[267] (Scheme 137).

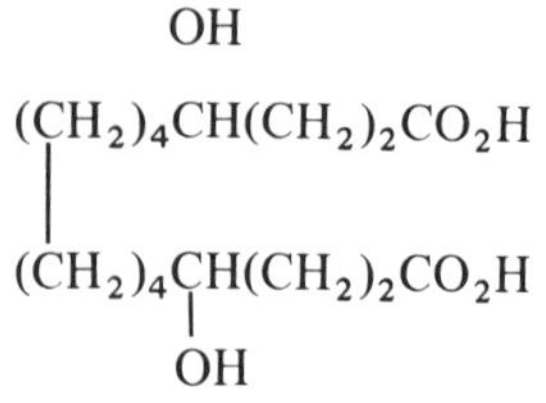
OH
(CH2)4CH(CH2)2CO2H
(CH2)4CH(CH2)2CO2H
OH

199

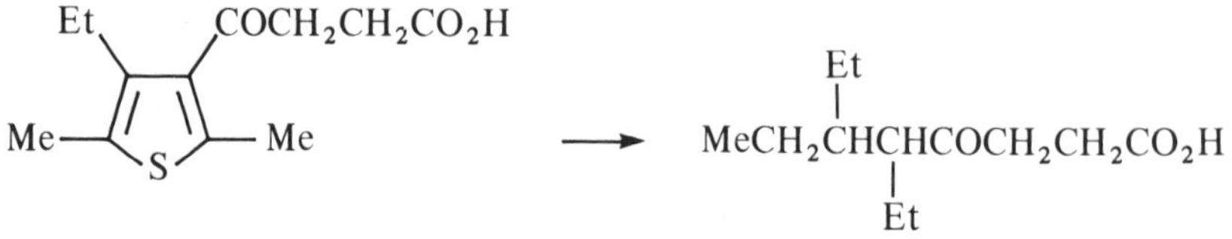
Et
COCH2CH2CO2H
Me
S
Me
Et
MeCH2CHCHCOCH2CH2CO2H
Et

200 **201**

Scheme 137

By the reductive desulfurization of thioindoxylic acid (**202**) Kao, Tilak, and Venkataraman[241] prepared 3-hydroxy-3-phenylpropionic acid with an admixture of 3-phenylpropionic acid (Scheme 138). It should be noted that Papa, Schwenk,

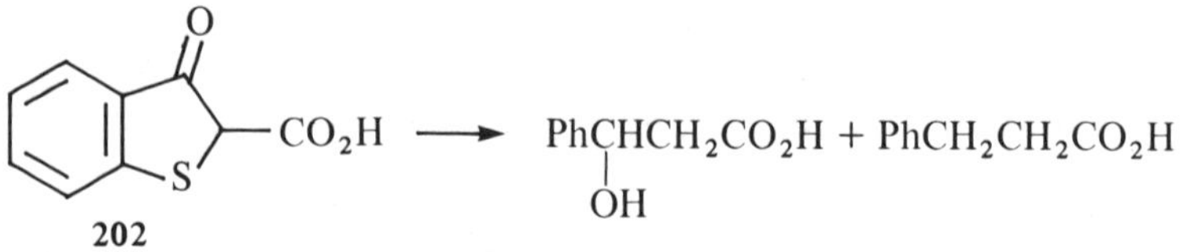
O
CO2H
S
PhCHCH2CO2H + PhCH2CH2CO2H
OH

202

Scheme 138

and Ginsberg[262] obtained mandelic acid and its meta-methylsubstituted derivative from thionaphthenequinone and 5-methylthionaphthenequinone respectively as a result of alkali hydrolysis followed by reductive desulfurization (Scheme 139).

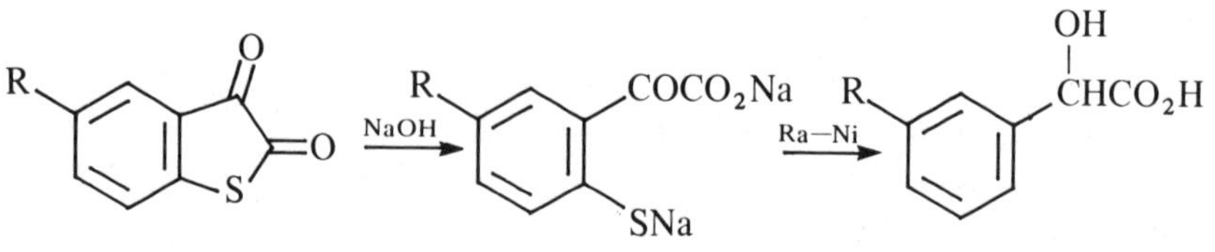
R
O
O
S
NaOH
R
COCO2Na
SNa
Ra–Ni
R
OH
CHCO2H

R = H, Me

Scheme 139

The paper of Grey, McGhie, and Ross[320] gives an idea about factors affecting the formation of aliphatic hydroxy- and ketoacids during the hydrogenolysis of ketoacids of thiophene series (203) and (204). In this publication as well as in the previous paper published by the same group[299] it was shown that the result of desulfurization is determined by the method of Raney nickel preparation. Keto-acids (205), (206) are obtained in 20–80% yields when reductive desulfurization is carried out by the Papa–Schwenk method or with Raney nickel W-6. However with nickel W-7 or nickel prepared according to Brown[223] there are formed hydroxy acids 207 and 208 (Schemes 140 and 141).

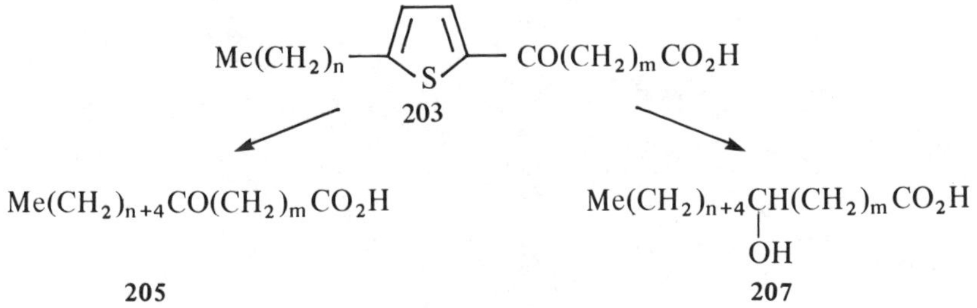

$$Me(CH_2)_n \underset{S}{\overset{}{\fbox{}}} CO(CH_2)_m CO_2H$$

203

$$Me(CH_2)_{n+4}CO(CH_2)_m CO_2H \qquad Me(CH_2)_{n+4}\underset{OH}{CH}(CH_2)_m CO_2H$$

205   207

(a) m = 7, n = 4; (b) m = 8, n = 3; (c) m = 4, n = 7; (d) m = 5, n = 6

Scheme 140

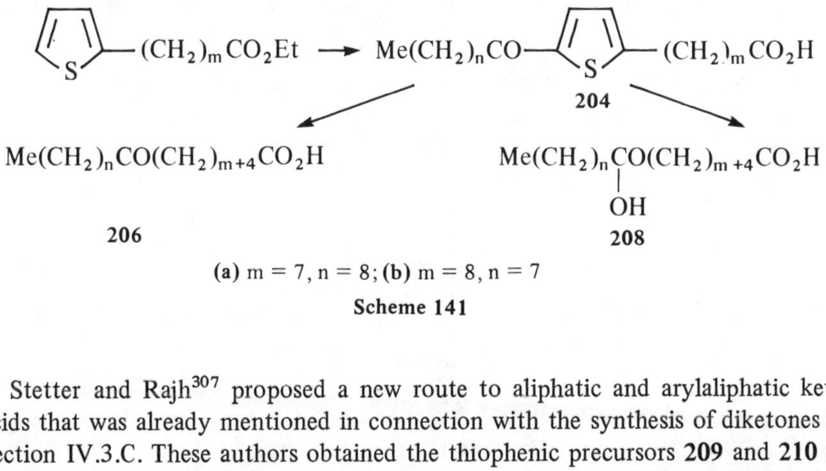

$$\underset{S}{\overset{}{\fbox{}}}(CH_2)_m CO_2Et \longrightarrow Me(CH_2)_n CO \underset{S}{\overset{}{\fbox{}}}(CH_2)_m CO_2H$$

204

$$Me(CH_2)_n CO(CH_2)_{m+4}CO_2H \qquad Me(CH_2)_n \underset{OH}{CO}(CH_2)_{m+4}CO_2H$$

206   208

(a) m = 7, n = 8; (b) m = 8, n = 7

Scheme 141

Stetter and Rajh[307] proposed a new route to aliphatic and arylaliphatic keto acids that was already mentioned in connection with the synthesis of diketones in Section IV.3.C. These authors obtained the thiophenic precursors 209 and 210 in yields of 75–80%, using a novel and interesting procedure, addition of aldehydes of the thiophene series to α,β-unsaturated ketones, esters, or nitriles in the presence of sodium cyanide. The keto esters (209) and keto nitriles (210) obtained in the last two cases are hydrolyzed into keto acids (211). These keto acids are subjected to reductive desulfurization as sodium salts in aqueous solution in the presence of methyl ethyl ketone (Scheme 142). By the use of 2,5-thiophenedicarboxaldehyde, diketodinitrile (212) was obtained which after hydrolysis gave diketodicarboxylic acid (213). Hydrogenolysis of the latter resulted in aliphatic diketodicarboxylic acid (214) (Scheme 143).

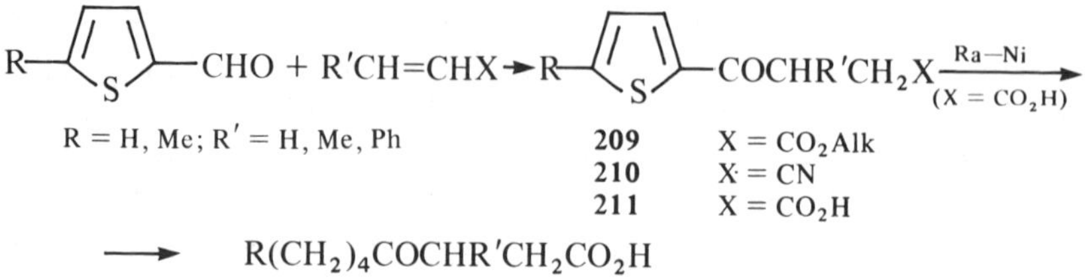

$$R-\underset{S}{\fbox{ }}-CHO + R'CH=CHX \rightarrow R-\underset{S}{\fbox{ }}-COCHR'CH_2X \xrightarrow[(X = CO_2H)]{Ra-Ni}$$

R = H, Me; R' = H, Me, Ph

209	X = CO$_2$Alk
210	X = CN
211	X = CO$_2$H

$$\longrightarrow R(CH_2)_4COCHR'CH_2CO_2H$$

Scheme 142

$$OHC-\underset{S}{\fbox{ }}-CHO + CH_2 = CHCN \longrightarrow$$

$$\longrightarrow XCH_2CH_2CO-\underset{S}{\fbox{ }}-COCH_2CH_2X \longrightarrow$$

| 212 | X = CN |
| 213 | X = CO$_2$H |

$$\xrightarrow[(X = CO_2H)]{Ra-Ni} HO_2CCH_2CH_2CO(CH_2)_4COCH_2CH_2CO_2H$$

214

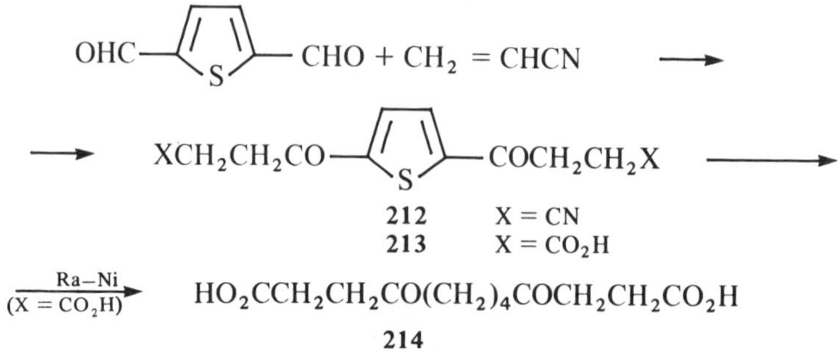

Scheme 143

E.   Synthesis of Aliphatic Amines and Some of Their Derivatives

The paper by Tilak et al.[321] may be considered to be the first publication in which reductive desulfurization of an amine is described. These authors have studied the hydrogenolysis of benzidine sulfone (215) under the action of Raney nickel in ethanol, which leads to N,N'-diethylbenzidine (Scheme 144).

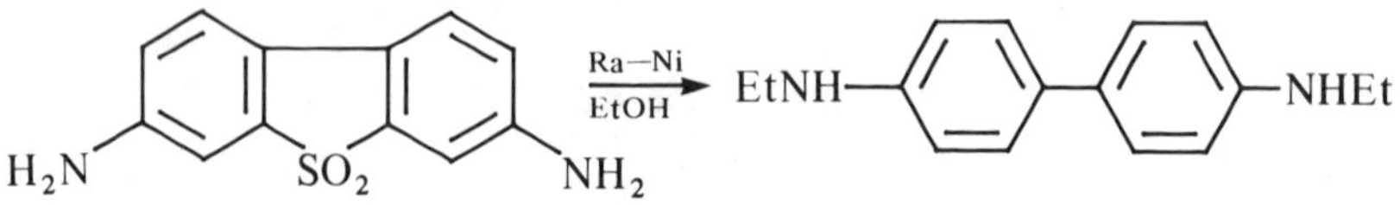

$$H_2N-\underset{SO_2}{\fbox{ }}-NH_2 \xrightarrow[EtOH]{Ra-Ni} EtNH-\fbox{ }-\fbox{ }-NHEt$$

215   **Scheme 144**

Alkylation of primary amines during reductive desulfurization is usually explained as a result of ethanol dehydrogenation under the action of Raney nickel followed by the formation of azomethine and its hydrogenation to a secondary amine. It is, as a rule, an undesirable process making the isolation of individual products difficult. These difficulties do not appear when the synthesis of tertiary amines proposed by Gol'dfarb and Ibragimova[322] is used. This method includes the synthesis of the tertiary amines 216 and 217 by the Leuckart reaction and subsequent desulfurization of these compounds gives tertiary aliphatic amines in 46–66% yields. By introducing the β-hydroxyethyl group into the thiophene ring of the tertiary amine mentioned, one can smoothly obtain the corresponding amino alcohols 218 and 219 after hydrogenolysis[323] (Schemes 145 and 146).

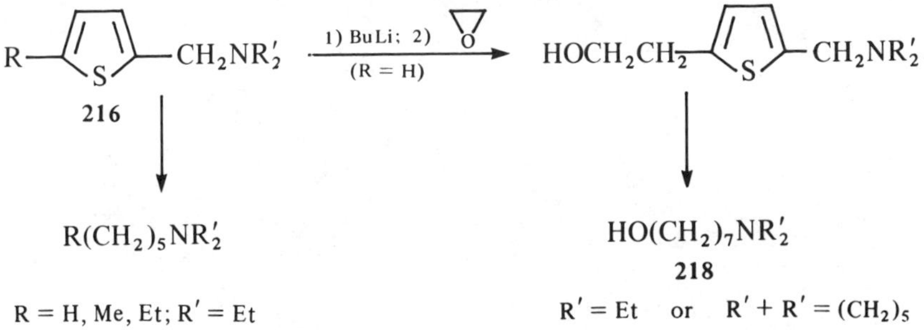

$$R-\underset{S}{[\text{thiophene}]}-CH_2NR_2' \xrightarrow[\text{(R = H)}]{\text{1) BuLi; 2)}\ \triangle O} HOCH_2CH_2-\underset{S}{[\text{thiophene}]}-CH_2NR_2'$$

216

$$R(CH_2)_5NR_2'$$

R = H, Me, Et; R' = Et

$$HO(CH_2)_7NR_2'$$

218

R' = Et   or   R' + R' = $(CH_2)_5$

Scheme 145

$$R''-\underset{S}{[\text{thiophene}]}-\underset{R'}{\overset{R}{\underset{|}{C}}}-\underset{S}{[\text{thiophene}]}-CH_2NEt_2$$

217

Ra—Ni   (R'' = H)   Ra—Ni   (R'' = CH_2CH_2OH)

$$CH_3(CH_2)_3\underset{R'}{\overset{R}{\underset{|}{C}}}(CH_2)_5NEt_2$$

R = Me, R' = Et

$$HO(CH_2)_6\underset{R'}{\overset{R}{\underset{|}{C}}}(CH_2)_5NEt_2$$

219   (a) R = R' = H; (b) R = Me, R' = Et

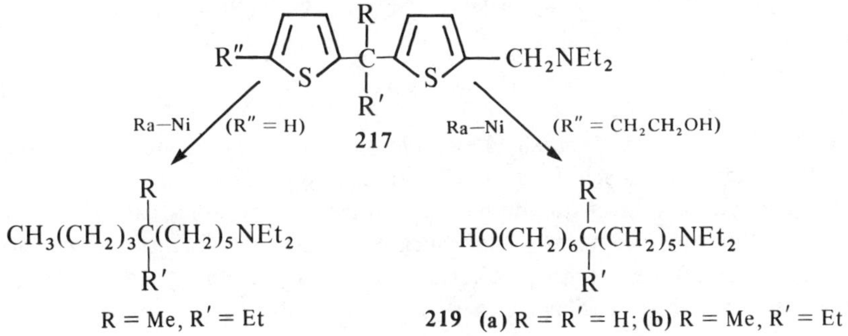

Scheme 146

Gol'dfarb, Fabrichnyi, and Rogovik[324] obtained without difficulty the aliphatic diamines with tertiary and secondary amino groups from diamines **220** and **221** (Scheme 147). The possibility of obtaining nonsymmetrical diamines is the peculiarity of this method. The synthetic routes to the corresponding thiophenic precursors such as **221** are demonstrated in Scheme 148.

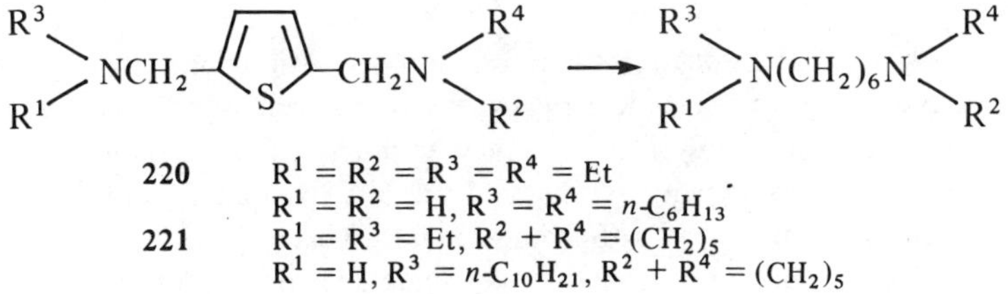

$$\underset{R^1}{\overset{R^3}{\underset{}{}}}NCH_2-\underset{S}{[\text{thiophene}]}-CH_2N\underset{R^2}{\overset{R^4}{}} \longrightarrow \underset{R^1}{\overset{R^3}{}}N(CH_2)_6N\underset{R^2}{\overset{R^4}{}}$$

220 $R^1 = R^2 = R^3 = R^4 = Et$
         $R^1 = R^2 = H, R^3 = R^4 = n\text{-}C_6H_{13}$

221 $R^1 = R^3 = Et, R^2 + R^4 = (CH_2)_5$
         $R^1 = H, R^3 = n\text{-}C_{10}H_{21}, R^2 + R^4 = (CH_2)_5$

Scheme 147

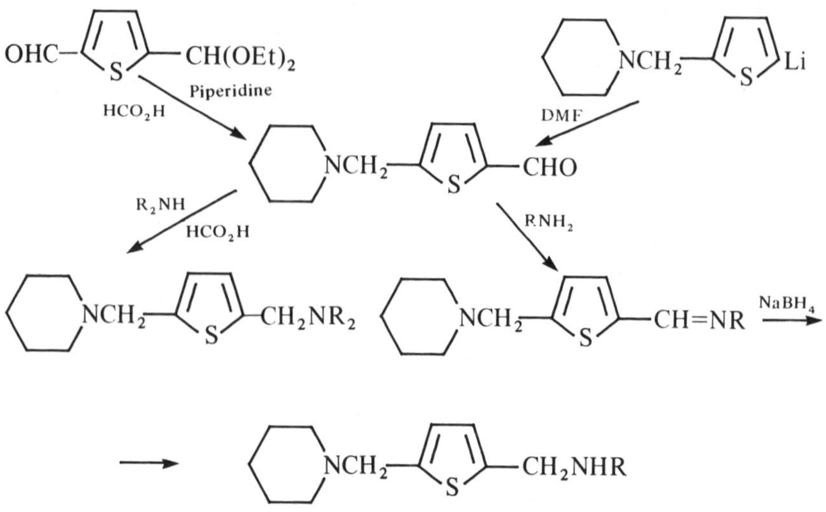

Scheme 148

Pastour and Barrat[325] obtained symmetrical secondary diamines by reductive desulfurization of the Schiff bases (222) that were formed from 2,5-thiophenedicarboxaldehyde and aromatic amines. It is interesting to note that in the case of the azomethine formed from α-naphthylamine, the hydrogenation of one of the benzene rings of the naphthalene residue was observed under the conditions of reductive desulfurization (Scheme 149).

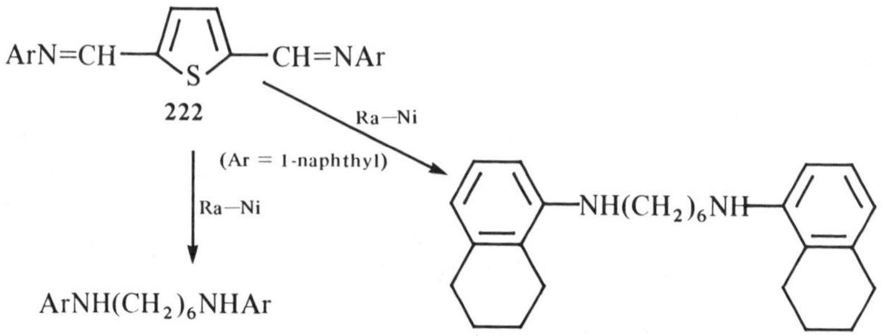

Ar = Ph, p-MeC$_6$H$_4$, m-MeC$_6$H$_4$, p-MeOC$_6$H$_4$, p-EtOC$_6$H$_4$, 2-naphthyl

Scheme 149

The possibilities of the synthesis of primary aliphatic amines from thiophenes have been studied by Gol'dfarb, Krasnyanskaya, and Fabrichnyi.[326] The undesirable alkylation in the hydrogenolysis process may be prevented by acylation of the thiophene amine,[324, 326] as shown in Scheme 150. The diacylated aliphatic diamines were obtained in a similar way[327] (Scheme 151). Červinka, Belovski, and Koralova[328] used desulfurization of the optically active amide 223 for the determination of the absolute configuration of 1-(2-thienyl)-ethylamine (Scheme 152).

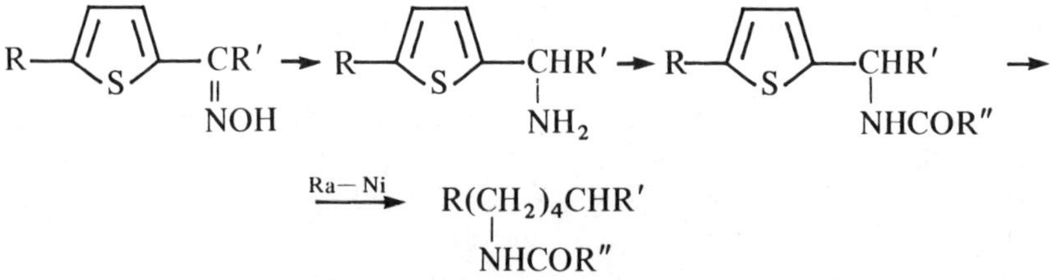

$$R-\underset{S}{\overset{}{\text{thiophene}}}-\underset{\underset{NOH}{\|}}{\overset{}{C}}R' \rightarrow R-\underset{S}{\overset{}{\text{thiophene}}}-\underset{\underset{NH_2}{|}}{\overset{}{C}}HR' \rightarrow R-\underset{S}{\overset{}{\text{thiophene}}}-\underset{\underset{NHCOR''}{|}}{\overset{}{C}}HR' \rightarrow$$

$$\xrightarrow{Ra-Ni} R(CH_2)_4\underset{\underset{NHCOR''}{|}}{\overset{}{C}}HR'$$

R and R' = H or Alk, R" = Me or Ph

Scheme 150

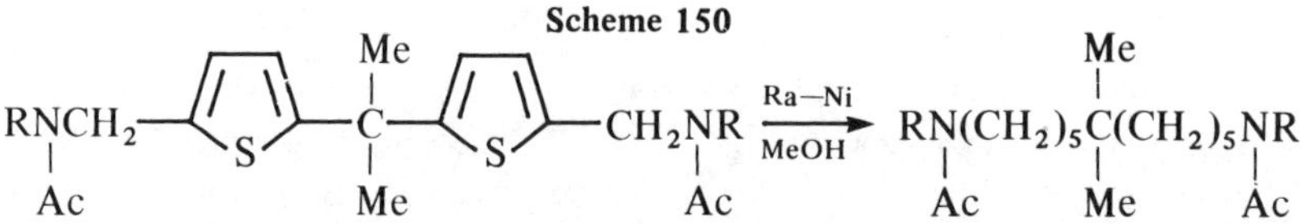

$$RNCH_2-\underset{S}{\overset{}{\text{thiophene}}}-\underset{\underset{Me}{|}}{\overset{\overset{Me}{|}}{C}}-\underset{S}{\overset{}{\text{thiophene}}}-CH_2NR \xrightarrow[MeOH]{Ra-Ni} RN(CH_2)_5\underset{\underset{Me}{|}}{\overset{\overset{Me}{|}}{C}}(CH_2)_5NR$$

with Ac below RNCH2, Me below central C, Ac below CH2NR; on right: Ac, Me, Ac below.

R = H, Me

Scheme 151

$$S(-)\ \underset{S}{\overset{}{\text{thiophene}}}-\underset{\underset{NHCOPh}{|}}{\overset{\overset{H}{|}}{C}}-Me \longrightarrow S(+)\ Me(CH_2)_3-\underset{\underset{NHCOPh}{|}}{\overset{\overset{H}{|}}{C}}-Me$$

223

Scheme 152

By carrying out the reductive desulfurization of thiophene ketoximes in the presence of ammonia,[326] it became possible to obtain also nonacylated primary amines (Scheme 153). Under these conditions, aldoximes undergo Beckmann rearrangement. They were transformed into primary aliphatic amines by the use of only Raney cobalt (as Section IV.3.K).

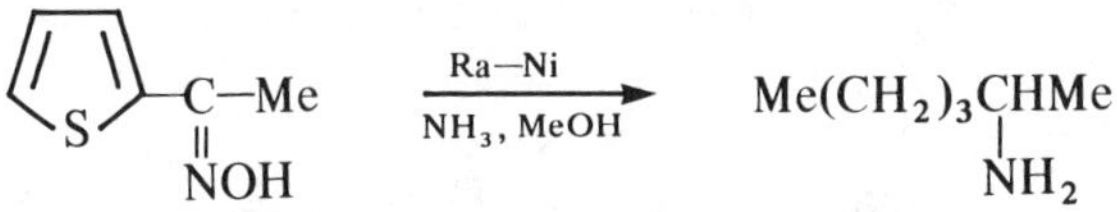

$$\underset{S}{\overset{}{\text{thiophene}}}-\underset{\underset{NOH}{\|}}{\overset{}{C}}-Me \xrightarrow[NH_3, MeOH]{Ra-Ni} Me(CH_2)_3\underset{\underset{NH_2}{|}}{\overset{}{C}}HMe$$

Scheme 153

Boshagen and Geiger[329] developed a novel method of the synthesis of 3-amino-2-acylbenzo[b]thiophenes (224) from the 3-chlorobenzisothiazolium salts (225) and used reductive desulfurization as one of the methods of structure elucidation of the products obtained (Scheme 154). The latter transformation leads to secondary enamino ketones in 75–92% yields.

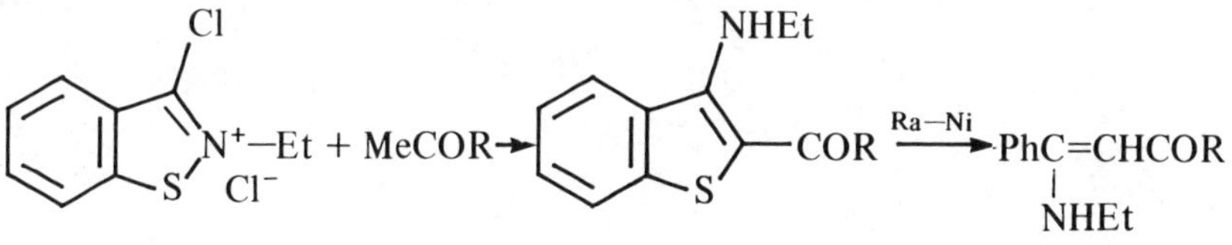

$$\underset{225}{\overset{}{\text{(Cl benzisothiazolium salt)}}} N^+-Et + MeCOR \rightarrow \underset{224}{\overset{NHEt}{\text{(benzothiophene)}}}-COR \xrightarrow{Ra-Ni} \underset{226}{\overset{}{PhC=CHCOR}}$$

with Cl⁻ at 225, NHEt at 226

R = Me, i-Pr, Ph

Scheme 154

F. Synthesis of Amino Acids

The data given above concerning the synthesis of carboxylic acids and amines make it evident that amino acids may also be obtained from thiophenes using reductive desulfurization. However, the investigations of the possibilities of this promising route preceded the study connected with the synthesis of amines from thiophenes. For instance, such important question as retaining the primary amino group under desulfurization conditions was solved first during the study of the synthesis of amino acids.

The routes to various aliphatic amino acids from thiophenes were elaborated mainly by Gol'dfarb, Fabrichnyi, and Shalavina (see Refs. 330 and 100c). Methods for the synthesis of amino acids with different mutual arrangement of amino and carboxy groups, of hydroxyamino acids, aminodicarboxylic acids, and diamino-monocarboxylic acids were developed. Among these products, there were obtained natural compounds and compounds not found in nature, but of interest for the study of their physiological activity or for the synthesis of physiologically active substances. Though all synthesized amino acids have been obtained as racemates, it is in principle possible to obtain optically active substances using the resolution at the stage of the aliphatic or the thiophene acid. Many synthesized amino acids are of interest for the study of polycondensation processes. For the sake of brevity, synthetic methods considered in this section are classified according to the type of thiophenic precursor, but not to the type of final product — the aliphatic amino acids.

a. SYNTHESIS OF α- AND β-AMINO ACIDS OF THE ALIPHATIC SERIES FROM THIOPHENECARBOXALDEHYDES

In the first study devoted to the synthesis of aliphatic amino acids,[331] Gol'dfarb and Fabrichnyi proposed the scheme based on reductive desulfurization of α-amino acids of the thiophene series, which are obtained by the Strecker method from the corresponding thiophenecarboxaldehydes. Using this method, two groups of investigators prepared α-amino acids, with both straight chain (227, R = H, alkyl)[331] (Scheme 155) and branched chain (228)[332,333] (Scheme 156), as well as α-aminodicarboxylic acids (227, R = $(CH_2)_nCO_2H$).[334]

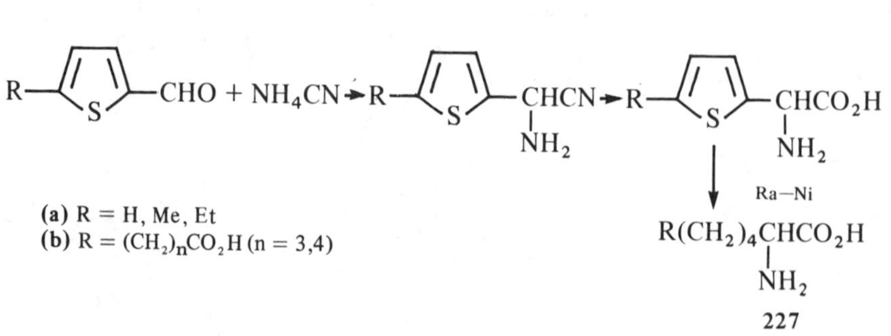

(a) R = H, Me, Et
(b) R = $(CH_2)_nCO_2H$ (n = 3,4)

227

Scheme 155

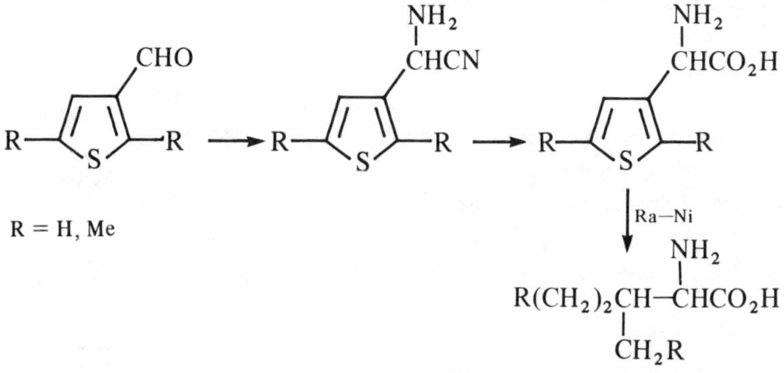

$$R(CH_2)_2CH{-}\underset{\underset{CH_2R}{|}}{\overset{\overset{NH_2}{|}}{C}}HCO_2H$$

228

Scheme 156

Since under the conditions of the Strecker reaction, the yields of α-amino acids of the thiophene series often are rather low, other methods for the preparation of aliphatic amino acids starting from thiophenecarboxaldehydes were developed. One includes the preparation and desulfurization of 2-phenyl-4-(2-thenylidene)-5-oxazolones (**229**). The heating of these oxazolones in methanolic solution in the presence of alkali with Raney nickel makes it possible to carry out in one operation the cleavage of the oxazolone ring, reduction of the C=C double bond in the acylamino acid formed, and reductive desulfurization of the latter[335,336] (Scheme 157). The yields of aliphatic benzoylamino acids (**230**) are up to 90%. Hydrogenolysis of oxazolones in methanol without alkali makes it possible to obtain methyl esters of benzoylamino acids smoothly.

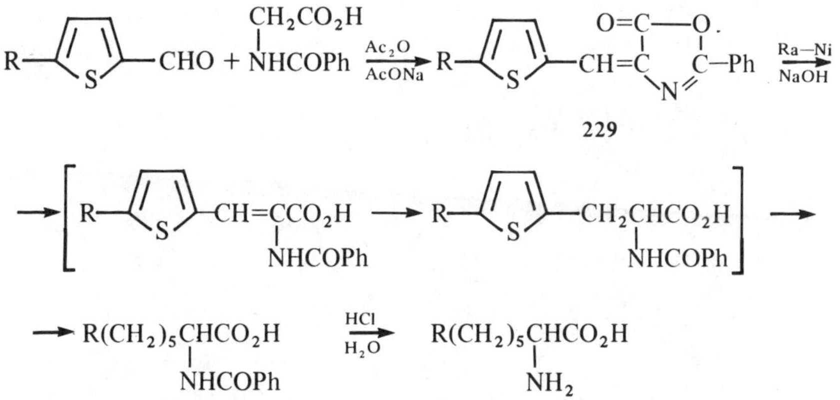

229

230   R = H, Alk
231   R = $(CH_2)_nCO_2H$
232   R = NHR'; $R' = H$, Ac, PhCO
233   R = $(CH_2)_nOR'$; $R' = Me$, n = 0; $R' = H$, Ac, n = 1, 2

Scheme 157

This method is especially convenient for obtaining long-chain amino acids. It should also be noted that, in contrast to the usual syntheses of such acids, the condensation of hippuric acid with aromatic aldehydes proceeds in high yields (60–70%). The method also allows aminodicarboxylic (231),[336] diaminocarboxylic (232),[336,337] and hydroxyamino acids (233)[338] to be obtained.

Another method for the synthesis of aliphatic hydroxyamino acids, proposed by Gol'dfarb, Fabrichnyi, and Rogovik,[338] was used for obtaining α-amino-β-hydroxyenanthic acid (234) from β-(2-thienyl)serine (235). The latter was prepared by the condensation of 2-thiophenecarboxaldehyde with N-benzylidene-glycine (Scheme 158).

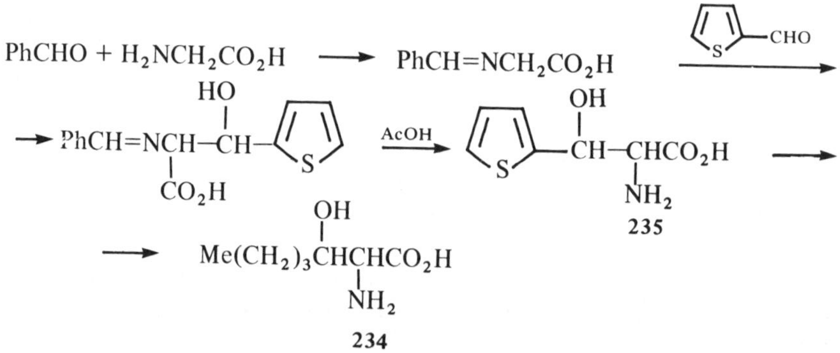

Scheme 158

Unfortunately, desulfurization of β-(2-thienyl)serine gives acid 234 in only 20% yield, so that the synthesis via oxazolones considered above has obvious advantages both in yields and in the variety of amino acids that can be prepared.

Starting from thiophenecarboxaldehydes, one can also obtain β-amino acids, the synthesis of which was developed by Gol'dfarb, Fabrichnyi, and Shalavina[339] (Scheme 159). Its first step consists in the transformation of thiophene aldehydes

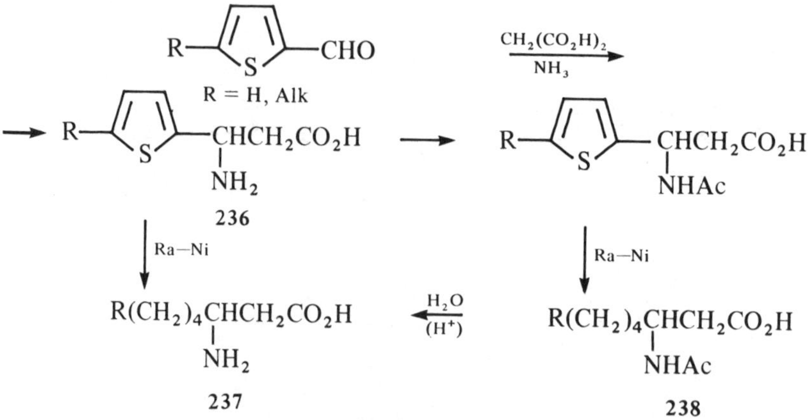

Scheme 159

into β-amino acids (236) using the Rodionov method. (β-Amino acids containing thiophene rings were first prepared by Mamaev, Suvorov, and Rokhlin.[340]) Then these acids are transformed under the action of Raney nickel into aliphatic β-amino acids (237).[339] Desulfurization in the presence of ammonia leads to variable yields (from 80% to zero), depending on the structure of the amino acid. Acylation of the amino group makes it possible in all cases to carry out the reductive desulfurization in good yields, since it not only prevents hydrogenolysis of the C—N bond, but also facilitates the isolation of the products (238) which are then easily hydrolyzed by hydrochloric acid.

b. SYNTHESES OF ALIPHATIC γ- AND δ-AMINO ACIDS FROM NITROTHIOPHENECARBOXYLIC ACIDS

The first paper devoted to the synthesis of aliphatic amino acids[331] proposed a scheme for preparing such acids by desulfurization of nitrothiophenecarboxylic acids and verified this possibility, using 5-nitro-2-thiophenecarboxylic acid, which was transformed into δ-aminovaleric acid in low yield (Scheme 160). Gol'dfarb, Fabrichnyi, and Shalavina[60] studied reductive desulfurization of some 4-nitro-2-thiophenecarboxylic acids and found that these acids are transformed into the corresponding aliphatic γ-amino acids in 50–80% yields (Scheme 161). Optimization of the conditions of reductive desulfurization in the case of 5-nitro-2-thiophenecarboxylic acid[341] made it possible to increase the yield of δ-aminovaleric acid to about 30%. Taking into account the accessibility of nitrothiophenecarboxylic acids and their one-step transformation into aliphatic amino acids, this route for the preparation of γ- and δ-amino acids of aliphatic series is of considerable interest.

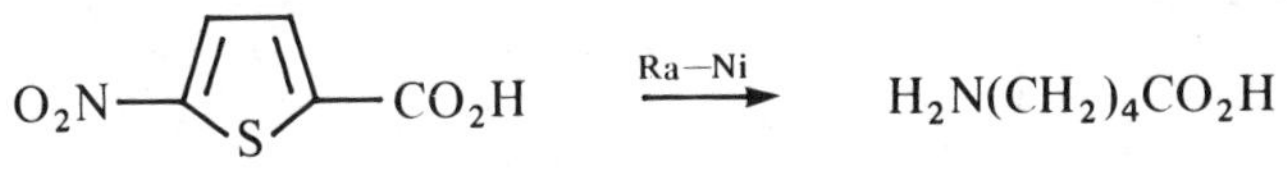

$$O_2N-\!\!\left\langle\!\!\sqrt{}_{S}\right\rangle\!\!-CO_2H \quad\xrightarrow{Ra-Ni}\quad H_2N(CH_2)_4CO_2H$$

Scheme 160

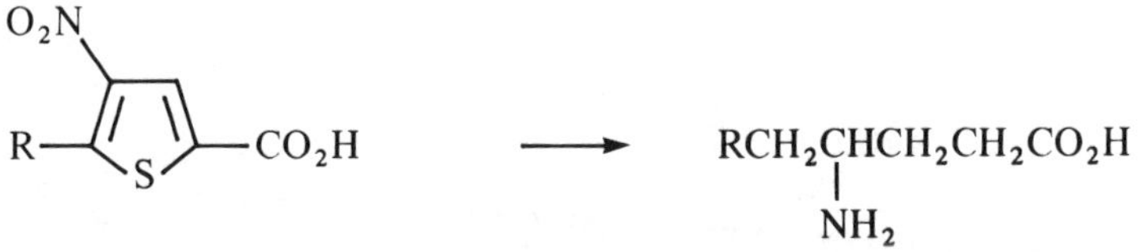

$$R-\!\!\left\langle\!\!\sqrt{}_{S}\right\rangle\!\!-CO_2H \quad\longrightarrow\quad RCH_2CHCH_2CH_2CO_2H$$
$$\underset{NH_2}{|}$$

Scheme 161

The transformation of nitrothiophenecarboxylic acids into aminothiophenecarboxylic acids and then into acylaminothiophenecarboxylic acids with subsequent hydrogenolysis of the latter may be regarded as a variant of the synthesis described above. Acylamino acids can be obtained in one step using reductive acylation of nitrothiophenecarboxylic acids.[59,60] Buu-Hoi and Sy[280] (see also Ref. 59) described the reductive desulfurization of 4- and 5-acetylamino-2-thiophenecarboxylic acids, though without indicating the yields of the formed γ- and δ-aminovaleric acids.

c. SYNTHESES OF ALIPHATIC AMINO ACIDS FROM OXIMINO ACIDS OF THIOPHENE SERIES

Owing to the accessibility of oximino acids and the possibility of their transformations into aliphatic amino acids of a different structure, the techniques based on the desulfurization of oximino acids are the most universal methods used. In many cases, it is advisable to introduce an intermediate stage of the reduction of oximino acid into the amino acid of the thiophene series. All transformations of oximino acids of the thiophene series into various aliphatic amino acids, with the exception of those considered in the following section, may in principle be demonstrated by two schemes.

The syntheses according to the first (Scheme 162) were described by Gol'dfarb, Fabrichnyi, and Shalavina in 1956.[61,332] The first stage of this synthesis is the acylation of thiophenes or its homologues with cyclic anhydrides or with chlorides of dicarboxylic acid halfesters. The keto acids formed are transformed by the action of hydroxylamine into oximino acids (239), which directly or after reduction into amino acids (240) using amalgamated aluminum, are subjected to reductive desulfurization with the formation of aliphatic amino acids (241). McGhie, Ross, and Lanny,[342] following the same scheme, prepared several amino acids from (5-alkyl-2-thienyl)oximinoalkanoic acids (239). Aliphatic amino acids with branched chain (242) were obtained as desulfurization products when position 5 of the thiophene ring was substituted by a branched alkyl.[342,343] Gol'dfarb et al.,[334,344] using as examples α-, γ-, and δ-amino acid syntheses, showed that the method under consideration also can be used for aminodicarboxylic acids (243). This is achieved when using esters of ω-(2-thienyl)alkanoic acids, instead of thiophene homologues, in the first step.

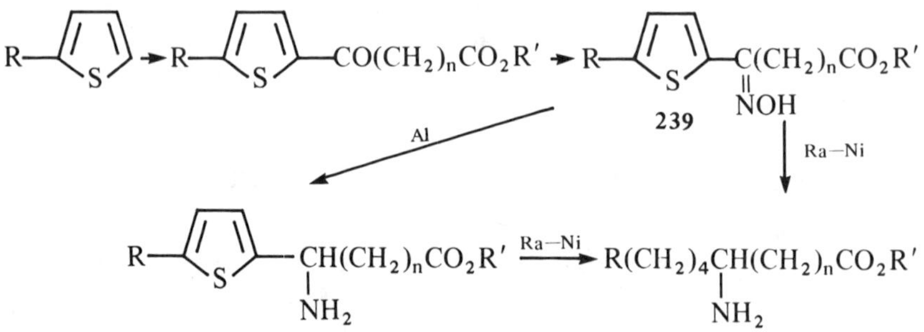

$$R\text{-}\underset{S}{\text{thienyl}} \rightarrow R\text{-}\underset{S}{\text{thienyl}}\text{-}CO(CH_2)_nCO_2R' \rightarrow R\text{-}\underset{S}{\text{thienyl}}\text{-}\underset{\underset{NOH}{\|}}{C}(CH_2)_nCO_2R'$$

239

Al

Ra—Ni

$$R\text{-}\underset{S}{\text{thienyl}}\text{-}\underset{\underset{NH_2}{|}}{C}H(CH_2)_nCO_2R' \xrightarrow{Ra-Ni} R(CH_2)_4\underset{\underset{NH_2}{|}}{C}H(CH_2)_nCO_2R'$$

240   R' = H, Alk

241   R = H, n-Bu
       n = 0, 2–5, 7, 8

242   R = i-Bu, t-Bu,
       EtCHCH₂
              |
              Me
       n = 2, 3, 8

243   R = (CH₂)ₘCO₂H
       m = 1, 3–5; n = 0, 2, 3

Scheme 162

The second route (Scheme 163) was also proposed and developed by Gol'dfarb and coworkers.[68,345,346] As in the synthesis of aminodicarboxylic acids, esters of ω-thienylalkanoic acids are used, which are then formylated or acylated with monocarboxylic acid chlorides. The following transformations do not differ from those shown in the previous scheme. The distinguishing feature of the second scheme is that it makes possible the preparation of amino acids in which amino and carboxy groups are far from each other, in particular, ω-amino acids (244). These acids are of interest for the study of polycondensation. The same scheme was used by McGhie et al.[342] for the synthesis of 10-aminoundecanoic acid. The preparation of the ω-amino acids with a branched chain was shown by Fabrichnyi, Shalavina, and Gol'dfarb, as examplified by the synthesis of 10-amino-3-methylcapric acid from the corresponding oximino ester[345] (Scheme 164).

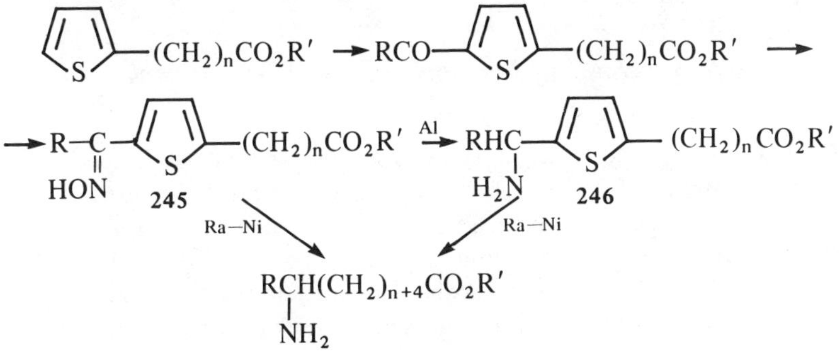

Scheme 163

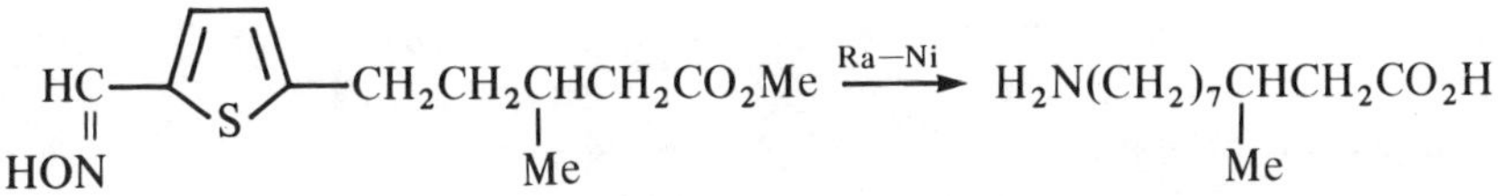

Scheme 164

In Ref. 343, the same authors compared the results of the synthesis of six aliphatic amino acids, which was carried out by the three methods: (A) direct desulfurization of oximino acids 239 and 245 ($R' = H$), (B) desulfurization of their esters 239 and 245 ($R' = Alk$), and (C) desulfurization of acetylamino acid esters 240 and 246 ($R' = Alk$) obtained by reduction of oximino acids followed by acetylation. The yields (calculated on the starting oximino acids) are given in Table 1. They show that for different cases there are different optimal ways of converting thiophene oximino acids into aliphatic amino acids. But none of the routes gives any acceptable results in the case of 10-amino-16-methylheptadecanoic and 12-aminotridecanoic acids.

TABLE 1.   PREPARATION OF ALIPHATIC AMINO ACIDS FROM OXIMINO ACIDS OF
            THE THIOPHENE SERIES 239 AND 245 USING DIFFERENT PROCEDURES[343]

| Amino acids | Yields (%)^a | | |
(241, 242, and 244; R' = H)	A	B	C
4-Aminooctanoic acid (241), n = 2, R = H	40	5	26
7-Aminoundecanoic acid (241), n = 5, R = H	51	0	0
5-Amino-11-methyldodecanoic acid (242),			
n = 3, R = i-Bu	7	28	0
10-Amino-16-methylheptadecanoic acid (242),			
n = 8, R = i-Bu	7	0	0
12-Aminododecanoic acid (244), n = 6, R = H	0	–	42
12-Aminotridecanoic acid (244), n = 6, R = Me	6	5	9

^{a} Calculated on starting oximino acids.

G.   Synthesis of Lactams and Their Transformations

Gol'dfarb et al. developed the syntheses of lactams of the thiophene series as well as their transformations into C-alkyl- and C-cycloalkyl-substituted ε-capro- and ζ-enantho-lactams, aminodicarboxylic and diaminocarboxylic acids, and derivatives of 2-oxohexahydropyrimidine and 2-oxoimidazolidine. These compounds proved to be of interest for the investigation of polymerization processes and synthesis of new polyamides as well as physiologically active compounds.

a.   SYNTHESES AND SOME PROPERTIES OF SUBSTITUTED
ε-CAPROLACTAMS AND ζ-ENANTHOLACTAMS

For the synthesis of capro- and enantholactams, use was made of reductive desulfurization of systems in which the lactam ring is fused with the thiophene ring. In most of the syntheses considered below, the key intermediates are thiophene lactams of two types (247 and 248) formed from the corresponding ketones (249) upon Schmidt rearrangement or from their oximes (250) by Beckmann rearrangement (Schemes 165 and 166). According to the first scheme proposed by Gol'dfarb et al.,[346-348] thiophene or 2-methylthiophene used as starting compounds are transformed by the usual methods (see Refs. 2 and 349) into bicyclic ketones (249), oximes of which undergo Beckmann rearrangement. When carrying out the latter, using the action of benzenesulfonyl chloride, there are obtained in high yields (75–95%) lactams (247) having the amino group directly attached to the β-position of the thiophene ring with a small admixture of the isomeric lactam (248), which is easily removed by recrystallization.[347,350,351] Reductive desulfurization of the lactams (247) leads to alkylsubstituted ε-capro- and ζ-enantholactams (251).

In this way, ε-ethyl-ε-caprolactam, ζ-ethyl-ζ-enantholactam, ε-propyl-ε-caprolactam, ζ-propyl-ζ-enantholactam, and β-methyl-ζ-ethyl-ζ-enantholactam have been obtained. Upon hydrolysis of these substances, the corresponding ε- and ζ-amino acids (252) were obtained[346,348] (Scheme 165).

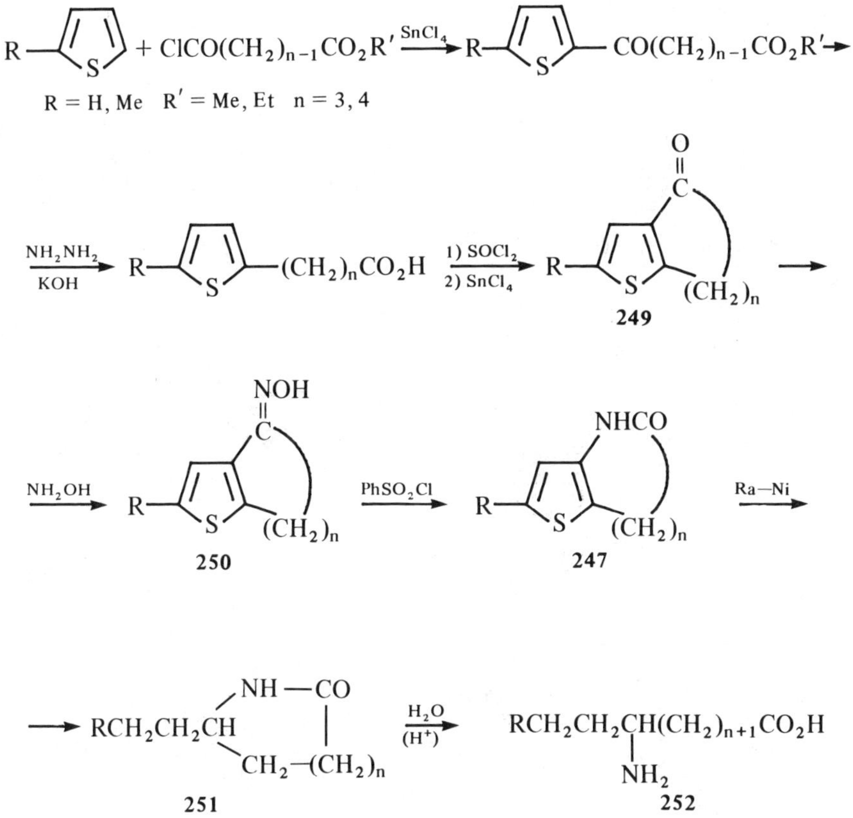

$$R-\underset{S}{\underset{||}{\bigcirc}} + ClCO(CH_2)_{n-1}CO_2R' \xrightarrow{SnCl_4} R-\underset{S}{\underset{||}{\bigcirc}}-CO(CH_2)_{n-1}CO_2R' \rightarrow$$

R = H, Me   R' = Me, Et   n = 3, 4

$$\xrightarrow[KOH]{NH_2NH_2} R-\underset{S}{\underset{||}{\bigcirc}}-(CH_2)_nCO_2H \xrightarrow[2)\ SnCl_4]{1)\ SOCl_2} R-\underset{S}{\overset{\overset{O}{\parallel}}{\bigcirc}}\underset{(CH_2)_n}{\bigcirc} \longrightarrow$$

249

$$\xrightarrow{NH_2OH} R-\underset{S}{\overset{\overset{NOH}{\parallel}}{\bigcirc}}\underset{(CH_2)_n}{\bigcirc} \xrightarrow{PhSO_2Cl} R-\underset{S}{\overset{NHCO}{\bigcirc}}\underset{(CH_2)_n}{\bigcirc} \xrightarrow{Ra-Ni}$$

250   **247**

$$\longrightarrow RCH_2CH_2CH\underset{CH_2-(CH_2)_n}{\overset{NH-CO}{\diagup}}\Big| \xrightarrow[(H^+)]{H_2O} RCH_2CH_2\underset{NH_2}{\overset{|}{CH}}(CH_2)_{n+1}CO_2H$$

251   **252**

Scheme 165

For the synthesis of other lactams (**253**) bearing an alkyl substituent attached to the α-carbon atom of the amino acid chain, the Beckmann rearrangement with the aid of polyphosphoric acid and the Schmidt rearrangement were used. In the case of thiophenocyclohexanone oximes, the first of these methods leads to mixtures of seven-membered lactams with predominance of the desired and easily isolated isomers (**248**).[348] The rearrangement of the seven-membered oxime, however, gives the same product (**247**) as that formed by the action of benzenesulfonyl chloride. By treating the ketone (**249**, n = 3, R = H) with sodium azide in polyphosphoric acid, the desired lactam of 2-(γ-aminopropyl)-3-thiophenecarboxylic acid (**248**, n = 3) was obtained in 70% yield, containing only a small amount of the undesirable isomer **247**.[348,352] However, in the case of the 5-methyl-substituted ketone **249**, (n = 3, R = Me), the percentage of the undesirable isomer is comparatively high and it is advisable to prepare the lactam (**248**, n = 3, R = Me) with the aid of the Beckmann rearrangement.[348] Significant difficulties arise when trying to prepare α-alkyl-ζ-enantholactams, since the action of sodium azide in polyphosphoric acid does not lead to lactams in the case of seven-membered ketones.[348] The necessary precursors (248, n = 4) have been isolated in small amounts either from the mixture of lactams formed by Beckmann rearrangement of the oxime (250, n = 4, R = H) or by the action of benzenesulfonyl chloride on the geometric isomer formed as a by-product during the oximation of the seven-membered ketone (249, n = 4, R = Me).[351] Reductive desulfurization of lactams (248) leads smoothly to α-ethyl- and α-propyl-substituted lactams (253), the hydrolysis of which results in the corresponding acids (254)[348] (Scheme 166).

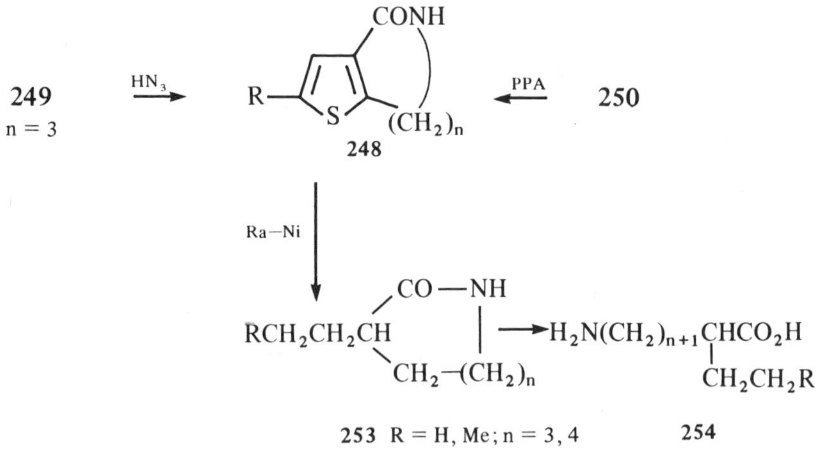

Scheme 166

The preparation of the other two types of lactams (255 and 256), which differ in the position of the alkyl substituent in the ring, is described in Ref. 353. The starting compound, 2,5-dichlorothiophene, is acylated with chlorides of succinic or glutaric acid half-ester. After selective dechlorination of the obtained keto acids (257) by the action of copper in propionic acid, reduction of the keto group and cyclization of 5-chloro-3-thienylalkanoic acid (258) is carried out. The obtained ketones (259) are directly, or through oximes (260), transformed into lactams (261 and 262). It should be noted that independently of the method used, only one of the two possible isomeric lactams (261) was obtained from the six-membered ketone (259) and, under conditions of Beckmann rearrangement with polyphosphoric acid, the oxime of the seven-membered ketone was transformed into a mixture of lactams (261 and 262), which were separated by crystallization. During reductive desulfurization of thiophene lactams, the hydrogenolysis of the C—Cl bond also occurs which leads to β-ethyl-ε-caprolactam (255, n = 3), β-ethyl-ζ-enantholactam (255, n = 4), and ε-ethyl-ζ-enantholactam (256)[353] (Scheme 167).

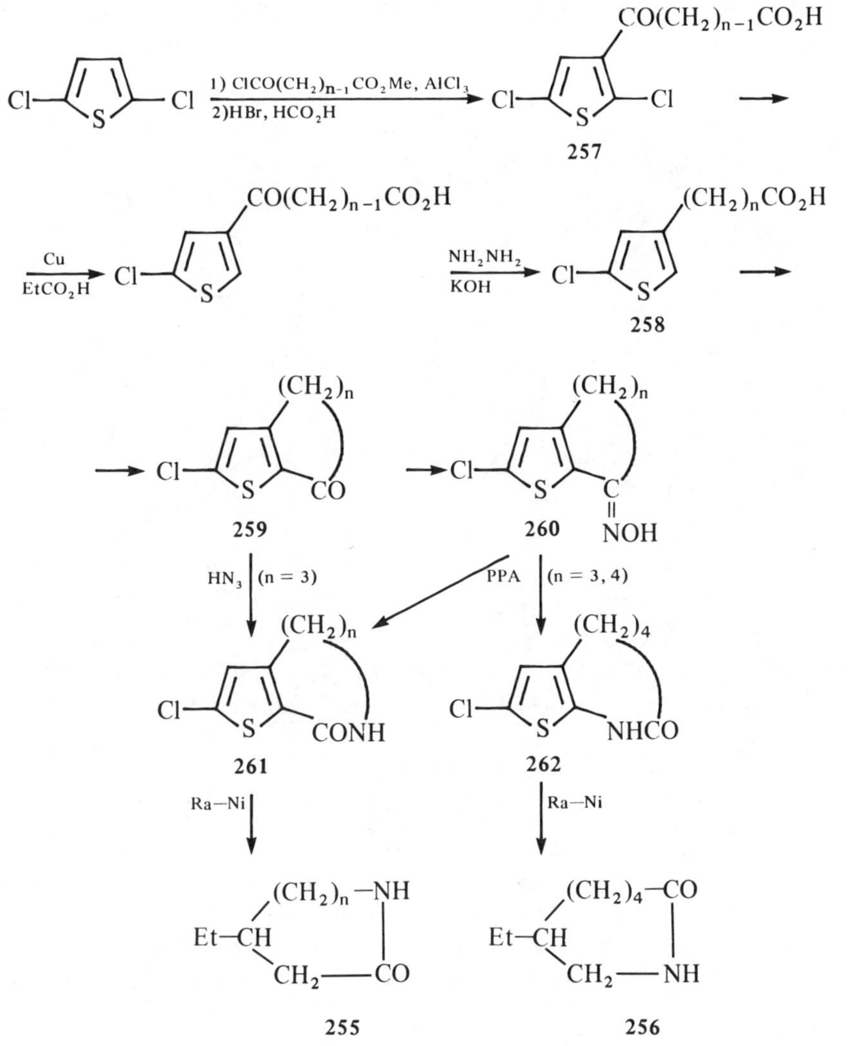

$CO(CH_2)_{n-1}CO_2H$

Cl—[thiophene]—Cl 1) $ClCO(CH_2)_{n-1}CO_2Me$, $AlCl_3$
2) HBr, HCO_2H → Cl—[thiophene]—Cl →

257

$CO(CH_2)_{n-1}CO_2H$

$\xrightarrow[EtCO_2H]{Cu}$ Cl—[thiophene]

$\xrightarrow[KOH]{NH_2NH_2}$ Cl—[thiophene]—$(CH_2)_nCO_2H$ →

258

$(CH_2)_n$

→ Cl—[thiophene]—CO

259

$(CH_2)_n$

→ Cl—[thiophene]—C
‖
NOH

260

HN_3 | (n = 3)   PPA | (n = 3, 4)

$(CH_2)_n$

Cl—[thiophene]—CONH

261

$(CH_2)_4$

Cl—[thiophene]—NHCO

262

Ra—Ni | | Ra—Ni

$(CH_2)_n$—NH

Et—CH
CH_2——CO

255

$(CH_2)_4$—CO

Et—CH
CH_2——NH

256

Scheme 167

The synthesis of C-cyclohexyl- and C-cycloheptyl-substituted ϵ-caprolactams (**263**) and ζ-enantholactams (**264**) was described in Ref. 354. The authors used as starting compounds the thiophenocycloalkanones described above, (**249**, n = 3, 4, R = H), which were reduced into cyclohexa- and cyclohepta[b]thiophenes (**265**), and the latter were subjected to a number of transformations similar to those described for the synthesis of bicyclic lactams and shown in Scheme 168. By the action of benzenesulfonyl chloride, oximes of the tricyclic ketones (**266**) are obtained and undergo Beckmann rearrangement leading to lactams (**267**), with the amino-group directly linked to the thiophene ring. The reductive desulfurization of these lactams gives, in yields of 70–90%, ε-cycloalkyl-ε-caprolactams and ζ-cycloalkyl-ζ-enantholactams (263), which are hydrolized into the corresponding amino acids (268).

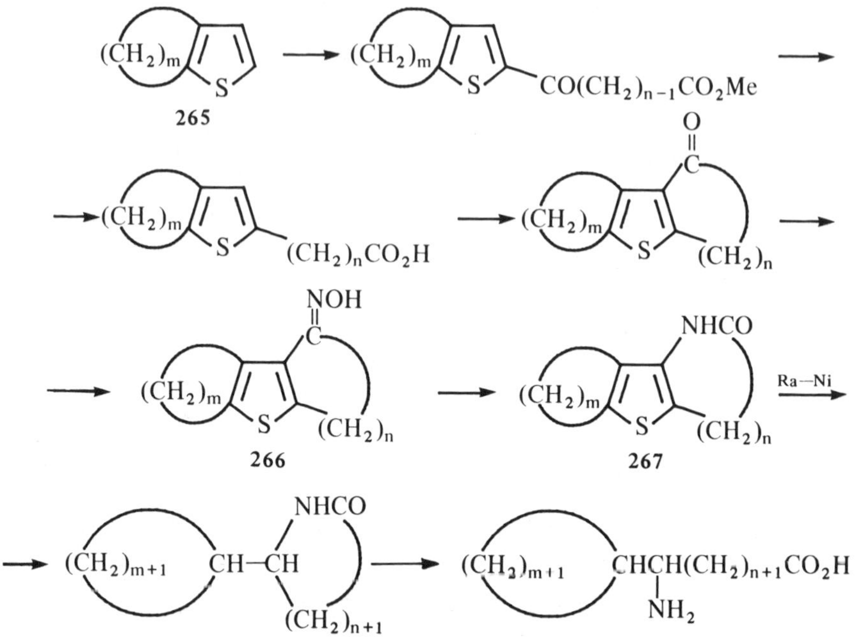

265

→ $CO(CH_2)_{n-1}CO_2Me$

→ $(CH_2)_n CO_2H$

266

267   Ra—Ni

263   m = 4, 5; n = 3, 4   268

Scheme 168

After the treatment of one of the oximes (266, m = 4, n = 3) with polyphosphoric acid, a mixture of lactams was obtained, from which the lactam of 2-(γ-aminopropyl)-4,5,6,7-tetrahydrobenzo[b]thiophene-3-carboxylic acid (269) was isolated and transformed upon hydrogenolysis into α-cyclohexyl-ε-caprolactam (264); the hydrolysis of the latter gave α-cyclohexyl-ε-aminocaproic acid (270) (Scheme 169).

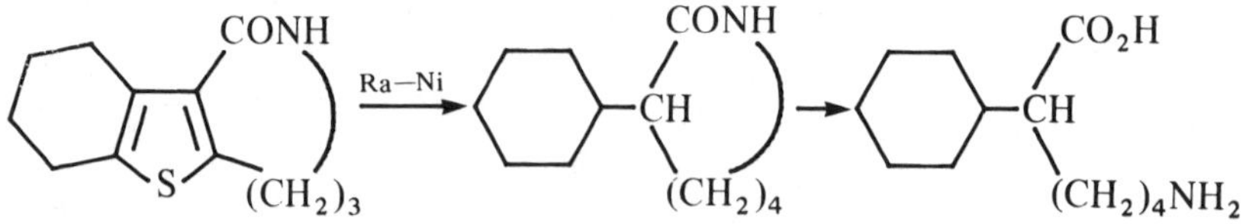

269   264   270

Scheme 169

The methods of lactam synthesis from thiophene derivatives considered above have certain advantages over other routes when obtaining individual compounds. This depends in particular on the definite directing effect of the thiophene ring in the course of Beckmann rearrangement, leading in a number of cases to a single isomer; in the case of alkylcycloalkanone oximes, Beckmann rearrrangement always proceeds in two possible directions. The indicated feature, as well as the formation of lactams with different positions of alkyl groups in the nuclei, allowed the kinetics to be studied and the thermodynamic parameters of polymerization for a large set of alkyl- and cycloalkyl-substituted ε-capro- and ζ-enantholactams to be determined.[355-360]

In particular, the data obtained show that the length of an alkyl chain (ethyl or propyl) affects very little the ability of lactams to be polymerized. Alkyl substituents reduce considerably the rate of polymerization of ε-caprolactams and cycloalkyl-substituted ε-caprolactams do not polymerize at all. At the same time, the presence of substituents in the ζ-enantholactam molecule has little effect on its polymerization ability.

The thiophene lactams and corresponding oximes are of interest not only for the preparation of C-substituted ε-capro- and ζ-enantholactams. They are also interesting as physiologically active substances, in particular those possessing both sedative and myorelaxant activity.[361] These substances, as demonstrated below, may also be used in the synthesis of other physiolocally active compounds.

b. SOME SYNTHESES BASED ON LACTAMS

Lactams of the thiophene series proved to be intermediates in the synthesis of lactams of aliphatic aminodicarboxylic and diaminomonocarboxylic acids, which can then be transformed smoothly into corresponding acids. The synthesis of some ε-aminodicarboxylic acids (271) through the corresponding lactams developed by Gol'dfarb et al.[362] is shown in Scheme 170. The yields at the stage of reductive

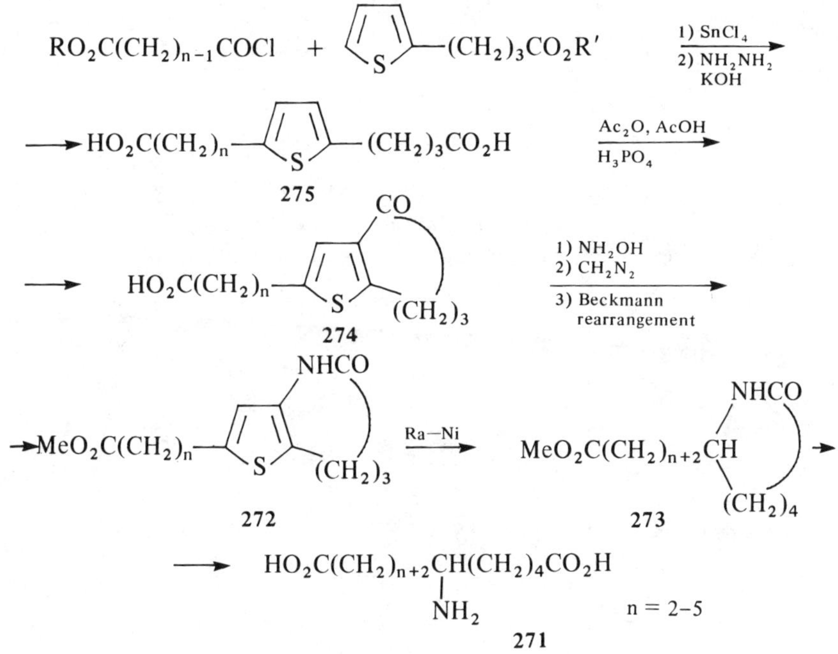

$RO_2C(CH_2)_{n-1}COCl$ + [thiophene]—$(CH_2)_3CO_2R'$ $\xrightarrow[\text{2) } NH_2NH_2]{\text{1) } SnCl_4}$
KOH

$\longrightarrow HO_2C(CH_2)_n$—[thiophene]—$(CH_2)_3CO_2H$ $\xrightarrow[H_3PO_4]{Ac_2O, AcOH}$

275

$\longrightarrow HO_2C(CH_2)_n$—[thiophene ring with CO and (CH$_2$)$_3$]

274

$\xrightarrow[\text{3) Beckmann rearrangement}]{\text{1) } NH_2OH}$ 2) CH_2N_2

$\longrightarrow MeO_2C(CH_2)_n$—[thiophene ring with NHCO and (CH$_2$)$_3$] $\xrightarrow{Ra-Ni}$ $MeO_2C(CH_2)_{n+2}\overset{NHCO}{\underset{(CH_2)_4}{CH}}$ $\longrightarrow$

272   **273**

$\longrightarrow HO_2C(CH_2)_{n+2}\underset{NH_2}{CH}(CH_2)_4CO_2H$ $n = 2-5$

271

Scheme 170

desulfurization of lactams (272) are 75–85%. The hydrolysis of aliphatic lactams (273) and the transformation of the obtained hydrochlorides into amino acids also proceed in high yields. The interesting feature of this scheme is that bicyclic keto acids (274) are prepared directly from the corresponding dicarboxylic acids (275) with the formation of only six-membered rings.

Preparation of diaminocarboxylic acids via the corresponding nitrolactams considerably expands the possibilities of their synthesis.[363] In contrast to the synthesis discussed above, through oxazolones leading to diamino acids with one amino group in the α-position to the carboxy group (Section IV.3.F.a), the route proposed by Fabrichnyi, Shalavina, and Gol'dfarb in Ref. 363, makes it possible to obtain diaminocarboxylic acids of various types. One of the variants of these syntheses leads to α-(aminoalkyl)-ε-caprolactams 276 and 277 and includes nitration of the thiophene lactam 278 in the free α-position of the thiophene ring or in the β-position if the α-position is blocked by the methyl group or the chlorine atom, the latter being eliminated in the course of reductive desulfurization (Scheme 171).[363]

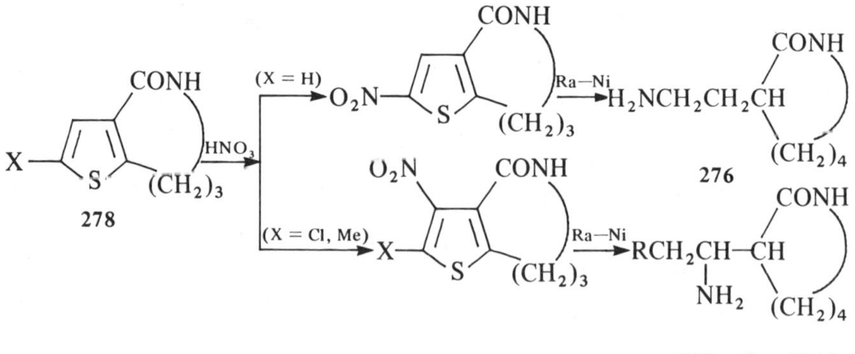

Scheme 171   277   R = H, Me

In another modification of the synthesis,[363,364] ε-(aminoalkyl)-ε-caprolactams (279, 280, n = 3) and also ζ-(aminoalkyl)-ζ-enantholactams (279, 280, n = 4) are obtained (Scheme 172). This variant also includes nitration of the thiophene

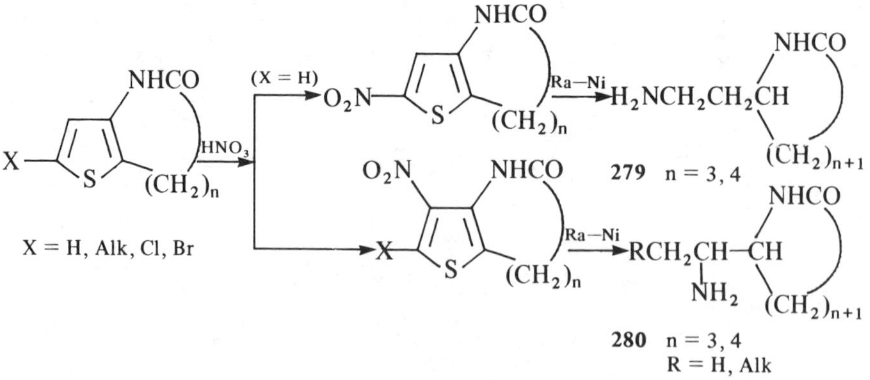

Scheme 172

lactams in the α- or β-position of the thiophene ring. In the second case, the α-position should be occupied by an alkyl group, chlorine, or bromine. In conditions of reductive desulfurization, the hydrogenolysis of C—Cl and C—Br bonds takes place, which results in the formation of lactams of aliphatic diaminocarboxylic acids that do not contain halogen.

Hydrolysis of diaminocarboxylic acid lactams leads smoothly to the corresponding amino acids. In the case of ε-(aminoalkyl)-ε-caprolactams and ζ-(aminoalkyl)-ζ-enantholactams, diaminocarboxylic acids **281** and **282** are formed in which the amino groups are separated by two- or three-carbon fragments. These structural peculiarities were used for the preparation of 2-oxoimidazolidine (**283**)[364] (Scheme 173) and 2-oxohexahydropyrimidine (**284**)[365] derivatives (Scheme 174), aided by the reaction of the above amino acids with urea.

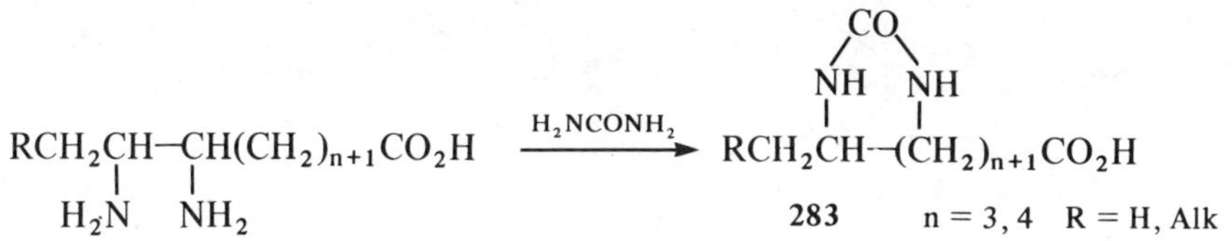

$$\underset{\underset{\textbf{281}}{\overset{|}{\underset{H_2N}{}}\ \overset{|}{\underset{NH_2}{}}}{RCH_2CH-CH(CH_2)_{n+1}CO_2H}\ \xrightarrow{H_2NCONH_2}\ \underset{\textbf{283}\quad n=3,4\quad R=H, Alk}{RCH_2CH-(CH_2)_{n+1}CO_2H}$$

$$\overset{\overset{CO}{\diagup\ \diagdown}}{\underset{\overset{|}{NH}\ \ \overset{|}{NH}}{}}$$

Scheme 173

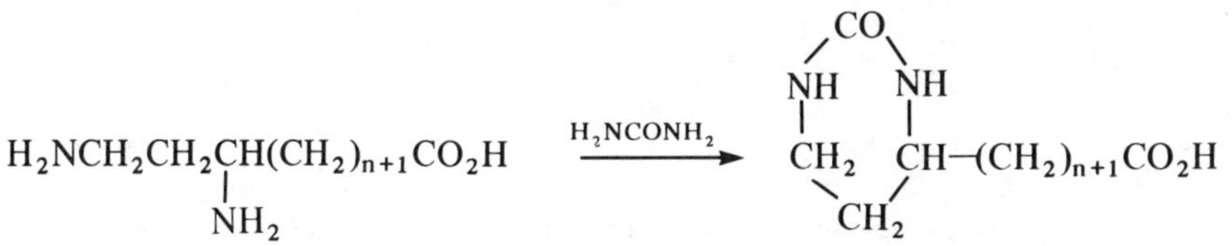

$$\underset{\underset{\textbf{282}}{\overset{|}{NH_2}}}{H_2NCH_2CH_2CH(CH_2)_{n+1}CO_2H}\ \xrightarrow{H_2NCONH_2}\ \underset{\textbf{284}\quad n=3,4}{\underset{\overset{\diagdown}{CH_2}}{CH_2\quad CH-(CH_2)_{n+1}CO_2H}}$$

$$\overset{\overset{CO}{\diagup\ \diagdown}}{\underset{\overset{|}{NH}\ \ \overset{|}{NH}}{}}$$

Scheme 174

In a similar way, (2-oxohexahydro-4-pyrimidinyl)alkanoic acids (**285**) with a branched chain were also obtained.

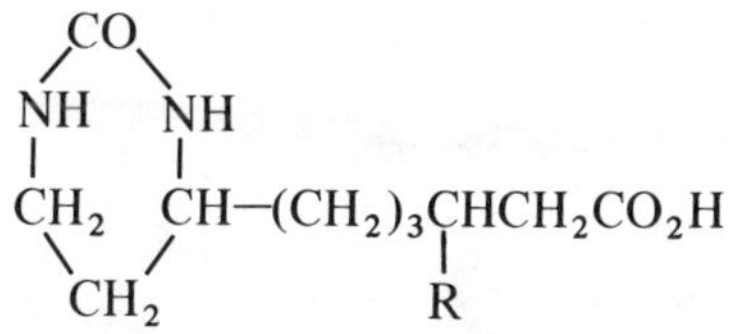

$$\overset{\overset{CO}{\diagup\ \diagdown}}{\underset{\overset{\diagdown}{CH_2}}{\underset{R}{CH_2\quad CH-(CH_2)_3CHCH_2CO_2H}}}$$

285   R = Me, Et

The 2-oxoimidazolidine and 2-oxohexahydropyrimidine derivatives are of interest for the study of their biological activity. In particular, dethiobiotin (**283**, R = H, n = 4) and dethionorbiotin (**283**, R = H, n = 3) were among the synthesized substances which, however, were obtained as mixtures of diastereoisomers. Dethiobiotin was also prepared with the help of reductive desulfurization of 2,3,4,5-

tetradehydrobiotin (286),[366] which in its turn was obtained from the lactam of
δ-(3-amino-2-thienyl)valeric acid (247, R = H, n = 4), as shown in Scheme 175.[48]
The obtained sample of dethiobiotin increased considerably the fodder yeast
growth.[367]

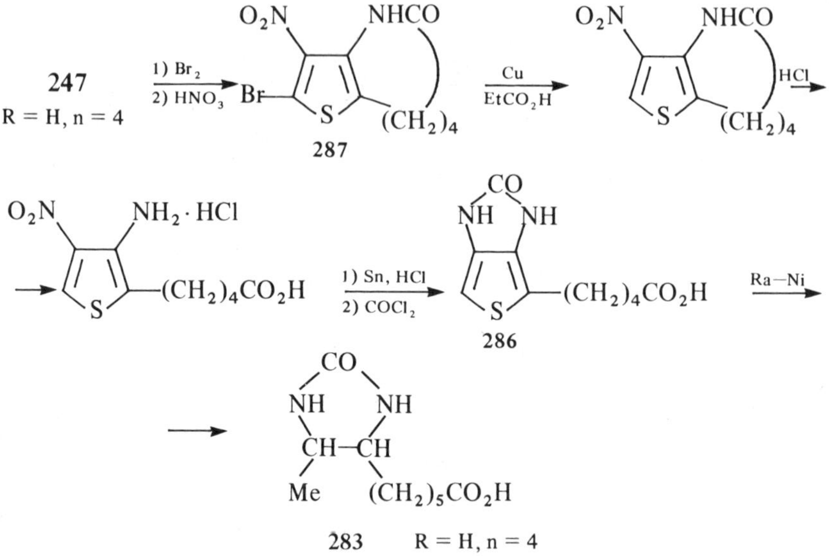

Scheme 175

A similar sample of dethiobiotin was also prepared directly from the lactam of
δ-(5-bromo-4-nitro-3-amino-2-thienyl)valeric acid (287), as shown in Scheme
176.[366]

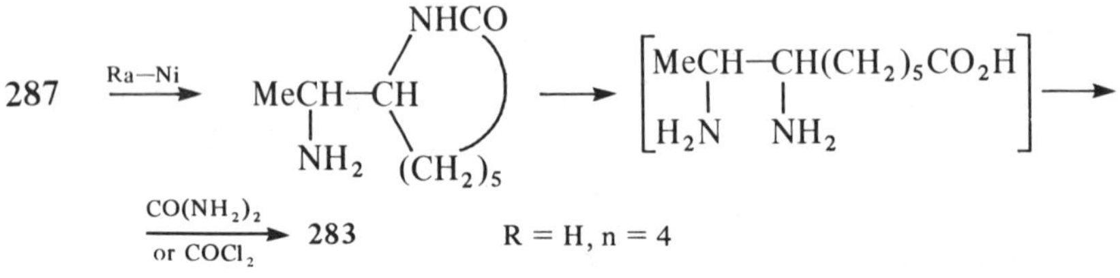

Scheme 176

H.   Synthesis of Some Alicyclic Compounds

There are two approaches to the synthesis of alicyclic compounds from thiophene derivatives. The first does not differ in the main from that used for the
preparation of aliphatic and arylaliphatic compounds: the starting substances are
alicyclic derivatives and thiophene is used both for lengthening the carbon chain
and for incorporating substituents. The main features of such an approach were
shown clearly in the paper by Buu-Hoi, Sy, and Xuong[368] concerning the syntheses

of dihydrohydnocarpic acid (288), of dihydrochaulmoogric acid (289), and also of an analogue of the former with a six-membered ring (290) (Scheme 177).

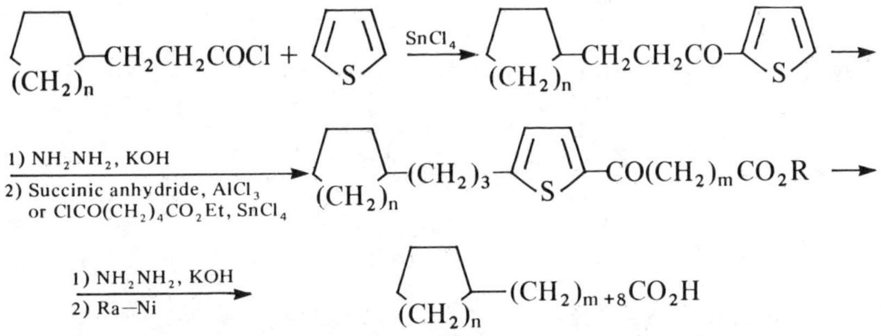

$$\text{(CH}_2)_n\text{-CH}_2\text{CH}_2\text{COCl} + \text{S} \xrightarrow{\text{SnCl}_4} \text{(CH}_2)_n\text{-CH}_2\text{CH}_2\text{CO-S} \longrightarrow$$

$$\xrightarrow[\substack{\text{2) Succinic anhydride, AlCl}_3 \\ \text{or ClCO(CH}_2)_4\text{CO}_2\text{Et, SnCl}_4}]{\text{1) NH}_2\text{NH}_2\text{, KOH}} \text{(CH}_2)_n\text{-(CH}_2)_3\text{-S-CO(CH}_2)_m\text{CO}_2\text{R} \longrightarrow$$

$$\xrightarrow[\text{2) Ra-Ni}]{\text{1) NH}_2\text{NH}_2\text{, KOH}} \text{(CH}_2)_n\text{-(CH}_2)_{m+8}\text{CO}_2\text{H}$$

288	n = 1, m = 2
289	n = 1, m = 4
290	n = 2, m = 2

Scheme 177

A synthetic route of the same kind was developed for the preparation of 1-alkyladamantanes (291 and 292), 2-alkyladamantanes (293 and 294), and also ω-functionalized derivatives of the first of them (295) by Hoeck, Strating, and Wynberg[369,370] (Schemes 178–180).

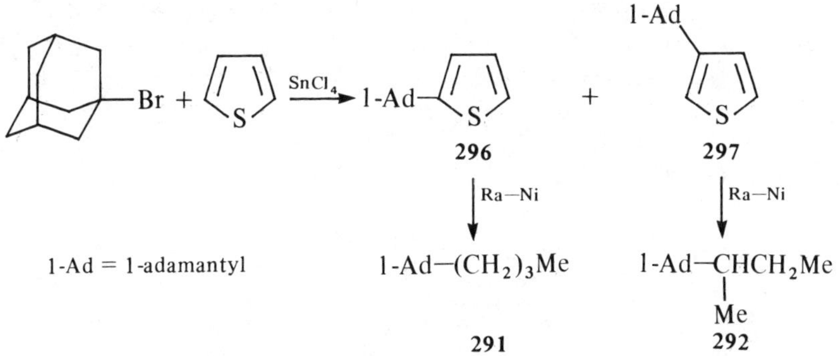

1-Ad = 1-adamantyl

1-Ad-(CH₂)₃Me → 1-Ad$-(CH_2)_3$Me **291**

1-Ad$-$CHCH$_2$Me | Me **292**

Scheme 178

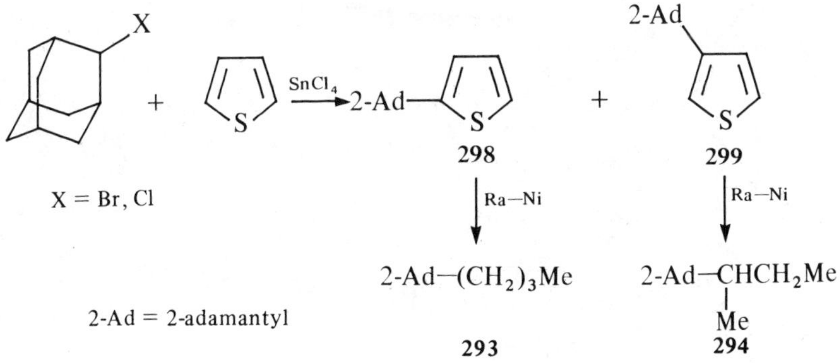

X = Br, Cl

2-Ad = 2-adamantyl

2-Ad$-(CH_2)_3$Me **293**

2-Ad$-$CHCH$_2$Me | Me **294**

Scheme 179

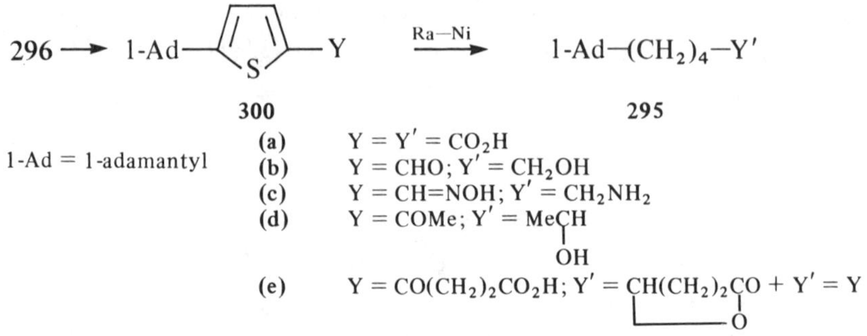

$$296 \longrightarrow 1\text{-Ad} \underset{S}{\overset{\text{Ra—Ni}}{\longrightarrow}} Y \quad 1\text{-Ad—}(CH_2)_4\text{—}Y'$$

300   **295**

1-Ad = 1-adamantyl

(a) $Y = Y' = CO_2H$
(b) $Y = CHO; Y' = CH_2OH$
(c) $Y = CH=NOH; Y' = CH_2NH_2$
(d) $Y = COMe; Y' = Me\underset{OH}{\overset{|}{C}}H$

(e) $Y = CO(CH_2)_2CO_2H; Y' = \underset{\underset{O}{|\_\_\_\_}}{C}H(CH_2)_2\overset{|}{C}O + Y' = Y$

Scheme 180

The reductive desulfurization of adamantylthiophenes (296–300) proceeds smoothly, leading to hydrocarbons (291–294) in 80–85% yields or to functionalized derivatives (295) in 60–75% yields. The method has a specific drawback: in the course of alkylation, a mixture of α- and β-substituted thiophenes is formed that were separated as chloromercuric derivatives. However, there are possibilities for selective synthesis of adamantylthiophenes. In this connection, we should mention a novel and comparatively simple synthesis of 2-(1-adamantyl)thiophene and 5-(1-adamantyl-2-thiophenecarboxylic acid from 1-acetyladamantane, including thiophene ring closure.[371]

When studying the absolute configuration of spiro[3,3]heptane-2,6-dicarboxylic acid (301), known as Fecht acid, Wynberg and Hubiers carried out some transformations of the racemic acid and its l-isomer into products bearing thiophene rings (302–304). One of them was subjected to reductive desulfurization[372] (Scheme 181).

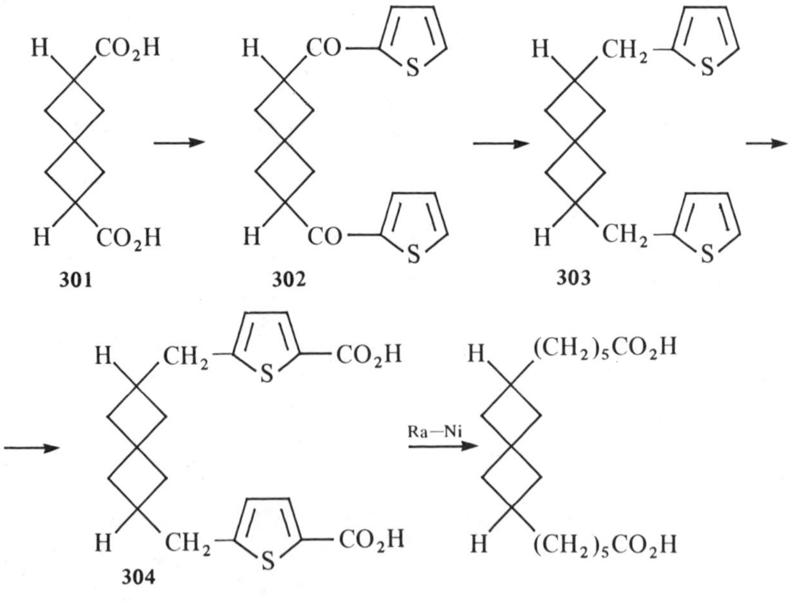

301   **302**   **303**

304

Scheme 181

The characteristic feature of the second approach to the synthesis of cyclo-aliphatic compounds is that carbon atoms of the thiophene ring are used to create the alicycle itself. This approach has found adequate development in the new method of the synthesis of macrocyclic compounds to which Section IV.3.I is devoted. The same principle was also used for the synthesis of some substances with common rings. Thus Gol'dfarb, Krasnyanskaya, and Fabrichnyi[326] described the synthesis of some alicyclic amines (305) and the corresponding acetamides (306) from the oximes of thienocycloalkanones (250) with mixtures of geometric isomers of 2-alkyl-1-aminocyclohexanes and -cycloheptanes obtained in the process. The hydrogenolysis products of the oximes (250) were obtained in ~ 50% yield, of the amine (307) in 30% yield, and of acylamines (308) in 80–90% yield, respectively (Scheme 182).

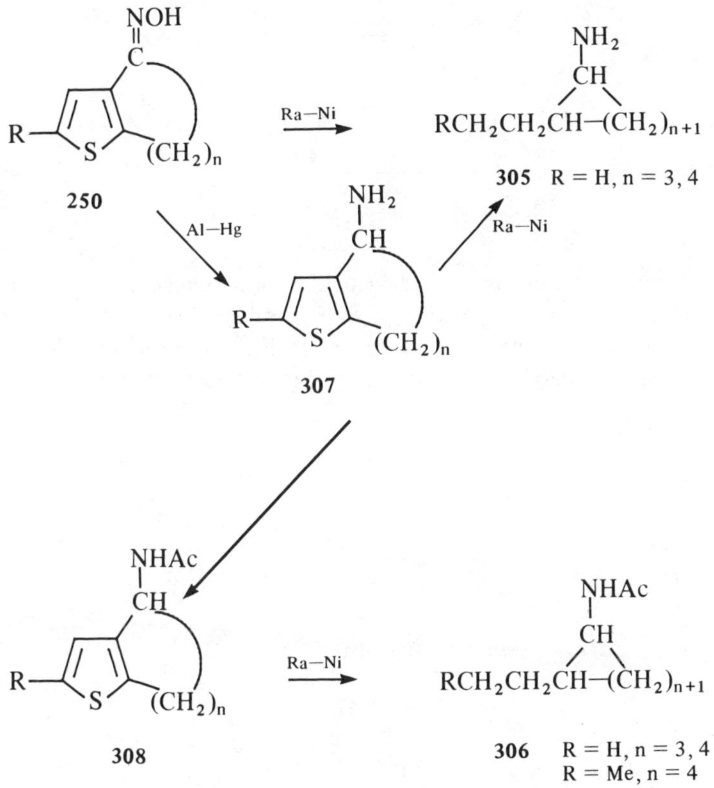

Scheme 182

Royer et al.[313,373] studied the synthesis of several benzosuberane derivatives (309–311) from 2-ethylbenzo[b]thiophene via the tricyclic system (312). The yield of propylbenzocycloheptene (310) is as high as 88% (Scheme 183).

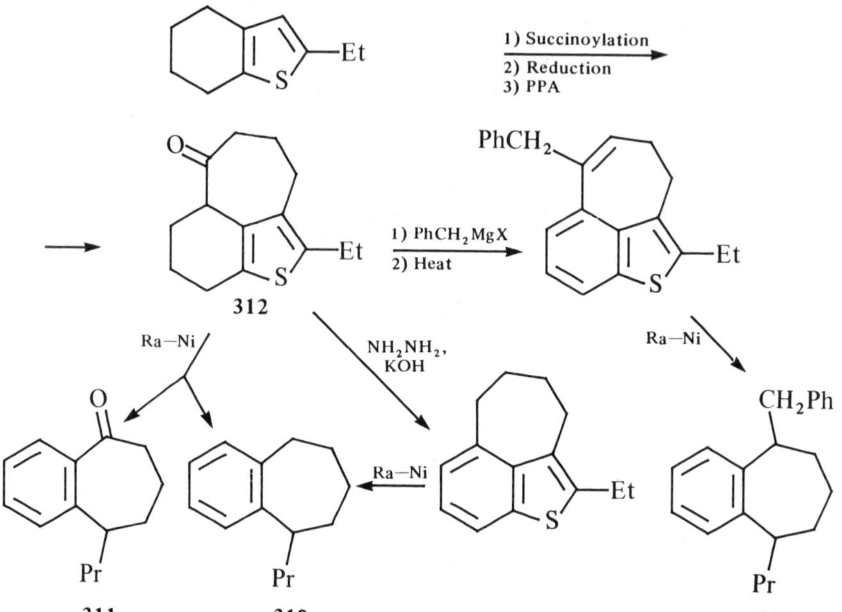

Scheme 183

The reductive desulfurization leading to the formation of alicyclic compounds was also used to prove the structure of some condensed systems containing thiophene or dihydrothiophene rings. A group of French authors[374] described the desulfurization of some bicyclic systems bearing *tert*-butyl groups (Schemes 184–186).

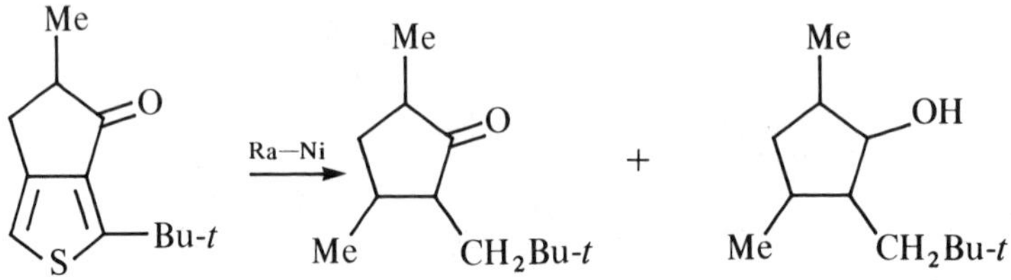

Scheme 184

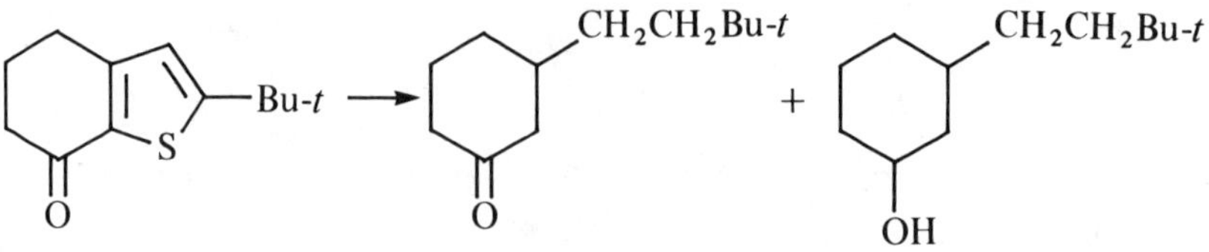

Scheme 185

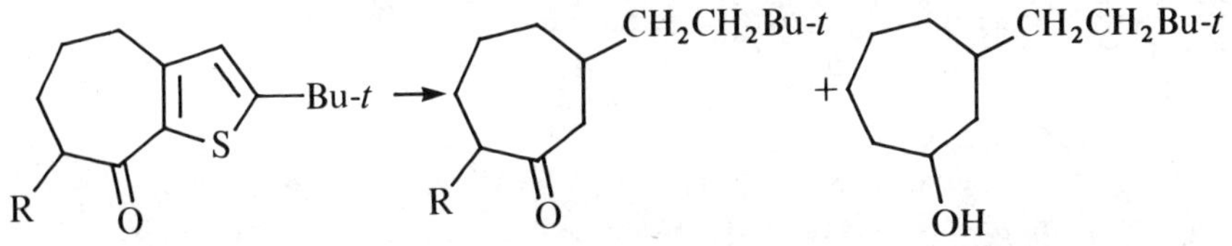

Scheme 186

Gol'dfarb et al.[375,376] prepared aralkylcyclohexanes and -cycloheptanes (**313** and **314**) in 60–80% yields from the dihydrothiophene derivatives **315** and **316**. The latter were obtained from the products formed during the interaction of so-called thiophenocycloalkanones[249] with benzene or chlorobenzene in the presence of aluminum chloride. In the course of reductive desulfurization, replacement of chlorine atoms in the benzene ring with hydrogen occurs (Scheme 187).

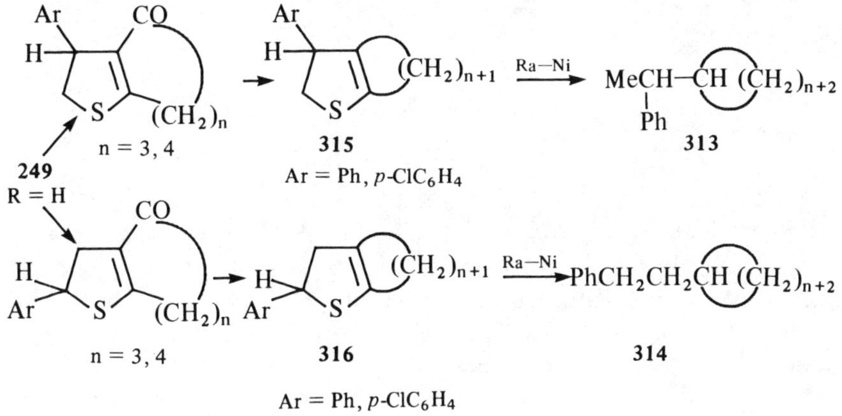

Scheme 187

Stork and Stotter[377] proposed a new approach to the synthesis of angular substituted polycyclic systems that included the reductive desulfurization of Diels–Alder adducts of substituted 2,5-dihydrothiophenes (**317**) (Scheme 188).

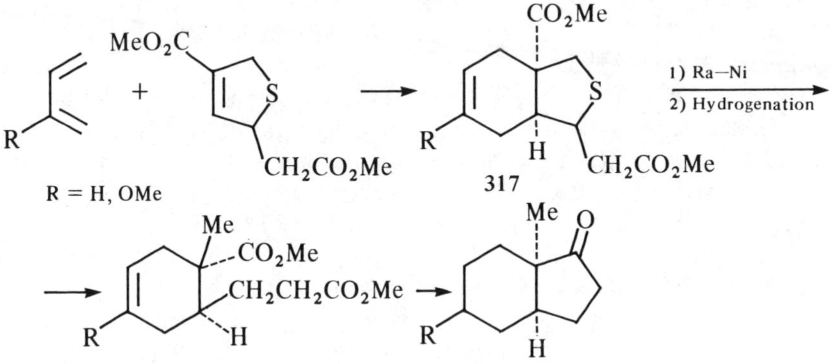

Scheme 188

I.   Synthesis of Alicyclic and Heterocyclic Many-Membered Compounds from Thiophene Derivatives

For the synthesis of macrocyclic compounds with the aid of reductive desulfurization of thiophene derivatives, two approaches are possible (Scheme 189):

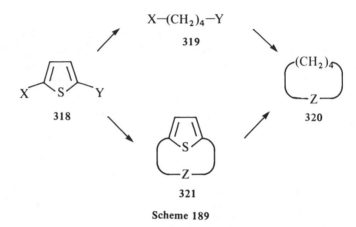

Scheme 189

1.   The preparation of long-chain bifunctional aliphatic compounds (**319**) from the thiophene derivatives (**318**) and their subsequent cyclization by one of the usual methods, which results in the products of type **320**. It should be noted that the thiophene ring greatly facilitates the incorporation of various substituents in the molecule of the long-chain compound.

2.   The preparation of bi- or polycyclic compounds that possess the thiophene ring (**321**) with the subsequent removal of the sulfur atoms serving as a bridge.

It is not necessary to discuss the first principle of macrocyclic system synthesis. It should be only noted that the synthesis of long-chain bifunctional aliphatic compounds, for instance, higher dicarboxylic acids and hydroxy acids using the reductive desulfurization of the corresponding thiophenic precursors, has been elaborated by Buu-Hoi[273,274] and Gol'dfarb.[282,284] These data are considered in Sections IV.3.B and IV.3.D.

The second principle has a number of advantages when compared with the first. Bicyclic thiophene derivatives obtained in the course of such syntheses are of interest themselves, and the second approach should in general give higher yields than the first of many-membered monocyclic compounds, when using the same cyclization reactions. According to the data published (see Ref. 378), cyclization may be facilitated by the presence of a "rigid group" (in this case, the thiophene ring). Furthermore, the thiophene ring can be utilized for lengthening the carbon chain and incorporating substituents; it is also important to note that because of its aromatic character, it can play the part of the second function, eliminating the necessity of incorporating the latter.

The principle given above for the synthesis of macrocyclic compounds via the bicyclic substances possessing the thiophene ring was proposed in 1957 by Gol'dfarb, Taits, and Belen'kii[379] and was later developed by the same authors (see Refs. 100d and 380). There were elaborated three routes to the precursors containing thiophene rings which are called cyclothienes[381] by analogy with cyclophanes (at present, such compounds are usually called thiophenophanes).

a.   SYNTHESIS OF MANY-MEMBERED CYCLOALKANONES

One method of the synthesis under consideration[379,381] is the intramolecular acylation of ω-(2-thienyl)alkanoyl chlorides (322) with subsequent conversion of the cyclization products (323 and 324) into alicyclic ketones (325) and diketones (326) by treatment with Raney nickel. Cyclization of the lower acid chlorides of this series was studied by Fieser and Kennely[2] as well as by P. and D. Cagniant.[382] It was directed into the β-position, leading to six- and seven-membered ketones (249). French authors[349] also described the cyclization of 6-(5-ethyl-2-thienyl)-caproyl chloride into the eight-membered cyclic ketone, though it proceeds in very low yield. As expected with longer aliphatic chains in ω-(2-thienyl)alkanoic acids used by Gol'dfarb et al., the reaction is directed into the more active α'-position (Scheme 190).

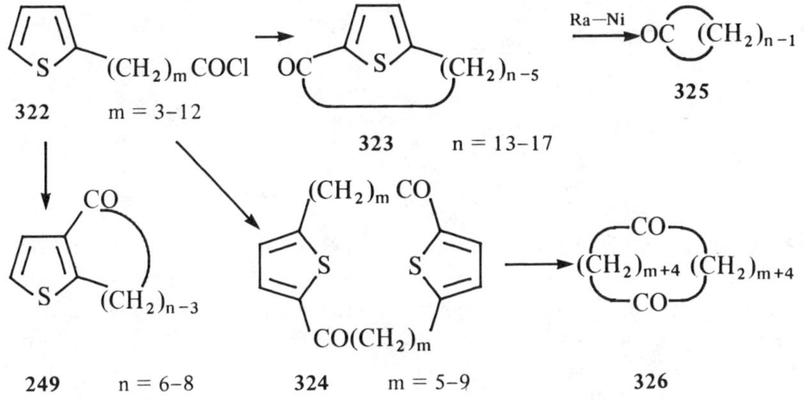

Scheme 190

This route is especially attractive because it does not require the preparation of a bifunctional thiophene compound (the role of the second function is played by the unsubstituted α'-position of the thiophene ring), which makes it easier to prepare the starting compounds. In other words, with this method, one can make full use of the advantages of the thiophene-based synthesis of macrocycles.

For the preparation of some higher ω-(2-thienyl)alkanoic acids, the ordinary scheme was used that included the acylation of thiophene by half-ester chlorides of dicarboxylic acids and subsequent reduction of the resulting keto esters. But the possibilities of this scheme are limited by the low accessibility of many higher dicarboxylic acids. In this connection, Belen'kii, Taits, and Gol'dfarb elaborated a convenient method for the preparation of the necessary acids through the ω-chloroalkyl-2-thienylketones (327), starting with the accessible ω-chloroalkanoic acids (328). One modification of this synthesis leads to the ω-(2-thienyl)alkanoic acids with an even number of carbon atoms in the aliphatic chain (329) and another leads to the acids with an odd number of carbon atoms in the chain (330) (Scheme 191).

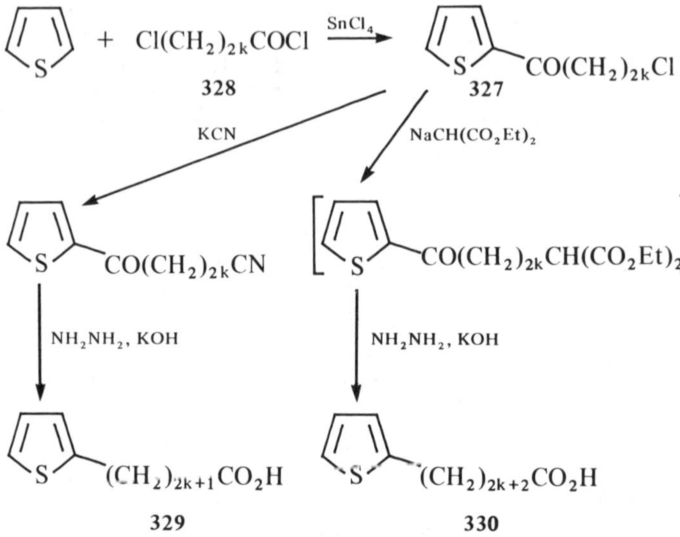

Scheme 191

Various conditions of the cyclization of ω-(2-thienyl)alkanoyl chlorides have been studied.[379,380,383,385] In particular, it was found that this reaction proceeds satisfactorily when using a high dilution technique in chloroform or carbon disulfide as a solvent and in the presence of partially hydrolyzed (by 1 mole water) aluminum chloride etherate as a condensing agent.[383] Under such conditions, the bicyclic ketones (323, n = 14–17) are obtained in yields that can be regarded as fairly high for such systems (53–63%). It was found that the presence of an inert sorbent, as well as an increasing concentration of the condensing agent in a homogeneous medium, leads to higher yields of bicyclic ketones possessing the thiophene ring.[385]

Under the conditions mentioned above, 9-(2-thienyl)nonanoyl chloride gives corresponding monoketone (323, n = 13) in only 8% yield. It proved impossible to isolate monoketones from cyclization products of 7-(2-thienyl)heptanoyl chloride and 8-(2-thienyl)octanoyl chloride. In these cases, only products of intermolecular cyclization (324) were formed. Such diketones, together with monoketones, were also obtained when the chlorides of 6-(2-thienyl)caproic, 9-(2-thienyl)pelargonic, and 10-(2-thienyl)capric acids were subjected to cyclization[384] (Scheme 190).

One can reduce the keto group in the cyclization products and introduce into the thiophene ring substituents retained after desulfurization[386] (Scheme 192).

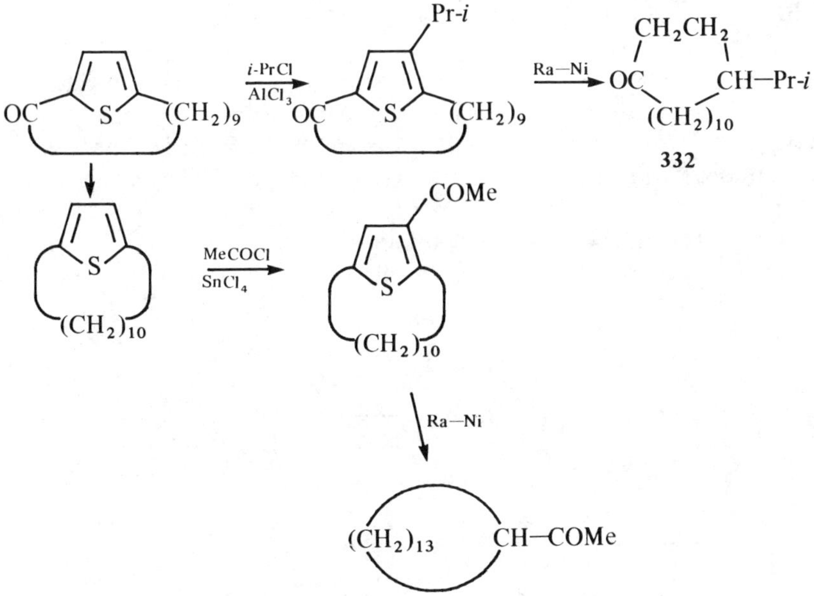

Scheme 192

The reductive desulfurization of macrocyclic compounds possessing the thiophene ring has some peculiarities. For instance, [10,10]-cyclodithiene (331), under the ordinary desulfurization conditions (by the action of Raney nickel in the ethanol–benzene mixture at 30–50°C), is transformed smoothly into cyclooctacosane in 71% yield[379,381] (Scheme 193), but in the case of ketones, reduction of the carbonyl group occurs. It was found that upon addition of acetone to the solvent mixture, the macrocyclic ketones are reduced without affecting the carbonyl group, and the yields of a number of cycloaliphatic ketones amounted to 70–90%. Among them were macrocyclic ketones with a musk odor, such as cyclotetradecanone, exaltone, cyclohexadecanone, dihydrocivetone (325), and several diketones (326)[379,381,386] as well as γ-isopropylcyclotetradecanone (332) and acetylcyclotetradecane (333).[386]

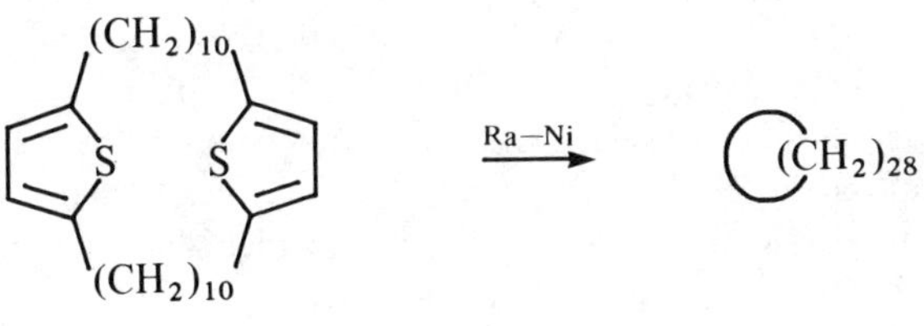

Scheme 193

Bicyclic systems with thiophene rings bearing methyl group were obtained by Taits et al., starting with 2- and 3-methylthiophenes.[387-389] Acylation of 2-methylthiophene with ω-chloralkanoyl chlorides or with half-ester chlorides of dicarboxylic acids is directed into position 5 and, after procedures similar to those considered above, leads to ω-(5-methyl-2-thienyl)alkanoic acids. When subjected to cyclization, chlorides of the latter (334) give the products of ring closure into position 4 (335, m = 9–11), with some contamination from ketones formed as a result of intermolecular acylation into position 3 (336).[387,388] On reductive desulfurization of these ketones by Raney nickel in the presence of acetone, α-ethyl

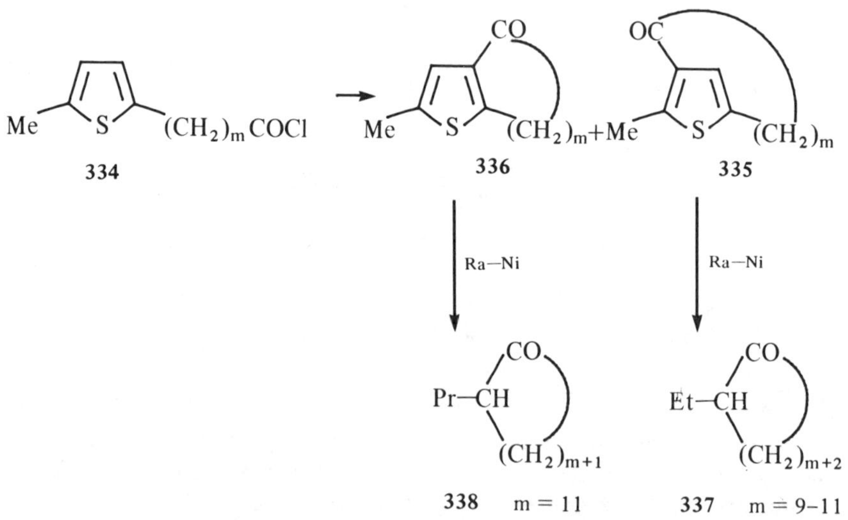

Scheme 194

(337) and α-propylcycloalkanones (338) were obtained (Scheme 194). It should be noted that the last stage of the synthesis proceeds very slowly (up to 30 h).

In the case of 3-methylthiophene, mixtures of the products of acylation in positions 2 (339) and 5 (340) in the ratio ∼ 2 : 1 are formed,[389] these products can be separated using chromatography on alumina. Subsequent transformations of these keto acids into acids (341 and 342) and then into chlorides (343 and 344) and finally intramolecular acylation of the latter leads to bicyclic ketones (345 and 346). By reductive desulfurization of the ketones in an ethanol–acetone mixture, the sulfur removal proceeds sufficiently rapidly, but in the reaction products, unconjugated double C=C bonds are retained, which demands additional hydrogenation over palladium–charcoal. As the result, β- (347) and γ-methylcycloalkanones (348), including racemic β-methylcyclopentadecanone (muscone),[389] are obtained (Scheme 195).

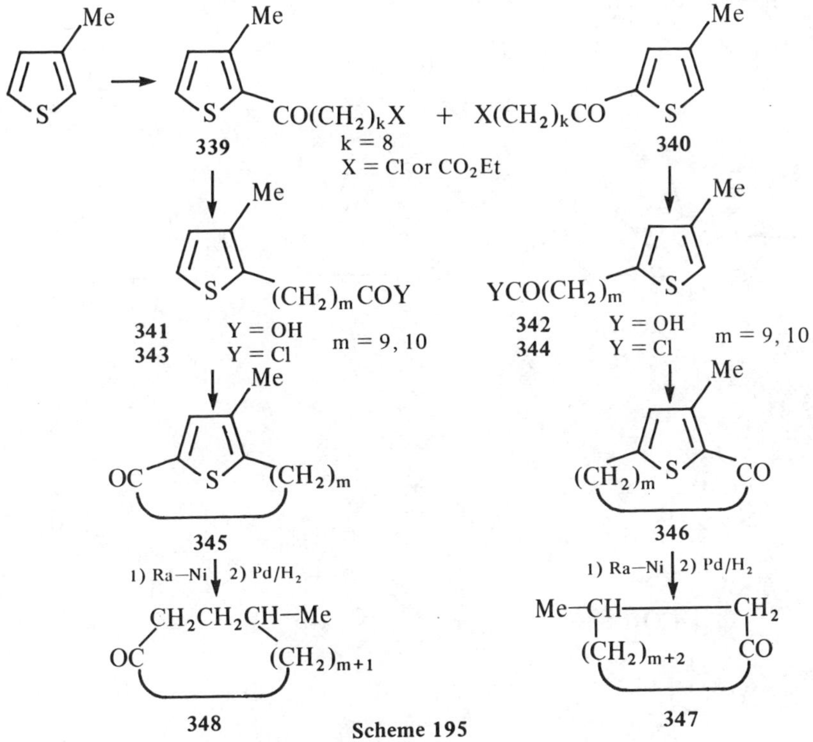

Scheme 195

b. SYNTHESES OF MACROCYCLIC β-KETOESTERS AND SOME OTHER SYSTEMS

Another route to the higher alicyclic compounds elaborated by Taits, Gol'dfarb, and coworkers[3,4] consists of the intramolecular alkylation of ω-halo-β-ketoesters such as **350** with subsequent "ketonic hydrolysis" and reductive desulfurization. For the preparation of starting compounds, thiophene, easily accessible chloroalkanoic acids (obtained from the products of ethylene and carbon tetrachloride telomerization), and monoethyl malonate chloride were used. On treating the corresponding chlorides (**349**) with sodium iodide in acetone or methyl ethyl ketone, 2-(ω-iodoalkyl)-5-(ethoxycarbonylacetyl)thiophenes (**350**) are easily formed in quantitiative yields. In some cases, there is no need to isolate them from the solution after the halogen exchange. On the cyclization of ketoesters (**350**) carried out using a high dilution technique in boiling methyl ethyl ketone over finely powdered potassium carbonate, the cyclic ketoesters (**351**, m = 8, 10) were obtained in 70–80% yields. The transformation of these ketoesters into ketones (**323**) and subsequent reductive desulfurization of the latter led to cyclopentadecanone (exaltone) and cycloheptadecanone (dihydrocivetone).[3,4] The alkylation of the unsubstituted ketoesters (**351**) gave (**352**) and then alicyclic alkyl-substituted ketoesters (**353**) and α-alkylcycloalkanones (**354**)[390] were obtained upon desulfurization and ketonic hydrolysis (Scheme 196).

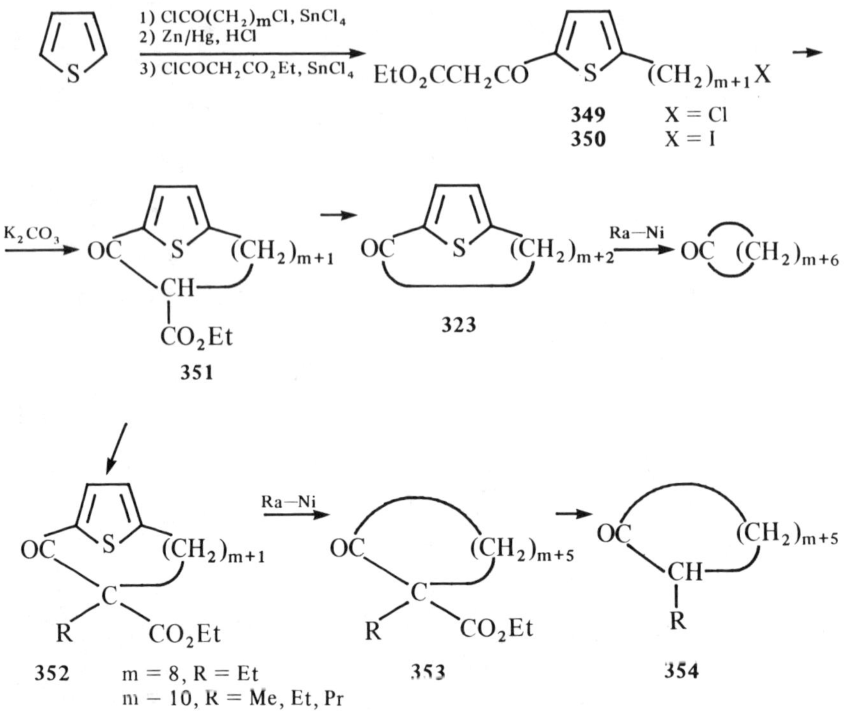

Scheme 196

In the case of 2-(7-iodoheptyl)-5-(ethoxycarbonylacetyl)thiophene (**350**, m = 6), the only product isolated (yield 47%) was **355**. On ketonic hydrolysis, this product was transformed into the corresponding diketone (**324**), the reductive desulfurization of which was considered above.

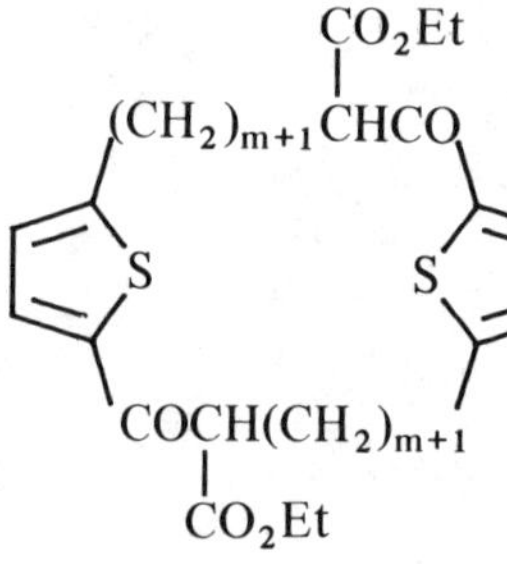

355

Taits, Krasnyanskaya, and Gol'dfarb, when studying the conditions[391] and kinetics[392] of 2-(9-iodononyl)-5-(ethoxycarbonylacetyl)thiophene cyclization over potassium carbonate as well as the effect of carbonates of other alkali metals,[391,393] demonstrated the role of the carbonate surface in this reaction. The study of the same cyclization in homogeneous medium[394] (in the presence of potassium *tert*-butylate) allowed the authors to determine the maximal concentration of the iodide at which intermolecular reaction leading to by-products does not take place; this is essential for the preparative use of the cyclization. Finally, the study of cyclization of iodides (**350**, n = 6, 8) in a wide range of concentrations[395] allowed the yield of bis-β-ketoester (**355**, m = 6) up to an increase of 58% and the conditions for the formation of its higher cyclic homolog (**355**, m = 8) in 37–45% yield to be found, with the monoester (**351**, m = 8) being obtained simultaneously in 56–37% yield.

The third route to bicyclic systems possessing thiophene rings, proposed by Taits and Gol'dfarb, is the acyloin condensation of the corresponding diesters.[42,379] The standard conditions of acyloin condensation (finely dispersed sodium metal in boiling xylene) were unsuitable for the cyclization of esters of the thiophene series. This can be explained on one hand by the degradative action of sodium on the thiophene ring under the reaction conditions and on the other by the deactivation of sodium by sulfur-containing degradation products. Lowering the reaction temperature to 50–60°C resulted in the desired acyloin (**356**), but in only 30% yield, based on the utilized ester; the latter is converted only 15%.[42] The low conversion occurs because at the temperature used, sodium was in the solid state so that its surface was not refreshed after deactivation. A somewhat higher yield of the mentioned acyloin (40%) was obtained with a new condensation reagent – the sodium–potassium alloy that is liquid at ordinary temperatures. However, even in this case, conversion of the ester did not exceed 25%. Only cyclization of esters whose thiophene ring has no unsubstituted positions proceeds in the presence of a sodium–potassium alloy with a 100% conversion and gives acyloins **357** and **358** in 70 and 40% yield, respectively.[42] Since the acyloin condensation in the thiophene series has a number of serious limitations, desulfurization of acyloins **356–358** was not studied.

356   357   358

Meth-Cohn[396] described the conditions of preparing in satisfactory yield (42%) the tetracyclic compound possessing three thiophene nuclei (**359**), which on boiling in benzene with an excess of Raney nickel was converted practically quantitatively into 1,2,4,5,7,8-hexaethylcyclononane (**360**) (Scheme 197).

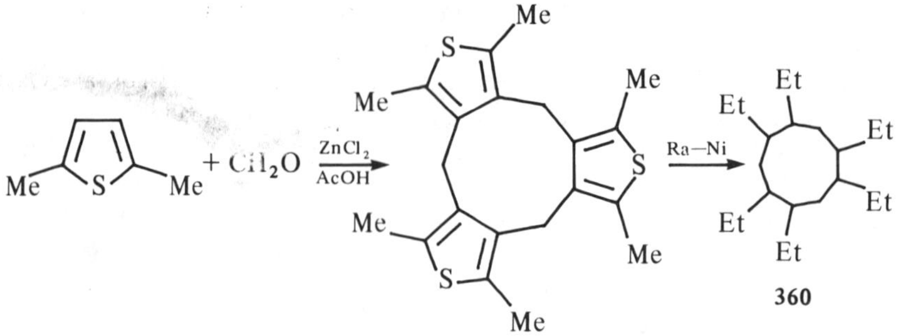

359

360

Scheme 197

Miyahara, Takahiko, and Yoshino[397] described a new interesting synthesis of many-membered cyclic systems possessing one thiophene and two benzene rings, **361** and **362**. The reductive desulfurization of the latter leads to [*n*,6]paracyclophanes (**363**) (Scheme 198).

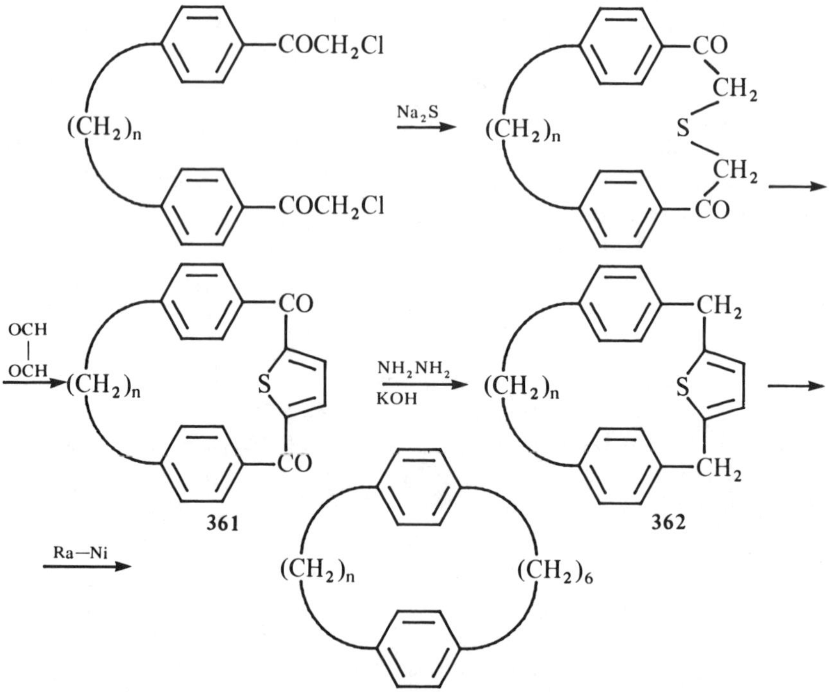

361

362

363 n = 3–8, 10

Scheme 198

c.   SYNTHESES OF SOME MANY-MEMBERED HETEROCYCLES

The above methods for the synthesis of macrocyclic systems possessing thiophene rings with subsequent reductive desulfurization may also be used for the preparation of many-membered heterocyclic compounds with nitrogen or oxygen as heteroatoms. Taits, Alashev, and Gol'dfarb[398] described the synthesis of macrocyclic ketolactones of types **364** and **365**, including intramolecular acylation of ω-(2-thienyl)alkyl alkanedicarboxylate chlorides (**366**) (Scheme 199).

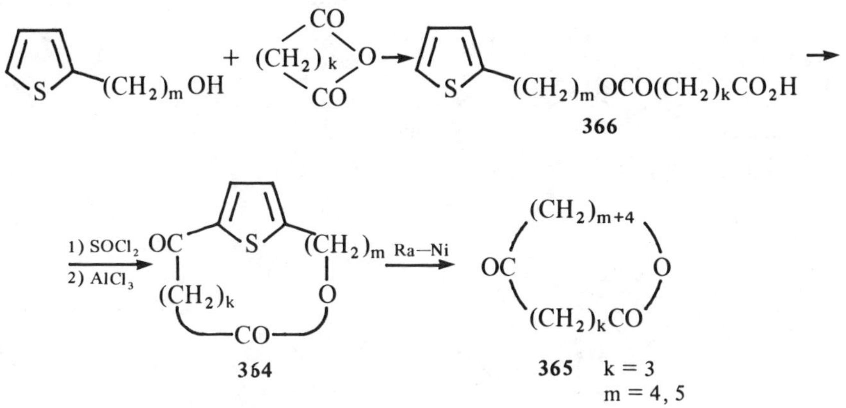

Scheme 199

The ω-(2-thienyl)alkanols were obtained using reduction of the corresponding ω-(2-thienyl)alkanoic acid esters with lithium aluminum hydride or by Kishner reduction of ω-acetoxyalkyl 2-thienyl ketones formed from ω-chloroalkyl 2-thienyl ketones.[32] Later, new methods for the synthesis of esters (**366**) with various values of m and k were elaborated.[81] These methods made it possible to obtain various macrocyclic ketolactones possessing thiophene rings, including isomeric lactones such as **364**, which differ in the position of the ester grouping in the ansa-system, and to investigate their stereochemical peculiarities.[399–403] However, detailed consideration of this material is not among the aims of this section, as desulfurization has been studied for only two ketolactones indicated in Scheme 199.

Alashev, Bulgakova, Taits, and Gol'dfarb[404,405] also described the synthesis of macrocyclic ketolactones that possess benzene and thiophene rings (**367**). Reductive desulfurization of these compounds leads to ketolactones (**368**) with skeletons related to that of the natural macrolide zearalenone, but to a great extent the products of partial (**369**) or complete (**370**) reduction of the keto group are also formed (Scheme 200).

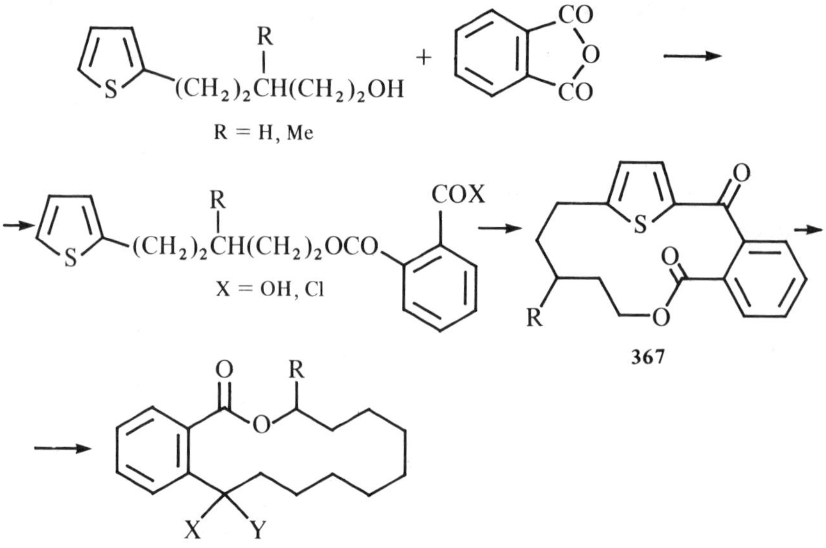

368 X + Y = O
369 X = H, Y = OH
370 X = Y = H

Scheme 200

Recently, Gol'dfarb, Taits, and Krasnyanskaya[406] described the first case of the synthesis of macrocyclic compound possessing the pyrazolone ring **(371)** using reductive desulfurization (Scheme 201).

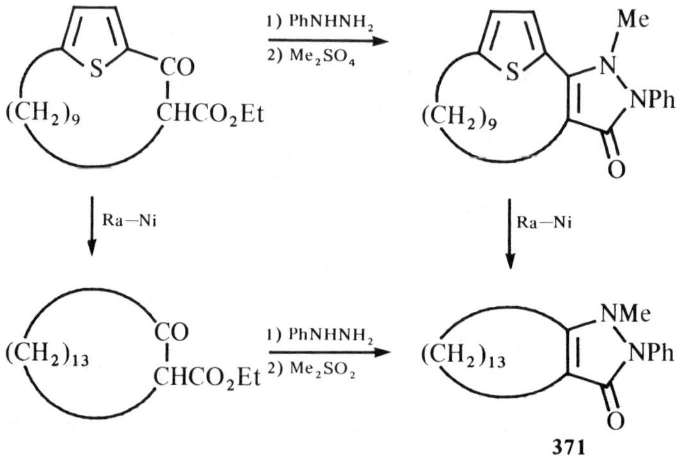

Scheme 201

J. Synthesis of Some Heterocyclic Compounds

Desulfurization of thiophene derivatives is widely used for the preparation of some oxygen- and nitrogen-containing as well as boron-nitrogen-containing heterocycles. Section IV.3.E discussed N-alkylpiperidine synthesis, which does not differ in principle from the synthesis of open-chain amines. The synthesis of substituted lactams from thiophene derivatives considered in detail in Section IV.3.F also calls to mind the methods used for the preparation of other heterocycles, as it will be clear from the following text. However, taking into account the genetic relation of lactams to amino acids, it was advisable to consider these types of compounds together. The same is also true for lactones, the synthesis of which was described in Section IV.3.D together with that of hydroxy acids. In Section IV.3.I, which is devoted to the macrocycle synthesis from thiophene derivatives, heterocycles are considered with carbocyclic compounds since the methods of their preparation are very close to the synthesis of many-membered carbocyclic compounds.

This section considers mainly the synthesis of heteroatomic compounds. The first such attempt was described as early as 1951 by Kotake and Sakan,[79] who studied the desulfurization of benzothienopyridine (372) leading to 4-phenylpyridine.

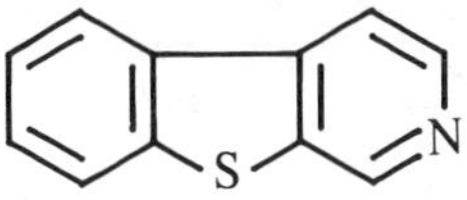

372

Gronowitz and Böler[407] carried out reductive desulfurization of 5-(4-pyrimidyl)-2-thiophenecarboxylic acid (373) into 5-(4-pyrimidyl)valeric acid (374). This reaction is an example of the synthesis of 4-substituted pyrimidines with a functional group in the side chain (Scheme 202). Myshkina, Stoyanovich, and

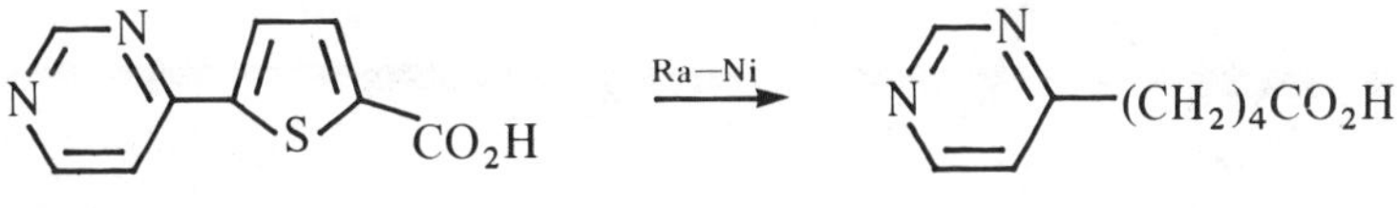

373   374

Scheme 202

Gol'dfarb[408] carried out hydrogenolysis of sterically crowded diamine (375) formed during azacycloalkenylation of thiophene with triacetoneamine and after additional hydrogenation obtained 1,4-(2,2,6,6-tetramethyl-4-piperidyl)butane (376) in 57% yield. The latter substance was used for the preparation of the stable iminoxyl radical (377) (Scheme 203).

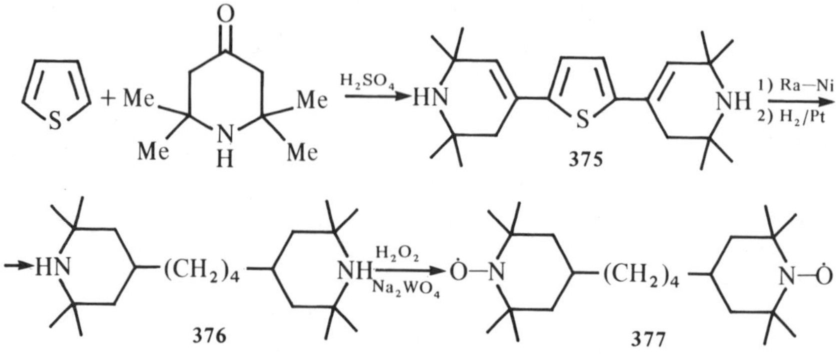

Scheme 203

An interesting route to isomeric bipyrroles (**378–380**), based on reductive desulfurization with subsequent hydrolysis and decarboxylation (Schemes 204–206), was proposed by Farnier, Soth, and Fournari.[409] The yields at the desulfurization stage are in the range of 70–80% and this method may be regarded just as a synthetic method, since the same authors elaborated a sufficiently simple way to prepare thiophenic precursors (**381–383**) in acceptable yields, starting with the corresponding thiophenecarboxaldehydes and azidoacetic ester.[410] This method is illustrated by the synthesis in which 2,5-thiophenedicarboxaldehyde is used as starting compound (Scheme 207).

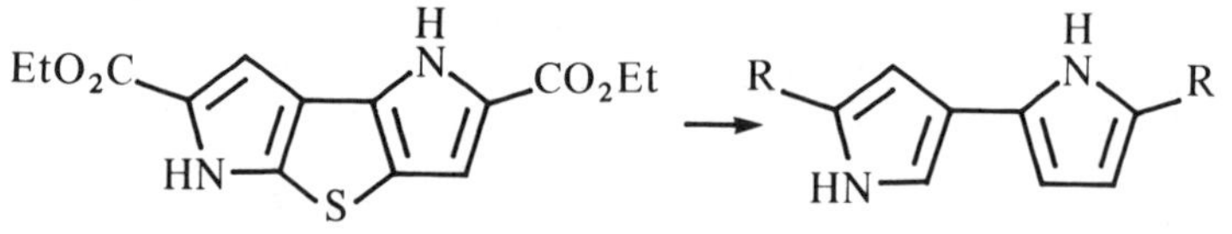

381

378   R = CO$_2$Et, H

Scheme 204

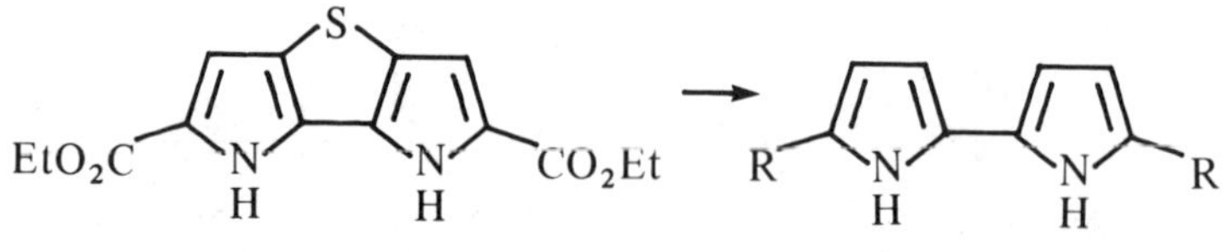

382

379   R = CO$_2$Et, H

Scheme 205

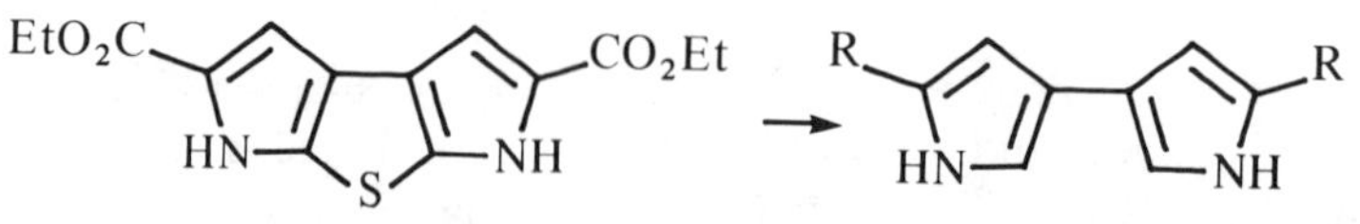

383

380   R = CO$_2$Et, H

Scheme 206

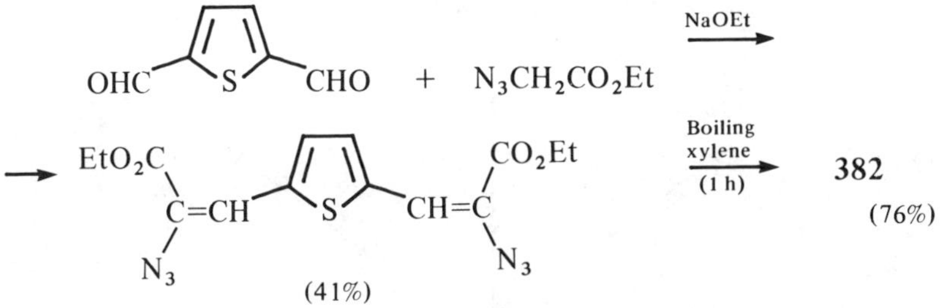

Scheme 207

The reductive desulfurization of 2,3-dimethylthieno[2',3'-g]benzofuran and its ethyl homologues results in dihydrobenzofuran derivatives. On heating with sulfur, these derivatives are transformed into corresponding benzofurans (**384**)[314] (Scheme 208).

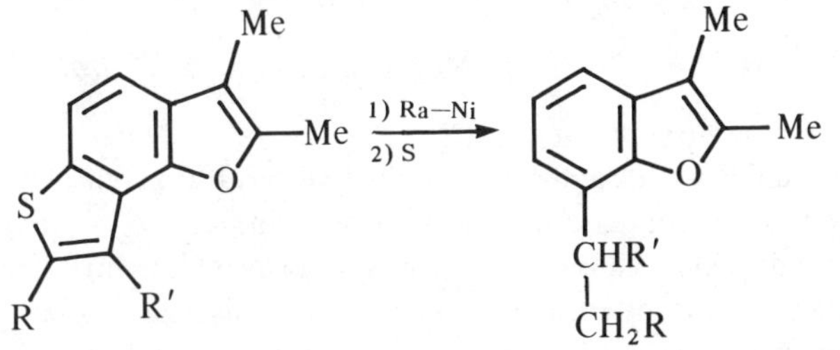

384 R and R' = H or Et

Scheme 208

There are interesting investigations devoted to the synthesis of six-membered heterocycles with boron and nitrogen as heteroatoms from condensed systems possessing the thiophene ring. In this connection, it should be noted that in conditions of reductive desulfurization, C—B and B—N bonds are not affected. In particular, the action of Raney nickel on 2,5-thiophenediboronic acid gives 1,4-butanediboronic acid as the sole product.[411] The synthesis of borazarene derivative **385**, using reductive desulfurization (Scheme 209), was described by Dewar and Marr.[412] Gronowitz et al.[413,414] prepared a number of borazaropyridines (**386**) (Scheme 210).

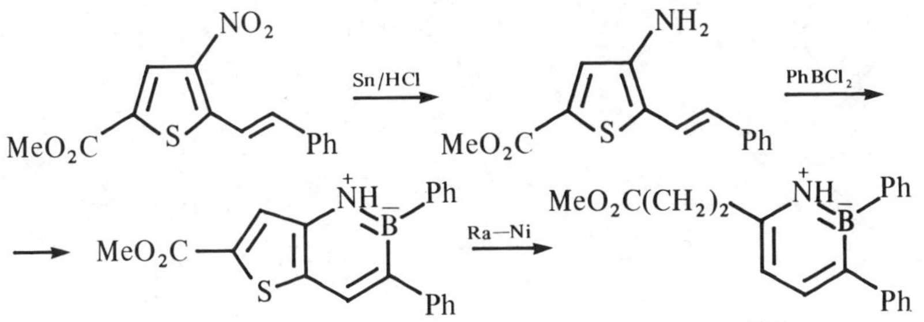

Scheme 209   **385**

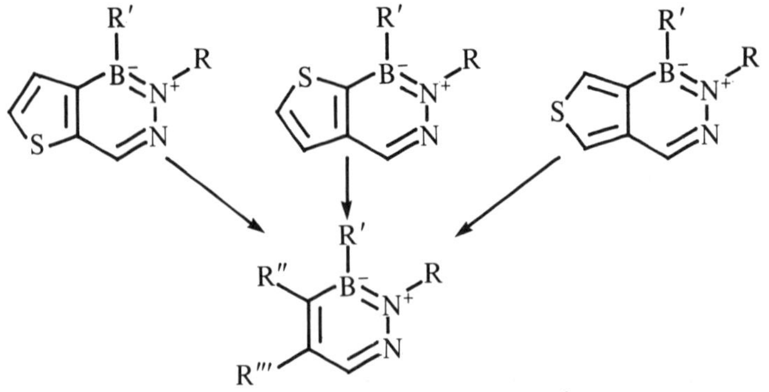

386   R = H, Me; R' = OH, Me
R" = H, Et; R''' = H, Me, Et

Scheme 210

K.   Use of Other Skeletal Metals and Related Catalysts

Raney nickel is unique in its activity for reductive desulfurization. Other skeletal metals give considerably worse results and find only limited use in the synthesis of aliphatic compounds from thiophene derivatives. Badger, Kovanko, and Sasse[415] used Raney cobalt for the desulfurization of dibenzothiophene, of some ketones and carboxylic acids of the thiophene series, and of thiazoles and thioamides. These authors estimated that in the case of 3-acetylbenzo[b]thiophene desulfurization, Raney cobalt is approximately 10 times less active than Raney nickel.

It is essential to note that skeletal cobalt differs not only in activity but also in the character of its action. Thus, when it is used, the formation of dimerization products considerably less than with Raney nickel.[415] Santalova, Konstantinov, and Gol'dfarb[416] obtained complex mixtures during the action of Raney nickel on some diamines (387) prepared from the product of chloromethylation of 2,2-di(2-thienyl)butane. The authors explain this result by the ability of Raney nickel to cleave the C—N bond. At the same time, the use of Raney cobalt made it possible to obtain the desired aliphatic diamines (388) (Scheme 211).[416]

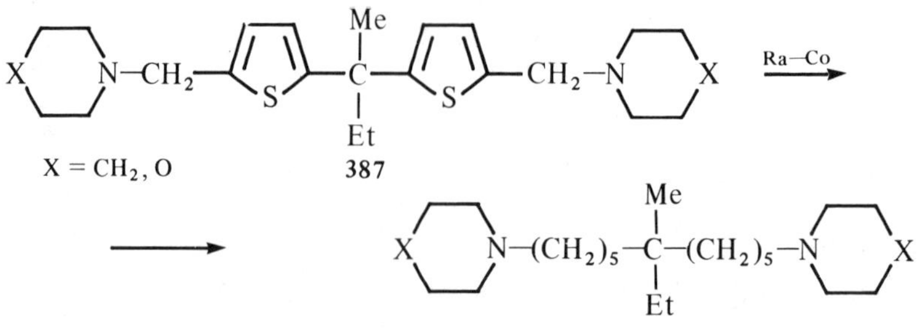

X = CH$_2$, O   387

388

Scheme 211

The specific feature of Raney cobalt action was noticed in the case of 3-carboxy-benzo[b]thiophene-1,1-dioxide:[417] considerable amounts of the products of sulfonyl group reduction and of phenoxyacetic acid (formed as a result of C—C bond hydrogenolysis) are obtained (Scheme 212).

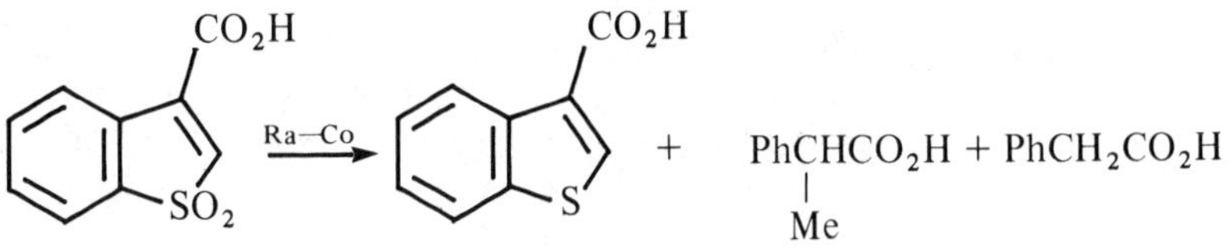

$$\text{+ PhCHCO}_2\text{H + PhCH}_2\text{CO}_2\text{H}$$
$$\quad\quad\quad | $$
$$\quad\quad\text{Me}$$

Scheme 212

Other skeletal metals possess much lower activity. Recently, Mircheva[418] studied the desulfurization ability of skeletal nickel, cobalt, copper, and iron with respect to thiophene. The very low activity of the last two metals was explained by the formation of oxides in the course of the preparation of Raney copper and iron from the corresponding alloys. Gol'dfarb and Zakharov studied reductive desulfurization of 6-(2-thienyl)caproic acid and its esters with skeletal iron and skeletal agents containing also small amounts of nickel (less than 14%) as well as iron.[419] In particular, the skeletal reagent prepared from the aluminum — stainless steel alloy appeared to be sufficiently effective (as compared with Raney nickel). But one should bear in mind that desulfurization conditions were unusual and quite drastic (250–300°C with hydrogen addition at initial pressure 100–110 atm).

Schut et al.[420] studied the desulfurization of 3,4-diphenylthiophene and methyl 7-(5-heptyl-2-thienyl)heptanoate using the excess of freshly prepared nickel boride catalyst obtained on the reduction of nickel dichloride with sodium borohydride in a methanol, ethanol, or ethanol–acetone mixture. The reaction is carried out without hydrogen addition and leads mainly to the products of incomplete reduction, possessing one C=C double bonds. Unfortunately, the reaction is not selective and results in the mixtures of cis- and trans-isomers (Schemes 213 and 214).

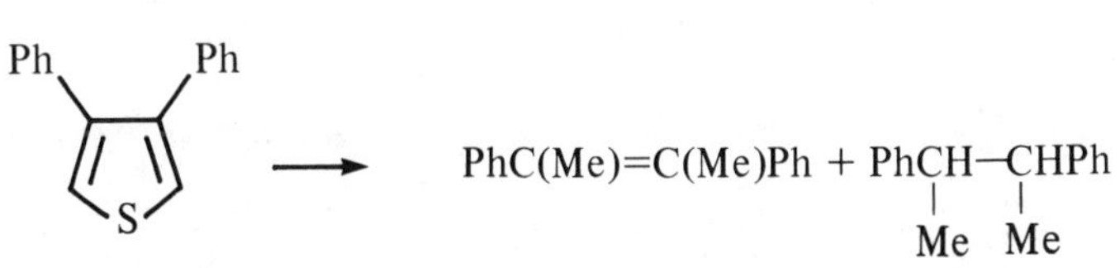

$$\text{PhC(Me)=C(Me)Ph + PhCH—CHPh}$$
$$\quad\quad\quad\quad\quad\quad\quad | \quad\; |$$
$$\quad\quad\quad\quad\quad\quad\text{Me   Me}$$

Scheme 213

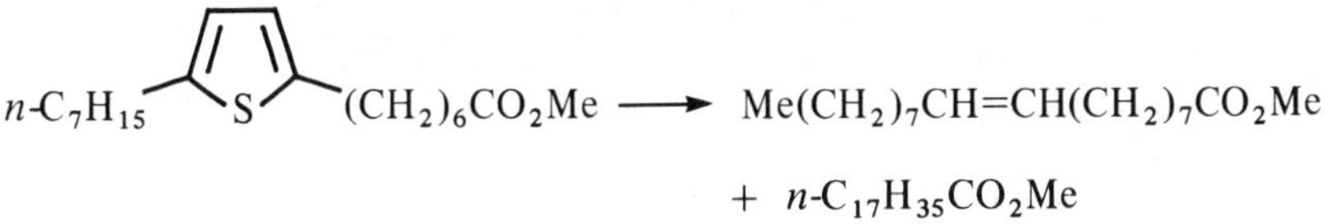

$$\text{Me(CH}_2)_7\text{CH=CH(CH}_2)_7\text{CO}_2\text{Me}$$
$$\text{+ } n\text{-C}_{17}\text{H}_{35}\text{CO}_2\text{Me}$$

Scheme 214

V. MISCELLANEOUS METHODS OF THIOPHENE RING OPENING

Besides reductive methods of cleavage and desulfurization of the thiophene ring, other transformations leading to the opening of the thiophene ring with the retention or removal of sulfur atom are also known. Though these processes differ in character from those considered in this chapter, some should be pointed out.

A method of thiophene desulfurization in glowing discharge was described. Under these conditions, dibenzothiophene-1,1-dioxide is transformed into biphenylene (yield 71%) and dibenzofuran (yield 25%);[421] thiophene gives a mixture of lower aliphatic hydrocarbons.[422] Thiophene desulfurization with iron carbonyls[423,424] or hydrogen atoms[425] does not proceed sufficiently selectively. Recently, Gronowitz and Frejd studied the ring cleavage of β-lithium-substituted thiophenes and Iddon et al. investigated similar benzo[b]thiophene derivatives (see Ref. 426). Janda and coworkers elaborated methods of electrochemical oxidation of thiophene, leading to the formation of functional aliphatic compounds that do not contain sulfur (for references to the original papers, see Ref. 1). Data concerning the reactions of thiophene with some carbenes, followed by sulfur elimination and the formation of polyene structures appeared in Refs. 427–429. Cases of ring cleavage products formation under the action of singlet oxygen on alkyl- and arylthiophenes are also known.[430,431]

When attempting to evaluate the importance of thiophene in modern synthetic chemistry, even taking into account the material considered above (which covers only a few aspects of its use), one can see a striking number of possibilities; and all this is achieved with the aid of thiophene. Hence, thiophene occupies a unique position among accessible heteroaromatic systems, because it is possible not only to obtain its own derivatives, but also to use thiophene itself as a powerful tool in the synthesis of various compounds belonging to other series.

REFERENCES

1. L. I. Belen'kii and V. P. Gul'tyai, *Khim. Geterotsikl. Soed.*, 723 (1981); *Chem. Abstr.*, 95, 97466 (1981).

2. L. F. Fieser and R. G. Kennely, *J. Am. Chem. Soc.*, 57, 1611 (1935).

3. S. Z. Taits and Ya. L. Gol'dfarb, *Izv. Akad. Nauk SSSR, Otd. Khim. Nauk*, 1698 (1960); *Chem. Abstr.*, 55, 8318 (1961).

4. Ya. L. Gol'dfarb, S. Z. Taits, and V. N. Bulgakova, *Izv. Akad. Nauk SSSR, Ser. Khim.*, 1299 (1963); *Chem. Abstr.*, 59, 13990 (1963).

5. W. J. King and F. F. Nord, *J. Org. Chem.*, 14, 638 (1949).

6. M. Sy, N. P. Buu-Hoi, and N. D. Xuong, *J. Chem. Soc.*, 1975 (1954).

7. Ya. L. Gol'dfarb and Ya. L. Danyushevskii, *Izv. Akad. Nauk SSSR, Otd. Khim. Nauk*, 1361 (1956); *Chem. Abstr.*, 51, 8065 (1957).

8. P. Cagniant and D. Cagniant, *Bull. Soc. Chim. Fr.*, 62 (1953).

9. L. I. Belen'kii, S. Z. Taits, and Ya. L. Gol'dfarb, *Izv. Akad. Nauk SSSR, Otd. Khim. Nauk*, 1706 (1961); *Chem. Abstr.*, 56, 3435 (1962).

10. B. P. Fabrichnyi, F. M. Stoyanovich, Yu. B. Vol'kenshtein, S. Z. Taits, Ya. L. Gol'dfarb, A. S. Mezentsev, M. A. Panina, I. B. Karmanova, and V. A. Koval'skaya, USSR Author Certificate 677,331 (1977); *Chem. Abstr.*, **95**, 150420 (1981).

11. Y. LeGuen and G. Thiault, French Demande 2,377,398 (1977); through *Chem. Abstr.*, **91**, 20313 (1979).

12. M. A. Ryashentseva, O. A. Kalinovskii, Kh. M. Minachev, and Ya. L. Gol'dfarb, *Khim. Geterotsikl. Soed.*, 694 (1966); *Chem. Abstr.*, **66**, 54772 (1967).

13. M. A. Ryashentseva, Kh. M. Minachev, E. P. Belanova, B. P. Fabrichnyi, and Yu. B. Vol'kenshtein, *Izv. Akad. Nauk SSSR, Ser. Khim.*, 2152 (1977); *Chem. Abstr.*, **88**, 22513 (1978).

14. I. Wender, R. Levine, and M. Orchin, *J. Am. Chem. Soc.*, **72**, 4375 (1950).

15. I. Wender, H. Greenfield, and M. Orchin, *J. Am. Chem. Soc.*, **73**, 2656 (1951).

16. E. Campaigne and J. L. Diedrich, *J. Am. Chem. Soc.*, **70**, 391 (1948).

17. O. Červinka, P. Maloň, and M. Prohazkova, *Coll. Czech. Chem. Comm.*, **39**, 1869 (1974).

18. F. Ernst, *Bericht*, **19**, 3278 (1886).

19. T. Mizutani, Y. Ume, and T. Matsuo, Japanese Patent No. 135,932 (1974); through *Chem. Abstr.*, **82**, 155744 (1975).

20. T. Mizutani, Y. Ume, and T. Matsuo, Japanese Patent No. 135,933 (1974); through *Chem. Abstr.*, **82**, 155745 (1975).

21. F. W. Dunn and K. J. Ditmer, *J. Am. Chem. Soc.*, **68**, 2561 (1946).

22. P. Foulatier and C. Caullet, *Compt. Rend. (C)*, **279**, 25 (1974).

23. P. Foulatier, J.-P. Salaün, and C. Caullet, *Compt. Rend. (C)*, **279**, 779 (1974).

24. M. R. Kegelman and E. V. Brown, *J. Am. Chem. Soc.*, **75**, 5961 (1953).

25. C. Caullet, M. Salaün, and H. Hebert, *Compt. Rend. (C)*, **264**, 228 (1967).

26. E. V. Kryukova and A. P. Tomilov, *Elektrokhimiya*, **5**, 869 (1969); *Chem. Abstr.*, **71**, 76739 (1969).

27. V. P. Gul'tyai, L. M. Korotaeva, A. S. Mendkovich, and I. V. Proskurovskaya, *Izv. Akad. Nauk SSSR, Ser. Khim.*, 834 (1981); *Chem. Abstr.*, **94**, 216563 (1981).

28. V. P. Gul'tyai and L. M. Korotaeva, *Izv. Akad. Nauk SSSR, Ser. Khim.*, 165 (1982).

29. V. P. Gul'tyai, L. M. Korotaeva, A. P. Rodionov, and A. M. Moiseenkov, *Izv. Akad. Nauk SSSR, Ser. Khim.*, 1150 (1981); *Chem. Abstr.*, **95**, 105310 (1981).

30. L. N. Nekrasov, L. N. Vykhodtseva, L. M. Korotaeva, and V. P. Gul'tyai, *J. Electroanal. Chem.*, **138**, 177 (1982).

31. D. W. H. MacDowell and A. W. Springsteen, *J. Org. Chem.*, **41**, 3046 (1976).

32. S. Z. Taits, F. D. Alashev, and Ya. L. Gol'dfarb, *Izv. Akad. Nauk SSSR, Ser. Khim.*, 402 (1968); *Chem. Abstr.*, **69**, 86736 (1968).

33. L. I. Belen'kii, I. B. Karmanova, Yu. B. Vol'kenshtein, and Ya. L. Gol'dfarb, *Izv. Akad. Nauk SSSR, Ser. Khim.*, 956 (1971); *Chem. Abstr.*, **75**, 88408 (1971).

34. J. Šrogl, M. Janda, I. Stibor, and H. Prohazka, *Z. Chem.*, **11**, 421 (1971).

35. S. Gronowitz, *Acta Chem. Scand.*, **13**, 1045 (1959).

36. A. Hallberg, S. Lilljefors, and P. Pedaja, *Synth. Comm.*, **11**, 25 (1981).

37. A. S. Alvarez-Insúa, S. Conde, and C. Corral, *J. Heterocycl. Chem.*, **19**, 713 (1982).

38. H. Neunhoeffer and G. Köhler, *Tetrahedron Lett.*, 4879 (1978).

39. I. B. Karmanova, Yu. B. Vol'kenshtein, and L. I. Belen'kii, *Khim. Geterotsikl. Soed.*, 490 (1973); *Chem. Abstr.*, **79**, 31770 (1973).

40. M. Martin-Smith and M. Gates, *J. Am. Chem. Soc.*, **78**, 6177 (1956).

41. R. Mozingo, S. A. Harris, D. E. Wolf, C. E. Hoffhine, W. R. Easton, and K. Folkers, *J. Am. Chem. Soc.*, **67**, 2092 (1945).

42.   S. Z. Taits and Ya. L. Gol'dfarb, *Izv. Akad. Nauk SSSR, Otd. Khim. Nauk,* 1289 (1963); *Chem. Abstr.,* **59,** 13990 (1963).

43.   V. Z. Sharf, S. Z. Taits, A. S. Gurovets, Yu. B. Vol'kenshtein, B. P. Fabrichnyi, and S. I. Shcherbakova, *Khim. Geterotsikl. Soed.,* 171 (1982); *Chem. Abstr.,* **96,** 181086 (1982).

44.   V. Z. Sharf, L. I. Belen'kii, A. S. Gurovets, and I. B. Karmanova, *Khim. Geterotsikl. Soed.,* 176 (1982); *Chem. Abstr.,* **96,** 181087 (1982).

45.   S.-O. Lawesson, *Ark. Kemi,* **11,** 317 (1957).

46.   I. O. Shapiro, L. I. Belen'kii, I. A. Romanskii, F. M. Stoyanovich, Ya. L. Gol'dfarb, and A. I. Shatenshtein, *Zh. Obshch. Khim.,* **38,** 1998 (1968); *Chem. Abstr.,* **70,** 19289 (1969).

47.   B. P. Fabrichnyi, I. F. Shalavina, and Ya. L. Gol'dfarb, USSR Author Certificate 170,521 (1965); *Chem. Abstr.,* **63,** 13215 (1965).

48.   B. P. Fabrichnyi, I. F. Shalavina, and Ya. L. Gol'dfarb, *Dokl. Akad. Nauk SSSR,* **162,** 120 (1965); *Chem. Abstr.,* **63,** 11538 (1965).

49.   B. P. Fabrichnyi, I. F. Shalavina, S. M. Kostrova, and Ya. L. Gol'dfarb, *Khim. Geterotsikl. Soed.,* 1358 (1971); *Chem. Abstr.,* **75,** 48799 (1971).

50.   J. C. Barker, J. F. C. Coutts, and P. R. Huddlestone, *Chem. Comm.,* 615 (1972).

51.   S. Gronowitz, A. Hallberg, and C. Glennow, *J. Heterocycl. Chem.,* **17,** 171 (1980).

52.   W. Steinkopf, *Lieb. Ann.,* **403,** 17 (1914).

53.   D. L. Eck and G. W. Stacy, *J. Heterocycl. Chem.,* **6,** 147 (1969).

54.   J. Cymmerman-Creig and D. Willis, *J. Chem. Soc.,* 1071 (1955).

55.   L. H. Klemm, R. Zell, I. T. Barnisch, R. A. Klemm, C. E. Klopfenstein, and D. R. McCoy, *J. Heterocycl. Chem.,* **7,** 373 (1970).

56.   L. H. Klemm and W. Hsin, *J. Heterocycl. Chem.,* **12,** 1183 (1975).

57.   Ya. L. Gol'dfarb, B. P. Fabrichnyi, and I. F. Shalavina, *Khim. Geterotsikl. Sord.,* 1323 (1982).

58.   C. Aretos and J. Vialle, in *Rhenium,* Elsevier, Amsterdam, 1962, p. 171; through *Ref. Zh. Khim.,* 2B631 (1965).

59.   Ya. L. Gol'dfarb, M. M. Polonskaya, B. P. Fabrichnyi, and I. F. Shalavina, *Dokl. Akad. Nauk SSSR,* **126,** 86 (1959); *Chem. Abstr.,* **53,** 21872 (1959).

60.   Ya. L. Gol'dfarb, B. P. Fabrichnyi, and I. F. Shalavina, *Zh. Obshch. Khim.,* **29,** 3636 (1959); *Chem. Abstr.,* **54,** 19638 (1960).

61.   H. Goldschmidt and W. Schulthess, *Bericht,* **20,** 1700 (1887).

62.   P. Chabrier, P. Tchoubar, and S. Le Tellier-Dupré, *Bull. Soc. Chim. Fr.,* 332 (1946).

63.   N. I. Putokhin and V. S. Egorova, *Zh. Obshch. Khim.,* **18,** 1866 (1948).

64.   N. I. Putokhin and V. S. Egorova, *Zh. Obshch. Khim.,* **10,** 1873 (1940); *Chem. Abstr.,* **35,** 4377 (1941).

65.   G. Barger and A. P. T. Easson, *J. Chem. Soc.,* 2100 (1938).

66.   J. M. Griffing and L. F. Salisbury, *J. Am. Chem. Soc.,* **70,** 3416 (1948).

67.   Ya. L. Gol'dfarb, B. P. Fabrichnyi, and I. F. Shalavina, *Dokl. Akad. Nauk SSSR,* **109,** 305 (1956); *Chem. Abstr.,* **51,** 1839 (1957).

68.   Ya. L. Gol'dfarb, B. P. Fabrichnyi, and I. F. Shalavina, *Zh. Obshch. Khim.,* **29,** 891 (1959); *Chem. Abstr.,* **54,** 1483 (1960).

69.   W. P. Bradley, *Bericht,* **19,** 2115 (1886).

70.   B. P. Fabrichnyi and I. F. Shalavina, USSR Author Certificate 771,100 (1980); *Chem. Abstr.,* **94,** 192124 (1981).

71.   B. P. Fedorov, G. I. Gorushkina, and Ya. L. Gol'dfarb, *Zh. Obshch. Khim.,* **31,** 3933 (1961); *Chem. Abstr.,* **57,** 8529 (1962).

72. B. P. Fedorov, G. I. Gorushkina, and Ya. L. Gol'dfarb, *Izv. Akad. Nauk SSSR, Ser. Khim.*, 2049 (1967); *Chem. Abstr.*, **68**, 108875 (1968).

73. M. A. Ryashentseva, Kh. M. Minachev, O. A. Kalinovskii, and Ya. L. Gol'dfarb, *Zh. Org. Khim.*, **1**, 1004 (1965); *Chem. Abstr.*, **63**, 11474 (1965).

74. O. Dann, *Bericht*, **76**, 419 (1943).

75. D. H. Jones and K. R. H. Wooldridge, *J. Chem. Soc.* (*C*), 550 (1968).

76. H. Burton and W. A. Davy, *J. Chem. Soc.*, 525 (1948).

77. M. Martin-Smith and M. Gates, *J. Am. Chem. Soc.*, **78**, 5351 (1956).

78. H. Gilman and G. R. Wilder, *J. Am. Chem. Soc.*, **76**, 2906 (1954).

79. H. Kotake and T. Sakan, *J. Inst. Polytech. Osaka City Univ., Ser. C*, **2**(1) 25 (1951); through *Chem. Abstr.*, **46**, 6121 (1952).

80. R. M. Kellog and J. Buter, *J. Org. Chem.*, **36**, 2236 (1971).

81. S. Z. Taits, A. A. Dudinov, F. D. Alashev, and Ya. L. Gol'dfarb, *Izv. Akad. Nauk SSSR, Ser. Khim.*, 148 (1974); *Chem. Abstr.*, **80**, 120672 (1974).

82. R. T. Gilsdorf and F. F. Nord, *J. Org. Chem.*, **15**, 807 (1950).

83. Ya. L. Gol'dfarb, M. B. Ibragimova, and O. A. Kalinovskii, *Izv. Akad. Nauk SSSR, Otd. Khim. Nauk*, 1098 (1962); *Chem. Abstr.*, **57**, 13710 (1962).

84. S. Conde, R. Madrofiero, M. P. Fernandez-Tomé, and J. del Rio, *J. Med. Chem.*, **21**, 978 (1978).

85. F. Ya. Perveev and N. I. Kudryashova, *Zh. Obshch. Khim.*, **23**, 976 (1953); *Chem. Abstr.*, **48**, 8219 (1954).

86. A.-B. Hörnfeldt, J. S. Gronowitz, and S. Gronowitz, *Acta Chem. Scand.*, **22**, 2725 (1968).

87. A. P. Stoll and R. Süess, *Helv. Chim. Acta*, **57**, 2487 (1974).

88. A. Biedermann, *Bericht*, **19**, 1615 (1886).

89. O. Hinsberg, *Bericht*, **48**, 1611 (1915).

90. F. G. Bordwell and W. H. McKellin, *J. Am. Chem. Soc.*, **73**, 2251 (1951).

91. F. M. Stoyanovich, G. I. Gorushkina, and Ya. L. Gol'dfarb, *Izv. Akad. Nauk SSSR, Ser. Khim.*, 387 (1968); *Chem. Abstr.*, **71**, 3198 (1969).

92. Z. V. Todres, F. M. Stoyanovich, Ya. L. Gol'dfarb, and D. N. Kursanov, *Khim. Geterotsikl. Soed.*, 632 (1973); *Chem. Abstr.*, **79**, 66117 (1973).

93. Ya. L. Gol'dfarb, G. P. Pokhil, and L. I. Belen'kii, *Dokl. Akad. Nauk SSSR*, **167**, 82 (1966); *Chem. Abstr.*, **65**, 2196 (1966).

94. M. Vafai and M. Renson, *Bull. Soc. Chim. Belges*, **75**, 145 (1966).

95. Ya. L. Gol'dfarb, M. A. Kalik, and M. L. Kirmalova, *Izv. Akad. Nauk SSSR, Otd. Khim. Nauk*, 1696 (1960); *Chem. Abstr.*, **55**, 8377 (1961).

96. Ya. L. Gol'dfarb and M. A. Kalik, *Khim. Geterotsikl. Soed.*, 788 (1968); *Chem. Abstr.*, **70**, 96512 (1969).

97. V. S. Bogdanov, M. A. Kalik, A. V. Kessenikh, and Ya. L. Gol'dfarb, *Khim. Geterotsikl. Soed.*, 793 (1968); *Chem. Abstr.*, **70**, 106286 (1969).

98. Ya. L. Gol'dfarb, M. A. Kalik, M. L. Kirmalova, and M. M. Polonskaya, *Izv. Akad. Nauk SSSR, Ser. Khim.*, 897 (1966); *Chem. Abstr.*, **65**, 13635 (1966).

99. Ya. L. Gol'dfarb and M. A. Kalik, *Usp. Khim.*, **41**, 679 (1972); *Chem. Abstr.*, **77**, 34205 (1972).

100. L. I. Belen'kii, E. P. Zakharov, M. A. Kalik, V. P. Litvinov, F. M. Stoyanovich, S. Z. Taits, and B. P. Fabrichnyi, in *Novye Napravleniya Khimii Tiofena* (Ya. L. Gol'dfarb, Ed.), Nauka, Moscow, 1976; (a) Chapter 3, pp. 155–189; (b) Chapter 5, pp. 256–285; (c) Chapter 6, pp. 286–320; (d) Chapter 7, pp. 321–404.

101.   V. S. Bogdanov, M. A. Kalik, I. P. Yakovlev, and Ya. L. Gol'dfarb, *Zh. Obshch. Khim.,*
        40, 2102 (1970); *Chem. Abstr.,* **74**, 140600 (1971).

102.   Ya. L. Gol'dfarb, M. A. Kalik, and M. L. Kirmalova, *Izv. Akad. Nauk SSSR, Otd.
        Khim. Nauk,* 701 (1962); *Chem. Abstr.,* **57**, 16529 (1962).

103.   Ya. L. Gol'dfarb and M. A. Kalik, *Khim. Geterotsikl. Soed.,* 475 (1969); *Chem. Abstr.,*
        72, 21549 (1970).

104.   Ya. L. Gol'dfarb, M. A. Kalik, and Z. G. Koslova, *Khim. Geterotsikl. Soed.,* 1331
        (1980); *Chem. Abstr.,* **94**, 208627 (1981).

105.   Ya. L. Gol'dfarb and M. A. Kalik, *Khim. Geterotsikl. Soed.,* 1323 (1970); *Chem. Abstr.,*
        74, 76240 (1971).

106.   Ya. L. Gol'dfarb, M. A. Kalik, and M. L. Kirmalova, *Izv. Akad. Nauk SSSR, Ser.
        Khim.,* 1801 (1963); *Chem. Abstr.,* **60**, 5434 (1964).

107.   Ya. L. Gol'dfarb, S. A. Ozolin', and V. P. Litvinov, *Zh. Obshch. Khim.,* **37**, 2220
        (1967); *Chem. Abstr.,* **68**, 87218 (1968).

108.   Ya. L. Gol'dfarb, S. A. Ozolin', and V. P. Litvinov, USSR Author Certificate 201,427
        (1966); *Chem. Abstr.,* **69**, 19135 (1968).

109.   Ya. L. Gol'dfarb, M. A. Kalik, and M. L. Kirmalova, *Izv. Akad. Nauk SSSR, Ser. Khim.,*
        1675 (1964); *Chem. Abstr.,* **61**, 16032 (1964).

110.   Ya. L. Gol'dfarb, M. A. Kalik, and M. L. Kirmalova, *Khim. Geterotsikl. Soed.,* 62
        (1967); *Chem. Abstr.,* **67**, 116772 (1967).

111.   Ya. L. Gol'dfarb, M. A. Kalik, and M. L. Kirmalova, *Khim. Geterotsikl. Soed.,* 71
        (1967); *Chem. Abstr.,* **67**, 116773 (1967).

112.   A. V. Mashkina, *Geterogennyi kataliz v khimii organicheskikh soedinenii sery,* Nauka,
        Sibir. Otdel., Novosibirsk, 1977.

113.   P. N. Confalone, G. Pizzoloto, and M. P. Uskokovič, *Helv. Chim. Acta,* **59**, 1005
        (1976).

114.   P. N. Confalone, G. Pizzoloto, and M. R. Uskokovič, *J. Org. Chem.,* **42**, 135 (1977).

115.   G. Claeson and H.-G. Jonsson, *Ark. Kemi,* **31**, 83 (1969).

116.   A. A. Balandin, M. L. Khidekel', and V. V. Patrikeev, *Zh. Obshch. Khim.,* **31**, 1876
        (1961); *Chem. Abstr.,* **55**, 27261 (1961).

117.   L. Rajca, A. Borowski, and A. Rajca, *Symp. Rhodium Homog. Cat. (Proc.),* 58 (1978);
        through *Chem. Abstr.,* **90**, 12835 (1979).

118.   H. Broadbent, L. H. Staush, and N. L. Jarvis, *J. Am. Chem. Soc.,* **76**, 1519 (1954).

119.   M. A. Ryashentseva and Kh. M. Minachev, *Usp. Khim.,* **38**, 2050 (1969); *Chem. Abstr.,*
        72, 47921 (1970).

120.   A. V. Mashkina and E. M. Vakurova, *Dokl. Akad. Nauk SSSR,* **168**, 821 (1966); *Chem.
        Abstr.,* **65**, 8851 (1966).

121.   M. A. Ryashentseva, Kh. M. Minachev, and E. P. Belanova, *Izv. Akad. Nauk SSSR, Ser.
        Khim.,* 1183 (1976); *Chem. Abstr.,* **85**, 123731 (1976).

122.   E. Campaigne and J. L. Diedrich, *J. Am. Chem. Soc.,* **73**, 5240 (1951).

123.   G. B. Hatch, U.S. Patent No. 2,648,675; through *Chem. Abstr.,* **48**, 8264 (1954).

124.   H. Greenfield, S. Metlin, M. Orchin, and I. Wender, *J. Org. Chem.,* **23**, 1054 (1958).

125.   I. Wender and M. Orchin, U.S. Patent No. 3,002,002 (1957); through *Chem. Abstr.,*
        56, 1430 (1962).

126.   D. N. Kursanov, Z. N. Parnes, G. I. Bolestova, and L. I. Belen'kii, USSR Author Certifi-
        cate 408,950 (1972); *Chem. Abstr.,* **80**, 133233 (1974).

127.   Z. N. Parnes, G. I. Bolestova, L. I. Belen'kii, and D. N. Kursanov, *Izv. Akad. Nauk
        SSSR, Ser. Khim.,* 1918 (1973); *Chem. Abstr.,* **80**, 14800 (1974).

128.   D. N. Kursanov, Z. N. Parnes, and N. M. Loim, *Synthesis,* 633 (1974).

129. D. N. Kursanov, Z. N. Parnes, M. I. Kalinkin, and N. M. Loim, *Ionnoe Gidrirovanie*, Khimiya, Moscow, 1979, pp. 61–74.

130. D. N. Kursanov, Z. N. Parnes, G. I. Bolestova, and L. I. Belen'kii, *Tetrahedron*, **31**, 311 (1975).

131. G. I. Bolestova, E. P. Zakharov, S. P. Dolgova, Z. N. Parnes, and D. N. Kursanov, *Khim. Geterotsikl. Soed.*, 1206 (1975); *Chem. Abstr.*, **84**, 17049 (1976).

132. Z. N. Parnes, G. I. Bolestova, S. P. Dolgova, V. E. Udre, M. G. Voronkov, and D. N. Kursanov, *Izv. Akad. Nauk SSSR, Ser. Khim.*, 1834 (1974); *Chem. Abstr.*, **82**, 16635 (1975).

133. E. S. Rudakov, Z. N. Parnes, A. M. Osipov, Yu. I. Lyakhovetskii, and D. N. Kursanov, *Izv. Akad. Nauk SSSR, Ser. Khim.*, 1173 (1976); *Chem. Abstr.*, **85**, 159796 (1976).

134. G. I. Bolestova, A. I. Korepanov, Z. N. Parnes, and D. N. Kursanov, *Izv. Akad. Nauk SSSR, Ser. Khim.*, 2547 (1974); *Chem. Abstr.*, **82**, 125212 (1975).

135. S. I. Zav'yalov, N. A. Rodionova, L. I. Zheleznaya, G. I. Bolestova, V. V. Filippov, Z. N. Parnes, and D. N. Kursanov, *Izv. Akad. Nauk SSSR, Ser. Khim.*, 1643 (1975); *Chem. Abstr.*, **83**, 164075 (1975).

136. S. I. Zav'yalov and O. V. Dorofeeva, *Izv. Akad. Nauk SSSR, Ser. Khim.*, 1634 (1980); *Chem. Abstr.*, **94**, 3963 (1981).

137. Z. N. Parnes, G. I. Bolestova, and D. N. Kursanov, *Izv. Akad. Nauk SSSR, Ser. Khim.*, 478 (1976); *Chem. Abstr.*, **85**, 20989 (1976).

138. Z. N. Parnes, G. I. Bolestova, and D. N. Kursanov, *Zh. Org. Khim.*, **13**, 476 (1977); *Chem. Abstr.*, **87**, 21664 (1977).

139. Z. N. Parnes, Yu. I. Lyakhovetskii, S. P. Dolgova, A. S. Pakhomov, and D. N. Kursanov, *Izv. Akad. Nauk SSSR, Ser. Khim.*, 2526 (1977); *Chem. Abstr.*, **88**, 62239 (1978).

140. L. I. Belen'kii, A. P. Yakubov, and Ya. L. Gol'dfarb, *Zh. Org. Khim.*, **11**, 424 (1975); *Chem. Abstr.*, **82**, 124371 (1975).

141. Z. N. Parnes, Yu. I. Lyakhovetskii, N. M. Loim, L. I. Belen'kii, P. V. Petrov, and D. N. Kursanov, *Izv. Akad. Nauk SSSR, Ser. Khim.*, 2145 (1976); *Chem. Abstr.*, **86**, 43481 (1977).

142. Z. N. Parnes, Yu. I. Lyakhovetskii, M. I. Kalinkin, L. I. Belen'kii, and D. N. Kursanov, *Tetrahedron*, **34**, 1703 (1978).

143. J. Wristers, *J. Am. Chem. Soc.*, **99**, 5051 (1977).

144. Yu. I. Lyakhovetskii, M. Kalinkin, Z. Parnes, F. Latypova, and D. Kursanov, *Chem. Comm.*, 766 (1980).

145. S. G. Mairanovskii, L. I. Kosychenko, and S. Z. Taits, *Elektrokhimiya*, **13**, 1250 (1977); *Chem. Abstr.*, **88**, 43133 (1978).

146. S. G. Mairanovskii, L. I. Kosychenko, and S. Z. Taits, *Izv. Akad. Nauk SSSR, Ser. Khim.*, 1382 (1980); *Chem. Abstr.*, **93**, 83354 (1980).

147. S. G. Mairanovskii, L. I. Kosychenko, *Elektrokhimiya*, **16**, 266 (1980); *Chem. Abstr.*, **92**, 137556 (1980).

148. V. S. Mikhailov, V. P. Gul'tyai, S. G. Mairanovskii, S. Z. Taits, I. V. Proskurovskaya, and Yu. G. Dubovik, *Izv. Akad. Nauk SSSR, Ser. Khim.*, 888 (1975); *Chem. Abstr.*, **83**, 87240 (1975).

149. V. P. Gul'tyai, I. V. Proskurovskaya, T. Ya. Rubinskaya, A. V. Lozanova, A. M. Moiseenkov, and A. V. Semenovskii, *Izv. Akad. Nauk SSSR, Ser. Khim.*, 1576 (1979); *Chem. Abstr.*, **91**, 157538 (1979).

150. R. M. Kellogg, *J. Am. Chem. Soc.*, **93**, 2344 (1971).

151. W. L. Mock, *J. Am. Chem. Soc.*, **88**, 2857 (1966).

152. S. D. McGregor and D. M. Lemal, *J. Am. Chem. Soc.*, **88**, 2858 (1966).

153. V. P. Gul'tyai, T. G. Konstantinova, and A. M. Moiseenkov, *Izv. Akad. Nauk SSSR, Ser. Khim.,* 687 (1981).

154. V. P. Gul'tyai, T. G. Konstantinova, A. M. Moiseenkov, V. P. Litvinov, and A. Konar, *Chem. Scr.,* **19,** 95 (1982).

155. V. P. Gul'tyai, *Abstr. Sandbjerg Meet. Org. Electrochem.,* 10–13 June 1982, p. 81.

156. V. P. Gul'tyai, L. M. Korotaeva, and T. Ya. Rubinskaya, *Dokl. Akad. Nauk SSSR,* **267,** 662 (1982).

157. S. F. Birch and D. T. McAllan, *Nature,* **165,** 899 (1950).

158. S. F. Birch and D. T. McAllan, *J. Chem. Soc.,* 2556 (1951).

159. S. F. Birch and D. T. McAllan, *J. Chem. Soc.,* 3411 (1951).

160. R. C. Krug and S. Tocker, *J. Org. Chem.,* **20,** 1 (1955).

161. W. Hückel and I. Nabih, *Chem. Ber.,* **89,** 2115 (1956).

162. A. J. Birch and D. Nasipuri, *Tetrahedron,* **6,** 148 (1959).

163. W. Hückel, S. Gupte, and M. Wartini, *Chem. Ber.,* **99,** 1388 (1966).

164. H. Gilman and A. L. Jacoby, *J. Org. Chem.,* **3,** 108 (1938).

165. A. J. Birch and J. Slobbe, *Heterocycles,* **5,** 905 (1976).

166. V. A. Pleskov, *Zh. Fiz. Khim.,* **9,** 12 (1937); *Acta Physicochim. URSS,* **6,** 1 (1937).

167. V. A. Pleskov and A. M. Monoszon, *Zh. Fiz. Khim.,* **6,** 1286 (1935); *Acta Physicochim. URSS,* **2,** 615 (1935).

168. Ya. L. Gol'dfarb, A. V. Semenovskii, E. P. Zakharov, G. V. Davydova, and F. M. Stoyanovich, *Izv. Akad. Nauk SSSR, Ser. Khim.,* 480 (1979); *Chem. Abstr.,* **90,** 168011 (1979).

169. W. G. Blenderman, M. M. Joullie, and G. Preti, *Tetrahedron Lett.,* 4985 (1979).

170. W. G. Blenderman and M. M. Joullie, *Synth. Comm.,* **11,** 881 (1981).

171. Ya. L. Gol'dfarb, E. P. Zakharov, A. S. Shashkov, and F. M. Stoyanovich, *Zh. Org. Khim.,* **16,** 1523 (1980); *Chem. Abstr.,* **94,** 30137 (1981).

172. K. Kosugi, A. V. Anisimov, H. Yamamoto, R. Yanashiro, K. Shirai, and T. Kumamoto, *Chem. Lett.,* 1341 (1981).

173. Ya. L. Gol'dfarb and E. P. Zakharov, *Izv. Akad. Nauk SSSR, Ser. Khim.,* 1909 (1975); *Chem. Abstr.,* **84,** 30786 (1976).

174. Ya. L. Gol'dfarb and E. P. Zakharov, *Khim. Geterotsikl. Soed.,* 337 (1977); *Chem. Abstr.,* **87,** 67792 (1977).

175. E. P. Zakharov and Ya. L. Gol'dfarb, *Izv. Akad. Nauk SSSR, Ser. Khim.,* 1877 (1978); *Chem. Abstr.,* **89,** 179476 (1978).

176. Ya. L. Gol'dfarb and E. P. Zakharov, *Khim. Geterotsokl. Soed.,* 1633 (1971); *Chem. Abstr.,* **76,** 139896 (1972).

177. Ya. L. Gol'dfarb and E. P. Zakharov, *Zh. Org. Khim.,* **6,** 1757 (1970); *Chem. Abstr.,* **73,** 109231 (1970).

178. Ya. L. Gol'dfarb and E. P. Zakharov, *Izv. Akad. Nauk SSSR, Ser. Khim.,* 2160 (1973); *Chem. Abstr.,* **80,** 36937 (1974).

179. Ya. L. Gol'dfarb and E. P. Zakharov, *Khim. Geterotsikl. Soed.,* 1499 (1975); *Chem. Abstr.,* **84,** 89924 (1976).

180. A. V. Semenovskii and M. M. Emel'yanov, *Izv. Akad. Nauk SSSR, Ser. Khim.,* 2578 (1980); *Chem. Abstr.,* **95,** 7463 (1981).

181. A. V. Semenovskii, M. M. Emel'yanov, *Izv. Akad. Nauk SSSR, Ser. Khim.,* 1359 (1981); *Chem. Abstr.,* **95,** 150923 (1981).

182. E. P. Zakharov, Ya. L. Gol'dfarb, and F. M. Stoyanovich, *Izv. Akad. Nauk SSSR, Ser. Khim.,* 1440 (1982).

183. (a) A. V. Lozanova, A. M. Moiseenkov, and A. V. Semenovskii, *Izv. Akad. Nauk SSSR, Ser. Khim.*, 958 (1980); *Chem. Abstr.*, **93**, 70917 (1980); (b) A. V. Lozanova, A. M. Moiseenkov, and A.V. Semenovskii, *Izv. Akad. Nauk SSSR, Ser. Khim.*, 1932 (1980); *Chem. Abstr.*, **94**, 30148 (1981).

184. A. V. Lozanova, A. M. Moiseenkov, and A. V. Semenovskii, *Izv. Akad. Nauk SSSR, Ser. Khim.*, 838 (1981); *Chem. Abstr.*, **95**, 80025 (1981).

185. M. M. Emel'yanov, A. V. Lozanova, A. M. Moiseenkov, V. A. Smit, and A. V. Semenovskii, *Izv. Akad. Nauk SSSR, Ser. Khim.*, 2788 (1982).

186. A. V. Lozanova, A. M. Moiseenkov, and A. V. Semenovskii, *Izv. Akad. Nauk SSSR, Ser. Khim.*, 2783 (1982).

187. J. van Schooten, J. Knotnerus, H. Boer, and P. M. Duinker, *Rec. Trav. Chim.*, **77**, 935 (1958).

188. R. Fricke and G. Spilker, *Bericht*, **58**, 24 (1925).

189. R. Fricke and G. Spilker, *Bericht*, **58**, 1589 (1925).

190. R. Fricke and G. Spilker, *Bericht*, **59**, 349 (1926).

191. F. Challenger and J. B. Harrison, *J. Inst. Petrol. Technol.*, **21**, 135 (1935); *Chem. Abstr.*, **29**, 4006 (1935).

192. H. Gilman and D. L. Esmay, *J. Am. Chem. Soc.*, **75**, 2947 (1953).

193. J. L. Gerlock, L. R. Mahoney, and T. M. Harvey, *Ind. Eng. Chem. Fundam.*, **17**, 23 (1978); through *Chem. Abstr.*, **88**, 91876 (1978).

194. Y. Kurauchi, S. Wada, K. Ohga, and Sh. Morita, *Aromatics*, **32**, 6 (1980); through *Ref. Zh. Khim.*, 22Zh120 (1980).

195. B. Moldavskii and N. Prokopchuk, *Zh. Prikl. Khim.*, **5**, 619 (1932); *Chem. Abstr.*, **27**, 274 (1933).

196. B. L. Moldavskii and Z. I. Kumari, *Zh. Obshch. Khim.*, **4**, 298 (1934); *Chem. Abstr.*, **29**, 1814 (1935).

197. S. R. Sergienko and V. N. Perchenko, *Dokl. Akad. Nauk SSSR*, **128**, 103 (1959); *Chem. Abstr.*, **54**, 1385 (1960).

198. R. D. Obolentsev and A. V. Mashkina, *Dokl. Akad. Nauk SSSR*, **131**, 1092 (1960); *Chem. Abstr.*, **54**, 19131 (1960).

199. N. K. Nag, A. V. Sapre, D. H. Broderick, and B. C. Gates, *J. Catal.*, **57**, 509 (1979); through *Chem. Abstr.*, **91**, 123135 (1979).

200. B. L. Lebedev, M. S. Knots, T. S. Kostromina, M. N. Kuznetsova, E. D. Radchenko, and A. V. Agafonov, *Neftekhimiya*, **20**, 890 (1980); *Chem. Abstr.*, **94**, 177641 (1981).

201. J. Maternova and M. Zdrazil, *Coll. Czech. Chem. Comm.*, **45**, 2532 (1980); through *Chem. Abstr.*, **94**, 102476 (1981).

202. A. R. Kuzyev, in *Organicheskie soedineniya sery, tom 2, Sintez, stroenie i reaktsionnaya sposobnost'*, Zinatne, Riga, 1980, p. 471; *Chem. Abstr.*, **94**, 65410 (1981).

203. R. Bartsch and C. Tanielian, *J. Catal.*, **35**, 353 (1974); through *Ref. Zh. Khim.*, 12N205 (1975).

204. D. R. Kilanovski, H. Teeuwen, V. H. J. De Beer, B. C. Gates, G. C. A. Schuit, and H. Kwart, *J. Catal.*, **55**, 129 (1978); through *Chem. Abstr.*, **90**, 57555 (1979).

205. S. C. Schuman and H. Shalit, *Catal. Rev.*, **4**, 245 (1970); through *Ref. Zh. Khim.*, 8B986 (1971).

206. M. Schmal and M. I. Pais da Silva, *Rev. Brasil Technol.*, **11**, 175 (1980); through *Chem. Abstr.*, **94**, 17477 (1981).

207. C. S. Brooks, *Surf. Technol.*, **10**, 379 (1980); through *Chem. Abstr.*, **93**, 138305 (1980).

208. O. M. Zakharova and B. V. Romanovskii, *Vestn. Moskov. Gosud. Univ., Ser. Khim.,* **21**(5), 438 (1980); *Chem. Abstr.,* **94,** 102740 (1981).

209. C. J. Thompson, H. J. Coleman, C. C. Ward, and H. T. Rall, *Anal. Chem.,* **32,** 214 (1960).

210. P. Truitt, E. H. Holst, and G. Sammons, *J. Org. Chem.,* **22,** 1107 (1957).

211. J. Bougault, E. Cattelain, and P. Chabrier, *Bull. Soc. Chim. Fr.,* **6,** 34 (1939).

212. J. Bougault, E. Cattelain, and P. Chabrier, *Bull. Soc. Chim. Fr.,* **7,** 781 (1940).

213. J. Bougault, E. Cattelain, and P. Chabrier, *Bull. Soc. Chim. Fr.,* **7,** 780 (1940).

214. V. duVigneaud, D. B. Melville, K. Folkers, D. E. Wolf, R. Mozingo, J. C. Kerestesy, and S. A. Harris, *J. Biol. Chem.,* **146,** 475 (1942); through *Chem. Abstr.,* **37,** 1716 (1943).

215. F. F. Blicke and D. G. Sheets, *J. Am. Chem. Soc.,* **70,** 3768 (1948).

216. F. Challenger, *Aspects of the Organic Chemistry of Sulphur,* Butterworths, London, 1959, Chapter 3.

217. M. Brouty and M. R. Paullaud, *Parfums, Cosmet., Savons,* **2,** 546 (1959); *Chem. Abstr.,* **54,** 853 (1960).

218. H. Hauptmann and W. F. Walter, *Chem. Rev.,* **62,** 347 (1962).

219. G. R. Pettit and E. E. van Tamelen, in *Organic Reactions,* Vol. 12, Wiley, New York, 1962, pp. 356–529.

220. W. A. Bonner and R. A. Grimm, in *The Chemistry of Organic Sulfur Compounds* Vol. 2 (N. Kharash, Ed.), Pergamon Press, Oxford, 1966, p. 35.

221. R. Mozingo, *Org. Synth. Coll.,* **3,** 181 (1955).

222. H. R. Billica and H. Adkins, *Org. Synth. Coll.,* **3,** 176 (1955).

223. D. J. Brown, *J. Soc. Chem. Ind.,* **69,** 353 (1950).

224. O. Kruber and A. Raeithel, *Chem. Ber.,* **87,** 1469 (1954).

225. W. Carruthers, *Nature,* **176,** 790 (1955).

226. L. Granatelli, *Anal. Chem.,* **31,** 434 (1959).

227. A. I. Meyers, *Heterocycles in Organic Synthesis,* Wiley-Interscience, New York, 1974.

228. F. F. Blicke and D. G. Sheets, *J. Am. Chem. Soc.,* **71,** 4010 (1949).

229. G. E. Wiseman and E. S. Gould, *J. Am. Chem. Soc.,* **76,** 1706 (1954).

230. G. M. Badger and W. H. F. Sasse, *J. Chem. Soc.,* 3862 (1957).

231. R. D. Chambers, J. A. Cunningham, and D. J. Spring, *J. Chem. Soc.,* (*C*), 1560 (1968).

232. W. Carruthers and A. G. Douglas, *J. Chem. Soc.,* 2813 (1959).

233. O. Kruber and A. Raeithel, *Chem. Ber.,* **86,** 366 (1953).

234. W. Carruthers, *J. Chem. Soc.,* 4186 (1953).

235. J. E. Banfield, W. Davies, B. C. Ennis, S. Middleton, and Q. N. Porter, *J. Chem. Soc.,* 2603 (1956).

236. O. Dann and M. Kokorudz, *Chem. Ber.,* **91,** 172 (1958); O. Dann, M. Kokorudz, and R. Gropper, *Chem. Ber.,* **87,** 140 (1954).

237. R. Gaertner, *J. Am. Chem. Soc.,* **74,** 4950 (1952).

238. J. E. Banfield, W. Davies, N. W. Gamble, and S. Middleton, *J. Chem. Soc.,* 4791 (1956).

239. P. D. Clark, K. C. Clarke, D. F. Ewing, and R. M. Scrowston, *J. Chem. Soc., Perkin 1,* 677 (1980); P. D. Clark and D. M. McKinnon, *Can. J. Chem.,* **59,** 227 (1981); P. D. Clark and D. M. McKinnon, *Can. J. Chem.,* **59,** 1297 (1981).

240. S. F. Birch, T. V. Cullum, R. A. Dean, and D. G. Redford, *Tetrahedron,* **7,** 311 (1959).

241. G. N. Kao, B. D. Tilak, and K. Venkataraman, *Proc. Indian Acad. Sci.,* **38A,** 244 (1953); through *Chem. Abstr.,* **49,** 1003 (1955).

242. H. Wynberg, A. Logothetis, and D. Ver Ploeg, *J. Am. Chem. Soc.,* **79,** 1472 (1957).

243. F. G. Bordwell, W. H. McKellin, and D. Babcock, *J. Am. Chem. Soc.*, **73**, 5566 (1951).

244. W. Davies and F. C. James, *J. Chem. Soc.*, 314 (1955).

245. W. Davies, F. C. James, S. Middleton, and Q. N. Porter, *J. Chem. Soc.*, 1565 (1955).

246. W. Davies, Q. N. Porter, and J. R. Wilmshurst, *J. Chem. Soc.*, 3366 (1957).

247. Ya. L. Gol'dfarb and M. S. Kondakova, *Izv. Akad. Nauk SSSR, Otd. Khim. Nauk*, 1131 (1952); *Chem. Abstr.*, **48**, 1330 (1954).

248. Ya. L. Gol'dfarb and I. S. Korsakova, *Dokl. Akad. Nauk SSSR*, **96**, 283 (1954); *Chem. Abstr.*, **49**, 5430 (1955).

249. W. Baker, A. S. El-Nawawy, and W. D. Ollis, *J. Chem. Soc.*, 3163 (1952).

250. G. M. Badger, N. Kowanko, and W. H. F. Sasse, *J. Chem. Soc.*, 2969 (1960).

251. B. B. Corson, H. E. Tiefenthal, G. R. Atwood, W. J. Heintzelman, and W. L. Reilly, *J. Org. Chem.*, **21**, 584 (1956).

252. O. Dann and G. Hauck, *Arch. Pharm.*, **293**, 187 (1960).

253. G. M. Badger, B. J. Christie, J. M. Pryne, and W. H. F. Sasse, *J. Chem. Soc.*, 459 (1957).

254. W. Davies and Q. N. Porter, *J. Chem. Soc.*, 459 (1957).

255. P. Pastour and C. Barrat, *Compt. Rend. (C)*, **263**, 1312 (1966).

256. G. Badger, P. Cheuychit, and W. H. F. Sasse, *J. Chem. Soc.*, 3235 (1962).

257. H. Wynberg and A. Logothetis, *J. Am. Chem. Soc.*, **78**, 1958 (1956).

258. H. Wynberg, G. L. Hekkert, J. P. M. Houbiers, and H. W. Bosch, *J. Am. Chem. Soc.*, **87**, 2635 (1965).

259. L. A. Hulshof and H. Wynberg, in *New Trends in Heterocyclic Chemistry* (R. B. Mitra, N. R. Ayyangar, V. N. Gogte, R. M. Acheson, N. Cromwell, Eds.), Elsevier, Amsterdam, 1979, p. 373.

260. M. Sy and M. Maillet, *Compt. Rend. (C)*, **262**, 151 (1966).

261. M. Sy, M. Maillet, and P. David, *Bull. Soc. Chim. Fr.*, 2609 (1967).

262. D. Papa, E. Schwenk, and H. F. Ginsberg, *J. Org. Chem.*, **14**, 723 (1949).

263. E. J. Modest and J. Szmuszkovicz, *J. Am. Chem. Soc.*, **72**, 577 (1950).

264. J. Szmuszkovicz and E. J. Modest, *J. Am. Chem. Soc.*, **72**, 571 (1950).

265. S. Hansen, *Acta Chem. Scand.*, **8**, 695 (1954).

266. G. M. Badger, H. J. Rodda, and W. H. F. Sasse, *Chem. Ind. (London)*, 308 (1954).

267. G. M. Badger, H. J. Rodda, and W. H. F. Sasse, *J. Chem. Soc.*, 4162 (1954).

268. M. Sy, N. P. Buu-Hoi, and N. D. Xuong, *Compt. Rend.*, **239**, 1813 (1954); through *Chem. Abstr.*, **50**, 296 (1956).

269. M. Sy, N. P. Buu-Hoi, and N. D. Xuong, *Compt. Rend.*, **239**, 1224 (1954); through *Chem. Abstr.*, **49**, 13211 (1955).

270. E. C. Spaeth and C. B. Germain, *J. Am. Chem. Soc.*, **77**, 4066 (1955).

271. J. F. McGhie, W. A. Ross, D. Evans, and J. E. Tomlin, *J. Chem. Soc.*, 350 (1962).

272. N. P. Buu-Hoi, O. Perin-Roussel, and P. Jacquignon, *J. Chem. Soc. (C)*, 942 (1969).

273. N. P. Buu-Hoi, M. Sy, and N. D. Xuong, *Compt. Rend.*, **240**, 442 (1955).

274. N. P. Buu-Hoi, M. Sy, and N. D. Xuong, *Bull. Soc. Chim. Fr.*, 1583 (1955).

275. N. P. Buu-Hoi, M. Sy, and N. D. Xuong, *Rec. Trav. Chim.*, **75**, 463 (1956); through *Chem. Abstr.*, **51**, 1138 (1957).

276. Ya. L. Gol'dfarb, M. A. Kalik, and M. L. Kirmalova, *Zh. Obshch. Khim.*, **29**, 2034 (1959); *Chem. Abstr.*, **54**, 8775 (1960).

277. Ya. L. Gol'dfarb and Yu. B. Volkenshtein, *Zh. Obshch. Khim.*, **31**, 616 (1961); *Chem. Abstr.*, **55**, 24712 (1961).

278. F. M. Stoyanovich, Ya. L. Gol'dfarb, and G. B. Chermanova, *Izv. Akad. Nauk SSSR, Ser. Khim.*, 2285 (1973); *Chem. Abstr.*, **80**, 37201 (1974).

279. M. Sy, *Bull. Soc. Chim. Fr.*, 1175 (1955).

280. N. P. Buu-Hoi and M. Sy, *Compt. Rend.*, **242**, 2011 (1956).

281. N. P. Buu-Hoi and M. Sy, *J. Org. Chem.*, **23**, 97 (1958).

282. Ya. L. Gol'dfarb and M. L. Kirmalova, *Izv. Akad. Nauk SSSR, Otd. Khim. Nauk*, 570 (1955); *Chem. Abstr.*, **50**, 6422 (1956).

283. Ya. L. Gol'dfarb and M. L. Kirmalova, *Zh. Obshch. Khim.*, **25**, 1373 (1955); *Chem. Abstr.*, **50**, 6422h (1956).

284. Ya. L. Gol'dfarb and M. L. Kirmalova, *Izv. Akad. Nauk SSSR, Otd. Khim. Nauk*, 479 (1957); *Chem. Abstr.*, **51**, 15490 (1957).

285. Ya. L. Gol'dfarb and M. L. Kirmalova, *Zh. Obshch. Khim.*, **29**, 897 (1959); *Chem. Abstr.*, **54**, 1485 (1960).

286. E. Lescot, N. P. Buu-Hoi, and N. D. Xuong, *J. Chem. Soc.*, 3234 (1959).

287. H. Wynberg and A. Bantjes, *J. Am. Chem. Soc.*, **82**, 1447 (1960).

288. V. P. Litvinov and Ya. L. Gol'dfarb, *Izv. Akad. Nauk SSSR, Ser. Khim.*, 2183 (1963); *Chem. Abstr.*, **60**, 9257 (1964).

289. H. Wynberg and J. Feijen, *Rec. Trav. Chim.*, **89**, 77 (1970).

290. F. M. Stoyanovich and B. P. Fedorov, *Zh. Org. Khim.*, **1**, 1282 (1965); *Chem. Abstr.*, **63**, 16288 (1965).

291. A. Fredga, *Ark. Kemi*, **6**, 277 (1953).

292. K. Pettersson, *Ark. Kemi*, **7**, 39 (1954).

293. J. D. Bu'Lock, G. N. Smith, and C. T. Bedford, *J. Chem. Soc.*, 1405 (1962).

294. N. P. Buu-Hoi, *Nature*, **180**, 385 (1957).

295. N. P. Buu-Hoi and N. D. Xuong, *Compt. Rend.*, **247**, 654 (1958).

296. N. V. Bac, N. P. Buu-Hoi, and N. D. Xuong, *Compt. Rend.*, **254**, 3555 (1962).

297. N. V. Bac, N. P. Buu-Hoi, and N. D. Xuong, *Bull. Soc. Chim. Fr.*, 1077 (1962).

298. N. P. Buu-Hoi, N. V. Bac, and N. D. Xuong, *Bull. Soc. Chim. Fr.*, 1747 (1962).

299. T. F. Grey, J. F. McGhie, M. K. Pradhan, and W. A. Ross, *Chem. Ind. (London)*, 578 (1954).

300. C. D. Hurd and B. Rudner, *J. Am. Chem. Soc.*, **73**, 5157 (1951).

301. Ya. L. Gol'dfarb and P. A. Konstantinov, *Izv. Akad. Nauk SSSR, Otd. Khim. Nauk*, 992 (1956); *Chem. Abstr.*, **51**, 5041 (1957).

302. Ya. L. Gol'dfarb and P. A. Konstantinov, *Izv. Akad. Nauk SSSR, Otd. Khim. Nauk*, 121 (1959); *Chem. Abstr.*, **53**, 16103 (1959).

303. G. N. Kao, B. D. Tilak, and K. Venkataraman, *Proc. Indian Acad. Sci.*, **32A**, 162 (1950); *Chem. Abstr.*, **45**, 6624 (1951).

304. Ya. L. Gol'dfarb and Yu. B. Vol'kenshtein, *Izv. Akad. Nauk SSSR, Otd. Khim. Nauk*, 737 (1963); *Chem. Abstr.*, **59**, 7459 (1963).

305. G. Kimura, Sh. Kaishi, M. Koshi, and Y. Inabi, *J. Soc. Org. Synth. Chem. Jpn.*, **22**, 461 (1964); through *Ref. Zh. Khim.*, 20Zh194 (1965).

306. S. Portnoy and H. Gisser, *J. Org. Chem.*, **27**, 3331 (1962).

307. H. Stetter and B. Rajh, *Chem. Ber.*, **109**, 534 (1976).

308. F. Challenger and J. L. Holmes, *J. Chem. Soc.*, 1837 (1953).

309. F. Challenger, *Sci. Prog.*, **41**, 593 (1953); *Chem. Abstr.*, **47**, 12344 (1953).

310. P. G. Gassman and D. R. Amick, *J. Am. Chem. Soc.*, **100**, 7611 (1978).

311. Ya. L. Gol'dfarb and P. A. Konstantinov, *Izv. Akad. Nauk SSSR, Otd. Khim. Nauk*, 217 (1957); *Chem. Abstr.*, **51**, 10474 (1957).

312. R. Royer, P. Demerseman, J.-P. Lechartier, and A. Cheutin, *J. Org. Chem.*, **27**, 3808 (1962).

313. R. Royer, P. Demerseman, and J.-P. Lechartier, *Compt. Rend.*, **254**, 2605 (1962).

314. P. D. Clark, D. F. Ewing, F. Kerrigan, and R. M. Scrowston, *J. Chem. Soc., Perkin 1*, 615 (1982).

315. K. S. Miller, C. Haymaker, and H. Gilman, *J. Org. Chem.*, **24**, 622 (1959); K. S. Miller, C. Haymaker, and H. Gilman, *J Org. Chem.*, **26**, 5217 (1961).

316. J. Sicé, *J. Am. Chem. Soc.*, **75**, 3697 (1953).

317. S. Gronowitz, *Ark. Kemi*, **12**, 239 (1958).

318. S. Gronowitz, *Ark. Kemi*, **13**, 87 (1958).

319. M. Hannoun, N. Blažević, D. Kolbah, M. Mihailić, and F. Kajfež, *J. Heterocycl. Chem.*, **19**, 1131 (1982).

320. T. F. Grey, J. F. McGhie, and W. A. Ross, *J. Chem. Soc.*, 1502 (1960).

321. K. H. Shah, B. D. Tilak, and K. Venkataraman, *Proc. Indian Acad. Sci.*, **28A**, 142 (1948); through *Chem. Abstr.*, **44**, 3958 (1950).

322. Ya. L. Gol'dfarb and M. B. Ibragimova, *Dokl. Akad. Nauk SSSR*, **106**, 469 (1956); *Chem. Abstr.*, **51**, 3439 (1957).

323. Ya. L. Gol'dfarb and M. B. Ibragimova, *Dokl. Akad. Nauk SSSR*, **113**, 594 (1957); *Chem. Abstr.*, **51**, 14720 (1957).

324. Ya. L. Gol'dfarb, B. P. Fabrichnyi, and V. I. Rogovik, *Izv. Akad. Nauk SSSR, Ser. Khim.*, 708 (1966); *Chem. Abstr.*, **65**, 10551 (1966).

325. P. Pastour and C. Barrat, *Compt. Rend.*, **250**, 3110 (1965).

326. Ya. L. Gol'dfarb, E. A. Krasnyanskaya, and B. P. Fabrichnyi, *Izv. Akad. Nauk SSSR, Otd. Khim. Nauk*, 1825 (1962); *Chem. Abstr.*, **58**, 8922 (1963).

327. P. A. Konstantinov, L. V. Semerenko, K. M. Suvorova, E. N. Bondar', and Ya. L. Gol'dfarb, *Khim. Geterotsikl. Soed.*, 230 (1968); *Chem. Abstr.*, **70**, 57554 (1969).

328. O. Červinka, O. Bélovsky, and L. Korálová, *Z. Chem.*, **9**, 448 (1969).

329. H. Böshagen and W. Geiger, *Lieb. Ann.*, **764**, 58 (1972).

330. Ya. L. Gol'dfarb, B. P. Fabrichnyi, and I. F. Shalavina, *Tetrahedron*, **18**, 21 (1962).

331. Ya. L. Gol'dfarb and B. P. Fabrichnyi, *Dokl. Akad. Nauk SSSR*, **100**, 461 (1955); *Chem. Abstr.*, **49**, 8244 (1955).

332. Ya. L. Gol'dfarb, B. P. Fabrichnyi, and I. F. Shalavina, *Zh. Obshch. Khim.*, **26**, 2595 (1956); *Chem. Abstr.*, **51**, 4943 (1957).

333. S. Nishimura, R. Motoyama, and E. Imoto, *Bull. Univ. Osaka Prefect., Ser. A*, **6**, 127 (1958); through *Chem. Abstr.*, **53**, 4248 (1959).

334. Ya. L. Gol'dfarb, B. P. Fabrichnyi, and I. F. Shalavina, *Izv. Akad. Nauk SSSR, Otd. Khim. Nauk*, 1276 (1956); *Chem. Abstr.*, **51**, 5702 (1957).

335. B. P. Fabrichnyi, E. A. Krasnyanskaya, and Ya. L. Gol'dfarb, *Dokl. Akad. Nauk SSSR*, **143**, 1370 (1962); *Chem. Abstr.*, **57**, 8426 (1962).

336. B. P. Fabrichnyi, E. A. Krasnyanskaya, I. F. Shalavina, and Ya. L. Gol'dfarb, *Zh. Obshch. Khim.*, **33**, 2697 (1963); *Chem. Abstr.*, **60**, 512 (1964).

337. B. M. de Malleray, *Helv. Chim. Acta*, **54**, 343 (1971).

338. Ya. L. Gol'dfarb, B. P. Fabrichnyi, and V. I. Rogovik, *Izv. Akad. Nauk SSSR, Otd. Khim. Nauk*, 2172 (1963); *Chem. Abstr.*, **60**, 9226 (1964).

339. Ya. L. Gol'dfarb, B. P. Fabrichnyi, and I. F. Shalavina, *Zh. Obshch. Khim.*, **28**, 213 (1958); *Chem. Abstr.*, **52**, 12838 (1958).

340. V. P. Mamaev, N. N. Suvorov, and E. M. Rokhlin, *Dokl. Akad. Nauk SSSR*, **101**, 209 (1955); *Chem. Abstr.*, **50**, 3387 (1956).

341.   B. P. Fabrichnyi, S. E. Zurabyan, I. F. Shalavina, and Ya. L. Gol'dfarb, *Izv. Akad. Nauk SSSR, Ser. Khim.*, 2102 (1967); *Chem. Abstr.*, **68**, 114333 (1968).

342.   J. F. McGhie, W. A. Ross, and D. H. Laney, *J. Chem. Soc.*, 2578 (1962).

343.   B. P. Fabrichnyi, I. F. Shalavina, and Ya. L. Gol'dfarb, *Zh. Obshch. Khim.*, **34**, 3878 (1964); *Chem. Abstr.*, **62**, 8997 (1965).

344.   B. P. Fabrichnyi, I. F. Shalavina, and Ya. L. Gol'dfarb, *Zh. Org. Khim.*, **15**, 1536 (1979).

345.   B. P. Fabrichnyi, I. F. Shalavina, and Ya. L. Gol'dfarb, *Zh. Obshch. Khim.*, **28**, 2520 (1958); *Chem. Abstr.*, **53**, 3052 (1959).

346.·   Ya. L. Gol'dfarb, B. P. Fabrichnyi, and I. F. Shalavina, *Zh. Obshch. Khim.*, **31**, 2057 (1961); *Chem. Abstr.*, **55**, 27050 (1961).

347.   B. P. Fabrichnyi, I. F. Shalavina, and Ya. L. Gol'dfarb, *Zh. Obshch. Khim.*, **31**, 1244 (1961); *Chem. Abstr.*, **55**, 23488 (1961).

348.   B. P. Fabrichnyi, I. F. Shalavina, and Ya. L. Gol'dfarb, *Zh. Org. Khim.*, **1**, 1507 (1965); *Chem. Abstr.*, **64**, 586 (1966).

349.   P. Cagniant and D. Cagniant, *Bull. Soc. Chim. Fr.*, 1152 (1956).

350.   B. P. Fabrichnyi, I. F. Shalavina, and Ya. L. Gol'dfarb, *Zh. Org. Khim.*, **3**, 2079 (1967); *Chem. Abstr.*, **68**, 87080 (1968).

351.   B. P. Fabrichnyi, I. F. Shalavina, and Ya. L. Gol'dfarb, *Zh. Org. Khim.*, **5**, 790 (1969); *Chem. Abstr.*, **71**, 21955 (1969).

352.   S. Nishimura, M. Nakamura, M. Suzuki, and E. Imoto, *Nippon Kagaku Zasshi*, **83**, 343 (1962); through *Chem. Abstr.*, **59**, 3862 (1963).

353.   B. P. Fabrichnyi, I. F. Shalavina, S. E. Zurabyan, Ya. L. Gol'dfarb, and C. M. Kostrova, *Zh. Org. Khim.*, **4**, 680 (1968); *Chem. Abstr.*, **69**, 18565 (1968).

354.   B. P. Fabrichnyi, I. F. Shalavina, Ya. L. Gol'dfarb, and S. M. Kostrova, *Zh. Org. Khim.*, **10**, 1956 (1974); *Chem. Abstr.*, **82**, 57495 (1975).

355.   A. V. Volokhina, B. P. Fabrichnyi, I. F. Shalavina, and Ya. L. Gol'dfarb, *Vysokomol. Soed.*, **4**, 1829 (1962); *Chem. Abstr.*, **59**, 771 (1963).

356.   O. B. Salamatina, A. K. Bonetskaya, S. M. Skuratov, B. P. Fabrichnyi, I. F. Shalavina, and Ya. L. Gol'dfarb, *Vysokomol. Soed.*, **7**, 485 (1965); *Chem. Abstr.*, **63**, 4395 (1965).

357.   O. B. Salamatina, A. K. Bonetskaya, S. M. Skuratov, B. P. Fabrichnyi, I. F. Shalavina, and Ya. L. Gol'dfarb, *Vysokomol. Soed.*, **10B**, 10 (1968); *Chem. Abstr.*, **68**, 69416 (1968).

358.   A. K. Bonetskaya, T. V. Sopova, O. B. Salamatina, S. M. Skuratov, B. P. Fabrichnyi, I. F. Shalavina, and Ya. L. Gol'dfarb, *Vysokomol. Soed.*, **11**, 894 (1969); *Chem. Abstr.*, **72**, 55937 (1970).

359.   Ya. Kondelikova, A. K. Bonetskaya, B. P. Fabrichnyi, I. F. Shalavina, S. M. Kostrova, and Ya. L. Gol'dfarb, *Vysokomol. Soed.*, **16B**, 694 (1974); *Chem. Abstr.*, **82**, 73537 (1975).

360.   A. K. Bonetskaya, in *Sovremennye Problemy Fizicheskoi Khimii*, Vol. 6, Moscow State University Press, Moscow, 1972, p. 194; *Chem. Abstr.*, **78**, 160159 (1973).

361.   Yu. I. Vikhlyaev, T. A. Klygul', E. I. Slyn'ko, Ya. L. Gol'dfarb, B. P. Fabrichnyi, I. F. Shalavina, and S. M. Kostrova, *Khim.-Farm. Z.*, No. 7, 8 (1974); *Chem. Abstr.*, **81**, 105342 (1974).

362.   Ya. L. Gol'dfarb, B. P. Fabrichnyi, I. F. Shalavina, and S. M. Kostrova, *Zh. Org. Khim.*, **11**, 2400 (1975); *Chem. Abstr.*, **84**, 121060 (1976).

363.   B. P. Fabrichnyi, I. F. Shalavina, and Ya. L. Gol'dfarb, *Zh. Org. Khim.*, **5**, 361 (1969); *Chem. Abstr.*, **70**, 114910 (1969).

364.   B. P. Fabrichnyi, I. F. Shalavina, S. M. Kostrova, and Ya. L. Gol'dfarb, *Zh. Org. Khim.*, **6**, 1091 (1970); *Chem. Abstr.*, **73**, 35280 (1970).

365. B. P. Fabrichnyi, I. F. Shalavina, S. M. Kostrova, and Ya. L. Gol'dfarb, *Zh. Org. Khim.*, 8, 187 (1972); *Chem. Abstr.*, 77, 114342 (1972).

366. B. P. Fabrichnyi, I. F. Shalavina, S. M. Kostrova, and Ya. L. Gol'dfarb, *Khim. Geterotsikl. Soed.*, 315 (1972); *Chem. Abstr.*, 77, 61878 (1972).

367. T. N. Semushina, B. I. Tokarev, V. V. Luk'yanova, N. M. Monakhova, N. G. Pavlova, and B. P. Fabrichnyi, *Gidroliz. Lesokhim. Pro.*, No. 3, 21 (1970); *Chem. Abstr.*, 73, 75625 (1970).

368. N. P. Buu-Hoi, M. Sy, and N. D. Xuong, *Compt. Rend.*, 240, 785 (1955).

369. W. Hoek, J. Strating, and H. Wynberg, *Rec. Trav. Chim.*, 85, 1045 (1966).

370. W. Hoek, H. Wynberg, and J. Strating, *Rec. Trav. Chim.*, 85, 1054 (1966).

371. V. P. Litvinov, V. I. Shvedov, and V. S. Dermugin, *Khim. Geterotsikl. Soed.*, 1128 (1982).

372. H. Wynberg and J. P. M. Houbiers, *J. Org. Chem.*, 36, 834 (1971).

373. R. Royer, P. Demerseman, J.-P. Lechartier, and A. Cheutin, *Bull. Soc. Chim. Fr.*, 1711 (1962).

374. G. Muraro, D. Cagniant, and P. Cagniant, *Bull. Soc. Chim. Fr.*, 343 (1973).

375. B. P. Fabrichnyi, S. M. Kostrova, V. S. Bogdanov, and Ya. L. Gol'dfarb, *Izv. Akad. Nauk SSSR, Ser. Khim.*, 2534 (1976); *Chem. Abstr.*, 86, 106270 (1977).

376. Ya. L. Gol'dfarb, B. P. Fabrichnyi, V. K. Zav'yalova, and V. S. Bogdanov, *Izv. Akad. Nauk SSSR, Ser. Khim.*, 1570 (1979); *Chem. Abstr.*, 91, 157537 (1979).

377. G. Stork and P. L. Stotter, *J. Am. Chem. Soc.*, 91, 7780 (1969).

378. L. I. Belen'kii, *Usp. khim.*, 33, 1265 (1964).

379. Ya. L. Gol'dfarb, S. Z. Taits, and L. I. Belen'kii, *Izv. Akad. Nauk SSSR, Otd. Khim. Nauk*, 1262 (1957); *Chem. Abstr.*, 52, 6310 (1958).

380. Ya. L. Gol'dfarb, S. Z. Taits, and L. I. Belen'kii, *Tetrahedron*, 19, 1851 (1963).

381. Ya. L. Gol'dfarb, S. Z. Taits, and L. I. Belen'kii, *Zh. Obshch. Khim.*, 29, 3564 (1959); *Chem. Abstr.*, 54, 19639 (1960).

382. P. Cagniant and D. Cagniant, *Bull. Soc. Chim. Fr.*, 680 (1955).

383. L. I. Belen'kii, S. Z. Taits, and Ya. L. Gol'dfarb, *Dokl. Akad. Nauk SSSR*, 139, 1356 (1961); *Chem. Abstr.*, 56, 450 (1962).

384. Ya. L. Gol'dfarb, S. Z. Taits, and L. I. Belen'kii, *Izv. Akad. Nauk SSSR, Ser. Khim.*, 1451 (1963); *Chem. Abstr.*, 59, 15243 (1963).

385. S. Z. Taits, L. I. Belen'kii, and Ya. L. Gol'dfarb, *Izv. Akad. Nauk SSSR, Ser. Khim.*, 1460 (1963); *Chem. Abstr.*, 59, 15244 (1963).

386. Ya. L. Gol'dfarb, S. Z. Taits, T. S. Chirkova, and L. I. Belen'kii, *Izv. Akad. Nauk SSSR, Ser. Khim.*, 2055 (1964); *Chem. Abstr.*, 62, 9104 (1965).

387. S. Z. Taits, O. A. Kalinovskii, V. S. Bogdanov, and Ya. L. Gol'dfarb, *Khim. Geterotsikl. Soed.*, 1467 (1970); *Chem. Abstr.*, 74, 53372 (1971).

388. O. A. Kalinovskii, S. Z. Taits, and Ya. L. Gol'dfarb, *Izv. Akad. Nauk SSSR, Ser. Khim.*, 2331 (1970); *Chem. Abstr.*, 75, 20078 (1971).

389. S. Z. Taits, O. A. Kalinovskii, V. S. Bogdanov, and Ya. L. Gol'dfarb, *Khim. Geterotsikl. Soed.*, 170 (1972); *Chem. Abstr.*, 76, 153469 (1972).

390. S. Z. Taits, V. N. Bulgakova, and Ya. L. Gol'dfarb, *Khim. Geterotsikl. Soed.*, 16 (1973); *Chem. Abstr.*, 78, 111037 (1973).

391. S. Z. Taits, E. A. Krasnyanskaya, and Ya. L. Gol'dfarb, *Izv. Akad. Nauk SSSR, Ser. Khim.*, 754 (1968); *Chem. Abstr.*, 70, 11445 (1960).

392. S. Z. Taits, E. A. Krasnyanskaya, and Ya. L. Gol'dfarb, *Izv. Akad. Nauk SSSR, Ser. Khim.*, 762 (1968); *Chem. Abstr.*, 70, 10759 (1969).

393. S. Z. Taits, E. A. Krasnyanskaya, A. L. Klyachko-Gurvich, and Ya. L. Gol'dfarb, *Izv. Akad. Nauk SSSR, Ser. Khim.*, 1807 (1973); *Chem. Abstr.*, 79, 145640 (1973).

394.   S. Z. Taits, E. A. Krasnyanskaya, and Ya. L. Gol'dfarb, *Izv. Akad. Nauk SSSR, Ser. Khim.*, 2228 (1970); *Chem. Abstr.*, **75**, 35685 (1971).

395.   S. Z. Taits, E. A. Krasnyanskaya, Ya. L. Gol'dfarb, N. F. Kononov, A. G. Pogorelov, and R. F. Merzhanova, *Izv. Akad. Nauk SSSR, Ser. Khim.*, 2536 (1975); *Chem. Abstr.*, **84**, 59421 (1976).

396.   O. Meth-Cohn, *Tetrahedron Lett.*, 91 (1973).

397.   Y. Miyahara, I. Takahiko, and T. Yoshino, *Chem. Lett.*, 563 (1978).

398.   S. Z. Taits, F. D. Alashev, and Ya. L. Gol'dfarb, *Izv. Akad. Nauk SSSR, Ser. Khim.*, 566 (1968); *Chem. Abstr.*, **69**, 76718 (1968).

399.   Ya. L. Gol'dfarb, S. Z. Taits, F. D. Alashev, A. A. Dudinov, and O. S. Chizhov, *Khim. Geterotsikl. Soed.*, 40 (1975); *Chem. Abstr.*, **83**, 10022 (1975).

400.   F. D. Alashev, A. V. Kessenikh, S. Z. Taits, and Ya. L. Gol'dfarb, *Izv. Akad. Nauk SSSR, Ser. Khim.*, 2022 (1974); *Chem. Abstr.*, **82**, 30869 (1975).

401.   B. Tashkhodzhaev, L. G. Vorontsova, and F. D. Alashev, *Izv. Akad. Nauk SSSR, Ser. Khim.*, 1287 (1976); *Chem. Abstr.*, **85**, 123793 (1976).

402.   B. Tashkhodzhaev, L. G. Vorontsova, and F. D. Alashev, *Izv. Akad. Nauk SSSR, Ser. Khim.*, 2246 (1976); *Chem. Abstr.*, **86**, 89649 (1977).

403.   B. Tashkhodzhaev, L. G. Vorontsova, and F. D. Alashev, *Izv. Akad. Nauk SSSR, Ser. Khim.*, 2475 (1976); *Chem. Abstr.*, **86**, 120546 (1977).

404.   F. D. Alashev, V. N. Bulgakova, Ya. L. Gol'dfarb, and S. Z. Taits, *Izv. Akad. Nauk USSR, Ser. Khim.*, 147 (1977); *Chem. Abstr.*, 171420 (1977).

405.   F. D. Alashev, V. N. Bulgakova, S. Z. Taits, and Ya. L. Gol'dfarb, *Izv. Akad. Nauk SSSR, Ser. Khim.*, 1377 (1980); *Chem. Abstr.*, **93**, 204612 (1980).

406.   Ya. L. Gol'dfarb, S. Z. Taits, and E. A. Krasnyanskaya, *Khim. Geterotsikl. Soed.*, 920 (1980); *Chem. Abstr.*, **93**, 204615 (1980).

407.   S. Gronowitz and J. Böler, *Ark. Kemi*, **28**, 587 (1968).

408.   L. A. Myshkina, F. M. Stoyanovich, and Ya. L. Gol'dfarb, *Izv. Akad. Nauk SSSR, Ser. Khim.*, 395 (1982).

409.   M. Farnier, S. Soth, and P. Fournari, *Can. J. Chem.*, **54**, 1083 (1976).

410.   M. Farnier, S. Soth, and P. Fournari, *Can. J. Chem.*, **54**, 1074 (1976).

411.   J. G. C. Coutts, H. R. Goldschmid, and O. C. Musgrave, *J. Chem. Soc.*, (*C*), 488 (1970).

412.   M. J. S. Dewar and P. A. Marr, *J. Am. Chem. Soc.*, **84**, 3782 (1962).

413.   J. Namtvedt and S. Gronowitz, *Acta Chem. Scand.*, **22**, 1373 (1968).

414.   S. Gronowitz and A. Maltesson, *Acta Chem. Scand.*, **25**, 2435 (1971).

415.   G. M. Badger, N. Kowanko, and W. H. F. Sasse, *J. Chem. Soc.*, 440 (1959).

416.   T. I. Santalova, P. A. Konstantinov, and Ya. L. Gol'dfarb, *Dokl. Akad. Nauk SSSR*, **131**, 1102 (1960); *Chem. Abstr.*, **54**, 21103 (1960).

417.   G. M. Badger, P. Cheuychit, and W. H. F. Sasse, *Aust. J. Chem.*, **17**, 371 (1964).

418.   V. Mircheva, *God. Vissh. Khim.-Tekhnol. Inst. Sofia*, **24**, 171 (1978); through *Chem. Abstr.*, **95**, 188412 (1981).

419.   Ya. L. Gol'dfarb and E. P. Zakharov, *Zh. Prikl. Khim.*, **45**, 1060 (1972); *Chem. Abstr.*, **77**, 101319 (1972).

420.   J. Schut, J. B. F. N. Engberts, and H. Wynberg, *Synth. Comm.*, **2**, 415 (1972).

421.   H. Suhr and P. Henne, *Lieb. Ann.*, 1610 (1977).

422.   H. Suhr, P. Henne, D. Iacocca, and M. J. Ropero, *Lieb. Ann.*, 441 (1980).

423.   H. D. Kaesz, R. B. King, T. A. Mannel, L. D. Nichols, and F. G. A. Stone, *J. Am. Chem. Soc.*, **82**, 4749 (1980).

424.   T. Chivers and R. L. Timms, *Can. J. Chem.*, **55**, 3509 (1977).

425. O. Horie, H. H. Nguyen, and A. Amano, *Chem. Lett.*, 1015 (1975).

426. S. Gronowitz and T. Frejd, *Khim. Geterosikl. Soed.*, 435 (1978); *Chem. Abstr.*, 89, 24187 (1978).

427. G. Canquis, B. Divisia, and G. Roverdy, *Bull. Soc. Chim. Fr.*, 3027 (1971).

428. L. N. Plekhanova, G. A. Nikiforov, V. V. Ershov, and E. P. Zakharov, *Izv. Akad. Nauk SSSR, Ser. Khim.*, 846 (1973); *Chem. Abstr.*, 79, 52946 (1973).

429. A. E. Vasil'vitskii, V. M. Shostakovskii, and O. M. Nefedov, *Khim. Geterotsikl. Soed.*, 723 (1982).

430. C. N. Skold and R. H. Schlessinger, *Tetrahedron Lett.*, 791 (1970).

431. W. van Tilbórg, *Rec. Trav. Chim.*, 95, 140 (1976).

CHAPTER VII

Thiophene 1,1-Dioxides, Sesquioxides, and 1-Oxides

MAYNARD S. RAASCH

Central Research and Development Department,[1] Experimental Station,
E. I. du Pont de Nemours and Company, Wilmington, Delaware

I.   THIOPHENE 1,1-DIOXIDES

The growing interest in thiophene 1,1-dioxides as reagents in synthesis makes timely a comprehensive review of their chemistry.

1.   Synthesis

A.   Removal of 2HX from Tetrahydrothiophene 1,1-Dioxide Derivatives

Thiophene 1,1-dioxide was first prepared by Bailey and Cummins[2] with a double Hofmann elimination (Scheme 1).

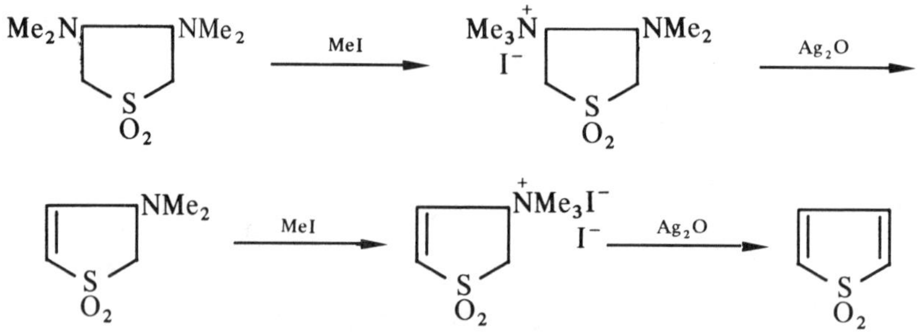

Scheme 1

Because of its tendency to dimerize with loss of sulfur dioxide, the compound could not be isolated in the pure state, but its chloroform solution could be stored at 5°C for several weeks without appreciable decomposition. This rather long synthesis has been replaced by a more convenient one, starting with commercially available "butadiene sulfone," adding bromine to it, and heterogeneously eliminating 2HBr with powdered sodium hydroxide in benzene[3] or tetrahydrofuran.[4]

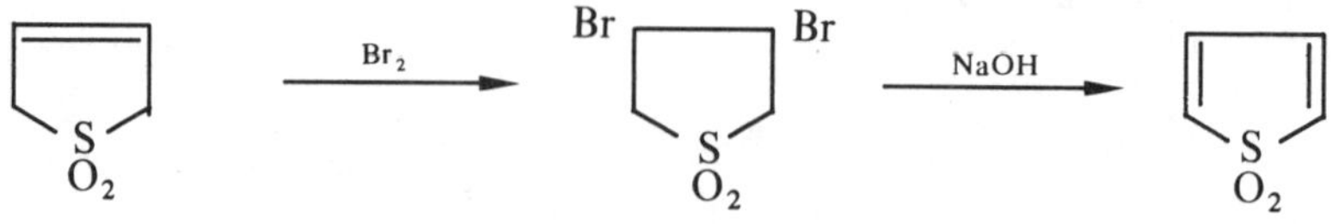

2-Methyl-, 3-methyl-,[5,6] 2-ethyl-,[6] 3-ethyl, 3-phenyl-,[5,6] 2,4-dimethyl-,[5] 3,4-dimethyl,[5,7,8] and 3,4-diphenylthiophene 1,1-dioxides[5] have been made by dehydrobromination of the appropriately substituted 3,4-dibromotetrahydrothiophene 1,1-dioxides. 3,4-Dichlorothiophene 1,1-dioxides has been prepared by the dehydrochlorination of 3,3,4,4-tetrachlorotetrahydrothiophene 1,1-dioxide with ammonia.[9] Bactericidal, fungicidal, herbicidal, nematocidal, insecticidal,[10,11] and gastropodicidal[13] activities have been recorded for this compound.

This method of synthesis may fail when an alternative pathway for the elimination of HX exists, or the conditions induce isomerization. For example, though removal of 2HBr from 3,4-dibromo-3,4-dimethyltetrahydrothiophene 1,1-dioxide with piperidine in benzene produced 3,4-dimethylthiophene 1,1-dioxide,[7] use of aqueous sodium hydroxide as the base generated an *exo* methylene group, probably by isomerization, and a thiophene 1,1-dioxide was not obtained.[14]

B. Oxidation of Substituted Thiophenes

The most common method of preparing substituted thiophene 1,1-dioxides is by direct oxidation of the thiophene. Hinsberg[15] synthesized the first thiophene 1,1-dioxides in 1915 by oxidizing tetraphenyl- and 3,4-diphenylthiophene with 30% hydrogen peroxide in acetic acid, though the latter dioxide was altered by his workup. The next work on a thiophene 1,1-dioxide was by Backer, Bolt, and Stevens in 1937, who introduced the use of perbenzoic acid as an oxidizing agent and prepared an authentic sample of 3,4-diphenylthiophene 1,1-dioxide.[16]

Melles and Backer[17] studied the influence of substituents on the possibility of oxidizing thiophenes to thiophene 1,1-dioxides with perbenzoic acid or hydrogen peroxide in acetic acid. Electron withdrawing substituents were considered harmful, since they make attack on the sulfur atom more difficult. No dioxide was obtained from 2-nitro-3,5-diphenyl-, 3-nitro-2,5-diphenyl-, 2-carboxy-3,4-diphenyl, 2,5-dicarboxy-3,4-diphenyl-, and 2,5-dicarboxymethyl-3,4-diphenylthiophene. 2-Acetothiophene was not oxidized by 3-chloroperbenzoic acid.[18] For reasons that are not clear, 2-phenyl-, 2,3-diphenyl-, and 2,4-diphenylthiophene gave undefined products, though 2,5- and 3,4-diphenylthiophene were operable. Electron donating groups favor oxidation, and 2,5- and 3,4-dimethylthiophenes were converted to 1,1-dioxides. 3,4-Dibromo-2,5-diphenylthiophene formed a dioxide, but its isomer, 2,5-dibromo-3,4-diphenylthiophene, did not. Other thiophenes that gave 1,1-dioxides are the 2-COOMe-3,4-Ph$_2$; 2-Bz-3,4-Ph$_2$; 2,5-Bz$_2$-3,4-Ph$_2$; 2-COOMe-3,4-Ph$_2$-5-Bz; 2-Br-3,4-Ph$_2$; 2,3-Br$_2$-4,5-Ph$_2$; and 2,4-Br$_2$-3,5-Ph$_2$ derivatives.[17]

Additional examples of thiophenes 1,1-dioxides prepared by oxidation of thiophenes may be seen by perusing Table 2.

The commercially available 3-chloroperbenzoic acid is used often as the oxidizing agent with dichloromethane, chloroform, or 1,2-dichloroethane as solvent. It will oxidize thiophenes that are unattacked by peracetic acid or hydrogen peroxide in acetic acid. Oxidation of alkylthiophenes is often plagued by low yields and tarry by-products. This situation has been improved by the use of an acid acceptor with 3-chloroperbenzoic acid. Thus, oxidation of 2-ethyl-5-methyl- and 2-isopropyl-5-methylthiophene in dichloromethane with 3-chloroperbenzoic acid in the presence of solid sodium bicarbonate gave 70% yields.[19]

Modification of the workup has also been recommended for increasing the yield from alkylthiophenes. Oxidation with 3-chloroperbenzoic acid in dichloromethane is carried out and the 3-chlorobenzoic acid is filtered off, with a further quantity being removed after cooling the solution to $-55°C$. The remaining 3-chlorobenzoic acid is then removed with the aid of macroreticular *tert*-amine resin of the type IMAC 22 NB or Amberlyst A21. Yields of 52–74% were obtained from 2,5-dimethyl-, 2,5- and 2,4-di-*tert*-butyl-, and 2,5-diphenylthiophene.[18,20]

The identification of other inexpensive, selective oxidizing agents that will convert thiophenes to their 1,1-dioxides and not attack the product, and whose reduced form is easily removable from the reaction mixture, is desirable.

C.   3,4-Dihydroxythiophene 1,1-Dioxide Derivatives

The condensation of diethyl oxalate with a series of bis(arylmethyl) sulfones and with bis(carbethoxymethyl) sulfone produced 2,5-substituted 4-hydroxy-3-oxo-2,3-dihydrothiophene 1,1-dioxides (**1**).[21]

When first reported, these compounds were represented as the tautomeric 3,4-dihydroxythiophene 1,1-dioxides.[22-24] The dibenzoates[23,24] prepared from them probably have the thiophene 1,1-dioxide structure. Recently, 4-hydroxy-3-oxo-2,5-diphenyl-2,3-dihydrothiophene 1,1-dioxide (**2**) has been reacted with tosyl chloride,[25] thiophosgene,[25] phosgene,[26,27] and oxalyl chloride[28] to form the thiophene 1,1-dioxide derivatives **3–6**.

4   5   6

Methylation of **2** and its bis(4-chlorophenyl) analogue with diazomethane produced 2,5-diaryl-3,4-dimethoxythiophene 1,1-dioxide.[30] However, attempts to repeat this resulted in the introduction of only one methyl group.[25] The use of methyl sulfate, methyl iodide, and ethyl iodide gave C- and O-alkylated products of the keto form.[29, 30]

The reaction of **2** with 2 moles cyclohexylamine resulted in the cyclohexyl-aminothiophene 1,1-dioxide salt **7**. With 1,2-ethanediamine, the cyclic diamino derivative **8** was formed.[25]

7   8

2,3-Dihydro-4-methyl-3-oxothiophene 1,1-dioxide (**9**) dissolved in alkali with an intense yellow color. The yellow needles precipitated from this solution by acid have been assigned the structure of a thiophene 1,1-dioxide (**10**).[14] Methylation of the ketone introduced two methyl groups to form 2,4-dimethyl-3-methoxythiophene 1,1-dioxide (**11**).[31]

9   10   11

D.   Special Syntheses

The addition of diethylpropynylamine to 2,3-diphenyl- and 2,3-bis(4-chlorophenyl)thiirene 1,1-dioxide in benzene generated diethylaminothiophene 1,1-dioxides (**13**),[32, 33] a reaction that may take place through a "Dewar thiophene" 1,1-dioxide (**12**). The compounds are coccidiostatic.

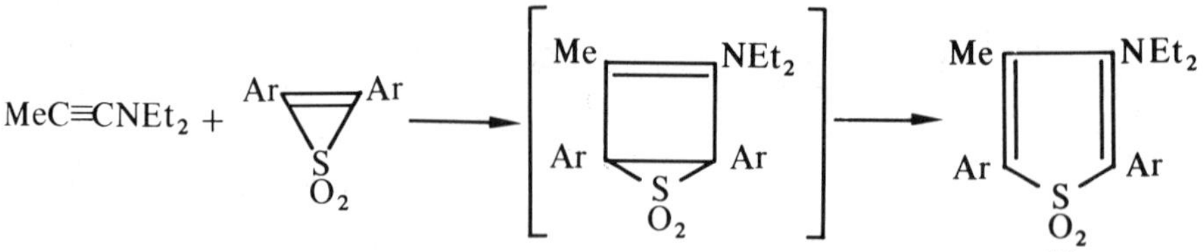

Similarly, the reaction of 2,3-diphenylthiirene 1,1-dioxide with 1-pyrrolidinoindene yielded a thiophene 1,1-dioxide (14) through the elimination of pyrrolidine.[34]

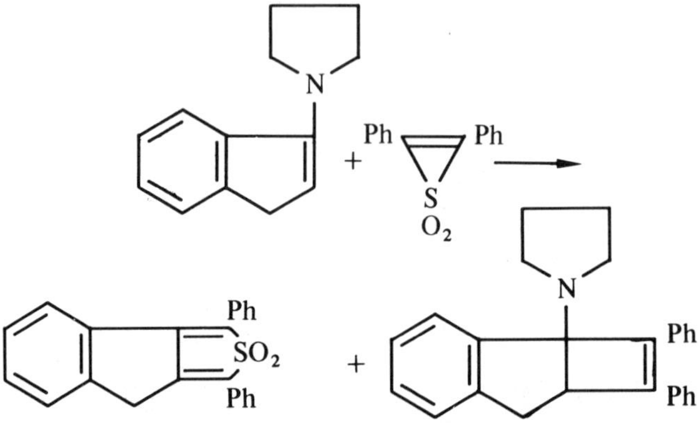

14   9.5%   55%

The rearrangement of a diallenic sulfone generated a thiophene 1,1-dioxide (Scheme 2).[35]

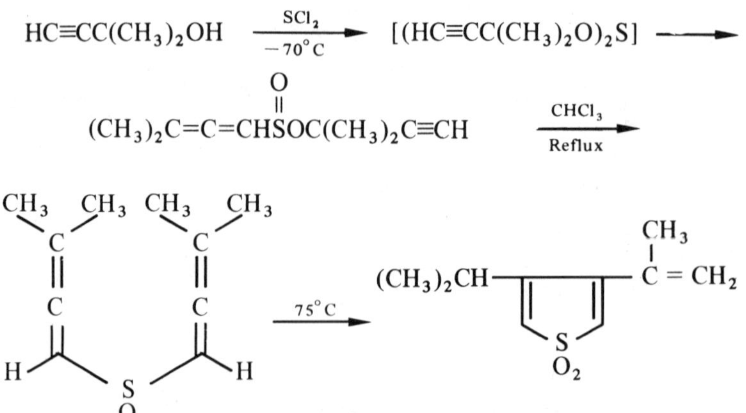

Scheme 2

Starting with 3-deuterio- and 3-bromo-2,2-dimethylpropargyl alcohols, the corresponding thiophene dioxides (15 and 16) were obtained.

Adding nitrogen tetroxide to 2,5-diisopropylidene-3-sulfolene produced 2,5-bis(1-methyl-1-nitroethyl)thiophene 1,1-dioxide (17).[36,37]

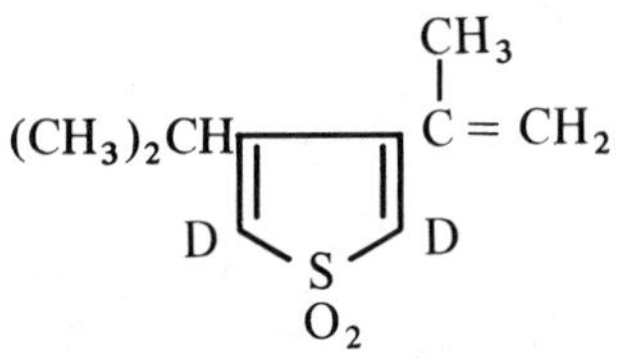

15

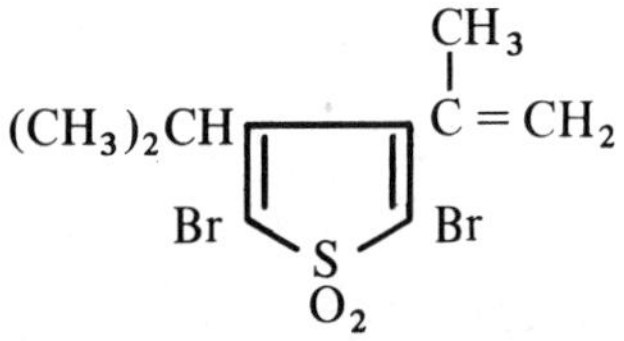

16

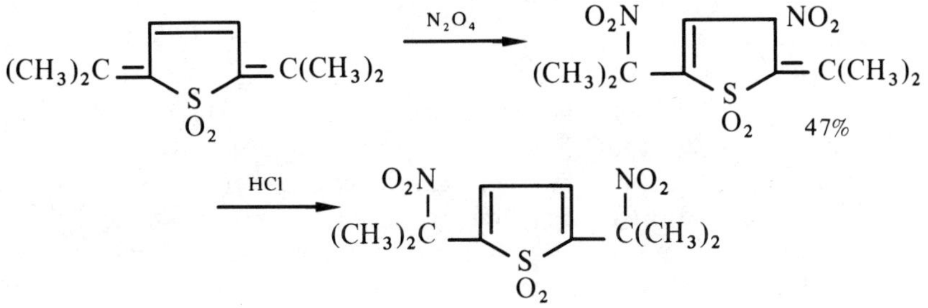

47%

17

Tetraphenylthiophene 1,1-dioxide has been made by the following reactions (Scheme 3).[38]

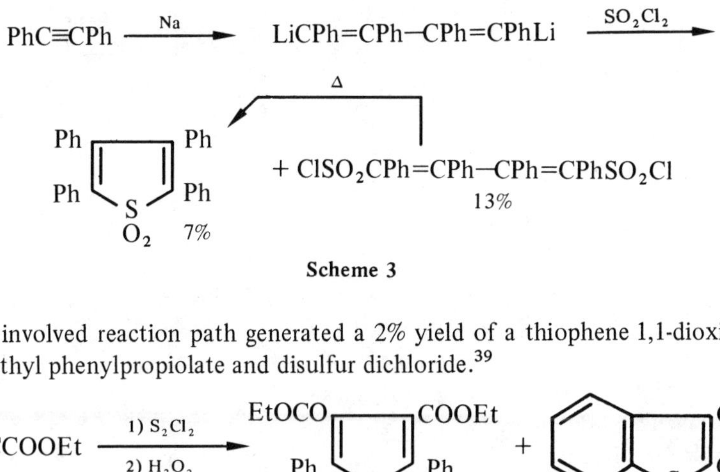

$$PhC \equiv CPh \xrightarrow{Na} LiCPh = CPh - CPh = CPhLi \xrightarrow{SO_2Cl_2}$$

$$+ ClSO_2CPh = CPh - CPh = CPhSO_2Cl$$
13%

Scheme 3

An involved reaction path generated a 2% yield of a thiophene 1,1-dioxide (18) from ethyl phenylpropiolate and disulfur dichloride.[39]

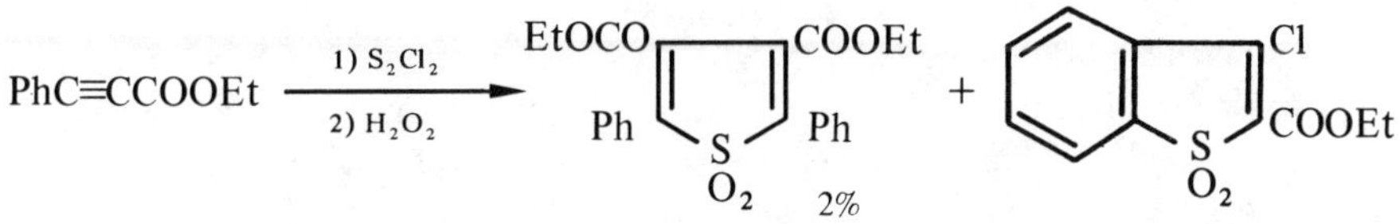

$$PhC \equiv CCOOEt \xrightarrow[\text{2) } H_2O_2]{\text{1) } S_2Cl_2}$$

18   2%   80%

Oxidation of a dithiane yielded a thiophene 1,1-dioxide.[165]

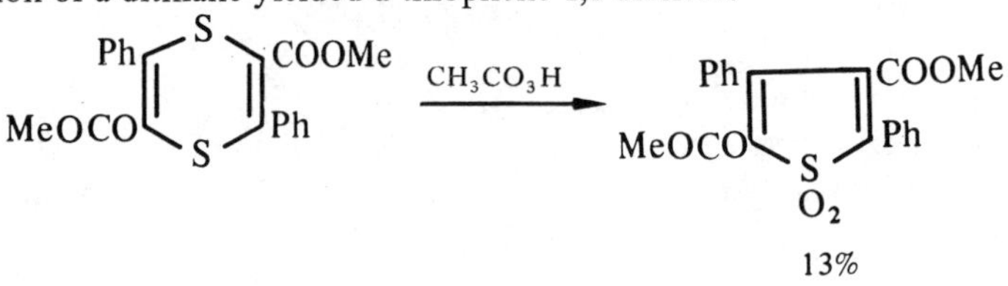

$$\xrightarrow{CH_3CO_3H}$$

13%

2. Reactions

A. Pyrolysis

Pyrolysis of tetraphenylthiophene 1,1-dioxide in boiling di-*n*-butylphthalate (bp 340°C) produced 1,2,3-triphenylnaphthalene and 1,2,3-triphenylazulene. The mechanism of these transformations was not established, but diphenylacetylene was judged not to be an intermediate, since it did not form any of the observed products under the same conditions. Pyrolysis of the molten compound at 500–550°C yielded, instead, tetraphenylthiophene (10%), the corresponding furan (17%), and an amorphous solid.[40] However, flash vapor-phase pyrolysis of tetraphenylthiophene 1,1-dioxide at 880°C gave diphenylacetylene (75%).[41] Presumably, sulfur dioxide is extruded to leave a diradical which undergoes fission to diphenylacetylene. Flash pyrolysis of 2,4-di-*tert*-butyl-, 2,5-di-*tert*-butyl-, 2,5-dimethyl-, and 2,5-diphenylthiophene 1,1-dioxide took a different course and yielded the corresponding furans. The dioxide may rearrange to a cyclic, six-membered sultine, which then extrudes sulfur monoxide to form the furan. 2,5-Dimethylthiophene 1,1-dioxide was quantitatively recovered when subjected to flash pyrolysis conditions at 400°C/0.1 mm Hg, which is an indication of its thermal stability.[41]

Pyrolysis of tetrachlorothiophene 1,1-dioxide at 660°C/1 mm gave perchlorovinylacetylene (67%).[42] Here, the diradical resulting from extrusion of sulfur dioxide undergoes rearrangement. This decomposition–rearrangement is mechanistically related to the pyrolysis of hexachlorobenzothiophene 1,1-dioxide to perchlorophenylacetylene and to the pyrolysis of octachlorobenzocyclobutene to perchlorostyrene.

B. Diels–Alder Reactions

Thiophene 1,1-dioxides undergo Diels–Alder reactions of many ramifications that can produce not only simple adducts, but also cycloheptatrienes, azulenes, and an array of complex polycyclic products. This is an area where thiophene 1,1-dioxides will find many applications in synthesis.

In thiophene 1,1-dioxides, the aromatic character of thiophene has been lost, since the unshared pairs of electrons of thiophene, which contribute to the resonance of the molecule, are now bonded with oxygen atoms. The thiophene 1,1-dioxides behave as cyclic 1,3-dienes. The Diels–Alder reaction of thiophene 1,1-dioxide has been the subject of a molecular orbital and mapping study.[57]

a. DIMERIZATION

Thiophene 1,1-dioxide undergoes dimerization even below room temperature, in which one molecule acts as a diene and another as a dienophile to produce a Diels–Alder reaction. The adduct (**19**) spontaneously extrudes sulfur dioxide to leave 3a,7a-dihydrobenzothiophene 1,1-dioxide (**20**) as the product (34%).[43] As a

1,3-diene, dihydrobenzothiophene 1,1-dioxide (20) can react further with another mole of thiophene 1,1-dioxide,[44] with maleic anhydride,[44] or with dimethyl acetylenedicarboxylate.[12]

Suggestions have been made that the ready dimerization of thiophene 1,1-dioxide may be conveniently understood in terms of second-order perturbation theory[45] and spiroconjugation.[46]

The rapid dimerization of thiophene 1,1-dioxide has resulted in the isolation, instead of this compound, of 20 and its higher condensation products in the removal of 2HX from 3,4-dichlorotetrahydrothiophene 1,1-dioxide[47,48] with ammonia, from 3,4-dibromotetrahydrothiophene 1,1-dioxide by treatment first with pyridine and then with piperidine,[44] and from 3,4-di(tosyloxy)-tetrahydrothiophene 1,1-dioxide with pyridine.[49]

Nine substituted thiophene 1,1-dioxides have been dimerized with loss of sulfur dioxide, including the 2-methyl, 3-methyl, 3-phenyl-,[12] and 2,4-dimethyl derivatives.[50,51] Indoline brought about the dimerization in water of 3,4-dimethylthiophene 1,1-dioxide to 21a.[52] The second-order reaction rate constant is $(4.2 \pm 0.5) \times 10^{-4}$ l. mole^{-1} s^{-1} at 130°C in diphenyl ether.[53] Oxidation of 2-(1-methylnonyl)-thiophene with 3-chloroperbenzoic acid gave a 4% yield of the oily product 22.[18] Kinetic data have been presented for the dimerization of 3,4-diphenylthiophene 1,1-dioxide in boiling phenol to 21b (62%).[54] The preparation of tetrachlorothiophene 1,1-dioxide was accompanied by the formation of 1.1% of 23.[42]

21a   R = Me
  b   R = Ph

The dimerization of 3,4-dichlorothiophene 1,1-dioxide in boiling xylene resulted in the loss of both sulfur dioxide and hydrogen chloride from the initial adduct to form 3,5,6-trichlorobenzothiophene 1,1-dioxide (24). A small amount of the Diels–Alder adduct of this product with 3,4-dichlorothiophene 1,1-dioxide was also produced (25). The isolation of a compound believed to be a true [2 + 2] dimer (26) of 3,4-dichlorothiophene 1,1-dioxide is interesting (Scheme 4).[9]

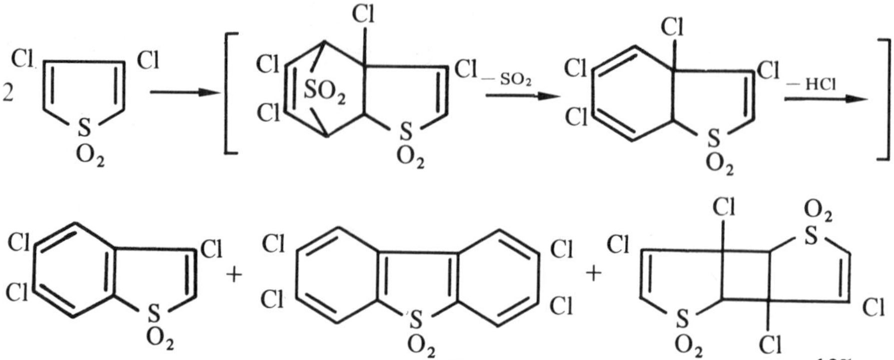

24 49% **25** Small amount **26** 12%

Scheme 4

3-Isopropyl-4-isopropenylthiophene 1,1-dioxide formed a true dimer of a different type. The double bond of the isopropenyl group, together with a double bond of the ring, functioned as a diene in a Diels–Alder cycloaddition to produce **27**.[50]

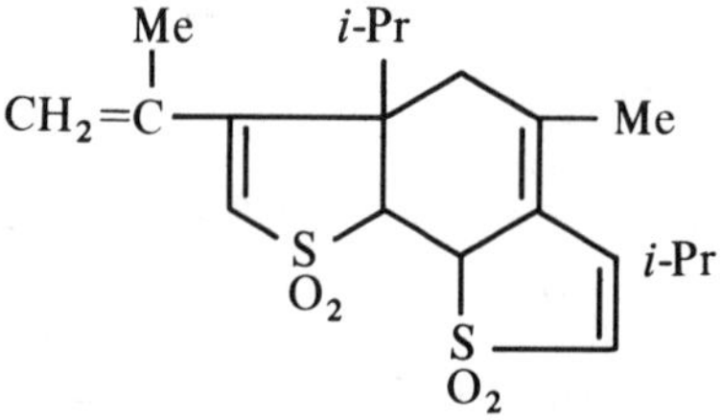

27

Benzo[3,4]cyclobuta[1,2-c]thiophene monoxide and dioxide do not undergo Diels–Alder dimerization, but the dioxide formed a 2+2 dimer of unestablished stereochemistry upon irradiation with a medium pressure lamp (**28**).[160]

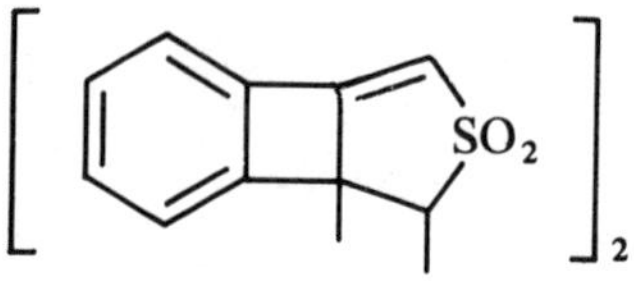

Tetraphenylthiophene 1,1-dioxide was unaffected by irradiation with UV light in Pyrex apparatus.[40] A photooxidation experiment with the same compound, using intense incandescent light, gave a resin.[56]

b. DIELS–ALDER REACTIONS OF THIOPHENE 1,1-DIOXIDE

Because of the rapidity with which it dimerizes, thiophene dioxide itself is of limited use as a diene in the Diels–Alder reaction. The reaction with diethyl acetylenedicarboxylate was unusual, as in several experiments the unstable primary adduct (28a) containing a sulfonyl bridge was isolated.[58] Later investigations,[12, 20] following the reaction by nmr, were unable to detect the presence of the primary adduct. Possibly, this failure was caused by the presence of base in their thiophene 1,1-dioxide, which was prepared by the Hofmann elimination method. Only two other cases of the isolation of the sulfonyl-bridge adduct are known. One is the adduct of 2,5-dimethylthiophene 1,1-dioxide with benzoquinone (28b).[59] The other is the adduct of 3,4-dichlorothiophene 1,1-dioxide with 2,5-dimethylfuran (Section 2.B.e).

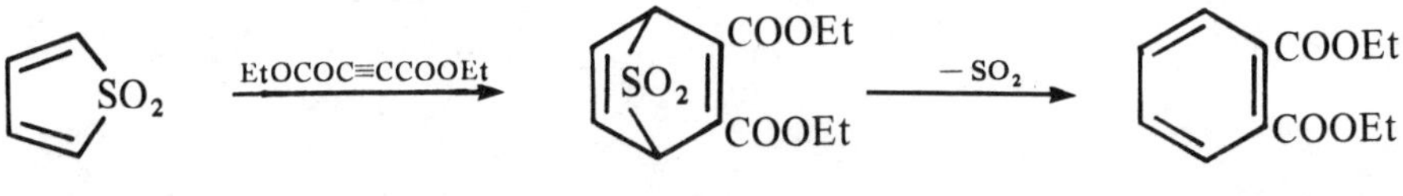

28a

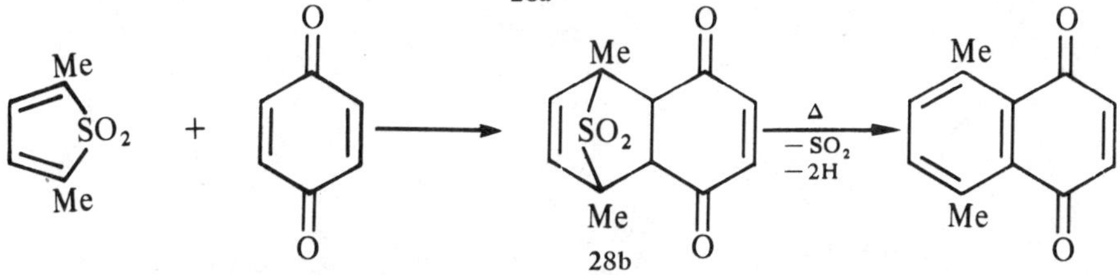

28b

With indene and thiophene 1,1-dioxide, a 3% yield was obtained and isolated as fluorene.[58] The use of thiophene 1,1-dioxide as a dienophile was more successful. It was added to 1,2-dimethylenecyclohexane,[58] cyclopentadiene, and spiro[2.4]hepta-4,6-diene[50] to form 29, 30, and 31. Compounds of type 30 were also obtained from cyclopentadiene and 2-methyl-, 3-methyl-, 3-phenyl-, 2,4-dimethyl-, 3,4-dimethyl-, 3-chloro-4-methoxy-, and 3-isopropyl-4-isopropenylthiophene 1,1-dioxide.[50]

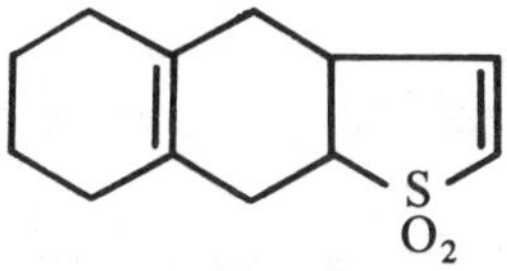

29 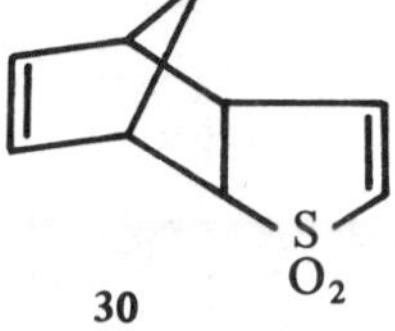   30 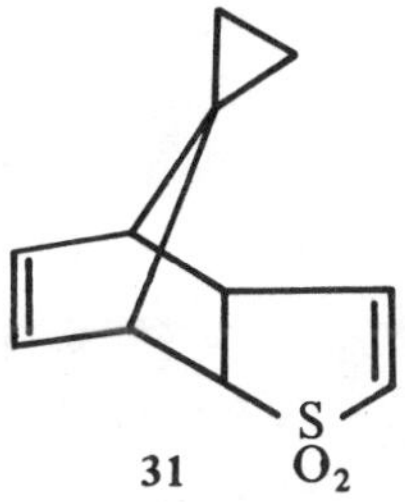   31

c.   DIELS–ALDER REACTIONS OF ALKYL- AND ARYLTHIOPHENE 1,1-DIOXIDES

The first Diels–Alder reactions of thiophene 1,1-dioxides were reported by Melles in 1952.[7] Tetraphenylthiophene 1,1-dioxide was reacted with maleic anhydride. As is characteristic in these reactions, sulfur dioxide was lost and the product was a dihydrophthalic anhydride (32).

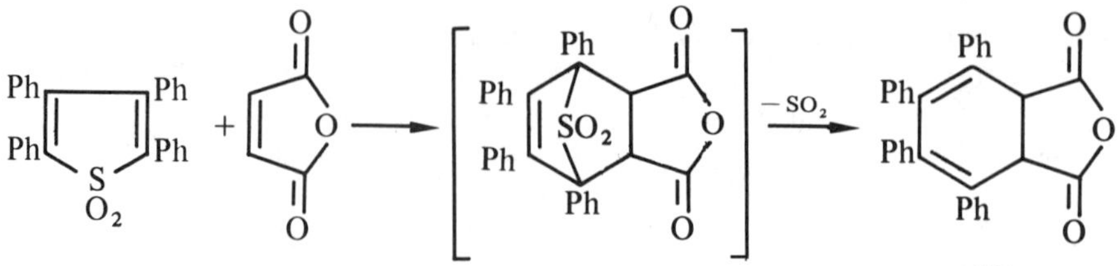

32

In less hindered examples, the product diene reacted with another molecule of maleic anhydride to give bicyclo[2.2.2]octene derivatives as the only isolated product (33). This was the case with 2,5-Me$_2$-; 2,5-Me$_2$-3,4-Br$_2$-; 3,4-Me$_2$-; 2,5-Ph$_2$-;

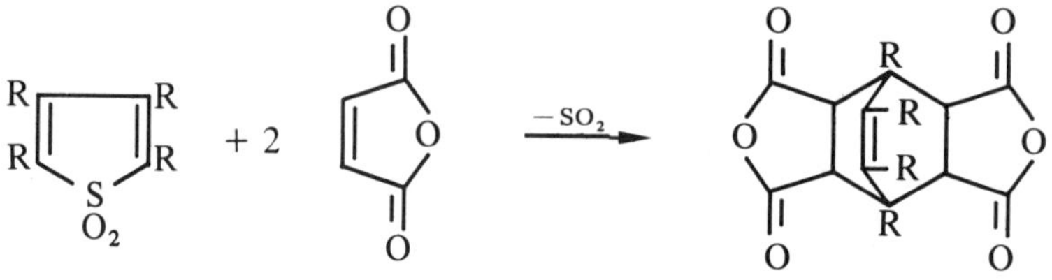

33

and 3,4-Ph$_2$-thiophene 1,1-dioxides.[7] Another example, illustrating the buildup of complex ring systems, is the addition of 3,4-dimethylthiophene 1,1-dioxide to 2 moles of benzothiophene 1,1-dioxide to give **34** or the *cis* isomer.[8]

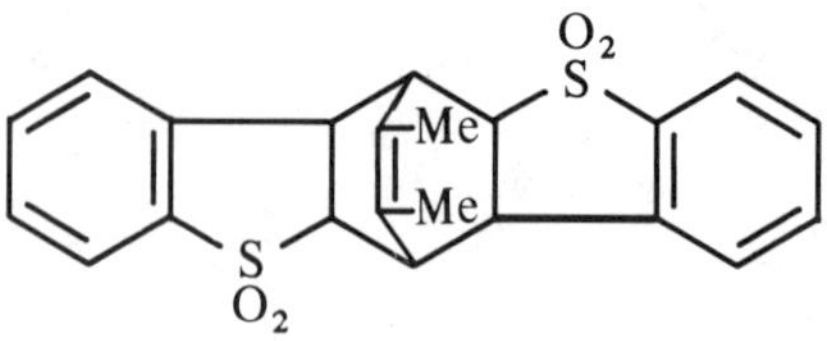

34

Successful 1:1 Diels–Alder additions, with loss of sulfur dioxide, have resulted from the following reactions: tetraphenylthiophene 1,1-dioxide with stilbene, phenylacetylene, and dimethyl acetylenedicarboxylate[60] and the free acid;[61] 2,5-diphenylthiophene 1,1-dioxide[62] with *N*-phenylmaleimide; and 2,5-dimethylthiophene 1,1-dioxide with *N*-phenylmaleimide.[20]

Although 2,5-diphenylthiophene 1,1-dioxide underwent Diels–Alder addition to phenylacetylene to form 1,2,4-triphenylbenzene (35) in 45% yield, the same

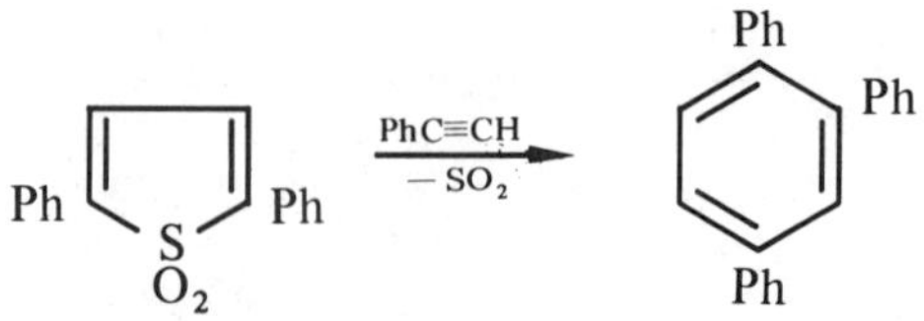

35

product was obtained from 3,4-diphenylthiophene in only 6.3% yield.[63] The use of α-acetoxystyrene was still less successful. The reaction between 3,4-diphenylthiophene 1,1-dioxide and phenylacetylene has been investigated in detail. The principal product (34% yield) arose from a Diels–Alder addition in which a double bond of the thiophene dioxide and a double bond of a phenyl group function together as the diene, which is reminiscent of the exo- and endocyclic double bonds functioning together in the dimerization of 3-isopropyl-4-isopropenylthiophene 1,1-dioxide. Consecutive cycloaddition, 1,5-sigmatropic rearrangement, and cheletropic elimination lead to **36** (Scheme 5).[64,65]

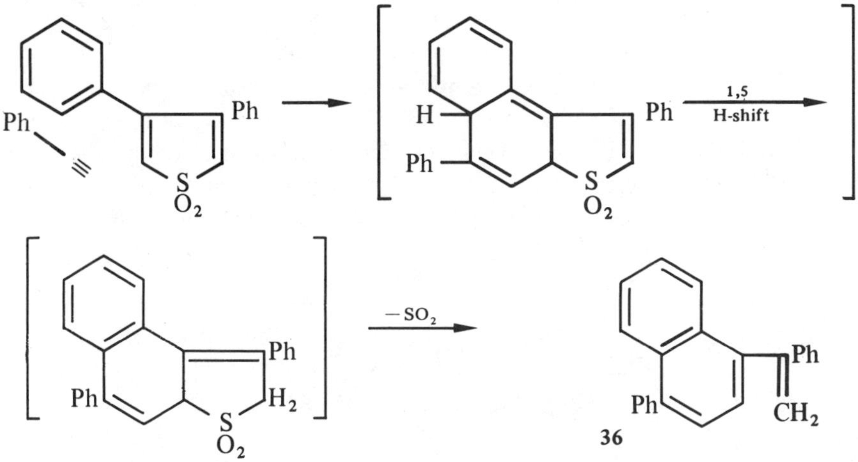

Scheme 5

Attempts to add the electron-rich olefins, ethyl vinyl ether and 1-pyrrolidino-2-methylpropene, to 3-ethyl- and 3,4-dimethylthiophene 1,1-dioxide, and 1-diethylamino-1,3-butadiene to 3,4-dimethylthiophene 1,1-dioxide, failed. These failures were attributed to the antibonding interaction of the sulfone oxygen orbital of the thiophene dioxide LUMO with the HOMO of the electron-rich olefin.[50] Other reported nonreactions are 2,4-dimethylthiophene 1,1-dioxide with dimethyl acetylenedicarboxylate, diphenylacetylene, 1-diethylaminopropyne, and 1,1-bis(dimethylamino)ethylene,[20] and 2,5-dibenzoyl-3,4-diphenyl- and 2-carbomethoxy-3,4-diphenyl-5-benzoylthiophene 1,1-dioxide with maleic anhydride.[7] Attempted Diels–Alder reactions of N-phenylmaleimide with 3-methyl- and 3-phenylthiophene 1,1-dioxides, and of dimethyl acetylenedicarboxylate with 3-methylthiophene 1,1-dioxide gave, instead, the Diels–Alder adducts of the 3a,7a-dihydrobenzothiophene 1,1-dioxides formed through dimerization of the thiophene 1,1-dioxides.[12]

d. DIELS–ALDER REACTIONS OF HALOGENATED THIOPHENE 1,1-DIOXIDES

The Diels–Alder addition of 3,4-dichlorothiophene 1,1-dioxide (**37**) with bicyclo[2,2,1]hepta-2,5-diene resulted in an unusual series of events leading to 1,2-dichlorobenzene and the adduct (**38**) of 3,4-dichlorothiophene 1,1-dioxide with cyclopentadiene (Scheme 6).[9]

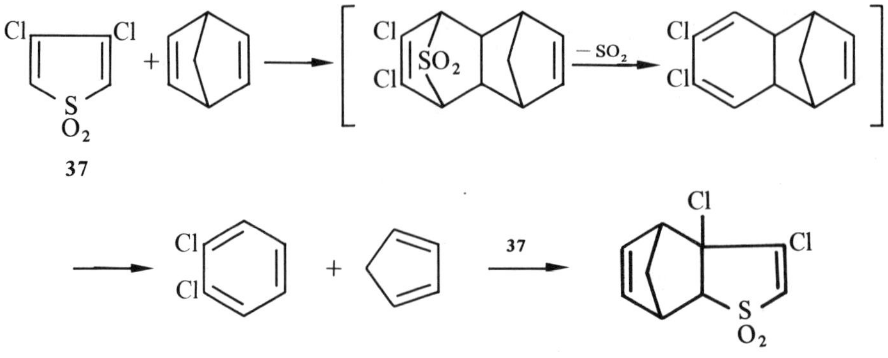

Scheme 6

37

38

In the reaction with cyclopentadiene, 3,4-dichlorothiophene 1,1-dioxide acted as a dienophile.[9,50] Butadiene and isoprene each yielded two reaction products, **39** and **40**, resulting from 3,4-dichlorothiophene 1,1-dioxide functioning both as a diene acceptor and as a diene.[9,66]

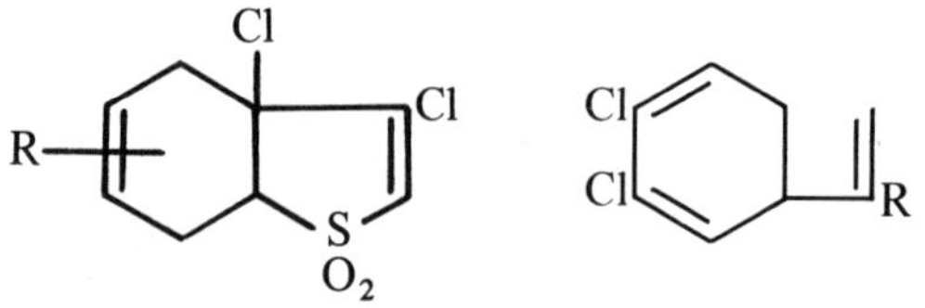

39a R = H
b R = Me

40a R = H
b R = Me

Addition of 3,4-dichlorothiophene 1,1-dioxide to benzoquinone formed 6,7-dichloronaphthoquinone. Under the reaction conditions, the intermediate dihydro compound or its 1,4-naphthalenediol tautomer was oxidized by the benzoquinone present (Scheme 7).[9,67]

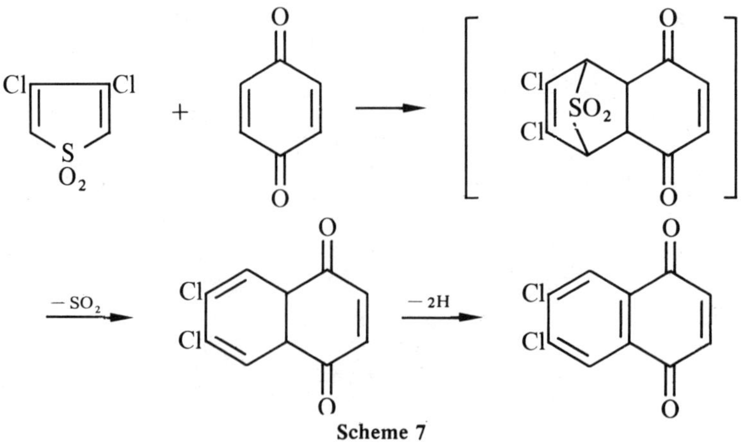

Scheme 7

Two moles of *N*-*n*-butyl- and *N*-(4-nitrophenyl)-maleimide added to 3,4-dichloro-thiophene 1,1-dioxide in the manner previously described to give **41**.[9]

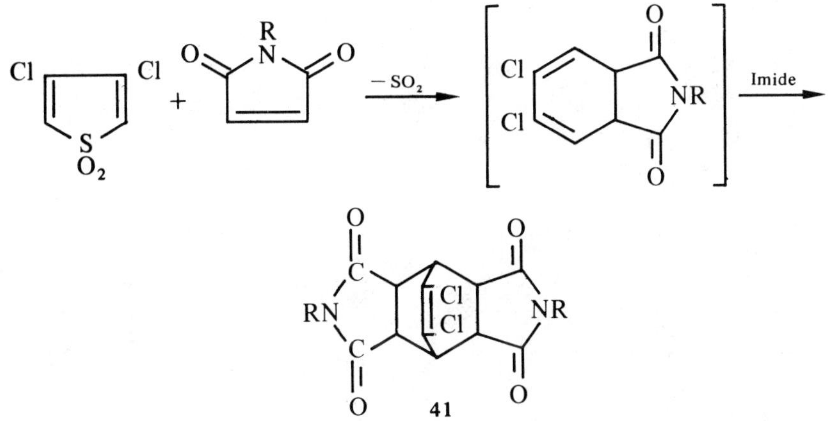

Tetrachlorothiophene 1,1-dioxide, prepared by oxidation of the thiophene with 3-chloroperbenzoic acid, has a combination of stability and reactivity that makes it a useful synthetic reagent. The compound has high additivity toward double bonds, which it annelates with a tetrachlorobutadienediyl group.

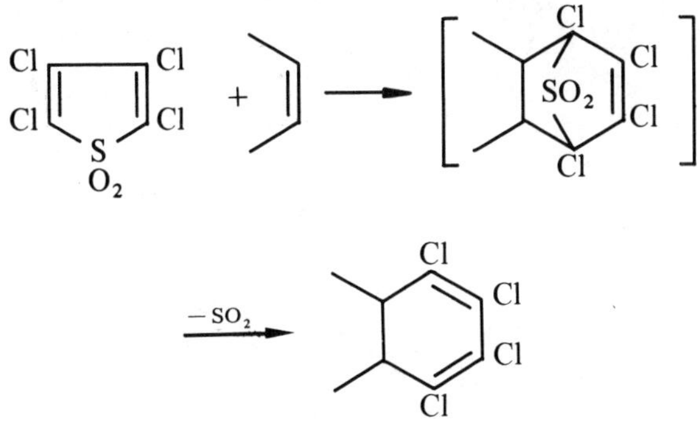

The products can be oxidatively aromatized, treated with base to form 1,2,4-trichloroaromatics, and reductively dechlorinated as part of the synthetic possibilities. Of particular interest is the facile assemblage of polycyclic carbocycles and heterocycles, by means of a double Diels–Alder reaction, through addition of tetrachlorothiophene 1,1-dioxide to 1,5- and 1,6-dienes. Tetrabromothiophene 1,1-dioxide reacts like its tetrachloro analogue.[42]

Tetrachlorothiophene 1,1-dioxide is reactive enough to combine slowly with ethylene at 28°C. Many unsaturated compounds react exothermically on slight warming. Forty-five C=C compounds of various types that have been reacted with the reagent are listed in Table 1.[42,68] Some of the products are shown here by way of illustration.

TABLE 1. ANNELATIONS WITH TETRACHLOROTHIOPHENE 1,1-DIOXIDE[42]

Reactant	Reaction Conditions Temperature, °C (time, h)	Product	Yield (%)
$CH_2=CH_2$	C_6H_6, 80 (200 psi), (4)		89
Cyclopentene	44, (2)		74
Cyclohexene	83, (1)		90
Cycloheptene	100, (1)		92
Cyclooctene	100, (1)		79

Substrate	Structure	Conditions	Yield
Cyclododecene		100, (16)	40
1,4-Cyclohexadiene		86, (1)	88
1,2-Cyclononadiene[68]		CCl$_4$, Reflux (1)	52
cis-1,trans-5-Cyclodecadiene		ClCH$_2$CH$_2$Cl, 80 (1)	63
Methylenecyclobutane		60, (3)	75
Methylenecyclohexane		100, (3)	73

TABLE 1. *Continued*

Reactant	Reaction Conditions Temperature, °C (time, h)	Product	Yield (%)
1,4-Dimethylenecyclohexane	100, (1)		70
(−)-β-Pinene	100, (3)		74
Methyleneadamantane	130, (4)		27
Indene	C_6H_6, 45, (1)		80
Acenaphthylene	C_6H_6, 100, (1)		91

endo-Dicyclopentadiene	C$_6$H$_6$, 80, (1)		65
PhC≡CH	Cooled in ice		86
1-Hexen-5-yne	(2)	HC≡C(CH$_2$)$_2$	70
1-Dodecene	100, (3)	CH$_3$(CH$_2$)$_9$	88
1,7-Octadiene	100, (1)	CH$_2$=CH(CH$_2$)$_4$	69
1,8-Nonadiene	100, (1)	CH$_2$=CH(CH$_2$)$_5$	49

589

TABLE 1. *Continued*

Reactant	Reaction Conditions Temperature, °C (time, h)	Product	Yield (%)
CH$_2$=CHCOOH	100, (1)	Cl-substituted cyclohexadiene with COOH	68
CH$_2$=CHCOOMe	80, (2)	Cl-substituted cyclohexadiene with COOMe	91
CH$_2$=CHCN	78, (17)	Cl-substituted cyclohexadiene with CN	83
CH$_2$=CHCH$_2$COOH	100, (16)	Cl-substituted cyclohexadiene with CH$_2$COOH	78
CH$_2$=CH(CH$_2$)$_8$COOH	100, (1)	Cl-substituted cyclohexadiene with (CH$_2$)$_8$COOH	78

Dienophile	Conditions	Product	Yield (%)
Dimethyl maleate	100, (16)		73
Maleic anhydride	100, (48)		61
Maleimide	$ClCH_2CH_2Cl$, 80, (1)		88
N-Methylmaleimide	100, (1)		90
$CH_2{=}C(Me)COPh$	100, (20)		65
$CH_2{=}CHCH_2Br$	70, (2)		88

591

TABLE 1. *Continued*

Reactant	Reaction Conditions Temperature, °C (time, h)	Product	Yield (%)
CH$_2$=CHCH$_2$NCS	100, (2)		55
Safrole	100, (1)		75
1,4-Benzoquinone	ClCH$_2$CH$_2$Cl, 80, (16)		74
1,4-Naphthoquinone	C$_6$H$_6$, 80, (16)		68
2-Vinylpyridine	(1) C$_6$H$_6$, 25 (2) HCl		75
N-Vinylpyrrolidone	C$_6$H$_6$, 25		84

N-Vinylsuccinimide	100, (0.5)	92
N-Methylpyrrole	ClCH$_2$CH$_2$Cl, 80, (3)	44
Indole	C$_6$H$_6$, 23, (15)	77
Diethyl 2,3-diazanorbornene-2,3-dicarboxylate	ClCH$_2$CH$_2$Cl, 80, (10)	58
Thiophene	85, (72)	61
Precocene II	C$_6$H$_6$, 80, (3)	74

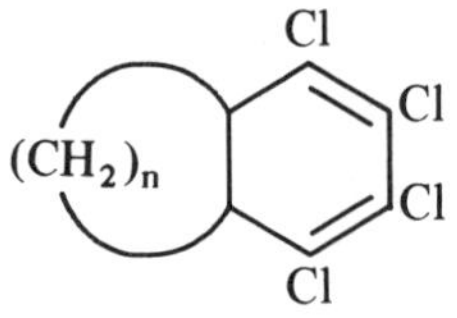

n = 3–6, 10

From cycloolefins

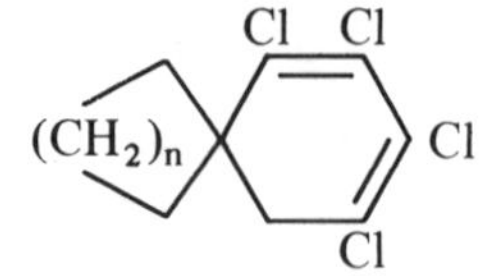

n = 1, 3

From methylenecycloalkanes

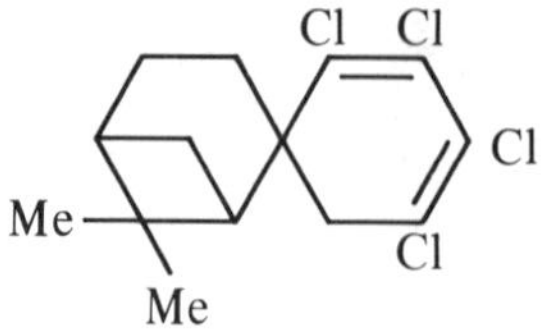

From (−)-β-pinene

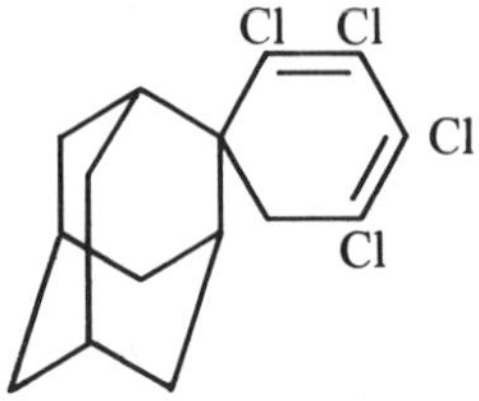

From methyleneadamantane 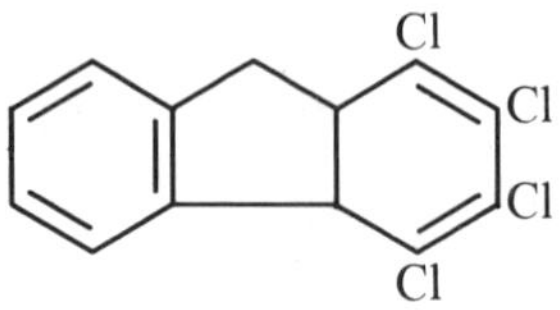   From indene 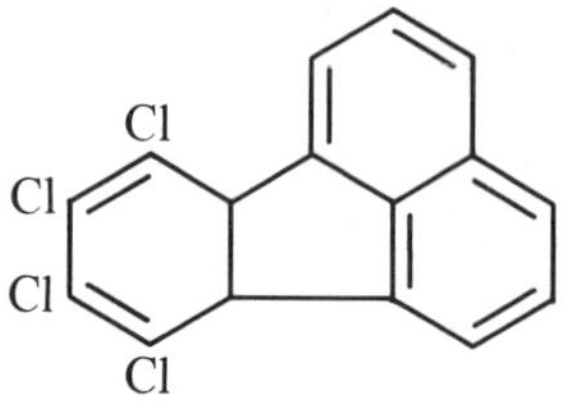   From acenaphthylene

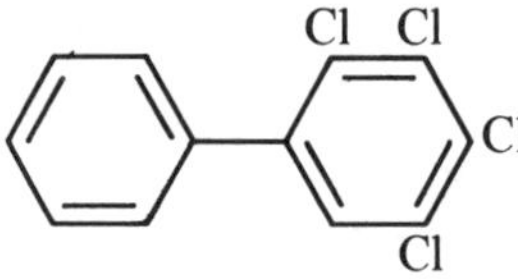

From PhC≡CH 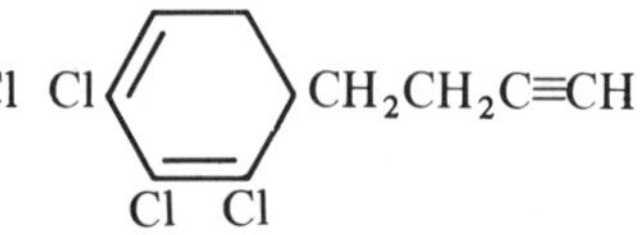   From 1-hexen-5-yne 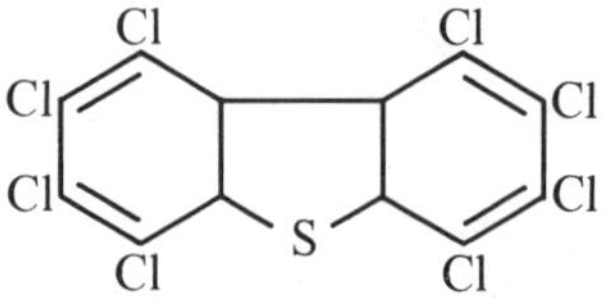   From thiophene

Addition of tetrachlorothiophene 1,1-dioxide to 1,3-dienes can occur in two ways. Thus, with excess cyclopentadiene, annelation of one double bond took place in 44% yield to form **42**, and Diels–Alder addition of the diene to one double bond of tetrachlorothiophene 1,1-dioxide occurred to give **43** in 25% yield. With 2,3-dimethylbutadiene, **44** was the isolated product.

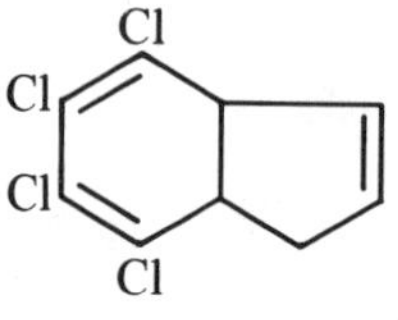

42

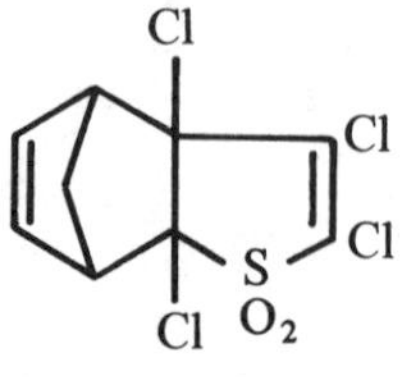

43

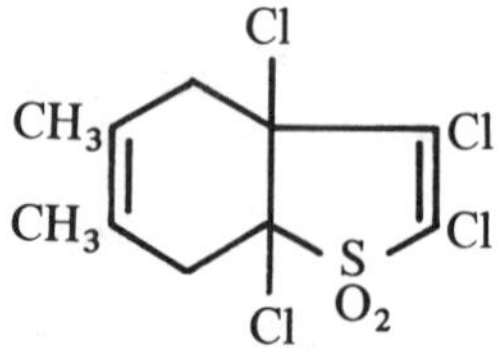

44

Besides the above [4 + 2] additions to a double bond of tetrachlorothiophene 1,1-dioxide, a [12 + 2] addition in 5% yield has been reported.[167]

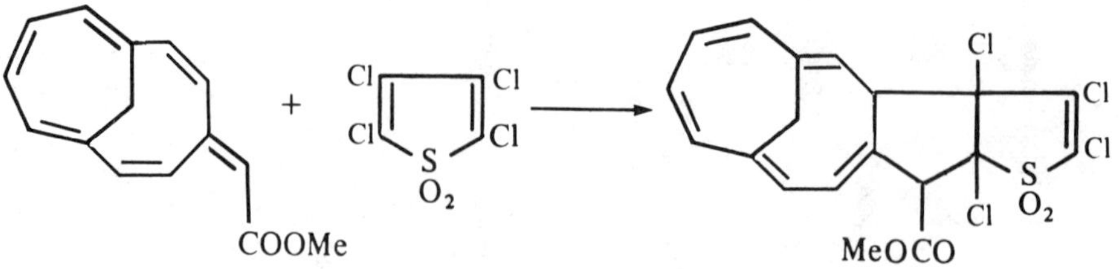

COOMe

MeOCO

The use of tetrachlorothiophene 1,1-dioxide provides facile routes to isotwistenes, homoisotwistenes, their hetero analogues, and higher carbocycles.[42]

Annelation of one double bond of a 1,5-acyclic diene with tetrachlorothiophene 1,1-dioxide produced a tetrachloro-1,3-cyclohexadienyl group in that position, which was then available for an intramolecular Diels–Alder reaction with the remaining double bond of the diene to form an isotwistene (45) in 50–60% yield.

The initial annelation took place at slightly elevated temperatures, but the intramolecular reaction, also exothermic, may require raising the temperature to 150–175°C.

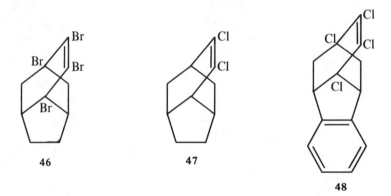

a, R = R′ = H; b, R = H, R′ = CH₃; c, R = R′ = CH₃   45

Similarly, tetrabromothiophene 1,1-dioxide and 3,4-dichlorothiophene 1,1-dioxide reacted with 1,5-hexadiene to form 46 and 47.

46   47

48

By the reaction of tetrachlorothiophene 1,1-dioxide with 1,2-divinylbenzene at 80°C, the more complex ring system represented by 48 was built up.

The 1,5-diene chains may be interrupted by a heteroatom. Reaction with tetrachlorothiophene 1,1-dioxide then yields heteroisotwistenes (49). Vinyl acrylates produced lactones (50).

By using acyclic 1,6-dienes instead of 1,5-dienes, homoisotwistenes were obtained. Thus, 1,6-heptadiene gave 51. The reaction can utilize readily available diallyl compounds to form heterohomoisotwistenes. Thus, diallyl ether, diallyl sulfide, N,N-diallylacetamide, N,N-diallylcyanamide, and N,N-diallylaminoacetonitrile form compounds 52.

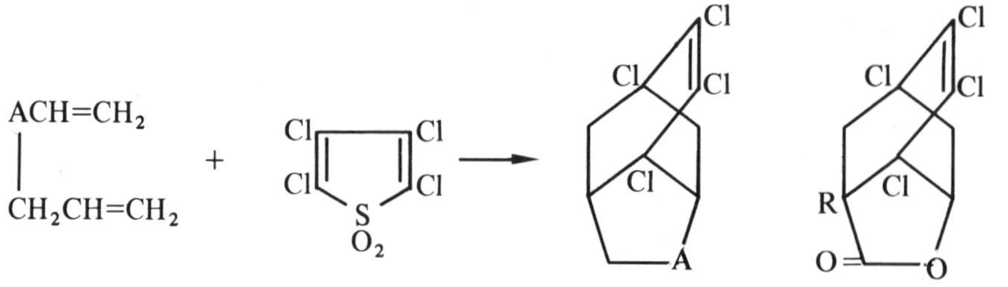

a A = O   b A = S

49

a   R = H
b   R = CH₃

50

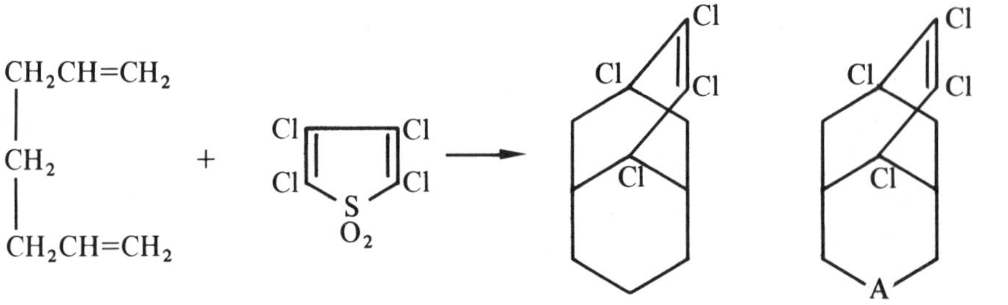

51   **52**

A = O, S, NHAc,
NCN, NCH₂CN

With 1,5-cyclooctadiene, tetrachloro-, tetrabromo-, and 3,4-dichlorothiophene 1,1-dioxides formed hexahydro-4,1,5-[1]propanyl[3]ylidene-1*H*-indenes (**53–55**) by intramolecular Diels–Alder reactions.[42] 1,5-Cyclononadiene yielded **56**.

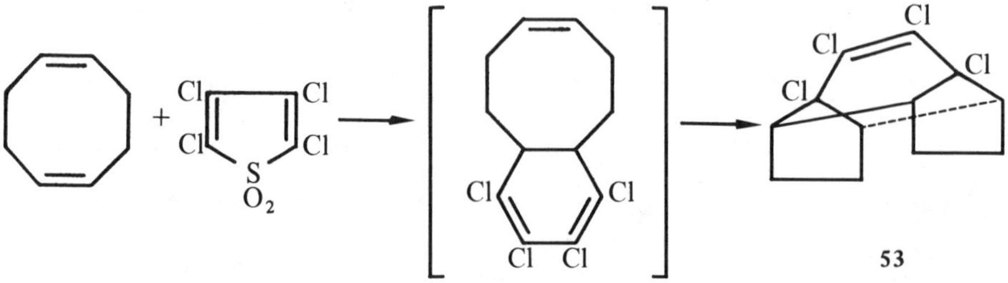

53

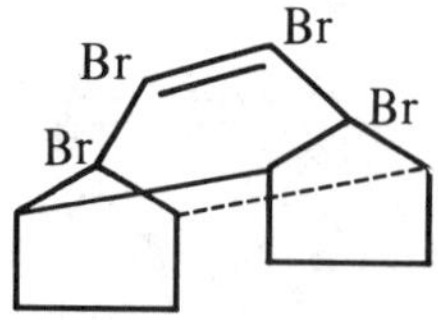

54 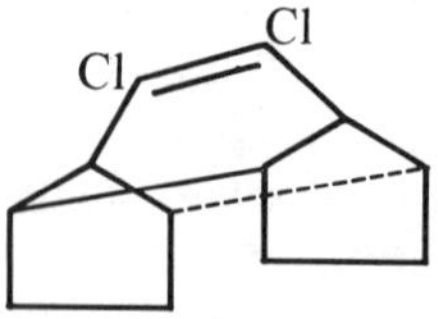   **55** 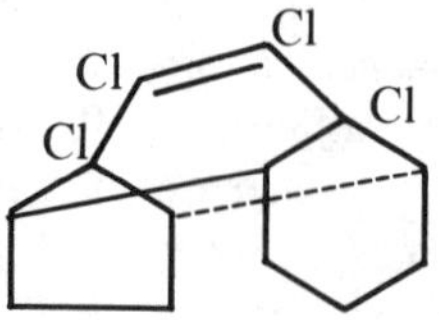   **56**

By using dibenzo[a,e]cyclooctene, a dibenzo analogue (57) of 53 was assembled.

57

The procedure has been adapted to the preparation of halogenated iceanes. Tetrachlorothiophene 1,1-dioxide was reacted with 9,9-dihalo-1,4,5,8-tetrahydro-4a,8a-methanonaphthalene to form two stereoisomeric trienes, one of which spontaneously closed by intramolecular Diels–Alder reaction to the iceane.[150,166]

$Z = CCl_2$, CBr_2, NHCO, CO_2CO, CH_2OCH_2, $OC(CH_3)_2O$

In the case of the compound where $Z = 0$, the adduct did not undergo an intramolecular Diels–Alder reaction. In this connection, the monoadduct of tetrachlorothiophene 1,1-dioxide and 1,4,5,8-tetrahydronaphthalene was also prepared.[166]

The Diels–Alder addition reactions of tetrachlorothiophene 1,1-dioxide shown in Scheme 8 have also been reported.[144] Kinetic studies of reactions of tetrachlorothiophene 1,1-dioxide with four-substituted styrenes (MeO, Me, Cl, H, Br, and CN) showed the expected faster reaction with electron-rich styrenes. Molecular orbital calculations (CNDO/2) indicated a 1.7 eV lowering of LUMO for tetrachlorothiophene 1,1-dioxide, compared with the value for 2,5-dimethylthiophene 1,1-dioxide.[144]

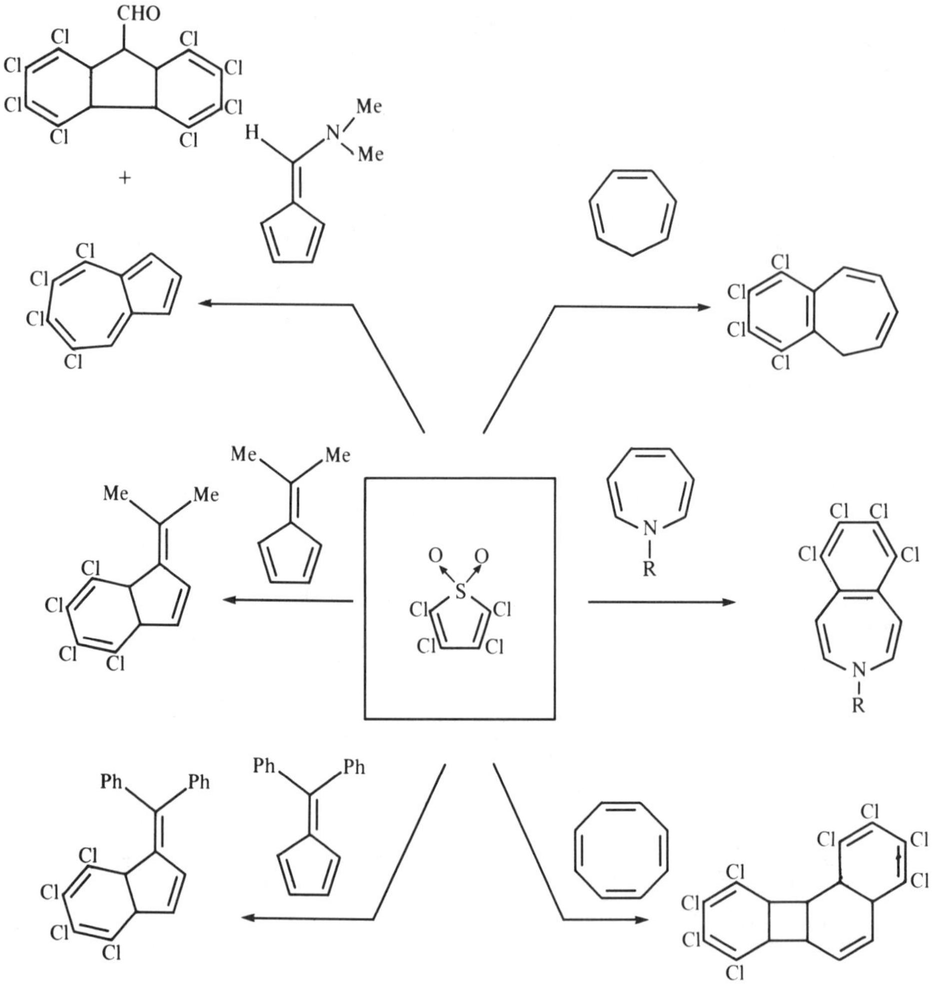

Scheme 8

In Scheme 8, the reaction product from tetrachlorothiophene 1,1-dioxide and cycloheptatriene was reported to be an oil. In an unpublished work,[146] this reaction yielded a crystalline compound (40%, mp 46–47°C, abs. EtOH) that had eliminated 1 mole of HCl. Assuming the usual mode of loss of HCl from these tetrachloro compounds,[42] this product should be:

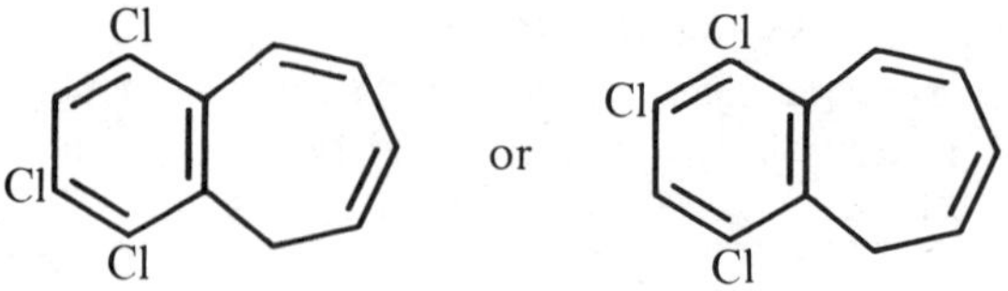

or

Annelation of 3-thiabicyclo[3.2.0]hept-6-ene 3,3-dioxide with tetrachlorothiophene 1,1-dioxide produced an adduct whose pyrolysis yielded 1,2,3,4-tetrachlorobenzene by the loss of both sulfur dioxide and butadiene.[156]

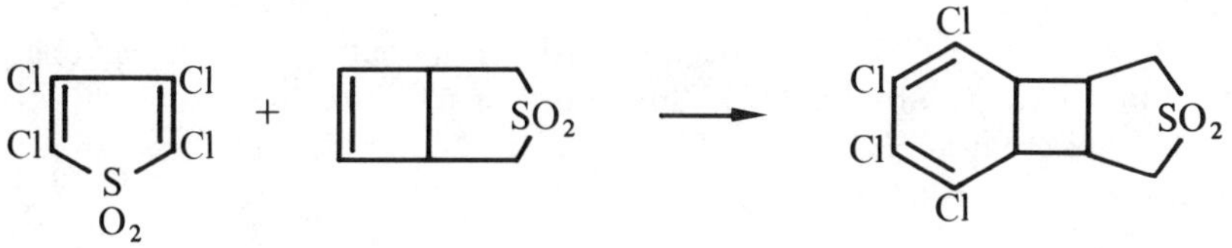

Tetrachlorothiophene 1,1-dioxide has also been used to annelate the tricarbonyl iron complexes of seven-membered ring compounds in 66–92% yields.[147]

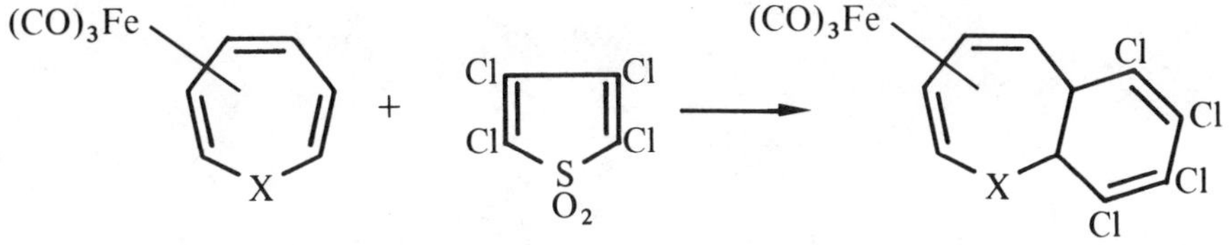

X = NCOOEt, CH₂, C=O

e.   ADDITION–REARRANGEMENT REACTIONS OF HALOGENATED THIOPHENE 1,1-DIOXIDES WITH FURANS

Halogenated thiophene 1,1-dioxides undergo rearrangement reactions with furans to form halobenzyl carbonyl compounds. In the case of 2,5-dimethylfuran and 3,4-dichlorothiophene 1,1-dioxide, the Diels–Alder primary adduct retaining the sulfonyl bridge (58) precipitated from the solution and could be isolated if the reaction was stopped after 1–2 min. Warming the reaction resulted in loss of sulfur dioxide and rearrangement of the annelation product to the benzyl ketone 59 (Scheme 9).[69]

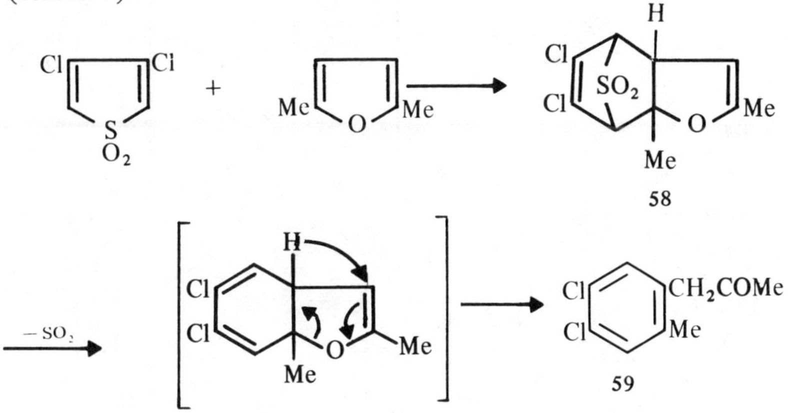

Scheme 9

The reaction also proceeded, without isolation of the primary adduct, between tetrachlorothiophene 1,1-dioxide and 2-Me-, 2,5-Me$_2$-, 2-n-C$_4$H$_9$-, 2-MeO, 2-AcOCH$_2$-, 2-AcNHCH$_2$-, 2-Ac-, and 2-MeOCO-furan. Tetrabromothiophene 1,1-dioxide was also operable. 2,5-Dimethoxy-2,5-dihydrofuran behaved like 2-methoxyfuran by losing methanol in the reaction. α-Angelica lactone (60) functioned like 2-hydroxy-5-methylfuran to form 61. With 2-vinylfuran, annelation of the vinyl group took place, and the product then underwent the furan reaction to yield 62.

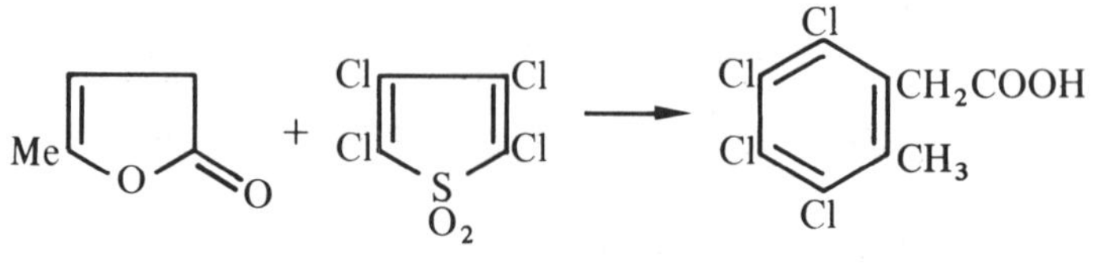

60   61

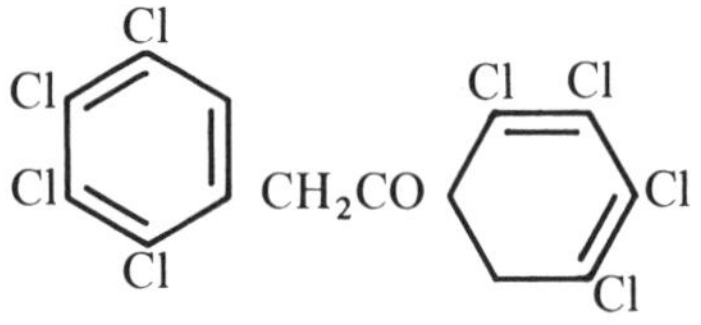

62

The reaction of furan itself with tetrachlorothiophene 1,1-dioxide differs from the examples given above. The rearrangement product, (2,3,4,5-tetrachlorophenyl)-acetaldehyde, was isolated in 8% yield, but a second product (63, 27%) was formed from furan acting as a 1,3-diene accepting one double bond of tetrachlorothiophene 1,1-dioxide and the resulting intermediate acting as a donor of one double bond to another molecule of the thiophene 1,1-dioxide (TCTD), which acted as a diene (Scheme 10).

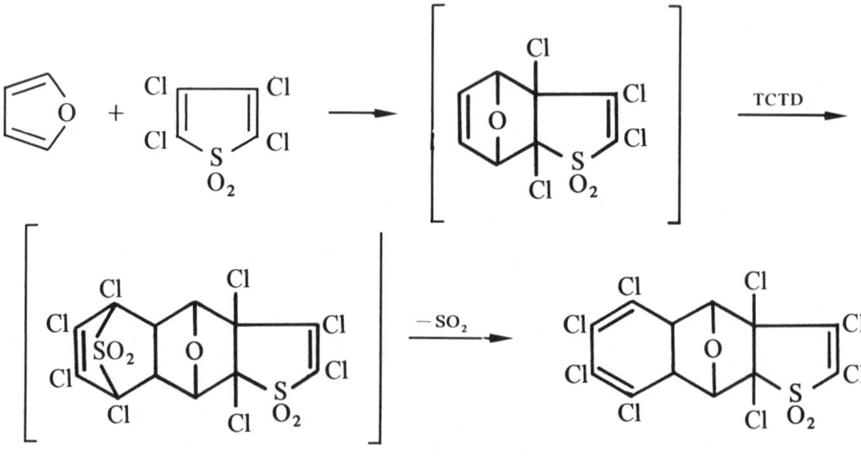

63

Scheme 10

f. SYNTHESIS OF CYCLOHEPTATRIENES

A facile synthesis of cycloheptatrienes in high yields utilizes the propensity of thiophene 1,1-dioxides for Diels–Alder reactions. Addition of thiophene 1,1-dioxide and its 2,5-dimethyl, 2,5-di-*tert*-butyl, 2,4-di-*tert*-butyl, and 2,5-diphenyl derivatives to cyclopropene and its methyl derivatives yielded cycloheptatrienes. Logically, the reaction proceeds through a Diels–Alder adduct. The expulsion of sulfur dioxide from the adduct may be synchronous with cyclopropyl cleavage. Thus, a norcaradiene may not be involved.[20, 70]

Cycloadditions of 2-methyl, 3-methyl-, 3-isopropyl-4-isopropenyl-, and 3-chloro-4-methoxythiophene 1,1-dioxides with 1-methylcyclopropene were carried out to study the ratios of regioisomers formed.[50]

In a program to prepare all the methylated tropylium ions, 3,4-dimethyl-, 2,3,4-trimethyl-, and tetramethylthiophene 1,1-dioxides were reacted with cyclopropene and its monomethyl derivatives to form methylated cycloheptatrienes.[71]

g. SYNTHESIS OF AZULENES

The [6+4] cycloaddition of 6-(dimethylamino)fulvenes and thiophene 1,1-dioxides to form azulenes was published nearly simultaneously by Copland, Leaver, and Menzies[4] and by Reiter, Dunn, and Houk.[5, 6] The yields from thiophene 1,1-dioxide and its 2-methyl, 3-methyl, 3-ethyl, 3-phenyl, 2,4-dimethyl, 3,4-dimethyl, 3,4-diphenyl, and 3,4-dichloro derivatives ranged from 1 to 60% for the latter. More recently, an 11% yield has been obtained from tetrachlorothiophene 1,1-dioxide.[144]

The reaction of 2-ethyl-5-methyl- and 2-isopropyl-5-methylthiophene 1,1-dioxide with 6-(dimethylamino)fulvene in refluxing pyridine gave azulenes that were converted to the natural products guaiazulene and chamazulene;[19] using pyridine as solvent doubled the yield. By reacting thiophene 1,1-dioxide with 6-(dimethylamino)fulvene-6-^{13}C and with 6-(dimethylamino)fulvene-6-d, azulene-4-^{13}C and azulene-4-d were prepared.[72a] Similarly, [4,7-$^{13}C_2$]-azulene was made from [2,5-$^{13}C_2$]-thiophene 1,1-dioxide.[72b]

C. 1,3-Dipolar Additions

The generation of thiophene 1,1-dioxide from 3,4-dibromotetrahydrothiophene 1,1-dioxide with pyridine or triethylamine in the presence of N-α-diphenylnitrone yielded mono- and diadducts.[143] Monoadduct formation was 10^3 times faster than diadduct formation.

Similarly, benzonitrile oxide and mesitonitrile oxide underwent regioselective cycloaddition to thiophene 1,1-dioxide to produce the following adducts.[143, 155]

Ar = Ph, 2,4,6-Me$_3$C$_6$H$_2$

Diazomethane has been added to 3,4-diphenyl- and 3,4-di-p-tolythiophene 1,1-dioxide to form a pyrazoline ring (64). By starting with 2,3-dihydro-4-methyl-3-oxothiophene 1,1-dioxide (65), a similar compound (66) resulted through a thiophene dioxide intermediate.[31]

Pyrolysis of the products yielded the expected cyclopropane ring.

Isoquinoline 2-oxide underwent a 1,3-dipolar addition with 3,4-dimethylthiophene 1,1-dioxide to form **67**.[73]

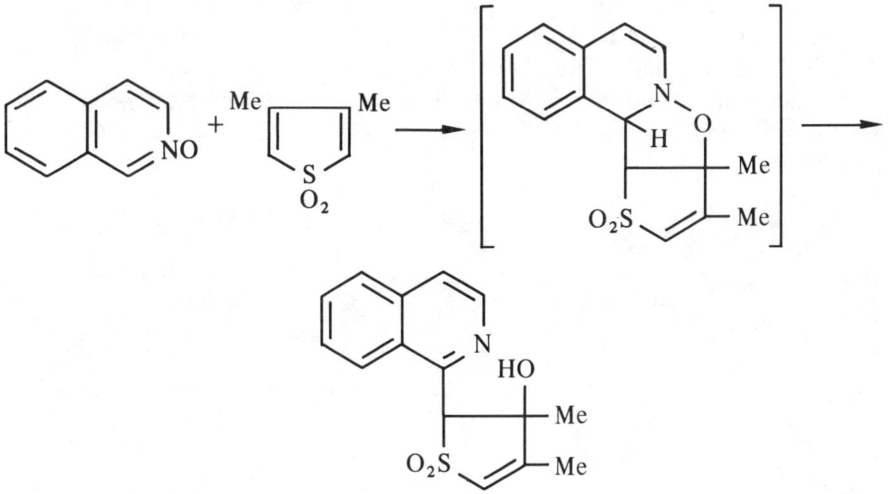

67

D. Nucleophilic Reactions

2,3-Addition took place between 3,4-dimethylthiophene 1,1-dioxide and piperidine or pyrrolidine in water to form **68**. 3-Methylthiophene 1,1-dioxide and piperidine yielded **69**.[52]

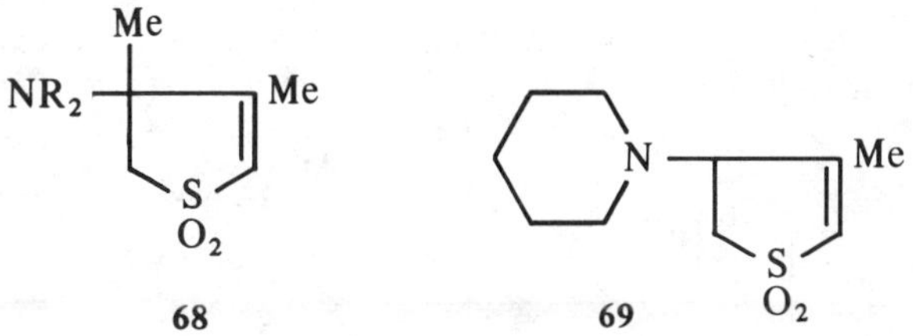

68   **69**

However, in piperidine or pyrrolidine alone, 3,4-dimethylthiophene 1,1-dioxide isomerized and the adducts formed under these conditions have the structure of **70**.[12,52]

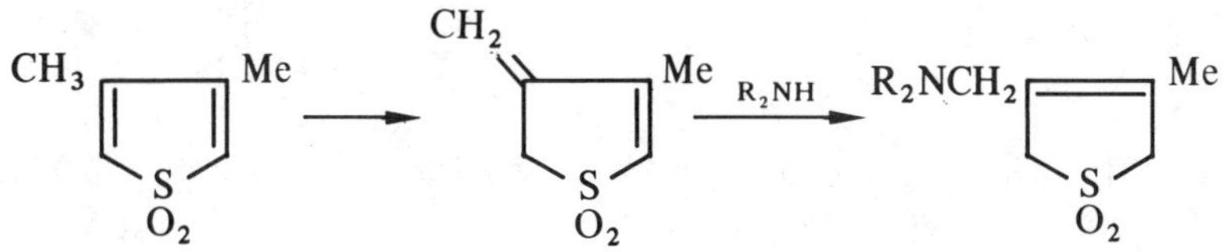

70

Dimethylamine has been added to thiophene 1,1-dioxide and its 3-methyl deriva-tive to form 3-dimethylamino-2,3-dihydrothiophene 1,1-dioxide[2] and its 4-methyl derivative,[12] respectively. 3,4-Dimethylthiophene 1,1-dioxide and ammonia gave a 6% yield of an adduct.[7] From 3,4-dibromo-2,5-dimethylthiophene 1,1-dioxide and piperidine, a compound believed to be 4-bromo-2,5-dihydro-2,5-dimethyl-2(or 5), 3-dipiperidinothiophene 1,1-dioxide was obtained.[7]

Sodium benzylmercaptide added to 3,4-dimethylthiophene 1,1-dioxide.[7] The reaction was represented as 2,5-addition to give 71, but the proof of structure does not rule out 72.

In the section on polymer chemistry, the addition of a polymeric thiol to 3,4-diphenylthiophene 1,1-dioxide is reported.[95]

This would indicate 2,5-addition, but no report of the addition of a monomeric thiol to 3,4-diphenylthiophene 1,1-dioxide to substantiate the reactions has appeared.

The reaction of 3-bromo-2-methyl-5-alkylthiophene 1,1-dioxides with organo-lithium reagents (Ph, n-Bu, t-Bu, Et, and Me) revealed two competing ring-opening reactions arising from nucleophilic organolithium attack at the 5-carbon and from bromine–lithium exchange:[151,152]

Substituting ethyl, isopropyl, n-butyl, and tert-butyl groups for the 5-methyl group altered the ratio of the two isomeric enynes in path a. The enyne with the larger group trans to the acetylene function was favored.

With 3-chloro-2,5-dimethylthiophene 1,1-dioxide and butyllithium as the reactants, chlorine–lithium exchange did not take place and the reaction followed path a.[152]

The reaction of 3-bromo-2,5-dimethyl-4-menthenylthiophene 1,1-dioxide with n-butyllithium, followed by methylation of the lithium sulfinate produced, proceeded by path b in 34% yield.[164]

E. Electrophilic Reactions

Bromine has been added to thiophene 1,1-dioxide in the 2,3-position.[43] Addition also took place with 2,5-di-$tert$-butyl-, 3,4-dibromo-2,5-di-$tert$-butyl-,[74] and 2,5-dibenzhydrylthiophene 1,1-dioxide.[75] The position of addition (2,3 or 2,5) was not established. Bromination of 3-thiabicyclo[3,2,0]hepta-1,4-diene 3,3-dioxide produced a tribromide.[158]

3,4-Dimethylthiophene 1,1-dioxide underwent 2,5-addition with nitrogen tetroxide.[76]

F. Reductions

The reduction of 3,4-dimethyl-,[7] 2,5-diphenyl-,[7] and tetraphenylthiophene 1,1-dioxides[7,15] with zinc in acetic and hydrochloric acids yielded the corresponding thiophenes. 3,4-Dibromo-2,5-diphenylthiophenes 1,1-dioxide gave 2,5-diphenylthiophene. However, the use of zinc and acetic acid alone on 3,4-dibromo-2,5-diphenyl-[7] and 3,4-dibromo-2,5-di-$tert$-butylthiophene 1,1-dioxide[74] replaced bromine with hydrogen without further reduction. Zinc and acetic acid or hydrochloric acid reduced 2,5-bis(4-nitrophenyl)-3,4-diphenylthiophene 1,1-dioxide to 2,5-bis(4-aminophenyl)-3,4-diphenylthiophene.[77] Zinc and acetic acid had no effect on tetraphenylthiophene 1,1-dioxide,[15] but reduced 3,4-diphenyl-[16] and 3,4-di-p-tolylthiophene 1,1-dioxides[78] to their 2,5-dihydro derivatives. The use of zinc in acetic and hydrochloric acids converted 2-carbomethoxy-3,4-diphenylthiophene 1,1-dioxide to its 2,5-dihydro derivative.[7] Zinc in acetic and hydrochloric acids did not reduce 2,5-di-benzhydrylthiophene 1,1-dioxide, but hydrogenation using Raney nickel as catalyst carried the reduction to 1,1,6,6-tetraphenylhex-ane.[75] Hydrogenation of 3,4-diphenylthiophene 1,1-dioxide using platinum oxide as catalyst produced, successively, the 2,5-dihydro- and 2,3,4,5-tetrahydro deriva-tives and 3,4-dicyclohexyltetrahydrothiophene 1,1-dioxide.[16] Tetraphenylthio-phene 1,1-dioxide was reduced to the thiophene by heating with selenium at 320°C.[79]

G.   Tricarbonyliron and Cyclopentadienylcobalt Complexes

In 1958, Weiss and Hübel found that 3,4-dimethyl- and tetraphenylthiophene 1,1-dioxide would react with iron pentacarbonyl at 140 and 170°C respectively, to form tricarbonyliron complexes (73).[80] An analogous complex has been made using 2,5-dimethylthiophene 1,1-dioxide and $Fe_3(CO)_{12}$.[41] The structure of the complex from 3,4-dimethylthiophene 1,1-dioxide has been established by a single crystal x-ray diffraction study.[55] From UV spectra and LCAO-MO theory, the π-electron systems of the ligands in these complexes appear to be only weakly perturbed by the interaction with the iron.[81]

73

These complexes can also be readily made by irradiating the reactants at room temperature in benzene solution. Indeed, this method was successful in preparing the stable tricarbonyliron complex of the unstable thiophene 1,1-dioxide itself.[3] Among the most intense ms peaks from this complex are those corresponding to the successive removal of the three CO and SO_2 groups, indicating that the iron π electron bonding system is very stable.[3,83] ESCA spectra of these complexes have also been reported.[82]

Irradiation of 3-chloro-2,5-dimethyl-, 3,4-dichloro-2,5-dimethyl- and 3,4-dibromo-2,5-dimethylthiophene 1,1-dioxide in the presence of iron pentacarbonyl formed tricarbonyliron complexes of these compounds plus complexes in which halogen was replaced by hydrogen. Thus, the tricarbonyliron complex of 2,5-dimethylthiophene 1,1-dioxide was produced in all three cases, and monohalo derivatives as well from dihalo compounds. The proton source may have been adventitious water.[83]

Photolysis in acetic acid of the tricarbonyliron complex of 2,5-dimethylthio-phene 1,1-dioxide resulted in regioselective reduction to give cis-2,5-dihydro-2,5-dimethylthiophene 1,1-dioxide.[84] Flash vacuum pyrolysis of the complex at 640–850° produced 2,5-dimethylthiophene, its dioxide, and 2,5-dimethylfuran.[41]

During decomposition of the complex by trimethylamine oxide in an aprotic solvent, an intermediate (74) in the decomplexation was isolated. Additional trimethylamine oxide released 2,5-dimethylthiophene 1,1-dioxide.[85]

$$(CO)_2Fe^- \overset{+}{-}\overset{}{N}HMe_2$$

74

The remarkable thermal stability of the cyclopentadienylcobalt unit led Drage and Vollhardt[153] to prepare complexes of this unit with thiophene 1,1-dioxides, and to flash pyrolyze them to cyclobutadiene complexes. Thus, cyclopentadienyl-cobalt 2,5-dimethylthiophene 1,1-dioxide was prepared in 84% yield by irradiating cyclopentadienyldicarbonylcobalt [CpCo(CO)$_2$] and 2,5-dimethylthiophene 1,1-dioxide in boiling benzene. This complex was flash pyrolyzed in vacuum to cyclopentadienylcobalt 1,2-dimethylcyclobutadiene (50%).

Analogous experiments were carried out with the methylcyclopentadienyl analogue and with 2,4-dimethyl-, 2-ethyl-5-methyl-, and 2-(2-butyl)-5-methylthiophene 1,1-dioxides.

H.   Other Reactions

The cyclic carbonate derivative 4 reacted with carboxylic acids in the presence of pyridine to produce effective acyl transferring agents (75) which can be adapted to the synthesis of peptides.[26,27,86] The product from pivalic acid formed an orange dicyclohexylamine salt whose structure might be represented as 76.[26] Compound 4 also underwent carbonate ring-opening additions to alcohols and amines. These products can be used to prepare urethanes, ureas, and isocyanates.[87] At 90°C, 4 dehydrated 4-chlorobenzylideneoxime to the nitrile.[27] Similar to the cyclic carbonate, the cyclic oxalate 6 can be employed in synthesizing 2-oxocarboxamides and peptides.[28]

$$\text{4} \xrightarrow{\text{RCOOH}} \text{75}$$

$$(C_6H_{11})_2\overset{\oplus}{N}H_2 \quad \text{76}$$

$$\text{6} \xrightarrow{\text{RCOOH}}$$

By employing the phenyl groups in styrene/divinylbenzene copolymer as one of the phenyl groups for structure **4**, a polymeric reagent has been made for carrying out the syntheses recorded for **4**[27] (see Section I.3).

The reduction of 2-(4-nitrophenyl)-3,4,5-triphenyl- and 2,5-bis(4-nitrophenyl)-3,4-diphenylthiophene 1,1-dioxides with sodium sulfide or stannous chloride to the amino compounds, followed by diazotization and coupling to 2-naphthols, yielded dyes[88] (see Table 2).

According to a Japanese patent, a catalytic amount of thiophene 1,1-dioxide brings about the decomposition of PhCH(Me)OOH in ethylbenzene to phenol in 90% yield at 150°C.[89]

3. Polymer Chemistry

The decomposition of 1-pentyne:sulfur dioxide 1:1 copolymer in refluxing dioxane produced a compound believed to be x,x-di-*n*-propylthiophene 1,1-dioxide.[90,91] This compound was copolymerized with isobutylene, using Al_2Br_5Cl as catalyst; no analysis of the product was given.[92]

The polymerization of styrene with benzoyl peroxide in the presence of 1–3% of 3,4-diphenylthiophene 1,1-dioxide or 3,4-bis(4-chlorophenyl)thiophene 1,1-dioxide

TABLE 2. THIOPHENE 1,1-DIOXIDES[a]

Substituents and Positions				Method of Synthesis	Yield (%)	mp (°C)	Reference[b]
2	3	4	5				
H	H	H	H	A			2, 3
H, ^{13}C	H	H	H, ^{13}C	A			72b
Me	H	H	H	A			5, 6
H	Me	H	H	D	Low		5, 6, 12
Me	Me	H	H	A			71
Me	H	Me	Me	E	52	89–90	18, 59
Me	H	Me	H	D	14	88.5–89.5	17
H	Me	Me	H	A	66	120–121	8, 52
Me	Me	Me	Me	D	63	114	17
Me	Me	H	Me	D	44	134.5–136	71
Me	Me	Me	H	D	Low		71
Et	H	H	H	D	61	112.8–113.8	71
H	Et	H	H	A			6
Et	H	H	i-Pr	A			5, 6
Me	H	H	sec-Bu	E	69	Oil	19, 153
Me	H	H	H	E	71	Oil	19
Me	H	H	H	E	25		153
H	—x,x-Di-n-propyl—		H	F	22	88	90, 91
Me	H	t-Bu	H	E	56	132–133	18
t-Bu	H	H	H	B		126–127	122
t-Bu	H	H	H	E	70	124–125	18
t-Bu	H	H	t-Bu	B		127.5–128.5	98
Me$_2$EtC	H	H	Me$_2$EtC	B		118.5–121	96, 97
t-Octyl	H	H	t-Octyl	E	0.4	Oil	98
MeOctylCH	H	H	H	B	65	127.5–128.5	157
H	—CH$_2$CH$_2$—		H	E	80	127.5–128.5	158

TABLE 2. *Continued*

Substituents and Positions				Method of Synthesis	Yield (%)	mp (°C)	Reference[b]
2	3	4	5				
Br		$-CH_2CH_2-$	Br	E	60	176.5–177.5	158
Ph		$-CH_2CH_2-$	Ph	E	40	244–246	161
H		$-CH(t\text{-}Bu)CH(t\text{-}Bu)-$	H	E	88	120–121	159
H		$-CH_2(o\text{-}C_6H_4)CHPh-$	H	E	30	82	159
H		$-(o\text{-}C_6H_4)-$	H	B	100	213–214	160
H	i-Propenyl	i-Pr	H	F		68–69	35
D	i-Propenyl	i-Pr	D	F			35
Br	i-Propenyl	i-Pr	Br	F			35
H	Cl	MeO	H				50
Me	Cl	H	Me	B	33	81–83	83
Me	Cl	Cl	Me	B	38	70–72	83
Me			Me	E	High		83
Me	Br	H	Me	E	53	102–104	151
Me	Br	H	Et	E	53	100–102	151
Me	Br	H	i-Pr	E	38	Oil	151
Me	Br	H	n-Bu	E	49	28–29	151
Me	Br	H	t-Bu	E	45	Oil	151
Me	Br	Menthenyl	Me	E	67	125–125.5	164
Br	Me	Br	Br	B	68	180.5–181	17
Me	Br	Me	Br	D			17
Et	Br	Br	Me	B	40	61–62	99
t-Bu	Br	Br	t-Bu	B	73	91	74
H	F	F	H	B			c
Cl	H	Cl	H	A	82	112–113	10
H	Cl	Cl	H	A			9,10
Cl	Cl	Cl	H	A			10
Cl	Cl	Cl	Cl	E	50	91–92	42
Br	Br	Br	Br	E	66	201–202	42
O_2NMe_2C	H	H	O_2NMe_2C	F	28	128–130	76

610

H	OH	Me	H	F	80	163	14
Me	OMe	Me	H	F		129	14
Ph_2CH	H	H	Ph_2CH	B	49	172–173	75
H	Ph	H	H	A			5, 6, 12
Ph	H	H	Ph	E	74	178–179	18
				D	40	178.5–179.5	17
				B	39		17
H	Ph	Ph	H	C	65	171.5–172.5	54
				B	32		16
				D	40	171	16
D	Ph	Ph	D	C	45	166–168	64
Ph	Ph	Ph	Ph	B	65	275–277	15, 123, 124
				F	20	286–287	38
Ph	$—1,2\text{-}C_6H_4CH_2—$		Ph	F	9.5	189–191	34
COOMe	$—1,2\text{-}C_6H_4C(O)—$		Ph	C	36	220	165
H	$4\text{-}MeC_6H_4$	$4\text{-}MeC_6H_4$	H	D	34	170	78
H	$4\text{-}ClC_6H_4$	$4\text{-}ClC_6H_4$	H	D	18	179	93
Ph	Cl	Cl	Ph	B		125	126
Ph	Cl	Cl	$4\text{-}MeC_6H_4$	B		123	126
Ph	Cl	Cl	$4\text{-}ClC_6H_4$	B		123–124	126
$4\text{-}MeC_6H_4$	Cl	Cl	$4\text{-}MeC_6H_4$	B		181–182	126
Br	Ph	Ph	H	D	11	157–157.5	17
Br	Br	Ph	Ph	B	14	185–186	17
Br	Ph	Br	Ph	B	33	140–141	17
$4\text{-}O_2NC_6H_4$	Ph	Ph	Ph	B		250	112, 113, 123
$4\text{-}O_2NC_6H_4$	$4\text{-}O_2NC_6H_4$	Ph	Ph	B	100	194	113
$4\text{-}O_2NC_6H_4$	Ph	Ph	$4\text{-}O_2NC_6H_4$	B	100	294	112, 113, 123
$4\text{-}BrC_6H_4$	Ph	Ph	$4\text{-}O_2NC_6H_4$	B		276–278	127
	2,3,4,5-Tetrakis($4\text{-}O_2NC_6H_4$)			B		315	127[d]
$4\text{-}H_2NC_6H_4$	Ph	Ph	Ph	F		256	77, 112
$4\text{-}H_2NC_6H_4$	Ph	Ph	$4\text{-}H_2NC_6H_4$	F		288	77, 112, 113
$4\text{-}MeOC_6H_4$	Ph	Ph	Ph	B		237	128
$4\text{-}BrC_6H_4$	Ph	Ph	$4\text{-}BrC_6H_4$	B		294–295	127

TABLE 2. *Continued*

	Substituents and Positions			Method of Synthesis	Yield (%)	mp (°C)	Reference[b]
2	3	4	5				
4-FC₆H₄	Ph	4-FC₆H₄	Ph	B	93	244.5–246.5	115
4-FC₆H₄	Ph	Ph	4-FC₆H₄	B		261–262.5	115
PhCO	Ph	Ph	H	D	49	198	17
PhCO	Ph	Ph	PhCO	D	43	244–245	17
COOMe	Ph	Ph	H	D	72	205–205.5	17
COOMe	Ph	Ph	PhCO	D	60	211.5–212.5	17
COOH	Ph	Ph	PhCO	D		186.5	17
COOMe	Ph	COOMe	Ph	C	16	166	165
Ph	COOEt	COOEt	Ph	F	2	234	39
Ph	NEt₂	Me	Ph	F		108–110	32, 33
4-ClC₆H₄	NEt₂	Me	4-ClC₆H₄	F		119–120	32
Ph	—NHCH₂CH₂NH—		Ph	F	45	224	25
Ph	O⁻B⁺	C₆H₁₁NH	Ph	F	55	204	25
Ph	O⁻B⁺	t-BuCOO	Ph	F		205–207	26
Ph	TsO	TsO	Ph	F	66	187	25
Ph	BzO	BzO	Ph	F	92	192.5–193.8	23
2-ClC₆H₄	BzO	BzO	2-ClC₆H₄	F		214–215	24
2,4-Cl₂C₆H₃	BzO	BzO	2,4-Cl₂C₆H₄	F	48	233–234	24
2,4,5-Cl₃C₆H₂	BzO	BzO	2,4,5-Cl₃C₆H₂	F		275–276	24
4-BrC₆H₄	BzO	BzO	4-BrC₆H₄	F		205–206	24
Ph		—OCOO—	Ph	F	73	> 250	26, 27
Ph		—OCSO—	Ph	F	44	200 dec	25
Ph		—OCOCOO—	Ph	F	96	247 dec	28
4-ClC₆H₄	MeO	MeO	Ph	F	76	148.3–149.3	30
4-RC₆H₄[e]	MeO	MeO	4-ClC₆H₄	F	79	207.5–208.5	30
4-R¹C₆H₄[e]	Ph	Ph	Ph	F		246	88
4-R²C₆H₄[e]	Ph	Ph	Ph	F		280	88
4-R²C₆H₄[e]	Ph	Ph	Ph	F		246	88
4-R³C₆H₄[e]	Ph	Ph	Ph	F		295	88

612

$4\text{-RC}_6\text{H}_4$[e]	Ph	$4\text{-RC}_6\text{H}_4$	F	> 330	88
$4\text{-R}^1\text{C}_6\text{H}_4$[e]	Ph	$4\text{-R}^1\text{C}_6\text{H}_4$	F	275	88
$4\text{-R}^2\text{C}_6\text{H}_4$[e]	Ph	$4\text{-R}^2\text{C}_6\text{H}_4$	F	235	88
$4\text{-R}^3\text{C}_6\text{H}_4$[e]	Ph	$4\text{-R}^3\text{C}_6\text{H}_4$	F	254	88
$4\text{-R}^4\text{C}_6\text{H}_4$[e]	Ph	$4\text{-R}^4\text{C}_6\text{H}_4$	F	224	88

[a] Methods of synthesis: (A) elimination of 2HX from tetrahydrothiophene 1,1-dioxide derivatives, (B) oxidation of thiophenes with hydrogen peroxide in acetic acid, (C) oxidation of thiophenes with peracetic acid, (D) oxidation of thiophenes with perbenzoic acid, (E) oxidation of thiophenes with 3-chloroperbenzoic acid, and (F) special synthesis (see text).

[b] References are to sources giving preparative information.

[c] This compound has Registry No. 695-69-2, but it does not appear in *Chemical Abstracts*, since the registry system started in 1965. An older patent contains an unexemplified disclosure of it (Ref. 10).

[d] Empirical formula and analysis are wrong.

[e]

R =

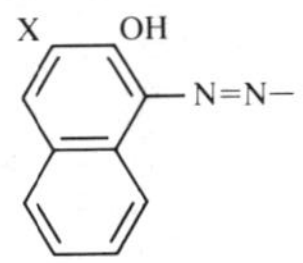

$X = H$; R^1, $X = PhNHCO$; R^2, $X = 4\text{-ClC}_6\text{H}_4\text{NHCO}$; R^3, $X = 1\text{-C}_{10}\text{H}_7\text{NHCO}$; R^4, $X = 4\text{-MeOC}_6\text{H}_4\text{NHCO}$.

caused a reduction in molecular weight and some retardation of the polymerization rate. Elemental analyses of an isolated polymer indicated a combined thiophene dioxide content of up to 7%.[93]

Polymers have been formed by a Diels–Alder reaction of 2,5-dimethyl- or 3,4-diphenylthiophene 1,1-dioxide with a bis-maleimide. The thiophene dioxide undergoes a Diels–Alder reaction with one of the maleimide moieties to produce an intermediate which can then self-condense in another Diels–Alder reaction to produce a polymer (Scheme 11).[94]

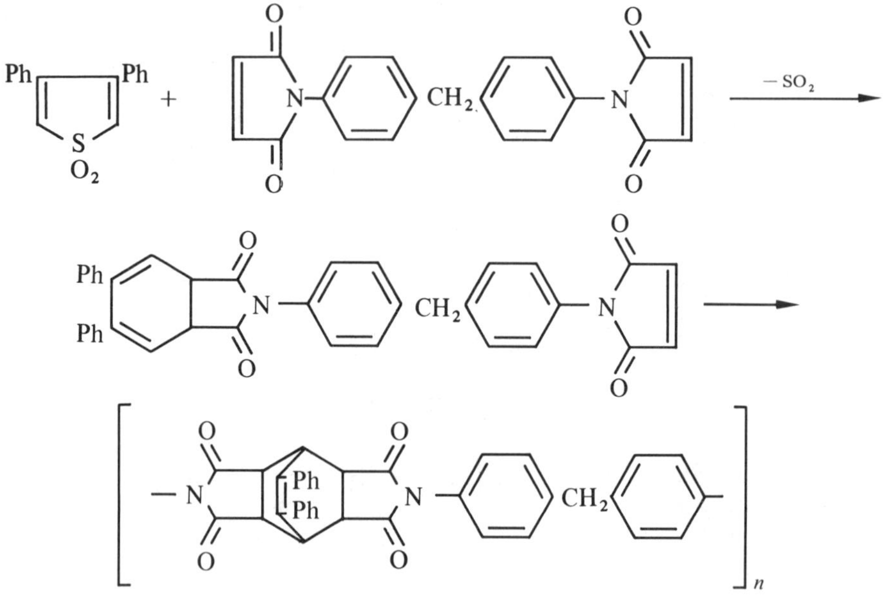

Scheme 11

A bis(thiophene 1,1-dioxide) was prepared in impure form, but dimerization made it unsuitable for use in Diels–Alder polymer reaction with diethynylbenzene.[65]

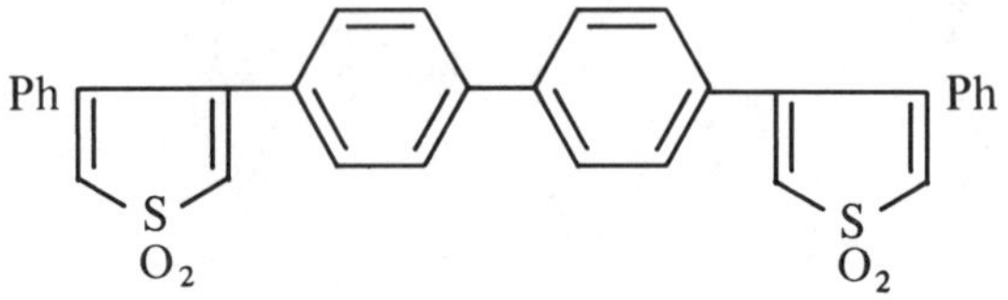

To prepare polymers with pendant 1,3-butadienyl groups that could serve as reversible acceptors for sulfur dioxide, divinylbenzene/styrene copolymer was first mercaptomethylated and the thiol groups were then added 2,5 to 3,4-dimethyl- and 3,4-diphenylthiophene 1,1-dioxides. Heat-induced or spontaneous (in the case of the diphenyl derivative) loss of sulfur dioxide from the products provided the 1,3-butadienyl groups (Scheme 12).[95]

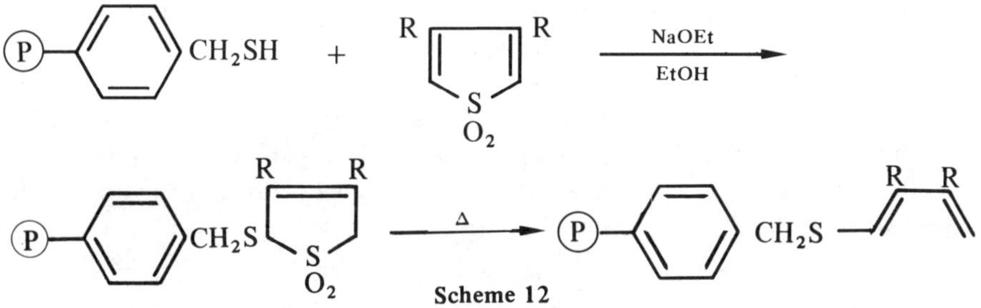

Scheme 12

Commercial Merrifield resin (chloromethylated styrene/2% divinylbenzene copolymer) has been converted to a 3,4-carbonyldioxy-1,5-diarylthiophene 1,1-dioxide (Scheme 13). This polymeric reagent reacts with carboxylic acids to produce acyl transferring agents useful in preparing peptides, as described for the monomeric compound[27] (see Section I.2.H).

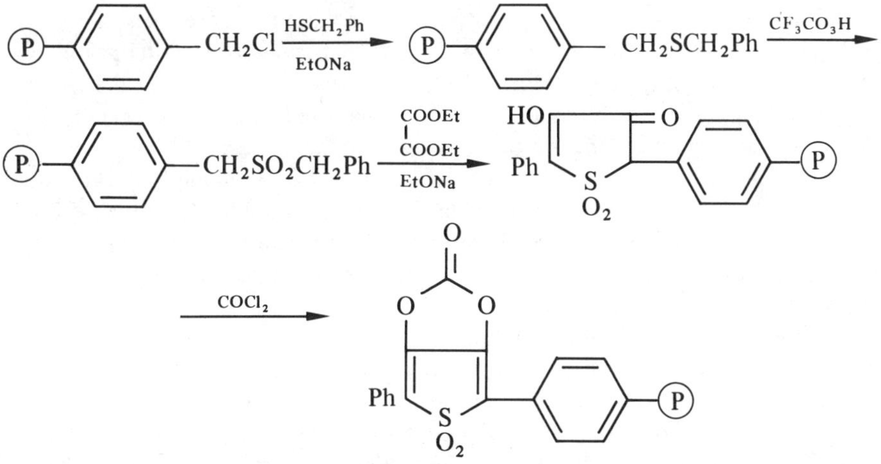

Scheme 13

4.   Thiophene 1,1-Dioxides in Petroleum Analysis

Thiophene dioxides have figured in proposed analytical procedures for petroleum to determine what thiophenes are present. The thiophene-containing fraction is oxidized and the thiophene dioxides are measured by polarographic and electron paramagnetic resonance methods. 2,5-Bis(1,1-dimethylpropyl)-, 2,5- and 2,4-di-*tert*-butyl-, and tetraphenylthiophene 1,1-dioxide were used in the study.[96,97] In an earlier evaluation of the method, measurements were made on 2,5-di-*tert*-butyl- and 2,5-di-*tert*-octylthiophene 1,1-dioxide and the polarographic wave was attributed to a 2-electron reduction.[98]

Gas-chromatographic separation of sulfones derived from petroleum has also been proposed, and 2,5-di-*tert*-butylthiophene 1,1-dioxide has been determined in mixtures of sulfones.[148] Substituted thiophene 1,1-dioxides have also been found in oxidized diesel fuel fractions by using mass spectroscopy.[154]

2-Ethyl-5-methylthiophene in petroleum was identified by conversion to the 3,4-dibromo derivative and oxidation to 3,4-dibromo-2-ethyl-5-methylthiophene 1,1-dioxide.[99]

5. Physical Studies

A. *Conjugation of the Sulfonyl Group*

The existence or nonexistence and extent of conjugation, hyperconjugation, and spiroconjugation involving the sulfonyl group in thiophene 1,1-dioxides have been the subject of many studies utilizing various techniques. The intent here is not to adjudicate the subject, but to call attention to the studies that have been made.

From an x-ray diffraction study of crystals of 2,5-di-*tert*-butylthiophene 1,1-dioxide, the conclusion was reached, from the observed alternation in the lengths of the ordinary and double bonds in the five-membered heterocycle, that there is no conjugation between the sulfonyl group and the system of short bonds.[100] However, photoelectron spectroscopy of 2,5-di-*tert*-butylthiophene 1,1-dioxide was interpreted to show that there is an unexpected, extraordinarily strong through-conjugation (by hyperconjugation) involving the sulfone group.[101] Raman and IR data for this compound reveal the loss in aromaticity, in comparison with the parent thiophene.[102]

Various molecular orbital calculations have been made on thiophene 1,1-dioxide itself.[103-108] In this molecule, the S=O bond was judged to be almost unaffected by mesomeric interactions.[103] Earlier calculations predicted strong conjugation with the sulfone group,[108, 109] but the citation of the acidity of 3,4-dihydroxy-thiophene 1,1-dioxide derivatives as experimental proof was infelicitous, as these compounds exist in the tautomeric, conjugated monoketo form. Diene character has been calculated to be greater in thiophene 1,1-dioxide than in thiophene 1-oxide,[107] but experimentally, thiophene 1-oxide is considered much more reactive as a diene than thiophene 1,1-dioxide.[59, 110]

Spiroconjugation has been suggested to be responsible for the high reactivity of thiophene 1,1-dioxide toward dimerization, as well as for the extraordinarily long wavelength absorption (λm 289 mμ).[46]

The IR and UV spectra of tetraphenylthiophene 1,1-dioxide and its 4-amino and 4-nitro derivatives were measured to identify chromophores. Conjugation was attributed to the S=O bond. The spectral evidence consisted of lowering the SO_2 stretching frequencies in the IR, and of shifting the absorption to longer wavelength in the UV.[111, 112] The compounds exhibited halochromism in concentrated sulfuric acid and in sodium methoxide solution, and the visible spectra were recorded.[111, 113]

Carbon-13 nmr data seem to indicate that the sulfonyl group does not take part in bond or electron delocalization[114] (see Section I.5.B).

B.   Nmr Data

Proton nmr data have been recorded for thiophene 1,1-dioxide[20] and its 2,5-dimethyl,[18] 3,4-dimethyl,[159] 2,3,4-trimethyl,[71] tetramethyl,[71] 2,5- and 2,4-di-tert-butyl,[18] 3-isopropenyl-4-isopropyl,[35] 2,5-diphenyl,[18] and 2-methyl-3-bromo-5-alkyl[151] derivatives. Values, including some ^{13}C data,[159, 163] have also been reported for a series of thiophene 1,1-dioxides in which positions 3 and 4 have been bridged by two carbon atoms.[157-162]

The ^{13}C chemical shifts and ^{13}C—H coupling for 2,4- and 3,4-dimethylthiophene 1,1-dioxide have been determined.[114, 159] A comparison between the data for thiophene derivatives and the thiophene dioxides seems to indicate that the sulfonyl group does not take part in bond or electron delocalization.[114]

C.   Mass Spectra

A mass spectral study of tetraphenyl-, 2,5-bis(4-fluorophenyl)-3,4-diphenyl-, and 2,4-bis(4-fluorophenyl)-3,5-diphenylthiophene 1,1-dioxide showed that a principal fragmentation was a loss of sulfur dioxide to form $Ar_4C_4^{+}$ which decomposed by halving to $C_2Ar_2^{+}$ and C_2Ar_2. The results from the fluorine-labeled molecules led to the conclusion that virtually no scrambling of groups takes place in the molecular ions, but does occur in the $Ar_4C_4^{+}$ ions.[115, 116]

Other aspects were considered in an investigation of the mass spectra of 2,5-dimethyl-, 2,5-diphenyl-, and 3,4-diphenylthiophene 1,1-dioxide. Elimination of SO presumably proceeds through rearrangement to the six-membered cyclic sultine. Loss of SO to form a furan ion is favored by the product ion stability. There was a low abundance of $M - SO_2$ ions.[117] In the case of 3-chloro-2,5-dimethyl-, 3,4-dichloro-2,5-dimethyl-, and 3,4-dibromo-2,5-dimethylthiophene 1,1-dioxide, no $M - SO_2$ ions were observed; initial ejection of SO took place.[83]

D.   Other Measurements

The luminescence spectrum of tetraphenylthiophene 1,1-dioxide has been determined under irradiation with UV light. Luminescence is more intense and occurs at a longer wavelength than for tetraphenylthiophene.[119]

3,4-Dimethylthiophene 1,1-dioxide exhibits solvent interaction (hydrogen bonding) with phenol and with chloroform.[118] Hydrogen bonding between thiophene 1,1-dioxide and 4-nitrophenol lowers the OH stretching frequency by 180 cm^{-1}.[104]

Dipole moments for tetraphenylthiophene 1,1-dioxide have been recorded: 5.3 ± 0.1 D (dioxane),[120] 5.33 D (dioxane), and 4.50 D (CHCl$_3$).[121]

II. THIOPHENE SESQUIOXIDES

Oxidation of thiophene with hydrogen peroxide in acetic acid provides, in 15% yield, a compound that has been given the trivial name thiophene sesquioxide (**77**), because it combines 2 moles of thiophene and 3 atoms of oxygen.[129-131] The compound may arise by the dimerization of thiophene 1-monoxide (this dimer has been reported[132]) followed by oxidation of the 1-sulfoxide group, or by a reaction between thiophene 1-monoxide and thiophene 1,1-dioxide. Either route involves a Diels–Alder reaction in which thiophene 1-monoxide functions as the diene. The structure of **77** was established as *syn,endo*-3a,4,7,7a-tetrahydro-4,7-epithiobenzo-[*b*]thiophene 1,1,8-trioxide by nmr analysis.[133]

The compound was also made using thiophene-2,5-d$_2$. In other work, structure **78** has been proposed for 2-methylthiophene sesquioxide.[18] Polarographic measurements have been made on thiophene sesquioxide.[97]

Perbenzoic acid has been used to prepare sesquioxides from thiophene (11%), 2-methylthiophene (17%), 3-methylthiophene (–), 3,4-dimethylthiophene (8%), and 3-phenylthiophene (32%).[134]

Chlorination of thiophene sesquioxide produced a compound (mp 51–52°C) that was assumed to be the tetrachloro adduct.[131]

III. THIOPHENE 1-MONOXIDES

1. Synthesis and Reactions

Thiophene 1-monoxides represent an insufficiently investigated class of compounds. Unstable thiophene 1-oxide (**80**) has been reported to be formed in 0.001 *M* solution by reacting *trans*-3,4-dimethylsulfonyloxytetrahydrothiophene 1-oxide (**79**) with sodium methoxide in methanol.[132]

Evidence adduced for the presence of **80** was UV absorption maxima at 215–216 and 263–265 nm, reduction to thiophene with sodium borohydride, and Diels–Alder dimerization to **81** in 15% yield when the reaction was run in concentrated solution. Evidence for the dimer was a UV maximum at 215 nm and the mass spectrum with peaks for the parent, $C_8H_8^+$, and $C_8H_7^+$. The extinction value for thiophene 1-oxide at 263 nm decreased about 53% in 14 min. This could be attributed to dimerization, unless reaction with methanol was also taking place. Despite the evidence, its adequacy for the presence of thiophene 1-oxide has been questioned, with the suggestion that the UV spectrum might be caused by 2,3-dihydro-3-methanesulfonyloxythiophene 1-oxide.[110]

When thiophene was oxidized at 0°C with 3-chloroperbenzoic acid in the presence of benzoquinone, the reactive thiophene 1-oxide generated *in situ* formed a Diels–Alder adduct (**82**) with the benzoquinone before the 1-oxide could be oxidized to the 1,1-dioxide. Naphthoquinone (**83**) and juglone (**84**) were also

formed. Similar results were obtained from 2-methyl-, 3-methyl-, 2,5-dimethyl-, 2,5-dichloro-, and 2,5-diphenylthiophene. 2,5-Dimethylthiophene gave a 33% yield of the adduct corresponding to **82** in the reaction given above at 0°C, but 2,5-dimethylthiophene 1,1-dioxide added only slowly to benzoquinone in refluxing chloroform. This illustrates the higher reactivity of the 1-oxide compared with the 1,1-dioxide. Reactions using napthoquinone instead of benzoquinone gave anthraquinones. Adducts with the sulfoxide bridge were formed, but could not be isolated in a pure state.[59]

Thiophene 1-oxide is presumed to have been formed by dehydrogenation of 2,3-dihydrothiophene 1-oxide, but was not isolated.[135]

The tricarbonyliron complex of thiophene 1-oxide has been reported without preparative information.[82]

As with thiophene 1,1-dioxide, substituted thiophene 1-oxides have greater stability than the parent molecule. By oxidation of 2,5-di-*tert*-butylthiophene with 1 eq. 3-chloroperbenzoic acid, a 5% yield of the stable 1-oxide was obtained along

with the 1,1-dioxide. 2,5-Bis(1,1,3,3-tetramethylbutyl)thiophene 1-oxide was prepared in the same way with unspecified yield.[110]

From the oxidation of benzo[3,4]cyclobuta[1,2-c]thiophene with 2 moles
hydrogen peroxide in acetic acid for 48 h, a 16% yield of the relatively unstable
sulfoxide was isolated (mp 143–145°C) and a 22% yield of the dioxide. The rate
of oxidation of the sulfoxide appeared to be similar to that of the thiophene.[156]

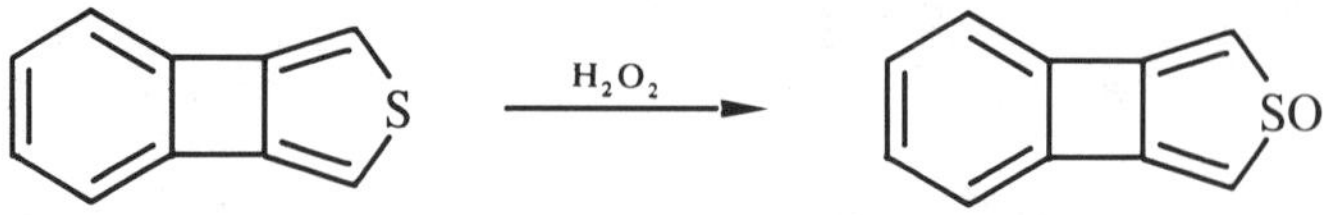

According to a patent assigned to Merck & Co., 2-chlorothiophene 1-oxide was
made by heating 0.01 M amounts of 2-chlorothiophene and 3-chloroperbenzoic
acids in 100 ml of refluxing chloroform for 30 min. The cold solution was filtered,
washed with aqueous solution of pH 8, and evaporated to leave the 1-oxide,
which was not characterized, but was used in the following reaction to prepare
a cephalosporin antibiotic (85)[136] (Scheme 14).

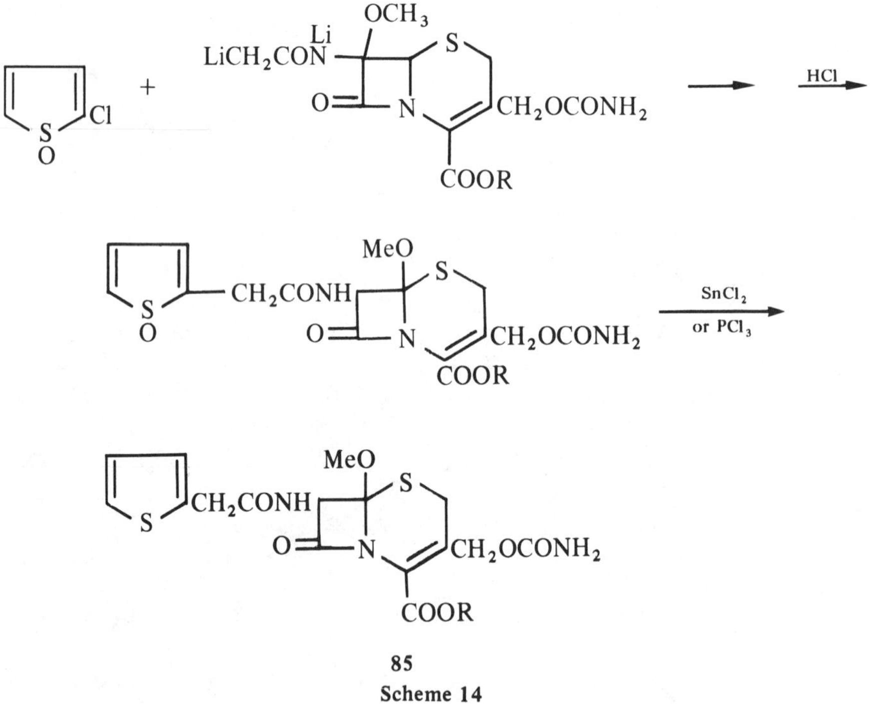

85

Scheme 14

A second patent relates to a Diels–Alder reaction between 2,3-dichloro-3-
methoxythiophene 1-oxide and 5-cyclopentyl-5-methyl-2-cyclopenten-1-one, but no
information on preparation or properties of the thiophene 1-oxide is divulged.[137]

Condensation of diethyl sulfinyldiacetate with diethyl oxalate has been reported to yield 2,5-dicarbethoxy-3,4-dihydroxythiophene 1-oxide (87).[22] Actually, the product probably has the structure of 88, but might be converted to derivatives of 87 by acylation or alkylation, as has been done for its sulfone analogue (Section I.1.C).

87

88

Oxidation of perfluorotetramethyl (Dewar thiophene) with peroxytrifluoroacetic yielded the 1-oxide (89). This compound is remarkable, because automerization causes all the trifluoromethyl groups to appear equivalent in the nmr spectrum above −100°. In effect, the SO group and the double bond migrate around the cyclobutene ring. The compound formed a Diels–Alder adduct (90) with furan, but attempts to oxidize 89 to the sulfone were unsuccessful.[138, 139, 147]

89

90

2. Physical Studies

Modified neglect of diatomic overlap (MNDO) calculations on thiophene 1-oxide gave figures for the optimized molecular geometry, and a value of 75.2 kcal/mol for the heat of formation. The calculated enthalpy of activation for inversion is greater than 30 kcal/mol, with the transition state for inversion regarded as planar.[140] A linear combination of Gaussian orbital calculations gave values of 3.83–10.5 kcal/mol for the pyrimidal inversion barrier for thiophene 1-oxide, with the higher range more likely. Diene character and tendency toward Diels–Alder additions was calculated to be less for thiophene 1-oxide than for thiophene 1,1-dioxide.[107] Experimentally, however, thiophene 1-oxide is much more reactive as a diene than thiophene 1,1-dioxide.[59, 110] The effect of steric factors should be considered in this comparison. From *ab initio* computations, a value of 19.6 kcal/mol was obtained for the inversion barrier for thiophene 1-oxide.[148] Using CNDO/2 calculations, barriers in kcal/mol for pyramidal inversion were found to be 24.5 for thiophene 1-oxide[106] and as indicated for the following molecules.[141, 142]

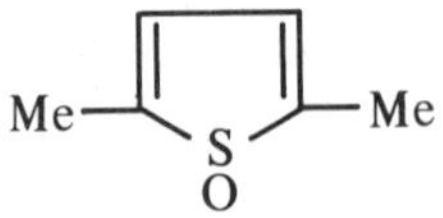

Me—⬠—Me
S
O
13.3

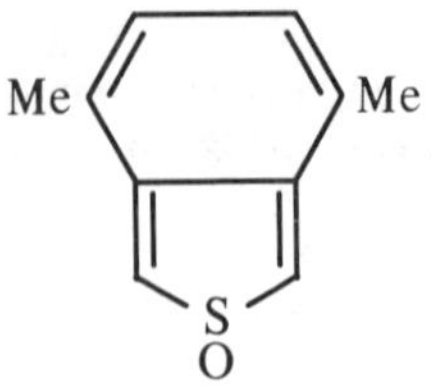

Me⬠Me
S
O
2.2

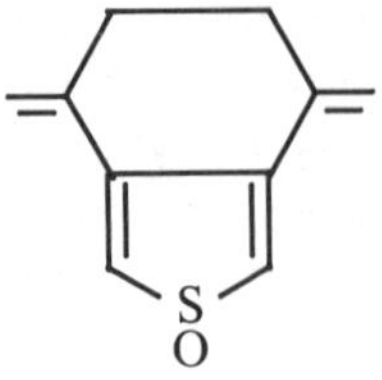

⬠
S
O
9.6

Spectroscopic evidence suggests that the sulfur atom in 2,5-di-*tert*-butylthiophene 1-oxide is also tetrahedrally hybridized. The sulfur–oxygen stretch in the infrared occurs at $9.5\,\mu$, a normal value. The nmr spectrum of 2,5-bis(1,1,3,3-tetramethyl-butyl)thiophene 1-oxide (**91**) is uniquely informative. At $-10°$ the geminal methylene protons in the side chain are magnetically nonequivalent. The source of the anisotropy is believed to be the sulfoxide functionality, which therefore must be pyramidal. The quartet in the nmr spectrum becomes a singlet above $60°$, and this is attributed to inversion of the sulfoxide function, which is proceeding through a planar transition state. From the nmr behavior, the derived free energy of activation for the inversion is 14.8 kcal/mol, which is also a reasonable estimate for the enthalpy of activation.[110]

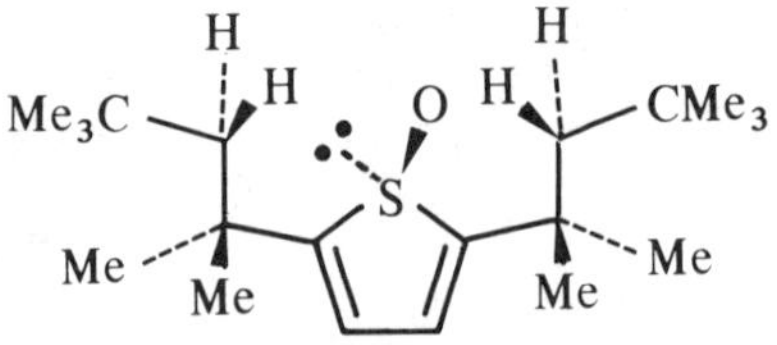

Me_3C —⟍⟋— H O H —⟍⟋— CMe_3
Me   S   Me
        Me   Me

91

ACKNOWLEDGMENT

I am indebted to the staff and facilities of the Technical Information Division, Du Pont Central Research and Development Department, for making this review possible.

REFERENCES

1. Contribution No. 4054.
2. W. J. Bailey and E. W. Cummins, *J. Am. Chem. Soc.*, 76, 1932 (1954).
3. Y. L. Chow, J. Fossey, and R. A. Perry, *J. Chem. Soc., Chem. Commun.*, 501 (1972).
4. D. Copland, D. Leaver, and W. B. Menzies, *Tetrahedron Lett.*, 639 (1977).
5. S. E. Reiter, L. C. Dunn, and K. N. Houk, *J. Am. Chem. Soc.*, 99, 4199 (1977).
6. S. E. Reiter, *Diss. Abstr. Int. B.*, 38, 5946 (1978) (with K. N. Houk).

7. J. L. Melles, *Recl. Trav. Chim. Pay-Bas,* **71,** 869 (1952).

8. W. Davies and Q. N. Porter, *J. Chem. Soc.,* 459 (1957).

9. H. Bluestone, R. Bimber, R. Berkey, and Z. Mandel, *J. Org. Chem.,* **26,** 346 (1961).

10. H. Bluestone, U.S. Patent No. 2,976,297 (1961); *Chem. Abstr.,* **55,** 16576d (1961); British Patent No. 885,252 (1961). These patents also contain unexemplified disclosures of 3,4-difluoro-; 3-chloro-; 2,3- and 2,5-dichloro-; tetrachloro-; 3-bromo-; 2,5- and 3,4-dibromo-; and 2,3,4-tribromothiophene 1,1-dioxides.

11. H. Bluestone, U.S. Patent No. 3,073,691 (1963); *Chem. Abstr.,* **59,** 576b (1963).

12. J. P. Walsh, *Diss. Abstr. Int. B.,* **31,** 1846 (1970) (with L. F. Hatch).

13. J. F. Alderman and P. H. Schuldt, U.S. Patent No. 3,806,598 (1974); *Chem. Abstr.,* **81,** 114885v (1974); French Patent No. 2,232,994 (1975); German Patent No. 2,248,467 (1975); Brazilian Patent No. 74/04791 (1975); South African Patent No. 74/3757 (1975).

14. H. J. Backer and J. Strating, *Recl. Trav. Chim. Pay-Bas,* **54,** 170 (1935).

15. O. Hinsberg, *Ber. Dtsch. Chem. Ges.,* **48,** 1611 (1915).

16. H. J. Backer, C. C. Bolt, and W. Stevens, *Recl. Trav. Chim. Pay-Bas,* **56,** 1063 (1937).

17. J. L. Melles and H. J. Backer, *Recl. Trav. Chim. Pay-Bas,* **72,** 314 (1953).

18. W. J. M. van Tilborg, *Synth. Commun.,* **6,** 583 (1976).

19. D. Mukherjee, L. C. Dunn, and K. N. Houk, *J. Am. Chem. Soc.,* **101,** 251 (1979).

20. W. J. M. van Tilborg, P. Smael, J. P. Visser, C. G. Kouwenhoven, and R. N. Reinhoudt, *Recl. Trav. Chim. Pay-Bas,* **94,** 85 (1975).

21. M. Chaykovsky, M. H. Lin, and A. Rosowsky, *J. Org. Chem.,* **37,** 2018 (1972).

22. R. H. Eastman and R. M. Wagner, *J. Am. Chem. Soc.,* **71,** 4089 (1949).

23. C. G. Overberger, S. P. Ligthelm, and E. A. Swire, *J. Am. Chem. Soc.,* **72,** 2856 (1950).

24. C. G. Overberger, R. A. Gadea, J. A. Smith, and I. C. Kogon, *J. Am. Chem. Soc.,* **75,** 2075 (1953).

25. W. Ried, O. Bellinger, and G. Oremek, *Chem. Ber.,* **113,** 750 (1980).

26. O. Hollitzer, A. Seewald, and W. Steglich, *Angew. Chem.,* **88,** 480 (1976); *Angew. Chem., Int. Ed. Engl.,* **15,** 444 (1976).

27. W. Steglich, O. Hollitzer, and A. Seewald, German Offen. Patent No. 2,625,539 (1977); *Chem. Abstr.,* **89,** 44240v (1978); U.S. Patent No. 4,122,090 (1978); British Patent No. 1,578,963 (1980); Canadian Patent No. 1,106,388 (1981); 1,110,385 (1981); *Chem. Abstr.,* **96,** 218227h (1982); French Patent No. 2,353,555 (1982); Swiss Patent No. 636,877 (1983).

28. W. Steglich, H. Schmidt, and O. Hollitzer, *Synthesis,* 622 (1978).

29. C. G. Overberger and J. M. Hoyt, *J. Am. Chem. Soc.,* **73,** 3957 (1951).

30. C. G. Overberger and J. M. Hoyt, *J. Am. Chem. Soc.,* **73,** 3305 (1951).

31. H. J. Backer, N. Dost, and J. Knotnerus, *Recl. Trav. Chim. Pay-Bas,* **68,** 237 (1949).

32. M. H. Rosen and H. M. Blatter, U.S. Patent No. 3,706,769 (1972); *Chem. Abstr.,* **78,** 97475e (1973).

33. M. H. Rosen and H. M. Blatter, U.S. Patent No. 3,629,437 (1971); *Chem. Abstr.,* **76,** 144849a (1972).

34. M. H. Rosen and G. Bonet, *J. Org. Chem.,* **39,** 3805 (1974).

35. S. Braverman and D. Segev, *J. Am. Chem. Soc.,* **96,** 1245 (1974).

36. V. M. Berestovitskaya, M. V. Titova, and V. V. Perekalin, *Zh. Org. Khim.,* **13,** 2454 (1977); *J. Org. Chem. USSR,* **13,** 2283 (1977).

37. M. V. Titova, V. M. Berestovitskaya, and V. V. Perekalin, *Metody Sint., Str. Khim. Prevrashch. Nitrosoedin., Gertsenovskie Chteniya, 31st,* 36 (1978) through *Chem. Abstr.,* **91,** 39231e (1979); **90,** 186705v (1979).

624 M. S. Raasch

38. E. H. Braye, W. Hübel, and I. Caplier, *J. Am. Chem. Soc.*, **83**, 4406 (1961).
39. W. Ried and W. Ochs, *Chem. Ber.*, **107**, 1334 (1974).
40. J. F. W. McOmie and B. K. Bullimore, *J. Chem. Soc., Chem. Commun.*, 63 (1965).
41. W. J. M. van Tilborg and R. Plomp, *Recl. Trav. Chim. Pay-Bas*, **96**, 282 (1977).
42. M. S. Raasch, *J. Org. Chem.*, **45**, 856 (1980).
43. W. J. Bailey and E. W. Cummins, *J. Am. Chem. Soc.*, **76**, 1936 (1954).
44. H. J. Backer and J. L. Melles, *Proc. Koninkl. Nederland Akad. Wetenschap.*, **54B**, 340 (1951) (in English); *Chem. Abstr.*, **47**, 6932b (1953).
45. E. W. Garbisch, Jr. and R. F. Sprecher, *J. Am. Chem. Soc.*, **88**, 3434 (1966).
46. H. E. Simmons and T. Fukunaga, *J. Am. Chem. Soc.*, **89**, 5208 (1967).
47. M. Prochazka and V. Horak, *Collect. Czech. Chem. Commun.*, **24**, 2278 (1957) (in German).
48. J. E. Mahan, U.S. Patent No. 2,786,851 (1957); *Chem. Abstr.*, **51**, 12146i (1957).
49. M. Prochazka and V. Horak, *Collect. Czech. Chem. Commun.*, **24**, 1509 (1959) (in German).
50. R. T. Patterson, *Diss. Abstr. Int. B*, **41**, 204 (1980) (with K. N. Houk).
51. K. N. Houk and R. T. Patterson, Abstracts, 181st National Meeting of the American Chemical Society, Atlanta, Ga., March 1981, No. O129.
52. J. T. Wrobel and K. Kabzinska, *Bull. Acad. Pol. Sci., Ser. Sci. Chim.*, **22**, 129 (1974) (in English).
53. W. J. M. van Tilborg and P. Smael, *Recl. Trav. Chim. Pay-Bas*, **95**, 132 (1976).
54. C. G. Overberger and J. M. Whelan, *J. Org. Chem.*, **26**, 4328 (1961).
55. K. Hoffmann and E. Weiss, *J. Organomet. Chem.*, **128**, 389 (1977).
56. J. Martel, *Compt. Rend.*, **244**, 626 (1957).
57. P. W. Lert and C. Trindle, *J. Am. Chem. Soc.*, **93**, 6392 (1971).
58. W. J. Bailey and E. W. Cummins, *J. Am. Chem. Soc.*, **76**, 1940 (1954).
59. K. Torssell, *Acta Chem. Scand.*, **B30**, 353 (1976).
60. E. W. Duck, *Res. Corresp., Suppl. Res. (London)*, **8**, 547 (1955).
61. J. M. Whelan, Jr., *Diss. Abstr.*, **20**, 1180 (1959) (with C. G. Overberger).
62. D. W. Jones, *J. Chem. Soc., Perkin I*, 1951 (1973).
63. C. G. Overberger and J. M. Whelan, *J. Org. Chem.*, **24**, 1155 (1959).
64. R. G. Nelb II and J. K. Stille, *J. Am. Chem. Soc.*, **98**, 2834 (1976).
65. J. K. Stille and R. G. Nelb, U.S. National Technical Information Service, Report No. AD-781546 (1974); *Chem. Abstr.*, **82**, 58165s (1975).
66. R. M. Bimber, U.S. Patent No. 3,110,739 (1963); *Chem. Abstr.*, **60**, 2870b (1964).
67. H. Bluestone, U.S. Patent No. 3,066,153 (1962); *Chem. Abstr.*, **58**, 9002e (1963).
68. M. S. Raasch and B. E. Smart, *J. Am. Chem. Soc.*, **101**, 7733 (1979).
69. M. S. Raasch, *J. Org. Chem.*, **45**, 867 (1980).
70. D. N. Reinhoudt, P. Smael, W. J. M. van Tilborg, and J. P. Visser, *Tetrahedron Lett.*, 3755 (1973).
71. K. Takeuchi, Y. Yokomichi, T. Kurosaki, Y. Kimura, and K. Okamoto, *Tetrahedron*, **35**, 949 (1979).
72. (a) J. Becker, C. Wentrup, E. Katz, and K.-P. Zeller, *J. Am. Chem. Soc.*, **102**, 5110 (1980); (b) K.-P. Zeller and S. Berger, *Z. Naturforsch.*, **36B**, 858 (1981).
73. K. Kabzinska and J. T. Wrobel, *Bull. Acad. Pol. Sci., Ser. Sci. Chim.*, **22**, 843 (1974) (in English).
74. Ya. L. Gol'dfarb and M. L. Kirmalova, *Dokl. Akad. Nauk. SSSR*, **91**, 539 (1953); *Chem. Abstr.*, **48**, 10723f (1954).

75. Ya. L. Gol'dfarb and M. S. Kondakova, *Izv. Akad. Nauk. SSSR, Ser. Khim.,* 1131 (1952); *Bull. Acad. Sci. USSR, Div. Chem. Sci.,* 993 (1952).

76. M. V. Titova, V. M. Berestovitskaya, and V. V. Perekalin, *Zh. Org. Khim.,* **15,** 877 (1979); *J. Org. Chem. USSR,* **15,** 786 (1979); *Metody Sint., Str. Khim. Prevrashch. Nitrosoedin.,* 37 (1980); *Chem. Abstr.,* **97,** 92065d (1982).

77. L. Fortina and G. Montaudo, *Ann. Chim. (Rome),* **50,** 451 (1960).

78. H. J. Backer, W. Stevens, and J. R. van der Bij, *Recl. Trav. Chim. Pay-Bas,* **59,** 1141 (1940).

79. L. Fortina and G. Montaudo, *Ann. Chim. (Rome),* **50,** 1401 (1960).

80. E. Weiss and W. Hübel, *J. Inorg. Nucl. Chem.,* **11,** 42 (1959); K. W. Hübel and E. L. Weiss, Belgian Patent No. 574,524 (1959); *Chem. Abstr.,* **54,** 8730a (1960); British Patent No. 913,763 (1962); *Chem. Abstr.,* **59,** 11568g (1963); U.S. Patent No. 3,260,730 (1966).

81. G. N. Schranzer and G. Kratel, *J. Organomet. Chem.,* **2,** 336 (1964).

82. J. H. Eekhof, H. Hogeveen, R. M. Kellog, and G. A. Sawatsky, *J. Organomet. Chem.,* **111,** 349 (1976).

83. V. Usieli, S. Gronowitz, and I. Andersson, *J. Organomet. Chem.,* **165,** 357 (1979).

84. M. Frank-Neumann, D. Martina, and F. Brion, *Angew. Chem.,* **90,** 736 (1978); *Angew. Chem., Int. Ed. Engl.,* **17,** 690 (1978).

85. J. H. Eekhof, H. Hogeveen, and R. M. Kellog, *J. Chem. Soc., Chem. Commun.,* 657 (1976).

86. G. Schnorrenberg and W. Steglich, *Angew. Chem.,* **91,** 326 (1979); *Angew. Chem., Int. Ed. Engl.,* **18,** 307 (1979).

87. H. Schmidt, O. Hollitzer, A. Seewald, and W. Steglich, *Chem. Ber.,* **112,** 727 (1979).

88. L. Fortina and G. Montaudo, *Ann. Chim. (Rome),* **51,** 95 (1961).

89. T. Yamahara, T. Takano, and M. Tamura, Japanese Kokai 52/113,928 (1977); *Chem. Abstr.,* **88,** 50482w (1978).

90. L. L. Ryden and C. S. Marvel, *J. Am. Chem. Soc.,* **58,** 2047 (1936).

91. C. S. Marvel and W. W. Williams, *J. Am. Chem. Soc.,* **61,** 2710 (1939).

92. D. W. Young, U.S. Patent No. 2,456,354 (1948); *Chem. Abstr.,* **43,** 3651b (1949).

93. C. G. Overberger, H. J. Mallon, and R. Fine, *J. Am. Chem. Soc.,* **72,** 4958 (1950).

94. S.-W. Chow and J. M. Whelan, Jr., U.S. Patent No. 2,971,944 (1961); *Chem. Abstr.,* **55,** 12941f (1961).

95. T. J. Nieuwstad, A. P. G. Kieboom, A. J. Breijer, J. van der Linden, and H. van Bekkum, *Recl. Trav. Chim. Pay-Bas,* **95,** 225 (1976).

96. F. N. Mazitova, N. A. Iglamova, and G. V. Dmitrieva, *Khim. Tekhnol. Topl. Masel,* 13 (1976); *Chem. Abstr.,* **85,** 200047n (1976).

97. N. A. Iglamova, F. N. Mazitova, A. A. Vatina, and A. V. Il'yasov, *Neftekhimiya,* **19,** 264 (1979); *Chem. Abstr.,* **91,** 76551e (1979).

98. H. V. Drushel and J. F. Miller, *Anal. Chem.,* **30,** 1271 (1958).

99. Ya. L. Gol'dfarb, G. I. Gorushkina, and B. P. Fedorov, *Izv. Akad. Nauk SSSR, Ser. Khim.,* 340 (1956); *Bull. Acad. Sci. USSR, Div. Chem. Sci.,* 327 (1956).

100. L. G. Vorontsova, *Zh. Strukt. Khim.,* **7,** 240 (1966); *J. Struct. Chem. USSR,* **7,** 234 (1966).

101. C. Müller, A. Schweig, and W. L. Mock, *J. Am. Chem. Soc.,* **96,** 280 (1974).

102. V. T. Aleksanyan, Ya. M. Kumel'fel'd, S. M. Shostakovskii, and A. I. L'vov, *Zh. Prikl. Spektrosc., Akad. Nauk. Belorussk. SSR,* **3,** 355 (1965); *Chem. Abstr.,* **64,** 10613a (1966).

103. F. de Jong and M. Janssen, *J. Chem. Soc., Perkin II,* 572 (1972).

104. F. de Jong and M. J. Janssen, *Recl. Trav. Chim. Pay-Bas,* **92,** 1073 (1973).

105. F. de Jong, A. J. Noorduin, T. Bouwan, and M. J. Janssen, *Tetrahedron Lett.,* 1209 (1974).

106. D. T. Clark, *Int. J. Sulfur Chem., Part C,* 7, 26 (1972).

107. M. H. Palmer and R. H. Findlay, *J. Chem. Soc., Perkin II,* 1223 (1975).

108. H. P. Koch and W. E. Moffitt, *Trans. Faraday Soc.,* 47, 7 (1951).

109. H. P. Koch, *J. Chem. Soc.,* 408 (1949).

110. W. L. Mock, *J. Am. Chem. Soc.,* **92,** 7610 (1970).

111. L. Fortina and G. Montaudo, *Gazz. Chim. Ital.,* **90,** 987 (1960).

112. N. Marziano and G. Montaudo, *Gazz. Chim. Ital.,* **91,** 587 (1961).

113. W. Dilthey and E. Graef, *J. Prakt. Chem.,* **151** (2), 257 (1938).

114. K. Kabzinska, *Bull. Acad. Pol. Sci., Ser. Sci. Chim.,* **24,** 363 (1976) (in English).

115. M. M. Bursey, T. A. Elwood, and P. F. Rogerson, *Tetrahedron,* **25,** 605 (1969).

116. M. K. Hoffman, T. A. Elwood, P. R. Rogerson, J. M. Tesarek, M. M. Bursey, and D. Rosenthal, *Org. Mass. Spectrom.,* 3, 891 (1970).

117. P. Vouros, *J. Heterocycl. Chem.,* **12,** 21 (1975).

118. K. Kabzinska and J. T. Wrobel, *Bull. Acad. Pol. Sci., Ser. Sci. Chim.,* **22,** 181 (1974) (in English).

119. Yu. M. Vinetskaya, M. G. Voronkov, B. M. Krasovitskii, and V. Udre, *Khim. Geterotsikl. Soedin.,* **4,** 180 (1968); *Chem. Heterocycl. Compd. USSR,* **4,** 139 (1968).

120. H. Lumbroso and G. Montaudo, *Bull. Soc. Chim. Fr.,* 2119 (1964).

121. M. G. Gruntfest, Yu. V. Kolodyazhnyi, V. Udre, M. G. Voronkov, and O. A. Osipov, *Khim. Geterosikl. Soedin.,* **6,** 448 (1970); *Chem. Heterocycl. Compd. USSR,* **6,** 413 (1970).

122. Ya. L. Gol'dfarb and I. S. Korsakova, *Dokl. Akad. Nauk. SSSR,* **89,** 301 (1953); *Chem. Abstr.,* **48,** 7598b (1954).

123. L. Fortina, *Ann. Chim. (Rome),* **49,** 2047 (1959).

124. M. G. Voronkov and V. Udre, *Khim. Geterotsikl. Soedin.,* **2,** 527 (1966); *Chem. Heterocycl. Compd. USSR,* **2,** 394 (1966).

125. M. G. Voronkov and V. Udre, *Khim. Geterosikl. Soedin.,* **1,** 683 (1965); *Chem. Heterocycl. Compd. USSR,* **1,** 458 (1965).

126. K. E. Schulte, H. Walker, and L. Rolf, *Tetrahedron Lett.,* 4819 (1967).

127. P. Maravigna and G. Montaudo, *Gazz. Chim. Ital.,* **94,** 146 (1964).

128. W. Dilthey, E. Graef, H. Diericks, and W. Josten, *J. Prakt. Chem.,* **151** (2), 185 (1938).

129. W. Davies, N. W. Gamble, F. C. James, and W. E. Savige, *Chem. Ind. (London),* 804 (1952).

130. W. Davies and F. C. James, *J. Chem. Soc.,* 15 (1954).

131. K. Okita and S. Kambara, *Kogyo Kagaku Zasshi,* **59,** 547 (1956); *Chem. Abstr.,* **52,** 3762g (1958).

132. M. Prochazka, *Collect. Czech. Chem. Commun.,* **30,** 1158 (1965) (in German).

133. R. E. Merrill and G. Sherwood, *J. Heterocycl. Chem.,* **14,** 1251 (1977).

134. J. L. Melles and H. J. Backer, *Recl. Trav. Chim. Pay-Bas,* **72,** 491 (1953).

135. K. Gollnick and S. Fries, *Angew. Chem.,* **92,** 849 (1980); *Angew. Chem., Int. Ed. Engl.,* **19,** 833 (1980).

136. R. A. Firestone, Canadian Patent No. 992,954 (1976); *Chem. Abstr.,* 87, 102355h (1977); Japanese Kokai 48/067292 (1973); East German Patent No. 102,703 (1973); Netherlands Patent No. 72/16270 (1973); Spanish Patent No. 409,591 (1976); Swiss Patent No. 578,580 (1976).

137. M. Sletzinger and G. G. Hazen, U.S. Patent No. 3,981,911 (1976); *Chem. Abstr.*, **86**, 77280q (1977).

138. J. A. Ross, R. P. Seiders, and D. M. Lemal, *J. Am. Chem. Soc.*, **98**, 4325 (1976).

139. C. H. Bushweiler, J. A. Ross, and D. M. Lemal, *J. Am. Chem. Soc.*, **99**, 629 (1977).

140. J. A. Hashmall, V. Horak, L. E. Khoo, C. O. Quicksall, and M. K. Sun, *J. Am. Chem. Soc.*, **103**, 289 (1981).

141. J. D. Andose, A. Rauk, R. Tang, and K. Mislow, *Int. J. Sulfur Chem.*, Part A, **1**, 66 (1971).

142. A. Rauk, J. D. Andose, W. G. Frick, R. Tang, and K. Mislow, *J. Am. Chem. Soc.*, **93**, 6507 (1971).

143. A. Bened, R. Durand, D. Pioch, P. Geneste, J. P. Declercq, G. Germain, J. Rambaud, and R. Roques, *J. Org. Chem.*, **46**, 3502 (1981).

144. K. Kanematsu, K. Harano, and H. Dantsuji, *Heterocycles*, **16**, 1145 (1981).

145. T. Ban, K. Nagai, Y. Miyamoto, K. Harano, M. Yasuda, and K. Kanematsu, *J. Org. Chem.*, **47**, 110 (1982).

146. M. S. Raasch, unpublished results.

147. J. P. Snyder and T. A. Halgren, *J. Am. Chem. Soc.*, **102**, 2861 (1980).

148. N. A. Iglamova, G. M. Loginova, and F. N. Mazitova, *Khim. Tekhnol. Topl. Masel*, 48 (1981); through *Chem. Abstr.*, **96**, 22126e (1982).

149. J. S. Amato, S. Karady, R. A. Reamer, H. B. Schlegel, J. P. Springer, and L. M. Weinstock, *J. Am. Chem. Soc.*, **104**, 1375 (1982).

150. D. P. G. Hamon and P. R. Spurr, *J. Chem. Soc., Chem. Commun.*, 372 (1982).

151. J. O. Karlsson, S. Gronowitz, and A. Hallberg, *Chem. Scripta*, **20**, 37 (1982).

152. J. O. Karlsson, S. Gronowitz, and A. Hallberg, *Acta Chem. Scand.*, **B36**, 341 (1982).

153. J. S. Drage and K. P. C. Vollhardt, *Organometallics*, **1**, 1545 (1982).

154. N. A. Iglamova, F. N. Mazitova, and E. S. Brodskii, *Neftekhimiya*, **22**, 407 (1982); *Chem. Abstr.*, **97**, 112131c (1982).

155. F. M. Albini, P. Ceva, A. Mascherpa, E. Albini, and P. Caramella, *Tetrahedron*, **38**, 3629 (1982).

156. R. A. Aitken, J. I. G. Cadogan, I. Gosney, B. J. Hamill, and L. M. McLaughlin, *J. Chem. Soc., Chem. Commun.*, 1164 (1982).

157. P. J. Garratt and D. N. Nicolaides, *J. Chem. Soc., Chem. Commun.*, 1014 (1972).

158. P. J. Garratt and D. N. Nicolaides, *J. Org. Chem.*, **39**, 2222 (1974).

159. P. J. Garratt and S. B. Neoh, *J. Org. Chem.*, **44**, 2667 (1979).

160. P. J. Garratt and K. P. C. Vollhardt, *J. Chem. Soc., Chem. Commun.*, 109 (1970); *J. Am. Chem. Soc.*, **94**, 7087 (1972).

161. P. J. Garratt and S. B. Neoh, *J. Org. Chem.*, **40**, 970 (1975).

162. P. J. Garratt, *Pure Appl. Chem.*, **44**, 783 (1975).

163. A. J. Jones, P. J. Garratt, and K. P. C. Vollhardt, *Angew. Chem.*, **85**, 260 (1973); *Angew. Chem., Int. Ed. Engl.*, **12**, 241 (1973).

164. A. Svensson, J. O. Karlsson, and A. Hallberg, *J. Heterocycl. Chem.*, **20**, 729 (1983).

165. F. Boberg, C.-F. Czogalla, and J. Schröder, *Justus Liebigs Ann. Chem.*, 1588 (1983).

166. P. R. Spurr and D. P. G. Hamon, *J. Am. Chem. Soc.*, **105**, 4734 (1983).

167. A. Beck, D. Hunkler, and H. Prinzbach, *Tetrahedron Lett.*, **24**, 2151 (1983).

CHAPTER VIII

Reactions at Sulphur

A. E. A. PORTER

Chemistry Department, University of Stirling, Stirling, Scotland

1. INTRODUCTION

The classical molecular orbital description of thiophene views the molecule as having a planar pentagonal ring with sp^2 hybridized carbon and sulphur atoms. The σ framework of the ring is derived by overlapping the sp^2 hybrid orbitals of carbon and sulphur, and this arrangement then formally leaves a single p_z electron associated with each carbon atom and an electron pair in the p_z orbital associated with the sulfur atom; these p_z orbitals are in the plane perpendicular to the plane of the ring and overlap results in an electron cloud above and below the plane of the ring. Since the electron cloud contains six electrons, a stable closed shell results, and this confers aromatic stability on the ring.

The remaining sp^2 hybrid orbital at sulphur contains an electron pair that is not formally involved in bonding and, in principle, this electron pair is a focus for electrophilic attack; however, in practice, it is known that electrophilic attack on thiophene takes place preferentially at the 2-position, and, in the case of nitration, the loss of the displaced proton is not rate-determining. These data are consistent with thiophene behaving as a typical aromatic system and in accord with valence–bond predictions that attack of the electrophile occurs preferentially at position 2 of the ring.[1]

II. S-ALKYLATION OF THIOPHENES

1. Preparation

In view of the observations discussed above, it was particularly interesting when Brumlick and coworkers[2] reported that thiophene could be methylated at sulphur, using trimethyloxonium fluoroborate or methyl iodide/silver perchlorate. The position of methylation was confirmed by cation exchange to yield the stable hexafluorophosphate salt (I, $X = PF_6^-$), which was characterized by elemental analysis and by catalytic hydrogenation to 2 which was independently prepared from thiophane.

It is unfortunate that the original work was published as a communication and no experimental conditions appeared subsequently, since we and others[3] have experienced considerable difficulty reproducing this work. In our hands, both trimethyloxonium fluoroborate and the triethyl analogue failed to react at all. Acheson and Harrison[3] demonstrated that the alkylation using methyl iodide/silver perchlorate, apart from giving poor yields, is positively hazardous and resulted in an explosion on at least one occasion; silver tetrafluoroborate appears to be safer. Even under carefully controlled conditions, the yields of the alkylated thiophenes rarely exceed 10%, and although many derivatives have been characterized spectroscopically, others are too labile for practical isolation (Table 1).

In a detailed examination of the factors affecting S-alkylation of thiophene derivatives,[3] it has been observed that substituents exerting negative inductive or mesomeric effects completely inhibit alkylation; thus, 2-halo- and 2-acylthiophenes have not been successfully alkylated. As might be expected, annulation of the thiophene ring facilitates alkylation at sulphur, and both benzo[b]thiophene derivatives and dibenzo[b,d]thiophenes are readily S-alkylated in high yields (Tables 2 and 3).

2. Spectroscopic Properties

UV spectroscopy represents an easy method of diagnosing S-alkylation of simple thiophenes, because it changes totally during the course of alkylation. Thiophene in isooctane solution shows a λ_{max} at 231 nm, and although lack of solubility of S-alkylthiophenium salts in isooctane precludes direct comparison, S-methyl thiophenum hexafluorophosphate exhibits two maxima at ~ 220 and 270 nm (Table 4) in aqueous solution, which appear to be largely independent of the counter ion.

TABLE 1

Entry Number	X	R	R_1	R_2	R_3	R_4	Yield (%)	mp (°C)
1	PF_6	Me	H	H	H	H	10^a	160–170(d)^a, 145–160^b
2	ClO_4	Me	H	H	H	H	5^a	112–114
3	PF_6	Me	H	Me	H	H	10^a	90–140
4	PF_6	Me	Me	H	Me	H	$12^{a,c}$	65–71
5	ClO_4	Me	Me	H	Me	H	5^a	75–78
6	PF_6	Et	Me	H	Me	H	3^a	78–80
7	PF_6	Me	Me	Me	Me	Me	95^c	94–94.5
8	PF_6	Et	Me	Me	Me	Me	$>95^c$	–
9	PF_6	Me	Me	$-CH_2D$	$-CH_2D$	Me	$>95^c$	–

^{a} See Ref. 3.
^{b} See Ref. 2.
^{c} The higher yields were obtained by applying the suggestion made by Acheson and Harrison.[3] Thus, the alkylations were carried out[4] using fluorosulphonate esters under anhydrous conditions and precipitating the salt by the addition of a saturated solution of $NaPF_6$. Although this procedure worked well for entries 7–9, it was not applicable to entries 1 and 4.

TABLE 2

R	R_1	R_2	R_3	R_4	R_5	Yield (%)	mp (°C)
Me	H	H	H	H	H	$73^{a,b}$	72–73
Et	H	H	H	H	H	68	75–76
Me	Me	H	H	H	H	97	85–87
Et	Me	H	H	H	H	55	54–60
Me	H	Me	H	H	H	89	97–98
Et	H	Me	H	H	H	75	68–69
Me	Me	Me	H	H	H	95	103–104
Et	Me	Me	H	H	H	95	105–106
Me	Me	Me	H	Me	H	96	130–132
Et	Me	Me	H	Me	H	80	132–133
Et	Br	H	H	H	H	25	87–89

^{a} See Ref. 3 for a fuller list of derivatives.
^{b} The quoted yield was the highest obtained using several preparative procedures.

TABLE 3

R	R_1	R_2	R_3	R_4	R_5	R_6	R_7	R_8	Yield (%)	mp (°C)
Me	H	H	H	H	H	H	H	H	93[a]	149–151
Me	H	H	Br	H	H	Br	H	H	31	224
Et	H	H	H	H	H	H	H	H	98	123–125
Et	H	H	Br	H	H	Br	H	H	18	224–225

[a] For a more comprehensive list of known salts, see Ref. 3.

Proton nmr spectra of *S*-alkyl thiophenium salts have been examined in detail, but solubility differences between a given thiophene and the corresponding *S*-alkylated salt precluded direct comparison of the spectra. Thiophene itself shows two multiplets centred on δ 7.00 (α,α' protons) and δ 7.2 (β,β' protons) in deuterio-chloroform. On alkylation, all protons in the ring appear as a single resonance at δ 7.53 and the S-CH$_3$ group resonates at δ 3.11. Among the simple *S*-methyl thiophenium salts, there appears to be a deshielding of the ring protons relative to the free thiophene (Table 5), and the *S*-methyl group usually resonates in the region δ 3.1–3.2 in the absence of further shielding or deshielding effects.

^{13}C Nmr data on the thiophenium salts are scant and the only comparison reported[4] is that between 2,3,4,5-tetramethylthiophene, which has resonances at 11.8 (2- and 5-methyl groups), 11.6 (3- and 4-methyl groups), 127 (α-ring

TABLE 4

Ring Substituent	Thiophene[a]	Salt[b]
None	231 (3.85)	220 (3.32)[c] 267 (2.845)[c] 225 (3.49)[d] 269 (2.75)[d]
3-Methyl	234 (3.72)	220 (3.45) 271 (3.08)
2,5-Dimethyl	236 (3.88)	209 (3.398) 285 (3.398)

[a] The values given are for λ$_{max}$ in isooctane solution; log ε is in parentheses.
[b] As aqueous solutions of the hexafluorophosphate salts.
[c] See Ref. 3.
[d] See Ref. 2.

TABLE 5

Compound[c]	S—CH$_3$	2H[a]	3H	4H	5H	Solvent[b]
Thiophene	–	7.00	7.2	7.2	7.00	CDCl$_3$
1-Methylthiophenium hexafluorophosphate	3.11	7.53	7.53	7.53	7.53	CH$_2$Cl$_2$
3-Methyl thiophene	–	6.85	–	6.85	7.15	CDCl$_3$
1,3-Dimethylthiophenium hexafluorophosphate	3.21	7.08	–	7.36	7.53	CH$_2$Cl$_2$
2,5-Dimethylthiophene	–	–	6.6	6.6	–	CDCl$_3$
1,2,4-Trimethylthiophenium hexafluorophosphate	3.21	–	6.98	–	6.98	CH$_2$Cl$_2$

[a] Values given for δ with TMS as an internal standard.
[b] See Ref. 3.
[c] See Table 1.

carbon) and 132.4 (β-ring carbon atom), and 1,2,3,4,5-pentamethylthiophene fluorosulphonate, where the equivalent carbon atoms resonate at 10.0, 12.8, 128.4, and 148.5 ppm (from TMS as an external standard). In addition, the methylated thiophene has a resonance at 26.0 ppm caused by the S-Me group.

Although the small difference in ^{13}C chemical shifts of the ring carbon atoms at positions 2 and 5 is negligible, there is, by comparison, a remarkable difference between the chemical shifts of the ring carbon atoms at positions 3 and 4 in the *S*-alkylated thiophene. It has been argued that this supports charge localization at these carbon atoms.

No systematic study of the IR spectra of *S*-alkylated thiophenes is available and the available mass spectral data[3] suggests that dealkylation occurs readily, frequently giving rise to mass spectra that are virtually identical with the starting thiophene. In some instances, a peak corresponding to (M$^+$-I) of the cation was observed, but the intensity of this peak was small and frequently unreproducible in successive runs.

3. Structure and Bonding in Thiophenium Salts

In the absence of hard experimental data, predictions on bonding are generally difficult and although no x-ray study has yet been undertaken on *S*-alkylthiophenium salts, the structures of *S*-methyldibenzothiophenium tetrafluoroborate (3) and 1-methylnaphtho[2,3-b]thiophenium tetrafluoroborate (4) have appeared.[5]

3   4

Comparisons of the structures of these compounds with the parent thiophenes have made meaningful assumptions on the structure and bonding in 1-methylthiophenium salts possible. The local geometry around the sulphur atom, and in particular the precise environment of the methyl group, is important and x-ray data confirms the pyramidal nature of the sulphur atom. Thus, the bond angle between the plane of the rings and the alkyl substituent on sulphur is 68.9° for **3** and 68.2° for **4**. Independent, although less direct, evidence is available[3] from a consideration of the proton nmr spectra of 1-ethylbenzo[b]thiophenium salts with substituents in the 2- or 3-positions of the ring. Thus, the appearance of the methylene protons of the ethyl group (**5**) as a twelve line multiplet is only consistent with these protons being in a diastereotopic environment, and this in turn is dependent upon the sulphur atom being an asymmetric center.

5

Since the sulphur atom in these systems is tetrahedral, implying sp^3 hybridization, and since sp^2 hybridization is essential to achieve an "aromatic sextet," it is to be expected that, at best, a reduction in aromatic stabilization energy should occur. The effect of d-orbital participation is impossible to assess; however, some charge delocalization is possible by $3d_\pi$-$2p_\pi$ bonding, and supporting evidence for this is available in both ^{13}C and ^1H nmr data.[3,4] Thus, the large chemical shift difference between the β-carbon atoms in 2,3,4,5-tetramethylthiophene (**6**) and the corresponding S-methylated derivative (**7**, $X = SO_3F^-$) have been interpreted in terms of a high proportion of resonance structure (**8**).

6   **7**   **8**

In comparing the downfield chemical shifts in simple thiophenes and the corresponding thiophenium salts, Acheson and Harrison observed that the 3-proton is usually subject to a greater downfield shift than the 2-proton, consistent with charge localization on the adjacent ring carbon atom. Similar proton shifts have been reported for the 3-proton in the benzo[b]thiophene series.

Although the molecular geometry of the 1-methyl thiophenium ion is not available directly, the bond lengths and angles (**9**) have been calculated using the MNDO method. Molecular parameters have also been calculated for thiophen-1-oxide

1.480 Å / 113.1° / 110.5° 92.5° / 1.353 Å / S 1.759 Å / 1.806 Å / CH$_3$

9

1.477 Å / 112.6° / 90.9° 111.8° / 1.351 Å / S 1.776 Å / 1.516 Å / O

10

1.455 Å / 111.6° / 111.8° 93.2° / 1.368 Å / S 1.692 Å

11

1.422 Å / 92° 112° / 1.369 Å / S 1.718 Å

12

(10) and thiophene (11). Though the values obtained for thiophene differ from the electron diffraction results (12), the method is considered to provide evidence in support of a tetrahedral sulphur atom in **9**.

In addition, there appears to be an increase in bond alternation consistent with a deterioration of the π-electron system, as evidenced by an increase in the carbon–sulphur bond lengths, a decrease in the C_α-C_β bond length, and an increase in the C_β-C_β bond length.

4. Reactions of *S*-Alkylthiophenium Salts

The apparent structural similarities between *S*-alkylthiophenium salts and thiophene sulphoxides has prompted several groups to suggest that both class of compounds are antiaromatic.[6] Thiophene sulphoxides are usually exceptionally reactive, undergoing spontaneous Diels–Alder dimerization unless bulky substituents are present at the 2- and 5-positions, and this type of reactivity is typical of anti-aromatic molecules (e.g., cyclobutadiene). Structural similarities are not, however, a measure of similarities in reactivity, and *S*-alkylthiophenium salts show no tendency to take part in Diels–Alder cycloaddition reactions[7] either in an intra- or intermolecular sense. Pentamethylthiophenium hexafluorophosphate failed to react with perfluorobut-2-yne and dicyanoacetylene under conditions that result in efficient cycloaddition reactions with thiophene sulphoxides.

The general instability of the thiophenium salts seems to be related to their alkylating properties; thus, attack at the alkyl substituent by a nucleophilic species (e.g., water) results in alcohol formation. Indeed, it would seem that the loss of thiophene as a stable molecule in an S_N2 reaction should be a very favorable process.

1-Methylbenzo[b]thiophenium tetrafluoroborate is readily prepared in high yields and is moderately stable. It is a particularly efficient methylating agent[3] and methylates dimethylsulphoxide on oxygen at room temperature. Pyridine is methylated instantaneously at room temperature and acridine, which normally requires more forcing conditions, is also readily N-methylated at room temperature.

In addition to the facile solvolysis of S-alkylthiophenium salts, they appear to be photochemically labile.[7] Irradiation of 1,2,3,4,5-pentamethylthiophenium fluorosulphonate in methylfluorosulphonate solution results in the migration of the S-methyl substituent to the α-position (Scheme 1).

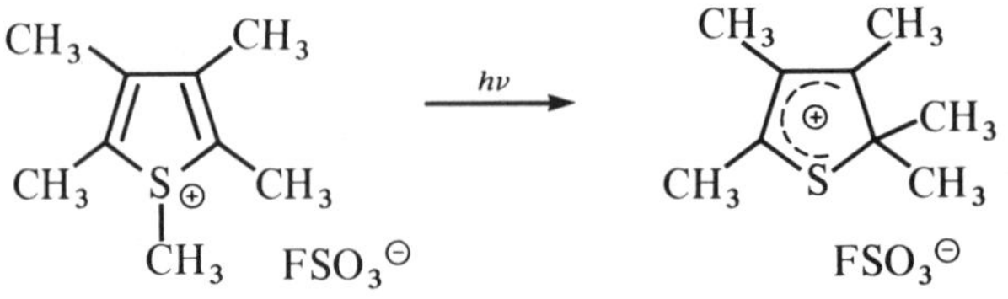

Scheme 1

The structure of the rearrangement product was confirmed by a consideration of the ^{1}H and ^{13}C nmr spectra of partially deuterated derivatives and it appears to be photochemically stable. All attempts to quench the reaction mixture failed to yield any identifiable products.

The precise mechanism of this transformation remains unclear, although a mechanism involving the homolytic cleavage of the carbon–sulphur bond (Scheme 2) and recombination of the methyl radical and radical cation appears an attractive possibility.

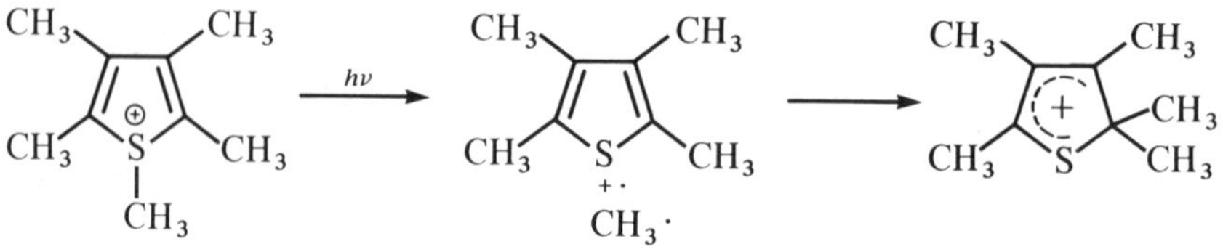

Scheme 2

III.   YLID FORMATION

1.   Preparation

The first reference to thiophenium ylids appeared[8] in 1972, when Durr and coworkers observed that the photolysis of tetraphenyldiazocyclopentadiene in thiophene produced a mixture of the carbene insertion product 13 and the ylid 14, although no information on the yield or physical properties of 14 was given. Subsequently, in 1974, Heldeweg and Hogeveen[4] demonstrated that 7 (X = PF$_6^-$)

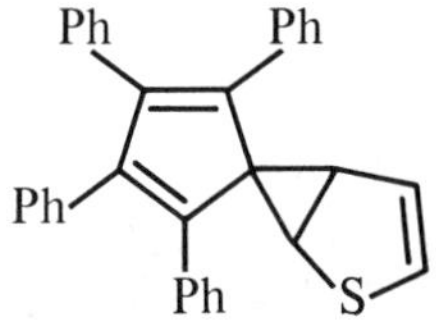

13 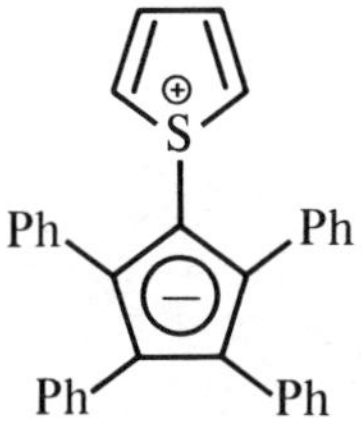   14

could be deprotonated with butyllithium at $-45°C$ in acetonitrile and the resultant ylid behaved as a typical sulphur ylid (Scheme 3) in its reaction with p-nitrobenzaldehyde to yield the epoxide (albeit in low yield).

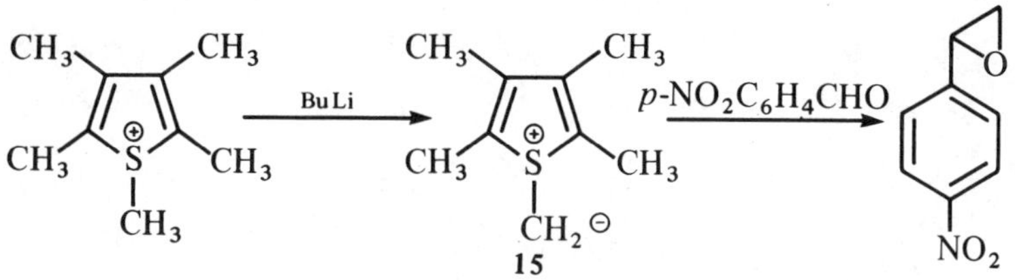

Scheme 3

The existence of two types of ylid, namely, stabilized ylids such as **14** and non-stability ylids such as **15**, offers direct evidence that thiophene is capable of behaving as a typical sulphide, e.g., diphenyl sulphide, in spite of the considerably reduced nucleophilicity of the sulphur atom.

No further interest was displayed in thiophenium ylids until 1977, when Ando and coworkers[9] reported that the photolysis of dimethyldiazomalonate in thiophene resulted in the formation of thiophenium bismethoxycarbonyl methylide (**16**) in low yield and that this ylid was a stable crystalline solid. Subsequently, my own group demonstrated[10] that modest to excellent yields of the bismethyloxycarbonyl methylides were available by the rhodium (II) acetate-catalyzed addition of diazomalonic esters to thiophene and its derivatives. Table 6 lists some ylids prepared by this method.

It is interesting to note that even when alternative reaction pathways are possible, ylid formation[11] is still favorable, as illustrated by the reaction of the alkene (**16**) with dimethyldiazomalonate (Scheme 4) to yield the ylid. In this reaction,

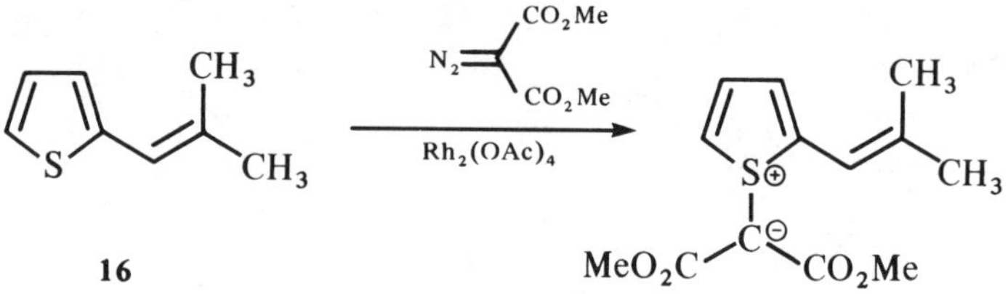

16

Scheme 4

TABLE 6

R_1	R_2	R_3	R_4	R_5	Yield (%)[a,b]	mp (°C)
Me	H	H	H	H	95	145–146
Et	H	H	H	H	90	111–111.5
CH_3	Cl	H	H	Cl	100	174.5–175
Et	Cl	H	H	Cl	60	82.5–83
CH_3	$-CH_2OH$	H	H	H	54	133–133.5
CH_3	CH_3	H	H	H	90	146–146.5
CH_3	Br	H	H	H	73	138–140
CH_3	Br	CH_3	H	H	86	128.5–129
CH_3	Br	H	H	Br	55	190–190.5
Me	H	H	—CH=CH—CH=CH—		97	177–177.5

[a] The yield is of the isolated and crystallized product.
[b] The reaction is very sensitive to the purity of the thiophene. Impurities poison the catalyst.

there is no evidence for cyclopropane formation from the reaction of the carbenoid with the side-chain double bond.

This type of behavior is also observed in 2-hydroxymethylthiophene (Table 6), in which the possible competing side reaction of insertion of the carbenoid into the —OH bond is not observed. 2-Thienylamine is exceptional in that it does not form ylids or give products derived from carbenoid insertion into the N—H bond, but that reaction is beyond the scope of this chapter.

A detailed analysis of the reactions of α-diazocarbonyl compounds with thiophene has revealed that ylid formation is not a general reaction.[12] Thus, it is known that diazoacetic esters[13] react with thiophene to yield the product of carbene insertion into the 2,3-bond (17, R = H, R_1 = CO_2Me), and similar behavior has been reported for diazomethane,[14] which yields 17 (R = R_1 = H).

Diazomalonic esters and diazoacetoacetic esters are generally considered to exhibit similar reactivities; however, diazoacetoacetic esters show no tendency to form stable thiophenium ylids (18, R = $COCH_3$, R_1 = CO_2Me). Instead, the major reaction product is the 2-insertion product 19 (R = $COCH_3$, R_1 = CO_2Me). A minor reaction product is the cyclopropane 17 (R = $COCH_3$, R_1 = CO_2Me). Similar behavior is observed with other diazoketones (e.g., diazoacetophenone).

17   18   19

Interestingly, Meldrum's diazo (**20**) fails to react with thiophene under a variety of conditions, and ylids such as **21** have not yet been synthesized.

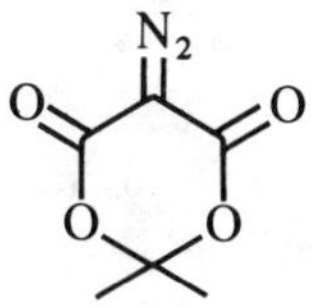

20

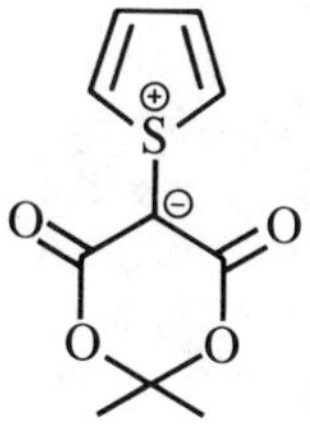

21

The known reactions of thiophene with diazoalkanes are readily rationalized in terms of a single reaction scheme (Scheme 5). If it is assumed that the first step in

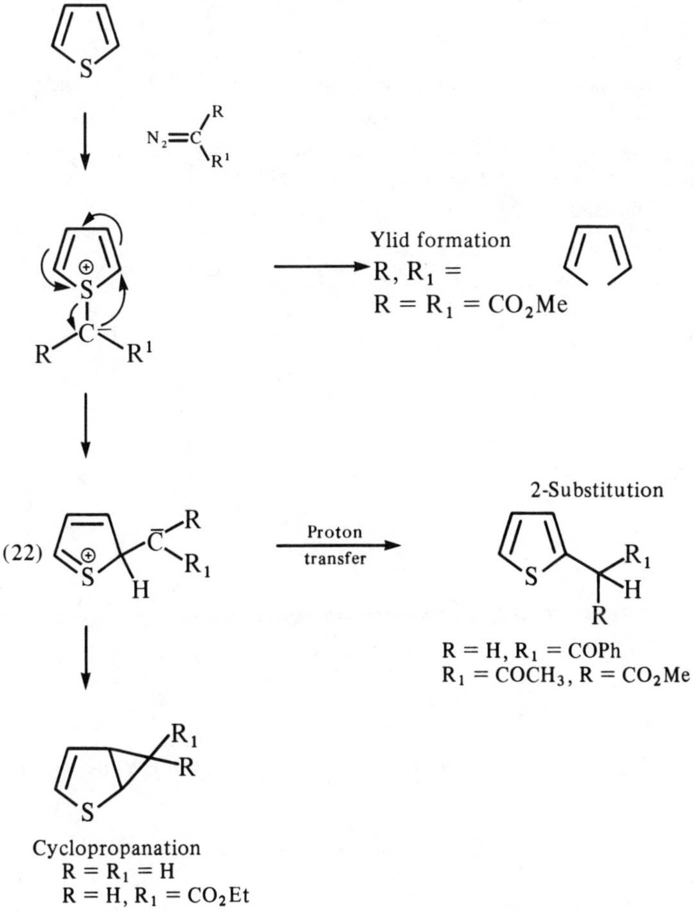

Ylid formation
R, R_1 =
R = R_1 = CO_2Me

(22)

Proton transfer

2-Substitution

R = H, R_1 = COPh
R_1 = $COCH_3$, R = CO_2Me

Cyclopropanation
R = R_1 = H
R = H, R_1 = CO_2Et

Scheme 5

all cases involves the reaction of the carbene or carbenoid derived from the diazo-alkane with the ring sulphur atom, then the ylid would be expected as the product. Indeed, in those cases where the ylid is stabilized by the delocalization of the insipient negative charge onto neighboring substituents, the resultant ylids are stable isolable entities. However, there appears to be a very delicate balance of electronic effects governing the ylid stability, and an intramolecular migration "vide infra" of the substituent from sulphur to C-1 of the ring appears to be facile. The resultant betaine (**22**) may then undergo two possible reactions, depending on the substituent; thus, proton transfer results in the 2-insertion product observed with, for example, diazoacetoacetic esters and diazoacetophenone. When the sub-stituents R and R_1 are not able to stabilize the betaine (**22**), spontaneous cycli-zation to the cyclopropane occurs.

The whole reaction sequence appears to be very sensitive to electronic effects and sometimes mixed products may be observed. In the case of diazoacetoacetic esters, there appears to be approximately a 10:1 ratio of 2-substitution:cyclo-propane formation.

There appears to be little supporting evidence in the literature for the reaction of thiophene and its derivatives with electrophiles at the sulphur atom, for example, during normal electrophilic substitution reactions; however, Del Mazza and Rein-ecke[15] suggested recently that benzyne is capable of reacting with the sulphur atom in thiophenes to give an intermediate betaine (Scheme 6) that reacts spontaneously to generate ylids such as **23**.

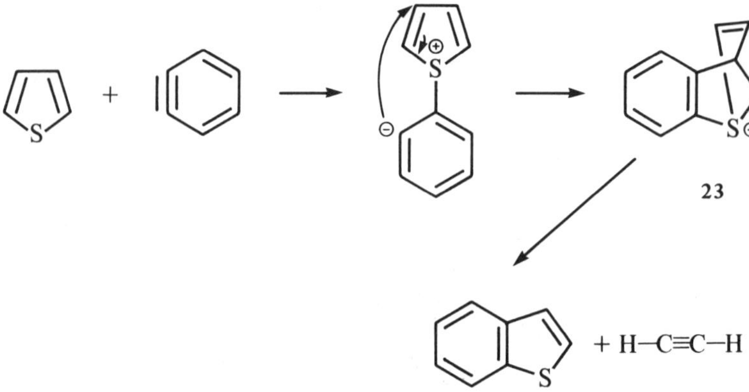

Scheme 6

2. Structure and Bonding in Thiophenium Ylids

In contrast to S-alkyl thiophenium salts, more structural data are available for thiophenium ylids. An x-ray crystal structure determination[10] of thiophenium bismethoxycarbonylmethylide has been carried out and accurate bond lengths and angles are available (**24**).

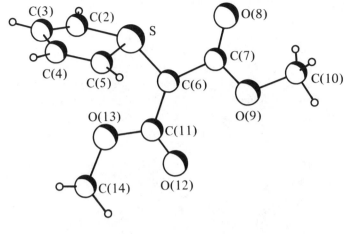

24

Bond lengths (Å): S—C(2), 1.746; S—C(5), 1.743; S—C(6), 1.709; C(2)—C(3), 1.324; C(3)—C(4), 1.439; C(4)—C(5), 1.319; C(6)—C(7), 1.442; C(6)—C(11), 1.432; C(7)—O(8), 1.215; C(7)—O(9), 1.350; O(9)—C(10), 1.442; C(11)—O(12), 1.212; C(11)—O(13), 1.353; O(13)—C(14), 1.441.

Bond angles (degrees): C(2)—S—C(5), 91.2; C(2)—S—C(6), 116.7; C(5)—S—C(6), 114.7; S—C(2)—C(3), 109.0; C(2)—C(3)—C(4), 114.2; C(3)—C(4)—C(5), 113.7; S—C(5)—C(4), 109.6; S—C(6)—C(7), 110.4; S—C(6)—C(11), 121.5; C(7)—C(6)—C(11), 128.1; C(6)—C(7)—O(8), 125.3; C(6)—C(7)—O(9), 112.6; O(8)—C(7)—O(9), 122.0; C(7)—O(9)—C(10), 114.7; C(6)—C(11)—O(12), 127.0; C(6)—C(11)—O(13), 111.7; O(12)—C(11)—O(13), 121.3; C(11)—O(13)—C(14), 116.6.

The molecule has an approximate mirror plane with the sulphur atom pyramidal, as expected. The angle (θ, structure **25**) between the plane of the ring and the ylid carbon atom is 49.7°, which is significantly less than in the corresponding sulphonium salts.

$$C_3-C_2-S \underset{\displaystyle C}{\overset{\displaystyle \theta}{\diagdown}}$$

25

The ylid bond [S—C(6)] is significantly shorter than the carbon–sulphur bond in aliphatic sulphides and shorter than the carbon–sulphur bond in thiophene itself.[16] The ring carbon–sulphur bonds [S—C(2)] are longer than those in thiophene and there seems to be a considerably reduced delocalization that results in a shortening of the [C(2)—C(3)] and [C(4)—C(5)] bonds and a lengthening of the [C(3)—C(4)] bond, relative to thiophene.

In contrast to the ^{13}C spectra of S-alkyl thiophenium salts, there do not appear to be anomalous chemical shift differences between the α- and β-carbon atoms of thiophene and the corresponding ylids. The ^{13}C nmr spectrum of thiophene shows two lines at 124.9 and 126.7 ppm from TMS, corresponding to the α- and β-carbon atoms, respectively. In thiophenium bismethoxycarbonyl methylide, these resonances are shifted to 131.01 and 133.64 ppm, respectively. This corresponds to a deshielding of the α-carbon atom by 6.11 ppm and the β-carbon atom by 6.94 ppm. These figures, which are broadly similar, are consistent with a more general depletion of electronic charge over the whole ring and seem to rule out significant stabilization by the type of d_{π}-p_{π} bonding (e.g., **26**) invoked for the thiophenium salts. These data, coupled with the short ylid S—C bond, would indicate that if d_{π}-p_{π} bonding is involved, it is almost certainly to the exocyclic ylid carbon atom (e.g., **27**).

The available proton nmr spectral data shows that chemical shift values for the α- and β-protons in thiophene and thiophenium bismethoxycarbonyl methylide are virtually identical at 7.0 and 7.2 ppm, respectively.

There is a small downfield shift of the ring protons in 2,5-dichlorothiophenium bismethoxycarbonylmethylide (0.28 ppm) relative to 2,5-dichlorothiophene. Similar small shifts are observed with other ylids, but since the chemical shift values appear to be concentration dependent, little meaningful information can be gained.

The signal at δ 3.72 caused by the methyl ester proton in 2,5-dichlorothiophenium bismethoxycarbonylmethylide appears as a broad singlet, and on decreasing the temperature, the signal eventually separates into four peaks.[17] Available data indicates that two separate processes are involved in this increase of complexity of both proton and ^{13}C nmr spectra. In principle, one might envisage three different processes which might be responsible: pyramidal inversion at sulphur, restricted rotation about the ylid carbon–sulphur bond, and restricted rotation about the ester-O—CH$_3$ bond.

In practice, it is recognized that barriers to pyramidal inversion at sulphur are high[18] and sulphur ylids, which are chiral owing to the pyramidal nature of the sulphur atom, have been isolated and shown to be configurationally stable. This suggests that the two observable barriers are probably rotational barriers. Since the ylid carbon–sulphur bond is short, consistent with a degree of double bond character, it would appear that the higher energy process is due to hindered rotation about this bond, whereas the lower energy barrier arises from hindered rotation about the ester bond.

By a detailed analysis of the coalescence temperatures (T_c) in a number of ylids, it has been possible to calculate the equilibrium constant (k_c) for the rotational processes. By substituting k_c into the Eyring equation, the free energies of activation of the rotational processes ($\Delta G^{\ddagger}$) become available (Table 7).

TABLE 7. BOND ROTATIONAL ENERGY BARRIERS IN THIOPHENIUM YLIDS

R	R$_1$	R$_2$	S—C Bond[a]			CO—O—CH$_2$-Bond[b]		
			T_c (°K)	$\delta\nu$ (Hz)	$\Delta G^{\ddagger}$ (kJ mole^{-1})	T_c (°K)	$\delta\nu$ (Hz)	$\Delta G^{\ddagger}$ (kJ mole^{-1})[c]
H	H	Me	190 (± 2)	—	—[d]	190 (± 2)	6.6	37.5 (± 0.4)
H	H	Et	198 (± 5)	—	—[e]	198 (± 5)	—	—
Cl	Cl	Me	315 (± 5)	34.1	60.6 (± 1)	193 (± 5)	20.8	37.2 (± 1)
H	Br	Me	263 (± 2)	7.62	51.8 (± 0.3)	—	—	—
H	Me	Me	240 (± 2)	7.22	47.3 (± 0.3)	200 (± 5)	37.83	37.0 (± 1)
Br	Br	Me	293 (± 2)	7.23	57.8 (± 0.4)	193 (± 5)	33.4	37.1 (± 1)
Me	Me	Me	257 (± 2)	7.03	50.8 (± 0.4)	200 (± 5)	36.9	38.4 (± 1)
Cl	Cl	Et	309 (± 2)	6.2	61.3 (± 0.4)	—	—	—

[a] These values were determined using solutions of the ylid in CDCl$_3$ at a concentration of 25 mg/ml.

[b] These values were obtained using CDFCl$_2$ as a solvent at concentrations of 4 mg/ml to avoid association effects.

[c] The value for k_c was determined using the approximation $k_c = \pi\delta\nu/2^{1/2} = 2.26\,\delta\nu$. Substituting k_c into the Eyring equation, $\Delta G^{\ddagger} = 4.57\,T_c[10.32 - \log(k_c/T_c)]$, we obtain $\Delta G^{\ddagger} = 19.14\,T_c[9.97 + (\log T_c/\delta\nu)]$ kJ mole^{-1}.

[d] These values could not be determined, since any slowing down in the rotation of the S—C bond is masked by the slowing down of the ester group rotation.

[e] This value was obtained from ^{13}C data.

A number of points from Table 7 are worthy of comment. First, the value of the barrier to rotation about the ester-OCH_3 bond appears to be constant at 37.5($\pm$ 1) kJ mole^{-1} and this is not unexpected. The barrier to rotation about the ylid sulphur–carbon bond varies, and in view of the differing sizes of the substituents at the α- and α'-positions of the thiophene ring, this would be expected. However, the lower barrier for the 2,5-dibromothiophenium bismethoxycarbonylmethylide (*cf.* the dichloro analogue) would suggest that steric factors alone are insufficient to explain the rotational barriers and electronic effects are probably also important.

3.   Reactions of Thiophenium Ylids

Thiophenium ylids and thiophene-1-oxides are isoelectronic and in principle one would expect a similar reactivity to be associated with both structural types. In practice, this is not borne out. Thiophenium ylids show no tendency to undergo dimerization, a reaction that is common among the sulphoxides and that has been used as a yardstick in assigning antiaromatic properties to this class of compounds.

Although no systematic study has been carried out on nonstabilized thiophenium ylids (e.g., **15**, Scheme 3), these ylids appear to be relatively stable and able to participate in reactions that are typical of sulphur ylids at $-45°C$. No report of dimers have appeared.

A more systematic study of stabilized thiophenium ylids such as **18** (R = R_1 = CO_2Me) has been carried out. Although thiophenium bismethoxycarbonyl methylides are thermally unstable above about 60°C, they show no tendency to undergo dimerization. In addition, they fail to react with electron-deficient dienophiles, such as maleic anhydride and acetylenic esters, or electron-rich dienophiles, such as vinyl ethers and vinyl esters in Diels–Alder reactions.[19]

On heating thiophenium bismethoxycarbonylmethylide above 60°C, it undergoes a smooth rearrangement[20] to produce thiophene-2-malonic ester (Scheme 7, R = H) in excellent yield.

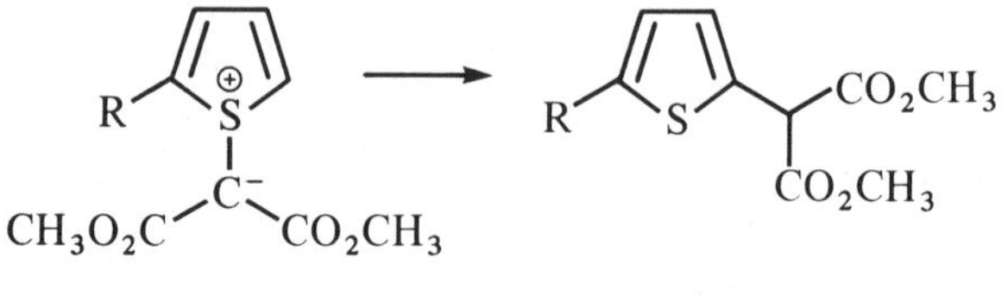

Scheme 7

Other reactions of this type are common; thus, 2-methylthiophenium bismethoxy-carbonylmethylide rearranges on heating to produce the corresponding 5-malonate (Scheme 7, R = CH_3).

Two mechanisms for this reaction are possible: an intramolecular 1,2-migration of the ylidic carbon from sulphur to C-2 of the ring (Scheme 8), or the initial dissociation of the ylid into the thiophene and a carbene, followed by electrophilic attack of the carbene at C-2 of the ring.

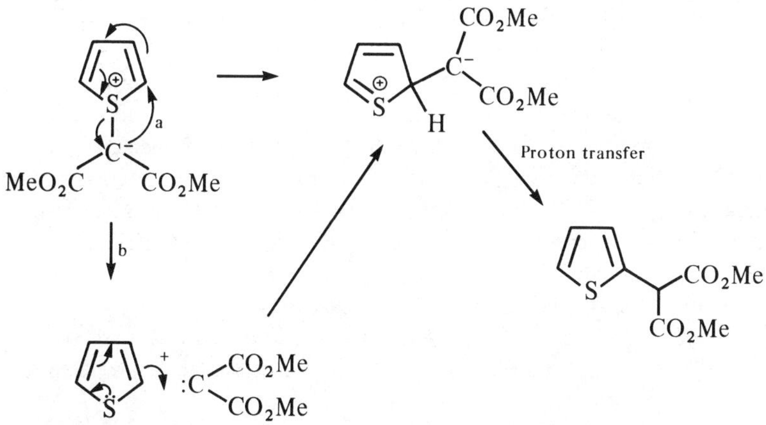

Scheme 8

Crossover experiments support the first mechanism (path a) in that when thiophenium bismethoxycarbonylmethylide is heated in 2-methylthiophene, only the product of intramolecular reaction (Scheme 7, R = H) is observed. Similarly, when 2-methylthiophenium bismethoxycarbonylmethylide is heated in thiophene, only the 2-methylthiophene-5-malonate is produced as the sole product.

In view of the known effect of an alkyl substituent on partial rate factors during electrophilic substitution reactions, 2-methylthiophene is expected to be between 10 and 100 times more reactive than thiophene; and if electrophilic substitution by a carbene were involved, then by carrying out the reaction (Scheme 7, R = H) in 2-methylthiophene, significant quantities of 2-methylthiophene-5-malonate would be expected to result.

When both α and α'-positions of the ring are blocked, rearrangement reactions are not clean and several products are formed, for example, 2,5-dichlorothiophenium bismethoxycarbonylmethylide (28) on heating at reflux[21] in anisole rearranges to the thieno[3,2-b]furan (29), presumably by an intramolecular mechanism (Scheme 9).

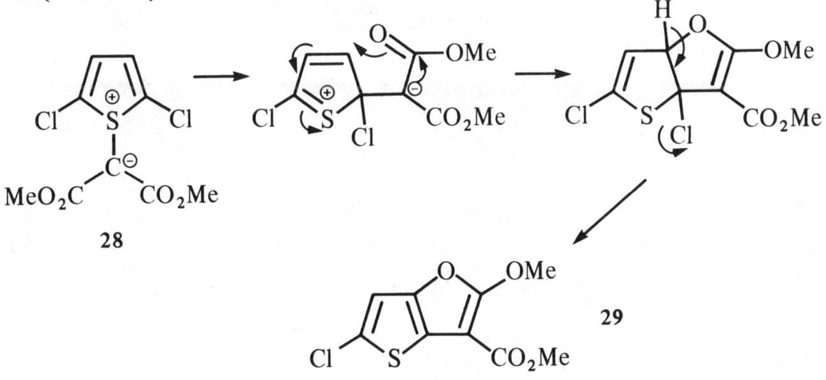

Scheme 9

Other 2,5-disubstituted thiophenium ylids appear to break down by less clearly defined mechanisms, which have not, as yet, been investigated.

In the presence of metal ions that stabilize carbenoid intermediates, 2,5-dichlorothiophenium bismethoxycarbonylmethylide (28) exhibits a different type of behavior and undergoes fragmentation to 2,5-dichlorothiophene and the metal carbenoid derived from bismethoxycarbonylcarbene.[22] It is of interest to note that the metal catalysts [RhII, Cu(I), Cu(II)] must be implicated in the cleavage of the ylid carbon–sulphur bond, rather than merely seeming to stabilize the carbene formed by thermal cleavage of the ylid, since in the absence of the metal ions, there is no evidence for the intermediacy of carbenic species.

A number of carbenoid reactions have been studied using 28 as the carbenoid precursor. Thus, it is possible to effect cyclopropanations of simple alkenes in high yield (Table 8).

Experimentally, the use of 28 as a carbene precursor has a number of advantages. The ylid is a stable crystalline solid with an apparently indefinite shelf life and the reactions are simply carried out by heating the ylid in an excess of the alkene at its boiling point.

Mesomerically activated arenes react to yield aryl malonates[23] in modest to excellent yields (Table 9).

TABLE 8

Reactant	Product	Yield (%)	Temperature (°C)	Time (h)
(cyclohexene)	(bicyclic) CO$_2$Me, CO$_2$Me	66	Reflux	16
(cyclooctene)	(bicyclic) CO$_2$Me, CO$_2$Me	86	120	0.5
CH$_2$=CH(CH$_2$)$_4$CH=CH$_2$	MeO$_2$C CO$_2$Me (cyclopropane)-(CH$_2$)$_4$CH=CH$_2$	83	120	1.5
CH$_2$=CH–OAc	OAc (cyclopropane) CO$_2$Me, CO$_2$Me	81	75	11
(CH$_2$)$_7$CO$_2$Me, (CH$_2$)$_7$CH$_3$	CO$_2$Me (CH$_2$)$_7$CO$_2$Me, CO$_2$Me (CH$_2$)$_7$CH$_3$	60	110	0.5

TABLE 9

Reactant	Product	Reaction Time	Yield (%)
Thiophene	Dimethyl thiophen-2-malonate	24 h	98
Pyrrole	Dimethyl pyrrole-2-malonate	5 min	79
Indole	Dimethyl indole-3-malonate	2–5 min	85
Anisole	Dimethyl 2(4'-methoxyphenyl)malonate	1.5 h	48
Methylenedioxybenzene	Dimethyl 2-(3',4'-methylenedioxyphenyl)malonate	1 min	20

Carbene insertion into active X—H bonds occurs readily; thus, aniline reacts to give dimethyl 2(aminophenyl) malonate (**30**, X = NH) in 95% yield and phenol gives **30** (X = 0) in 40% yield. These reactions have some potential. Thus, the

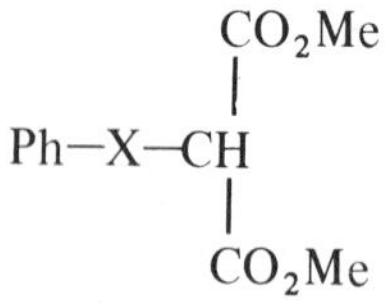
CO2Me
Ph—X—CH
CO2Me

30

carbene generated from **20** reacts readily with the amine (**31**) to give the insertion product (Scheme 10), whereas dimethyl diazomalonate fails to yield the insertion product.[24]

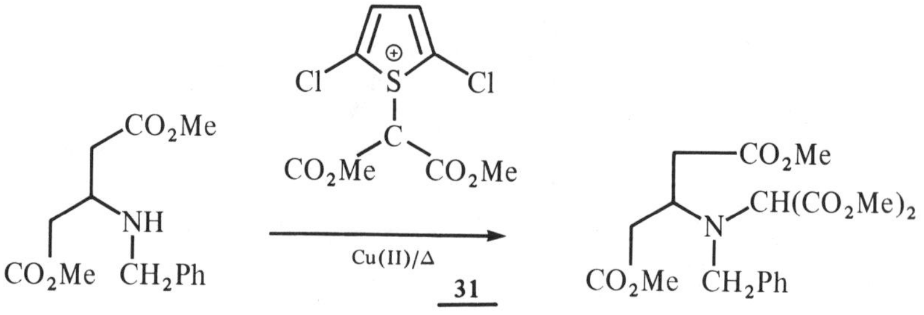
Cl
S
Cl
C
CO2Me
CO2Me
CO2Me
NH
CO2Me
CH2Ph
Cu(II)/Δ
31
CO2Me
CH(CO2Me)2
N
CO2Me
CH2Ph

Scheme 10

Thiophenium ylids also appear to become protonated with acids, and when the anion derived from the acid is sufficiently nucleophilic, the thiophene residue is displaced. Acetic acid reacts with **28** to yield dimethyl 2-acetoxymalonate (**32**).

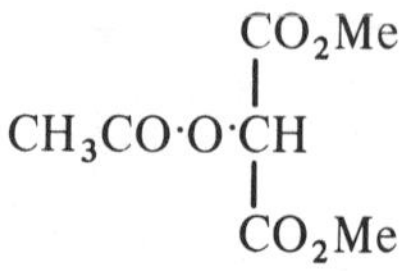
CO2Me
CH3CO·O·CH
CO2Me

32

In conclusion, the structure of, and bonding in, thiophenium ylids and *S*-alkyl thiophenium salts remain unclear and await the development of adequate calculations to rationalize the known structural features and chemical reactivity of these systems. It seems probable that a wealth of new and interesting chemistry will be uncovered in any systematic study of their properties and reactivities.

REFERENCES

1. See for example R. M. Acheson, in *Introduction to the Chemistry of Heterocyclic Compounds,* Wiley, New York, 1967.
2. G. C. Brumlick, A. I. Kosak, and R. Pitcher, *J. Am. Chem. Soc.,* **86**, 5360 (1964).

3. R. M. Acheson and D. R. Harrison, *J. Chem. Soc. (C)*, 1764 (1970).
4. R. F. Heldeweg and H. Hogeveen, *Tetrahedron Lett.*, 75 (1974).
5. J. A. Hashmall, V. Horak, L. E. Khoo, C. O. Quicksall, and M. K. Sun, *J. Am. Chem. Soc.*, **103**, 289 (1981).
6. W. L. Mock, *J. Am. Chem. Soc.*, **92**, 7610 (1970).
7. H. Hogeveen, R. M. Kellogg, and K. A. Kuindersma, *Tetrahedron Lett.*, 3929 (1973).
8. H. Dürr, B. Hen, B. Ruge, and G. Schweppes, *Tetrahedron Lett.*, 1257 (1972).
9. W. Ando, H. Higuchi, and T. Migita, *J. Org. Chem.*, **42**, 3365 (1977).
10. R. J. Gillespie, J. Murray-Rust, P. Murray-Rust, and A. E. A. Porter, *J.C.S. Chem. Comm.*, 83 (1978).
11. V. M. Shostakovskii, A. E. Vasil'vitskii, V. L. Zlatkina, and O. M. Neftov, *Izv. Akad. Nauk. SSSR, Ser. Khim.*, **9**, 2180 (1980).
12. R. J. Gillespie and A. E. A. Porter, *J. Chem. Soc. Perkin I*, 2624 (1979).
13. W. Steinkopf and H. Augestad-Jensen, *Annalen*, **428**, 154 (1922).
14. E. Muller, H. Kesler, H. Fricke, and A. Sühr, *Tetrahedron Lett.*, 1047 (1963).
15. D. Del Mazza and M. G. Reinecke, *J.C.S. Chem. Comm.*, 124 (1981).
16. W. R. Harshbarger and S. H. Bauer, *Acta Cryst.*, **B36**, 1010 (1970).
17. A. E. A. Porter, J. A. Rechka, and F. G. Riddell, unpublished observations.
18. B. M. Trost and L. S. Melvin, *Sulphur Ylids*, Academic Press, New York, 1975.
19. R. J. Gillespie and A. E. A. Porter, unpublished observations.
20. R. J. Gillespie, A. E. A. Porter, and W. E. Willmott, *J.C.S. Chem. Comm.*, 85 (1978).
21. R. J. Gillespie, J. Murray-Rust, P. Murray-Rust, and A. E. A. Porter, *J.C.S. Chem. Comm.*, 366 (1979).
22. J. Cuffe, R. J. Gillespie, and A. E. A. Porter, *J.C.S. Chem. Comm.*, 641 (1978).
23. R. J. Gillespie and A. E. A. Porter, *J.C.S. Chem. Comm.*, 50 (1979).
24. J. Cuffe, S. Husinec, and A. E. A. Porter, unpublished observations.

CHAPTER IX

Radical Reactions of Thiophene

A. E. A. PORTER

Chemistry Department, University of Stirling, Stirling, Scotland

I. INTRODUCTION

In preparing this chapter, I specifically restricted the scope of the contents to a detailed consideration of two general headings, namely, the generation and reactions of 2- and 3-thienyl radicals and the homolytic substitution of thiophene rings. In both cases, a direct change in the nature of the substituents attached to the thiophene ring occurs.

Thiophene derivatives also undergo radical substitution reactions at benzylic positions (e.g., in the bromination of 3-methylthiophene with NBS in the presence of a radical initiator[1]), but such reactions usually result in side chain rather than ring substitution, and as such fall outside of the scope of this chapter.

II.   PRODUCTION OF THIENYL RADICALS

1.   Radicals from Peroxides

Following the initial observation that the thermal decomposition products of dibenzoyl peroxide result in the formation of benzoyloxy radicals[2] that subsequently fragment to phenyl radicals and CO_2, Ford and MacKay[3] and Schuetz and Teller[4] investigated the thermal decomposition of 2-thenoyl peroxides (1).

1   2

The peroxides are readily prepared from the corresponding thenoyl chloride and sodium peroxide and proved to be relatively stable. The rate constants[4] for their decomposition in CCl_4 solution, in the presence of 3,4-dichlorostyrene as a radical scavenger, at 75°C, were shown to be broadly similar to that of benzoyl peroxide (Table 1).

TABLE 1.   RATES OF DECOMPOSITION OF THIENOYL
                    PEROXIDES

Peroxide	$k \times 10^3$ (min^{-1})
Bisbenzoyl	2.52
Bis(5-methyl-2-thenoyl)	2.54
Bis(5-t-butyl-2-thenoyl)	2.42
5-Methyl-bis(2-thenoyl)	1.79
Bis(4-methyl-2-thenoyl)	1.76
Bis(2-thenoyl)	1.33
Bis(5-chloro-2-thenoyl)	0.95
Bis(5-bromo-2-thenoyl)	0.92
Bis(4-bromo-2-thenoyl)	0.69

In the presence of the radical scavenger, most of the peroxides studied followed first-order decomposition rates; however, in the absence of the scavenger, the rates of decomposition were found to be of a higher order. 2-Thenoyl peroxide decomposes at 75°C in CCl_4 solution in the absence of the scavenger to yield 2-thenoic acid (20 mole %), carbon dioxide (20 mole %) and a polymer (46% by weight of the original peroxide).

If the decomposition results in a rapid decarboxylation to yield 2-thienyl radicals (2), then by analogy with the phenyl radicals generated by similar procedures, reaction with aromatic substrates should result in thienylated products. Decomposition of 2-thienoyl peroxide in benzene, nitrobenzene, chlorobenzene, bromobenzene, and iodobenzene produced little evidence of thienylation. The major observable product (0.6–0.9 mole/mole of peroxide) was 2-thenoic acid, presumably formed by hydrogen radical abstraction by the 2-thenoyl radical (3). This observation is supported by the marked increase in acid formation (1.4 and 2.0

moles, respectively) observed when the decomposition was carried out in toluene and cumene.

Phenyl-2-thenoate was among the minor products formed in benzene, which indicates that the 2-thenoyl radical appears to react with aromatic substrates, rather than lose CO_2 to form 2. In addition, low yields of arylated products were observed (e.g., in thiophene), and MacKay and Ford have suggested that the intermediate thenoyl radicals were more stable than their benzenoid counterparts by virtue of resonance contributions such as 4.

MacKay[5] has also presented results of a study of the thermal decomposition of 3-thenoylperoxides (5) in a number of solvents, and in contrast to the scant evidence for the production of 2-thienyl radicals, clear-cut evidence was obtained for the formation of 3-thienyl radicals (6). Thus, in thiophene, benzene, and

chlorobenzene, low yields of the arylated products were obtained; and in chlorobenzene, the *ortho* isomer (7) predominated, a common occurrence in free-radical arylation reactions.

Although these reactions constitute direct evidence for the involvement of 3-thienyl radicals, it is clear that decomposition of the peroxides fails to provide a practical source of 2- and 3-thienyl radicals and more useful methods are obviously necessary if radical thienylations are to be effected.

2.   Radicals from Thienylamines

The classical Gomberg–Bachmann reaction[6] involves arylation using phenyl radicals generated by the decomposition of diazonium salts, under basic conditions, in the presence of an aromatic substrate (Scheme 1).

$$ArNH_2 \xrightarrow{NaNO_2/HCl} ArN_2^+Cl^- \xrightarrow[Ar']{OH^-} Ar-Ar' + N_2 + H_2O + Cl^-$$

Scheme 1

In principal, this reaction should be applicable in the thiophene series, but aminothiophenes are renowned for their instability. Even the salts undergo rapid autoxidation in the solid state and solution, unless electron-withdrawing groups are present in the ring to effect stabilization.

Putokhin and Yakovlev[7] claimed that 2-thienylamines may be obtained as stable "double salts" by the reduction of 2-nitrothiophene with $SnCl_2/HCl$. These "double salts" $[C_4H_3S\overset{+}{N}H_3Cl^-]_2SnCl_4$ are claimed to be stable in aqueous solution and to undergo normal diazotization reactions. The resultant diazonium salts have been trapped as azo-dyes with various amines, phenols, and naphthols.

These claims were subsequently refuted when Shishkin and Mamaev[8] reported that the double salts of the thienylamines survived the diazotization reaction essentially unchanged. Since no subsequent work supports the initial claims, they must be regarded as suspect.

There is, however, little doubt that mesomerically deactivating groups, such as a carboxyl group in the 5-position of the thiophene ring, stabilize the 2-thienylamines and these derivatives undergo normal diazotization reactions. The resultant diazonium salts undergo the Gomberg–Bachmann reaction to produce the expected radical, which has been trapped in benzene solution,[5] albeit in low yield (Scheme 2).

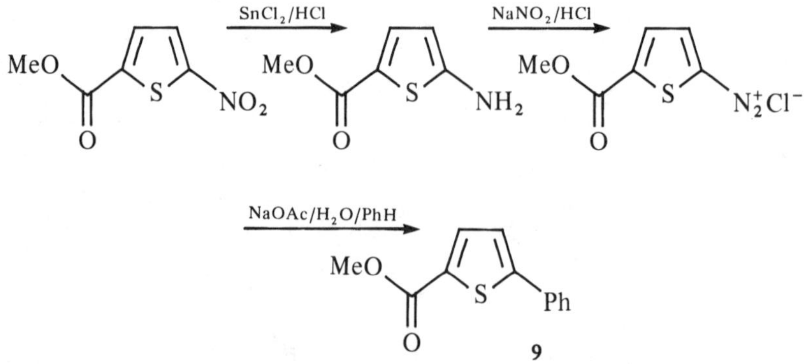

Scheme 2

A more successful variant of this reaction has been developed; thus, the reduction of the diazonium chloride with $SnCl_2/HCl$ produces the hydrazinothiophene (8) as the hydrochloride salt, which is relatively stable. When 8 was subjected to silver oxide oxidation in dry benzene, the phenylated product (9) was isolated in 85% yield. The full potential of this reaction does not appear to have been exploited in that no further examples in the thiophene series have been studied.

3. Radicals from Iodothiophenes

The most synthetically useful method of producing thienyl radicals is by the photolysis of iodothiophenes. The facile homolytic cleavage of the carbon–iodine bond is well-established, largely because of the extensive studies of Kharasch[9] and coworkers. Thus, the photolysis of iodobenzene derivatives in aromatic solvents results in phenylation by the initially formed radical (Scheme 3).

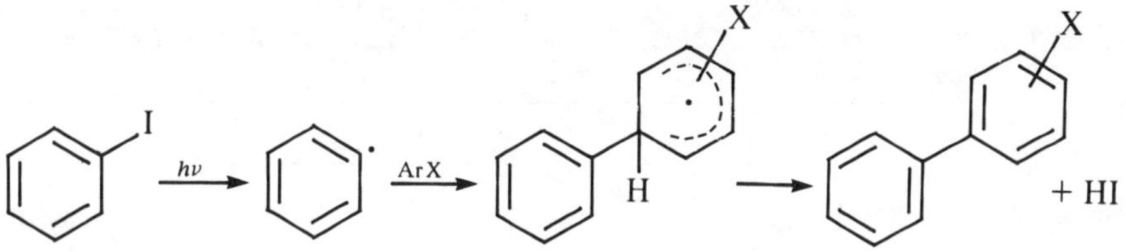

Scheme 3

Similarly, 2-[10] and 3-iodothiophenes[11,12] have been subjected to photolysis, using a low-pressure mercury lamp, in a number of aromatic substrates. Generally, 2-iodothiophene gives moderate to good yields of the phenylated products (60–70%); however, 3-iodothiophene gives significantly lower yields ($\sim 20\%$). In addition, when substituted benzenes are subjected to thienylation, product mixing resulting from o-, m-, and p-substitution are observed (Table 2).

TABLE 2. ISOMER RATIOS DURING HOMOLYTIC SUBSTITUTION[a]

Substrate	2-Iodothiophene[b]			3-Iodothiophene[c]			Iodobenzene[b]		
	o	m	p	o	m	p	o	m	p
Anisole	63.3	13.5	22.9	67.8	15.6	16.6	67.7	17.5	14.8
Cumene	39.3	36.4	24.2	39.8	39.1	21.1	36.0	41.2	22.8
Methyl benzoate	49.3	24.2	25.5	59.3	17.4	23.3	45.8	17.0	37.2

[a] Determined by gas–liquid chromatography using a 10% silicone oil column at 180°C.
[b] See Ref. 10.
[c] See Ref. 11.

The isomer ratios observed during the thienylation reactions follow the usual trend for homolytic aromatic substitution reactions and show no great differences from those obtained during phenylation. A comparison of the product distribution observed during 2-thienylation with that of phenylation does, however, reveal that there is a small decrease in the *meta*-substitution product in anisole and cumene and a corresponding increase in methyl benzoate, and that the reactivity at the *meta*-positions is generally considered to be more meaningful as a determinant of the polar characters of a radical species.[13] These variations, when considered alone, are too small to demonstrate any significant difference in the behavior of 2- and 3-thienyl radicals. However, Tiecco and Tundo have pointed out that when these results are combined with the relative reactivities of the substrates under the thienylation conditions (Table 3), the evidence for a significant effect of the hetero-atom on the reactivity of the radical becomes apparent. The 2-thienyl radical shows some selectivity and behaves as a weak electrophile, whereas the 3-thienyl radical is not appreciably different in its reactivity from the phenyl radical or the benzo[b]-thienyl radical[15] generated under similar photochemical conditions.

TABLE 3.   RATES OF REACTION OF 2- AND 3-THIENYL RADICALS RELATIVE TO PHENYL RADICALS

Substrate	Relative Rates		
	$\frac{X}{H} K_{Ph}$	$\frac{X}{H} K_{2\text{-Thienyl}}$	$\frac{X}{H} K_{3\text{-Thienyl}}$
Anisole	1.82	2.83	1.91
Cumene	0.84	1.20	0.94
Methylbenzoate	2.40	0.88	1.75

This effect is perhaps best rationalized in terms of an inductive effect by the heteroatom, since the unpaired electron is an orbital such that mesomeric inter-action with the heteroatom in either the 2- or 3-thienyl is improbable. As such, it would be expected to be smaller in the 3-thienyl than in the 2-thienyl radical.

Little practical use has been made of thienylation using 2- and 3-thienyl radicals. Kellogg and Wynberg[12] used the photolysis of 2- and 3-iodo thiophenes in benzene and hexadeuteriobenzene solution to prepare the corresponding phenyl and penta-deuterophenyl thiophenes in an investigation of the photochemical rearrangement of 2-phenylthiophene to 3-phenylthiophene. It is interesting to note that under the photolysis conditions employed to prepare the phenylated thiophenes, the rearrangement was not observed. Tiecco and Tundo[14] indicated[16,17] that the photolysis of certain 2- and 3-iodothiophene derivatives (**10** and **11**) in benzene solution gives high yields (70–90%) of the phenylated products. In all cases, the observed products were free from isomeric impurities.

10   **11**

X = 3-CHO, 4-NO$_2$, 4-CO$_2$Me, 5-Cl, 5-NO$_2$ X = 2-NO$_2$, 4-CO$_2$Me, 5-NO$_2$, 5-Cl

III. HOMOLYTIC AROMATIC SUBSTITUTION OF THIOPHENES

1. Introduction

Although homolytic substitution reactions have been extensively investigated, it is only during the last 30 years that five-membered heterocycles have received attention. After the initial research of Gomberg and Bachmann,[6] the subject lay dormant until 1950 and 1951, when Buu-Hoi and Hoan[18] and Smith and Boyer,[19] published details of preparative arylations carried out under Gomberg conditions. Thus, arylation of thiophene with o-nitrobenzene diazonium chloride gave the 2-(2'-nitrophenyl)thiophene (**12**) in a yield of 51%. As partial proof of its structure, **12** was cyclized to the thienoindole **13**. It is particularly interesting, in view of subsequent work discussed below, that the 3-substitution product (**14**) was not observed.

| 12 | 13 | 14 |

Homolytic phenylation of thiophene has since furnished a plethora of interesting results, and it is worthwhile to consider these results in more detail.

Tedder and coworkers[20, 21] noted that thiophene may be phenylated, even under acidic conditions, using dinitrobenzene diazonium salts. This behavior is striking in that anisole, which is broadly comparable with thiophene in its reactivity toward electrophiles (Table 4), under similar reaction conditions, undergoes coupling reactions with no evidence of phenylation.

In comparable experiments, thiophene and anisole were treated with 2,4-dinitrobenzene diazonium sulphate in glacial acetic acid at $0°C$; from the anisole reaction, the diazo dye was isolated in 26% yield, whereas the reaction with thiophene resulted in the formation of the phenylated derivatives **15–17** in yields of 17, 2.8, and 2%, respectively.

TABLE 4. PARTIAL RATE FACTORS FOR BROMINATION, CHLORINATION, PROTODESILYLATION, AND NITRATION OF ANISOLE (4-POSITION) AND THIOPHENE (2-POSITION)

Reaction	Thiophene	Anisole
Bromination	1.7×10^9	1.1×10^{10}
Chlorination	1.3×10^7	9.7×10^6
Protodesilylation	5×10^3	1.01×10^3
Nitration	1.5×10^2	1.7×10^2

In the case of substituted thiophenes, three different types of behavior are observed — phenylation, side-chain oxidation, and diazo coupling. 2,5-Dimethylthiophene yields a 1:1 mixture of the 2,4-dinitrophenylhydrazone (**18**) and the coupled product (**19**). Similar reactions have been observed in a number of thiophene derivatives (Table 5).

TABLE 5.　REACTIONS OF 2,4-DINITROBENZENE DIAZONIUM SALTS WITH THIOPHENE DERIVATIVES

Thiophene	Arylation (%)	Azodye Formation (%)	2,4-Dinitrophenyl Hydrazone Formation (%)
Thiophene	20	—	—
2-Methylthiophene	27	—	—
3-Methylthiophene	30	—	—
2-t-Butylthiophene	—	20	—
2-Phenylthiophene	—	13	—
2,4-Dimethylthiophene	—	60	—
2,5-Dimethylthiophene	—	13.5	13.5
2,3,5-Trimethylthiophene	—	—	93
Tetramethylthiophene	—	—	93

Under control conditions, the diazonium salts were shown to be stable, but on addition of thiophene, rapid evolution of nitrogen ensues. Little precedent exists for aromatic substrates initiating decomposition of diazonium salts in acidic media, although ferrocene[22] exhibited similar reactivity, undergoing concomitant phenylation.

A number of mechanistic possibilities exist for this reaction; however, it seems probable that 2,4-dinitrophenyl radicals are involved, since the corresponding aryl carbonium ions would be unlikely to attack a highly deactivated thiophene nucleus (e.g., compounds 15 and 16). Perhaps the simplest conceivable mechanism for arylation involves electron transfer from the thiophene to the diazonium salt, followed by elimination of nitrogen; such a reaction would lead to thiophene radical cations and phenyl radicals (Scheme 4), which could combine with the expulsion of a proton. However, if electron transfer were the initial step, then polyalkyl thiophenes should also be arylated, since the initial electron transfer should occur at a significantly faster rate.

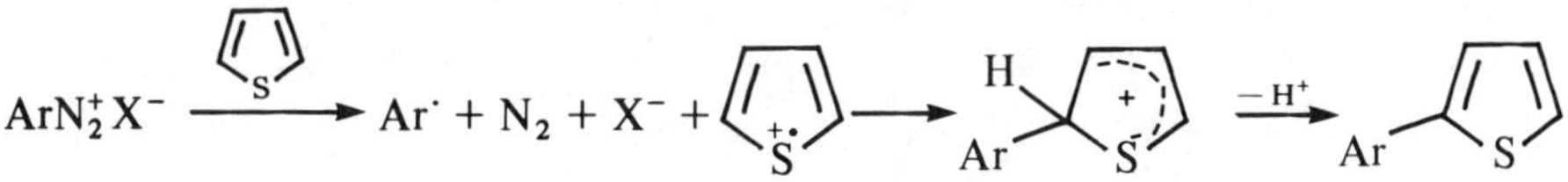

Scheme 4

Diazo coupling, rather than arylation, occurs with the polyalkyl thiophenes, which suggests that a competition between arylation and coupling may exist, and if the nucleus is sufficiently activated, coupling may be preferred. It does seem difficult however, to accept that a *t*-butyl or phenyl group is more activating than a methyl group; yet both 2-*t*-butyl and 2-phenylthiophene (Table 5) undergo coupling, whereas the 2-methyl derivative is arylated. It has been suggested that a mechanism involving the initial reaction of the diazonium salt with the sulphur atom in the thiophene ring may be involved (Scheme 5).

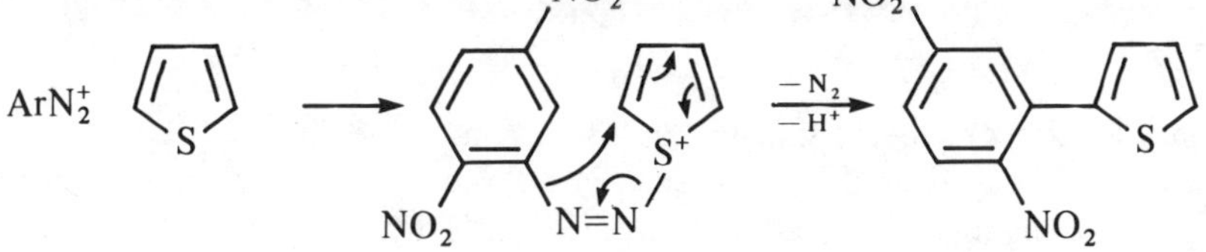

Scheme 5

Reactions at sulphur in thiophene are known to be very dependent on small changes in steric and electronic effects, and it seems possible that this reaction provides a further example of this phenomenon.

2.   Studies with Benzoyl Peroxide

Benzoyl peroxide has found use as a source of phenyl radicals during homolytic aromatic phenylation reactions through decarboxylation of the intermediate benzoyloxy radicals (Scheme 6). Under normal circumstances, decarboxylation of

$$PhCO\!-\!O\!-\!O\!-\!COPh \longrightarrow 2\,PhCOO^{\cdot} \longrightarrow 2\,Ph^{\cdot} + 2\,CO_2$$

Scheme 6

the benzoyloxy radicals is sufficiently fast to preclude their reaction with aromatic substrates; however, depending on the nature of the substrate and/or the presence of oxidants in the reaction medium, reaction with activated substrates such as naphthalene[23] or anisole[24] may result in substitution by benzoyloxy radicals. Systems such as furan,[25] with a known propensity toward 1,4-addition, yield mixtures of *cis*- and *trans*-2,5-dibenzoyloxy-2,5-dihydrofuran (**20**). In view of these results, it seemed strange when MacKay and Ford[3] reported that the only observable product on reaction of thiophene with benzoyl peroxide was 2-benzoylthiophene (**21**).

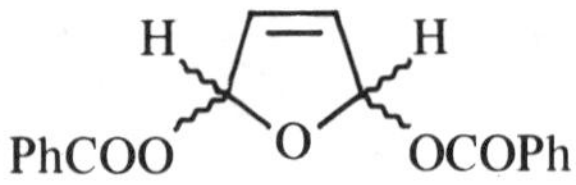

20 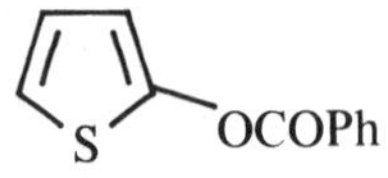   **21**

In a reinvestigation of this reaction,[26] it was shown that a complex mixture containing 2- and 3-phenylthiophenes, together with the isomeric dithienyls, 2-benzoyloxy-3,2-thienyl (**22**), and 2-thienoylbenzoate is formed. Griffin and

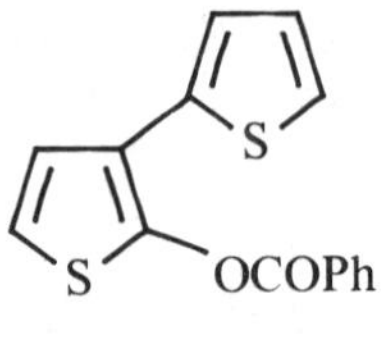

22

Martin[27] advanced a mechanism for the formation of the bithienyls by the initial formation of 2-thienyl radicals (Scheme 7) followed by reaction of the 2-thienyl radicals with thiophene to generate the isomeric bithienyls.

Hydrogen abstraction at the α-position of the thiophene ring was also observed during the photolysis of iodobenzene in thiophene and this was offered as evidence to support the intermediacy of 2-thienyl radicals in the reaction with benzoyl peroxide.

Camaggi and coworkers,[28] however, found little evidence for the involvement of 2-thienyl radicals in this reaction. They advanced an alternative mechanism

$$PhCOO{-}OCOPh \xrightarrow{\Delta} 2PhCO_2. \quad (i)$$

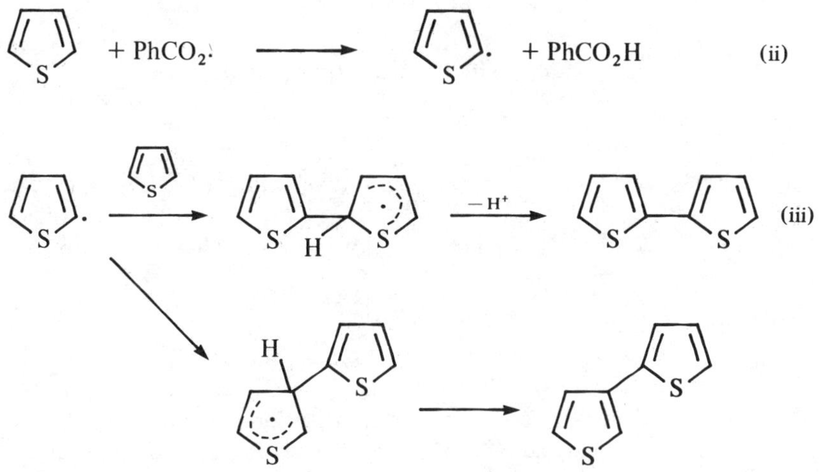

(ii)

(iii)

Scheme 7

involving a competition between decarboxylation (ultimately resulting in phenylation) and attack of the benzoyloxy radicals on the thiophene ring to produce the intermediate (**23**). This compound then undergoes dimerization at positions 3 or 5 of the ring to give **24** (Scheme 8), which subsequently eliminates benzoic acid to yield the bithienyls.

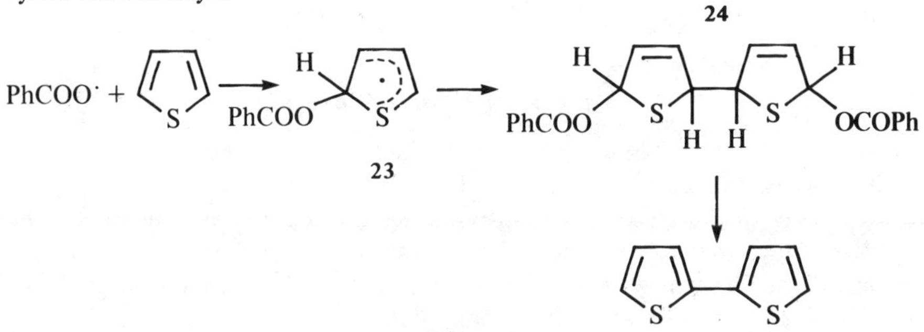

Scheme 8

Indirect supporting evidence for this reaction has been provided in that when benzoyl peroxide was allowed to react with thiophene in the presence of nitrosobenzene, the yield of 2-benzoylthiophene (**21**) formed by oxidation of the intermediate (**23**) was considerably increased, and formation of the bithienyls was suppressed. In addition, when benzoyl peroxide was decomposed in thiophene in the presence of monosubstituted benzene derivatives, no evidence of thienylation of these substrates was observed, offering further confirmation that the thienyl radicals are not directly involved.

In view of the large number of products formed by decomposition of benzoyl peroxide in thiophene, this method is clearly unsuitable for homolytic phenylation of the thiophene ring.

3. Studies with Phenylazotriphenyl Methane

Decomposition of phenylazotriphenylmethane has served as a source of phenyl radicals in the phenylation of benzene derivatives (Scheme 9). However, as in the

$$Ph-N=N-CPh_3 \longrightarrow Ph^{\cdot} + N_2 + Ph_3C^{\cdot}$$

Scheme 9

case of benzoyl peroxide, the decomposition of phenylazotriphenylmethane in thiophene does not offer a practical method of phenylation of thiophenes, owing to low yields of products and varying isomer ratios.[28] Both 2- and 3-phenylthiophenes appear to be formed and the ratio of these products is very dependent upon reaction conditions. In addition, triphenyl methane, 2-triphenylmethylthiophene, and *cis* and *trans* addition products (**25** and **26**) are observed. These addition products are readily dehydrogenated to **27** using chloranil.

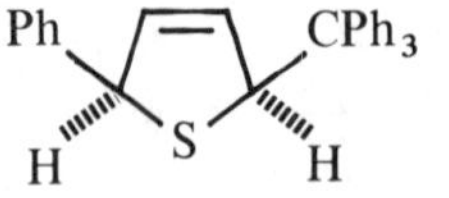

$$\begin{array}{ccc} \mathbf{25} & \mathbf{26} & \mathbf{27} \end{array}$$

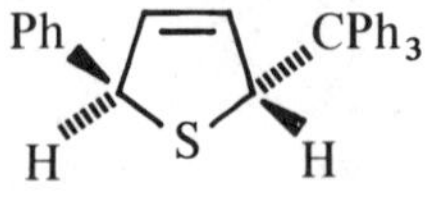

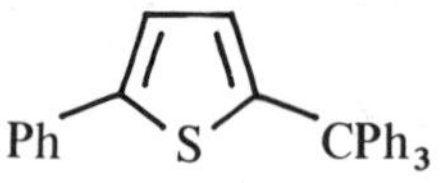

4. Practical Arylation of Thiophene

In view of the foregoing discussion, it would seem that arylation of thiophene is little more than a chemical curiosity, but of course this is not the case, and a number of practical arylation methods exist. Table 6 indicates some of the sources of phenyl radicals that have found use, along with the isomer ratios of products. From this table, it is evident that the ratio of isomers is fairly constant under comparable conditions. The sharp increase in the percentage of the 3-isomer at 150°C is consistent with other observations on the reactivity of thiophene at the 3-position at elevated temperatures.

Although isomer ratios are quoted clearly in most publications, it is sometimes more difficult to obtain details of the yields during the reactions; however, figures of 30–80% seem to be the norm.[29]

In a direct comparison of the reactivity of thiophene and benzene toward phenyl radicals, it was demonstrated that thiophene is 2.6 times more reactive than benzene

TABLE 6.   PHENYLATING AGENTS AND THEIR CORRESPONDING ISOMER RATIOS[a] ON REACTION WITH THIOPHENE.

Phenylating Agent	Yield (%)	2-Phenylthiophene	3-Phenylthiophene	Temperature (°C)
$PhN_2^+X^- + NaOH$	42.1	93	7	0
$Ph \cdot NH \cdot NH_2 + Ag_2O$	–	92.7	7.3	0
$Ph \cdot N(NO)COCH_3$	30	94.8	5.2	20
$PhNH_2 + C_5H_{11}ONO$ (in air)	48.5	93.1	6.9	30
$PhNH_2 + C_5H_{11}ONO$ (in oxygen)	–	92.3	7.7	30
$PhNH_2 + C_5H_{11}ONO$ (in nitrogen)	–	93.4	6.6	30
$Ph \cdot N = N - NHPh$	40	88.7	11.3	150

[a] Determined by gas–liquid chromatography.

and the partial rate factors for the 2- and 3-positions were 7.25 and 0.5, respectively. Thus, the orders of reactivity are: 2-position of thiophene > benzene > 3-position of the thiophene ring.

Arylation using aryl radicals generated by the aprotic diazotization of the corresponding aniline derivatives has also been examined and a remarkably constant ratio of 2:3 substitution in the thiophene ring was observed (Table 7). In each case, the

TABLE 7.   ARYLATION OF THIOPHENES USING APROTIC DIAZOTIZATION OF $R-C_6H_4NH_2$

R	2-Arylthiophene	3-Arylthiophene	Partial Rate Factors	
			k_2	k_3
H	93.1	6.9	7.25	0.5
p-OMe	91	9	4.6	0.5
p-Me	93	7	6.7	0.5
p-Cl	95	5	8.27	0.4
p-NO$_2$	96	4	10.4	0.4
m-Me	94	6	6.7	0.5
m-Cl	94	6	8.8	0.5
m-NO$_2$	96	4	10	0.4

presence of electron-withdrawing substituents in the phenyl radical results in an increase in the reactivity of the thiophene ring relative to benzene, and of course the converse is true for electron-releasing substituents.

Substituents in the thiophene ring appear to affect the position of further substitution by phenyl radicals. 2-Substituents generally favor the entry of a phenyl radical into positions 3 and 5 of the thiophene ring, with position 4 remaining relatively inert (Table 8).[16]

TABLE 8.   ISOMER RATIOS IN THE PHENYLATION[a] OF 2-SUBSTITUTED THIOPHENES

Substituent	3-Substitution (%)	4-Substitution (%)	5-Substitution (%)
Me	16.6	4.4	79.0
Br	31.7	3.6	64.7
Cl	33.7	7.3	59.0
-CO$_2$Me	49.7	2.0	48.3
NO$_2$	76.8	-	23.2

[a] Phenyl radicals generated by the aprotic diazotization of aniline.

These results are of course not unanticipated, since the presence of a strong electron-withdrawing group such as nitro or carboxylate is expected to stabilize the intermediate radical species **28** to a greater extent than **29**, whereas when R is an electron-releasing substituent, the converse should hold. Generally, the yields of these arylation reactions are very low[30] (< 25%), and in cases where simple alkyl substituents are present in the thiophene ring, benzylic coupling is observed. Thus, for example, attempts at arylation of 2-methylthiophene invariably result in the formation of the dithienylethane **30** as a by-product.

Arylation studies on 3-substituted thiophenes have produced results in accord with expectations[17] in that there is a predominance of 2-arylation observed, very little arylation at position 4, and varying degrees of arylation at position 5, depending on the substituents. For example, 3-methylthiophene gives 2-phenyl-3-methyl-thiophene and 2-phenyl-4-methylthiophene in a ratio of 2.5:1, whereas 3-nitro-thiophene gives 2-phenyl-3-nitrothiophene exclusively.

5. Heteroarylation Studies

Most of the studies of arylation *vide supra* have been carried out to determine the mechanisms and electronic effects during arylation. In contrast, most of the studies relating to heteroarylation have been carried out specifically to synthesize a given heteroaryl-substituted thiophene, and as such are of more interest to the synthetic organic chemist. Two approaches, which give moderate to excellent yields of substituted thiophenes, seem to have found the greatest use: the aprotic diazotization of heterocyclic amines under pseudo-Gomberg conditions and the photolysis of iodoheteroarenes[31,32] in thiophene. Both methods offer advantages and disadvantages; the heterocyclic amines[33] are generally readily available and inexpensive. The heteroaryl iodo-compounds usually give higher yields. In both instances, mixtures of 2- and 3-substituted products are isolated, although the ratio of 2:3 substitution appears to vary depending on the conditions and the substrate. Table 9 lists some of the heteroarylthiophenes prepared in this way.

In the case of the aprotic diazotizations, the intermediate diazonium species was decomposed at 70–80°C and the products were isolated by evaporation of the solvent. Product ratios were then determined by gas–liquid chromatography, and in all cases, a ratio of 87% of the 2-isomer to 13% of the 3-isomer was observed (the experimental error on the method was estimated to be ± 2%).

Heteroarylation of thiophene by the photolysis of 3-iodopyridine resulted in a mixture of 2-(3-pyridyl)thiophene and 3-(3-pyridyl)thiophene where the ratio of the two isomers was 7:1. The 5-iodopyrimidines gave predominently the 2-substitution product. Thus, on photolysis in thiophene, 2-chloro-5-iodopyrimidine resulted in a mixture that contained 97% of the 2-substituted product. The reasons for this apparent variation remain unclear.

TABLE 9.   HETEROARYLATION OF THIOPHENE

Heteroaryl Radical	Yield[a]	Method[b]
2-Thiazolyl	25	A
4-Methyl-5-acetyl-2-thiazolyl	35	A
4-Methyl-5-carboethoxy-2-thiazolyl	35	A
5-Bromo-2-thiazolyl	20	A
2-Benzothiazolyl	40	A
3-Methyl-5-isothiazolyl	30	A
3,4-Dimethyl-5-isoxazolyl	40	A
2-Pyridyl	20	A
3-Pyridyl	47, 42	A, B
3-Methyl-2-pyridyl	15	A
3-Quinolyl	45	A
8-Quinolyl	50	A
2-Pyrazinyl	25	A
2-Pyrimidyl	30	A
5-Pyrimidyl	58	B
2-Chloro-5-pyrimidyl	58	B
2,4-Dichloro-5-pyrimidyl	62	B
4-Chloro-5-pyrimidyl	10	B

[a] The quoted yields refer to isolated mixtures of 2- and 3-substituted thiophenes. The relative ratio of 2:3 substitution was not quoted in all cases, but appeared to range from 7:1[30] to almost exclusive 2-substitution.[31]

[b] Method A — aprotic diazotization of the corresponding heterocyclic amine with isoamyl nitrite in thiophene solution under pseudo-Gomberg conditions. Method B — photolysis of 25% w/v solutions of the corresponding iodoheteroarene in thiophene solution with a low-pressure Hanovia mercury arc lamp (quartz filter) until all traces of the iodo compound disappeared.

6.   Miscellaneous Radical Substituents of the Thiophene Ring

A.   Alkylation

Ford and MacKay[34] first observed that benzyl radicals have a high affinity for the thiophene ring. Thus, benzyl radicals generated[35] by the decomposition of di-*t*-butyl peroxide in toluene/thiophene mixtures resulted in benzylation of the thiophene ring. There was no evidence for the formation of 1,2-diphenylethane in this reaction, and it was concluded that the intermediate benzyl radicals react selectively with the thiophene ring.[36] Although the trapping of the radicals appears to be relatively efficient, their generation is inefficient and yields of the 2- and 3-benzylated thiophenes using this procedure are very low, making the reaction of little more than theoretical interest. Cyclohexylation of thiophene carried out[37] using cyclohexyl radicals generated by an analogous procedure has also been reported, but this reaction appears to have little practical value.

Thermal decomposition of diacetyl peroxide in substituted thiophene derivatives results in methylation of the ring (Scheme 10).[38] Again, this reaction has little

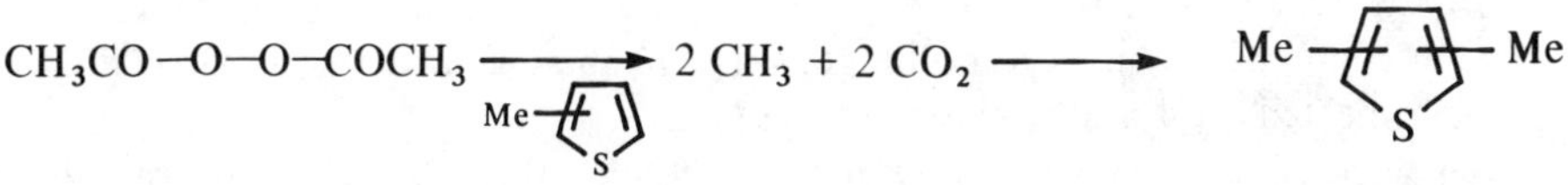

$$CH_3CO\!-\!O\!-\!O\!-\!COCH_3 \longrightarrow 2\ CH_3^{\cdot} + 2\ CO_2 \longrightarrow$$

Me—⬡—Me (thiophene with Me substituents)

Scheme 10

preparative value. Thiophene itself gives a 10% yield of 2-methylthiophene, and the yields and ratios of products for a number of substituted thiophenes are shown in Table 10.

B. Thiylation

In a series of studies,[39] Gol'dfarb and coworkers described the homolytic thiylation of thiophene derivatives by thiyl radicals generated, using Fenton's reagent, by the oxidation of the corresponding thiols. Somewhat unexpectedly, the yields from these reactions are good and the reaction appears to be of preparative value. Since the generation of the Fenton's reagent involves the use of strong hydrogen peroxide, oxidation of the ring sulphur atom is expected to be a competing reaction. However, no evidence for the formation of thiophene sulphoxides or their dimers was reported, and it would appear that the more reactive thiol yields the thiyl radical in preference to any oxidation of the thiophene ring.

Using this procedure, 2-methylthiophene was reacted with ethanethiol, propanethiol, and benzylmercaptan to yield the corresponding S-alkylthiothiophenes (**31**) in yields of 54.5, 42.0, and 70%, respectively. Attempts to identify any isomeric

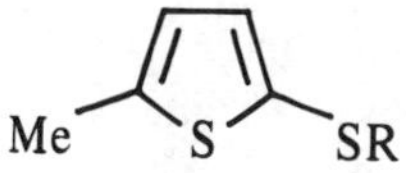

Me—⟨S⟩—SR

31

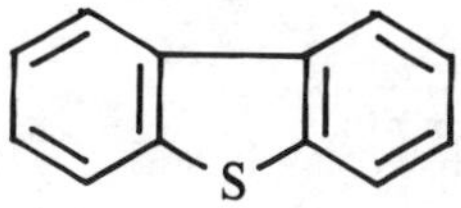

(dibenzothiophene structure)

32

TABLE 10. HOMOLYTIC METHYLATION OF THIOPHENE DERIVATIVES

Thiophene	Yield (%)	Reaction Time (h)	Position of Substitution
Thiophen	10	6	2
2-Methylthiophene	4	4	5
3-Methylthiophene	10	6	2 (major)
		6	5 (trace)
2,5-Dimethylthiophene	trace	4	3
2,4-Dimethylthiophene	21	4	5

contaminants by gas–liquid chromatography proved impractical; however, a comparison of the *S*-alkylthiothiophenes prepared by this method with known standards confirmed their isomeric purity.

In the reactions of alkylthio radicals with 2-chloro- and 2-bromo-thiophenes, large amounts of the bisalkylthiothiophenes were obtained (Scheme 11) in addition to the expected alkylthiohalothiophenes (33).

These reactions are particularly significant because they clearly demonstrate that the thiophene ring is stable under the fairly drastic oxidation conditions used and they suggest that other reactions applied to the more inert π-deficient heterocycles may be directly applicable in the thiophene series.

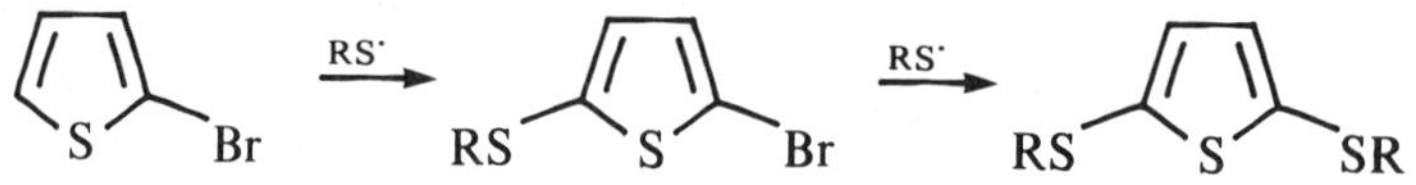

33

Scheme 11

C. Amination

Amination and amidation by radicals generated by the oxidation of formamide has been applied extensively in the pyridine series. Except for a brief reference to unpublished observations, no report of such reactions in the thiophene series has been reported.[40]

7. Condensed Thiophenes

Homolytic substitution of condensed thiophenes is expected to give rise to complex mixtures of products, because of the increased reactivity of all positions in the molecules with respect to an attacking radical species. Indeed, the first example studied confirmed this expectation.[41] Phenylation of dibenzo[b,d]thiophene (32) resulted in substitution at all four ring sites, with a relative ratio of products of 31:12:21:28 for positions 1–4 of the ring. This lack of selectivity clearly rules out homolytic substitution as a practical method for introducing substituents into this ring system. Other such examples have been quoted by Tiecco and Tundo.[14]

8. Conclusions

There is little doubt that most of the work on homolytic substitution on thiophene and its derivatives has had a strong bias toward mechanistic studies, where useful theoretical correlations have been possible. In recent years, however, it has proven possible in many instances to utilize homolytic substitution as a practical method of synthesis, and it appears that there is still considerable scope in such reactions for the preparation of specifically substituted thiophenes.

REFERENCES

1. L. Horner and E. H. Winkelman, *Newer Methods of Preparative Organic Chemistry*, Academic Press, New York, 1964, p. 182.

2. G. S. Hammond and L. M. Soffer, *J. Am. Chem. Soc.*, **72**, 4711 (1950).

3. M. C. Ford and D. MacKay, *J. Am. Chem. Soc.*, **72**, 4620 (1957).

4. R. D. Schuetz and D. M. Teller, *J. Org. Chem.*, **27**, 410 (1962).

5. D. MacKay, *Can. J. Chem.*, **44**, 2881 (1966).

6. M. Gomberg and W. E. Bachmann, *J. Am. Chem. Soc.*, **46**, 2339 (1924).

7. N. I. Putokhin and V. I. Yakovlev, *Dokl. Akad. Nauk. SSSR*, **98**, 89 (1954).

8. G. V. Shishkin and V. P. Mamaev, *Izv. Sibirsk. Otd. Akad. Nauk. SSSR*, **2**, 112 (1962) (cf. *CA*, **58**, 2421 (1963)).

9. N. Kharash, *Intra-sci. Chem. Rep.*, **3**, 203 (1969).

10. L. Benatti and M. Tiecco, *Boll. Sci. Fac. Chim. Ind. Bologna Suppl.*, **24**, 45 (1966).

11. G. Martelli, P. Spagnolo, and M. Tiecco, *J. Chem. Soc. B.*, 902 (1968).

12. R. M. Kellogg and H. Wynberg, *J. Am. Chem. Soc.*, **89**, 3495 (1967).

13. R. Ito, T. Migita, N. Morikawa, and O. Simamura, *Tetrahedron*, **21**, 955 (1966).

14. M. Tiecco and A. Tundo, *Int. J. Sulphur Chem.*, **8**, 295 (1973).

15. L. Benatti, G. Martelli, P. Spagnolo, and M. Tiecco, *J. Chem. Soc. B.*, 472 (1969).

16. C. M. Camaggi, G. De Luca, and A. Tundo, *J. Chem. Soc. Perkin II*, 412 (1972).

17. C. M. Camaggi, G. De Luca, and A. Tundo, *J. Chem. Soc. Perkin II*, 1594 (1972).

18. Ng Ph. Buu-Hoi and Ng Hoan, *Rec. Trav. Chim.*, **69**, 1455 (1950).

19. P. A. S. Smith and J. H. Boyer, *J. Am. Chem. Soc.*, **73**, 2626 (1951).

20. M. Bartle, S. T. Gore, R. K. Mackie, and J. M. Tedder, *J. Chem. Soc. Perkin I*, 1636 (1976).

21. S. T. Gore, R. K. Mackie, and J. M. Tedder, *J. Chem. Soc. Perkin I*, 1639 (1976).

22. G. D. Broadhead and P. L. Pausen, *J. Chem. Soc.*, 367 (1955).

23. R. L. Dannley and M. Gippin, *J. Am. Chem. Soc.*, **74**, 339 (1952).

24. B. M. Lynch and R. B. Moore, *Can. J. Chem.*, **40**, 1461 (1962).

25. K. E. Kolb and W. A. Black, *Chem. Comm.*, 1119 (1969).

26. C. M. Camaggi, R. Leardini, A. Tundo, and M. Tiecco, *J. Chem. Soc. Perkin I*, 271 (1974).

27. C. E. Griffin and K. R. Martin, *Chem. Comm.*, 154 (1965).

28. C. M. Camaggi, R. Leardini, M. Tiecco, and A. Tundo, *J. Chem. Soc. B.*, 1251 (1969).

29. C. M. Camaggi, R. Leardini, M. Tiecco, and A. Tundo, *J. Chem. Soc. B*, 1683 (1970).

30. R. Frimm L. Fisera, and J. Kovac, *Coll. Czech. Chem. Comm.*, **38**, 1809 (1973).

31. H-S. Ryang and H. Sakurai, *J. Chem. Soc. Chem. Comm.*, 594 (1972).

32. D. W. Allen, D. J. Buckland, B. G. Hutley, A. C. Oades, and J. B. Turner, *J. Chem. Soc. Perkin I*, 621 (1977).

33. G. Vernin and J. Metzger, *J. Org. Chem.*, **40**, 3183 (1975).

34. M. C. Ford and D. MacKay, *J. Chem. Soc.*, 4620 (1957).

35. A. L. J. Beckwith and W. A. Waters, *J. Chem. Soc.*, 1001 (1957).

36. J. I. G. Cadogan, D. H. Hay, and W. A. Sanderson, *J. Chem. Soc.*, 3203 (1960).
37. J. R. Shelton and A. L. Lipman, *J. Org. Chem.*, **39**, 2386 (1974).
38. M. J. Srogl, I. Stibor, M. Nemec, and P. Vopatrna, *Tetrahedron Lett.*, 637 (1973).
39. Y. L. Gol'dfarb, G. P. Pokhil, and L. I. Belen'kii, *Zh. Obshch. Khim.* (English translation), 37, 2541 (1967).
40. F. Minisci and O. Porta, *Adv. Heterocycl. Chem.*, 16, 123 (1974).
41. E. B. McCall, A. J. Neale, and T. J. Rawlings, *J. Chem. Soc.*, 5288 (1962).

CHAPTER X

Cycloaddition Reactions of Thiophenes, Thiophene 1-Oxides, and 1,1-Dioxides

P. H. BENDERS, D. N. REINHOUDT, and W. P. TROMPENAARS

Laboratory of Organic Chemistry, Twente University of Technology, Enschede, The Netherlands

I.   INTRODUCTION

Cycloaddition reactions of thiophenes, thiophene 1-oxides, and 1,1-dioxides are rather recent, compared with other reactions of these classes of compounds. In Hartough's book, *Thiophene and its Derivatives*, published in 1952, these reactions are dealt with in less than two pages and less than 10 relevant references are cited.[1] In a 1963 review by Gronowitz[2] the situation is not very different. The entire class of reactions is covered in less than four pages. The (2+2)-cycloadditions of thiophenes have more recently been reviewed, together with the same reactions of other heterocycles.[3] In this chapter, the literature has been covered up to the end of 1981 and results published early in 1982 have also been included, incidentally.

Various systems have been used to classify cycloaddition reactions. For reasons discussed previously,[3] in this chapter, Huisgen's proposal to classify them according to the number of atoms that each of the reactants contributes to the resulting cycloadduct has been adopted. The number of atoms contributed by the thiophene or the corresponding 1-oxide or 1,1-dioxide is given first.

The quantum chemical treatment of cycloaddition reactions of the thiophenes, 1-oxides, and 1,1-dioxides is rather limited. In relation with 1,3-dipolar cyclo-addition reactions of thiophene with benzonitrile oxides and the regioselectivity that is observed, the frontier molecular orbitals of thiophene have been calculated by Beltrame et al.[4,5] and Caramella et al.[6] These cycloadditions are predominantly governed by the HOMO(thiophene)-LUMO(1,3-dipole) orbital interaction. According to Lert and Trindle,[7] in comparison with furan, for example, it is difficult to predict the behavior of thiophene toward dienophiles because of the possible d-orbital participation of the sulfur atom in the bonding. Kanematsu et al.[8] applied the concept of continuity–discontinuity of cyclic conjugation for predicting the reactivity of thiophenes as a 4π-component in cycloaddition reactions. A quantitative argument supporting their theory is that electron-donating substituents increase the energy of the HOMO (e.g., in the case of 2,5-diaminothiophene with 2.6 eV) as compared with thiophene. This suggests that it will render these thiophenes more reactive with electron-deficient dienophiles. Thiophene 1,1-dioxides may be regarded as localized π-electron systems with a very low LUMO energy level. Therefore, they should react easily with electron-rich dienophiles. Substitution with halogen atoms will futher lower the energy level of the LUMO. Mock[9] discussed the ready dimerization of thiophene oxides in terms of a relatively small gap between the HOMO and LUMO energy levels. The effect of the sulfone moiety on the conjugation of the π-system in thiophene 1,1-dioxides is described in detail by Müller et al.[10]

II. CYCLOADDITION REACTIONS OF THIOPHENES

1. (2+1)-Cycloaddition Reactions of Thiophenes

The (2+1)-cycloaddition of thiophenes has been reported both with carbenes and with nitrenes, although in the latter case, the actual cycloadducts have never been isolated. Steinkopf and Augestad-Jensen[11] were the first to study reactions of thiophenes with diazo compounds. By heating a mixture of thiophene and ethyl diazoacetate at 126–128°C for 16 h, they obtained a small amount of a crystalline compound (mp 36.5°C) that was assigned the ethyl 2-thiabicyclo[3.1.0]hex-3-ene-6-carboxylate structure 1. The product was further characterized by conversion into the corresponding amide 2 and acid 3. Pettit[12] prepared 1 using a modified procedure and converted 1 via the isocyanate 4 into thiopyrylium halides (5). Photolysis of ethyl diazoacetate in thiophene for 50 h gives 1 in 23% yield.[13,14] Treatment of 1 with acid gives the 3-substituted thiophene (see also Ref. 17). Porter et al.[15–18] studied the rhodium(II)- and copper(I)-catalyzed addition of diazo compounds to thiophenes. They reported the formation of three types of products (6–8), the ratio of which is dependent on the nature of the diazo compound. The (2+1)-cycloadducts 1 and 6 are formed, although not exclusively, with ethyl and n-butyl diazoacetate and with ethyl diazoacetoacetate (Table 1). From the reaction of diazoacetophenone and thiophene at room temperature, only phenyl (2-thienylmethyl) ketone (7) could be isolated in 22.6% yield.[17] Under the same conditions, dimethyl diazomalonate reacts with thiophene itself and with thiophenes with +I or +M effects to give thiophenium bis(methoxycarbonyl)methylides (8) in high yields.[15,17] Except for the 2,5-disubstituted derivatives, the latter rearrange thermally to dimethyl (2-thienyl)malonates.[16] Gillespie and Porter[17] investigated

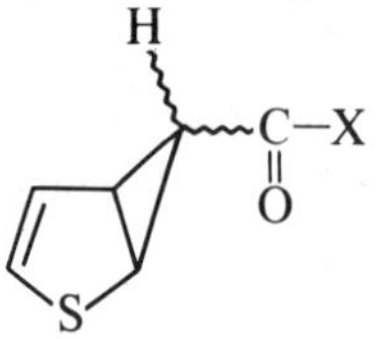

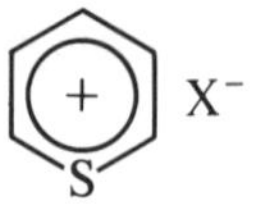

1 $X = OC_2H_5$
2 $X = NH_2$
3 $X = OH$
4 $X = NCO$

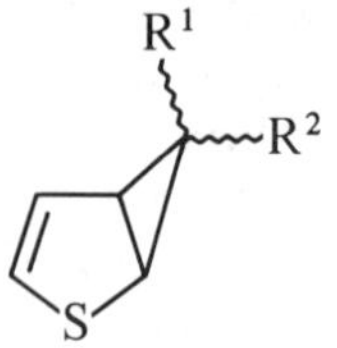

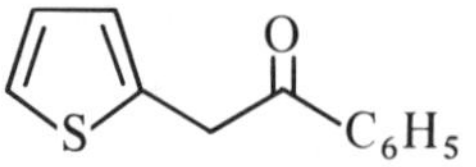

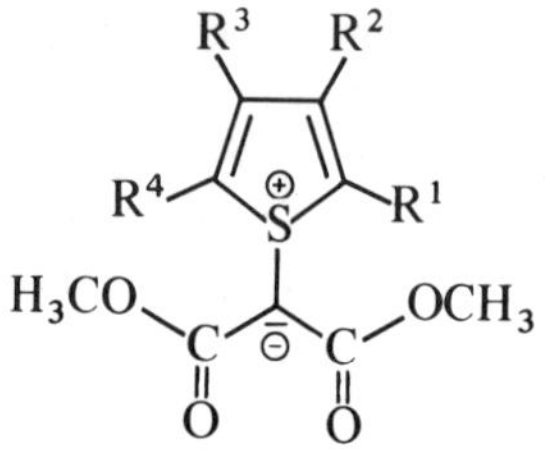

the stereochemistry of the (2+1)-cycloadducts of thiophene and carbenes by ^{1}H nmr spectroscopy. They concluded that the carboxyl group in **3**, which was obtained by hydrolysis of the ester, occupies the *exo*-position and that the reaction product of thiophene and ethyl diazoacetoacetate (**6**, $R^1 = COCH_3$, $R^2 = COOC_2H_5$) is a mixture of *endo*- and *exo*-isomers. 2,5-Dichlorothiophene reacted in the presence of rhodium acetate with ethyl diazoacetoacetate to give the (2+1)-cycloadduct **9** in 67% yield. This compound is not stable thermally and rearranges upon refluxing in toluene for 38 h to give the benzene derivative **10**.[18]

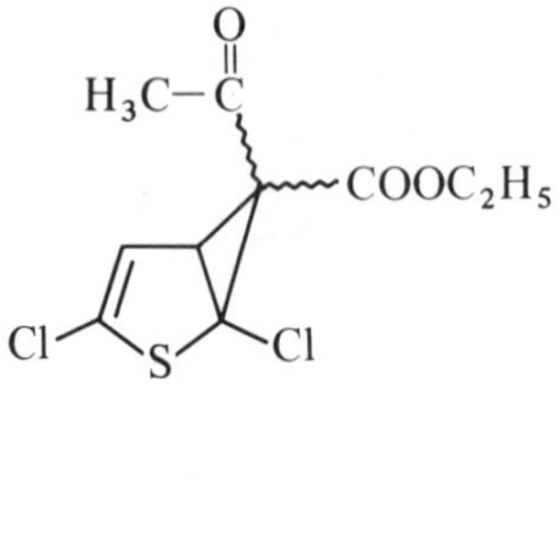

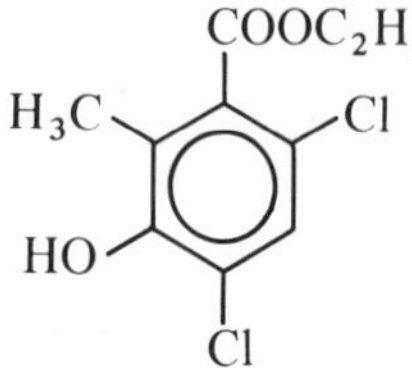

TABLE 1. Rh(II)-CATALYZED REACTIONS OF THIOPHENE WITH DIAZO COMPOUNDS[17]

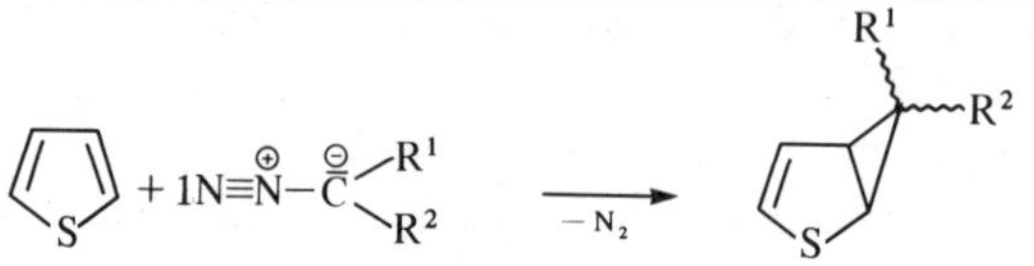

R^1	R^2	Temperature (°C)	Reaction Time (h)	Yield of 6 (%)
H	COOn-C$_4$H$_9$	32	73	60
H	COOn-C$_4$H$_9$	42	3	58
H	COOn-C$_4$H$_9$	52	2.5	62
H	COOn-C$_4$H$_9$	62	0.5	62
H	COOn-C$_4$H$_9$	85	0.25	71
COCH$_3$	COOC$_2$H$_5$	85	0.5	13^a
COCH$_3$	COOC$_2$H$_5$	25	20	5^a

^{a} The major product in this reaction was ethyl 3-oxo-2-(2-thienyl)butanoate.

More complex carbenoids, such as diazocyclopentadienes,[19] diazoanthrone,[20] and 4-diazo-2,6-di(*tert* butyl) 2,5-cyclohexadien-1-one,[21-23] also react with thiophenes. Irradiation of tetraphenyldiazocyclopentadiene in the presence of 2,5-dimethylthiophene gave a (2+1)-cycloadduct (**11**) in 1.5% yield. The major product, however, was **12** (30% yield), which is formed by a photochemical (di-π-methane) rearrangement of **11**.[19] Cauquis et al.[20] found a sulfur-free reaction product (**14**)

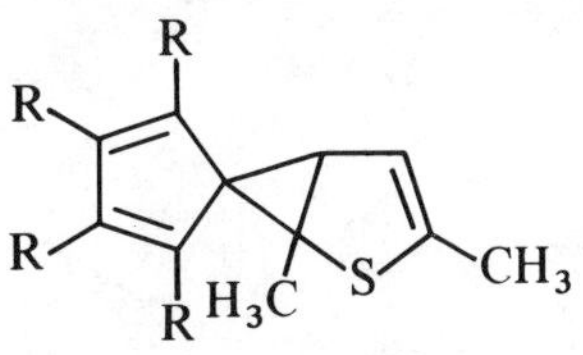

| 11 | R = C₆H₅ |

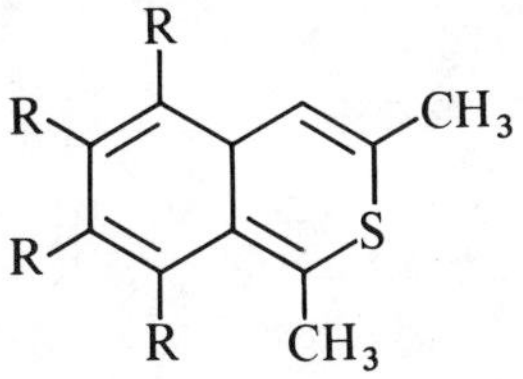

11   R = C_6H_5 **12**   R = C_6H_5

instead of a (2+1)-cycloadduct when diazoanthrone was reacted with thiophene photochemically. They have discussed the formation of **14** both in terms of desulfuration of the bis(2+1)-cycloadduct **13** and via rearrangement of the (2+1)-cycloadduct **15** to the thioaldehyde **16**, followed by reaction of the thioaldehyde group in **16** with a second molecule of diazoanthrone or with the corresponding carbene, and subsequent desulfuration of the thiirane **17** (Scheme 1). Similar results

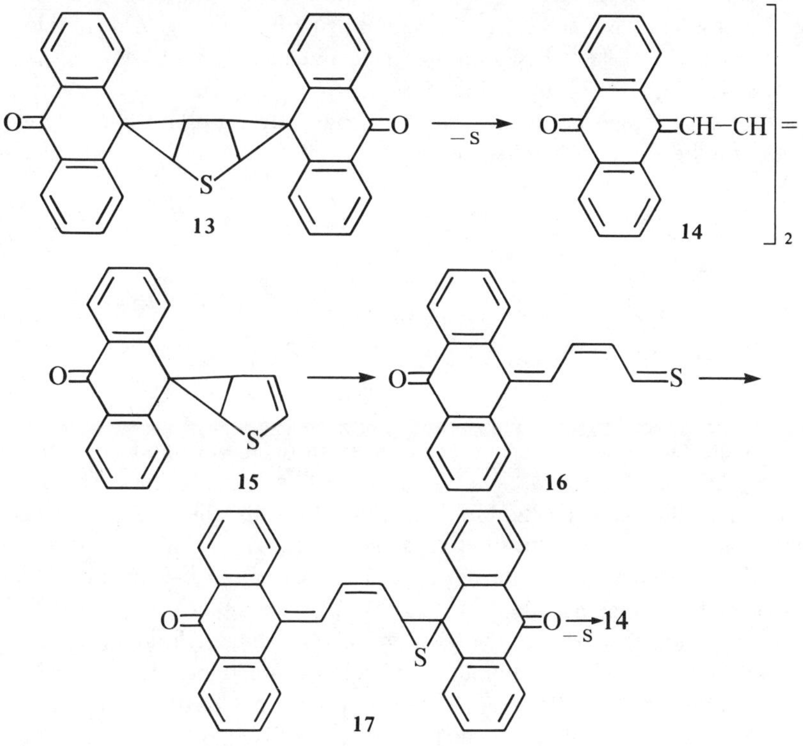

Scheme 1

were found by Plekhanova et al.[21-23] in the reaction of thiophene with 2,6-di(*tert*-butyl)-2,5-cyclohexadien-1-one carbene.

2-Methylthiophene reacts with dichlorocarbene to give a 2*H*-thiopyran (**20**), most likely by (2+1)-cycloaddition to give **18**, which may undergo ring opening with subsequent deprotonation to **19**. Reaction of the exocyclic double bond in **19** with dichlorocarbene yields **20**[24] (Scheme 2).

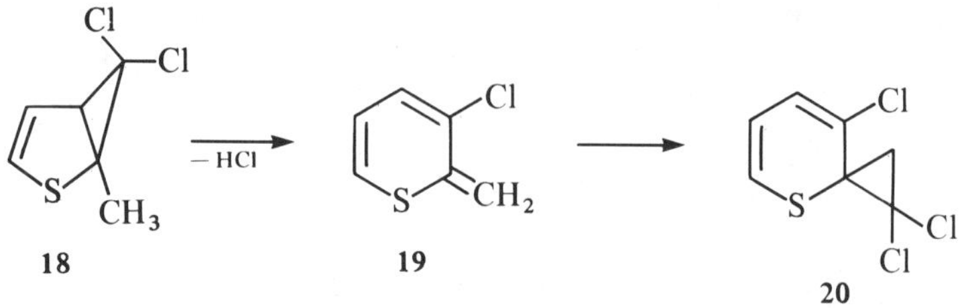

Scheme 2

Hitherto, 6-aza-2-thiabicyclo[3.1.0]hex-3-enes (e.g., **21**) have not been isolated from reactions of thiophenes and nitrenes, although they have been proposed as intermediates. Thermolysis at 130°C of ethyl azidoformate in thiophene or 2,5-dimethylthiophene gave *N*-(ethoxycarbonyl)pyrrole and the 2,5-dimethyl derivative in 21 and 18% yield, respectively.[25] Both (2+1)- and (4+1)-cycloaddition have been proposed as the first step. Irradiation of pentafluorophenyl azide in thiophene gave **22** in 11.5% yield.[26] Based on the results obtained with 2-azidophenyl thienyl sulfides, Lindley, Meth-Cohn, and Suschitzky[27] proposed that an electrophilic attack of the nitrene at the 2-position followed by a ring opening-ring closure process is more likely in these reactions.

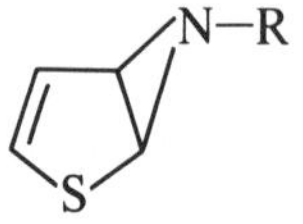

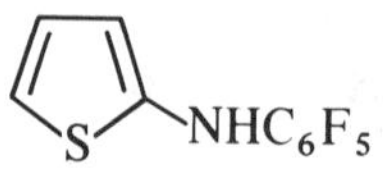

Intramolecular nitrene insertion reactions of nitrenes and thiophenes have been proposed by Jones and coworkers[28,29] to yield transient aziridines. The products isolated from the decomposition of 2-(2-azidobenzyl)thiophene (**23a**) in trichloro-benzene at 190°C were thieno[3,2-*b*]quinoline (**24a**) and 1,2-dihydrophyrrolo-[1,2-*a*]indole-3-thione (**25**) in yields of 5 and 6%, respectively. A similar reaction of 2-(2-azidobenzyl)-5-methylthiophene (**23b**) gave only 2-methylthieno[3,2-*b*]-quinoline (**24b**). Reaction of 3-(2-azidobenzyl)-2,5-dimethylthiophene (**26**) was more selective and gave 2,4-dimethylthieno[3,2-*c*]quinoline (**27**) in 34% yield. The formation of **30** in the thermal decomposition of (2-azidophenyl)-dithienyl-methanes, via the ring opened product **29**[30,31] (Scheme 3), is experimental evidence that the (2+1)-cycloadducts (e.g., **28**) are indeed intermediates in these reactions.

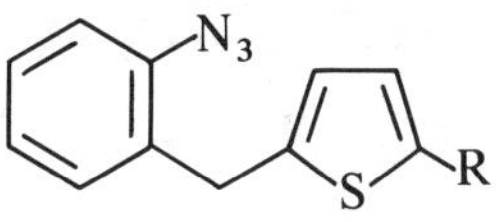

23
a   R = H
b   R = CH$_3$

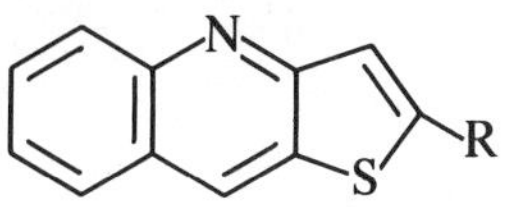

24
a   R = H
b   R = CH$_3$

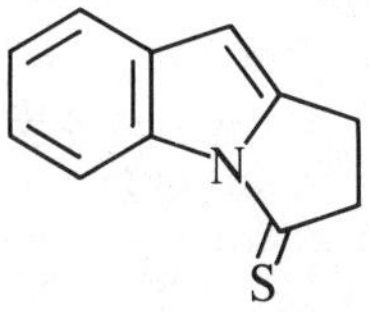

25

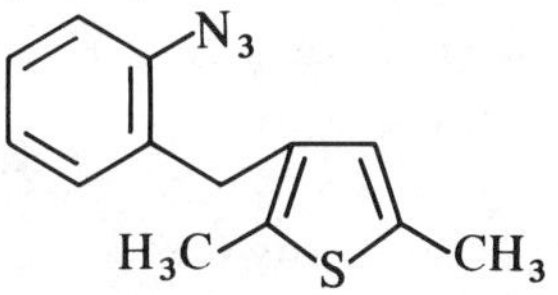

26

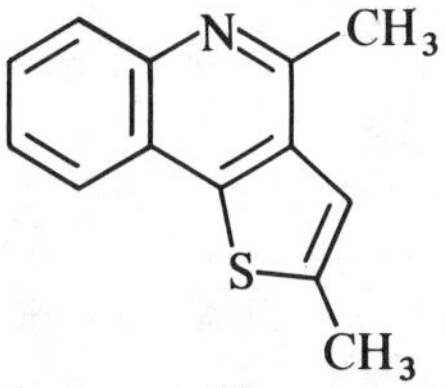

27

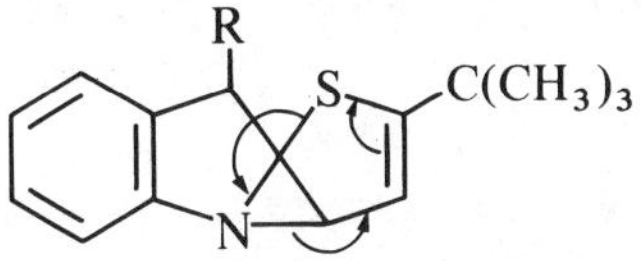

28

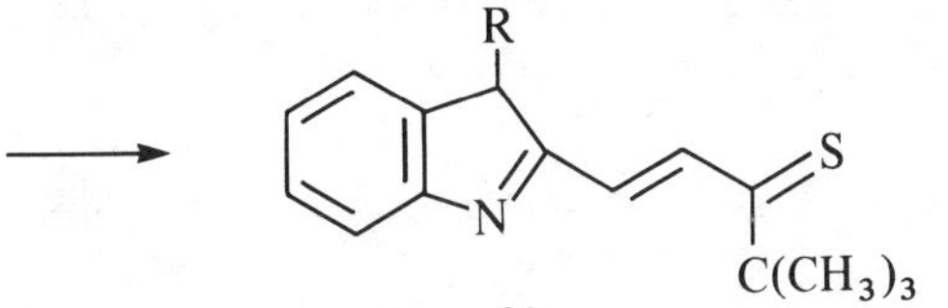

29

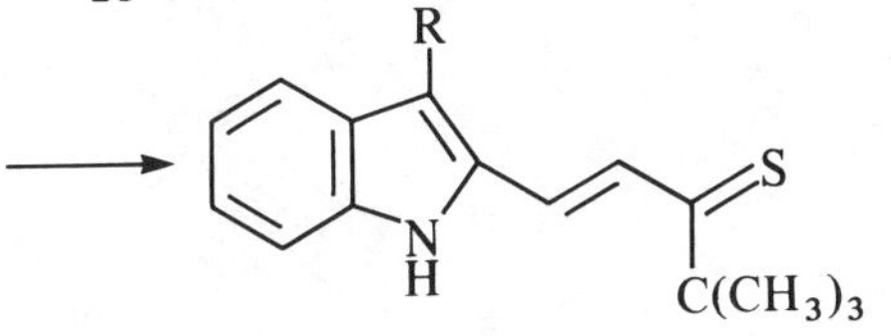

30

R = 5-(*tert*-Butyl)-2-thienyl

Scheme 3

2.   (2+2)-Cycloaddition Reactions of Thiophenes

A.   *Introduction*

The thermal (2+2)-cycloaddition of thiophenes has been observed in the aluminum chloride catalyzed reactions of alkyl-substituted thiophenes with dicyanoacetylene and in reactions of (*N,N*-dialkyl)aminothiophenes with dimethyl acetylenedicarboxylate (DMAD). In addition to the (4+2)-, and in the case of benzyne the (3+2)-cycloaddition mode, (2+2)-cycloaddition also takes place in the reactions of thiophenes with thiophyne and with benzyne.

Photochemical (2+2)-cycloaddition has been observed in reactions of dimethylthiophenes with carbonyl compounds and in reactions, in which an enone moiety constitutes part either of the olefinic reagent or of the thiophene. Also a photochemical (2+2)-cycloaddition of thiophene to a (heterocyclic) C=N bond has been reported.

B.   (2+2)-Cycloaddition in Reactions of Alkyl-Substituted Thiophenes with Dicyanoacetylene

Tetra- and trialkylthiophenes **31** react with dicyanoacetylene in the presence of 1 eq. aluminum chloride in dichloromethane at 0°C to 2-thiabicyclo[3.2.0]hepta-3,6-diene-6,7-dicarbonitriles (**32**)* as the major products.[32-35] With the trialkyl-substituted thiophene, **31d**, (2+2)-cycloaddition takes place both to the 2,3 and to the 4,5 position to yield **32d** and **32e** in a 2:1 ratio. 4,5,6,7-Tetrahydro-1,3-dimethylbenzo[c]thiophene (**31a**) reacts with dicyanoacetylene to give an "ene" adduct (**33**, 5.5%) in addition to the (2+2)-cycloadduct (**32a**). The thiophenes **31c-d** give as by-products small amounts (4 and 6%) of the 1,2-benzenedicarbo-nitriles **38c-d**, which result from (4+2)-cycloaddition and subsequent extrusion of sulfur. 2,5-Dimethylthiophene gives a 2:1 reaction product (**34**), the formation of which can be explained via Friedel-Crafts alkylation of the thiophene by the initially formed (2+2)-cycloadduct.[35]

31

32

a	$R^1, R^2 = (CH_2)_4$	a	$R^1, R^2 = (CH_2)_4$	
b	$R^1 = R^2 = CH_2D$	b	$R^1 = R^2 = CH_2D$	
c	$R^1 = R^2 = CH_3$	c	$R^1 = R^2 = CH_3$	
d	$R^1 = C(CH_3)_3, R^2 = H$	d	$R^1 = C(CH_3)_3, R^2 = H$	
		e	$R^1 = H, R^2 = C(CH_3)_3$	

33

34

On heating at 110–140°C, the (2+2)-cycloadducts **32b–e** rearrange to the corresponding 4,5-dicarbonitrile isomers **35b–e** in yields of 82–84%.[32-34, 38] Kinetic studies point to a concerted symmetry-allowed antarafacial, antarafacial Cope rearrangement.[38] Under these conditions, the tricyclic (2+2)-cycloadduct **32a**, which was designed not to undergo a Cope rearrangement, "disproportionates"

---

* Initially, the products of reaction of dicyanoacetylene with tetramethylthiophene (**31c**) and with the 3,4-bis(monodeuteriomethyl)-derivative (**31b**), obtained under similar conditions, were erroneously assigned a thiepine structure.[36] Unfortunately, this incorrect assignment has persisted in a comprehensive treatment of organic chemistry.[37]

to the desulfurated and sulfurated products **36a** and **39a** in yields of 88 and 70%, respectively.[39] On irradiating with UV light, the (2+2)-cycloadducts **32b–e** rearrange via cleavage of the C-1-S bond to the 1,7-dicarbonitrile isomers **37b–e** in yields of 67–74%.[34,38] Prolonged irradiation of **32d** or **37d** gives a different isomer (**37f**), which is formed by a rearrangement reaction of **37d**.[38]

| | 35 | | 36 | | 37 | | 38 |

| a | $R^1, R^2 = (CH_2)_4$ |

b	$R^1 = R^2 = CH_2D$		b	$R^1 = R^2 = CH_2D, R^3 = CH_3$
c	$R^1 = R^2 = CH_3$		c	$R^1 = R^2 = R^3 = CH_3$
d	$R^1 = C(CH_3)_3, R^2 = H$		d	$R^1 = C(CH_3)_3, R^2 = H, R^3 = CH_3$
e	$R^1 = H, R^2 = C(CH_3)_3$		e	$R^1 = H, R^2 = C(CH_3)_3, R^3 = CH_3$
			f	$R^1 = CH_3, R^2 = H, R^3 = C(CH_3)_3$

| | 39 | | 40 |

| a | $R^1, R^2 = (CH_2)_4, R^3 = CH_3, R^4 = CN$ |
| d | $R^1 = C(CH_3)_3, R^2 = H, R^3 = CN, R^4 = CH_3$ |

Apart from the isomer **37d** with the *tert*-butyl group on the bridgehead C-5 atom, both the 2-thiabicyclo[3.2.0]hepta-3,6-diene-4,5-dicarbonitriles (**35**), as well as the 1,7-dicarbonitrile isomers (**37**), are thermally rather stable up to 140°C. 5-(*tert*-Butyl)-3,6-dimethyl-2-thiabicyclo[3.2.0]hepta-3,6-diene-1,7-dicarbonitrile (**37d**) reacts at 140°C to give a mixture of the Cope rearranged isomer **40**, the desulfurated product (the 1,2-benzenedicarbonitrile **38d**), and possibly the sulfurated derivative of **37d** (namely, **39d**). Pyrolysis of the 2-thiabicyclo[3.2.0]hepta-3,6-dienes **35c–e** and **37c–f** at 285°C results in the extrusion of sulfur with the formation of the 1,2-benzenedicarbonitriles **36c–e** and **38c–f**, respectively, in yields of 42–56%. This reaction is assumed to take place via isomerization of the 2-thiabicyclo[3.2.0]hepta-3,6-diene to a thiepin formed by ring opening of the *cis*-fused cyclobutene ring, and subsequent electrocyclization to a thianorcaradiene, from which sulfur is rapidly eliminated.[39] However, other pathways may be followed at least in part, as was demonstrated in the pyrolysis of **37b** in the presence of triphenylphosphine at 275°C, which gave, according to ^{1}H nmr spectroscopy, an approximately equimolar mixture of the 1,2-benzenedicarbonitriles **36b** and **38b**, respectively.[34]

C. (2+2)-Cycloaddition in Reactions of (N,N-Dialkyl)Aminothiophenes with Dimethyl Acetylenedicarboxylate

Most examples of a (2+2)-cycloaddition in reactions of (N,N-dialkyl)aminothiophenes with DMAD to give 2-thiabicyclo[3.2.0]hepta-3,6-dienes concern the reaction of 3-(1-pyrrolidinyl)thiophenes (**41**). 3-(1-Pyrrolidinyl)thiophenes actually behave as "pseudo" enamines. The reactivity depends on the substitution pattern of the thiophene nucleus, which governs the electron density at C-2. It is interesting to note the relationship between the substituents attached to the thiophene ring, the electron density at the 2-position, as reflected in the chemical shift of H-2 (Table 2), and the reactivity in (2+2)-cycloaddition. Whereas DMAD reacts with 3-(1-pyrrolidinyl)- (**41a**) and with 3-methyl-4-(1-pyrrolidinyl)thiophene (**41b**) in deuteriochloroform even at −30°C,[33,40] with dimethyl 4-(1-pyrrolidinyl)-2,3-thiophenedicarboxylate (**41i**), a reaction was only observed at 100°C.[41,42] The presence of a substituent at C-2 lowers the "enamine" type of reactivity of **41**, because of steric interference with the pyrrolidinyl group. For example, 2-methyl-3-(1-pyrrolidinyl)thiophene (**41j**) reacts only at room temperature, whereas 3-

TABLE 2.   CHEMICAL SHIFTS OF H-2 IN 3-(1-PYRROLIDINYL)THIOPHENES (**41a–i**)

41

Compound	R^1	R^2	R^3	δ (H-2)^a	Reference
41a	H	H	H	5.5–5.7(m)	44
41b	H	CH$_3$	H	6.04, 6.15(d)	33, 42, 45
41c	H	H	CH$_3$	5.51 (brd s)	42, 45
41d	H	H	i-C$_3$H$_7$	5.56 (brd s)	42
41e	H	H	t-C$_4$H$_9$	5.51(d)	42
41f	H	H	C$_6$H$_5$	5.74(d)	42, 45
41g	H	H	2-CH$_3$O—C$_6$H$_4$	5.84 (brd s)	42
41h	H	C$_6$H$_5$	C$_6$H$_5$	6.21(s)	42, 45
41i	H	COOCH$_3$	COOCH$_3$	6.38(s)	42, 45
41j	CH$_3$	H	H	—	

^{a} **41a** in CCl$_4$, **41b–i** in CDCl$_3$. Abbreviations used: m, multiplet; d, doublet; and brd s, broad singlet.

TABLE 3. BENZENE DERIVATIVES (45) AND 6,7,7a,8-TETRAHYDRO-5H-THIENO-[2,3-b]PYRROLIZINES (53) ISOLATED FROM THE REACTIONS OF 3-(1-PYRROLIDINYL)THIOPHENES (41) WITH DMAD

E = COOCH$_3$

Compound	R^1	R^2	R^3	Yield (%)	Reference
45a	H	H	H	a	33
45b	H	CH$_3$	H	a	33
45e	H	H	t-C$_4$H$_9$	77	42
45f	H	H	C$_6$H$_5$	42	42
45i	H	E	E	35	42
45j	CH$_3$	H	H	a	33
53b		CH$_3$	H	50	42
53c		H	CH$_3$	52	42
53d		H	i-C$_3$H$_7$	61	42
53e		H	t-C$_4$H$_9$	69	42
53f		H	C$_6$H$_5$	63	42
53g		H	2-CH$_3$O–C$_6$H$_4$	64	42
53h		C$_6$H$_5$	C$_6$H$_5$	31	42
53i		E	E	46	42

a Yields of 50–80% have been reported for 45a, 45b, and 45j.

methyl-4-(1-pyrrolidinyl)thiophene (41b) reacts at $-30°$C.[33,40] The products of the reaction of 3-(1-pyrrolidinyl)thiophenes with DMAD strongly depend on the solvent used. In apolar solvents, benzene derivatives (45) are the isolated products. In protic polar solvents, 6,7,7a,8-tetrahydro-5H-thieno[2,3-b]pyrrolizines (53) are formed (Table 3). In aprotic polar solvents, the two modes of reaction compete.[33,40–43]

A plausible reaction pathway for the formation of the benzene derivatives (45) comprises the following: (1) (2+2)-cycloaddition of DMAD to the 2,3-position of the 3-(1-pyrrolidinyl)thiophene (41) to give a dimethyl 5-(1-pyrrolidinyl)-2-thiabicyclo[3.2.0]hepta-3,6-diene-6,7-dicarboxylate (42), (2) opening of the cis-annulated cyclobutene ring in 42, and (3) desulfuration of the resulting thiepin (43), presumably via the isomeric thianorcaradiene (44) (Scheme 4). The validity of the proposed reaction pathway has been proven by monitoring the course of the reaction by ^{1}H nmr spectroscopy, using 3-methyl-4-(1-pyrrolidinyl)thiophene (41b) and 2-(tert-butyl)-4-(1-pyrrolidinyl)thiophene (41e) as the substrates at -30 and $25°$C, respectively.[33,40,42] In aprotic polar solvents, the intermediate thiepin (43) (or thianorcaradiene, 44) could be intercepted when dimethyl 4-(1-pyrrolidinyl)-2,3-thiophenedicarboxylate (41i) and 2-phenyl-4-(1-pyrrolidinyl)thiophene (41f)

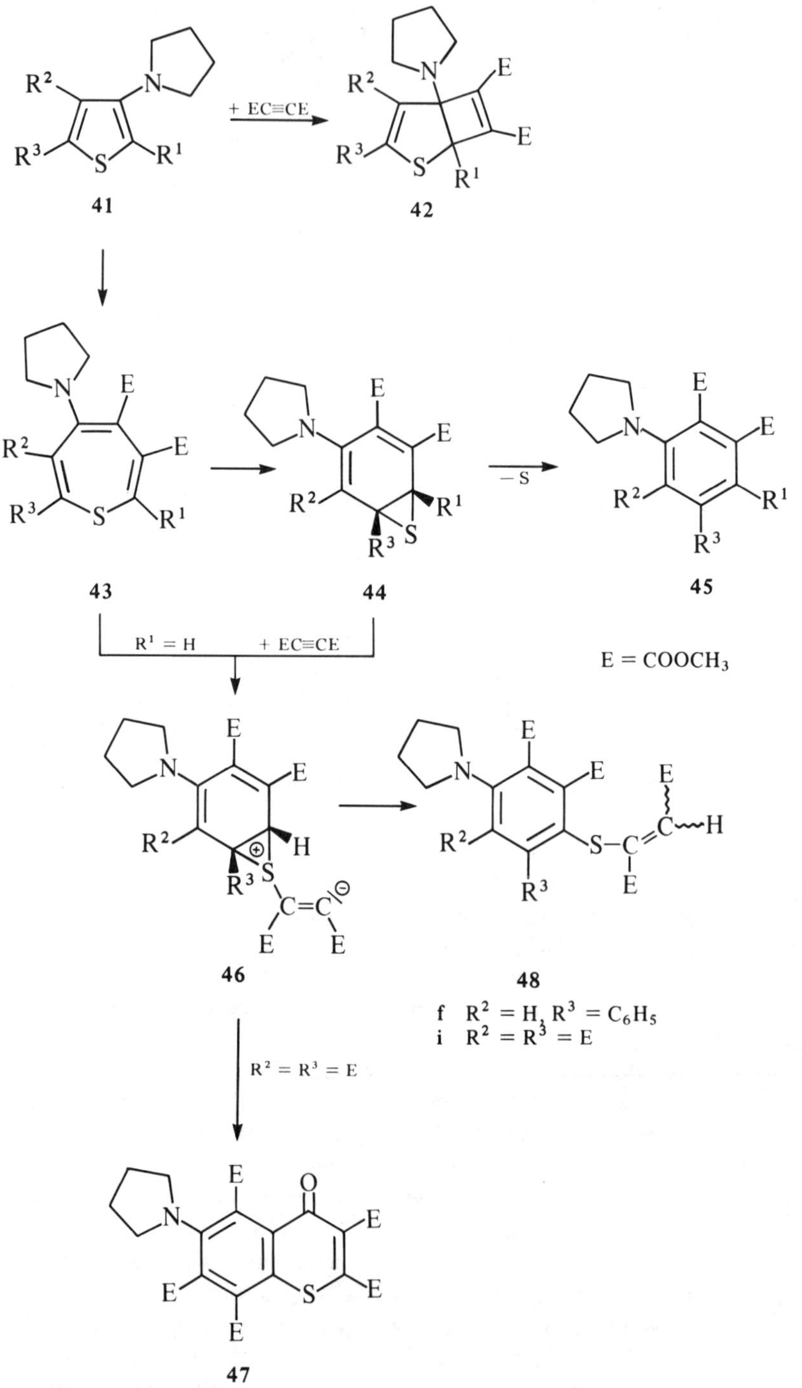

Scheme 4

were reacted with an excess of DMAD. In these cases, a thiopyran-4-one derivative (47) and/or dimethyl (*E*)- and (*Z*)-(phenylthio)-2-butenedioates (48i, f) have been isolated in addition to a 5*H*-thieno[2,3-*b*]pyrrolizine (53i, f). The formation of 47 has been rationalized by *S*-alkylation of the initially formed thiepin (43) (or thianorcaradiene, 44) by DMAD, followed by an intramolecular nucleophilic attack of the carbanion at the adjacent ester function. The methoxy group of the ester acts as the leaving group and will subsequently abstract a proton. The formation of 48 has been accounted for by a proton transfer to the carbanionic center and aromatization[41,42] (Scheme 4).

A possible mechanism for the formation of 6,7,7a,8-tetrahydro-5*H*-thieno-[2,3-*b*]pyrrolizines (53) is as follows. The initial step is an electrophilic attack of DMAD at C-2 of the thiophene 41 to give an intermediate (49), which is the tied ion pair form of the 1,4-dipole. This step will be the rate-determining one, because it involves the loss of the thiophene aromaticity. Differentiation in the reaction pathway occurs in the second solvent-depending step. In apolar solvents, a second σ-bond is formed to give the (2+2)-cycloadduct, whereas in polar solvents, the solvent-stabilized charge-separated 1,4-dipolar species 50 originates. Intramolecular abstraction of one of the α-aminomethylene protons by the carbanionic center yields the azomethinylide 51. Tautomerization (i.e., a 1,3 hydrogen shift probably occurring via the solvent) gives the 1,5 dipolar intermediate 52, which finally undergoes a symmetry-allowed disrotatory electrocyclization to 53[42] (Scheme 5). The formation of pyrrolizines turned out to be a general reaction of pyrrolidinyl enamines and DMAD.[46,47] With pyrrolidinyl enamines of cyclic ketones, the (2+2)-cycloadduct (compare 42) obtained from the reaction of the enamine and DMAD in an apolar solvent could be converted into a pyrrolizine (compare 53) by reacting them in methanol. This observation supports strongly the mechanism given above.[48] On the contrary, Michael adducts, for example, dimethyl (*E*)- and (*Z*)-[2-(1-pyrrolidinyl)benzo[*b*]thien-3-yl]-2-butenedioate (55) have been shown to give pyrrolizine derivatives (56) on heating.[49] This indicates that a Michael adduct has to be considered also as a possible intermediate in the formation of the 5*H*-thieno[2,3-*b*]pyrrolizines (53). The formation of the Michael adduct 54 (see Scheme 5) can be rationalized by intermolecular proton transfer, most probably via the solvent, in the 1,4-dipolar charge-separated ion pair 50.[47]

55   56   E = COOCH₃

With dicyanoacetylene as the electron-deficient acetylene, 3-methyl-4-(1-pyrrolidinyl)thiophene (41b) reacts instantaneously at − 60°C to give the linear Michael adduct 57, as was concluded on the basis of ¹H nmr spectroscopic data; whereas from the reaction of the less reactive dimethyl 4-(1-pyrrolidinyl)-2,3-

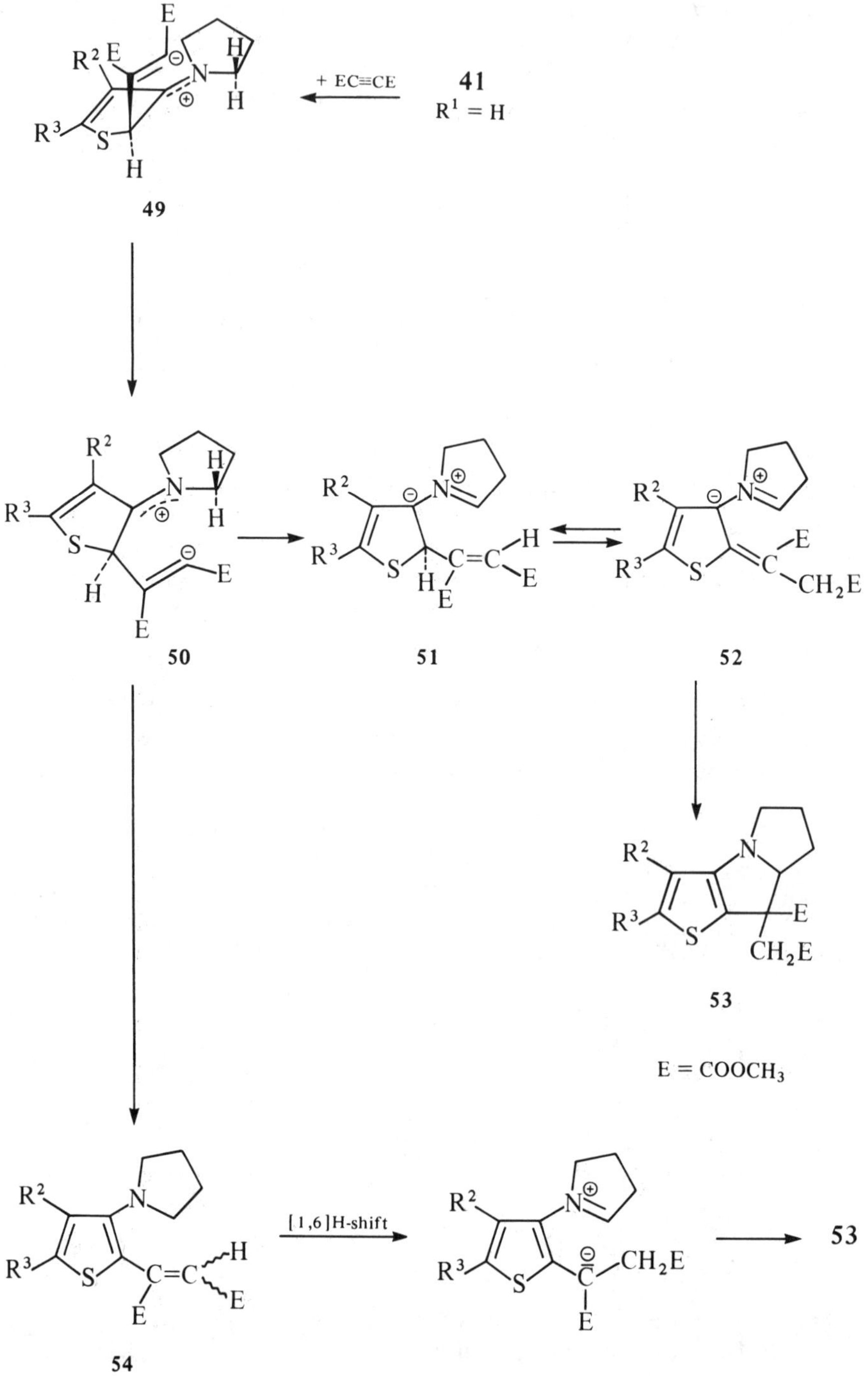

Scheme 5

thiophenedicarboxylate (**41i**), at $-60°C$, the 6,7,7a,8-tetrahydro-5*H*-thieno[2,3-*b*]-pyrrolizine **58** was isolated.[33,42]

A (2+2)-cycloaddition to the unsubstituted side of the ring of an aminothiophene has been observed when ethyl 2-amino-3-thiophenecarboxylate reacts with DMAD in DMSO. The main product of this reaction is the Michael adduct **59**. In addition, the (2+2)-cycloadduct **60**, the 1:2 addition product **61**, and the corresponding aromatic derivative **62** have been obtained.[50]

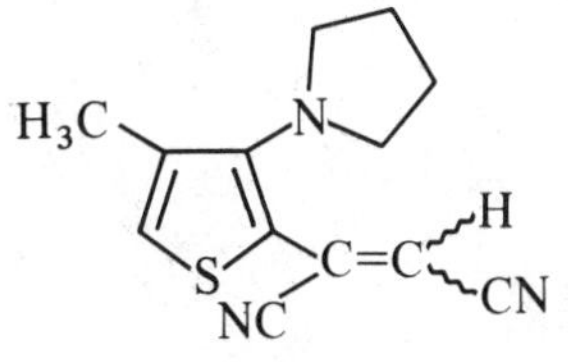

57

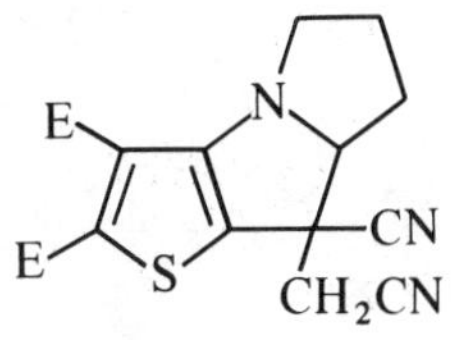

58

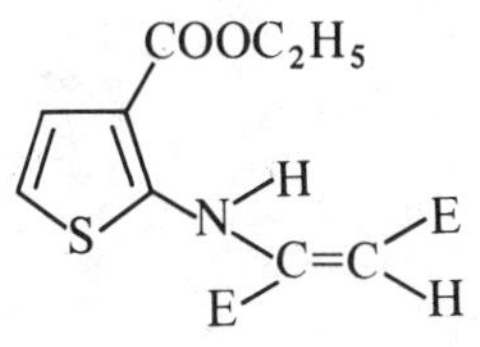

59

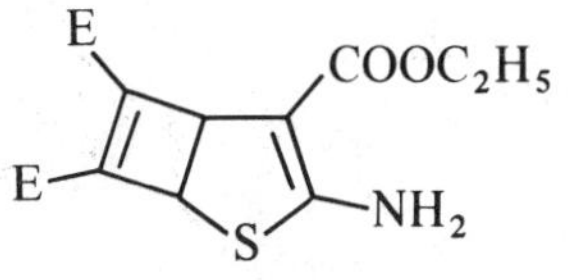

60

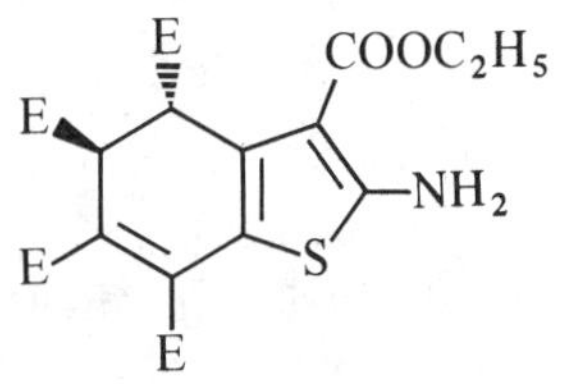

61

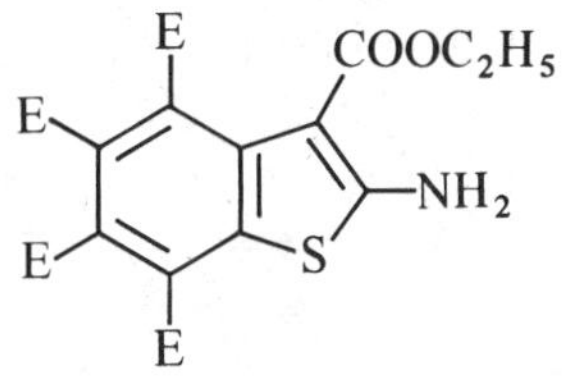

62   E = COOCH$_3$

D.   (2+2)-Cycloaddition in Reactions of Thiophenes with (het)-Arynes

(2+2)-Cycloaddition concurrent with (4+2)-cycloaddition of thiophene to 2,3-didehydrothiophene (thiophyne, **63**) was suggested to account for the formation of thieno[2,3-*b*]thiophene (thiophthene, **64**) on the one hand and of benzo[*b*]thiophene (**65**) and phenylthiophene (**66**) on the other hand, as the minor products in the pyrolysis of thiophene. The major product is a mixture of all three possible bithienyls (**67**)[51,52] (Scheme 6) (see also Section II.6.B).

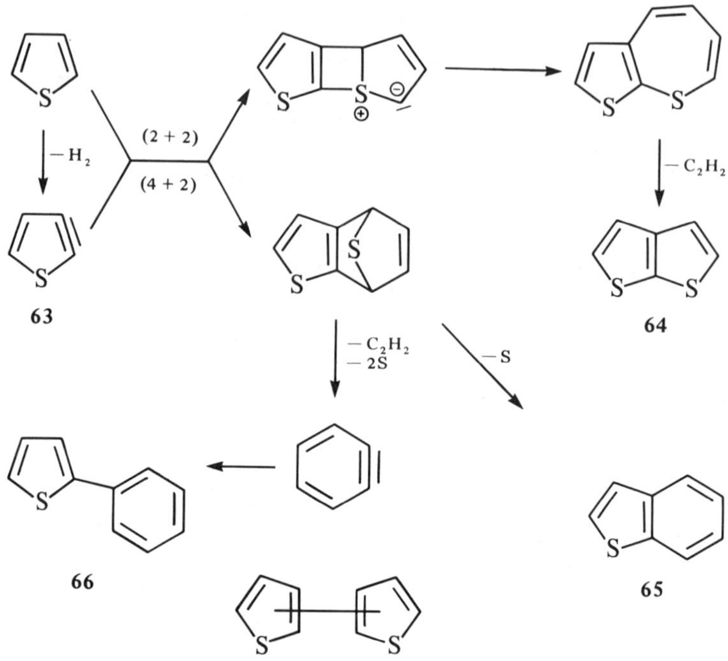

Scheme 6

However, the intermediacy of the aryne **63** in this reaction, as well as in the copyrolysis of thiophene and (hetero)aromatic dicarboxylic anhydrides, has been questioned in a recent comprehensive review on five-membered hetarynes by Reinecke.[53] Until now, the only methods suitable for the generation of 2,3-didehydrothiophene* (thiophyne, **63**) appear to be flash- or flow-vacuum thermolysis (FVT) and plasmolysis of 2,3-thiophenedicarboxylic anhydride (**68**) in the presence of aryne trapping agents.[53-55] The work done in this field has been surveyed in the review mentioned above. The most interesting result in this context is the formation of 4,5- (**71**) and 6,7-dimethylbenzo[b]thiophene (**72**) in preference to 4,7- (**70a**) or 5,6-dimethylbenzo[b]thiophene (**70b**), respectively, on FVT of 2,3-thiophenedicarboxylic anhydride (**68**) in the presence of 2,5- (**69a**) or 3,4-dimethylthiophene (**69b**). The former benzo[b]thiophenes (**71** and **72**) can be rationalized as arising from a (2+2)-cycloaddition of the thiophene to thiophyne, the latter ones (**70a** and **b**) as arising from a (4+2)-cycloaddition, followed in both cases by desulfuration[53,56,57] (Scheme 7) (see Section II.6.B).

(2+2)-Cycloaddition combined with (4+2)-cycloaddition of thiophene to six-membered (het)arynes (**73**) was assumed to explain the formation of the major

---

* By way of contrast, no convincing evidence is as yet available to support the existence of the isomeric 3,4-didehydrothiophene.

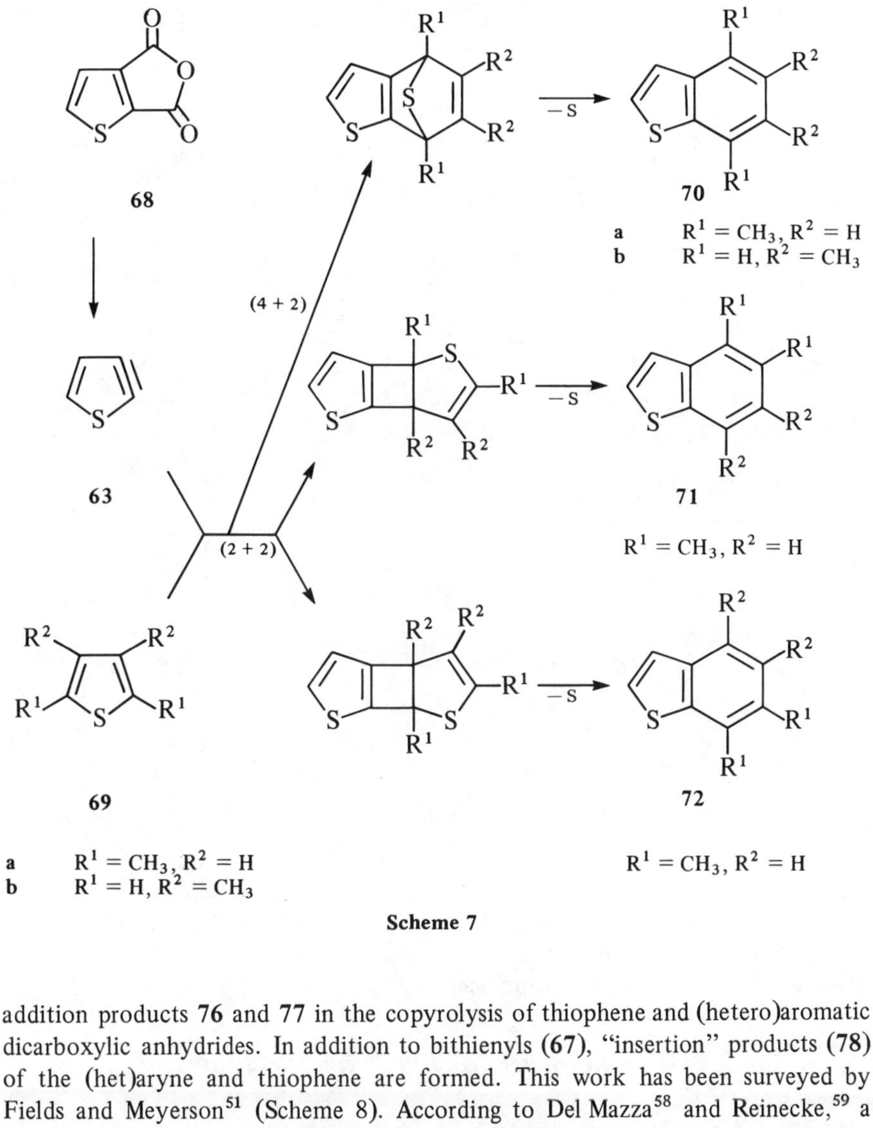

68

63

(4 + 2)

(2 + 2)

69

a $R^1 = CH_3, R^2 = H$
b $R^1 = H, R^2 = CH_3$

70

a $R^1 = CH_3, R^2 = H$
b $R^1 = H, R^2 = CH_3$

71

$R^1 = CH_3, R^2 = H$

72

$R^1 = CH_3, R^2 = H$

Scheme 7

addition products **76** and **77** in the copyrolysis of thiophene and (hetero)aromatic dicarboxylic anhydrides. In addition to bithienyls (**67**), "insertion" products (**78**) of the (het)aryne and thiophene are formed. This work has been surveyed by Fields and Meyerson[51] (Scheme 8). According to Del Mazza[58] and Reinecke,[59] a questionable point in this mechanism is the ring contraction of the 1-benzothiepins (**74**) to cyclobutenes (**75**) under thermal conditions. Thermally, 1-benzothiepins either extrude sulfur or rearrange to give naphthalenes.[60] Alternatively, a (3+2)-cycloaddition of thiophene to the (het)aryne has been proposed[58, 59] (see Section II.5). Until recently, the reported reactions of (halogenated) benzynes with thiophenes in solution failed to provide evidence for a (2+2)-cycloaddition mode (see Section II.6.B). During a study of the reaction of mono- and disubstituted thiophenes with benzyne (**80**), generated from diphenyliodonium-2-carboxylate (**79**), Del Mazza observed, *inter alia*, at least two different isomeric naphthalenes (**81** and/ or **82**, and **83**) and one or two benzo[*b*]thiophenes. The formation of the aromatic

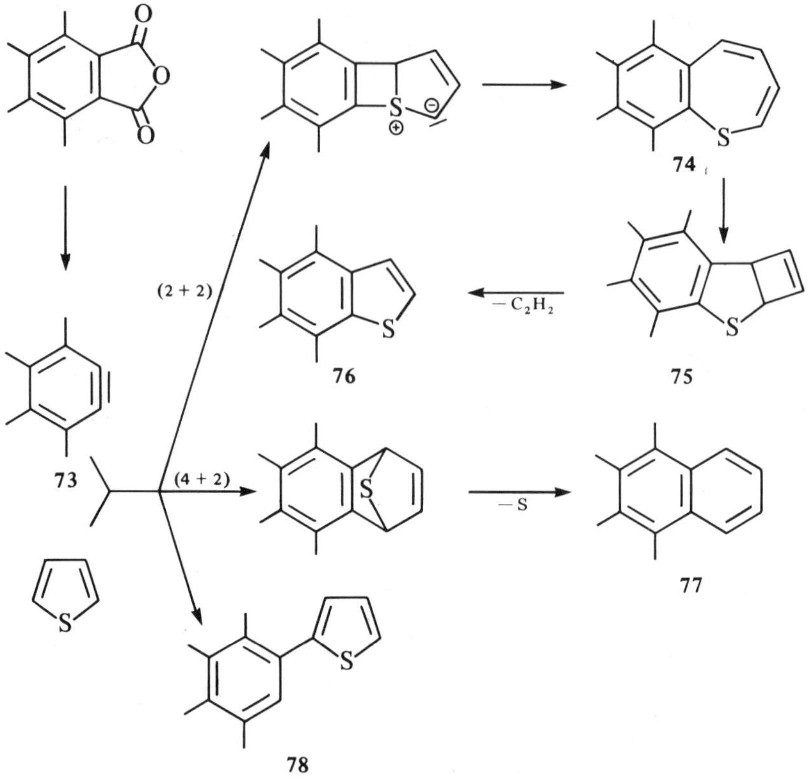

Scheme 8

hydrocarbons was accounted for by (2+2)- and (4+2)-cycloaddition of the thiophene to benzyne with subsequent loss of sulfur[58] (Scheme 9), and the formation of the benzo[*b*]thiophenes by (3+2)-cycloaddition followed by extrusion of an acetylene[58, 59] (see Section II.5 and Scheme 13).

E. *Photochemical* (2+2)-*Cycloadditions*

2,5-Dimethylthiophene undergoes a photochemical (2+2)-cycloaddition with carbonyl compounds.[61–64] The reactions are regioselective and give 7-oxa-2-thia-bicyclo[3.2.0]hept-3-enes (**84**) in 50–62% yield.[61,62] In all cases (except when $R^1 = R^2 = C_6H_5$), two isomers are formed that differ in the relative positions of the two substituents (R^1 and R^2) in the 6-position.[62] Only one other thiophene has been reported to give an oxetane (**85**), namely 2,3-dimethylthiophene. As a possible reason why other thiophenes do not give this type of (2+2)-cycloaddition, it has been suggested that the thiophene acts as a quencher of the triplet state of the ketones, The effect of the methyl group(s) might be the raising of the lowest $n\pi^*$ triplet energy level of the thiophene.[64]

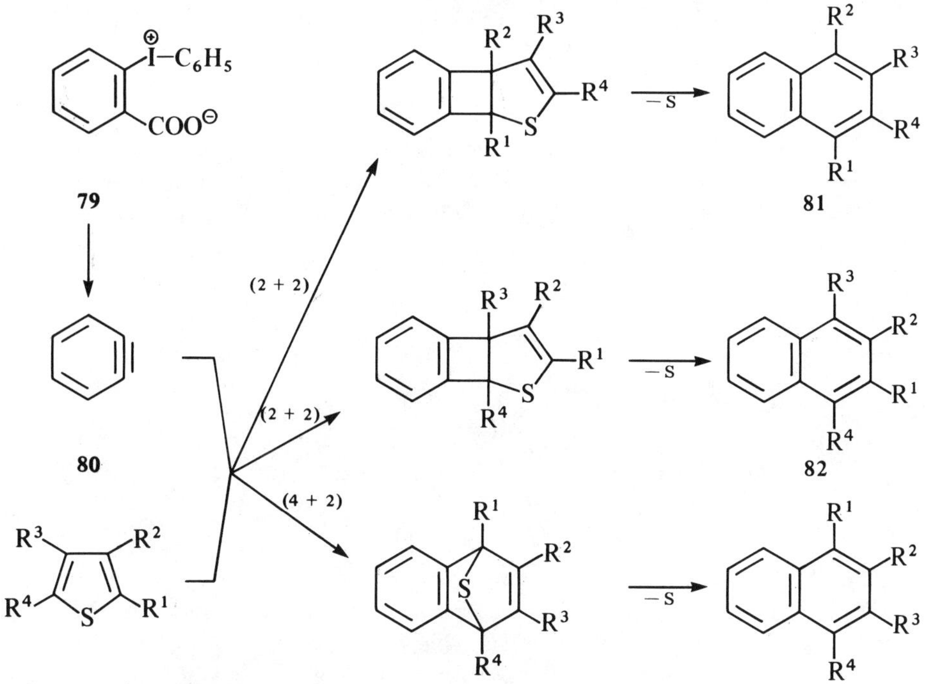

79

80

(2 + 2)

(2 + 2)

(4 + 2)

−S

−S

−S

81

82

83

a $R^1 = R^2 = R^3 = R^4 = H$
b $R^1 = CH_3, CH_3O, Br; R^2 = R^3 = R^4 = H$
c $R^2 = CH_3, Br; R^1 = R^3 = R^4 = H$
d $R^1 = R^4 = CH_3, CH_3O, Cl, Br; R^2 = R^3 = H$
e $R^1 = R^4 = H; R^2 = R^3 = Br$
f $R^1 = CH_3, R^2 = R^3 = H, R^4 = OCH_3$

Scheme 9

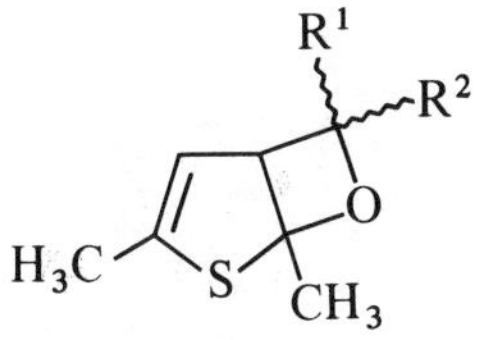

84

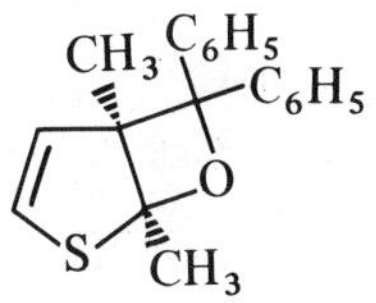

85

a $R^1 = C_6H_5, R^2 = C_6H_5$
b $R^1 = C_6H_5, R^2 = 2\text{-Pyridyl}$
c $R^1 = C_6H_5, R^2 = 3\text{-Pyridyl}$
d $R^1 = C_6H_5, R^2 = 4\text{-Pyridyl}$
e $R^1 = C_6H_5, R^2 = 2\text{-Thienyl}$
f $R^1 = H, R^2 = 1\text{-Naphthyl}$

Various photochemical (2+2)-cycloadditions of thiophenes have been reported in which an enone moiety is incorporated either in the olefinic reagent or in the thiophene. Thiophene reacts with 2,3-dimethyl- and with 2-methylmaleic anhydride upon irradiation in the presence of benzophenone.[65,66] The (2+2)-cycloadducts

86a and **b** were obtained as the major reaction products, whereas in the case of 2,5-dimethylthiophene, **86c** and **86d** were only minor products, since oxetane formation (*vide supra*) predominates.[61] Nakano et al.[63] obtained the diacids **87** by irradiation of the corresponding thiophenes with 2,3-dimethyl- or with 2-methyl-maleic anhydride in the presence of benzophenone as a triplet sensitizer. Only the yield of **87b** (14%) was given. It has not been established whether opening of the anhydride ring occurred during the irradiation or the work-up procedure. The

86

a	$R^1 = R^2 = H, R^3 = CH_3$
b	$R^1 = R^2 = R^3 = H$
c	$R^1 = R^2 = R^3 = CH_3$
d	$R^1 = R^2 = CH_3, R^3 = H$

87

a	$R^1 = R^2 = R^3 = H$
b	$R^1 = R^3 = H, R^2 = CH_3$
c	$R^1 = CH_3, R^2 = R^3 = H$
d	$R^1 = R^3 = CH_3, R^2 = H$

stereochemistry of **86** was proven to be *exo* with respect to the dihydrothiophene and the anhydride rings by ^1H nmr spectroscopy. (2+2)-Cycloadducts of 2,3-dimethylmaleimide, 2,3-dichloro-*N*-methylmaleimide, and 2-methylmaleimide with thiophene (**88a–c**) are mentioned in a paper by Rivas et al.[67] However, experimental details such as yields and physical constants were not reported. Compounds were characterized by ^1H nmr spectroscopy. Wamhoff and Hupe[68] reported that irradiation of thiophene and of 2-methylthiophene with 2,3-dihalomaleimides in acetone yielded 2-halo-3-(2-thienyl)maleimides (**89**) in low yields (15–< 1%). Irradiation

88

a	$R^1 = R^2 = CH_3, X = NH$
b	$R^1 = R^2 = Cl, X = NCH_3$
c	$R^1 = CH_3, R^2 = H, X = NH$

89

a	$R^1 = R^2 = H, X = Cl$
b	$R^1 = H, R^2 = CH_3, X = Br$
c	$R^1 = R^2 = H, X = I$
d	$R^1 = CH_3, R^2 = H, X = Cl$
e	$R^1 = R^2 = CH_3, X = Br$
f	$R^1 = CH_3, R^2 = H, X = I$

90

91

92

a	$R = H$
b	$R = CH_3$

of 2,3-dibromo-*N*-methylmaleimide in neat thiophene gave a photoproduct formed by reaction with 2 molecules of thiophene, for which structure **90** has been proposed with the (2+2)-cycloadduct **91** as an intermediate. Cantrell[69] found that irradiation of 2-cyclohexenone in thiophene and 2,5-dimethylthiophene gives **92a** (8%) and **92b** (51%), respectively. The stereochemistry of the (2+2)-cycloadducts was not further investigated. When a carbonyl group is attached to the thiophene nucleus, photochemical (2+2)-cycloadditions with alkenes[70-72] and with acetylenes[73,74] take place. Irradiation of 2-acetylthiophene in the presence of excess of 2,3-dimethyl-2-butene gave a mixture of three products. The major product was proven to be the (4+2)-cycloadduct **93a** (38%), whereas the minor products were the (2+2)-cycloadduct **94** and the oxetane **95a** in yields of 10 and 11%, respectively. However, irradiation of 2-acetylthiophene and 2-methylpropene afforded only the oxetane **95b** (6%) in addition to the two isomeric (4+2)-cycloadducts **93b** and **93c** (23 and 18%, respectively). 2-Benzoylthiophene reacted under similar conditions with 2,3-dimethyl-2-butene to give no ring-addition products; only the oxetane **95c** could be isolated in 76% yield.[70,71] Cantrell[71] proposed that the reactive excited states of these 2-acylthiophenes are predominantly of the π,π^*-type, a hypothesis supported by the long-lived phosphorescence of the ketone. Similar results have been reported by Arnold et al.[72] who also found oxetane

93

a   R^1 = R^2 = CH$_3$
b   R^1 = H, R^2 = CH$_3$
c   R^1 = CH$_3$, R^2 = H

94

95

a   R^1 = R^2 = R^3 = CH$_3$
b   R^1 = H, R^2 = R^3 = CH$_3$
c   R^1 = R^2 = CH$_3$, R^3 = C$_6$H$_5$

96

R = 4-NC—C$_6$H$_4$

formation when 2- and 3-benzoylthiophene and their 4-cyano or 4-methoxy derivatives were irradiated in the presence of 2-methylpropene. Irradiation of mixtures of 3-aroylthiophenes and DMAD gave complex mixtures of products. It was suggested that (2+2)-cycloaddition across the C-2—C-3 bond was one of the reaction pathways, but experimental proof has not been presented.[73] Sensitized and

TABLE 4.   REACTIONS OF THIOPHENES WITH TETRACYANOETHYLENE OXIDE

R¹	R²	Temperature (°C)	Reaction Time (h)	Yield of 97 (%)	Reference
H	H	130	16	50	76
H	H	150	28	70	78
Cl	H	a	16	48	76
H	Br	a	–	50	79
I	H	a	–	50	79
H	I	a	–	50	79
CH₃	H	150	–	36	79
H	CH₃	145	–	39	79

a Refluxing 1,2-dibromoethane.

nonsensitized photoaddition of the same acetylene to thiophene and 2,5-dimethyl-thiophene yielded dimethyl phthalate and the 3,6-dimethyl-substituted derivative, respectively.[74] It is unlikely that a (2+2)-cycloadduct is the intermediate in these reactions (see Section II.2.B).

Photochemical (2+2)-cycloaddition of thiophene with the C=N bond of 3-(4-cyanophenyl)isoxazoline gives **96**.[75]

3.   (2+3)-Cycloaddition Reactions of Thiophenes

Thiophenes react only with very reactive 1,3-dipoles. The first recorded reaction of this type is the reaction of (substituted) thiophenes with tetracyanoethylene oxide at temperatures of 130–150°C. The substituted 4,4,6,6-tetracyano-3a,4,6,6a-tetrahydrothieno[2,3-c]furans (**97**) were obtained in yields between 36 and 70% (Table 4).[76-79] The formation of compounds **97** is in agreement with a carbonyl ylide as an intermediate, since the reactions of similar cyanoethylene oxides with olefins show all the characteristics of concerted 1,3-dipolar cycloaddition reactions.[77, 80, 81] Gronowitz and Uppström[79] measured the rates of these cycloadditions at 150°C in 1,2-dichloroethane and found a decrease in rates when electron-withdrawing groups are present at the thiophene ring: k(2-CH₃): k(H): k(2-Cl) = 9.66 : 1 : 0.64. Several substituted thiophenes show other types of reaction. 2- and 3-Methoxythiophene and 2-(methylthio)thiophene react with tetracyanoethylene oxide to give the corresponding thenoyl cyanides **99**. The formation of these compounds was explained by a two-step process beginning with the formation of a dicyanomethylide (**98**) and carbonyl cyanide. In the second step, excess of thiophene reacts with carbonyl cyanide (Scheme 10). Independent experiments revealed

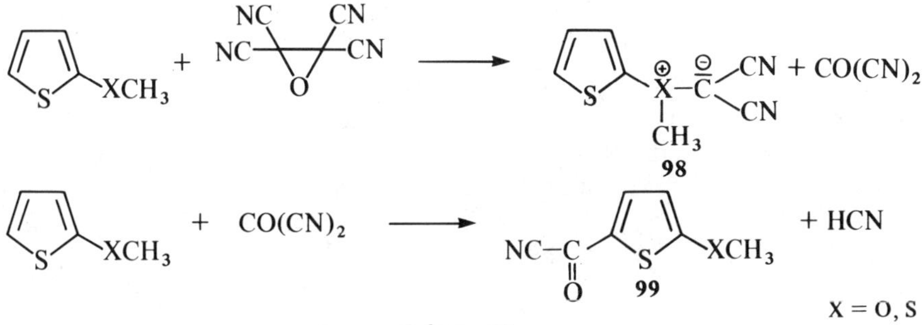

Scheme 10

the formation of thenoyl cyanides by reaction of thiophenes with carbonyl cyanide at 130°C.[82] 2,5-Dibromothiophene reacts with tetracyanoethylene oxide at 160°C to give 2,5-bis(dicyanomethylene)-2,5-dihydrothiophene (**100a**) together with carbonyl cyanide and bromine; the yield of **100a** was 70%. The molecular bromine reacted immediately with the excess of 2,5-dibromothiophene to yield 2,3,5-tribromothiophene and tetrabromothiophene. 2,3,5-Tribromothiophene and 2,3,5-trichlorothiophene reacted in a similar mode and gave compounds **100b** and **100c** in yields of 28 and 16%, respectively. A possible mechanism is given in Scheme 11, assuming that tetracyanoethylene oxide can react both as a nucleophile and as an electrophile. As an alternative, the thermal decomposition of tetracyanoethylene oxide to give dicyanomethylene and carbonyl cyanide, followed by a (2+1)-cycloaddition to give **101** and subsequent rearrangement, was discussed[83] (Scheme 11). Strongly electron-deficient thiophenes like 2-cyano- and 2-nitrothiophene did not react.[79]

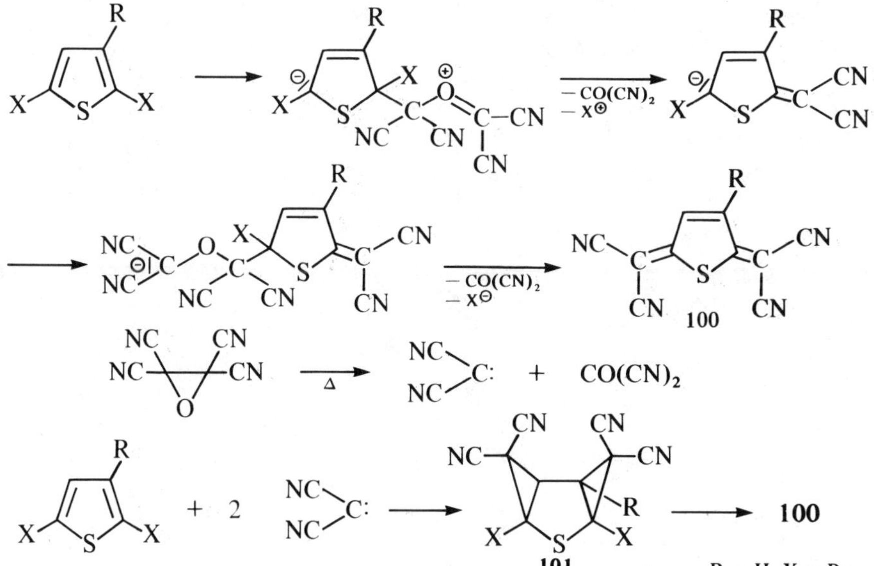

a	R = H, X = Br
b	R = X = Br
c	R = X = Cl

Scheme 11

3-(1-Pyrrolidinyl)thiophene reacts with 2,4,6-trimethylbenzonitrile oxide to give the oxime of 2,4,6-trimethylphenyl 2-[3-(1-pyrrolidinyl)thienyl] ketone (102). Although the initially formed thieno[2,3-*d*]isoxazoline 103 was not detected as an intermediate, a 1,3-dipolar addition and subsequent isomerization involving the aromatization of the thiophene ring is supported by analogous reactions of other 3-(1-pyrrolidinyl)heteroaromatics.[84] Stable 1,3-dipolar adducts of thiophene and

102   103

benzonitrile oxides have been reported by Beltrame et al.[4] and Caramella et al.[6] 3,5-Dichloro-2,4,6-trimethylbenzonitrile oxide reacts with thiophene, used as the solvent, at reflux temperature. After 48 h, the 1:1-cycloadduct 104a was isolated in 25% yield. As one of the by-products, the bis(1,3-dipolar) cycloadduct of thiophene (105a) was obtained in a yield of 8%. The structure of 104a was proven by acid-catalyzed hydrolysis to 106a and reaction with bromine, which afforded 107a. Furthermore, reaction of 104a with 3,5-dichloro-2,4,6-trimethylbenzonitrile oxide gave a mixture of two isomeric bis(1,3-dipolar) cycloadducts. The predominant isomer was identified as 105a and the other, minor isomer was shown to be 108a.[4] The unsubstituted benzonitrile oxide, generated *in situ* in thiophene at 0°C, gave 1.9% of 104b and 2.5% of 105b together with a 3:1 reaction product. Acid hydrolysis of 104b gave 2-benzoylthiophene and further reaction of 104b with benzonitrile oxide gave 105b exclusively.[6] Similar cycloadducts (104c and 105c)

104   105   106

107   108   109

a   R =   [structure: H₃C, Cl, H₃C, Cl substituted benzene ring]—CH₃   b   R = C₆H₅   c   R =   [structure: H₃C, H₃C substituted benzene ring]—CH₃

were obtained when 2,4,6-trimethylbenzonitrile oxide was reacted with thiophene at room temperature. After 7 months, **104c** and **105c** were obtained in 1.6 and 40% yield, respectively, along with the symmetrical bisadducts **108c** (2%) and **109c** (3%). An independent experiment revealed that **104c** added the nitrile oxide yielding **105c** and **108c** in the same ratio (ca. 20:1) as found in the cycloaddition to thiophene.[6] The observed regiochemistry in the 1,3-dipolar additions of thiophene is in agreement with the predictions made by qualitative FMO considerations. The HOMO(thiophene)-LUMO(benzonitrile oxide) interaction predominantly governs the cycloaddition.[5,6] Grünanger and coworkers[6] suggested that the bis(1,3-dipolar) cycloadducts **105** and **108** have the *anti*-stereochemistry.

4.   (2+4)-Cycloaddition Reactions of Thiophenes

Hurd et al.[85] proposed that in the pyrolysis of thiophene, benzo[b]thiophene is formed by a (2+4)-cycloaddition of thiophene and a C₄H₄ diradical from pyrolytic cleavage of thiophene and subsequent dehydrogenation. More solid evidence that thiophene can react as a 2π-component in Diels–Alder reactions was obtained in reactions with hexachlorocyclopentadiene and tetrachlorothiophene 1,1-dioxide. Reaction of hexachlorocyclopentadiene and thiophene at 150–160°C for 14 days gave the bis[(4+2)-cycloadduct] **110**.[86] Raasch[87] reported a similar type of cyclo-adduct (**111**) by reacting thiophene with tetrachlorothiophene 1,1-dioxide in a yield of 61%. The reaction conditions (85°C, 72 h) are considerably milder than in the case of the hexachlorocyclopentadiene (see Section IV.6.L). Suitable substituted thiophenes also react with dimethyl 1,2,4,5-tetrazine-3,6-dicarboxylate in

[structures 110 and 111]

110   111

a (2+4)-cycloaddition fashion. In a consecutive reaction, nitrogen is eliminated from the initially formed cycloadduct to give **112** ⇌ **113**. Starting from 2,5-dimethylthiophene, **113d** is the isolated product. In the case of thiophene itself and the monosubstituted derivatives, **112** or the tautomer **113** is oxidized under reaction conditions and **114a–c** are isolated (Scheme 12).[88]

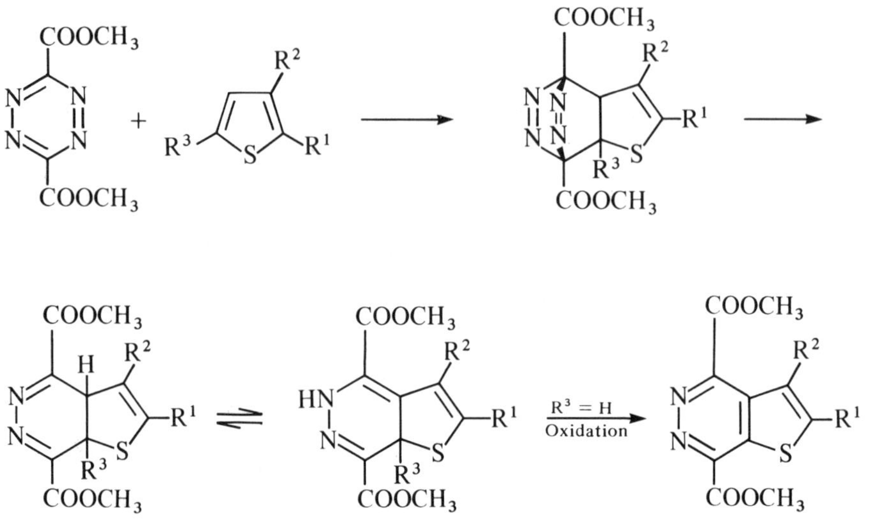

112   **113**   **114**

a	$R^1 = R^2 = R^3 = H$
b	$R^1 = R^3 = H, R^2 = CH_3$
c	$R^1 = Cl, R^2 = R^3 = H$
d	$R^1 = R^3 = CH_3, R^2 = H$

Scheme 12

5. (3+2)-Cycloaddition Reactions of Thiophenes

In studying the solution-phase reaction of thiophenes with benzyne **(80)**, generated from diphenyliodonium-2-carboxylate **(79)**, Del Mazza[58] and Reinecke[59] observed the formation of benzo[b]thiophenes (**115, 116**) in low but reproducible yields in addition to, *inter alia*, naphthalenes. The naphthalenes were rationalized as arising from (2+2)- and (4+2)-cycloaddition of the thiophene to benzyne and subsequent extrusion of sulfur (see Section II.2.D and Scheme 9). For the formation of the benzo[b]thiophenes **115** and **116**, (3+2)-cycloaddition with subsequent expulsion of an acetylene moiety was proposed (Scheme 13). With monosubstituted thiophenes, the unsubstituted side is preferentially attacked, as shown in the reactions of 2-bromo-, 2-methoxy-, and 3-methylthiophene, but for 2-methylthiophene, either side may react.

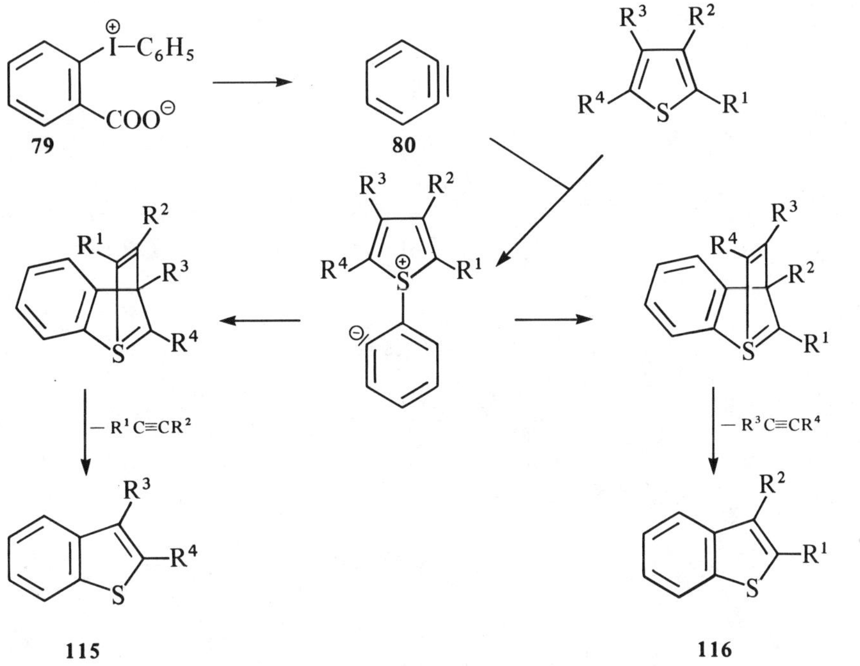

a $R^1 = R^2 = R^3 = R^4 = H$
b $R^1 = CH_3, CH_3O, Br; R^2 = R^3 = R^4 = H$
c $R^2 = CH_3, R^1 = R^3 = R^4 = H$
d $R^1 = R^4 = CH_3, R^2 = R^3 = H$
e $R^1 = R^4 = H, R^2 = R^3 = Br$

Scheme 13

6. (4+2)-Cycloaddition Reactions of Thiophenes

The thermal reaction of a thiophene, acting as a 4π-component, with a dienophile in a suprafacial, ~~suprafacial~~ manner is an orbital symmetry-allowed process. However, examples of such a (4+2)-cycloaddition are generally restricted to (1) reactions of thiophenes and reactive dienophiles having a triple bond [electron-deficient acetylenes, (het)arynes], (2) reactions of thiophenes with electron-releasing substituents and dienophiles containing an electron-poor double bond, and (3) reactions of alkyl- and phenyl-substituted thiophenes with singlet oxygen. Furthermore, at very high pressure, thiophene reacts with maleic anhydride to give the (4+2)-cycloadduct. The unfavorable Diels–Alder equilibrium has been explained by Lert and Trindle[7] by stabilization of the antisymmetric π MO of the butadiene moiety in the thiophene by the antisymmetric 3d AO of the sulfur; such a stabilization is not possible in the product. Photochemically, the reaction discussed above is a symmetry-forbidden process, whereas the symmetry-allowed suprafacial, antarafacial mode of reaction is not possible for steric reasons. The only photochemical

TABLE 5.   BENZENE DERIVATIVES (118) OBTAINED FROM THE REACTIONS OF THIOPHENES WITH ACETYLENES

118

Compound	R^1	R^2	R^3	R^4	R^5	R^6	Yield (%)	Reference
118a	H	H	H	H	CH_3	CH_3	a	74
118b	CH_3	H	H	CH_3	CH_3	CH_3	$(2)^b$	74
118c	H	H	H	H	C_6H_5	C_6H_5	$(38)^b$	74
118d	H	H	H	H	E^c	E	$(56)^b$	74
118e	CH_3	H	H	CH_3	E^c	E	$(42)^b$	74
118f	C_6H_5	H	C_6H_5	H	E^c	E	12	90
118g	H	H	H	H	CN	CN	$8/(36)^b$	34, 89
118h	CH_3	H	H	H	CN	CN	9	34, 89
118i	CH_3	H	H	CH_3	CN	CN	49	34, 89
118j	CH_3	CH_3	H	CH_3	CN	CN	24	34
118k	CH_3	$t\text{-}C_4H_9$	H	CH_3	CN	CN	51	34, 89
118l	C_2H_5	C_2H_5	H	C_2H_5	CN	CN	40	34
118m	CH_3	CH_3	CH_3	CH_3	CN	CN	$56/(82)^b$	34
118n	CH_3	CH_2D	CH_2D	CH_3	CN	CN	60	34
118o	H	C_6H_5	C_6H_5	H	CN	CN	18	34, 89
118p	C_6H_5	C_6H_5	C_6H_5	C_6H_5	CN	CN	9	34

a Not determined.
b Determined by gas–liquid chromatography.
c $E = COOCH_3$.

(4+2)-cycloaddition was observed on irradiation of mixtures of 2-acetylthiophene and 2,3-dimethyl-2-butene and 2-methylpropene, respectively (see Section II.2.E).

A. (4+2)-*Cycloaddition in Reactions of Thiophenes with Acetylenes*

Thiophene and its alkyl- and phenyl-substituted derivatives react with acetylenes to produce benzene derivatives **118** as the isolated products (Table 5). The yields and reaction conditions depend strongly on the electrophilic character of the acetylene used and on the degree and kind of substitution of the thiophenes. Reacting 2,5-dimethylthiophene with dicyanoacetylene for 12 h at 100°C afforded 3,6-dimethyl-1,2-benzenedicarbonitrile (**118i**) in 49% yield, whereas with 2-butyne as the dienophile, the presence of only 2% of 1,2,3,4-tetramethylbenzene (**118b**) could be detected by gas–liquid chromatography after 24 h at 300°C. The reaction of thiophene with dicyanoacetylene took 48 h at 120°C to yield 8% of 1,2-benzene-dicarbonitrile (**118g**) and the corresponding reaction of tetramethylthiophene took 12 h at 55°C to give 56% of tetramethyl-1,2-benzenedicarbonitrile (**118m**), whereas tetraphenylthiophene afforded only 9% of 3,4,5,6-tetraphenyl-1,2-benzenedicarbo-nitrile (**118p**) after 65 h at 130°C.[34, 74, 89] The course of the reaction is generally explained by a (4+2)-cycloaddition of the thiophene to the acetylene as the initial step, followed by extrusion of sulfur. In some cases, a 1:1 adduct could be detected by means of mass spectrometry. In other cases, products that could arise from a *retro* Diels–Alder reaction of **117** were identified[43, 74] (Scheme 14). From the reaction of 2,4-diphenylthiophene with DMAD in addition to the benzene derivative **118f**, two other products were isolated. The major product was 2,5-diphenylthio-phene and the minor product dimethyl 2-phenylnaphtho[2,1-*b*]thiophene-4,5-dicarboxylate.[90] A rationale for the formation of the latter compound is given in Section II.7.

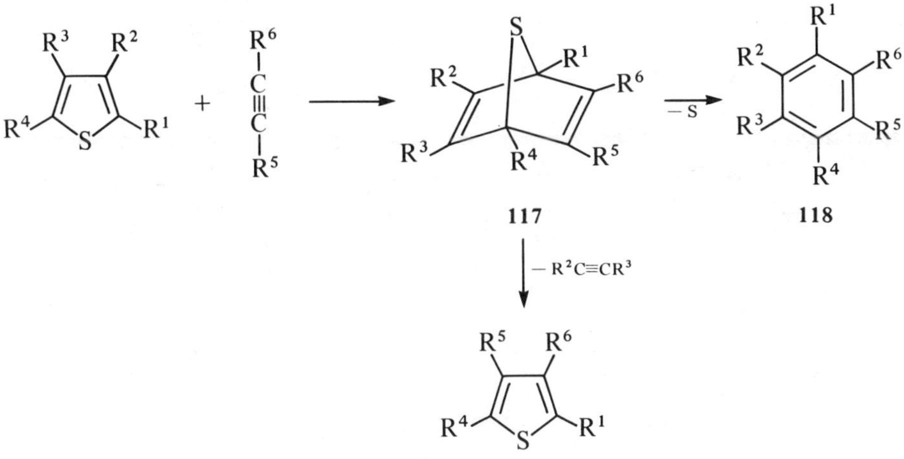

117   **118**

$$- R^2C{\equiv}CR^3$$

Scheme 14

B.   (4+2)-*Cycloaddition in Reactions of Thiophenes* *with (het)-Arynes*

Combined with (2+2)-cycloaddition, (4+2)-cycloaddition of thiophene to 2,3-didehydrothiophene (thiophyne, **63**) was proposed to account for the formation of the minor pyrolysis products of thiophene[51,52] (see Section II.2.D and Scheme 6). However, the intermediacy of a thiophyne in this reaction has been disputed by Reinecke.[53] Hitherto, examples of a (4+2)-cycloaddition of thiophenes to thiophyne **(63)** are restricted to FVT of 2,3-thiophenedicarboxylic anhydride **(68)** in the presence of thiophene[53–55,57] and of 2,5- **(69a)** or 3,4-dimethylthiophene **(69b)**.[53,56,57] With thiophene, the only other identifiable product besides sulfur was benzo[*b*]thiophene **(65**, 59%); but with the two latter substrates in addition to 4,7- **(70a)** and 5,6-dimethylbenzo[*b*]thiophene **(70b)**, respectively, 4,5- **(71)** and 6,7-dimethylbenzo[*b*]thiophene **(72)** were obtained. The formation of the benzo[*b*]thiophenes **65** and **70** has been explained by desulfuration of the initially formed (4+2)-cycloadducts of the thiophene to thiophyne, the formation of **71** and **72** by desulfuration of the initially formed (2+2)-cycloadducts (see Section II.2.D and Schemes 6 and 7).

(4+2)-Cycloaddition together with (2+2)-cycloaddition of thiophene to six-membered (het)arynes **(73)** generated in the gas phase have been proposed to account for the major addition products **77** and **76**, respectively, in the copyrolysis of thiophene and (hetero)aromatic dicarboxylic anhydrides[51] (see Section II.2.D

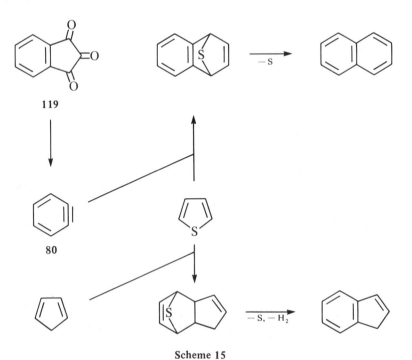

Scheme 15

and Scheme 8). The diene reactivity of thiophene in the gas phase was also demonstrated by the formation of naphthalene from its reaction with benzyne (80), generated by FVT of indanetrione (119), and by the formation of indene from thiophene plus cyclopentadiene under FVT conditions[55] (Scheme 15).

Until recently, in reactions of thiophenes with (halogenated) benzynes in the liquid phase, only (4+2)-cycloaddition was observed. Reacting 2-fluorophenyllithium or 2-fluorophenyl Grignard compounds as well as tetrafluorobenzenediazonium-2-carboxylate with thiophenes yields the naphthalenes 121a-i[91-95] (Table 6). By monitoring the course of the reaction of pentafluorophenyllithium with thiophene by ^{1}H nmr spectroscopy, evidence could be provided that the Diels–Alder adduct 120 ($R^1 - R^4$ = H and $R^5 - R^8$ = F) was an intermediate in this reaction[92] (Scheme 16). Thiophenes exhibit a diene reactivity not only to the more

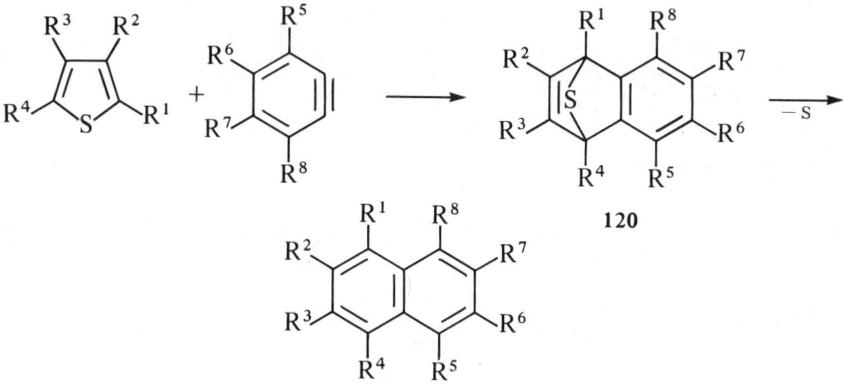

Scheme 16

reactive halogenated benzynes, but also to benzyne itself. This was demonstrated in the reactions with benzenediazonium-2-carboxylate[34] (Table 6) as well as diphenyliodonium-2-carboxylate (79) as the benzyne (80) precursors.[58,96] In the latter case, and with mono- and disubstituted thiophenes, except the naphthalenes (83) derived from (4+2)-cycloaddition and subsequent desulfuration, isomeric naphthalenes (81 and/or 82) were observed. These are presumed to arise from a (2+2)-cycloaddition of the thiophene to benzyne with subsequent loss of sulfur[58] (see Section II.2.D and Scheme 9). Furthermore, benzo[b]thiophenes (115 and 116) were observed whose formation was explained by a (3+2)-cycloaddition followed by loss of an acetylene[58,59] (see Section II.5 and Scheme 13).

C. (4+2)-Cycloaddition in Reactions of Thiophenes with Dienophiles Containing a Double Bond

Diels–Alder addition of thiophene to another molecule of thiophene was assumed by Wynberg and Bantjes[97] to explain the formation of benzo[b]thiophene (65) and

TABLE 6. NAPHTHALENES (121) OBTAINED FROM THE REACTIONS OF THIOPHENES WITH HALOGENATED BENZYNES AND OF TETRAALKYLTHIOPHENES WITH BENZYNE

121

Compound	R^1	R^2	R^3	R^4	R^5	R^6	R^7	R^8	Yield (%)	Reference
121a	H	H	H	H	F	F	F	F	35	91–93
121b	H	H	H	H	F	Cl	F	Cl	39	94
121c	H	H	H	H	F	Cl	OCH₃	Cl	–	94
121d	H	H	H	H	F	OCH₃	F	F	40	95
121e	H	H	H	H	F	F	H	F	31	95
121f	CH₃	H	H	H	F	F	F	F	–	92
121g	Br	Br	H	H	F	F	F	F	39	95
121h	H	H	H	H	F	F	F	F	63	95
121i	Cl	Cl	Cl	Cl	F	F	F	F	14	92
121j	CH₃	CH₃	CH₃	CH₃	H	H	H	H	12	34
121k	C₂H₅	C₂H₅	C₂H₅	C₂H₅	H	H	H	H	17	34

phenylthiophene (66) in the pyrolysis of thiophene. Thieno[2,3-*b*]thiophene (thiophthene, 64) was presumed to arise as a known product from the reaction of acetylene and sulfur (Scheme 17). As an alternative, Fields and Meyerson[51, 52]

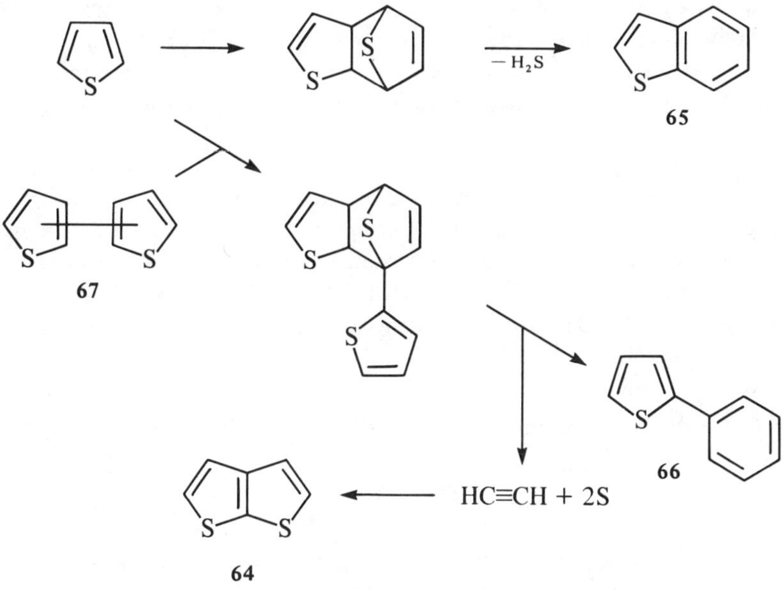

Scheme 17

proposed a mechanism with 2,3-didehydrothiophene (thiophyne, 63) as the key intermediate in the formation of the minor pyrolysis products 64–66 of thiophene (see Section II.2.D and Scheme 6). Apart from the question of thiophyne intermediacy in the latter mechanism, both mechanisms suffer from the requirement that acetylene is an expected by-product that no one has actually found from thiophene.[85]

Gaertner and Tonkyn[98] failed to add maleic anhydride to tetramethylthiophene, even when the reaction was carried out in boiling nitrobenzene (bp 211°C). Negative results have also been reported by Clapp,[99] who attempted to bring about a reaction between tetraphenylthiophene and maleic anhydride. Similarly, thiophene does not react in a (4+2)-cycloaddition mode with dienophiles such as dimethyl maleate, dimethyl fumarate, methyl acrylate, acrylonitrile or acrylaldehyde, even under very high pressure. Under these circumstances (3 h, 100°C, 15 kbar), the Diels–Alder adduct 122 could be obtained only with maleic anhydride, in yields of 37–47%. On the basis of the spectroscopic data and chemical evidence, it has been suggested that the adduct has the *exo*-configuration.[100]

At atmospheric pressure, a diene reactivity toward compounds containing a double bond has been observed with 2,5-dimethoxythiophene, with 2,4-bis(*N*-isopropyl-*N*-phenylamino)thiophene, and with 2,3,4,5-di(1,8-naphthylene)thiophene. The last compound reacts with maleic anhydride at 225°C and with diphenylethene at 310–320°C to give 3,4,5,6-di(1,8-naphthylene)phthalic anhydride (acenaphtho[1,2-*j*]fluoranthene-4,5-dicarboxylic anhydride, **123a**, 60%) and 1,2-diphenyl-3,4,5,6-di(1,8-naphthylene)benzene (4,5-diphenylacenaphtho[1,2-*j*]fluoranthene, **123b**, 14%), respectively, which result from a Diels–Alder reaction and subsequent elimination of hydrogen sulfide.[99] The two other recorded thiophenes with Diels–Alder reactivity toward dienophiles having a double bond are mutually related by the presence of electron-releasing substituents attached to the thiophene nucleus. Reaction of 2,5-dimethoxythiophene with an equivalent amount of maleic anhydride in refluxing xylene for 3.5 h affords 1,4-dimethoxybicyclo[2.2.2]oct-7-ene-2,3,5,6-tetracarboxylic dianhydride (**124**) in 42% yield. The course of the reaction has been envisaged as a (4+2)-cycloaddition of the thiophene to the dienophile, followed by extrusion of sulfur and a second Diels–Alder reaction of maleic anhydride with the 1,3-cyclohexadiene formed.[101] 2,4-Bis(*N*-isopropyl-*N*-phenylamino)thiophene undergoes a Michael-type addition at the 5-position with iso(thio)cyanates,[102] β-nitrostyrene, (ethoxymethylene)malonitrile, diethyl azodicarboxylate, and dimethyl acetylenedicarboxylate; but with acrylonitrile, 4-phenyl-1,2,4-triazoline-3,5-dione, and *N*-phenylmaleimide, the 1,3-cyclohexadiene derivatives **125**, **126**, and **127**, respectively, are obtained. The formation of these products

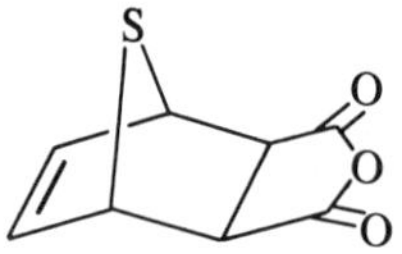

122

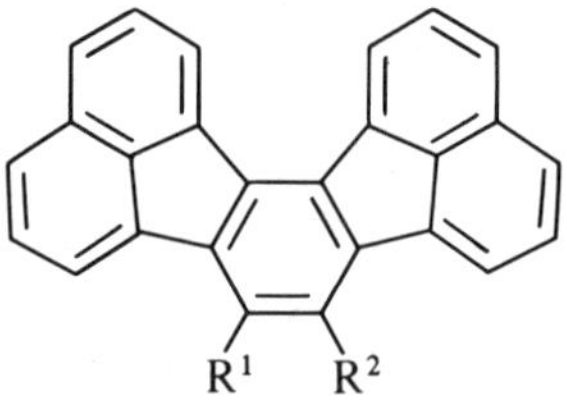

123

a $R^1, R^2 = -\overset{O}{\underset{\|}{C}}-O-\overset{O}{\underset{\|}{C}}-$

b $R^1 = R^2 = C_6H_5$

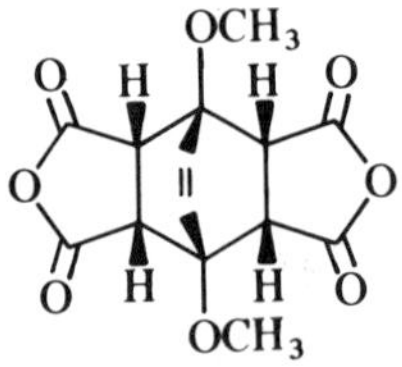

124

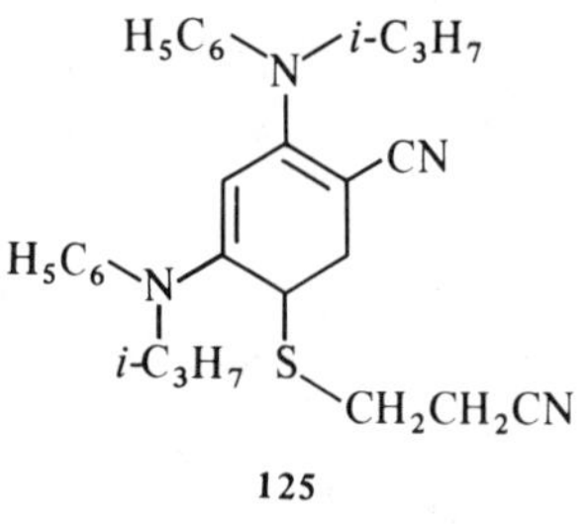

125

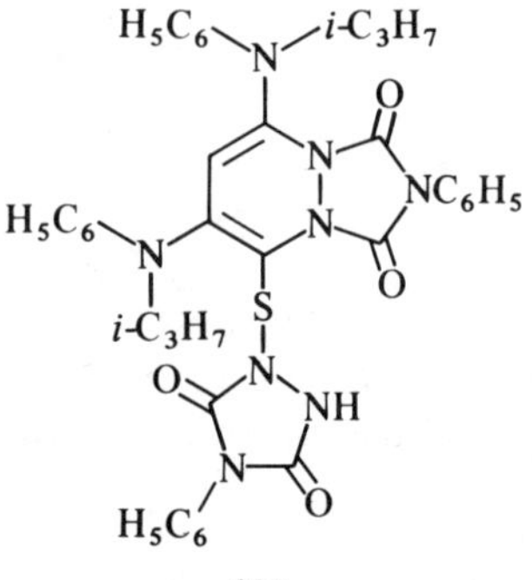

126

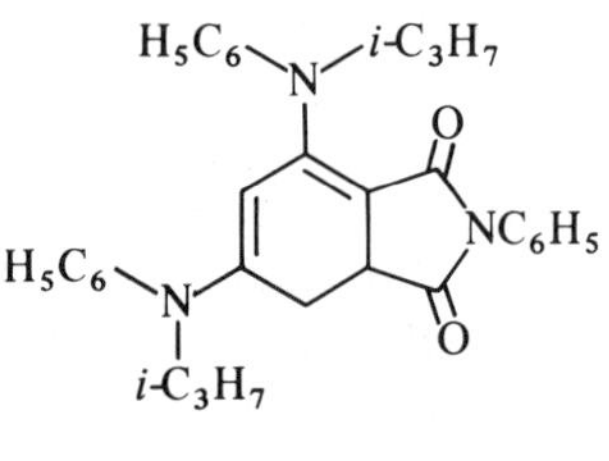

127

was explained by a concerted (path a) or stepwise (path b, c) Diels–Alder addition of the dienophile A=B to the 2,4-thiophenediamine **128**, followed by carbon–sulfur cleavage. With acrylonitrile and with 4-phenyl-1,2,4-triazoline-3,5-dione, the intermediate that arises from one carbon–sulfur rupture (path d) was intercepted to yield the ultimate products **125** and **126**, whereas in the reaction with N-phenyl-maleimide, extrusion of sulfur led to the final product **127**. Alternatively, the products **125–127** could arise without going through an initially formed Diels–Alder adduct via path b, g, and e or f[103] (Scheme 18).

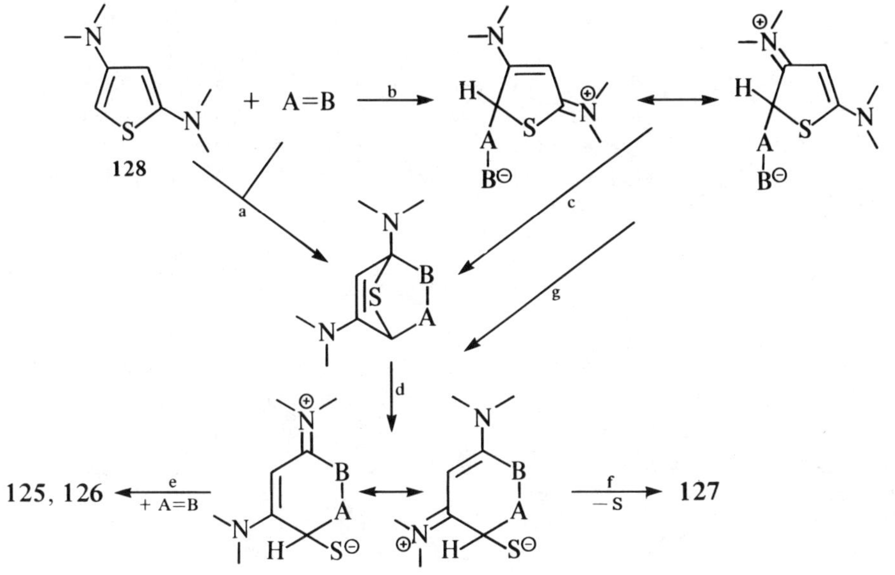

Scheme 18

D.   (4+2)-Cycloaddition in Reactions of Thiophenes with Singlet Oxygen

Oxidation of alkyl- and phenyl-substituted thiophenes with photochemically generated singlet oxygen affords 2-buten-1,4-diones **(130)** as the products reported in nearly all cases. With 2,5-dimethylthiophene as the substrate, the main product is (Z)-3-hexen-2-one-5-sulfine **(133a)**; the yield of the sulfine **(133a)** increases and that of the diketone **(130a)** decreases with increasing solvent polarity. After oxygenation of 4,5,6,7-tetrahydro-2,3-dimethylbenzo[b]thiophene in methanol, only one product was isolated, namely the sulfoxide **134e**, which can be derived from the corresponding sulfine **133e** by incorporation of a solvent molecule. Apart from a small amount of the ketoaldehyde **130d**, photooxidation of 2,4-di(tert-butyl)thiophene gives 3,5-di(tert-butyl)-3-hydroxy-3H-1,2-oxathiole 2-oxide **(136d)**.[104–106] The formation of all the observed reaction products was explained by a (4+2)-

cycloaddition of the thiophene to singlet oxygen as the initial step. Cleavage of either a carbon–sulfur or an oxygen–oxygen bond in the intermediate peroxide **129**, followed by extrusion of sulfur, would give the diketone **130**, which might also arise from the oxathiirane **135** by desulfuration. The peroxide **129** may also be envisaged as an intermediate in the formation of the sulfine **133**, the sulfoxide **134e**, and the 1,2-oxathiole 2-oxide **136d**. Possible steps in some of the reaction pathways are depicted in Scheme 19.[104-106] Evidence for the peroxide intermediate was obtained by Adam and Eggelte.[107] Photooxygenation of 2,5-dimethyl- and 2,5-di(*tert*-butyl)thiophene in dichloromethane at − 78°C, followed by reduction with diimine at the same temperature, afforded the 2,3-dioxa-7-thiabicyclo[2.2.1]-heptanes **137a** and **137b** as an oil and white needles in 46 and 50% yield, respectively. An alternative pathway for sulfine formation involves attack of singlet oxygen at the thiophene sulfur, followed by ring closure of the resulting 1,3-dipolar intermediate **131**. Ring opening of the cycloadduct **132**, thus formed, would give the sulfine **133**.[104] (The reaction of vinylthiophenes with singlet oxygen is discussed in Section II.7.)

7.   (4+2)-Cycloaddition Reactions of Vinylthiophenes

In this section, (4+2)-cycloaddition reactions of the (hetero)diene system composed of one π-bond of the thiophene nucleus and an exocyclic double bond are discussed. This mode of reaction has been observed in reactions of vinylthiophenes with (electron-deficient) dienophiles, such as maleic anhydride, 1,4-benzoquinone, 4-phenyl-1,2,4-triazoline-3,5-dione, azodicarboxylates, singlet oxygen, and DMAD, and in the reaction of 2,4-diphenylthiophene with DMAD. Even an intramolecular Diels–Alder reaction of a vinylthiophene has been reported, the dienophilic part of which is an acetylenic ester moiety. A (4+2)-cycloaddition has also been encountered in reactions of aryl 2-thienyl thiones with maleic anhydride and with bicyclo[2.2.1]hept-2-ene. In these compounds, the reacting 4π-electron system is composed of a π-bond of the thiophene ring and of a conjugated thione double bond.

Generally, 2-vinylthiophenes (**138**) have been reacted with maleic anhydride, without a solvent, at 100°C[108] or in refluxing benzene.[109,110] With a number of 2-vinylthiophenes (**138**), 1:1 addition products (**139**) have been isolated (Table 7), but with 1-(2-thienyl)cyclohexene, 1-(2-thienyl)-4-methylcyclohexene, and 1-(2-thienyl)cycloheptene, 1:2 addition products of unknown structure have been obtained. To effect addition to 1-(2-thienyl)-2-methyl- and to 1-(2-thienyl)-2,5-dimethylcyclohexene, a reaction temperature of 220°C was necessary. In these cases, the crude initial addition products were dehydrogenated directly with sulfur to give **140**. To explain their formation, a shift of the cyclohexene double bond to the alternative position, in conjugation with the thiophene, was assumed prior to the reaction.[108] Under similar conditions, as in the reaction with maleic anhydride, 2-vinylthiophene reacts with 1,4-benzoquinone in acetic acid to give a 43% yield of **141**. The initial cycloadduct is dehydrogenated by the excess of quinone.[110]

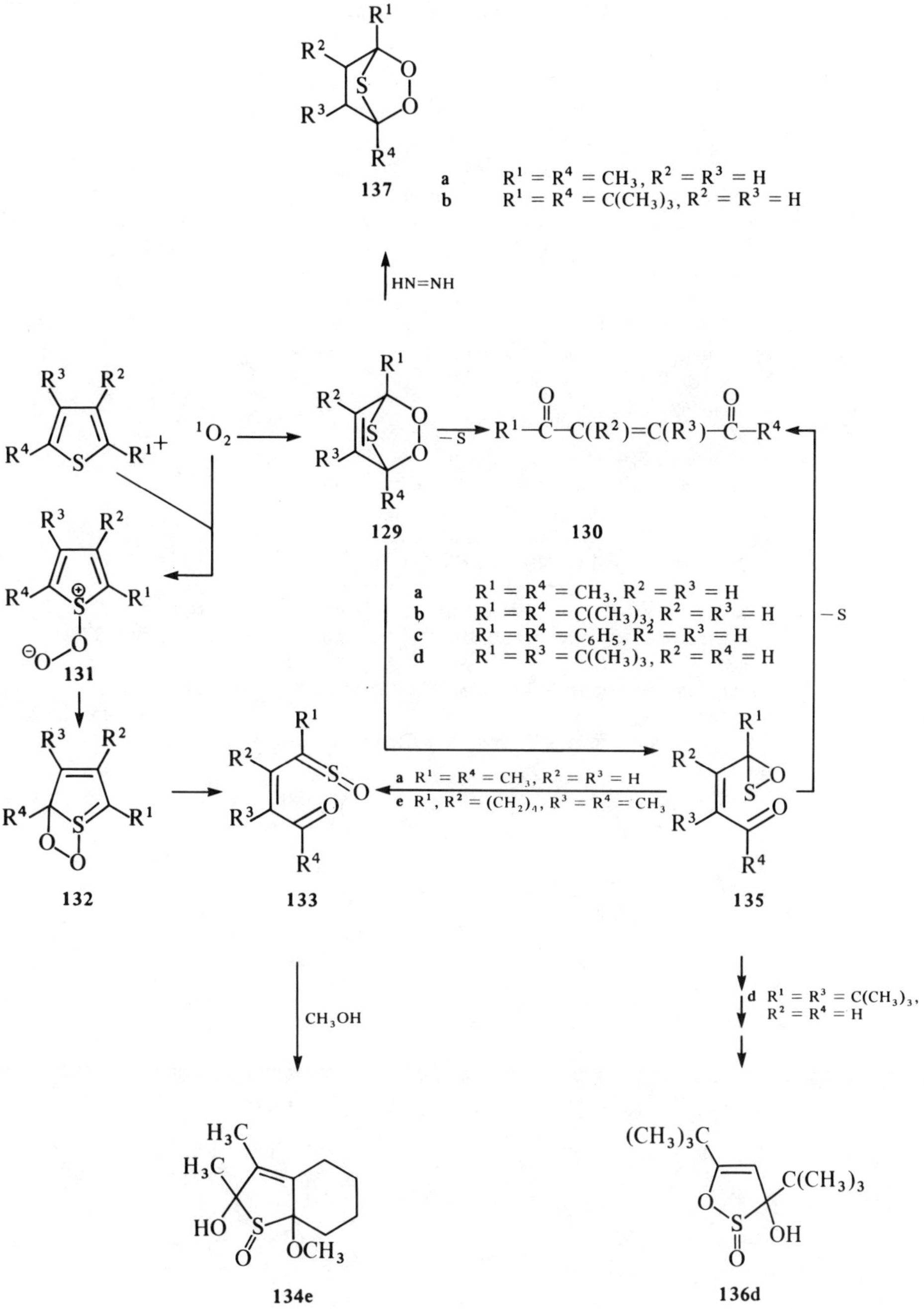

a $R^1 = R^4 = CH_3, R^2 = R^3 = H$
b $R^1 = R^4 = C(CH_3)_3, R^2 = R^3 = H$

137

$R^1 = R^4 = CH_3, R^2 = R^3 = H$
$R^1 = R^4 = C(CH_3)_3, R^2 = R^3 = H$

129   **130**

a $R^1 = R^4 = CH_3, R^2 = R^3 = H$
b $R^1 = R^4 = C(CH_3)_3, R^2 = R^3 = H$
c $R^1 = R^4 = C_6H_5, R^2 = R^3 = H$
d $R^1 = R^3 = C(CH_3)_3, R^2 = R^4 = H$

a $R^1 = R^4 = CH_3, R^2 = R^3 = H$
e $R^1, R^2 = (CH_2)_4, R^3 = R^4 = CH_3$

131

132   **133**   **135**

d $R^1 = R^3 = C(CH_3)_3,$
    $R^2 = R^4 = H$

134e   **136d**

Scheme 19

707

TABLE 7.   2-VINYLTHIOPHENES (138) FROM WHICH DIELS–ALDER
ADDUCTS WITH MALEIC ANHYDRIDE HAVE BEEN
OBTAINED[a, b]

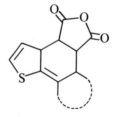

138

	Reference
(2-Thienyl)ethene (2-vinylthiophene)	109, 110
1-(2-Thienyl)-2-methylcyclohexene[c]	108
1-(2-Thienyl)-5-methylcyclohexene[c]	108
1-(2-Thienyl)-2,5-dimethylcyclohexene[c]	108
1-(2-Thienyl)cyclooctene	108
3-(2-Thienyl)indene	108
1-(2-Thienyl)-3,4-dihydronaphthalene	108
1-(2-Thienyl)-6-methoxy-3,4-dihydronaphthalene	108
2-(2-Thienyl)-1,4,5,6,7,8,9,10-octahydronaphthalene[d]	108
1-(2-Thienyl)-3,4-dihydroacephenanthrene	108
4-(2-Thienyl)-6,7-dihydrobenzo[b]thiophene	108

[a] Reactions carried out at steam-bath temperature, unless otherwise stated.
[b] Yields are essentially quantitative, except with the parent compound
(49%).
[c] Reaction conditions – 3 h at 220°C; the adduct was not isolated in a
pure form.
[d] Reaction conditions – 24 h at 120°C; the adduct was not isolated in a
pure form.

138

139

140

141

a   R = H
b   R = CH₃

With the more reactive 4-phenyl-1,2,4-triazoline-3,5-dione, 2-vinylthiophene affords the Diels–Alder adduct **142** at − 20°C in dichloromethane.[111] After refluxing a solution of 2-(2-propenyl)thiophene and di(*tert*-butyl) azodicarboxylate in benzene for 10–14 h, two products were isolated, namely, the dihydrothieno-[3,2-*c*]pyridazine **143b** (10%) and the "ene" addition product **146** (23.5%). From

	a	R = H
	b	R = CH₃

142   **143**   **144**

145   **146**

the reaction with 2-vinylthiophene, an inseparable mixture of products was obtained which showed, in the mass spectrum, molecular ions corresponding to the dihydrothieno[3,2-*c*]pyridazine **143a** and to the tetrahydro derivative **144**. Treatment of the mixture with trifluoroacetic acid to remove the ester functions, followed by oxidation, led to the known thieno[3,2-*c*]pyridazine **145**.[112]

On sensitized photooxygenation of 2-vinylthiophenes, a (4+2)-cycloaddition of the diene moiety consisting of the exocyclic and the neighboring thiophene π-bond to singlet oxygen occurs to give the thermally stable 1,4-endoperoxides **147** (Table 8). The reaction is stereospecific, as was demonstrated with (*E*)-2-(1-propenyl)thiophene, with (*E*)-2-styrylthiophene, and with (*E*)-1-(1-naphthyl)-2-(2-thienyl)ethene as the substrates, which yield exclusively the endoperoxides **147c**, **147d**, and **147e**, respectively. When there is a methyl group in the α-position of the vinyl side chain, in addition to the cycloadducts **147f** and **g**, products resulting from an "ene" reaction (**148f** and **g**) have been obtained.[113] (For the reaction of alkyl- and phenyl-substituted thiophenes with singlet oxygen, see Section II.6.D.)

	f	R² = R³ = CH₃
	g	R² = R³ = H

147   **148**

TABLE 8.   1,4-ENDOPEROXIDES (147) OBTAINED FROM THE REACTIONS OF 2-VINYLTHIOPHENES WITH SINGLET OXYGEN[113]

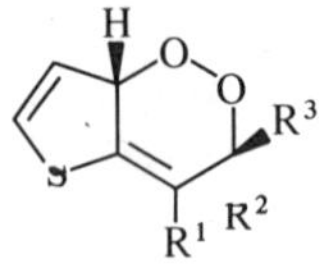

147

Compound	R^1	R^2	R^3	Yield (%)
147a	H	H	H	10
147b	H	CH_3	CH_3	75
147c	H	CH_3	H	51
147d	H	C_6H_5	H	72
147e	H	α-$C_{10}H_7$	H	77
147f	CH_3	CH_3	CH_3	25
147g	CH_3	H	H	27
147h	C_6H_5	H	H	15

If the vinyl double bond constitutes part of an annulated cyclopentenyl moiety, the possibility of a sigmatropic rearrangement (149 ⇌ 150) exists. Keeping a mixture of 4,6-diphenyl-4*H*-cyclopenta[*b*]thiophene (149, $R^1 = R^3 = C_6H_5$, $R^2 = H$) and an excess of DMAD for 3 h at 10°C and 3 h at room temperature, the adduct 151a was isolated in nearly quantitative yield. When, however, the reaction was carried out in refluxing benzene, the cycloadduct 152 was obtained in nearly quantitative yield. A similar result (compound 153) has been effected with maleic anhydride as the dienophile. On the contrary, under these conditions, 5-methyl-6-phenyl-4*H*-cyclopenta[*b*]thiophene (149, $R^1 = H$, $R^2 = CH_3$, $R^3 = C_6H_5$) and DMAD afford the adduct 151b exclusively.[114]

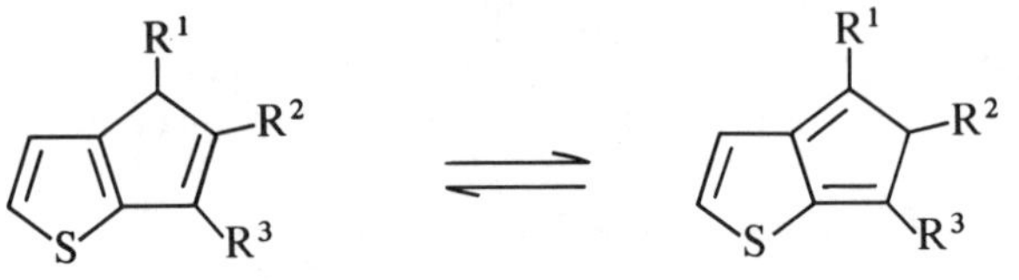

149

150

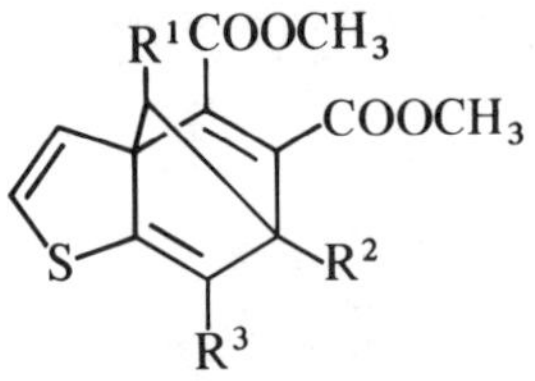

151

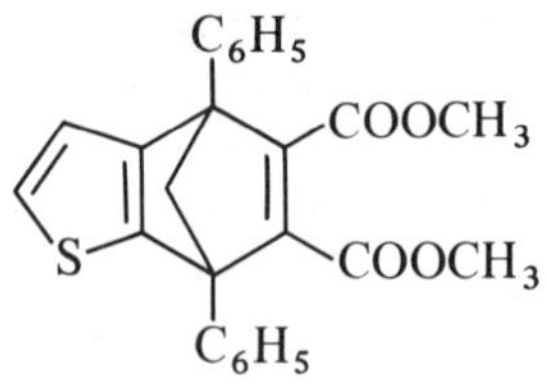

152

a $R^1 = R^3 = C_6H_5$, $R^2 = H$
b $R^1 = H$, $R^2 = CH_3$, $R^3 = C_6H_5$

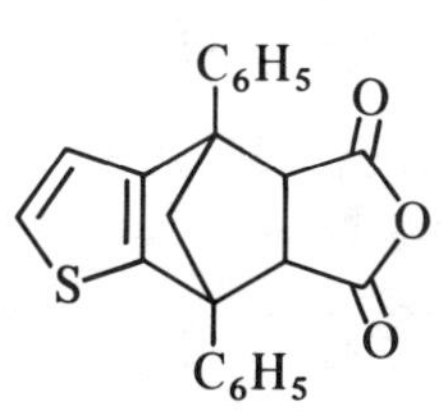

153

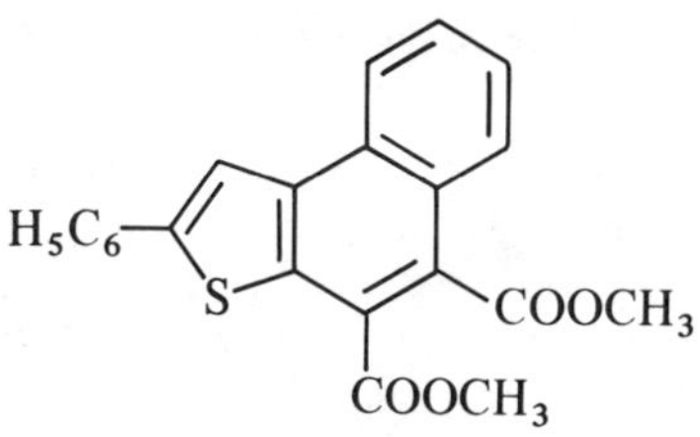

154

After refluxing a mixture of 2,4-diphenylthiophene and an excess of DMAD for 10 h in 1,2-dichlorobenzene, three compounds were isolated, namely, 2,5-diphenylthiophene, dimethyl 3,5-diphenyl-1,2-benzenedicarboxylate (118f), and dimethyl 2-phenylnaphtho[2,1-*b*]thiophene-4,5-dicarboxylate (154). The formation of the benzene derivative 118f has been accounted for by (4+2)-cycloaddition of the thiophene to the acetylene and subsequent extrusion of sulfur (see Section II.6.A). The naphtho[2,1-*b*]thiophene 154 is believed to arise from an unusual Diels–Alder addition of DMAD to the diene comprised of one unsaturated bond of the thiophene and one "Kekulé-bond" of the 4-phenyl substituent. The initial cycloadduct would readily undergo dehydrogenation to afford 154.[90]

If the exocyclic carbon–carbon double bond constitutes part of a diene system, both with one of the π-bonds of the thiophene nucleus as well as with another olefinic double bond (as in 155 and 157), Diels–Alder addition takes place exclusively to the diene moiety, in which the thiophene is not involved. This can be concluded from reactions of 155 with maleic anhydride and of 157 with maleic anhydride and with 1,4-benzoquinone, which afford 158, 159, and 160, respectively. The formation of 158, for which no proof of structure has been given, was explained by isomerization of 155 to 156 prior to the cycloaddition reaction.[115, 116]

155

156

157

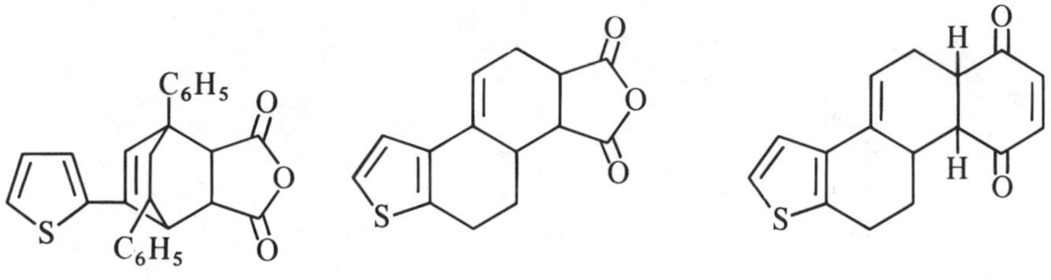

158

159

160

Finally, an intramolecular Diels–Alder reaction to a vinylthiophene 4π-electron moiety has been observed on refluxing a solution of (E)-3-(2-thienyl)propenyl phenylpropiolate (**161a**) or the corresponding 3-thienyl isomer (**161b**) in acetic anhydride to give the dihydrobenzo[*b*]thiophene lactones **162** and **163**, respectively.[117]

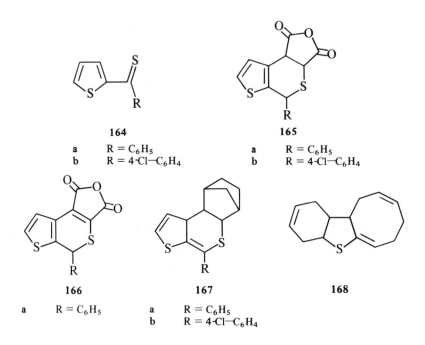

161	162	163

a   2-Thienyl
b   3-Thienyl

Not only an exocyclic olefinic double bond can form a reactive diene system with a thiophene π-bond but also a carbon sulfur double bond. This was demonstrated in the reactions of aryl 2-thienyl thiones (**164**) with dienophiles. Refluxing a solution of phenyl 2-thienyl thione (**164a**) and maleic anhydride in xylene afforded both an aromatized (**165a**) and an aromatized oxidized cycloadduct (**166a**) in low yields. Under comparable conditions, the 4-chloro derivative **164b** yielded only the corresponding aromatized cycloadduct **165b** in 18% yield. The reactions of phenyl (**164a**) and 4-chlorophenyl 2-thienyl thione (**164b**) with bicyclo[2.2.1]hept-2-ene gave the initial cycloadducts **167**.[118]

164

a   R = C$_6$H$_5$
b   R = 4-Cl—C$_6$H$_4$

165

a   R = C$_6$H$_5$
b   R = 4-Cl—C$_6$H$_4$

166

a   R = C$_6$H$_5$

167

a   R = C$_6$H$_5$
b   R = 4-Cl—C$_6$H$_4$

168

8. (4+4)-Cycloaddition Reaction of 2-Vinylthiophene

A (4+4)-cycloaddition of 2-vinylthiophene to butadiene is probably involved in the formation of the benzo[*b*]cycloocta[*d*]thiophene derivative **168** in the nickel acetylacetonate-triphenylphosphite-triethylaluminum catalyzed reaction of both reactants.[119]

III. CYCLOADDITION REACTIONS OF THIOPHENE 1-OXIDES

1. Introduction

Procházka[120] claimed to have prepared the parent thiophene 1-oxide in solution by eliminating two molecules of methanesulfonic acid from tetrahydro-3,4-bis(methylsulfonyloxy)thiophene 1-oxide. The parent compound could not be isolated, but dimerized via a Diels–Alder mode of reaction. The formation of thiophene 1-oxides as intermediates in the oxidation of thiophene and some of its derivatives was deduced from the structure of the Diels–Alder adducts, the so-called "sesquioxides," formed in a consecutive reaction.[121–124] An application of this common feature of the thiophene 1-oxides to undergo Diels–Alder reactions has been described by Torssell,[125] who oxidized thiophene and some of its derivatives in the presence of the 1,4-benzoquinone type of dienophiles. Mock[9] prepared two sterically hindered thiophene 1-oxides, namely, 2,5-di(*tert*-butyl)- and 2,5-di(*tert*-octyl)thiophene 1-oxide, in low yields by oxidation of the corresponding thiophenes. Another approach to a substituted thiophene 1-oxide was described by Eastman and Wagner,[126] who condensed diethyl sulfinyldiacetate with diethyl oxalate in a basic medium and obtained a stable product in 13% yield, which was described as diethyl 3,4-dihydroxythiophene-2,5-dicarboxylate 1-oxide. The cycloaddition reactions of these transient thiophene 1-oxides are described in this section.

2. Dimerization of Thiophene 1-Oxides

The dimerization of thiophene 1-oxide (**169**), as described by Procháska,[120] is a Diels–Alder type of cycloaddition in which the thiophene 1-oxide acts both as a dienophile and as a diene, yielding 3a,4,7,7a-tetrahydro-4,7-epithiobenzo[*b*]thiophene 1,8-dioxide (**170**) (Scheme 20). Evidence for the intermediate thiophene

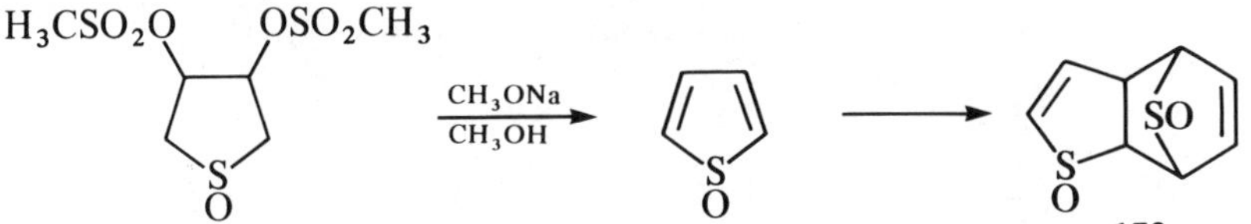

Scheme 20

1-oxide was found in its UV spectrum and in the reduction with sodium borohydride to thiophene. The formation of thiophene 1-oxide as a transient intermediate in the oxidation of thiophene had already been postulated in earlier work. Stevens,[121] Melles and Backer,[122] Davies and James,[123] and Okita and Kambara[124] reported the isolation of a "sesquioxide," 3a,4,7,7a-tetrahydro-4,7-epithiobenzo[b]thiophene 1,1,8-trioxide (172a). Davies and James[123] gave two possible mechanisms for its formation (Scheme 21). Melles and Backer[127] oxidized 3,4-dimethylthiophene with

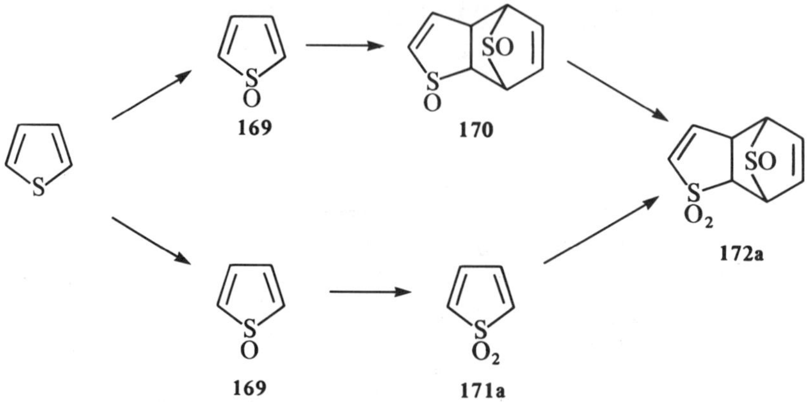

Scheme 21

2 eq. perbenzoic acid (PBA) in chloroform and obtained 3,4-dimethylthiophene 1,1-dioxide (171b) in 62.5% yield. With 1 eq. PBA, 3a,4,7,7a-tetrahydro-3,3a,5,6-tetramethyl-4,7-epithiobenzo[b]thiophene 1,1,8-trioxide (172b) was isolated as the only product in 8% yield (Scheme 22). 2-Methylthiophene, 3-methylthiophene, and 3-phenylthiophene react similarly, yielding the corresponding "sesquioxides." Which one of the possible regioisomers was formed could not be established.

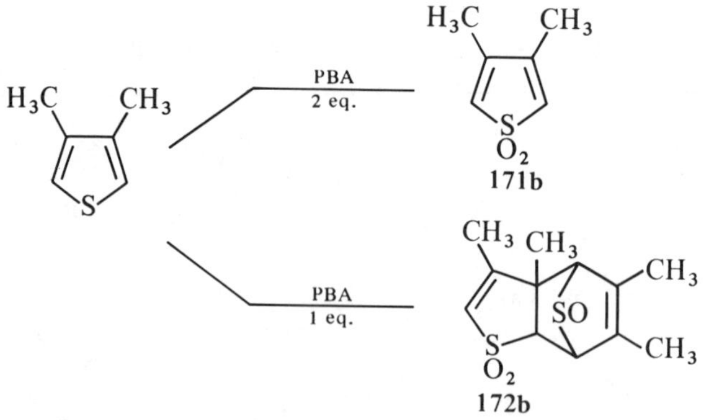

Scheme 22

3. (4+2)-Cycloaddition Reactions of Thiophene 1-Oxides

The marked propensity of thiophene 1-oxides for (4+2)-cycloaddition reactions has been demonstrated by Helder.[34] Oxidation of tetramethylthiophene with *m*-chloroperbenzoic acid (MCPBA) in the presence of dicyanoacetylene yielded tetramethyl-1,2-benzenedicarbonitrile (**118m**, 70%) (Scheme 23). An alternative route,

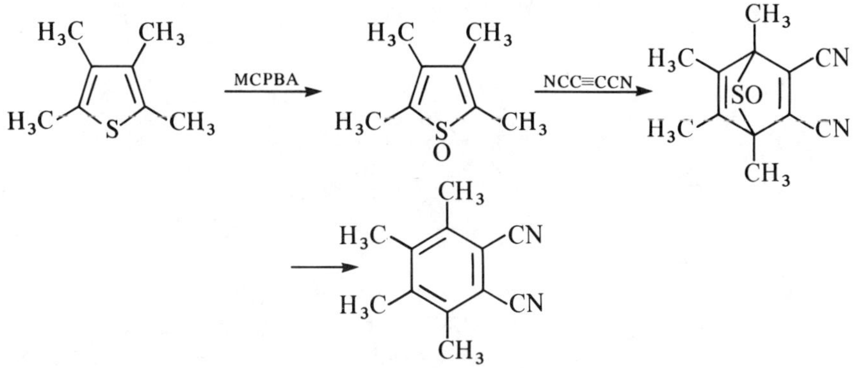

Scheme 23

the addition of dicyanoacetylene to tetramethylthiophene (see Section II.6.A) and subsequent oxidation of the primary Diels–Alder adduct could not definitely be ruled out. Torssell[125] generated thiophene 1-oxide by oxidizing thiophene with MCPBA in the presence of 1,4-benzoquinone in dichloromethane. The presence of a (4+2)-cycloadduct, 1,4,4a,8a-tetrahydro-1,4-epithio-5,8-naphthoquinone 9-oxide (**173a**), in the crude reaction product was proved spectroscopically; 5-hydroxy-1,4-naphthoquinone (juglone, **174a**) and 1,4-naphthoquinone (**175a**) were isolated in yields of 21 and 7%, respectively, as the products of further oxidation. When the reaction was performed in chloroform, **173a** could be isolated in 21% yield. Under similar conditions, 2-methyl- and 3-methylthiophene give the adducts **173b** and **c** in yields of 24 and 18%, respectively, together with 5-methyl- (**175b**) and 8-hydroxy-5-methyl-1,4-naphthoquinone (**174b**) and 6-methyl- (**175c**) and 5-hydroxy-6-methyl-1,4-naphthoquinone (**174c**), respectively. 2,5-Dimethyl- and 2,5-dichlorothiophene were also oxidized with MCPBA in the presence of 1,4-benzoquinone, yielding mixtures of **173d** and **175d** and of **173e** and **175e**, respectively. Under the same reaction conditions, 2,5-dimethylthiophene 1,1-dioxide hardly reacts with 1,4-benzoquinone; therefore, it was concluded that 2,5-dimethylthiophene 1-oxide is a much more reactive diene as the corresponding 1,1-dioxide. Torssell could prove that under the same reaction conditions, **173d** was much more stable than the adduct of the corresponding thiophene 1,1-dioxide. This led to the conclusion that an SO bridge is more stable than an SO₂ bridge in this kind of compound (see Section IV.6.K). A variety of substituted thiophenes were treated in the

173 174 175

a	$R^1 = R^2 = R^3 = R^4 = H$
b	$R^1 = CH_3, R^2 = R^3 = R^4 = H$
c	$R^3 = CH_3, R^1 = R^2 = R^4 = H$
d	$R^1 = R^4 = CH_3, R^2 = R^3 = H$
e	$R^1 = R^4 = Cl, R^2 = R^3 = H$

same way. In these cases, the corresponding adducts were not obtained, but at best small amounts of the corresponding 1,4-naphthoquinones were secured. Thiophene, 2-methyl-, and 3-methylthiophene were oxidized also in the presence of 1,4-naphthoquinone, and although the presence of Diels–Alder adducts was shown by ^{1}H nmr spectroscopy, only the corresponding 9,10-anthraquinones were isolated.

IV. CYCLOADDITION REACTIONS OF THIOPHENE 1,1-DIOXIDES

1. Introduction

Bailey and Cummins[128] reported the first preparation of the parent thiophene 1,1-dioxide (171a) by a six-step synthesis starting from 2,5-dihydrothiophene 1,1-dioxide. Thiophene 1,1-dioxide could not be isolated in a pure state, but a dilute chloroform solution of the compound could be stored at 5°C for a long time. On evaporation of the solvent, dimerization took place with evolution of sulfur dioxide. A similar approach to prepare the parent compound had been used previously by Backer and Melles.[122, 129] However, sulfur dioxide was evolved from a benzene solution of thiophene 1,1-dioxide, and after 1 day at room temperature, a brown viscous mass was separated, from which only a trimeric product could be isolated (see Section IV.4). Direct oxidation of thiophene iteself does not result in the formation of thiophene 1,1-dioxide; the only product isolated is the sesquioxide 172a, a Diels–Alder adduct of thiophene 1-oxide and 1,1-dioxide (see Section III.2 and Scheme 21). Substituted thiophene 1,1-dioxides are well-characterized substances. The first examples, 3,4-diphenyl- and tetraphenylthiophene 1,1-dioxide, were prepared in 1915 by Hinsberg,[130] by direct oxidation of the corresponding thiophenes.

In this section, the cycloaddition reactions of thiophene 1,1-dioxides are analyzed.

2.   (2+2)-Cycloaddition Reactions of Thiophene 1,1-Dioxides

Bluestone et al.[131] described the dimerization of 3,4-dichlorothiophene 1,1-dioxide (171d), in which 171d behaves both as a diene and as a dienophile. As usual (see Section IV.4), the Diels–Alder adduct eliminated sulfur dioxide and, in this special case, hydrogen chloride to yield a 3,5,6-trichlorobenzo[b]thiophene 1,1-dioxide (176) as the major product. In a consecutive reaction, 176 reacted further as a dienophile with 171d to give the adduct 177 after loss of sulfur dioxide and hydrogen chloride. Surprisingly, a third product was isolated with the molecular formula of a dimer. Based on ^{1}H nmr data and because the crystals exhibited zero piezoelectricity, it was concluded that the structure was completely centrosymmetric and, therefore, the compound was described as the (2+2)-cycloadduct 178 (Scheme 24).

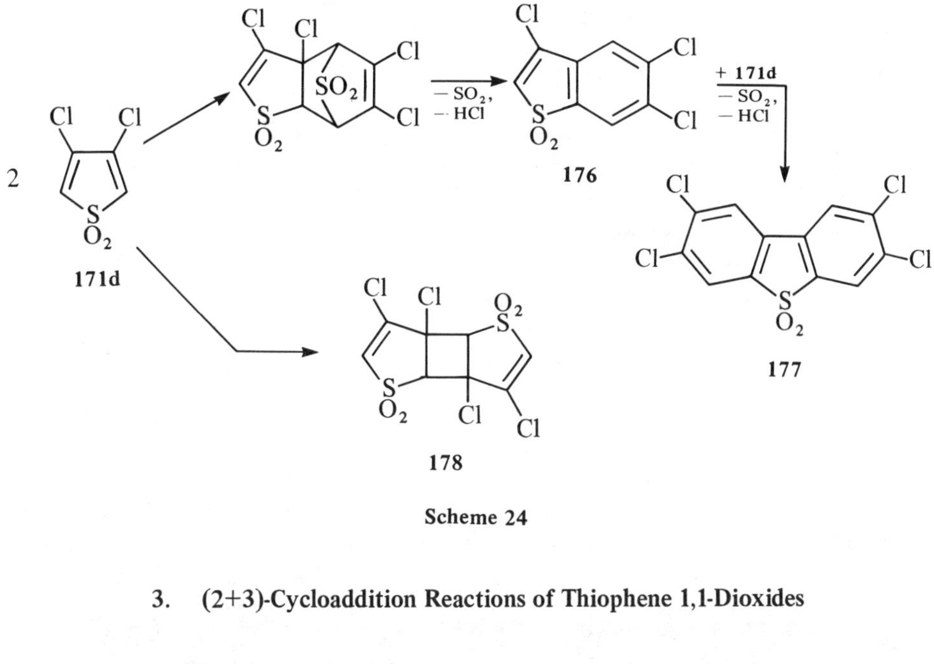

Scheme 24

3.   (2+3)-Cycloaddition Reactions of Thiophene 1,1-Dioxides

Kabzińska[132] claimed the formation of an intermediate (2+3)-cycloaddition product (179) on reacting 3,4-dimethylthiophene 1,1-dioxide (171b) with iso-quinoline N-oxide. Rearomatization of the isoquinoline nucleus by cleavage of the weak N—O bond and formation of an O—H bond yielded 2,3-dihydro-2-(isoquinolinyl)-3,4-dimethylthiophene-3-ol 1,1-dioxide (180). Geneste and co-workers[133] reacted thiophene 1,1-dioxide (171a), generated in situ by treatment of 3,4-dibromotetrahydrothiophene 1,1-dioxide with pyridine, with two types of 1,3-dipoles — an acyclic nitrone and a nitrile oxide. Both cycloadducts are formed regioselectively. Reaction of thiophene 1,1-dioxide with N,α-diphenyl-nitrone afforded 181 in a yield of 70%, and reaction with 2,4,6-trimethylbenzo-

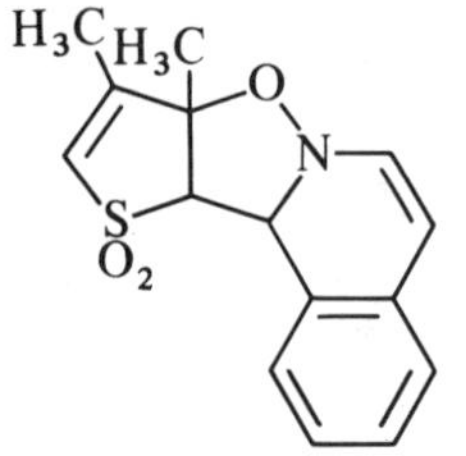

P. H. Benders, D. N. Reinhoudt and W. P. Trompenaars

179 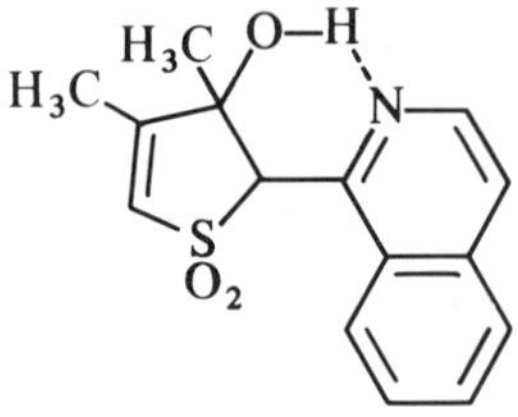　　　　　　　　　　　180

nitrile oxide gave compound **182** (67%). Both **181** and **182** can react further as dipolarophiles to give the corresponding diadducts. Kinetic studies of the cyclo-addition of thiophene 1,1-dioxide and the nitrone showed that the monoadduct formation is 10^3 times faster than the diadduct formation.

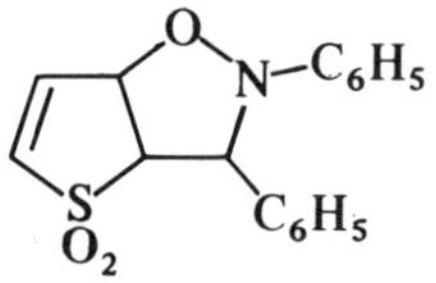

181 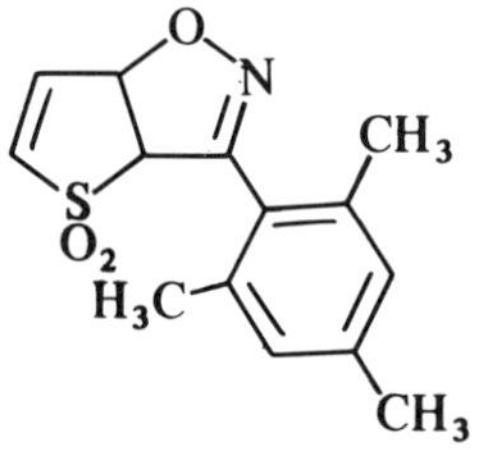　　　　　　　　　　　182

4.　"Dimerization" of Thiophene 1,1-Dioxides

Bailey and Cummins,[134] who reported the synthesis of thiophene 1,1-dioxide in solution,[128] found that on evaporation of the solvent, even below room tempera-ture, sulfur dioxide was evolved and a polymeric material was formed together with a crystalline compound, which was shown to be the 3a,7a-dihydrobenzo[b]thio-phene 1,1-dioxide (**184a**). In this reaction, thiophene 1,1-dioxide behaves both as a 2π- and as a 4π-component; the initial Diels–Alder adduct (**183**) liberates sulfur dioxide to give **184a** (Scheme 25). The same type of dimer (**184**) as that described

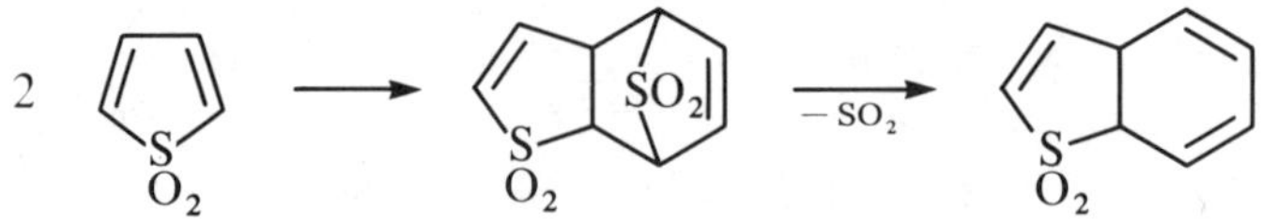

171a　　　　　　　　183　　　　　　　184a

Scheme 25

for the parent thiophene 1,1-dioxide (171a) has been recorded for 3,4-dimethylthio-phene 1,1-dioxide (171b → 184b) by Wróbel,[137] for 3,4-diphenylthiophene 1,1-dioxide (171e → 184e) by Whelan,[135] and for tetrachlorothiophene 1,1-dioxide (171f → 184f) by Raasch[87] (Table 9). Dimerization of 3,4-dichlorothiophene 1,1-dioxide (171d) afforded 3,5,6-trichlorobenzo[b]thiophene 1,1-dioxide (176) (see Section IV.2 and Scheme 24). Unsymmetrically substituted thiophene 1,1-dioxides might give rise to the formation of regioisomers. Patterson,[136] who has "dimerized" 2-methyl-, 3-methyl-, 3-phenyl-, and 2,4-dimethylthiophene 1,1-dioxide (171g–j), found, however, only one adduct in each case (Table 9). An interesting variation was found in the dimerization of 3-isopropenyl-4-isopropylthiophene 1,1-dioxide (171k). In this case, the C-2-C-3 double bond of the thiophene ring together with the double bond of the isopropenyl group form the diene component in the Diels–Alder reaction with a second molecule of 171k. The benzo[2,1-b:3,4-b']dithiophene derivative 185 was formed in a 60% yield[136] (Scheme 26).

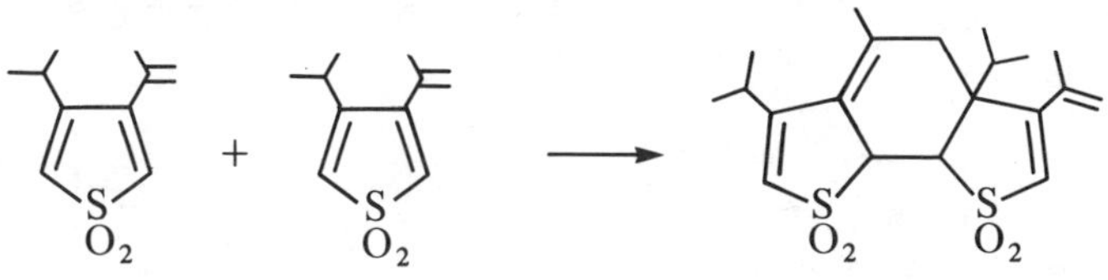

171k   171k   185

Scheme 26

Backer and Melles[129] attempted to prepare thiophene 1,1-dioxide by dehydro-bromination of 3-bromo-2,3-dihydrothiophene 1,1-dioxide with piperidine in benzene. During the preparation, sulfur dioxide was evolved and a brown viscous mass was separated from the benzene solution. Extraction with acetic acid afforded a crystalline product to which, on the basis of elemental analysis, a "trimeric" structure 186 or 187 has been assigned.

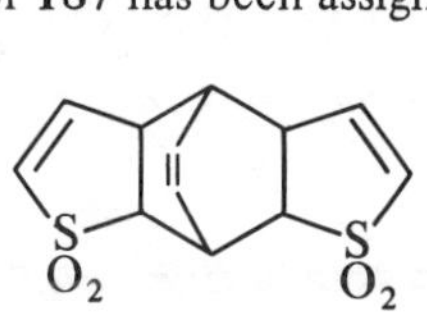

186 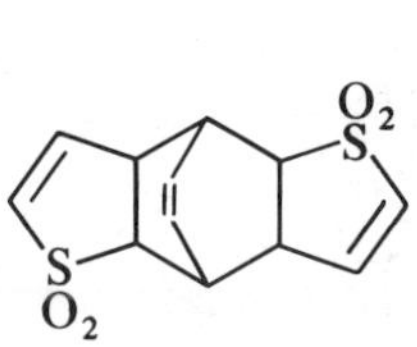   187

5.   (2+4)-Cycloaddition Reactions of Thiophene 1,1-Dioxides

In this section, reactions in which the thiophene 1,1-dioxide acts as a dienophile in Diels–Alder reactions are described.

Oxidation of thiophene with peracids yields a sesquioxide (172a), the for-mation of which was explained by a Diels–Alder reaction of the intermediate

TABLE 9. DIMERIZATION OF THIOPHENE 1,1-DIOXIDES (171)

$$2 \quad 171 \quad \xrightarrow{-SO_2} \quad 184$$

Thiophene 1,1-Dioxide	Dimer	R^1	R^2	R^3	R^4	R^5	R^6	R^7	R^8	Reference
171a	184a	H	H	H	H	H	H	H	H	134
171b	184b	H	CH₃	CH₃	H	H	CH₃	CH₃	H	137
171d	176	H	Cl	Cl	H	—	—	Cl	H	131
171e	184e	H	C₆H₅	C₆H₅	H	H	C₆H₅	C₆H₅	H	135
171f	184f	Cl	Cl	Cl	Cl	Cl	Cl	Cl	Cl	87
171g	184g	CH₃	H	H	H	H	H	H	CH₃	136
171h	184h	H	CH₃	H	H	H	H	CH₃	H	136
171i	184i	H	C₆H₅	H	H	H	H	C₆H₅	Cl	136
171j	184j	H	CH₃	H	CH₃	H	CH₃	H	CH₃	136

thiophene 1,1-dioxide (171a, dienophile) and thiophene 1-oxide (169, diene).[122-124] Similar reactions have been observed with substituted thiophenes[122] (see Section III.2 and Schemes 21 and 22). An analogous mechanism has been proposed for the dimerization of thiophene 1,1-dioxides (see Section IV.4 and Scheme 25).

Bailey and Cummins[138] reported the reaction of thiophene 1,1-dioxide (171a) as a reactive dienophile with 1,2-bis(methylene)cyclohexane. After 42 h at room temperature, the octahydronaphtho[2,3-b]thiophene 1,1-dioxide 188 was isolated in 50% yield. Bluestone et al.[131] reacted 3,4-dichlorothiophene 1,1-dioxide (171d) with monomeric cyclopentadiene in acetone. The reaction was strongly exothermic. The reaction was started at 5°C, but it is not clear from the experimental data if the temperature was kept constant. This may be an important detail, as will be discussed later. Two products (189d and 191d) were isolated, the formation of which was explained by the reaction of 171d both as a dienophile (giving the major product 189d, 61%) and as a diene (giving the minor product 191d, 16%) (Scheme 27). Similar sets of products (192 and 193) were found by reacting 171d with 1,3-butadiene and 2-methyl-1,3-butadiene.[131] Patterson[136] repeated the reaction of

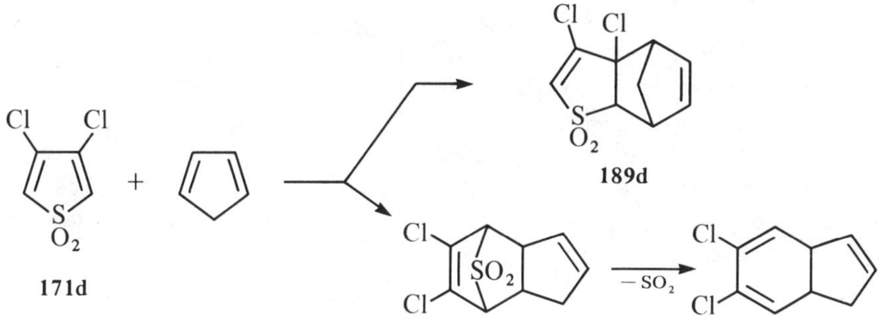

Scheme 27

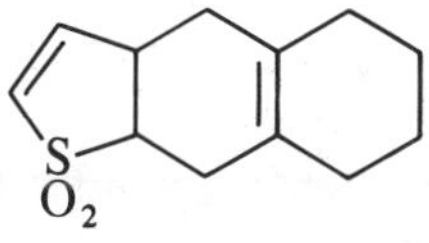

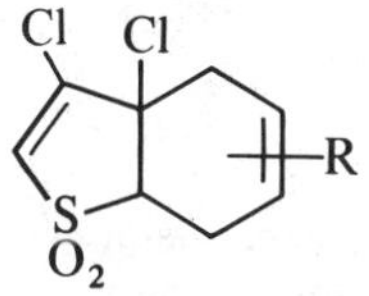

a	R = H	a	R = H
b	R = CH$_3$	b	R = CH$_3$

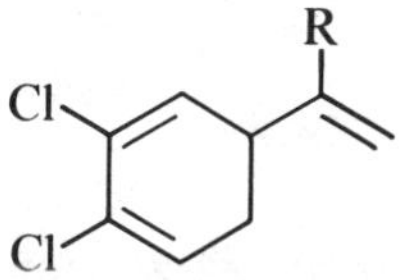

3,4-dichlorothiophene 1,1-dioxide (171d) with cyclopentadiene at a well-controlled temperature (25°C). Only compound 189d was isolated in a 68% yield, and he concluded that under the conditions described by Bluestone,[131] 189d undergoes a Cope rearrangement to give 190d, which loses sulfur dioxide, yielding 191d (Scheme 28). Under similar controlled-reaction conditions Patterson,[136] also found only one

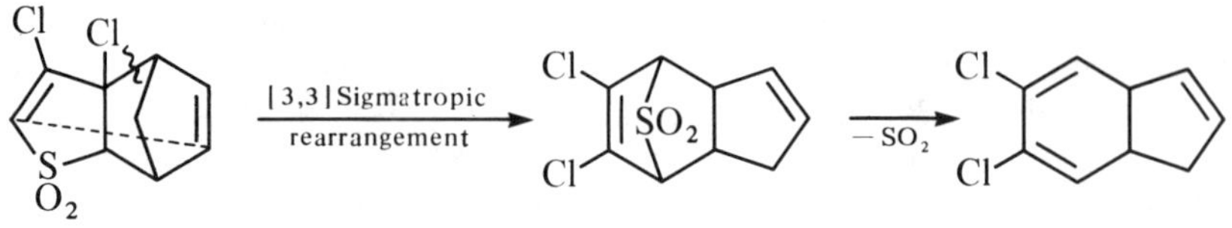

Scheme 28

cycloadduct in the reactions of cyclopentadiene with the parent thiophene 1,1-dioxide (171a) and with 3,4-dimethyl- (171b), 2-methyl- (171g) 3-methyl- (171h), 3-phenyl- (171i), 2,4-dimethyl- (171j), 3-isopropenyl-4-isopropyl- (171k), and 3-chloro-4-methoxythiophene 1,1-dioxide (171l), respectively. In all these reactions, the thiophene 1,1-dioxide reacts as a dienophile (Table 10). The regioselectivity of the Diels–Alder reaction is predominantly determined by the steric effects of the substituents of the thiophene 1,1-dioxide, and the least sterically crowded Diels–Alder adducts are formed. The periselectivity of the reaction with cyclopentadiene was further demonstrated in the reaction of thiophene 1,1-dioxide (171a) with spiro[2.4]hepta-4,6-diene, in which 171a acts only as the dienophile, yielding 194 in 60% yield. Raasch[87] reported an exothermic reaction of tetrachlorothiophene 1,1-dioxide (171f) with cyclopentadiene in dichloromethane. Two adducts (189f and 191f) were isolated, the formation of which can also be explained by the Patterson mechanism (Table 10). From the reaction of tetrachlorothiophene 1,1-dioxide (171f) and 2,3-dimethylbutadiene, only the (2+4)-cycloadduct 195 was

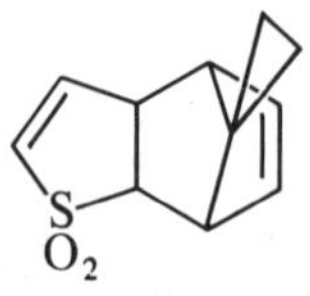

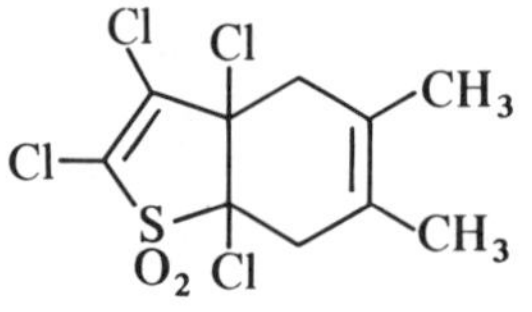

isolated in a 50.5% yield, even though the reaction was performed under reflux.[87] The high reactivity of 171f was further demonstrated by its reaction with furan.[139] The first step is a (2+4)-cycloaddition reaction and the transient intermediate 196 reacts with a second molecule of 171f to yield 197. (Scheme 29). From the same

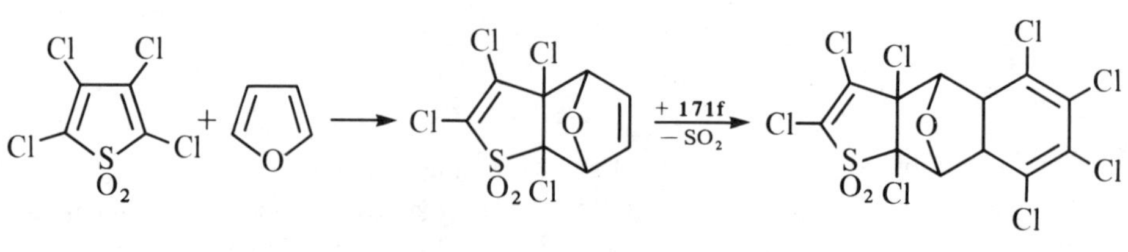

Scheme 29

TABLE 10. CYCLOADDITION OF THIOPHENE 1,1-DIOXIDES (171) AND CYCLOPENTADIENE

| | Thiophene 1,1-Dioxide | | | | Reaction Temperature (°C) | Compound | Yield (%) | | | | | Compound | Yield (%) | Reference |
Compound	R^1	R^2	R^3	R^4				R^5	R^6	R^7	R^8			
171a	H	H	H	H	25	189a	65	H	H	H	H	—	—	136
171b	H	CH_3	CH_3	H	25	189b	45	H	CH_3	CH_3	H	—	—	136
171d	H	Cl	Cl	H	a	189d	61	H	Cl	Cl	H	191d	16	131
171d	H	Cl	Cl	Cl	25	189d	68	Cl	Cl	Cl	Cl	—	—	136
171f	Cl	Cl	Cl	Cl	a	189f	25	H	H	Cl	CH_3	191f	44	87
171g	CH_3	H	H	H	25	189g	67	H	H	H	CH_3	—	—	136
171h	H	CH_3	H	H	25	189h	60	H	H	CH_3	H	—	—	136
171i	CH_3	C_6H_5	H	H	25	189i	40	CH_3	H	C_6H_5	H	—	—	136
171j	CH_3	H	CH_3	H	25	189j	40	H	H	CH_3	H	—	—	136
171k	H	$i\text{-}C_3H_5$	$i\text{-}C_3H_7$	H	25	189k	60	H	$i\text{-}C_3H_7$	$i\text{-}C_3H_5$	H	—	—	136
171l	H	Cl	OCH_3	H	25	189l	65	H	Cl	OCH_3	H	—	—	136

^{a} Exothermic reaction.

723

reaction, a second product was isolated, formed by a (4+2)-cycloaddition of **171f** and furan, followed by a rearrangement (see Section IV.6.L).

6.   (4+2)-Cycloaddition Reactions of Thiophene 1,1-Dioxides

A.   Introduction

In this section, reactions in which the thiophene 1,1-dioxide acts as a diene with a large variety of dienophiles are surveyed. Sulfur dioxide is spontaneously eliminated from the intermediate adducts. In several cases, consecutive reactions have been observed.

B.   Acyclic Alkenes as Dienophiles in (4+2)-Cycloaddition Reactions of Thiophene 1,1-Dioxides

Tetrachlorothiophene 1,1-dioxide (**171f**) is such a reactive diene that, even with ethene, it reacts slowly at room temperature to give 1,2,3,4-tetrachloro-1,3-cyclohexadiene (**198a**); at 80°C in benzene (1.4 bar), it is converted almost quantitatively in 4 h. The reaction of **171f** with 1-dodecene affords **198b**. With 1-hexen-5-yne, reaction takes place exclusively at the double bond to give **198c**[87] (Scheme 30). Raasch[87,139] also reported the reaction of tetrachlorothiophene 1,1-dioxide (**171f**) with a great number of functionalized alkenes, all reacting in a similar way to give the 1,2,3,4-tetrachloro-1,3-cyclohexadiene derivatives **198** in yields of 50–90% (Scheme 30). Apart from the reactions of the stable and reactive tetrachloro-

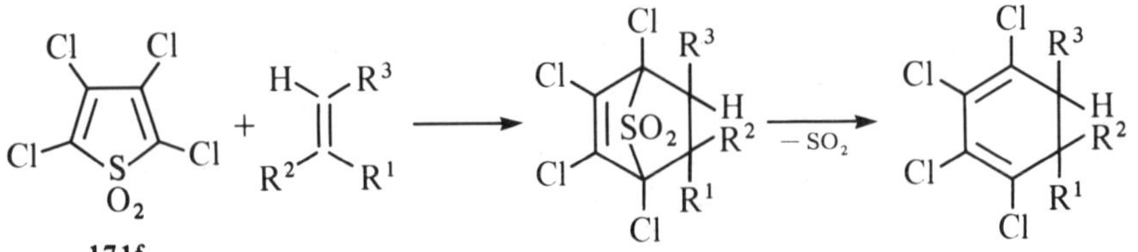

171f   **198**

a $R^1 = R^2 = R^3 = H$
b $R^1 = (CH_2)_9CH_3$, $R^2 = R^3 = H$
c $R^1 = (CH_2)_2C{\equiv}CH$, $R^2 = R^3 = H$
d $R^1 = (CH_2)_8COOH$, $R^2 = R^3 = H$
e $R^1 = CH_2Br$, $R^2 = R^3 = H$
f $R^1 = CH_2NCS$, $R^2 = R^3 = H$
g $R^1 = CH_2COOH$, $R^2 = R^3 = H$

h $R^1 = CH_2-$ [structure], $R^2 = R^3 = H$

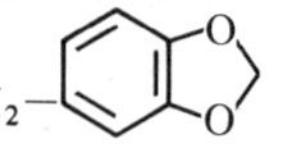

i $R^1 = -N$ [structure], $R^2 = R^3 = H$

j $R^1 = -N$ [structure], $R^2 = R^3 = H$

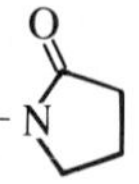

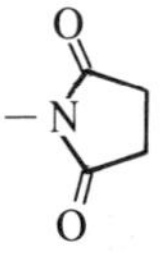

k $R^1 = $ 2-Furyl, $R^2 = R^3 = H$
l $R^1 = $ 2-Pyridyl, $R^2 = R^3 = H$
m $R^1 = COOH$, $R^2 = R^3 = H$
n $R^1 = COOCH_3$, $R^2 = R^3 = H$
o $R^1 = CN$, $R^2 = R^3 = H$
p $R^1 = R^3 = COOCH_3$, $R^2 = H$
q $R^1 = CH_3$, $R^2 = C_6H_5CO$, $R^3 = H$
r $R^1, R^3 = CH_2CH{=}CHCH_2$, $R^2 = H$
s $R^1 = (CH_2)_4CH{=}CH_2$, $R^2 = R^3 = H$
t $R^1 = (CH_2)_5CH{=}CH_2$, $R^2 = R^3 = H$

Scheme 30

thiophene 1,1-dioxide (**171f**) with simple and functionalized alkenes, as described above, only one other example of this type of reaction has been recorded. Whelan[135] reported the reaction of 3,4-diphenylthiophene 1,1-dioxide (**171e**) with 1-acetoxy-styrene to yield 1,2,4-triphenylbenzene (Scheme 31).

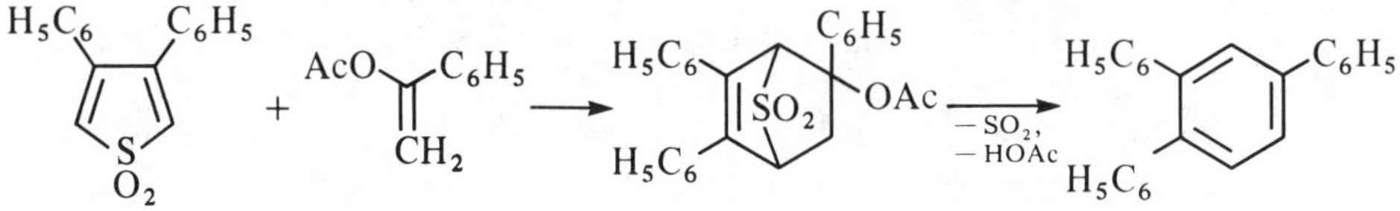

171e

Scheme 31

C. Cycloalkenes as Dienophiles in (4+2)-Cycloaddition Reactions of Thiophene 1,1-Dioxides

Reinhoudt, van Tilborg, et al.[140,141] described the high-yield synthesis of substituted 1,3,5-cycloheptatrienes (**200**) by reaction of thiophene 1,1-dioxides (**171**) with suitable substituted cyclopropenes (**199**). Sulfur dioxide is expelled very quickly from the intermediate Diels–Alder adduct, synchronous with cleavage of the cyclopropane ring (Scheme 32). The cycloaddition is greatly influenced by steric effects, as can be seen from the reaction rates that have been determined.[141]

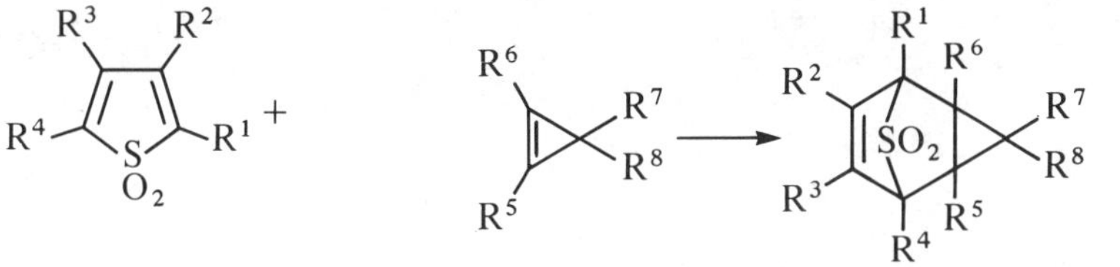

171   **199**

a $R^1 = R^2 = R^3 = R^4 = H$ a $R^5 = R^6 = R^7 = R^8 = H$
c $R^1 = R^4 = CH_3, R^2 = R^3 = H$ b $R^7 = CH_3, R^5 = R^6 = R^8 = H$
n $R^1 = R^4 = t\text{-}C_4H_9, R^2 = R^3 = H$ c $R^5 = CH_3, R^6 = R^7 = R^8 = H$
o $R^1 = R^3 = t\text{-}C_4H_9, R^2 = R^4 = H$ d $R^7 = R^8 = CH_3, R^5 = R^6 = H$
p $R^1 = R^4 = C_6H_5, R^2 = R^3 = H$

$$-SO_2$$

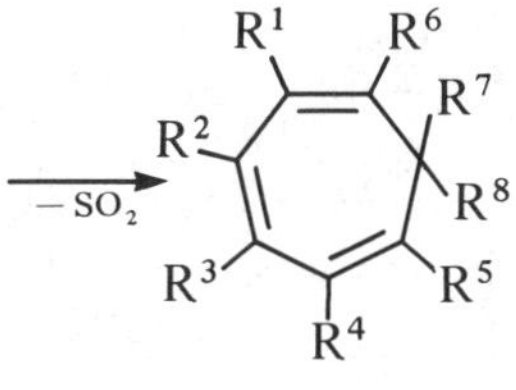

200

171a + 199c → 200a
171c + 199a → 200b
171c + 199b → 200c
171c + 199c → 200d
171n + 199b → 200e
171o + 199a → 200f
171p + 199a → 200g

Scheme 32

Patterson[136] reacted 2-methylthiophene 1,1-dioxide (171g) with 1-methylcyclo-propene to give approximately a 1:1 mixture of the two regioisomeric dimethyl-1,3,5-cycloheptatrienes 200h ($R^1 = R^5 = CH_3$) and 200i ($R^1 = R^6 = CH_3$). The same ratio was found for 3-methylthiophene 1,1-dioxide (171h), yielding 200j ($R^2 = R^5 = CH_3$) and 200k ($R^2 = R^6 = CH_3$). The cycloaddition of 3-isopropenyl-4-isopropylthiophene 1,1-dioxide (171k) with 1-methylcyclopropene was found to be more regioselective: 200l ($R^2 = i\text{-}C_3H_7$, $R^3 = i\text{-}C_3H_5$, and $R^6 = CH_3$) was isolated in 40% yield and 200m ($R^2 = i\text{-}C_3H_7$, $R^3 = i\text{-}C_3H_5$, and $R^5 = CH_3$) in 20% yield. An even higher regioselectivity was found in the reaction of 3-chloro-4-methoxythiophene 1,1-dioxide (171l) with 1-methylcyclopropene: 200n ($R^2 = Cl$, $R^3 = OCH_3$, and $R^6 = CH_3$) and 200o ($R^2 = Cl$, $R^3 = OCH_3$, and $R^5 = CH_3$) were isolated in yields of 58 and 12%, respectively.

Cyclopentene, -hexene, -heptene, -octene, and -dodecene react with tetrachloro-thiophene 1,1-dioxide (171f) in a manner analogous to ethene (see Section IV.6.B), to yield the cycloadducts 201.[87] Tetrachlorothiophene 1,1-dioxide (171f) reacts in a similar way with indene, acenaphthylene, and endo-dicyclopentadiene to give 202, 203, and 204, respectively.[87] Bailey and Cummins[138] described the reaction of the parent thiophene 1,1-dioxide (171a) with indene; its disproportionation product, fluorene (instead of dihydrofluorene), was isolated in a low yield (3%).

201

a   n = 1
b   n = 2
c   n = 3
d   n = 4
e   n = 8

202

203   204

D. Cycloalkanes with Exocyclic Double Bonds as Dienophiles in (4+2)-Cycloaddition Reactions of Thiophene 1,1-Dioxides

Raasch[87] reacted tetrachlorothiophene 1,1-dioxide (171f) with methylenecyclo-butane, methylenecyclohexane, 1,4-dimethylenecyclohexane, β-pinene, and methyl-eneadamantane to yield the corresponding spirocompounds 205–209.

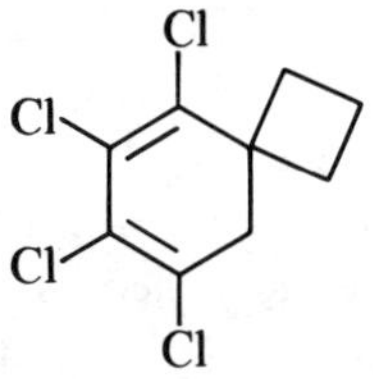

205 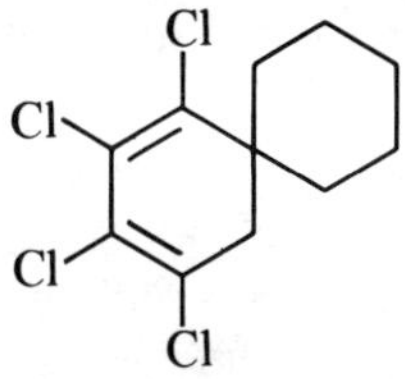   **206** 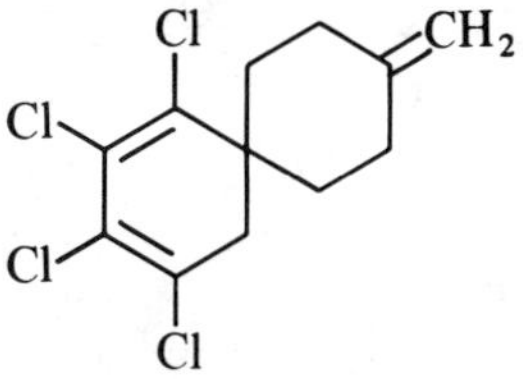   **207**

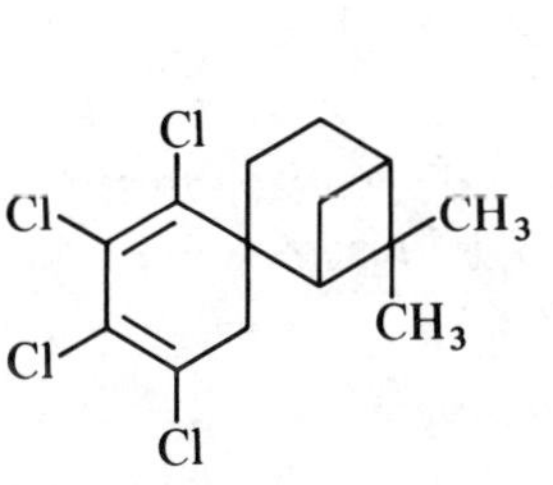

208 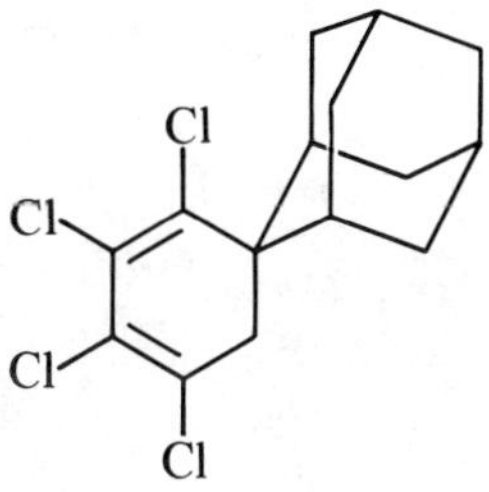   **209**

E.   1,2-Dienes as Dienophiles in (4+2)-Cycloaddition Reactions of Thiophene 1,1-Dioxides

Raasch and Smart[142] reacted tetrachlorothiophene 1,1-dioxide (**171f**) with 1,2-cyclononadiene (**210**) in refluxing tetrachloromethane and obtained a mixture of the triene **211** (52%) and the sulfinic acid **212** (9%). Compound **211** could be transformed into **212** by passing sulfur dioxide into a dichloromethane solution of **211**. By performing the reaction of **171f** and **210** at room temperature in dichloromethane, **212** was isolated in 79% yield (Scheme 33).

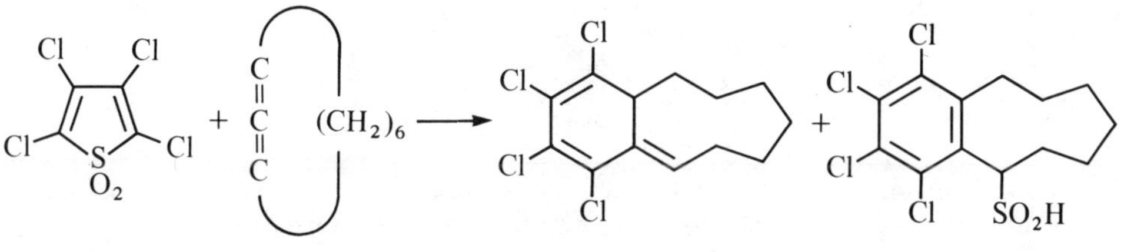

171f   **210**   **211**   **212**

Scheme 33

F.   1,3-Dienes as Dienophiles in (4+2)-Cycloaddition Reactions of Thiophene 1,1-Dioxides

Bluestone et al.[131] described the reaction of 3,4-dichlorothiophene 1,1-dioxide (**171d**) and Raasch[87] discussed the reaction of tetrachlorothiophene 1,1-dioxide

(171f) with cyclopentadiene. They explained the formation of the two types of products (189d + 191d and 189f + 191f) by the competition of a (2+4)- and a (4+2)-cycloaddition reaction of the thiophene 1,1-dioxides. Patterson[136] investigated the reaction of a variety of substituted thiophene 1,1-dioxides with cyclopentadiene under well-controlled reaction conditions, and he concluded that the thiophene 1,1-dioxides react only as dienophiles. The formation of the so-called (4+2)-cycloadducts was explained as resulting from a Cope-rearrangement of the initially formed (2+4)-cycloadducts. Therefore, these reactions have been dealt with as (2+4)-cycloaddition reactions of thiophene 1,1-dioxides (see Section IV.5).

G.   1,4-*Dienes as Dienophiles in* (4+2)-*Cycloaddition Reactions of Thiophene* 1,1-*Dioxides*

Bluestone et al.[131] reacted 3,4-dichlorothiophene 1,1-dioxide (171d) with bicyclo[2.2.1]hepta-2,5-diene in refluxing chloroform and isolated 189d in a moderate yield. The intermediate Diels–Alder adduct 213 loses sulfur dioxide as usual, and the resulting 214 undergoes a retro Diels–Alder reaction to form 1,2-dichlorobenzene and cyclopentadiene. Subsequently, 171d reacts as a dienophile with cyclopentadiene to give 189d (Scheme 34). It is interesting to note

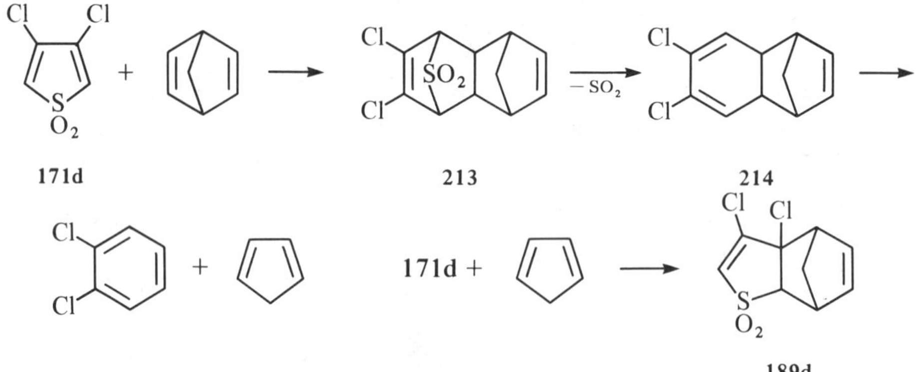

Scheme 34

that compound 191d was not found in the reaction mixture (see Section IV.5 and Scheme 27). Tetrachlorothiophene 1,1-dioxide (171f) reacts with only one of the two double bonds of 1,4-cyclohexadiene to give 1,2,3,4-tetrachloro-4a,5,8,8a-tetrahydronaphthalene (198r, see Scheme 30) in 88% yield.[87]

H.   1,5-*Dienes as Dienophiles in* (4+2)-*Cycloaddition Reactions of Thiophene* 1,1-*Dioxides*

Raasch[87] found that one double bond of an acyclic 1,5-diene system reacts with a thiophene 1,1-dioxide to give an intermediate (4+2)-cycloadduct (215) with the remaining double bond in such a position that it is available for an intramolecular Diels–Alder reaction yielding an isotwistene (216) (Scheme 35). 1,2-Divinylbenzene

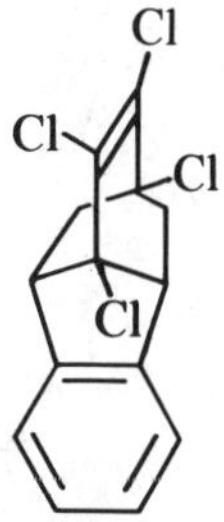

	217		218		219

a	X = O	a	R = H
b	X = S	b	R = CH$_3$

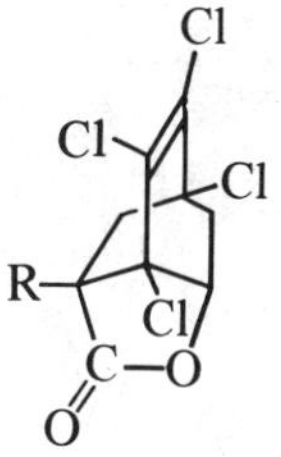

reacts similarly with tetrachlorothiophene 1,1-dioxide (171f) to give the isotwistene 217. The heterotwistenes 218a, 218b, and 219 are formed if the two double bonds of the 1,5-diene system are connected by a —CH$_2$—O—, a —CH$_2$—S—, or a —C(O)—O-linkage, respectively. 1,5-Cyclooctadiene reacts with halogenated thiophene 1,1-

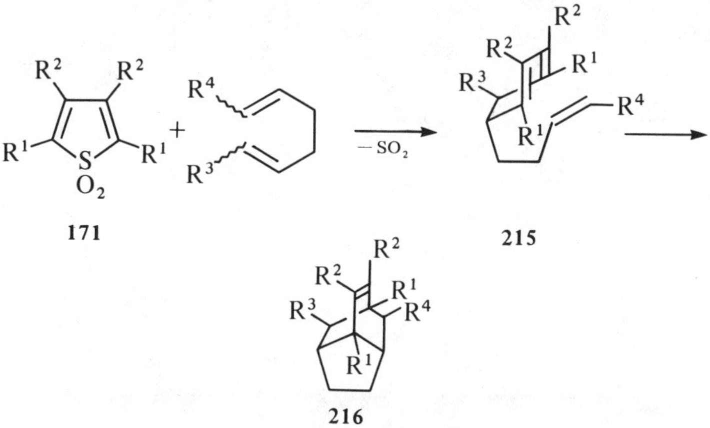

a	R^1 = R^2 = Cl, R^3 = R^4 = H
b	R^1 = R^2 = Cl, R^3 = H, R^4 = CH$_3$
c	R^1 = R^2 = Cl, R^3 = R^4 = CH$_3$
d	R^1 = R^2 = Br, R^3 = R^4 = H
e	R^1 = H, R^2 = Cl, R^3 = R^4 = H

Scheme 35

dioxides, yielding the tetracyclic compounds 220. The reaction proceeds very smoothly and products are isolated in high yields (Scheme 36). Similar complicated ring systems (221 and 222) have been obtained by reacting tetrachlorothiophene 1,1-dioxide (171f) with dibenzo[a,e]cyclooctene and with 1,5-cyclononadiene, in

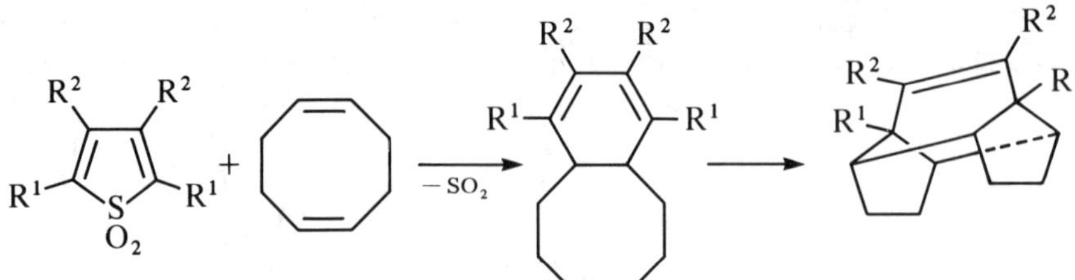

Scheme 36

a $R^1 = R^2 = Cl$
b $R^1 = H, R^2 = Cl$
c $R^1 = R^2 = Br$

57 and 58% yield, respectively. A prerequisite for the formation of these reaction products is the *Z-Z* configuration of the reacting cyclic 1,5-diene. Although the reaction of (*Z,Z*)-1,5-cyclodecadiene and tetrachlorothiophene 1,1-dioxide (171f) has not been described, Raasch[87] found only a single Diels–Alder reaction with (*Z,E*)-1,5-cyclodecadiene yielding 223.

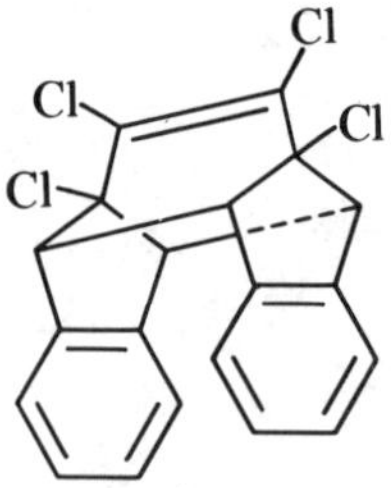

221 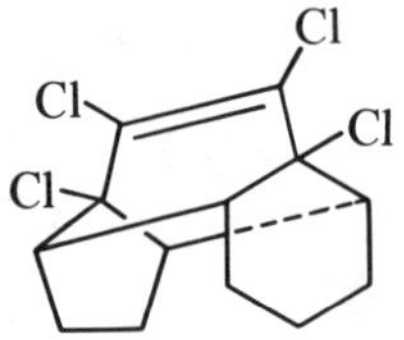   222

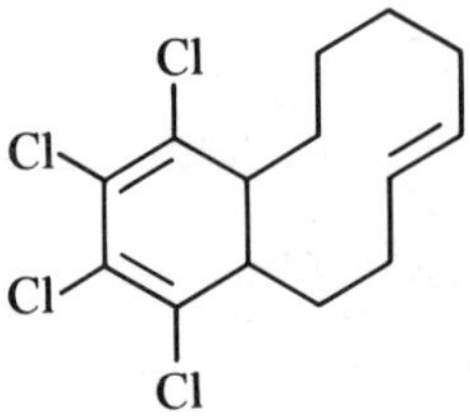

223 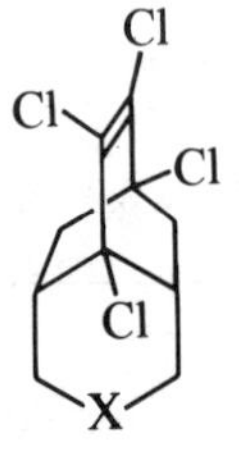   224

a $X = CH_2$
b $X = O$
c $X = S$
d $X = NCOCH_3$
e $X = NCN$
f $X = NCH_2CN$

I.   1,6-Dienes as Dienophiles in (4+2)-Cycloaddition Reactions of Thiophene 1,1-Dioxides

1,6-Heptadiene and 4-hetero-1,6-heptadienes have been reacted with tetrachloro-thiophene 1,1-dioxide (171f) to yield the homotwistenes 224a–f.[87]

J.   1,7- and 1,8-Dienes as Dienophiles in (4+2)-Cycloaddition Reactions of Thiophene 1,1-Dioxides

Tetrachlorothiophene 1,1-dioxide reacted with 1,7-octadiene and with 1,8-nonadiene at 100°C to give the monoadducts 198s and 198t, respectively (see Scheme 30). The reactivity of the remaining double bond of the diene system has not been discussed.[87]

K.   Cyclic Functionalized Alkenes as Dienophiles in (4+2)-Cycloaddition Reactions of Thiophene 1,1-Dioxides

Melles[143] reported the (4+2)-cycloaddition reaction of methyl- and phenyl-substituted thiophene 1,1-dioxides with maleic anhydride at 150–200°C. Sulfur dioxide was formed under the reaction conditions and the resulting cyclic 1,3-diene 225 readily added a second molecule of maleic anhydride to yield substituted bicyclo[2.2.2]-oct-7-ene-2,3,5,6-tetracarboxylic dianhydrides (226) (Scheme 37).

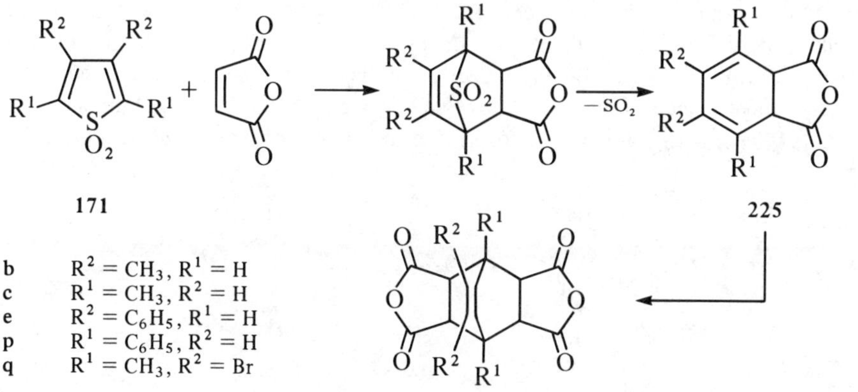

171

b	$R^2 = CH_3$, $R^1 = H$
c	$R^1 = CH_3$, $R^2 = H$
e	$R^2 = C_6H_5$, $R^1 = H$
p	$R^1 = C_6H_5$, $R^2 = H$
q	$R^1 = CH_3$, $R^2 = Br$

226

Scheme 37

With tetraphenylthiophene 1,1-dioxide (171r), the reaction stopped after the second step and 225 ($R^1 = R^2 = C_6H_5$) was isolated. A number of other sub-stituted thiophene 1,1-dioxides investigated by Melles[143] decomposed under the reaction conditions without the formation of a well-defined product, whereas others were inert. The parent thiophene 1,1-dioxide dimerizes much faster than it could add maleic anhydride.[138] 3,4-Dichlorothiophene 1,1-dioxide (171d) appeared to react with maleic anhydride to give a similar 1:2 adduct (226, R^1 = H, R^2 = Cl), as described by Melles for the methyl- and phenyl-substituted derivatives (*vide supra*).[131] The more reactive tetrachlorothiophene 1,1-dioxide (171f) afforded the adduct 225, at 100°C, (R^1 = R^2 = Cl) in 61% yield.[87]

Bluestone[131] reacted 3,4-dichlorothiophene 1,1-dioxide (171d) with N-substituted maleimides and obtained the 1:2 adducts 227. This reaction is similar to the one described for maleic anhydride. Van Tilborg et al.[141] reported the reaction of 2,5-

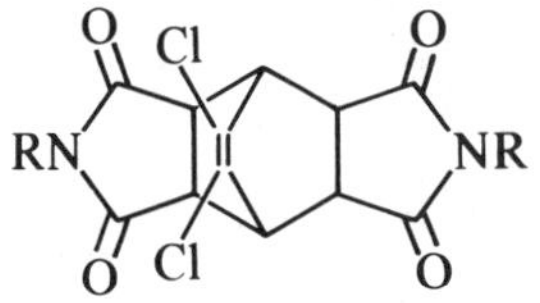

227			**228**
a	R = n-C₄H₉	a	R = H
b	R = 4-O₂N—C₆H₄	b	R = CH₃

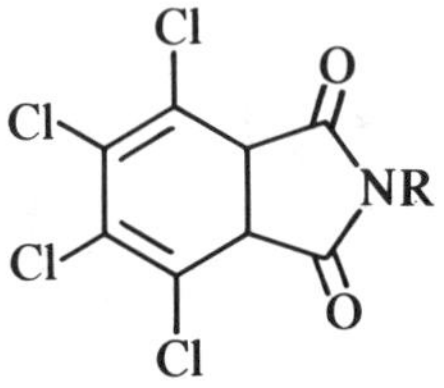

dimethylthiophene 1,1-dioxide (171c) with N-phenylmaleimide to give the expected Diels–Alder adduct. Tetrachlorothiophene 1,1-dioxide (171f) reacted with maleimide and N-methylmaleimide to give 228a and 228b, respectively, in high yields.

Bluestone et al.[131] reacted 3,4-dichlorothiophene 1,1-dioxide (171d) with an excess of 1,4-benzoquinone in benzene (70°C, 88 h) and isolated in high yield a product described as 6,7-dichloro-1,4-naphthoquinone (229). Quinhydrone was isolated during the work-up of the reaction, indicating that an oxidation–reduction reaction takes place (Scheme 38). Torssell[125] reacted 2,5-dimethylthiophene 1,1-dioxide (171c) with 1,4-benzoquinone in refluxing chloroform for 20 h and isolated

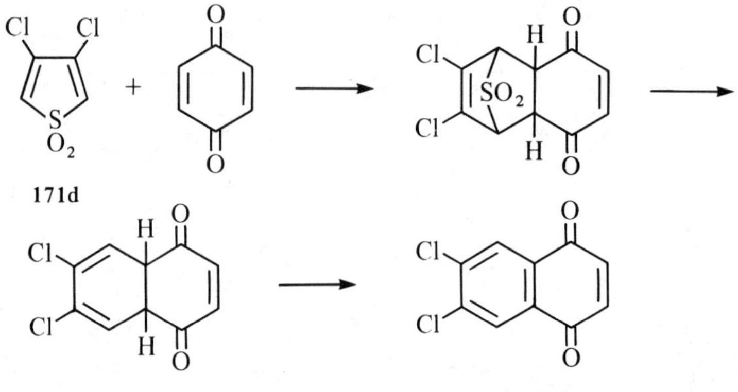

Scheme 38

the primary Diels–Alder adduct **230**. This surprisingly stable adduct lost sulfur dioxide on prolonged refluxing, yielding the disproportionation product 5,8-dimethyl-1,4-naphthoquinone (**175d**) (see Section III.3). Raasch[87] reported the reaction of tetrachlorothiophene 1,1-dioxide (**171f**) with 1,4-benzoquinone and 1,4-naphthoquinone, performed at approximately the same temperature, and isolated the expected dihydroquinones **231** and **232**, respectively.

230

231

232

233

Tetrachlorothiophene 1,1-dioxide (**171f**) and diethyl 2,3-diazabicyclo[2.2.1]-hept-5-ene-2,3-dicarboxylate afforded **233** in 58% yield.[87]

L. Heteroaromatic Compounds as Dienophiles in (4+2)-Cycloaddition Reactions of Thiophene 1,1-Dioxides

Tetrachlorothiophene 1,1-dioxide (**171f**) acts as a dienophile in the primary reaction with furan and as a diene in the consecutive reaction with the transient intermediate[139] (see Section IV.5 and Scheme 29). A second pathway of the reaction of **171f** and furan afforded 2,3,4,5-tetrachlorophenylacetaldehyde (**234a**) as a minor product, the formation of which was accounted for by a (4+2)-cyclo-addition followed by a rearrangement (Scheme 39). This addition–rearrangement reaction appeared to be a general reaction for a variety of 2- and 2,5-substituted furans; the products **234b–m** have been isolated in good yields (Table 11). In the case of monosubstituted furans, the less sterically crowded Diels–Alder adducts are formed. Surprisingly, the Diels–Alder reaction of 3,4-dichlorothiophene 1,1-dioxide (**171d**) and 2,5-dimethylfuran afforded 5,6-dichloro-3a,4,7,7a-tetrahydro-2,7a-dimethyl-4,7-epithiobenzofuran 8,8-dioxide (**235**), isolated after 2 min reaction time in a yield of 52%. Heating the reactants caused the evolution of sulfur

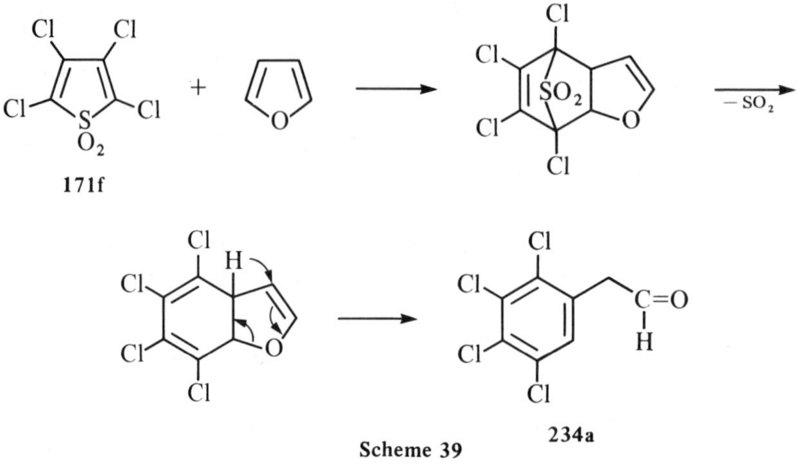

Scheme 39

dioxide and the rearranged product **234b** was isolated. Compound **171f** also reacted with 2,5-dihydro-2,5-dimethoxyfuran, yielding the same product (**234j**) as that obtained with 2-methoxyfuran. Under the reaction conditions, an equilibrium probably exists between the two furan derivatives.

TABLE 11.   REACTIONS OF THIOPHENE 1,1-DIOXIDES (171) WITH 2- AND 2,5-SUBSTITUTED FURANS[139]

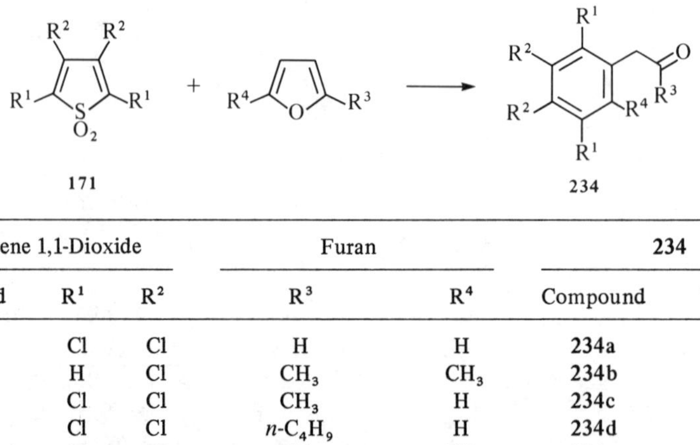

Thiophene 1,1-Dioxide			Furan		234	
Compound	R^1	R^2	R^3	R^4	Compound	Yield (%)
171f	Cl	Cl	H	H	234a	8
171d	H	Cl	CH_3	CH_3	234b	77
171f	Cl	Cl	CH_3	H	234c	92
171f	Cl	Cl	n-C_4H_9	H	234d	93
171f	Cl	Cl	CH_3	CH_3	234e	90
171f	Cl	Cl	CH_2OCOCH_3	H	234f	74
171f	Cl	Cl	$CH_2NHCOCH_3$	H	234g	84
171f	Cl	Cl	$COOCH_3$	H	234h[a]	62
171f	Cl	Cl	$COCH_3$	H	234i[a]	37
171f	Cl	Cl	OCH_3	H	234j	77
171f	Cl	Cl	OH	CH_3	234k	62
171f	Cl	Cl	2,3,4,5-$Cl_4C_6H_3$	H	234l	69
171m	Br	Br	CH_3	CH_3	234m	90

[a] The product obtained exists in the enolic form.

235   236   237

Tetrachlorothiophene 1,1-dioxide (**171f**) reacts successively with both double bonds of *N*-methylpyrrole to give **236** in 44% yield. Indole reacts at room temperature to give **237** in 77% yield.[87]

The high reactivity of tetrachlorothiophene 1,1-dioxide (**171f**) was again demonstrated by the reaction with thiophene. Under relatively mild conditions, the diadduct **111** was formed in 61% yield[87] (see Section II.4).

The (4+2)-cycloaddition reactions of thiophene 1,1-dioxides with themselves are surveyed in Section IV.4. In this section, the corresponding mode of reaction with other (annulated) thiophene 1,1-dioxides is described, although the latter ones hardly can be regarded as heteroaromatics. Davies and Porter[144] reacted 3,4-dimethylthiophene 1,1-dioxide (**171b**) with benzo[*b*]thiophene 1,1-dioxide and they isolated only the 1:2 adduct **238** (or **239**). Bluestone et al.[131] reacted 3,4-dichlorothiophene 1,1-dioxide (**171d**) with 3,5,6-trichlorobenzo[*b*]thiophene 1,1-dioxide (**176**), yielding the adduct **177** (see Section IV.2 and Scheme 24).

238   239

M.   Alkynes as Dienophiles in (4+2)-Cycloaddition Reactions of Thiophene 1,1-Dioxides

Bailey and Cummins[138] reported the (4+2)-cycloaddition reaction of the parent thiophene 1,1-dioxide (**171a**) with diethyl acetylenedicarboxylate. The Diels–Alder adduct **240** was isolated in 16% yield, although the reaction was not reproducible every time. The structure was proven by elemental analysis and by its chemical behavior. On heating **240**, sulfur dioxide was evolved to form diethyl phthalate (Scheme 40). Van Tilborg et al.[141] repeated the reaction with DMAD, but all attempts to obtain the adduct failed. Overberger and Whelan[135, 145] reacted 3,4-diphenyl- (**171e**) and 2,5-diphenylthiophene 1,1-dioxide (**171p**) with phenylacetylene and obtained 1,2,4-triphenylbenzene in both cases. Reaction of **171e**

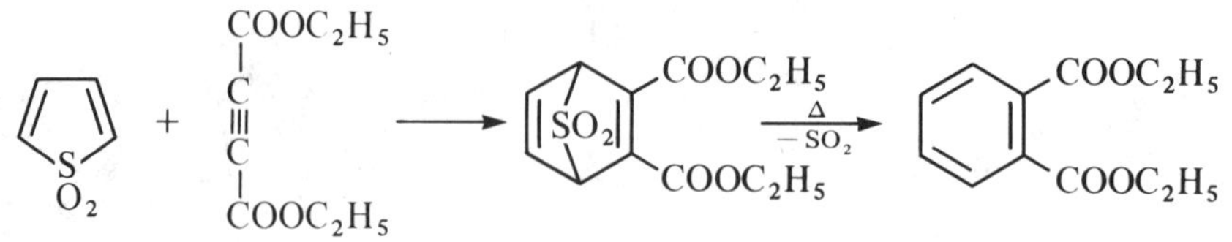

Scheme 40

with acetylenedicarboxylic acid produced 4,5-diphenylphthalic anhydride. Nelb and Stille[146] repeated the reaction of **171e** and phenylacetylene in refluxing toluene and, although they isolated 1,2,4-triphenylbenzene in very low yield, the major product was 1-phenyl-4-(1-styryl)naphthalene (**241**). A mechanism was proposed (as depicted in Scheme 41), based on deuterium labeling studies involving 2,5-dideuterio-3,4-diphenylthiophene 1,1-dioxide. Helder[34] reported the reaction of

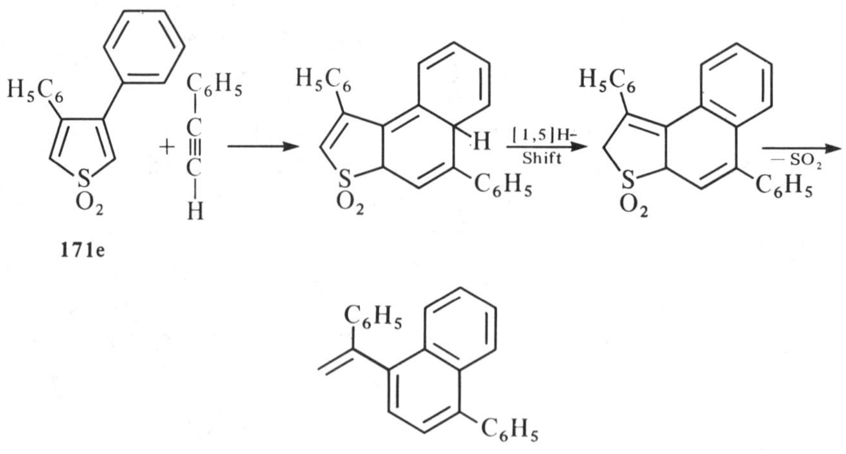

Scheme 41

3,4-diphenylthiophene 1,1-dioxide (**171e**) with the electron-deficient dicyano-acetylene (100°C, 1 h) and isolated the expected 4,5-diphenyl-1,2-benzenedi-carbonitrile (**118o**) in 18% yield. 3-(*tert*-Butyl)-2,5-dimethylthiophene 1,1-dioxide (**171s**), with the electron-donating alkyl groups, reacted with dicyanoacetylene (100°C, 16 h) to give 4-(*tert*-butyl)-3,6-dimethyl-1,2-benzenedicarbonitrile (**118k**) even in 86% yield. When the polarity of the reactants is reversed, this type of (4+2)-cycloaddition also takes place. Raasch[87] found a relatively high reactivity of tetrachlorothiophene 1,1-dioxide (**171f**) toward phenylacetylene. The reaction was performed with ice cooling, affording 1,2,3,4-tetrachloro-5-phenylbenzene in 86% yield.

7.   (4+6)-Cycloaddition Reactions of Thiophene 1,1-Dioxides

Leaver and coworkers[147] and Houk and associates[148, 149] independently observed the formation of azulenes by reacting thiophene 1,1-dioxides and 6-(dimethyl-amino)fulvenes. The transient (4+6)-cycloadducts eliminate sulfur dioxide and dimethylamine, yielding the azulenes (Scheme 42). The yields are often low,

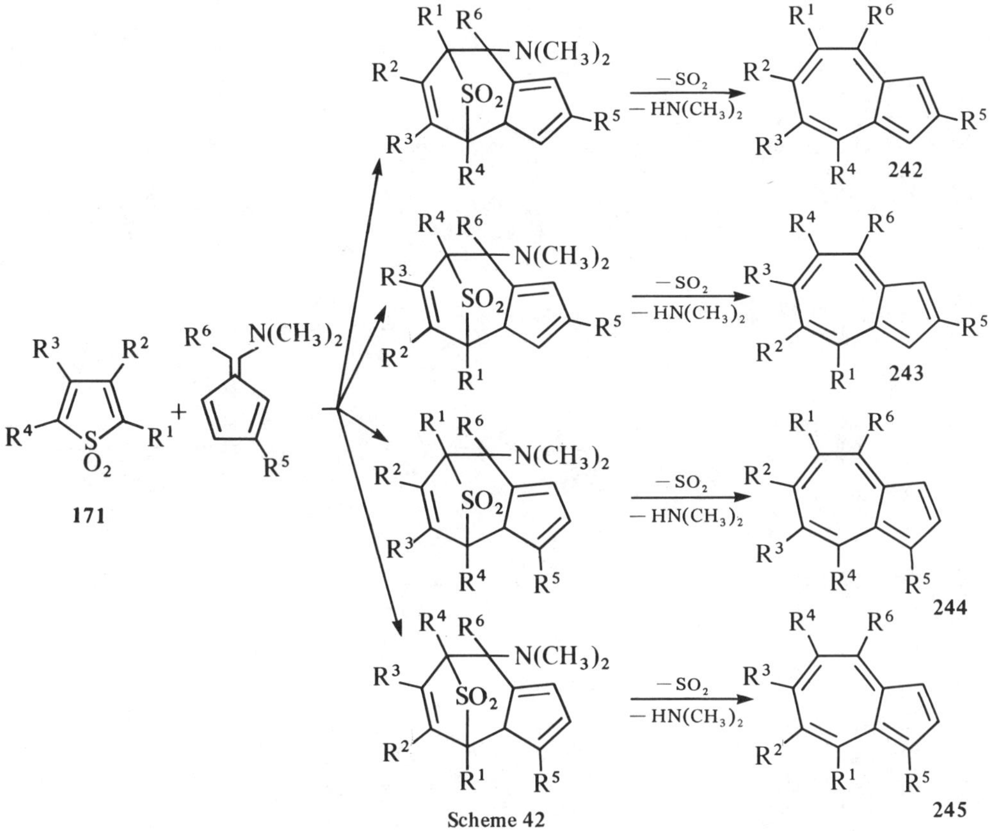

Scheme 42

because of the tendency of many thiophene 1,1-dioxides to dimerize or to react with amines. Nevertheless, the reaction provides a synthetic route to azulenes, because thiophene 1,1-dioxides are readily available and the azulenes are simply isolated after chromatography. Leaver and coworkers[147] reported the reactions of the parent thiophene 1,1-dioxide (171a) and of 3,4-dichlorothiophene 1,1-dioxide (171d) with 6-(dimethylamino)fulvene, yielding the parent azulene (33%) and 5,6-dichloroazulene (46%), respectively. Houk and coworkers[148, 149] also studied reactions with unsymmetrically substituted thiophene 1,1-dioxides, which proceed in general with high regioselectivity (Table 12). An application of this (4+6)-cycloaddition reaction to the synthesis of guaiazulene and chamazulene has also been described.[150]

TABLE 12. REACTIONS OF THIOPHENE 1,1-DIOXIDES (171) WITH 6-(DIMETHYLAMINO)FULVENES[a]

Compound	Thiophene 1,1-Dioxide R^1	R^2	R^3	R^4	Fulvene R^5	R^6	Yield of Azulene (%) 242	243	244	245	Reference
171a	H	H	H	H	H	H	10[b]				148, 149
171a	H	H	H	H	H	CH_3	33[b]				147
171a	H	H	H	H	H	CH_3	4[b]				148, 149
171a	H	H	H	H	H	$N(CH_3)_2$	10[b]				148, 149
171a	H	H	H	H	CH_3	CH_3	–	–	13[c]		149
171b	H	CH_3	CH_3	H	H	H	8[b]				148, 149
171b	H	CH_3	CH_3	H	H	CH_3	7[b]				148, 149
171b	H	CH_3	CH_3	H	H	$N(CH_3)_2$	4[b]				149
171b	H	CH_3	CH_3	H	CH_3	CH_3	Trace[d]		5[c]		149
171c	CH_3	H	H	CH_3	CH_3	H	5[b]				149
171d	H	Cl	Cl	H	H	H	60[b]				148, 149
171d	H	Cl	Cl	H	H	H	46[b]				147
171d	H	Cl	Cl	H	H	CH_3	15[b]				148, 149
171d	H	Cl	Cl	H	H	$N(CH_3)_2$	Trace[b]				149
171d	H	Cl	Cl	H	CH_3	CH_3	–	–	16.5[c]		149
171e	H	C_6H_5	C_6H_5	H	H	H	15[b]				149
171g	CH_3	H	H	H	H	H	5[e]				148, 149
171g	CH_3	H	H	H	H	CH_3	Trace[e]			—	148, 149
171g	CH_3	H	H	H	CH_3	CH_3	–		2	—	149
171h	H	CH_3	H	H	H	H	25[e]			—	149
171h	H	CH_3	H	H	H	CH_3	15[e]			—	148, 149
171h	H	CH_3	H	H	H	$N(CH_3)_2$	5[e]			—	148, 149
171h	H	CH_3	H	H	CH_3	CH_3	–		10	—	148, 149
171i	H	C_6H_5	H	H	H	H	30[e,g]	30[f,g]			149
171i	H	C_6H_5	H	H	H	CH_3	10[e]	Trace[f]			148, 149
171i	H	C_6H_5	H	H	H	$N(CH_3)_2$	6[e]	Trace[f]			148, 149
171i	H	C_6H_5	H	H	CH_3	CH_3	–		10	Trace	148, 149
171j	CH_3	H	CH_3	H	H	H	–	10[f]	–		148, 149

738

										Ref.
171j	CH$_3$	H	CH$_3$	H	H	CH$_3$	–	3^f	–	149
171j	CH$_3$	H	CH$_3$	H	H	N(CH$_3$)$_2$	–	Trace^f	–	149
171j	CH$_3$	H	CH$_3$	H	CH$_3$	CH$_3$	–	10	–	149
171t	H	C$_2$H$_5$	H	H	H	H	12^e	–	–	148, 149
171t	H	C$_2$H$_5$	H	H	H	CH$_3$	12^e	–	–	148, 149
171t	H	C$_2$H$_5$	H	H	H	N(CH$_3$)$_2$	12^e	–	–	148, 149
171t	H	C$_2$H$_5$	H	H	CH$_3$	CH$_3$	–	–	12.5	149
171u	C$_2$H$_5$	H	H	H	H	H	2^e	–	–	149

^{a} See Scheme 42.
^{b} **242 = 243 = 244 = 245.**
^{c} **244 = 245.**
^{d} **242 = 243.**
^{e} **242 = 244.**
^{f} **243 = 245.**
^{g} Total yield of **242** and **243.**

739

REFERENCES

1.   H. D. Hartough, *The Chemistry of Heterocyclic Compounds,* Vol. 3 (A. Weissberger, Ed.), Interscience, New York, 1952.

2.   S. Gronowitz, in *Advances in Heterocyclic Chemistry,* Vol. 1 (A. R. Katritzky, Ed.), Academic Press, New York, 1963, p. 1.

3.   D. N. Reinhoudt, in *Advances in Heterocyclic Chemistry,* Vol. 21 (A. R. Katritzky and A. J. Boulton, Eds.), Academic Press, New York, 1977, p. 253.

4.   P. L. Beltrame, M. G. Cattania, V. Redaelli, and G. Zecchi, *J. Chem. Soc., Perkin Trans. 2,* 706 (1977).

5.   P. L. Beltrame, M. G. Cattania, and G. Zecchi, *Croat. Chem. Acta,* **51,** 285 (1978); *Chem. Abstr.,* **90,** 203385k (1979).

6.   P. Caramella, G. Cellerino, P. Grünanger, F. Marinone Albini, and M. R. Re Cellerino, *Tetrahedron,* **34,** 3545 (1978).

7.   P. W. Lert and C. Trindle, *J. Am. Chem. Soc.,* **93,** 6392 (1971).

8.   K. Kanematsu, K. Harano, and H. Dantsuji, *Heterocycles,* **16,** 1145 (1981).

9.   W. L. Mock, *J. Am. Chem. Soc.,* **92,** 7610 (1970).

10. C. Müller, A. Schweig, and W. L. Mock, *J. Am. Chem. Soc.,* **96,** 280 (1974).

11. W. Steinkopf and H. Augestad-Jensen, *Justus Liebigs Ann. Chem.,* **428,** 154 (1922).

12. R. Pettit, *Tetrahedron Lett.,* (23) 11 (1960).

13. G. O. Schenck and R. Steinmetz, *Angew. Chem.,* **70,** 504 (1958).

14. G. O. Schenck and R. Steinmetz, *Justus Liebigs Ann. Chem.,* **668,** 19 (1963).

15. R. J. Gillespie, J. Murray-Rust, P. Murray-Rust, and A. E. A. Porter, *J. Chem. Soc., Chem. Commun.,* 83 (1978).

16. R. J. Gillespie, A. E. A. Porter, and W. E. Willmott, *J. Chem. Soc., Chem. Commun.,* 85 (1978).

17. R. J. Gillespie and A. E. A. Porter, *J. Chem. Soc., Perkin Trans. 1,* 2624 (1979).

18. R. J. Gillespie, J. Murray-Rust, P. Murray-Rust, and A. E. A. Porter, *Tetrahedron,* **37,** 743 (1981).

19. H. Dürr, B. Heu, B. Ruge, and G. Scheppers, *J. Chem. Soc., Chem. Commun.,* 1257 (1972).

20. G. Cauquis, B. Divisia, and G. Reverdy, *Bull. Soc. Chim. Fr.,* 3027 (1971).

21. L. G. Plekhanova, G. A. Nikiforov, V. V. Ershov, and E. P. Zakharov, *Izv. Akad. Nauk SSSR, Ser. Khim.,* 846 (1973); through *Chem. Abstr.,* **79,** 52946p (1973).

22. L. G. Plekhanova, G. A. Nikiforov, V. V. Ershov, and E. P. Zakharov, *Nov. Khim. Karbenov, Mater. Vses. Soveshch. Khim. Karbenov Ikh Analogov, 1st,* 237 (1972); through *Chem. Abstr.,* **82,** 72695w (1975).

23. G. A. Nikiforov, L. G. Plekhanova, and V. V. Ershov, *Tezisy Dokl.-Simp. Khim. Tekhnol. Geterotsikl. Soedin. Goryuch. Iskop., 2nd,* 104 (1973); through *Chem. Abstr.,* **86,** 72325h (1977).

24. P. Weyerstahl and G. Blume, *Tetrahedron,* **28,** 5281 (1972).

25. K. Hafner and W. Kaiser, *Tetrahedron Lett.,* 2185 (1964).

26. R. A. Abramovitch, S. R. Challand, and Y. Yamada, *J. Org. Chem.,* **40,** 1541 (1975).

27. J. M. Lindley, O. Meth-Cohn, and H. Suschitzky, *J. Chem. Soc., Perkin Trans. 1,* 1198 (1978).

28. G. R. Cliff, G. Jones, and J. McK. Woollard, *Tetrahedron Lett.,* 2401 (1973).

29. G. R. Cliff, G. Jones, and J. McK. Woollard, *J. Chem. Soc., Perkin Trans. 1,* 2072 (1974).

30. G. Jones, C. Keates, I. Kladko, and P. Radley, *Tetrahedron Lett.*, 1445 (1979).

31. P. C. Hayes, G. Jones, C. Keates, I. Kladko, and P. Radley, *J. Chem. Res., Synop.*, 288 (1980).

32. D. N. Reinhoudt, H. C. Volger, C. G. Kouwenhoven, H. Wynberg, and R. Helder, *Tetrahedron Lett.*, 5269 (1972).

33. D. N. Reinhoudt and C. G. Kouwenhoven, *Tetrahedron*, **30**, 2093 (1974).

34. R. Helder, Thesis, University of Groningen, 1974.

35. R. H. Hall, H. J. den Hertog, Jr., and D. N. Reinhoudt, *J. Org. Chem.*, **47**, 967 (1982).

36. H. Wynberg and R. Helder, *Tetrahedron Lett.*, 3647 (1972).

37. N. D. A. Walshe, in *Heterocyclic Compounds*, Vol. 4 (P. G. Sammes, Ed.), Pergamon Press, Oxford, 1979, p. 866.

38. R. H. Hall, H. J. den Hertog, Jr., and D. N. Reinhoudt, *J. Org. Chem.*, **47**, 972 (1982).

39. R. H. Hall, H. J. den Hertog, Jr., D. N. Reinhoudt, S. Harkema, G. J. van Hummel, and J. W. H. M. Uiterwijk, *J. Org. Chem.*, **47**, 977 (1982).

40. D. N. Reinhoudt and C. G. Kouwenhoven, *J. Chem. Soc., Chem. Commun.*, 1233 (1972).

41. D. N. Reinhoudt, G. Okay, W. P. Trompenaars, S. Harkema, D. M. W. van den Ham, and G. J. van Hummel, *Tetrahedron Lett.*, 1529 (1979).

42. D. N. Reinhoudt, J. Geevers, W. P. Trompenaars, S. Harkema, and G. J. van Hummel, *J. Org. Chem.*, **46**, 424 (1981).

43. D. N. Reinhoudt, W. P. Trompenaars, and J. Geevers, *Tetrahedron Lett.*, 4777 (1976).

44. F. A. Buiter, J. H. Sperna Weiland, and H. Wynberg, *Recl. Trav. Chim. Pays-Bas*, **83**, 1160 (1964).

45. D. N. Reinhoudt, W. P. Trompenaars, and J. Geevers, *Synthesis*, 368 (1978).

46. D. N. Reinhoudt, J. Geevers, and W. P. Trompenaars, *Tetrahedron Lett.*, 1351 (1978).

47. W. Verboom, G. W. Visser, W. P. Trompenaars, D. N. Reinhoudt, S. Harkema, and G. J. van Hummel, *Tetrahedron*, **37**, 3525 (1981).

48. G. W. Visser, W. Verboom, W. P. Trompenaars, and D. N. Reinhoudt, *Tetrahedron Lett.*, 1217 (1982).

49. G. W. Visser, W. Verboom, P. H. Benders, and D. N. Reinhoudt, *J. Chem. Soc., Chem. Commun.*, 669 (1982).

50. H. Biere, C. Herrmann, and G.-A. Hoyer, *Chem. Ber.*, **111**, 770 (1978).

51. E. K. Fields and S. Meyerson, in *Advances in Physical Organic Chemistry*, Vol. 6 (V. Gold, Ed.), Academic Press, New York, 1968, p. 1.

52. E. K. Fields, in *Organic Reactive Intermediates* (S. P. McManus, Ed.), Academic Press, New York, 1973, p. 449.

53. M. G. Reinecke, in *Reactive Intermediates*, Vol. 2 (R. A. Abramovitch, Ed.), Plenum Press, New York, 1982, p. 367.

54. M. G. Reinecke and J. G. Newsom, *J. Am. Chem. Soc.*, **98**, 3021 (1976).

55. M. G. Reinecke, J. G. Newsom, and L.-J. Chen, *J. Am. Chem. Soc.*, **103**, 2760 (1981).

56. M. G. Reinecke, J. G. Newsom, and K. A. Almqvist, *Tetrahedron*, **37**, 4151 (1981).

57. M. G. Reinecke, *Tetrahedron*, **38**, 427 (1982).

58. D. Del Mazza, Thesis, Texas Christian University, Fort Worth, Texas, 1980.

59. D. Del Mazza and M. G. Reinecke, *J. Chem. Soc., Chem. Commun.*, 124 (1981).

60. D. N. Reinhoudt and C. G. Kouwenhoven, *Tetrahedron*, **30**, 2431 (1974).

61. C. Rivas, M. Vélez, and O. Crescente, *J. Chem. Soc., Chem. Commun.*, 1474 (1970).

62. C. Rivas and R. A. Bolivar, *J. Heterocycl. Chem.*, **10**, 967 (1973).

63. T. Nakano, C. Rivas, C. Perez, and K. Tori, *J. Chem. Soc., Perkin Trans. 1*, 2322 (1973).

742   P. H. Benders, D. N. Reinhoudt and W. P. Trompenaars

64. C. Rivas, D. Pacheco, F. Vargas, and J. Ascanio, *J. Heterocycl. Chem.*, **18**, 1065 (1981).
65. G. O. Schenck, W. Hartmann, and R. Steinmetz, *Chem. Ber.*, **96**, 498 (1963).
66. C. Rivas, S. Krestonosich, E. Payo-Subiza, and L. Cortes, *Acta Cient. Venez.*, **21**, 28 (1970); through *Chem. Abstr.*, **73**, 24636v (1970).
67. C. Rivas, C. Pérez, and T. Nakano, *Rev. Latinoam. Quim.*, **6**, 166 (1975); *Chem. Abstr.*, **84**, 120785r (1976).
68. H. Wamhoff and H.-J. Hupe, *Tetrahedron Lett.*, 125 (1978).
69. T. S. Cantrell, *J. Org. Chem.*, **39**, 3063 (1974).
70. T. S. Cantrell, *J. Chem. Soc., Chem. Commun.*, 155 (1972).
71. T. S. Cantrell, *J. Org. Chem.*, **39**, 2242 (1974).
72. D. R. Arnold, R. J. Birtwell, and B. M. Clarke, Jr., *Can. J. Chem.*, **52**, 1681 (1974).
73. D. R. Arnold and C. P. Hadjiantoniou, *Can. J. Chem.*, **56**, 1970 (1978).
74. H. J. Kuhn and K. Gollnick, *Tetrahedron Lett.*, 1909 (1972); H. J. Kuhn and K. Gollnick, *Chem. Ber.*, **106**, 674 (1973).
75. T. Kumagai, Y. Kawamura, K. Shimizu, and T. Mukai, *Koen Yoshishu – Hibenzenkei Hokozoku Kagaku Toronkai [oyobi] Kozo Yuki Kagaku Toronkai, 12th*, 317 (1979); through *Chem. Abstr.*, **92**, 197557r (1980).
76. W. J. Linn and R. E. Benson, *J. Am. Chem. Soc.*, **87**, 3657 (1965).
77. W. J. Linn, *J. Am. Chem. Soc.*, **87**, 3665 (1965).
78. P. Brown and R. C. Cookson, *Tetrahedron*, **24**, 2551 (1968).
79. S. Gronowitz and B. Uppström, *Acta Chem. Scand., Ser. B*, **29**, 441 (1975).
80. H. Hamberger and R. Huisgen, *J. Chem. Soc., Chem. Commun.*, 1190 (1971).
81. A. Dahmen, H. Hamberger, R. Huisgen, and V. Markowski, *J. Chem. Soc., Chem. Commun.*, 1192 (1971).
82. S. Gronowitz and B. Uppström, *Acta Chem. Scand., Ser. B*, **28**, 339 (1974).
83. S. Gronowitz and B. Uppström, *Acta Chem. Scand., Ser. B*, **28**, 981 (1974).
84. D. N. Reinhoudt and C. G. Kouwenhoven, *Recl. Trav. Chim. Pays-Bas*, **93**, 321 (1974).
85. C. D. Hurd, A. R. Macon, J. I. Simon, and R. V. Levetan, *J. Am. Chem. Soc.*, **84**, 4509 (1962).
86. H. Hamadait and M. Neeman, Israeli Patent No. 9749, June 6, 1957; through *Chem. Abstr.*, **52**, 1263 (1958).
87. M. S. Raasch, *J. Org. Chem.*, **45**, 856 (1980).
88. G. Seitz and T. Kämpchen, *Chem. Ztg.*, **99**, 292 (1975); G. Seitz and T. Kämpchen, *Arch. Pharm. (Weinheim, Ger.)*, **311**, 728 (1978); *Chem. Abstr.*, **90**, 6328t (1979).
89. R. Helder and H. Wynberg, *Tetrahedron Lett.*, 605 (1972).
90. K. Kobayashi and K. Mutai, *Chem. Lett.*, 1149 (1977).
91. D. D. Callander, P. L. Coe, and J. C. Tatlow, *J. Chem. Soc., Chem. Commun.*, 143 (1966).
92. D. D. Callander, P. L. Coe, J. C. Tatlow, and A. J. Uff, *Tetrahedron*, **25**, 25 (1969).
93. S. Hayashi and N. Ishikawa, *Nippon Kagaku Zasshi*, **91**, 1000 (1970); through *Chem. Abstr.*, **74**, 75787y (1971).
94. S. Hayashi and N. Ishikawa, *Yuki Gosei Kagaku Kyokaishi*, **28**, 533 (1970); through *Chem. Abstr.*, **73**, 45241c (1970).
95. P. L. Coe, G. M. Pearl, and J. C. Tatlow, *J. Chem. Soc. C*, 604 (1971).
96. D. Del Mazza and M. G. Reinecke, *Heterocycles*, **14**, 647 (1980).
97. H. Wynberg and A. Bantjes, *J. Org. Chem.*, **24**, 1421 (1959).
98. R. Gaertner and R. G. Tonkyn, *J. Am. Chem. Soc.*, **73**, 5872 (1951).
99. D. B. Clapp, *J. Am. Chem. Soc.*, **61**, 2733 (1939).

100. H. Kotsuki, S. Kitagawa, H. Nishizawa, and T. Tokoroyama, *J. Org. Chem.*, **43**, 1471 (1978); H. Kotsuki, H. Nishizawa, S. Kitagawa, M. Ochi, N. Yamasaki, K. Matsuoka, and T. Tokoroyama, *Bull. Chem. Soc. Jpn.*, **52**, 544 (1979).

101. J. M. Barker, P. R. Huddleston, and S. W. Shutler, *J. Chem. Soc., Perkin Trans. 1*, 2483 (1975).

102. J. P. Chupp, *J. Heterocycl. Chem.*, **7**, 285 (1970).

103. J. P. Chupp, *J. Heterocycl. Chem.*, **9**, 1033 (1972).

104. C. N. Skold and R. H. Schlessinger, *Tetrahedron Lett.*, 791 (1970).

105. H. H. Wasserman and W. Strehlow, *Tetrahedron Lett.*, 795 (1970).

106. W. J. M. van Tilborg, *Recl. Trav. Chim. Pays-Bas*, **95**, 140 (1976).

107. W. Adam and H. J. Eggelte, *Angew. Chem.*, **90**, 811 (1978).

108. J. Szmuszkovicz and E. J. Modest, *J. Am. Chem. Soc.*, **72**, 571 (1950).

109. J. F. Scully and E. V. Brown, *J. Am. Chem. Soc.*, **75**, 6329 (1953).

110. W. Davies and Q. N. Porter, *J. Chem. Soc.*, 4958 (1957).

111. W. A. Pryor, J. H. Coco, W. H. Daly, and K. N. Houk, *J. Am. Chem. Soc.*, **96**, 5591 (1974).

112. G. Jones and P. Rafferty, *Tetrahedron*, **35**, 2027 (1979).

113. M. Matsumoto, S. Dobashi, and K. Kondo, *Tetrahedron Lett.*, 4471 (1975).

114. J. Skramstad and T. Midthaug, *Acta Chem. Scand., Ser. B*, **32**, 413 (1978).

115. A. Sammour and H. H. Zoorob, *Acta Chim. Acad. Sci. Hung.*, **86**, 53 (1975); *Chem. Abstr.*, **84**, 17216g (1976).

116. I. R. Trehan, R. Inder, and D. V. L. Rewal, *Indian J. Chem., Sect. B*, **14**, 210 (1976); *Chem. Abstr.*, **85**, 94588c (1976).

117. L. H. Klemm and K. W. Gopinath, *J. Heterocycl. Chem.*, **2**, 225 (1965).

118. H. Ohmura and S. Motoki, *Chem. Lett.*, 235 (1981).

119. U. M. Dzhemilev, L. Y. Gubaidullin, and G. A. Tolstikov, *Izv. Akad. Nauk SSSR, Ser. Khim.*, 1469 (1978); through *Chem. Abstr.*, **89**, 108895x (1978).

120. M. Procházka, *Collect. Czech. Chem. Commun.*, **30**, 1158 (1965).

121. W. Stevens, Thesis, University of Groningen, 1940.

122. J. L. Melles and H. J. Backer, *Recl. Trav. Chim. Pays-Bas*, **72**, 491 (1953).

123. W. Davies and F. C. James, *J. Chem. Soc.*, 15 (1954).

124. K. Okita and S. Kambara, *Kogyo Kagaku Zasshi*, **59**, 547 (1956); through *Chem. Abstr.*, **52**, 3762 (1958).

125. K. Torssell, *Acta Chem. Scand., Ser. B*, **30**, 353 (1976).

126. R. H. Eastman and R. M. Wagner, *J. Am. Chem. Soc.*, **71**, 4089 (1949).

127. J. L. Melles and H. J. Backer, *Recl. Trav. Chim. Bays-Bas*, **72**, 314 (1953).

128. W. J. Bailey and E. W. Cummins, *J. Am. Chem. Soc.*, **76**, 1932 (1954).

129. H. J. Backer and J. L. Melles, *Proc. K. Ned. Akad. Wet., Ser. B*, **54**, 340 (1951); *Chem. Abstr.*, **47**, 6932 (1953).

130. O. Hinsberg, *Ber. Dtsch. Chem. Ges.*, **48**, 1611 (1915).

131. H. Bluestone, R. Bimber, R. Berkey, and Z. Mandel, *J. Org. Chem.*, **26**, 346 (1961).

132. K. Kabzińska and J. T. Wróbel, *Bull. Acad. Pol. Sci., Ser. Sci. Chim.*, **22**, 843 (1974); *Chem. Abstr.*, **84**, 16441q (1976).

133. A. Bened, R. Durand, D. Pioch, P. Geneste, J. P. Declercq, G. Germain, J. Rambaud, and R. Roques, *J. Org. Chem.*, **46**, 3502 (1981).

134. W. J. Bailey and E. W. Cummins, *J. Am. Chem. Soc.*, **76**, 1936 (1954).

135. J. M. Whelan, Jr., Thesis, Polytechnic Institute of Brooklyn, 1959; through *Diss. Abstr.*, **20**, 1180 (1959).

136.   R. T. Patterson, Thesis, Louisiana State University, Baton Rouge, 1979; *Diss. Abstr., Int. B,* **41**, 204 (1980).

137.   J. T. Wróbel and K. Kabzińska, *Bull. Acad. Pol. Sci., Ser. Sci. Chim.,* **22**, 129 (1974); *Chem. Abstr.,* **80**, 133161b (1974).

138.   W. J. Bailey and E. W. Cummins, *J. Am. Chem. Soc.,* **76**, 1940 (1954).

139.   M. S. Raasch, *J. Org. Chem.,* **45**, 867 (1980).

140.   D. N. Reinhoudt, P. Smael, W. J. M. van Tilborg, and J. P. Visser, *Tetrahedron Lett.,* 3755 (1973).

141.   W. J. M. van Tilborg, P. Smael, J. P. Visser, C. G. Kouwenhoven, and D. N. Reinhoudt, *Recl. Trav. Chim. Pays-Bas,* **94**, 85 (1975).

142.   M. S. Raasch and B. E. Smart, *J. Am. Chem. Soc.,* **101**, 7733 (1979).

143.   J. L. Melles, *Recl. Trav. Chim. Pays-Bas,* **71**, 869 (1952).

144.   W. Davies and Q. N. Porter, *J. Chem. Soc.,* 459 (1957).

145.   C. G. Overberger and J. M. Whelan, *J. Org. Chem.,* **24**, 1155 (1959).

146.   R. G. Nelb, II and J. K. Stille, *J. Am. Chem. Soc.,* **98**, 2834 (1976).

147.   D. Copland, D. Leaver, and W. B. Menzies, *Tetrahedron Lett.,* 639 (1977).

148.   S. E. Reiter, L. C. Dunn, and K. N. Houk, *J. Am. Chem. Soc.,* **99**, 4199 (1977).

149.   S. E. Reiter, Thesis, Louisiana State University, Baton Rouge, 1977; *Diss. Abstr. Int. B,* **38**, 5946 (1978).

150.   D. Mukherjee, L. C. Dunn, and K. N. Houk, *J. Am. Chem. Soc.,* **101**, 251 (1979).

CHAPTER XI

Photochemical Reactions of Thiophenes

ALAIN LABLACHE-COMBIER

Laboratorie de Chimie Organique Physique, Université des Sciences et Techniques de Lille, Villeneuve d'Asq Cedex, France

I. PHOTOREARRANGEMENTS

1. Introduction

Considerable attention has been focused in recent years on the photorearrangement of thiophenes in solution, as well as that of other five-membered aromatic heterocycles.[1] In the gas phase at 2139 or 2288 Å, thiophene is photodecomposed into CH_2H_2, CH_2CCH_2, CH_3CCH, CS_2, and CH_2CHCCH; a polymer is also formed.[2]

2.   Thiophene, Alkylthiophenes, and Arylthiophenes

The pioneer and the most extensive studies in this field have been performed by Wynberg's group.[3] If the mechanism of the photoisomerization of some compounds (e.g., some furans or isoaxazoles) has been proved unambiguously, only the stable photoisomers of thiophene derivatives substituted by electron withdrawing groups – tetrakis (trifluoromethyl) thiophene – and cyanothiophene has been isolated[3-7] and, therefore, the mechanism of photoisomerization of most of the thiophenes described remains speculative, even if intermediates have been trapped.[8,9]

A.   Experimental Data

The photoisomerization of 2-arylthiophene to 3-arylthiophene seemed a priori to be a rather simple reaction:[10]

The photorearrangement occurs within the thiophene ring and not the aryl ring.[11]

The aryl group remains linked to the same carbon atom of the thiophene ring during the photorearrangement. This has been proved by a ^{14}C labeling experiment on 2-phenylthiophene.[12]

However, the data obtained from the photoisomerization of deuterium-labeled 2-phenylthiophene and from ethyl-phenylthiophenes[15] show that the thiophene ring photoisomerization is a rather complex reaction. Scrambling occurs at all possible ring positions of the isomerized 3-phenylthiophene and the recovered starting material is also found to have undergone deuterium scrambling.[13,14]

The major path of photorearrangement of phenyl-, deuterio-, and methyl-substituted 2-phenylthiophenes involves an interchange of the C_2–C_3 atoms, without concomitant inversion of the C_4–C_5 carbon atoms.

Phenyl-, deuterio-, and methyl-substituted 3-phenylthiophenes exhibit considerable specificity of rearrangement, as shown in Scheme 2.

Some 2-alkylthiophenes also photoisomerize[16] (Scheme 2).

B.   Mechanisms

Numerous mechanisms have been postulated to rationalize the rearrangement patterns. The reactive excited state is the singlet in the case of 2-phenyl thiophene.[17] The proposed mechanisms include:

Initial formation of a thiophene valence bond isomer,[18] analogous Dewar (1-1′), prismane (Ladenburg) (2-2′), or benzvalene (3-3′) bond isomers or benzene.[19] This is not in accord with experimental data (Scheme 3).

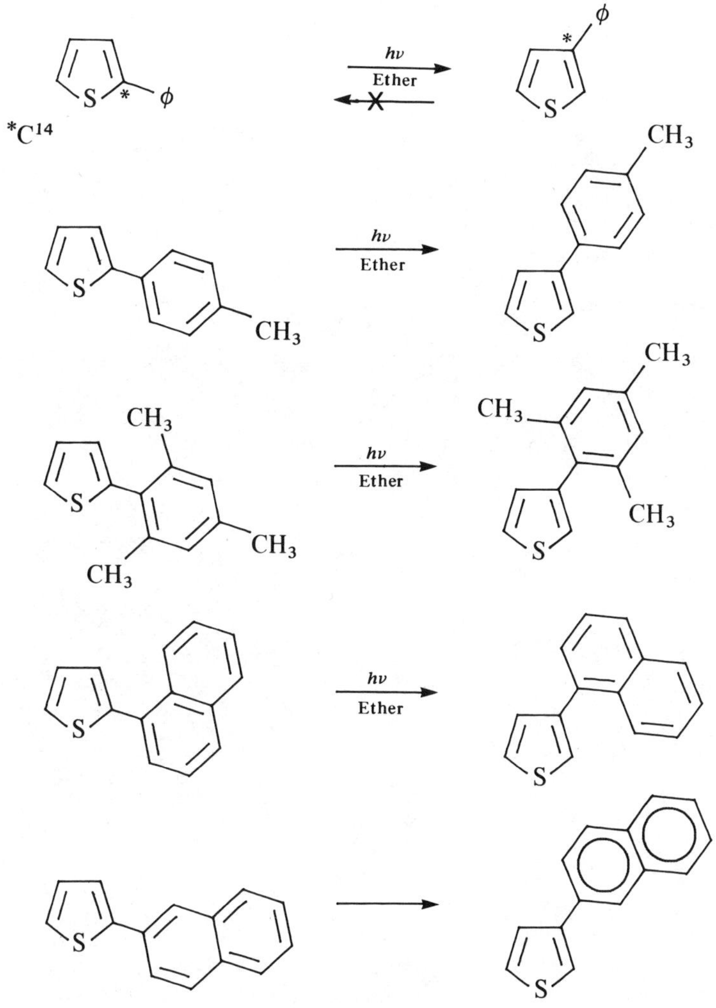

Scheme 1

A ring contraction-ring expansion mechanism route, analogous to the one by which 3,5-diphenylisoxazole photoisomerizes to 2,5-diphenyloxazole[20] (Scheme 4). This mechanism can satisfactorily explain the major rearrangement paths of phenyl-, deuterio-, and methyl-substituted arylthiophenes. Wynberg rejected it on the assumption that, in the photoarrangement of 2-phenylthiophene to 3-phenylthiophene, the intermediate would be thioaldehyde 4 and not thioketone 5, which was, a priori, expected to be more stable than 4.[18] π-Bond order calculations lead, in fact, to the opposite conclusion.[9]

A zwitterionic intermediate[18] (Scheme 5a). This mechanism, supported by Wynberg, fits well with the observed data on 2-arylthiophenes, but not for 3-

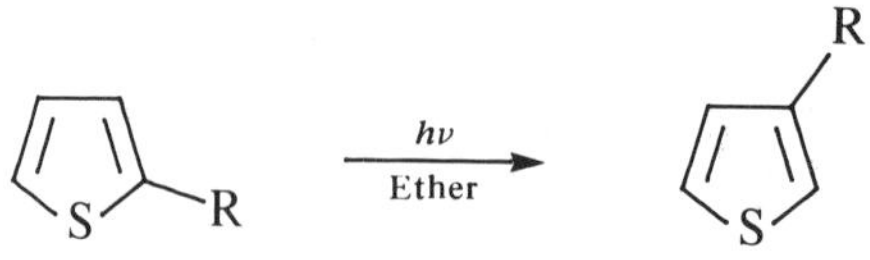

A. Lablache-Combier

$$\text{(structure)} \xrightarrow[\text{Ether}]{h\nu} \text{(structure)}$$

Yield (%)

R = CH$_3$	8–9
R = CH$_2$C$_6$H$_5$	15
R = CH$_2$C(CH$_3$)$_3$	10
R = C(CH$_3$)$_3$	27

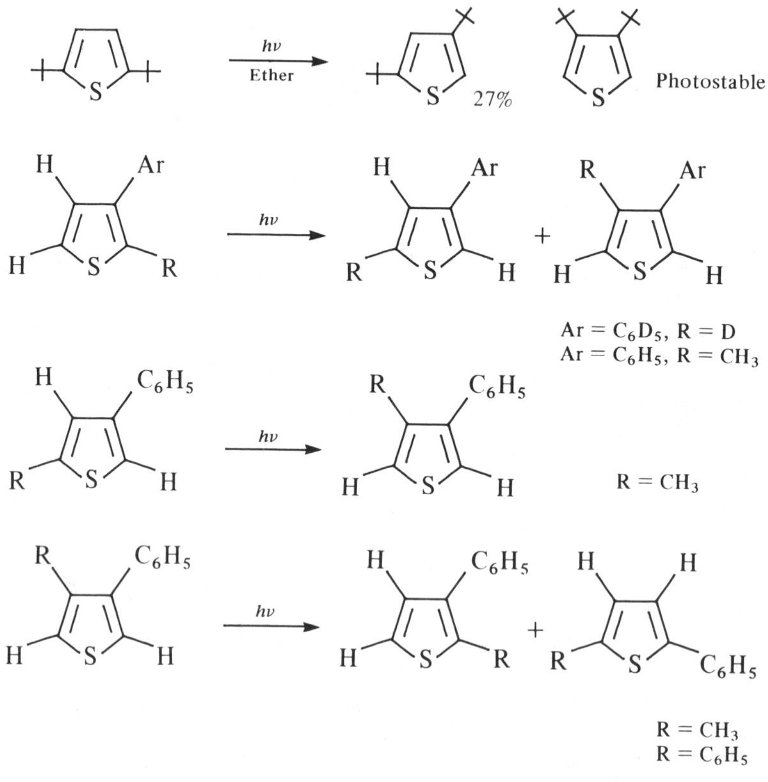

$$\text{(structure)} \xrightarrow[\text{Ether}]{h\nu} \text{(structure)} \; 27\% \qquad \text{(structure)} \quad \text{Photostable}$$

$$\text{(structure)} \xrightarrow{h\nu} \text{(structure)} \; + \; \text{(structure)}$$

Ar = C$_6$D$_5$, R = D
Ar = C$_6$H$_5$, R = CH$_3$

$$\text{(structure)} \xrightarrow{h\nu} \text{(structure)} \qquad R = CH_3$$

$$\text{(structure)} \xrightarrow{h\nu} \text{(structure)} \; + \; \text{(structure)}$$

R = CH$_3$
R = C$_6$H$_5$

Scheme 2

arylthiophenes; 3-phenylthiophene itself does not photoisomerize, but slowly decomposes. Wynberg proposed the intermediate, Scheme 5b, to account for the experimental data on 3-arylthiophenes.

Kellog proposed another mechanism in which in the intermediate, a two-atom fragment, is twisted 90° out of the plane formed by the remaining three atoms (5').[21] This might indeed represent a stable geometrical situation on the singlet excited-state potention surface, from which isomerization is supposed to take

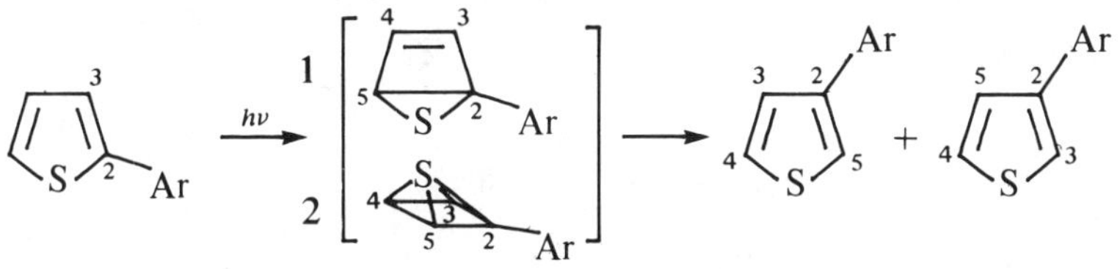

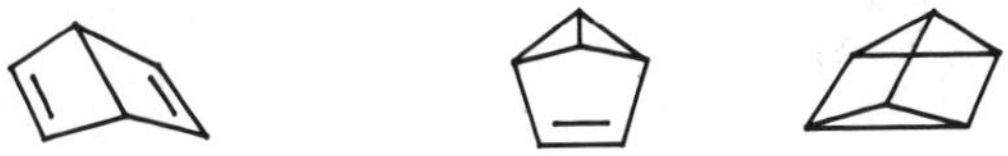

Scheme 3

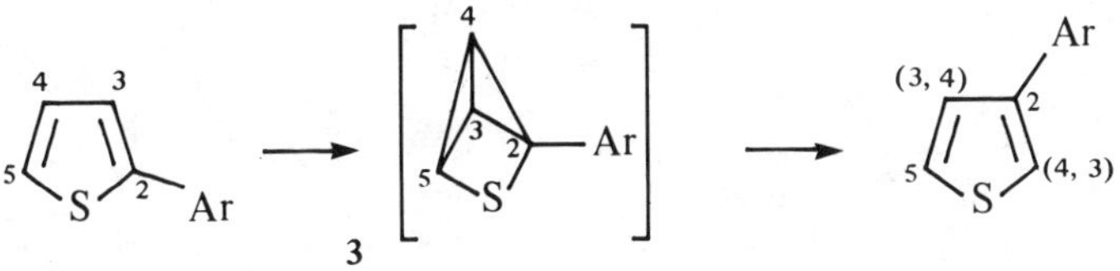

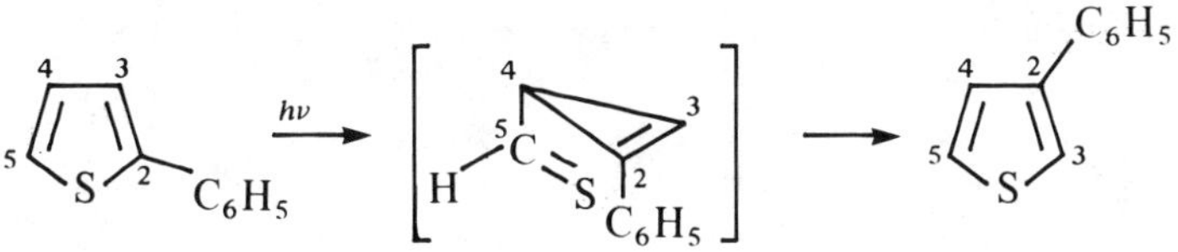

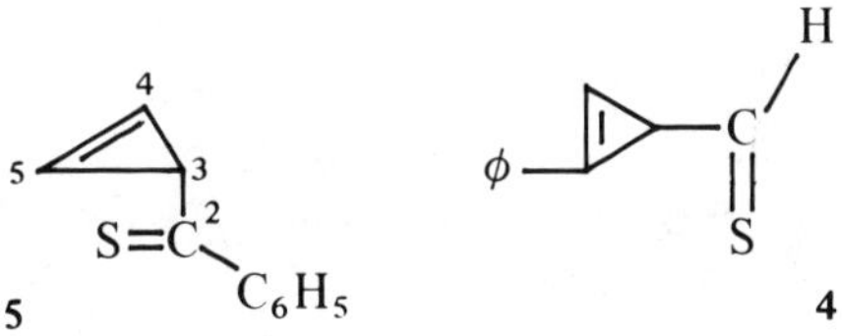

Scheme 4

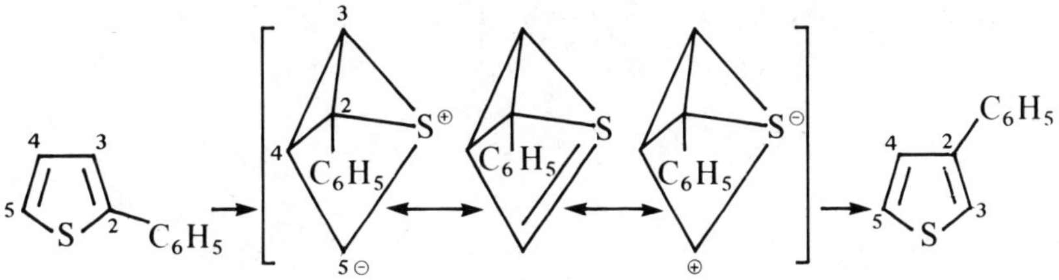

Scheme 5a

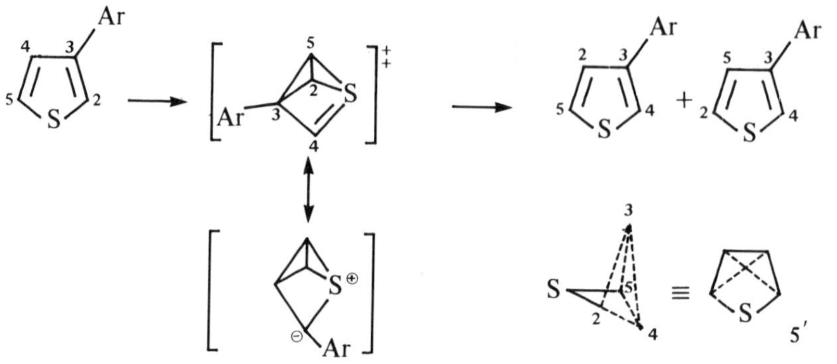

Scheme 5b

place.[22] This mechanism also explains the observed phenylmethyl photorearrangements, but it has no experimental or theoretical support.

It has recently been suggested that the difference in photochemical behavior between 2-phenylthiophene and 3-phenylthiophene is related to the electronic structure of the initial photoreactive states.[23] For 2-phenylthiophene, the lowest singlet excited state is delocalized over the whole π system, whereas it is mainly localized on the phenyl moiety for 3-phenylthiophene. For the first compound, there is a crossing of $\pi\pi^*$ electronic states and photoisomerization becomes irreversible, whereas for the latter, no crossing occurs and a "no-reaction reaction" is preferred.

Van Tamelen and Whitesides[24] proposed the mechanism in Scheme 6, which explains the majority of the results encountered, including those for 3-arylthiophenes, if the following assumptions are made:

Dewar structures and cyclopropenylthiocarbonyls are in equilibrium.

In the formation of a given cyclopropene, the ring contractions occur in such a fashion as to place the phenyl group on the double bond of the cyclopropene. Although the requirements of the second assumption are not always adhered to, it does account for the major products of the photoreactions.

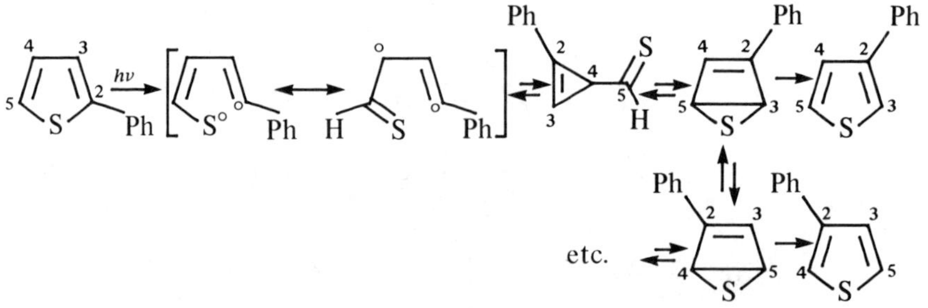

Scheme 6

In two cases, a stable, nonaromatic photorearranged thiophenic derivative has been isolated: by photolysis, tetra(trifluoromethyl)thiophene is converted to tetrakis(trifluoromethyl) Dewar thiophene 6[4-7a,25] (Scheme 7a). 3-Cyano-2 methylthiophene and 3-cyano-4 methylthiophene lead by 254-nm UV irradiation in cyclohexane to two Dewar thiophenes, which are in equilibrium. The more stable is 2-cyano-3 methyl-5 thiabicyclo[2.1.0]pentene-2 6'.[7c] These Dewar derivatives give Diels–Alder adducts with furan or with 2,5-diphenyl[3,4] benzofuran.[6b,7c] The cyanothiophene photorearrangement is a singlet state reaction[75] (Scheme 7b).

As cyanothiophenes, cyanopyrroles lead to Dewar derivatives.[7d] In general, electron-withdrawing substituents stabilize the intermediate(s) in the photoisomerization of aromatic compounds.[1]

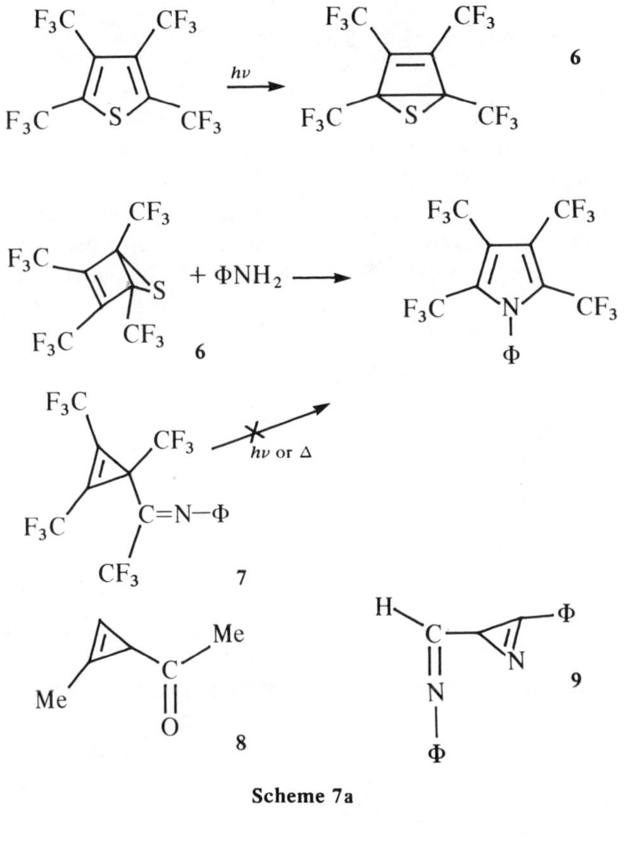

Scheme 7a

C. Intermediate(s) Trapping by a Primary Amine

The photorearrangement intermediates of thiophenes have been trapped by primary amines; irradiated in a primary amine, thiophene, the methylthiophenes, 2-phenylthiophene, and the dimethylthiophenes are converted into pyrroles.[8]

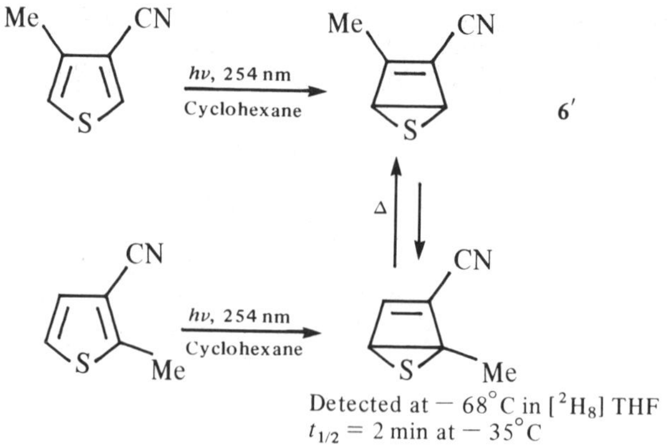

Detected at $-68°C$ in $[^2H_8]$ THF
$t_{1/2} = 2$ min at $-35°C$

Scheme 7b

Corresponding furans are photoconverted to the same pyrrole(s). This suggests that the intermediates, in both cases, have a similar structure.[9, 26]

The structure of the pyrroles obtained from the thiophenes irradiated in propyl-amine cannot be explained, if it is assumed that the amine reacts on a zwitterionic intermediate, whatever its structure.[27] On the other hand, cyclopropenylthioketone or -thioaldehyde intermediates and Dewar thiophene intermediates can account for some of the experimental data[27] (Scheme 8). If it is assumed that the structure of

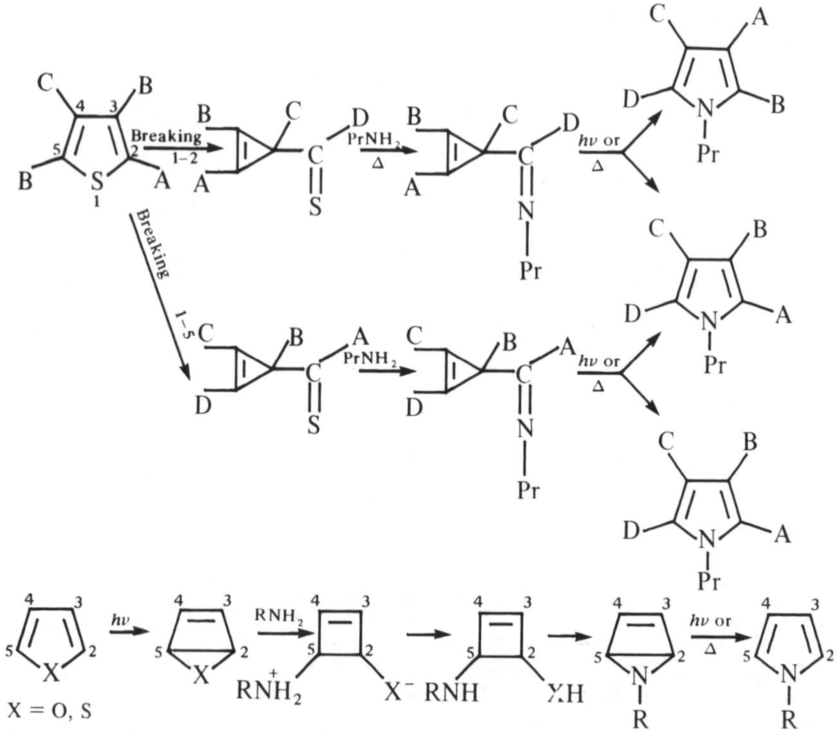

Scheme 8

the intermediate is either a cyclopropenylthioketone or -thioaldehyde, or a Dewar thiophene, depending upon the substituent(s) of the thiophenic ring, all the experimental data can be explained.

D.   Tetrakis Trifluoromethyl – Dewar Thiophene Reactivity

Kobayashi and coworkers investigated the reactivity of tetrakis(trifluoromethyl) Dewar thiophene 6.[7a, 28] They showed:

that 6 heated with aniline is converted to N-phenyl-2,3,4,5-tetrakis (trifluoromethyl)pyrrole;

that there is an interconversion between this pyrrole and its Dewar isomer.

These data support the hypotheses of Couture and Lablache-Combier about a possible reaction between a Dewar thiophene and a primary amine, leading to a pyrrole.[27] However, N-phenyl-cyclopropenylimine (7) is not converted either thermally or photochemically to a pyrrole.[28] Does this mean that such a cyclopropenylimine cannot be an intermediate in the phototransformation of a thiophene into a pyrrole, as proposed in Scheme 8? It certainly does not:

Substitution of a hydrogen by a trifluoromethyl group drastically changes the behavior and the stability of a molecule.[7a]

Cyclopropenylketone 8 irradiated in n-butylamine is converted to a 2:1 mixture of N-butyl-2,5 dimethylpyrrole and N-butyl-2,4-dimethylpyrrole. The first step of this reaction is certainly imine formation. Some other cyclopropenylketones behave similarly.[29]

Imine 9 is converted by a pyrazole by heating or to an imidazole by UV irradiation.[30]

E.   Conclusion

To conclude the discussion about the photoisomerization of thiophene and its alkyl or aryl derivatives, it can be said that the mechanism of this reaction is not yet clearly proved and may differ from one compound to the other. For certain derivatives, a ground-state cyclopropene–thioaldehyde or –thioketone, or a Dewar thiophene may be a true intermediate (Scheme 9a). On the other hand, for some other derivatives there is perhaps no ground-state intermediate. In this case, a cyclopropene–thioketone or –thioaldehyde and a Dewar thiophene may be two equivalent graphical representations of the same excited-state hypersurface (Scheme 9b). Similar questions arose when looking for the photoisomerization of six-membered aromatic N-oxide derivatives. Is a ground state oxaziridine the intermediate in this rearrangement or not?[31]

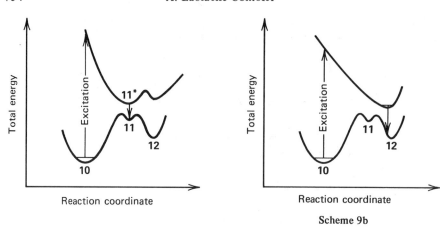

Scheme 9a

Scheme 9b

(a) Cross section of the hypothetical hypersurface illustrating the thermal (nonobserved) rearrangement of **10** via **11** to **12**. There is an intermediate (**11**) in the photoisomerization of **10** to **12**. (b) Cross section of the excited state leading to the reaction. There is no intermediate in the photoisomerization of **10** to **12**.

3.   Heteroarylthiophenes

Photoisomerization of dithienyl compounds has been reported[16] (see Scheme 1a). The mechanism suggested by Wynberg to take into account the structure of the product formed is completely different from those suggested for alkyl- or arylthiophene photorearrangements (see Scheme 10b). However, 2-(2-thienyl)furan behaves as 2-phenylthiophene: the thiophenic ring with the weakest C—X bond isomerizes.[32]

4.   Thiophenium Ions

Thiophenium ions of general formula 13 are photostable, whereas S-methylated thiophenium ions photoisomerize (see Scheme 11). The selective shift of an S-methyl substituent from sulfur to an adjacent carbon atom, without apparent further effect, on the ring substituents is best explained by a homolytic scission of the C—S bond.

5.   Photo-Fries Rearrangements

The phenyl ester of thiophen-2-carboxylic acid is photoisomerized to the corresponding o- and p-hydroxyphenylketones.[34] This is a classical photo-Fries rearrangement.[35]

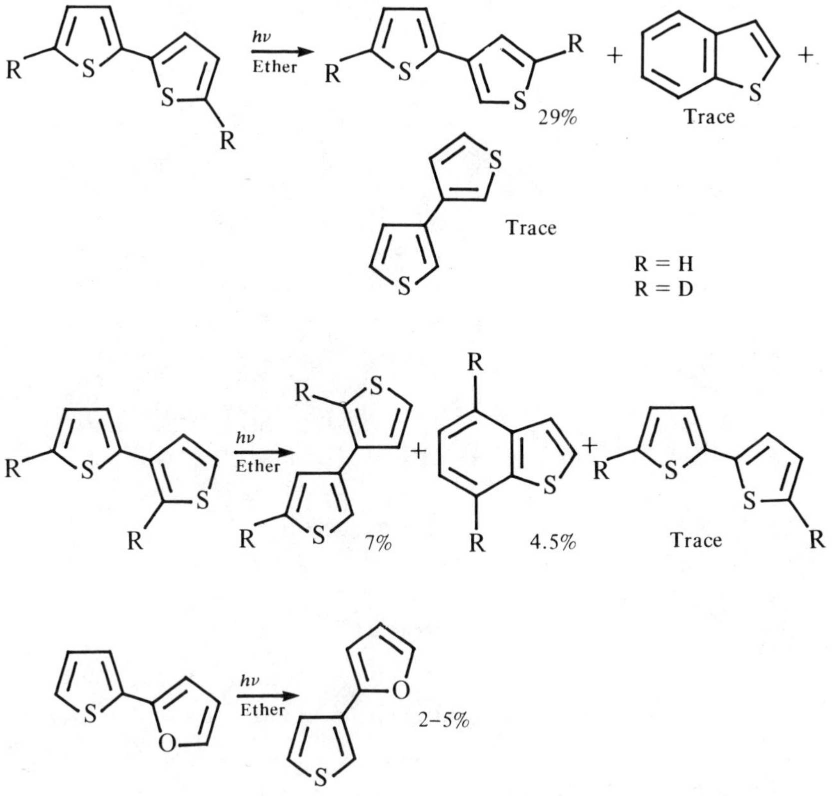

Scheme 10a

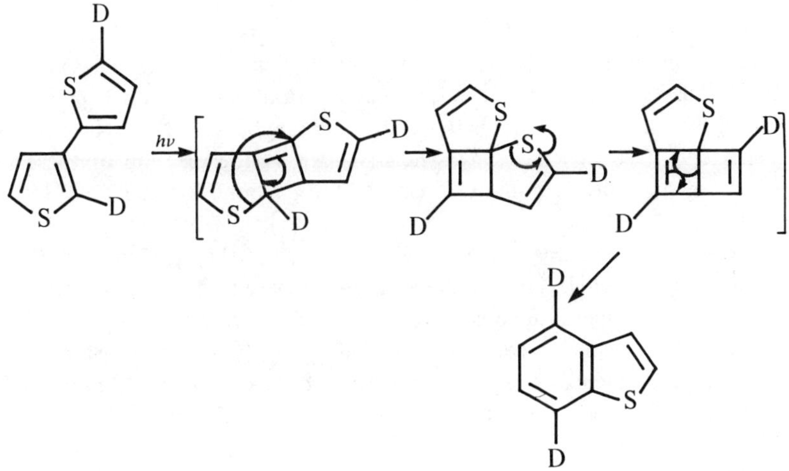

Scheme 10b

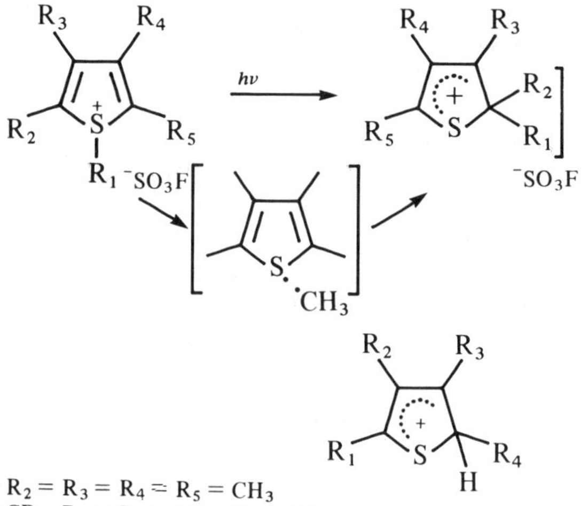

$$R_1 = R_2 = R_3 = R_4 = R_5 = CH_3$$

a $R_1 = R_2 = R_3 = R_4 = R_5 = CH_3$
b $R_1 = CD_3, R_2 = R_3 = R_4 = R_5 = CH$
c $R_1 = R_2 = R_5 = CH_3, R_2 = R_4 = CH_2D$
d $R_1 = CD_3, R_2 = R_5 = CH_3, R_3 = R_4 = CH_2D$

Scheme 11

II. PHOTOCYCLIZATIONS

2,3-Diphenylthiophene photocyclizes to phenanthro[9,10-b]thiophene.[11] This is a general reaction,[36] but 3,4-diphenylthiophene is not converted to phenanthro-[9,10-c]thiophene, it photoisomerizes as shown in Scheme 12.[11]

Some thienyl and furylethenes, which are stilbene analogues, photocyclize[37] (see Scheme 13a). Similar derivatives of thieno[2,3-b]thiophene photocyclize also, leading to heterohelicenes.[38] An example is given in Scheme 13b. The photocyclization of 1-phenyl-4 thienyl and 1,4-dithienyl-1,3 butadienes has been reported[39] (see Scheme 13c). The photocyclization of some cis-1 phenyl-2(3-thienyl)ethylene and some iodophenyl and iodothienyl-(thienyl) ethylene has also been reported[40] (see Scheme 13d). The first step of the reaction of these iodo compounds is probably a homolytic carbon–iodine bond cleavage.[40a] This is a general reaction.[40b]

The ESR spectra of isomeric diethienylethylenes, 1-(2 thienyl)-2 phenylethylene and 1-(2-thienyl)-2 phenylacetylene anion radicals, have also been analyzed.[41]

α-Diformyl thiophenes undergo cyclization when irradiated in CCl$_4$[42] (see Scheme 13e). Irradiation of compound 14 leads only to a cis-trans isomerization of the exocyclic ethylenic bond.[43] Compound 15 behaves similarly.[44] Fluorescence characteristics of some heteroatomic derivatives of 2,2′ dithienyl have been published.[45]

The anilide of thiophen-2 carboxylic acid photocyclizes[46] (see Scheme 14). Enamide photocyclization is a general reaction.[47]

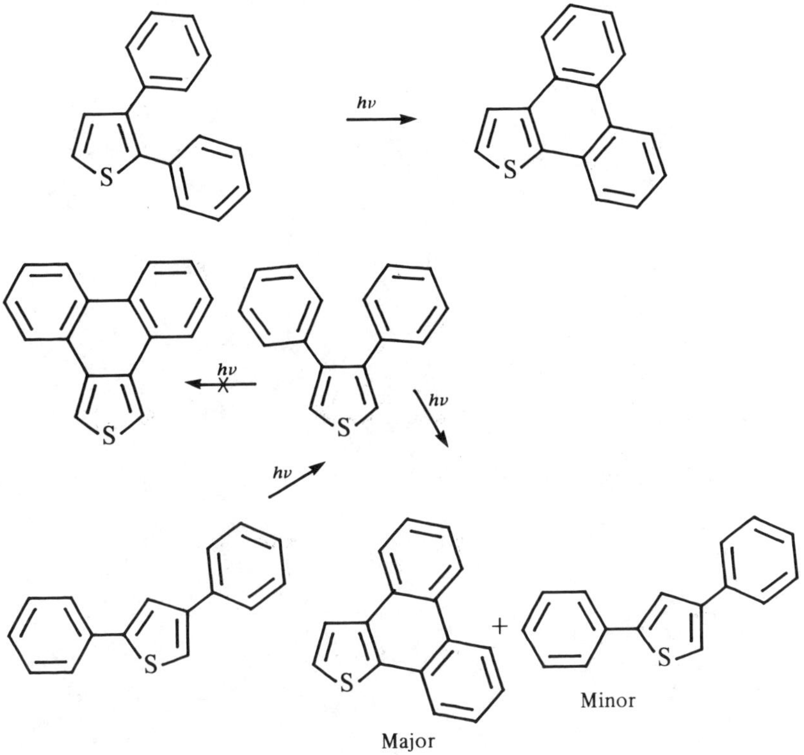

Scheme 12

UV irradiation of 5-(3'-thienyl)-2 chlorothionicotinate in benzene leads, via deshydrohalogenation and cyclization, to the formation of a thiolactone. Cleavage of the carbonyl–sulfur bond is a competitive process[48a] (see Scheme 15). On UV irradiation in benzene, some S-aryl esters of thiophen-2 and thiophen-3 carbothioic acid yield thienobenzothiopyranones or pyranopyridones. A thiophene carbaldehyde is generally also formed. It can even be the main product of the photoreaction[48b] (see, for example, Scheme 15).

III. PHOTODIMERIZATIONS, PHOTOADDITIONS

1. Photodimerization

Photodimerization of thiophene analogues of chalcone[49] and of styril thiophenes[50] have been reported (for examples, see Scheme 16). Such [2+2] photodimerizations are very general photoreactions.[51] In the case of the styrilthiophenes studied, the structure of the dimer formed is not the same in benzene solution and

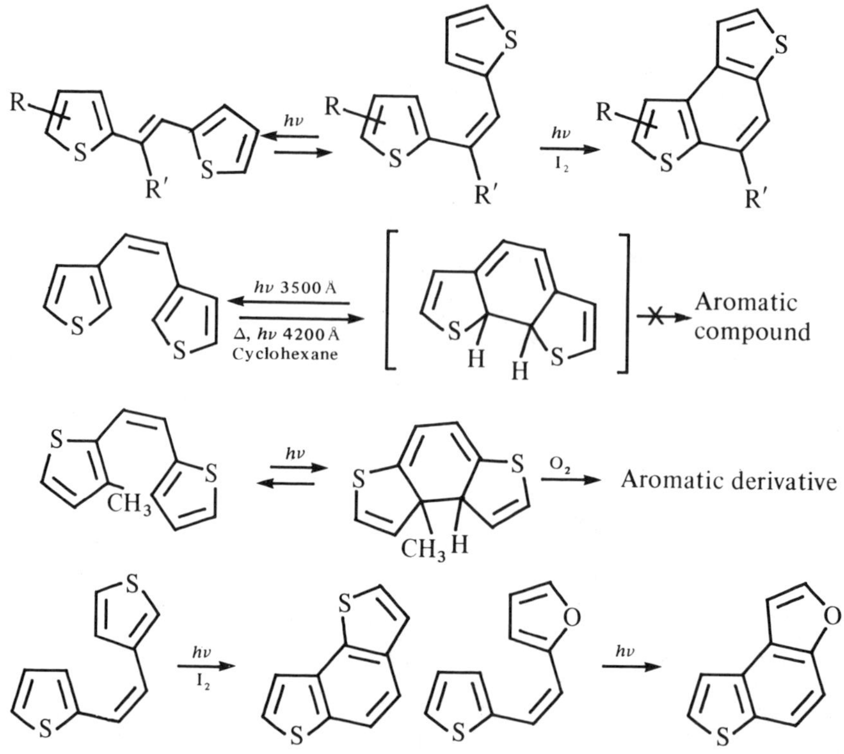

Scheme 13a

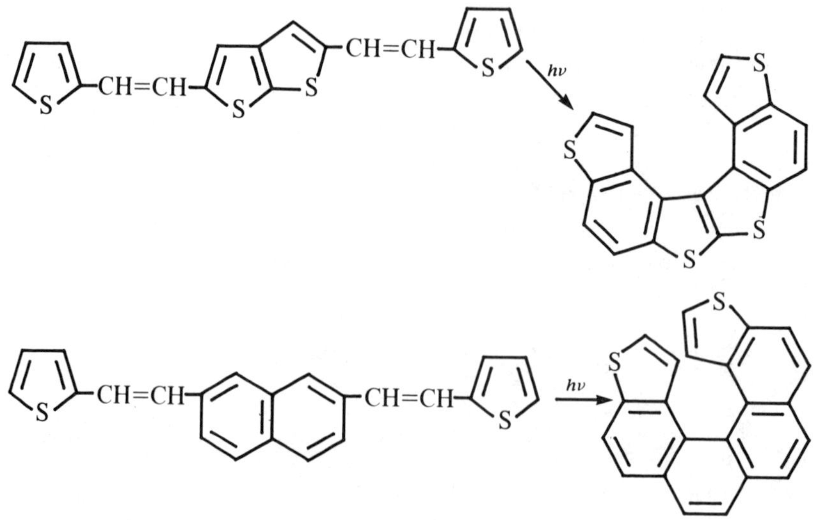

Scheme 13b

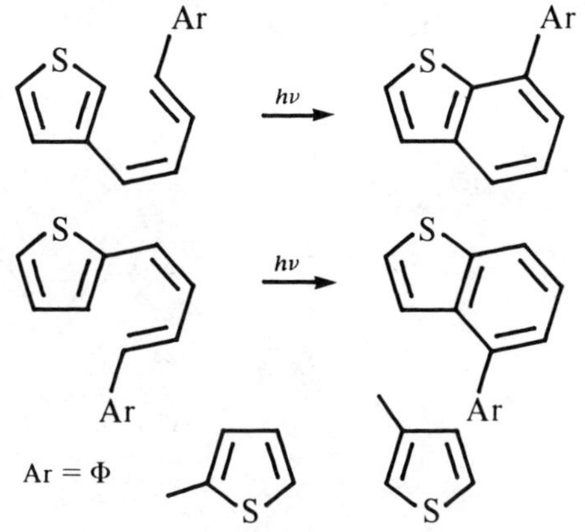

Scheme 13c

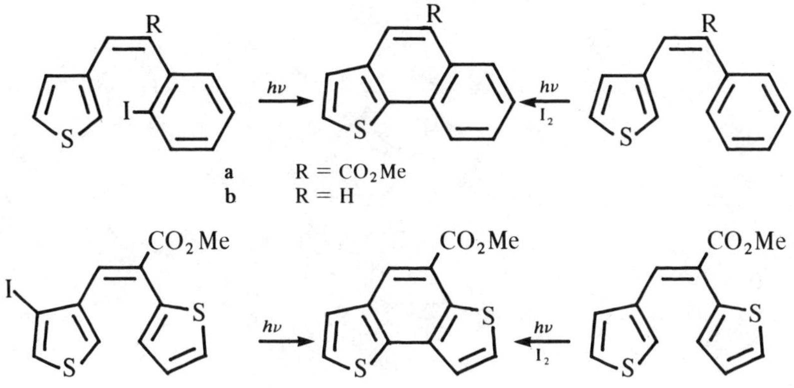

a	R = CO$_2$Me
b	R = H

Scheme 13d

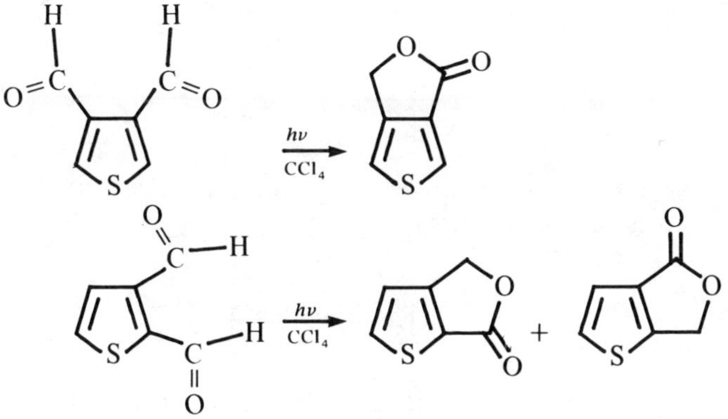

Scheme 13e

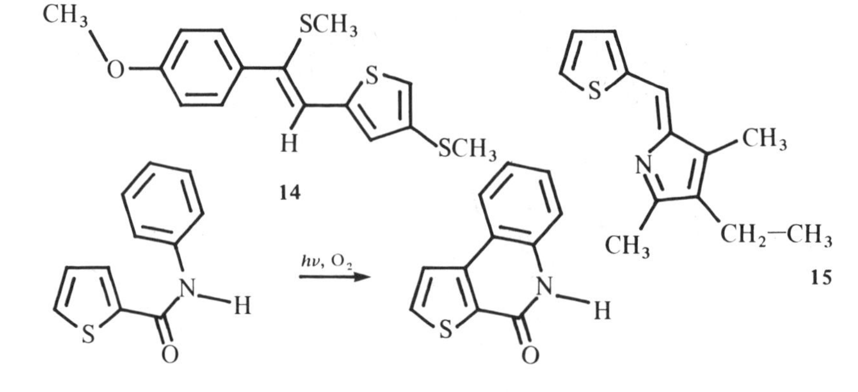

Scheme 14

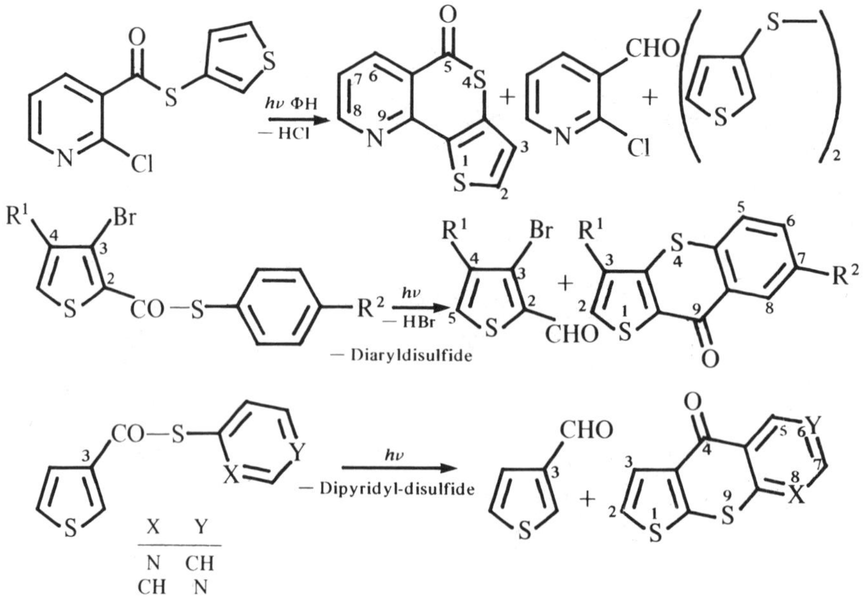

Scheme 15

the solid state, where photodimerization does not occur in every case. The monomer crystal structure is a major factor.[50]

2.   Photoaddition

We deal here only with photoreaction in which a thiophene is photoexcited, but not with reactions in which another photoexcited molecule adds to a ground-state thiophene.

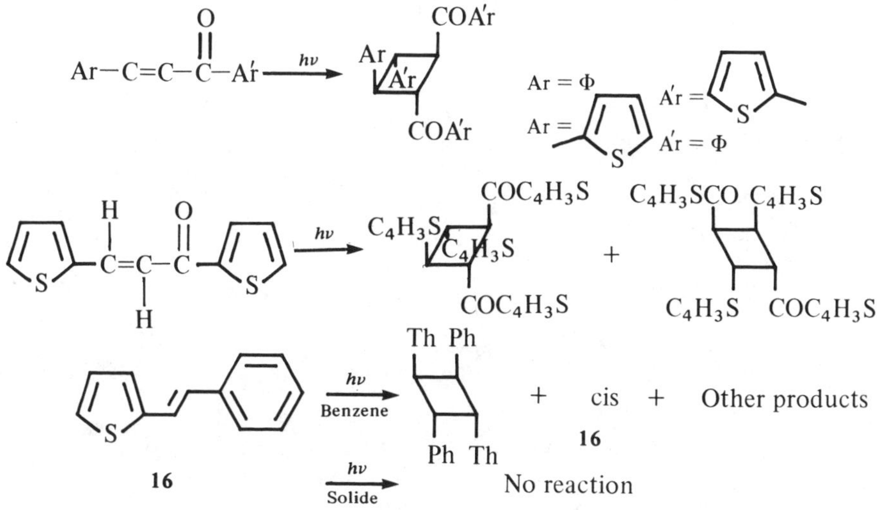

$$Ar-C=C-\overset{\overset{O}{\|}}{C}-A\acute{r} \xrightarrow{h\nu}$$

COA'r — Ar, A'r, COA'r structure

Ar = Φ   Ar = (thiophene)   A'r = (thiophene)   A'r = Φ

$$C_4H_3S-\overset{H}{\underset{H}{C}}=C-\overset{O}{\overset{\|}{C}}-C_4H_3S \xrightarrow{h\nu} C_4H_3S, C_4H_3S, COC_4H_3S \quad + \quad C_4H_3SCO\ C_4H_3S / C_4H_3S\ COC_4H_3S$$

Th Ph structure:

(thiophene)–CH=CH–(benzene) **16** $\xrightarrow[\text{Benzene}]{h\nu}$ [Ph Th / Th Ph cyclobutane] + cis + Other products **16**

$\xrightarrow[\text{Solide}]{h\nu}$ No reaction

16

Scheme 16

A. Thiophenes

The benzophenone photosensitized addition of 2,5-dimethylthiophene to maleic anhydride derivatives leads to a small amount of the expected [2+2] adduct and to an oxetane (Scheme 17).[51-53] Thiophene and the other methyl-substituted thiophenes do not react. 2,5-Dimethylthiophene also gives oxetanes when irradiated with 1-naphthaldehyde, 2-, 3-, and 4-benzoylpyridine (for examples, see Scheme 17).[53] The photoreaction of thiophene on dihalogenomaleimide leads in most of the cases studied to the corresponding 2-thienyl maleimide. The first step of this reaction is probably homolytic cleavage of a maleimide C-halogen bond.[54] The formation of a bis adduct has been reported. The first step is, in this case, the formation of a [2+2] photoadduct[54] (see Scheme 17). In contrast to 2-methylfuran, thiophene does not add to 1,1-diphenyl ethylene in a reaction photoinitiated by 1-cyanonaphthalene.[55]

2.5-Dimethylthiophene and thiophene give an unstable adduct with dimethyl acetylenedicarboxylate, which splits off sulfur to give a dimethyl phthalate derivative. The same reaction can be performed thermally.[56] The unstable adduct is probably not a [4+2] adduct, as initially postulated,[56] but a [2+2] adduct,[57] as shown in Scheme 18. 3-Benzoylthiophene irradiated in the presence of dimethylacetylene dicarboxylate gives a [2+2] adduct on the thiophene ring[58] (see Scheme 18). Whereas irradiation of 254 nm of a furan–benzene mixture leads to several adducts formed in good yield, only a trace of a 1-1 photoadduct of thiophene and benzene is obtained when a mixture of these two compounds is irradiated at 254 nm. It is only obtained in the presence of a proton donor.[59]

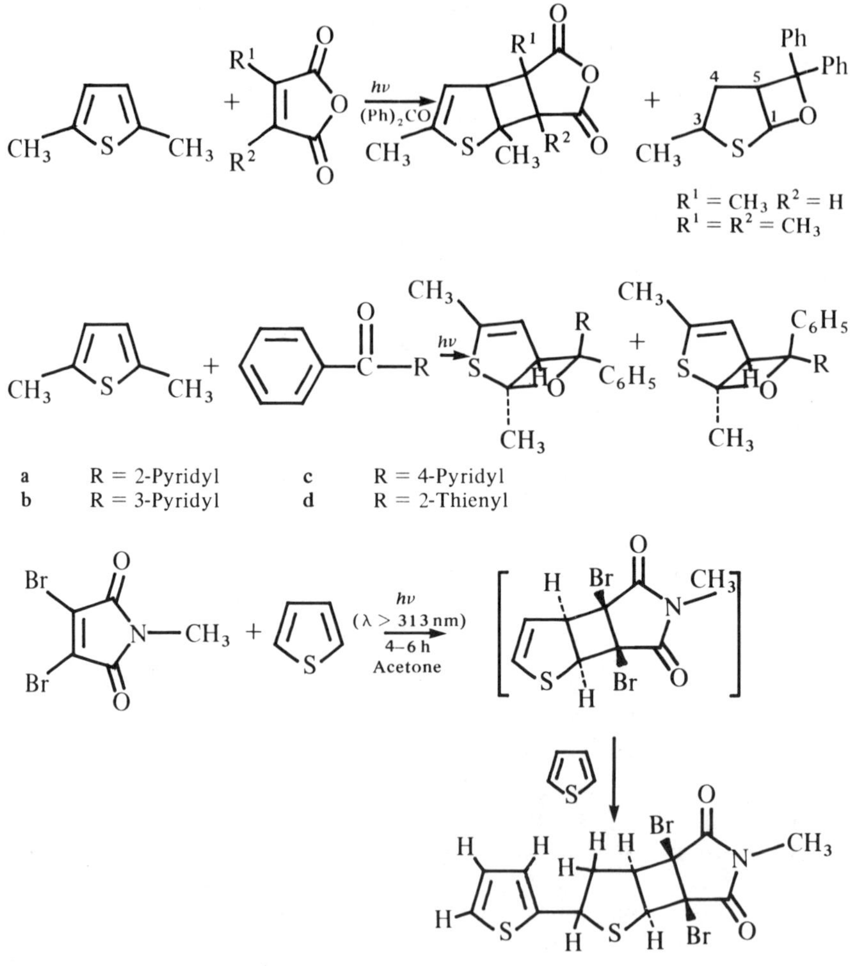
R¹ = CH₃ R² = H
R¹ = R² = CH₃
a R = 2-Pyridyl
b R = 3-Pyridyl
c R = 4-Pyridyl
d R = 2-Thienyl
hν
(Ph)₂CO
(λ > 313 nm)
4–6 h
Acetone

Scheme 17

Photostimulated S_{RN_1} (Bunnett's reaction) of acetone enolate ion with 2-chloro-, 2-bromo, and 3-bromothiophenes in liquid ammonia has been reported. Similar reactions can be stimulated by solvated electrons[60] (see Scheme 19). Irradiation of the charge transfer complex formed by 2,3-dichloro-1,4 naphthoquinone with a thiophene leads to 2-thienyl substituted naphthoquinone. The reaction proceeds by a radical pair intermediate, as shown by CIDNP experiments[61] (see Scheme 19). 2-Bromo-3 methoxy-1,4-naphthoquinone irradiated with 1-phenyl-2-(2-thienyl)-ethylene leads to condensed quinones,[62] as shown in Scheme 19.

3-(2-Thienyl)-2,2 dimethyl-2H azirine is photoisomerized to a nitrile isopropylide, which acts as a 1-3 dipole and adds regiospecifically to the activated C—C and C—O double bond (Scheme 20).[63] This is a very general reaction of [2H]azirine.[64]

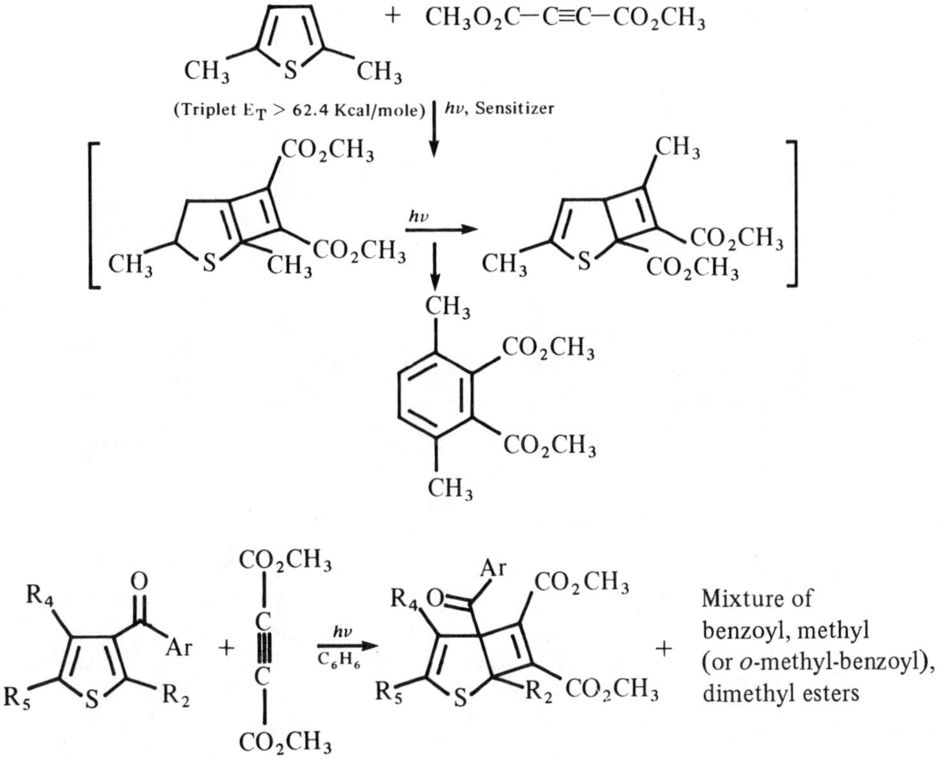

$$+ \quad CH_3O_2C-C\equiv C-CO_2CH_3$$

(Triplet $E_T > 62.4$ Kcal/mole) | $h\nu$, Sensitizer

Scheme 18

B. Thienylketones or Aldehydes

The photochemical reactivity of some benzoylthiophenes has been studied in great detail. From absorption and emission spectra of 2- and 3-benzoyl-thiophene, their p-cyano and p-methoxy-derivatives,[65] 2-(2-methyl-benzoyl), -2 benzoyl-3 methyl, 17, and 2-benzoyl-4 methyl, 18, -thiophene,[66] -2 and -4 methyl-3 benzoyl-thiophene, 2,5-dimethyl-3 benzoylthiophene and 3(-2 methylbenzoyl)thiophene,[58] their partial energy diagrams have been constructed. The lowest triplet state of 2-benzoyl thiophenes derivatives has a $\pi\pi^*$ character. However, for some 3-methyl-benzoyl thiophene, T_1 can be $\pi\pi^*$.[58] All of these compounds, except 17 and 18, give an oxetane when irradiated with isobutene. These oxetanes are thermally unstable; they eliminate formaldehyde and yield 1-aryl thienyl-2 methylpropene[66,67] (see Scheme 21). Such reactions are very general.[68] No photoenolization occurs,

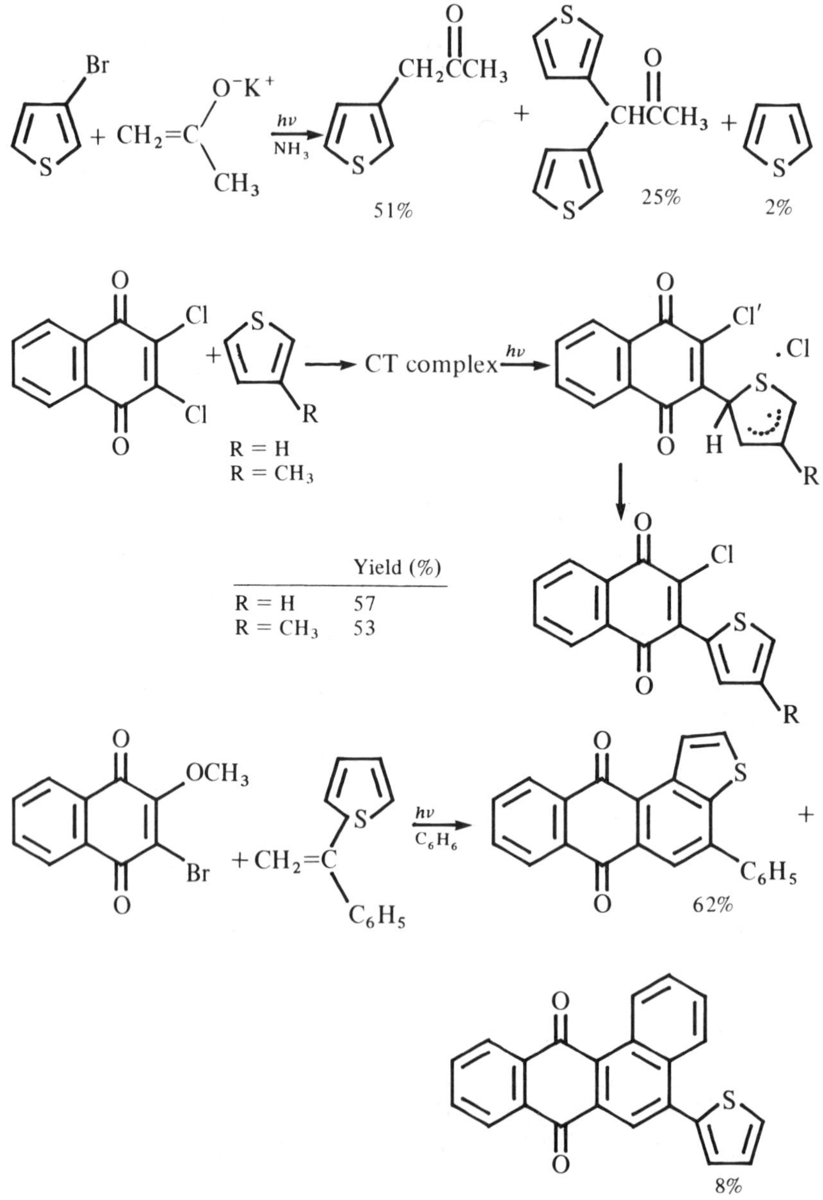

Scheme 19

even when there is a hydrogen γ to the carbonyl group, except for the methylated
3-benzoyl thiophenes studied.[58]

The oxetane formed by photoaddition of 3-benzoylthiophene on tetramethyl
ethylene is thermally stable.[69b,c] Under the same conditions, 2-acetylthiophene

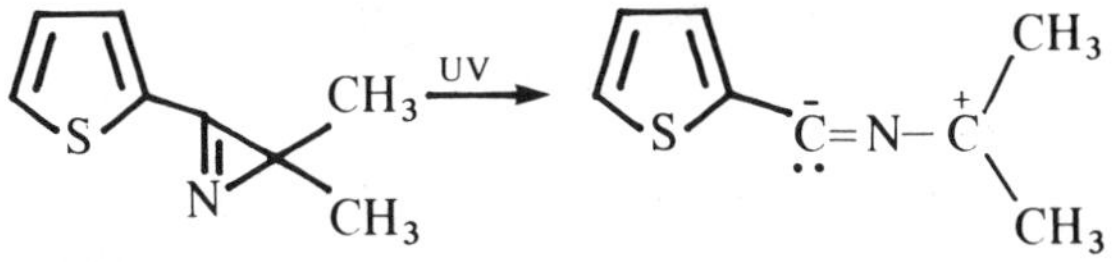

Scheme 20

gives mainly a [4 + 2] adduct,[69a,b] whereas the oxetane formed from 3-acetylthio-
phene is thermally unstable[69c] (see Scheme 21). 2- and 3-Formylthiophene also
give [2 + 2] adducts with tetramethylethylene, but dithienylketone is only photo-
reduced to the pinacol when irradiated in the presence of this compound[69c] (see
Scheme 21). 3-Benzoylthiophene gives a [2 + 2] adduct on the thiophene ring when
irradiated in the presence of dimethyl acetylenedicarboxylate in benzene[58] (see
Scheme 18). It was not possible to trap a photoenol from a methylated-3 benzoyl-
thiophene.

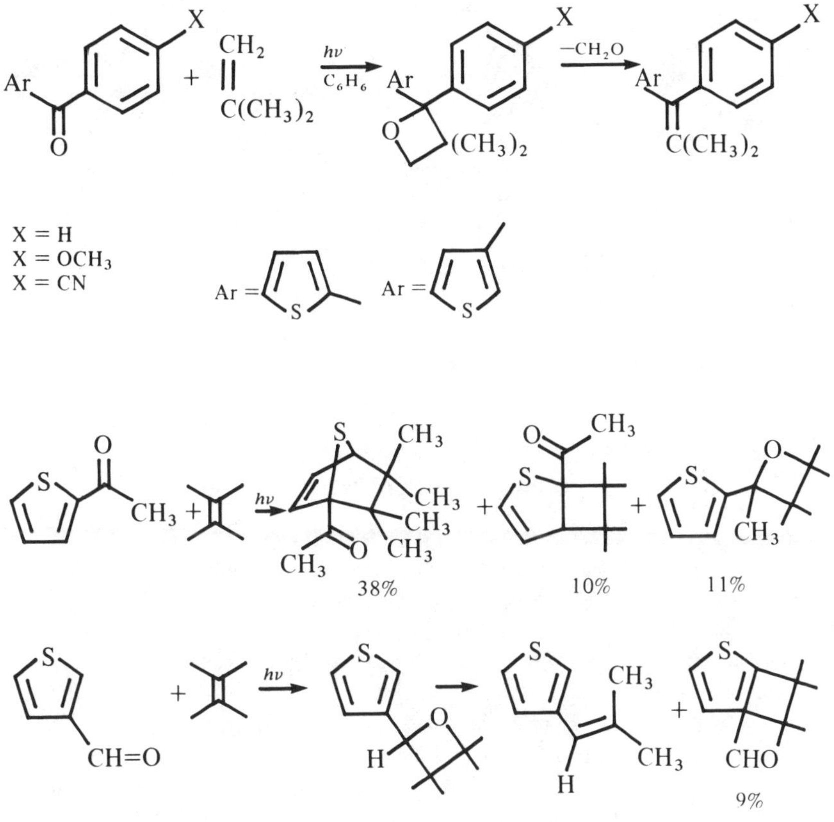

X = H
X = OCH$_3$
X = CN

Scheme 21

IV. PHOTOSUBSTITUTIONS

1. Photodeshalogenation of Halothiophenes

Thiophene is formed by irradiation of either 2- or 3-bromothiophene in a solvent such as a cyclohexane or ether. It is a singlet-state reaction.[70, 71] The reaction pathway depicted in Scheme 22 is most likely radical in nature, but oxygen has no effect on the reaction quantum yield. Consequently, it has been proposed that the reaction involves a π complex (19) between the thienyl radical and the bromine atom. The formation of a π complex between a phenyl radical and a chlorine atom has been suggested to occur during the photolysis of chlorobenzene in cyclohexane.[72]

2- or 3-Iodothiophene are often used to generate the corresponding thienyl radical by photolysis.[73] The reactivity of thienyl radicals has been reviewed.[74]

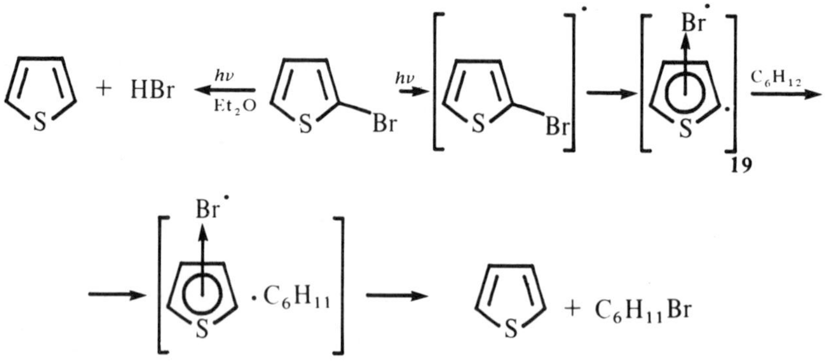

Scheme 22

2. Photosubstitution of Nitrothiophenes

Aromatic photosubstitution is a general reaction rationalized by Havinga's and Cornelisse's group.[75] Only one paper on this subject involving thiophene derivatives has been published so far.[76] 2-Nitrothiophene and 2-nitrobromothiophene are converted to the corresponding 2-cyanothiophene by irradiation in the presence of CN^-. The reaction is thought to proceed via a triplet state through an intermediate σ complex, and can be classified as a S_{N_2} (Ar^*) substitution — a counterpart of ground state aromatic substitution (see Scheme 23). Methoxide and cyanate ion also displace the nitro group of 2-nitrothiophene. Under the same conditions, 3-nitrothiophene is photodecomposed. The expected σ complex (20) formed as an intermediate in the photoreaction may undergo ring opening, as indicated in Scheme 23.

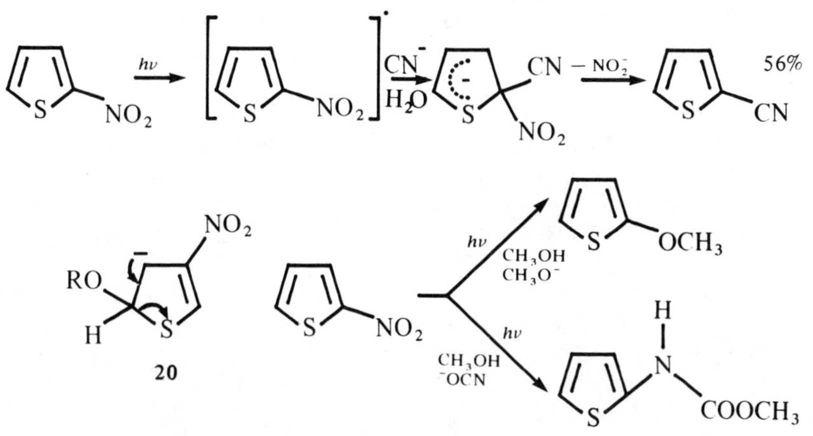

Scheme 23

3.   Synthesis of Grignard Reagents

The formation of Grignard reagents from 2-bromothiophene is enhanced by irradiation of the mixture of this organic halide and magnesium turnings in diethyl ether.[77]

REFERENCES

1.   For reviews see: P. Beak and W. Messer, in *Organic Photochemistry,* Vol. 2 (O. Chapman, Ed.), Marcel Dekker, New York, 1969, p. 117; A. Lablache-Combier and M. A. Rémy, *Bull. Soc. Chim. Fr.,* 679 (1971); A. Lablache-Combier, *l'Actual. Chim.,* 9 (December 1973); A. Lablache-Combier, in *Photochemistry of Heterocyclic Compounds* (O. Buchardt, Ed.), Wiley, New York, 1976, p. 123; A. Padwa, in *Rearrangements in Ground and Excited States,* Vol. 3 (P. De Mayo, Ed.), Academic Press, New York, 1980, p. 123.

2.   H. A. Wiebe and J. Heicklen, *Can. J. Chem.,* **47,** 2965 (1969).

3.   H. Wynberg, *Acc. Chem. Res.,* **4,** 65 (1971).

4.   H. Wiebe, S. Braslovsky, and J. Heicklen, *Can. J. Chem.,* **50,** 2721 (1972).

5.   Y. Kobayashi, I. Kumadaki, A. Ohsawa, and Y. Sekine, *Tetrahedron Lett.,* 2841 (1974).

6.   Y. Kobayashi, I. Kumadaki, O. Ohsawa, Y. Sekine, and H. Mochizuki, *Chem. Pharm. Bull.,* **23,** 2773 (1975).

7.   (a) Y. Kobayashi and I. Kumadaki, *Acc. Chem. Res.,* **14,** 76 (1981); J. A. Barltrop, A. C. Day, and E. Irving, *Chem. Commun.,* 880 (1979); (c) J. A. Barltrop, A. C. Day, and E. Irving, *Chem. Commun.,* 966 (1979); (d) J. A. Barltrop, A. C. Day, I. D. Moxon, and R. W. Ward, *Chem. Commun.,* 786 (1975); J. A. Barltrop, A. C. Day, and R. W. Ward, *Chem. Commun.,* 131 (1978).

8.   A. Couture and A. Lablache-Combier, *Chem. Commun.,* 524 (1969); *Tetrahedron,* **27,** 1059 (1971).

9.   A. Couture, A. Delevallée, A. Lablache-Combier, and C. Parkanyi, *Tetrahedron,* **31,** 785 (1975).

10. H. Wynberg and H. Van Driel, *J. Am. Chem. Soc.*, 87, 3998 (1965).

11. H. Wynberg, H. Van Driel, R. M. Kellog, and J. Buter, *J. Am. Chem. Soc.*, 89, 3487 (1967).

12. H. Wynberg and H. Van Driel, *Chem. Commun.*, 204 (1966).

13. H. Wynberg, R. M. Kellog, H. Van Driel, and G. E. Beekhuis, *J. Am. Chem. Soc.*, 88, 5047 (1966).

14. R. M. Kellog and H. Wynberg, *J. Am. Chem. Soc.*, 89, 3495 (1967).

15. H. Wynberg, G. E. Beekhuis, H. Van Driel, and R. M. Kellog, *J. Am. Chem. Soc.*, 89, 3498 (1967).

16. R. M. Kellog, J. K. Dik, H. Van Driel, and H. Wynberg, *J. Org. Chem.*, 35, 2737 (1970).

17. R. M. Kellog and H. Wynberg, *Tetrahedron Lett.*, 5895 (1968).

18. H. Wynberg, R. M. Kellog, H. Van Driel, and G. E. Beeckhuis, *J. Am. Chem. Soc.*, 89, 3501 (1967).

19. D. Bryce-Smith and A. Gilbert, *Tetrahedron*, 32, 1309 (1976).

20. E. F. Ullman and B. Singh, *J. Am. Chem. Soc.*, 88, 1844 (1966).

21. R. M. Kellog, *Tetrahedron Lett.*, 1429 (1972).

22. N. D. Epiotis, *The Theory of Organic Reactions*, Springer-Verlag, Berlin, 1978.

23. A. Mehlhorn, F. Fratev, and V. Monev, *Tetrahedron*, 37, 3627 (1981).

24. E. E. Van Tamelen and T. H. Whitesides, *J. Am. Chem. Soc.*, 93, 6129 (1971).

25. Y. Kobayashi, I. Kumadaki, A. Ohsawa, Y. Sekine, and A. Ando, *Heterocycles*, 6, 1587 (1977).

26. A. Couture and A. Lablache-Combier, *Chem. Commun.*, 891 (1971).

27. A. Couture, A. Delevallée, A. Lablache-Combier, and C. Parkanyi, *Tetrahedron*, 31, 785 (1975).

28. Y. Kobayashi, A. Ando, K. Kawada, and I. Kumadaki, *J. Org. Chem.*, 45, 2968 (1980).

29. T. Tsuchiya, H. Arai, and M. Igeta, *Chem. Commun.*, 550 (1972); *Chem. Pharm. Bull.*, 29, 1516 (1973); *Tetrahedron*, 29, 2747 (1973).

30. A. Padwa, J. Smolanoff, and A. Tremper, *Tetrahedron Lett.*, 29 (1974).

31. A. Lablache-Combier, in *Photochemistry of Heterocyclic Compounds* (O. Buchardt, Ed.), Wiley, New York, 1976, p. 207; K. B. Tomer, N. Harrit, I. Rosenthal, O. Buchardt, P. L. Kumler, and D. Creed, *J. Am. Chem. Soc.*, 95, 7402 (1973).

32. H. Wynberg, H. J. M. Sinnige, and H. M. J. C. Creemers, *J. Org. Chem.*, 36, 1011 (1971).

33. H. Hogeveen, R. M. Kellog, and K. A. Kuidersma, *Tetrahedron Lett.*, 3929 (1973).

34. Y. Kanacka and Y. Hatanaka, *Heterocycles*, 2, 423 (1974).

35. D. Bellus, in *Advances in Photochemistry*, Vol. 8 (J. N. Pitts Jr., G. S. Hammond, and W. A. Noyes Jr., Ed.), Wiley Interscience, New York, 1971, p. 109.

36. For reviews see: E. V. Blackburn and C. J. Timmons, *Q. Rev. Chem. Soc.*, 23, 482 (1969); F. R. Stermitz, in *Organic Photochemistry*, Vol. 1 (O. L. Chapman, Ed.), Marcel Dekker, New York, 1967, p. 247.

37. R. M. Kellog, M. B. Groen, and H. Wynberg, *J. Org. Chem.*, 32, 3093 (1967).

38. J. H. Dopper, D. Oudman, and H. Wynberg, *J. Am. Chem. Soc.*, 95, 3692 (1973).

39. C. C. Leznoff, W. Lilie, and C. Manning, *Can. J. Chem.*, 52, 132 (1974).

40. (a) D. De Luca, G. Martelli, P. Spagnolo, and M. Tiecco, *J. Chem. Soc.*, C, 2504 (1970); (b) J. Grimshaw and A. P. de Silva, *Chem. Soc. Rev.*, 10, 181 (1981).

41. L. Lunazzi, A. Mangini, C. Placucci, P. Spagnolo, and M. Tiecco, *J. Chem. Soc. Perkin II*, 192 (1972).

42. C. Paulmier, J. Bourgignon, J. Morel, and P. Pastour, *C. R. Acad. Sci. Paris, Série C*, 270, 494 (1970).

43.   N. Lozac'h and M. Stavaux, unpublished results.

44.   J. A. Van Koeveringe and J. Lugtenberg, *Rec. J. R. Neth. Chem. Soc.,* **96,** 55 (1977).

45.   R. E. Atkinson and F. E. Hardy, *J. Chem. Soc. Perkin II,* 27 (1972).

46.   Y. Kanaoka and K. Itoh, *Synthesis,* 36 (1972).

47.   For a review see I. Ninomiya and T. Naito, *Heterocycles,* **15,** 1433 (1981).

48.   (a) K. Beelitz, K. Praefcke, and S. Gronowitz, *Z. Naturforsch.,* **34b,** 1573 (1979); (b) K. Beelitz, G. Buchholz, and K. Praefcke, *Liebigs Ann. Chem.,* 2043 (1979).

49.   H. Wynberg, M. B. Groen, and R. M. Kellog, *J. Org. Chem.,* **35,** 2828 (1970).

50.   B. S. Green and L. Heller, *J. Org. Chem.,* **39,** 196 (1974).

51.   C. Rivas, M. Velez, and O. Crescente, *J. Chem. Comm.,* 1474 (1970).

52.   S. E. Flores, M. Velez, and C. Rivas, *Org. Mass. Spectrom.,* **5,** 1049 (1971).

53.   C. Rivas and R. A. Bolivar, *J. Heterocycl. Chem.,* **10,** 967 (1973).

54.   H. Wamhoff and H. J. Hupe, *Tetrahedron Lett.,* 125 (1978).

55.   K. Mizuno, M. Ishii, and Y. Otsuji, *J. Am. Chem. Soc.,* **103,** 5570 (1981).

56.   H. J. Kuhn and K. Gollnick, *Tetrahedron Lett.,* 1909 (1972).

57.   H. J. Kuhn and K. Gollwick, *Chem. Ber.,* **106,** 674 (1973).

58.   D. R. Arnold and C. P. Hadjiantoniou, *Can. J. Chem.,* **56,** 1970 (1978).

59.   J. C. Berridge, A. Gilbert, and G. N. Taylor, *J. Chem. Soc. Perkin I,* 2175 (1980).

60.   J. F. Bunnett and B. F. Gloor, *Heterocycles,* **5,** 377 (1976); J. F. Bunnett, *Acc. Chem. Res.,* **11,** 413 (1878).

61.   K. Maruyama and T. Otsuki, *Chem. Letters,* 851 (1977).

62.   K. Mazuyama, K. Mitsui, and T. Otsuki, *Chem. Letters,* 853 (1977).

63.   K. H. Pfoertner and R. Zell, *Helv. Chem. Acta,* **63,** 645 (1980).

64.   P. Gilgen, H. Heimgartner, and H. Schmid, *Heterocycles,* **6,** 143 (1977); A. Padwa, *Acc. Chem. Res.,* **9,** 371 (1976); M. Nastasi and J. Streith, in *Rearrangements of Ground and Excited States,* Vol. 3 (P. de Mayo, Ed.), Academic Press, New York, 1980, p. 445.

65.   D. R. Arnold and R. J. Birtwell, *J. Am. Chem. Soc.,* **95,** 4599 (1973); **96,** 6818 (1974).

66.   D. R. Arnold and B. M. Clarke Jr., *Can. J. Chem.,* **53,** 1 (1975).

67.   D. R. Arnold, R. J. Birtwell, and B. M. Clarke Jr., *Can. J. Chem.,* **52,** 1681 (1974).

68.   N. J. Turro, *Modern Molecular Photochemistry,* Benjamin, New York, 1978, p. 442.

69.   (a) T. S. Cantrell, *Chem. Commun.,* 155 (1972); (b) T. S. Cantrell, *J. Org. Chem.,* **39,** 2242 (1974); (c) T. S. Cantrell, *J. Org. Chem.,* **42,** 3774 (1977).

70.   A. T. Jeffries, III and C. Parkanyi, *Z. Naturforsch.,* **31b,** 345 (1976).

71.   C. Parkanyi, *Bull. Soc. Chim. Belg.,* **90,** 599 (1981).

72.   M. A. Fox, W. C. Nichols, Jr., and D. M. Lemal, *J. Am. Chem. Soc.,* **95,** 8164 (1973).

73.   N. Kharusch, *Intra Sci. Chem. Rep.,* **3,** 203 (1969); N. Kharasch, W. Wolf, T. Erpelding, P. G. Naylor, and L. Tokes, *Chem. Ind. (London),* 1720 (1962); W. Wolf and N. Kharasch, *J. Org. Chem.,* **26,** 283 (1961); L. Benatti and M. Tiecco, *Boll. Sci. Fac. Chim. Ind. Bologna,* **24,** 45 (1966); G. Martelli, P. Spagnolo, and M. Tiecco, *J. Chem. Soc. B,* 901 (1968).

74.   M. Tiecco and A. Tundo, *Int. J. Sulfur Chem.,* **8,** 295 (1973).

75.   J. Cornelisse, *Pure Appl. Chem.,* **41,** 433 (1975); E. Havinga and J. Cornelisse, *Chem. Rev.,* **75,** 353 (1975); J. Cornelisse, G. P. de Gunst, and E. Havinga, *Adv. Phys. Org. Chem.,* **11,** 225 (1975); E. Havinga and J. Cornelisse, *Pure Appl. Chem.,* **47,** 1 (1976); J. Cornelisse, G. Lodder, and E. Havinga, *Rev. Chem. Intermed.,* **2,** 231 (1979).

76.   M. B. Groen and E. Havinga, *Mol. Photochem.,* **6,** 9 (1974).

77.   B. Gandha and J. K. Sugden, *Synth. Commun.,* **9,** 845 (1979).

Author Index

Numbers in parentheses are reference numbers and indicate that the author's work is referred to although his name is not mentioned in the text. Numbers in *italics* show the pages on which complete references are listed.

Aadahl, G., 392(252), *444*
Abbieu, B. E., 394(279), *445*
Abdel-Fattah, B., 398(304), *446*
Abdel-Megeid, F. M. E., 183(589), *212*
Abd El-Salam, A. M., 435(670), *456*
Abdulla, W. A., 398(299), *445*
Abdurakhmanov, M. A., 338(69), *349*
Abegaz, B., 83(306a), 84(306a), *203*
Abiko, Y., 401(316), *446*
Abood, L. G., 374(146–148), *441*
Aboul Enein, H. Y., 379(187), *442*
Aboul-Enein, M. N., 435(671), *456*
Abou Ouf, A. A., 398(299, 300, 302), *445*
Abraham, W. R., 268(38), 273(38), 286(65), 309(38, 65, 93, 94), 315(143), *320, 321, 323*
Abramov, I. A., 27(175a), 591(175a), *199*
Abramovich, V. B., 4(14, 16, 23–25, 27), 5(30, 32), *194*
Abramovich, Z. I., 331(22), *347*
Abramovitch, R. A., 676(26), *740*
Abreu, B. E., 373(138), *441*
Abronim, I. A., 226(84), 229(84), 232(84), 249(84), *257*
Acheson, R. M., 629–635(1, 3), 648, *649*
Actor, P., 415(467), *450*
Adam, W., 706(107), *743*
Adams, C. R., 5(42), *195*
Adamson, D. W., 369(99, 101), 374(140), *439, 441*
Adkins, H., 484(222), 485(222), *562*
Adlerová, E., 395(285), *445*
Advani, B. G., 79(293, 294, 296, 297), 80(300), 81(294, 296, 297), 82(300), *202, 203*
Afavas'eva, Yu. A., 4(22), 5(33, 34, 44, 45, 49), *194, 195*
Agadzhanova, N. V., 335(40), 336(40), *348*
Agafonov, A. V., 482(200), *561*
Agawa, T., 66(265), *202*
Agten, J. T. M., 369(96), *439*
Ahmed, F. S. M., 435(670), *456*
Ahmed, M., 311(110), *322*
Aihara, J.-I., 226(50), 228(50), *256*
Aisaka, A., 381(194), *442*
Aitken, R. A., 599(156), 620(156), *627*

Akhmetshin, M. I., 4(14), *194*
Akonyan, N. E., 50(267), *202*, 378(174), *442*
Akonyan, R. A., 46(239), 47(239), 50(267), *201, 202*, 378(174), *442*
Alaimo, R. J., 422(529, 530), *452*
Alashev, F. D., 461(32), 465(81), 547(32, 81, 398–405), *555, 557, 568*
Al Badr, A. A., 379(187), *442*
Albanus, L., 379(149), *441*
Albini, E., 602(155), *627*
Albini, F. M., 602(155), *627*
Albrecht, R., 415(468, 469), *450, 451*
Albrechtsen, R., 429(640), *455*
Alderman, J. F., 572(13), 601(13), *623*
Aleksanyan, V. T., 616(102), *625*
Alekseev, V. V., 27(175a), 59(175), *199*
Alemagna, A., 28(181), 29(181), *199, 213*
Alexander, W., 159(529), *210*, 394(279), *445*
Alfonzo, L. M., 188(611), *212*
Alieva, R. B., 335(40), 336(40), *348*
Allais, A., 366(83), *439*
Allan, C. J., 237(144), *258*
Allen, D. W., 665(32), *669*
Allison, D. A., 237(144), *258*
Allner, K., 426(601), *454*
Almqvist, K. A., 686(56), 700(56), *741*
Alperman, H. G., 152(504), 103(504), *209*, 387(224, 225), *445*
Altgelt, K. H., 346(108), *351*
Alvarez, M., 167(540), *210*, 368(90, 91), *439*
Alvarez-Insúa, A. S., 462(37), *555*
Amano, A., 554(425), *569*
Amaranath, L., 428(619), *455*
Amato, J. S., (149), *627*
Amick, D. R., 507(310), *564*
Amin, S. G., 388(227), 401(326), *443, 446*
Amirov, Ya. S., 4(24), *194*
Amosova, S. V., 11(81a), 14(89, 89a), 127(380), 144(462), *196, 205, 208*
Amrein, B. J., 419(509), *452*
An, V. V., 148(477), *208*
Ananthan, S., 432(655), *456*
Andermann, G., 226(87), 229(87), 237(87), *257*
Anderson, D., 429(628), *455*

771

Toyoshima, H., 411(427), *449*
Trakhtenberg, P. L., 411(428), *449*
Traverso, G., 183(590), *212*
Traynelis, V. J., 188(609, 610, 612), 192(609, 610, 612), *212*
Trebaul, C., 20(115), *197*
Trehan, I. R., 392(255, 256), *444*, 711(116), *743*
Treibs, W., 140(447), *207*
Tremper, A., 753(30), *768*
Treuner, U. D., 406(368), *448*
Treuner, W. D., 406(367), *447*
Trieu, N. D., 111(356), 113–116(356), *205*
Trinajstic, N., 226(72, 73, 134, 138), 229(72, 73), 231(72, 73), 232(134), 233(138), 242(72), 245(72, 73), 248(72), 249(134), 250(138), *256, 258*
Trindle, C., 578(57), *624*, 673(7), 697(7), *740*
Trinh, S., 417(479), *451*
Tripathi, H. N., 409(408), 416(408), *449*
Tripathi, K., 409(408), 416(408), *449*
Trippler, S., 144(459, 461), 145(463), *208*
Trofimov, B. A., 11(81a), 14(89, 89a), *196*, 127(380), 144(462), *205, 208*
Trofimova, A. G., 127(380), *205*
Trolin, S., 374(144), *441*
Trompen, W. P., 424(557), *453*
Trompenars, W. P., 153(515, 516), 154(515, 516), 157(516), 159(516), 160(515), 161(515, 516), *209*, 680(41, 42, 45), 681(41, 42, 43), 683(41, 42, 46–48), 685(42), 699(43), *741*
Tronche, P., 394(278), *445*
Tronchet, J. M. J., 11(84a), *196*
Tron-Loisel, H., 375(161, 162), *441*
Trost, B. M., 126(372), *205*, 642(18), *649*
Trouiller, G., 372(129), *440*
Trška, P., 226(120), 231(120), 243(120), 248(120), 249(120), *257*
Truitt, P., 483(210), 487(210), *562*
Tsubokawa, M., 387(221), *443*
Tsuchiya, T., 753(29), *768*
Tsuji, M., 363(65), *438*
Tsujikawa, T., 427(609), *454*
Tsukada, M., 127(375), *205*
Tsukada, T., 381(196), *442*
Tsukada, W., 387(321), *443*
Tsukuma, S., 411(427), *449*
Tsumagari, T., 48(249), 49(249), *201*, 360(35, 36), 362(56–58), *438*
Tsurumi, K., 386(215), *443*
Tsvetkova, I. V., 426(575), *453*
Tuchmann-Duplessis, H., 415(466), *450*
Tullar, B. F., 27(139), *198*
Tundo, A., 5(68), *196*, 656(14, 16, 17), 660(26, 28), 662(28, 29), 664(16), 665(17), *669*, 766(74), *769*

Tupper, D. E., 161(532), 163(532), *210*, 357(23), 358(24), 430(23), *473*
Turk, C. F., 419(504), *452*
Turner, J. B., 665(32), *669*
Turro, N. J., 763(68), *769*

Udre, V., 5(59), 6(59), 129(397, 398), 132(419–421), *195, 206, 207*, 471(132), *554*, 611(124), 617(119, 121), *626*
Ueno, K., 387(221, 222), *443*
Uhlenbrock, J. H., 272(43), *320*
Ulendeeva, A. D., 331(19), *347*
Ullar, I. A., 161(532), 163(532), *210*
Ullas, G. V., 432(655), *456*
Ullman, E. F., 747(20), *768*
Ume, Y., 460(19, 20), *555*
Umio, S., 134(439), *207*
Undheim, K., 187(607), *212*
Ung, S. N., 372(129), *440*
Uppström, B., 692(79), 693(79, 82, 83), *742*
Uriu, S. A., 426(586), *454*
Usdin, E., 355(16), *437*
Usieli, V., 606(83), 610(83), 617(83), *625*
Ushirogochi, A., 67(280a), *202*
Uskokovič, M. P., 468(113), *558*
Uskokovič, M. R., 153(513), 157(513), 159(513, 526), *209, 210*, 468(114), *558*
Usolzeva, M. V., 339(78), *349*

Vachkov, K. V., 34(158, 158a), *198*
Vadzis, M., 404(344), *447*
Vafai, M., 466(94), *557*
Vafina, A. A., 338(71), *349*
Vagabov, M. B., 148(476), *208*
Vagabov, M. G., 337(43), *348*
Vagabov, M. V., 148(478a), *208*, 337(44, 45), *348*
Vaidya, N. K., 415(471), *451*
Vaidya, S. R., 188(614), 189(614), *212*
Vaisberg, K. M., 4(16), *194*
Vaitiekunas, A., 354(3), 355(3), 426(3), *346*
Vakhreeva, K. I., 411(429, 430), *449*
Vakurova, E. M., 469(120), *558*
Valtsova, A. A., 331(17), *347*
Van Abbe, N. J., 412(439), *450*
van Bekkum, H., 604(95), 614(95), *625*
Van Bergen, T. J., 166(539), 167(539), *210*
Van Bever, W. F. M., 369(97), *439*
Van Brussel, E., 402(336), *446*
Van Campen, M. G., 373(131), *440*
Van Cuong Pham, T., 269(36), 274(36), 315(36), *320*
Van Daele, P. G. H., 369(96), 385(213), *439, 443*
van den Ham, D. M. W., 680(41), 681(41), 683(41), *741*

Subject Index